THE SEA

Ideas and Observations on Progress in the Study of the Seas

Volume 1 PHYSICAL OCEANOGRAPHY

Volume 2 COMPOSITION OF SEA-WATER

COMPARATIVE AND DESCRIPTIVE OCEANOGRAPHY

Volume 3 THE EARTH BENEATH THE SEA

HISTORY

THE SEA

Ideas and Observations on Progress in the Study of the Seas

General Editor

M. N. HILL

VOLUME 3

The Earth Beneath the Sea

History

INTERSCIENCE PUBLISHERS

a division of

John Wiley & Sons · New York · London · Sydney

First published 1963 by John Wiley & Sons, Inc.

Library of Congress Catalog Card Number 62-18366

ISBN 0 470 39621 0

PRINTED IN THE UNITED STATES OF AMERICA

10 9 8 7 6 5 4

CONTRIBUTORS TO VOLUME 3

G. ARRHENIUS, Scripps Institution of Oceanography, University of California, La Jolla, California

R. A. BAGNOLD, Rickwoods, Mark Beech, Edenbridge, Kent, England

E. C. BULLARD, Department of Geodesy and Geophysics, Madingley Rise, Cambridge, England

J. DORMAN, Lamont Geological Observatory, Columbia University, Palisades, New York

C. L. DRAKE, Lamont Geological Observatory, Columbia University, Palisades, New York

H. E. EDGERTON, Massachusetts Institute of Technology, Cambridge, Massachusetts

K. O. EMERY, University of Southern California, Los Angeles, California. Now at Woods Hole Oceanographic Institution, Woods Hole, Massachusetts

C. EMILIANI, Institute of Marine Science, University of Miami, Virginia Key, Miami, Florida

D. B. ERICSON, Lamont Geological Observatory, Columbia University, Palisades, New York

J. I. EWING, Lamont Geological Observatory, Columbia University, Palisades, New York

M. EWING, Lamont Geological Observatory, Columbia University, Palisades, New York

R. L. FISHER, Department of Earth Sciences, University of California, San Diego, La Jolla, California

RICHARD F. FLINT, Department of Geology, Yale University, New Haven, Connecticut

R. N. GINSBURG, Shell Development Company, A Division of Shell Oil Company, Exploration and Production Research Laboratory, Houston, Texas

EDWARD D. GOLDBERG, University of California, La Jolla, California

J. J. GRIFFIN, Scripps Institution of Oceanography, University of California, La Jolla, California

A. GUILCHER, Institut de Géographie, Université de Paris, Paris, France

J. C. HARRISON, Hughes Research Laboratories, Malibu, California

BRUCE C. HEEZEN, Department of Geology and Lamont Geological Observatory, Columbia University, Palisades, New York

J. B. HERSEY, Woods Hole Oceanographic Institution, Woods Hole, Massachusetts

H. H. HESS, Department of Geology, Princeton University, Princeton, New Jersey

M. N. HILL, Department of Geodesy and Geophysics, Madingley Rise, Cambridge, England

D. L. INMAN, Scripps Institution of Oceanography, University of California, La Jolla, California

I. R. KAPLAN, Laboratory for Microbiological Chemistry, Habassah Medical School, Hebrew University, Jerusalem, Israel

F. F. KOCZY, The Marine Laboratory, Institute of Marine Science, University of Miami, Miami, Florida

H. S. LADD, U. S. Geological Survey, U. S. National Museum, Washington 25, D.C.

A. S. LAUGHTON, National Institute of Oceanography, Wormley, Godalming, Surrey, England

R. MICHAEL LLOYD, Shell Development Company, A Division of Shell Oil Company, Exploration and Production Research Laboratory, Houston, Texas

J. S. MCCALLUM, Shell Oil Company, Roswell, New Mexico

R. G. MASON, Scripps Institution of Oceanography, University of California, La Jolla, California, and Geophysics Department, Imperial College of Science and Technology, London, England

H. W. MENARD, Scripps Institution of Oceanography, University of California, La Jolla, California

STANLEY L. MILLER, Department of Chemistry, University of California, San Diego, La Jolla, California

JOHN E. NAFE, Lamont Geological Observatory, Columbia University, Palisades, New York

J. OLIVER, Lamont Geological Observatory, Columbia University, Palisades, New York

R. W. RAITT, Scripps Institution of Oceanography, University of California, La Jolla, California

W. R. RIEDEL, Scripps Institution of Oceanography, University of California, La Jolla, California

S. C. RITTENBERG, Department of Bacteriology, University of California, Los Angeles, California

F. P. SHEPARD, Scripps Institution of Oceanography, University of California, La Jolla, California

G. G. SHOR, JR., Scripps Institution of Oceanography, University of California, San Diego, La Jolla, California

K. W. STOCKMAN, Shell Development Company, A Division of Shell Oil Company, Exploration and Production Research Laboratory, Houston, Texas

J. LAMAR WORZEL, Lamont Geological Observatory, Columbia University, Palisades, New York

PREFACE

Oceanography has surged forward as a subject for research during the past twenty years, and great progress has been made in our understanding of the structure of the water-masses, of the crust of the earth beneath the oceans, and of the processes which are involved in creating these structures. This progress has been particularly rapid in the recent past for two prime reasons. Firstly, techniques of investigation have become available which have made many hitherto intractable problems capable of solution. For example, it has only recently become possible to measure the value of gravity in a surface ship, to determine the deep currents of the oceans by direct methods, or to analyze wave spectra with the detail which modern electronic computers allow. The second reason for the present rapidity of progress lies in the increasing availability of the resources for both theoretical and practical marine investigations; the international character of oceanography and the growing world-wide interest in fundamental research have resulted in the provision of opportunities hitherto impossible.

Some years ago Dr. Roger Revelle of the Scripps Institution of Oceanography suggested to us that perhaps the production of treatises on the new developments in oceanography were not keeping pace with the progress in the subject. It was true that papers were appearing in healthy numbers throughout the journals of the world, but there were no recent comprehensive works such as *The Oceans*, written by Sverdrup, Johnson and Fleming and published in 1942. This work had been, and still is, of value to many, if not all, oceanographers; Revelle suggested that we produce another such volume containing ideas and observations concerning the work accomplished during the twenty years since this masterpiece. It was suggested that this new work should not attempt to be a textbook but a balanced account of how oceanography, and the thoughts of oceanographers, were moving.

It was early apparent that a work of this nature would have to be from the pens of many authors; the subject had become too broad, the oceanographers perhaps too specialized, to allow a small number of contributors to cover a vast field of study. It was also apparent that we could do no more than include biology in so far as it was directly related to physical, chemical and geological processes in the ocean and on the ocean floor. Marine biology could alone fill that space which we thought was the maximum we should allow for these volumes. Biology is, therefore, somewhat scattered through this work; so it has to be, since the contributions of biologists extend throughout the disciplines embraced by the contents of these volumes.

It has been said in the past that oceanography as a subject did not, or even should not, include the study of the mode of formation and the structure of the muds and hard rocks forming the floor of the oceans and the seas. This exclusion is no longer possible, and we have included in these volumes geological and geophysical ideas and observations concerning the earth beneath the seas. In the great oceanographical (or oceanological) establishments of the world the earth sciences relating to the oceanic environment are pursued alongside one another without discrimination as to whether the air over the sea, the water, or the solids beneath the sea could claim prior importance. We have adopted the same breadth of outlook in these volumes.

With a composite work such as we have tried to produce, it has proved difficult to ensure a precise balance in the emphasis given to our wide variety of topics. Sometimes, maybe, there will be too much written about a narrow field; sometimes we shall perhaps be criticized for the converse, too little written about a broad field. At least, however, we hope we have covered most of the topics suitable for a work on the new developments in our subject.

When we drew up the first list of contents of this work we had supposed that we should be producing one volume. We soon found, however, that we had omitted a number of topics, and the contents list consequently increased. We also found that our contributors could not possibly have provided, within the limits we originally proposed, adequate discourses on their various topics. It was apparent that we should have to subdivide *The Sea* into three volumes. The first consists of new thoughts and ideas on Physical Oceanography; the second on the Composition of Sea-Water and Comparative and Descriptive Oceanography; and the third on the Earth Beneath the Sea and History. The three volumes, therefore, are divided from one another by major divisions of oceanography, and we believe for this reason that our readers will not be inconvenienced by constant cross-referencing from one volume to another. This clear division has also allowed us to index the volumes separately; it should be possible to select the volume containing information on any particular topic without difficulty. We have no overall index, since for a work of this magnitude it would be cumbersome.

We considered that for ease of production we should start a new numbering sequence at the beginning of each chapter for figures, equations and tables. In order to avoid confusion in referring back, we have printed the chapter number and section number on alternate pages.

In collecting the material for these volumes we have had most helpful cooperation from our contributors. In some topics, however, we have failed to find authors, or potential authors have withdrawn. We must admit, therefore, to omissions, some of which are conspicuous and important. By further effort we could doubtlessly have repaired these gaps in subject matter, but this would have delayed publication, and we believe the contributions we already have make a valuable collection. It is possible that in the future a fourth volume might be produced to deal with topics omitted or new since we first started collecting material for *The Sea*.

We are greatly indebted to the secretarial assistance provided by Mrs. Joyce

Nightingale and Mrs. Joyce Day. Mr. Cameron D. Ovey has provided editorial assistance of outstanding value, and Dr. J. E. Holmstrom has produced an index of great merit. To them and to the publishers and printers we extend our thanks.

May, 1962 M.N.H.

CONTENTS

THE EARTH BENEATH THE SEA

HISTORY

The Earth
Beneath the Sea

I. GEOPHYSICAL EXPLORATION

1. ELEMENTARY THEORY OF SEISMIC REFRACTION AND REFLECTION MEASUREMENTS

J. I. Ewing

1. Introduction

Seismic techniques have been among the most useful in furthering our knowledge of the structure and thickness of the earth's crust under the sea. The techniques are based on the measurement of the propagation of sonic energy, usually generated by the explosion of TNT or dynamite charges. The sound energy is received at varying distances from the explosion site by hydrophones or geophones and recorded on oscillograph tapes with an accurate time scale. Although useful information can sometimes be obtained from studying the character of the transmitted energy, the basic measurement taken from the oscillograms is the time of arrival of sound waves for which the path from explosion to detector is known or can be ascertained. If the path (distance) and the travel time are known, the speed of propagation can be computed. By recording either a line of shots at a fixed receiving position or a single shot at a line of receiving positions, a relationship of travel time versus distance is established, from which the number and thickness of the sub-surface strata and the speed of sound in each can be determined. Knowing the speed of sound in a particular layer enables us to make at least an intelligent guess about the type of material constituting the layer; or if it is known, from drilling or other means, that a certain layer is present at a particular depth beneath the surface, this allows us to determine the continuation of the layer into areas where the other information is unavailable.

2. Reflected Waves

If we know the speed of sound in various sub-surface strata and if the interfaces between the strata are good reflectors, it is simple to measure the thickness of each by measuring the time required for a pulse of sonic energy to travel through the layer and back by reflection. Standard echo-sounders measure water depth in just this way, and some sounders under certain conditions will measure the depth to sub-bottom interfaces within a few tens of meters of the bottom. The limitations of this usage arise from the fact that relatively high sonic frequencies are required to give good resolution of sea-floor topography, and high frequencies, because of absorption, cannot penetrate far into the unconsolidated sediments or rock layers. Low-frequency energy from

[*MS received July, 1960*]

explosions is capable of penetrating to depths of thousands of meters, and the seismic-reflection technique using explosions has been one of the major tools of sub-surface exploration in the past few decades.

The propagation of seismic energy can be considered adequately for most purposes to behave according to ray theory. The shot point is the origin of ray paths which extend in all directions, and each ray can be considered to be a bundle of energy travelling that particular path. Only the rays that leave the origin and eventually arrive at a detector are of interest. In a multi-layered medium, there are several possible paths, most of them involving reflection or refraction at the interfaces between the layers.

The simplest case to consider is that involving reflections from horizontal plane interfaces. The lower part of Fig. 1 shows the ray paths of the direct

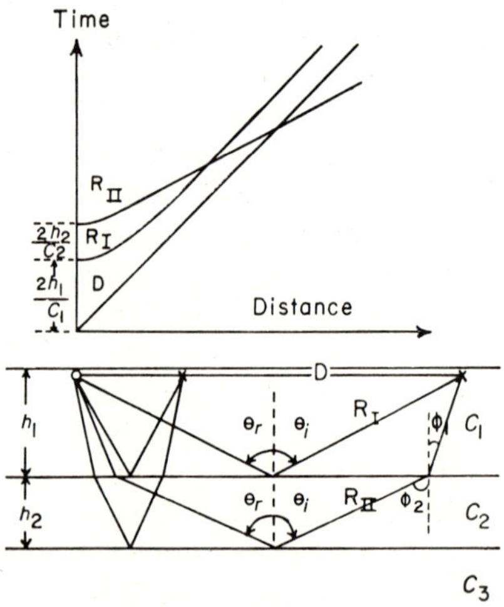

Fig. 1. Ray diagram and time–distance graph for reflections in a three-layer model, $C_1 < C_2 < C_3$.

○ = detector
× = shot
$\theta_i = \theta_r$

waves and of the first-order reflected waves from two shots to a detector. The shots and detector are in Layer 1, which has a seismic velocity C_1 and thickness h_1. Underneath are two layers with velocities C_2 and C_3. The upper part of the figure shows the travel time *vs.* distance graph for the arrivals. The direct waves determine the straight line, D. The reflected waves determine the hyperbolic curves, R_I and R_{II}.[1] The inverse slope of D, $\Delta D/\Delta T$, gives the seismic velocity in Layer 1. The R_I curve is asymptotic to D. Its intercept on the ordinate, $2h_1/C_1$, gives the travel time for a reflected wave from the top of

[1] In this and succeeding discussions, R_I, R_{II}, R_{III}...designate reflected waves from the 1st, 2nd, 3rd...interfaces below the surface. Arabic subscripts R_2, R_3...designate second, third...order reflections from the sea floor.

Layer 2 at zero distance between shot and detector. Similarly, the R_{II} intercept represents the travel time for a normal incidence reflection from the top of Layer 3. Higher order reflections from these interfaces determine hyperbolic curves which intercept the ordinate at times $2nh_1/C_1$ and $2n[(h_1+h_2)/C_1C_2]$ and are asymptotic to the first-order curves. The number and strength of reflections received from each shot will depend on absorption, scattering and spreading losses in Layers 1 and 2 and on reflectivity and scattering at the interfaces. In the cases we consider here, the 1–2 interface is the water–sediment boundary and the coefficient of reflectivity for it can be expressed by Rayleigh's equation for plane waves incident on a boundary between two non-rigid compressible media with velocities and densities C_1, ρ_1 and C_2, ρ_2 and angle of incidence θ_i:

$$K_i = \frac{(\rho_2C_2/\rho_1C_1)-[1+\tan^2\theta_i(1-C_2^2/C_1^2)]^{1/2}}{(\rho_2C_2/\rho_1C_1)+[1+\tan^2\theta_i(1-C_2^2/C_1^2)]^{1/2}}. \tag{1}$$

This shows that the coefficient of reflectivity is a function of the angle of incidence, θ_i, and of the ratio of acoustic impedances, ρC, between the two media.

In many oceanic areas where $C_1 < C_2$ and $\rho_1C_1 < \rho_2C_2$, K increases from some finite value at $\theta_i = 0$ to unity at $\theta_i = \sin^{-1}(C_1/C_2)$, where total reflection occurs. In other areas there is evidence that the velocity in the uppermost sediments is lower than in the water, in which case total reflection occurs only at grazing incidence. The reader is referred to Officer (1958) for a treatment of reflectivity which considers various ratios of C_1 and C_2, and of ρ_1C_1 and ρ_2C_2. For the deeper interfaces, where rigidity becomes an important factor, other complications are introduced into the consideration of reflectivity, owing to the fact that some of the energy in the incident wave can now go into a refracted compressional wave, as before, and in addition, into reflected and refracted shear waves. Reference is made to Nafe (1957) who investigated reflection and transmission coefficients at an interface between two solids for several values of velocity and density appropriate to crustal layering.

In the time–distance graph, it is obvious that if the intercepts of R_I and R_{II} are measured and C_1 and C_2 are known, the thickness of Layer 1 and of Layer 2 can be computed directly. In marine operations, C_1 is the speed of sound in the water and can be computed from hydrographic data. In some cases C_2 can be measured directly by refracted waves, but, particularly in deep ocean areas, the velocity contrast between Layers 1 and 2 is small and it is usually necessary to derive the value of C_2 by comparison of the R_{II} curve with a family of theoretical curves computed for various assumed velocities and thicknesses of Layer 2.

Fig. 2, taken from Katz and Ewing (1956), shows theoretical curves for the relationship of three different values of h_2/h_1 and five different values of C_2/C_1 to R_{In} (reflection from sea bottom) and $R_{IIn}-R_{In}$ (time difference between bottom and sub-bottom reflection). Observed values of R_{In} and $R_{IIn}-R_{In}$ plotted on the same scale will determine a curve similar to one of these. The theoretical curve which most closely fits the observed values thus indicates the velocity and thickness of Layer 2. These curves were computed assuming uniform velocity in each layer. A similar method of comparing observed with theoretical curves is used to handle models with velocity gradients. This is discussed in detail in Chapter 5.

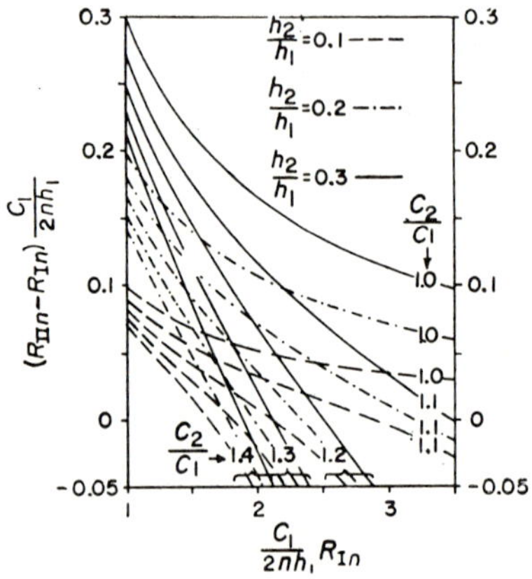

Fig. 2. Relationship of bottom reflected and sub-bottom reflected waves for different ratios of layer thicknesses, h_2/h_1, and velocities, C_2/C_1. (After Katz and Ewing, 1956.)

3. Refracted Waves

In seismic exploration, sound waves refracted at an interface between two media with different velocities are of particular interest. In Fig. 1 the R_{II} ray is refracted by the velocity change across the 1–2 interface such that it propagates in Layer 2 along a path more nearly horizontal than that in Layer 1. The refraction is described by Snell's Law,

$$\frac{\sin \varphi_1}{C_1} = \frac{\sin \varphi_2}{C_2}. \tag{2}$$

When φ_2 is 90°, equation (2) reads

$$\sin \varphi_1 = C_1/C_2. \tag{3}$$

This defines the *critical angle* of incidence, φ_c, , for which the ray is refracted parallel to the 1–2 interface. Similarly, there is a value of φ_2 for which a ray travelling in Layer 2 will be refracted along the 2–3 interface. The sound wave refracted at the critical angle travels along the interface at the speed of sound in the higher velocity layer, and part of its energy is continually re-emergent into the lower velocity medium at an angle equal to that of the incident ray.

Fig. 3 shows the ray paths for direct, refracted and reflected waves for a three-layer model and their relationship to each other on a time–distance graph.

Since we are dealing with seismic measurements in the sea, here and in future discussions Layer 1 is the water layer. The direct wave in Layer 1, D, and the refracted waves in Layers 2 and 3, G_1 and G_2, determine straight line segments, and the reflected waves determine hyperbolic curves. There is a critical distance at which a refracted wave from either of the deeper layers can be returned to the surface. This distance is dependent on the velocity contrast between the layer in question and Layer 1, and on the velocity and thickness

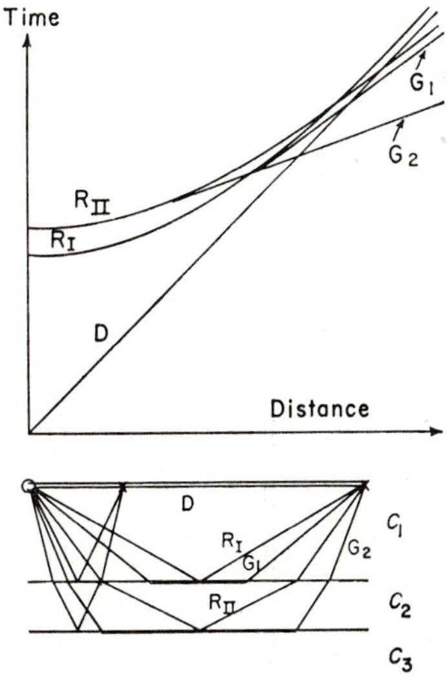

Fig. 3. Ray diagram and time–distance graph for reflected and refracted waves in a three-layer model, $C_1 < C_2 < C_3$.

of the layers between. Notice that the refraction line G for any layer is tangent to the reflection curve associated with the top of that layer. The point of tangency occurs at critical distance. In addition to the above relationship, i.e. G_n is tangent to R_n, notice also that R_n is asymptotic to G_{n-1}. Since both refracted and reflected waves are recorded in the normal method of operation, these relationships are of great importance in the analysis of seismic data, imposing, as they do, strict conditions upon the identification and interpretation of arrivals. Fig. 4 shows sample oscillograph records with various reflected and refracted arrivals indicated. Notice that the waterborne waves, D and R_1, contain energy in a wide range of frequencies, whereas the refracted and reflected ground waves contain only low-frequency energy. In general it is not possible to identify the path of a ground wave arrival by its appearance on one record. Such identification is made principally from the relationship of the arrival to others on the time–distance graph.

For simplicity of discussion we have been considering structures in which all of the interfaces are horizontal. In such cases the inverse slope of each refraction line $G_1, G_2, G_3 \ldots G_n$ gives the true seismic velocity in that layer. If

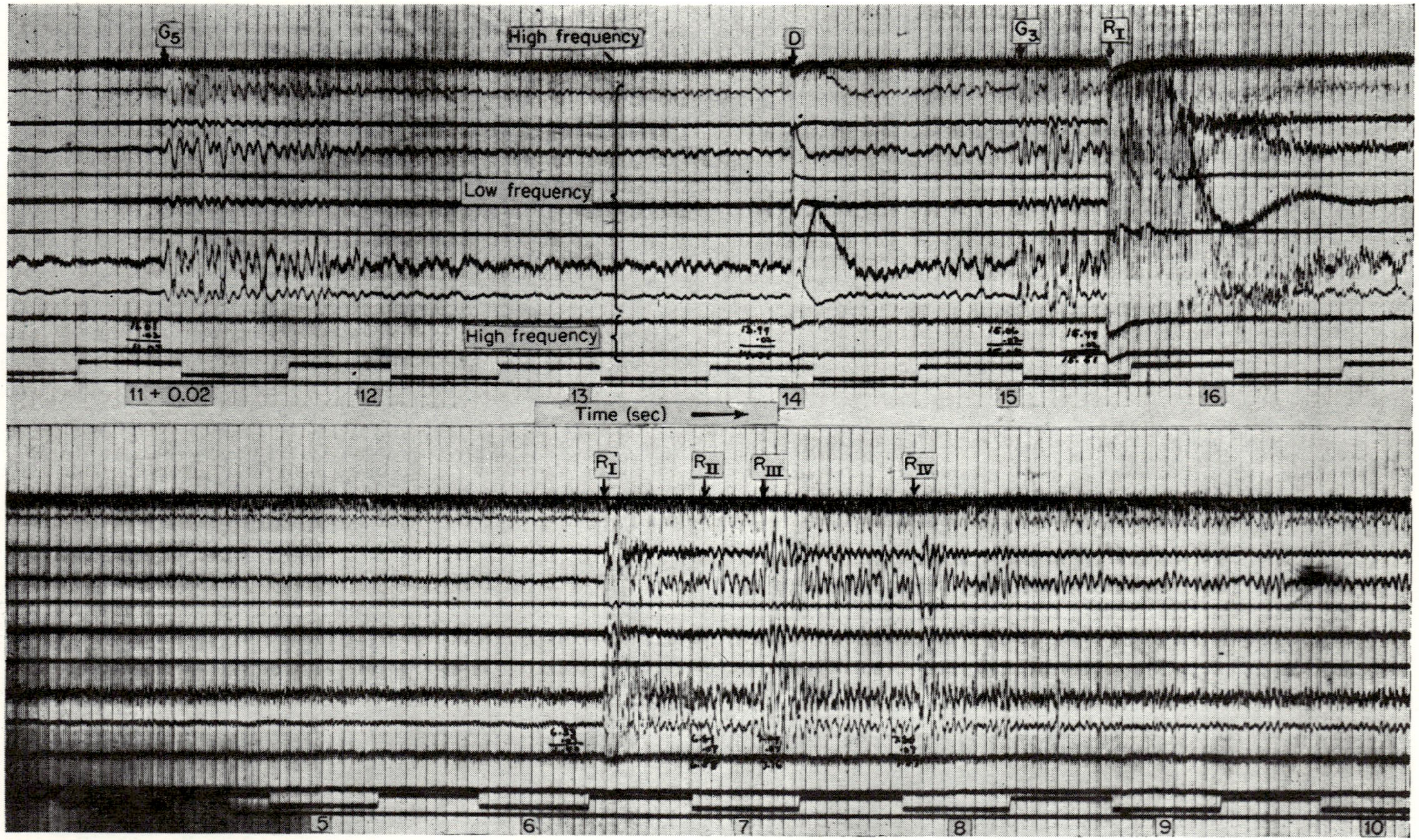

Fig. 4. Sample oscillograph records. The upper record shows refracted ground waves, G_5 and G_3, a direct water wave, D, and a bottom reflected wave, R_I, for a shot–detector distance of 11.5 nautical miles. The lower record shows bottom reflected, R_I, and sub-bottom reflected, R_{II}, R_{III} and R_{IV}, arrivals at a range of 1.5 miles. Water depth is 2650 fathoms.

the layers are dipping, the inverse slopes give apparent velocities which are too high if the shots are up-dip from the receiving position or too low if the shots are down-dip. In order to determine the existence of dips, and their direction and magnitude, it is necessary to shoot end-to-end profiles or reversed profiles. The former consists of two lines of shots usually fired in opposite directions away from a common receiving position. The latter consists of a line of shots received at both ends of the line. Certain conditions are imposed upon the time–distance data by the end-to-end arrangement and others are imposed by the reversed one; hence it is most advantageous to combine the two and shoot end-to-end reversed profiles. Fig. 5 shows time–distance graphs of two such profiles and the structure section for a three-layer case with dipping beds.

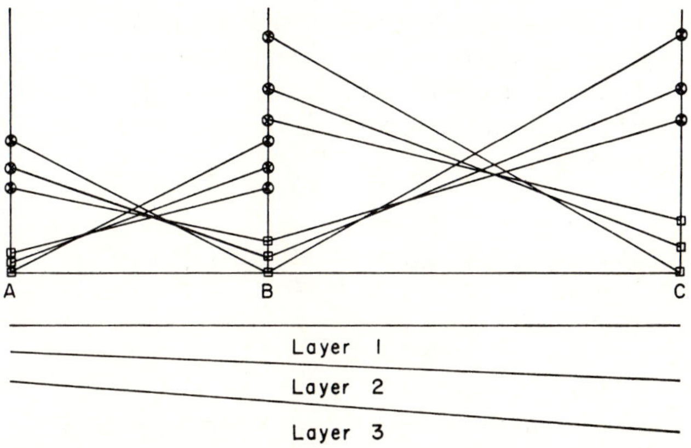

Fig. 5. Time–distance graphs for end-to-end reversed profiles across dipping beds, showing relationship of reverse points and intercepts. Only refracted waves are considered here.

The conditions imposed by the end-to-end arrangement are that the intercepts (squares) of the lines in Profile 1 must be the same as those in Profile 2 at point B. The conditions imposed by a reversed profile are that the reverse points (circles) for each layer be the same at each end of the profile. This latter results from the reciprocity theorem, which states that, at any distance, the shot and receiving positions may be interchanged without affecting the travel time. These conditions, both for end-to-end and reversed profiles, provide a valuable check on the interpretation of travel-time data. If two or more end-to-end reversed profiles are shot and the data can be shown to satisfy these conditions, the results may be viewed with considerable confidence. In addition to these, it is also frequently possible to provide a further test of the data and of the interpretation with the conditions imposed by the required relationship of the refraction lines and reflection curves, as discussed earlier. The importance of these various means of cross-checking the results is obvious when one bears in mind that we are dealing with geological layering and not with models.

4. Computation of Layer Velocities and Thicknesses

The foregoing has shown the relationships between ray paths and time–distance graphs for various structural models. The following, after Ewing *et al.* (1939), describes the "slope-intercept" method of computing the velocity structure from a time–distance graph. In these computations, the assumption is made that each layer is homogeneous and bounded above and below by smooth planes, and that the seismic velocity in each layer is higher than that in the layer above. In addition, we know that the seismic waves are bent at interfaces between two media according to Snell's Law (2) and that a wave travelling in a layer with velocity C and incident upon the surface of the layer at an angle α with the normal has an apparent velocity $C/\sin\alpha$ along the surface.

Fig. 6 shows a four-layer structural model with dipping interfaces. From the time–distance graph we can measure the apparent velocity (inverse slope) for each layer at each end of the profile, and we can measure the intercept times for the refraction lines corresponding to each layer.

The following equations,

$$\left.\begin{aligned}
&\frac{C_1}{C_{2a}} = \sin(i_{12}-\omega_{12}) \qquad \frac{C_1}{C_{2b}} = \sin(i_{12}+\omega_{12})\\
&\qquad\qquad \frac{C_1}{C_2} = \sin i_{12}\\
&\frac{C_1}{C_{3a}} = \sin(\alpha_{13}-\omega_{12}) \qquad \frac{C_1}{C_{3b}} = \sin(\beta_{13}+\omega_{12})\\
&\frac{C_1}{C_2} = \frac{\sin\alpha_{13}}{\sin(i_{23}-\omega_{23})} \qquad \frac{C_1}{C_2} = \frac{\sin\beta_{13}}{\sin(i_{23}+\omega_{23})}\\
&\qquad\qquad \frac{C_2}{C_3} = \sin i_{23}\\
&\frac{C_1}{C_{4a}} = \sin(\alpha_{14}-\omega_{12}) \qquad \frac{C_1}{C_{4b}} = \sin(\beta_{14}+\omega_{12})\\
&\frac{C_1}{C_2} = \frac{\sin\alpha_{14}}{\sin(\alpha_{24}-\omega_{23})} \qquad \frac{C_1}{C_2} = \frac{\sin\beta_{14}}{\sin(\beta_{24}+\omega_{23})}\\
&\frac{C_2}{C_3} = \frac{\sin\alpha_{24}}{\sin(i_{34}-\omega_{34})} \qquad \frac{C_2}{C_3} = \frac{\sin\beta_{24}}{\sin(i_{34}+\omega_{34})}\\
&\qquad\qquad \frac{C_3}{C_4} = \sin i_{34},
\end{aligned}\right\} \tag{4}$$

can be used to compute the true seismic velocity, C, of each layer and the angle of inclination, ω, of each interface by a stepwise process. The velocities C_{na} and C_{nb} are apparent velocities which can be read directly from the time–distance graph. Thus, being given C_1 (directly measurable), C_{2a}, and C_{2b}, we

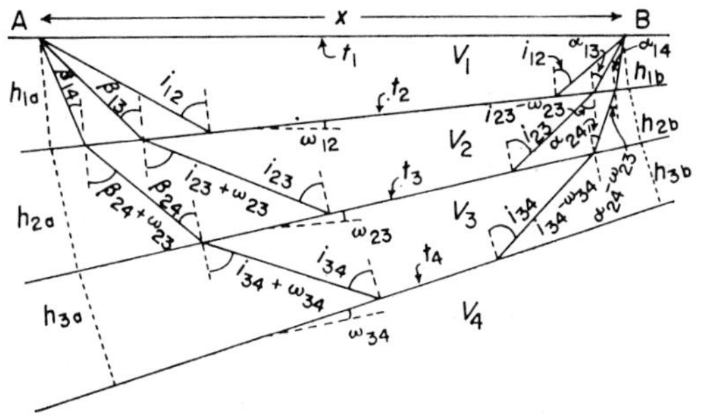

Fig. 6. Ray diagrams defining angles used in computing velocity and thickness for sloping layers.

can compute C_2 and ω_{12}. Using these additional values in the next step allows us to compute C_3 and ω_{23}, and so on for each layer.

For computing the thickness of each layer we again follow a stepwise process using the intercept times taken from the time–distance graph. Fig. 7 shows the fictitious ray paths associated with the intercept times, i.e. the travel times for zero shot–detector distance. This figure is derived from Fig. 6 by simply moving point B to point A and considering that the rays originating

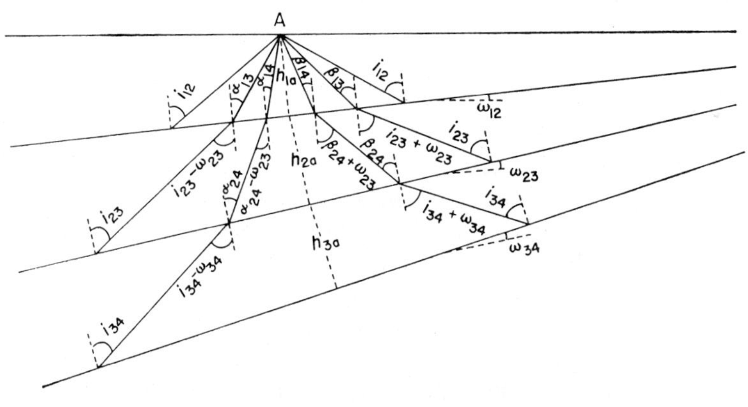

Fig. 7. Diagram of fictitious ray paths associated with intercepts.

at B move also. Obviously these paths are not real, since a finite range exists at which a refracted wave can be returned to the surface; but they have a mathematical significance in these computations. When point B is moved a short distance toward point A, the total path is shortened principally in the portion where the ray follows an interface. In Fig. 7 these portions of the ray paths are, in effect, negative distances. Thus the intercept time associated with a particular interface amounts to the sum of the travel times along each segment of the path above the interface minus the travel time along the interface. For example the intercept time for Layer 2 at A is

$$t = \frac{2h_1}{C_1 \cos i_{12}} - \frac{2h_1 \tan i_{12}}{C_2},$$

reducing to

$$t = \frac{2h_1}{C_1} \cos i_{12}.$$

Similar relationships can be derived from the deeper interfaces giving the following set of equations:

$$\begin{aligned} t_{2a} &= \frac{2h_{1a}}{C_1} \cos i_{12} \\ t_{3a} &= \frac{2h_{2a}}{C_2} \cos i_{23} + \frac{h_{1a}}{C_1} (\cos \alpha_{13} + \cos \beta_{13}) \\ t_{4a} &= \frac{2h_{3a}}{C_3} \cos i_{34} + \frac{h_{2a}}{C_2} (\cos \alpha_{24} + \cos \beta_{24}) + \frac{h_{1a}}{C_1} (\cos \alpha_{14} + \cos \beta_{14}). \end{aligned} \tag{5}$$

These equations give the time-intercept formulae for point A. A similar set exists for point B. With (4) and (5) and the slopes and intercepts taken from the time–distance graph, true velocities, inclinations and thicknesses can be computed.

5. Shear Waves and Complex Refracted Waves

In Fig. 8 are shown the ray paths and refraction lines for the most commonly observed arrivals in seismic work at sea. This model is typical of the structure on the continental rise of the east coast of the United States. The M discontinuity and the mantle are not shown because doing so would have unduly complicated the picture with additional lines.

Of the compressional wave arrivals shown, the least commonly observed is that in Layer 2. In fact, it is seldom seen in deep-water areas and is shown here mainly to indicate its relationship to other arrivals. Layer 4 (see Raitt, Chapter 6) is observed in all but some of the deepest areas, where the arrivals are masked and are difficult to see as second arrivals.

Of the shear-wave arrivals, i.e. those which propagate as compressional waves in Layers 1–3 and as vertically polarized shear waves (S_v) in the high-velocity layers, the one which travels in Layer 5 (designated by open circles) is most commonly observed. Though relatively common, its strength is markedly variable from one profile to another, indicating an appreciable variation in the conditions essential to efficient transformation from compressional to shear propagation.

Note that the shear waves in both Layers 4 and 5 are transformed at the 3–4 interface. Transformation from compressional to shear waves at the 4–5 interface results in downward refraction owing to the fact that the compressional wave velocity, α_4, in Layer 4 is higher than the shear-wave velocity, β_5, in Layer 5; hence there can be no arrivals at the surface in this case. It is

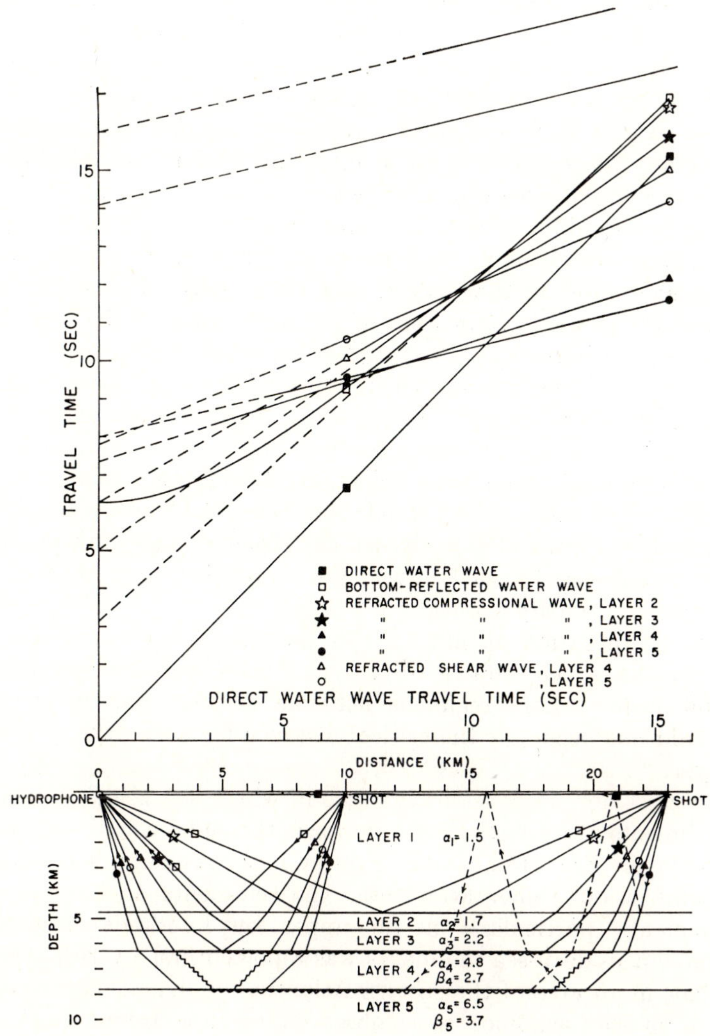

Fig. 8. Ray diagram and time–distance graph for commonly observed arrivals in oceanic areas.

theoretically possible to receive an arrival which was first transformed from compression to shear at the 3–4 interface, then was re-transformed to compressional and refracted horizontally at the 4–5 interface. Such an arrival has not been identified, however, very possibly owing to insufficient energy transfer associated with the transformations. S_v waves in the low-velocity sediments have been observed occasionally in shallow water (Drake *et al.*, 1959) but have not been reported in deep-water measurements.

The lines whose intercepts are 14.5 and 16.0 are determined by reflected-refracted and doubly refracted compressional waves from Layer 5. Ray paths associated with these arrivals are shown by dashed lines. The reflected-refracted arrival has only two possible paths for each shot, i.e. the reflection through the water layer can occur at either the shot end, as shown, at the detector end, or both. The extra trip to the surface taken by the doubly refracted arrival can occur at any place between the point of initial horizontal refraction and the point of re-emergence leading directly to the detector. For uniform, horizontal layering, the doubly refracted arrival can be very strong, because the energy travelling any of the possible paths will arrive at the detector in phase and at the same time as that taking any other path. The strength of this arrival, therefore, gives an indication of the uniformity of the layering. Both of the arrivals just discussed propagate horizontally at the speed of the compressional wave in Layer 5, hence their slopes on the time–distance graph are the same as that of the singly refracted wave in that layer. The intercept of the reflected-refracted line is equal to that of the once refracted line plus the delay time corresponding to a reflection through the water layer at the angle indicated. The intercept of the doubly refracted line is obviously twice that of the singly refracted line. Reflected-refracted and multiply refracted arrivals are often seen for each layer. For simplicity, only those associated with the Layer 5 wave are shown.

Note that in the time–distance graph for the case just considered, the refraction lines have been drawn solid over the region where arrivals can be received and dashed in the region of fictitious arrivals, i.e. the change from dashed to solid line occurs at the critical distance for each type of arrival and for each layer.

It is of interest now to consider the case in which the velocity in the upper layer of sediments (Layer 2) is not uniform throughout but increases with depth in the layer. From the results of many measurements, it is known that, owing to compaction, cementation, etc., the sediments do have appreciable velocity gradients, particularly in the upper few hundreds of meters. The effect of gradient in deep-sea sediments was reported first by Hill (1952) when he found that, in an area of the eastern Atlantic, no arrivals corresponding to reflections from the sea floor were observed (at low frequencies), whereas arrivals, somewhat delayed, could be accounted for as having followed paths refracted by a velocity gradient in the sediments. Velocity–depth relationships have since been determined empirically for both shallow-water and deep-water sediments (Nafe and Drake, 1957), and studies have been made on artificial compaction (Laughton, 1954), indicating velocity gradients of 0.5–2 $\sec^{-1}$ in a variety of sediment types. Although it is well known that velocity gradients are present in the water layer, they are small compared with those in sediments, and for simplicity we will assume constant velocity in the water.

Fig. 9 shows a two-layer case in which there is assumed to be a small velocity discontinuity at the water–sediment interface, and in which the velocity in the upper sediments is approximately equal to that of water at the sea floor, increasing linearly with depth beneath the bottom. To demonstrate the full effect of gradient, we first assume that the sediments are infinitely thick and that velocity increases with depth at a constant rate K sec^{-1}:

$$C = C_0 + KZ. \tag{6}$$

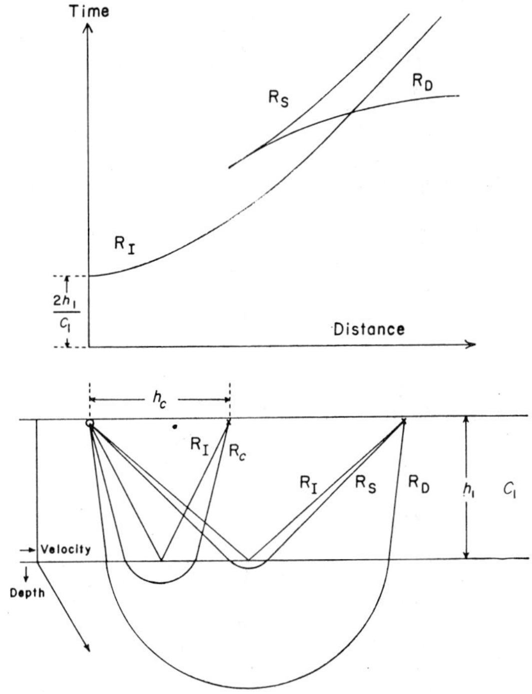

Fig. 9. Ray diagram and time–distance graph showing effect of velocity gradient in thick sediments.

With this velocity–depth structure, ray paths in the sediments are segments of circles whose centers lie at a height of C_0/K above the water–sediment interface. For a particular depth of water and for a particular value of K, there exists a critical distance, D_c, between shot and detector for which a ray originating at the surface and penetrating into the sediments can be returned to the surface. This ray is designated R_c in Fig. 9. At ranges shorter than D_c, only reflected waves from the sea floor can return to the surface. At any range greater than D_c two arrivals are returned to the surface, one which penetrates more deeply and one which penetrates less deeply into the sediments than the arrival following the path of R_c. Since much of its path is through the high-velocity sediments, the deep ray, R_D, reaches the detector earlier than the shallow ray, R_S. As shown in the time–distance graph in Fig. 9, the R_D curve approaches zero slope (infinite velocity) at large distances and the R_S curve approaches R_I. (The R_S curve will be asymptotic to R_I, asymptotic to a line parallel to R_I or will cross R_I depending on whether the velocity in the upper sediments is equal to, less than or greater than the water velocity.)

In most instances where gradients in oceanic sediments have been studied, it has been found that velocity–depth relationships of the type

$$C = C_0 + K\sqrt{Z} \tag{7}$$

or

$$C = C_0 + KZ + K'(1 - e^{-10Z}) \tag{8}$$

fit the observational data better than the linear relationship given by (6).

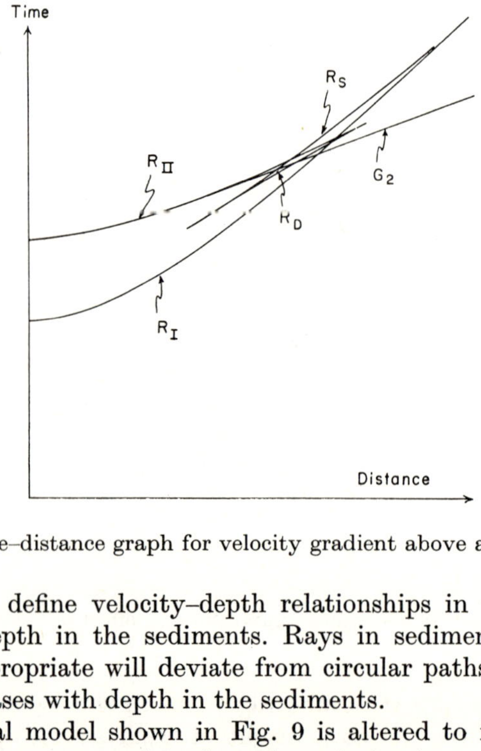

Fig. 10. Time–distance graph for velocity gradient above a discontinuity.

Both (7) and (8) define velocity–depth relationships in which the gradient decreases with depth in the sediments. Rays in sediments for which these equations are appropriate will deviate from circular paths, i.e. the curvature of the path decreases with depth in the sediments.

If the structural model shown in Fig. 9 is altered to include a boundary within the sedimentary material where the velocity increase is discontinuous, we have a model which closely approximates the upper sediments in many ocean-basin areas. Layer 2 would have a parabolic or exponential gradient and Layers 1 and 3 negligible gradients by comparison. The time–distance graph in Fig. 10 shows the relationship of the reflected and refracted waves for such a model. The lower part of the R_{II} curve is determined by waves reflected from the first sub-bottom interface. Beyond a certain range, determined by the thickness of Layer 2 and by the velocity gradient, the R_{II} curve ends. If Layer 2 is thick enough or has a sufficiently high velocity gradient, R_{II} terminates in a cusp where it joins the R_D curve, as shown in the diagram. In the absence of sufficient thickness or velocity gradient, R_{II} simply merges with R_S at the range which produces a grazing ray at the 2–3 interface.

6. Some Complications and Deviations from Simple Theory

Although geologic structures which are encountered in exploration can often be approximated by models which are relatively simple, there are usually certain complications. For example, the interfaces are seldom found to be plane boundaries. They are more apt to be undulating or broken by folding or faulting. This has the effect of making the refraction lines and reflection curves undulating or broken. Other complications arise from the fact that some layers have lateral velocity variations, or have a velocity dependence on direction of propagation. This latter effect is particularly common in laminated rocks such as shales or schists. Many other problems associated with structural or compositional inhomogeneities can easily be imagined.

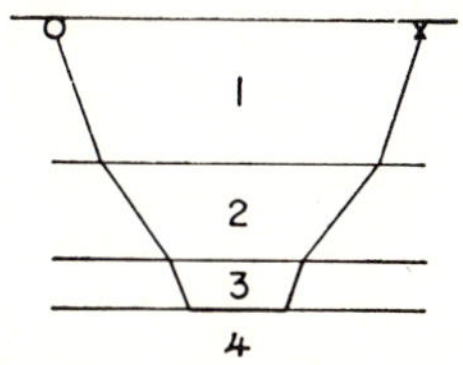

Fig. 11. Ray path for a model in which the velocity in Layer 3 is less than that in Layers 2 and 4.

Complications also arise from difficulties in identifying arrivals owing to low signal strength or to coincidence of arrival times of two or more phases. Unless there is an appreciable difference in the characteristic frequency of two arrivals which reach a detector at almost the same time, it is usually difficult to pick the beginning of the second one. Hence, the most reliable arrivals are first ones, or those which come long after the motion from earlier ones has appreciably died out (G_3, for example, in Fig. 4). In certain structures there may be layers for which the refracted waves are never the first arrivals. The reliability of detecting these "masked" layers by refracted arrivals is particularly dependent upon the closeness of shot spacing, since a single record cannot be counted on for a precise pick.

Thin layers, even if they are not masked, may be missed, or their velocity and thickness inaccurately determined by profiles in which the shot spacing is too wide. Field observations and model experiments have shown that, in a layer which is thin compared with a wavelength of the seismic energy, reliable refracted arrivals are obtained only over relatively short distances. Hence a thin layer will produce reliable arrivals over only a short portion of a profile, and close shot spacing is essential for detecting it.

A low-velocity layer between two media of higher velocity is not detectable by normal refraction techniques. Fortunately, occurrences of this type of layering are believed to be uncommon. Fig. 11 is a ray diagram for a hypothetical example. The time–distance graph for this model would give no indication of Layer 3 from refracted waves, and computation of the structure with standard formulae for refraction data would give an excessive value of thickness for Layer 2.

Owing to velocity gradients in the ocean, the R_I arrivals do not follow straight paths as shown in the preceding figures, but travel the curved paths shown in Fig. 12. For shot 1 the paths are slightly bent, but can be closely approximated by straight lines. The two rays emerging from shot 2 indicate the limits of a bundle of rays observable at that range. The deep ray grazing the bottom (RSR ray in Ewing and Worzel, 1948) marks the maximum range of a first-order reflected wave. The R_I ray paths for shot 3 are combined surface sound-channel paths and deep refracted paths (the leakage arrivals described by Hersey *et al.*, 1952). Beyond the range of the grazing ray, the R_I arrivals differ from shot to shot only in the length of the path in the surface channel. Hence, they plot on a line on the time–distance graph which is parallel, rather than asymptotic, to D.

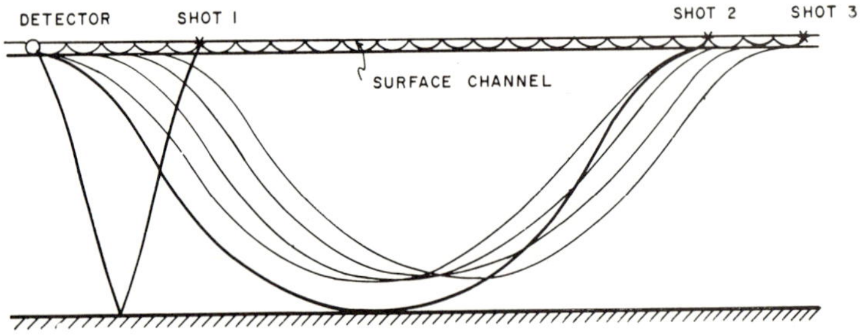

Fig. 12. Ray paths in deep water for shots and detector located in the surface sound channel. For average North Atlantic areas, shot 2 would correspond to a shot at 40–45 miles range.

The ray diagram shown in Fig. 12 is appropriate for most deep-water areas. In areas of shallower water it is possible for the long range R_I arrivals to re-enter the surface channel after having been reflected from the sea floor. In this case the R_I curve is also asymptotic to a line parallel to D. Whether the asymptote is above or below D on the time–distance graph depends upon the speed of sound in the surface channel relative to the average speed in the water beneath the channel and upon the depth of water. In high latitudes particularly, the speed in the surface channel may be lower than the mean vertical velocity, causing the R_I curve to cross line D. The existence of the surface channel depends upon the presence at the surface of a layer of isothermal water or of one in which the normal thermocline is inverted, i.e. temperature increasing with depth. Without one or the other of these temperature structures, there is no channel and the direct wave, which is so useful for measuring distance in seismic work, cannot travel for any appreciable distance between two near-surface points. In such cases, it is possible to compute distance from the reflected arrivals, but the process is time-consuming and less accurate.

Lamont Geological Observatory (Columbia University) Palisades, New York. Contribution No. 594.

References

In preparing this contribution, reference was made to the work of such a large number of authors that the text would have been unduly cumbersome had each been listed in the usual fashion. Reference has been made in a few cases to call attention to specific points concerning theory or technique. The following is a list of contributions in which most of the basic theory was developed or summarized.

Drake, C. L., C. Gaibar-Puertas, J. E. Nafe and M. Langseth, 1959. Estudios de prospeccion sismica por refraccion en el Golfo de Cadiz; Revista de Geofisica. *Inst. Nac. Geofis.*, 1–20.

Ewing, J. and M. Ewing, 1959. Seismic refraction measurements in the Atlantic Ocean basins in the Mediterranean, on the Mid-Atlantic Ridge, and in the Norwegian Sea. *Bull. Geol. Soc. Amer.*, **70**, 291–318.

Ewing, M., G. H. Sutton and C. B. Officer, 1954. Seismic refraction measurements in the Atlantic Ocean, Part VI: Typical deep stations, North America basin. *Bull. Seism. Soc. Amer.*, **44**, 21–38.

Ewing, M., G. P. Woollard and A. C. Vine, 1939. Geophysical investigations in the emerged and submerged Atlantic coastal plain, Part III: Barnegat Bay, New Jersey section. *Bull. Geol. Soc. Amer.*, **50**, 257–296.

Ewing, M. and J. L. Worzel, 1948. Long range sound transmission, in "Propagation of Sound in the Ocean". *Geol. Soc. Amer. Mem.*, 27.

Ewing, M., J. L. Worzel, N. C. Steenland and F. Press, 1950. Geophysical investigations in the emerged and submerged Atlantic coastal plain, Part V: Woods Hole, New York and Cape May sections. *Bull. Geol. Soc. Amer.*, **61**, 877–892.

Gaskell, T. F., M. N. Hill and J. C. Swallow, 1958. Seismic measurements made by H.M.S. *Challenger* in the Atlantic, Pacific and Indian Oceans and in the Mediterranean Sea, 1950–53. *Phil. Trans. Roy. Soc. London*, **A251**, 23–83.

Hamilton, E. L., 1959. Thickness and consolidation of deep-sea sediments. *Bull. Geol. Soc. Amer.*, **70**, 1399–1424.

Hersey, J. B. and M. Ewing, 1949. Seismic reflections from beneath the ocean floor. *Trans. Amer. Geophys. Un.*, **30**, 5–14.

Hersey, J. B., C. B. Officer, H. R. Johnson and S. Bergstrom, 1952. Seismic refraction observation north of the Brownson Deep. *Bull. Seism. Soc. Amer.*, **42**, 291–306.

Hill, M. N., 1952. Seismic refraction shooting in an area of the eastern Atlantic. *Phil. Trans. Roy. Soc. London*, **A244**, 561–596.

Katz, S. and M. Ewing, 1956. Seismic refraction measurements in the Atlantic Ocean, Part VII: Atlantic Ocean basin, west of Bermuda. *Bull. Geol. Soc. Amer.*, **67**, 475–510.

Laughton, A. S., 1954. Laboratory measurements of seismic velocities in ocean sediments. *Proc. Roy. Soc. London*, **A222**, 336–341.

Nafe, J. E., 1957. Reflection and transmission coefficients at a solid–solid interface of high velocity contrast. *Bull. Seism. Soc. Amer.*, **47**, 205–219.

Nafe, J. E. and C. L. Drake, 1957. Variations with depth in shallow and deep water marine sediments of porosity, density, and the velocities of compressional and shear waves. *Geophysics*, **22**, 523–552.

Nettleton, L. L., 1940. *Geophysical prospecting for oil*. McGraw-Hill, New York.

Officer, C. B., 1955. A deep-sea seismic reflection profile. *Geophysics*, **20**, 270.

Officer, C. B., 1958. *Introduction to the theory of sound transmission*. McGraw-Hill, New York.

Raitt, R. W., 1956. Seismic-refraction studies of the Pacific Ocean basin, Part I: Crustal thickness of the central equatorial Pacific. *Bull. Geol. Soc. Amer.*, **67**, 1623–1640.

2. REFRACTION AND REFLECTION TECHNIQUES AND PROCEDURE

G. G. SHOR, JR.

1. Introduction

In basic essentials, seismic refraction work at sea is quite simple and uniform in practice among the various institutions participating. One to four hydrophones (pressure detectors) are placed in the water below the depth of disturbance by wave action, and charges of high explosive are fired by a moving ship at varying ranges along a line radial to the hydrophone group. The shots are normally fired by means of a time fuse at a sufficient depth such that a gas bubble is created which oscillates at low frequency and does not blow out to the water surface. The sound of the detonation is picked up on a towed or hull-mounted detector on the firing ship and is used, with appropriate corrections for distance of the shot from the detector, as the zero time for recording. Small shots at close spacing are fired near the receiving hydrophones; as distance away increases, the charge size and distance between shots is rapidly increased.

Receiving hydrophones, generally high-sensitivity crystal or ceramic units, are supported by floats in such a manner as to be either neutrally buoyant or very slightly heavy, depending on the mode of handling. Signals from the hydrophones are pre-amplified, transmitted to the recording equipment by wire in the case of a two-ship operation or by radio on a one-ship operation, and then separated into multiple-frequency bands, amplified and recorded on an oscillograph as a "wiggly line" record. Additional information recorded on the oscillograph includes the firing mark from the shooting ship, fork-controlled timing lines and time-ticks from a break-circuit chronometer.

When the details of field procedure are examined more closely than in the brief summary above, variations in technique between institutions become evident. Many of the variations have originated from the efforts of different groups to solve the same problems independently, and when two equally successful methods have been obtained they have continued to exist. In many cases superior methods of operation originated at one institution have been generally adopted by all. Some of the differences are explicable because of differences in ship type or availability.

Published information on instrumentation and techniques has been scattered through articles on data analysis or published as technical reports. The principal reference on methods and instrumentation for seismic refraction used at the Scripps Institution of Oceanography (SIO) is by Raitt (1952); instrumentation at Woods Hole Oceanographic Institution (WHOI) is described by Dow (1952); instrumentation at Lamont Geological Observatory (LGO) by Sutton (1952) and by Luskin and Koelsch (1958). A summary of field techniques presently used at Lamont and Woods Hole appears in Officer *et al.* (1959); earlier information can be found in Ewing *et al.* (1937, 1939, 1940, 1946, 1950), Officer *et al.* (1952) and Hersey *et al.* (1952). Work at Cambridge University is described

[*MS received July, 1960*]

by Hill (1952); Russian methods are given in Galperin and Kosminskaya (1958), Andreyeva and Udintsev (1958) and Galperin, Goryachev and Zverev (1958).

2. Shooting Techniques

A great impetus to refraction work at sea occurred in 1946 when large quantities of surplus high explosives from World War II became available in many countries. These included such material as aerial bombs, mines, depth charges, cast TNT blocks and diverse types of material intended for demolition work. As most such material deteriorates when stored for extended periods, over-age material is generally dumped at sea. A sizable quantity of this surplus material was made available for scientific work instead; as a result the cost of operation was materially reduced and the techniques were generally fitted to the types of material available. These explosives have usually been varying types and packaging arrangements of TNT, Tetryl, Tetrytol and PETN.

Because all explosives have to be carried long distances aboard small ships, a great deal of work has been devoted to the dual problem of increasing signal strength and decreasing noise in the low-frequency spectrum, in which the best propagation of the refracted waves occurs. The best method found so far for obtaining the maximum signal from a given size of charge is to fire it at a depth equal to a quarter of the wavelength for the bubble-pulse frequency.

The occurrence of the bubble-pulse phenomenon was observed as early as 1925 by Weibull (1954) and has since been described by a number of investigators. Detailed study by a group at WHOI resulted in data such as those given in Fig. 1 (from Raitt, 1952, based on work by Arons and Yennie, 1948), in which it is apparent that the highest intensity occurs at the bubble-pulse frequency. The value of this frequency is determined by the equation (for TNT):

$$f = \frac{(H+33)^{5/6}}{4.36W^{1/3}},$$

where H is in feet, W in pounds and f in cycles/sec, from which one can determine the depth of detonation of a charge of given weight for an observed bubble-pulse frequency. In combination with the requirement of quarter-wave depth, this gives

$$\begin{aligned} H &= C_0/4f \\ H(H+33)^{5/6} &= 1.09C_0W^{1/3}, \end{aligned}$$

where C_0 is the velocity of sound in sea-water, for optimum shooting. A curve for f and H as functions of W for normal sound conditions is given in Fig. 2.

In some cases it has been necessary to place charges at depths other than the optimum. Certain types of explosives such as 300-lb and larger depth charges sink at such a rapid rate that they reach their optimum depth before a

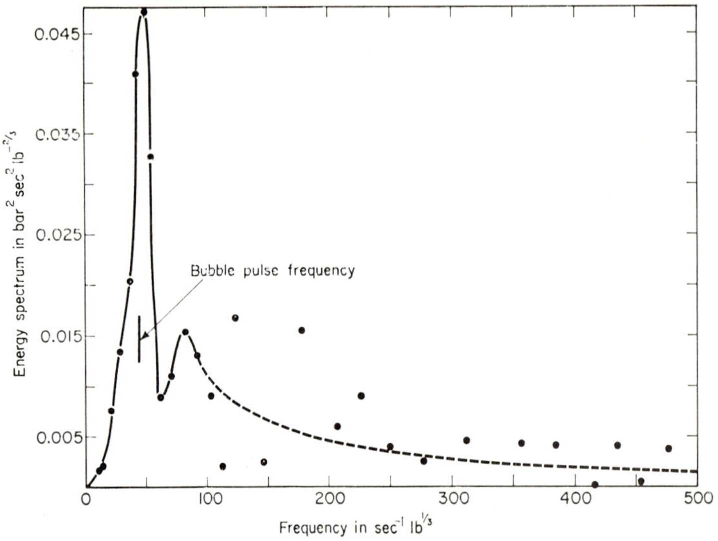

Fig. 1. Low-frequency part of the Fourier energy spectrum of an explosion at 500-ft depth. (After Raitt, 1952.)

slowly moving ship can reach a safe distance away. The damaging effects are accentuated in shallow water, and the larger charges must often be fired at greater depths than is desirable. At ranges short of that at which the first refracted wave is detected, a series of small charges is frequently fired to get a wide-angle reflection profile; in this case the amount of energy is usually more than adequate, even with quite small charges, and efficiency is sacrificed so as to obtain a less complex arrival by firing the charges so near the surface that

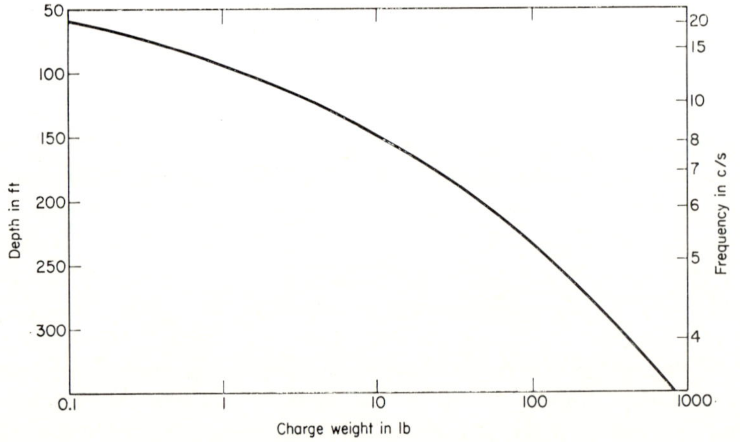

Fig. 2. Optimum charge depth and frequency as functions of charge weight. (After Raitt, 1952.)

the bubble blows out. This is usually done by fastening the charge to a balloon or air-filled plastic bag.

In the early stages of post-war work and in most of the pre-war operations, charges were detonated by electric caps by a firing impulse from the shooting ship, or in the case of depth charges, by depth-sensitive detonators. Since about 1949 almost all operations have been conducted using non-electric caps and slow-burning time fuses for detonation; this eliminates problems of retrieving the firing line between shots or slowing the shooting ship to fire. Reliability of the method is generally better than 97% if properly waterproofed fuse is used. The lower accuracy of determination of firing time in the fuse method (0.01 sec instead of 0.001 sec or better) has been found to be unimportant in deep-sea work.

When electric detonation is used, the firing impulse can be recorded or sent by radio from the shooting box to the receiving ship directly. When time fuses or pressure detonators are used, a different technique is required. Shots of sizes varying from $\frac{1}{2}$ to 300 lb are fired on a pre-determined schedule set by the operator on the receiving ship. The shooting crew consists of an operator and a shooter. The operator announces the shot number, time, weight, fuse length and whether it is to float or sink. On the scheduled minute he instructs the shooter to light the fuse. The shooter lights the fuse, drops the charge over the side (with the aid of a tilt-table for the largest charges), and calls, "Mark". Operators at both ends of the radio circuit start stop-watches at the mark, and shortly before the expected time of detonation start their recording equipment. The operator on the shooting ship switches the radio circuit to put the signal from the fathometer audio output or a special hull-mounted hydrophone or a towed hydrophone on the radio in place of the microphone; he also records the same signal along with time-ticks from a marking chronometer on a pen recorder. When the charge fires, it produces a record aboard both ships of the direct wave from the shot, and under favorable conditions also records the bubble pulse, one or more bottom reflections, and the wave reflected once at the water surface and once at the bottom. When the fathometer output is used, the details other than firing mark and bottom reflection are usually lost because of the narrow pass band of the fathometer receiver. Both operators stop their watches at the time of detonation, and the stop-watch time multiplied by the ratio of ship speed to the velocity of sound in the surface layer (the ship's Mach number!) is used as the firing time correction. If the sinking rate of the charge approaches the ship speed, an additional correction for the slant path is necessary. Usually this correction is less than 0.01 sec and can be neglected. In case radio transmission is missed, shot time can be recovered from the shooting ship's firing record using chronometer corrections determined from the shots immediately before and after. Chronometers are normally checked against WWV time signals at least daily; when radio communicatious are unavailable or poor, profiles are sometimes shot on a fixed time schedule with firing marks recorded against chronometer only, with more frequent checks of WWV signals.

Data recording on the shooting ship is relatively simple. The operator keeps a log of shot number, time of day, weight, explosive type, fuse length, whether flotation balloons are used, sinking time, water depth at the time the shot is dropped, ship course and speed, and time of passing abeam of the receiving ship. If necessary he marks the fathometer tape at the time the shot is dropped; in some systems a mark is created when the fathometer ping is automatically turned off as he shifts his radio circuit, and the time of detonation marks itself as a black line. The operator also identifies and retains all firing mark records. Ship's bridge personnel provide a shooting track with positions of first and last shots, all course and speed changes and navigational fixes, and the range and bearing of the receiving ship when passing. This track usually has to be reconciled with that provided independently by the receiving ship to make the abeam positions agree.

Explosive need depends markedly on the length of line to be shot. In current practice a deep-sea station with penetration to the mantle (two profiles, each about 40 miles long) requires about 1600 lb of TNT in about 80 shots. This total includes many small shots fired at short range for a reflection profile; the major weight of explosive is used in the few distant shots fired for mantle arrivals. These latter shots are normally between 50 and 100 lb of high explosives each; in bad weather or when signals are weak, 300-lb shots are not uncommon. The equation

$$\text{Maximum range in kilometers} = 7\sqrt{(\text{charge weight in pounds})}$$

given by Raitt (1952) for operations in the tropical Pacific continues to be valid in most areas. For work on the continental shelf as much as 4000 lb per station may be required if mantle arrivals are sought. Short lines in shallow water for studies of the shallower layers can be fired with a very few pounds of explosives.

3. Receiving Techniques

Receiving techniques have followed three separate patterns. Although many persons have contributed their share, the principal developers of these methods have been Maurice Ewing of Woods Hole and the Lamont Geological Observatory, Russell Raitt of the Marine Physical Laboratory of Scripps Institution of Oceanography, and Maurice Hill of Cambridge University. The WHOI–LGO methods have been used by Lamont, Woods Hole, Hudson Laboratories, and collaborating groups; the SIO methods have been followed by Scripps, Pacific Naval Laboratory, and the U.S.S.R. Institute of Oceanology and Institute of Physics of the Earth. Hill's method has been followed by Cambridge and the National Institute of Oceanography. It is described separately in Chapter 3. The methods discussed here are necessarily those which the author knows best from personal experience, discussions with the users, and the published literature.

In the SIO method, the receiving ship lies to or steams slowly into the wind, and three or four hydrophones on neutrally buoyant suspensions are laid out at distances of 300 to 2000 ft from the ship, as shown in Fig. 3a. Each hydrophone has been previously balanced by use of an aluminum float and small lead weights so that it is very slightly buoyant at the surface and will be as nearly as possible neutral at operating depth. A 50-ft leader cable without built-in buoyancy is supported to neutral buoyancy by small aluminum floats distributed along its length. In field use the cable hangs in loops between the small floats, creating a mechanical filter to eliminate disturbances propagated along the cable. The end of the leader is fastened to a heavy weight. From the

(a)

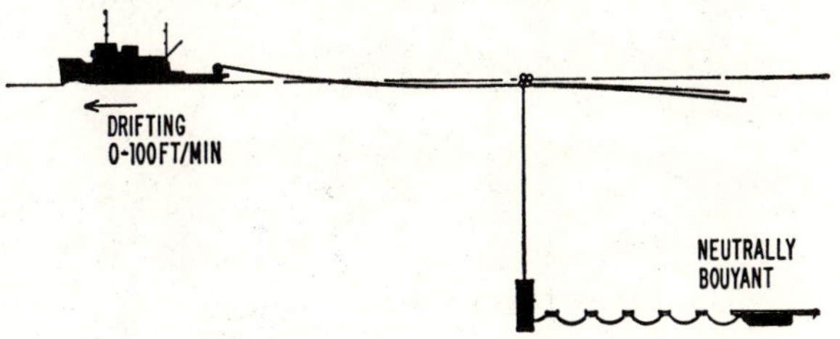

(b)

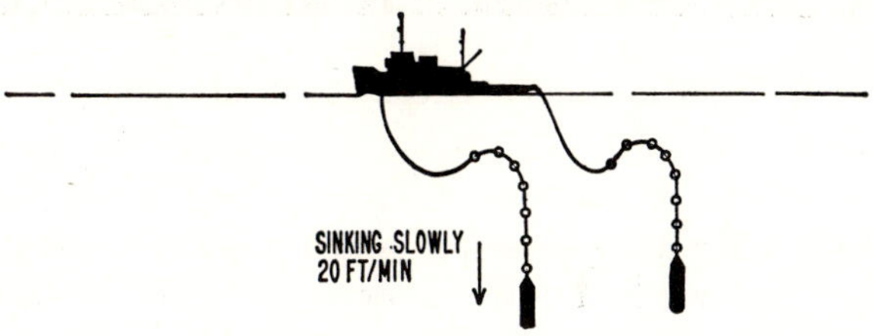

Fig. 3. Hydrophone suspensions: (a) SIO method; (b) LGO–WHOI method.

weight a neutrally buoyant cable leads up 200 ft to the water surface where it is fastened to a group of small plastic floats and then back to a cable-puller or a winch on the ship. The hydrophone system is shown in Fig. 4.

During reception of a seismic station other operations such as coring, heat-flow measurements, underwater photography, hydrographic casts, bathythermograph observations, and normal laboratory and ship work are continued so that in all cases the ship's auxiliary power is kept on. In windy conditions the main propulsion engines are on, and the ship heads into the wind with the screw turning over slowly to keep the hydrographic wire vertical and the hydrophone cables slack. If ship drift becomes too great and the hydrophone cables pull tight the mechanical filtering action is lost, the hydrophones rise toward the surface, and wave and flow noise develop. In such cases, if steaming

Fig. 4. Photograph of SIO hydrophone system.

is not adequate or if for some reason it is not possible to steam (as in the case where the dredge is down and currents are pulling the hydrophones), cable is paid out from the winches into a heap on deck and paid out rapidly at shot iime for the most distant shots only.

A block diagram of the SIO shooting and receiving system is shown in Fig. 5. The hydrophones used are normally either Brush Electronics Model AX-58C or Research Manufacturing Co. Model R-100. The former uses Rochelle salt crystals immersed in castor oil, the latter barium titanate ceramic cylinders cast in plastic. Both types produce a signal of approximately 100 μV per dyne/cm^2 for signals between 2 c/s and 20 kc/s. The output is at high impedance and must be matched to the cable with a cathode follower pre-amplifier (current amplifier) at the hydrophone; the SIO, LGO and WHOI pre-amplifier circuits are shown in Fig. 6. Batteries to operate the pre-amplifier are located not in the hydrophone but at the weight (actually a brass box filled with batteries) 50 ft away; this eliminates the need to open the hydrophone case or to re-balance at times of battery change. Hydrophones turn on when the battery case is right-side-up. The least signal from sea noise in calm weather in the low-frequency (2 to 20 c/s) range of interest for seismic refraction is of the order of 10 μV hydrophone output; first stage noise must therefore be

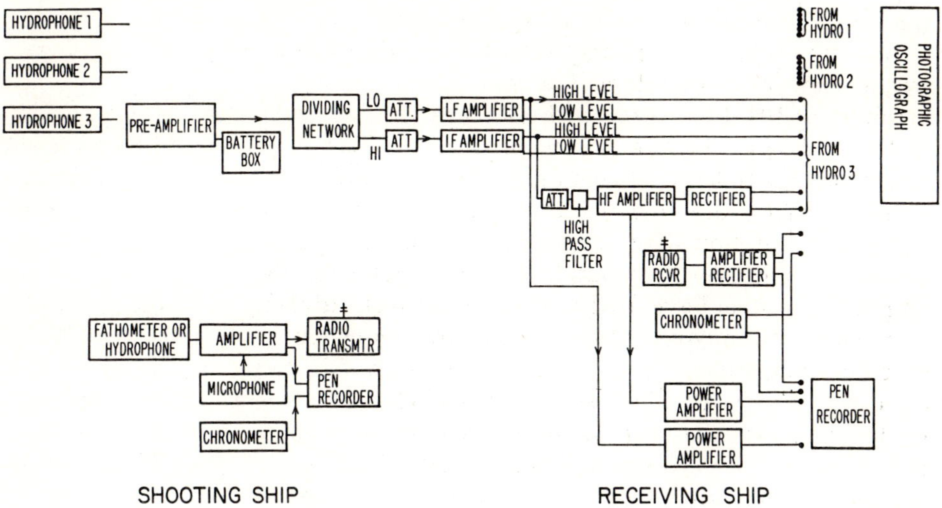

Fig. 5. Block diagram; SIO shooting and recording system.

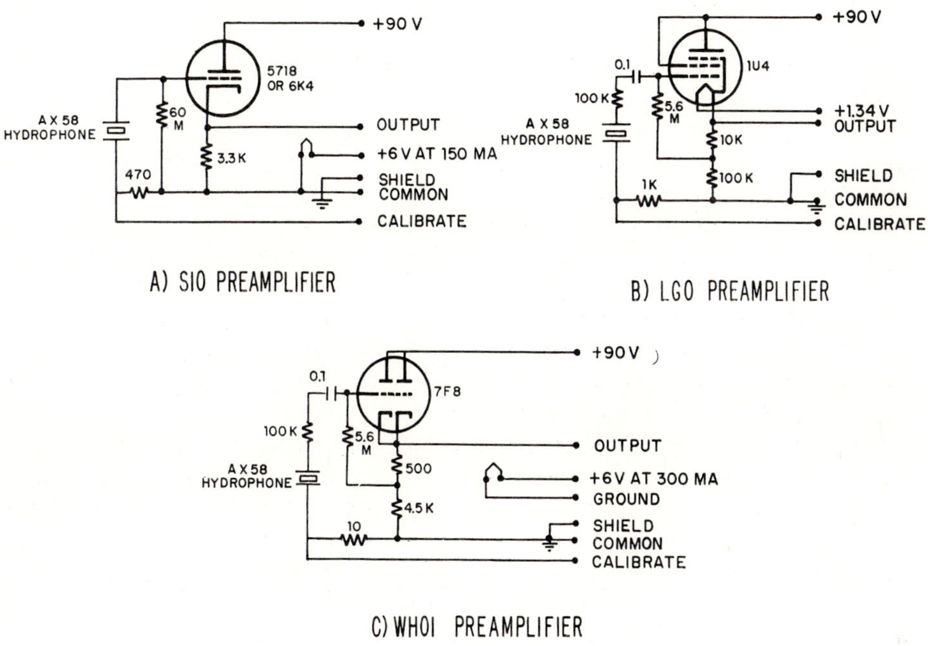

Fig. 6. Hydrophone pre-amplifiers.

kept well below this level. Direct waves through the water from nearby shots (and even from distant shots in areas of good sound propagation conditions) frequently create hydrophone output voltage in excess of 1 V in the higher frequencies; it is, therefore, desirable to have a dynamic range in the pre-amplifier of 100 dB or more with as nearly linear response as possible.

At the shipboard end of the cables the signal is fed to an input panel which uses blocking condensers to eliminate the d.c. component from the pre-amplifier signal and provides test points for checking pre-amplifier voltage and for inserting calibration signals. The signals from all hydrophones are then distributed through step attenuators to the amplifiers. Various versions of the amplifying system have used cross-over networks dividing at 20 c/s, paralleled amplifiers with low-pass filters for the low-frequency trace, and narrow-band amplifiers in parallel. The present system used at SIO feeds the signals through a cross-over network which sends signals below 20 c/s to the "LF" amplifiers in the block diagram (Fig. 5) and signals above 20 c/s to the "IF" amplifiers. Signals near 20 c/s are strongly suppressed. The choice of 20 c/s for the cross-over stems from two causes: first, the energy in the refracted arrivals lies in greater part below this frequency, and secondly, this is the mechanical frequency of most of the water-borne sound created by rotating machinery on the ships that have been used. If ships with machinery of other speeds are used, the cross-over frequency must be changed to keep the maximum attenuation at the noise frequency. Signals are separated before amplification to reduce the range of linear amplification needed to prevent intermodulation, which, if present, will cause the direct water wave (strongest in high frequencies) to produce a spurious signal on the low-frequency traces, masking the bottom refracted arrival on shots at certain ranges. The low-frequency signals are given up to 40 dB of amplification, and then two outputs at levels 26 dB apart are carried out for recording. The amplifiers are battery powered to reduce hum; great care must be taken to avoid ground loops and nearby sources of electrical noise. High dynamic range, linearity, freedom from blocking after overload, low internal noise, stability, good low-frequency response and reproducibility of gain settings are features of these amplifiers. IF amplifiers are in most respects identical with those used for the LF channels, except for the addition of rejection filters for 60 c/s electrical pick-up. The outputs of the IF's go both to the recorders at two amplitude levels and to high-frequency (HF) amplifiers that pass only frequencies above 600 c/s. The HF output is then rectified and recorded at two levels.

Outputs of the amplifiers are recorded with 230 c/s galvanometers on a 17-channel photographic oscillograph along with the radio-transmitted signal from the shooting ship and second marks from a chronometer. A secondary time base is provided by time lines controlled by a 100 c/s fork. In addition, a monitor record is taken with a pen-and-ink recorder using two pens operated by galvanometers and one operated by a relay. This requires the use of a set of auxiliary power amplifiers. With proper switching the operator can record on this the chronometer ticks, the firing mark from the radio, the low-frequency outputs of two of the hydrophones on the first arrivals, and high- and low-frequency outputs from one of the hydrophones on the water-borne arrivals. Information from the monitor records is used by the recording operator to check for instrument trouble or noise, and to decide on changes in charge size to be called to the shooting ship. A running plot of rough readings from the monitor records is maintained during the run for preliminary interpretation. Whenever a roll of recording paper has been run through the oscillograph, it is removed and developed to check on the quality of the photographic recording on all channels and to permit the beginning of data reduction aboard ship.

The recording system used at SIO is completely transistorized except for the pre-amplifiers.

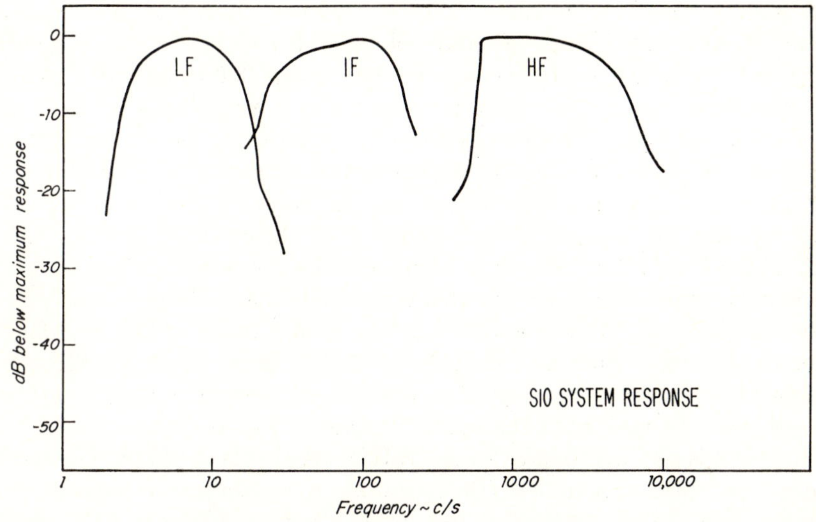

Fig. 7. Response of SIO system from hydrophone output to oscillograph trace; dB relative to maximum response of each channel.

Auxiliary information that must be noted includes all of the information listed for the shooting ship plus attenuator settings, depth at the receiving ship and a running record of instrument connections and settings. At the conclusion of each station, electrical calibrations are made on one more of the hydrophones by inserting a series of calibration signals (from 2 c/s to 20 kc/s) to the calibration resistor in the pre-amplifier and taking a brief record. This gives the total response of the system excluding the hydrophone element; an example of the response of the system from hydrophone output to oscillograph trace is given in Fig. 7.

In the system of recording developed at Lamont and Woods Hole, most of the differences in procedure stem from the practice of silencing the ship during receiving periods. The receiving ship lies to, securing all machinery possible and operating essential receiving equipment and radios from the ship's batteries and from battery-driven a.c. converters located near the equipment they supply. With the ship thus silenced the need for streaming the hydrophones away from the ship and for filtering out the 20 c/s mechanical noise and 60 c/s electrical pick-up disappears. This was the first method used; the SIO system was developed to meet the problem of ships that are difficult to silence and the presence of other scientific programs requiring power during stations.

In the LGO–WHOI system, hydrophones are identical with those at SIO; the pre-amplifiers are similar but not identical, as shown in Fig. 6, B and C. In a recent version of the LGO pre-amplifiers the filament batteries have been put into the hydrophone and power is turned on when the plate voltage is sent down the cable. The hydrophones are not balanced; instead, a series of five or six floats is fastened to the cable (which is heavier than water) at increasing distances of 10 to 60 ft from the hydrophone. The floats are so arranged that the system has slight positive buoyancy when about 200 to 300 ft of cable is in the water. Prior to arrival of the refracted wave, extra cable is paid out to make the system heavy and the hydrophone then sinks slowly during arrival of the shot, as shown in Fig. 3b. Submergence of the floats eliminates wave noise; the slack bight of wire prevents tugging from the ship, and flow noise is greatly reduced by the slow steady sinking of the hydrophone. In order to determine the hydrophone depth, a bourdon-tube depth gauge is placed on the hydrophone or near it on the cable in a separate housing. Two hydrophones are usually used, one at the bow and one at the stern; more would be difficult to handle. Cables are pulled and slacked by hand. Systems of amplifying and recording have varied. Broad-band amplifiers of a number of different types have been used in the past; the common feature has been the use of several frequency bands and amplitude levels. WHOI has in recent years standardized on their "Suitcase" amplifier (Dow, 1952), a battery-powered, accurately calibrated, wide-band, high-gain, low-noise, dual-channel amplifier designed for general underwater sound use. It has a pass band between 6 dB points of 8 c/s to 50 kc/s; interstage plug-in filtering permits selection of any desired response band. A companion power amplifier drives a hot-wire (Sanborn) recorder. All frequency bands used in the SIO system can be covered by the WHOI "Suitcase" system. In addition, on some operations, a set of standard land-type multiple-channel refraction amplifiers with variable filtering has been used on the output of the "Suitcase" to provide as many as 12 different frequency and amplitude channels per hydrophone. Tape recording of the seismic signals has been done with the "Suitcase" system, primarily for analysis of the reflected and direct water arrivals.

Work at Lamont since about 1952 has been done with an amplifier system with three frequency bands (3.5–135 c/s, 30–250 c/s and 620–14,000 c/s rectified), each at five output levels, described by Sutton (1952). The most recent LGO instrumentation has been a completely transistorized set of portable amplifiers providing three frequency bands at two output levels for each of two hydrophones. The relative frequency response of the LGO transistor amplifiers is shown in Fig. 8, as given by Luskin and Koelsch (1958).

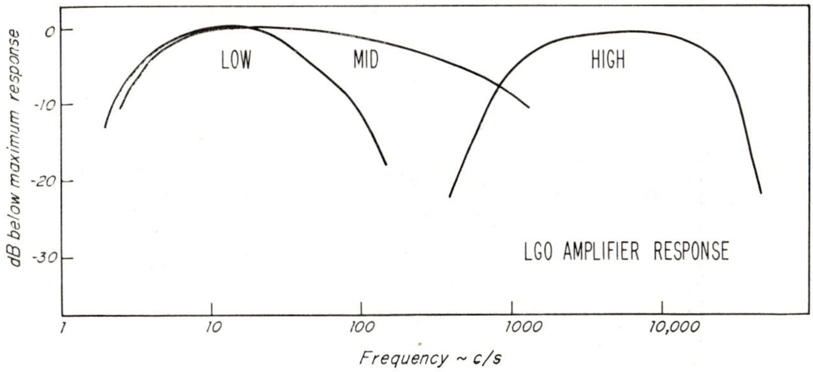

Fig. 8. Response of LGO amplifiers relative to maximum for each channel.

4. Plan of Operations

The lines of shots described earlier can be combined in a number of ways to make a seismic station. Combinations of linear profiles in common use are shown in Fig. 9 and may be classified as:

1. Single profiles
2. Split profiles
3. Reversed profiles
4. Overlapping profiles
5. Compound profiles
6. Fan profiles
7. Area surveys

A. Single Profile

The single profile is, as its name implies, a single line of shots extending radially from the receiving ship. Variations in layer thickness along the line of shooting cannot be detected in this method and will give false velocity and depth determinations; this method is, therefore, avoided if possible.

B. Split Profile

The split profile consists of two single profiles forming a straight line with the receiving ship in the middle. If there are no large variations in dip within the length of the profile, this method will give velocity, layer thickness, and an indication of dip. It is most applicable to work in areas of broad structure such as the normal ocean basins and the great rises. In areas of complexity, such as the continental shelves, trenches and seamounts, changes may take place within the length of the profile that will make interpretation difficult or impossible. Despite the apparent disadvantages, the advantages are many. The method requires only a single set of personnel and equipment for receiving, and only one ship need carry explosives. Safe and legal handling and stowage of large quantities of explosives at sea is not simple, and on many ships nearly impossible. Similarly, the maintenance of receiving equipment in the marine environment presents many problems. The two ships never cover the same track, making for better bathymetric coverage in reconnaissance areas (of

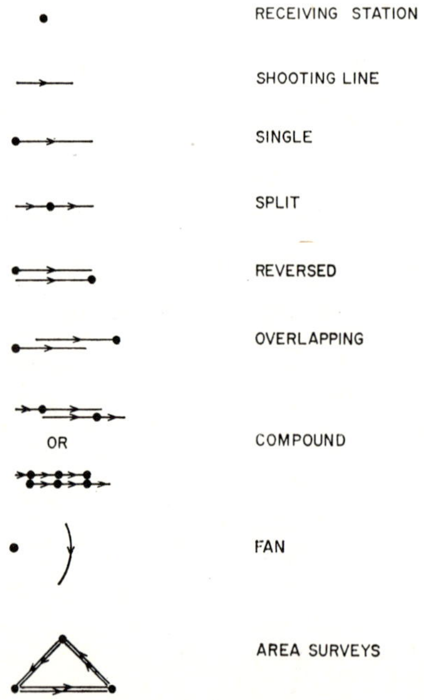

Fig. 9. Station arrangements.

which the oceans have many). Programs requiring time on station can be concentrated on the receiving ship while the shooting ship carries equipment used underway such as a magnetometer and a gravimeter. Split profiling has been done primarily by the SIO group.

C. Reversed Profile

Reversed profiling requires either that two ships be used each of which is capable of both shooting and receiving or that the ships interchange positions at the end of each run, losing considerable time. Determination of dip and elimination of errors due to structural changes along the line of profiling is improved, especially if second arrivals can be read from the records. If only first arrivals can be used, as is too often the case, true reverse coverage of arrivals from each layer is not normally obtained. As used extensively by LGO and WHOI, one ship lies to on station while the second shoots a profile, usually about 40 miles long on deep-sea lines where arrivals from the mantle are being sought. As soon as the final shot of the profile is fired, the shooting ship stops and launches hydrophones; the receiving ship gets underway and shoots on the same course up to the new receiving position. Ship drift on station normally amounts to several miles and is rarely known in amount until after the fact; profiles are, therefore, rarely truly reversed, forming instead a parallelogram. It can be assumed if the shooting lines are parallel that the velocities obtained will be true, and any mis-match in the reverse times is normally blamed on drift. On very long profiles in complex areas lateral changes in velocity may be pronounced in the shallow layers so that the velocities and thicknesses obtained are at best averages; in such cases, where time allows, resort is sometimes made to compound profiling.

D. Overlapping Profiles

Overlapping profiles are essentially incomplete reversed profiles so arranged that refraction arrivals from the deepest layer are received at both ends of the line from shots with the same geographic position. This permits determination of mantle velocity from plots of arrival-time differences to eliminate effects of variation in thickness of the shallower layers. If carried out to full length, these would be very long reversed profiles; for a given amount of time and explosives, reversed control is sacrificed to get a better velocity determination on the mantle.

E. Compound Profiles

Compound profiling involves combinations of the foregoing; in one form, used at SIO on expedition MUKLUK, it involved shooting reversed profiles on the ends of which were attached short unreversed lines to get a more positive check on the sediment and second-layer velocities. Each station then consisted of four shooting lines, two of which were reversed. A more elaborate version of this was the combined LGO–WHOI–Hudson Labs.–Texas A & M MOHOLE expedition of 1959, in which four ships participated, with three receiving along a straight line at any given time and the fourth ship shooting from a point short of the first ship to a point beyond the third. At the conclusion of each run the shooting ship stopped and became a receiver, while the first receiving ship in line picked up its hydrophones and became a shooting ship. This system can give reversals on all layers; it has as its only drawback the expense and difficulty of co-ordinating a large number of ships and personnel.

F. Fan Profiles

In fan profiles the shooting line never passes the receiving position. This is normally done in conjunction with a profile of another type and is a correlation method to determine variations in structure over a larger area. The shooting ship maintains constant distance from the receiving ship and varies its azimuth.

G. Detailed Area Surveys

Detailed area surveys require continuous profiling. In this case reversed profiles are placed end-to-end either along a straight course or in a series of closed loops.

5. Miscellaneous Variations

In addition to the methods described above, a number of variants have been used or are under development for special purposes. One of the most important of these is shooting and recording on the sea floor in deep water. This was attempted with some success at Woods Hole in 1937–40 (Ewing *et al.*, 1946) using a self-contained shooting and recording system that was dropped to the bottom, fired its charges and took its own records, and floated back to the surface when it was done. The project was plagued with instrumental problems and was halted by World War II. The data to be gained from such a system are of great importance, for the layers beneath the ocean on which we have the least information are those in the first kilometer below the bottom, refracted arrivals from which are infrequently received. A bottom-shooting and/or recording system would improve the information on these layers. Recent development work on such a system has been in progress at LGO.

Systems for performing refraction work from a single ship have been used almost from the beginning. LGO and WHOI have used the "whaleboat" method, in which shots were fired on fixed schedule from a small boat and the firing time recorded against a chronometer mark. Such a boat has been used at SIO using radio communications with the ship. Such operations are practicable only in areas of calm weather. The most successful single-ship operation is that described in this volume by Hill (Chapter 3), in which sonobuoys have been used for the receiving station.

A radically different method of operation, adapting the methods used in oil-exploration seismology to deep-sea problems, has been developed by Savit *et al.* (1962) of the Western Geophysical Company. In this method, the shooting ship stays at a fixed position, and the receiving ship tows a long streamer with 24 groups of hydrophones. Shots are fired each time the streamer has been moved ahead its own length. Low-impedance hydrophones without preamplifiers are used. The signals are passed through wide-band amplifiers and are tape recorded on a 24-channel tape recorder for later playback with filtering on to a variable-area record. This method of operation gives more data points per mile along each profile and permits the use of automatic data-handling systems. The disadvantages are the high initial cost of the equipment and the difficulty of reducing noise in the towed hydrophone system to a level that will permit recording arrivals from the mantle at long range.

6. Reflection Methods

Two major types of reflection work have been done in the deep sea: variable-angle reflection shooting and vertical reflection work. The former has been done generally in conjunction with refraction lines, although a few special reflection profiles have been performed, primarily by Officer (1955, 1955a). The latter has been of two types, continuous profiling (described by Hersey, Chapter 4) and point recording. Reflection shooting in the deep sea has attained neither the degree of sophistication apparent in shallow-water oil exploration work nor the general use of the deep-sea refraction work, for reasons that will become apparent below.

The problems of reflection work are primarily detection and classification. Detection of sub-bottom reflections requires first an adequate signal source to overcome the background. This is easily obtained with small explosive charges using a quiet hydrophone from a motionless ship; it becomes more difficult when a moving ship and/or non-explosive source is used. The second requirement is a method by which reflections from below the sea floor may be identified amid the signals created by reflections from various points on an irregular sea floor. This second requirement impinges on the problem of classification. The sub-bottom reflection must be distinguished from the bottom reflection, and one sub-bottom reflection must be distinguished from another where interference exists. The earliest method of detection of seismic reflections on land depended solely on amplitude; every large new arrival was identified as a reflection. In a few extremely favorable areas such methods gave usable results (see Weatherby, 1949). Similarly, at sea such methods can be used occasionally. The land work has shifted over to identification on the basis of correlation of arrival time between closely spaced recordings, both at various angles from a single shot and between nearby shots at vertical incidence; the trend in the work at sea is the same.

In any deep-sea refraction line, reflections from the sea floor are recorded and reflections from below the bottom may be present. Under favorable circumstances many useful data can be obtained by firing a closely spaced series of small charges at ranges short of that at which the first refracted arrival can be detected. Two methods of firing are used; to avoid complexity of the arrival caused by the bubble pulses and the surface reflections, charges are frequently fired at depths so shallow that the bubble blows out to the surface before any bubble pulses can occur. To provide low-frequency energy for deep penetration into the bottom, charges must be fired at greater depth. Frequently charges at the surface and at depth are alternated (see Hersey *et al.*, 1959). From variable-angle reflection profiles such as these, the depth of the reflecting layer and the velocity above the reflector can be obtained. Recording is normally by the same method as the refraction profile, with three frequency bands recorded on a photographic oscillograph. Distinction of the sub-bottom arrival from topographic echoes is done primarily by frequency content; echoes appearing on the high-frequency as well as the low-frequency traces are topographic; echoes on the low-frequency traces only are sub-bottom. Officer (1955, 1955a) has tape recorded reflection arrivals and played them back through narrow-band filters for this same purpose. For such a purpose the tape recorder used must have good response in the low-frequency bands.

Reflection records at vertical incidence can be taken from a single ship with little effort. The first extensive program of this sort was performed by Weibull (1954) on the Swedish Deep-Sea Expedition, 1947–1948; his method was unique and has not been copied extensively. He distinguished between sub-bottom reflections and topographic reflections by firing several shots at different depths below the surface; only for the sub-bottom reflection did the time-lag after the bottom reflection remain a constant for shots at different depths. Interference and complication of the record by bubble pulses was eliminated by firing the shots at depths so great that the bubble-pulse interval was much shorter than the bottom-to-sub-bottom interval for which discrimination was desired. Concurrently, the surface reflection was made so late that it normally occurred after all readable sub-bottom reflections were past. This method was performed with pipe-encased charges with a weighted nose. They were fired by means of a special pressure-sensitive detonator. When it was desired to fire charges on bottom a delay disk of sintered copper was placed between the pressure detonator and the sea, delaying build-up of pressure in the detonator. Detection was with a wide variety of relatively insensitive near-surface hydrophones. Recording was done by photographing the trace on an oscilloscope and by means of a wire recorder with subsequent playback.

Independently and during approximately the same period, Hersey and Ewing (1949) developed a method more akin to that used in the refraction work. Shots of 1 lb of TNT were fired at depths of 100 to 300 ft in the earlier part of the work and at 2 or 3 ft later. A hydrophone at depths varying from 150 to 1000 ft fed an audio-amplifier and an octave-band filter. Traces were photographed from an oscilloscope. The principal energy recorded was in the frequencies above 100 c/s. Since that time further development has led to an underway shooting method in which the hydrophone is towed behind the ship and the cable is slacked at the time of each shot. Many thousands of reflection shots have been taken in this manner. On-station reflection shots with charges fired at depths of 150 ft and hydrophones at the same depth were used by Shor (1959) in an area of the east Pacific where a strong reflection occurs late enough to be unaffected by the bubble pulses and surface reflections.

Development of improved reflection methods is progressing rapidly; major changes have occurred between the writing of this paper and its return in proof. Ewing and Tirey (1961) have developed a system of recording in which the firing of each shot starts the rotation of a recording drum on which are recorded each peak and trough of the returning signal as a black dot. The hydrophone and amplifiers used are the same as those of refraction work; the hydrophone is towed behind the ship, and the towing wire is slacked at each shot to allow the hydrophone to sink slowly through the water. With many closely spaced shots, one obtains a record, similar to that of a fathometer, which permits rapid correlation. This system gives consistent penetration to the base of the low-velocity sediments; sometimes horizons within the sedimentary column are also recorded.

Extensive work has been done with non-explosive systems for under-way reflection work. One electromechanical system is discussed by Hersey in this volume. Several versions of underwater arc discharge systems have been used with varying success. One described by Shor *et al.* (in press) has produced echoes from within the sedimentary column and from the base of the sediments in deep water. All these systems are limited by the noise produced by the motion of the hydrophones through the water; attempts to overcome this have included slacking a single hydrophone, or the use of multiple-hydrophone streamers to cancel noise, or systems by which the echoes from several consecutive signals are added before recording. Future reflection work in the deep sea will undoubtedly be done with high-repetition-rate systems which record in the style of a fathometer whether the source of the energy be high explosives, gas, compressed air, or electricity.

References

Andreyeva, I. B. and G. B. Udintsev, 1958. The structure of the bottom of the Japan Sea according to data from investigations by the expedition on the VITYAZ. *Izvest. Akad. Nauk S.S.S.R., Ser. Geol.*, No. 10, 3–20.

Arons, A. B. and D. R. Yennie, 1948. Energy partition in underwater explosion phenomena. *Revs. Modern Physics*, **20**, 519–536.

Dow, W., 1952. Three channel amplifier for use in refraction and sound measurements at sea. (Unpublished manuscript) Woods Hole Oceanographic Inst. Ref. 52–10.

Ewing, M., A. P. Crary and H. M. Rutherford, 1937. Geophysical investigations in the emerged and submerged Atlantic coastal plain, Part I: Methods and results. *Bull. Geol. Soc. Amer.*, **48**, 753–802.

Ewing, J. I. and G. B. Tirey, 1961. Seismic profiler. *J. Geophys. Res.*, **66**, 2917–2927.

Ewing, M., G. P. Woollard and A. C. Vine, 1939. Geophysical investigations in the emerged and submerged Atlantic coastal plain, Part III: Barnegat Bay, New Jersey, section. *Bull. Geol. Soc. Amer.*, **50**, 257–296.

Ewing, M., G. P. Woollard and A. C. Vine, 1940. Geophysical investigations in the emerged and submerged Atlantic coastal plain, Part IV: Cape May, New Jersey, section. *Bull. Geol. Soc. Amer.*, **51**, 1821–1840.

Ewing, M., G. P. Woollard, A. C. Vine and J. L. Worzel, 1946. Recent results in submarine geophysics. *Bull. Geol. Soc. Amer.*, **57**, 909–934.

Ewing, M., J. L. Worzel, J. B. Hersey, F. Press and G. R. Hamilton, 1950. Seismic refraction measurements in the Atlantic Ocean basin, Part 1. *Bull. Seism. Soc. Amer.*, **40**, 233–242.

Ewing, M., J. L. Worzel, N. C. Steenland and F. Press, 1950. Geophysical investigations in the emerged and submerged Atlantic coastal plain, Part V: Woods Hole, New York and Cape May sections. *Bull. Geol. Soc. Amer.*, **61**, 877–892.

Galperin, E. I., A. V. Goryachev and S. M. Zverev, 1958. Crustal structure researches in the transition region from the Asiatic continent to the Pacific. The Pacific Geologo-geophysical Expedition U.S.S.R. Academy of Sciences. (XII Seismology) No. 1. Publishing House of Akademia Nauk, U.S.S.R., Moscow.

Galperin, E. I. and I. P. Kosminskaya, 1958. Characteristics of the methods of deep seismic sounding on the sea. *Bull. Acad. Sci. U.S.S.R., Geophys. ser.*, No. 7, 475–483.

Hersey, J. B., E. T. Bunce, R. F. Wyrick and F. T. Dietz, 1959. Geophysical investigation of the continental margin between Cape Henry, Virginia, and Jacksonville, Florida. *Bull. Geol. Soc. Amer.*, **70**, 437–466.

Hersey, J. B. and M. Ewing, 1949. Seismic reflections from beneath the ocean floor. *Trans. Amer. Geophys. Un.*, **30**, 5–14.

Hersey, J. B., C. B. Officer, H. R. Johnson and S. Bergstrom, 1952. Seismic refraction observation north of the Brownson Deep. *Bull. Seism. Soc. Amer.*, **42**, 291–306.

Hill, M. N., 1952. Seismic refraction shooting in an area of the eastern Atlantic. *Phil. Trans. Roy. Soc. London*, **A244**, 561–596.

Luskin, B. and D. E. Koelsch, 1958. Transistorized seismic equipment for refraction measurements at sea. (Unpublished manuscript) Lamont Geological Observatory, Tech. Rep. No. 17.

Officer, C. B., 1955. Geologic interpretation of a series of seismic reflection profiles from Bermuda to the continental margin. *Deep-Sea Res.*, **2**, 253–260.

Officer, C. B., 1955a. A deep-sea seismic reflection profile. *Geophysics*, **20**, 270.

Officer, C. B., J. I. Ewing, J. F. Hennion, D. G. Harkrider and D. E. Miller, 1959. Geophysical investigations in the eastern Caribbean: Summary of 1955 and 1956 cruises. In *Physics and Chemistry of the Earth*, **3**. Pergamon Press, London.

Officer, C. B., M. Ewing and P. C. Wuenschel, 1952. Seismic refraction measurements in the Atlantic Ocean, Part IV: Bermuda, Bermuda Rise and Nares Basin. *Bull. Geol. Soc. Amer.*, **63**, 777–808.

Raitt, R. W., 1952. Oceanographic instrumentation. *Nat. Res. Council Pub.*, **309**, 70.

Savit, C. H., D. M. Blue and J. G. Smith, 1962. Exploration seismic techniques applied to oceanic crustal studies. *J. Geophys. Res.*, **67**, 1919–1931.

Shor, G. G., Jr., 1959. Reflection studies in the eastern equatorial Pacific. *Deep-Sea Res.*, **5**, 283–289.

Shor, G. G., Jr., D. G. Moore and W. B. Huckabay, in press. Deep-sea tests of a new non-explosive reflection profiler.

Sutton, G., 1952. Three channel amplifier for use in refraction and sound measurements at sea. (Unpublished manuscript) Lamont Geological Observatory, Tech. Rep. No. 2.

Weatherby, B. B., 1949. The history and development of seismic prospecting. In *Geophysical Case Histories*, **1** (ed. L. L. Nettleton). Society of Exploration Geophysicists, Tulsa, Okla.

Weibull, W., 1954. Sound explorations. *Reps. Swed. Deep-Sea Exped., 1947–48*, **4**, 1–34.

3. SINGLE-SHIP SEISMIC REFRACTION SHOOTING

M. N. Hill

1. Introduction

Shortly after the end of the Second World War it was realized that the difficulties of obtaining in Britain two ships for refraction shooting at sea would prohibit the use of the more conventional methods described in Chapter 2. It was, therefore, decided in the Department of Geodesy and Geophysics at Cambridge that a single-ship method must be developed.

The first system tried was to fire charges suspended beneath buoys. These charges were electrically detonated by clocks in the buoys and a preliminary warning before firing, and the shot instant, were radio-transmitted to the ship. The receiving hydrophones and their associated amplifiers were connected to the recording equipment in the ship. This method was soon abandoned as laborious and unsafe. At long ranges there would have been danger of the buoys being picked up by unknown ships before the charges had fired.

The next system tried was first used in shallow water and is that which is still in use, namely the sono-radio buoy system. The first experimental area was in the western approaches to the English Channel and the results were published by Hill and King (1953) and with more observations included by Day *et al.* (1956). The first deep water experiments were made in 1949 and the results published by Hill in 1952; these experiments were followed by further work in the Atlantic (Hill and Laughton, 1954) and by the work of H.M.S. *Challenger* (Gaskell *et al.*, 1958). This last paper contains a description of the method of analysis of results. With the method used in all these experiments hydrophones are suspended below buoys which transmit by radio the signals received to the ship from which the charges are fired. With this system (Fig. 1) a number of receiving points can be obtained for each shot; it is usual to lay four buoys at appropriate spacings. Further, the ship is in control of the charge-firing arrangements and the danger is, therefore, greatly reduced from that of the first system.

The disadvantages of the system compared with the two-ship system are as follows:

(*a*) The radio range between the buoys and the ship is limited to about 45 km because of the restricted radio power of the buoys.

(*b*) The buoys are not expendable and have to be recovered. This means that, having shot a line, it is necessary to steam back to the buoys, thereby increasing the total steaming distance in comparison with some of the two-ship methods. It further results in weather restrictions; recovery becomes difficult in winds higher than Force 5 to 6.

(*c*) Once launched, the buoys are no longer under control and, for example, the amplifier gain settings cannot be altered. Further, it is not possible to quieten the hydrophones by cable streaming methods such as are described in the previous chapter.

[*MS received April, 1962*]

(*d*) At present there is only one channel from each buoy. Multi-channel radio transmission from each buoy could greatly improve the amount of information received at the ship.

These disadvantages are, however, to some extent outweighed by some obvious advantages apart from that of being able to operate with a single ship:

(*a*) Multiple receiving points can easily be obtained; this not only results in economy in the use of explosives but also, for example, in information concerning the dip of the strata being obtained without reversal of the line.

(*b*) Ship-made noise is not picked up by the hydrophones. There are no moving parts in the buoys and quieter operation should, therefore, be possible than with hydrophones near ships.

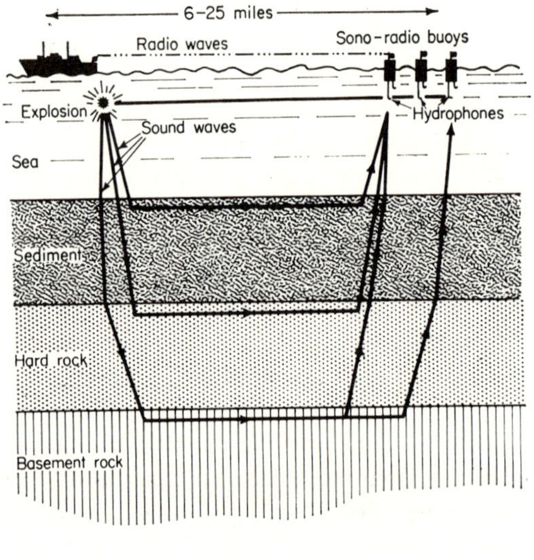

Fig. 1. The sono-radio buoy system of seismic refraction shooting. Explosions are fired from the ship and the signals received by hydrophones below the buoy are transmitted back to the ship for recording.

The disadvantages of the limited radio range of the sono-radio buoys could be overcome by increasing the radio power transmitted, but the increase required, say, to double the range to 90 km would be of the order of twenty times. This results in a cumbersome buoy. During early 1962, however, an alternative system for long-range seismic refraction shooting has been developed and successfully tried by the Department of Geodesy and Geophysics. A film recorder and radio receiver are installed in the sono-radio buoy. This recorder is switched on and off by the ship's radio transmitter and the shot instant transmitted to the buoy. This greatly increases the range of operation since the ship's transmitter has relatively great power. The disadvantage of the method lies in the fact that once radio contact is lost from the buoy transmitter the quality of the seismic signals cannot be judged until the buoys are recovered. However, the quality of the recording is improved by eliminating the radio link, and multi-channel recording is possible.

Apart from single-ship operation, sono-radio buoys can also be used with advantage where two ships are available. The receiving ship lies amidst an array of buoys spaced to conform to any pattern suitable for the particular problem.

2. Instrumentation

A. Hydrophones

In the system at present used by the Cambridge group the sensing head of the hydrophone consists of a barium titanate tube radially polarized and silvered in four longitudinal strips inside and outside. The tubes are 1 in. O.D. and approximately $\frac{3}{4}$ in. I.D. and 1 in. in length. Each of the longitudinal strips can be considered as a separate hydrophone element and the four are electrically connected in series. By this means the voltage sensitivity is increased by a factor of approximately four times from the value it would have if the tube were fully silvered.

After the electrical connections have been made the tube is completely potted in epoxy resin to form a cylinder $1\frac{1}{8}$ in. in diameter. This cylinder is longer than the barium titanate tube and is machined at one end to take an "O" ring for sealing into the preamplifier tube. At the other end the cylinder has a domed end; the intention being that this dome should prevent eddy formation if the hydrophone has water moving past it. The domed end has an axial hole in it: this is loaded with an appropriate amount of lead shot to provide horizontal trim for the complete assembly. A photograph of the sensing head is shown in Fig. 2.

The preamplifier tube is made from a corrosion resistant aluminium alloy of wall thickness approximately $\frac{1}{16}$ in. and of outside diameter equal to that of the sensing head. The length of this tube is selected to make the whole assembly neutrally buoyant. The tube contains a four-stage thermionic valve (DF66) amplifier, a gain control, and the battery supply consisting of one $1\frac{1}{2}$-V cell and one $22\frac{1}{2}$-V battery. The frequency response of the hydrophone and preamplifier is flat from 5 c/s to 5000 c/s.

The coaxial hydrophone cable is 300 ft in length, rubber covered, and approximately $\frac{1}{4}$ in. in outside diameter. For the first 100 ft from the hydrophone the cable is buoyed with tubular floats approximately every 10 ft to make it overall neutrally buoyant. At the 100-ft position a 10 lb weight is attached and thence the cable is connected directly to the buoy. It is thus expected that the hydrophone and 100 ft of cable will stream out horizontally at a depth of 200 ft below the buoy (Fig. 3).

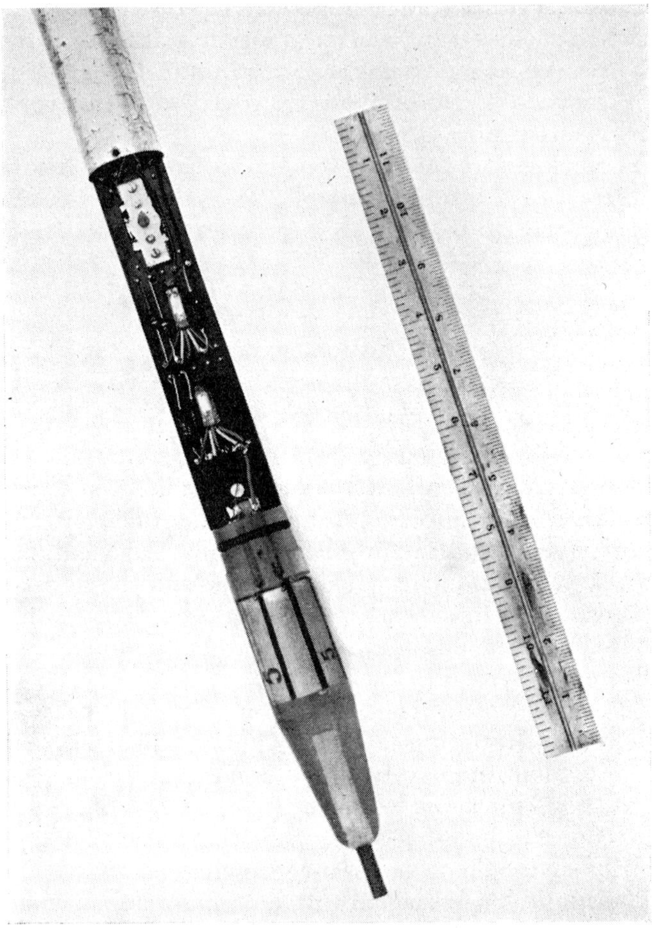

Fig. 2. The barium titanate pressure-sensitive element and part of the preamplifier of the hydrophone used below sono-radio buoys. Scale in inches.

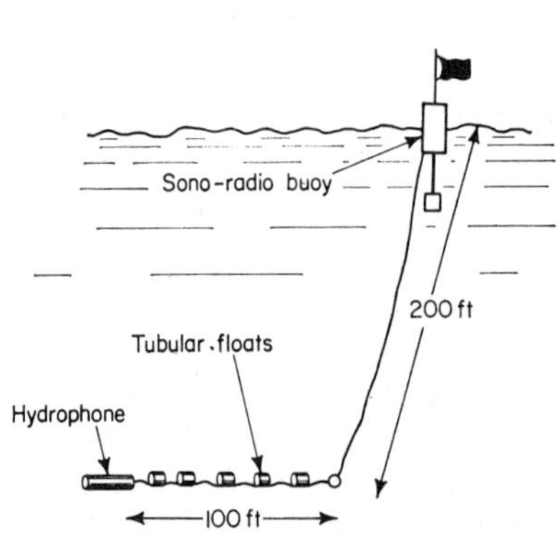

Fig. 3. The rigging of the hydrophone below a sono-radio buoy.

B. The Sono-Radio Buoys

There are two types of buoy at present in use. The first contains the radio transmitter and its associated dry batteries. The second, which is considerably larger, contains a transmitter, a radio receiver, the dry battery supply and an eight-channel 35-mm photographic recorder, associated filters, and amplifiers. A photograph of one of the smaller buoys is shown in Fig. 4.

Fig. 4. A photograph of a sono-radio buoy and hydrophone. The 8-ft-long bottom pole, made from galvanized iron tubing, is not shown.

The bottom poles are 8 ft in length with lead weights at their bases to maintain the buoy-can vertical. The top poles are 11-ft bamboos with heavy gauge copper wires strapped to them; these wires act as aerials. The poles are mounted on the aerial insulators on the tops of the buoys. A small flag is tied to the top of the pole.

C. The Radio Transmitters

The radio transmitters operate at preset frequencies between 40 and 47 mc/s and radiate approximately 3 W r.f. F.M. is used as the modulation system and the maximum deviation is ± 125 kc/s. Thermionic valves are used throughout. Amplitude modulation is unsatisfactory because of the movement of the buoy antenna.

The signals from the hydrophones are connected directly to the reactance modulator; this alters the oscillator frequency by an amount proportional to the signal voltage. The oscillator frequency is doubled in the anode circuit of the oscillator valve and doubled a second time in the drive valve for the push-pull output valves. The output from the transmitter is matched to the half-wave aerial on top of the buoy.

The battery supply for the transmitters consists of two composite $1\frac{1}{2}$-V + 90-V dry batteries each weighing about 5 lb. The life of these two batteries is approximately 24 h for continuous operation.

With advances in transistor design, it is expected that the transmitters could in future be transistorized. This would allow increase in the radiated power for a given weight of batteries, or alternatively a longer life. The signal-to-noise ratio could be improved by using phase modulation by an amplitude-modulated sub-carrier of, say, 10 kc/s.

D. The Long-Range Buoy-Radio Receivers

Commercial transistorized receivers are installed in the recording buoys. These receive signals from the ship's transmitter in the frequency range 1.5 to 2.5 mc/s and are connected to the transmitting aerial; suitable filtering to remove the 40–45 mc/s signal is required at the input. At the audio output there is a narrow band filter centred at 800 c/s and suitable relays for switching the recorder on and off.

In order to avoid spurious signals operating the switching system, the 800 c/s amplitude modulated transmission from the ship has to persist for 2 sec before the recorder is switched on. Thereafter the recorder will continue to operate so long as the modulation persists and for 10 sec after it is switched off. Thereby short breaks in the radio transmissions will not affect its operation. The modulation can be keyed in the ship and the pattern is recorded both in the buoy and in the ship. This pattern can be related to the shot instant, which is not directly transmitted to the buoys; the recorders are not switched on until after the shot fires.

E. The Long-Range Buoy-Recorders

The miniature 35-mm photographic recorders in the buoys contain 200 ft of film. A recording speed of 2 cm/sec is usually employed; this gives a total recording time of 50 min. There are up to 12 recording channels available.

F. The Ship's Radio Receivers

There are two receiving aerials in the ship, one mounted on the forecastle and the other astern. These aerials are commercial dipole directional television aerials; one has its maximum response ahead and the other astern thus enabling good reception to be obtained both on the way out and on the way back to the buoys. Relatively broad band aerial amplifiers are installed in the ship near the aerials and the high-level r.f. signals are thence fed to the receivers in the laboratory. Each receiver is tuned to a particular buoy.

The audio output from each receiver is filtered to provide one band from 5–500 c/s and a second from 5–50 c/s.

Each receiver has a tuning indicator and a radio frequency signal level indicator.

G. The Recorder

The output from the receivers is recorded on an ultraviolet-light 12-channel galvanometer recorder. This type of recording allows inspection of the record before processing. Time marks are produced by a phonic motor driven from a crystal-controlled frequency source.

The methods of analysis of the results are as described in Chapter 1.

3. Operation

The number of buoys normally used is four. These, in deep-water areas, are spaced along the line of shots at intervals of about 1 mile. The first shot is fired about 4 miles from the nearest buoy and thereafter at fixed intervals of time, usually between 4 and 8 an hour. The ship's speed is approximately 8 knots. Between 10 and 20 shots will be fired for a full profile.

The explosive used is "Geophex", which does not require watertight containers or primers when initiated with aluminium detonators. Slow-burning fuse, specially covered in order to stop penetration of sea-water, allows the charge to sink and the ship to get clear before detonation. A length of fuse giving a burning time of 80 sec is standard; the charge depth is then approximately 300 ft.

The charge size is varied from about 5 lb for the close shots to a maximum of 300 lb for the longest range shots.

It is usual to keep the observation roughly plotted as the information comes in; thus, on the passage back to the buoys, any gaps can be filled in.

To avoid unnecessary searching for the buoys on returning to them, sound-ranging shots of $2\frac{1}{2}$ lb weight are fired when the range is estimated as being about 3 miles from the nearest buoy. In good visibility the buoys can be seen at about 1 mile distance.

If the opportunity exists, shots are fired from both ends of the line of buoys.

In shallow-water work the technique is unchanged apart from the spacing between shots and buoys.

References

Day, A. A., M. N. Hill, A. S. Laughton and J. C. Swallow, 1956. Seismic prospecting in the Western Approaches of the English Channel. *Q. J. Geol. Soc. London*, **112**, 15–44.

Gaskell, T. F., M. N. Hill and J. C. Swallow, 1958. Seismic measurements made by H.M.S. *Challenger* in the Atlantic, Pacific and Indian Oceans and in the Mediterranean Sea, 1950–53. *Phil. Trans. Roy. Soc. London*, **A251**, 23–83.

Hill, M. N., 1952. Seismic refraction shooting in an area of the Eastern Atlantic. *Phil. Trans. Roy. Soc. London*, **A244**, 561–596.

Hill, M. N. and W. B. R. King, 1953. Seismic prospecting in the Western Approaches of the English Channel. *Q. J. Geol. Soc. London*, **109**, 1–20.

Hill, M. N. and A. S. Laughton, 1954. Seismic observations in the Eastern Atlantic. *Proc. Roy. Soc. London*, **A222**, 348–356.

4. CONTINUOUS REFLECTION PROFILING

J. B. Hersey

1. Introduction

A continuous seismic-reflection apparatus consists of a sound source and a nearby hydrophone receiver which are towed through the water while operating exactly as the conventional echo-sounding instrument does; the source sends out sound pulses at regular intervals; the receiver receives subsequent echoes reflected from the bottom of the ocean and from reflectors below the bottom.

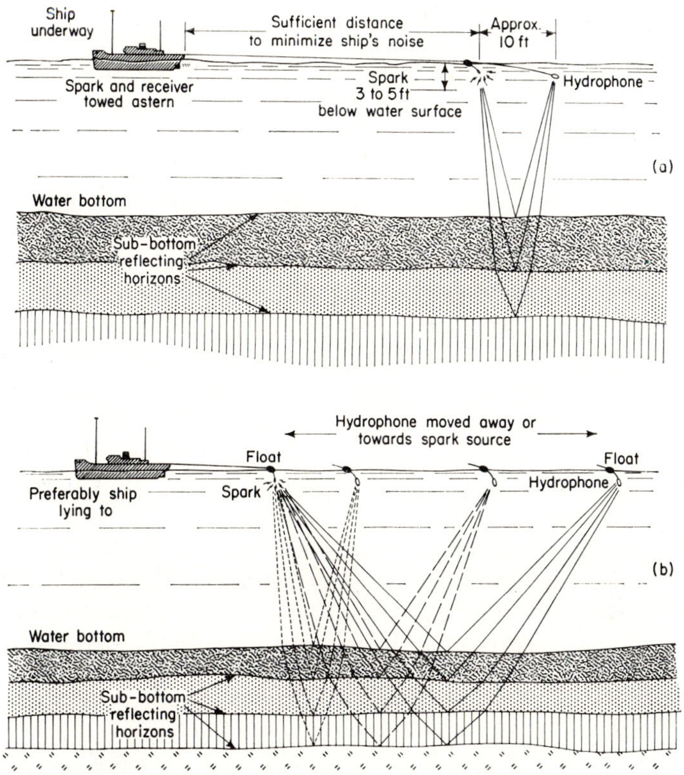

Fig. 1. (a) Continuous profiling for normal incidence reflections. (b) Procedure for oblique reflection profiles.

The echoes are then converted to electrical signals and fed ultimately to a graphic recorder like those commonly used in echo-sounding. (The basic operation is illustrated in Fig. 1.) Both single-frequency and broad-spectrum sound sources have been used. Either are short pulses of sound radiated on a precise schedule that is controlled by the graphic recorder. Since different

[*MS received July, 1960*]

frequency bands are effective for different studies the broad spectrum pulses are the more commonly used. Previously, seismic-reflection methods for exploring buried structures have depended principally on explosive sound sources. The difficulties associated with explosives have limited the techniques of exploration severely. Aside from problems of safety, perhaps the impracticality of shooting many shots in close sequence has been the biggest drawback. Despite these drawbacks very successful reflection profiling has been carried out with explosives during cruises of *Vema* of the Lamont Geological Observatory (Ewing and Tirey, 1961). Several sources have been developed that can be actuated rapidly and continually, and, further, have the requisite broad spectrum. Any one of these that can be triggered precisely is useable, though some are more limited than others.

The method depends critically on the automatic correlation of successive recordings provided by the graphic recorder. The first graphic recorders we are aware of were used for echo-sounding by Marti (1922), and during the 1930's the graphic recorder gained rapid acceptance for recording water depth continuously in shallow water. Prior to the use of graphic recorders echo-sounding information was presented on scanning discs or other metering devices which provided no permanent record and especially no record which permitted easy, rapid correlation of successive soundings. Then observers often misinterpreted echo returns as bottom soundings when they may equally well have been echoes from objects in the water, sub-bottom echoes, or echoes from rough topography. Shortly after graphic recorders came into common use, observations were made in shallow-water areas where the sound pulse had penetrated the bottom to be reflected from a surface below the bottom. A well known early example is a recording during trials of T.S.S. *Awatea* (see Veatch and Smith, 1939, pl. vi). Murray (1947) likewise recorded sub-bottom echoes with a high-frequency echo-sounder in the Gulf of Maine. Other early observations are known which have not been published. One of these was made by the writer and others during 1942 in the inner parts of a branch of Pearl Harbor where channel dredgings had been dumped in large quantities. The dredgings were known from diving operations to be very soft, watery mud and it did not seem surprising that ultrasonic pulses could penetrate short distances, of the order of 20 to 30 ft, through this watery mud to reflect from the rock surface below.

Since echo-sounders all use high frequencies one must infer from the paucity of reported observations of sub-bottom echoes that sediments having low absorption for high-frequency sound are rare in shallow water. For reliable penetration sounds containing low-frequency energy were needed even for work in shallow water. In deep water the sound pulses commonly used in early sounding operations were so long as to obscure echoes from layers only a few tens of feet below the bottom. Hence it is not surprising that there are no reports prior to 1950 of bottom penetration in deep water from echo-sounders. Since that time modest sub-bottom penetration in deep water has been observed much more often when short sound pulses have been used (Heezen *et al.*, 1959;

Hersey and Rutstein, 1958; Worzel, 1959). Again in deep water low-frequency energy was needed.

The bottom reflection studies in deep water, reported by Hersey and Ewing (1949), showed that sound pulses of considerably lower frequency, of the order of a few hundred cycles, would usefully penetrate several hundred feet of bottom sediments in the open ocean. These frequencies are well above the band from 20 to 150 c/s commonly used in seismic prospecting. However, at that time it was necessary to use the high energy of explosions to obtain these echoes. The explosions were not repeated rapidly enough over long periods to correlate echoes from sub-bottom reflecting layers. Consequently, although many recordings were taken in the western Atlantic between 1946 and 1950, it was never possible to learn very much about deep-water sediment structures from them.

We needed a sound source which could be actuated repeatedly on a rapid and regular schedule to correlate echoes from sub-oceanic reflectors over any considerable distance. Early in the 1950's several groups, academic and commercial, became interested in developing such seismic apparatus. The Magnolia Petroleum Company laboratory developed an instrument, the Sonoprobe, which was based on an electro-acoustic transducer generating a low-frequency pulse; it was used with a graphic recorder (McClure, Nelson and Huckaby, 1958). This instrument has been used both commercially and by academic research groups. The United States Geological Survey, starting with some studies of silting behind Hoover Dam in Lake Meade in 1947, used echo-sounders employing 14.2 kc/s to measure the rate of silting there. Subsequently, they made additional studies with the same technique in Lake Michigan and Passamaquoddy Bay (Smith, Cummings and Upson, 1954; Smith, 1958). They reduced their operating frequency to 11, and finally 6 kc/s as their work progressed. At the Woods Hole Oceanographic Institution in 1952 the writer and P. F. Smith made some observations in Great Harbor at Woods Hole of sub-bottom echoes of the sound generated from an electric spark discharged under water. This sound source was copied from the design developed by Anderson (1953). We recorded on magnetic tape and at the same time observed the echo sequences on a cathode ray oscilloscope. Later the writer and S. T. Knott made a series of observations also in Great Harbor and later in Vineyard Sound (Knott and Hersey, 1956). In all of these the spark sound source was used with a graphic recorder. At the Naval Electronics Laboratory at about the same time Padberg (1958) developed a directional transducer for generating high-energy low-frequency sound pulses from which he recorded sub-bottom echoes on a graphic recorder. Beckmann *et al.* (1959) have reported successful work with a repeating gas exploder as sound source (see below). Several commercial manufacturers of echo-sounding equipment have been called upon to modify their equipment to radiate lower frequency pulses in systems very similar to those mentioned above. Thus, considerable interest has been shown in this method of studying comparatively shallow structures in water-covered areas both at sea and in inland waters.

2. Design

Methods adapted by the various separate groups using the continuous profiling technique are similar. A schematic diagram of a typical instrument is shown in Fig. 2. The elements of design are quite simple. Sound absorption per unit distance is known to increase rapidly with frequency both in water and sediment. Further, absorption in sediment is much higher than in water at all frequencies of interest. Thus low frequencies will be required to penetrate sediment (or rock). The question *how low?* we have had to decide empirically. We know that absorption is small enough in some water-filled sediments to permit ultrasonic echoes (*ca.* 24 kc) to be recorded through at least 30 to 40 ft of soft mud. 6–14 kc/s have been used successfully to penetrate bottom sediments to well over 100 ft. Hersey and Ewing (1949) recorded apparent sub-

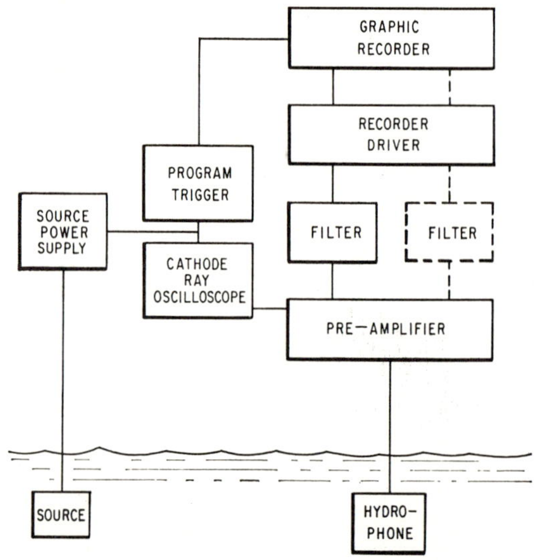

Fig. 2. Schematic diagram of continuous profiling electronics.

bottom echoes from reflectors down to 2400 ft in a 300–600 c/s band. The experience of many years of commercial exploration supports the estimate that 20 to 150 c/s is the most practical band for deep penetration. Hence for the needs of one problem or another there is a very wide frequency range of proven utility so far as penetration is concerned. However, when fine detail is needed, as in studying thin-bedded sediments, there is a lower limit to the frequency band that will serve. Therefore both low and high frequencies may be needed; a good design will provide a wide choice of frequency bands, or several discrete frequencies.

Except for adaptations of conventional echo-sounders, which do use discrete frequencies, all designers have chosen a short, broad-band pulse as their sound signal. Frequency selection is accomplished with filters following the hydrophone pre-amplifier, as shown in Fig. 2. Broad-band filters have been used; obviously at any frequency a band as wide as the order of an octave will be required if the response time of the filter (taken as the reciprocal band-width) is not to limit resolution.

All designers have used graphic recorders which record intensity changes as relative darkening of an electro-sensitive recording paper. Cathode-ray-scope visual presentations are included in all the systems, and, at least experimentally, long sequences of oscillograms have been recorded photographically. In the case of the new Marine Seismic System of the Field Research Laboratory, Socony Mobil Oil Company, photographic recording with the much-used galvanometer camera has been employed to compare with the variable density recording. It seems obvious that oscillographic recordings will be needed for

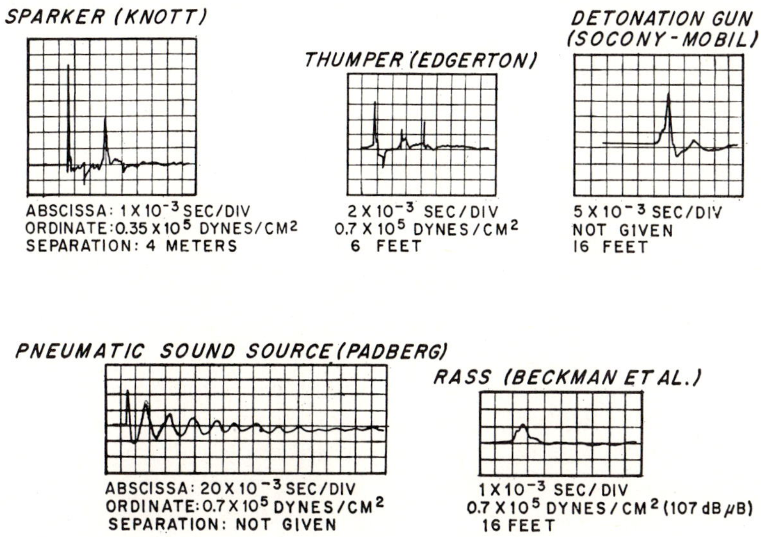

Fig. 3. Pressure–time curves of continuous profiling sources.

completely quantitative study of transmission and reflection processes, but it seems equally obvious that the automatic correlation provided by the graphic recorder together with the ease of measuring precise travel time with it will be useful for some time to come.

The following is a summary of the principal individual characteristics of the several instruments.

(a) *The Sonoprobe:* a directional magnetostrictive transducer. Acoustical pulse described as "one complete cycle of a 3800-cycle sine wave" (McClure *et al.*, 1958).

(b) *Padberg* (*in litt.*)*:* short-pulsed strongly directional magnetostrictive transducer. Padberg has also experimented with spark and pneumatic sources (see Fig. 3).

(c) *The Seismic Profiler:* a high-tension spark discharged under water (Knott and Hersey, 1956). Recently the boomer, an eddy-current-actuated sound-pulse generator, was developed by Harold Edgerton primarily for use with the seismic profiler but also for other applications in acoustical oceanography. The design lends itself to inexpensive alteration of the radiated pressure-pulse shape and duration. The name *seismic profiler* has also been given to the method employing explosives by Ewing and Tirey (1961). Possibly this is a suitable generic name for this kind of seismic instrument system, whether the source is high explosives or an electro-mechanical transducer.

(d) *The Sub-bottom Depth Recorder:* similar to the seismic profiler. A spark source is used for high frequencies and fine resolution; a repeating gas explosion (propane–oxygen) is used for lower frequencies and deeper penetration (Beckmann *et al.*, 1959).

The Field Research Laboratory, Socony Mobil Oil Corporation, has developed a gas exploder in the form of a vertical shock tube with one end under water. Starting with spark ignition near the top a shock wave develops as gases (propane, oxygen) burn. The shock wave strikes the water surface at the bottom, generating the sound pulse. The pressure–time wave-form (see Fig. 3) contains both short and long duration pulses (*in litt.*, manuscript in preparation).

A. Sound Sources

Fig. 3 contains the pressure–time curves available for the above sound sources. They were all taken with broad-band pressure gauges at distances from 6 to 60 ft from the source. The directional properties of these pulse generators (excepting Padberg's) have not as far as we know been investigated in any detail. All electrically powered sources employ a large condenser for energy storage and various methods of triggering its discharge. Stored energy varies from 4 to over 25,000 joules in models that have been extensively used. Acoustical energy is from 0.5 to 10% of the electrical energy. [For a detailed discussion of power supplies and transducers see Hersey *et al.* (1962) and Caulfield (1962).]

B. Receiving Hydrophones

Piezoelectric and magnetostriction hydrophones have been used more or less indiscriminately by various investigators, who have developed habits depending on preference, convenience and estimates of performance in towing. Of the published accounts of investigations at sea the Sonoprobe system employs a receiver fixed by a rigid mounting to the rail of the ship, whereas both the seismic profiler and the sub-bottom depth recorder employ single hydrophones towed well astern. Mostly, hydrophones are chosen for broad-band, flat response (at least 30 to 10,000 c/s) and for best towing. Since noise, whether ambient, self noise, or flow noise, is almost always more intense at low frequencies than at high, hydrophones having a high-pass frequency characteristic response are sometimes selected. Sensitive hydrophones are required but there is no need in the present state of the art for especially low noise properties compared with the quietest ambient levels for underway observations. Recently, especially in the work of John Ewing and his co-workers at Lamont, the so-called "slacking technique" has been used successfully to obtain records of high quality in deep water from a ship proceeding at 10 knots. The slacking technique, long familiar in oceanic seismic-refraction research, is to slack the hydrophone towing cable seconds before the desired wave train is to be received, thus placing the hydrophone nearly still in the water. The Lamont procedure while profiling is to shoot small charges of explosive on a two-minute schedule, slacking the hydrophone cable anew for each shot and then hauling in cable before the next.

A long array of hydrophones designed to be uniformly neutrally buoyant has also been employed by several groups as a means of improving the signal-to-noise ratio at high towing speeds. The recordings in Figs. 14 and 15 were made with a 40-ft array of five hydrophones. By such arrays useful observations have been made in the deep ocean at towing speeds of 8 knots. This speed does not appear to be an upper limit.

C. Pre-amplifiers

Conventional broad-band amplifier designs have been used; they are being altered and improved almost continually. Hersey (1957) describes an early design by Willard Dow.

D. Filters

Commercially available passive and electronic filters have usually been used. For good transient response it is well to avoid filters relying on sharply resonant circuits for sharp cut-off and, of course, care must be taken to match the filters to surrounding circuitry. In using the seismic profiler we have preferred, whenever possible, to split the pre-amplifier output feeding it to one high-frequency and one low-frequency filter. The high-frequency band is chosen either to discriminate between side-echoes from a rough bottom and sub-bottom reflections, or more rarely for high resolution of shallow echoes. The low-frequency band is chosen for maximum penetration. The two bands are recorded on a dual-channel graphic recorder.

E. Recorder Drivers

Little detailed information has been published about these units. The characteristic problem of seismic exploration recording is the very large contrast in amplitude between high-energy arrivals travelling only a short distance and the faintest reflections, some of which may have short travel times as well. Some kind of compression is almost mandatory (otherwise the recording paper is burned and torn to shreds!). McClure *et al.* (1958) report logarithmic compression for the Sonoprobe; Beckmann *et al.* (1959) report time-variable gain (TVG) for the sub-bottom depth recorder, and recent forms of compression circuitry in the seismic profiler include an adjustable clipper or, alternatively, approximately logarithmic output.

F. Special Data-Processing Techniques

The Field Research Laboratory of Socony Mobil Oil Company reports success with a magnetic drum compositor which retains and delays up to 16 successive between-pulse wave trains allowing them to add together before recording. The wave trains are automatically matched at the trigger pulse of the sound source. Correlated echoes are enhanced much more strongly than random noise components. The electro-sensitive papers have a somewhat similar action if the paper is moved so slowly through the recorder that successive stylus-sweeps overlap. Other correlation techniques that have been studied recently are gradually being introduced into seismography; the basic method of regular repetition of pulse-and-recording while the system moves slowly over the structure under study is well designed for their profitable application.

G. Recorders

The graphic recorders in use divide into dry paper and wet paper, and into stylus- and helix-recorders. There have been several produced commercially that will serve, but those most in use today are the Alden[1] (wet paper and helix-recorder), the Westrex[2] (dry paper and stylus) and the Muirhead[3] (wet paper and helix-recorder). All are facsimile recorders which have been adapted. The Alden and Westrex are described by Knott & Hersey (1956) and Luskin *et al.* (1954) respectively.

Multiple-channel recording is sometimes useful as in the simultaneous recording of low- and high-frequency bands discussed above (see Filters).

In conventional graphic recording of echo-soundings, individual oscillations are not resolved; the correlation presented is that of envelope matching (Fig. 4). Pulses filtered to emphasize low frequencies during continuous reflection profiling have been recorded to resolve individual oscillations either by half-wave rectification or high time resolution. Thus phase correlations can be made. Phase correlation recordings are illustrated in Figs. 5 and 19 by half-wave rectification, and in Fig. 21(b) by full-wave rectification and high resolution. In Fig. 6 full-wave and half-wave rectification are compared during a profile. Here echo sequences stand out more clearly when half-wave rectification is

[1] Alden Products Company, Brockton, Massachusetts. Product known as "The Precision Graphic Recorder".

[2] Westrex Company. Product known as the "Precision Depth Recorder".

[3] Muirhead Instruments Inc. Product known as the "Facsimile Depth Recorder".

employed. Thus, recordings which display phase relations are often more surely correlated in complex wave trains; also they can be used to study phase changes in reflection and other transmission processes.

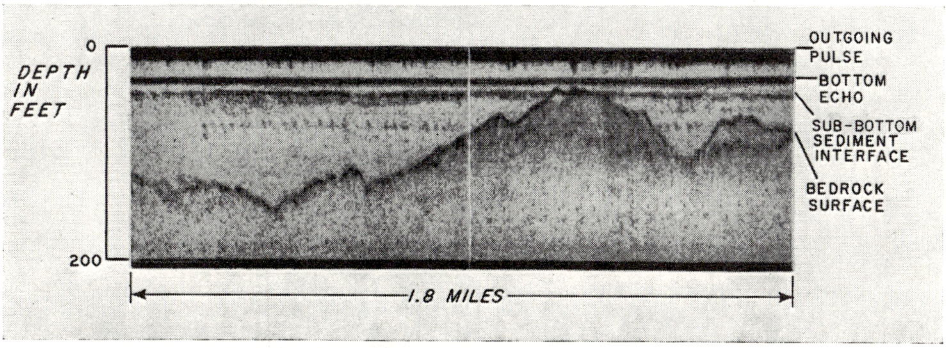

Fig. 4. A representative continuous reflection profile over an uneven bedrock surface buried in sediment in Long Island Sound, New York. (After Beckmann.)

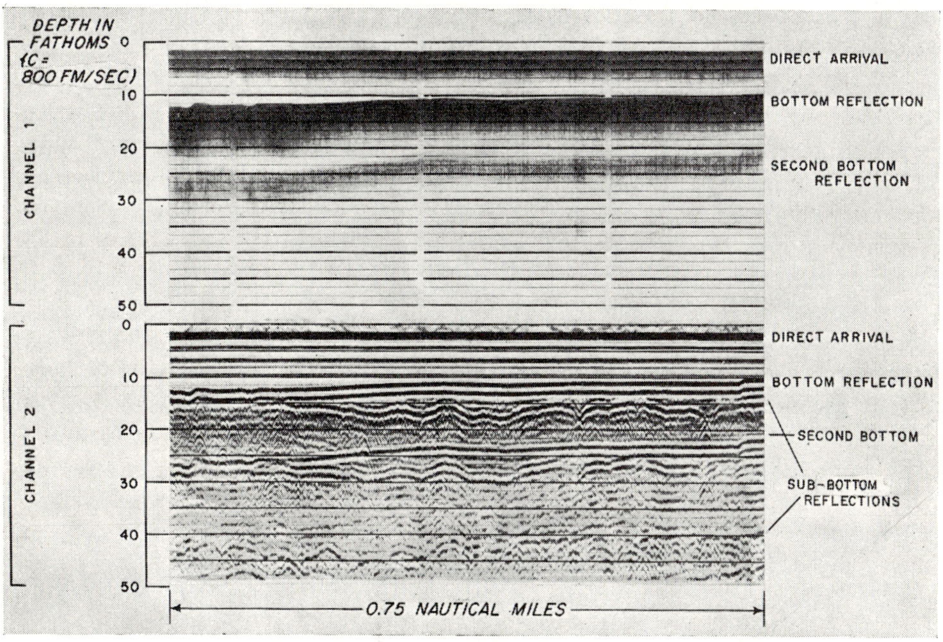

Fig. 5. Dual-channel recording made in Vineyard Sound, Massachusetts. Channel 1, filtered 2400 c/s high-pass; channel 2, filtered 75–600 c/s band-pass.

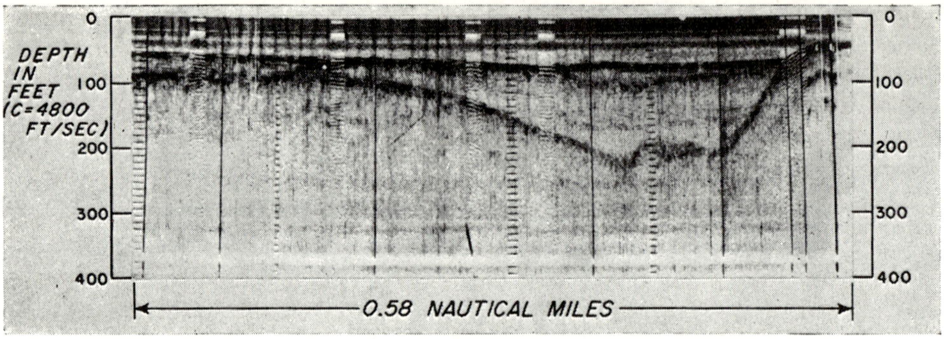

Fig. 6. Reflection profile across a buried channel in Narragansett Bay, Rhode Island. Brief intervals of half-wave rectification illustrate the enhanced clarity of recordings that display phase correlations over those displaying envelope correlation.

3. Operations at Sea

An effective continuous profiler can be mounted in almost any craft, excepting possibly a canoe. Low-power profilers have proven very convenient for inshore, shallow-water work; on seagoing ships one has room for more power supply and other complications which result in deeper sub-bottom penetration, which is usually needed offshore, becoming possible. The research ship or boat commonly tows the sound source and hydrophone (or hydrophone array) a considerable distance astern. This is done partly to clear the ship's wake, whose bubbles will absorb the sound, and partly to place the hydrophone farther from the acoustic noise sources on the ship. Recently, successful experiments have been carried out at the Woods Hole Oceanographic Institution with small hydrophones towed from a boom ahead of the bow. Towing speeds are commonly 3 to 6 knots. Source and hydrophone depth vary, but are commonly adjusted to be about a quarter wavelength for the lowest principal frequency in use. (That is, an arbitrary frequency near the low cut-off of the lowest filter.)

For sounding sub-bottom structures, source and receiver are set close together; say, 10 ft. For measuring seismic velocities, source and receiver are gradually separated while operating. Oblique reflections, refractions, or both, may be recorded. Often the drift of the observing ship is enough to carry it from the cabled hydrophone, or the hydrophone may be deliberately towed away from the ship (see Fig. 1b). Obviously, considerable variety is possible, including placing source, receiver or both remote from the observing ship without connecting cables. Then information can be recorded remotely or relayed by radio or underwater sound.

4. Results

Continuous profiling has been used for about five or six years. Applications were made first to shallow water and shallow unconsolidated sediment structures. Later on, deeper parts of the continental shelf were studied, and, at present writing, successful studies have been carried out both in oceanic depths of 5 to 6 km and in the Puerto Rico Trench. During these studies, reflections over 1 km below the bottom have been recorded.

Even in so young a development it is already difficult to sketch, much less to describe completely, the investigations which have been undertaken. However, the following are representative.

A. *Inland Waters*

Early emphasis was placed both on obtaining sub-bottom echoes and on comparing them with geologic evidence from bore holes and by charting reflecting horizons to outcrops. In 1956 and 1957, J. B. Hersey, D. Fink and Miss E. T. Bunce conducted two surveys in different parts of Narragansett

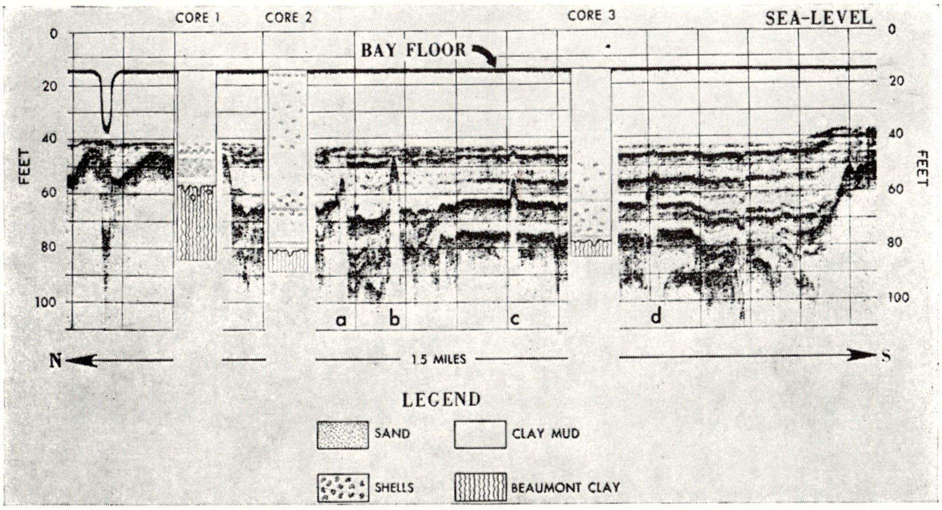

Fig. 7. Marine Sonoprobe profile across Corpus Christi Bay showing buried oyster reefs (see a, b, c and d). (After McClure, Nelson and Huckaby, 1958, fig. 9.)

Bay, one for the First Naval District Office of Public Works and one for the Army Corps of Engineers. The principal objectives were to map bedrock under glacial deposits in parts of the Bay. Fig. 6 is representative of recordings obtained. Bedrock reflections and other shallower horizons were traceable over most of the areas studied. Beckmann *et al.* (1959) report surveys in Long Island Sound and Chesapeake Bay. Fig. 4 shows a beautifully clear recording of bedrock obtained at shallow depth with a spark sound source. In these surveys continuous profiles were successfully correlated with core samples and penetration tests measured by the number of blows per foot required to drive drilling tools into the sediment. Reflection depths agreed well with abrupt changes in the number of blows per foot.

McClure, Nelson and Huckaby (1958) report comparisons of reflections with borings in several structures in shallow waters of the Gulf of Mexico. They obtained reflections from well known formations, and from deposits of shells (Fig. 7). Other similar correlations with bottom cores have been reported privately from work by Knott and Officer in the Gulf of Mexico.

B. Continental Shelves

These intriguing morphological features are receiving detailed attention off California from D. G. Moore of the U.S. Naval Electronics Laboratory (see Fig. 8)

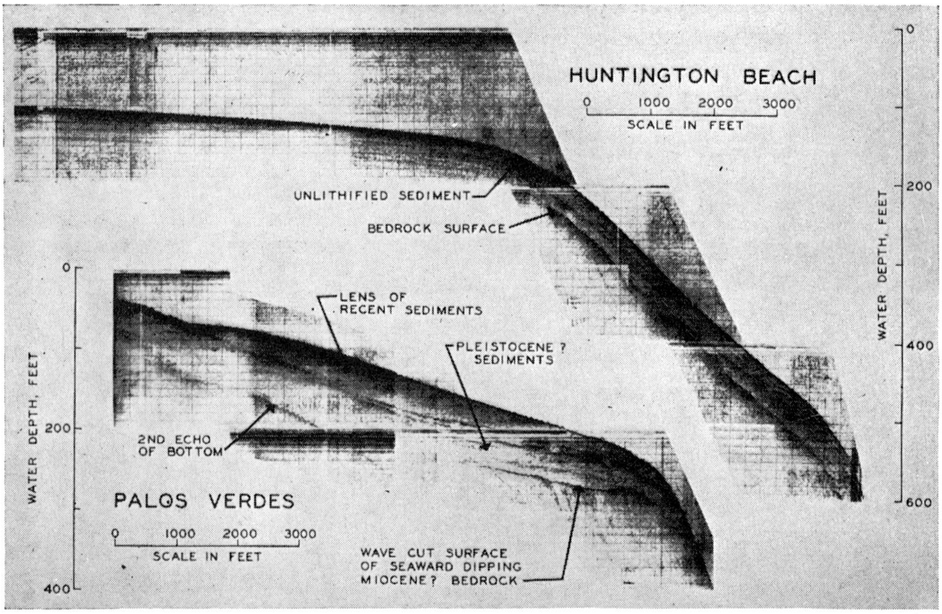

Fig. 8. Sonoprobe recording of the continental shelf off California. (After Moore, 1960, Plate 3.)

and C. B. Officer (Fig. 9), off the coasts of New Jersey, New York, and southern New England separately by John Ewing and S. T. Knott, and in Cape Cod Bay by Hoskins and Knott (1961). Profiles taken independently by Ewing and Knott across the shelf east of New Jersey are shown in Fig. 10. Recordings were made by the writer and R. Weller over the edge of the Spanish shelf south of Cartagena which are similar to those of Ewing and Knott. Thus far, little or no velocity information is available to support these profiles. However, it seems apparent that the present method can be used to increase our understanding of the processes that form the continental shelves.

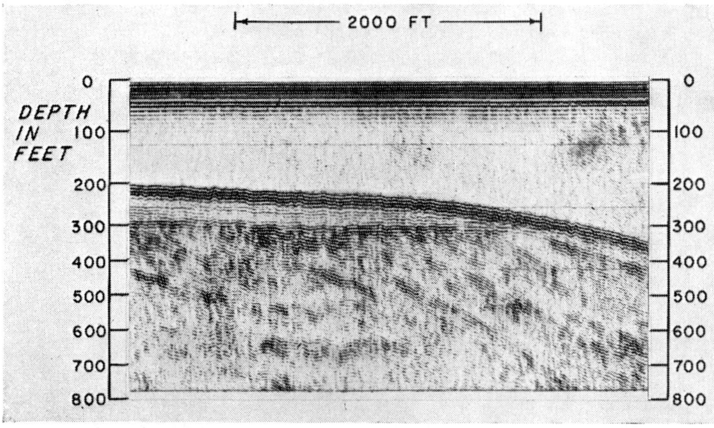

Fig. 9. The continental shelf from a commercial survey off California. (After Officer.)

C. Intermediate Water Depths

Reconnaissance profiles have been made by Woods Hole groups in the Mediterranean Sea to the northwest of the Strait of Messina, and in the North Atlantic on the Blake Plateau. Near the Strait of Messina the results suggest sediment-filled channels trending to the northwest downslope from the Strait (see Fig. 11). In these channels sediments attain a thickness of a thousand feet or more in places where the water depth is over 1200 ft. On the Blake Plateau flat-lying beds at shallow depths below the bottom were indicated. Profiles from the shelf-edge to abyssal depths have revealed outcrops on the continental slope of Europe and North America and layering over 1 km thick beneath the continental rise off eastern North America.

D. Deep-Ocean Studies

During some early tests in 1956, a Woods Hole group recorded apparent sub-bottom reflections with a spark source in water over 2640 fm deep. This could be done only from a ship lying to and on silent ship routine. More recently, Ewing and Tirey (1961) have reported extensive reflection profiling in the deep-sea in 1960 with explosives, while Hersey (1962) reported similar observations with a boomer in 1961. Since that time many thousands of miles of reflection profiling (with 0.5 lb of explosives) have been recorded from *Vema* of Lamont while, from *Chain* of Woods Hole, both boomer (5000, 9000, and 13,000 joules electrical storage) and sparker (25,000 joules) have been used in selected locations in the North Atlantic Ocean and the Mediterranean Sea.

Examples of somewhat contrasting structures are shown in Figs. 12–15, taken from various studies on *Chain*. A thick section, presumed to be horizontally stratified sediments in the abyssal plain west of Gibraltar, is shown in Fig. 12. Fig. 13 shows a long section across the low rise surrounding Madeira

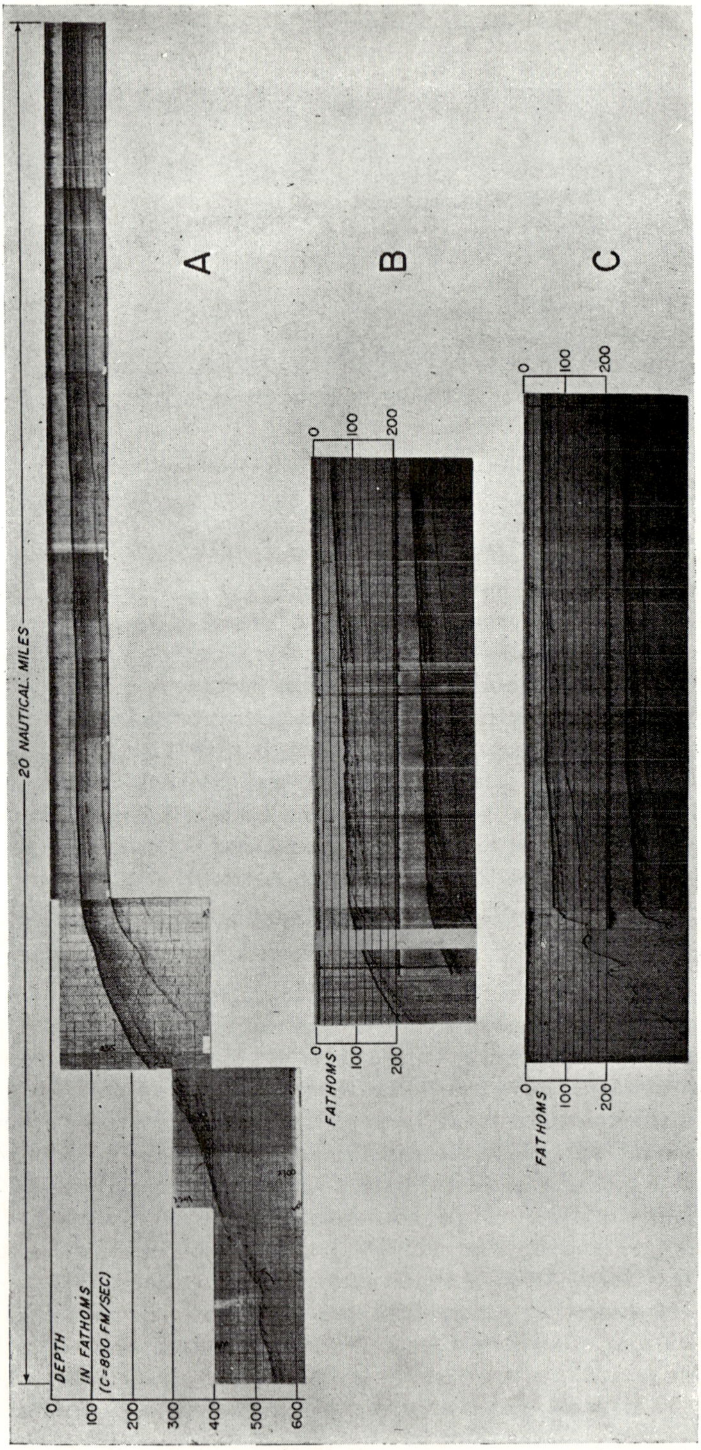

Fig. 10. Profiles across the edge of the continental shelf east of New Jersey. A, Knott; B and C, Ewing.

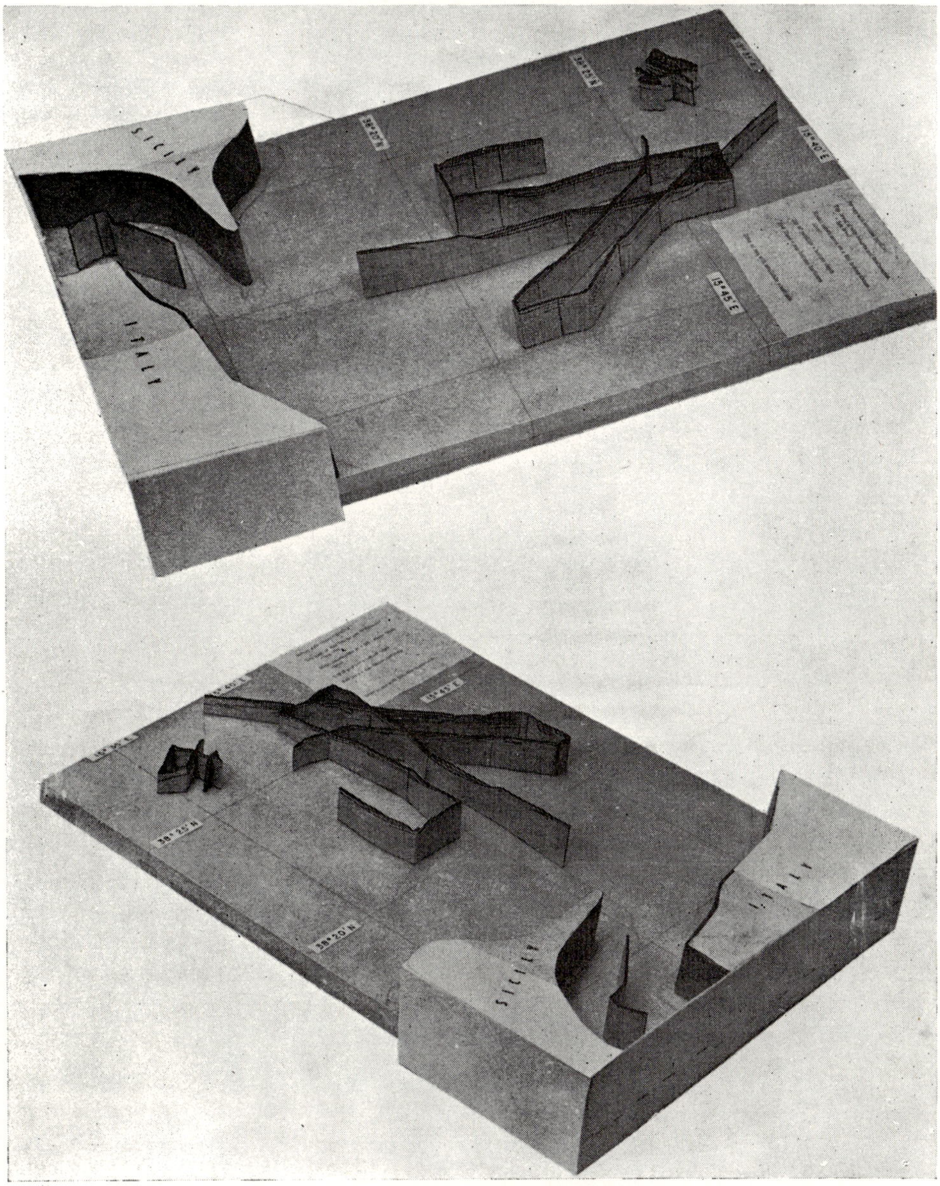

Fig. 11. A model of a continuous profiler survey in the Tyrrhenian Sea northwest of the Strait of Messina. (Woods Hole Oceanographic Institution.)

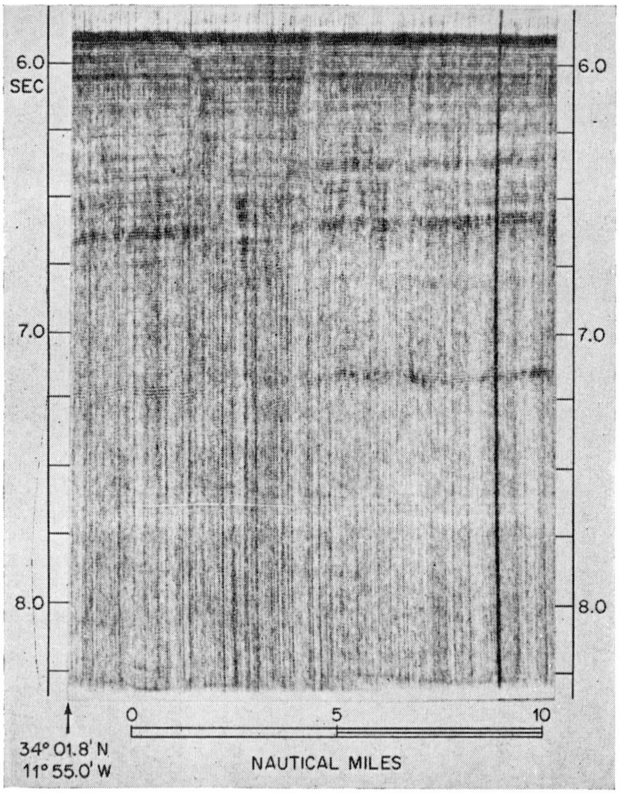

Fig. 12. A profile over the abyssal plain east of the Strait of Gibraltar. Bottom echo at 5.85 sec. Deepest sub-bottom echo about 7.15 sec.

Source: 5000 joule boomer
Receiver: AX 58 hydrophone
Towing speed: 3 knots
Analysis filter: 75–150 c/s band-pass

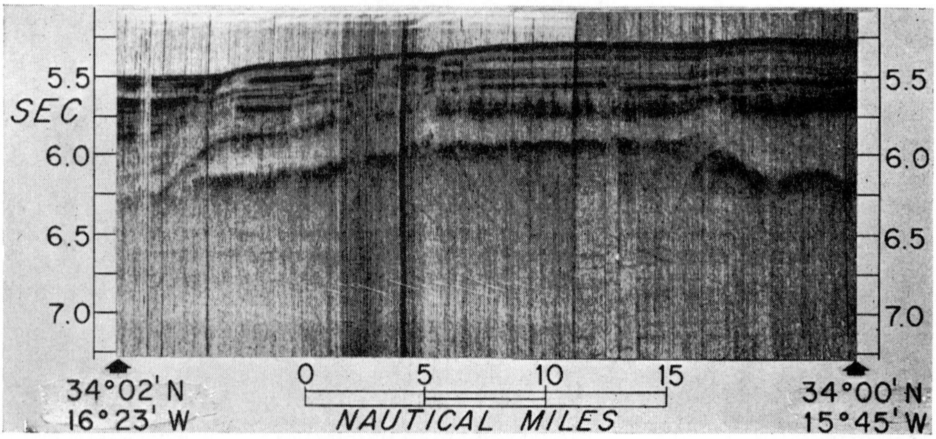

Fig. 13. An east–west profile about 100 miles north of Madeira. Bottom echo at 5.0 to 5.5 sec. Deepest sub-bottom echo 6.0 to 6.4 sec.

Source: 5000 joule boomer
Receiver: AX 58 hydrophone
Towing speed: 3 knots
Analysis filter: 75–150 c/s band-pass

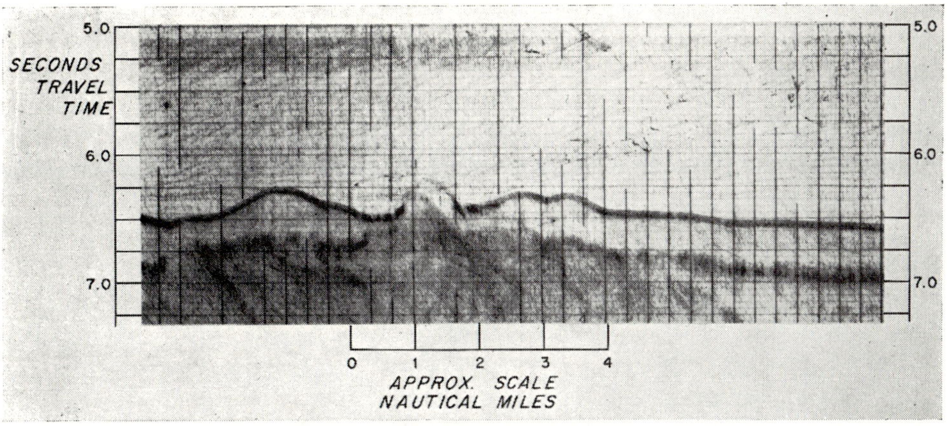

Fig. 14. Portion of a profile about 90 miles bearing 256°T from Bermuda, on the Bermuda Rise.

Source: 25,000 joule sparker
Receiver: 40 ft, five-element hydrophone array
Towing speed: about 5 knots
Analysis filter: 100–200 c/s band-pass

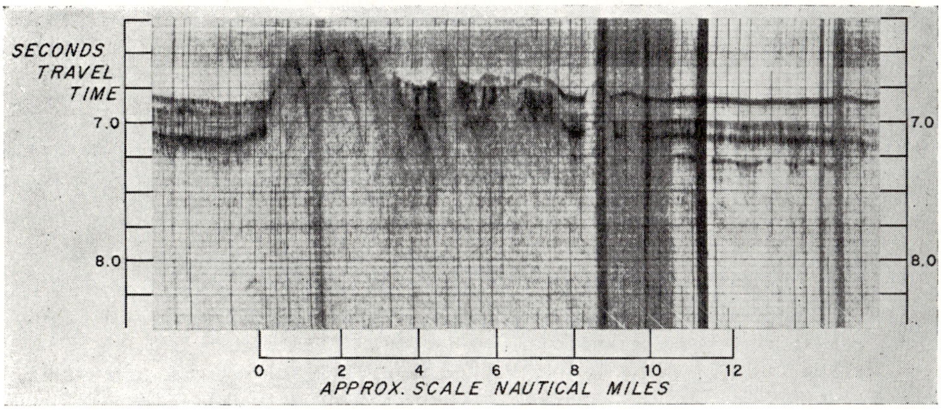

Fig. 15. A portion of a profile about 150 miles bearing 256°T from Bermuda, over abyssal hills on the Bermuda Rise.

Source: 25,000 joule sparker
Receiver: 40 ft, 5-element hydrophone array
Towing speeds: vary from 4 knots to 7 knots
Analysis filter: 150–300 c/s band-pass

and the several seamounts to the north. The shallower layers are probably sediments, but the shapes of the deeper reflectors together with their association with the seamounts suggest a volcanic origin. At this writing there has been no opportunity to investigate these structures in any detail. Thus we can only speculate about the significance of the obvious interruptions in the horizontal reflectors in both profiles. Some of the interruptions on both profiles

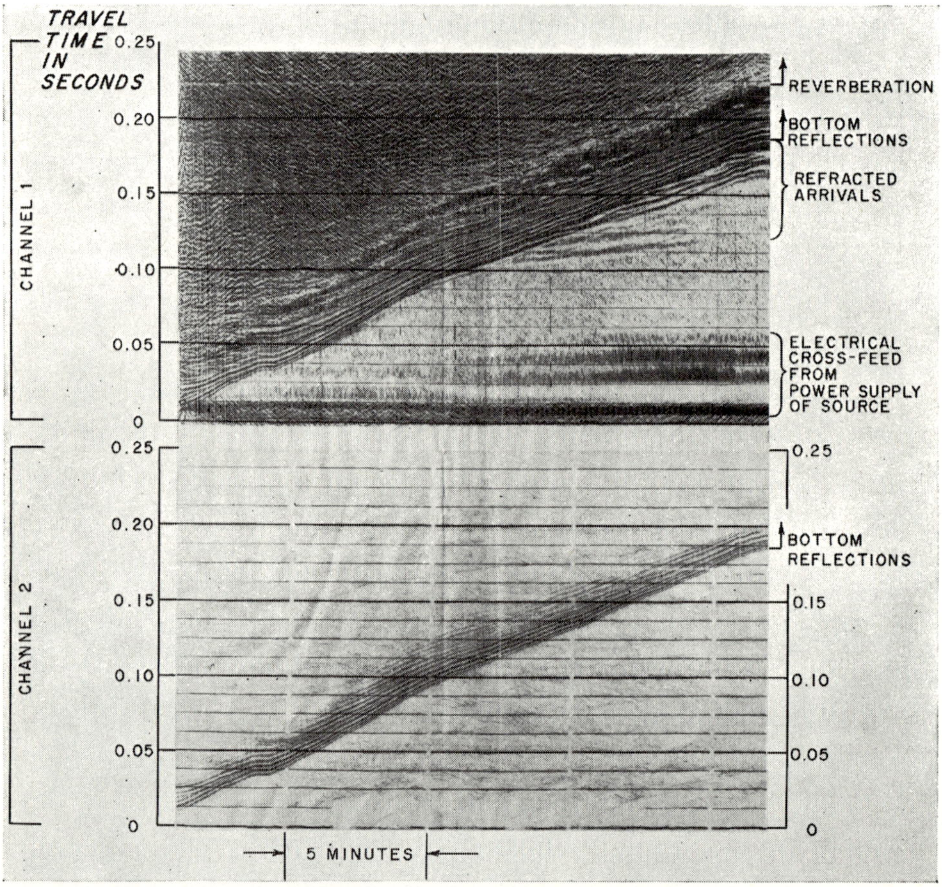

Fig. 16. A dual-channel recording of a refraction profile in Buzzards Bay, Massachusetts. Channel 1, filtered 75 c/s hi-pass.
Channel 2, filtered 1200 c/s hi-pass. Gain of channel 1, 30 dB higher than channel 2.

correspond through several successive reflections in depth suggesting either buried seamounts or intrusives.

Figs. 14 and 15 were recorded over the Bermuda rise. The deeper reflector of Fig. 14 is characteristically rough and has large relief. This recording was taken within 90 miles of Bermuda; 60 to 70 miles farther to the southwest, as in Fig. 15, a similar reflector is overlain by several nearly horizontal, strong reflectors except where it protrudes to form abyssal hills. The identity of the rough surface and the shallower layers has not been firmly established, but it seems certain that the rough surface is either the top of the deepest crustal layer having a velocity close to 6.5 km/sec (sometimes called layer 3) or the top of one of the layers of intermediate velocity, 3.0 to 5.5 km/sec, sometimes called layer 2. In the short time we have been recording reflections continuously in the deep ocean several recordings, widely distributed in location, have shown multiple layering, as in Figs. 12, 13 and 15. One is led to wonder whether generalizations like layers 1, 2 and 3 are not now over-simplified as means of classifying oceanic structures.

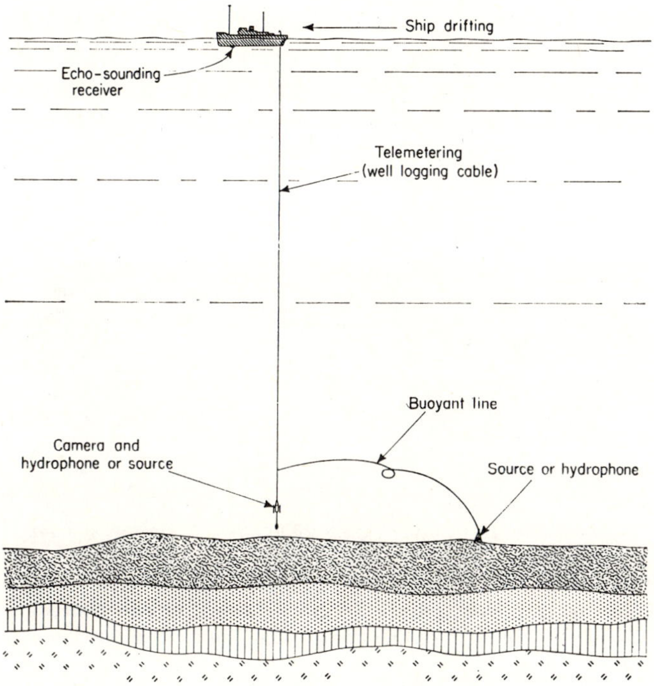

Fig. 17. Bottom seismic gear for deep-water use.

E. Seismic Refraction Observations

Scientists from both the Lamont Geological Observatory and Woods Hole have recorded refraction arrivals during oblique reflection profiles. Fig. 16 is representative; refraction arrivals recorded in this manner with high enough signal-to-noise ratio will be invaluable in supplementing present refraction techniques with explosives. Continuous refraction profiling should prove especially valuable in the study of unconsolidated sediments in deep water where shallow explosions and receivers have proven inadequate because the sediment arrivals are commonly masked. The outstanding requirement is a source that can be lowered to the bottom, there to operate long enough to enable the observing ship to drift away with a deep-suspended receiver while recording the profile. It is not difficult to invent such a system; possibly the next few years will see one put into successful operation. One untried possibility is sketched in Fig. 17. While the scheme itself is speculative, the component parts either exist or are under construction at the time of writing.

5. Interpretations

A. Geophysical

Continuous-reflection-profiling data provide a means of determining the geometric relations between the reflection record and the structure; from this description of structure and other independent knowledge, geologic inferences can be drawn in the manner usual for such problems. It is not proposed to review the subject of geological interpretation of geophysical data, but a

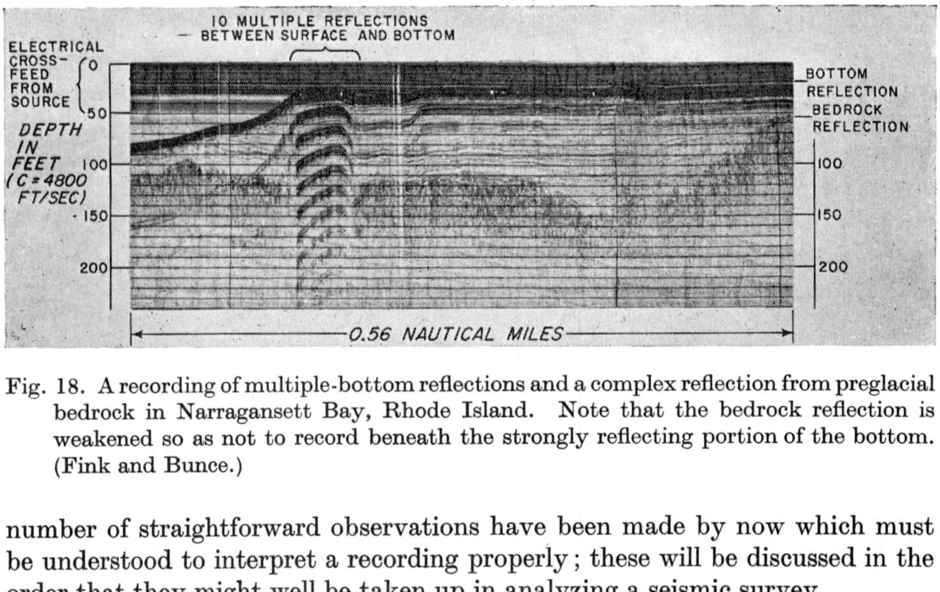

Fig. 18. A recording of multiple-bottom reflections and a complex reflection from preglacial bedrock in Narragansett Bay, Rhode Island. Note that the bedrock reflection is weakened so as not to record beneath the strongly reflecting portion of the bottom. (Fink and Bunce.)

number of straightforward observations have been made by now which must be understood to interpret a recording properly; these will be discussed in the order that they might well be taken up in analyzing a seismic survey.

A graphic recorder will bring out correlated signal sequences which have nothing to do with the profiling (except to degrade it). Typical occurrences are power-line cross-feed and rhythmic noise from ship's engines or propellers. These are best identified when the recording is made. It is usually a simple matter to shut off the sound pulse long enough to distinguish such noises. They should, of course, be eliminated or reduced if possible.

In shallow-water surveying an echo-sounder record often contains echo sequences of several multiple reflections between bottom and surface. Fig. 18

contains a striking example in which 10 multiples were recorded across a small area; unfortunately we have no clue about the reason for such high reflectivity. In continuous sub-bottom profiling, surface-bottom multiples are often superimposed on the sub-bottom echoes; in addition, there may be other multiple reflection sequences involving many different combinations of reflectors. In Fig. 19 not only are there three bottom-surface multiples but a complex pattern of sub-bottom echoes are repeated as well. In Fig. 20 a similar observation was made where the sub-bottom reflectors were so deep that bottom-surface multiples and the sub-bottom–surface-bottom multiples are completely separated. These examples clearly illustrate one of the principal difficulties of

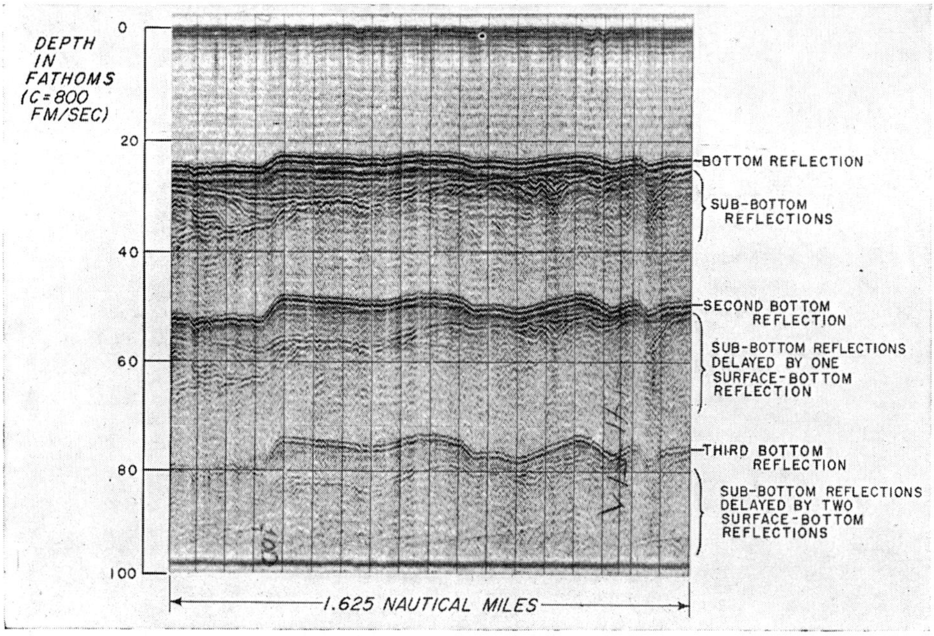

Fig. 19. Multiple-bottom-surface reflections recorded 10 miles northeast of Provincetown, Massachusetts. (After Knott.)

interpretation; painstaking identification of probable multiple echoes is a crucial step in analyzing continuous profiles.

In addition to providing "false" echoes the multiples may mask deeper reflections. Furthermore, several instances have been observed where strong reflectors seem to have returned so much energy that there was little left to be transmitted on to deeper horizons. McClure *et al.* (1958) observed such strong scattering in buried shell beds in the Gulf of Mexico (Fig. 7). Similar apparent masking occurred in Narragansett Bay (Fig. 18). Possibly both scattering and masking could be circumvented to some degree by proper choice of pulse shape (i.e. lower frequencies more dominant).

Amongst the echo sequences some represent echoes from extended interfaces between materials of different acoustic impedance, while others may be echoes scattered by smaller prominences (or concavities) in the bottom topography. These latter echo sequences are commonly called side echoes; usually they can be recognized by their crescent form. Such echo sequences are thought to be from reflectors small compared with the distance the ship moves during the sequence. If so they should approximate hyperbolae. These reflectors are located along the track of the ship at the top of the crescent, but a single profile near one of them does not locate it athwartships. Crescent-form echoes are familiar in echo-sounding (Fig. 21a); also they have been recorded as

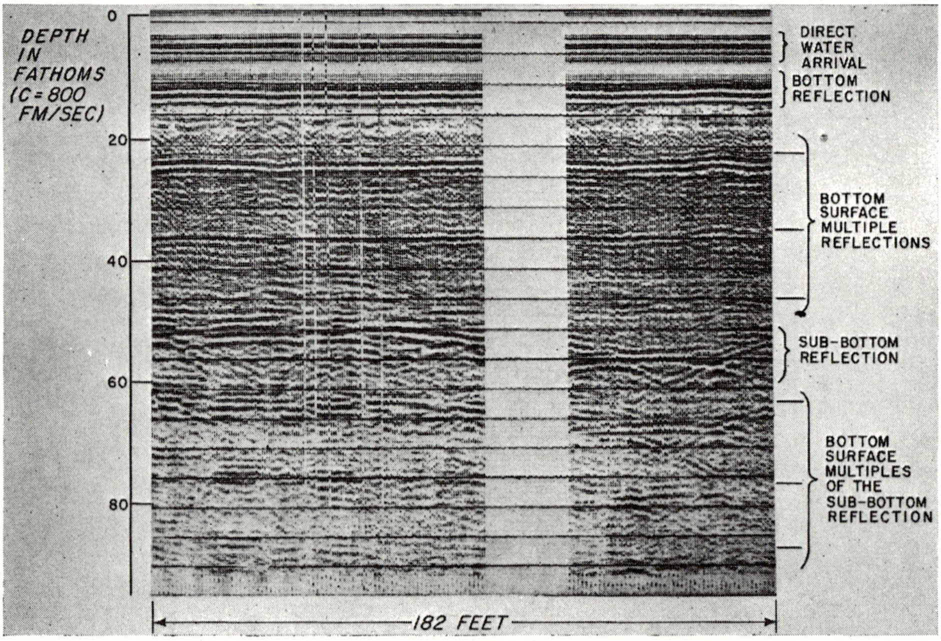

Fig. 20. Multiple-bottom and sub-bottom reflections in Vineyard Sound, Massachusetts.

apparent sub-bottom echoes (Fig. 21b). The latter record was taken in Vineyard Sound, Massachusetts, by the writer and others. At the time we did not prove that the crescent echo sequences came from sub-bottom reflectors, but we saw large numbers similar to those shown here, none of which rose above the bottom echo. This fact suggests that the reflectors are sub-bottom, possibly glacial boulders, which are well known on land nearby.

Sloping echo sequences require individual interpretation before a structure-profile section can be interpreted geologically. The echo-sounding literature covers extensively the interpretation of single echo sequences (for a review, see Elmendorf and Heezen, 1957). Thus, sloping echo sequences known from other control to have been recorded exactly normal to the strike of a slope can be interpreted to give the true depth and attitude of its interface by the well known arc-swinging technique. Where a sub-bottom slope is concerned, the geophysicist must determine whether he is required to consider the velocity structure in constructing arcs. This technique in echo-sounding is based on the assumption that the water has constant velocity. These rather tedious slope corrections are frequently by-passed unless the exact attitude and location of the reflecting surface is critically important.

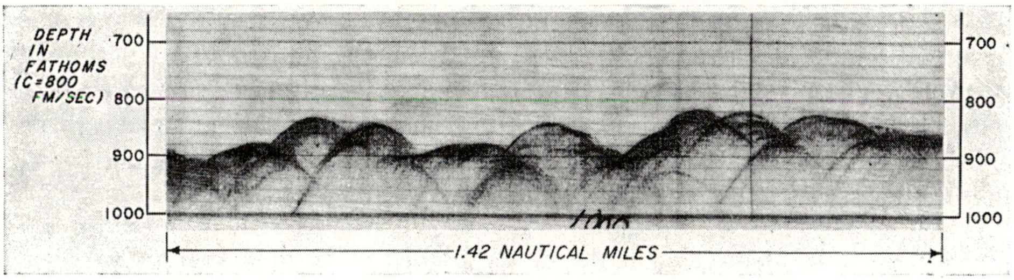

(a)

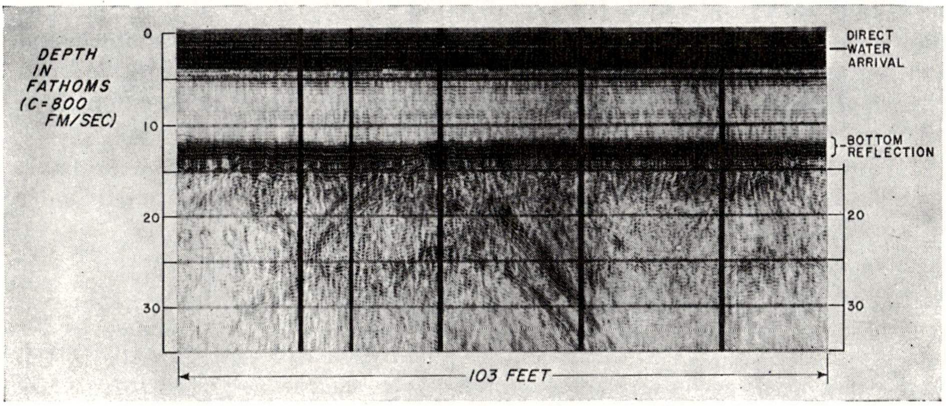

(b)

Fig. 21. (a) Crescent-form echo sequences from a rough bottom on the continental rise, western North Atlantic Ocean. (b) Sub-bottom crescent-form echo sequences from Vineyard Sound, Massachusetts.

After multiple reflections and side echoes have been identified and slope corrections made, the primary echo sequences are plotted on a structure profile. If velocity information is available it may be worthwhile to convert travel times to distances at this point. However, the common practice has been to preserve the travel times through this step. With either choice, reflecting horizons may be traced and charted by studying the whole pattern of profiles. Model-making has proven helpful in several investigations. Fig. 11 from the reconnaissance survey of the northwestern approach to the Strait of Messina is an example.

B. Geological Interpretation

The strata and other more complex structures delineated by the reflection method must be identified; this is usually best accomplished by associating them with known rocks or sediments, as by bottom cores. Inshore studies with small survey boats often permit one the luxury of tracing a sub-bottom reflector to an outcrop on shore. As mentioned previously there have been successful comparisons of profiles with core samples and with ease of penetration measured by the number of blows per foot required to penetrate bottom materials. Such a correlation is reassuring, but core samples are more useful for geologic interpretation. Control by core samples is now mostly limited to a few tens of feet in non-commercial research. The limitation is largely economic in shallow water; it is to be hoped that in the future the coring of deeper holes will be financed in areas of critical interest in shallow water as well as through such deep-water projects such as the Mohole.

Velocity and acoustic impedance (velocity × density) by themselves are not reliable means of identifying rocks or sediments. Thus it is well known that calcareous rock can have velocities ranging from close to 1.5 to over 7 km/sec. Nevertheless, considerable progress is being made in appreciating the characteristics of rocks and soft sediments affecting velocity. Some of these, such as the relationship between velocity and density, velocity and water content, and velocity and hydrostatic pressure, are discussed elsewhere in this work. We can hope that more and more reliable geological deductions can be based on these relationships.

Where regional or nearby local geology is well understood the probable geological structure can be deduced, as in the several examples cited under "Results".

6. Geophysical Measurements

Exactly as this recording method helps to correlate echo sequences for interpreting structure, so it can help to identify specific travel-paths in the water and in the bottom, thus permitting the scientist to measure processes which have previously been only vaguely identified in most work. Often the oblique reflection method described by Officer (1955) has seemed inadequate when applied to shot sequences because of small uncertainties of correlation. This difficulty virtually disappears in continuous profiling. Continuous refraction profiling is just coming into use, but early indications are: (1) we will find first arrival identification more certain, including ability to pick the first motion, (2) "second" arrivals can be picked with much more confidence, and (3) the effect of irregular topography on travel times will be displayed in much greater detail.

So far as the writer has been able to discover no measurements of acoustic impedance have been attempted during continuous profiling researches. The graphic records of multiple reflections, as in Figs. 18 and 19, suggest that such quantitative studies would be fruitful. Furthermore, it should be possible to use continuous profiling techniques to measure effective absorption coefficients.

7. Concluding Remarks

1. Several reflection surveys by continuous profiling methods have shown good agreement with evidence in cores and outcrops.

2. Sediment penetrations of several thousand feet have been achieved with the more advanced instruments, while even the earliest instruments recorded fine detail.

3. The principal obstacles to widespread use of the method appear to be: (a) masking by multiple reflections between shallow layers and between the sea surface and bottom, and (b) the probable rapid attenuation or scattering of down-going energy through certain strata such as beds of gravel or shell.

4. The method appears to offer a great improvement on well known reflection and refraction methods of measuring compressional wave velocities.

5. Acoustic impedance measurements can probably be made with considerable assurance because travel paths of arrivals can be identified and their energy measured.

6. Absorption in sediments can be measured in some places.

7. Obvious extensions of the method can be made to study seismic properties and the structure of sediments in deep water with source and receiver at the bottom.

References

Anderson, V. C., 1953. Wide band sound scattering in the deep scattering layer. *Scripps Inst. Oceanog. Ref.* 53–36, 35 pp. (unpublished manuscript).

Beckmann, W. C., A. C. Roberts and B. Luskin, 1959. Sub-bottom depth recorder. *Geophysics*, **24**, 749–760.

Caulfield, David, 1962. Predicting sonic pulse shapes of underwater spark discharges. *Deep-Sea Res.*, **9**, 339–348.

Elmendorf, C. H. and B. C. Heezen, 1957. Oceanographic information for engineering submarine cable systems. *Bell Syst. Tech. J.*, **36**, 1047–1093.

Ewing, J. I. and G. B. Tirey, 1961. Seismic profiler. *J. Geophys. Res.*, **6**, 2917–2927.

Heezen, B. C., Marie Tharp and M. Ewing, 1959. The floors of the oceans. I. The North Atlantic. *Geol. Soc. Amer. Spec. Paper* 65, 113 pp.

Hersey, J. B., 1957. Electronics in oceanography. *Electronics and Electron Physics*, **9**, 239–295. Academic Press, New York.

Hersey, J. B., 1962. Findings made during the June 1961 cruise of *Chain* to the Puerto Rico Trench and Caryn Sea Mount. *J. Geophys. Res.*, **67**, 1109–1116.

Hersey, J. B., H. E. Edgerton, S. O. Raymond and G. Hayward, 1961. Pingers and thumpers advance deep-sea exploration. *J. Instrument Soc. Amer.*, **8**, 72–77.

Hersey, J. B. and M. Ewing, 1949. Seismic reflections from beneath the ocean floor. *Trans. Amer. Geophys. Un.*, **30**, 5–14.

Hersey, J. B. and M. S. Rutstein, 1958. Reconnaissance survey of Oriente Deep (Caribbean Sea) with a precision echo sounder. *Bull. Geol. Soc. Amer.*, **69**, 1297–1304.

Hoskins, Hartley and S. T. Knott, 1961. Geophysical investigation of Cape Cod Bay, Massachusetts, using the continuous seismic profiler. *J. Geol.*, **69**, 330–340.

Knott, S. T. and J. B. Hersey, 1956. Interpretation of high-resolution echo-sounding techniques and their use in bathymetry, marine geophysics, and biology. *Deep-Sea Res.*, **4**, 36–44.

Luskin, B., B. C. Heezen, M. Ewing and M. Landisman, 1954. Precision measurement of ocean depth. *Deep-Sea Res.*, **1**, 131–140.

McClure, C. D., H. F. Nelson and W. B. Huckaby, 1958. Marine Sonoprobe system, new tool for geologic mapping. *Bull. Amer. Assoc. Petrol. Geol.*, **42**, 701–716.

Marti, M., 1922. Oscillographe enregistreur continu à grande sensibilité et son application au sondage continu par le son (7 Sept. 1922). Brevet d'Invention XII 3 No. 568–075.

Moore, David G., 1960. Acoustic-reflection studies of the continental shelf and slope off Southern California. *Bull. Geol. Soc. Amer.*, **71**, 1121–1136.

Murray, H. W., 1947. Topography of the Gulf of Maine. Field season of 1940. *Bull. Geol. Soc. Amer.*, **58**, 153–196.

Officer, C. B., 1955. A deep-sea seismic reflection profile. *Geophysics*, **20**, 270–282.

Padberg, L. R., 1958. Method of subsurface exploration by sonic means. U.S. Patent No. 2,866,512, 7 pp.

Smith, W. O., 1958. Recent underwater surveys using low-frequency to locate shallow bedrock. *Bull. Geol. Soc. Amer.*, **69**, 69–98.

Smith, W. O., C. B. Cummings and J. E. Upson, 1954. Mapping bedrocks with low-frequency sound. (Abs.). *Bull. Geol. Soc. Amer.*, **65**, 1307.

Veatch, A. C. and P. A. Smith, 1939. Atlantic submarine valleys of the United States and the Congo Submarine Valley. *Geol. Soc. Amer. Spec. Paper* 7, 101.

Worzel, J. L., 1959. Extensive deep-sea sub-bottom reflections identified as white ash. *Proc. Nat. Acad. Sci.*, **45**, 349–355.

5. THE UNCONSOLIDATED SEDIMENTS

J. I. Ewing and J. E. Nafe

1. Introduction

In this discussion the unconsolidated sediments are taken to be those in which the speed of compressional waves is less than about 4 km/sec. The upper limit is entirely arbitrary and has been chosen to include all the material with velocities lower than that of "Layer 2", which is discussed by Raitt in the next chapter. Layers in which the velocity is between 2 and 4 km/sec have been described by various investigators as semi-consolidated sediments. Other authors have designated layers as unconsolidated, semi-consolidated and consolidated sediments and rocks. In almost every case the designation has been made solely on the basis of seismic wave velocity, since identification according to sediment type, age or stratigraphic position has not been possible except in certain shallow-water areas where wells have been drilled nearby.

Most of the marine seismic measurements reported to date have been made in the Atlantic and Pacific Oceans and in the shallower inter-continental seas. In all areas investigated, the thickness of the unconsolidated sediments varies widely from zero on some of the topographic highs to several kilometers in some of the deeps and in some continental rise areas. The relationship of sediment thickness and topographic setting has become increasingly more understandable, and even predictable, in recent years since Kuenen introduced the concept of transport of sediments by density or turbidity currents. This concept has been thoroughly verified by sediment coring and seismic measurements. It is now well known that even the coarser sediments derived from continental erosion can be carried great distances from land, flowing out into the ocean basins and being empounded in the deepest accessible areas (see Chapter 14 of this volume). It is also known from underwater photography, coring, dredging, and seismic measurements that many topographic highs are devoid of even pelagic sediments, presumably being kept clean by deep currents or by slumping.

In addition to the variation of sediment thickness with topographic setting, there is a marked difference in average thickness of sediments in the Atlantic Ocean and in the Pacific. The Atlantic has an average of more than 1 km (Ewing and Ewing, 1959), compared with less than $\frac{1}{2}$ km in the Pacific (Raitt, 1956). This difference is generally attributed to the facts that the Atlantic is much smaller and that more large, sediment-bearing rivers flow into it.

The thickness of sediments cannot always be measured exactly by seismic methods, owing to the difficulty of measuring the velocity, particularly in the uppermost part of the section in the deep ocean. Refracted waves give accurate measurements of the higher velocity layers[1]; but because of various factors such as poor velocity or acoustic impedance contrast between the water and upper sediments, or between two sediment layers, clear reflected or refracted waves from those layers are sometimes not observed. Fig. 1 shows the time–distance graph and structure section for a profile recorded in the South Atlantic

[*MS received July, 1960*]

in an area of thick sedimentary cover. (For a general description of time–distance graphs and of their relationship to various structures, the reader is referred to Chapter 1.) This is typical of the seismic data from which sedimentary velocities and thicknesses are computed. The R_I, R_{II} and R_{III} curves are determined by reflected waves from the first, second and fourth subsurface interfaces respectively. The lines G_3 and G_4 are determined by refracted arrivals associated with the tops of Layers 4 and 5. Note that no G_1 line (corresponding to refracted arrivals along the sea floor) is observed. This is typical of practically all, if not all, deep-ocean areas. The significance of the absence of

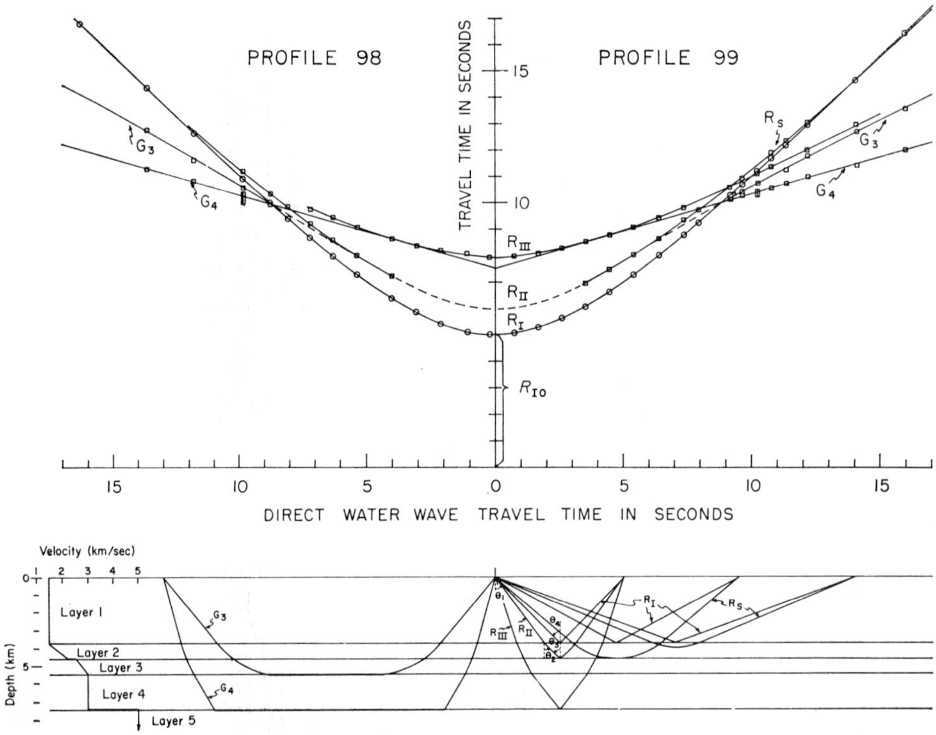

Fig. 1. Time–distance graph, structure section and ray diagram for end-to-end profiles S1-98 and 99. Velocity versus depth relationship shown at left of section. Ray paths at left of receiving position are for waves refracted at interfaces. At the right are ray paths for reflected waves and waves refracted by the velocity gradient.

such arrivals is that, in most places, the seismic wave velocity in the upper part of the deep-ocean sediments must be equal to or less than that in the water at the sea floor. The fact that reflected arrivals are received from this interface indicates that the acoustic impedance, ρC (density × velocity), is different above and below the interface. The water and sediment mixture must be more dense than the water alone, hence the following relationships can be written as typical of the water–sediment interface

$$C_1 \geqq C_2 \qquad \rho_1 C_1 \neq \rho_2 C_2 \qquad \rho_1 < \rho_2. \tag{1}$$

2. Evidence for Gradients and Low-Velocity Sediments

The evidence for the velocity structure shown by the velocity versus depth curve in Fig. 1 is derived from both refracted and reflected waves. In a time–distance graph, there is, theoretically, a reflection curve for each interface and a refraction line tangent to it, the inverse slope of which gives the velocity in the material below. If there is a velocity discontinuity at each interface and if velocity does not vary with depth in any layer, the time–distance graph will consist of straight line segments (for refracted arrivals) and hyperbolic curves (for reflections). If the velocity increases in each successively deeper layer, R_N will cross R_{N-1}. In the case shown by Fig. 1, the average velocity, $\bar{C}_2$, in Layer 2 is higher than that in Layer 1, but R_{II} does not cross R_I. This indicates that the longer range arrivals on the R_{II} curve were not reflected from the second interface, as were those at shorter range, but were bent upward by a velocity gradient before penetrating to the depth of the interface. This behavior of reflected arrivals is typical of all ocean-basin areas where detailed seismic studies have been reported (Hill, 1952; Officer, 1955; Katz and Ewing, 1956; Nafe and Drake, 1957; Ewing and Ewing, 1959). It is convincing evidence that appreciable velocity gradients exist in the deep ocean sediments. Evidence for velocity gradients is found also in refraction data and has been summarized by Nafe and Drake (1957). Refracted arrivals from various depths in the sediments taken from a number of profiles throughout the Atlantic Ocean show a systematic increase of velocity with depth. The average value of gradient is in agreement with that computed from reflection data.

As mentioned before, the fact that refracted arrivals are not received from the water–sediment interface is interpreted to mean that either there is no velocity discontinuity across the interface or the velocity below is lower than that above. We have evidence for both cases. In certain areas the first sub-bottom reflection curve, R_{II}, is observed to approach a line parallel to and above R_I on the time–distance graph. This can occur only if the velocity in the upper sediments is lower than water velocity. In other areas the R_{II} curve approaches R_I, indicating no velocity discontinuity across the water–sediment interface.

Further evidence for low-velocity sediments has been shown by Officer (1955) and Katz and Ewing (1956) from the study of guided waves traveling in the bottom just beneath the water–sediment interface. These waves have been recorded in many areas of the Atlantic. They appear at a constant frequency at ranges beyond that of the water-borne refracted wave which grazes the sea floor (see Fig. 5, p. 79). To account for the constant frequency and travel time of these arrivals, it is required that they be coupled waves, excited by the grazing ray in the water, and that they travel in a wave guide in the sediments. The wave guide might be a low-velocity layer, as described by Katz and Ewing, or it might be bounded above by a discontinuity and below by a velocity gradient. In either case the coupling would require that the upper sediments have a lower velocity than the bottom water.

The evidence cited above indicates two significant features of deep-ocean sediments: (1) they have appreciable velocity gradients, particularly in the upper few hundreds of meters, and (2) the velocity in the uppermost part of the sedimentary column is equal to or lower than that in the bottom water in most places. Direct measurement of sound velocities in sediments (Laughton, 1954; Hamilton, 1956; Hamilton *et al.*, 1956; Shumway, 1956; Sutton, Berckhemer and Nafe, 1957) has shown velocities generally ranging between 1.45 to 1.80 km/sec. Most values for deep ocean sediments are in the lower range, in good agreement with the seismic evidence. In the intermediate and shallow areas, higher velocities are found by both types of measurement. A comparison of velocity versus depth relationships in shallow-water and deep-water sediments is given by Nafe and Drake in Chapter 29.

3. Variable Angle Reflections

The method used to determine the amount of gradient by seismic reflection profiles is outlined briefly in Chapter 1 and has been described in more detail by Hill (1952), Officer (1955) and Katz and Ewing (1956). It is a process in which a family of theoretical curves is compared with the observational data

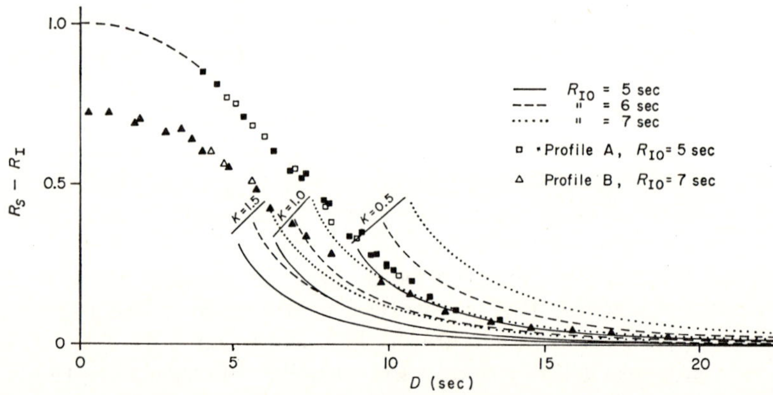

Fig. 2. Theoretical curves for $R_S - R_I$ versus D. Observational data from two profiles are plotted. Solid symbols represent arrivals observed in first-order reflections and refractions; open symbols represent second-order observations.

and the best fit determined. The curves shown in Fig. 2 (from Nafe and Ewing, unpublished manuscript) give theoretical values of $R_S - R_I$ versus range, D, for three different values of water depth and gradient, where R_S is the travel time of the wave refracted in the sediments, R_I is that of the bottom-reflected wave and D is that of the direct water wave. These curves were computed for linear velocity gradients, for which the formulae are

$$R_S = R_{I0} \sec \theta_1 + \frac{2}{K} (\sinh^{-1} \cot \theta_2 - \sinh^{-1} \cot \theta_3)$$

$$R_I = R_{I0} \sec \theta_4$$

$$D = \frac{C_1}{C_0} \left[R_{I0} \tan \theta_1 + \frac{2C_2}{KC_1} \left(\frac{\cos \theta_2 - \cos \theta_3}{\sin \theta_2} \right) \right], \tag{2}$$

where R_{I0} is the time for the bottom-reflected wave at zero shot-receiver distance, K is the gradient, C_0 is the velocity in the surface channel, C_1 is the mean velocity in the water, and C_2 is the velocity in the upper part of the sediments. The ray paths, with angles θ_1–θ_4 indicated, are shown in Fig. 1. The theoretical curves of Fig. 2 are those corresponding to the upper branch, R_S, of the refraction curve shown in Fig. 9, Chapter 1. Each curve begins at the "critical range" corresponding to the minimum distance at which a ray can be returned to the surface by refraction under the conditions of velocity gradient and water depth indicated. The R_D branches of the curves, corresponding to the deeper penetrating rays, are not shown. They join the R_S branches in cusps at the critical range and are curved in the opposite direction. For simplicity it was assumed that the water has uniform velocity equal to that in the upper sediments. Hence, in equations (2), $C_0 = C_1 = C_2$, $\theta_1 = \theta_2$ and, since we are dealing only with refracted waves, $\theta_3 = \pi/2$. For the ranges involved, it can be shown that errors introduced by the assumption that $C_0 = C_1$ are small, and as discussed earlier, we have much evidence that the velocity in the upper sediments is approximately equal to that in the water.

These same formulae can also be used for the cases in which the upper sediment velocity is either lower than or higher than water velocity. For $C_1 > C_2$, the curves corresponding to those in Fig. 2 would be displaced down and to the right. For $C_1 < C_2$, they would be displaced upward and toward the left. In the first case there is no critical angle for reflection, even at grazing incidence, hence the R_S arrival is always later than R_I. In the second case, R_S will cross R_I. In the case illustrated, i.e. $C_1 = C_2$, R_S joins R_I at the range corresponding to a grazing ray.

On these theoretical curves are plotted squares and triangles representing the observational data from two reflection profiles, A and B, with water depths corresponding to $R_{I0} = 5$ and 7 sec respectively. The data are plotted for ranges from 0 to 20 sec, hence both reflected, R_{II}, and refracted, R_S, arrivals are shown. In profile A, the outer points best fit the refraction curve for $R_{I0} =$ 5 sec and $K \simeq 0.45$ sec^{-1}. At ranges beyond about 9.3 sec, the observed arrivals are purely refracted. Since 9.3 sec is the *minimum* range at which a refracted arrival can be returned to the surface under the conditions $R_{I0} = 5$ sec and $K = 0.45$ sec^{-1}, the arrivals at shorter range must be reflected. The fact that the reflected arrivals join smoothly to the refracted ones indicates that the reflecting interface must lie at approximately the depth penetrated by the critical ray, i.e. the ray at range 9.3 sec. If the interface had been deeper, the reflection curve would have crossed the R_S–R_I curve above the cusp and joined the R_D–R_I curve as indicated diagrammatically in Fig. 3.

If a reflecting interface is shallower than the depth of penetration of the critical ray, the R_{II}–R_I curve approaches and joins the R_S–R_I curve from below as demonstrated by the arrivals of profile B in Fig. 2. For this profile $R_{I0} =$ 7 sec and the best fit to the data is with the curve for $K = 1.0$ sec^{-1}.

Fig. 4 shows the results from another profile for which $R_{I0} = 6$ sec and $K = 0.9$ sec^{-1}. There are two reflecting interfaces, both shallower than the depth

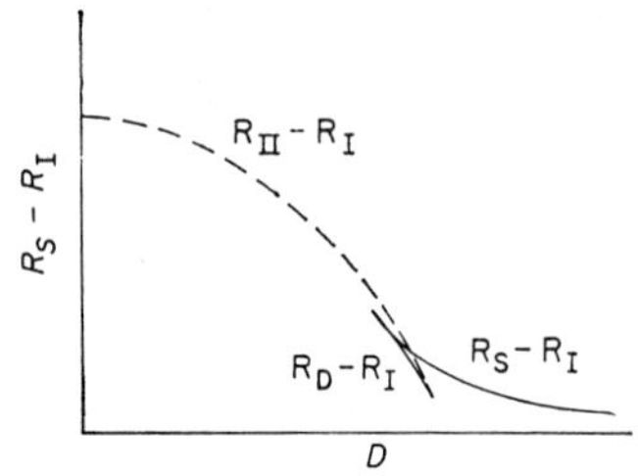

Fig. 3. $R_S - R_I$ versus D for case where 2–3 interface is deeper than depths penetrated by purely refracted ray.

of penetration of the critical ray, as evidenced by the fact that both reflection curves pass below the cusp of the refraction curve for $K = 0.9$ sec^{-1}. From the fact that both curves indicate approximately the same value of K, it would appear that the upper reflector is a thin layer, above and below which the velocity and velocity gradient are not greatly different. This can be stated only qualitatively because the precision of picking the arrivals in the region where the curves come together is not high.

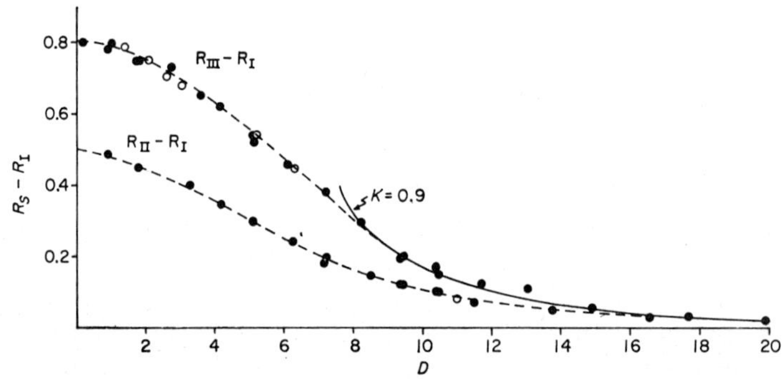

Fig. 4. $R_S - R_I$ versus D for gradient 0.9 sec^{-1}, $R_{I0} = 6$ sec and two reflecting interfaces.

In the profiles shown in Fig. 1, the gradient in Layer 2 is found by the methods just described to be approximately 0.5 sec^{-1}. There is some evidence that the velocity continues to increase with depth below the 2–3 interface. This comes from the relationship of the R_{II} curves and the G_3 lines. Although nearly so, the observed portions of the refraction lines are not tangent to the reflection curves as would be the case if they are associated with the same interface. Certainly it is possible that two interfaces are present; the upper one producing reflections but no refractions and the lower one producing refractions but no reflections, but this seems an unlikely interpretation. In order for the extension of the refraction lines to be tangent to the R_{II} curve, they would have to curve downward in the manner indicated by the dashed lines, indicating a gradient below the 2–3 interface. The fact that the outer

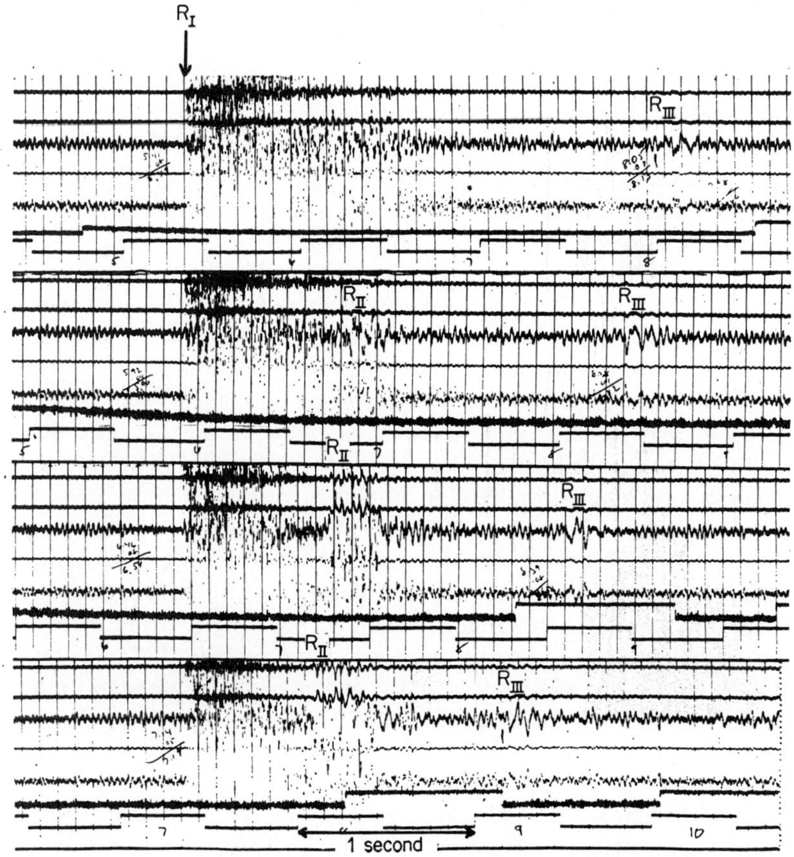

Fig. 5. Seismic records showing critical angle reflections for R_{II}. R_{III} arrivals are from base of low-velocity sediments. Water depth = 3.75 km. Thickness of low-velocity sediments $\simeq$ 4 km.

portions of the G_2 lines are observed as straight lines indicates that at some depth below the 2–3 interface, the velocity becomes constant or nearly so.

The 2–3 interface is judged to be a discontinuity with the higher velocity below on the basis of the observation of a critical angle for reflections from it.

Fig. 5 shows four records of the reflections at ranges of 2.6, 3.5, 4.4 and 5.4 sec from Profile 99. Note the sharp change in character and strength of R_{II} in this relatively short range. The same behavior was noticed in Profile 98 and is a common feature of most sub-bottom reflection profiles in deep-water areas.

The 4–5 interface is determined by the refracted arrivals corresponding to lines G_4 and by reflected arrivals R_{III}. Samples of the R_{III} signals are shown in Fig. 5 in addition to the R_{II} arrivals. In the case of the 4–5 interface, note that there is perfect agreement between reflected and refracted arrivals (Fig. 1) in that the G_4 lines are tangent to the R_{III} curves.

As mentioned earlier, it is not unusual to record strong reflections from within the sedimentary column without receiving the corresponding refracted arrivals. This can be an indication that the reflector is a thin layer, or it could in some cases be explained if the velocity discontinuity is to a lower velocity below. Still another explanation can be that the discontinuity is one of density rather than of velocity, i.e. $C_1 = C_2$; $\rho_1 C_1 \neq \rho_2 C_2$.

The foregoing has been an attempt to describe the methods of obtaining velocity structure in the sediments in deep-ocean areas, and some specific results have been shown. Although the cases discussed were confined to comparison with linear gradients, comparisons have been satisfactorily made with parabolic and exponential curves. It is improbable that any one type of gradient is applicable everywhere, and, indeed, in most cases the best fit might be obtained with a combination, i.e. linear to a certain depth, and parabolic below that. Previously reported results on gradients measured by seismic techniques have been given by Hill (1952), Officer (1955), Katz and Ewing (1956) and Nafe and Drake (1957), in addition to laboratory measurements on artificially compacted sediments by Laughton (1954). These results and others unreported have shown average gradients, principally from Atlantic Ocean profiles, to vary between 0.4 and 2.5 sec^{-1} for the upper 0.3 to 1.0 km of sediments. The variations are undoubtedly due in part to observational or interpretational differences. Some profiles are shot with more appropriate charge sizes or shot spacing than others, and some have much simpler interpretations than others. However, a considerable part of the variations is probably real, indicating differences in sediment types, rates of sedimentation, porosity, grain size, lithification, and other factors affecting seismic velocity (Sutton, Berckhemer and Nafe, 1957). At present there has not been a sufficient number of determinations to permit the classification of areas by sediment velocity structure except possibly, as suggested by Nafe and Drake (1957), to predict that in areas where calcium carbonate deposition is high, sediment velocities will also tend to be high. Most of the better determinations have been made in abyssal plains or other flat-lying areas in order to avoid topographic complications, and results from areas which receive no turbidity-current deposits, if available, might show significant differences.

Table I gives values of K from several profiles determined by fitting the theoretical curves for linear gradients. The highest value is that measured by

Hill (1952). The lowest was measured on cruise VEMA-15, Profile A of Fig. 2. The low value of the gradient in this case is not unreasonable, because it is the average gradient in approximately 1 km of sediments. The higher values come from profiles in which only the upper 0.2–0.4 km was measured, and higher gradients are expected near the sea floor. Hill's value of 2.5 sec^{-1} was also measured over a depth of approximately 0.4 km. A possible alternative interpretation of his results, allowing that the closer points represent reflections instead of refractions, could be made which would give a lower value of gradient. As shown in Figs. 2 and 4, if there is a sub-bottom reflector at a depth less than that reached by the ray corresponding to the cusp, the reflection curve R_{II} associated with it will join smoothly to the R_S curve, and it would be difficult to distinguish between the two types of arrival.

TABLE I

Velocity Gradients in the Sedimentary Layer

Measurements	Latitude	Longitude	Water depth (R_{I0}), sec	Velocity gradient (K), sec^{-1}
Sl-98 & 99	52° 25′S	40° 35′W	5.0	0.45
H-8	23° 32′N	71° 05′W	7.2	1.0
H-11	26° 10′N	75° 02′W	6.1	0.9
H-19	33° 20′N	71° 30′W	7.0	1.4
H-20	33° 05′N	73° 45′W	6.5	1.4
H-15	36° 15′N	67° 10′W	6.5	1.5
H-33	38° 05′N	70° 00′W	5.1	1.0
H-4	21° 26′N	67° 29′W	6.8	1.2
O-55	35° 15′N	64° 30′W	6.7	1.0
MH-52	53° 50′N	18° 40′W	3.2	2.5

4. Normal-Incidence Reflections

In addition to variable-angle reflection and refraction measurements, a large number of normal-incidence reflection measurements have been made. Only a small percentage of these have been published (Hersey and Ewing, 1949; Shor, 1959). Hersey and Ewing made certain classifications of reflection records on the basis of the character of the bottom reflection and of the number and type of sub-bottom reflections. They were able to show some correlation between reflection types and physiographic provinces for certain areas of the western North Atlantic. Shor's measurements in the Pacific covered the boundary between the present-day clay deposition and carbonate deposition areas and showed markedly thicker sedimentary cover in the latter (southern) region. His results also showed greater accumulation of material in valleys than on hills.

Sub-bottom reflections on standard echo-sounding records have been discussed by Hersey and Rutstein (1958), Heezen, Tharp and Ewing (1959), Worzel (1959) and Ewing, Luskin, Roberts and Hirshman (1960). These studies have shown that, in many areas, penetration of the sediments to depths greater than 100 ft is achieved, even at the relatively high frequencies employed in the sounding equipment (12 kc/s). The sub-bottom echoes are often stronger than the bottom reflection. In certain areas, particularly on moderate topographic highs, these sub-bottom interfaces can be traced for many miles. As in Shor's results, the depth to the sub-bottom reflector is usually correlated with topography; the upper layer thins or pinches out on hills and thickens in valleys. In abyssal plains, the sub-bottom reflectors are usually not as continuous as on the rises, as might be expected when one considers that the plains are subjected to turbidity current deposition and the rises only to pelagic sedimentation.

Efforts to sample specific sub-bottom interfaces with coring apparatus have been made on two occasions. Worzel identified a prominent reflector in the southeastern Pacific as a widespread layer of white volcanic ash. Ewing *et al.* (1960) found that strong reflectors on the outer rise of the Puerto Rico Trench correlated with increases in rigidity, presumably associated with increased carbonate content.

Continuous profiling of sub-bottom reflections in shallow-water areas is described by Hersey in Chapter 4. This recent development has made it possible to obtain detailed surveys of sediment structure to depths of thousands of feet in some areas and offers great advances in the study of marine stratigraphy.

5. Summary

Measurements of velocity gradients by variable-range reflection studies give average values between 0.9 and 1.4 sec^{-1} for the upper 0.2 to 0.4 km of sediments in deep-water areas. With the evidence cited previously for low velocities immediately below the water–sediment interface, we can consider that the velocity versus depth relationship most common in deep-water sediments is approximately described by a parabolic or exponential function in which the upper sediments have velocities equal to or slightly lower than that in the water. As greater depth below the sea floor is reached, the high initial gradient diminishes rapidly and may become insignificant at depth in some areas. There is some indication that the high gradients persist to some depth at which de-watering is largely achieved. Below this depth compaction would be retarded, resulting in lower gradients. Such behavior would be consistent with observations of porosity versus pressure of the kind summarized by Hamilton (1959) on the consolidation of sediments. Porosity decreases rapidly at first, then more slowly with increasing pressure, and would be expected to result in the variable gradients of the type observed.

As shown by Nafe and Drake (1957), gradients in shallow-water sediments generally do not vary as markedly with depth as do those in deep water.

Reference is suggested to this paper and to Chapter 29 for an analysis of the relationships between compressional- and shear-wave velocities, porosity, density and other physical properties. The evidence they presented for differences between the velocity–depth dependence in deep and shallow water was statistical and based on tabulations of reported velocities and depths. For accurate information on the depth variation of velocity in the uppermost portion of the sedimentary column more sensitive methods are required. Seismic records obtained by shooting and recording on the ocean bottom should provide the necessary additional precision.

Thicknesses of low-velocity sediments measured by reflection and refraction techniques range from an average of 1 km in the Atlantic to $\frac{1}{2}$ km in the Pacific. As suggested by Hamilton (1959) in a comprehensive review of compaction and lithification processes and of measured velocities and thicknesses, there is good reason to believe that the underlying "Layer 2", described by Raitt in Chapter 6, is also sedimentary material, presumably well consolidated or lithified. This interpretation has been cited by Hamilton and others to establish agreement between the amount of material estimated to have been eroded from continents and that found by seismic measurements in the oceans.

Lamont Geological Observatory (Columbia University), Palisades, New York. Contribution no. 593.

References

Ewing, J. and M. Ewing, 1959. Seismic refraction measurements in the Atlantic Ocean basins, in the Mediterranean Sea, on the Mid-Atlantic Ridge, and in the Norwegian Sea. *Bull. Geol. Soc. Amer.*, **70**, 291–317.

Ewing, M. and B. C. Heezen, 1955. Puerto Rico Trench topographic and geophysical data. *Geol. Soc. Amer. Spec. Paper* 62, 255–286.

Ewing, J., B. Luskin, A. Roberts and J. Hirshman, 1960. Sub-bottom reflection measurements on the continental shelf, Bermuda banks, West Indies arc, and in the west Atlantic basins. *J. Geophys. Res.*, **65**, 2849–2859.

Ewing, M., G. H. Sutton and C. B. Officer, 1954. Seismic refraction measurements in the Atlantic Ocean. Part VI: Typical deep stations, North America Basin. *Bull. Seism. Soc. Amer.*, **44**, 21–38.

Hamilton, E. L., 1956. Low sound velocities in high-porosity sediments. *J. Acoust. Soc. Amer.*, **28**, 16–19.

Hamilton, E. L., 1959. Thickness and consolidation of deep-sea sediments. *Bull. Geol. Soc. Amer.*, **70**, 1399–1424.

Hamilton, E. L., G. Shumway, H. W. Menard and C. J. Shipek, 1956. Acoustic and other physical properties of shallow water sediments off San Diego. *J. Acoust. Soc. Amer.*, **28**, 1–15.

Heezen, B. C., M. Tharp and M. Ewing, 1959. The floors of the oceans. I. The North Atlantic. *Geol. Soc. Amer. Spec. Paper* 65, 122 pp.

Hersey, J. B. and M. Ewing, 1949. Seismic reflections from beneath the ocean floor. *Trans. Amer. Geophys. Un.*, **30**, 5–14.

Hersey, J. B. and M. Rutstein, 1958. A reconnaissance survey of the Oriente Deep with a precision echo sounder. *Bull. Geol. Soc. Amer.*, **69**, 283–289.

Hill, M. N., 1952. Seismic refraction shooting in an area of the eastern Atlantic. *Phil. Trans. Roy. Soc. London*, **A244**, 561–596.

Katz, S. and M. Ewing, 1956. Seismic refraction measurements in the Atlantic Ocean. Part VII: Atlantic Ocean Basin, west of Bermuda. *Bull. Geol. Soc. Amer.*, **67**, 475–510.

Kuenen, P. H., 1948. Turbidity currents of high density. *Intern. Geol. Cong., Rep. 18th Session, Great Britain, 1948*, Pt. VIII, 44–52.

Laughton, A. S., 1954. Laboratory measurements of seismic velocities in ocean sediments. *Proc. Roy. Soc. London*, **A222**, 336–341.

Nafe, J. E. and C. L. Drake, 1957. Variation with depth in shallow and deep water marine sediments of porosity, density and the velocities of compressional and shear waves. *Geophysics*, **22**, 523–552.

Officer, C. B., 1955. A deep sea reflection profile. *Geophysics*, **20**, 270–282.

Raitt, R. W., 1956. Seismic refraction studies of the Pacific Ocean basin. Part I. *Bull. Geol. Soc. Amer.*, **67**, 1623–1640.

Shor, G. G., Jr., 1959. Reflexion studies in the eastern equatorial Pacific. *Deep-Sea Res.*, **5**, 283–289.

Shumway, G., 1956. Resonance chamber sound velocity measurements. *Geophysics*, **21**, 305–319.

Sutton, G. H., H. Berckhemer and J. E. Nafe, 1957. Physical analysis of deep-sea sediments. *Geophysics*, **22**, 779–812.

Worzel, 1959. Extensive deep sea sub-bottom reflections identified as white ash. *Proc. Nat. Acad. Sci.*, **45**, 349–355.

6. THE CRUSTAL ROCKS

R. W. RAITT

1. Introduction

Knowledge of the depth, velocity and thickness of the rock layers beneath the unconsolidated sediment has been gained mostly from observations of the first arrival times of waves refracted through these layers. Later arrivals sometimes supply additional useful information, but on the whole our present ideas regarding oceanic crustal structure depend almost entirely on the interpretation of the travel times of first arrivals.

The minimum distance at which a bottom-refracted wave is observed to arrive ahead of the bottom reflection is about 6 or 8 km in the typical deep-sea location. At this distance its intensity is strong, even for a small explosive charge of only a few pounds weight. There is usually no difficulty in measuring its time of arrival to a precision of about 0.01 sec.

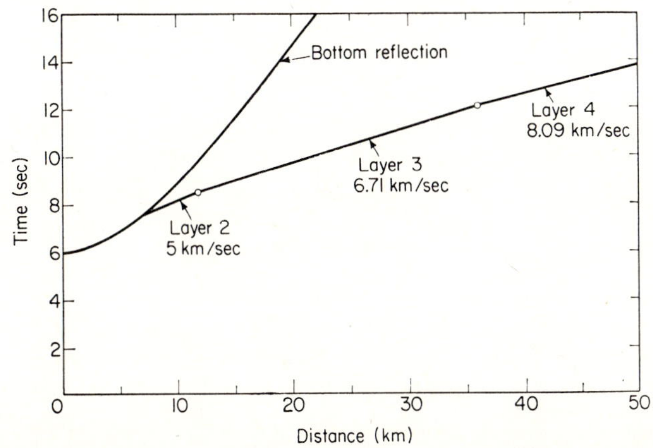

Fig. 1. Average travel-time–distance plot of an oceanic refraction profile.

Fig. 1 is a travel-time–distance plot which illustrates the main features of oceanic refraction profiles. It represents the calculated travel times for a model of the average oceanic structure as summarized by Hill (1957), consisting of the following layers:

Layer	Thickness, km	Velocity, km/sec
Sea-water	4.5	1.5
1	0.45	2
2	1.75	5
3	4.7	6.71
4	–	8.09

The refracted arrivals have three segments representing the three layers 2, 3 and 4 respectively. The distance intervals within which a given layer is observed

[*MS received July, 1960*]

as a first arrival are: Layer 2, 7 km to 12 km; Layer 3, 12 km to 36 km; Layer 4, > 36 km.

In actual practice the time–distance plots differ from the idealized model of Fig. 1 for various reasons:

1. Layer velocities and thicknesses differ somewhat from the assumed values.
2. Layering is not horizontal.
3. The velocity within one layer may vary.
4. A layer may be prevented from observation as a first arrival by prior arrivals from another layer.

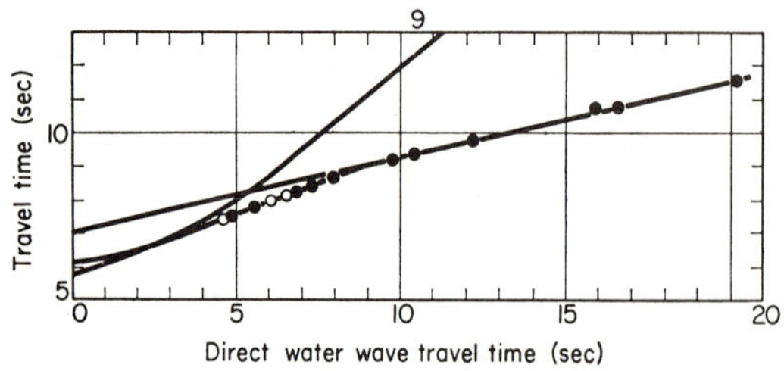

Fig. 2. Travel-time plot for *Challenger* Station CR9 near the Hawaiian Islands. Layer 2 velocity = 4.30 km/sec; Layer 3 velocity = 6.58 km/sec. (After Gaskell, Hill and Swallow, 1958.)

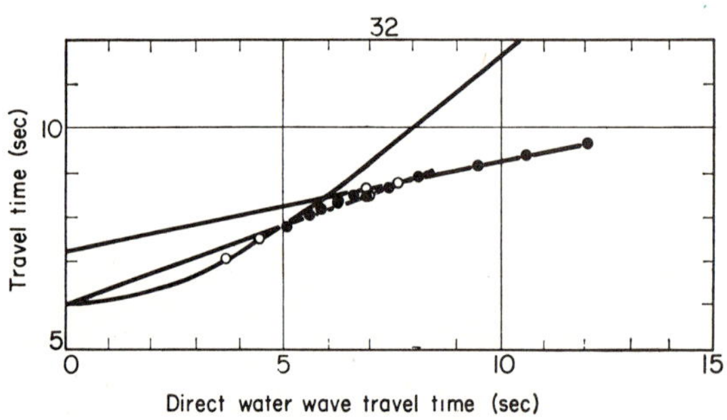

Fig. 3. Travel-time plot of *Challenger* Station 32 in the Atlantic near the British Isles. Layer 2 velocity = 4.36 km/sec; Layer 3 velocity = 7.15 km/sec. (After Gaskell, Hill and Swallow, 1958.)

Nevertheless, widely spaced deep-sea seismic-refraction stations usually have a strong similarity to the model. Studies at widely separated stations in the Atlantic and Pacific Oceans have shown the presence of all of the four layers. In the Indian Ocean the few observations of Gaskell, Hill and Swallow (1958) on H.M.S. *Challenger* did not reach Layer 4, but Layers 1, 2 and 3 were observed.

These seismic-refraction studies are not yet as widely distributed over the world oceans as might be desired. They tend to be concentrated in areas of good weather, or in regions close to the operating bases of the investigating laboratories. There are large areas in which there are no observations. For example, there are no published data on the Arctic Ocean and the South Atlantic, and there are only five stations in the Indian Ocean concentrated in a small region near the equator.

Present results certainly cannot be regarded as an adequate sample of all oceans. There is still the possibility that some of the areas not yet investigated will differ significantly from the average of those already studied. This is a continuing study, and exploration of many of the open areas is being planned or is in progress. As new data become available, modification of our present ideas about oceanic structure is expected.

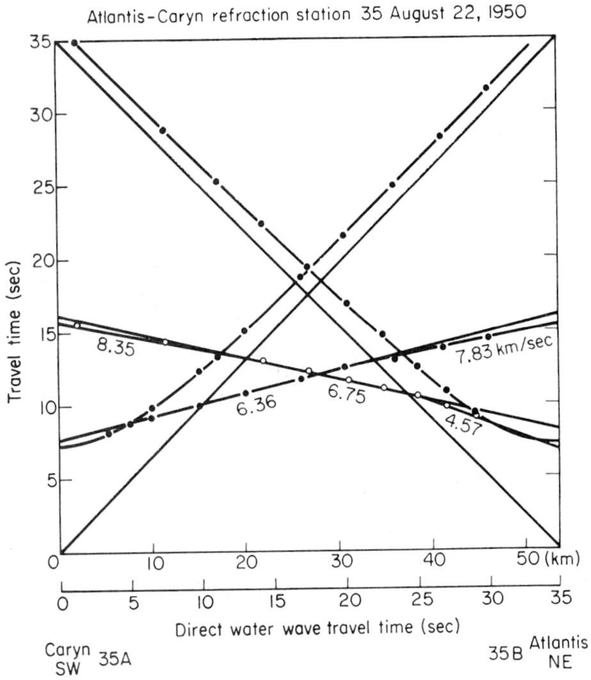

Fig. 4. Travel-time plot of *Atlantis–Caryn* Station 35, southeast of Bermuda. Layer 2 velocity = 4.57 km/sec; Layer 3 velocity = 6.54 km/sec; Layer 4 velocity = 8.08 km/sec. (After Officer, Ewing and Wuenschel, 1952.)

Seismic-refraction observations of the oceanic crust have been obtained by investigations at a relatively small number of laboratories. In the United States work has been done by Woods Hole Oceanographic Institution, Lamont Geological Observatory and Scripps Institution of Oceanography; in England by the Department of Geodesy and Geophysics of the University of Cambridge; and in the U.S.S.R. by the Institute of Geophysics and the Institute of Oceanology.

Examples of travel-time plots taken from published work of these laboratories are illustrated in Figs. 2–8. Fig. 2 represents a radio-sonobuoy station of the Cambridge University Department of Geodesy and Geophysics (Gaskell, Hill and Swallow, 1958). It illustrates a very clear-cut example of the second layer at a Pacific station which is more conspicuous than the average station because of the thin sediment, 0.05 km, and the unusually thick Layer 2, 2.7 km. Layer 3 is well observed, but penetration to Layer 4 was not achieved.

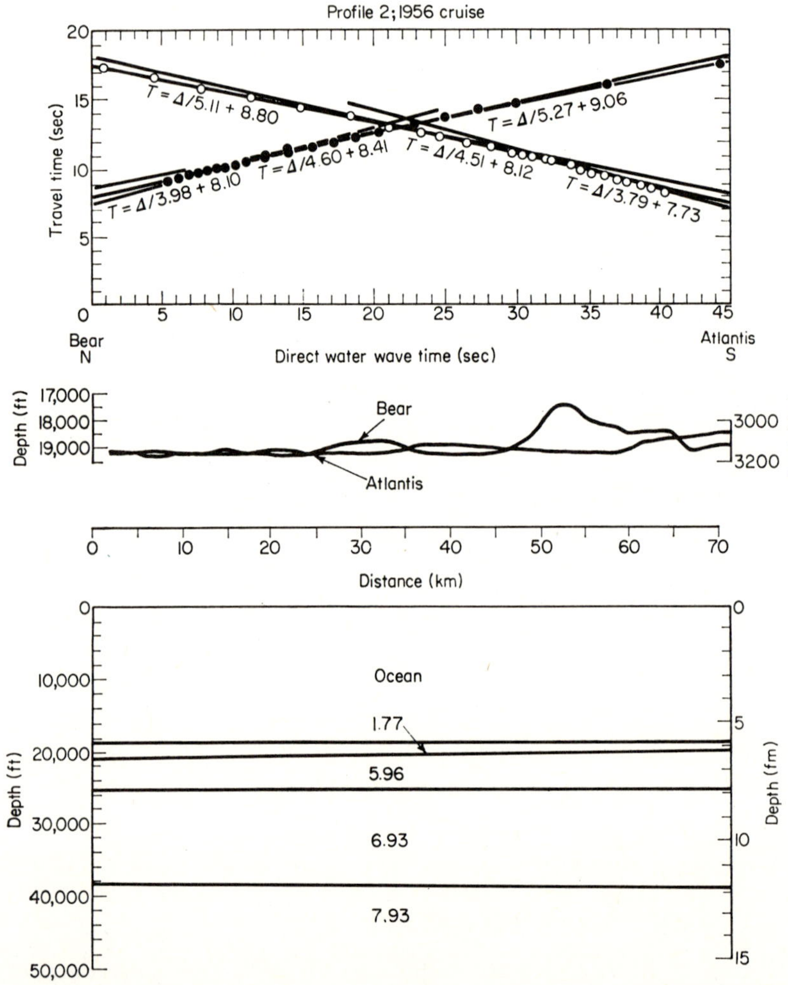

Fig. 5. Travel-time plot of *Atlantis–Bear* profile 2, 1956 cruise. Layer 2 velocity = 5.96 km/sec; Layer 3 velocity = 6.93 km/sec; Layer 4 velocity = 7.93 km/sec. (After Officer *et al.*, 1959.)

Fig. 3 represents another station of the Cambridge group, this one also taken by H.M.S. *Challenger* in the Atlantic. It also gives a good example of Layer 2 because of its fairly great thickness, 2.3 km. The length of the Layer 3 segment is shorter than in Fig. 2 but long enough to give a good demonstration of the Layer 3 velocity, 7.10 km/sec, and to determine the Layer 2 thickness.

Fig. 4 shows an example of a station recorded by Lamont Geological Observatory (Officer, Ewing and Wuenschel, 1952) using Woods Hole Oceanographic Institution vessels *Atlantis* and *Caryn*. It gives an example of all three layers 2, 3 and 4 depicted in Fig. 1.

Fig. 5 shows a station recorded about 300 nautical miles northeast of Puerto Rico. It is one of a large number of stations observed in a combined operation of Woods Hole Oceanographic Institution and Lamont Geological Observatory (Officer *et al.*, 1959) in a study of the structure of the eastern Caribbean. The structure here is nearly horizontal and yielded good determinations of all three layers beneath the sediment.

Fig. 6 is an example of one of the very earliest successful deep-sea refraction

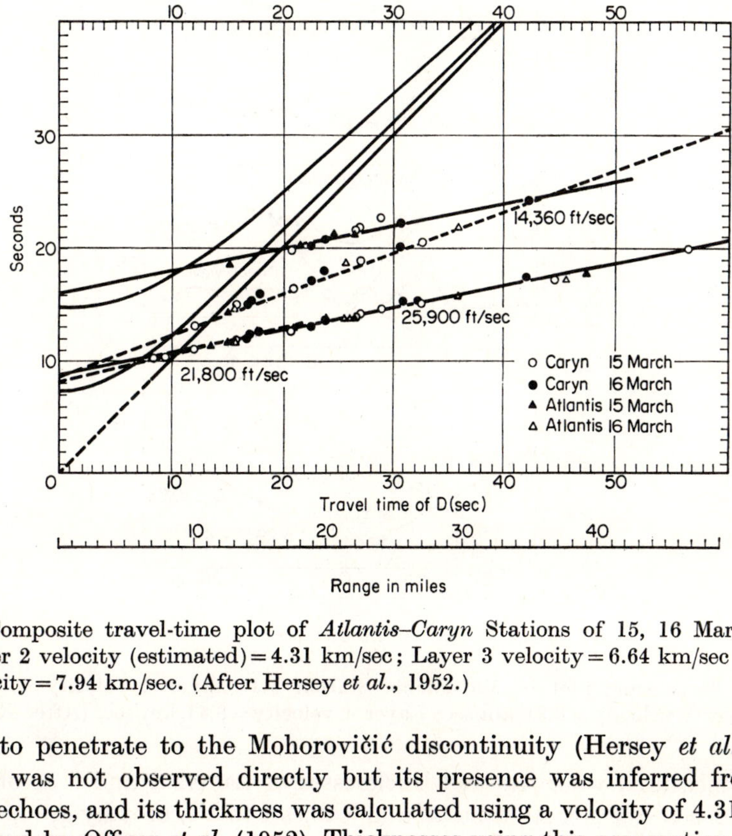

Fig. 6. Composite travel-time plot of *Atlantis–Caryn* Stations of 15, 16 March, 1949. Layer 2 velocity (estimated) = 4.31 km/sec; Layer 3 velocity = 6.64 km/sec; Layer 4 velocity = 7.94 km/sec. (After Hersey *et al.*, 1952.)

profiles to penetrate to the Mohorovičić discontinuity (Hersey *et al.*, 1952). Layer 2 was not observed directly but its presence was inferred from sub-bottom echoes, and its thickness was calculated using a velocity of 4.31 km/sec determined by Officer *et al.* (1952). Thicknesses using this assumption and the intercepts of Layers 3 and 4 are: Layer 1 = 0.42 km; Layer 2 = 2.26 km; Layer 3 = 2.49 km.

Fig. 7 is an example of a Pacific station in the region between the Hawaiian Islands and the coast of North America (Raitt, 1956). Three layers were observed. The second layer has the characteristic high velocity of the region. Because of this comparatively small velocity contrast with Layer 3 and the scatter of points caused by the rough bottom, Layer 2 is not prominent. However, the systematic decrease in velocity for the innermost shots on all profiles is unmistakable evidence for its presence. Calculated thicknesses are: Layer 1 = 0.26 km; Layer 2 = 0.93 km; Layer 3 = 6.24 km.

Fig. 8 is an unusual case of a very thick Layer 2 recorded near the equator in the Central Pacific on the "Capricorn" expedition (Raitt, 1956). The second layer arrivals are observed out to 20 km on either profile. Calculated thicknesses are: Layer 1 = 0.20 km; Layer 2 = 5.40 km; Layer 3 = 7.41 km.

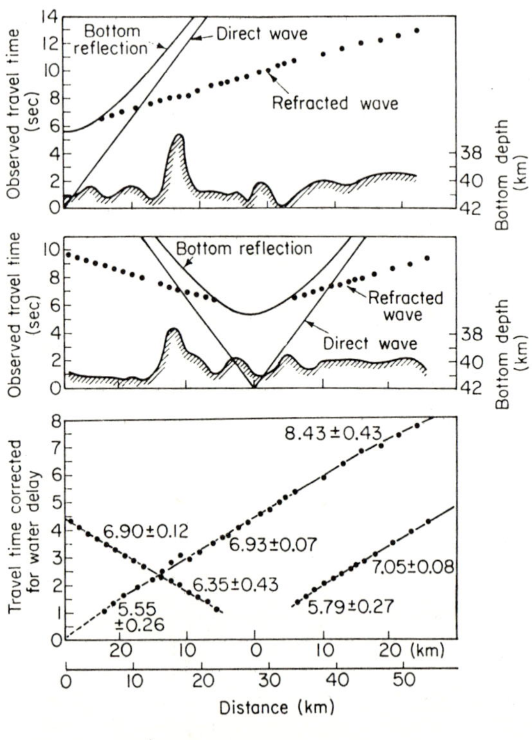

Fig. 7. Travel-time plot of Mid-Pacific Station M1. Layer 2 velocity = 5.88 km/sec; Layer 3 velocity = 6.91 km/sec; Layer 4 velocity = 8.41 km/sec. (After Raitt, 1956.)

The measured velocities and thicknesses of Layers 2 and 3 as observed at 134 deep-sea stations are tabulated in Tables I to VI, broadly grouped into six different oceanic areas: I. Western North Atlantic Ocean; II. Eastern North Atlantic Ocean; III. Eastern North Pacific Ocean; IV. Western North Pacific Ocean; V. South Pacific Ocean; VI. Indian Ocean. These results were taken from articles sometimes containing stations obviously atypical of a deep-ocean basin. Such stations have been excluded from this tabulation. The stations excluded have been those in ocean depths less than 3 km, in deep trenches such as the Tonga Trench, on the continental slope, on the flanks of islands, and on oceanic rises such as the Mid-Atlantic Ridge or Easter Island Rise known to be anomalous regions (Ewing and Ewing, 1959; Raitt, 1956).

As was indicated above, present results are but a fragment of a continuing investigation. We can expect our conclusions to be modified, or perhaps even changed radically, as new results are unfolded. However, present world-wide results as summarized in the tables support a picture of oceanic structure that has not changed greatly in the last few years. The summary of results presented by Hill in 1957 describes the average oceanic structure about as well as today.

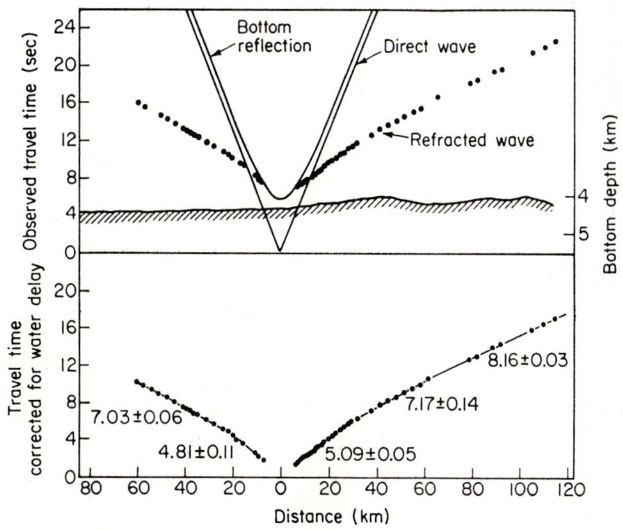

Fig. 8. Travel-time plot of Capricorn Station C2. Layer 2 velocity = 4.98 km/sec; Layer 3 velocity = 7.10 km/sec; Layer 4 velocity = 8.16 km/sec. (After Raitt, 1956.)

2. Layer 2

Surprisingly, the layer found at the base of the unconsolidated sediments, called Layer 2 for lack of a more descriptive name, has proved to be more difficult to measure than the deeper Layer 3 forming the main constituent of the crust. The reasons lie mostly in the short distance over which it is observed as a first arrival. Furthermore, this distance is sensitive to variation in the thicknesses of Layer 1 and Layer 2. If Layer 1 were thicker than in the example of Fig. 1 or if Layer 2 were thinner, the distance over which Layer 2 brings in a first arrival would be shortened, and it would not require a drastic change to eliminate Layer 2 entirely as a first arrival. It is occasionally observed as a second arrival but interference with the refracted wave of Layer 3 usually makes this observation unreliable.

TABLE I

Western North Atlantic Ocean

Station number	Ocean depth, km	Layer 2 Velocity, km/sec	Layer 2 Thickness, km	Layer 3 Velocity, km/sec	Layer 3 Thickness, km	Layer 4 Velocity, km/sec	Reference
—	5.30	(4.31)	2.26	6.64	2.49	7.94	Hersey *et al.* (1952)
A154	5.12			7.05	2.81	7.80	Ewing *et al.* (1950)[a]
A160-4	5.79			6.05			Tolstoy, Edwards and Ewing (1953)
A160-5	6.04			7.07			Tolstoy, Edwards and Ewing (1953)
A160-6	4.43	4.12	2.36	6.0			Tolstoy, Edwards and Ewing (1953)
A164-32	4.73	4.71	2.48	(7.10)			Officer, Ewing and Wuenschel (1952)
A164-33	4.72	(4.71)	2.25	7.10			Officer, Ewing and Wuenschel (1952)
A164-34	5.34	(4.51)	1.16	6.81	4.42	8.50	Officer, Ewing and Wuenschel (1952)
A164-35	5.71	4.57	1.72	6.54	2.63	8.08	Officer, Ewing and Wuenschel (1952)
A172-19	5.30	(4.51)	1.55	6.53	1.71	7.49[b]	Officer, Ewing and Wuenschel (1952)
A172-20	4.85	4.47	1.98			7.52[b]	Officer, Ewing and Wuenschel (1952)
16	5.84			6.51	2.35	7.56[b]	Ewing, Sutton and Officer (1954)
17	5.84			6.31	4.26	8.09	Ewing, Sutton and Officer (1954)
18	5.58			6.54	4.30	7.79	Ewing, Sutton and Officer (1954)
27	5.09			6.26	3.64	8.21	Ewing, Sutton and Officer (1954)
28	5.03			6.36	5.97	8.27	Ewing, Sutton and Officer (1954)
29	4.49			6.82	8.00	(8.00)	Ewing, Sutton and Officer (1954)
A164-9	5.05			6.55			Katz and Ewing (1956)
A164-11	5.29			6.46			Katz and Ewing (1956)
A164-12	5.40			6.87			Katz and Ewing (1956)
A164-13	4.15			6.24			Katz and Ewing (1956)
A164-14	4.23			6.25			Katz and Ewing (1956)
G-3	5.01			6.93	4.08	7.84	Ewing and Ewing (1959)
G-4	5.65			6.60	5.05	8.14	Ewing and Ewing (1959)
G-5	5.49			6.82	6.68	8.23	Ewing and Ewing (1959)
G-10	5.49			6.57	5.98	8.29	Ewing and Ewing (1959)
G-11	5.76			6.76	5.15	8.04	Ewing and Ewing (1959)
G-12	5.49			6.90	6.28	8.14	Ewing and Ewing (1959)
G-13	5.43			6.85	6.19	8.32	Ewing and Ewing (1959)

TABLE I (*contd.*)

Station number	Ocean depth, km	Layer 2 Velocity, km/sec	Layer 2 Thickness, km	Layer 3 Velocity, km/sec	Layer 3 Thickness, km	Layer 4 Velocity, km/sec	Reference
G-14	5.06			6.58	5.72	7.75	Ewing and Ewing (1959)
D-2	4.36			6.62	3.11	7.94	Ewing and Ewing (1959)
D-15	5.40			6.82	3.65	8.25	Ewing and Ewing (1959)
D-16	4.87			6.85	4.73	8.06	Ewing and Ewing (1959)
F-1	3.39			6.94	5.50	(8.00)	Ewing and Ewing (1959)
A156-28	5.02			6.59			Ewing and Ewing (1959)
A173-4	4.73			6.62	4.35	7.70	Ewing and Ewing (1959)
A157-29	5.05			6.82			Ewing and Ewing (1959)
A157-30	4.85			6.66	5.48	(8.00)	Ewing and Ewing (1959)
A157-31	5.05			6.65			Ewing and Ewing (1959)
A157-32	5.02			6.21	4.41	8.10	Ewing and Ewing (1959)
A157-33	5.05			6.63			Ewing and Ewing (1959)
G-15	4.57			6.81	6.71	8.12	Ewing and Ewing (1959)
E-1	4.79			6.69	5.25	8.22	Ewing and Ewing (1959)
A150-4	5.63			(6.50)			Ewing and Ewing (1959)
A150-5	4.67			(6.50)			Ewing and Ewing (1959)
A164-14	4.23			6.25			Katz and Ewing (1956)
A164-15	4.85			6.85			Katz and Ewing (1956)
A164-25	4.99			7.06			Katz and Ewing (1956)
A164-26	5.24			6.82			Katz and Ewing (1956)
A164-27	5.17			6.94			Katz and Ewing (1956)
A179-5	4.77	5.02	3.94	7.12			Katz and Ewing (1956)
A179-6	4.61			6.73	5.87	7.87	Katz and Ewing (1956)
A179-7	4.47			6.75	6.23	8.50	Katz and Ewing (1956)
A179-8	5.43			6.88			Katz and Ewing (1956)
A179-9	5.28			6.59			Katz and Ewing (1956)
A179-10	(5.14)	5.62	5.18[b]			8.05	Katz and Ewing (1956)
A179-11	5.24			6.32	4.06	7.68	Katz and Ewing (1956)
A179-12	5.28			6.36	3.29	7.83	Katz and Ewing (1956)

TABLE I (*contd.*)

Station number	Ocean depth, km	Layer 2 Velocity, km/sec	Layer 2 Thickness, km	Layer 3 Velocity, km/sec	Layer 3 Thickness, km	Layer 4 Velocity, km/sec	Reference
A179-13	5.18	4.47	2.93	6.82			Katz and Ewing (1956)
A179-14	5.22	4.60	2.04	6.73			Katz and Ewing (1956)
A179-15	4.57			6.65			Katz and Ewing (1956)
1956–1	5.87	4.22	0.74	6.87	6.45	8.21	Officer *et al.* (1959)
2	5.80	5.96	1.47	6.93	3.96	7.96	Officer *et al.* (1959)
3	5.49	5.70	2.34	7.11	4.12	8.44	Officer *et al.* (1959)
4	6.02	5.66	2.43	6.92	5.32	8.36	Officer *et al.* (1959)
5	6.41	5.44	2.53	6.98	5.01	8.64	Officer *et al.* (1959)
1955–18	4.95			6.68	5.62	8.13	Officer *et al.* (1959)
CR3	5.07	4.4	2.5	7.2			Gaskell, Hill and Swallow (1958)
CR4	5.32	4.84	1.7	5.58			Gaskell, Hill and Swallow (1958)
CR5	5.25	4.5	1.7	7.16			Gaskell, Hill and Swallow (1958)
Number		16	19	65	38	37	
Average		4.89	2.11	6.70	4.76	8.10	
Standard deviation		0.59	0.70	0.27	1.41	0.24	

( ) Assumed velocity
[a] Recalculated by Ewing, Sutton and Officer (1954).
[b] Anomalous value omitted from averages and standard deviations.

Much of the seismic-refraction work at sea emphasizes attaining maximum penetration in the crust and mantle. For this, large shots are needed and large shot spacing is usually adequate for delineation of first arrivals of Layer 3 and Layer 4. However, a shot spacing adequate for the long segments of first arrivals of Layers 3 and 4 is not adequate to reveal Layer 2, which requires close spacing of shots at a distance near the critical distance of first emergence of the refracted wave.

In the Pacific (Tables III, IV and V) Layer 2 is present nearly everywhere. In the Atlantic (Tables I and II), however, Layer 2 is found at less than one-third of the stations. It is not clear that these results prove a significant difference between the Atlantic and Pacific Oceans in the occurrence of Layer 2, for it may be masked more frequently in the Atlantic by a greater thickness of sediment. It is also possible that Layer 2 has escaped detection in the Atlantic by other causes, such as an unfavorable distribution of shots. It is likely that more observations of Layer 2 would have been made if shot patterns had been fired specifically to observe Layer 2, or better yet, if it had been possible to place hydrophones and shots on the bottom in the manner described by J. I. Ewing in Chapter 1. The Cambridge method, using radio-sonobuoys and a very close spacing of shot distance in the critical range, favors the observation of Layer 2. The Cambridge group finds Layer 2 about as frequently

TABLE II

Eastern North Atlantic

Station number	Ocean depth, km	Layer 2 Velocity, km/sec	Layer 2 Thickness, km	Layer 3 Velocity, km/sec	Layer 3 Thickness, km	Layer 4 Velocity, km/sec	Reference
E-7	5.49			6.46	4.31	7.68	Ewing and Ewing (1959)
E-8	4.85			6.28	3.81	7.77	Ewing and Ewing (1959)
E-9	4.39			6.26	3.62	7.65	Ewing and Ewing (1959)
D-12	4.50			6.14	6.49	7.77	Ewing and Ewing (1959)
D-5	5.17			6.49	4.61	8.04	Ewing and Ewing (1959)
D-6	3.45			6.64			Ewing and Ewing (1959)
CR1	4.64	(4.5)	2.7	6.18			Gaskell, Hill and Swallow (1958)
CR32	4.5	4.36	2.3	7.15			Gaskell, Hill and Swallow (1958)
CR38	4.37	4.52					Gaskell, Hill and Swallow (1958)
CR39	4.21	5.4					Gaskell, Hill and Swallow (1958)
CR41	4.11	5.33	1.7?	6.7?			Gaskell, Hill and Swallow (1958)
HL9	3.75	5.40	8.1[a]			7.94	Hill and Laughton (1954)
HL16	4.45			7.13	4.6	8.18	Hill and Laughton (1954)
HL18	4.75	5.80					Hill and Laughton (1954)
HL10	4.80			6.56			Hill and Laughton (1954)
HL17	4.60			6.49	3.4	7.80	Hill and Laughton (1954)
HL11	4.45			6.75			Hill and Laughton (1954)
Number		6	3	13	7	8	
Average		5.14	2.23	6.56	4.41	7.85	
Standard deviation		0.57	0.50	0.32	1.03	0.18	

[a] Anomalous value omitted from averages and standard deviations.

in the Atlantic as in the Pacific, and Hill (1957) believes that there is no convincing evidence of its absence. Ewing and Press (1955) are doubtful of its universal presence.

At some of the stations at which Layer 2 was not directly observed by bottom refraction, there was indirect evidence for its existence. One piece of evidence comes from the reflection times of strong sub-bottom reflections. If these reflections are assumed to come from the base of the unconsolidated sediment, the sediment thickness estimated thereby is often less than that calculated from the refraction data in the absence of Layer 2 (Ewing and

TABLE III

Eastern North Pacific

Station number	Ocean depth, km	Layer 2 Velocity, km/sec	Layer 2 Thickness, km	Layer 3 Velocity, km/sec	Layer 3 Thickness, km	Layer 4 Velocity, km/sec	Reference
CR8	5.30	(4.3)	0.6	6.36			Gaskell, Hill and Swallow (1958)
M1	4.18	5.88	0.93	6.96	6.24	8.41	Raitt (1956)
M2	4.83	5.97	1.42	6.88	4.31	8.05	Raitt (1956)
M3	4.91	(5.99)	0.70	6.70			Raitt (1956)
M4	4.80	5.76	0.35	6.74			Raitt (1956)
M5	4.81	6.32	1.16	6.79			Raitt (1956)
M6	5.25	(5.99)	0.72	6.58	4.42	8.24	Raitt (1956)
M7	5.80	6.04	0.81	6.73	4.14	8.15	Raitt (1956)
C24	4.46	4.92	1.18	6.84	4.03	8.21	Raitt (1956)
C25	4.26	5.78	0.66	6.90	4.19	8.16	Raitt (1956)
C26	4.39	4.52	1.26	6.78	5.12	8.46	Raitt (1956)
Number		8	11	11	7	7	
Average		5.65	0.89	6.75	4.64	8.24	
Standard deviation		0.61	0.33	0.17	0.79	0.15	

Ewing, 1959). The discrepancy is removed by postulating Layer 2. Another piece of evidence comes from the crustal shear waves occasionally observed as second arrivals. It was observed by Ewing, Sutton and Officer (1954), by Katz and Ewing (1956) and by Gaskell, Hill and Swallow (1958) that the travel times of these shear waves were greater than expected if there was no Layer 2 and the conversion of compressional to shear waves took place at the base of the sediments. This discrepancy is removed if Layer 2 exists and the conversion takes place at the boundary between Layer 1 and Layer 2.

Unfortunately, these indirect indications of the presence of Layer 2 do not measure its velocity or thickness and are often ignored in the structural interpretation. For example, some of the stations of Table I which have no Layer 2 in their interpretation are reported by Ewing and Ewing (1959) to have strong indirect evidence for its presence.

The same reasons which make its detection difficult also make the measurement of the velocity of Layer 2 inaccurate. The measured values range between the rough limits of 4 to 6 km/sec, and it is not certain how much of this variability is error of measurement. Most of this error results from the short distance over which Layer 2 can be seen as a first arrival. Hence the velocity is very sensitive to local variation in sediment thickness. As the topography of the base of the sediment may be rough, even where the ocean floor is smooth, this gives a source of error very difficult to control.

There is some internal evidence that there are real differences in the velocities of Layer 2. For example, in the Pacific Ocean, southeast of Hawaii, the average value of 5.99 km/sec was systematically greater than the average value of

TABLE IV

Western North Pacific

Station number	Ocean depth, km	Layer 2 Velocity, km/sec	Layer 2 Thickness, km	Layer 3 Velocity, km/sec	Layer 3 Thickness, km	Layer 4 Velocity, km/sec	Reference
CR9	4.63	4.30	2.70	6.85			Gaskell, Hill and Swallow (1958)
CR10	4.78	(4.3)	1.4	6.93			Gaskell, Hill and Swallow (1958)
CR11	5.17	(4.3)	2.0	6.69			Gaskell, Hill and Swallow (1958)
CR12	4.88			6.4	2.1	8.5	Gaskell, Hill and Swallow (1958)
CR17	4.41	4.3	1.1	6.40			Gaskell, Hill and Swallow (1958)
CR20	5.52	5.75					Gaskell, Hill and Swallow (1958)
CR21	5.82	5.70					Gaskell, Hill and Swallow (1958)
1, 2, 3, 4,	3.6			6.2	7.0	8.1	Andreyeva and Udintsev (1958)
M9	5.20	4.26	2.27	6.57	4.70	7.92	Raitt (1956)
M10	4.83	4.81	2.21	6.92	5.66	8.28	Raitt (1956)
M11	4.93	4.86	2.18	7.02			Raitt (1956)
M12	4.20	4.92	2.59	6.63			Raitt (1956)
M13	5.55	6.24	1.57	6.83			Raitt (1956)
M14	4.93	4.39	2.60	6.92	3.14	8.42	Raitt (1956)
M15	4.46	5.16	1.90	6.56	5.69	8.28	Raitt (1956)
Cl	3.86	4.87	5.84[a]	6.90	5.76	8.09	Raitt (1956)
C2	4.41	4.98	5.40[a]	7.10	7.41	8.16	Raitt (1956)
C3	5.17	5.73	2.10	6.72	4.23	8.14	Raitt (1956)
Number		14	12	16	9	9	
Average		5.02	2.05	6.73	5.08	8.21	
Standard deviation		0.63	0.50	0.25	1.72	0.18	

[a] Anomalous value omitted from averages and standard deviations.

5.09 km/sec observed elsewhere on the "Mid Pacific" and "Capricorn" expeditions (Raitt, 1956). Similar observations were made by Gaskell, Hill and Swallow (1958) in the Philippine Sea.

The observed range of velocities belongs to a wide variety of rock types, and is of little help in identifying the nature of the material (Birch, Schairer and Spicer, 1942; Birch, 1960). Rocks of nearly all types have velocities in this range. There are many lithified sedimentary types and metamorphic rocks of sedimentary origin. Also these velocities are characteristic of basaltic lavas, as indicated by laboratory experiments (Hill, 1957) and seismic-refraction measurements of the volcanic core of Eniwetok and Bikini Atolls (Raitt, 1954, 1957). Most plutonic rocks, especially firm specimens of dense basic rocks such as gabbro, diabase, norite and dunite, have velocities well above the range of Layer 2. However, many granites and granodiorites have compressional wave velocities of about 5 km/sec (Birch, 1960), a characteristic Layer 2 velocity.

TABLE V

South Pacific

Station number	Ocean depth, km	Layer 2 Velocity, km/sec	Layer 2 Thickness, km	Layer 3 Velocity, km/sec	Layer 3 Thickness, km	Layer 4 Velocity, km/sec	Reference
CR16	4.88	4.48	1.5	6.84			Gaskell, Hill and Swallow (1958)
C7	4.13	5.00	2.02	7.04	7.41	8.42	Raitt (1956)
C9	6.10	(5.09)	0.45	6.42	5.27	8.25	Raitt (1956)
C12	5.14	4.49	1.51	6.68	9.58	8.77	Raitt (1956)
C13	4.85	5.8	0.86	6.75	4.90	8.17	Raitt (1956)
C14	5.21	(5.09)	1.16	6.45	4.62	8.21	Raitt (1956)
C15	4.62	4.48	1.54	6.81	5.37	8.43	Raitt (1956)
C16	4.59	5.51	1.31	6.69	5.90	8.34	Raitt (1956)
C18	4.19	5.04	1.59	6.91	4.57	8.14	Raitt (1956)
C19	4.07	5.22	0.81	6.69	4.60	8.00	Raitt (1956)
C20	3.58	(5.09)	0.73	6.48	4.24	8.12	Raitt (1956)
C23	4.33	6.02	1.20	6.90	3.31	8.30	Raitt (1956)
Number		9	12	12	11	11	
Average		5.12	1.22	6.72	5.43	8.29	
Standard deviation		0.58	0.45	0.20	1.72	0.21	

TABLE VI

Indian Ocean

Station number	Ocean depth, km	Layer 2 Velocity, km/sec	Layer 2 Thickness, km	Layer 3 Velocity, km/sec	Layer 3 Thickness, km	Reference
CR22	5.19			6.70		Gaskell, Hill and Swallow (1958)
CR23	4.97	5.09	2.9	7.1		Gaskell, Hill and Swallow (1958)
CR24	3.03	3.92	2.7	6.7		Gaskell, Hill and Swallow (1958)
CR25	4.86	5.09	1.1	6.74		Gaskell, Hill and Swallow (1958)
CR26	4.70	4.4	1.6	6.22		Gaskell, Hill and Swallow (1958)
Number		4	4	5		
Average		4.62	2.08	6.69		
Standard deviation		0.57	0.87	0.31		

There has been much speculation about the nature of Layer 2. Hamilton (1959) has presented a thorough discussion of the evidence for its sedimentary origin. An attractive feature of this hypothesis is that it indicates there is a much greater total amount of sediment than the meager quantity contained in the unconsolidated layer. It would then agree with the estimate of Kuenen (1950) that the total volume of sediment deposited in the deep sea in the history of the ocean is about 10^9 km^3, a value several times as large as the seismic estimates of unconsolidated sediment in Layer 1.

Other evidence suggests an igneous origin. For example, it appears that Layer 2 is thicker near some groups of Pacific islands and seamounts (Raitt, 1957), giving support to a volcanic origin at least in some areas. The high values, 5.99 km/sec, east of Hawaii are characteristic of a granitic continental crust and may be intrusive, plutonic rocks. It has been suggested by Ewing and Ewing (1959) that Layer 2 is the upper part of Layer 3, forming a transition layer of increasing velocity with depth, presumably the result of the sealing of holes and fissures by the increasing pressure of the overburden.

It would be surprising, in view of the great variety of rocks forming similar layers beneath the continent, to find that Layer 2 is monotonously uniform under the oceans. It seems likely that it is a variety of rocks. The wide range of velocities supports this conclusion.

3. Layer 3

Observation of Layer 3 is much more satisfactory than that of Layer 2. It is nearly universally observed by all investigators at truly oceanic stations. It customarily occurs as a long linear segment of the travel-time plot extending from about 12 km shot distance to about 35 km (see Figs. 1–8). Complications sometimes occur in regions of rough bottom topography, partly because of large corrections for water-depth changes, but also because of complications in the crustal structure associated with large topographic features.

Generally the stations chosen for Tables I to VI were placed as much as possible in regions of flat or subdued topography in order to reduce these complications. The success of this procedure is testified by the internal consistency of the velocities. For example, in Table III the standard deviation of the Layer 3 velocities in the eastern North Pacific is only 0.17 km/sec; and much of this was contributed by a low value, CR8, which is well to the north of the region covered by these stations. If this station is not included in the average and standard deviation, the average is 6.79 ± 0.11 km/sec for the remaining stations of this area. This standard deviation contains the real variation of the velocity and hence can be regarded as an upper limit to the random standard error of measurement for this group of stations.

It is probably too optimistic to assume that this estimate of the upper limit to the random error of measurement also applies to the other groups tabulated in Tables I, II, IV, V and VI. If this assumption is valid, then a large part of the variability observed for these stations represents a real variation of velocity. Unfortunately, the quality of the travel-time data is also variable; and it is not possible to say with confidence that a given variation is real even though it may be several times the standard deviation determined from the homogeneous group of Table III.

In addition to the random error of measurement there can be systematic errors. One of these arises from failure to recognize Layer 2 arrivals at short range so that one includes them with the Layer 3 solution. This will cause the Layer 3 velocity to be systematically low. A test for this was made by averaging Layer 3 velocities of Table I in two separate groups: (1) the group of 17 stations in which Layer 2 was observed; (2) the group of 48 stations in which Layer 2 was not identified. Group (1) had a mean of 6.83 km/sec with a standard deviation of 0.31 km/sec. Group (2) had a mean of 6.65 km/sec with a standard deviation of 0.25 km/sec. Hence Group (2) is significantly lower, as expected. This is additional evidence for the widespread occurrence of Layer 2. It indicates that velocity measurement of Layer 3 is more reliable at stations where Layer 2 has been detected.

The thickness of Layer 3 depends on the intercept of the travel-time segment for Layer 4. It has been pointed out by Hill (1957) that this intercept is sensitive to the value assumed for the Layer 4 velocity. If the estimate of the velocity is too high, Layer 3 will be estimated too thick. Hence the degree of positive correlation between Layer 3 thickness and Layer 4 velocity will indicate the magnitude of the error in the Layer 4 velocity. Hill showed that there was a positive correlation, which indicated that some of the variability of Layer 4 velocity was random error of measurement.

As a conclusion of this argument, one might obtain more estimates of Layer 3 thickness by using a standard average value of Layer 4 velocity instead of the locally determined velocity for each station. This procedure might be recommended for those stations in which the quality of the Layer 4 arrival is obviously poor. However, it cannot be applied universally for there is good evidence that there are real variations in the velocity of Layer 4. For example, certain areas, such as the Mid-Atlantic Ridge (Ewing and Ewing, 1959), parts of the eastern Caribbean (Officer *et al.*, 1959), and the Easter Island Rise (Raitt, 1956), have Layer 4 velocities significantly lower than normal oceanic values.

The widespread occurrence of Layer 3 and the remarkable stability of its velocity in widely different parts of the world ocean demonstrate that it is a characteristic feature of the oceanic crusts. The 126 tabulated observations of Layer 3 velocity are all well above the velocity of about 6.0 km/sec characteristic of the principal refracting layer on the continents (Gutenberg, 1955).

There were only 8 stations of the 134 in Tables I to VI at which Layer 3 was not identified. Four of these stations, CR38, CR39, CR20 and CR21, were short-range sonobuoy stations in the eastern North Atlantic and western North Pacific and did not reach velocities greater than 5.75 km/sec. This is probably because of an unusual thickness of Layer 2 at these stations. A similar result was found at station HL18 (Hill and Laughton, 1954), where the maximum velocity attained was 5.80 km/sec. The other three stations have anomalous Layer 2 thicknesses or abnormally low Layer 4 velocities. Hence, there is no evidence at all for the absence of Layer 3, and its existence throughout all oceanic regions is well established.

This characteristic oceanic property of Layer 3 has caused Ewing and Ewing (1959) to call it the "oceanic" layer. It has also been called the "basaltic" layer, which implies that there is some knowledge of its chemical composition. Until we have identified it by direct sampling with a drill, this description of its nature is purely speculative. Its velocity puts it at the upper end of the range for rocks of sedimentary origin; but if it is sediment, it would probably be a very hard limestone, which seems hardly likely in view of the extraordinary supply of $CaCO_3$ required.

Comparing it with laboratory measurements by Birch (1960), it falls in the range expected for a rock of composition intermediate between acidic granite and ultrabasic dunite. Velocities of 6.7 km/sec are commonly observed in gabbro and rocks of similar composition. Whatever it is, its uniformity throughout the world indicates a similar nature. The variability of the measured velocities seems larger than experimental error and indicates a small-scale variability. It is significant that averages of large regions show such little systematic differences.

4. Average Oceanic Crust

In summary, the results of seismic-refraction measurements of the velocity and thickness of the crustal layers beneath the unconsolidated sediment show little significant difference between major oceanic areas. The average picture of the oceanic crust can be represented as an average of all available oceanic stations. The current averages and standard deviations of published values, including all of the data tabulated above, are:

	Velocity, km/sec	Thickness, km
Layer 2	5.07 ± 0.63	1.71 ± 0.75
3	6.69 ± 0.26	4.86 ± 1.42
4	8.13 ± 0.24	—

References

Andreyeva, I. B. and G. B. Udintsev, 1958. The structure of the bottom of the Japan Sea according to data from investigations by the expedition on the "*Vityaz*". *Izvest. Akad. Nauk S.S.S.R., Ser. Geol.*, **10**, 3–20.

Birch, F., 1960. The velocity of compressional waves in rocks to 10 kilobars, Part I. *J. Geophys. Res.*, **65**, 1083–1102.

Birch, F., J. F. Schairer and H. C. Spicer, Editors, 1942. Handbook of physical constants. *Geol. Soc. Amer. Spec. Paper* 36, 325 pp.

Ewing, J. and M. Ewing, 1959. Seismic refraction measurements in the Atlantic Ocean basins in the Mediterranean, on the Mid-Atlantic Ridge, and in the Norwegian Sea. *Bull. Geol. Soc. Amer.*, **70**, 291–318.

Ewing, J. I., C. B. Officer, N. R. Johnson and R. S. Edwards, 1957. Geophysical investigations in the eastern Caribbean: Trinidad Shelf, Tobago Trough, Barbados Ridge, Atlantic Ocean. *Bull. Geol. Soc. Amer.*, **68**, 897–912.

Ewing, M. and F. Press, 1955. Geophysical contrasts between continents and ocean basins. *Geol. Soc. Amer. Spec. Paper* 62, 1–6.

Ewing, M., G. H. Sutton and C. B. Officer, 1954. Seismic refraction measurements in the Atlantic Ocean, Part VI: Typical deep stations, N. America Basin. *Bull. Seism. Soc. Amer.*, **44**, 21–38.

Ewing, M., J. L. Worzel, J. B. Hersey, F. Press and G. R. Hamilton, 1950. Seismic refraction measurements in the Atlantic Ocean basin, Part 1. *Bull. Seism. Soc. Amer.*, **40**, 233–242.

Galperin, E. I., A. V. Goryachev and S. M. Zverev, 1958. Crustal structure. Researches in the transition region from the Asiatic continent to the Pacific. The Pacific Geologo-Geophysical Expedition. USSR Academy of Sciences (XII Seismology), No. 1. Publishing house of Akademia Nauk, U.S.S.R., Moscow, 28 pp.

Gaskell, T. F., M. N. Hill and J. C. Swallow, 1958. Seismic measurements made by H.M.S. "*Challenger*" in the Atlantic, Pacific and Indian Oceans and in the Mediterranean Sea, 1950–53. *Phil. Trans. Roy. Soc. London*, **A251**, 23–83.

Gutenberg, B., 1955. Wave velocities in the earth's crust. *Geol. Soc. Amer. Spec. Paper* 62, 19–34.

Hamilton, E. L., 1959. Thickness and consolidation of deep-sea sediments. *Bull. Geol. Soc. Amer.*, **70**, 1399–1424.

Hersey, J. B., C. B. Officer, H. R. Johnson and S. Bergstrom, 1952. Seismic refraction observations north of the Brownson Deep. *Bull. Seism. Soc. Amer.*, **42**, 291–306.

Hess, H. H., 1959. The AMSOC hole to the earth's mantle. *Trans. Amer. Geophys. Un.*, **40**, 340–395.

Hill, M. N., 1952. Seismic refraction shooting in an area of the eastern Atlantic. *Phil. Trans. Roy. Soc. London*, **A244**, 561–596.

Hill, M. N., 1957. Recent geophysical exploration of the ocean floor. *Physics and Chemistry of the Earth*, **2**, 129–163. Pergamon Press, London.

Hill, M. N. and A. S. Laughton, 1954. Seismic observations in the eastern Atlantic, 1952. *Proc. Roy. Soc. London*, **A222**, 348–356.

Katz, S. and M. Ewing, 1956. Seismic-refraction measurements in the Atlantic Ocean, Part VII: Atlantic Ocean basin, west of Bermuda. *Bull. Geol. Soc. Amer.*, **67**, 475–510.

Kuenen, Ph. H., 1950. *Marine Geology*. John Wiley and Sons, Inc., New York, 568 pp.

Officer, C. B., J. I. Ewing, R. S. Edwards and H. R. Johnson, 1957. Geophysical investigations in the eastern Caribbean: Venezuelan Basin, Antilles Island Arc and Puerto Rico Trench. *Bull. Geol. Soc. Amer.*, **68**, 359–378.

Officer, C. B., J. I. Ewing, J. F. Hennion, D. G. Harkrider and D. E. Miller, 1959. Geophysical investigations in the Eastern Caribbean: Summary of 1955 and 1956 cruises. *Physics and Chemistry of the Earth*, **3**, 17–109. Pergamon Press, London.

Officer, C. B., M. Ewing and P. C. Wuenschel, 1952. Seismic refraction measurements in the Atlantic Ocean, Part IV: Bermuda, Bermuda Rise and Nares Basin. *Bull. Geol. Soc. Amer.*, **63**, 777–808.

Raitt, R. W., 1954. Seismic-refraction studies of Bikini and Kwajalein atolls. *U.S. Geol. Surv. Prof. Paper* 260-K, 507–526.

Raitt, R. W., 1956. Seismic-refraction studies of the Pacific Ocean basin, Part I: Crustal thickness of the central equatorial Pacific. *Bull. Geol. Soc. Amer.*, **67**, 1623–1640.

Raitt, R. W., 1957. Seismic-refraction studies of Eniwetok Atoll. *U.S. Geol. Surv. Prof. Paper* 260-S, 685–698.

Tolstoy, I., R. S. Edwards and M. Ewing, 1953. Seismic refraction measurements in the Atlantic Ocean, Part 3. *Bull. Seism. Soc. Amer.*, **43**, 35–48.

7. THE MANTLE ROCKS

J. I. Ewing

1. Introduction

The Earth's mantle is larger in volume by a factor of 10 than the core and crust combined. Its outer surface is found at an average depth of 12 km under the ocean basins and at about 35–50 km under the continents. The depth is intermediate in intermediate areas. The inner surface (the core-mantle discontinuity) is about halfway between the surface and the center of the earth. The composition of the mantle is thought to be principally olivine, on the basis of seismic wave velocity, density and examination of materials brought up in oceanic volcanoes. For a review of the variation of seismic velocity with depth, based on the study of earthquake waves, the reader is referred to Chapter 8.

The present discussion is concerned with the information about mantle structure under the oceans derived from seismic refraction studies (see Chapters 1 and 2). To date, the length of marine seismic profiles has not exceeded 200 km and has generally been less than 100 km; hence the velocity measurements have not extended to any appreciable depth in the mantle as in earthquake seismology, but rather have sampled only the upper part. This offers at least one advantage in studying the nature of the crust-mantle (M) discontinuity. Even with the detail obtained by surface-wave dispersion analyses, the results from earthquake seismology show the average structure over a path that is usually thousands of kilometers long. Seismic refraction measurements give results in considerable detail from a local area, hence comparisons can be made from one place to another and certain conclusions about the homogeneity can be drawn from the observed differences or similarities.

2. Compressional-Wave Measurements

In Fig. 1 are plotted approximately 180 recorded seismic compressional-wave velocities higher than 7 km/sec from profiles in the Atlantic, Pacific and Indian Oceans, the Caribbean Sea and the Gulf of Mexico. Only those velocities were plotted which were measured at profiles where an overlying layer of high-velocity crustal rock (6–7 km/sec velocity) was present. This eliminated Mid-Atlantic Ridge profiles and some others on continental margins. These values come from Officer *et al.* (1952), Hersey *et al.* (1952), Ewing *et al.* (1954), Hill and Laughton (1954), Ewing *et al.* (1955), Katz and Ewing (1956), Raitt (1956), Gaskell *et al.* (1958), Hersey *et al.* (1959), Ewing and Ewing (1959) and unpublished results by Ewing *et al.* and Hennion and Ewing (MS). A quick look at Fig. 1 indicates the problem of defining the mantle rocks by seismic velocity. There is no doubt that a peak in the frequency plot occurs at about 8.1–8.2 km/sec, and this is in good agreement with the generally accepted value of P_n from earthquake seismology. As Hill (1957) pointed out in a review of recent explorations of the ocean floor, this would at first sight seem to indicate that the rocks with compressional-wave velocity 8.1 km/sec and

[*MS received August, 1960*]

lying at an average depth of 12 km below sea-level are of the same material as those underneath the continental M discontinuity, which is at an average depth of 30–40 km. He further notes that it is somewhat surprising to expect that a material would have the same compressional-wave velocity under the widely different conditions of temperature and pressure found at 12 km depth under oceans and 35 km under continents, despite the fact that increasing temperature and pressure change the seismic velocity in opposite directions. Still further uncertainties about the mantle composition are apparent in the light of recent hypotheses that the velocity change at the M discontinuity is

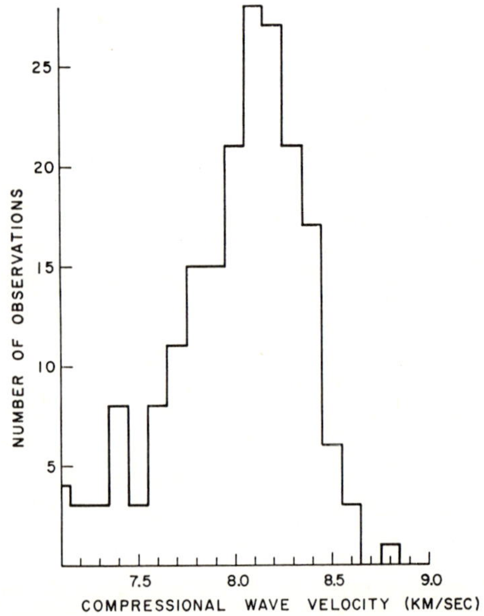

Fig. 1. Graph of observed sub-crustal velocities versus number of observations.

caused by a phase change rather than by a change in material. In view of the foregoing, therefore, the definition of mantle rocks here employed is based on 8.1 km/sec as typical of the compressional-wave velocity in them. Thus, a lower velocity is assumed to represent mantle rocks which are either partially contaminated by crustal material, or to represent a zone where fractional phase change has occurred. Velocities higher than 8.1 km/sec then must represent mantle material which has been more thoroughly differentiated from crustal material or in which, possibly as a result of mineral inhomogeneity, the phase change has produced an unusually high-velocity material. It is possible, of course, that all velocities different from 8.1 km/sec are the result of inaccurate measurement, but most of those who have analyzed any appreciable number of profiles would not accept that explanation. The relatively wide variation of velocities around 8.1 km/sec is entirely consistent with earthquake-wave observations. As mentioned earlier, these tend to average the velocities over long paths and would most likely give the result as 8.1 km/sec.

The velocities plotted in Fig. 1 are from profiles which were selected only insofar as none was used unless a crustal velocity was measured in the same profile. Fig. 2 shows similar plots for Atlantic and Pacific profiles which were selected as probably unaffected by proximity to prominent topographic features such as oceanic ridges, trenches, large seamounts and continental margins. The principal difference between the two oceans shown by these plots is that none of the Pacific profiles shows a velocity below 8.0 km/sec. These data are from Raitt (1956), and of 31 determinations of sub-crustal velocity he found only four to be below 8 km/sec. These four were from profiles

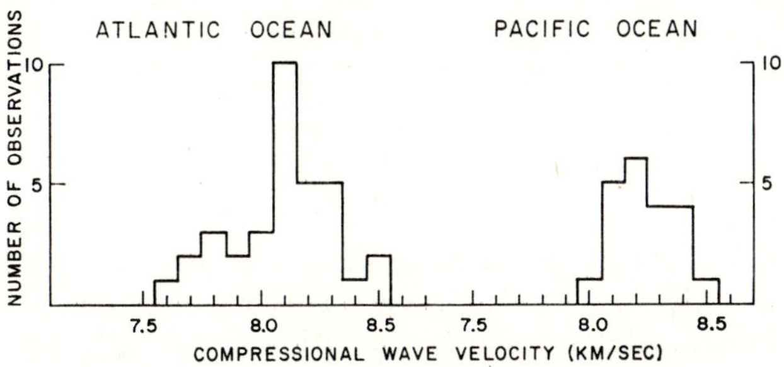

Fig. 2. Comparision of sub-crustal velocities versus number of observations for Atlantic and Pacific Oceans.

near Hawaii, Tonga and on the Easter Island Rise. These have been omitted in the graph. In the Atlantic, there are several ocean-basin profiles in which the sub-crustal velocity is lower than 8 km/sec. In addition to those plotted, there are a few low values in the eastern basin which were omitted because of obvious complication by large-scale topography. With the exception of two values which came from weakly shot profiles in the Nares Basin, the low velocities plotted were measured either on the Bermuda Rise, within 200 miles of the islands, or near the continental rise of eastern North America. Thus it is a fair statement that the open basin areas of the Atlantic are not different from those in the Pacific. Being much smaller, the Atlantic obviously has a noticeably higher percentage of its area influenced by large-scale topographic features. In these areas where low sub-crustal velocities are found, it is most probable that longer profiles would show an increase of velocity with depth, reaching a value of 8 km/sec or higher at a depth not far below the bottom of the crustal layer. Structure of this type has been measured by profiles in the Venezuelan Basin of the Caribbean Sea (Officer *et al.*, 1959).

3. Shear-Wave Measurements

Measurements of shear-wave velocity in the mantle by seismic-refraction profiles have been reported for profile 26 (1956 cruise) in the Caribbean Sea by

Officer *et al.* (1959). One other such measurement is reported by Luskin and Ewing (*in litt.*) for a profile midway between New York and Bermuda, and two measurements are known for the area north of the Puerto Rico Trench (Drake, *in litt.*; Hersey *et al.*, 1952). Table I lists the data for these four measurements.

TABLE I

Compressional and Shear-Wave Velocities and Poisson's Ratio for Mantle Rocks

Profile	Water depth, km	V_p, km/sec	V_s, km/sec	Poisson's ratio
Carib 26–56	5.00	8.16	4.62	0.27
[a]V–16 (N.Y.–Bda.)	4.74	8.06	4.27	0.31
[a]V–15 North of Puerto	5.36	8.08	4.24	0.31
[a]V–15 Rico Trench	5.36	8.06	4.27	0.31

[a] Unreversed measurement.

It is not understood why transformed shear waves from the mantle are observed so seldom. Although crustal shear waves are observed relatively much more often, they too exhibit a marked variation in strength from one area to another which is not obviously associated with geographical location, seismic velocities or other observed factors (Ewing and Ewing, 1959). Whether the propagation of shear waves is through the crust or through the mantle, the transformation from compressional waves occurs at the base of the low-velocity sediments, or possibly at some interface within the sediments. Therefore, the strength of any shear-wave arrivals from the high-velocity layers must be dependent on the character of this interface. It may be possible to account for the relative strength of the mantle shear waves versus the crustal shear waves by the dependence of transformed energy on the angle of incidence, which is different for the two paths.

Whatever the cause, few measurements of mantle shear waves have been obtained from marine seismic-refraction studies. The average determination of Poisson's ratio for the mantle from these measurements is higher than the value 0.25 typically found for dense, high-velocity rocks. The discrepancy, indicating a high value of the ratio V_p/V_s, may not have any significance in view of the scarcity of measurements, although two of the high values are from particularly good profiles and it would be difficult to interpret them differently. If future refraction studies give similar results, the low shear-wave velocity would seem to indicate a significantly lower velocity immediately below the crust-mantle interface than at a few kilometers deeper. This would not be inconsistent with the compressional-wave data from refraction studies; because the mantle p waves are always masked for a considerable distance beyond the range associated with the critical angle, and an increase of velocity with depth near the M discontinuity would be difficult to detect. The three lowest values of V_s in Table I were determined by refractions very near critical angle and thus must have measured the uppermost part of the mantle. The other measurement (4.6 km/sec) was based on arrivals appreciably beyond the critical range, and this is in agreement with the velocity determined by the study of surface-wave dispersion (Dorman, Ewing and Oliver, 1960).

To test for a relationship between sub-crustal compressional-wave velocity and depth, the graph shown in Fig. 3 was plotted. The data were taken from Atlantic and Pacific profiles in which the water depth was between 4 and 6 km, trenches being avoided. There appears to be a trend of the data in the direction of increasing velocity with increasing depth. Some of this trend may be the effect of lines having been drawn on time–distance graphs which, although

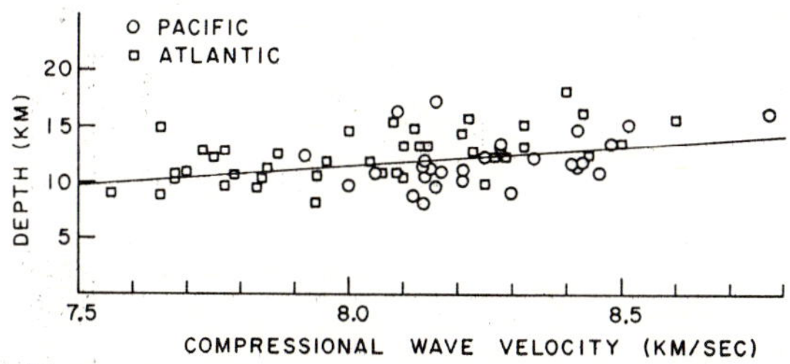

Fig. 3. Graph of mantle compressional wave velocities versus depth of the M discontinuity.

passing through the centers of gravity of the points, had incorrect slopes. For typical oceanic profiles errors produced by misalignment will cause the data to trend approximately in the direction indicated by the line in Fig. 3. This line fits the points about as well as any other, hence one could deduce that both the high and low velocities are improperly determined values and that there is no relationship of velocity with depth. The observed trend in Fig. 3 would, however, be expected if there is a velocity gradient in the upper mantle. Evidence from shear waves was cited previously as possibly indicative of a gradient.

In Fig. 4 are plotted crustal thicknesses versus elevation of the sea floor from about 90 of the better published profiles in the Pacific and Atlantic Oceans, the Caribbean Sea and the Gulf of Mexico. For this graph the data were selected if the sub-crustal velocity was 7.7 km/sec or higher, and, as before, trench profiles were avoided. Although there is a great amount of scatter, a trend is indicated which could extend smoothly to continental crustal thicknesses and appropriate depths of the M discontinuity (30–40 km crust for elevations of 0–2 km).

4. Summary

Diagrams showing observed velocities higher than 7 km/sec have peaks at 8.1–8.2 km/sec indicating this as the average value of compressional waves in the upper mantle in deep-ocean areas. Some deviations above and below the average value undoubtedly represent experimental errors, but some probably indicate inhomogeneity in the mantle material. Low velocities are consistently found in certain regions near large topographic features such as islands, oceanic ridges and continental margins. It is difficult to find a pattern of occurrence of the high velocities.

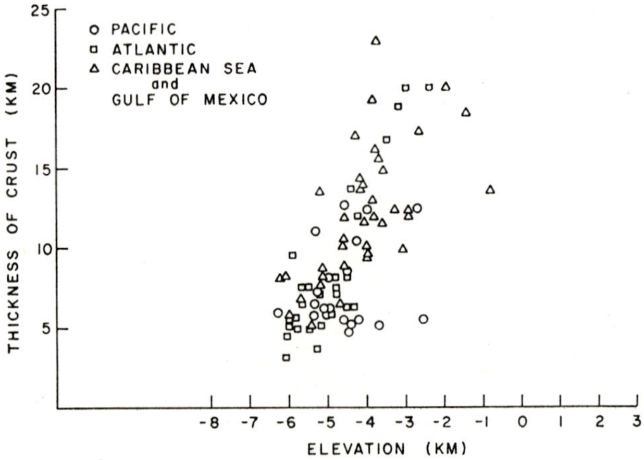

Fig. 4. Crustal thicknesses in the Atlantic and Pacific Oceans, the Caribbean Sea and the Gulf of Mexico as a function of sea-floor elevation.

Measurement of the velocity structure in the immediate vicinity of the M discontinuity is difficult because the critical angle arrivals are masked by preceding crustal arrivals, but results from a few shear-wave observations give some indication of a velocity gradient in the uppermost part of the mantle. Possible evidence for gradient is given also by a graph of mantle velocity versus depth of the M discontinuity.

References

Dorman, J., M. Ewing and J. Oliver, 1960. Study of shear-velocity distribution in the upper mantle by mantle Rayleigh waves. *Bull. Seism. Soc. Amer.*, **50**, 87–115.

Ewing, J. and M. Ewing, 1959. Seismic refraction measurements in the Atlantic Ocean basins, in the Mediterranean Sea, on the Mid-Atlantic Ridge, and in the Norwegian Sea. *Bull. Geol. Soc. Amer.*, **70**, 291–318.

Ewing, M. and B. C. Heezen, 1955. Puerto Rico Trench topographic and geophysical data. *Geol. Soc. Amer. Spec. Paper* 62, 255–268.

Ewing, M., G. H. Sutton and C. B. Officer, 1954. Seismic refraction measurements in the Atlantic Ocean. Part VI: Typical deep stations. *Bull. Seism. Soc. Amer.*, **44**, 21–38.

Ewing, M. and J. L. Worzel, 1954. Gravity anomalies and structure of the West Indies, Part I. *Bull. Geol. Soc. Amer.*, **65**, 165.

Ewing, M., J. L. Worzel, D. B. Ericson and B. C. Heezen, 1955. Geophysical and geological investigations in the Gulf of Mexico, Part I. *Geophysics*, **20**, 1–18.

Gaskell, T. F., M. N. Hill and J. C. Swallow, 1958. Seismic measurements made by H.M.S. *Challenger* in the Atlantic, Pacific and Indian oceans and in the Mediterranean Sea, 1950–53. *Phil. Trans. Roy. Soc. London*, **A251**, 23–83.

Hersey, J. B., C. B. Officer, H. R. Johnson and S. Bergstrom, 1952. Seismic refraction observation north of the Brownson Deep. *Bull. Seism. Soc. Amer.*, **42**, 291–306.

Hersey, J. B., E. T. Bunce, R. F. Wyrick and F. T. Dietz, 1959. Geophysical investigations of the continental margin between Cape Henry, Va., and Jacksonville, Fla. *Bull. Geol. Soc. Amer.*, **70**, 437–466.

Hill, M. N., 1957. Recent geophysical exploration of the ocean floor. *Physics and Chemistry of the Earth*, **2**. Pergamon Press, London.

Hill, M. N. and A. S. Laughton, 1954. Seismic observations in the eastern Atlantic. *Proc. Roy. Soc. London*, **A222**, 348–356.

Katz, S. and M. Ewing, 1956. Seismic refraction measurements in the Atlantic Ocean. Part VII: Atlantic Ocean basin, west of Bermuda. *Bull. Geol. Soc. Amer.*, **67**, 475–510.

Officer, C. B., M. Ewing and P. C. Wuenschel, 1952. Seismic refraction measurements in the Atlantic Ocean. Part IV: Bermuda, Bermuda Rise, and Nares Basin. *Bull. Geol. Soc. Amer.*, **63**, 777–808.

Raitt, R. W., 1956. Seismic refraction studies of the Pacific Ocean basin, Part I: Crustal thickness of the central equatorial Pacific. *Bull. Geol. Soc. Amer.*, **67**, 1623–1640.

8. EXPLORATION OF SUB-OCEANIC STRUCTURE BY THE USE OF SEISMIC SURFACE WAVES

J. OLIVER and J. DORMAN

1. Introduction

The study of seismic surface waves generated by an earthquake or a large nuclear explosion offers one of the two principal seismic methods for exploration beneath the bottom of the sea. Surface-wave techniques, which utilize waves of long periods, i.e. greater than about 5 sec, determine average features of structures of the order of tens or hundreds of kilometers in depth and hundreds or thousands of kilometers in length. The other principal seismic method, the seismic refraction or the travel-time technique, which is treated elsewhere in this volume, provides relatively detailed information on earth structures of the order of 10 km in depth and some tens of kilometers in length. The lengths of waves used in the surface-wave method range from about 10 to about 2000 km. Resolution of structural features is surprisingly good in spite of the great length of these waves, although local details are not as well resolved as in the case of the seismic-refraction technique, which utilizes waves only fractions of a kilometer in length. Data for surface-wave studies are abundant, at least for certain areas, whereas seismic-refraction data are difficult and expensive to obtain.

Up to the present, most information obtained from surface waves has come from analysis of their dispersion, i.e. the dependence of velocity on wave period. Valuable data on dispersion may be obtained from the analysis of a single seismogram and, since this dispersion is sensitive to and therefore diagnostic of the layered structure of the earth, a mean structure along the path may be deduced from such data.

Surface-wave studies have established the relative uniformity of crustal thickness in all major oceanic basins of the world, indicating that the mantle rises to shallow depth in these regions. Average sediment thicknesses of the order of 0.5 to 1.0 km and geographical variations in this thickness have been determined by surface-wave methods. In the case of deeper structure, primarily the velocity distribution in the upper few hundred kilometers of the sub-oceanic mantle, surface-wave data are a most important source of information. This is so, partly because the travel-time data based on both explosive and earthquake sources are meager, and partly because of the inherent weakness of the travel-time method for the study of structures in which the velocity decreases appreciably with depth over certain depth ranges. Surface waves of both the Love and Rayleigh types show that the top of the low-velocity zone of the upper mantle is shallower than it is beneath the continents. Our knowledge of earth structure based on surface-wave techniques can be markedly supplemented in the future by analysis of existing data using conventional techniques. In addition, there remain many unexplained or partially explained features of surface-wave trains traversing oceanic paths, and these features, once interpreted, will offer further sources of information on earth structure.

[*MS received July, 1960*]

2. The General Nature of Seismic Surface Waves

The data for the study of the ocean floor through the application of surface-wave techniques consist of seismograms from a large number of stations for a large number of earthquakes.

Each seismogram is a record relating to some particular component of ground motion as a function of time. Three factors affect the character of this record. They are (1) the nature of the source; (2) the nature of the propagation of the waves; and (3) the response of the instrument. Generally speaking, the third factor, instrument response, is understood well enough to be compensated for, and thus removed.

Of the other two factors, excitation and propagation, the greatest effect upon the character of the seismogram clearly results from propagation. The seismic-wave train from a natural earthquake with a duration measured in seconds, or fractions of a second, at the source commonly lasts many tens of minutes at a distant station. It is to the topic of propagation effects that most of this chapter will be devoted. Excitation is of considerable importance, however, and should become increasingly important in the future as our methods for studying this phenomenon improve. Some observations relating to various kinds of excitation will be discussed later.

Providing instrumentation is adequate and the source is of the proper size, the seismograms from a seismograph station at some distance from the source will include waves of a number of identifiable types. Body waves of the conventional P (compressional) and S (shear) types as well as various reflected and/or refracted combinations of these types will be registered largely during the early portion of the seismic recording. In general, these waves will have penetrated deep within the earth, and it is from such waves that most of our knowledge of the earth's deep interior, the core and the mantle, has been derived. When the source is not too deep, the early body waves will be followed by trains of dispersed long-period waves of relatively large amplitudes and long durations (Fig. 1). These are the surface waves. Their amplitudes are large because the waves are confined to the surface and near-surface media, and the frequency content, the duration and the dispersion result largely from wave-guide propagation within the earth's upper layers. Measurement of such dispersion provides us with much of our information on the earth's crust and upper mantle.

The seismic surface-wave trains may consist of waves of two general types. These are named after their discoverers, A. E. H. Love and Lord Rayleigh.

Love waves have velocities in general slightly greater than Rayleigh waves of the same period, and are characterized by surface-particle motion which is transverse to the direction of propagation and in the horizontal plane. This particle motion is that of the component of shear waves usually labelled SH, and, in fact, Love waves may be thought of as SH waves trapped within the surface wave guide.

Rayleigh waves have particle motions which are retrograde elliptical and in the longitudinal-vertical plane. In the general case of the layered earth, both Love and Rayleigh waves are dispersive as a result of wave-guide propagation.

For studies of phenomena involving dispersion, it is convenient to use the concepts of phase and group velocity. Both phase and group velocity are, in general, functions of wave period. Group velocity (the velocity with which the energy travels) is measured in the conventional way, i.e. once the arrival time of waves of a given period has been determined for a station, the group velocity for waves of that period is simply the distance from the epicenter of the shock divided by the travel time. Phase velocity is somewhat more difficult to measure, requiring either information about the nature of the source or data from two or more recording stations. Phase velocity refers to the velocity of the

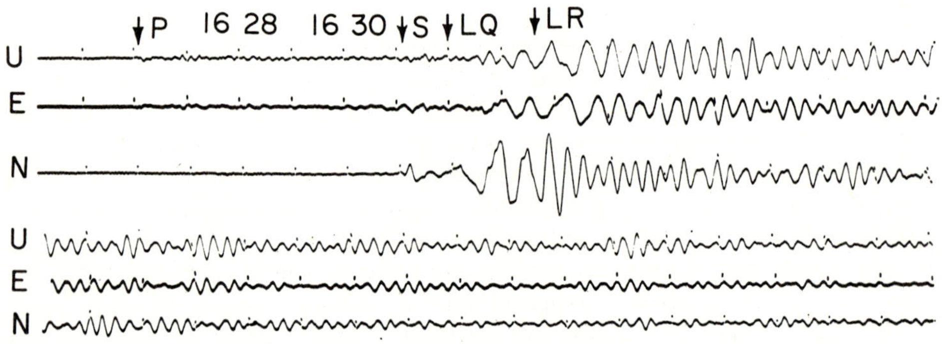

Fig. 1. Three-component Palisades long-period seismograms of the Mid-Atlantic Ridge shock of 8 June, 1960, at 16-19-48 G.C.T. The epicenter at 35°N, 35°W is 31.1° or 3460 km east of Palisades. The body-wave phases, P and S, are followed by Love (LQ) waves recorded on the N–S (transverse) component and Rayleigh (LR) waves on the vertical and E–W (longitudinal) components. The long duration of the oceanic Rayleigh-wave train is best seen on the vertical component. Much of the dispersion of the Love-wave train is due to the continental portion of the path between the edge of the continental shelf and Palisades. Clock correction is + 2 sec. Short-period waves preceding LR and LQ and following S may correspond to higher modes. Symbol at left indicates direction of ground motion corresponding to upward trace motion.

individual Fourier component wave of a given period. In a dispersive wave train, it is the velocity associated with a certain feature (say a peak or trough) at the particular time that that feature is associated with a particular wave period. Thus, as the dispersive wave packet propagates, peaks and troughs move through the packet at a velocity different from that of the packet. This velocity varies with position in the packet, i.e. in association with waves of different periods within the packet.

Once the wave trains on a seismogram are identified, usually on the basis of particle motion and velocity, it is possible to obtain the period as a function of arrival time and in this way, to determine precisely group velocities for each kind of wave. If adequate data are available, phase velocities may also be determined.

Unfortunately, with existing techniques, it is not possible to deduce earth structure directly from a measurement of the dispersion, i.e. a plot of group and/or phase velocity as a function of period. Instead, dispersion curves are computed for a variety of theoretical models, and then compared with the experimental data. This process is repeated until a good fit is obtained. In selecting models for computation it is frequently possible to use information from other sources, for example, seismic refraction data, and thereby to narrow the number of possible solutions, and hence to increase the accuracy of the surface-wave technique.

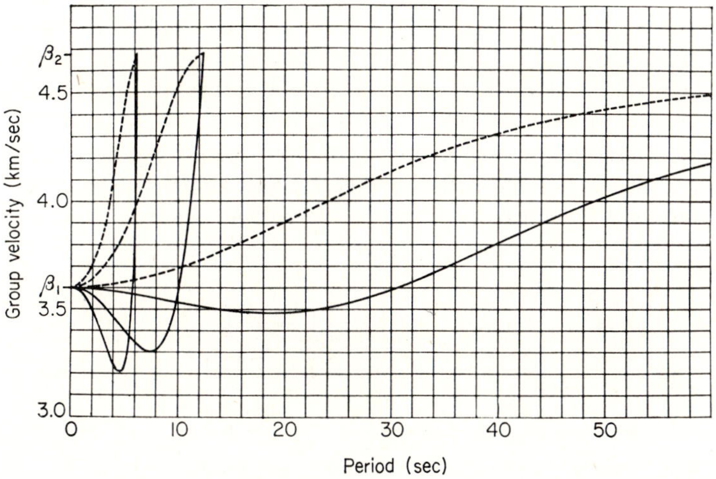

Fig. 2. Theoretical group-velocity (solid lines) and phase-velocity (dashed lines) dispersion curves for the first three modes of Love waves on a half-space with single surface layer. The shear velocities are $\beta_1 = 3.6$ km/sec for the layer and $\beta_2 = 4.68$ km/sec for the underlying material. The thickness of the layer is 35 km. Density ratio is $\rho_2/\rho_1 = 1.11$. The fundamental mode curves cover the entire range of frequencies reaching the velocity β_2 at infinite period. The second and third modes have a cut-off or maximum possible period at which the group and phase velocities equal β_2. This structural configuration is a simple approximation of the continental crust overlying the upper mantle. Waves corresponding to these three modes of propagation have been observed from natural earthquakes, the fundamental mode being very commonly observed.

Theoretical considerations of surface-wave propagation in a horizontally stratified wave guide produce, among other things, a most important relation called the period equation, i.e. the relation between phase, and indirectly group, velocity and wave period. A brief discussion of this theory and of methods of finding and solving a period equation are given by Haskell (1953) and Dorman (1962). Solution of this period equation gives various "allowed" values of velocity and period. These values, frequently called normal modes, may be thought of in much the same way as are the normal modes of an organ pipe or a banjo string. There are a fundamental mode and an infinity of harmonics or higher modes. The great difference, however, is that, because of the propagating nature of the waves in the surface wave guide of the earth, a normal mode is not represented by a single period as in the case of the banjo string, but rather by a continuous set of allowed values of velocity as a function of period. In practice, only the fundamental and first few higher modes are important, largely because of greater attenuation with distance of the shorter period waves which, in general, are associated with the higher modes.

For a simple model, consisting, for example, of a single solid layer overlying a sub-stratum, it is easy to describe and predict the general configuration of the normal-mode dispersion curves. As a guide to discussion of more complex models, let us consider this simple case first in some detail. In the case of Love waves, as illustrated in Fig. 2, the phase-velocity curves for all modes approach the shear velocity in the upper layer as the wave period approaches zero, and the shear velocity in the sub-stratum as the periods become long. The fundamental mode includes the entire spectrum from zero to infinity, but the higher modes have long-period cut-offs at successively shorter periods as the mode number increases. The corresponding group-velocity curves have similar limits,

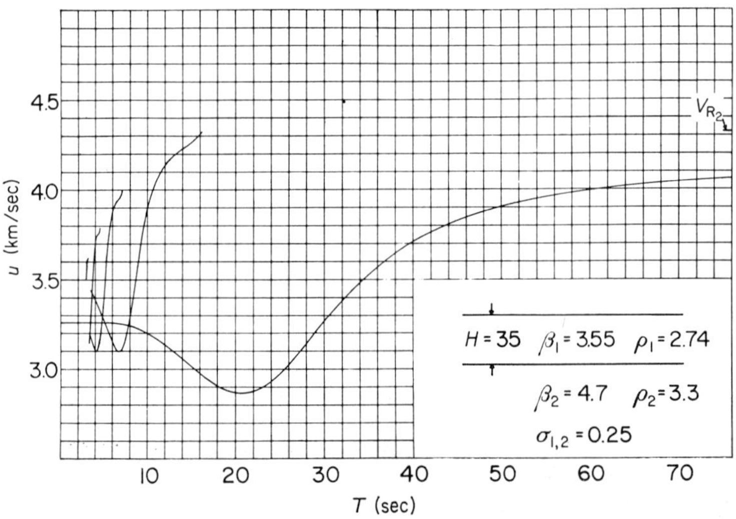

Fig. 3. Theoretical group velocity versus period for Rayleigh waves and higher modes on a model of the continental crust and mantle. Phase-velocity curves, not shown, have the same general relations to their respective group-velocity curves as in the case of Love waves. The curves are all based on roots of the same period equation. However, as a point of difference with the Love-wave curves, the asymptotic velocity of the fundamental or Rayleigh mode as the period increases is $V_{R_2} = 0.9194\beta_2$, whereas higher or shear modes reach β_2 at their cut-off period. The higher-mode curves are not complete at their short- and long-period ends because of computing difficulties. (See Case 8008 of Oliver, Kovach and Dorman (1961) for further data on this theoretical model.)

but differ in that a minimum of the group velocity exists at some intermediate period. Love waves are unaffected by a liquid layer, and so the above remarks apply equally to the case of a water-covered or non-water-covered region if the solid layers are the same in each case.

In the case of Rayleigh waves, as illustrated in Fig. 3, the fundamental or Rayleigh mode differs from all the higher modes in that it not only includes the entire spectrum of periods, but also in that the phase velocity approaches the Rayleigh-wave velocity in the upper medium as the wave periods decrease, and the Rayleigh velocity in the lower medium as the periods increase. Higher

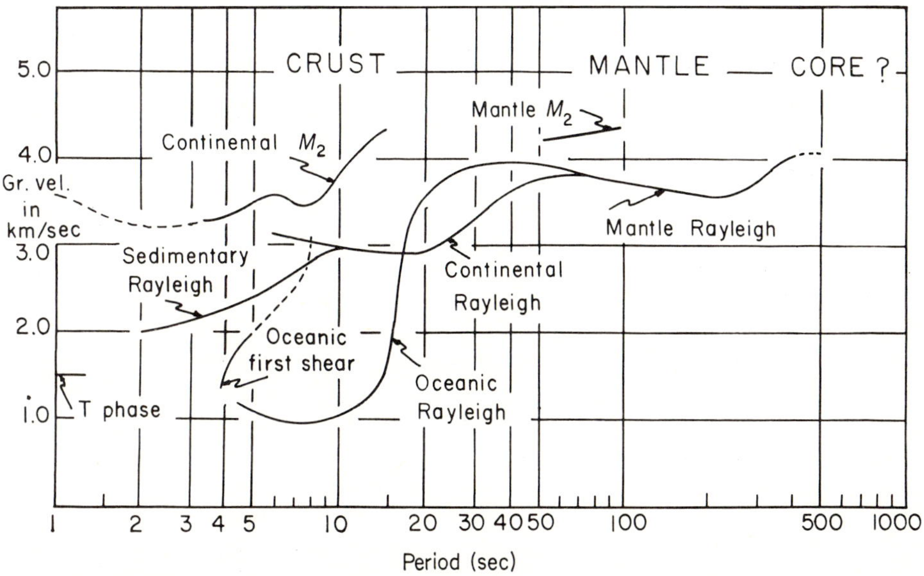

Fig. 4. Summary of Rayleigh-wave group-velocity observations. Effective depth of penetration of the surface waves increases with period, leading to the terms crustal and mantle Rayleigh waves. These curves are intended to represent average continental and oceanic conditions. However, for periods less than about 50 or 75 sec, regional differences within continents and ocean basins correspond to observed data, which depart significantly from these average curves. The configuration of the mantle-oceanic Rayleigh-wave curve and of the oceanic first-shear-mode curve are discussed in detail in this paper.

modes of the Rayleigh type, here termed shear modes, have the same limits as the Love modes, and also have similar long-period cut-offs. When the upper layer is liquid, waves of the Rayleigh type are strongly affected and the very short waves have velocities approximately, but not exactly, equal to the velocity of sound in water (Biot, 1952). As in the Love-wave case, the group-velocity curves all have a minimum at an intermediate value of the period.

In practice, it is not possible to approximate the earth adequately through the use of such simple models. The observed curves, as well as the theoretical curves for multi-layered models, are much more complex. However, certain features of the complex models and curves which we shall consider can be understood if the results for the simple two-media case are borne in mind.

In order to place in the proper perspective the detailed studies of surface-wave dispersion for oceanic paths, let us consider first a general summary of observed surface-wave dispersion, i.e. in this case, group velocity as a function of period. The data are summarized in Figs. 4 and 5 but various portions of the

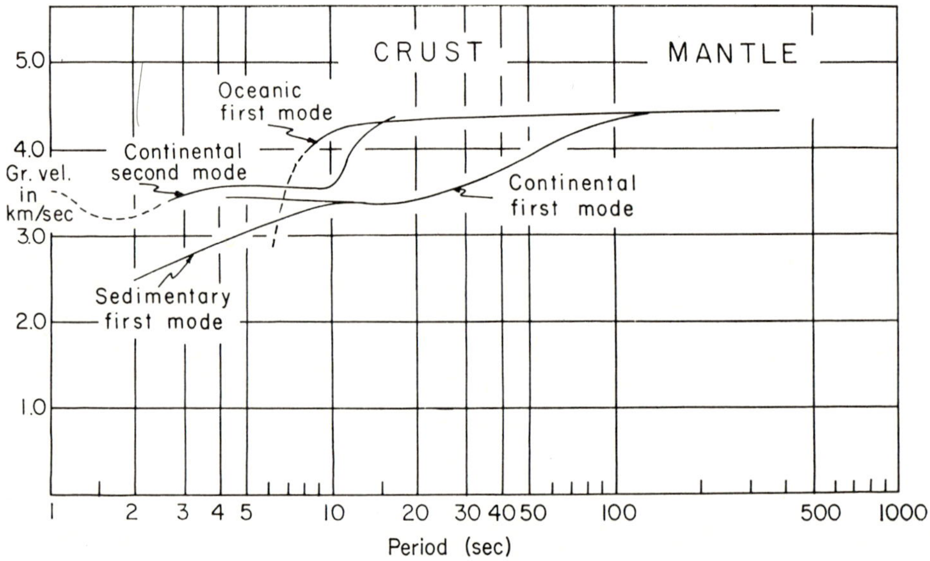

Fig. 5. Summary of Love-wave group-velocity observations. Regional differences within continents and oceans are associated with observed departures from these average curves out to periods probably as great as 100 sec. The relatively high group velocity of the oceanic first mode is due to the shallow depth of high-velocity rock beneath the oceans. The absence of observations of the oceanic second mode is probably due to attenuation of the very short period waves which would necessarily prevail in this mode.

curves in these figures will be discussed in more detail later. Consider the observations of waves of the Rayleigh type first (see Ewing and Press, 1956). For periods greater than about 75 sec, i.e. for waves whose lengths are greater than about 300 km, differences between continents and oceans are very small, and the waves are represented by a single line in Fig. 4. The minimum value of the group velocity at a period of about 225 sec results primarily from the effect of the velocity increase at depths of the order of hundreds of kilometers in the earth's mantle.

For periods less than 75 sec, the curve splits and the distinction between the continental and oceanic types of crust becomes readily apparent. In the period range 70 to 17 sec, the ocean curve falls well above the continental curve. This is largely due to the shallow depth (about 11 km) of the high-velocity mantle beneath oceans as compared to the relatively great depth (about 40 km) of the mantle beneath the continents.

At a period of about 20 sec, the effect of the water and low-velocity sedimentary layer becomes important, the ocean curve drops steeply, and, for periods shorter than about 17 sec, falls well below that for the continents.

From earthquake sources, no waves corresponding to the fundamental Rayleigh mode for periods less than about 13–15 sec are observed. However, in the case of certain special sources, i.e. nuclear explosions, some waves corresponding to the shorter-period portions of the fundamental curve shown in Fig. 4 were observed.

At very short periods in the continental case, the effect of low-velocity sedimentary layers becomes important, and the curve may fall to smaller velocities, as indicated in the figure. Short-period waves associated with the first higher mode, and in some cases still higher modes, have been identified for the continental crust, and these waves account for much of the seismic disturbance recorded by long-period instruments under certain conditions.

The identification of higher-mode waves in the ocean is somewhat more tenuous. Such waves appear to have been generated by a submarine nuclear explosion as indicated in Fig. 4. Earthquakes sometimes generate a train of short-period waves which have on occasion been identified as higher-mode waves, but quantitative verification of this point has been difficult to obtain. These two cases are discussed in considerably more detail in a later section.

Fig. 5 shows similar data for Love waves. In general the curves are similar to those of the Rayleigh type, with certain exceptions in addition to the generally higher velocities. The minimum in the mantle-wave range is not so pronounced in the case of Love waves, and may, in fact, be absent. The pronounced effect which the water layer has on Rayleigh waves is, of course, lacking in the Love-wave case, although the low-velocity sediments in ocean basins have an important effect on very short-period Love waves (about 8 sec and less). This effect is also discussed in detail in a later section. Let us now consider various sections of these curves for the oceanic case in considerably more detail.

3. Evidence from Mantle Surface Waves

The dispersive characteristics of oceanic surface waves of period greater than about 30 sec are governed primarily by the properties of the mantle. In general, studies of these waves, particularly of waves with periods greater than about one minute, have been based on data from very large earthquakes, although with modern instrumentation mantle waves may be detected in some cases from much smaller shocks.

The first identification of mantle Rayleigh waves and a study of their group velocities was carried out by Ewing and Press (1954), using data obtained by Benioff from three large earthquakes. A later study by Ewing and Press (1954a) of the great Kamchatka shock of 4 November, 1952, included data on waves with periods as long as about 500 sec. Their data are shown in Fig. 6. Later, Benioff and Press (1958) extended this range to waves with periods greater than 600 sec. The large Chilean shock of 22 May, 1960, produced waves of even longer periods, perhaps as great as about one hour, but analysis is not yet complete. Other data on mantle Rayleigh-wave group velocities for purely oceanic paths were given by Sutton (*in litt.*; see data in Fig. 6), Benioff and Press (1958) and Kovach (1959).

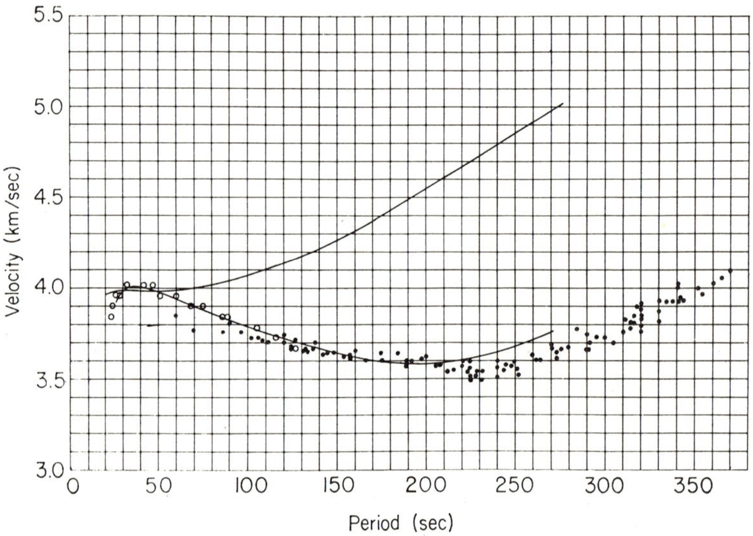

Fig. 6. Observed and theoretical Rayleigh-wave dispersion in the range of periods influenced primarily by the upper mantle. Solid dots are group-velocity data of Ewing and Press derived from multiple paths around the earth. Open circles are data of Sutton for paths across deep basins of the Pacific. The solid curves are computations of Dorman, Ewing and Oliver on the basis of a layered model of the oceanic crust and mantle (curve B, Fig. 7). The upper curve is phase velocity, the lower curve group velocity. Good agreement between theoretical and observed dispersion is evidence supporting the existence of a shear-velocity distribution beneath the oceans very similar to curve B.

Data on very long-period waves are usually obtained for very long paths which include both oceanic and continental segments. Ewing and Press, as well as subsequent investigators, detected no systematic differences in group velocities for these two kinds of paths. Preliminary measurements of phase velocity likewise showed no differences between continents and oceans in this respect (Nafe and Brune, 1960). More recent and more precise measurements of phase velocity by Brune (*in litt.*) do indicate a small but measurable difference in such cases, extending to periods at least as great as about 400 sec. However, the analysis of these data in terms of earth structure is not yet complete, and, in any case, is not likely to affect significantly the results presented here.

Data on long-period Love waves affected primarily by the mantle are available from a variety of sources. The term "G-wave" frequently appears in this connection and is used to describe a long-period Love wave of large amplitude and little dispersion propagating over oceanic paths. There is no intent, however, in using the term G-wave to imply a type of propagation mechanism other than that of the Love wave.

Satô (1958) analyzed multiple long-period Love-wave arrivals from several large earthquakes recorded on the Pasadena strain seismometer and found Fourier components with periods between 50 and 300 sec, all with group velocities of about 4.4 km/sec. Satô's results and other studies strongly suggest that Love waves with periods between 150 and 300 sec are dispersed very little in the cases of both oceanic and continental paths. Press (1959), Kovach (1959), Bäth and Vogel (1957), and Ewing, Jardetzky and Press (1957, p. 212) published data on continental Love waves indicating that oceanic and continental Love-wave dispersion curves generally differ at all periods less than at least 100 sec.

Long-period oceanic Love-wave dispersion was analyzed by Landisman, Satô and Ewing (1958) using a theoretical model consisting of two homogeneous surface layers underlain by a layer with constant gradient in rigidity. The upper layer was taken to represent the oceanic crust and the intermediate layer the upper mantle. The model giving the best fit to Satô's data and other data is shown in Fig. 7 as curve C. The indicated decrease in shear velocity at a depth of 50 km is shallower and of greater magnitude than that derived by Gutenberg and by Lehmann (Fig. 7, curve A) for various continental areas on the basis of body-wave travel times.

Results on the existence and the great extent of the low-velocity layer under oceans were confirmed and elaborated by Dorman, Ewing and Oliver (1960), who found a structural configuration consistent with the observed Rayleigh-wave group-velocity dispersion curve. This work was based on machine calculations according to the multilayer method (see Dorman, 1962). The crust and the mantle to a depth of 2800 km were represented by about 35 homogeneous layers in these calculations. The experimental study was limited to obtaining a solution for the shear-velocity distribution in the mantle. The effect of density variations was fixed by using Bullen's model A for density and the effects of variations in crustal structure were eliminated by using the same crustal layers based on seismic refraction results throughout the calculations. The oceanic Rayleigh-wave dispersion curve of Sutton (*in litt.*) for periods less than 140 sec and the mantle Rayleigh-wave dispersion data of Ewing and Press for longer periods were used as data for comparison with theoretical calculations. The final approximation of mantle shear-velocity distribution is shown in Fig. 7, curve B. The dispersion curve computed from this model is shown with the observed data in Fig. 6. The quality of agreement between the experimental and theoretical curves over a broad spectrum lends weight to this solution. Deviation of the theoretical curve from the experimental curve for periods greater than about 200 sec is of the sort expected from the use of a flat-earth calculation. Fig. 7 shows by comparison of the oceanic solution (curve B) with a shear-velocity distribution for the continental case (curve A, based on unpublished data of Lehmann) that the region of low shear velocity extends to much shallower depths beneath the oceans and also that somewhat lower shear velocities prevail down to depths of about 400 km. The sensitivity of this method appears to be such that estimates of shear velocity are accurate to about 0.1 km/sec, providing the assumed density distribution and crustal layering are correct.

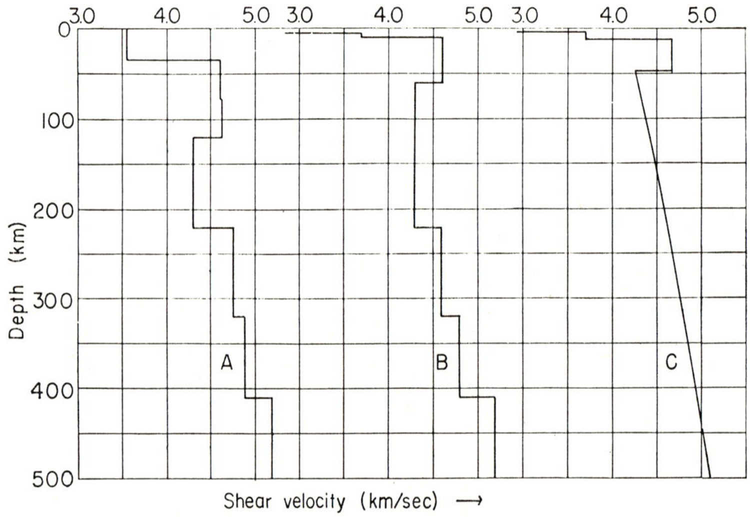

Fig. 7. Oceanic and continental shear-velocity distributions for the crust and mantle. Curve A, based on data of Lehmann derived from body-wave studies, gives a computed dispersion curve which agrees well with continental crust-mantle Rayleigh-wave dispersion data. Curve B gives a computed group-velocity curve, shown in Fig. 6, which fits oceanic crust-mantle Rayleigh-wave dispersion data. Curve B was chosen experimentally as the minimum departure from curve A which is consistent with observed oceanic Rayleigh-wave data. Curve C was chosen by Landisman, Satô and Ewing (1958) as the best experimental fit to oceanic Love-wave dispersion data. The oceanic water and mud layers used in computation are not represented in curves B and C. Curves B and C indicate that relatively low shear velocities are found at very shallow depths in the sub-oceanic mantle as compared with the continental case represented by curve A.

Takeuchi, Press and Kobayashi (1959) performed computations on mantle Rayleigh-wave dispersion based on two models of the mantle using a variational method. This did not differentiate between continental and oceanic mantle but explained the general features of the mantle Rayleigh-wave dispersion curve beyond 100 sec period, including the effect of the world-wide existence of the low-velocity region of the upper mantle. Although the accuracy of those calculations was questioned in a subsequent paper (Press and Takeuchi, 1960), the derived structural features still appear to be correct.

Clearly, all factors affecting the results have not been taken into account in the studies described above on oceanic mantle structure; yet, despite the preliminary nature of these results, it is clear that sub-oceanic and sub-continental mantle structure are not the same. Additional data are necessary to ascribe the observations either to regional differences in temperature distribution or in composition in the mantle. Further analysis of long-period surface-wave data, particularly of phase-velocity measurements for purely oceanic paths, are certain to provide additional information on the sub-oceanic mantle.

4. Evidence from Crustal Surface Waves

Crustal surface waves are so named because their group-velocity curves have a normally dispersed branch (i.e. a segment for which group velocity increases with period) as a result of lower seismic velocities of the crustal layers, and, in the case of oceanic Rayleigh waves, also because of the water layer. The study of dispersed surface-wave trains, such as those of crustal surface waves, was handicapped for a long time by methods of reading which apply best to ray theory analysis of body waves, i.e. only the arrival time of the first motion or of the maximum amplitude was read. This arrival time was related to the period of the first readable oscillation or the period of the maximum amplitude. This method necessitated reading large numbers of seismograms to obtain limited dispersion data with great scatter. The lower group velocities of late waves of long trains were never considered. Differences in velocities of surface waves for different parts of the earth were known from the early data (see Macelwane, 1951, and Gutenberg, 1951, for summary). However, few precise interpretations were attempted partly because of the scatter of the data. Moreover, in the case of Rayleigh waves, the correct theory was not applied until later. An important improvement in technique was made by Bullen (1939) who measured period versus arrival time for a number of successive cycles in a Rayleigh-wave train, smoothed the values and converted arrival time to group velocity. Wilson (1940) read the several swings on oceanic Love-wave seismograms and related each period to its proper group velocity, thus making use of data from the entire train of surface waves from a single shock. Following this principle, Ewing and Press (1952) plotted the number of each wave crest against its arrival time throughout an entire oceanic Rayleigh-wave train and, from this, obtained a smooth relation between period and group velocity, clearly showing the possibility of obtaining a large number of dispersion data from a single seismogram, as in Fig. 9 (which appeared in their paper). This method, which is now widely used, resulted in such a striking improvement in the quality of observations that most experimental surface-wave dispersion data of present-day significance have been derived by it. Almost all of the data reported here have been analyzed by this technique. An earlier summary of results was given by Press and Ewing (1955) and a detailed review of theory and results was given by Ewing, Jardetzky and Press (1957, chap. 4) and by Ewing and Press (1956a).

The crustal Love-wave train for oceanic paths is usually short in duration and normally dispersed, arriving before the beginning of the Rayleigh-wave train. Dispersion is small as indicated in Fig. 8. Occasionally this train continues into the "short-period train" (see Fig. 13) as discussed below under that title. Wilson (1940) based two important conclusions on Love-wave dispersion data: (a) that the Pacific, Atlantic and Indian Ocean basins were similar in structure and (b) that the high-velocity mantle was much closer to the surface beneath the oceans than beneath the continents. Coulomb (1952) published data

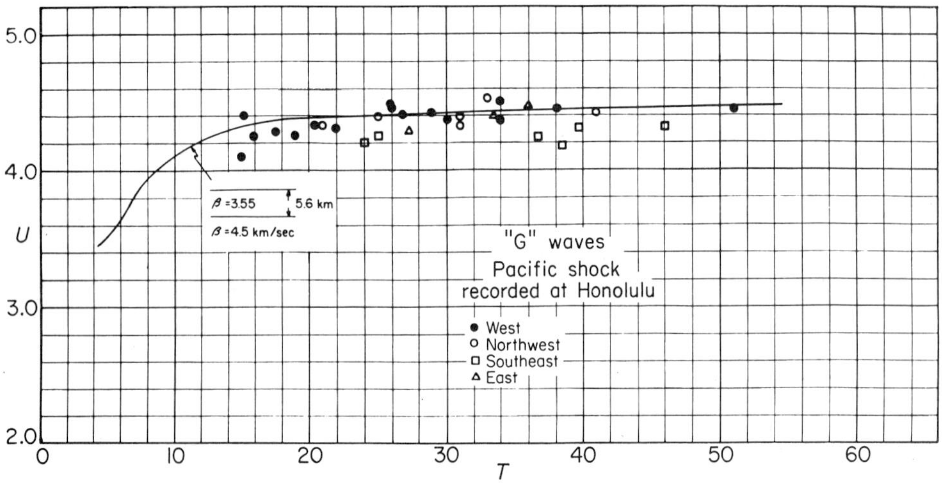

Fig. 8. Love-wave dispersion observations (group velocity versus period) for paths across the Pacific to Honolulu (Oliver, Ewing and Press, 1955). Agreement with group velocity for the single-layer theoretical case shown by the solid curve indicates the relative thinness of the oceanic crust. These data reveal significant variations in velocity depending on azimuth from Hawaii. In particular, the low velocity of waves from the southeast indicates lower shear velocities or a thicker crust, or both, which may be localized in the Easter Island Rise.

derived from the great Queen Charlotte Islands earthquake which were consistent with Wilson's data, as were the data of Caloi and Marcelli (1952) for the Atlantic. Oliver, Ewing and Press (1955) studied dispersion of Love waves recorded at Hawaii from a number of Pacific shocks and at Atlantic stations from several Atlantic shocks. Their results confirm Wilson's conclusions. Some scatter in Love-wave velocities of given periods was observed, which appears to relate to variations in the crustal structure. Low velocities were recorded at Honolulu from shocks southeast of Hawaii, as shown in Fig. 8, indicating that a somewhat different structure exists in the vicinity of the Easter Island rise. Data from the Queen Charlotte Islands recorded at Honolulu give the highest group velocity for any Pacific path with equally high velocities for Love waves recorded at Bermuda from Mid-Atlantic Ridge shocks. Other studies by Evernden (1954) and by DeNoyer (1959) based on Berkeley seismograms of

Pacific shocks give slightly lower velocities for oceanic Love-wave propagation in certain period ranges. Most of these differences can be ascribed to differences in path.

From both Love- and Rayleigh-wave observations at Christchurch, New Zealand, Officer (1955) concluded that the crust of the South Pacific Basin is somewhat thicker than that of the North Pacific Basin by comparison with the Hawaiian data of Oliver, Ewing and Press. Officer also found the Tasman

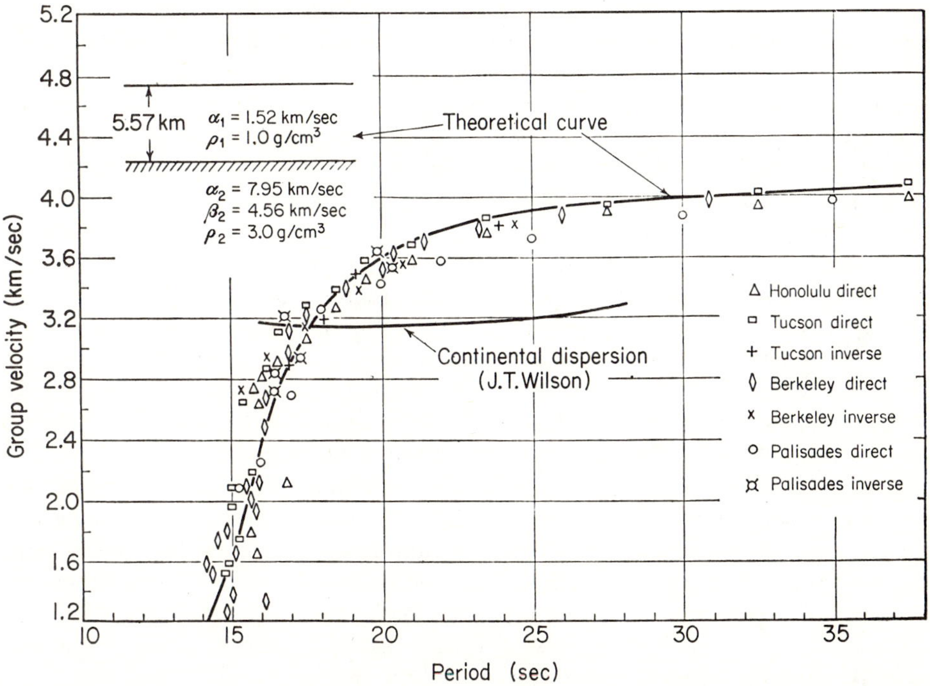

Fig. 9. Dispersion observations of oceanic Rayleigh waves from the Solomon Islands earthquake of 29 July, 1950, plotted with a theoretical oceanic dispersion curve (Ewing and Press, 1952). Waves over both the direct and antipodal paths were used at Berkeley, Tucson and Palisades, and over the direct path only at Honolulu. Corrections were made for the continental portions of the paths. Within the precision of the experiment no significant difference exists between data for the antipodal path through the Indian and Atlantic Oceans and data for the direct path through the Pacific Ocean. This is indicative of the uniformity of water depth and bottom structure in all of the oceans.

Basin to have an oceanic crust and found the crustal thickness of New Zealand and the nearby shallow submarine ridges to be about 20 km. Bäth and Vogel (1958) found dispersion curves of oceanic character for Love and Rayleigh waves across the Norwegian sea.

In general, for the deep basins of the Atlantic and Pacific, the higher-velocity Love-wave observations are consistent with a basaltic crust about 6 km thick overlying the higher-velocity mantle while the lower velocity observations indicate about twice this thickness with the upper part having a considerably lower shear velocity than basalt. It is, perhaps, significant that the observations made at the oceanic island stations, except for the Hawaiian data from the direction of the Easter Island rise, give consistently higher group velocities than those made at Berkeley. The effect of the portion of continental path near Berkeley, although its length is short, would be to reduce the group velocity of the waves observed there. Real differences in the data, however, undoubtedly exist, and probably represent real differences in sub-bottom structure. However, because of the large dispersion in the case of continental paths for Love waves in the crustal period range as compared with small dispersion for waves of the same period range for oceanic paths, there is always a danger that the effects of a small portion of continental or near-continental structure will obscure the effect of the oceanic portion of the path. Furthermore, mantle heterogeneities are of considerable importance in this problem.

Probably the most striking feature of seismic wave trains which have traversed paths which are at least partly oceanic is the long, nearly sinusoidal train of Rayleigh waves with periods slightly greater than about 15 sec (Fig. 1). This appearance results largely from the effect of the water layer on Rayleigh waves propagating in the crust–upper-mantle rock system. Because much of the earth is covered by deep water and because this water layer is remarkably uniform in depth over much of the earth, the predominant period in this wave train varies little with different oceanic propagation paths.

The theory of Rayleigh-wave propagation in rock overlain by a compressible ocean was considered by Stoneley (1926). He derived the pertinent period equation but did not obtain numerical solutions in the period range of significant dispersion. In 1950, Ewing and Press showed that such a theory could account for the main features of the observed Rayleigh-wave train (Fig. 9). The first theoretical model consisted simply of a water layer overlying a solid half-space with the average properties of the crust and mantle. Later, Jardetzky and Press (1953) showed that the effect on the dispersion curves of considering the crust and mantle separately in such a problem was small, as is also shown by the theoretical curves in Fig. 13. The sedimentary layer in both cases was included as part of the liquid layer. The properties of the Rayleigh waves of short period, which travel at a velocity less than the sound velocity in water, were clarified by Biot (1952).

Once the propagation of oceanic Rayleigh waves was understood, they were used for exploration of the various ocean basins. Ewing and Press (1952) in a study of waves propagating directly, and also through the antipodal path to Honolulu, Berkeley, Tuscon and Palisades from a shock in the Solomon Islands, found no measurable difference in the basement structure of the Pacific, the Indian and the Atlantic Oceans. This result brought Rayleigh-wave evidence into agreement with the conclusions Wilson had made on the basis of Love-wave data.

Oliver, Ewing and Press (1955) confirmed these broad conclusions for much of the Pacific and parts of the Atlantic Ocean. They also found mean sediment thicknesses in the Pacific ranging from 0.4 km north of Hawaii to 1.1 km southwest of Hawaii. These values are calculated by subtracting the mean water depth from the liquid layer thickness obtained by comparison of experimental data with theoretical models for different thicknesses of this layer. The values generally agree well with those obtained from refraction shooting. Oliver, Ewing and Press also pointed out that their data suggest somewhat greater depths for the mantle in the western Pacific than for the eastern Pacific. Along certain paths (for example, those from Hawaii to the Philippines, Formosa and Chile) the data, by virtue of slightly lower group velocities for waves of period greater than about 20 sec, suggest structures differing somewhat from that typical of ocean basins. These differences might lie in somewhat lower crustal or mantle velocities, or in greater crustal thickness, or a combination of these.

Berckhemer (1956) showed the close coincidence between Rayleigh-wave dispersion data for paths across basins of the eastern Atlantic Ocean with dispersion observed by Oliver, Ewing and Press for basins of the western Atlantic, clearly establishing the essential similarity of crustal structure on both sides of the Mid-Atlantic Ridge. Bäth (1959) presented group-velocity dispersion data for Love and Rayleigh waves for a wide variety of azimuths from Uppsala. With one exception, data for the ocean basins, though distinctly oceanic in character, could be obtained only by applying large corrections for continental portions of the paths. However, very favorable paths across the Arctic basin yielded well-determined dispersion curves clearly showing oceanic crustal structure.

Although surface waves suggest anomalies in such areas as the Easter Island Rise and Polynesia, the contrast between dispersion observed over all oceanic paths with that for continents overshadows the relatively small differences cited above, and is evidence for the relative uniformity of ocean structure.

Regardless of the crustal model adopted, all theoretical Rayleigh-mode dispersion curves have a minimum group velocity less than 1 km/sec in the 12–14 sec period range (see Fig. 13). This minimum is equivalent to the one which gives rise to the prominent Airy-phase arrival observed in some explosions in shallow water, suggesting that the corresponding Airy phase might be observed when the source is a natural earthquake. No clear-cut identification of such Airy phases has been made in the case of earthquake sources, however, and recent indications are that the reason for the absence of the waves corresponding to the Airy phase and to the shorter-period branch of the Rayleigh-mode dispersion curve is the depth of natural earthquakes.

However, waves corresponding to the short-period inversely dispersed branch of the Rayleigh-wave curve and the minimum group velocity have been observed from artificial sources. Oliver and Ewing (1958) discussed such arrivals at Pasadena and Berkeley, and they associated these with the atomic explosion WIGWAM detonated deep in the Pacific Ocean off lower California. Similar waves recorded at Honolulu were generated by the high-altitude explosions TEAK and ORANGE (see Pomeroy and Oliver, 1960), further indicating that the reason for the absence of these waves in the case of earthquakes is not an unknown structural feature, but rather the location of the source.

5. Evidence from Short-Period Surface Waves

It is to be expected that oceanic surface-wave trains include waves with periods shorter than those of the crustal waves discussed above, which were explained entirely as the fundamental modes of the Love and Rayleigh types. In addition to the portions of the fundamental-mode dispersion curves which continue to high frequencies, it was noted previously that propagation in an infinity of higher modes of both wave types is theoretically possible. A practical limit to the frequency of coherent signals which can be propagated through the

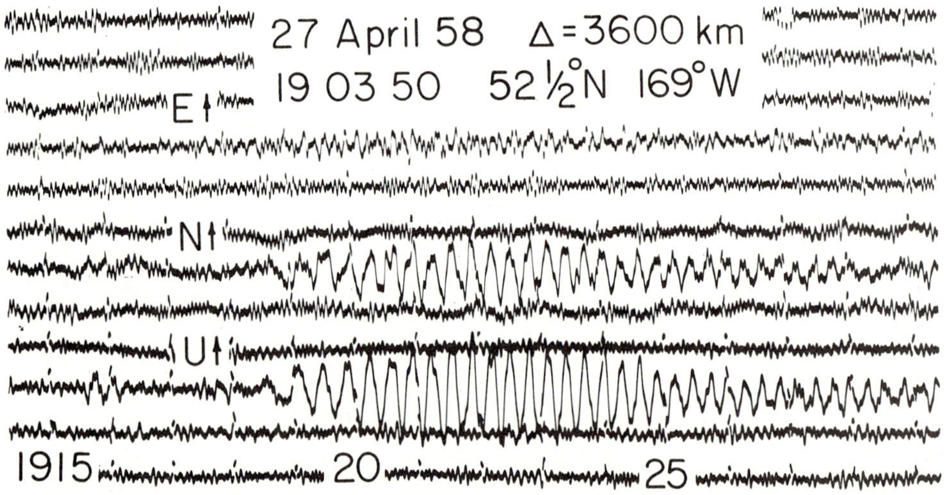

Fig. 10. Honolulu seismograms of a shock in the Aleutians. Because of the direction, almost directly north, from which the waves approached, the north–south component gives a record of longitudinal motion while the east–west component gives the transverse motion. Fundamental-mode Rayleigh waves are clearly visible on the vertical and longitudinal instruments, whereas long-period Love waves are absent on the transverse component. Short-period waves, however, are present on the transverse.

sub-bottom and to the distance at which they can be detected is imposed by attenuation. On the other hand, theory shows that propagation of short-period waves is confined to shallow depths and, therefore, crustal structure must be more important in determining the properties of short-period waves.

A train of short-period (6 to 9 sec) waves, which is often found on seismograms for oceanic paths, was described by Coulomb (1952) and by Oliver, Ewing and Press (1955). Examples of this wave train are shown on the Honolulu seismograms of the shocks of 27 April, 1958, 23 September, 1958, and 13 July, 1958, which all occurred in the Aleutian–Gulf of Alaska region (see Figs. 10, 11 and

12). The short-period wave train on the horizontal transverse component frequently occurs as a continuation of the long-period Love-wave train toward shorter periods and lower group velocities. The short-period waves, when recorded, are frequently found in all three components of ground motion, sometimes with approximately equal amplitudes in the later portions of the train (see seismograms of Fig. 11) or sometimes predominantly on the trans-

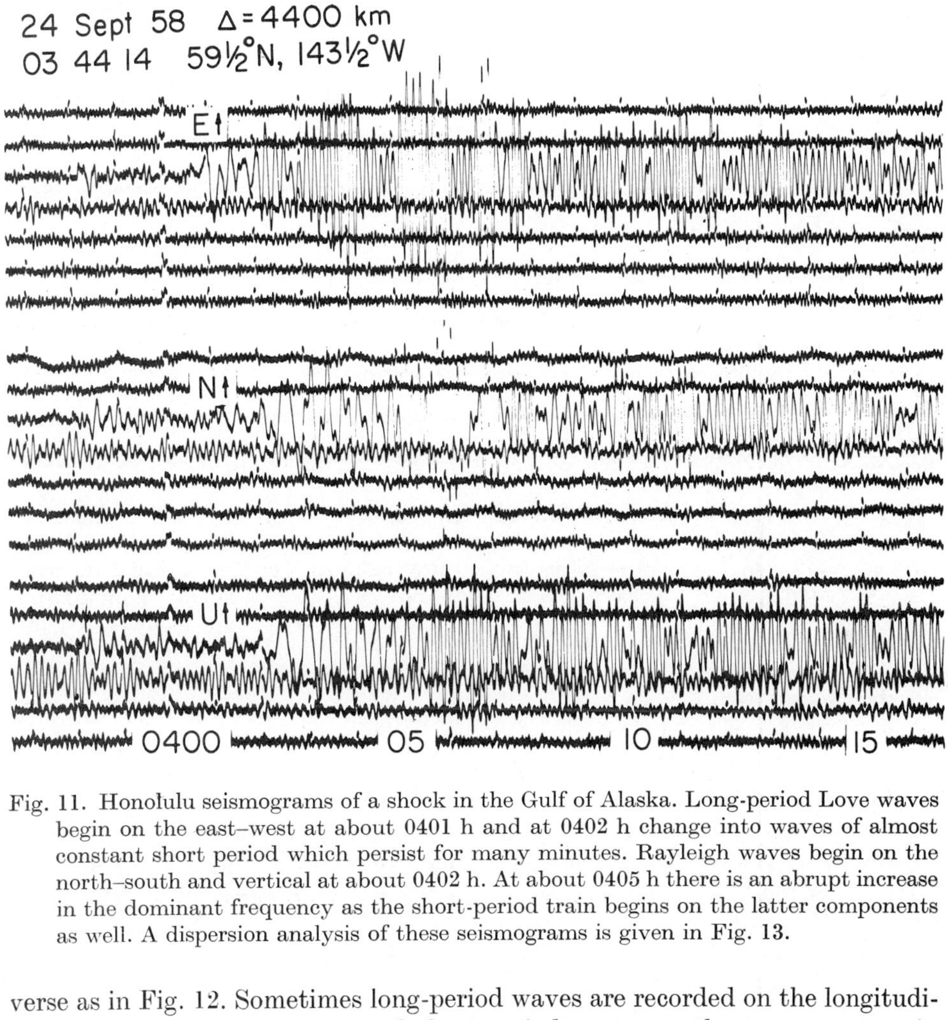

Fig. 11. Honolulu seismograms of a shock in the Gulf of Alaska. Long-period Love waves begin on the east–west at about 0401 h and at 0402 h change into waves of almost constant short period which persist for many minutes. Rayleigh waves begin on the north–south and vertical at about 0402 h. At about 0405 h there is an abrupt increase in the dominant frequency as the short-period train begins on the latter components as well. A dispersion analysis of these seismograms is given in Fig. 13.

verse as in Fig. 12. Sometimes long-period waves are recorded on the longitudinal and vertical components and short-period waves on the transverse, as in the seismograms of Fig. 10. Furthermore, the beginnings of the wave trains on the transverse component usually have longer periods and precede the short-period train on the other two components, as in Fig. 11. When observable, the periods are always about the same on all three components at the same time, but particle motion is apparently not coherent, i.e. the relative phases between the components change throughout the train. The dispersion data plotted in Fig. 13, which are derived from the seismograms in Fig. 11, show that, following the long-period Love-wave train, motion of about 9-sec period was recorded on all components for group velocities between 4 and about 1 km/sec. There is a tendency for the period to decrease slightly with time.

The beginning of the short-period train which has transverse particle motion,

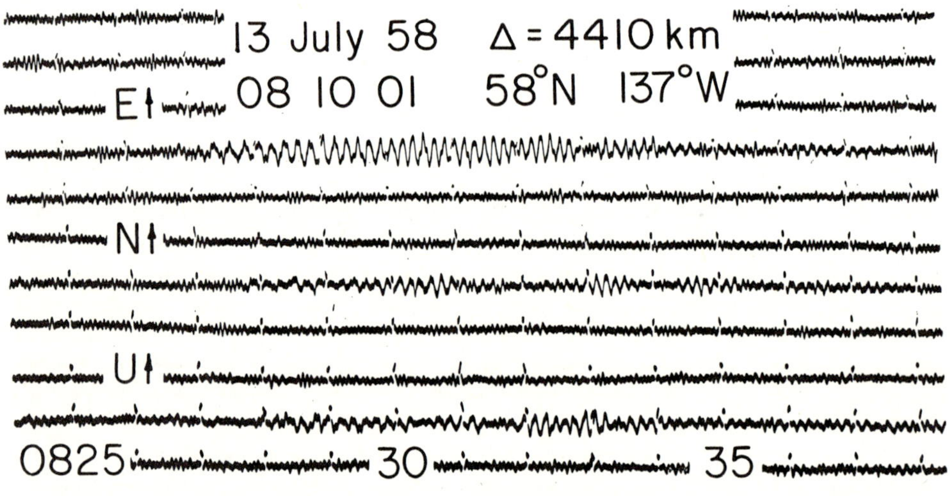

Fig. 12. Honolulu seismograms of a shock in the Gulf of Alaska. These seismograms are characterized by relatively strong Love waves, beginning at about 0827 h on the east–west, compared with weakly recorded Rayleigh waves on the vertical and north–south components. The period of the Love waves decreases rapidly at first to an almost constant period of about 9 sec, which persists for several minutes. Vertical motion at about 0833 h is a part of the short-period train. The seismograms of Figs. 10, 11 and 12 illustrate the long-duration, constant-frequency characteristics of the short-period train.

as shown in Fig. 11 or Fig. 12, has been shown previously to correspond to a portion of the fundamental Love-mode dispersion curve (see DeNoyer, 1959; Oliver, Dorman and Ewing, 1959). The remainder of the short-period train on all three components has not been explained previously, and it has been suggested that the waves correspond to higher-mode propagation of the Love and/or Rayleigh types. The authors have recently compared data of the type described above with dispersion computations based on layered models of oceanic structure derived from seismic refraction measurements.

An attempt to explain the short-period data by including a mud layer with low shear velocity in the theoretical model was made as shown in Fig. 13. The data of Nafe and Drake (1957, and in Chapter 29) on the physical properties of marine sediments provided a useful basis for these investigations. The dispersion curves in Fig. 13 show that the first Love mode and first shear mode are closely coincident and reach very low group velocities. Both of these features seem to be required by the data. However, the period of the strongly dispersed waves in both these modes is too short in this computation. Additional recent study of the dispersive effects of the mud parameters by the authors, as well as the latest data quoted by Nafe and Drake, indicate that a mud layer of lower shear velocity than that used in Fig. 13 may provide a model of bottom structure consistent with dispersion data as well as data derived from other sources. However, conclusive results are not yet available.

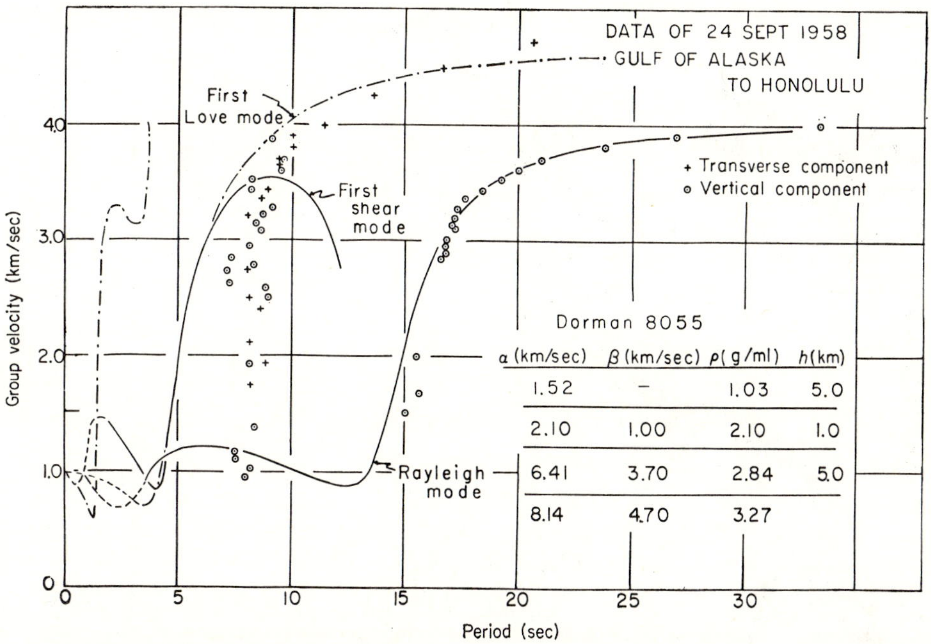

Fig. 13. Typical oceanic dispersion data, derived from the seismograms of Fig. 11, are plotted with theoretical surface-wave group-velocity dispersion data. The theoretical curves were computed on the basis of the layered structure indicated. Agreement between theoretical and experimental data is good in the Rayleigh mode owing to proper choice of the thickness of water plus mud (layers one and two), which determines the period of the steep portion of the curve. However, poor agreement in the first Love and first shear modes is due mainly to an incorrect choice of parameters to represent the mud.

Factors related to the presence or absence of various surface-wave types on seismograms were discussed by Oliver, Ewing and Press (1955) who compiled the geographical data of Fig. 14. Primarily, on seismograms written at Hawaii short-period waves predominate for shocks in the area including the Easter Island Rise, the west coast of North America and the Aleutians. For shocks elsewhere on the Pacific margins, long-period waves predominate at Hawaii. When two stations are on the same azimuth from a shock, the appearance of long-period waves is favored at the more distant station while short-period

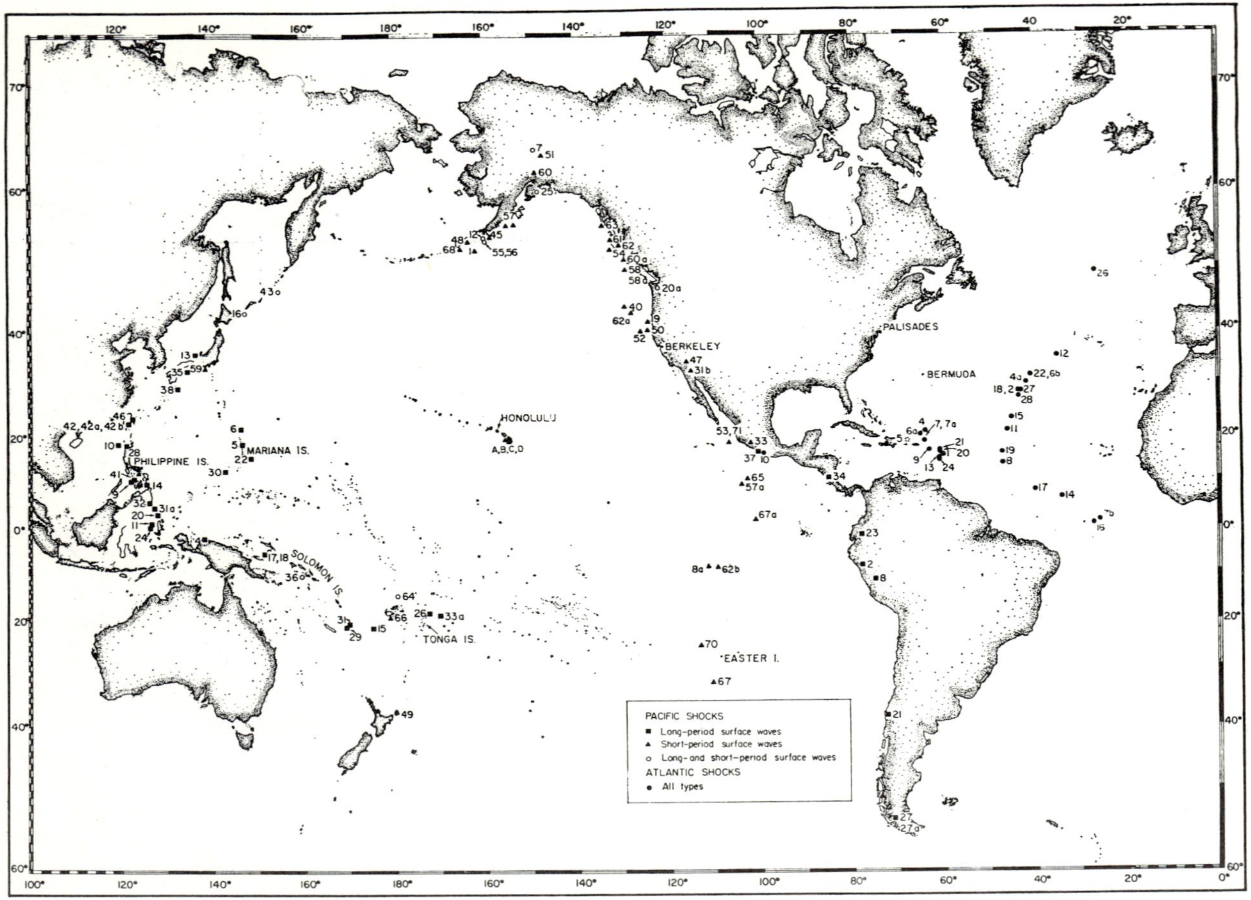

Fig. 14. The epicenters of earthquakes studied by Oliver, Ewing and Press (1955) identified according to the types of surface waves observed from them. Long path length appears to favor the observation of long-period waves, whereas short-period waves are observed over all of the shorter paths. Crustal structure is also a factor, since long-period waves only are observed for paths crossing Polynesia and the Easter Island Rise.

waves are more likely to predominate at the closer station. Also, short-period waves seem to be greatly attenuated whenever they cross a continental margin to reach the recording station.

The T-phase is a high-frequency (of the order of a few cycles per second), late-arriving wave generated by some submarine or coastal earthquake and recorded at large distances from the source, usually by stations near a coast. The T-phase is due to propagation of energy as sound waves in the ocean with a velocity of 1.5 km/sec, the sound velocity of water. It appears that energy transfer between the solid bottom and the water takes place primarily along submarine slopes. In some cases the energy is transmitted to the water at a submarine slope immediately in the epicentral region (see Ewing, Press and Worzel, 1952). In other cases it may travel as far as 51° as a body P wave before striking a slope and entering the water (Shurbet, 1955). The duration of the phase may be several minutes and the average velocity observed varies according to the proportions of the path segments through rock and water.

Tsunamis are earthquake-generated sea waves. The dominant wavelengths are usually much greater than the depth of the ocean (h) so that the waves propagate as shallow-water gravity waves with velocity $\sqrt{(gh)}$. The generation of these waves necessarily involves a sudden change in configuration of the sea bottom associated with the earthquake. Either vertical movements along submarine fault outcrops or slumps or turbidity currents triggered by the earthquake are considered possible mechanisms.

References

Bäth, M., 1959. Seismic surface-wave dispersion: A world-wide survey. *Geofis. pur. appl.*, **43**, 131–147.

Bäth, M. and A. Vogel, 1957. Continental dispersion of seismic surface waves. *Geofis. pur. appl.*, **38**, 10–18.

Bäth, M. and A. Vogel, 1958. Surface waves from earthquakes in northern Atlantic-Arctic Ocean. *Geofis. pur. appl.*, **39**, 35–54.

Benioff, H. and F. Press, 1958. Progress report on long period seismographs. *Geophys. J. Roy. Astr. Soc.*, **1**, 208–215.

Berckhemer, H., 1956. Rayleigh-wave dispersion and crustal structure in the East Atlantic Ocean Basin. *Bull. Seism. Soc. Amer.*, **46**, 83–86.

Biot, M. A., 1952. The interaction of Rayleigh waves and Stoneley waves in the ocean bottom. *Bull. Seism. Soc. Amer.*, **42**, 81–92.

Bullen, K. E., 1939. On Rayleigh waves across the Pacific ocean. *Mon. Not. Roy. Astr. Soc., Geophys. Suppl.*, **4**, 579–582.

Caloi, P. and L. Marcelli, 1952. Onde superficiali attraverso il bacino dell'Atlantico. *Ann. Geofis.*, **5**, 397–407.

Coulomb, J., 1952. Love waves of the Queen Charlotte Islands earthquake of August 22, 1949. *Bull. Seism. Soc. Amer.*, **42**, 29–36.

DeNoyer, J., 1959. Crustal structure of the North Pacific from Love-wave dispersion. *Bull. Seism. Soc. Amer.*, **49**, 331–336.

Dorman, J., 1962. Period equation for waves of Rayleigh type on a layered, liquid-solid half space. *Bull. Seism. Soc. Amer.*, **52**, 389–397.

Dorman, J., M. Ewing and J. Oliver, 1960. Study of the shear-velocity distribution in the upper mantle by mantle Rayleigh waves. *Bull. Seism. Soc. Amer.*, **50**, 87–116.

Evernden, J. F., 1954. Love-wave dispersion and the structure of the Pacific basin. *Bull. Seism. Soc. Amer.*, **44**, 1–5.

Ewing, M., W. S. Jardetzky and F. Press, 1957. *Elastic Waves in Layered Media.* McGraw-Hill Book Co., New York.

Ewing, M. and F. Press, 1950. Crustal structure and surface wave dispersion. *Bull. Seism. Soc. Amer.*, **40**, 271–280.

Ewing, M. and F. Press, 1952. Crustal structure and surface wave dispersion, Part II: Solomon Islands earthquake of July 29, 1950. *Bull. Seism. Soc. Amer.*, **42**, 315–325.

Ewing, M. and F. Press, 1954. An investigation of mantle Rayleigh waves. *Bull. Seism. Soc. Amer.*, **44**, 121–147.

Ewing, M. and F. Press, 1954a. Mantle Rayleigh waves from the Kamchatka earthquake of November 4, 1952. *Bull. Seism. Soc. Amer.*, **44**, 471–479.

Ewing, M. and F. Press, 1956. Rayleigh wave dispersion in the period range 10 to 500 seconds. *Trans. Amer. Geophys. Un.*, **37**, 213–215.

Ewing, M. and F. Press, 1956a. Surface waves and guided waves. *Encyclopedia of Physics, XLVII,* Geophysics I. Springer-Verlag, Berlin.

Ewing, M., F. Press and J. L. Worzel, 1952. Further study of the T phase. *Bull. Seism. Soc. Amer.*, **42**, 37–51.

Gutenberg, B., 1951. Structure of the crust. Continents and Oceans. In *Internal constitution of the earth,* B. Gutenberg, ed. Dover Publications.

Haskell, N. A., 1953. Dispersion of surface waves on multilayered media. *Bull. Seism. Soc. Amer.*, **43**, 17–34.

Jardetzky, W. S. and F. Press, 1953. Crustal structure and surface-wave dispersion, Part III: Theoretical dispersion curves for sub-oceanic Rayleigh waves. *Bull. Seism. Soc. Amer.*, **43**, 137–144.

Kovach, R. L., 1959. Surface wave dispersion for an Asio-African and a Eurasian path. *J. Geophys. Res.*, **64**, 805–813.

Landisman, M., Y. Satô and M. Ewing, 1958. Shear-wave velocities in the upper mantle. (Abstract) *Trans. Amer. Geophys. Un.*, **39**, 522.

Macelwane, J. B., S. J., 1951. Evidence on the interior of the earth derived from seismic sources. In *Internal constitution of the earth,* B. Gutenberg, ed. Dover Publications.

Nafe, J. and J. Brune, 1960. Observations of phase velocity for Rayleigh waves in the period range 100–400 seconds. *Bull. Seism. Soc. Amer.*, **50**, 427–439.

Nafe, J. E. and C. L. Drake, 1957. Variation with depth in shallow and deep water marine sediments of porosity, density and the velocities of compressional and shear waves. *Geophysics*, **22**, 523–552.

Officer, C. B., Jr., 1955. Southwest Pacific crustal structure. *Trans. Amer. Geophys. Un.*, **36**, 3.

Oliver, J., J. Dorman and M. Ewing, 1959. Surface wave dispersion for oceanic paths. (Abstract) *Bull. Geol. Soc. Amer.*, **70**, 1738.

Oliver, J. and M. Ewing, 1958. Short-period oceanic surface waves of the Rayleigh and first shear modes. *Trans. Amer. Geophys. Un.*, **39**, 482–485.

Oliver, J., M. Ewing and F. Press, 1955. Crustal structure and surface wave dispersion, Part IV, Atlantic and Pacific Ocean basins. *Bull. Geol. Soc. Amer.*, **66**, 913–946.

Oliver, J., R. Kovach and J. Dorman, 1961. Crustal structure of the New York–Pennsylvania Area. *J. Geophys. Res.*, **66**, 215–225.

Pomeroy, P. and J. Oliver, 1960. Seismic waves from high altitude nuclear explosions. *J. Geophys. Res.*, **65**, 3445–3457.

Press, F., 1959. Some implications on mantle and crustal structure from G waves and Love waves. *J. Geophys. Res.*, **64**, 565–568.

Press, F. and M. Ewing, 1955. Earthquake surface waves and crustal structure. *Geol. Soc. Amer. Spec. Paper* 62, 51–60.

Press, F. and H. Takeuchi, 1960. Note on the variational and homogeneous layer approximations for the computation of Rayleigh wave dispersion. *Bull. Seism. Soc. Amer.*, **50**, 81–85.

Satô, Y., 1958. Attenuation, dispersion and the wave guide of the G-wave. *Bull. Seism. Soc. Amer.*, **48**, 231–251.

Shurbet, D. H., 1955. Bermuda T phases with large continental paths. *Bull. Seism. Soc. Amer.*, **45**, 23–35.

Stoneley, R., 1926. The effect of the ocean on Rayleigh waves. *Mon. Not. Roy. Astr. Soc., Geophys. Suppl.*, **1**, 349–356.

Takeuchi, H., F. Press and N. Kobayashi, 1959. Rayleigh-wave evidence for the low-velocity zone in the mantle. *Bull. Seism. Soc. Amer.*, **49**, 355–364.

Wilson, J. T., 1940. The Love waves of the South Atlantic earthquake of August 28, 1933. *Bull. Seism. Soc. Amer.*, **30**, 273–301.

Lamont Geological Observatory Contribution No. 605.

9. GRAVITY AT SEA[1]

J. LAMAR WORZEL and J. C. HARRISON

1. Introduction

The technique of gravity surveying at sea was revolutionized during the International Geophysical Year by the development of gravity meters which can operate on surface ships under reasonably favorable sea conditions. It is now possible to make continuous gravity measurements on ships of 1000 or more tons displacement for a useful proportion of the time in many of the world's oceans and seas. The oceanographic survey ship engaged on bathymetric surveys should in the future reckon on being able to take concurrent continuous gravity and magnetic measurements. These types of survey are conveniently made together as all three methods require the ship to be underway continuously at constant speed; they all require precise navigation and the cost of installing and operating the magnetometer and gravity meter is small compared with that of keeping a fair-sized ship at sea.

The gravitational method is based on measuring the variation in the acceleration due to gravity at the sea-level surface. This variation is caused by changes in the distribution of mass beneath the ship. Most important of these changes is usually that caused by variations in the depth of water, but these can be corrected for provided that depth soundings are taken during the gravity measurements and that the bathymetry of the surrounding areas is fairly well known. The variations in the corrected gravity anomaly are caused by density variations beneath the sea-bed.

The method has the advantage that it is not limited with regard to depth. Anomalous masses can be detected however deep they may be, provided that their attraction at the surface is large enough to be measured. The disadvantage is that there is generally no single solution to the problem of finding a mass distribution that will produce the measured gravity field, so that a number of assumptions must be made before a structure can be deduced from a map of gravity anomalies. In these features the method differs somewhat from seismic surveying in which reasonably definite information may be learned about the shallow layers although the depth of penetration is limited by the difficulties of shooting long lines. The two methods thus complement one another, and more is gained by surveying one area in the two ways than the sum of the information from the two surveys taken individually. Now that continuous measurements can be taken from a surface ship underway, gravity survey is an extremely useful method of reconnaissance for locating interesting features which do not show topographically. The detailed analysis of a survey often requires other kinds of geological and geophysical data before a satisfactory interpretation can be made.

2. Methods and Theory

A. Development of Surface-Ship Gravity Meters

Gravity measurements at sea need to be accurate to at least 10 mgal to be of any use at all, and an accuracy of 5 mgal is often required. An instrumental

[1] Lamont Geological Observatory Contribution No. 596.

[*MS received August, 1960*]

accuracy better than 2 mgal would be useful only in a limited number of cases where exceptionally accurate navigation is possible. The amplitudes of the horizontal- and vertical-wave accelerations experienced on a surface ship usually lie between 10,000 and 100,000 mgal depending on the state of the sea and the size of the ship, but these accelerations decrease with depth below the surface and are typically less than 5000 mgal at 250 ft depth.

The first attempts to measure gravity at sea were made by O. Hecker, and later by W. G. Duffield, using a barometric apparatus on board a surface ship. Although an apparatus of this type constructed by H. Haalck eventually gave results of useful accuracy in extremely calm water in 1938, this method has never been widely used.

From the first cruise of the Dutch submarine, K II, in 1923, up to 1955, all reliable measurements of gravity at sea had been made with the Vening Meinesz pendulum apparatus operating in a submarine. This apparatus has been extensively described in the literature (Vening Meinesz, 1929, 1941). It is a thoroughly reliable apparatus, if handled with due care, and the accuracy of a measurement is between 1 and 2 mgal when the wave accelerations are less than about 6000 mgal and when there is little energy present with the same period as the pendulums. These conditions can normally be realized with a submarine able to dive to 250 ft. A total of some 4214 measurements have been taken with this apparatus, of which 2943 were taken by the Lamont Geological Observatory of Columbia University. The apparatus has the disadvantage that the computation of the gravity values from the photographic records is laborious, taking about one hour per record. It is also not suited for development into a surface-ship instrument.

Gravity meters have replaced the pendulum apparatus for measurements on land with the exception of absolute determinations of gravity and the establishment of calibration ranges. The first gravity meter used at sea was the vibrating string instrument constructed by R. L. G. Gilbert, which was successfully tested on board a submarine in 1948 (Gilbert, 1949). Unfortunately, it has not been used successfully at sea since these tests, although a somewhat similar meter, developed from the Vibrotron pressure gauge by the Borg-Warner Corporation, was successfully tested on a surface ship in 1959.

In the meantime L. J. B. LaCoste and Anton Graf had independently developed spring-type gravity meters capable of functioning on submarines. The LaCoste meter was tested by F. N. Spiess and G. L. Brown on two cruises in 1955 (Spiess and Brown, 1958). Sixteen previously established pendulum stations were reoccupied on the first cruise and on the second they made 15 simultaneous pendulum and gravity-meter measurements. The root-mean-square difference between the pendulum and meter observations was 3.3 mgal in both tests, which indicates that the standard error of a gravity-meter reading is about the same as that of a pendulum observation. The first tests of the Graf meter were made on a surface ship on Lake Starnberg in 1955 (Graf, 1956). Worzel and Graf (see Worzel and Graf, 1957) tested Graf's instrument by comparison with simultaneous pendulum observations at 53 sea stations during April, 1956. Thirty-nine of the gravity-meter measurements lay within 3 mgal of the pendulum results and all but three lay within 9 mgal.

These meters possess several advantages over the pendulum apparatus; they are less fragile and easier to install and very much less labor is required in the computation of the results; but most important of all, they are suitable for development for use on surface ships. The Graf meter was the first to be used on a surface ship, in November, 1957 (Worzel, 1959), and tests of the LaCoste meter followed in 1958 (LaCoste, 1959; Harrison, 1959).

B. Some Theoretical Conclusions

Surface-ship operation of gravity meters possesses the advantage that the measurements are made effectively at sea-level and that, apart from the wave movements, there can be no changes in vertical velocity of the instrument during a measurement. There is, therefore, no need for the accurate leveling survey which may be more trouble than the actual gravity measurements in land surveys and, provided the wave periods are removed from the gravity record, no account need be taken of accelerations caused by changes in the vertical velocity of the vessel, which is an important factor in a submarine or airplane survey. It is, however, necessary to make a correction (the Eötvös correction which may be large) for the ship's speed, and the gravity measurement must be taken in the presence of large vertical and horizontal accelerations. The ship may be at sea for several weeks, or even months, between ports, so that the meter must have a small or linear drift rate and be free from sudden jumps in reading (tares).

The beam of a spring gravity meter must be heavily damped so that the meter can withstand vertical accelerations of amplitude as high as 50,000 or 100,000 mgal without the beam hitting its stops. The spring and reading system must be linear or symmetrical so that the mean beam position, determined by averaging the beam record, is independent of the magnitude of the wave accelerations. Alternatively, in a non-linear instrument such as Gilbert's vibrating string meter, it is necessary to apply a correction depending on the magnitude of the vertical accelerations. Apart from these effects of non-linearity, the vertical accelerations can be averaged out during a measurement provided they are not large enough to cause the meter beam to hit its stops. The same is not true for the horizontal accelerations.

The problem of the horizontal accelerations may be met in two ways: first, the meter may be mounted on a stabilized platform which keeps its axis vertical. In this case, the effect of the horizontal accelerations is eliminated provided the gravity meter is constructed in such a way that it is unaffected by accelerations perpendicular to its measuring axis. Alternatively, the meter may be hung in a gimbal suspension and allowed to swing with the wave accelerations. In this case a correction must be applied for the horizontal accelerations. The Graf meter contains no method for correcting for horizontal accelerations and must, therefore, be used on a stabilized platform. The LaCoste meter includes a system for measuring and correcting for the horizontal accelerations and may be used without a stabilized platform. It could also be used on a stable platform. The theory of operation of meters in these two cases has been discussed elsewhere (Harrison, 1960) and this reference should be consulted for the derivation of the results quoted below.

A meter used on land need be leveled with an accuracy of only 5′ in order to attain an accuracy of 1 mgal. However, a meter used at sea on a stabilized platform is very sensitive to stabilization errors that have the same period as the wave accelerations, because, in this case, a part of the horizontal acceleration is added to the vertical component. The error introduced depends on the phase angle between the tilt of the platform and the horizontal acceleration. For the most unfavorable phase angle, an error of 1 mgal is caused by a tilting of 4″ amplitude when the amplitude of the horizontal accelerations is 100,000 mgal, and by a tilting of 44″ when their amplitude is 10,000 mgal. Stable platforms can be made to satisfy the necessary requirements.

LaCoste has pointed out a further source of error in gravity measurements made with a beam gravity meter on a stable platform, and that is the cross-coupling effect. The beam is deflected by the vertical accelerations, and the horizontal accelerations, acting on the deflected beam, produce a couple which depends on the product of the horizontal and vertical accelerations and the phase angle between the beam deflection and the horizontal accelerations. This effect is at a maximum for circular motion when the movement of the meter beam lags $\pi/2$ radians on the vertical acceleration, as in a heavily damped system whose free period is not very different from the wave period. The error introduced by this effect can easily exceed 500 mgal for horizontal and vertical accelerations of 100,000 mgal amplitude. It can largely be eliminated if the meter is slowly rotated about a vertical axis, which has the effect of altering the phase relation between the horizontal and vertical accelerations, and it is, in any case, only important where there is a persistent phase relationship between the vertical and horizontal accelerations.

A meter suspended in gimbals swings in response to the horizontal accelerations. It tends to hang in the direction of the total resultant acceleration formed by the vectorial addition of the horizontal acceleration with the sum of the vertical acceleration and the acceleration due to gravity. This sum is always greater than the vertical component alone, no matter what the direction of the horizontal acceleration, so that a correction must be subtracted from the measured acceleration to compensate for the effect of the horizontal accelerations. This effect was first pointed out by Browne (1937) in connection with the correction of measurements taken with the Vening Meinesz pendulum apparatus for horizontal accelerations of the submarine. The theory has been extended by LaCoste (see Harrison, 1960) for gravity meters on surface ships. LaCoste treats the gravity-meter unit in gimbals as a compound pendulum performing oscillations forced by the wave accelerations in addition to swinging with its own free period. The problem is simplified greatly if the sensing element of the meter is positioned a distance below the gimbal support equal to the length of a simple pendulum with the same period as the compound pendulum formed by the meter unit. In this case the correction for the horizontal accelerations amounts to the time average of $\frac{1}{2}g\theta^2$, where θ is the angle between the axis of the gravity meter and the true vertical. The direction of the true vertical must be established using long-period pendulums or a gyroscopically controlled device. The forces perpendicular to the measuring axis of the meter are very small, so the cross-coupling effect is not important. The theory holds to an accuracy of 1 mgal for horizontal accelerations up to about 60,000 mgal. For larger accelerations, fourth-order terms must be considered.

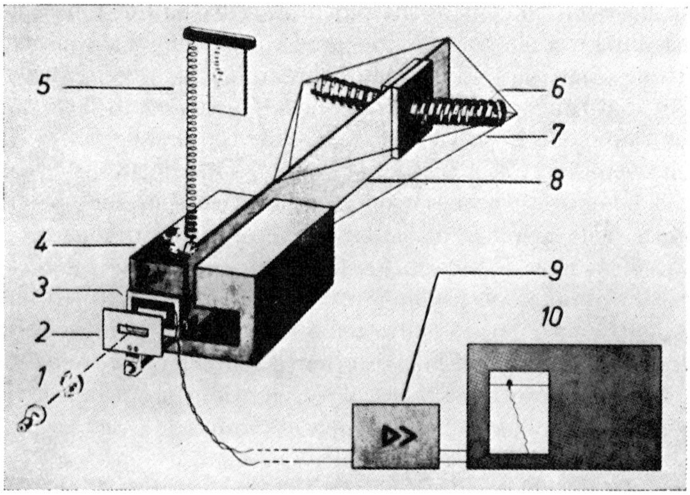

Fig. 1. Schematic diagram of the Graf Sea Gravimeter. 1. Photocell lamp and optics. 2. Diaphragm. 3. Photo-electric cell. 4. Damping magnet. 5. Measuring spring and scale. 6. Torsion spring. 7. Fibers for constraining weight beam motion. 8. Weight beam. 9. Amplifier. 10. Recorder.

C. The Graf Sea Gravimeter

The main features of the Graf Sea Gravimeter are shown in Fig. 1. It consists of a horizontal beam of aluminum supported on nearly horizontal helical ribbon springs. These springs act as the main springs and the pivots for the beam motion. Eight ligatures are arranged to restrain all beam motions except rotations about the main axis. The main springs are set under sufficient tension that the center of gravity of the beam would be raised above the axis of rotation of the beam at all values of gravity likely to be encountered on the surface of the seas. A weak reading spring is attached to a lever arm on the beam and to a nut whose position is controlled by a dial reading screw. By adjusting the reading screw, sufficient tension can be applied by the reading spring to bring the center of gravity of the beam to the same horizontal plane as the axis of rotation of the beam. A dial, attached to the reading screw, permits a reading of the tension of the spring which can be related to the value of gravity by suitably calibrating the instrument. The position of the beam is determined by the differential output of two photocells mounted behind a slit attached to the end of the beam. These photocells are so mounted and adjusted that they give a null reading when the center of gravity of the beam is in the same horizontal plane as the axis of rotation of the beam, and give equal readings for the same upward and downward deflections. A large permanent magnet is placed so that the beam moves in the gap between the pole faces, thereby generating eddy currents in the aluminum beam which strongly damp the beam motion. A pressure compensator and temperature compensator are

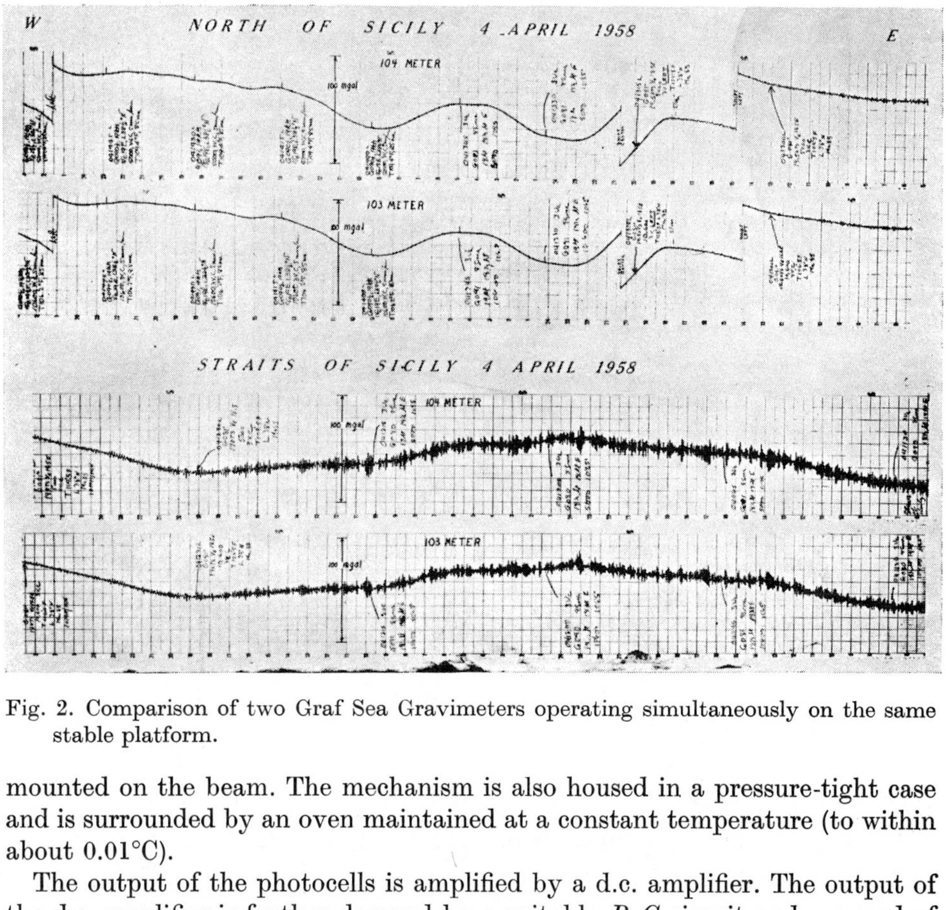

Fig. 2. Comparison of two Graf Sea Gravimeters operating simultaneously on the same stable platform.

mounted on the beam. The mechanism is also housed in a pressure-tight case and is surrounded by an oven maintained at a constant temperature (to within about 0.01°C).

The output of the photocells is amplified by a d.c. amplifier. The output of the d.c. amplifier is further damped by a suitable R–C circuit and a record of this output is made on a potentiometer-type recorder. The record allows a static variation of ± 50 mgals from the true null position, which is arranged to fall at the center of the recorder paper.

The method of operation, then, is to adjust the dial reading until the recorder pen is near the center of the record. The highly damped wave motions appear as small oscillations on the record (see Fig. 2). The changes of gravity appear as changes of the axial position of the trace on the record. When the trace approaches the margin of the record, as gravity changes, a change of the dial reading will return the axis of the trace to an appropriate place on the record.

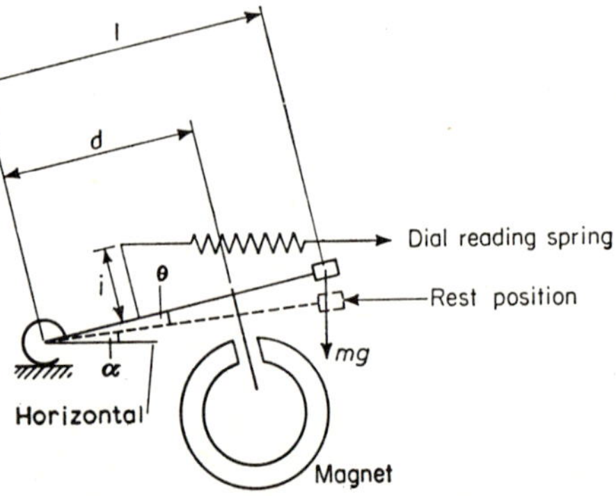

Fig. 3. Diagram of the Graf Sea Gravimeter delineating quantities used in the equations. θ = angular deflection of main springs when beam is horizontal, K = main spring constant (torque/unit angle), S = damping coefficient, $\rho(h-h_0)$ = force produced by reading spring, $A \cos \omega t$ = horizontal acceleration of waves, and $A \sin \omega t$ = vertical acceleration of waves.

From Fig. 3, one may see that the differential equation of motion of the beam, where θ is small (less than $1°$ in this instrument), is

$$\ddot{\theta}+2\mu\dot{\theta}+p^2\theta = X, \tag{1}$$

where

$$2\mu = Sd/I,$$

$$p^2 = \frac{K}{I}+\frac{mAl \sin \omega t \sin \alpha}{I}-\frac{mAl \cos \omega t \cos \alpha}{I}-\frac{mgl \sin \alpha}{I},$$

$$X = \frac{K}{I}(\theta-\alpha)-\frac{\rho(h-h_0)i}{I}+\frac{mAl \sin \omega t \cos \alpha}{I}+\frac{mAl \cos \omega t \sin \alpha}{I}-\frac{mgl \cos \alpha}{I}.$$

The solution of (1) is

$$\theta = \frac{X}{p^2}+e^{-\mu t}(Be^{ct}+De^{-ct}), \tag{2}$$

where $c = (\mu^2-p^2)^{1/2}$.

In this meter $\mu > p$ so that the solution is that of a forced oscillation superimposed on an exponential approach to (X/p^2).

Since α is very small ($< 11''$),

$$\frac{X}{p^2} = \frac{K(\theta-\alpha)-\rho(h-h_0)i+mAl \sin \omega t-mgl}{K-mAl \cos \omega t}, \tag{3}$$

the average value of which, over a 5-min interval, is

$$\left(\frac{\overline{X}}{p^2}\right) = \frac{K(\theta - \alpha) - \rho(h - h_0)i - mgl}{K}, \tag{4}$$

the same as for a deflection in the absence of the wave motion.

In practice one computes the gravity value from the dial setting of the instrument and adds to it the correction read from the axial position of the record. The 5-min averaging time is imposed by the time constant of the complete mechanical-electrical system. Graf (1958) gives more details about the construction and testing of the instrument.

The first tests of the Graf Sea Gravimeter were made in August, 1955, on Lake Starnberg (Graf, 1956) using a gimbal mounting. Later tests were made

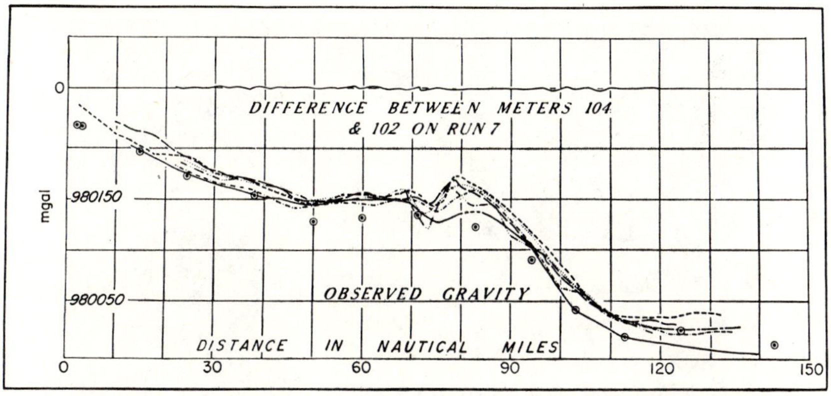

Fig. 4. Comparison of observed gravity on eight runs over as nearly the same course as possible and the differences observed by two meters on the same stable platform on the same run.

in the Adriatic Sea in May, 1956 (Graf, 1957). Comparisons were made with the Vening Meinesz pendulum apparatus on board U.S.S. *Becuna* in August, 1956 (Worzel and Graf, 1957). Of the 59 observations 3 had large discrepancies, 17 fell within the range of 3–9 mgal and 39 within the range 0–3 mgal, showing a close approach to a statistical distribution. The first successful surface-ship gravity observations were made with this instrument mounted on a gyro-stabilized platform on an I.G.Y. cruise on U.S.S. *Compass Island* in November, 1957 (*I.G.Y. Bull.*, 1958). More complete comparisons, as well as a number of continuous profiles, were reported by Worzel (1959). Seven continuous profiles, 150 nautical miles long, agreed with themselves and with previous submarine gravity measurements within a spread of 20 mgal (Fig. 4). The spread is partially caused by navigation errors, and partially by the inability to recover the same course and positions. At three other crossings of former submarine profiles differences of the order of 6 mgal were observed. The continuous track of U.S.S. *Compass Island* passed within 5 miles of 11 other submarine gravity

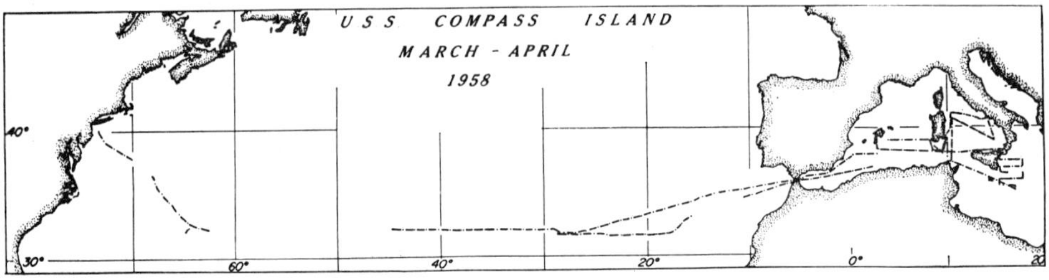

Fig. 5. 6500 miles of track of U.S.S. *Compass Island* on which gravity measurements were obtained.

Fig. 6. LaCoste–Romberg surface-ship gravity meter installed on M.V. *Horizon*.

observations. Differences of less than 10 mgal were observed in nine cases and in two cases 50 mgal, probably due to errors in the submarine values. At six crossings of the ship's own track a root-mean-square difference of 5 mgal was observed. Two meters observing on the same stable platform showed differences less than 1 mgal. It is concluded that the gravity-meter values are reliable to ± 1 mgal but that the gravity values measured on a surface vessel are reliable to ± 5 mgal, due primarily to the navigational limitations. Fig. 5 shows the 6500 miles of track of a five-week cruise of *Compass Island* on which continuous gravity observations were obtained. In the past two years successful observations have been made on R.V. *Vema*, a 700-ton auxiliary schooner, about 20% of the time on her world-wide cruises.

D. The LaCoste–Romberg Surface-Ship Gravity Meter

The LaCoste–Romberg instrument (Fig. 6) utilizes an adaption of the suspension used in their land meters with a beam of increased moment of inertia, damping and a photo-electric reading system. The instrument is

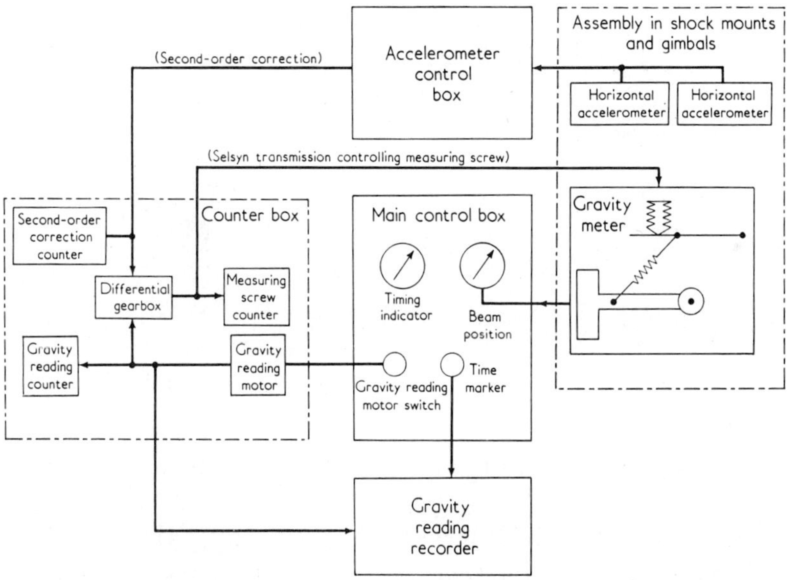

Fig. 7. Block diagram of LaCoste–Romberg surface-ship meter.

suspended in gimbals and long-period pendulums are used to establish the direction of the true vertical. An electrical voltage, proportional to the square of the angle between the instrument in its gimbal suspension and the reference pendulums, is developed by the horizontal accelerometers. This voltage is also proportional to the correction for the horizontal accelerations of the instrument. It is averaged and converted into a shaft rotation which is continuously added to the measuring-screw position through a differential gear-box (see Fig. 7). The ratio of shaft rotation to accelerometer output is adjusted in the laboratory before each trip during tests on a wave-simulating machine. In these tests the meter is adjusted to give the same reading with the machine in motion as it does when still.

The meter itself is very stable as regards drift and other spurious changes of reading. The reading of meter #3 at the Marine Physical Laboratory, San Diego, changed by only 2.3 mgal between 21 February and 21 December, 1959, during which period it was used on five separate surveys on four different ships. The reading method depends on the equation of motion of the beam (see

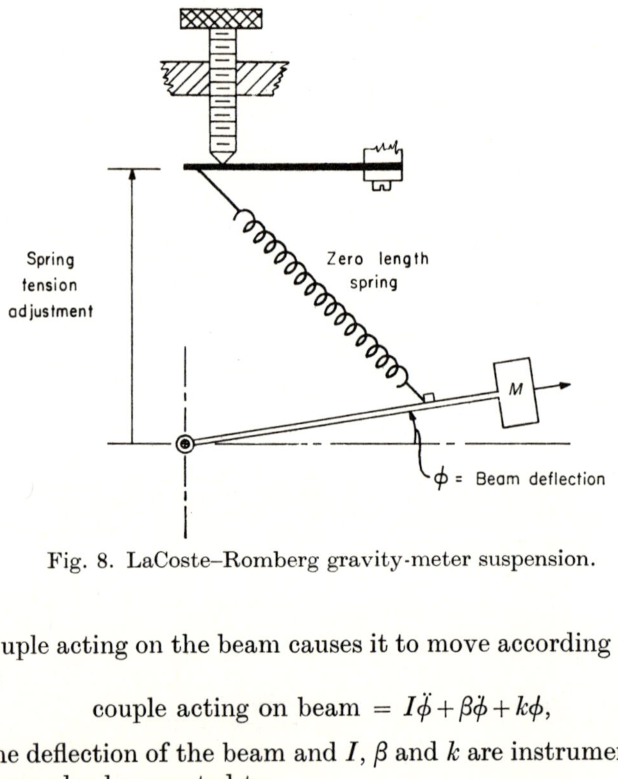

Fig. 8. LaCoste–Romberg gravity-meter suspension.

Fig. 8). A couple acting on the beam causes it to move according to an equation of the type:

$$\text{couple acting on beam} = I\ddot{\phi}+\beta\dot{\phi}+k\phi, \tag{5}$$

where ϕ is the deflection of the beam and I, β and k are instrumental constants. This couple can also be equated to

$$L(g+\ddot{z}+B-C), \tag{6}$$

where $\ddot{z}$ is the vertical acceleration of the meter, B is the difference between the total acceleration acting on the meter and the vertical component (i.e. the instantaneous Browne correction), and L is an instrumental constant. C is proportional to the spring tension set by the measuring screw and would be the gravity reading in the case of a static instrument with the beam set on the reading line.

Equating (5) and (6) and integrating both with respect to time over the interval T_1 to T_2:

$$\int_{T_1}^{T_2} (I\ddot{\phi}+B\dot{\phi}+k\phi)\,dt = L\int_{T_1}^{T_2} (g+\ddot{z}+B-C)\,dt. \qquad (7)$$

Of the terms on the right-hand side of (7), $\int_{T}^{T_2} (g+\ddot{z})\,dt=\bar{g}(T_2-T_1)$, where $\bar{g}$ is the average value of g. (T_2-T_1) must be long enough to average out the vertical acceleration $\ddot{z}$. $\int_{T_1}^{T_2} B\,dt=\bar{B}(T_2-T_1)$ and $\int_{T_1}^{T_2} C\,dt=\bar{R}(T_2-T_1)+\bar{B}(T_2-T_1)$, where R is the gravity setting. (B computed in the accelerometer control box is added to R in the differential gear box and the sum positions the measuring screw.) The right-hand side of (7) thus becomes

$$L(\bar{g}+\bar{B}-\bar{R}-\bar{B})(T_2-T_1) = L(\bar{g}-\bar{R})(T_2-T_1), \qquad (8)$$

so that

$$\bar{g} = \bar{R}-1/L(T_2-T_1)\int_{T_1}^{T_2} (I\ddot{\phi}+B\dot{\phi}+k\phi)\,dt. \qquad (9)$$

The beam deflection, ϕ, is measured by a photo-electric system and the quantity $(I\ddot{\phi}+\beta\dot{\phi}+k\phi)\,dt$ is computed by an analogue computer and recorded on a pen recorder. $(I\ddot{\phi}+\beta\dot{\phi}+k\phi)\,dt$ is read from the graph and $\bar{g}$ computed from equation (9). The various settings of R are recorded in digital form on an Esterline Angus 20-pen recorder and $\bar{R}$ can be determined from this record. A 12-min averaging time has been found to be generally convenient under normal sea conditions, although shorter intervals can be used under very calm conditions.

In reading a spring-type gravity meter on land, $g=\bar{g}$ and R is normally adjusted to bring the beam to the zero point ($\phi=0$). In this case the beam is still, so that $\dot{\phi}=\ddot{\phi}=0$ and $(I\ddot{\phi}+\beta\dot{\phi}+k\phi)\,dt=0$. Hence $g=R$. It is also possible to read the meter with the beam off the null position and to make an appropriate correction, in which case $g=R-L^{-1}k\phi$. When reading the meter on *terra firma*, $\dot{\phi}$ and $\ddot{\phi}$ are always zero, but on a moving ship or in an airplane this is not true because of the disturbing accelerations and the constantly changing value of gravity. Corrections need to be applied for the velocity and acceleration of the beam, in addition to the correction for the departure of the beam from the null position. If these corrections are not made, it is necessary to wait until the mean values of $\dot{\phi}$ and $\ddot{\phi}$ are small before making a valid gravity reading, with the result that the meter reading lags behind the changes of gravity and that rapid variations are not read. This feature makes it possible to use the LaCoste–Romberg meter in an aircraft, where the changes of gravity are very rapid.

The meter was first used on the I.G.Y. cruise of the Texas A and M College ship *Hidalgo* (a 136-ft converted mine-sweeper) in the spring of 1958 (LaCoste, 1959) with encouraging results at a number of check stations. Further tests were made on board the University of California Research Vessel *Horizon* (150 ft length, 505 gross tons) in the fall of 1958. *Horizon*'s track passed close to 12 submarine stations and the root-mean-square difference between the surface and submarine measurements was 4.4 mgal (Harrison, 1959). This is satisfactory considering the navigational uncertainties. The first full-scale survey made with the meter was that in the Gulf of California from February to April, 1959, during the Scripps Institute of Oceanography's Vermilion Sea Expedition. Some 5000 miles of continuous data were obtained, the coverage in the central part of the gulf being shown in Fig. 9 to indicate the detail obtainable

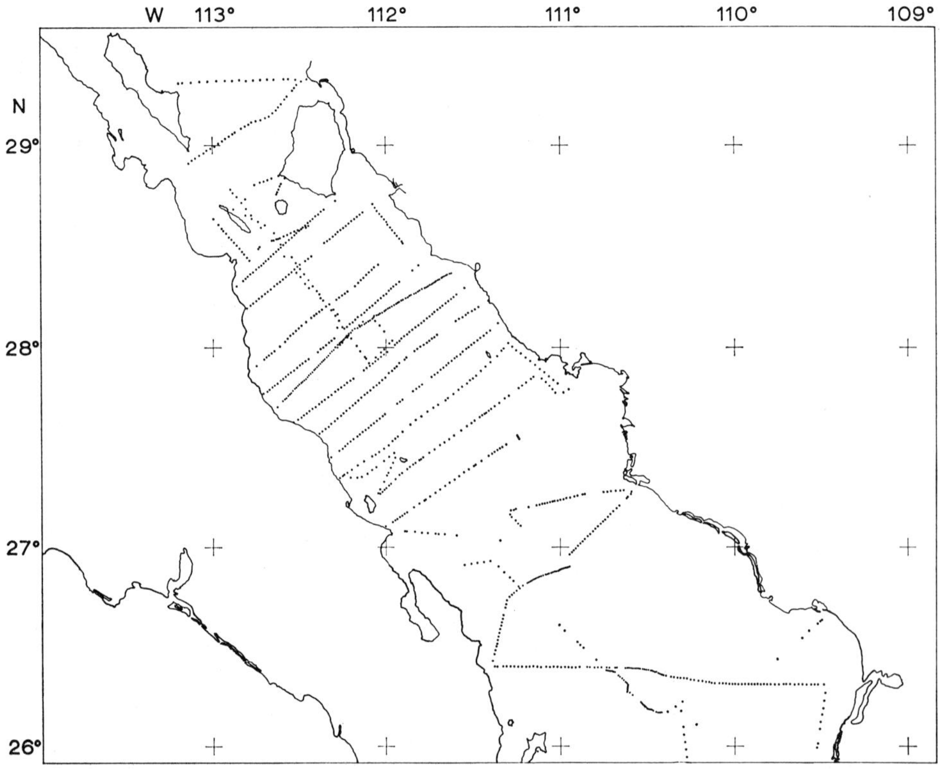

Fig. 9. Gravity coverage in central section of Gulf of California obtained on Vermilion Sea Expedition in the spring of 1959.

with this instrument under favorable conditions. Simultaneous measurements were taken with a LaCoste–Romberg underwater meter on the sea floor and with the surface meter at 27 anchored stations. Surface measurements were also made with *Horizon* anchored as close as possible (normally within 400–1000 yds) to the shore, and a landing party sent ashore to make a simultaneous measurement with a geodetic land meter. Navigational uncertainties are eliminated from these comparisons and the results, shown in Table I, indicate an accuracy of 2–3 mgal for the surface measurements. The meter has also been

TABLE I

Gravity Comparisons in the Gulf of California

Station	Ship – land, mg	Ship – underwater, mg	Underwater – land, mg
Mazatlan	+2.5		
La Paz	+0.2		
88/89	+4.1		
123	+1.1		
137		−1.1	
Guaymas	+0.9		0
138		−7.1	
141		−5.3	
141a	−5.3	−3.3	−2.0
142		−2.4	
143	−0.2		
145		−1.0	
146		−2.6	
147	−0.1		
154		−2.7	
155		−1.5	
156		−2.2	
157		−0.4	
158		−3.1	
162	−3.6	−3.1	−0.5
175		−2.8	
180	−2.0	−3.1	+1.1
181		−2.5	
182		−2.7	
187		−1.2	
188		−1.2	
189		−1.6	
210	−1.2	−2.9	+1.7
211		−5.1	
212		−4.7	
213		−2.7	
216		−4.0	
217		−0.4	
San Felipe	+2.8	−3.2	+0.4
	−0.1 ± 2.6	−2.7 ± 1.5	+0.1 ± 1.3

extensively used in the open ocean with a success that may be gauged from Fig. 10, which shows a survey in the Pacific Ocean off Southern California.

E. Airborne Gravity Measurements

The method of reading the LaCoste–Romberg gravity meter, in which the couple on the beam is computed from the full equation of motion (see 2-D), allows gravity measurements to be taken extremely rapidly. Aircraft flying in quiet air are much steadier than a surface ship of moderate size under normal

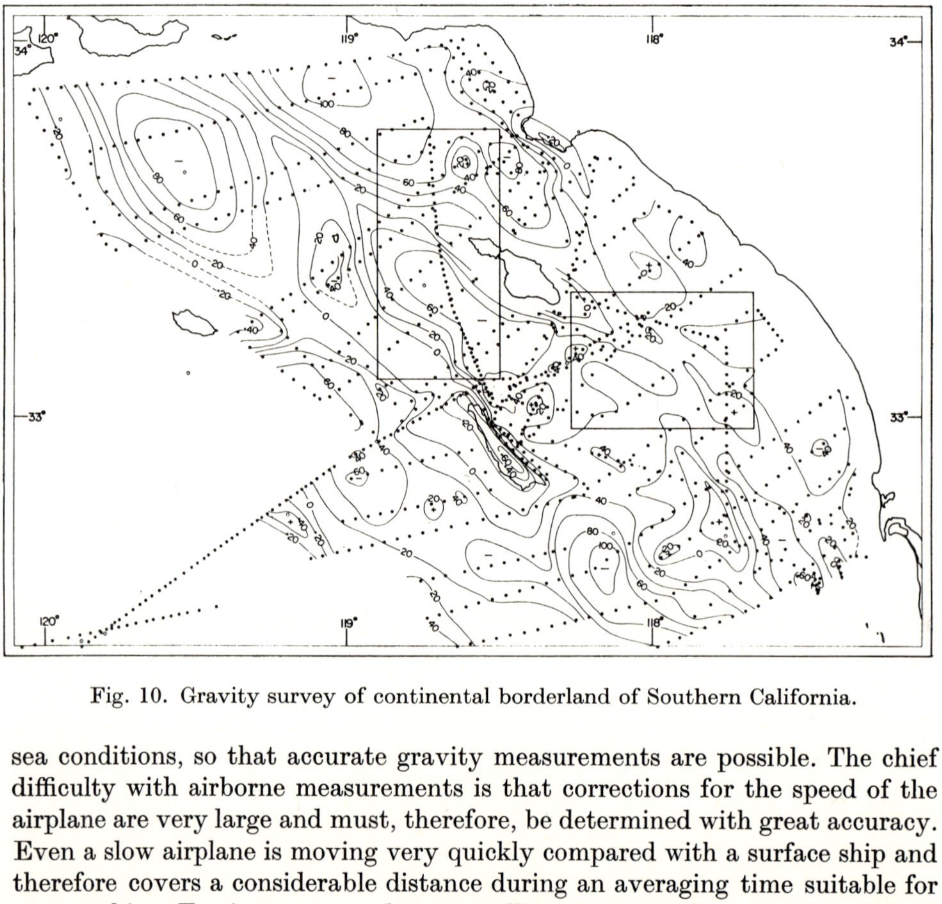

Fig. 10. Gravity survey of continental borderland of Southern California.

sea conditions, so that accurate gravity measurements are possible. The chief difficulty with airborne measurements is that corrections for the speed of the airplane are very large and must, therefore, be determined with great accuracy. Even a slow airplane is moving very quickly compared with a surface ship and therefore covers a considerable distance during an averaging time suitable for use on ships. For instance, a plane travelling at 180 mph covers 36 miles in the 12-min averaging time found convenient for shipboard measurements. While this averaging time can be reduced considerably, it is clear that airborne surveys cannot yield great detail. On the other hand, for geodetic and some geophysical purposes, measurements averaged over, say, 10 miles would be quite adequate. The measurements could be made at a height of, say, 20,000 ft, which means that it would be possible to map gravity over practically the entire earth on a single equipotential surface, thus eliminating the theoretical difficulty raised by masses between a measurement on land and the sea-level reference surface. Further, it would be possible to carry out the mapping in a matter of decades rather than generations.

The first airborne tests of the LaCoste–Romberg meter were made by Thompson and LaCoste on board a United States Air Force KC-135 (a military version of the Boeing 707) over Edwards Air Force Base, California (Thompson, 1959; Thompson and LaCoste, 1960). Precise navigational data were provided from the airplane by a Doppler radar system with barometric measurements for elevation changes which were also obtained from the ground by phototheodolites. An overall accuracy of about 10 mgal was obtained.

Further tests were made in a joint venture by the Gravity Meter Exploration Company, Fairchild Aerial Surveys, LaCoste and Romberg, and the Institute of Geophysics of the University of California (Nettleton, LaCoste and Harrison, 1960). A B-17 airplane was used; aerial photographs were used for horizontal positioning and altitude, and vertical velocities were determined by a combination of data from a hypsometer (pressure measuring device) and from a precision radar altimeter. Again it was found possible to take airborne gravity measurements and, despite some Eötvös corrections in excess of 1000 mgal, the corrected gravity values at the intersection of flight lines usually agreed to within 10 mgal. Repeated flights along the same line gave agreeing results and the contour map, constructed at 12,000 ft elevation from the airborne results, was similar to the map of free-air anomalies on the ground. The actual gravity values at 12,000 ft agreed to within 10 mgal with the ground values projected upward, except where the ground stations were situated on very high terrain and there were insufficient data to project the ground values upward with any confidence. A 3-min averaging time was used for these measurements, during which the airplane covered about 10 miles over the ground.

These tests show that it is possible to measure gravity with useful accuracy from an airplane and open up the possibility of making a regional survey of the oceans in a few years. However, it is necessary to develop a navigational system, or combination of systems, to enable the airplane to obtain its position, elevation, true course and speed with the required accuracy when over the sea and beyond the range of visual aids to navigation. The error in speed must not greatly exceed one knot, and the airplane's true course must be known with an accuracy of $\pm\frac{1}{2}^{\circ}$. A change in vertical velocity during a measuring interval is equivalent to a net vertical acceleration of the meter and can only be distinguished from a change in gravity by accurate measurement of variations in the airplane's height. The hypsometer and radar altimeter used in the B-17 tests were accurate to about 2 ft. A sinusoidal height variation of 1-ft amplitude (2-ft range) and 6-min period produces a corresponding acceleration of nearly 10 mgal.

These requirements are stringent but can probably be met. Certainly the method is sufficiently promising to warrant further tests and a careful investigation into the navigational problems. Probably all the relevant data should be recorded directly in digital form on punched paper tape and computed on a high-speed computer using numerical rather than analogue filters.

F. Reduction of Observations and Methods of Interpretation

The instruments described above yield a gravity value on the moving ship or submarine. Coriolis force due to the vessel's motion on the rotating earth alters the observed value of gravity and means that a correction—the Eötvös correction—must be applied to the observed value to reduce it to that which would have been measured on a stationary vessel. The correction to be added to the observed gravity value is given (Worzel, 1959) by the formula:

$$\delta g = 7.487S \sin C \cos \varphi + (S^2/240.8).$$

C is the true course made good, φ the latitude and S the ship's speed in knots.

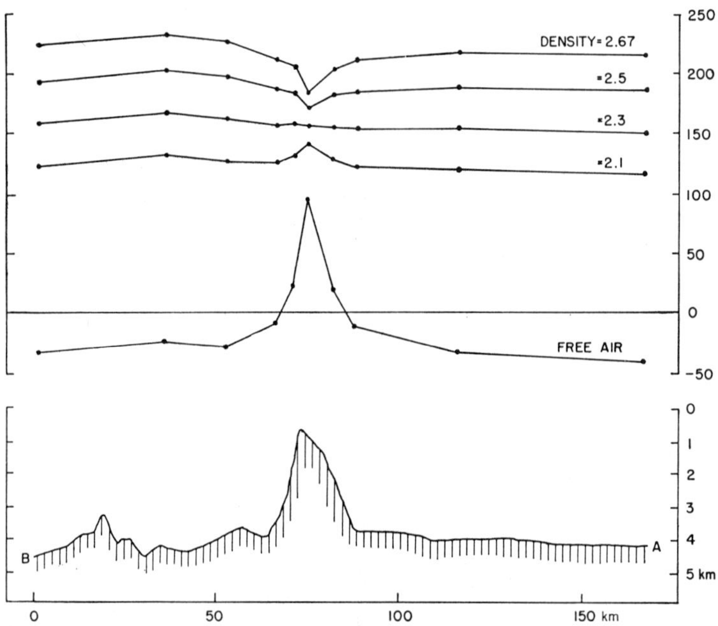

Fig. 11. Gravity-anomaly profiles across Seamount Jasper. (After Harrison and Brisbin, 1959.)

Measured gravity corrected for the Eötvös effect is next reduced to sea-level. A shipboard measurement is taken close to sea-level and only a small correction for the free-air gradient is needed. The value of gravity at the latitude of observation according to the International Formula is subtracted from this finally corrected observed value. This difference is known as the free-air anomaly by analogy with the free-air anomaly on land. In fact the correction to sea-level for a sea observation is trivial, and it is not necessary to make any of the assumptions about the mass between the gravity meter and the sea-level surface that are made for measurements on land. The free-air anomaly is expressed in milligals (1 mgal = 0.001 cm/sec^2).

Free-air anomalies are caused by differences between the distribution of masses on the real earth and on the earth model assumed in the International Formula. The most rapid variations of free-air anomaly, though usually not the most important on a large scale, are due to the changing effect of the bottom topography. Small-scale variations in the free-air anomaly are often closely correlated with changes in the depth of water, as is plain from Figs. 11, 12 and 20, and these changes must be removed or allowed for before the other reasons for variation, such as geological structures beneath the sea floor, can be investigated.

The depth of water in the immediate vicinity of the ship is the most influential in affecting the gravity anomaly, and the effect of a given topographic feature decreases rapidly with increasing distance. As the ship moves into shallower water, there is an increase in the mass immediately beneath it and a consequent increase in free-air anomaly. One method for removing this direct correlation with topography is to compute the increase in gravity at the point of observation that would be caused by substituting rock of a certain density for the sea-water. This "topographic correction" is added to the free-air anomaly and the new anomaly so produced is called the Bouguer anomaly. The gravity profile will show no correlation with bottom topography if the correct density is chosen for the added rock and there are no concealed masses. This is illustrated in Fig. 11, which shows profiles of the free-air anomaly and Bouguer anomalies computed for densities of 2.1, 2.3, 2.5 and 2.67 g/cm^3 across Seamount Jasper (Harrison and Brisbin, 1959). The free-air profile shows a sharp maximum over the seamount; this maximum is not quite removed when a density of 2.1 is adopted. The profile for 2.3 g/cm^3 density shows no indication of the presence of the seamount while those profiles for higher assumed densities show a minimum over the peak. Replacing the sea-water by rock of 2.3 g/cm^3 density eliminates the gravitational effect of the seamount altogether—it is (hypothetically) buried in material of its own density and is no longer detectable gravitationally.

The computation of Bouguer anomalies at sea involves the hypothetical addition of mass to the ocean and so these anomalies do not give a true picture of the variation of mass per unit area over the earth's surface. They are always large and positive over the deep sea, because of the added mass, and large and negative over elevated land areas, because here mass is subtracted in the computation. Fig. 11 shows the increase in size of the Bouguer anomaly with increase of the assumed density; although a density of 2.3 g/cm^3 is the correct one to use in this profile, because it is the one that best eliminates the gravitational effect of the seamount, it certainly is not the correct one to use for comparing the mean level of anomaly in the Pacific with the level in shallow water close to the shore. In the latter case it is proper to use the mean density of the continental block between sea-level and the level of the sea floor in the deep ocean. The uncertainty about which density to use in the Bouguer reduction may make it preferable to compute the attraction of trial structures, including the assumed sub-bottom mass distribution and the varying water depth, and to compare this attraction directly with the free-air anomalies. The methods are ultimately equivalent as the attraction of the varying depth of water is taken into account in both. In general, the Bouguer anomaly is more useful in fairly small-scale problems and the free-air anomaly in larger-scale problems, though there is a wide scope for personal preference.

Computation of the topographic correction is troublesome. When submarine measurements were being accumulated at a relatively slow rate, it was possible to consider computing this correction manually at each station, using the classical methods inaugurated by J. F. Harford and W. Bowie. A template

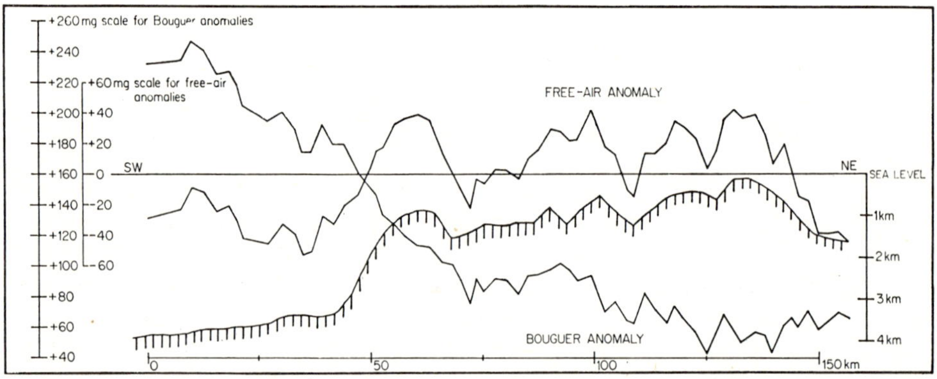

Fig. 12. Continental slope profile off Southern California. (After Harrison, 1959.)

divided into zones by concentric circles, each zone being divided into compartments by radial lines, was used and the average height or depth in each compartment estimated. Tables (see Heiskanen and Vening Meinesz, 1958, §6-2, pp. 159 *et seq.*) are available from which the contribution of each compartment at the station positioned at the central point of the template can be read, and the total correction determined by summing all these contributions. This method is laborious and it is no longer practicable to attempt to keep pace with the rate of accumulation of gravity data, now that thousands of miles of continuous gravity measurements can be taken yearly. The advent of high-speed digital computers provides the ultimate answer to this problem, but it will take several years of experience to discover and standardize the best procedures. Unfortunately, the correction depends mainly on the topography very close to the station, and it must be described for the machine in considerable detail. However, the depth of each unit area need only be read off once. Talwani, Worzel and Landisman (1959) have described a method suitable for two-dimensional structures. Quite a number of problems can be treated two-dimensionally to a good approximation, and the method may also be as satisfactory as any when the surveying ship makes a single pass through a relatively unsurveyed area. In this case the only precise bathymetric information is that along the ship's track, and a two-dimensional assumption is as good as any other.

Kukkamaki (1954) has derived an expression for the attraction of a vertical line mass at a point as a function of the heights of the upper and lower limits of the mass and distance from the point. The topography may be put on cards as the average height of squares or as spot heights on a regular grid and Kukkamaki's expression may be integrated numerically over any required area with the aid of some method such as Simpson's rule. Alternatively, Talwani and Ewing (1960) have described a method based on an expression for the attraction of a horizontal lamina with a polygonal boundary at an external point. The topographic contours are approximated by polygons and the attraction of the water-mass computed by integration over the laminas bounded by the actual and interpolated contour lines. A modification can be made to take account of the earth's curvature.

Most major topographic features are compensated isostatically; that is to say the mass per unit area over the Earth's surface remains constant, the surface excesses of mountain ranges and high plateaus and the deficits of ocean basins being cancelled by compensating masses at depth. Several hypotheses have been proposed to describe the compensating masses in detail. The variation of gravity at sea-level can be computed from the topography and the compensating masses distributed according to any one of these hypotheses. The difference between the actual value and this computed value is known as the isostatic anomaly. There are as many different types of isostatic anomaly as there are isostatic hypotheses. None of the hypotheses is particularly realistic in view of present knowledge, and the chief use of the isostatic anomaly is that it usually varies more slowly than the free-air or Bouguer one. Hence, in an area of very sparse data, the isostatic anomalies are more likely to be typical of the whole area than any other. They also make it very easy to see quickly whether an area is in isostatic equilibrium and it should be easy to compute by a high-speed computer once the topography is in a form suitable for computation of the topographic corrections. The most realistic isostatic hypothesis, and that which generally gives the smallest and most slowly varying anomalies, is the Airy–Heiskanen system for a normal crustal thickness of 30 km. This is now the most generally adopted system.

A map of topographically corrected gravity anomalies shows the presence of hidden mass excesses and deficits. It is not usually possible to make a unique interpretation of this anomaly pattern, though it is always possible to compute the gravitational attraction of an assumed structure and to compare this with the observed anomalies. An otherwise acceptable solution to a structural problem may be rejected on gravitational grounds, but the structure deduced from a gravity profile usually contains a number of additional assumptions. This ambiguity is usual in geophysical methods and, for this reason, interpretations are very much more trustworthy when based on several different types of survey. The surveys complement one another, each providing information helpful in the interpretation of the others, and it is important to survey interesting areas in as many ways as possible. In particular, gravity, magnetic and topographic surveys can very conveniently be made concurrently.

Two definite pieces of information can be learned from the anomaly field produced by a buried mass. It is possible to assign a maximum permissible depth to the mass and to estimate the total mass excess or deficit. It is further possible to calculate the exact distribution of mass, if this mass can be assumed to be a surface distribution on a horizontal plane at known depth. Otherwise the interpretive process consists of computing the attraction of trial models and modifying, accepting or rejecting the model on the basis of the fit with the observed anomalies. These interpretive methods are described in more detail in many of the standard geophysical texts (see Garland, 1956).

3. Gravity Observations and Geological Interpretations

Early studies of the seas were hampered by a lack of data and a concentration of the data over anomalous features. The important main conclusion was drawn that the ocean basins were closely in isostatic equilibrium. The geological conclusion that the Mohorovičić discontinuity (crust–mantle interface) lay at a depth of 10 to 15 km below sea-level (depending on the choice of density differential between the crust and the mantle) was not drawn until after the early measurements of seismic refraction work indicated the interface at these depths (see Chapter 7).

Worzel and Shurbet (1955) derived a "standard" section from the comparison of six sea stations with seven land stations. Seismic, gravity, elevation and water-depth data were available for each station. They concluded that a column composed of 5 km of water of density 1.03, 1 km of sediment of average density 2.30, 4.5 km of crustal rocks of average density 2.84 and a mantle of average density 3.27 beneath the crustal rocks was most representative of the average condition in typical ocean basins; a column composed of 33 km of crustal rocks of average density 2.84 overlying a mantle of average density 3.27 was most representative of the average conditions of continents. These two columns are in isostatic equilibrium.

The oceans cover about 70% of the surface of the Earth. There are, however, many features in the oceans that cannot be called typically oceanic such as continental margins, islands, seamounts, mid-ocean ridges, etc. Worzel and Talwani (1959) studied the 4214 pendulum-gravity stations at sea to determine how well the International Formula of gravity fitted the observations. Sea observations in typically oceanic areas are especially good for such a study, as, first, the observations are made directly on the geoid to be studied and, secondly, there are no anomalous masses near the point of observation.

Measurements within the area of influence of the features clearly not typically oceanic were eliminated from consideration. The criteria for eliminating observations were determined from gravity sections across such features (see, for instance, Figs. 16, 17, 18, 20, 23, 24 and 27). The following observations were eliminated:

1. Measurements in the shallow-water area inside the 2-km depth curve and within $1\frac{1}{2}°$ (about 161 km) of the 2-km depth curve on continental and island margins.

2. Observations within trenches and within $1\frac{1}{2}°$ of the lip of the trenches (the lip was considered to be the change from a steep to a gentle slope).

3. Observations on ridges within the 4-km curves.

About a quarter of the observations (numbering 1013) remained. These were plotted on the Prince of Monaco charts. No simple pattern could be discerned

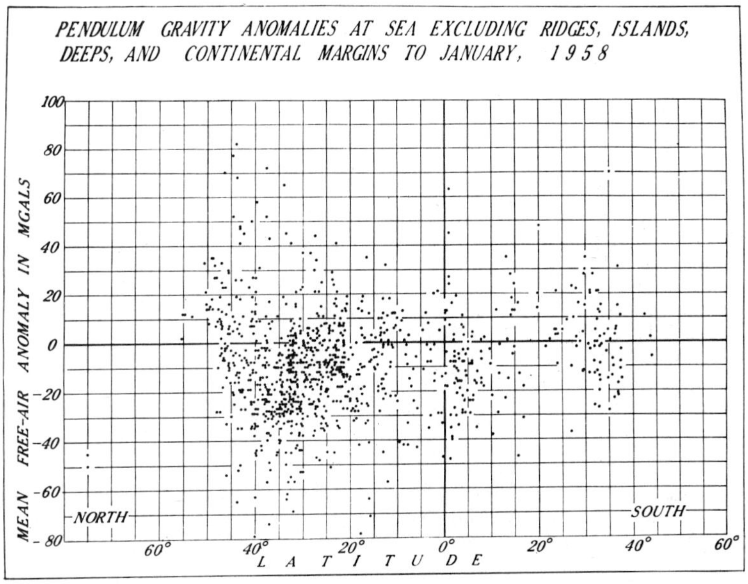

Fig. 13. Anomalies from pendulum measurements for "typical" ocean basins. (After Worzel and Talwani, 1959.)

other than a latitude relationship. Fig. 13 shows the points plotted in relation to the latitude. Fig. 14 shows the same data plotted as average values for 5° intervals of latitude against latitude. The number of observations included in each average is indicated in the middle graph, and the percentage of the Earth's surface area between each 5° of latitude that is oceanic is plotted at the top.

For the oceanic part of the Earth it was concluded:

1. From the average of all the values, the International Gravity Formula is 8.8 mgal too large.

2. There is a definite asymmetry between the Northern and Southern Hemispheres.

3. The geoid is depressed below the ellipsoid from about 50°N to about 15°S

being depressed most between 30 and 35°. From 15°S to about 40°S the geoid is raised slightly above the ellipsoid.

4. The values in the Arctic from 73° (although difficult to analyze because the topography is so poorly known) indicate that the mean anomaly between 75° and 90°N is about 0.

It is considered suggestive that the latitude belt of the greatest geoidal down-warp in the oceans coincides with the latitude of the highest mountain and, hence, the greatest geoidal up-warp on continents.

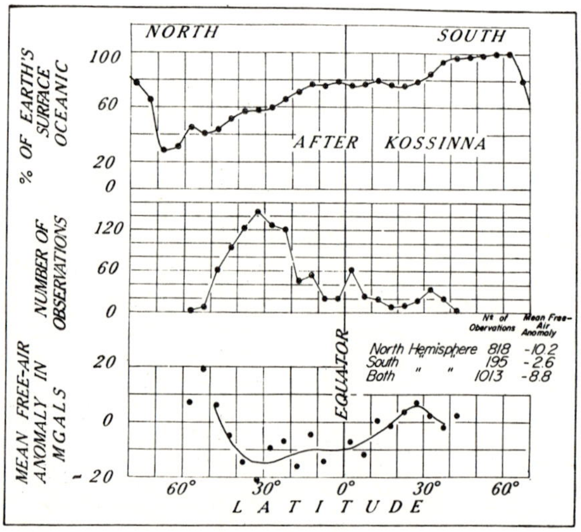

Fig. 14. Five-degree average anomalies for "typical" ocean basins compared with latitude. At the top, the percentage of the Earth's surface which is oceanic for each latitude is plotted. (After Worzel and Talwani, 1959.)

Koryakin (1958) has made a number of sections across the Atlantic Ocean. These sections are based on some of the published submarine gravity measurements and some surface-ship measurements reported to have been made by Gajnanov and Koryakin in 1954–55. They estimate that the latter values have a root-mean-square error of $\pm$16–17 mgal. Unfortunately, only one of the five profiles using these values is illustrated, and that on such a small scale that it would be difficult to add these observations to the world data.

Fig. 15 shows Koryakin's section from Puerto Rico to the shoreline of Africa, near the Canary Islands. His sections are computed on the following assumptions: (1) a water density of 1.03; (2) a sediment density of 2.30; (3) a crustal density of 2.8; (4) a mantle density of 3.3; (5) Helmert's 1909 formula of world gravity; and (6) the mantle interface in the Nares Basin is 10 km below sea-level. As a section generalized from limited data, it gives a good first-order approximation of the crustal section across the ocean provided that one is willing to assume that the crustal and mantle densities are constant and of the above magnitude.

A. Continental Margins

Heiskanen and Vening Meinesz (1958) conclude that of the 26 coastal margins studied, in all but two cases the regions were most closely in local isostatic equilibrium for an Airy isostatic assumption and $T = 30$. Near the Canary Islands and the west coast of Africa, Australia's northwest cape and Socotra, a value of $T = 20$ seemed better but they thought the data inconclusive.

Worzel and Shurbet (1955a) have derived seven sections across the continental margin of the northeast coast of the United States from gravity and

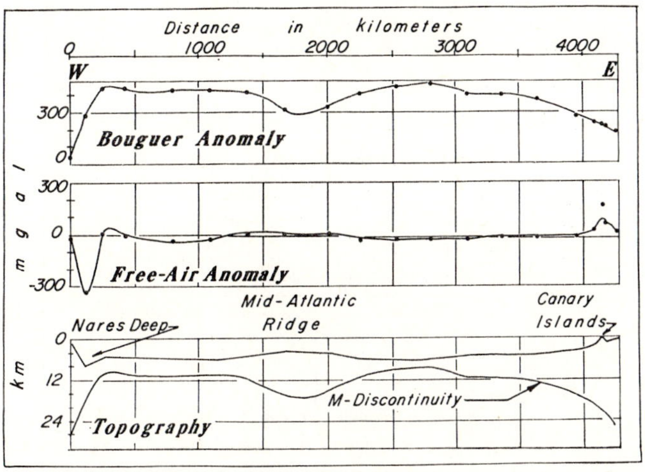

Fig. 15. Crustal profile from Puerto Rico to the African coast near the Canary Islands. (After Koryakin, 1958.) This section was computed from the gravity anomalies assuming a depth to the M-discontinuity just north of the Nares Deep.

seismic data. These sections show a considerable variation of sedimentary cover. The crust–mantle boundary is computed on the basis of the following assumptions: (1) a water density of 1.03; (2) a mean density of the sedimentary layers of 2.30; (3) a mean density of the crustal layers of 2.84; (4) a mean density of the mantle of 3.27; and (5) the standard sea-crustal section of Worzel and Shurbet (1955). Fig. 16 shows three sections representing the variation encountered.

It was concluded that the true edge of the continent occurs at about the 1000-fm curve; the maximum sedimentary thickness occurs at the base of the continental slope and a considerable thickness persists across the continental rise; the distribution of sediments is probably brought about by the action of turbidity currents; sedimentation, mostly derived from the continents, must have modified the basic structure with the various sections representing inter-

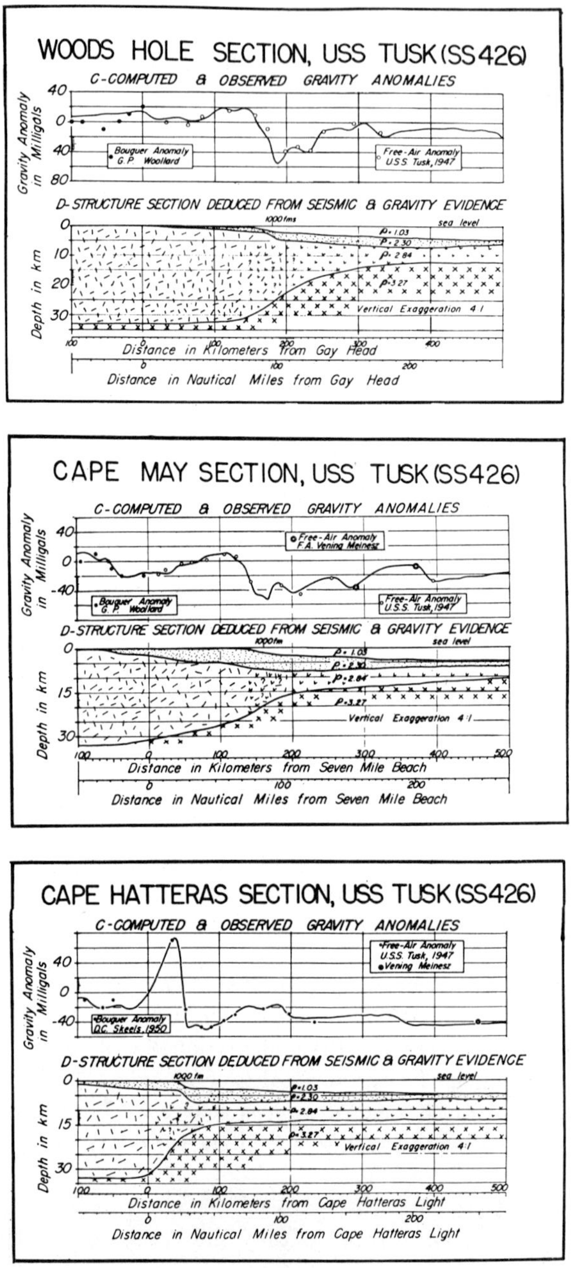

Fig. 16. Typical continental margin sections computed from seismic and gravity data for the east coast of the United States. (After Worzel and Shurbet, 1955a.)

mediate stages; all of the sections are closely in isostatic balance; and the M-discontinuity varies considerably in steepness but continental crustal thickness changes to ocean crustal thickness within 200 km.

Drake *et al.* (1959) computed various structures for the Portland, Maine section on various assumptions of crustal layering and densities covering the gamut of seismological and geological possibilities. They found that none of the

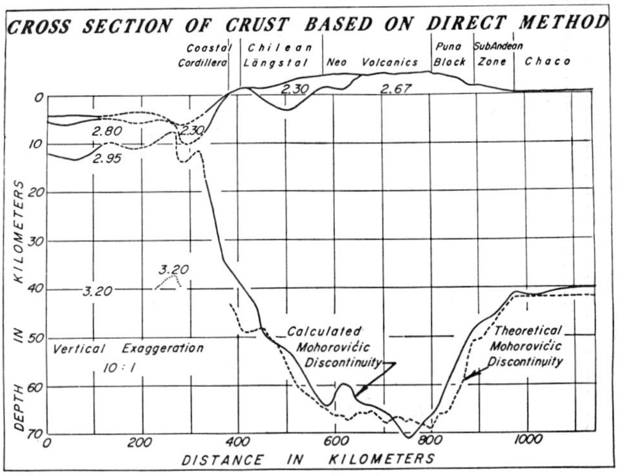

Fig. 17. Crustal section across continental margin and trench near Antofagasta, Chile, computed from gravity data. (After Wuenschel, unpublished thesis.)

possibilities that they considered reasonable appreciably modified the section from those computed in the earlier work of Worzel and Shurbet except for a different choice of density for the mantle.

Wuenschel, in his unpublished Ph.D. thesis, derived three sections across the western continental margin of southern South America. Fig. 17 shows one of his profiles, which is quite similar to the other two. The analysis was complicated by the deep trench near the continental slope and the high mountains on the shore side, and the lack of detailed geological knowledge in the region. The densities he used are shown on the figure. Once again one sees the abrupt change from continental crustal thickness (mountainous here) to oceanic thicknesses in a distance of less than 200 km.

Fig. 18 shows a first approximation crustal section along the west coast of the United States near San Diego. No seismic data were available so all the crust was considered to have a density of 2.84. It is known that there are a number of basins with some sedimentary fill in the region but so far no estimate of the density or the sediment thickness can be made. It is not expected that the section will be appreciably modified when these sediment accumulations are known. Once again, one is struck by the abrupt shallowing of the mantle

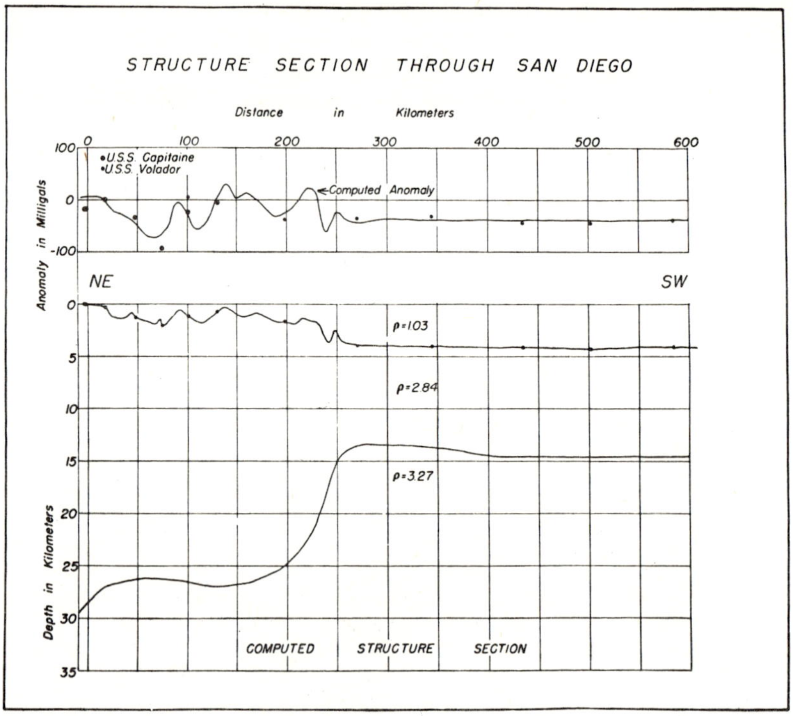

Fig. 18. Crustal section across continental margin near San Diego, California. (After Worzel and Talwani, unpublished.)

at the continental edge, which here occurs about 200 km from the shoreline at about the 1000-fm curve.

Thompson and Talwani (1959) have computed a section across the continental margin along the west coast of the United States, near San Francisco. The densities used are shown in the section illustrated in Fig. 19. The section is complicated by the various mountain ranges in that coastal region. The abrupt change of crustal thickness at about the 1000-fm curve is again evident.

B. Mid-Ocean Ridges

The only mid-ocean ridge examined in detail is the Mid-Atlantic Ridge at about latitude 30°N. Fig. 20 shows, at the top, a continuous gravity curve obtained with the Graf Sea Gravimeter in crossing this feature, and, at the bottom, the topography. After the effect of the topography (Talwani, Heezen and Worzel, 1961) has been removed, an anomaly (Bouguer) is obtained which must be explained by densities within the crust or the mantle. There were some seismic measurements available to provide some control for the upper part of the crust (Ewing and Ewing, 1959). Fig. 21a shows the resulting section for the conventional view that the variation of the thickness of the crust is primarily responsible for the remaining anomaly. Even with the abrupt change of depth

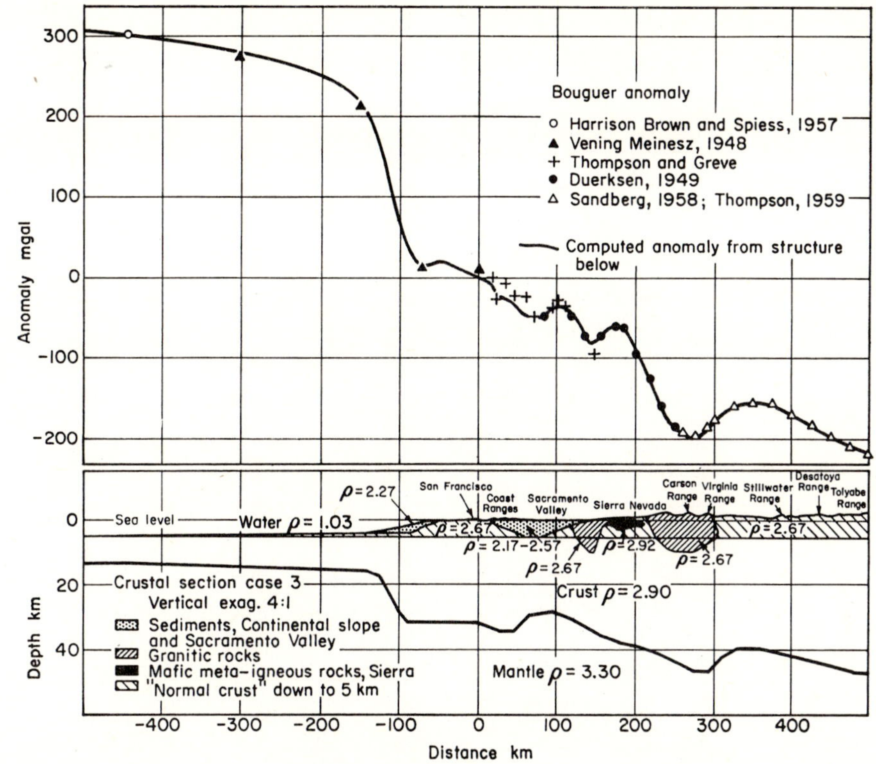

Fig. 19. Crustal section across continental margin near San Francisco computed from surface geology and gravity data. (After Thompson and Talwani, 1959.)

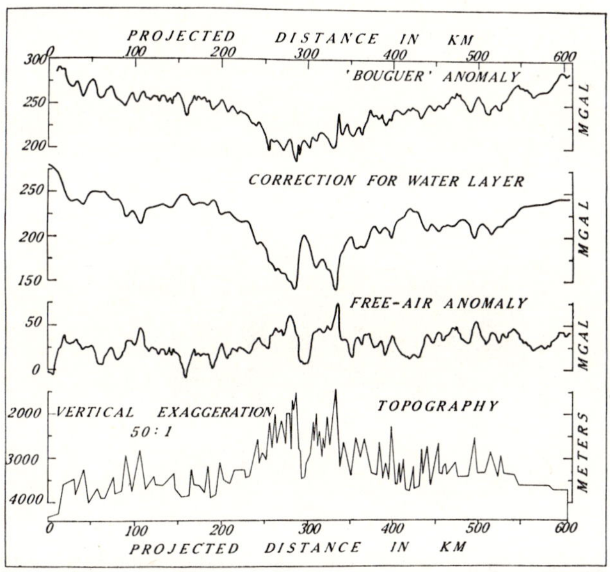

Fig. 20. Gravity and topographic sections across the Mid-Atlantic Ridge. (After Talwani, Heezen and Worzel, 1961.)

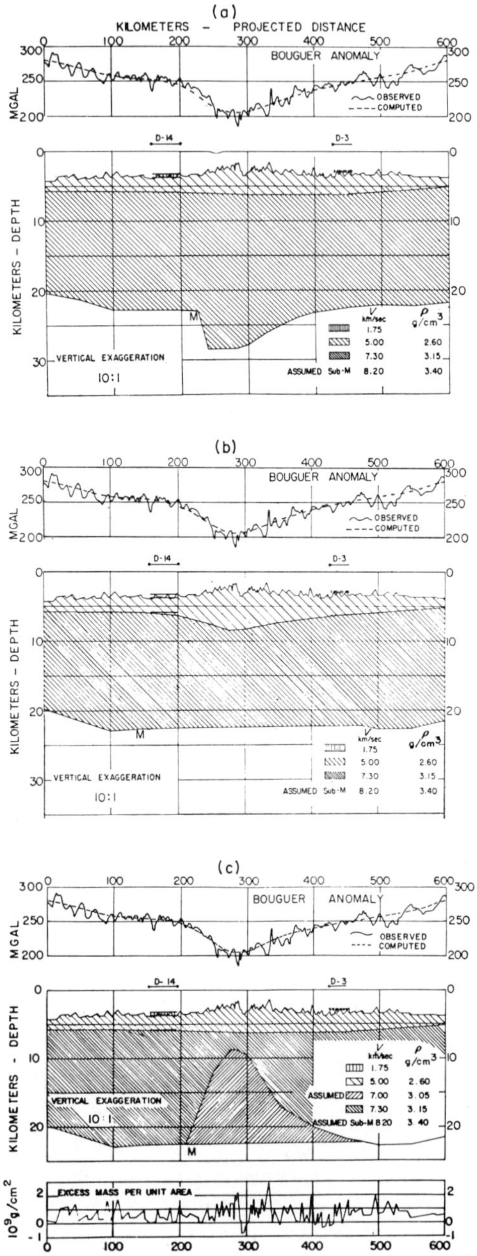

Fig. 21. Crustal sections computed to fit the seismic and gravity data for the Mid-Atlantic Ridge. (After Talwani, Heezen and Worzel, 1961.)

of the M-discontinuity at about 230 km, the computed curve is not steep enough to fit the observed curve. This indicates that most, or all, of the anomalous mass is at this depth or shallower.

Fig. 21b shows an alternative mass distribution where most of the anomaly is caused by a thickening of the topmost layer. This is the 5.0 km/sec layer found in the seismic stations on either flank. A good reversed seismic station near the peak could establish whether this type of distribution is most likely.

Fig. 21c shows a third alternative mass distribution where most of the anomaly is caused by low density in the lower part of the crust beneath the central part of the ridge. Seismic refraction measurements probably could not delineate this structure as the low-density material could be expected to have lower velocities and would thus refract the seismic waves downward rather than upward. Perhaps earthquake surface-wave studies could detect this type of mass distribution. This type of section shows much greater promise of explaining the magnetic anomalies often associated with the rift valley of the ridge.

Combinations of the mass distributions shown in Fig. 21 can also be found to explain the gravity anomalies over the ridge. It would be difficult to explain the major part of the anomaly, however, by masses deeper than 25 km.

C. Islands and Seamounts

Heiskanen and Vening Meinesz (*loc. cit*) have summarized their results of isostatic studies for volcanic islands in Table II. These are carried out for densities of the island materials of 2.937 and 3.07, and they give comparisons for a value on the island and for a deep-sea station clear of the island influence. The computations are carried out for various degrees of regionality indicated by radius R. They conclude that the best results (closest to zero anomaly) are obtained for a regionality of 232.4 km and for a density between the two extremes for an Airy isostatic assumption, $T = 30$. They further conclude, on the basis of estimates of the mechanical properties of the crust and the mantle, that the rigid crust is about 35 km thick. In this concept the upper part of the mantle, although showing no density contrast with the mantle beneath it, has different elastic properties, the upper part resisting viscous flow, while the lower part is capable of viscous flow.

Woollard (1954) has computed a density distribution to account for the gravity anomaly of the islands of Hawaii and Bermuda. Fig. 22 has been adapted from this study. He finds that a density of 2.30 for the Hawaiian Island mass above the sea floor gives the smoothest Bouguer anomaly and accounts for a regional Bouguer anomaly by a gentle depression of the M-discontinuity. For Bermuda he finds the density 2.7 for the island mass gives the smoothest Bouguer anomaly curve. The study by Harrison and Brisbin (1959) of Seamount Jasper (see Fig. 11) indicates a density of 2.3 assuming no underlying density contrasts.

Worzel and Talwani in a paper in preparation have made computations for

TABLE II

Mean Isostatic Gravity Anomalies in Milligals for $T = 30$ km in Volcanic-Island Areas

Radius R:	0 km		58.1 km		116.2 km		174.3 km		232.4 km		Number of stations
Density:	2.937	3.07	2.937	3.07	2.937	3.07	2.937	3.07	2.937	3.07	
Hawaii	+188	+174	+163	+148	+108	+91	+52	+32	+10	−13	6
Oahu	+106	+98	+86	+77	+47	+36	+16	+3	−4	−17	6
Deep sea	−16	−16	−9	−9	+1	+2	+11	+12	+15	+16	4
Madeira	+107	+96	+81	+69	+43	+28	+15	−1	−5	−24	3
Deep sea	+14	+14	+16	+17	+16	+16	+14	+14	+12	+11	10
Sao Vicente, Cape Verde Island	+128	+120	+107	+98	+70	+58	+53	+40	+28	+12	4
Deep sea	+5	+4	+8	+8	+10	+10	+7	+7	+4	+2	4
Hamilton, Bermuda	+149	+139	+120	+109	+75	+61	+45	+30	+24	+7	1
Deep sea	−11	−11	−8	−8	−6	−6	−7	−7	−10	−10	4
Las Palmas, Grand Canary	+72	+68	+64	+61	+46	+40	+33	+26	+22	+13	2
Deep sea	−8	−8	−4	−4	+1	+1	+4	+4	+5	+4	8
Port Louis, Mauritius	+140	+133	+119	+111	+85	+75	+59	+47	+37	+23	1
Deep sea	+18	+18	+23	+24	+22	+22	+20	+20	+17	+16	4
Fayal	+24	+19	+17	+12	+12	+2	−3	−10	−18	−29	3
Ponta Delgada, Sao Miguel	+43	+37	+29	+17	+9	+0	−8	−18	−27	−40	1
Sea depth 1000–2500 m	+39	+37	+39	+37	+34	+31	+26	+23	+14	+8	14
Depth 2500 m	+37	+35	+38	+38	+42	+42	+47	+46	+47	+45	8

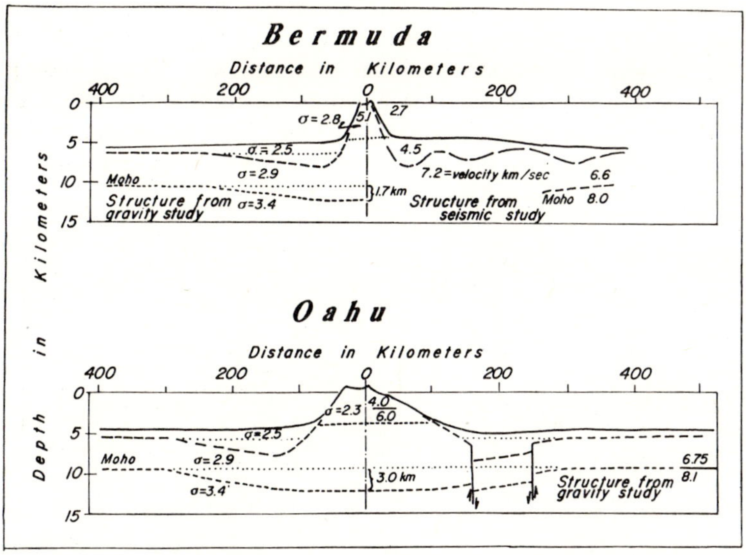

Fig. 22. Crustal sections computed to fit the seismic and gravity data for Bermuda and Oahu. (After Woollard, 1954.)

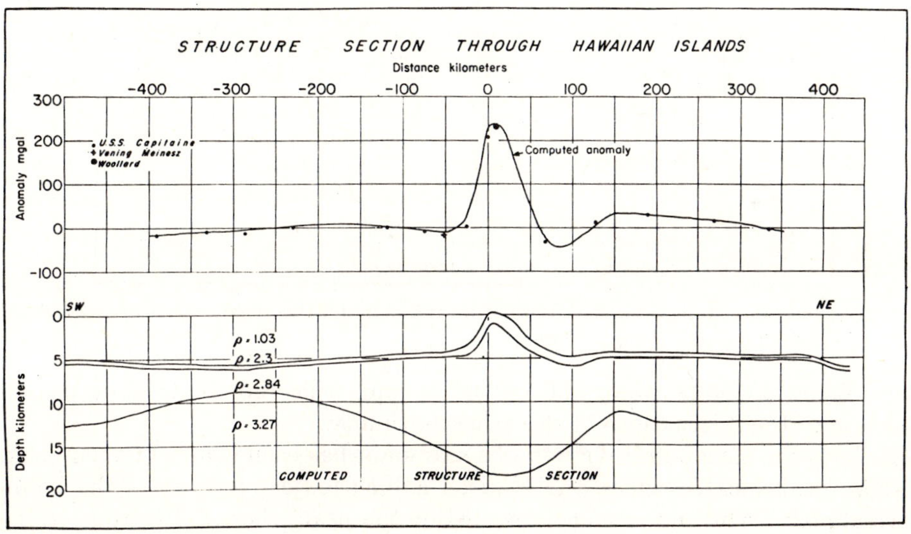

Fig. 23. Crustal section computed for Hawaii. (After Worzel and Talwani, unpublished.)

another mass distribution to account for the gravity anomalies for Hawaii. In this case there were considerably more gravity data available. Fig. 23 shows the resulting section. The asymmetry indicated in the section would be more exaggerated if the section were made farther to the southeast and less exaggerated if it were made farther to the northwest.

A continuous gravity profile has been obtained on an east to west crossing of Cruiser Bank. Some seismic data (Bentley *et al.*, unpublished) exist on the seamount and in the nearby flanking regions to provide control on the upper parts of the sections. A two-dimensional analysis was used to compute the variations of the M-discontinuity which would be required to account for the gravity anomaly (Worzel and Talwani, 1959a). Fig. 24 shows the resulting section. Since this feature is not perfectly represented by a two-dimensional

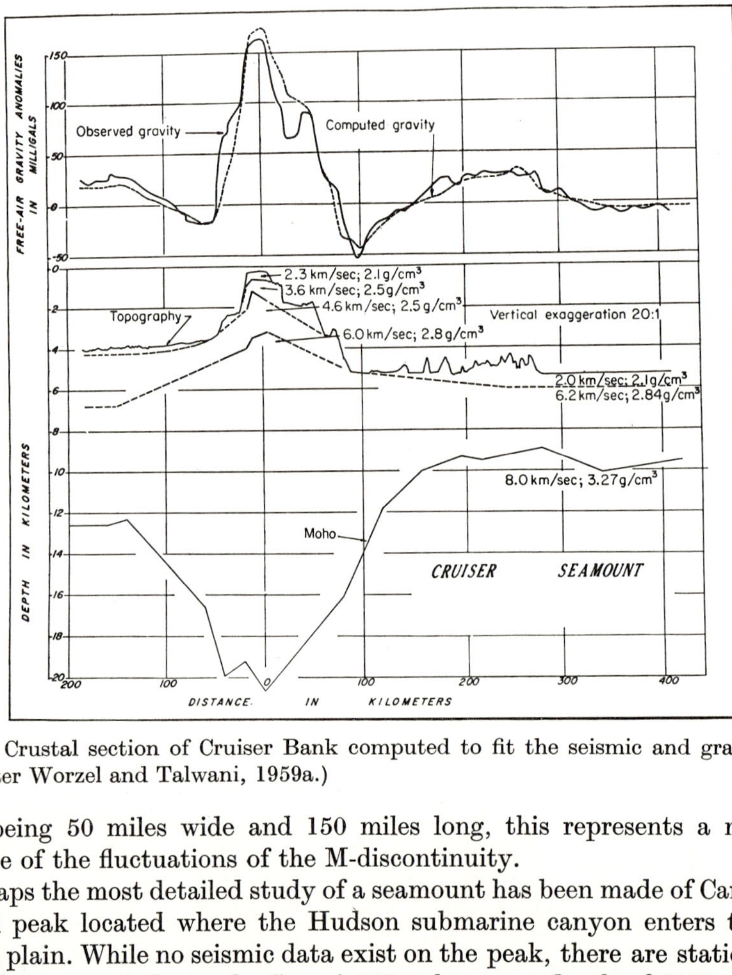

Fig. 24. Crustal section of Cruiser Bank computed to fit the seismic and gravity data. (After Worzel and Talwani, 1959a.)

form, being 50 miles wide and 150 miles long, this represents a maximum estimate of the fluctuations of the M-discontinuity.

Perhaps the most detailed study of a seamount has been made of Caryn Peak, a small peak located where the Hudson submarine canyon enters the Sohm abyssal plain. While no seismic data exist on the peak, there are stations in the deep areas around the peak. Core A-158 taken at a depth of 3970 m was on the flank of Caryn Seamount. Upper Cretaceous fossils, altered by replacements or recrystallization of the original $CaCO_3$ in a way considered due to vulcanism, were found in the core (Miller and Ewing, 1956). This makes it most probable that the vulcanism responsible for this feature must have ended toward the end of the Upper Cretaceous.

Fig. 25 shows the computations for nine different mass distributions for the north–south and east–west crossings of this feature (Worzel and Talwani,

1959a). At the top right of each cross-section is shown the continuous free-air anomaly curve observed with the Graf Sea Gravimeter. Gravity anomalies in the region indicated that the normal anomaly in the vicinity of this seamount would be about -38 mgal. In Case A it is assumed that the seamount is superimposed on the sea crust and sediments, and the three computations represent density assumptions of 2.30, 2.50 and 2.84 for the cross-hatched mass shown in the diagram to the left. In Case B, the more logical view that the seamount is based at or near the base of the sedimentary layers is computed for the same density assumptions. In Case C, a local isostatic compensation is assumed at

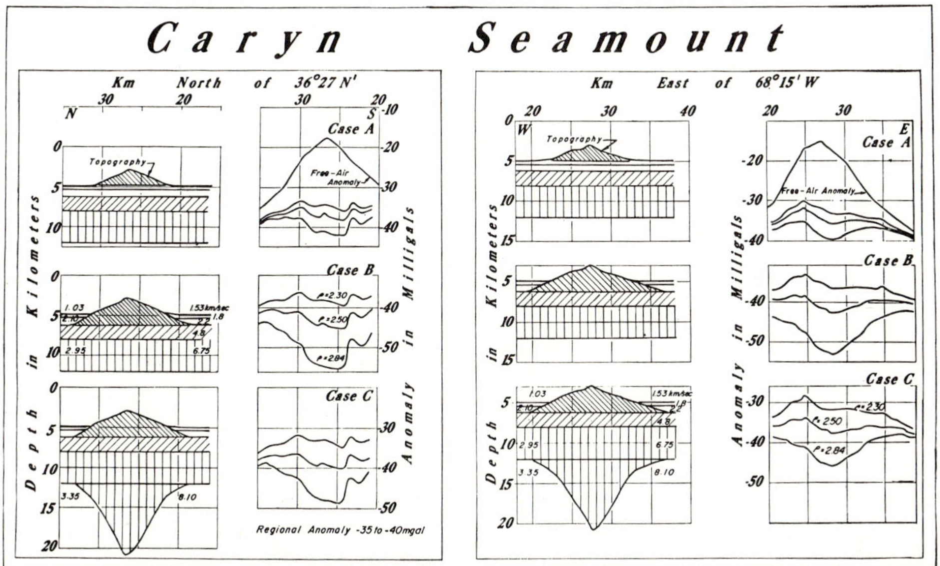

Fig. 25. Computed curves for various structure sections of Caryn Seamount. (After Worzel and Talwani, 1959a.)

the base of the crust with a density the same as that for the 6.75 km/sec crustal layer which overlies the M-discontinuity. The densities for the various layers determined by refraction seismology were chosen according to the Nafe–Drake velocity–density relation (Talwani *et al.*, 1959).

Using the criterion of the removal of the anomalous feature, one would choose a solution of a density near 2.50 for Case A, the solution of a density near 2.30 for Case B, or the solution of a density between 2.30 and 2.50 for Case C. The remaining fluctuations of the curves in each case are probably due to an imperfect knowledge of the shape of the body, or due to the fact that the density may not be well represented by a mean density, or to the errors of calculation; or perhaps due to the errors of measurement caused by an imperfect knowledge of the ship's velocity.

Most workers have accepted a Case A type of solution. In view of the evidence of the core on the flank of the seamount a solution of type B or type C is more probable. The type B solution would imply that the volcanic material came from deep within the mantle. The type C solution would imply that either the volcanic material which forms the seamount originated in the region of the compensation or that the vulcanism originated deep within the mantle depositing some material as the island and some as the root.

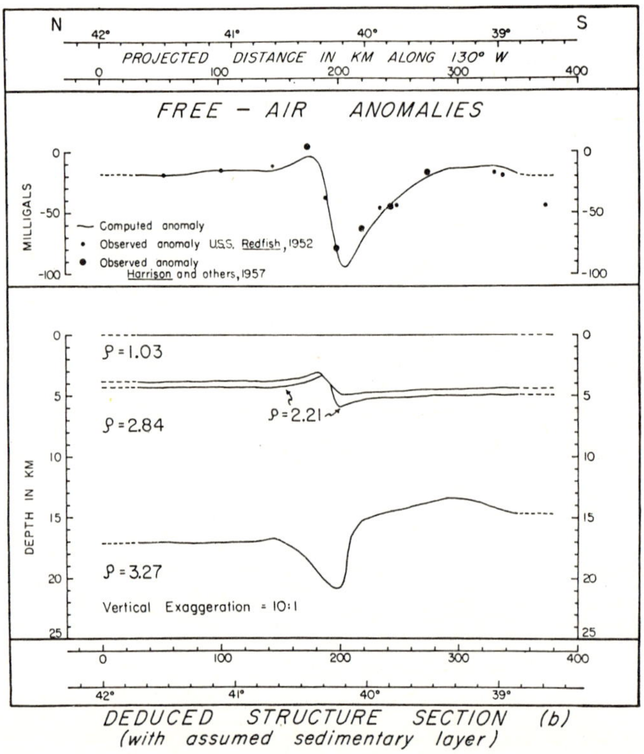

Fig. 26. Mendocino Escarpment. (After Talwani *et al.*, 1959.)

D. Mendocino Escarpment

A mass distribution which accounts for the observed gravity anomaly over the Mendocino Escarpment has been given by Talwani *et al.* (1959). Fig. 26 shows the crustal structure deduced assuming a sedimentary layer as shown and assuming that the residual anomaly is all due to the fluctuations of the M-discontinuity. As this section is improbable on geological and seismic grounds, a solution of the type shown above for the Mid-Atlantic Ridge, i.e. a low density mass within the crust, would seem more likely.

E. Ocean Trenches

The discovery by Vening Meinesz of the belt of large negative-gravity anomalies associated with the oceanic trenches is one of the greatest discoveries of contemporary geology and geophysics. Since their discovery, many geophysical investigations have been made by Vening Meinesz and others to try to delineate the structure related to these belts.

In the East Indies, Vening Meinesz has evolved a detailed and complicated theory to account for the anomalies. In summary he invokes a compression which causes a plastic downbuckle of the earth's crust causing the earthquakes, negative gravity anomalies, and shear weaknesses, which provide lanes for vulcanism in the island arcs, and uplift of the islands. The reader is referred to chapter 10C of Heiskanen and Vening Meinesz (1958) for a comprehensive treatment of this subject.

Tsuboi (1957) has calculated a mass distribution for the East Indies to account for the gravity anomalies. Fig. 27 is adapted from his work in which he

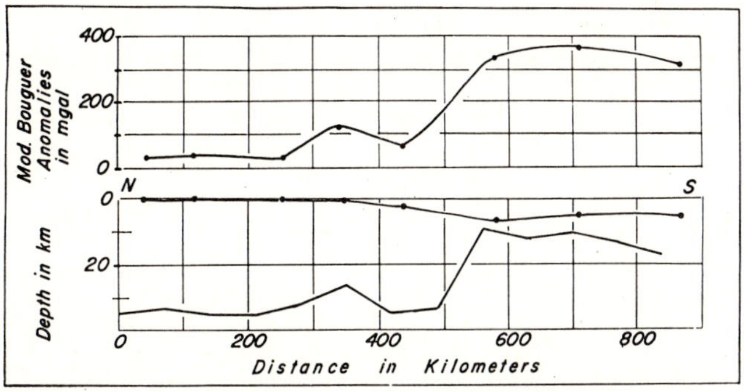

Fig. 27. Structure section through the Dutch East Indies. (Adapted from Tsuboi, 1957.)

used a density differential of 0.4 g/cm^3 at the M-discontinuity. It is readily seen that the thinnest part of the crust is beneath the trench.

Probably the most detailed investigations have been made of the Puerto Rico Trench. The latest work is reported by Talwani *et al.* (1959) and the reader is referred to it if he desires references to the earlier work. Fig. 28 shows the mass distribution computed to fit the detailed anomaly curve. Many seismic data are available to control the mass distribution of the upper crustal layers. The densities are indicated and were chosen to fit the Nafe–Drake velocity–density curve. The depths of the layers determined seismically are indicated by the circled points. The M-discontinuity was chosen to obtain the best fit of the gravity data using a depth to the mantle of 11 km at a point 310 km north of 18°N. Here, in the trench, the crust is intermediate in thickness between an oceanic and a continental crust. The free-air anomaly of -346 mgal measured here is the largest negative anomaly yet found.

Recently a continuous gravity profile has been observed with the Graf Sea Gravimeter from well out into the Atlantic Ocean to well inside the Caribbean Sea, passing close to Barbados. Many seismic refraction stations were available from the work of Officer *et al.* (1959) to provide density estimates from the Nafe–Drake curve and depths of the upper layers for control of the computations. The mass distribution required to fit the gravity anomaly is shown in Fig. 29 (Sutton, Talwani and Worzel, 1960). Here there is no trench, but there is a large negative-gravity anomaly. The crustal thickness is again intermediate in the region of the negative-anomaly belt and there is a big accumulation of sediment.

Raitt *et al.* (1955) have made seismic measurements in the Tonga Trench

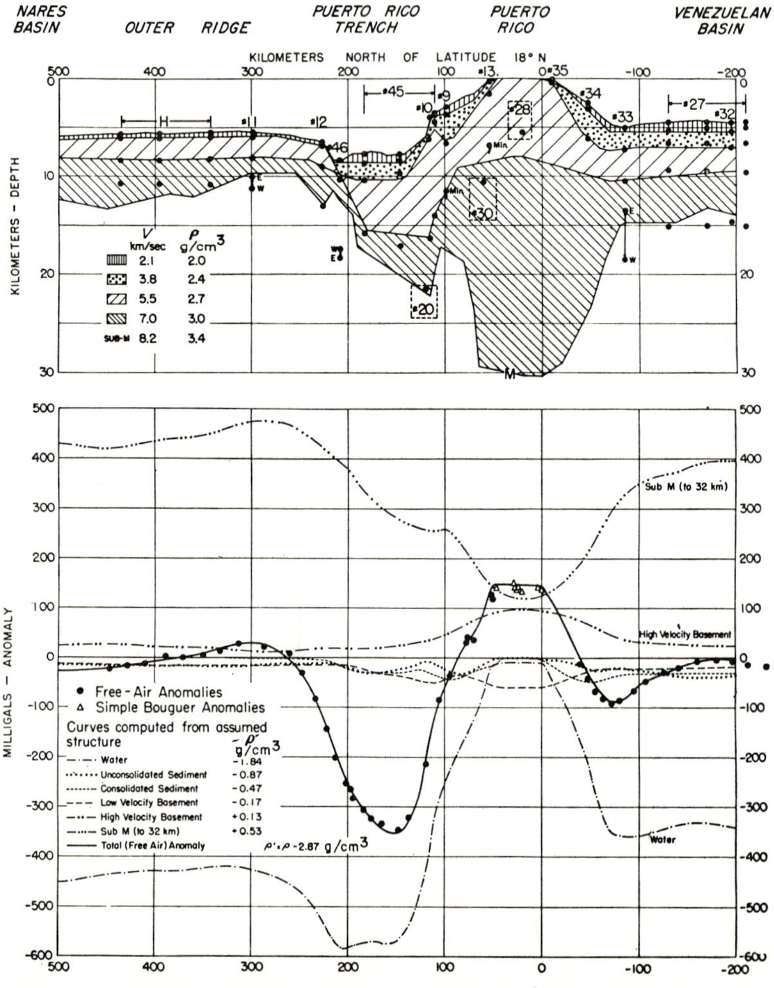

Fig. 28. Structure section across Puerto Rico Trench computed to fit the seismic and gravity data. (After Talwani, Sutton and Worzel, 1959.)

and in the nearby regions. More recently, gravity measurements have been made as a part of the International Geophysical Year cooperation between the Lamont Geological Observatory and the Bureau of Mineral Resources of Australia. Fig. 30 shows the section deduced by Raitt and the computed gravity anomalies for this section, using the Nafe–Drake density curve, compared with the observed anomalies (Talwani, Worzel and Ewing, 1961). The computed anomaly curve falls at least 100 mgal below the observed points in the region from 800 km to 900 km east of Suva, Fiji. Thus there seems to be a conflict between the seismic and the gravity data. In Fig. 31 the boundaries in the regions where no seismic data exist have been adjusted and an almost masked (seismically) interface is assumed in the trench region in order to compute a gravity curve which fits the observed data. In this way the conflict between the seismic and gravity data can be resolved.

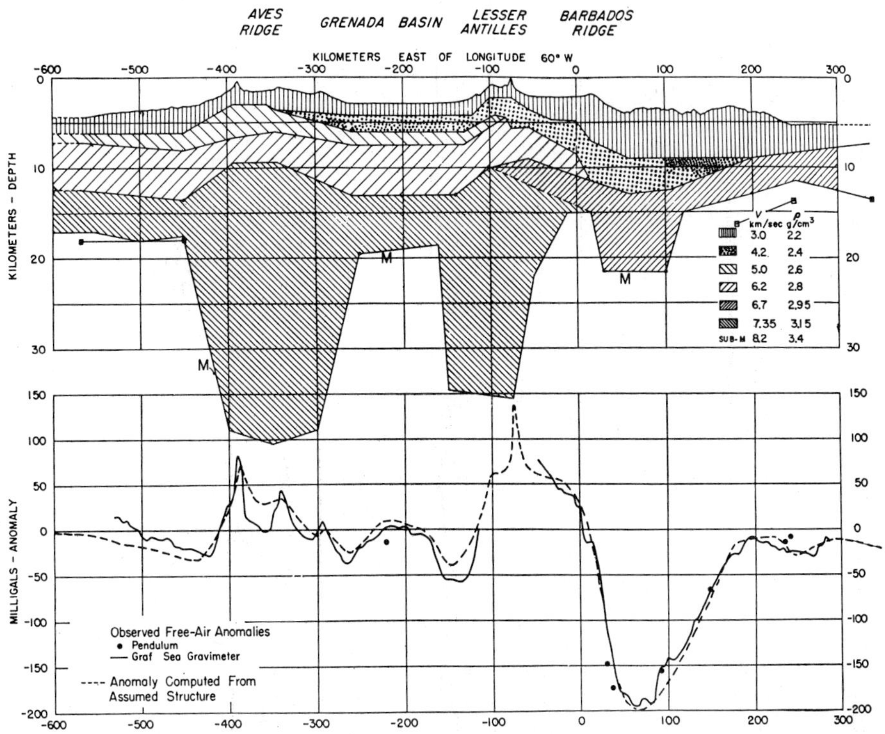

Fig. 29. Structure section across the gravity-anomaly belt of the Lesser Antilles near Barbados computed to fit the seismic and gravity data. (After Sutton, Talwani and Worzel, 1960.)

The reader is again referred to Fig. 17 for a view of the structure of a trench complicated by a continental margin and an adjacent mountain range.

F. Summary

The advent of gravity meters capable of making gravity measurements on surface ships, even though at present they require moderate seas in which to operate, ushers in a new era in the measurement of gravity at sea. We can expect a vast increase in the data of gravity at sea and a great deal better insight into the most "normal" part of our globe.

Most authors have computed a mass distribution to account for the observed gravity anomalies on the assumption that fluctuations of the M-discontinuity can account for the anomalies, at least after one has accounted for the sedi-

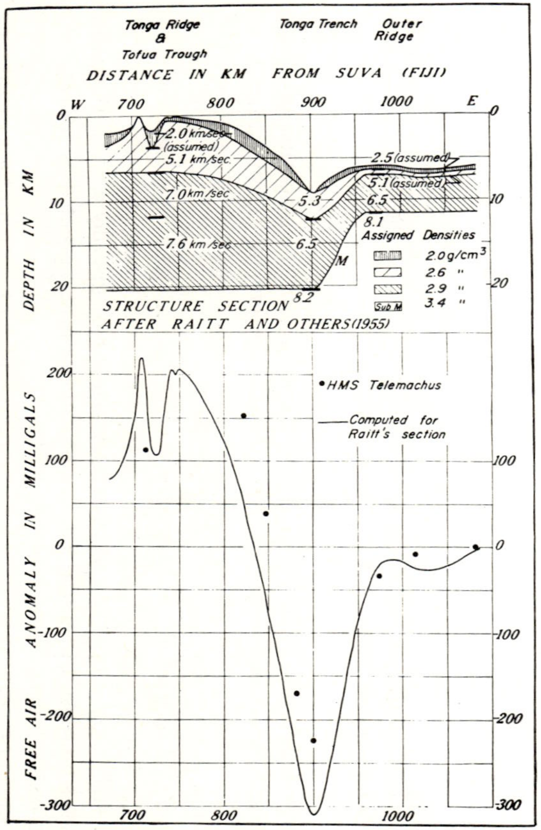

Fig. 30. Structure section across Tonga Trench computed by Raitt to fit the seismic data and comparison of the computed gravity curve for this section with the observed data. (After Talwani, Worzel and Ewing, 1961.)

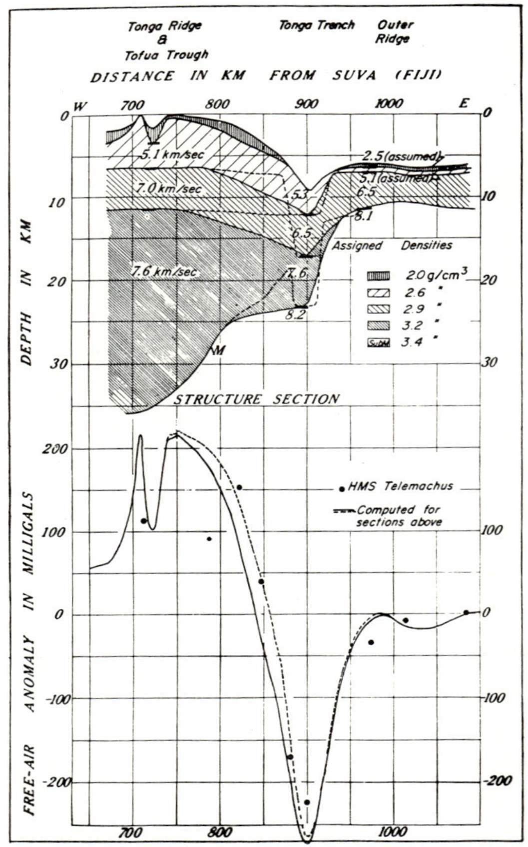

Fig. 31. Structure sections across Tonga Trench computed to fit the seismic and gravity data. (After Talwani, Worzel and Ewing, 1961.)

mentary accumulations in the region near the ocean bottom. Probably, in the regions of continental margins, this view will prove to be substantially correct. In regions such as the mid-ocean ridges and volcanic-island arcs, it may be necessary to consider some or all of the deficiency of mass represented by the down-bulge of the M-discontinuity to occur within the crustal layers. Only more definitive data can decide. Since the low-density matter of these layers is probably of lower velocity than the overlying material, it will be masked to refraction seismology. It may be necessary to detect the layers from surface-wave studies or from studies of shadow zones, both of which look difficult at present.

It now appears that the crust beneath the trenches is only slightly different in thickness from an oceanic crust and that, in some trenches, there is a large accumulation of sedimentary material while in others there is only a small amount.

References

Browne, B. C., 1937. The measurement of gravity at sea. *Mon. Not. Roy. Astr. Soc., Geophys. Suppl.*, **4**, 271–279.

Drake, C. L., M. Ewing and G. H. Sutton, 1959. Continental margins and geosynclines: The East Coast of North America, North of Cape Hatteras. *Physics and Chemistry of the Earth,* Vol. III. Pergamon Press, New York–London–Paris–Los Angeles.

Ewing, John and Maurice Ewing, 1959. Seismic-refraction measurements in the Atlantic Ocean basins, in the Mediterranean Sea, on the Mid-Atlantic Ridge, and in the Norwegian Sea. *Bull. Geol. Soc. Amer.*, **70**, 291–318.

Ewing, Maurice, J. L. Worzel, J. B. Hersey, Frank Press and G. R. Hamilton, 1950. Seismic refraction measurements in the Atlantic Ocean Basin, Pt. 1. *Bull. Seism. Soc. Amer.*, **40** [misprinted **4** on offprints], 233–242.

Garland, G. D., 1956. "Gravity and Isostasy," in *Encyclopedia of Physics,* Vol. XLVII, 202–245. Springer-Verlag, Berlin.

Gilbert, R. L. G., 1949. A dynamic gravimeter of novel design. *Proc. Phys. Soc. London,* **B62**, 445–454.

Graf, A., 1956. Beschreibung eines neuentwickelten Seegravimeters und Ergebnisse der ersten Messfahrt auf dem Starnberger See an Bord der *Seeshaupt. Bayer. Akad. Wiss., Math-Nat. Kl.*, New Series, **75**, 5–16.

Graf, A., 1957. Über die bisherigen Erfahrungen und Messergebnisse mit dem Seegravimeter. *Z. Geophys.*, **23**, 4–25.

Graf, A., 1958. Das Seegravimeter. *Z. Instrumentenk.*, **66**, no. 8, 151–161.

Harrison, J. C., 1959. Tests of the LaCoste-Romberg Surface-Ship gravitymeter, I. *J. Geophys. Res.*, **64**, 1875–1881.

Harrison, J. C., 1960. "The Measurement of Gravity at Sea" in *Methods and Techniques in Geophysics*. Interscience Publishers, New York–London.

Harrison, J. C. and W. C. Brisbin, 1959. Gravity anomalies off the West Coast of North America, 1: Seamount Jasper. *Bull. Geol. Soc. Amer.*, **70**, 929–934.

Heiskanen, W. A. and F. A. Vening Meinesz, 1958. *The Earth and Its Gravity Field.* McGraw-Hill Book Co., Inc., New York–Toronto–London.

IGY Bulletin, 1958. First sea surface gravimeter. *Trans. Amer. Geophys. Un.*, **39**, No. 8, 175–178.

Koryakin, E. D., 1958. On the deep structure of the Earth's crust within the confines of the Atlantic Ocean. *Izvest. Akad. Nauk U.S.S.R., Ser. Geofiz.* 1477–1484; English edition, translated and published by Amer. Geophys. Union, January, 1960. (Dec. 1958, 860–865.)

Kukkamaki, T. J. 1955. Gravimetric reductions with electronic computers. *Ann. Acad. Sci. Fennicae, Ser. A.*, III, Geol.-Geog., **42**.

LaCoste, L. J. B., 1959. Surface ship gravity measurements on the Texas A. and M. College ship, the 'Hidalgo'. *Geophysics*, **24**, 309–322.

Miller, E. T. and M. Ewing, 1956. Geomagnetic measurements in the Gulf of Mexico and in the vicinity of Caryn Peak. *Geophysics*, **21**, 406–432.

Nettleton, L. L., Lucien LaCoste and J. C. Harrison, 1960. Tests of an Airborne gravity meter. *Geophysics*, **25**, 181–202.

Officer, C. B., J. I. Ewing, J. F. Hennion, D. G. Harkrider and D. E. Miller, 1959. Geophysical investigations in the Eastern Caribbean: Summary of 1955 and 1956 cruises. *Physics and Chemistry of the Earth*, Vol. III. Pergamon Press, New York–London–Paris–Los Angeles.

Raitt, Russell W., Robert L. Fisher and Ronald G. Mason, 1955. Tonga Trench. In "Crust of the Earth". *Geol. Soc. Amer. Spec. Paper* 62, 237–254.

Spiess, F. N. and G. L. Brown, 1958. Tests of a new submarine gravity meter. *Trans. Amer. Geophys. Un.*, **39**, 391–396.

Sutton, G. H., M. Talwani and J. L. Worzel, 1960. West–East crustal section through the gravity anomaly belt of the Lesser Antilles along 14° 20′N. *J. Geophys. Res.*, **65**, 2526 (abstract).

Talwani, M. and M. Ewing, 1960. Rapid computation of gravitational attraction of three-dimensional bodies of arbitrary shape. *Geophysics*, **25**, 203–225.

Talwani, Manik, George H. Sutton and J. Lamar Worzel, 1959. Crustal section across the Puerto Rico Trench. *J. Geophys. Res.*, **64**, 1545–1555.

Talwani, Manik, J. Lamar Worzel and Maurice Ewing, 1961. Gravity anomalies and crustal section across the Tonga Trench. *J. Geophys. Res.*, **66**, 1265–1278.

Talwani, M., J. L. Worzel and M. Landisman, 1959. Rapid gravity computations for two-dimensional bodies with application to the Mendocino Submarine fraction zone. *J. Geophys. Res.*, **64**, 49–59.

Talwani, Manik, B. C. Heezen and J. Lamar Worzel, 1961. Gravity anomalies, physiography, and crustal structure of the Mid-Atlantic Ridge. *Trav. Sci. Bur. Centr. Seism. U.G.G.I.*, Int. Serie A, fasc. 22, 81–111.

Thompson, George A. and Manik Talwani, 1959. Crustal section across California and Sierra Nevada. *Bull. Geol. Soc. Amer.*, **70**, 1688 (abstract).

Thompson, L. G. D., 1959. Airborne gravity meter test. *J. Geophys. Res.*, **66**, 488.

Thompson, L. G. D. and L. J. B. LaCoste, 1960. Aerial gravity measurements. *J. Geophys. Res.*, **65**, 305–322.

Tsuboi, Chuji, 1957. Crustal structure along a certain profile across the East Indies as deduced by a new calculation method. Gedenboek for F. A. Vening Meinesz. *Verhandel. Koninkl. Ned. Geol.-Mijnb. Gen., Geol. Ser.*, **78**, 287–294.

Vening Meinesz, F. A., 1929. *Theory and Practice of Pendulum Observations at Sea.* Waltman, Delft.

Vening Meinesz, F. A. 1941. *Theory and Practice of Pendulum Observations at Sea*, Part II. Waltman, Delft.

Woollard, G. P., 1954. The crustal structure beneath oceanic islands. *Proc. Roy. Soc. London*, **A222**, 361–387.

Worzel, J. Lamar, 1959. Continuous gravity measurements on a surface ship with the Graf Sea Gravimeter. *J. Geophys. Res.*, **64**, 1299–1315.

Worzel, J. L. and A. Graf, 1957. Comparison of the Graf Sea Gravimeter with the Vening Meinesz apparatus on board the submarine U.S.S. *Becuna*. *Bull. Geod. Int.*, **45**, 38–53.

Worzel, J. Lamar and G. L. Shurbet, 1955. Gravity interpretations from standard oceanic and continental crustal sections. *Geol. Soc. Amer. Spec. Paper* 62, 87–100.

Worzel, J. Lamar and G. L. Shurbet, 1955a. Gravity anomalies at continental margins. *Proc. Nat. Acad. Sci.*, **41**, 458–469.

Worzel, J. L. and Manik Talwani, 1959. The Geoid in ocean areas. *Intern. Oceanog. Cong., Preprints.*, Amer. Assoc. Adv. Sci., 63.

Worzel, J. L. and Manik Talwani, 1959a. Gravity anomalies on seamounts. *Bull. Geol. Soc. Amer.*, **70**, 1702–1703 (abstract).

10. THE MAGNETIC FIELD OVER THE OCEANS

E. C. Bullard and R. G. Mason

1. Historical

It is believed that the compass was first discovered in China (Needham, 1962). The earliest certain references to it are in the *Wu Ching Tsung Yao*, edited by Tsêng Kung-Liang and finished in 1044 and in the *Mêng Chhi Pi Than* of Shen Kua written about 1088. Needham considers it likely that it had been used for fortune telling for some hundreds of years before this. The first mention of its use in a ship is by Chu Yii in a work written between 1111 and 1117.

No connection is known between these Chinese discoveries and the first use of the compass in Europe. The evidence has been discussed at length by Crichton Mitchell (1932) who concludes that the first certain reference to it in European literature is in two books written about 1187 by Alexander Neckam, a monk of St. Albans, who treats the compass as a device well known in his time. In the 13th and 14th centuries there are many references, particularly by poets, to whom the idea of a needle that points the way in storm and darkness has always appealed. For example, in 1375 the Scottish poet Barbour (Skeat, 1870) wrote of the journey of Robert the Bruce from Arran to Carrick in 1306:

"Thai rowit fast with all thar micht
Till that apon thame fell the nycht
That It wox myrk on gret maner
Swa that thai wist nocht quhar thai wer
For thai na nedil had na stane."

In China in 1088 Shen Kua knew that the compass needle does not point to the true north (or, as he would have said, to the true south). In Europe this appears to have been unknown till late in the 15th century (Needham, 1962; Crichton Mitchell, 1937).

The first systematic observations at sea were made during the voyage of João de Castro in 1538 to 1541 (Hellmann, 1897). He sailed around the Cape, up the west coast of India and into the Red Sea; during this voyage he made 43 determinations of declination. These and other early observations have been collected by van Bemmelen (1899) and by Gaibar-Puertas (1953). From these van Bemmelen has compiled maps showing the declination at 50-year intervals from 1550 to 1700. As the observations at sea were made in wooden ships, they were usually not seriously disturbed by neighbouring masses of iron, and it is believed that the maps do give a good general representation of the declination over a large part of the earth.

It is not possible to give any account here of the voyages and magnetic observations of the 16th- and 17th-century navigators. The voyages of Foxe and James to Hudson's Bay at the beginning of the 17th century and of Edmond Halley in the Atlantic in 1698 to 1700 are of particular significance. The results of the former show that the north magnetic pole was then, as it is today, to the north-west of Hudson's Bay. Halley went from England to

[*MS received October, 1960*]

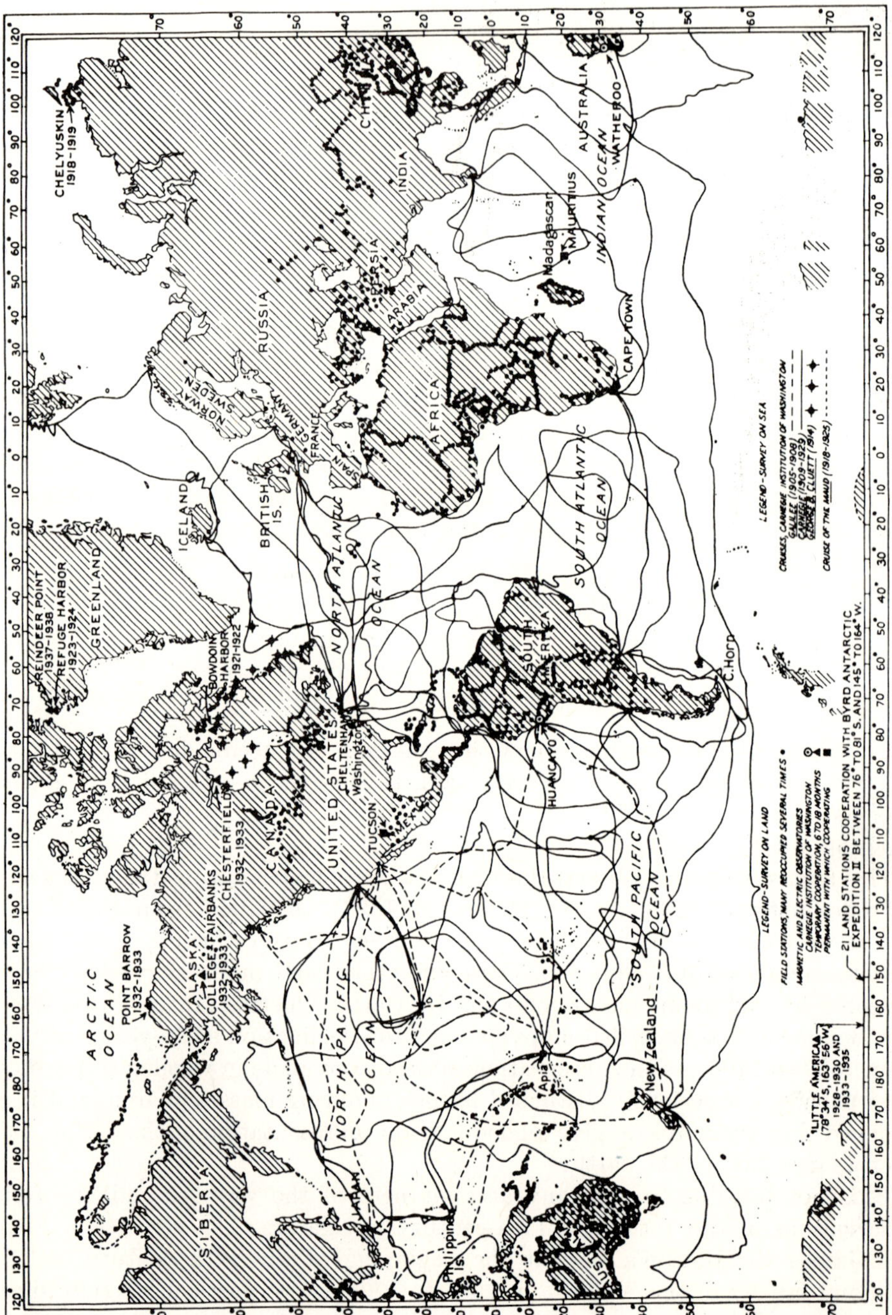

Fig. 1. Tracks of the ships of the Carnegie Institution of Washington. (After Fleming, 1937. By courtesy of the Carnegie Institution.)

latitude 52° 24′S and returned along the coast of North America. He obtained observations almost daily, of which van Bemmelen's collection gives only a selection. On his return he published the first chart showing lines of equal declination (isogonic lines); this was for 1700 and covered the Atlantic Ocean. It was followed in 1702 by a chart covering the whole world, but with no lines of equal declination over most of the Pacific (Chapman, 1941; Bullard, 1956).

Observations of declination, and occasionally of inclination, continued through the 18th century and led to a gradual improvement of the magnetic charts. The first charts of the horizontal component of the field and of the total force were those of Hansteen published in 1819.

After the change from wooden to iron and steel ships, it became impracticable to make accurate magnetic measurements at sea, except in specially constructed non-magnetic vessels. The most successful of these were the *Galilee* (1905–1908) and the *Carnegie* (1909–1929). Their tracks are shown in Fig. 1; the results from these cruises still constitute the main evidence for the behaviour of the magnetic field over the oceans. Tracks of other ships making magnetic surveys between 1839 and 1916 are given by Fleming (1937). After the destruction of the *Carnegie* by fire in 1929, no new ocean-going non-magnetic ship was constructed and no systematic measurements were made at sea for many years. Since 1956 this work has been taken up again by the Russian ship *Zarya*, which has made a number of traverses of the Atlantic, Pacific and Indian Oceans. A description of this ship and its work has been given by Ivanov (1960, 1961) and Matveyev (1962); she is a schooner 37.5 m long and of 600 tons displacement.

Extensive measurements of all three components of the field have also been made in Canadian and United States aircraft (Vestine, 1960), and a large programme, "Project Magnet", covering almost the whole of the oceans is at present being carried out by the United States Navy. The intended tracks of the project are shown in Fig. 2.

The surveys from these special aircraft and from non-magnetic ships give measurements of all three components of the field. Such measurements are essential to a knowledge of the field as a whole, and are necessary if it is to be divided into parts of internal and external origin (Chapman and Bartels, 1940; Vestine *et al.*, 1947). They do, however, require elaborate equipment and special ships or specially modified aircraft. Fortunately, much geological information can be obtained from the much simpler measurement of total force. This can be made with an instrument towed behind a ship or aircraft at a sufficient distance to avoid the magnetic disturbance due to the engines and hull.

2. Measurement of the Magnetic Field at Sea

Three quantities are required to define the magnetic field at a point. At least one of these must be the field strength or the strength of a component of the field; the other two can be field components, or angles giving the direction of the field relative to axes fixed relative to the earth. There are thus four operations involved in the complete determination of the field at a point: (1) The measurement of the field strength; (2) the establishment of the direction of the magnetic field; (3) the establishment of the true vertical; and (4) the establishment of the direction of the true north.

Until about 1945 all magnetic measurements at sea were made with adaptations of the instruments used on land. The instruments were mounted on gimbals whose period was short compared to that of the ship's motion; they must, therefore, have maintained a fixed attitude relative to the apparent vertical. It is to be supposed that the effect of the difference between the true and apparent verticals was eliminated or reduced by averaging by eye while

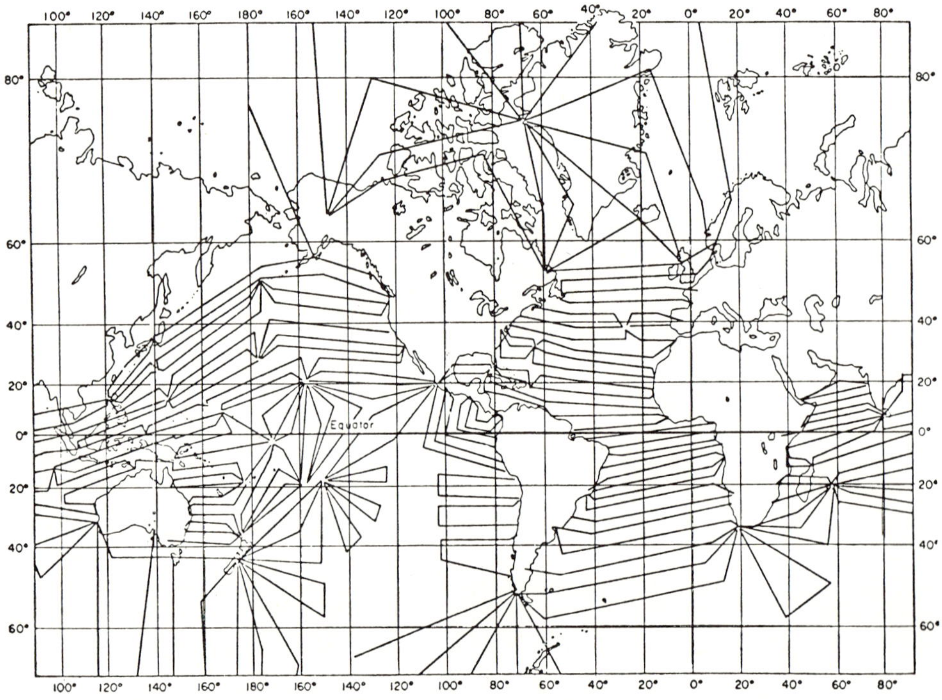

Fig. 2. Tracks of Project Magnet. (After Vestine, 1960, fig. 4.)

taking a reading. The published accounts are curiously reticent about this important point; even the period of the *Carnegie*'s gimbals is not stated. Azimuth was then, as now, determined by astronomical observation; this involves not only the observation of the direction of the sun or of a star, but also the determination of the vertical. It is not necessary to describe here the compasses, dip circles, earth inductors, deflectors and other instruments used, since they are now only of historical interest. A very detailed description of their final forms has been given by Bauer *et al.* (1917).

In recent years two basic improvements have been made. First, gyro-stabilized platforms have become available and, secondly, electronic techniques have been devised for measuring the field. A description of a stabilized platform used in an aircraft has been given by Serson, Mack and Whitham (1957). This provides a vertical reference direction which is within a few minutes of arc of the true vertical and an azimuth reference direction which is stable to a similar accuracy for periods of up to an hour; the azimuth of this reference is checked at intervals of about 15 min by astronomical observations. This device was made some years ago and probably does not represent the limit of what would now be possible; in particular, a gyro-stabilized sextant would greatly improve the astronomical azimuth determinations. Such stabilized platforms are easier to use in an aircraft than in a ship, since the rolling and pitching of an aircraft is usually much less than that of a ship. So far as the writers know, no magnetic measurements have been made in a ship using a stabilized platform. This is probably because the magnetic disturbances in a steel ship make accurate measurements impossible and no ocean-going non-magnetic ship was in service between the loss of the *Carnegie* and the recent commissioning of the *Zarya*. An airborne instrument without a gyro-stabilized platform, which measures both the strength and the direction of the field, has been described by Schonstedt and Irons (1955).

In deep water the magnetic field may have a gradient of over 100 γ/km ($1\ \gamma = 10^{-5}$ gauss) and on the continental shelf it may, exceptionally, reach ten times this value. To obtain meaningful contours of features with such gradients and a total amplitude of the order of 1000 γ requires positions to be known to a few hundred metres. Within the range of precise radio-navigational aids, such as Decca, there is no difficulty, but in the open ocean the accuracy required is beyond that which can ordinarily be attained by astronomical observations interpolated by dead reckoning. The usual procedure in surveying features far from shore is to establish a system of buoys moored on light wires and carrying radar reflectors, and to conduct the survey relative to these. A ship has the great advantage over an aircraft that it is possible to observe the field and the depth simultaneously, so that the magnetic survey is correctly related to the topography, even if the whole survey is displaced and distorted by errors in fixing the position of the ship.

In a survey aimed at improving the general magnetic charts and not at the study of geological detail, a lesser accuracy both of position and field is sufficient, or at any rate must be tolerated. Here the main need for accurate positions comes from the desire to determine the secular variation. In ten years the field components change by an amount of the order of 500 γ whilst the field gradient, after smoothing out local anomalies, is of the order of 5 γ/km. Thus the mean of a number of repeat observations will give a useful determination of the secular variation even if their positions are subject to errors of 20 km, as is easily possible in aircraft flights in remote regions. The density of observations needed is discussed by Serson and Hannaford (1957).

In all recent investigations one of two methods has been used for the measurement of the strength of the field. The most convenient for use from a ship is one depending on proton magnetic resonance, though fluxgate methods, devised originally for the detection of submarines, are still widely employed, particularly in the United States. The first measurements with a towed magnetometer were made by Miller and Ewing (1956) with a fluxgate instrument.

The fluxgate method will not be described in detail here. It depends on applying an alternating field to a rod of high-permeability alloy. The alternating field is sufficient to drive the material to saturation in both directions. If the steady field along the rod is zero, the wave form will be symmetrical and free from even harmonics; otherwise even harmonics are generated in the energizing coil which are proportional in amplitude to the value of the steady field acting along the rod. It is, therefore, possible to determine when the rod is set perpendicular to the earth's field.

The fluxgate magnetometer can be used to measure specific components of the field, but because of the need to operate it in a towed fish, and the consequent difficulty of maintaining stable reference directions, it is generally arranged as a self-orienting total-field instrument. This is accomplished by mounting three fluxgate units perpendicular to one another in a motor-driven gimbal. Two of them are made to orient the third along the earth's field by means of servo-loops, which control the orienting motors in such a way as to maintain zero field along their axes. The even harmonic signal from the third unit is then a measure of the total force of the earth's magnetic field.

The accuracy and sensitivity of the method can be increased by reducing the field acting along the third rod to zero by passing a measured direct current through the energizing coil, using the fluxgate as a null detector. The accuracy of measurement then depends on the stability of the coil and the accuracy with which the current in it can be controlled and measured. There is no difficulty in controlling the current to one part in 10^5 for periods of a few hours, representing better than 1 γ; the main source of error lies in the difficulty of stabilizing the current and the dimensions of the coil over a long period. For a few hours a stability of the order of 1 γ is possible, but it is doubtful whether, even with frequent comparison with a proton magnetometer, an absolute accuracy much better than 10 γ can be expected. In practice, it is usual to adjust the current in discrete steps, corresponding to uniform increments of field, the signal due to the residual field being applied to a recording instrument.

The basic disadvantage of the fluxgate, compared with the proton precession method, is that it is not absolute, and, because of the instability of the coil system, it requires frequent calibration. It possesses one advantage in that it is more suitable for measuring vector components of the field.

Instruments developed in the United States have been described by Rumbaugh and Alldredge (1949) and Schonstedt and Irons (1955) and Russian instruments by Logachev (1955), Finger *et al.* (1961) and Kudrevsky (1961).

The proton magnetic resonance method was first used to measure the earth's field by Packard and Varian (1954). An instrument for use on land has been described by Waters and Phillips (1956) and by Waters and Francis (1958), and one used in a rocket by Heppner, Stolarik and Meredith (1958). Hill (1959) and others have used the instrument at sea.

The method depends on the interaction of the magnetic moment of protons with a magnetic field. A magnetic field of about 100 gauss is applied to a sample of water. This causes about 10^{-7} of the protons in the water to become oriented along the field. If the applied field is removed the protons precess and produce an alternating e.m.f. in a coil surrounding the sample.

The phenomenon can be described from the point of view of either classical or quantum mechanics (Andrew, 1956; Abragam, 1961). Both give the frequency, f, of the proton precession as

$$f = \gamma H / 2\pi,$$

where γ is the gyromagnetic ratio of the proton. This has been determined by the measurement of the frequency in a known field. The result obtained by Driscoll and Bender (1958, 1958a) and adopted by the International Union of Geodesy and Geophysics (Nelson, 1960) is

$$\gamma = 26{,}751.3 \pm 0.2 \text{ gauss}^{-1} \text{ sec}^{-1}.$$

For another determination see Vigoureux (1962). The value of γ required is the observed value and not that corrected for screening by the extra-nuclear electrons, which is what is usually given in tables of physical constants.

It is necessary to leave the polarizing field on for a few seconds to allow the orientation of the protons to reach thermal equilibrium. On turning off the field, the protons will precess about the earth's field and will produce an external alternating field for a time depending partly on the time taken to reach a new thermal equilibrium and partly on a tendency for the protons to exchange energy and get out of step. In practice, the signal decays to $1/e$ in a few seconds. If the field is not constant over the sample, the rate of precession of the protons will differ in different parts of it, and the signal will decay owing to their getting out of step. An inhomogeneity of 1 in 10,000 would cause the protons to get out of step in 5 sec and would have an appreciable effect on the decay time; it is, therefore, desirable that the field should be uniform over the sample to within a few γ. This condition is easily satisfied in a towed instrument, but usually prevents the instrument working with the sample aboard ship or in a building.

The proton magnetometer requires no calibration. The relation (1) between frequency and field depends only on the gyromagnetic ratio of the proton and is independent of all instrumental and environmental factors. Also, the frequency is independent of the orientation of the coil, and no stabilization is necessary. The signal is a maximum if the applied field and the pick-up coil are at right angles to the Earth's field, and decreases as $\sin^2 \theta$ if this angle is reduced. This is not a serious limitation in practice, since, over most of the earth, the field makes a large angle with the axis of a horizontal coil and in the equatorial regions a vertical coil can be used.

Abragam *et al.* (1957) have described a method by which the proton resonance may be made to control the frequency of a continuously running oscillator, but this method has not been used at sea.

A method depending on the optical excitation of rubidium vapour has been described by Skillman and Bender (1958). A somewhat similar, and probably superior, device depending on the optical excitation of helium has been described by Keyser *et al.* (1961). Both instruments are more sensitive than the proton magnetometer; they are, however, less convenient, and neither has been used at sea.

Hill has used a proton magnetometer in a fish towed behind a ship. In his instrument the protons in the hydrogen atoms of water are used. The instrument now in use is shown in Figs. 3 and 4; in it the water is in a bottle containing

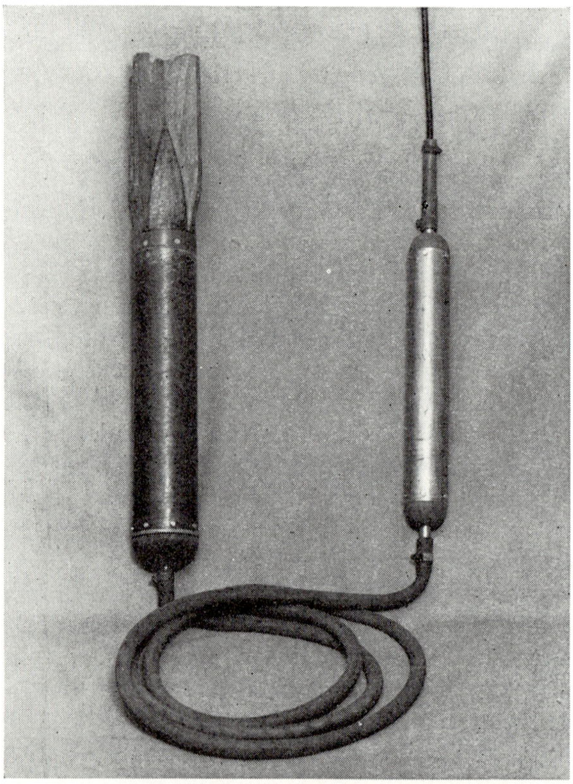

Fig. 3. Proton magnetometer for towing behind a ship.

550 cm^3 and is surrounded by a coil having 4220 turns of copper wire 0.6 mm in diameter. The fish is a cylindrical container with fins to render it stable. It is made of fibre-glass impregnated with an epoxy resin and is 25 cm long and 10 cm in external diameter. A current of 1 amp through the coil gives a polarizing field of 320 gauss.

The fibre-glass fish contains only the bottle and the coil. It is joined by a 3-m length of screened cable to a second fish which contains the pre-amplifier and the relay to break the polarizing current and connect the coil to the pre-amplifier. This second fish is of brass and is rather smaller than the first. By separating the relays from the bottle, their magnetic effect is made negligible. The coil is tuned to the proton precession frequency by a capacitor whose value can be altered from the ship by a stepping switch. The steps in resonant frequency are about 50 cycles. The Q of the coil is about 90.

Precautions must be taken to ensure that the polarizing current is reduced to a very small value when it is nominally turned off. If the field is to be measured to 1 γ, a polarizing field of 100 gauss must be reduced to 10^{-7} of its initial value; a single set of relay contacts cannot be relied on to do this; a second relay is, therefore, provided in the ship, and as an additional precaution the leads carrying the polarizing current are short-circuited by contacts in the fish when the current is off. If the polarizing current is broken sufficiently rapidly, considerable sparking takes place at the relay contacts and leads to a rapid deterioration in them. This may be avoided by placing a resistance

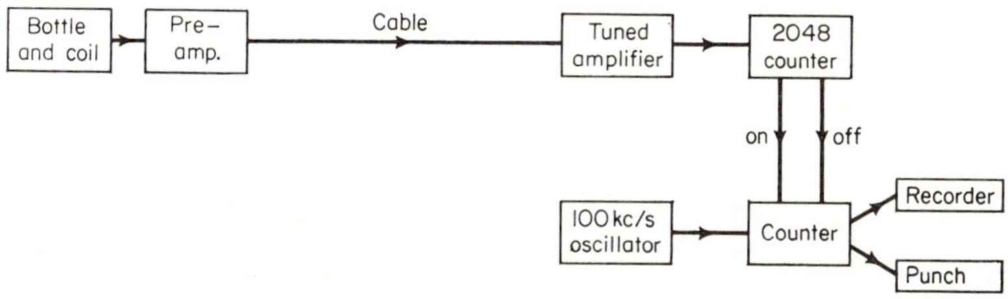

Fig. 4. Block diagram of proton magnetometer.

across the ship end of the cable leading to the coil or by providing a relay on the ship which reduces the current relatively slowly to about 10% of its initial value.

The pre-amplifier in the fish feeds into a cable running to the ship. This cable has six copper cores and one steel core to provide for the stress of towing. If the bottle is more than two ship's lengths astern of the ship the magnetic disturbance produced by most ships is less than 3 γ and can usually be neglected. The variation of the ship's field with course and distance has been discussed by Bullard and Mason (1961).

In the ship the signal is amplified by an amplifier with a band-width of about 100 c/s and fed to a binary counter which counts 2048 cycles of the proton precession. In order to allow the switching transients to subside, the counter is allowed to count for about a quarter of a second before the 2048 count starts. The beginning and end of the 2048 count start and stop a second binary coded decimal counter which counts the oscillations of an e.m.f. derived from a 100-kc/s quartz oscillator. If N be the number of cycles of the 100-kc/s oscillations counted in 2048 proton periods, the precession frequency is $2048 \times 10^5/N$ and the field, F, is given by:

$$F = 2048 \times 10^5 \times 2\pi/\gamma N = (48{,}102.2 \pm 0.4)/N.$$

In a field of 0.5 gauss the time of observation is about one second and the count.

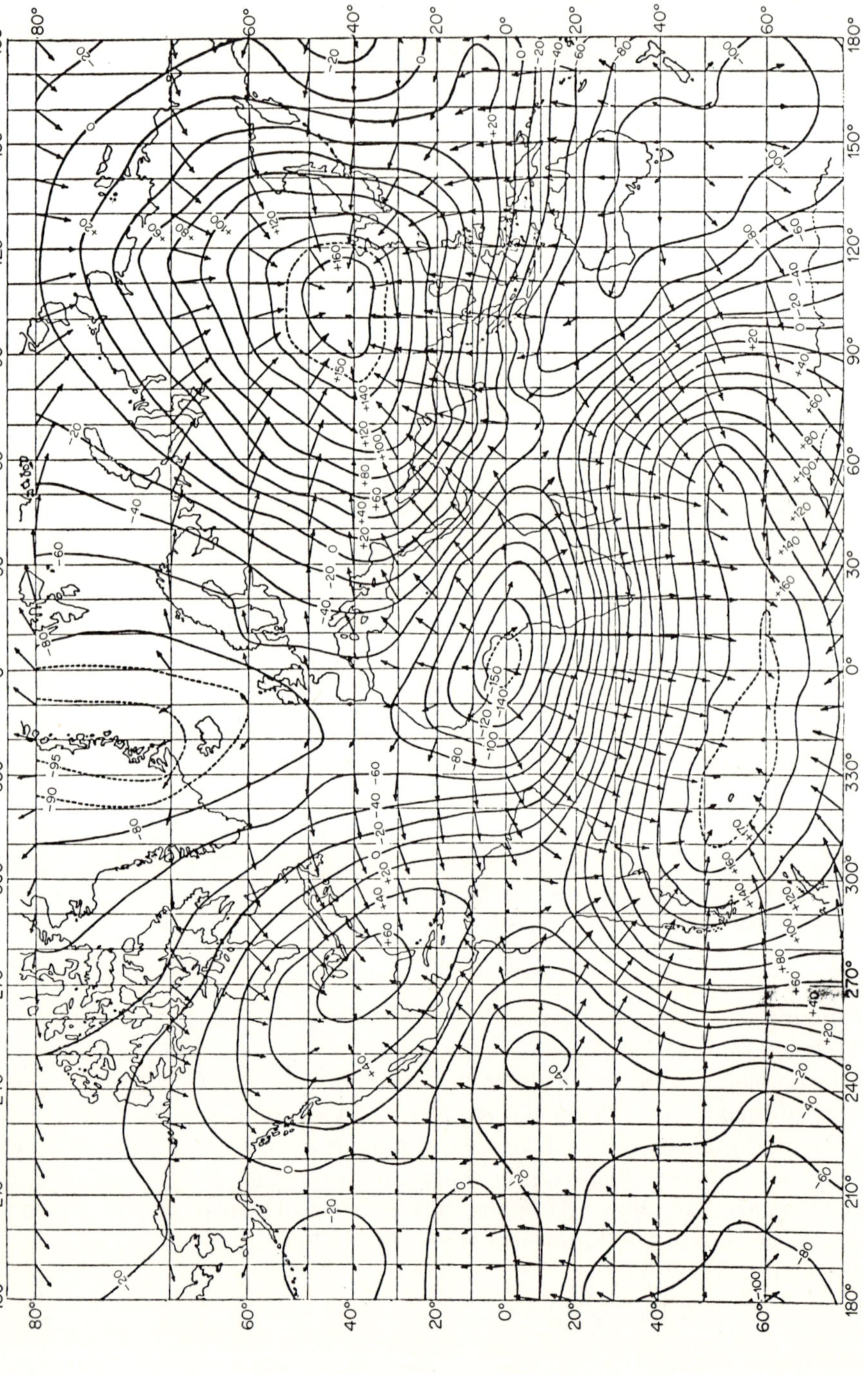

Fig. 5. The non-dipole field in 1945. The contours give the vertical field at intervals of 0.02 gauss. The arrows give the horizontal component on a scale of 70 mm/gauss. (After Bullard *et al.*, 1950.)

N, is about 100,000. A change of unity in the count then corresponds to a change of $\frac{1}{2}\gamma$ in F. If the possible accuracy is to be achieved, the gate that starts and stops the 100-kc/s count may be opened and closed consistently to 10^{-5} sec; that is, to 1/50 of a cycle of the proton signal. As the signal decays to about 30% during the count, the gate must be designed so that its times of opening and closing are not affected by the amplitude of the signal.

The last three digits of the count are recorded on a potentiometer recorder and all six digits on punched tape. A description of the punching arrangements has been given by Bullard (1961). The whole equipment is cycled once every minute, or once every half minute. The cycle is initiated by a clock driven by a precision oscillator.

3. General Features of the Field

Over most of the earth's surface the field is within 20% of that which would be produced by a dipole at the centre of the Earth. No analysis has been made for the land and the oceans separately, but there is no reason to suppose that there is any significant difference in the moment or direction of the dipole between land and sea. If the field due to the best-fitting dipole is removed, the residue is known as the *non-dipole field*. The three components of this for the year 1945 are shown in Fig. 5. The rate of change of the vertical component in 1942 is shown in Fig. 6.

Such diagrams use data collected over a long period and are the result of extensive smoothing and extrapolation both in time and space; they do not represent the immediate results of observation in the way that the magnetic survey of a single country does. The extent of their deficiencies is illustrated by measurements made by recent Japanese Antarctic expeditions (Nagata, Oguti and Kakinuma, 1958; Oguti and Kakinuma, 1959). They found that in the region between South Africa and the Antarctic continent the field is decreasing by 200 γ/year, which is about three times the rate of decrease given by current magnetic charts.

In spite of these uncertainties, Figs. 5 and 6 do demonstrate that the great geographical and geological differences between the continents and the oceans are not reflected in the world-wide pattern of the magnetic field. This is further demonstrated by Fig. 7, which shows a section from a chart of the total field in one of the areas where it is best known. The lines representing the smoothed field run from continent to ocean without any conspicuous disturbance. This broad similarity of the field over oceans and continents is to be expected if the main features of the field have their origin in the core of the earth, the outer surface of which is at a depth of 2900 km.

A possible exception to this general similarity has been suggested by Runcorn, who has pointed out that the centres of rapid change, shown in Fig. 6, tend to avoid the Pacific Ocean. It is possible that this is due to the temperatures beneath the Pacific being higher than elsewhere and producing a higher electrical conductivity in the mantle and a greater screening of the secular variation (Bullard, Maxwell and Revelle, 1956).

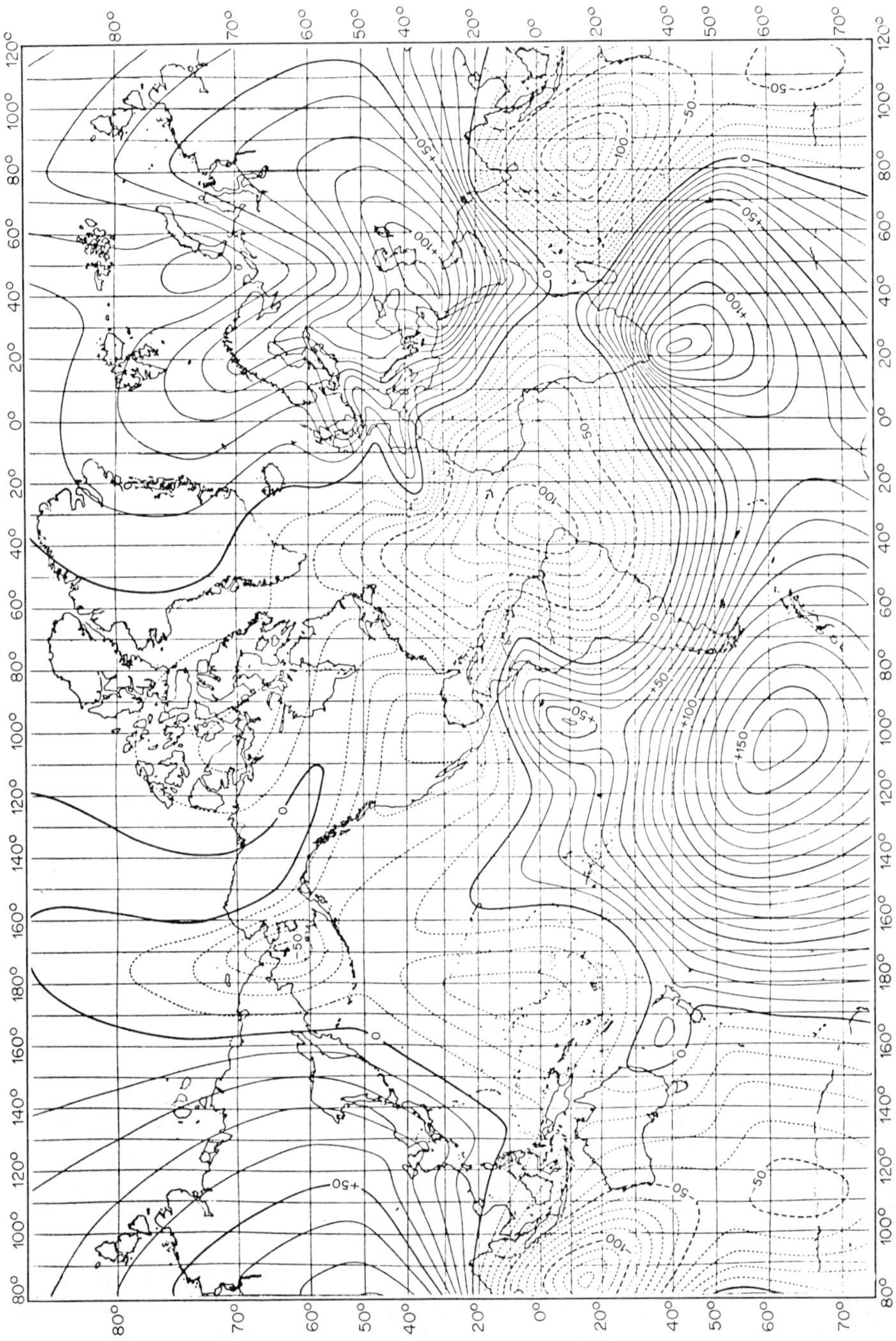

Fig. 6. The rate of change of the vertical component of the field in 1942. (After Vestine *et al.*, 1947, fig. 135a.)

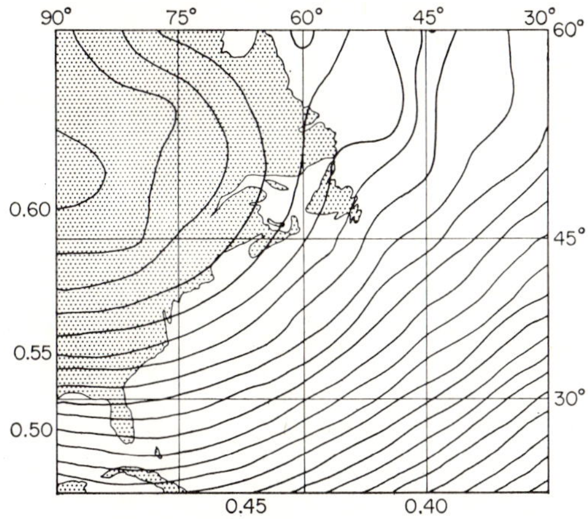

Fig. 7. Section from the U.S. Hydrographic Office chart of the total field in 1955 (No. 1703).

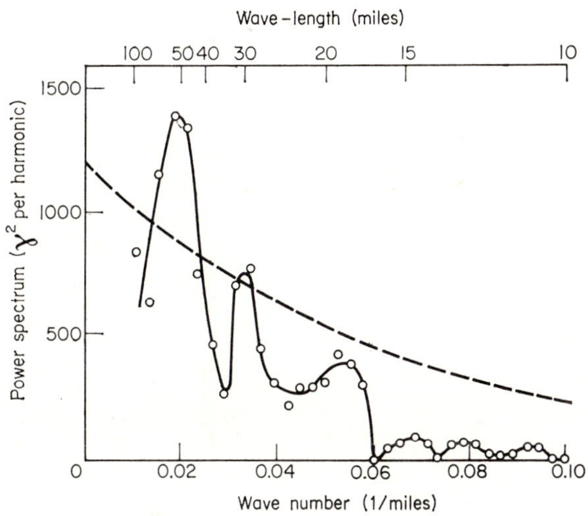

Fig. 8. Power spectrum of the magnetic field on a profile in the north-western Atlantic. The dashed line represents the spectrum which would be observed if the spectrum at the sea bottom were "white".

Since measurements at sea are taken at or above the surface of the sea and many thousands of metres above the ocean bottom, local anomalies will be smoothed and attenuated. If the anomalies are represented by a *space spectrum*, a component of wave-length λ will be reduced in the ratio exp $(-2\pi h/\lambda)$ if it is measured at a height h above the bottom. In water of 5000-m depth the attenuation is by a factor of over 2 for all wave-lengths below 14 km. There is, therefore, a very substantial loss of detail as compared to a survey conducted on land. A smoothed power spectrum taken by a ship along an east–west profile in a depth of about 2500 fm in the eastern basin of the North Atlantic is shown in Fig. 8. The profile is 379 miles in length and runs along latitude

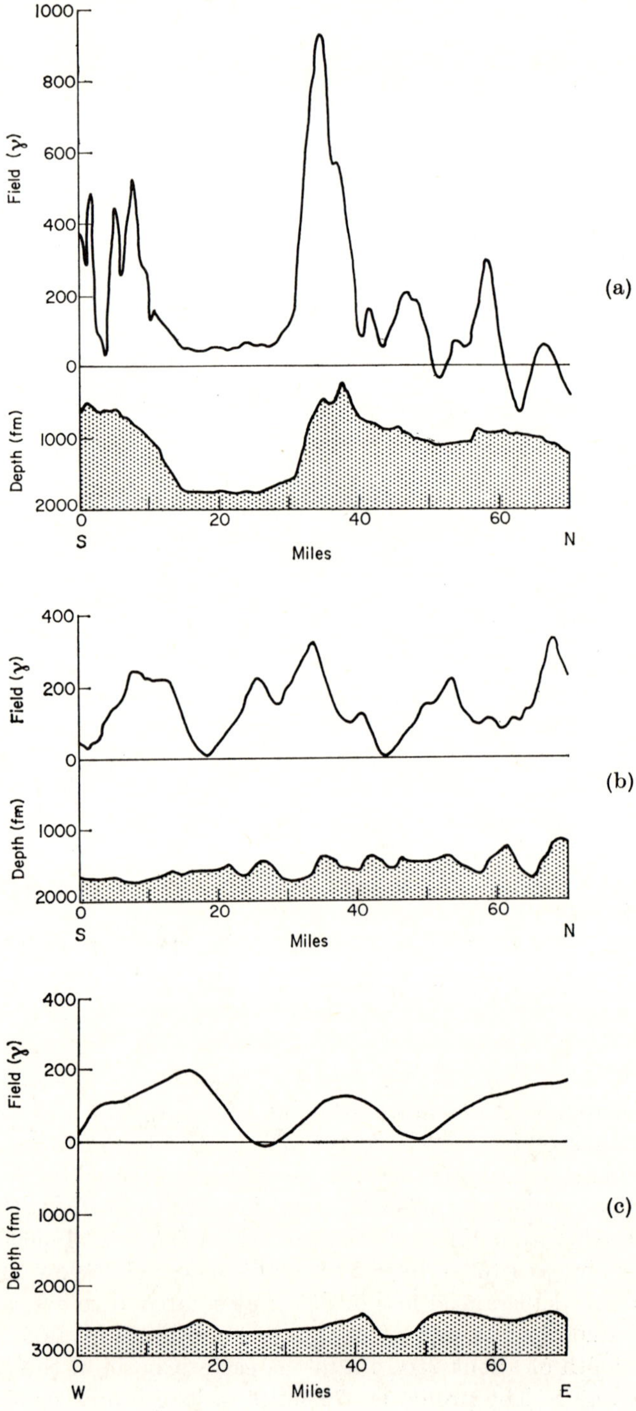
1000
800
600
400
200
0
Field (γ)
Depth (fm)
1000
2000
0
20
40
60
S
N
Miles
(a)
400
200
0
Field (γ)
Depth (fm)
1000
2000
0
20
40
60
S
N
Miles
(b)
400
200
0
Field (γ)
Depth (fm)
1000
2000
3000
0
20
40
60
W
E
Miles
(c)

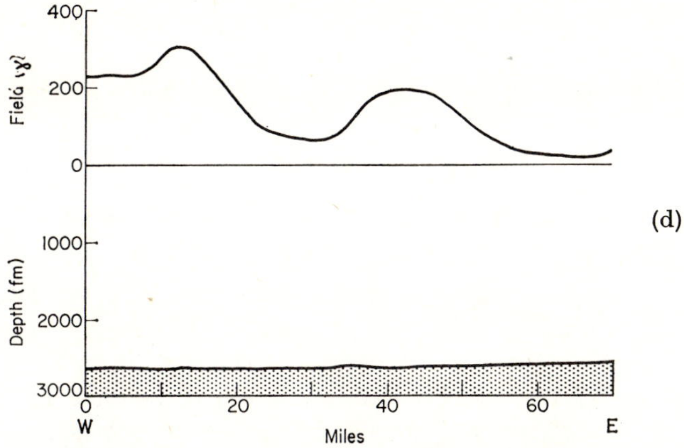

Fig. 9. Magnetic profiles and bottom topography in the North Atlantic.
(a) 37° 48′N, 25° 57′W to 38° 50′N, 26° 15′W.
(b) 44° 52′N, 27° 00′W to 45° 58′N, 27° 28′W.
(c) 46° 20′N, 20° 24′W to 46° 25′N, 19° 12′W.
(d) 45° 40′N, 06° 45′W to 45° 20′N, 05° 40′W.

$46\frac{1}{2}$°N from 17°W to 23°W. The dashed curve shows the spectrum that would be obtained if all wave numbers were equally represented at the sea bottom, that is, if the spectrum at the sea bottom were "white". The observed spectrum is, on the average, steeper than the dashed curve; thus measurements at the sea bottom would show a spectrum that falls towards the shorter wave-lengths. This might be due to the sources of the disturbance being buried beneath the ocean floor, or to the short wave-lengths being absent in the source.

The loss of detail is also shown in Figs. 9a, b, c and d, which show the total field over 70-mile sections of ship's track after the removal of a regional trend. Fig. 9a shows a section of track in comparatively shallow water over the Mid-Atlantic Ridge just north of the Azores; the field changes by nearly 1000 γ in a few miles and the changes are correlated with the bottom topography. Fig. 9b shows the field over the ridge in deeper water further to the north; the fluctuations are smaller and not so sharp. Fig. 9c shows the field over the approaches to the ridge in a depth of about 2500 fm (this is part of the line from which the spectrum shown in Fig. 8 was obtained). Fig. 9d shows a section over the Biscay abyssal plain where the bottom is flat and the sources are presumably buried beneath the sediment.

If the irregularities in the field are partly due to the inhomogeneities in the crust and partly to motions in the fluid core of the earth, it is to be expected that the spectrum would fall into two parts, a part due to the core with wave-lengths of a few thousand kilometres and a part due to the crust with wave-lengths up to some tens of kilometres. The intermediate wave-lengths might be expected to be of small amplitude since the mantle at depths below about a hundred kilometres will be above the Curie point and therefore non-magnetic. The fall at long wave-lengths seen in the spectrum in Fig. 8 may show the short-wave limit of this gap (Bullard *et al.*, 1962). The existence of the gap and its long-wave limit at about 2000 km is also suggested by a study of some profiles from Project Magnet by Alldredge *et al.* (1961), though the method of analysis does not allow any detailed conclusions to be drawn.

4. Reduction of Magnetic Observations

The magnetic observations may be affected by the magnetism of the ship and by time variations of the earth's magnetic field, as well as by the geology. If the magnetometer is more than two ship's lengths astern of the ship, the observed field will require no correction for the magnetization of the ship. This condition is easily realized in practice. Three types of time variations must be considered, the normal daily variation, magnetic storms, and the long-term secular variation.

The daily variation of the total field on land is usually of the order of 40 γ, and a correction for this is desirable. A correction depending on records from the nearest observatory on land may be used, but it is doubtful if this really improves the consistency of the observations. The induction of electric currents in the ocean will tend to reduce the daily variation at sea below that observed on land. The calculations of Chapman and Whitehead (1923) suggest that the reduction will be to about two-thirds but Rikitake (1961, 1962) has pointed out that the effect will be reduced if the mantle is a good conductor of electricity at a depth of a few hundred kilometres. Observations on oceanic islands sometimes show a small daily variation; for example, the mean daily variation of the total field at San Miguel in the Azores during the months of May, June and July, 1958, was 30 γ, whilst at Coimbra in Portugal, on almost the same latitude, it was 53 γ. A few measurements with a towed magnetometer, at a point between the two, suggested that the daily variation in the open ocean was nearer that in the Azores. However, data from oceanic islands must be regarded with caution because of local effects arising from the induced currents flowing in the ocean round them. Observations by R. G. Mason (unpublished) on the island of Oahu show that at points only 20 miles from the Honolulu magnetic observatory the amplitude of the daily variation can differ from that at the observatory by as much as 10%; on averagely disturbed days hour-by-hour differences of more than 20 γ are not uncommon. Anomalous effects might also be expected near the boundaries of oceans and continents. Observations by Hill and Mason (1962) suggest that in deep water off the end of the English Channel the daily variation is greater than it is in western Europe. The theory has been investigated by R. B. Roden (unpublished) who has shown that such an effect is to be expected, especially on the east side of an ocean. The increase is caused by the same current system as produces the decrease in mid-ocean.

A daily variation of up to 200 γ has been found at sea near the equator (Oguti and Kakinuma, 1959). This high value is related to the equatorial electrojet system of the ionosphere. The effect of the electrojet is at a maximum near the magnetic equator and dies away within a few hundred kilometres on either side. The rapid changes of field are confined to the daylight hours. Within an area affected by the electrojet, short-period fluctuations in times of the order of one hour and with amplitudes of up to 100 γ may occur at any time, but particularly during the hours of daylight. There is no way of distinguishing these time variations from spatial variations, other than by a resurvey or by having a second magnetometer in a moored buoy. Similar complications arise in the auroral zones. Experience suggests that, except in these special areas, the error incurred by neglecting the normal daily variation may be acceptable.

Magnetic storms present a more serious problem. It is doubtful if a useful correction can be made using data from a land station and the only way of eliminating the uncertainties is to reject observations taken when the level of disturbance at a land station exceeds some arbitrary limit, or to provide a magnetometer in a moored buoy to determine the correction. Electric currents induced in the ocean will modify the storm-time fluctuations, particularly the high-frequency components, in much the same way as they modify the daily variation, but the extent of this modification is unknown. A number of magnetic storms occurred during the survey of the north-east Pacific described in a later section, but without any significant effect on the contours (which were plotted at an interval of 50 γ). The effects of magnetic storms and daily variation would be more serious in more detailed surveys.

A correction for secular variation may be necessary if the survey of an area is spread over more than a few months. The rate of change of the field can be taken from neighbouring land stations or from magnetic charts. A single value can usually be used for an area up to 1000 km across.

In order to plot anomalies in an intelligible form it is necessary to remove the regional trend. Since the removal of a complicated or irregular trend may introduce spurious features into the pattern of anomalies, it is desirable that a simple and easily understood method should be used. A linear variation with distance on a Mercator chart is often satisfactory over areas a few hundred miles across. In the magnetic survey of the north-east Pacific a second-order polynomial in latitude and longitude was fitted to the observed field, averaged within 20′ rectangles; a third-order polynomial has been successfully used in an area $40° \times 40°$ in the north-western Indian Ocean. A linear variation plus a two-dimensional Fourier series was satisfactory in a $20° \times 20°$ area in the Atlantic (Bullard *et al.*, 1962) but cannot be generally recommended since the Fourier series necessarily has the same values on two opposite edges of the area, which often prevents a good fit being obtained with a few terms. The determination of trend from observations within an area is in a large degree arbitrary and it is to be hoped that Project Magnet will so improve our knowledge of the field that a spherical harmonic expansion of the whole earth can be used to give the trend in the areas covered by detailed surveys.

When the ship is in range of accurate navigational aids the measurements at the crossing points of tracks usually agree to within 20 γ, but after a few hours' steaming in the open ocean, discrepancies of 50 γ are not uncommon.

5. The Magnetic Properties of Submarine Rocks

The geological interpretation of magnetic measurements depends on the large differences that exist between the magnetic properties of different types of rock. Basic rocks in particular tend to be more magnetic than others, and any situation involving local excesses or deficiencies of basic rocks will lead to anomalies in the earth's magnetic field. Little information has been published on the magnetic properties of submarine rocks; this is partly because of the lack of representative material to study.

In the interpretation of marine surveys it is often found necessary to take larger values for susceptibility than those given for continental basalts (by, for example, L. B. Slichter, in Birch *et al.*, 1942). This is justified on the grounds that basic igneous rocks commonly possess a permanent, or remanent, magnetization several times greater in intensity than that induced in the earth's field; if the remanent magnetization is approximately aligned in the direction of the field, then the rock possesses an apparent susceptibility several times greater than the measured value. Thus susceptibilities ranging up to 0.015 e.m.u. have been used for basalts. At the other extreme, the unconsolidated sediments of the sea floor are generally thought to be too weakly magnetized to produce measurable anomalies at the surface.

A recent study of some basalts dredged from the escarpment of the Mendocino Fault in the north-east Pacific (R. G. Mason, unpublished) has revealed exceptionally high intensities of remanent magnetization, ranging up to 0.3 e.m.u./cm^3, and correspondingly large ratios of remanent to induced magnetization. The former, expressed in terms of apparent susceptibility, range from 0.003 to 0.6 with a median value of 0.035, the latter from 2 to 400, with a median value of 20. The results are presented in Fig. 10 as a histogram which shows, for comparison, the results of an investigation of more than 600 lavas from Mull. The corresponding figures for the Mull lavas are 0.001 to 0.5, with a median value of 0.010, for apparent susceptibility, and 0 to 130, with a median value of 3, for the ratio of remanent to induced magnetization. The lavas from the Mendocino Fault are thus, on an average, more than three times as magnetic as those from Mull.

Not all submarine lavas are so highly magnetic. Much of the material obtained in dredge hauls is badly weathered and, as might be expected, is considerably less magnetic than the average unweathered rock. Matthews (1961) has found apparent susceptibilities ranging from 0.002 to 0.025, with a median value of 0.010, for weathered basalts from the eastern North Atlantic; similar results have been obtained for weathered basalts from the Pacific.

The relation between the magnetic properties of the Mendocino Fault basalts and their mineralogy has been investigated by J. M. Hall (unpublished). In general, intensities of remanent magnetization increase with decrease in grain size of the magnetic minerals, particularly in the range 50 μ–5 μ. With the exception of the most magnetic of the specimens examined, which is rich in maghemite (γ-Fe_2O_3), these are mostly impure magnetites less than 50 μ in

diameter; it is probably because of the small grain size of the magnetic minerals that these rocks are so magnetic. Since rapid cooling encourages fine grain size, submarine lavas may be generally more magnetic than their continental counterparts. The occurrence of maghemite is interesting. Maghemite is a highly magnetic mineral believed to be generally formed from magnetite, either by direct oxidation in a low oxygen environment or via hydrated compounds, possibly by hydrothermal solutions at relatively low temperatures and pressures (Mason, 1943). It is metastable and at higher temperatures and pressures it is changed irreversibly to haematite (α-Fe_2O_3); it is possible that a deep-ocean environment favours oxidation of magnetite to maghemite rather than to the less magnetic haematite.

In common with other ferromagnetic materials, a rock becomes non-magnetic when heated to a temperature exceeding its Curie point. The latter varies from rock to rock, depending on the magnetic minerals involved, but is generally

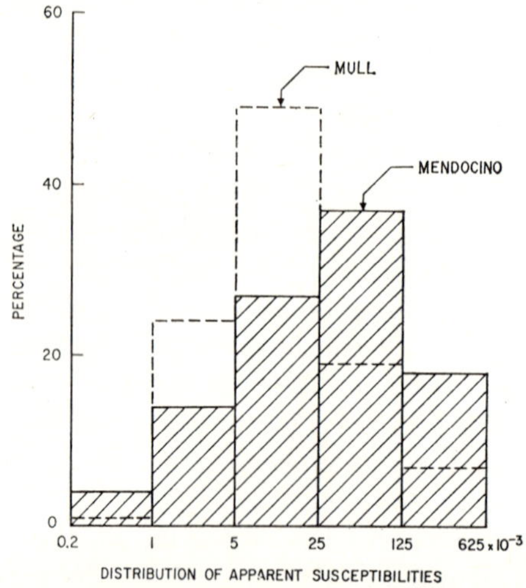

Fig. 10. The distribution of apparent susceptibilities of (a) basalts dredged from the Mendocino escarpment and (b) basaltic lavas from Mull. (From R. G. Mason, unpublished.)

less than 575°C. Since temperatures within the earth increase with depth, there will exist everywhere a surface of magnetic discontinuity corresponding to the Curie point isotherm. There is some uncertainty as to the depth at which the Curie point might be reached; on the basis of present information it seems reasonable to assume that it will be between 20 and 50 km below the sea floor, and, therefore, well below the base of the oceanic crust. Lateral variations in its depth will lead to anomalies in the magnetic field, which, therefore, provide a means for studying the thermal state of the upper mantle.

6. The Magnetic Surveys of the North-east Pacific

Much of the magnetic data obtained at sea is in the form of long unrelated profiles recorded during passage of a ship. These are of limited usefulness

since they give no indication of the lateral extent or direction of anomalous magnetic trends, and, therefore, provide no basis for quantitative geological interpretation. For example, isolated lines across the area covered by the magnetic map of the north-east Pacific (Fig. 11) would give no information about the magnetic trends; the general impression of the magnetic character of the area that would be gained from a study of such lines would depend entirely on their orientation. If the geological pattern is to be revealed by a magnetic survey, the lines along which measurements are made must be close enough together to allow a contoured map of the field to be drawn.

Experience shows that if the sharpest anomalies, arising from sources close to the ocean floor, are to be adequately resolved, the separation between adjacent lines must not exceed about half the depth of the water. Satisfactory contouring requires a relative positional accuracy between adjacent lines of about 20% of the spacing, i.e. about 10% of the depth, or about 500 m over the deep ocean. Over a small area buoys can be moored and the survey conducted relative to them. In surveys covering a larger area, elaborate radio-navigation systems are necessary. Even then, the operational range is restricted to a few hundred kilometres from land, though systems are available which give ocean-wide coverage to an accuracy of within a few kilometres.

The survey conducted by the Scripps Institution of Oceanography, in conjunction with the U.S. Coast and Geodetic Survey, off the west coast of the United States (Mason, 1958; Mason and Raff, 1961; Raff and Mason, 1961) represents the first attempt to make a detailed magnetic map of an extensive area of the oceans. The results are of exceptional interest in that they reveal major structural trends of which there is little or no indication in the topography, and they provide evidence for unsuspected horizontal displacements along some of the faults of the north-east Pacific greater than any that have so far been observed over the continents.

The area covered by the survey lies off the foot of the continental slope along the Pacific coast of the United States; it is 400–500 km in width and extends off-shore from the north Mexican border in the south to the southern end of Queen Charlotte Island in the north. It was completely covered with a regular grid of lines, mostly about 8 km apart, which were positioned with a probable error of 100 m by means of a radio-navigation system. The survey itself was made with a highly stabilized fluxgate magnetometer which, by frequent comparison with a proton precession magnetometer, provided a measure of the total field with a standard deviation from the absolute value of about 15 γ; over periods of a few hours the stability was better than 1 γ.

The magnetic anomaly map, Fig. 11, is a plot of the observed field from which the regional field has been subtracted. The latter was derived from the map of the observed field by a smoothing process. No attempt was made to remove the effect of diurnal and other time variations of the earth's magnetic field as these were thought to have no significant effect on the final plot.

The general features of the bathymetry have been described by Menard (1955). The area is crossed by two of the four great, almost east–west, fault

Fig. 11a. Map of the total magnetic intensity from 32°N to 42°N after removal of the regional field. The contour interval is 50 γ, and spot values are rounded off to the nearest 5 γ beyond the extreme value recorded. (From Mason and Raff, 1961.)

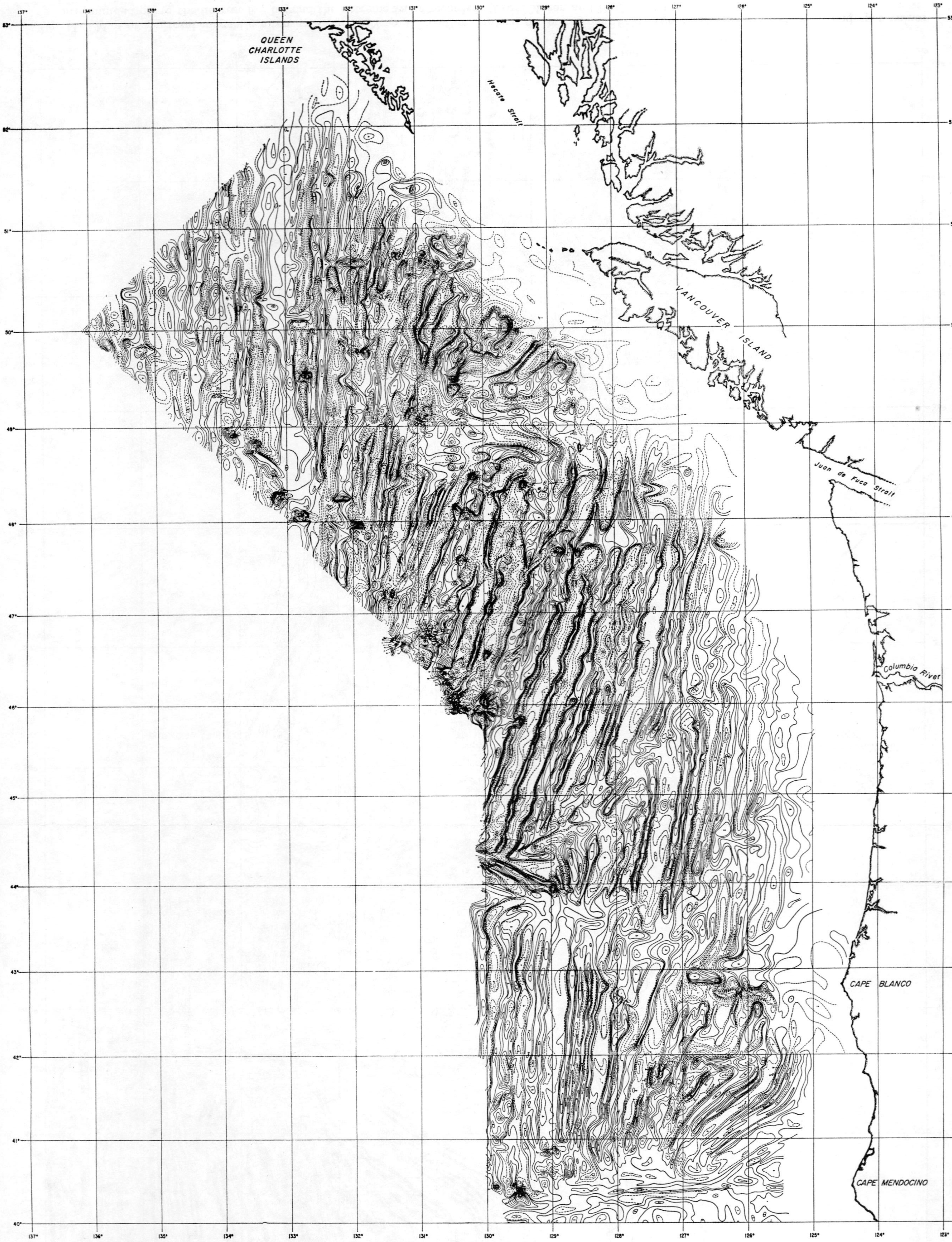

Fig. 11b. Map of the total magnetic intensity from 40°N to 52½°N after removal of the regional field. The contour interval is 100 γ, and spot values are rounded off to the nearest 5 γ beyond the extreme value recorded. (From Raff and Mason, 1961.)

zones of the north-east Pacific, the Mendocino and the Murray. The former is by far the most spectacular topographic feature in the area; the bottom rises about 1 km from south to north across it and over most of the map it forms an asymmetrical ridge with a south-facing escarpment 5000–10,000 ft ($1\frac{1}{2}$–3 km) high. It marks the boundary between the deep plain of the north-east Pacific, a comparatively smooth abyssal plain 2100–2600 fm (4–5 km) deep, and the shallower ridge and trough province, an area of narrow north- or north-east-trending ridges and troughs, to the north. The Murray Fault, about 6° to the south, divides the deep plain from the slightly shallower Baja California Seamount Province. A relatively minor topographic feature, the Pioneer Ridge, crosses the deep plain about 180 km south of the Mendocino Fault. Fig. 12 shows the relation of these features to the area covered by the map, Fig. 11.

The change in elevation across the Mendocino Fault suggests a major change in crustal structure, and this is confirmed by seismic refraction measurements. On the south side of the Mendocino Fault the seismic results are consistent with a normal oceanic crust, consisting of up to a few hundred metres of unconsolidated sediments, a "second layer" 1–2 km thick and generally thought to consist of volcanic rocks and consolidated sediments, and the main, presumably basaltic, crustal layer about 5 km thick. The Mohorovičić discontinuity lies at a depth of about 11 km below the surface of the sea. On the north side, where the bottom is shallower, the crust appears to be thinner than it is to the south, which is contrary to what would be expected if opposite sides of the fault are in isostatic equilibrium; however, since velocities beneath it are rather less than what is considered normal for the mantle, there is some doubt as to whether the Mohorovičić discontinuity has in fact been reached. It may be noted that the line of the Mendocino Fault marks the approximate boundary between various physiographic provinces of the neighbouring continent (Eardley, 1951, p. 430).

One of the most impressive features of the map is the vastness of the area covered by what is essentially a single, homogeneous, north–south-lineated magnetic pattern. This pattern is known to extend for more than 2500 km in the north–south direction and although its western limit has not yet been determined, it probably extends for a like distance to the west of the coast. It is, therefore, a major feature of the earth's surface.

At the known faults, and at other places for which there is otherwise no evidence for faulting, the pattern of the anomalies, though generally not their direction, changes abruptly. At the Murray Fault a 200-km section of the pattern on the north side can be matched, with a fair degree of certainty, against a similar section on the south side, in a way suggesting a right-lateral displacement of about 150 km across it. There is no obvious match of the pattern across either the Pioneer or the Mendocino Faults, within the limits of the map. However, the fortunate fact that the lineations of the field are approximately normal to the faults makes it possible, with comparatively little effort, to extend the map in the vicinity of these faults for the purpose of looking for such a match. This has been done (Vacquier, 1959; Vacquier *et al.*, 1961; Raff,

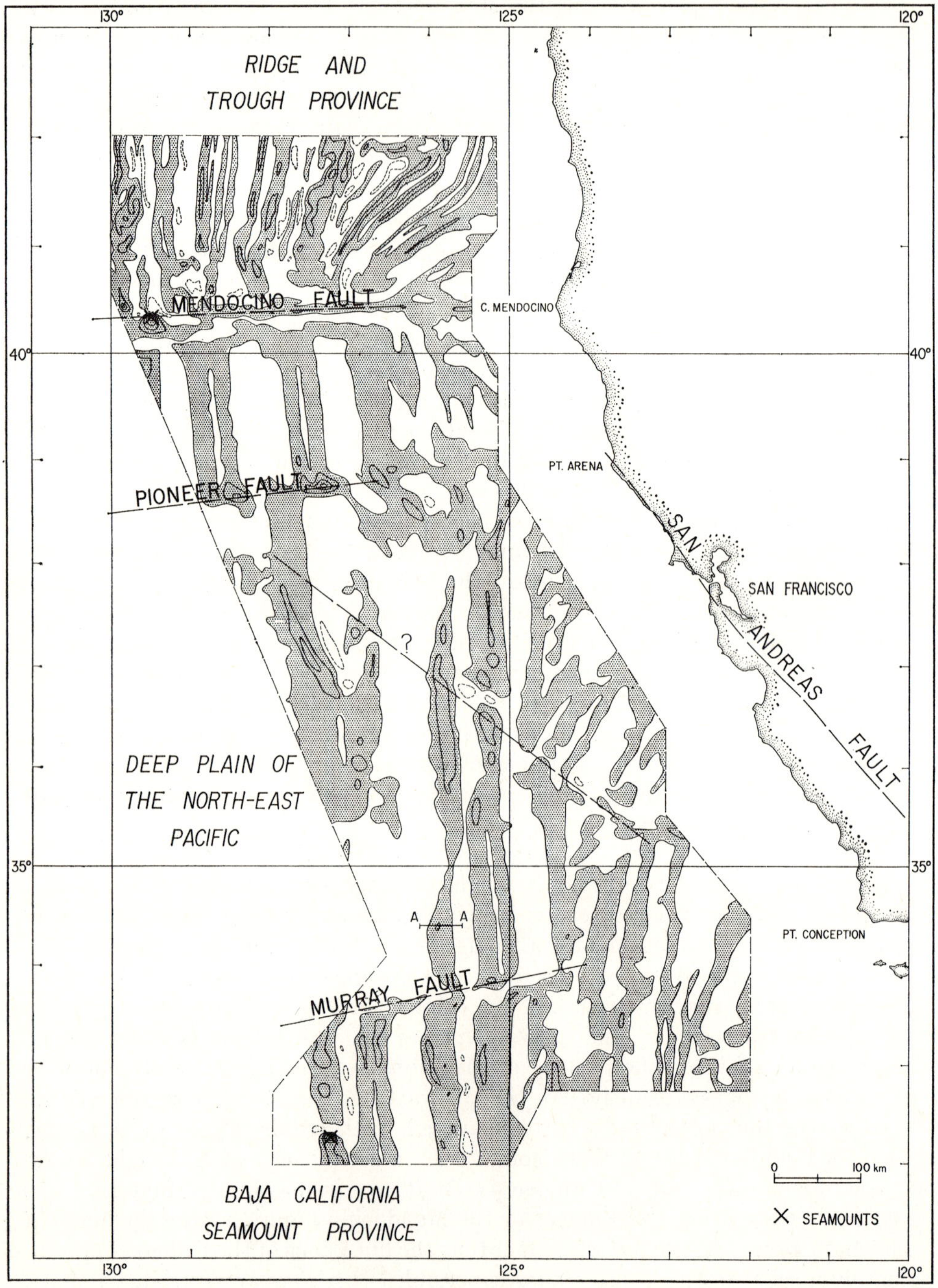

Fig. 12a. Index map to Fig. 11a. The shaded areas are positive and the contour interval is 250 γ. (After Mason and Raff, 1961, fig. 1.)

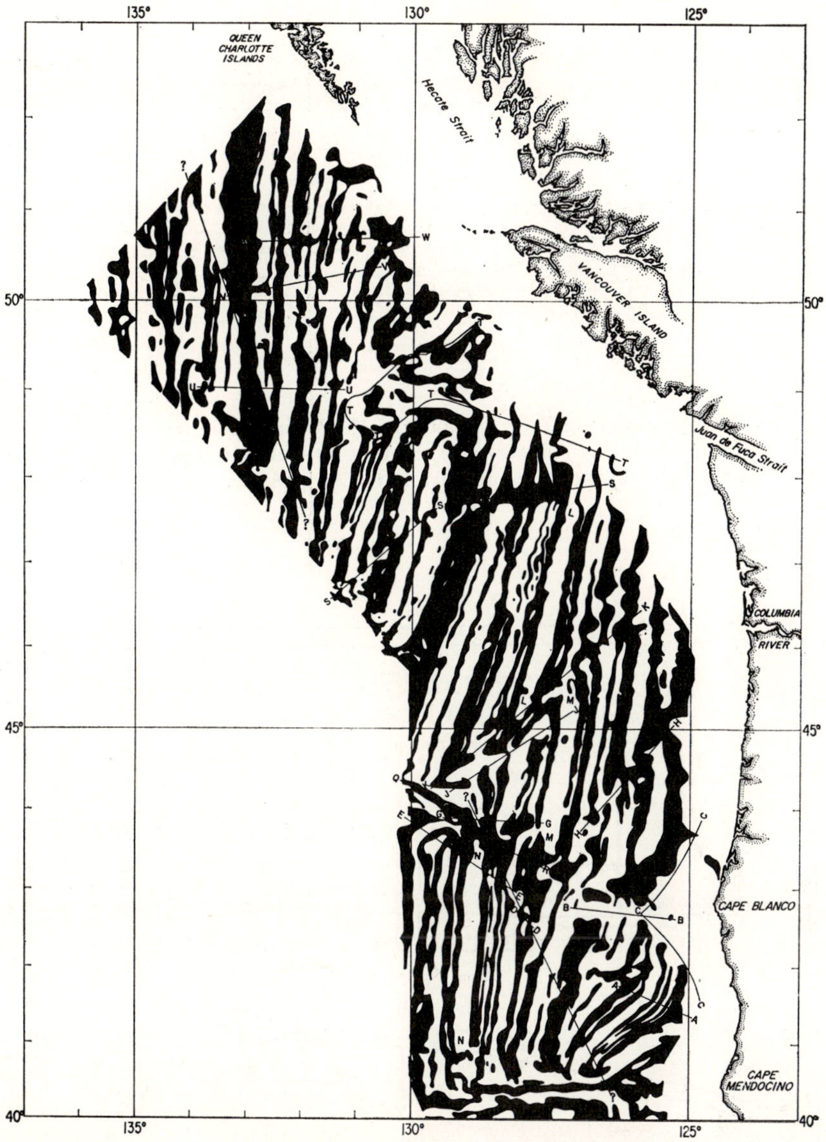

Fig. 12b. Index map to Fig. 11b. The shaded areas are positive and the contour interval is 500 γ. (After Raff and Mason, 1961, fig. 1.)

1962). Fig. 13 shows profiles from both sides of both faults; these have been displaced relative to one another to give what is judged to be the best overall match. They are interpreted as indicating a left-lateral slip of about 265 km across the Pioneer Fault and about 1185 km across the Mendocino, making an overall left-lateral displacement of about 1450 km. The quality of the match

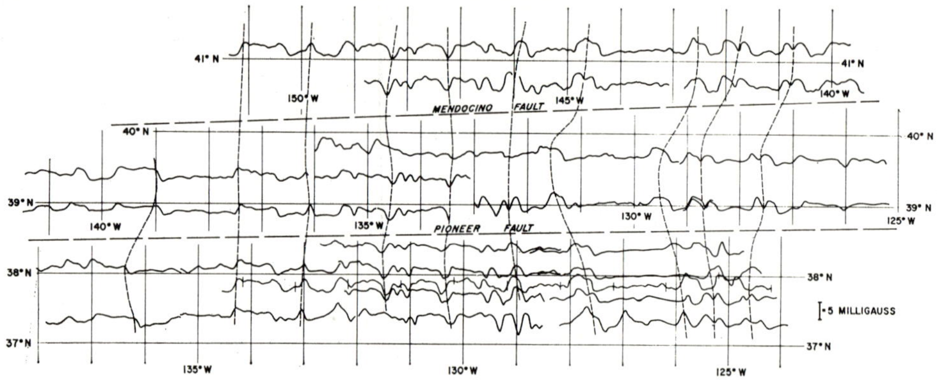

Fig. 13. Magnetic profiles from both sides of the Mendocino and Pioneer faults, displaced laterally to give what is judged to be the best overall match. (After Vacquier *et al.*, 1961, fig. 4. By courtesy of the Bulletin and the Scripps Institution of Oceanography.)

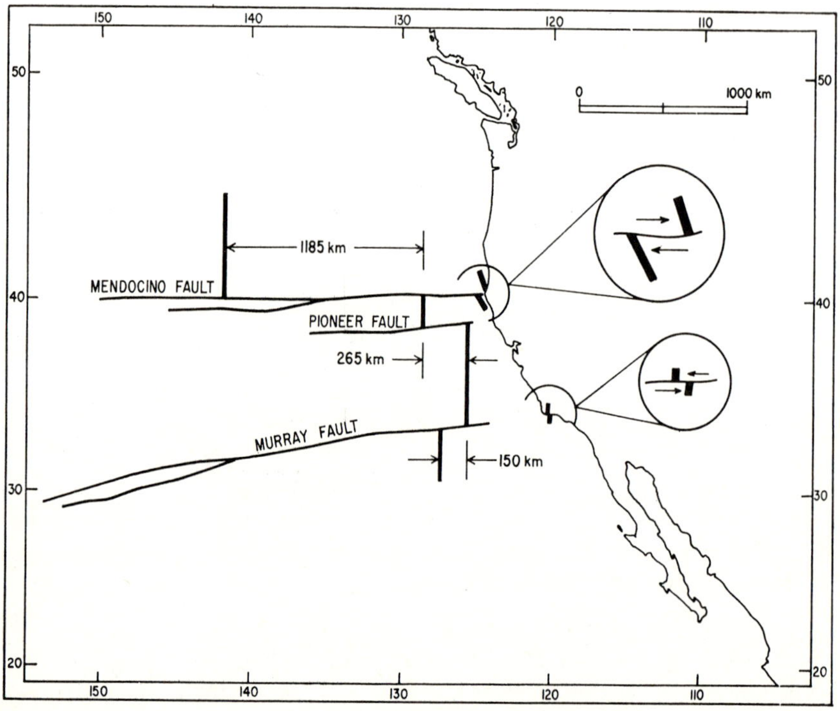

Fig. 14. Summary of the lateral displacements across the Mendocino, Pioneer and Murray faults. (After Menard, 1960.)

leaves little doubt that these displacements represent real displacements of the crust, or at least of those layers with which anomalies are associated. Fig. 14 summarizes the displacements across the three faults. A summary of displacements on large wrench faults on land and at sea has been given by Menard (1962).

A subject almost as intriguing as the lateral faults is that of the origin and significance of the magnetic pattern. The lineations of the field are remarkably straight over very large distances. The principal group of north–south-trending anomalies, for example, strikes almost exactly 5°W of N from the southern edge of the map to the Mendocino Fault, a distance of 900 km. They are also

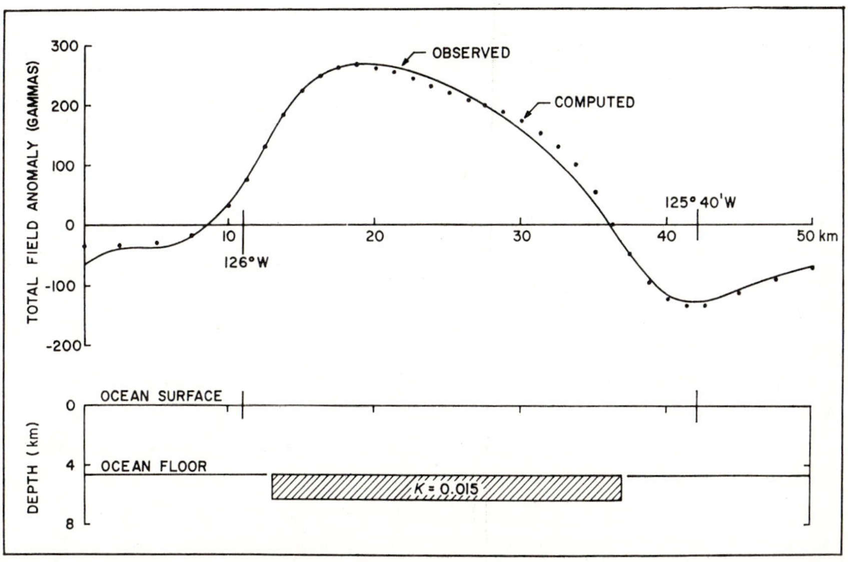

Fig. 15. Magnetic profile along line A–A of Fig. 12a and its interpretation in terms of an infinite north–south slab.

remarkably uniform in magnitude and in width. As far as can be judged from the best available topographic maps, over most of the area the pattern shows little correlation with topography, and there is no indication of any systematic differences between seismic velocities in the anomalous areas and in the areas between, though the amount of data is not yet sufficient to demonstrate this point conclusively.

As regards their immediate cause, the anomalies are such as might be expected of flat, slab-like structures approximately underlying the areas of positive anomaly. Fig. 15 shows the observed anomaly along line A–A of Fig. 12a and superimposed on it the computed anomaly for an east–west crossing of an infinite north–south slab immediately underlying the ocean floor. The agreement is striking.

Although magnetic data are not capable of unique interpretation, some general conclusions can be reached about these slabs. In the first place, even if they were at the shallowest possible depth in the crust, and if they were of a highly magnetic material such as basalt, they would have to be almost 2 km thick in order to account for the magnitudes of the anomalies. At the most, their upper surfaces could lie no more than about 1 km below the sea floor, otherwise they would not produce anomalies of the observed sharpness. They would have to lie at least partly in the "second layer" of the crust. They could be thicker slabs, extending to the base of the crust; in that case their upper surfaces would have to be even shallower.

The geological possibilities fall into three categories: (1) isolated bodies of magnetically anomalous material, for example basic lava flows, within the "second layer", (2) elevated folds or fault blocks of the main crustal layer, and (3) zones of intrusion of highly magnetic material from the mantle. The first would fit in well with current views about the somewhat enigmatic "second layer"; for the second, one could draw a parallel with the block-fault topography of the neighbouring Basin and Range Province. Objection could be raised to all three on the grounds of lack of topographic and seismic expression; these objections would be removed if the rocks involved were as highly magnetic as those from the Mendocino Fault discussed earlier, in which case the amount of rock involved would be correspondingly less. Fig. 16 illustrates the three cases in terms of two-dimensional structures each of which has been adjusted by trial and error to fit the anomaly exactly (Mason, 1958).

A more fundamental question is, how did such an orderly pattern come about in the first place? There seems to be little doubt that the magnetic structures are related in some way to lines of weakness in the oceanic crust; in view of the approximately orthogonal relationship between them, one might suspect that the pattern and the faults had a common origin. However, from the closeness with which the pattern on opposite sides of the fault can be matched, and in the absence of any magnetic structures cutting across them (except towards the foot of the continental slope, where there is evidence for complex faulting with north–south components of slip), it must be concluded that the faults are more recent than the pattern; in that case, there seems to be no good reason why they should be related in any way.

Irrespective of how the displacements came about, the oceanic crust seems to be sufficiently rigid for sections of it covering hundreds of thousands of square kilometres to be capable of moving distances of the order of hundreds of kilometres relative to one another with comparatively little distortion. Although some blocks have suffered rotation, the magnetic-anomaly map shows little evidence of distortion within individual blocks, except in the fault zones and near the eastern edge of the map. However, Menard (1959) has drawn attention to a distinctive pattern of minor lineations, characteristic of tension, which borders the faults, suggesting that all the crustal blocks have been under tension at one time or another. The systematic deflection of the lineations in a clockwise sense near the eastern edge of the map south of the Mendocino Fault, and the sharp flexures of the pattern immediately north of the fault, suggest a right-lateral movement of the continent relative to the oceanic crust. This is in the same sense as the movement along the San Andreas Fault, and along the tentative NW–SE fault indicated in Fig. 12a.

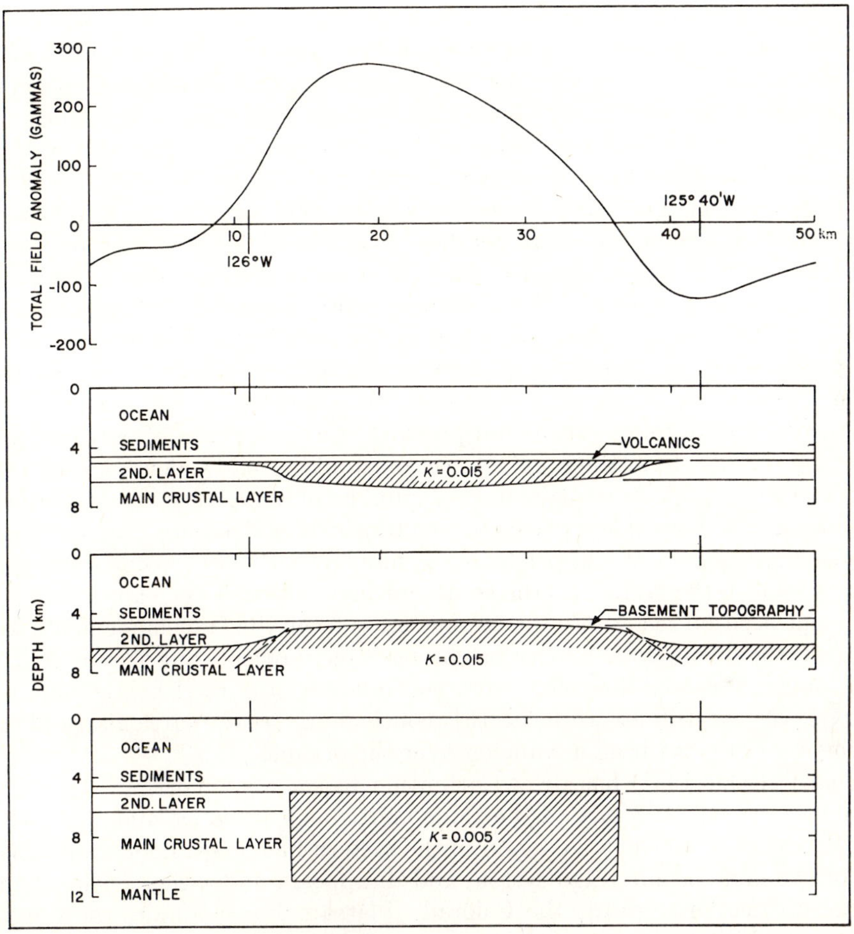

Fig. 16. Three possible interpretations of the magnetic profile along line A–A of Fig. 12a, in terms of two-dimensional sections adjusted to fit the anomaly exactly.

Two further facts must be taken into account in any hypothesis about the origin of the crustal displacements. The first is that they are not matched on the adjacent continent. Where the Mendocino Fault intersects the continental slope there is a right-lateral offset of about 100 km, but this is in the opposite direction to the displacement of the magnetic pattern. There is no offset of the continental slope along the line of the Murray Fault, but small east–west strike-slip faults occur in the Santa Barbara Channel islands and in the transverse ranges of California which are thought to mark an extension of the fault zone on land. However, again the displacements are generally in the opposite sense to the magnetic displacements. From what is known of the geology of North America, it can be said with some certainty that no east–west strike-slip displacements of the required magnitude can have occurred anywhere at any time since late Pre-Cambrian.

The second fact is that there is, on the north side of the Mendocino Fault, a strip of the oceanic crust over 1000 km wide from the foot of the continental slope outwards for which there is no corresponding pattern on the south side of the fault. Further, although there is some evidence for a general deepening of the magnetic structures towards the continental slope, over such parts of the continental shelf as have been mapped the field appears to be relatively featureless and, although the east–west anomaly associated with the Mendocino Fault can be followed shoreward, there is no evidence that the north–south structures continue under it. Even if their depth increased several-fold, they should still produce an observable effect at the surface. An explanation of the disappearance of the magnetic pattern under the continent might be that although the structures continue under it, the pattern has been erased by the combined effects of temperature and pressure; the conversion of maghemite to haematite referred to in the previous section would provide such a mechanism.

These facts might be explained in a number of ways: (1) the major fault movements may have taken place at some time earlier than late Pre-Cambrian; the displacements in the opposite sense may reflect more recent movement along the line of the faults; (2) the continent may not have been anywhere near its present position relative to the oceanic crust when the fault movements took place; (3) the displacements may be comparatively recent, but they may have taken place so slowly that they were continuously absorbed by tectonic and other processes still at work in the continental crust, particularly if the continental crust has been advancing over the oceanic.

Menard (1960, 1961) has drawn attention to some similarities between the anomalous crustal structure of the ridge and trough province and that of the East Pacific Rise, an elevated region of the eastern Pacific sea floor which forms part of the mid-oceanic ridge system and disappears under the continent in a northerly direction towards the Colorado Plateau. For example, the crust is apparently thin and seismic velocities abnormal in both areas, and along the crest of the East Pacific Rise and in parts of the ridge and trough province heat-flow values are abnormally high. Menard seeks to link these facts by the hypothesis of a rising convection current which produces high heat flow, uplift, tension cracks, lateral movement and thinning of the crust by normal faulting, and differential stresses, leading to the strike-slip faults. The tension cracks and normal faults might have controlled the magnetic structures.

If the East Pacific Rise and the ridge and trough province were once part of a single continuous structure, then the strike-slip displacements must continue under the continents, presumably within the deeper layers. The explanation of the change in crustal structure across the Mendocino Fault might then be that transcurrent faulting has brought dissimilar parts of the crust opposite one another. Study of the way in which the more than 1000 km of unmatched magnetic pattern on the south side of the Pioneer Fault between about 135° and 145°W is absorbed in the westward direction, and the discovery of similar patterns in other parts of the oceans, would do much towards narrowing down the range of tenable hypotheses.

7. The Continental Shelves

The continental shelves are structurally more closely related to the continents than to the deep ocean. Magnetic surveys over them may be expected to show the same features as do those on land and to lead to the same ambiguities in interpretation. Plots of the field frequently show well-marked linear trends, such as those shown in Fig. 17, but it is impossible to say from magnetic measurements alone whether the regions of high magnetic intensity are regions where the basement is shallower than it is elsewhere, or whether the magnetic trends represent the trends of intrusions of highly magnetic material within the basement. Sometimes the "roughness" of the magnetic plot may give a clue that the basement is shallow, but numerical estimates of depth are not to be relied on. In fact, seismic work suggests that the area of large anomalies shown in Fig. 17 to the west of Brittany is an area where the basement is shallower than it is farther to the north, where the magnetic plot is relatively featureless (Day *et al.*, 1956).

Even if the cause of the anomalies is not known in detail, a magnetic map may show trends that run parallel to other features of the area; Fig. 17, for example, shows persistent ENE–WSW trends parallel to the centre line of the English Channel, the folding of south-west Ireland and the line of bathyliths from Dartmoor to the Scillies. The west coast of Europe is intersected or approached by folding of three periods, the Caledonian, the Hercynian and the Alpine; when the area of survey has been extended, it is to be hoped that their course and direction across the continental shelf can be plotted. It would be of particular interest to know whether the Caledonian trend can be traced to the north and west of Ireland and whether the Alpine folding in the south of Spain sends a branch to meet the rather vague ridge and line of earthquake epicentres that runs from the Azores towards Gibraltar.

Perhaps the most important use for the magnetometer on and near the continental shelf is to determine whether the structural lines of the continents run out into the deep ocean. This can best be investigated in an area such as western Europe where powerful structures strike out more or less perpendicular to the continental margin. Many other areas, such as that to the north of Newfoundland, should also be investigated. Whether the structural lines from the continents persist into the oceans or are truncated at the continental margin is one of the fundamental questions of geology; unfortunately, the magnetic evidence is at present insufficient to decide it.

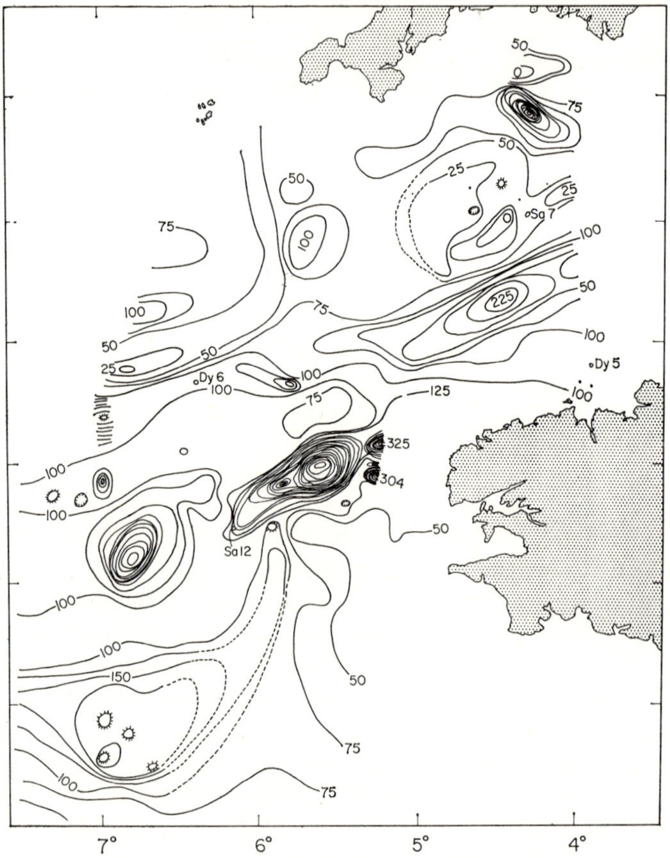

Fig. 17. Magnetic anomalies to the west of the English Channel. Contour interval 25 γ. A linear trend has been removed. (From T. D. Allan, unpublished.)

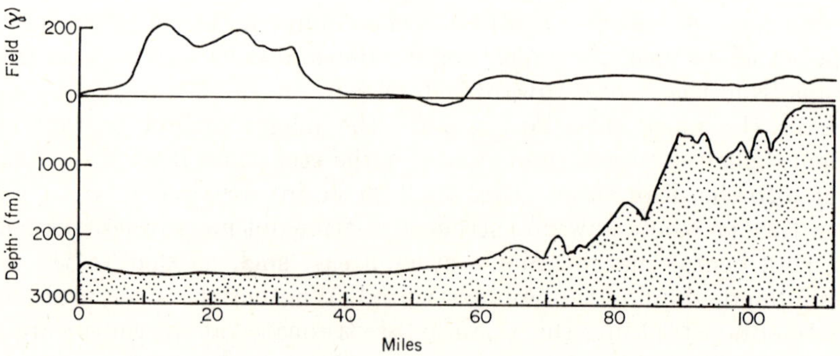

Fig. 18. Section off the mouth of the English Channel.

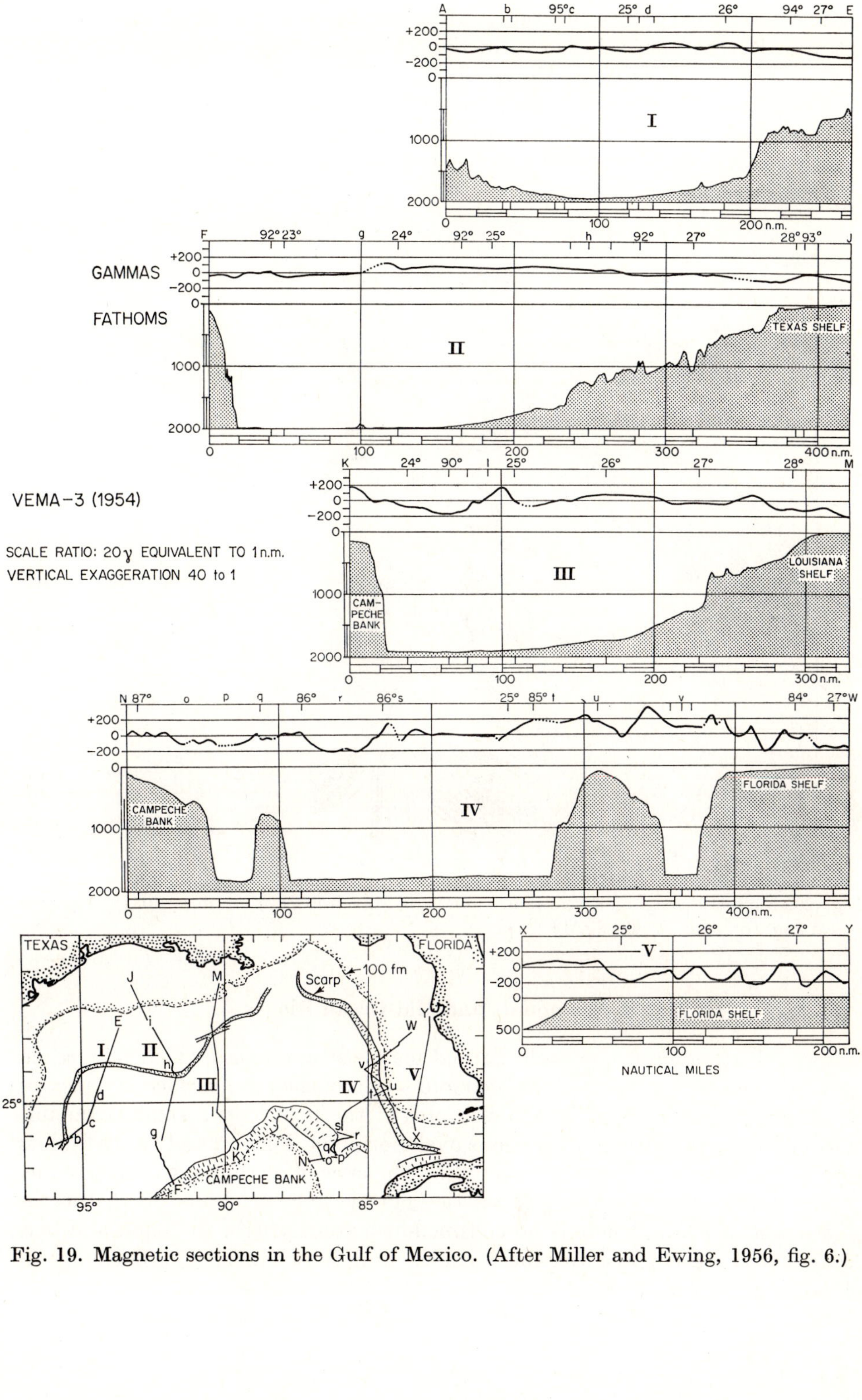

Fig. 19. Magnetic sections in the Gulf of Mexico. (After Miller and Ewing, 1956, fig. 6.)

The continental slope and the rise leading up to it frequently show no significant anomaly. A profile taken while running south along longitude $5\frac{3}{4}$°W into the Bay of Biscay is shown in Fig. 18 and sections taken by Miller and Ewing (1956) in the Gulf of Mexico in Fig. 19. Such smooth profiles might be expected in places where the outer part of the shelf is composed of sediments. They are not, however, always found; for example, Keller *et al.* (1954) find a positive anomaly of about 500 γ at the outer edge of the shelf off the east coast of the United States.

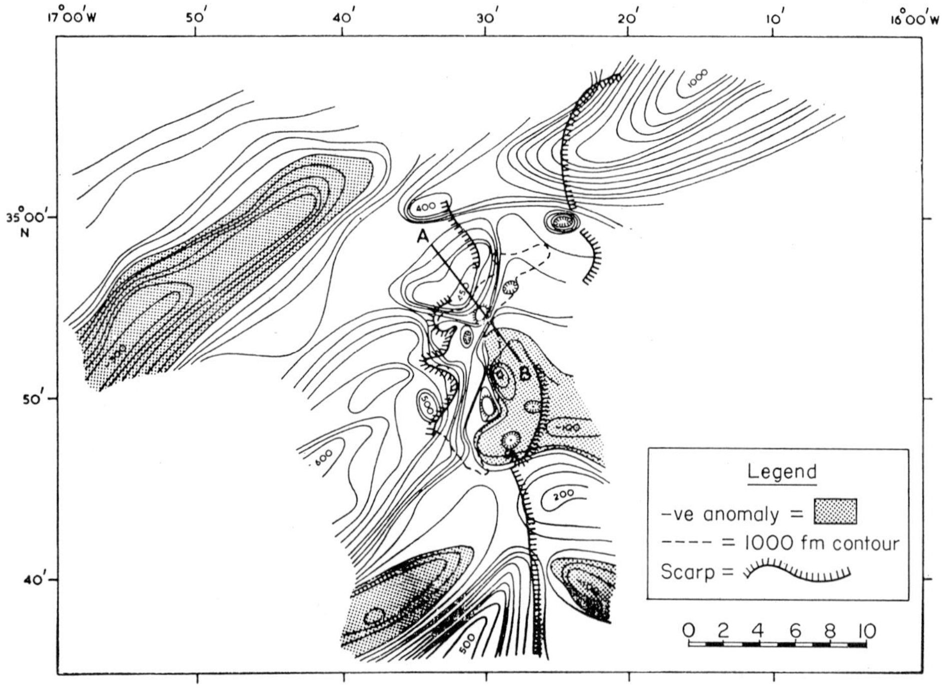

Fig. 20. Magnetic map of a seamount north of Madeira. Contour interval 50 γ. (After Laughton *et al.*, 1960, fig. 14.)

8. Seamounts and Mid-Ocean Ridges

Typical seamounts show well marked magnetic anomalies which may reach several thousand gammas if the seamount approaches the surface or emerges as a volcanic island. Fig. 20 shows a plot of the field over a small seamount north of Madeira and Fig. 21 sections over Bermuda. Figs. 22 and 23 show plots of two seamounts in the area covered by the survey of the north-east Pacific.

Fig. 24 shows a plot of the magnetic field over Galicia Bank off the north-west coast of Spain. There is no conspicuous anomaly over the bank and it is clear that it is not a basaltic volcano of the kind shown in Fig. 20. Dredging on

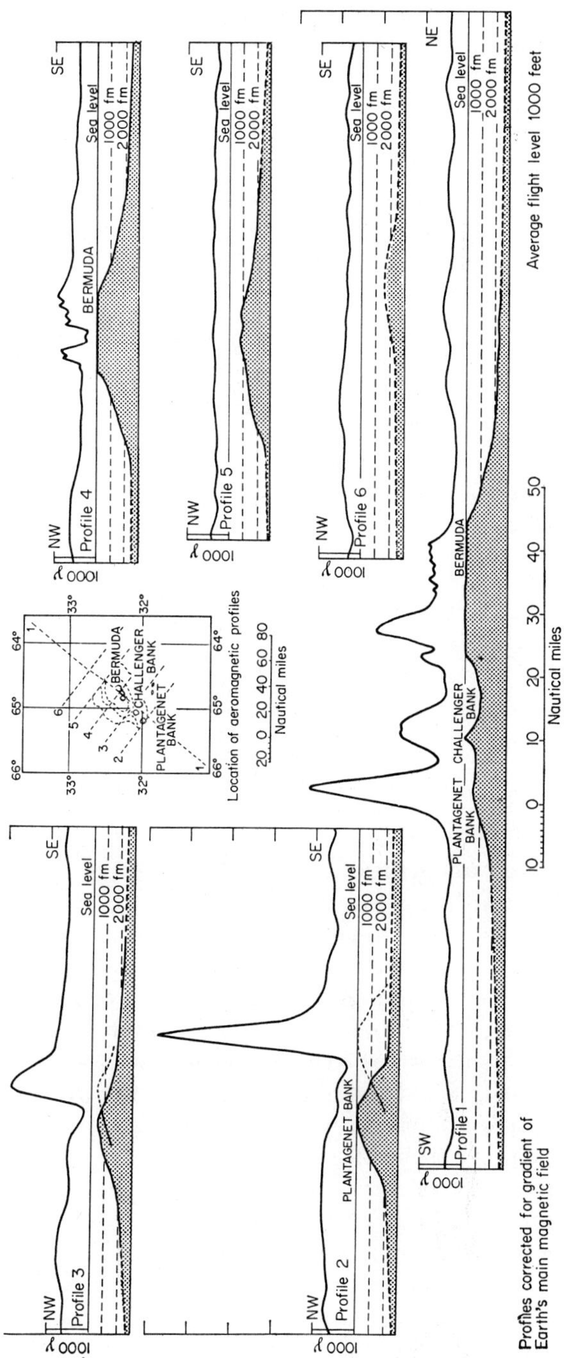

Fig. 21. Magnetic profiles across Bermuda. (After Keller *et al.*, 1954, fig. 8.)

this bank has yielded a variety of metamorphic and sedimentary rocks; many of these are erratics, but it is believed that some of the limestones are in place and it seems probable that the bank is to be regarded as a part of the continent separated from the mainland by faulting. Islands with great thicknesses of coral also show less marked anomalies than those composed throughout of basalt. Fig. 25 shows a map of Eniwetok Island which has a range of anomaly of only a few hundred gammas.

It is rare that sufficient is known about the form and susceptibility of the rocks of a seamount for any meaningful comparison to be made between the

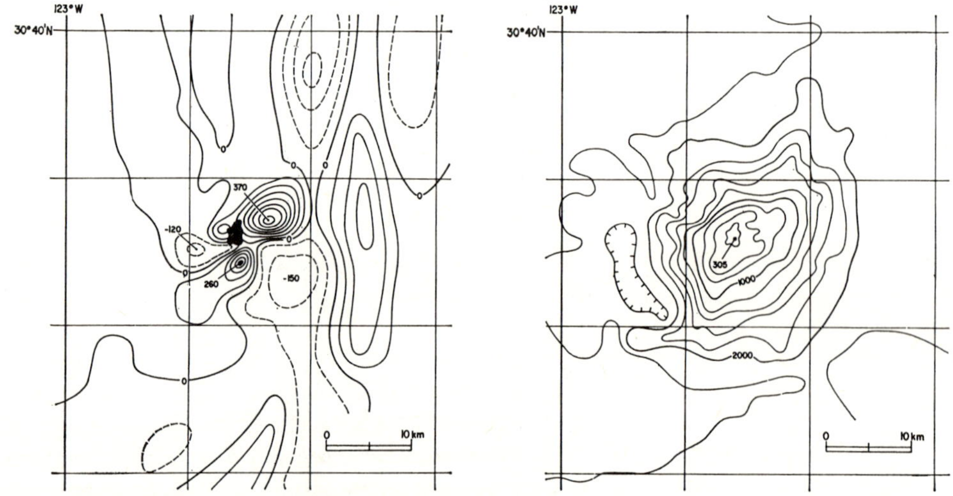

Fig. 22. Magnetic and topographic maps of Jasper Seamount. Magnetic anomalies in gammas. Contour interval 50 γ. Depths in fathoms. Contour interval 500 fm. (From R. G. Mason, unpublished.)

observed and expected anomalies. The positive peak of the anomaly is frequently displaced towards the magnetic equator, as it should be (see, for example, Fig. 21), and a rough quantitative agreement with theory has been obtained by Miller and Ewing (1956) for Caryn peak and by Keller *et al.* (1954) for Bermuda. On the other hand, the positive peak is often displaced in quite a different direction. Fig. 22 shows a plot of the field over Jasper seamount in the Pacific in which the major positive peak is displaced 4 km to the ENE. In such cases it is not easily decided whether this is due to discordant permanent magnetization of the seamount as a whole or to inhomogeneity of its petrology, and hence of its magnetic properties.

In some cases the effect of the visible seamount may be overshadowed by the effect of its roots on the normal crustal structure. Fig. 23 shows a plot of the anomaly over a row of four seamounts in the north-east Pacific. It is dominated by the magnetic lows to the south of the topographic peaks and the highs to the north. At first sight it might appear that the seamounts are negatively magnetized, but, when the observed anomaly is compared with the theoretical anomaly computed for normal magnetization of the topography, it is seen that

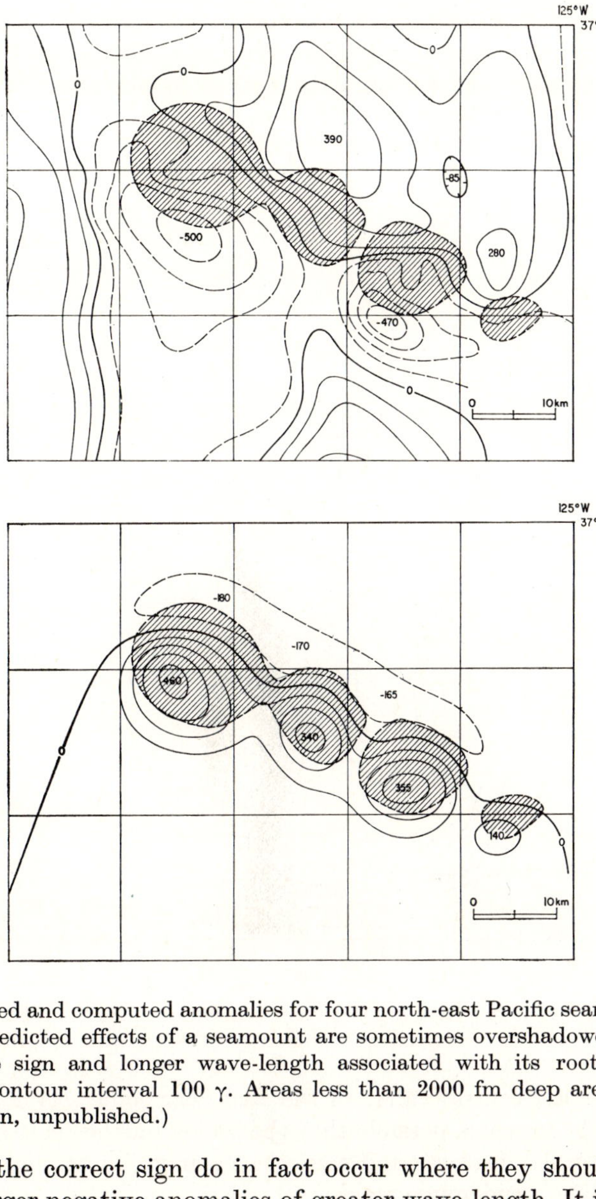

Fig. 23. Observed and computed anomalies for four north-east Pacific seamounts, showing how the predicted effects of a seamount are sometimes overshadowed by anomalies of opposite sign and longer wave-length associated with its roots. Anomalies in gammas. Contour interval 100 γ. Areas less than 2000 fm deep are shaded. (From R. G. Mason, unpublished.)

anomalies of the correct sign do in fact occur where they should be, but are masked by larger negative anomalies of greater wave-length. It is evident that the latter anomalies represent a deficiency of magnetic material in the roots of the seamounts as compared to the normal crust. There is no seamount for which the magnetization has been conclusively shown to be reversed, though Laughton *et al.* (1960) have suspected that the one shown in Fig. 20 may be.

When a number of volcanic masses are joined together, as on the Mid-Atlantic Ridge, a very complicated pattern is produced in which it is difficult to correlate topography and magnetic field. Heezen, Tharp and Ewing (1959) have found a systematically positive anomaly over the central valley on the crest of the ridge, as is shown in Fig. 26. When the field is varying so rapidly, it is not easy to be sure that the anomaly is really associated with the valley and Hill (1960) failed to find any anomaly associated with it. Girdler (1958), C. L. Drake (unpublished) and L. T. D. Allan (unpublished) have found striking

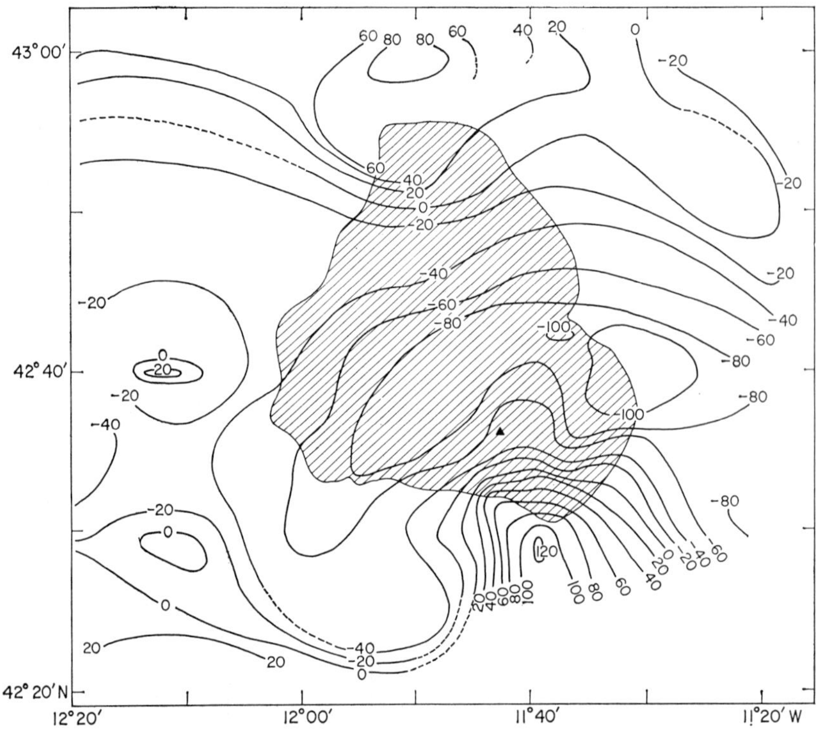

Fig. 24. Magnetic map of Galicia Bank. Contour interval 20 γ.

anomalies over the central valley of the Red Sea; some sections are shown in Figs. 27a and b. It seems possible that the valley on the ridge and that in the Red Sea are similar features and that the magnetic anomalies are associated with dykes filling a widening crack.

9. The Deep Trenches

Magnetic profiles have been published for the Aleutian (Keller *et al.*, 1954), Tonga (Raitt *et al.*, 1955) and Kuril-Kamchatka (Galperin *et al.*, 1958) trenches. Fig. 28 shows eight crossings of the Aleutian Trench and Fig. 29 two profiles across the Tonga Trench.

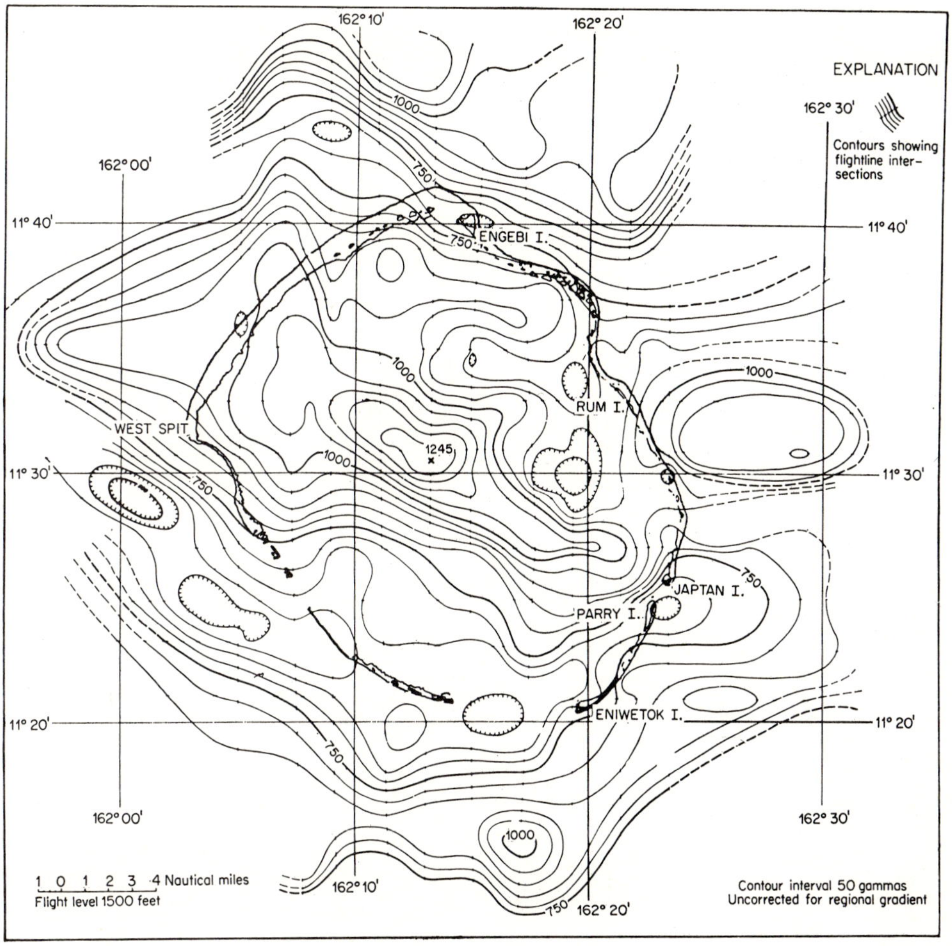

Fig. 25. Magnetic map of Eniwetok. (After Keller *et al.*, 1954, fig. 7.)

Keller *et al.* (1954) conclude that the anomalies observed across the Aleutian Trench are produced by susceptibility contrasts within the rocks of the sea floor and bear no direct relation to the trench. However, they could also arise from variations in the depth of the basement rocks under the sea floor. In the absence of information from other sources, for example seismic measurements, it is impossible to decide between these two alternatives.

In the case of the Tonga Trench the magnetic measurements were supported by echo-sounder profiles and by several seismic sections. Fig. 29 shows a possible interpretation of the magnetic profiles in terms of a magnetic basement, taking into account these other sources of information. Although this approach cannot be expected to lead to precise values for the thickness of the sediments, it does give a qualitative picture of how the thickness might vary from place to place. The discrepancy between the observed and computed curves cannot be

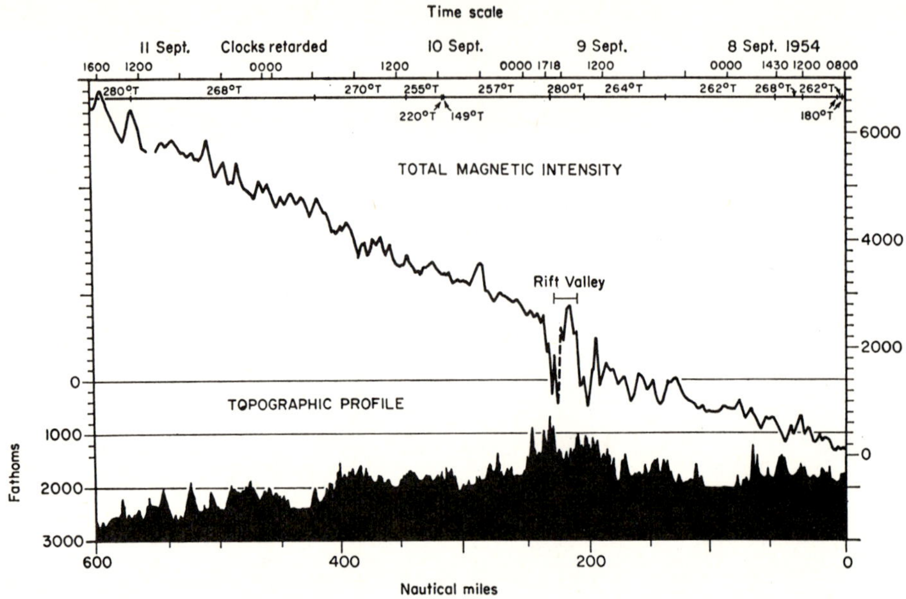

Fig. 26. Magnetic sections across the central valley of the Mid-Atlantic Ridge. (After Heezen *et al.*, 1959, fig. 48.)

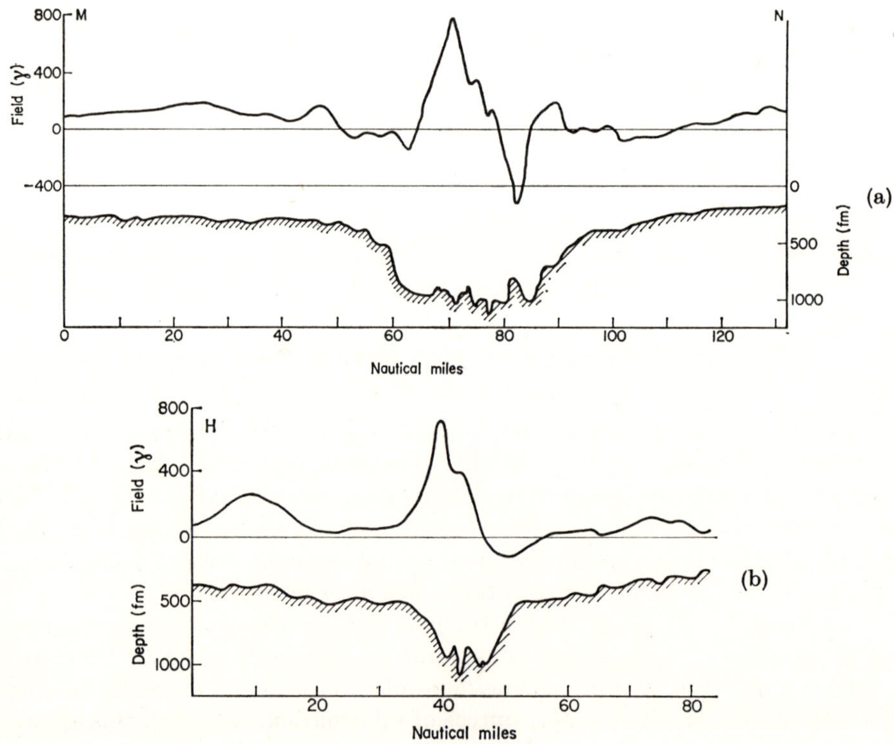

Figs. 27a and b. Magnetic profiles across the Red Sea. (T. D. Allan, unpublished.)

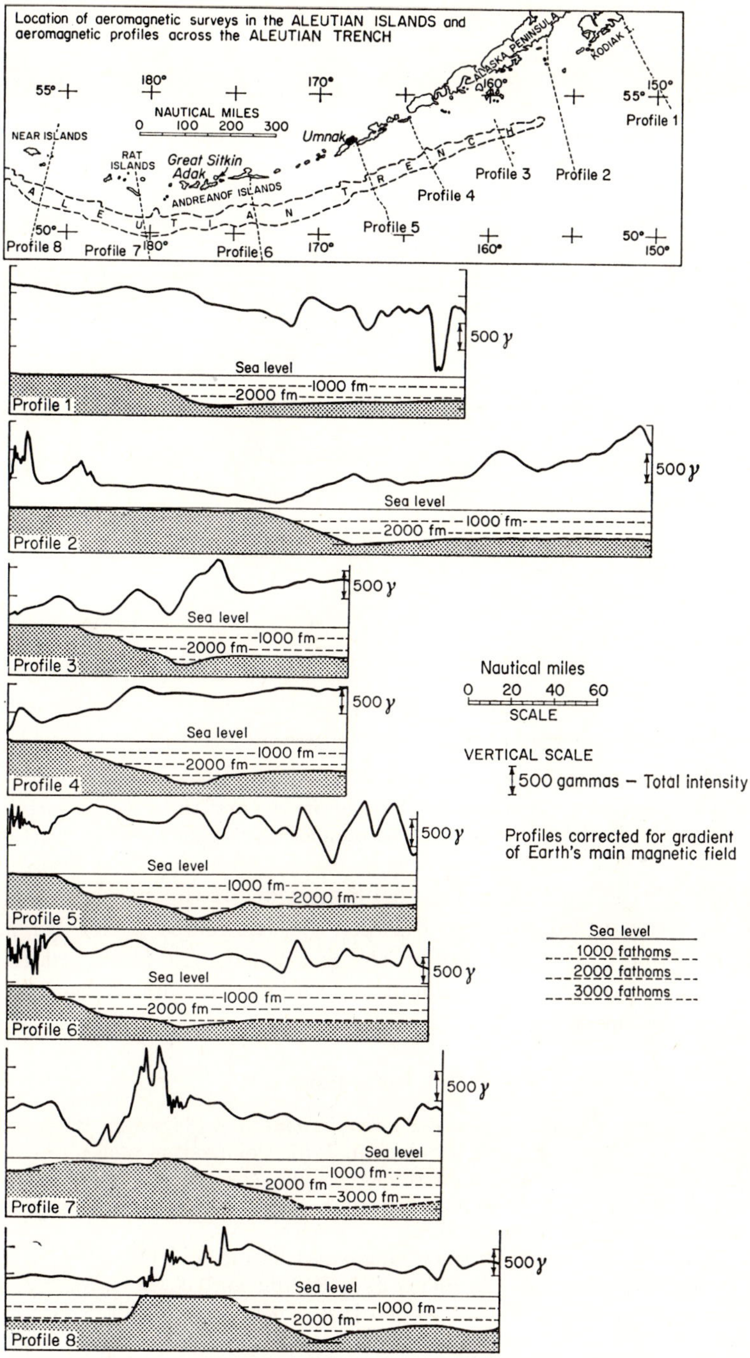

Fig. 28. Magnetic and bathymetric profiles across the Aleutian Trench. (After Keller *et al.*, 1954, fig. 10.)

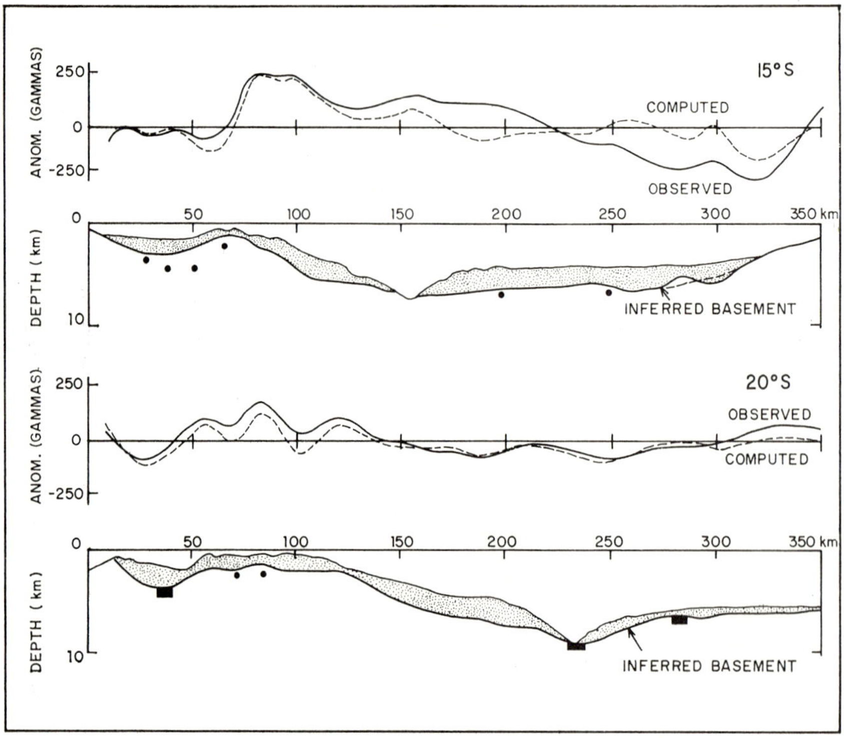

Fig. 29. Magnetic profiles across the Tonga Trench and their interpretation in terms of a magnetic basement. The bars are seismic depths and the dots are estimated depths based on the "half-width" of suitable magnetic anomalies. (After Raitt *et al.*, 1955.)

accounted for by any reasonable configuration of the basement. It could be caused by structural differences between opposite sides of the trench; it could also be explained by susceptibility changes within the basement rocks, or by their possessing a remanent magnetization aligned in some direction other than that of the present field.

References

Abragam, A., 1961. *The Principles of Nuclear Magnetism*. Clarendon Press, Oxford.

Abragam, A., J. Combrisson and I. Solomon, 1957. Polarisation nucléaire par effet Overhauser dans les solutions d'ion paramagnétiques. *C.R. Acad. Sci. Paris*, **245**, 157–160.

Alldredge, L. R. and G. D. Van Voorhis, 1961. Depth to source of magnetic anomalies. *J. Geophys. Res.*, **66**, 3793–3800.

Andrew, E. R., 1956. *Nuclear Magnetic Resonance*. Cambridge Univ. Press.

Bauer, L. A., W. J. Peters, J. A. Fleming, J. P. Ault and W. F. G. Swann, 1917. Ocean magnetic observations. *Publ. Carnegie Inst. Wash.*, No. 175, Pt. 3.

Bemmelen, W. van, 1899. Die Abweichung der Magnetnadel . . . bis zur Mitte XVIII Jahrhunderts. *Observations Batavia Roy. Mag. Met. Obs.*, Suppl. to vol. 21, 109 pp.

Birch, F., J. F. Schairer and H. C. Spicer (Eds.), 1942. Handbook of physical constants. *Geol. Soc. Amer. Spec. Paper* 36, 296–297.

Bullard, E. C., 1956. Edmond Halley (1655–1742). *Endeavour*, **15**, 189–199.

Bullard, E. C., 1961. The automatic reduction of geophysical data. *Geophys. J.*, **3**, 237–243.

Bullard, E. C., C. Freedman, H. Gellman and J. Nixon, 1950. The westward drift of the earth's magnetic field. *Phil. Trans. Roy. Soc. London*, **A243**, 67–92.

Bullard, E. C., M. N. Hill and C. S. Mason, 1962. Chart of the total force of the earth's magnetic field for the North-Eastern Atlantic Ocean. *Geomagnetica*, pp. 185–191. Serviço Meteorológico Nacional, Lisbon.

Bullard, E. C. and R. G. Mason, 1961. The magnetic field astern of a ship. *Deep-Sea Res.*, **8**, 20–27.

Bullard, E. C., A. E. Maxwell and R. Revelle, 1956. Heat flow through the deep sea floor. *Advances in Geophys.*, **3**, 153–181.

Cahill, L. J. and J. S. van Allen, 1956. High altitude measurements of the earth's magnetic field with a proton precession magnetometer. *J. Geophys. Res.*, **61**, 547–559.

Chapman, S., 1941. Edmond Halley as physical geographer, and the story of his charts. *Occas. Notes Roy. Astr. Soc.*, No. 9, 117–134.

Chapman, S. and J. Bartels, 1940. *Geomagnetism.* Clarendon Press, Oxford.

Chapman, S. and T. T. Whitehead, 1923. The influence of electrically conducting material within the earth on various phenomena of terrestrial magnetism. *Trans. Camb. Phil. Soc.*, **22**, 463–482.

Crichton Mitchell, A., 1932. Chapters in the history of terrestrial magnetism, Chapter I. On the directive property of a magnet. *Terr. Mag.*, **37**, 105–146.

Crichton Mitchell, A., 1937. Chapters in the history of terrestrial magnetism, Chapter II. The discovery of the magnetic declination. *Terr. Mag.*, **42**, 241–280.

Day, A. A., M. N. Hill, A. S. Laughton and J. C. Swallow, 1956. Seismic prospecting in the western approaches of the English Channel. *Q. J. Geol. Soc. London*, **112**, 15–44.

Driscoll, R. L. and P. L. Bender, 1958. Proton gyromagnetic ratio. *Phys. Rev. Let.*, **1**, 413–414.

Driscoll, R. L. and P. L. Bender, 1958a. A free precession determination of the proton gyromagnetic ratio. *I.R.E. Trans.*, **1** (7), 176.

Eardley, A. J., 1951. *Structural Geology of North America.* Harper, New York, p. 430.

Finger, D. L., G. I. Kolomiytseva, V. V. Noyysh and G. M. Priyezzhev, 1961. Experimental survey of terrestrial magnetism with magnetometers towed behind an iron ship. *Geomagnetizm i Aeronomiya*, **1**, 421–425.

Fleming, J. A., 1937. Magnetic survey of the oceans. In *International Aspects of Oceanography* by Vaughan, T. W. National Academy of Sciences, Washington.

Gaiber-Puertas, C., 1953. *Cons. Sup. Inves. Cient., Memoria* no. 11. 475 pp.

Galperin, E. I., A. V. Goryachev and S. M. Zverev, 1958. *The Pacific Geologo-Geophysical Expedition, U.S.S.R. Academy of Sciences, 1957* (XII Seismology) no. 1. Akad. Nauk S.S.S.R., Moscow.

Girdler, R. W., 1958. The relationship of the Red Sea to the East African Rift system. *Q. J. Geol. Soc. London*, **114**, 78–105.

Girdler, R. W. and G. Peter, 1960. An example of the importance of natural remanent magnetization in the interpretation of magnetic anomalies. *Geophys. Prospecting*, **8**, 474–483.

Harradon, H. D., 1944. Some early contributions to the history of geomagnetism—VII. *Terr. Mag.*, **49**, 185–198.

Heezen, B. C., M. Tharp and M. Ewing, 1959. The floors of the oceans, I, The North Atlantic. *Geol. Soc. Amer. Spec. Paper* 65, 122 pp.

Hellmann, G., 1897. Die Anfange der magnetischen Beobachtungen. *Z. Ges. Erdk.*, **32**, 112–136.

Heppner, J. P., J. D. Stolarik and L. H. Meredith, 1958. The earth's magnetic field above WSPG, New Mexico, from rocket measurements. *J. Geophys. Res.*, **63**, 277–288.

Hill, M. N., 1959. A ship-borne nuclear-spin magnetometer. *Deep-Sea Res.*, **5**, 309–311.

Hill, M. N., 1960. A median valley of the Mid-Atlantic Ridge. *Deep-Sea Res.*, **6**, 193–205.

Hill, M. N. and C. S. Mason, 1962. Diurnal variation of the earth's magnetic field at sea. *Nature*, **195**, 365–366.

Ivanov, M. M., 1960. Sea magnetic survey on board the non-magnetic ship *Zarya*. *I.A.G.A. Bull. No. 16. Trans. Toronto Meeting*, 365–7.

Ivanov, M. M., 1961. Work of the non-magnetic ship *Zarya* in the Pacific Ocean. *Okeanologiya*, **1**, 920–922.

Keller, F., J. L. Meuschke and L. R. Alldredge, 1954. Aeromagnetic surveys in the Aleutian, Marshall and Bermuda Islands, *Trans. Amer. Geophys. Un.*, **35**, 558–572.

Keyser, A. R., J. A. Rice and L. D. Schearer, 1961. A metastable helium magnetometer for observing small geomagnetic fluctuations. *J. Geophys. Res.*, **66**, 4163–4175.

Kudrevsky, A. I., 1961. The series nuclear magnetometer PM–1. *Geomagnetizm i Aeronomiya*, **1**, 436–440.

Laughton, A. S., M. N. Hill and T. D. Allan, 1960. Geophysical investigations of a seamount 150 miles north of Madeira. *Deep-Sea Res.*, **7**, 117–141.

Logachev, A. A., 1955. *Kurs magnitorazvedki* (Textbook of magnetic exploration). Gosgeoltekhizdat, Moscow, 302 pp.

Mason, B., 1943. Mineralogical aspects of the system FeO–Fe_2O_3–MnO–Mn_2O_3. *Geol. Fören. Stockholm Forhandl.*, **65**, 97–180.

Mason, R. G., 1958. A magnetic survey off the west coast of the United States between latitudes 32° and 36°N, longitudes 121° and 128°W. *Geophys. J.*, **1**, 320–329.

Mason, R. G. and A. D. Raff, 1961. A magnetic survey off the west coast of North America, 32°N to 42°N. *Bull. Geol. Soc. Amer.*, **72**, 1259–1265.

Matthews, D. H., 1961. Lavas from an abyssal hill on the floor of the North Atlantic Ocean. *Nature*, **190**, 158–159.

Matveyev, B. M., 1962. Magnetic research on the schooner *Zarya*. *Vestn. Akad. Nauk S.S.S.R.*, No. 6.

Menard, H. W., 1955. Deformation of the northeastern Pacific basin and the west coast of North America. *Bull. Geol. Soc. Amer.*, **66**, 1149–1198.

Menard, H. W., 1959. Minor lineations in the Pacific Basin. *Bull. Geol. Soc. Amer.*, **70**, 1491–1496.

Menard, H. W., 1960. The East Pacific Rise. *Science*, **132**, 1737–1746.

Menard, H. W., 1961. The East Pacific Rise. *Sci. Amer.*, **205**, 52–61.

Menard, H. W., 1962. Correlation between length and offset on very large wrench faults. *J. Geophys. Res.*, **67**, 4096–4101.

Miller, E. T. and M. Ewing, 1956. Geomagnetic measurements in the Gulf of Mexico and in the vicinity of Caryn Peak. *Geophysics*, **21**, 406–432.

Nagata, T., 1953. *Rock Magnetism*. Maruzen, Tokyo.

Nagata, T., T. Oguti and S. Kakinuma, 1958. Geomagnetic anomaly between South Africa and Antarctic continent roughly along 40°E meridian line. *Proc. Japan Acad.*, **34**, 427–431.

Needham, J., 1962. *Science and Civilization in China*, Vol. IV: 1 (Physics). Cambridge Univ. Press.

Nelson, J. H., 1960. The gyromagnetic ratio of the proton. *J. Geophys. Res.*, **65**, 3826.

Oguti, T. and S. Kakinuma, 1959. Preliminary report of geomagnetic survey during JARE the second. *Ant. Rec.*, No. 7, 17–25.

Packard, M. and R. Varian, 1954. Proton gyromagnetic ratio. *Phys. Rev.*, **93**, 941.

Raff, A. D., 1962. Further magnetic measurements along the Murray fault. *J. Geophys. Res.*, **67**, 417–418.

Raff, A. D. and R. G. Mason, 1961. A magnetic survey off the west coast of North America, 40°N to 52½°N. *Bull. Geol. Soc. Amer.*, **72**, 1259–1265.

Raitt, R. W., R. L. Fisher and R. G. Mason, 1955. Tonga Trench. In "Crust of the Earth" (Edited by Poldervaart, A.). *Geol. Soc. Amer. Spec. Paper* 62, 237–254.

Rikitake, T., 1961. S_q and Ocean. *J. Geophys. Res.*, **66**, 3245–3254.

Rikitake, T., 1962. Supplement to paper "S_q and Ocean". *J. Geophys. Res.*, **67**, 2588–2591.

Rumbaugh, L. H. and L. R. Alldredge, 1949. Airborne equipment for geomagnetic measurements. *Trans. Amer. Geophys. Un.*, **30**, 836–848.

Schonstedt, E. O. and H. R. Irons, 1955. NOL vector airborne magnetometer type 2A. *Trans. Amer. Geophys. Un.*, **36**, 25–41.

Serson, P. H. and W. L. W. Hannaford, 1957. A statistical analysis of magnetic profiles. *J. Geophys. Res.*, **62**, 1–18.

Serson, P. H., S. Z. Mack and K. Whitham, 1957. A three component airborne magnetometer. *Pub. Dom. Obs. Ottawa*, **19**, 1–97.

Skeat, W. W. (Ed.), 1870. *The Bruce* (by Barbour, J.), Book 5, lines 19–23. Early English Text Society, London.

Skillman, T. L. and P. L. Bender, 1958. Measurement of the earth's magnetic field with a rubidium vapor magnetometer. *J. Geophys. Res.*, **63**, 513–515.

Vacquier, V., 1959. Measurement of horizontal displacement along faults in the ocean floor. *Nature*, **183**, 452–453.

Vacquier, V., A. D. Raff and R. E. Warren, 1961. Horizontal displacements in the floor of the Pacific Ocean. *Bull. Geol. Soc. Amer.*, **72**, 1267–1270.

Vestine, E. H., 1960. The survey of geomagnetic field in space. *Trans. Amer. Geophys. Un.*, **41**, 4–21.

Vestine, E. H., L. Laporte, C. Cooper, I. Lange and W. C. Hendrix, 1947. Description of the earth's main magnetic field and its secular change, 1905–1945. *Publ. Carnegie Inst. Wash.*, No. 578.

Vigoureux, P., 1962. A determination of the gyromagnetic ratio of the proton. *Proc. Roy. Soc.* **A270**, 72–89.

Waters, G. S. and P. D. Francis, 1958. A nuclear magnetometer. *J. Sci. Instrum.*, **35**, 88–93.

Waters, G. S. and G. A. Phillips, 1956. A new method of measuring the earth's magnetic field. *Geophys. Prosp.*, **4**, 1–9.

11. THE FLOW OF HEAT THROUGH THE FLOOR OF THE OCEAN

E. C. BULLARD

1. Method

The rate of flow of heat through unit area of the floor of the ocean is determined as the product of a temperature gradient and a thermal conductivity. The temperature gradient is measured by forcing a probe into the ocean floor. The probe contains two or more temperature-sensing elements whose readings are recorded by a recorder housed in a pressure-tight container at the top of the probe.

The thermal conductivity is determined by measurement on samples brought up by a coring tube.

2. The Temperature Probe

Most work published up to the present time has been done with cylindrical probes. Those used in the Pacific from the ships of the Scripps Institution have the junctions 1.4 to 2.7 m apart and are 1.9 to 4.2 cm in diameter; the one used in *Discovery II* is 4.6 m long and has external and internal diameters of 2.7 and 1.1 cm. A detailed description of the latter probe has been given by Bullard (1954). The recorder is contained in a pressure-tight case, to the bottom of which the probe is attached as shown in Fig. 1. The case carries fins to ensure stability when running down through the water; instability might occur after the point has entered the sediment, but records from an inclinometer in the recording case show that, in practice, the probe is within a few degrees of the vertical after penetration. A similar apparatus has been described by Uyeda *et al.* (1961) and by Von Herzen *et al.* (1962).

Recently, Gerard, Langseth and Ewing (1962) at the Lamont Observatory have constructed an apparatus in which the temperature gradient is measured by an attachment to a corer. This has the great advantage that the total time taken is greatly reduced and the core is obtained at the exact point where the temperature measurement is made. To avoid the temperature disturbance produced by the corer the thermistors used for measuring the temperature gradient are carried on fins projecting from the coring tube. The thermistors are in tubes about 3 mm in diameter projecting 2.5 cm below the fins and about 7 cm from the coring tube.

The temperature-sensitive element in a probe may be a thermocouple or a thermistor. As the temperature gradient is usually about 50°C/km, the temperature difference between two points 4 m apart is only 0.2°C; a pair of copper–constantan thermo-junctions gives about 30 μV/°C and would, therefore, give 6 μV for this temperature difference. Even if several junctions are used in series this is inconveniently small and to get a deflection of a few millimetres requires a galvanometer with a sensitivity of a few tenths of a millimetre per microvolt. The galvanometer must be specially balanced to make it insensitive

[*MS received August, 1960*]

to imperfect levelling and must withstand the shock it experiences when the probe runs into the sediment. The zero position on the record is not known, but the deflection may be determined by reversing the thermocouple connections to the galvanometer. Electronic amplification before recording is not practicable for such small e.m.f.'s. The advantages of the thermocouple are that it is itself simple and robust, and that it measures a temperature difference directly and does not require it to be found as the difference of two temperatures measured separately.

Fig. 1. Probe for measuring temperature gradient. (By courtesy of British Petroleum Co.)

A thermistor measures a single temperature and therefore needs careful calibration or matching if it is one of a pair used to determine a small temperature difference. The resistance change of a thermistor is typically 5%/°C. The e.m.f. that can be obtained from a bridge of which it forms one arm is limited by the heating produced by the measuring current, but is of the order of 3 mV for 0.2°C. This e.m.f. may be used directly to drive an insensitive and robust galvanometer of the type used in seismic prospecting or the bridge may be made self-balancing by an amplifier connected in a servo-loop; a pen recorder may then be used. If an amplifier is to be used the bridge may be fed by a.c., or a vibrating reed may be used between the bridge and the amplifier. A system of relays or a rotary switch is necessary to connect the thermistors successively to the bridge; calibrating resistors may be included in the cycle so that the resistance of the thermistors is obtained by interpolation and not by assuming the zero and sensitivity of the system to be known. A check on the resistance of the thermistors when at the same temperature may be obtained by noting the readings when the instrument is suspended a short distance above the bottom.

When a probe weighing 160 kg in air and 130 kg in water strikes the bottom at a speed of 3 m/sec and penetrates 5 m, 1700 cal of heat are produced. This heat appears partly in the probe and partly in the sediment; if it all appeared

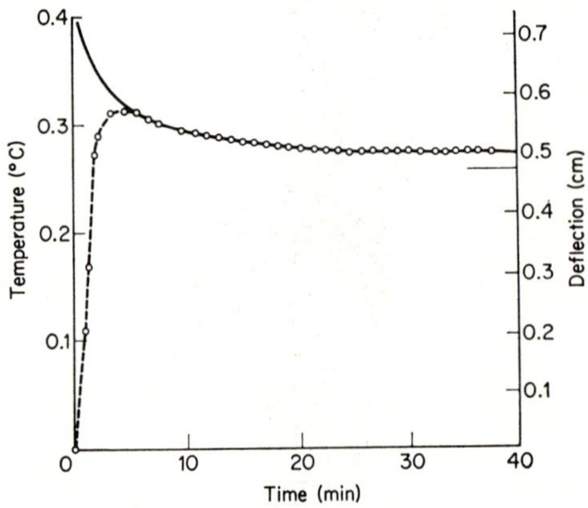

Fig. 2. Variation with time of the temperature difference between the top and the bottom of a probe 2.7 cm in diameter. The full-line curve represents the theoretical expression (1). (After Bullard *et al.*, 1956, fig. 2.)

in the probe used in *Discovery II* it would raise its temperature by 0.8°C, which is considerably larger than the temperature difference to be measured. In practice part of the heat goes into the sediment.

The time taken for heat to spread from the surface of the probe to the centre is only a few minutes and is much shorter than the time needed for the probe to reach temperature equilibrium with the sediment. If the lower end of the probe is heated above the temperature of the sediment into which it has penetrated the temperature will rise for about 3 min, pass through a maximum and then fall towards its final value as shown in Fig. 2.

It is not practicable to leave a probe in the bottom for more than about 40 min; after this time the temperature gradient in the probe usually exceeds that in the sediment by 10 to 15%. It is, therefore, desirable to extrapolate the observed temperature differences to get the value which would be obtained after a very long time. In making this correction it may be assumed that any cross-section of the probe is at a uniform temperature and that conduction is purely radial. The temperature, θ, of the probe at a time t after entry may then be shown to be (Bullard, 1954)

$$\theta = \theta_1 + \theta_0 F(\alpha, \tau), \tag{1}$$

where θ_1 and θ_0 are constants and

$$F(\alpha, \tau) = \frac{4\alpha}{\pi^2} \int_0^\infty \frac{\exp(-\tau x^2)\, dx}{x[\{xY_0(x) - \alpha Y_1(x)\}^2 + \{xJ_0(x) - \alpha J_1(x)\}^2]}. \tag{2}$$

Here τ is a non-dimensional time given by

$$\tau = \kappa t / a^2,$$

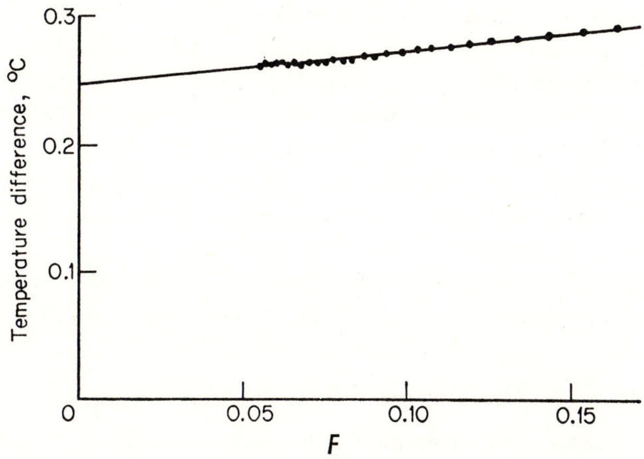

Fig. 3. Correction for lack of equilibrium. The temperature difference between the top and bottom of the probe is plotted against the theoretical function F of (1). The intercept of the straight line fitted to the points gives the equilibrium temperature that would be attained if the probe were left in the sediment for an indefinitely long time. (After Bullard *et al.*, 1956, fig. 3.)

where κ is the thermometric conductivity of the sediment and a is the radius of the probe; τ is about $0.1t$ if t is in minutes and a is 1.3 cm. α is given by

$$\alpha = 2\pi a^2 \rho\sigma / m,$$

where ρ and σ are the density and specific heat of the sediment and m is the water equivalent of a 1-cm length of the probe (that is, the mass of water that would require the same amount of heat to raise its temperature 1°C as a 1-cm length of the probe does). α is usually about 2. The Y's and J's in (2) are Bessel functions.

The function $F(\alpha, \tau)$ has been tabulated by Bullard (1954) and by Jaeger (1956). It is 1 for t zero, decreases rapidly at first as t increases and for large t approaches zero as $m/4\pi k t$. The observed temperatures may be fitted to (1) by least squares. The fit is usually excellent as is shown in Fig. 3; the constants of the best fitting line give estimates of θ_1 and θ_0. $(\theta_1 + \theta_0)$ is the temperature that would be attained if the initial frictional heat spread instantaneously through the probe; in fact, the thermocouple never reaches this temperature since some heat is lost to the sediment before it has had time to penetrate to the centre of the probe. For example, in Fig. 3, the observed maximum temperature difference is 0.31°C and occurs after 3 min, whilst $(\theta_1 + \theta_0)$ is 0.51°C. The equilibrium temperature difference, θ_1, is 0.251°C, whilst the last observed temperature difference was 0.266°C; the correction is, therefore, 6%. Experiments with a probe in which the temperature gradient could be measured both over the whole length and over the bottom half have shown the latter to be 8% greater than the former (Bullard and Day, 1961). The reason for this is unknown; bacterial action in the surface layers and the increase of conductivity with depth would produce a difference in the opposite direction. It may be that in the upper 2 m some of the heat is carried by movement of water. In view of the large variation in the observed heat flow from place to place these uncertainties in measuring the gradient and in the extrapolation to find the equilibrium gradient are relatively unimportant.

The miniature probes attached to a corer reach equilibrium in under a minute and no extrapolation is needed to find the steady state; their thermal properties have been studied in the laboratory by Uyeda *et al.* (1961).

The reliability of temperature measurements made with probes has recently been confirmed by measurements made in the 600-ft trial boring for the Mohole.

3. The Thermal Properties of Sediments

It would be possible to measure the thermal conductivity of the sediments of the ocean floor by generating a known amount of heat in a probe and following the subsequent change in temperature; this experiment has never been carried out and all measurements have been made on samples obtained with a coring tube.

Two methods of measurement have been used to obtain the thermal conductivity of samples. Butler and Ratcliffe at the National Physical Laboratory have used a static method with disc-shaped specimens (Ratcliffe, 1960). This method is the one normally employed for precise absolute measurements of the conductivity of poor conductors and is believed to be free from serious systematic error. It is, however, tedious as it is necessary to wait for the apparatus to attain temperature equilibrium. A transient method has, therefore, been developed (Von Herzen and Maxwell, 1959) in which a hypodermic needle 0.086 cm in diameter and 6 cm long is inserted into a piece of core and its temperature measured as a function of time while heat is dissipated at a constant rate, Q, by a heater within the needle. If r is the radius of the needle and k the conductivity of the sediment, the temperature, θ, at time t is

$$\theta = \frac{Q}{4\pi k} \ln \frac{2.246\kappa t}{r^2},$$

where κ is the thermometric conductivity of the sediment. This expression is a good approximation for times large compared to r^2/κ. For the needle used, r^2/κ is about 0.8 sec and a straight line is obtained when θ is plotted against $\ln t$ for all times between 10 sec and 10 min.

The results show that the conductivity of a sediment depends on its water content and varies very little with the nature of the mineral particles. It is, therefore, possible to calculate the conductivity if the water content is known. The observations can be represented satisfactorily by the theoretical expression for the conductivity of an assembly of spheres distributed in water (Bullard, Maxwell and Revelle, 1956; Hôrai and Uyeda, 1960). The comparison for 44 specimens measured by Butler and Ratcliffe is shown in Fig. 4. Empirically it

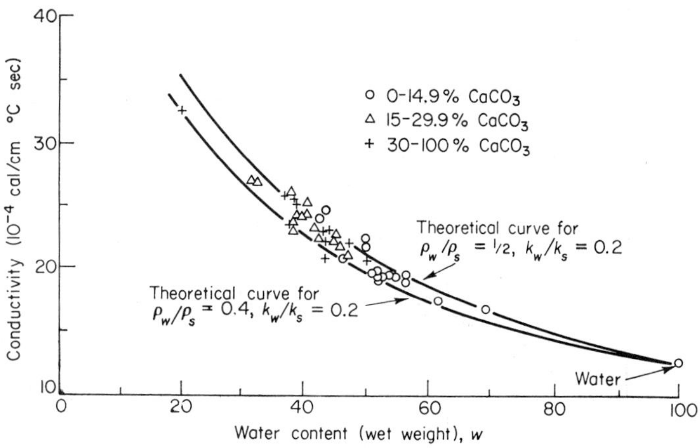

Fig. 4. The variation of the thermal conductivity of ocean-bottom sediment with water content. The curves are theoretical expressions for spheres of density ρ_s and conductivity k_s dispersed in water of density ρ_w and conductivity k_w. (After Bullard *et al.*, 1956, fig. 4.)

is found that the thermal resistivity (the reciprocal of the conductivity) is linearly related to water content as is shown in Fig. 5 (Bullard and Day, 1961). The best straight line is

$$R = (161 \pm 14) + (651 \pm 30)w, \tag{3}$$

where R is in cm sec °C/cal and w is the water content expressed as a proportion of the wet weight. There is no systematic difference between the conductivities of red clay and *Globigerina* ooze of the same water content.

The measurements were made at a room temperature of about 25°C and at atmospheric pressure. The values required are those at the temperature and pressure of the sea floor, which may be taken as 3°C and 500 kg/cm². Butler has shown that the decrease in temperature will reduce the conductivity by 6% (Ratcliffe, 1960). If the pressure coefficient is the same as that for pure water the increase in pressure will raise the conductivity by 2 to 3%. The corrections are only roughly known but a uniform reduction in conductivity by 4% should avoid serious systematic error. The relation (3) then becomes

$$R = 168 + 678w. \tag{4}$$

The conductivity that is required in the calculation of heat flow is the reciprocal of the mean resistivity. Since the resistivity is linearly related to the water content it may be found from the mean water content. The water content at a point in a core may vary from about 25 to 60%, but the mean down a core is much more uniform. For example, 10 cores of *Globigerina* ooze from the north-eastern basin of the Atlantic all had water contents between 48 and 40% and mean conductivities between 0.0020 and 0.0023 cal/cm sec °C (Bullard, 1960).

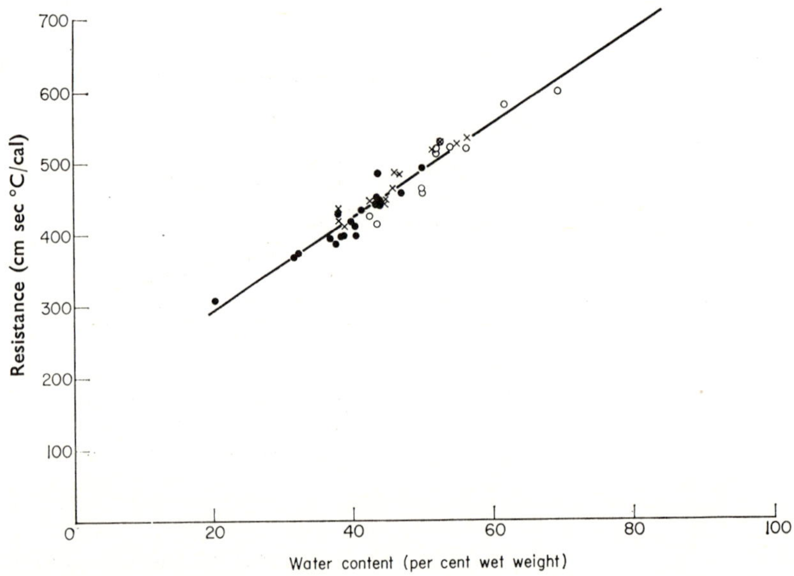

Fig. 5. Relation between thermal resistivity and water content. ○ red clay, ● *Globigerina* ooze, × mud.

To calculate the thermometric conductivity, $k/\sigma\rho$, the density, ρ, and the specific heat, σ, of the sediment are required. If the sediment can be regarded as a mechanical mixture of particles of mean density ρ_s, the spaces between which are filled with water of density ρ_w, its density is given by

$$\rho^{-1} = \rho_s{}^{-1} + (\rho_w{}^{-1} - \rho_s{}^{-1})w.$$

Naturally, the best value for ρ_s varies with the type of sediment. A least squares reduction of the 44 densities given in Bullard (1956) and Bullard and Day (1961) gave 2.56 g/cm³ for *Globigerina* ooze, 2.64 for red clay and 2.34 for mud; all the data combined gave 2.51 with a standard deviation of 3%. If this accuracy is sufficient, ρ_s may be taken as 2.5 g/cm³. If the water content is determined by drying, it is the content of pure water and not sea-water that is given by w; ρ_w should, therefore, be taken as 1 g/cm³, and

$$\rho^{-1} = 0.40 + 0.60w. \tag{5}$$

Similarly, the specific heat of a sediment may be calculated from the specific heats of its constituents; though there is a small discrepancy between the specific heat of sea-water and that calculated for a mixture of salt and pure water (Cox and Smith, 1959), it seems sufficient to take the specific heat of the solids as 0.18 cal/g °C and of the liquid removed on drying as 1.00 cal/g °C. The specific heat of the sediment is then given by

$$\sigma = 0.18 + 0.82w. \tag{6}$$

Typical values of the thermal properties of ocean-bottom sediment as a function of water content are given in Table I. These are calculated from the expressions (4), (5) and (6) above. Ratcliffe (1960) gives nomograms connecting density, water content and conductivity.

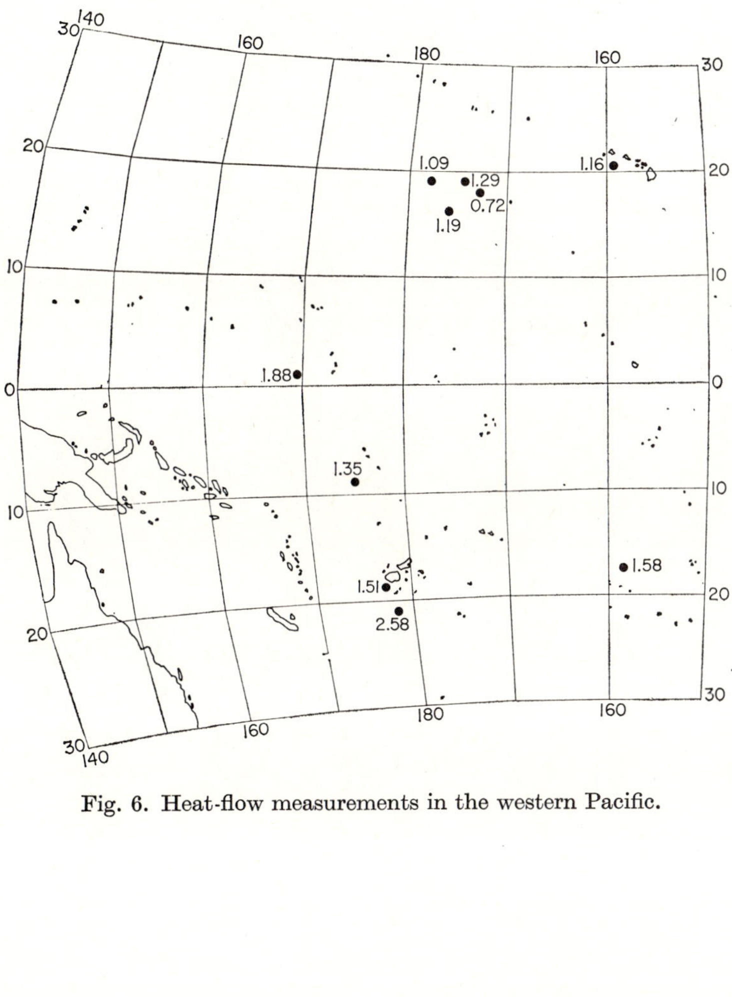

Fig. 6. Heat-flow measurements in the western Pacific.

Table I

Typical Thermal Properties of Ocean Sediments

Water content, % wet wt.	Density, g/cm^3	Specific heat, cal/g °C	Thermal resistance, cm sec °C/cal	Thermal conductivity, cal/cm sec °C $\times 10^{-4}$	Thermometric conductivity, cm^2/sec $\times 10^{-4}$
20	1.92	0.34	304	33	50
30	1.72	0.43	371	27	36
40	1.56	0.51	439	23	29
50	1.43	0.59	507	20	23
60	1.32	0.67	574	17	20
70	1.22	0.75	643	16	17

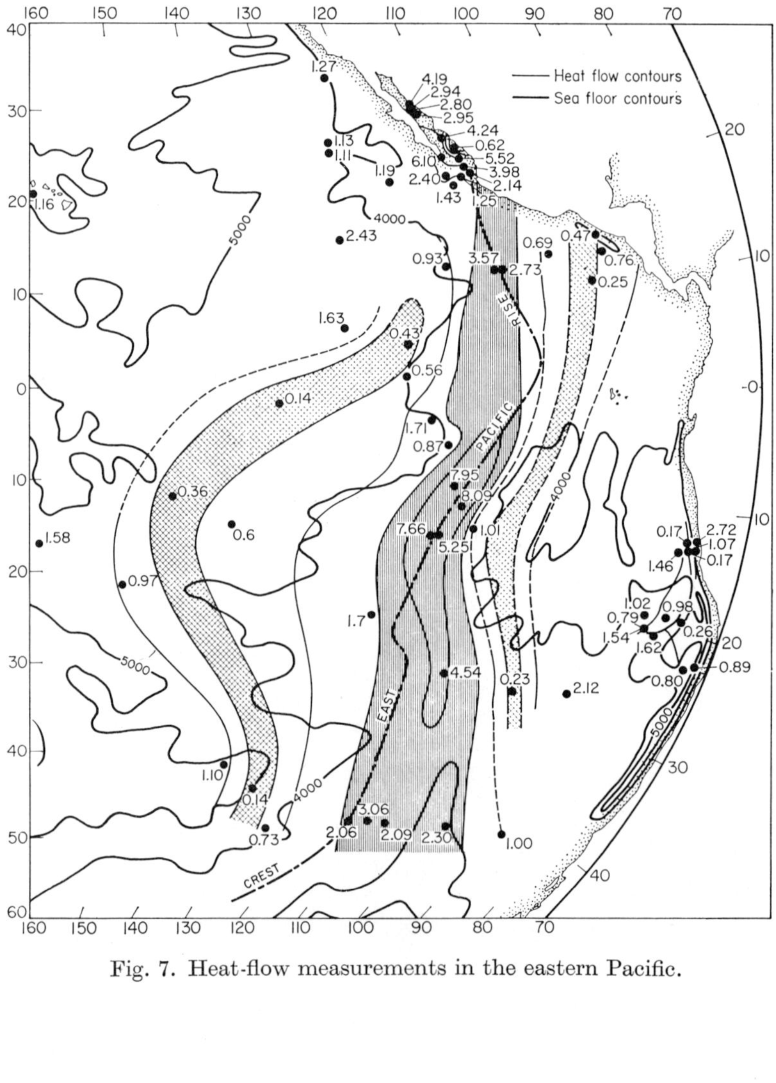

Fig. 7. Heat-flow measurements in the eastern Pacific.

4. The Results of Heat-Flow Measurements

Measurements have been made in the Pacific by workers in ships of the Scripps Institution (Revelle and Maxwell, 1952; Von Herzen, 1959; Foster, 1962) and by Uyeda *et al.* (1962). In the Atlantic, measurements have been made by Bullard (1954), Bullard and Day (1961), Reitzel (1961), Gerard *et al.* (1962), Nason and Lee (1963) and Lister (unpublished). The earlier Pacific results are plotted in Figs. 6 and 7 and the Atlantic results in Fig. 8. These figures include unpublished results by R. P. Von Herzen and C. Lister which

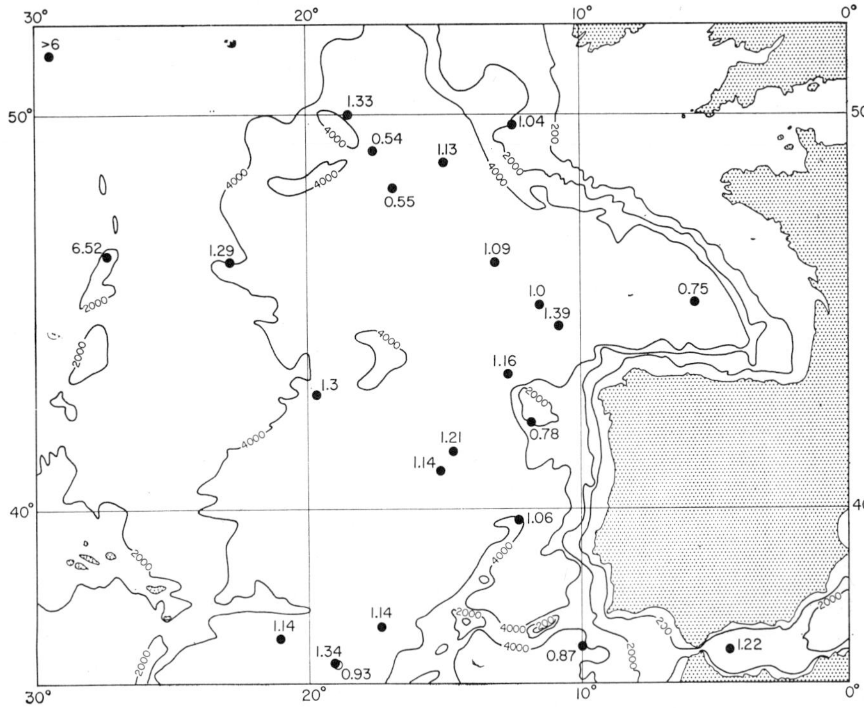

Fig. 8. Heat-flow measurements in the Atlantic; the value in the north-west corner of the figure is an unpublished result by J. Reitzel in which the recorder went off the scale.

they have kindly allowed me to use. Lister has made a number of measurements to the south of the area shown in Fig. 8 which lie in the range 0.8 to 2.2 μcal/cm^2 sec. Reitzel has obtained 1.2 μcal/cm^2 sec 500 km north-west of Bermuda and Gerard *et al.* 1.0 to 1.8 μcal/cm^2 sec in the western basin of the North Atlantic.

The values on the East Pacific Rise and on the Mid-Atlantic Ridge show a large spread, the heat flow on the west of the rises being usually greater than elsewhere and occasionally above 8 μcal/cm^2 sec. This is shown by the histograms in Fig. 9 and in the section in Fig. 10; it is also confirmed by further measurements on the Mid-Atlantic Ridge by Gerard *et al.* and by Nason and Lee. Unpublished work by Von Herzen has shown that the heat flows on the crest of the East Pacific Rise occur in two strips 20 to 40 km wide, and that the curve of Fig. 10 should consist of two peaks and not of a single broad maximum. The high heat flows also seem to be similarly distributed on the Mid-Atlantic Ridge. Nason and Lee found a narrow peak of high values not far from a place where Lister found nothing abnormal.

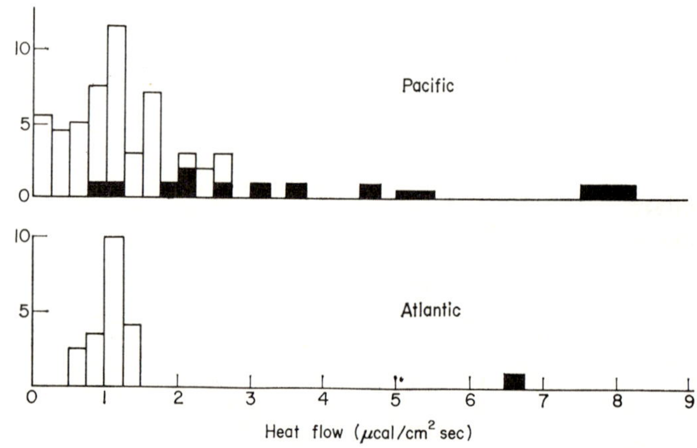

Fig. 9. Histogram of heat-flow values in the Atlantic and Pacific Oceans. The black bars are values on the Mid-Atlantic Ridge and the East Pacific Rise. Values in the Gulf of California are excluded from this figure.

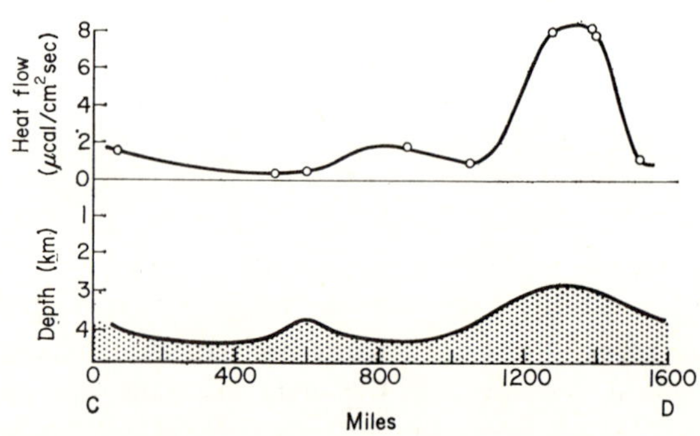

Fig. 10. Section across East Pacific Rise showing heat flow and generalized topography.

Heat flows measured in the Gulf of California are high and appear to form an extension of the line of high values on the crest of the East Pacific Rise. It is possible that the same belt of high heat flow runs out to sea again to the north of California (Menard, 1961).

Values in the Acapulco and South American trenches tend to be below those elsewhere, as is shown in Fig. 11, which shows a section off the coast of Peru in latitude 14°S. These low values are probably not a general feature of deep ocean trenches, since they have not been found in the Japan Trench (Uyeda *et al.*, 1962), the Puerto Rico Trench (Gerard *et al.*, 1962) or the Aleutian Trench (Foster, 1962).

Excluding the high values on the ridges and in the Gulf of California, the values give a roughly normal distribution, the means and standard deviations of the values available in late 1960 being:

Atlantic	1.06 ± 0.055 μcal/cm² sec	S.D. = 0.25.
Pacific	1.09 ± 0.094 μcal/cm² sec	S.D. = 0.65.

The means agree closely and it seems probable that their mean,

$$1.08 \pm 0.054 \ \mu\text{cal/cm}^2 \text{ sec},$$

represents the mean oceanic heat flow away from the ridges with an uncertainty of 5 or 10%. This is close to the continental mean, which is less well determined

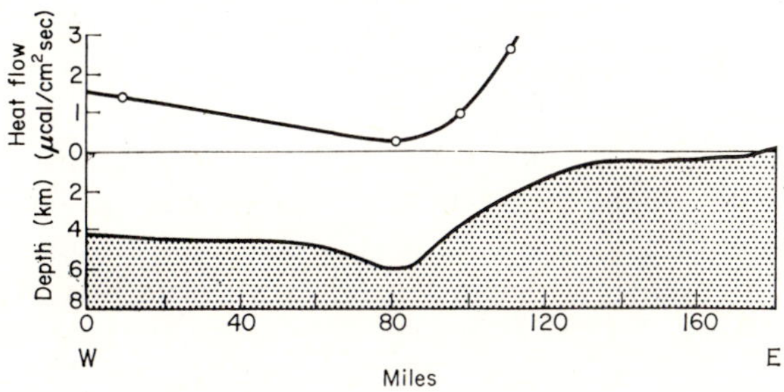

Fig. 11. Section across the South American trench showing heat flow and generalized topography.

but probably lies between 1.1 and 1.5 μcal/cm² sec. The scatter of the Pacific results is greater than that of those from the Atlantic. This is probably due to the former being drawn from a much wider area.

To maintain a heat flow as large as that observed requires very large amounts of heat if it is to be continued through geological time. In 10^7 years the total heat flowing through 1 cm² amounts to 3×10^8 cal, which is the heat generated by the complete combustion of a layer of coal 270 m thick. The contribution to the heat flow from chemical and biological processes in the sediments is negligible compared to this, as also is that from radioactivity in the sediments and from compaction. These and other sources of heat have been discussed by Bullard *et al.* (1956).

The only adequate source of heat that has been suggested is radioactivity within the earth. The observed heat flow is about that which would be expected from the uranium, thorium and potassium contained in an earth built up from stony and iron meteorites. If the earth were in fact built up from meteorites, the radioactivity would initially be distributed uniformly through the mantle. This would lead to melting and probably to an upward migration of the radioactive elements, which tend to be concentrated in the materials crystallizing late in the process of solidification. On the continents it is probable that a large part of the radioactivity is now concentrated in the granites and gneisses of the crust. The oceanic basalts are less radioactive than continental rocks and produce about three times less heat per unit volume; the bulk of the observed oceanic heat flow must, therefore, come from beneath the "Moho". If the rocks beneath the "Moho" had as low a radioactivity as ultra-basic rocks or meteorites, the heat would have to come from great depths in the mantle and the temperature gradients needed to convey the heat to the surface by conduction would lead to melting. There are two ways of resolving this difficulty: either the radioactivity must be concentrated towards the top of the mantle so that most of the heat is generated in the upper 200 km, or there must exist some more efficient process than conduction to bring heat to the surface from great depths. Convection currents in the mantle might provide such a mechanism.

The high heat flows over the ocean ridges can hardly be accounted for by local concentrations of radioactive material and it seems most probable that their immediate cause is the intrusion of igneous rocks beneath the ocean floor. Vulcanism occurs today at several points on the Mid-Atlantic ridge and there seems nothing improbable in suggesting that dykes may be injected along the line of the ridge. Since a dyke 1 km thick cools in a time of the order of 10^5 years, a series of intrusions would be required. The temperature beneath the ridges must rise very rapidly with depth. The observed temperature gradient at the station on the Mid-Atlantic Ridge in Fig. 8 is 315°C/km. Beneath the sediments this will fall to about 130°C/km owing to the conductivity of basalts being greater than that of the sediments. The temperature would, therefore, reach the melting point of basalt at a depth of 5 to 10 km below the ocean floor. Intrusion and vulcanism seems likely to occur under such circumstances.

It is perhaps significant that the high heat flow on the Mid-Atlantic Ridge was measured in the central valley. This valley is a frequently observed feature of the ridge and is believed to run more or less continuously through the North and South Atlantic, round South Africa and up the ridges of the Indian Ocean, one of which runs into the Red Sea. In the Red Sea there is a trough similar to that on the ridge. From gravity and magnetic surveys, Girdler (1958) has concluded that this valley marks the site of a dyke, or series of dykes, 50 km in width. It would be of great interest to know if the heat flow is high in this valley also.

The relation of the Red Sea to the ridges of the Indian Ocean and the Atlantic appears similar to that of the Gulf of California to the East Pacific Rise. Here again, where the ridge runs into the narrow sea there are high heat flows. There is, however, a less well-marked trough and gravity high.

If high heat flows prove to be characteristic of mid-ocean ridges and to be due to intrusion, it will be necessary to explain why intrusion should occur on such a scale over such great distances. It has been suggested that the pattern of heat flow might indicate the locus of an ascending convection current in the mantle, though there is no compelling reason to suppose that this is so; if it were, the low heat flows on each side of the East Pacific Rise could be interpreted as the places where the current sinks again. The horizontal limbs joining the rising and sinking currents would then produce a horizontal force on the crust and perhaps tensional cracks, along which the dykes postulated to explain the high heat flow could rise. The irregularity of the heat flow on the ridges is consistent with its being associated with a swarm of dykes.

The reality of this picture and its connection with a possible widening of the oceans and with continental drift must await a more thorough survey of many features of the ridges.

The low heat flows observed in the Acapulco and South American trenches have no obvious explanation. A large thickness of poorly conducting sediments in the bottom of a trench would present a large thermal resistance and would tend to divert the heat flow and produce a reduced flow in the trench and an increased flow on each side. However, seismic studies do not show an exceptional thickness of sediments in these trenches and it seems that this explanation cannot be the correct one. If South America were overriding the Pacific, cold surface rocks from the ocean floor would be forced beneath the continental margin and the heat flow would be reduced. With a rate of drift of 0.2° of arc in 10^6 years (2.2 cm/year), which is of the order suggested by palaeomagnetism, the width of the trench (60 km) would be traversed in 3×10^6 years. The thermal time constant for a sheet of rock 5 km thick is of the order of 2×10^6 years, so that this rather fanciful explanation cannot be immediately rejected as thermally impossible. The high value of 2.72 μcal/cm^2 sec shown in Fig. 7 at the station on the landward slope of the trench in latitude 13°S would be difficult to explain on this theory.

5. Future Work

If the distortion of the rocks of the earth represents the working of a heat engine, the measurement of the variations of heat flow from place to place are of great importance. The present methods are tedious and do not always yield satisfactory results. More reliable and more easily used equipment should be developed.

The most urgent programme of measurement is to determine how generally high heat flows occur along ridges and to make measurements on closely spaced sections across them. Measurements in the Red Sea and along its junction with the ridges of the Indian Ocean would also be particularly interesting.

References

Bullard, E. C., 1954. The flow of heat through the floor of the Atlantic Ocean. *Proc. Roy. Soc. London*, **A222**, 408–429.

Bullard, E. C. and A. A. Day, 1961. The flow of heat through the floor of the Atlantic Ocean. *Geophys. J.*, **4** (*The Earth Today*), 282–292.

Bullard, E. C., A. E. Maxwell and R. Revelle, 1956. Heat flow through the deep sea floor. *Advances in Geophys.*, **3**, 153–181.

Cox, R. A. and N. D. Smith, 1959. The specific heat of sea water. *Proc. Roy. Soc. London*, **A252**, 51–62.

Foster, T. D., 1962. Heat flow measurements in the Northeast Pacific and in the Bering Sea. *J. Geophys. Res.*, **67**, 2991–2993.

Gerard, R., M. G. Langseth and M. Ewing, 1962. Thermal gradient measurements in the water and bottom sediment of the Western Atlantic. *J. Geophys. Res.*, **67**, 785–803.

Girdler, R., 1958. The relationship of the Red Sea to the East African rift system. *Q. J. Geol. Soc. London*, **114**, 79–105.

Hôrai, K. and S. Uyeda, 1960. Studies in the thermal state of the earth (V). Relation between thermal conductivity of sedimentary rocks and water content. *Bull. Earthq. Res. Inst.*, **38**, 199–206.

Jaeger, J. C., 1956. Conduction of heat in an infinite region bounded internally by a circular cylinder of a perfect conductor. *Austral. J. Phys.*, **9**, 167–179.

Maxwell, A. E. and R. Revelle, 1956. Heat flow through the Pacific Ocean basin. *Pubs. Bur. Cent. Seism. Intern. Trav. Sci.*, Fasc. 19, 395–405.

Menard, H. W., 1961. The East Pacific Rise. *Sci. Amer.*, **205**, 52–61.

Nason, R. D. and W. H. K. Lee. 1962. *Nature*, **196**, 975.

Ratcliffe, E. H., 1960. The thermal conductivity of ocean sediments. *J. Geophys. Res.*, **65**, 1535–1541.

Reitzel, J., 1961. Some heat flow measurements in the North Atlantic. *J. Geophys. Res.*, **66**, 2267-2268.

Revelle, R. and A. E. Maxwell, 1952. Heat flow through the floor of the Eastern North Pacific Ocean. *Nature*, **170**, 199–200.

Uyeda, S., Y. Tomoda, K. Hôrai, H. Kanamori and H. Futi, 1961. Studies of the thermal state of the earth (VII). A sea bottom thermogradmeter. *Bull. Earthq. Res. Inst.*, **39**, 115–131.

Uyeda, S., K. Hôrai, M. Yasni and H. Akamatsu, 1962. Heat flow measurements over the Japan Trench. *J. Geophys. Res.*, **67**, 1186–1188.

Von Herzen, R., 1959. Heat flow values from the Southeastern Pacific. *Nature*, **183**, 882–883.

Von Herzen, R. and A. E. Maxwell, 1959. The measurement of thermal conductivity of deep-sea sediments by a needle probe method. *J. Geophys. Res.*, **64**, 1557–1563.

Von Herzen, R. P., A. E. Maxwell and J. M. Snodgrass, 1962. *Temperature—Its Measurement and Control in Science and Industry*, **3**, pt. 1, 769–777. Reinhold Publ. Co., New York.

Note added in Proof

A very large number of measurements has been made recently in the Pacific; papers by Von Herzen, Von Herzen and Uyeda, and by Northrop, Rhea and Von Herzen are in press. W. H. K. Lee has collected all available information for 587 oceanic stations all over the world. He gets a mode of 1.1 μcal/cm^{2}sec and a mean of 1.6; the mean is too large owing to over-representation in the sample of the crests of the mid-ocean ridges. The mean away from the ridges is probably near 1.1 μcal/cm^{2}sec. The modes and the means for the oceans and the continents do not differ significantly.

Measurements by Von Herzen in the Indian Ocean show a pattern similar to that in the Atlantic and Pacific with high values on the crest of the Carlsberg Ridge and in the Gulf of Aden. The latter may be compared with the high values in the Gulf of California.

A Russian measurement of 1.9 μcal/cm^{2}sec in the Black Sea was published by Sisoev, N. N., 1961, *Okeanologiya*, **1**, 886–887.

II. TOPOGRAPHY AND STRUCTURE

12. TOPOGRAPHY OF THE DEEP-SEA FLOOR[1]

Bruce C. Heezen and H. W. Menard

1. Introduction

In *The Oceans*, Sverdrup, Johnson and Fleming (1942) state that, "from the [purely physical] oceanographic point of view, the chief interest in the topography of the sea floor is that it forms the lower and lateral boundaries of the water. The presence of land barriers or submarine ridges that impede a free flow of water introduces special characteristics in the pattern of circulation and in the distribution of properties and organisms." On the other hand, the geomorphologist is concerned primarily with the distribution and dimensions of the topographic features that occur on the submerged portions of the Earth's crust. The physical oceanographer views the sea floor as simply a bottom to the receptacle within which the ocean water rests; in contrast, the geologist's interest in topography is in what it can tell him concerning the geological structure, the erosional and depositional and tectonic history, and the origin of the ocean basins and their relief. Thus, attitudes, objectives and the systems of nomenclature used by these two groups differ widely.

The physical oceanographer divides the ocean into a series of broad basins separated by ridges. These basins are generally defined by the contours shown on the General Bathymetric Chart of the Oceans at a scale of 1 : 10,000,000. A map of ocean basins is essentially a map showing boundaries which lie along the crests of all major ocean ridges and along the margins of all land-masses (Wüst, 1940; see also Fig. 1). All the water contained in the depression between the land and the crest of a prominent ridge is considered to lie in a certain basin. One cannot deny the need and the usefulness of such a system, particularly in regard to problems of physical oceanography (Wiseman and Ovey, 1954, 1955). But this crude "bathymetric system" of classification is poorly suited to studies of submarine physiography, for the generalized bathymetric contours, on which the bathymetric system is based, cut across physiographic units with little regard for physiographic boundaries determined by a detailed study of the relief of the sea floor (Heezen *et al.*, 1959; see also Fig. 2).

The development of a new physiographic outlook toward submarine relief is the major advance in submarine topography since *The Oceans* was written. Physiographic studies of sea-floor relief were made possible by the development and wide use of continuous-recording echo-sounders. With the exception of the magnificent work done by the *Meteor* Expedition (Maurer and Stocks, 1933), no detailed profiles of the deep-sea floor were available prior to World War II.

[1] Lamont Geological Observatory (Columbia University) Contribution No. 592.

[*MS received July, 1960*]

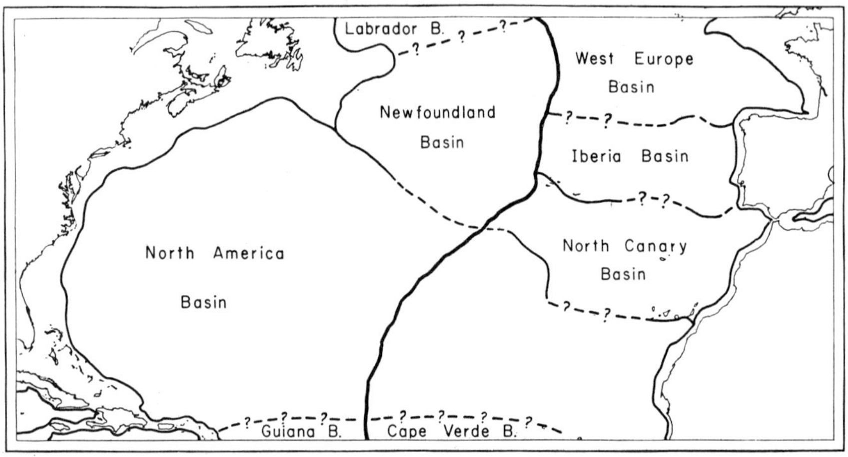

Fig. 1. Major basins of the North Atlantic Ocean. Heavy solid lines indicate boundaries formed by the axis of the Mid-Atlantic Ridge. Light solid lines indicate boundaries formed by shelf break and submarine ridges. Dashed lines indicate arbitrary boundaries. (After Wüst, 1940.)

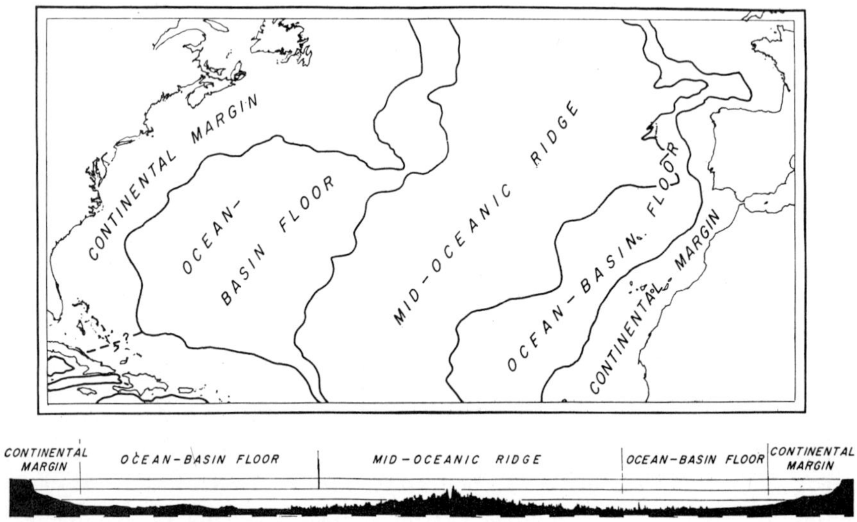

Fig. 2. Major morphological divisions of the North Atlantic Ocean. The profile is a representative profile from New England to the Sahara Coast. (After Heezen *et al.*, 1959.)

The *Meteor* profiles were constructed from closely spaced echo-soundings taken by extremely diligent workers. Continuous-recording echo-sounders were developed in the 1920's but none of these early instruments was suitable for use in the deep sea. It was not until World War II that continuously recording deep-sea echo-sounders were developed. Although the information shown on such

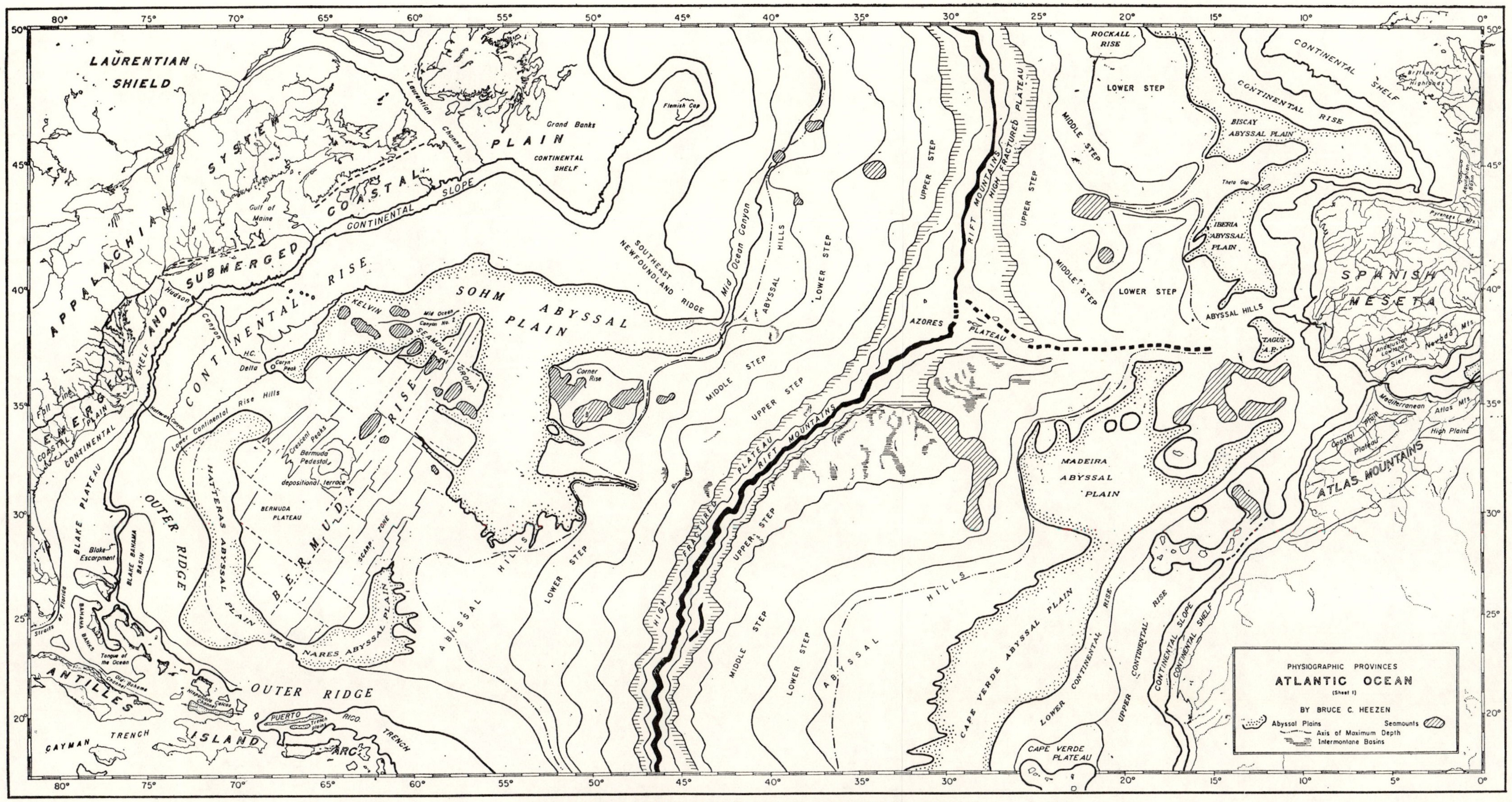

PHYSIOGRAPHIC PROVINCES, ATLANTIC OCEAN

Fig. 3. Physiographic provinces of the North Atlantic Ocean. (After Heezen *et al.*, 1959.)

records was exciting and absolutely new, it soon became obvious that much critical detail was being lost due to the small size of the record. In addition, precision of timing was not guaranteed and large errors in depth were common.

Between 1946 and 1952 several oceanographic institutions modified sounders to give larger records and attached constant frequency units. Suitable for more advanced studies of physiography than previously possible, these makeshift models were still inadequate.

The Precision Depth Recorder and the Precision Graphic Recorder developed in 1953 and 1954 now achieve an accuracy better than 1 fm in 3000 when operated at recording speeds of 18 inches = 1 sec (400 fm).

In summary, then, the systematic study of the physiography of the deep-sea floor was not made possible until the continuously recording echo-sounder was put into widespread use following World War II; and it has been possible to study the smaller details of topography only since the development of the first precision sounder in 1953 (Luskin *et al.*, 1954).

The study of the geomorphology of the deep-sea floor is in a very exciting stage. So few areas have been sounded adequately that it is impossible to make a cruise with a precision sounder and not discover entirely new physiographic features. The next decade should be one of the most exciting in geomorphology, for in this period enough data will have been collected to produce a fairly comprehensive geomorphology of the world oceans.

Although it is the genetic side of physiography that separates it from "topography" and lends it an absorbing interest, physiographic classifications need not be purely genetic ones. Indeed, the state of knowledge of sea-floor topography rather prohibits the use of an entirely genetic classification. Thus, the general classification used in this chapter is largely a descriptive one.

In writing this chapter, the authors have drawn heavily upon papers written in collaboration with their co-workers. Particular reference in this regard should be made to the following papers: Heezen, Tharp and Ewing (1959), Menard (1959), Menard and Dietz (1951), Dietz and Menard (1953), Menard (1955), Heezen, Ericson and Ewing (1954), and Ewing and Heezen (1955).

2. Outline of Submarine Topography

The relief of the Earth lies at two dominant levels: one, within a few hundred meters of sea-level, represents the normal surface of the continental blocks; the other, between 4000 and 5000 m below sea-level, comprising over 50% of the Earth's surface, represents the deep-sea floor. The topographic provinces beneath the sea can be included under three major morphologic divisions: continental margin, ocean-basin floor, and mid-oceanic ridge. These are indicated on a typical trans-oceanic profile taken from the North Atlantic in Fig. 2. Each of these major divisions can be further divided into categories of provinces and the categories divided into individual physiographic provinces (Fig. 3).

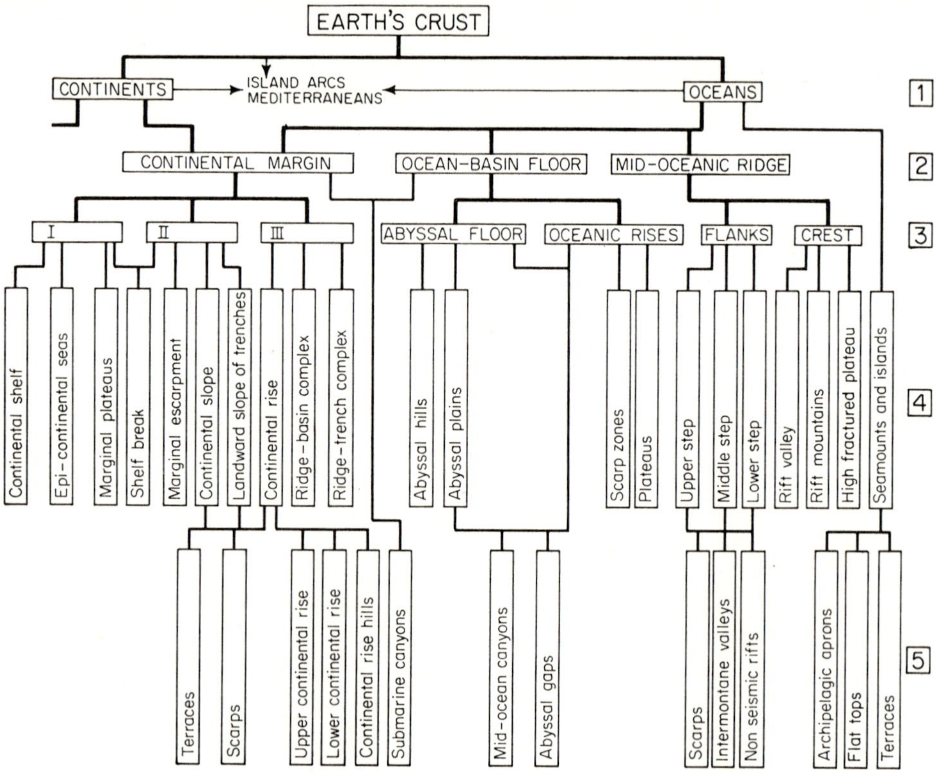

Fig. 4. Outline of submarine topography. Line 1, first-order features of the crust; line 2, major topographic features of the ocean; line 3, categories of provinces and super-provinces; line 4, provinces; line 5, sub-provinces and other important features. (After Heezen *et al.*, 1959.)

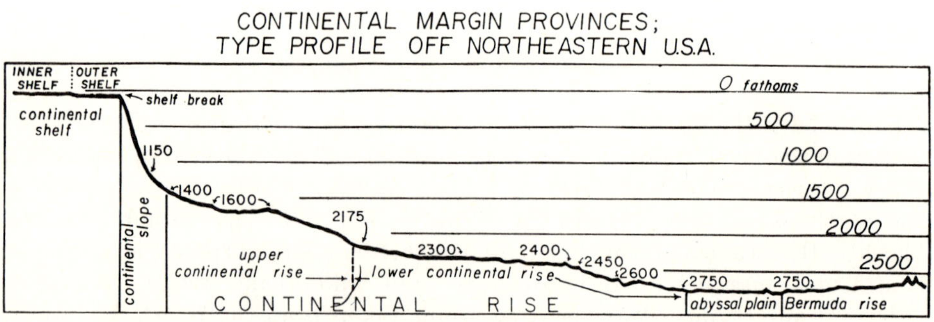

Fig. 5. Continental margin provinces: type profile off northeastern United States. Profile plotted from P.D.R. records. This profile is representative of the sector from Georges Banks to Cape Hatteras, eastern United States. (After Heezen *et al.*, 1959.)

A. Continental Margins

a. Physiographic provinces

The continental margin includes those provinces associated with the transition from continent to ocean floor. The continental margin in the Atlantic and Indian Oceans is generally composed of continental shelf, continental slope, and continental rise. A typical profile off northeastern United States is shown in Fig. 5. Gradients on the continental shelf average 1 : 1000, while on the continental slope, gradients range from 1 : 40 to 1 : 6, and occasionally local slopes approach the vertical. The continental rise (or continental apron) lies at the base of the continental slope. Continental rise gradients average 1 : 300, but individual slope segments may be as low as 1 : 700, or as steep as 1 : 50.

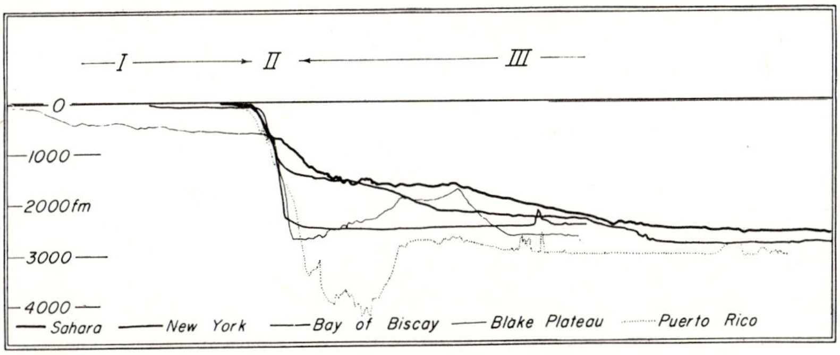

Fig. 6. Three categories of continental margin provinces. Category I provinces lie on the continental block, Category II provinces form the side of the continental block, and Category III provinces are the upturned or depressed margins of the oceanic depression. (After Heezen *et al.*, 1959.)

The continental slopes are cut by many submarine canyons. Some of the larger canyons, such as the Hudson, Monterey and Congo, extend across the continental rise (Fig. 7). Submarine alluvial fans extend out from the seaward ends of the larger canyons. The continental shelf and continental slope form the upper and lateral surfaces of the continental terrace (Dietz and Menard, 1951).

The continental margin can be divided into three categories of provinces (Fig. 6). Category I includes the continental shelf, marginal plateaus and shallow epicontinental seas, all slightly submerged portions of the continental block. Category II includes the continental slope, marginal escarpments and the landward slopes of marginal trenches, all expressions of the outer edge of the continental block. Category III includes the continental rise, marginal basin–outer ridge complex and the marginal trench–outer ridge complex. The continental slope of northeastern United States can be traced directly into the marginal escarpment (Blake Escarpment) off southeastern United States (Fig. 5) and the landward slope of the Antilles marginal trench (Puerto Rico Trench). The lower continental rise off New England can be traced into the Antilles Outer Ridge. Seismic-refraction studies show that a trench filled with

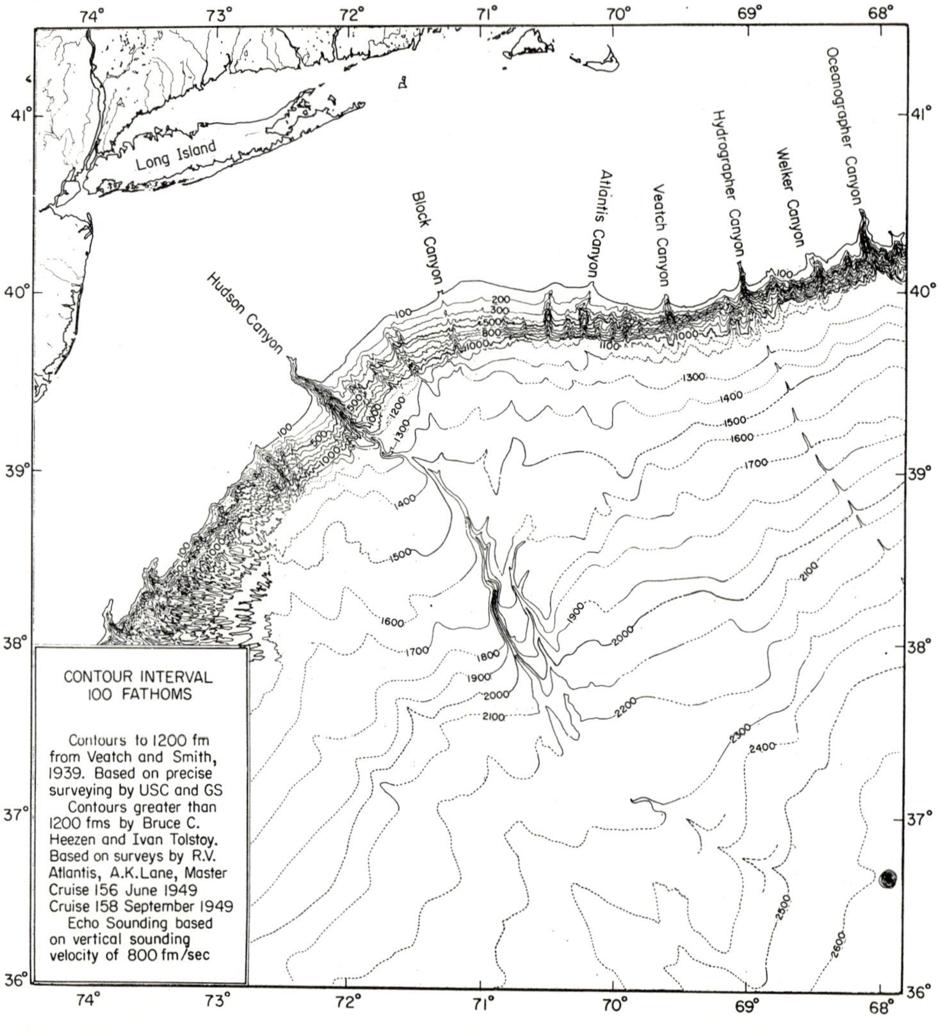

Fig. 7. Preliminary chart of Hudson submarine canyon. (After Heezen *et al.*, 1959.)

sediments and sedimentary rocks lies at the base of the continental slope off New England (Fig. 8). Thus the main difference in morphology in many places between the trenchless continental margins and those with a marginal trench appears to be that, in the former, the trench has been filled with sediments. Elsewhere, as in the Gulf of Alaska, a buried marginal trench has not yet been revealed by seismic stations.

In the continental margins of the Atlantic, Indian, Arctic and Antarctic Oceans, and the Mediterranean Sea, the continental rise generally represents the Category III provinces. The Pacific, however, is bounded by an almost continuous line of marginal trenches. The high seismicity, vulcanism and

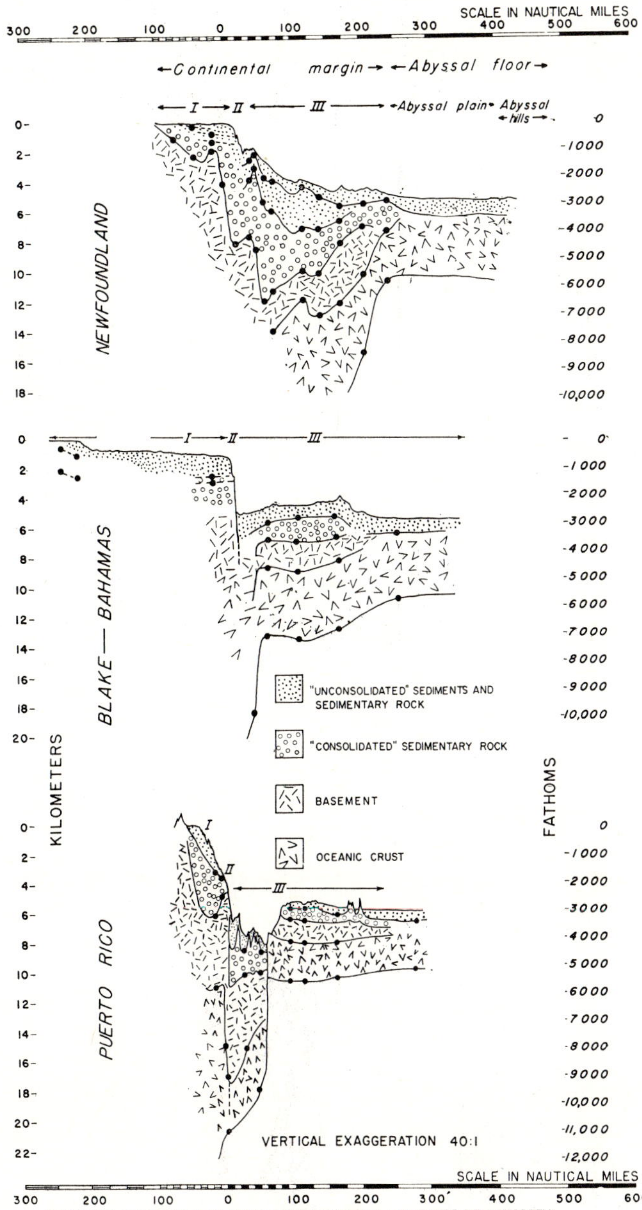

Fig. 8. Crustal structure and continental margin provinces. The three profiles represent typical areas of the continental margin of North America. They may represent a genetic sequence, starting with an unfilled trench, such as Puerto Rico, and ending with a filled marginal trench, as illustrated by the Newfoundland profile. (After Heezen *et al.*, 1959.)

youthful relief of the Pacific borders all suggest a recent origin. In contrast, the non-seismic, non-volcanic character, as well as the lower relief of the Indian and Atlantic margins, suggests a greater age. Thus, on the old, stable continental margins, the deposition of sediments derived from the land has filled the marginal trenches and produced the continental rise. The local relative relief on the continental margin rarely exceeds 20 fm, with the major exceptions of submarine canyons and occasional seamounts.

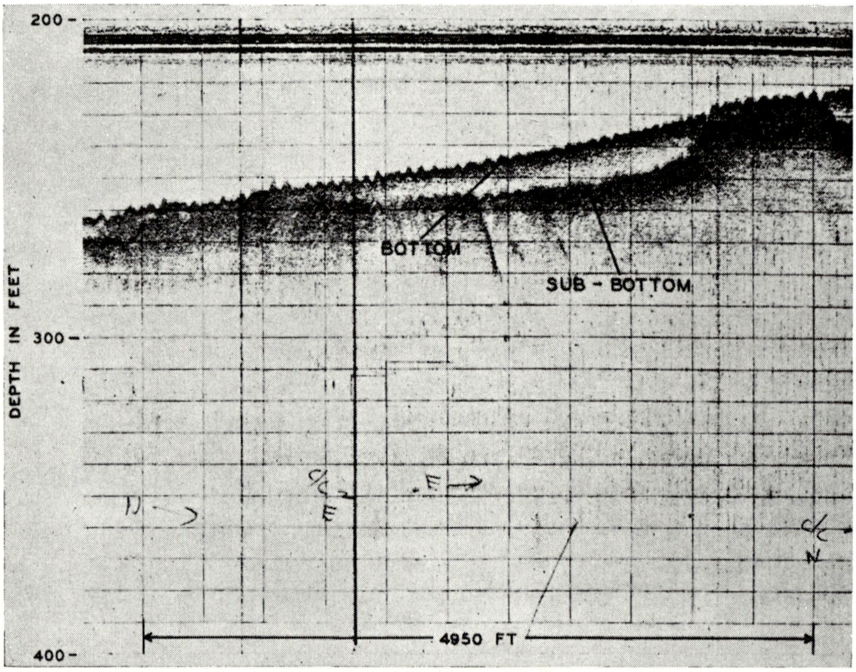

Fig. 9. Sonoprobe record on the continental shelf off San Diego, showing buried bedrock surfaces. Note the smooth bottom-sediment profile and the easily detected, irregular, sub-bottom rock profile, illustrating a buried beach on the continental shelf. (After Moore, 1960.)

b. Topographic forms of the continental shelf (see Chapter 13)

Submerged marine beach terraces have been identified on the continental shelf throughout the world. These terraces represent various Pleistocene low levels and are preserved in varying degrees of completeness. In some areas subsequent deposition has badly obscured these ancient beaches (Fig. 9). In general, the continental shelf has a very low monotonous relief. But off major rivers, such as the Hudson, partially buried ancient river channels can be identified, and in certain well-surveyed areas, what appear to be ancient beach ridges can be recognized.

"The redistribution of sediments by tidal streams has been demonstrated in a number of ways. For example, the numerous banks in the North Sea are elongate, parallel to the streams; while the depth and position of certain channels are subject to such important changes in position that resurveys have to be made each year or so. In the same region, the streams, reaching about a knot or more at the surface, have brought a large area of sand into ridges normal to their path. Off the Dutch coast, the grade of the sand decreases in their inferred direction of advance. Similar relief is found even at 90 fm near the edge of the continental shelf at the western approaches to the English Channel" (Stride, 1959). Sand ridges on the continental shelf were first recognized on records made by normal echo-sounders. But a recently developed, horizontally beamed echo-sounder has provided a much better method of studying small-scale linearities on the shelf. In addition to studying in more detail the previously recognized sand ridges, Stride (1960) has found that in some areas, notably in the area south of the Dorset coast, the curving linear patterns represent the outcrop of sedimentary beds. In areas totally devoid of recent sediment, small cuestas a few feet high mark the outcrop pattern of beds of varying resistance. In this manner, he has traced anticlines, synclines and fault lines from near the shore out to a depth of 40 or 50 fm. There is a misleading tendency to assume that sediment covers the sea floor everywhere. Now that we realize that strong currents have swept various areas of the continental shelf clean of sediments, it is not at all surprising that the topographic expression of outcrop patterns long recognized in the emerged coastal plain are also seen in the submerged coastal plain. Thus, the topography of the continental shelf differs only slightly from that which can be observed on the emerged adjacent coastal plains.

Structural benches, the topographic expression of outcropping beds, have been identified on the continental slope (Heezen *et al.*, 1959). Near Cape Hatteras, Virginia, structural benches have been dated by extrapolating data obtained in several test borings near the coastline (Fig. 10). Benches on Georges Bank have been dated by dredging, and the dips of beds have been deduced by correlation of structural benches along the walls of submarine canyons. It appears that, through the action of slumps, bottom currents and turbidity currents, sediments are continually being removed from the continental slope. There is little cover of recent sediments to obscure the outcrops of the ancient formations.

c. Shelf break

An abrupt change in slope known as the shelf break marks the boundary between the continental shelf and continental slope (Fig. 11). The slope change has been variously attributed to: (1) sedimentation at wave base and in equilibrium with present sea-level; (2) sedimentation at wave base but related to some past lower sea-level; (3) wave abrasion in equilibrium with present sea-level; and (4) wave abrasion related to some past lowered sea-level.

Modern knowledge of the activity of waves on the continental shelves has resulted in the conclusion that the break in slope is caused by wave abrasion

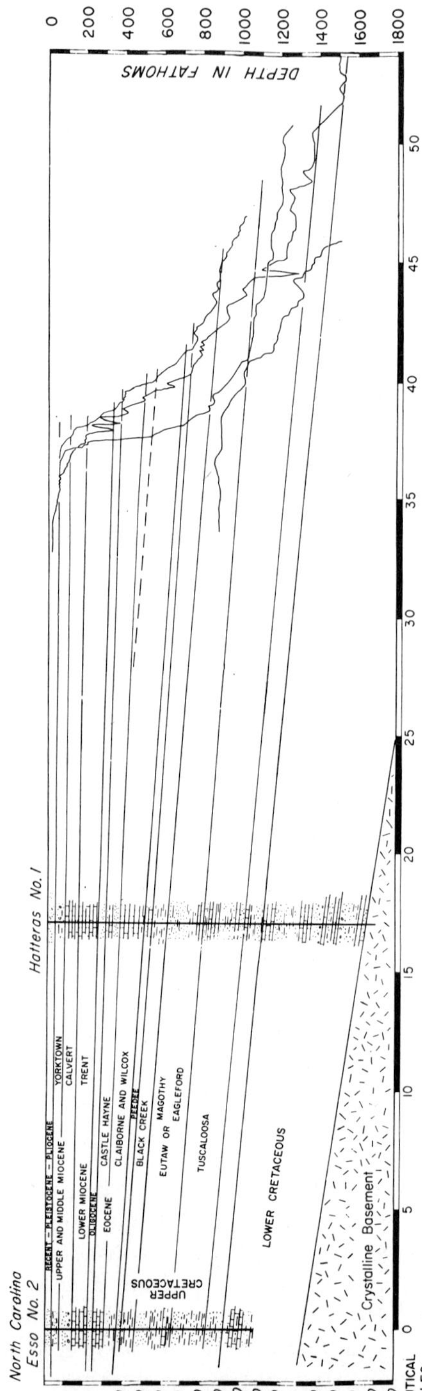

Fig. 10. Geological section at Cape Hatteras, Virginia, showing prominent outcrop benches on the continental slope. The four sounding profiles are projected to the line of the profile. Note the resistant formations on prominent structural benches on the continental slope. (After Heezen *et al.*, 1959.)

related to a sea-level lowered to within about 5 fm of the break (Dietz and Menard, 1951). It is clear, from several lines of reasoning, that the sea-level has recently risen eustatically, and that, in general, this recent rise has taken place more rapidly than changes of land level affected by erosion, sedimentation, or diastrophism. However, the shelf break falls at different depths in different areas, ranging from 12 fm to 250 fm. The mean depth of the shelf break is approximately 73 fm. Heezen *et al.* (1959) suggested that the lack of agreement in the depth of the shelf break is due to the fact that shelf break does not everywhere represent a single ancient eustatic level. Thus, the most prominent break may locally be a Pliocene or Miocene structural bench, but elsewhere, late Pleistocene or Recent strata may form the shelf break. In some areas the

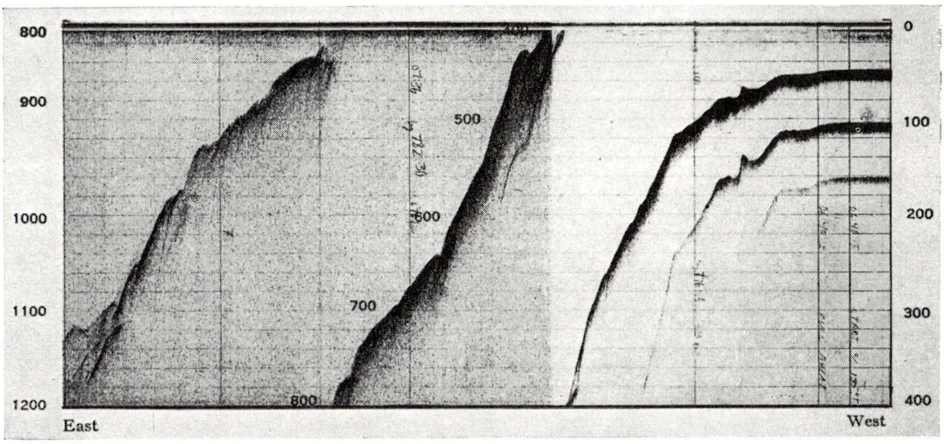

Fig. 11. Shelf break in the continental slope off New York. The record was made with the Precision Depth Recorder. (After Heezen *et al.*, 1959.)

late Pleistocene eustatic low level may truncate the continental slope and there the shelf break may truly represent a late-stage eustatic sea-level.

The deeper structure of the continental margin indicates a fundamental structural discontinuity at the base of the continental slope (Fig. 8). It would seem a small extrapolation to attribute a fault origin to the continental slope. Although faulting may have played a part in the earliest history of the Category II provinces, alternate periods of sedimentation and marine planation on the continental shelf, and long-continued erosion by slumps, turbidity currents and deep-sea currents on the continental slope, together with a general subsidence of the area, could alone have produced the characteristic form of the continental terrace.

d. Submarine canyons

Submarine canyons were early recognized and have captured the imagination of several generations of geologists and hydrographers (see Chapter 20).

These deep gorges, cutting all the continental slopes of the world, vary from less than a mile to ten miles in width, from a few fathoms to several hundred fathoms in depth, and from a few miles to a few hundred miles in length (Figs. 7 and 12). The slopes of the walls of submarine canyons are generally

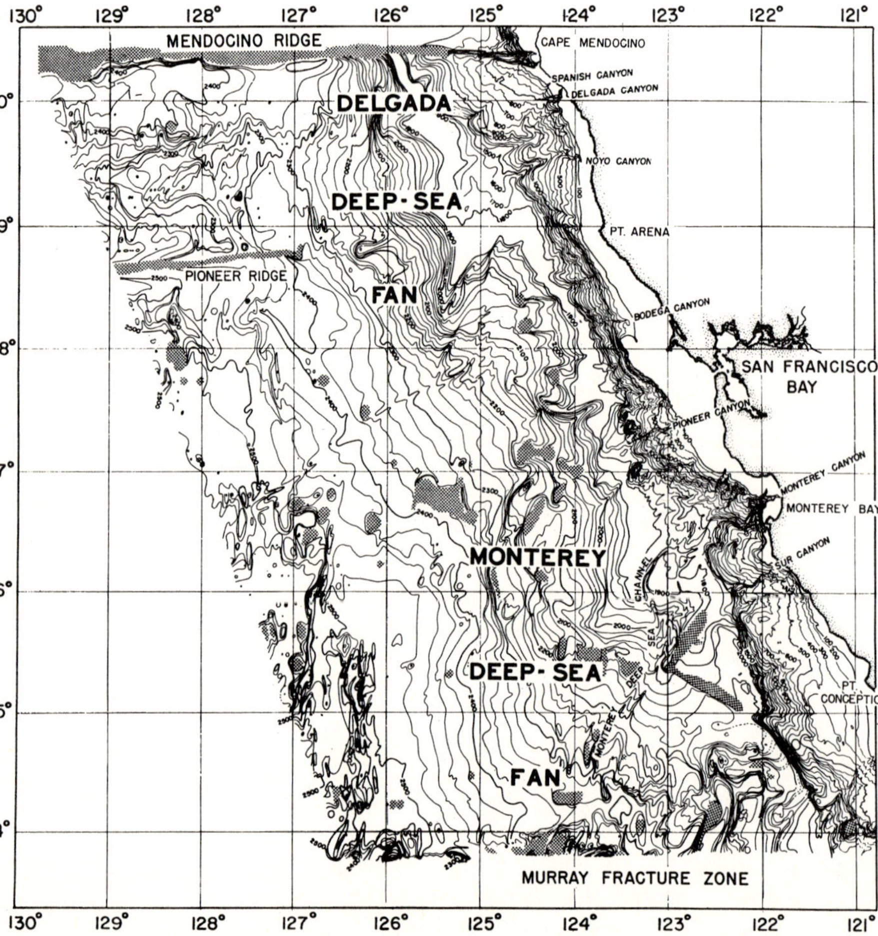

Fig. 12. Topography of Central California. Counter interval 20 fm below 1900 fm. Local elevations indicated by pattern of dots. Note ranges of hills west of the Monterey fan. (After Menard, 1960.)

greater than the gradient of the continental slope. Some investigators (Menard, 1955) prefer to limit the term "submarine canyon" to that portion of each feature which is incised into the continental slope, and to refer to that portion of the same feature which cuts into the continental rise as a deep-sea channel. Other investigators (Heezen *et al.*, 1959; Ericson *et al.*, 1951) prefer to refer to the entire system as a "submarine canyon", regardless of the fact that the size of the feature sharply decreases on the lower gradients of the continental rise.

The structural benches which are present on the continental slope are somewhat more clearly defined in the walls of the submarine canyons.

e. Marginal plateaus

A marginal plateau is a shelf-like feature having a greater depth than the continental shelf and separated from the continental shelf by a low continental slope. The surfaces of the marginal plateaus off Pernambuco Province, Brazil, and west of Angola are irregular in profile. In contrast, the Blake Plateau is superficially similar to the continental shelf in slope and in the frequency and magnitude of relief features. On closer examination, however, it is found to be much rougher than the continental shelf and is dominated by relief which appears to be primarily tectonic in character.

The Blake Plateau is a well-developed representative of this morphologic type. In the area of the Blake Plateau, the shelf break lies parallel to the coast about 60 miles offshore (Figs. 3, 13). The continental slope extends at a gradient of 1:40 only to depths of 300 to 400 fm, where the lower gradients, i.e. about 1 : 1000, of the Blake Plateau are found. The main Blake Plateau is 170 miles wide (east–west) and extends from the Grand Bahama Island to 30°N. The Blake Escarpment forms a precipitous drop to abyssal depths along the eastern edge of the plateau. The top of the Blake Escarpment lies at about 550 fm and its base at 2600 fm. The escarpment is typically formed by two or three distinct slope segments. The outer part of the Blake Plateau is typically smooth, with very low and rolling topography. The inner part, however, lying at the base of the continental slope, is broken by many small, steep-sided valleys 1 to 3 miles wide, and from 5 to 20 fm deep. The trend of this topography appears to parallel the continental margin and one trough has been traced for a considerable distance by Hersey.

The outer edge of the Blake Plateau breaks off abruptly at about 550 fm. Here gradients increase to 1 : 30. This segment continues with a few minor breaks to a depth of 1200–1500 fm, where a narrow bench occurs. Below this bench the escarpment drops so steeply that only a few side echoes are recorded on the echogram. The gradient here exceeds 1 : 2 in several profiles. At 2400 fm there is another narrow bench in the southern part of the area, but toward the north the abyssal plain of the Blake–Bahama basin lies directly at the foot of the steepest segment. A peculiar fact is that along many east–west cross-sections, the deepest point in the basin lies directly at the foot of the escarpment. A similar deepening, adjacent to the Campeche and West Florida escarpments in the Gulf of Mexico, has also been found.

Marginal plateaus and marginal escarpments are found off Recife, Brazil, Rio de Janeiro, Brazil, Angola and the Congo (Heezen *et al.*, 1959). A very similar feature extends for several hundred miles eastward from the Falkland Islands.

Much of the great Melanesian subcontinent resembles in depth and in topography the smaller marginal plateaus mentioned above. The area off Southern

California, known as the continental borderland, might be considered a highly fractured marginal plateau. Here, basins adjacent to shore in the continental margin are being filled by turbidity currents while the outer basins receive only pelagic sediments. The ancient Los Angeles basin is completely filled and the modern basins lying seaward of it are rapidly filling (Gorsline and Emery,

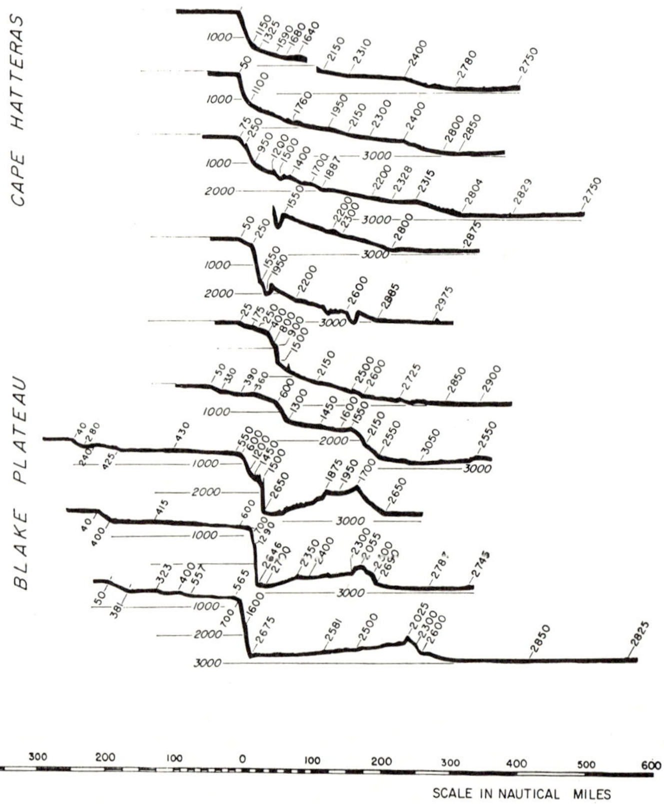

Fig. 13. Ten topographic profiles east of Cape Hatteras, Virginia, and the Blake Plateau. The sequence of profiles shown in this figure illustrates the transition from the typical continental shelf, continental slope, continental rise profiles of the northeastern United States to the more complex profiles across the Blake Plateau and marginal basin of the southeastern United States. Note the well-developed marginal escarpment and outer ridge on the lower three profiles. (After Heezen *et al.*, 1959.)

1959). It can be safely predicted that if tectonic activity in the area diminishes, sedimentation will eventually completely fill the offshore basins of the continental borderland one by one and produce a shelf-like feature which may eventually resemble a marginal plateau very similar to the Blake Plateau. The analogy seems rather good when one compares the deeper structure revealed by results of seismic-refraction work on the Blake Plateau with the modern topo-

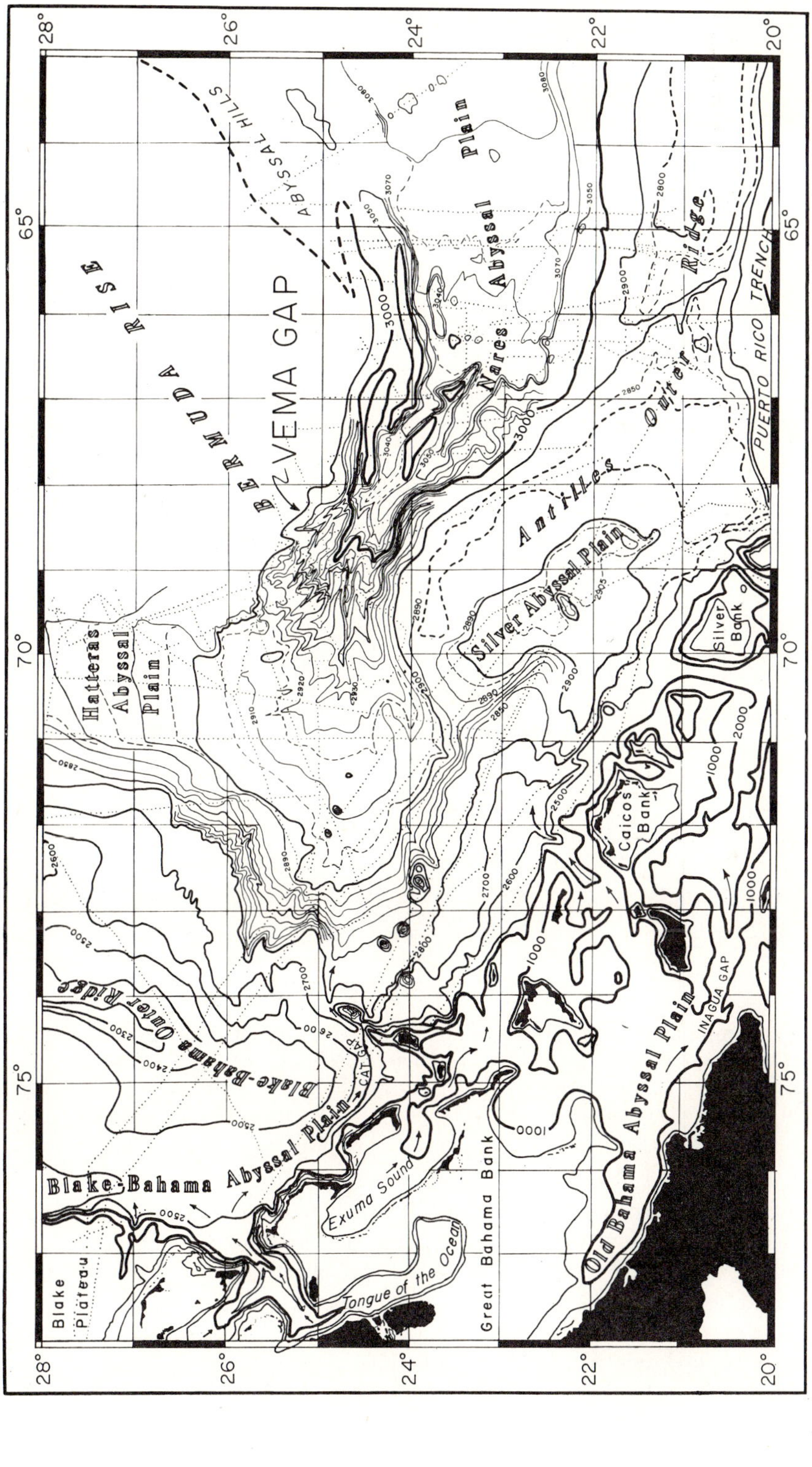

Fig. 14. Outer ridge, abyssal plains and abyssal gaps east of the Bahama Islands, Contoured from Lamont Geological Observatory precision soundings (lines of dots) by Bruce C. Heezen.

graphy of the Southern California continental borderland. The Blake Plateau apparently consists of many basins filled to overflowing.

f. Continental rise

The topography of the continental rise (continental apron), which lies at the base of the continental slope, is generally smooth (Fig. 14). Although minor irregularities are not uncommon, these irregularities are rarely more than

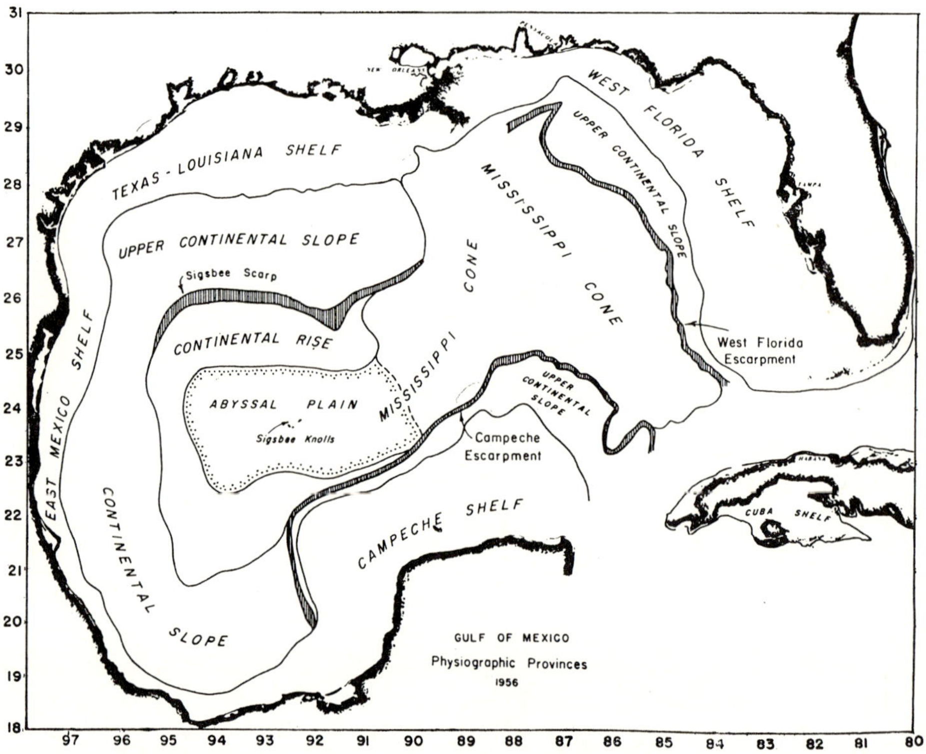

Fig. 15. Physiographic provinces of the Gulf of Mexico, showing the prominent Mississippi Cone. The Mississippi Cone dominates the floor of the Gulf of Mexico and turbidity currents traversing this cone have built the Sigsbee Abyssal Plain. (After Ewing, Ericson and Heezen, 1958.)

10 fm in amplitude. Occasionally, the continental rise is punctuated by numbers of rather large seamounts often linear in pattern (e.g. off the eastern coast of Brazil, the northwest coast of Africa, the northeast coast of the United States, Peru, the west coast of Central America, the coast of Baja California and the Gulf of Alaska). The continental rise may often be divided into two parts: the upper continental rise and the lower continental rise. A typical example is shown in Fig. 5. Although in several areas, notably central

California, the continental rise is bounded on the seaward side by abyssal hills, in most areas an abyssal plain intervenes. In general there is a fairly abrupt change in gradient from a gradient steeper than 1 : 1000 on the continental rise to one flatter than 1 : 1000 on the abyssal plain. In some areas, such as off northeastern United States, a line of lower continental-rise hills falls along this boundary.

g. Deep-sea fans and abyssal cones

A turbidity current emerging from a submarine canyon onto the relatively flat sea floor of the lower continental rise or the abyssal plain will be checked in velocity and will deposit some of its sediment. Successive deposits at the mouth of a single canyon will ultimately build a deep-sea fan. Well developed fans occur off the west coast of North America from the Gulf of Alaska to Southern California (Menard, 1955). If the submarine canyon remains relatively stable in position for a long period of time, one would expect a family of overlapping deep-sea fans gradually to build up a larger feature. Some investigators (Menard) do not distinguish between the smaller fans and the resulting larger composite feature (Fig. 11). However, Ewing, Ericson and Heezen (1958) make a distinction between the smaller features known as deep-sea fans and the larger composite ones which they term abyssal cones (Fig. 15). Abyssal cones are found off Mississippi (Ewing *et al.*, 1958) and the Congo (Heezen *et al.*, 1957) and recent investigations show that both the Ganges and Indus have abyssal cones.

After building in one area for a long time, a deep-sea fan can be expected to reach a stage at which the active turbidity-current channel breaks through one of its natural levees and begins to build a new distributory channel and new deep-sea fan. Distributory channels extend across the abyssal cone (or large deep-sea fan), functioning in much the same way as the distributory channels in the delta of a large river such as the Mississippi.

In most areas where the continental slope or marginal escarpment drops directly into a marginal basin, there is a marginal trench and associated outer ridge. An exception occurs off Southern California where abyssal hills lie at the base of the continental slope, and a trench is absent. The topography of the outer ridge is quite similar to that of the continental rise, with the exception of being slightly more irregular (Fig. 16). The marginal trenches, such as the Puerto Rico Trench, Japan Trench, the Kurile–Kamchatka Trench, etc., are bounded by very steep walls. These walls are broken into small but discernible steps that are generally interpreted as fault splinters along the major faults bounding the trenches.

A small abyssal plain lies in the floor of most marginal basins. This characteristic is well illustrated by the Blake–Bahama Abyssal Plain, lying at the foot of the Blake Escarpment. In the Puerto Rico Trench, the trench plain reaches a maximum width of 12 miles. The Acapulco Trench Abyssal Plain, at its maximum development, is 6 miles in width. Some trench plains are often less than a mile in width and appear only as a series of later echoes on the echogram, arriving after the echoes from both sides of the trench (Fisher, 1954). In some areas where detailed surveys have been made of the continental margin, it has been discovered that much of the fine-scale relief is subparallel to the adjacent margin. This seems to suggest that the tectonic relief throughout the continental margin province is controlled by a single structural system, which lies predominantly parallel to the continental slope.

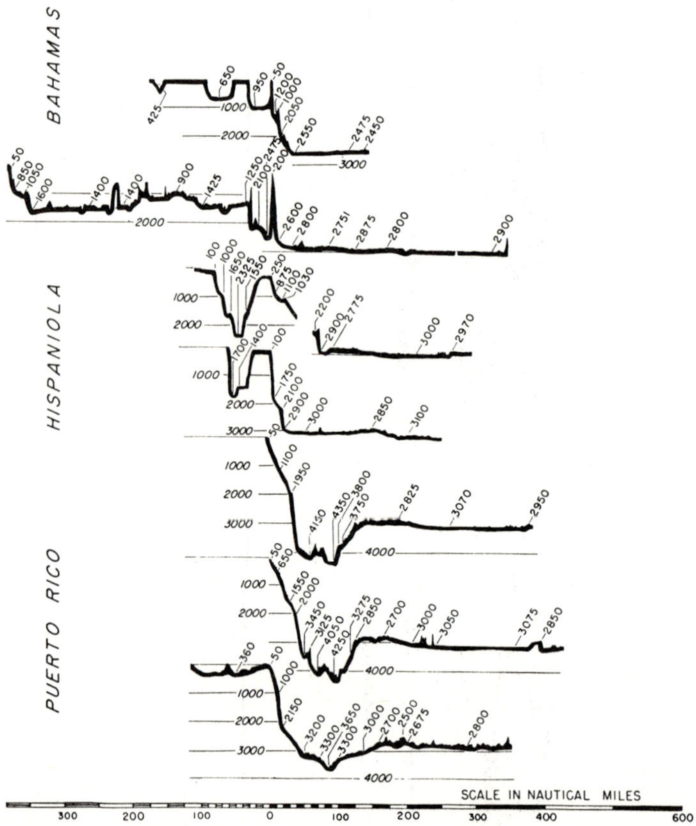

Fig. 16. Profiles illustrating the marginal trench and outer ridge in the northern Caribbean. (After Heezen *et al.*, 1959.)

h. Island arcs

Lying on the northwestern margin of the Pacific Ocean are a series of great island arcs—the Aleutians, the Kuriles, the Japanese and the Marianas. The outer convex side of these island arcs is considered to be the seaward limit of the Asian continent by some petrologists because of the occurrence of andesite. However, consideration of other criteria, such as the depth of water or crustal thickness, makes it clear that most of the great island arcs lie within the Pacific Basin rather than at the margin of the continent. The Philippine Sea is roughly

1000 miles wide between the continent and the Marianas Islands and it is deeper than normal oceanic depth. The few seismic stations available suggest a relatively normal oceanic crust. Likewise, the Aleutian Islands are separated from the continent by an area of normal oceanic crust which appears shallow because of a thick fill of sediments in the Bering Sea Basin. Nevertheless, the island arc trenches lying along the convex sides of these arcs may be considered analagous to the marginal trenches observed along the Pacific coasts of Central and South America and off the north coast of Puerto Rico (see Chapter 17).

Trenches and island arcs were among the first surveyed features of oceanic topography because they were barriers to submarine cables. The island arcs include lines of active volcanoes and they are associated with more large earthquakes and more pronounced gravity anomalies than any other part of the Earth's crust. The more or less arcuate trenches bordering them are the deepest places in the ocean, which is particularly noteworthy because they are near the margins of the ocean basins rather than in the middle. At present, the greatest depth known is about 10,600 m in the Marianas Trench,[1] but very similar depths have been found in the Kurile and the Tonga Trenches.

A second class of arcs, the Antillean Arcs of the Atlantic, are generally considered as extensions of the Pacific tectonic belt into the Atlantic Ocean. The West Indies and Scotia Arcs are more complex than the North Pacific Island Arcs, and are more sharply curving; and the deep-sea trenches associated with them are more limited in extent, either due to the fact that they were originally smaller or that they have been filled. The Indonesian area is similar to both the great island arcs of the North Pacific and the more contorted Antillean Arcs. The Banda Sea appears to be quite similar to the Antillean Seas, whereas the Java–Sumatra area has many similarities to the Japan and Kurile–Kamchatka area. Again, the Philippine Arc seems to fall in a class very close to the Japan Arc but lies landward from the Marianas Arc.

The typical Antillean Arcs and typical North Pacific Island Arcs are associated with deep-sea trenches. However, there is a third general occurrence of deep-sea trenches that has been mentioned before, i.e. along the continental margins of Central America and South America and along the eastern margin of the Melanesian sub-continent.

The epicontinental marginal seas of northeast Asia, the Bering, Okhotsk and Japan, lying as they do between a volcanic island arc and the continent, are rapidly filling with sediments supplied both from vulcanism and from erosion of the adjacent stable land-masses. If subsidence is arrested and sedimentation goes on, the Bering Sea may eventually reach a stage similar to the present Okhotsk Sea, and the Okhotsk Sea may eventually become a shallow sea similar to the Japan Sea. Further, the Japan Sea may eventually be filled to such a degree that it greatly resembles the East China Sea, which lies landward of the Ryukyu Arc. Kay (1951) has pointed out that the Paleozoic Appalachian geosyncline was probably similar to modern island arcs of the North

[1] In 1962 a new record depth was measured from H.M.S. *Cook*.

Pacific. Thus, we may anticipate, just as the Appalachian geosyncline is now part of the North American continent and as the East China Sea is now virtually part of the Chinese continental shelf, that the Japan, Okhotsk and Bering Seas will also become incorporated in the Asian continent in some future geologic period.

The Bering Sea is composed of two parts, the broad continental shelf extending from the mainland of Alaska, and the deeper southwestern part lying north of the Aleutian Islands between Dutch Harbor and Attu Island. An abyssal plain with sediment several kilometers thick lies in the deepest part of the Bering Sea, bearing evidence that the sea is being filled by turbidity currents. The floor of the Okhotsk Sea is similar to that of the Bering Sea except that the deeper portion covers a much smaller area. It, too, is occupied by an abyssal plain. The minor relief features of the Okhotsk Sea have been described in some detail by Udintsev (1957), who has proposed a complex genetic classification of the minor relief forms of the floor of the Okhotsk Sea.

i. Margins of a subcontinent

The area east of Australia and north of New Zealand, extending to a few degrees north of the equator, is part continental and part oceanic in character. It lies from a few hundred to a few thousand meters below sea-level; the deeps have an oceanic crust and the banks have a continental one. Although the region differs markedly in surface relief from the continents, its margins with the ocean basins are equally sharp.

To the east lie the Tonga and Kermadec Trenches, and the block-faulted region is modified into a double island arc very similar to the Marianas Islands except for curvature. To the north the edge of the region is marked by a dense group of submerged atolls such as Turpie and Alexa Bank rising from a region of relatively shallow depths. An apparently normal, smooth continental rise exists along the east part of the northern boundary of the region, but to the west, toward the Santa Cruz Islands, the base of the slope is lined with deep troughs such as the Cape Johnson and Vitiaz Troughs. This little known region may provide a perfect example of a transition between an oceanic trench and a continental rise.

j. The Mediterranean

The Mediterranean Sea, which separates Europe and Africa, has many of the characteristics of an ocean basin. The continental shelves are moderately well developed in the western Mediterranean. The continental slopes, which seem to be of the same form as those of the Atlantic, are cut by submarine canyons that appear to have the same characteristics as submarine canyons found elsewhere in the world. The floor of the western Mediterranean is occupied by the Balearic Abyssal Plain. Judging from the gradients in depth of the basin, the plain has been dominated by deposition from the Rhône and the submarine canyons of Alpes-Maritimes. However, deposition from submarine canyons along the Algerian coast has also occurred.

The Tyrrhenian Sea is floored by a circular abyssal plain and in many ways is similar to the western Mediterranean basin. The fractured continental shelf extending from Tunisia through Sicily to Italy divides the eastern and western Mediterranean into two separate basins. In the eastern Mediterranean, the continental shelves and slopes, particularly in the Levant, are similar to those in other parts of the world. However, in the region of Crete and in the Aegean Sea, complex faulting and apparently a great deal of subsidence have produced the unique irregular topography of the Aegean Sea. The floor of the eastern Mediterranean is also radically different from that of the western Mediterranean. Although there is some smoothing of the topography seaward of the Nile Delta, and a small abyssal plain south of the Straits of Messina, the remainder of the eastern Mediterranean is complexly broken up by a fine-scale relief having few counterparts in other areas of the ocean. We can only conclude that the balance between tectonic activity and sedimentation in the eastern Mediterranean is strongly in favor of tectonics. This is true despite the sizable volume of sediments which must be distributed in the eastern Mediterranean by turbidity currents and surface turbidity originating at the mouth of the Nile. Thus, we must conclude that the eastern Mediterranean has recently been drastically deformed and that this deformation is still in progress.

B. Ocean-Basin Floor (see Fig. 4)

The second of the three basic divisions of the oceans includes the broad floors of the ocean basins. Approximately one-third of the Atlantic, one-third of the Indian, and three-fourths of the Pacific Ocean fall in this major division. The ocean-basin floor can be divided into two categories of provinces: (1) abyssal floor; (2) oceanic rises and asymmetrical, aseismic ridges. Although seamounts, seamount groups and linear archipelagos are found throughout the oceans, they are discussed in this section, since the greatest number occur on the ocean-basin floor.

a. Abyssal floor

In the Atlantic, Indian and northeast Pacific Oceans, abyssal plains occupy a large part of the abyssal floor (Fig. 17). An abyssal plain is a smooth portion of the deep-sea floor where the gradient of the bottom does not exceed 1 : 1000 (Fig. 14). Gradients of abyssal plains range from 1 : 1000 to 1 : 5000 although some areas have been found where the gradients are less than 1 : 7000. Abyssal plains adjoin all continental rises and can be distinguished from the continental rises by distinct changes in bottom gradient. At their seaward edge most of the abyssal plains gradually give way to abyssal hills. Individual abyssal hills are 50–200 fm high and 2–6 miles wide. In the Atlantic, the abyssal hill provinces only locally exceed 50 miles in width, whereas they cover most of the Pacific floor. Abyssal plains in the Atlantic area range from 100–200 miles in width (Figs. 18, 19 and 20). Core samples of sediment obtained from an abyssal plain

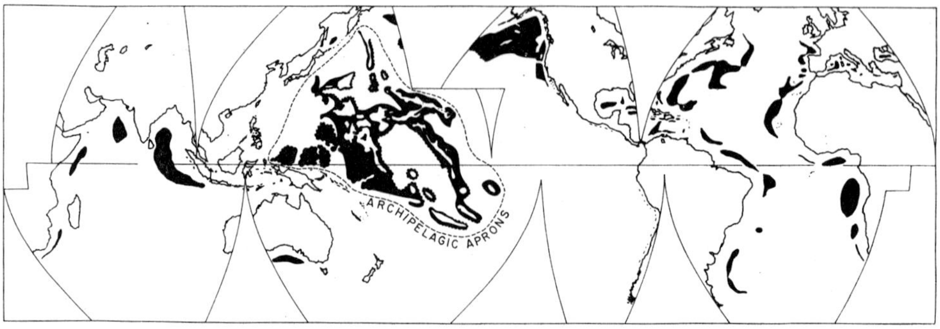

Fig. 17. Abyssal plains and archipelagic plains of the world. (Modified from Menard, 1959.)

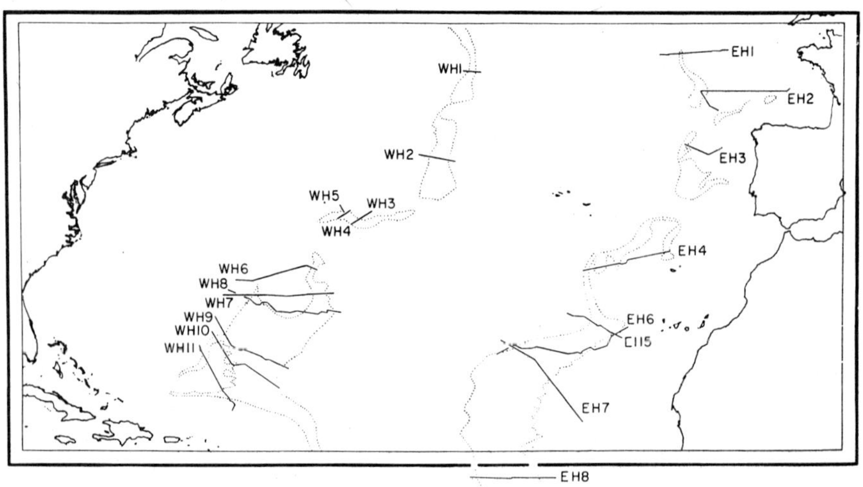

Fig. 18. Index of abyssal hills–abyssal plains profiles of the North Atlantic shown in Figs. 19 and 20. (After Heezen *et al.*, 1959.)

invariably contain beds of sand, silt and gray clay intercalated in the red or gray pelagic clay that is generally characteristic of the deep-oceanic environment.

b. Abyssal plains (see Chapter 14)

Abyssal plains were discovered in 1947 when continuously recording deep-sea echo-sounders first became generally available. Their origin was disputed for several years, but at present every line of evidence available suggests that they were produced by turbidity currents spreading out on the sea floor from the continents: (1) the plains contain coarse sand and anomalous shallow-water fossils that can have been transported only by turbidity currents; (2) throughout the world abyssal plains occur only where turbidity currents spreading out from continents are not barred by bottom topography. Thus, they are wide-

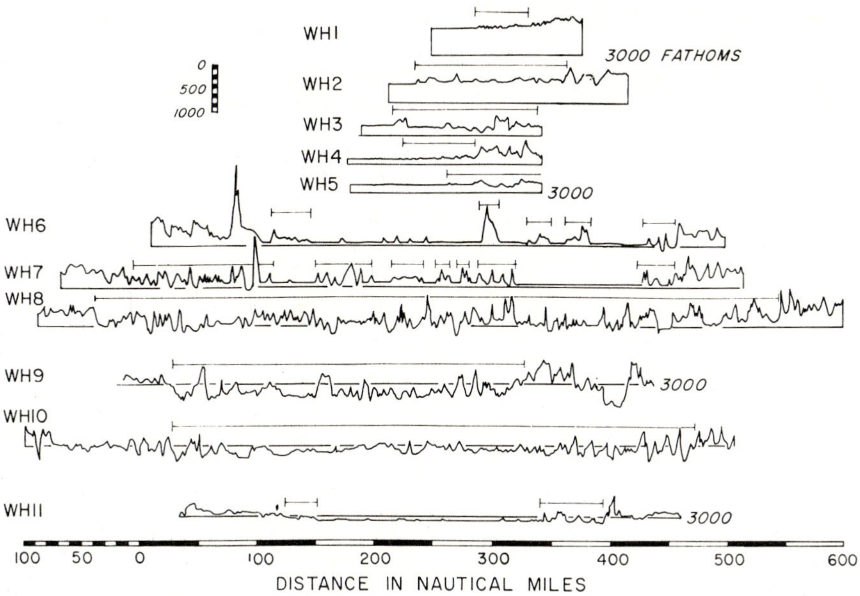

Fig. 19. Profiles of abyssal plains and abyssal hills in the western North Atlantic. Profiles are indexed in Fig. 18. (After Heezen *et al.*, 1959.)

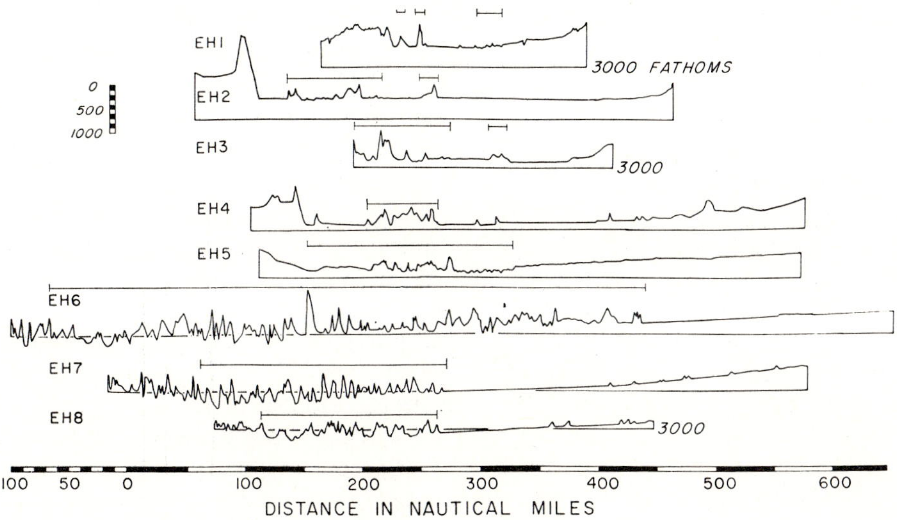

Fig. 20. Profiles of abyssal plains and abyssal hills in the eastern North Atlantic. Profiles are indexed in Fig. 18. (After Heezen *et al.*, 1959.)

spread in the Atlantic but relatively rare in the Pacific, because turbidity currents cannot flow past island arcs and trenches. (3) The small-scale relief of plains and their approaches can be explained only by the action of bottom-moving turbidity currents (Heezen *et al.*, 1954; Menard, 1955). Submarine canyons cut in the continental slope and continental rise lead out to deep-sea fans, which in turn are crossed by channels extending for some distance on the abyssal plains. In short, these are all elements of a drainage system reaching to a depth of more than $2\frac{1}{2}$ miles. An enormous turbidity current swept down from the Grand Bank area off Newfoundland and broke a large number of submarine cables in 1929 (Heezen and Ewing, 1952). Ancient turbidity currents of this scale have been reported as existing in the geological record and they may be relatively common in some places and at some times. However, much of the deposition by modern turbidity currents has been in great fans at the mouths of submarine canyons. Turbidity currents arising from slumps on the continental slope between the canyons cannot be expected to play a large role in shaping the topography, since their deposits are not concentrated into individual large features but spread over a broad front. Hence, their effect is limited to smoothing the topography. In many regions the sediment available to turbidity currents is dumped into the heads of submarine canyons by longshore drift. Hence, the sediment slumped into the canyon heads emerges on the fans in turbidity currents. In other parts of the world, rivers, such as the Congo and the Magdalena, lead directly to a submarine canyon. These submarine canyons carry sediment from the river delta directly to the deep sea. At the mouths of the Congo and Magdalena Canyons huge bulging abyssal cones blanket the sea floor. Seaward of the abyssal cones, turbidity currents follow distributory channels onto the abyssal plains (see Chapter 27).

In addition to their occurrence on the ocean-basin floor, abyssal plains are also found in the marginal trenches, in marginal basins, and in epicontinental marginal seas, and features of exactly the same morphology and origin are found in certain lakes.

c. Abyssal hills

An abyssal hill is a small hill that rises from the ocean-basin floor and is from a few fathoms to a few hundred fathoms in height, and from a few hundred feet to a few miles in width (Figs. 18, 19 and 20). The term "abyssal-hills province" is applied to those areas of the ocean-basin floor in which nearly the entire area is occupied by hills, i.e. the province lies at approximately the depth of the adjacent abyssal plain but lacks a smooth floor. Isolated abyssal hills and groups of abyssal hills also occur in the abyssal plains.

Abyssal hills are found along the seaward margin of most abyssal plains and occur in profusion in basins isolated from adjacent land areas by ridges, rises, or trenches. In the North Atlantic the abyssal hills form two strips parallel to the Mid-Atlantic Ridge for virtually its entire length. The Bermuda Rise is bordered on the southeast by abyssal hills which join with the strips adjoining the Mid-Atlantic Ridge. Southeast of the Bermuda Rise, the abyssal plain is absent; consequently, the western Atlantic Abyssal-Hills Province exceeds 500 miles in width.

In the North Atlantic the axes of maximum depth on the eastern and western sides of the Mid-Atlantic Ridge lie in the abyssal-hills province (Fig. 3). This pattern probably continued through the South Atlantic, Indian and South Pacific Oceans. Individual abyssal hills are identical to the smaller hills rising from the steps of the mid-oceanic ridge and are probably of the same origin. The abyssal hills and the lower step of the Mid-Atlantic Ridge can be distinguished topographically only by the contrast in level.

Abyssal hills cover a much larger proportion of the Pacific basin than in any other ocean. In certain areas where the relief is in general similar to abyssal hills but of somewhat larger amplitude (i.e. many of the individual peaks exceed 500 fm in height), the term "seamount province" has been used. A good example of such an area is the Baja California seamount province (Menard, 1955), where 220 seamounts have been found in an area of about 1,000,000 square miles, which is hilly between seamounts. Echograms have been obtained along approximately 50,000 miles of track in this province and no extensive smooth bottom has been found except for a very limited area at the base of Patton Escarpment. Detailed surveys off the United States show groups of ridges averaging 20×100 miles and a few hundred fathoms of relief. These low, linear mountains trend north–south. In contrast, about 50 of the higher seamounts have been surveyed and all appear to be volcanoes although many are elongate like the Hawaiian volcanoes. Detailed surveys around taut buoys or with precision positioning have been made of abyssal hills in several places in the Pacific. Although they are linear off California, the hills in the equatorial Pacific are low, irregular domes and are probably formed by laccolithic intrusions. Statistical analysis of echograms in abyssal-hill areas also indicates that low domes are extremely common.

d. Abyssal gaps

If two adjacent, but distinct, abyssal plains have no through passage at or below the level of the higher plain, they are said to be separated by a sill, a ridge or a rise, depending on the dimensions of the features involved. However, several plains are connected by constricted passages. An abyssal gap is a passage connecting two abyssal plains lying at different levels in the vicinity of the gap. The sea floor slopes down continuously through the gap from the higher to the lower abyssal plain at a gradient considerably greater than that of either adjacent abyssal plain. Prominent abyssal gaps are known in the western Atlantic (Vema Gap) and in the eastern North Atlantic (Theta Gap) and an abyssal gap connects Colombia and the Venezuela abyssal plains of the Caribbean. Investigations of both Vema Gap and Theta Gap have shown that small mid-ocean canyons (deep-sea channels) connect the upper plain with the lower plain. Sediments collected in abyssal gaps have suggested that turbidity currents have transported material through the gaps.

e. Mid-ocean canyons

Cascadia Deep-Sea Channel has been traced as a continuous feature from the submarine canyon at the mouth of the Columbia River to a point 600 miles to the west. A few sounding lines indicate that its total length is about 1000 miles.

It is 1–5 miles wide and has 10–100 fm relief below an abyssal plain. Except where it passes through a narrow gorge in a submarine mountain range, it is steep-walled, flat-floored, and generally flanked by levees. A very similar feature, the Equatorial Atlantic Mid-Ocean Canyon, has been subject to a rather detailed survey which has confirmed the continuity of the canyon for several hundred miles. Although incomplete, the study suggests that the mid-ocean canyon disappears somewhere on the continental rise and does not connect with a submarine canyon. Some question exists, therefore, as to whether what appear to be identical morphological forms are actually the same. All known deep-sea channels extend without interruption from the continental slope portions of submarine canyons to deep-sea fans or cones to abyssal plains, and they are clearly the channels followed by turbidity currents

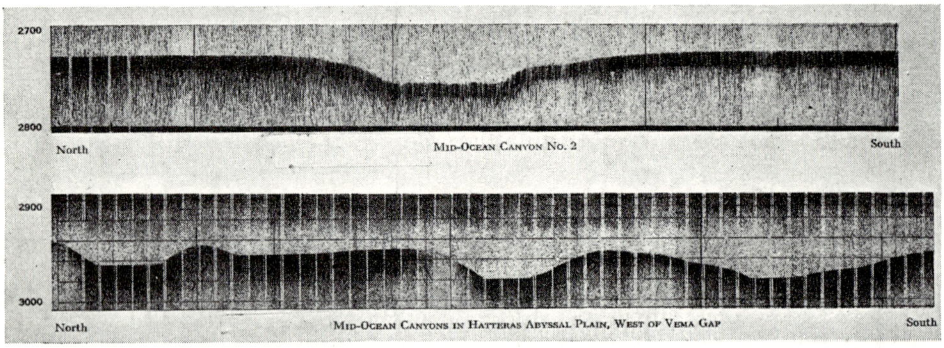

Fig. 21. P.D.R. records of Mid-Ocean Canyon No. 2 and mid-ocean canyons at Vema Gap. Position of Vema Gap and Mid-Ocean Canyon No. 2 as indicated in Fig. 3. Depth in fathoms. (After Heezen *et al.*, 1959.)

originating at the heads of the canyons. Each mid-ocean canyon discovered in the North Atlantic, on the other hand, leads to an abyssal gap in a manner suggesting a genetic relationship. It is not known whether these mid-ocean canyons connect to submarine canyons.

Mid-ocean canyons or deep-sea channels (if they are not the same thing) have been found by random crossings on several abyssal plains and it would appear that they are widespread (Fig. 21). All recognized mid-ocean canyons in the Atlantic parallel the adjacent continental margin, suggesting a tectonic control. Although all have a continuous downward slope, they cut across the regional slope of the abyssal plain. This latter characteristic is prominently displayed in the northeastern Pacific where deep-sea channels hook to the left across the contours of fans.

f. Oceanic rises

Heezen, Tharp and Ewing (1959) define oceanic rises as "aseismic areas slightly elevated above the abyssal floor which do not form parts of the continental margin or the mid-oceanic ridge" (Fig. 4). Menard (1960) and others use a broader, less restricted definition.

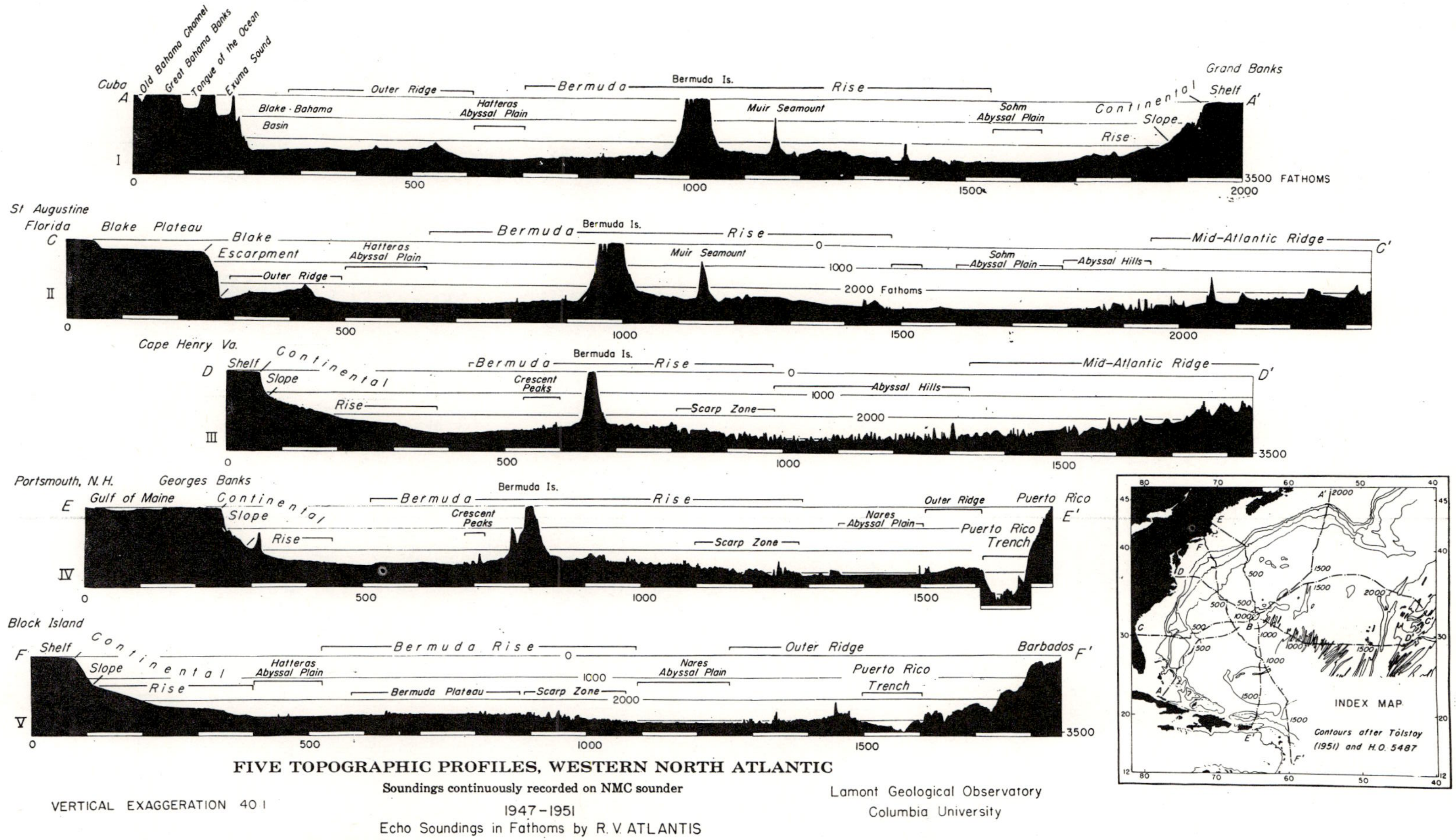

Fig. 22. Five topographic profiles across the Bermuda Rise, western North Atlantic. (After Heezen *et al.*, 1959.)

In the North Atlantic, the Bermuda Rise is the best known example (Figs. 3 and 22). In contrast to the mid-oceanic ridges, oceanic rises are non-seismic; their relief is more subdued and they are asymmetrical in cross-section. The western and central parts of the Bermuda Rise are characterized by gentle, rolling relief. The average depth gradually decreases toward the east. In the eastern third, the rise is cut by a series of 500–1000-fm-high scarps which drop the sea floor to the level of the abyssal plain on the east. The series of eastward-facing scarps suggest block faulting. Situated approximately in the center of the Bermuda Rise is the volcanic pedestal of Bermuda. A small archipelagic apron surrounds the pedestal. The turbidity-current origin of the smooth surface of the apron is supported by cores containing shallow-water carbonate clastic sediments in depths of 2300 fm.

g. Fracture zones

Fracture zones are long, thin bands that are conspicuously more mountainous than the sea floor in general and ordinarily separate regions with different depths. The first wholly submarine fracture zones were discovered only a few

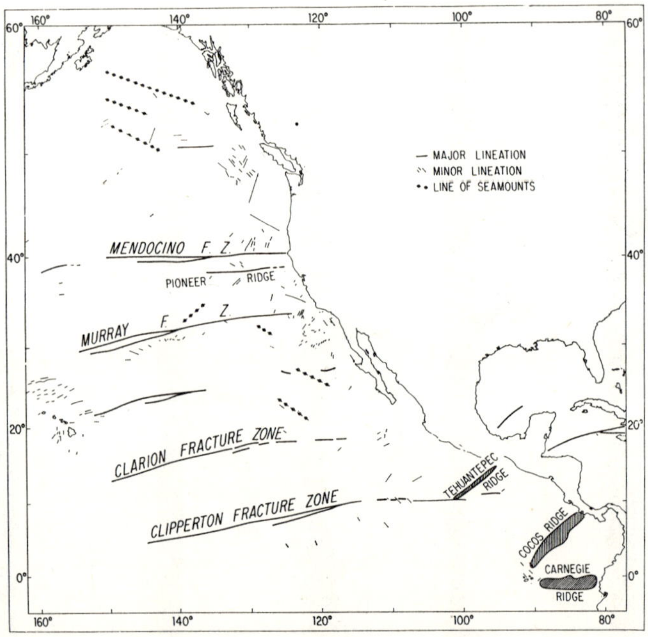

Fig. 23. Major and minor lineations in the northeastern Pacific. (After Menard, 1959.)

years ago in the northeastern Pacific basin. Now scores are known in the Pacific and Atlantic. They form a pattern of east–west parallel lines, relatively evenly spaced down the whole eastern Pacific (Fig. 23). Three follow great circles and all are remarkably straight. Each is about 100 km wide and more than 2000 km long. Within the zones are asymmetrical ridges and narrow troughs parallel to the general trend. Large volcanic seamounts are also common. The Murray Fracture Zone may be taken as an example. It extends along the sea floor in an easterly direction from near the Hawaiian Islands to Southern California, or about 3000 km. The western section of the zone is composed of two asymmetrical elevations with the steep sides facing and a band of narrow troughs and ridges between (Fig. 24). The total relief in a cross-section 100 km wide is

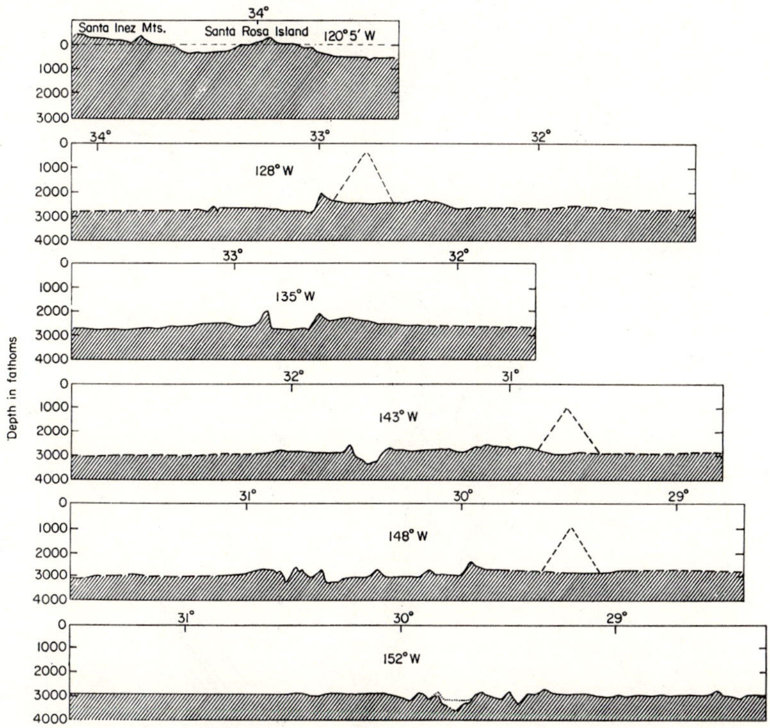

Fig. 24. North–south profiles of the Murray Fracture Zone between Hawaii and Southern California. (After Menard, 1955.)

about 1600 m, and most of the changes in depth are abrupt. South of the regular ridges and troughs, and parallel to them, is a great range of submarine volcanoes called the Moonless Mountains because none of them has ever been above sea-level, although several are more than 3000 m high. In the central section of the Murray Fracture Zone, the northern asymmetrical irregular ridge disappears, and the relief of the southern ridge becomes greater. In addition, a regional change in depth appears, such that the sea floor, for hundreds of kilometers north of the fracture zone, is about 400 m deeper than the sea floor for an equal distance to the south (Fig. 25). A single seamount rises above the crest of the ridge to give a local relief of 7450 m. This seamount, Erban Guyot, has a flat top from which beach gravels and fossils have been dredged. Although it is now 450 m beneath the sea surface, it apparently was an island some time

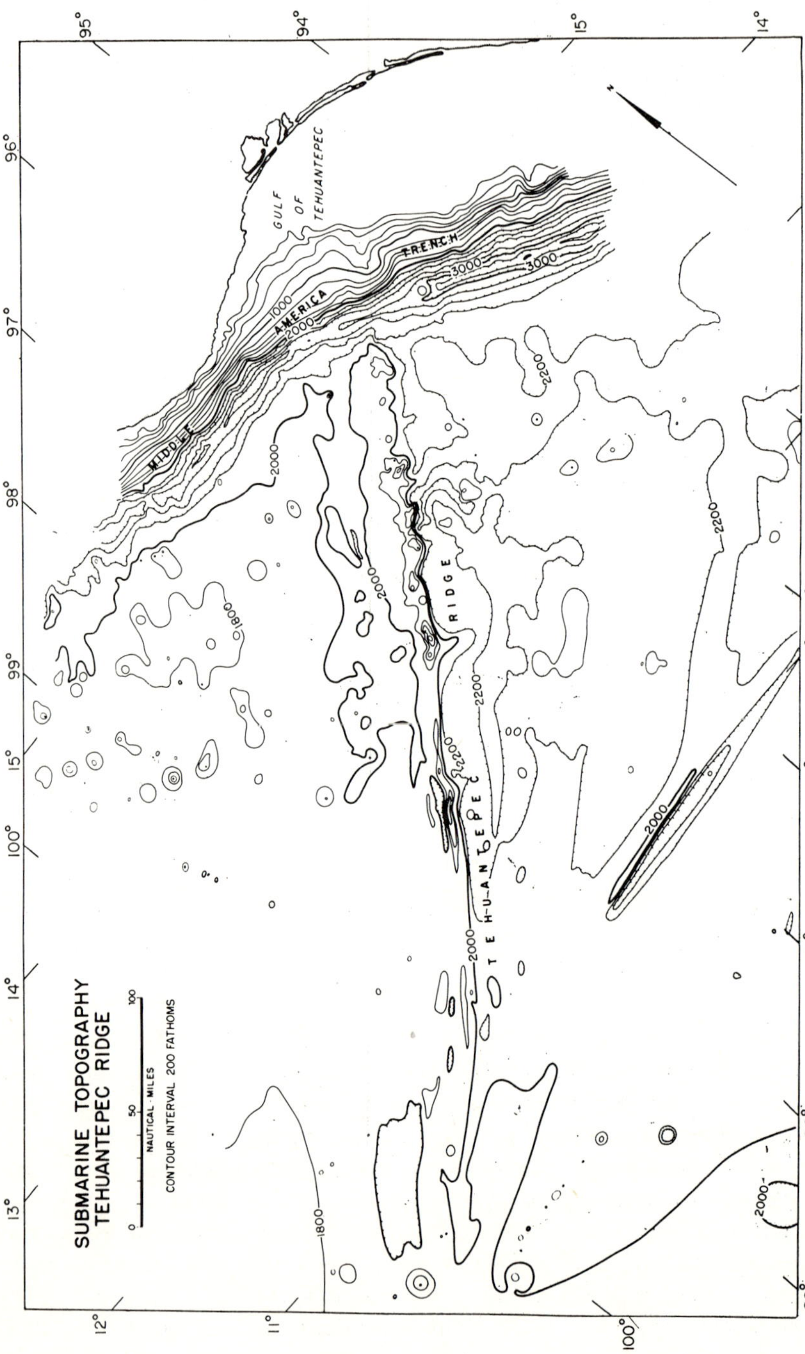

Fig. 25. Tehuantepec Ridge, a branch of the Clipperton Fracture Zone. This figure illustrates the typical fracture-zone profile of an asymmetrical ridge and a regional change in depth. (After Menard and Fisher, 1958.)

more than 20,000,000 years ago. In the section of the Murray Fracture Zone nearest the continent, the relief becomes more subdued, but there are several short troughs more than 500 m deep and the regional change in depth still exists. A sea-borne magnetic survey (Vacquier *et al.*, 1961) of this region shows a north-trending pattern of anomalies with magnitudes of about 500 γ. The anomaly pattern is disturbed along the east-trending line of the Murray Fracture Zone. The pattern can be matched on opposite sides of the disturbance and it appears that it has been offset 150 km, which would place it among the largest fault displacements known. Even so, the displacements on the Pioneer and Mendocino fracture zones are much larger (Fig. 26). The transverse ridges of California lie on the trend of the Murray Fracture Zone and, at least super-

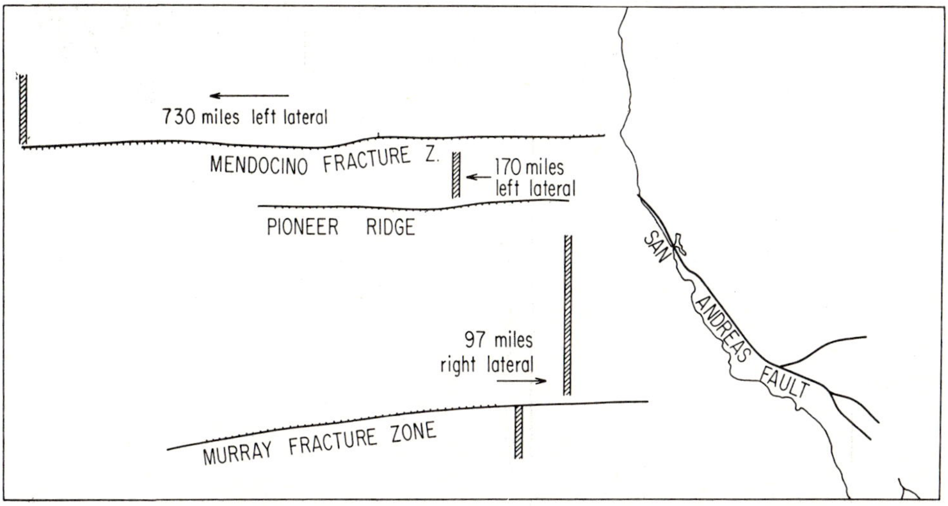

Fig. 26. Displacement of a prominent magnetic anomaly by three fracture zones. Distances in statute miles. (After Vacquier *et al.*, 1961.)

ficially, they seem to be an extension of the zone. Fracture zones are difficult to detect without continuously recording echo-sounders, and it is probable that many more will be found (see Chapter 16).

h. Minor lineations

Widely distributed minor lineations on the deep-sea floor frequently lie in pinnate patterns relative to the major lineations such as island arcs, archipelagos and fracture zones. Individual minor lineations are generally 20 to 200 km long and consist of: (1) elongate submarine volcanoes, atolls on volcanic platforms and volcanic islands; (2) lines of closely spaced circular submarine volcanoes; (3) asymmetrical ridges; (4) narrow troughs ordinarily associated with parallel asymmetrical ridges; and (5) straight troughs resembling submarine canyons on continental and insular slopes.

Compared with faults on land, these "minor" lineations are rather large features. But despite their size, only a few areas of the oceans have been sufficiently well sounded that such features can be delineated. It is only possible to map minor lineations when precision sounding equipment is employed on ships positioned by precision electronic navigation systems. A few such areas (Fig. 23) have been surveyed in the Pacific (Menard, 1959).

i. Aseismic ridges

A significant discovery of the *Meteor* Expedition was that a continuous ridge runs in a southwesterly direction from the vicinity of Walvis Bay, Southwest

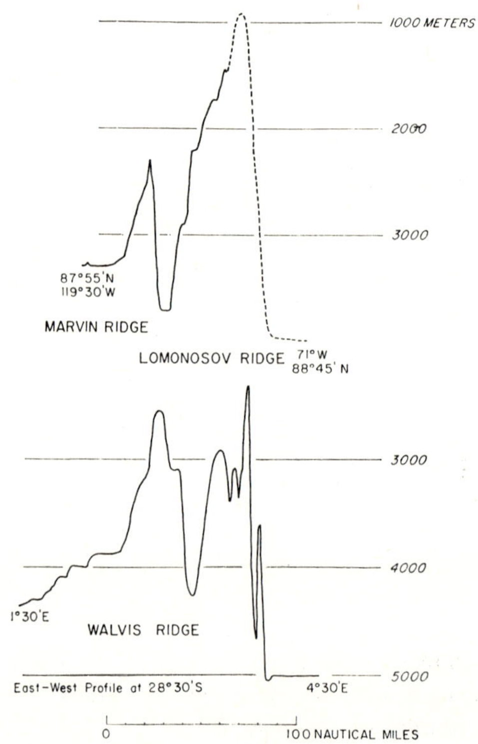

Fig. 27. Profiles of the Lomonosov Ridge in the Arctic Ocean and the Walvis Ridge in the South Atlantic. These two ridges are excellent examples of the asymmetrical, aseismic ridges. (After Heezen and Ewing, 1960.)

Africa (Fig. 27). The discovery of this feature strongly influenced the German investigators (Stocks and Wüst, 1935) and a now-familiar herringbone pattern of ridges in the whole Atlantic was postulated. Not many of these postulated ridges survived later investigation. The Walvis Ridge strikes out from the continent seaward of the Angola uplift. The ridge is strongly asymmetrical, narrow and steep-sided on the south. The main ridge typically has two crests separated by a central asymmetrical depression, and a smaller frontal scarp is found 50 miles southeast of the main Walvis Ridge Scarp. Walvis Ridge can be traced southwestward as a large feature over 1000 fm high for 1500 miles and its tectonic trend continues for 2000 miles further in a line of similar but lower asymmetrical scarps and ridges. It is in many ways similar to the Mendocino Escarpment of the North Pacific. Other such ridges are now known in the Indian Ocean. There are several similar ridges in the vicinity of Madagascar and Madagascar itself may be a feature very similar to the Walvis Ridge, with its characteristic asymmetrical shape.

In the Arctic Ocean the Lomonosov Ridge (Fig. 27) is of similar form and in common with the other ridges of this group is devoid of seismic activity (Heezen and Ewing, 1960; Dietz and Shumway, 1960). Although both the Lomonosov and Walvis Ridges are flanked by huge escarpments over 2000 fm high, the surface of each ridge is relatively smooth and stands in strong contrast to the intensely disrupted relief of the mid-oceanic ridges or of the abyssal hills. This smoothness of the relief and the lack of earthquakes indicates that these aseismic ridges, now tectonically stable, have probably been stable for some time. A core taken from the southern scarp of the Walvis Ridge penetrated a condensed section of Tertiary oozes and ended in Upper-Cretaceous *Globigerina* ooze (D. B. Ericson, *in litt.*). This would seem to suggest that the Walvis Ridge has existed in essentially the same form since the Cretaceous and that the thinness of the condensed section resulted in slumping of material from the southern side of the Walvis Ridge; this slumping has continued since late Cretaceous and hence it suggests that the scarp has existed since late Cretaceous time.

The Muscarene Ridge in the Indian Ocean lies parallel to the mid-oceanic ridge mid-way between Madagascar and the mid-oceanic ridge. Granite rocks outcrop in the Seychelles Islands. This bank, flat-topped at about 100 fm and over 100 miles wide, drops abruptly to abyssal depths on either side. The topography of the Muscarene Ridge is massive and blocky, and it, too, is devoid of the intensely disrupted small-scale relief typical of the mid-oceanic ridges. However, dredging from the crest of the Mendocino Ridge and a core taken from the eastern Bermuda Rise, coupled with the results of seismic-refraction studies, indicate that these features are composed of basic rocks. Thus, the Muscarene Ridge may, with Madagascar, form an independent class of aseismic ridges. They may, in fact, be fragments of continents.

j. Seamounts, seamount groups and linear archipelagos (see Chapter 15)

A seamount is by definition any submerged peak over 500 fm high. Seamounts are distributed through all the physiographic provinces of the oceans. Seamounts sometimes occur randomly scattered, but more often lie in linear rows (Fig. 28). It seems safe to conclude that virtually all conical seamounts are extinct or active volcanoes. The Kelvin Seamount Group, an 800-mile-long line of seamounts, stretches out from the vicinity of the Gulf of Maine toward the Mid-Atlantic Ridge (Fig. 3). The Atlantis–Great Meteor Seamount Group extends for 400 miles along a north–south line, south of the Azores (Fig. 3). In the southwest Pacific, many lines of islands and seamounts cross the ocean. In the mid-Pacific southwest of Hawaii is a large area of seamounts whose flat summits range from 450–850 fm beneath sea-level. These seamounts have been termed "Guyots". From the flat summits, shallow-water fossils of Cretaceous age have been dredged (Hamilton, 1956). Such sunken islands are not entirely limited to the Pacific; in the Atlantic, several of the Kelvin Seamounts are

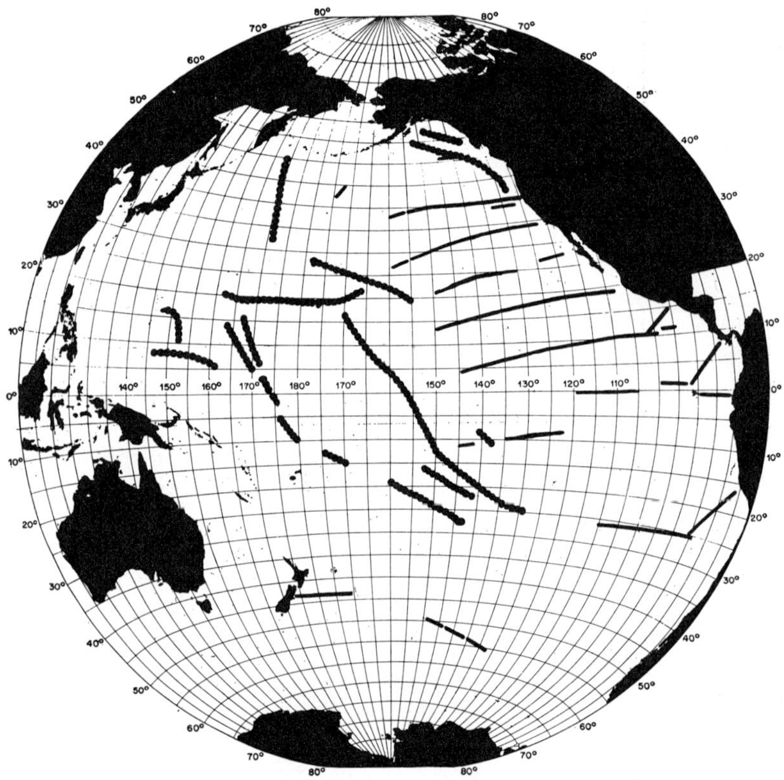

Fig. 28. Fracture zones of the eastern Pacific compared with trends of archipelagos of the western Pacific. Fracture zones are shown by solid lines and archipelagos by chains of dots. (After Menard, 1959.)

flat-topped at 650 fathoms, and the seamounts of the Atlantis–Great Meteor group have flat summits at 150-250 fm.

The number of seamounts in the sea is of interest in view of tectonics and it can be estimated in the Pacific by several methods converging on the number 10^4 (see Chapter 15). The distribution of large seamounts and volcanic islands is not uniform. Instead, they are clustered in groups of 10 to 100. Abyssal hills are randomly distributed, however, so that all the sea floor has been the site of vulcanism, intrusions or faulting. The vulcanism has ranged in geological time at least from late Cretaceous to the present day, and it seems to have been concentrated in different areas at different times. This is indicated by distribution maps of guyots and existing islands. At present, oceanic islands are concentrated in the southwestern Pacific, whereas in the past they were also common in the northeastern part.

That island groups lie along straight lines has long been recognized as geologically significant. Modern oceanographic exploration has added little information about the islands, except to show that there are similar groups of marine volcanoes which are wholly submerged, and that some of the seamounts formerly were islands. The deep-sea floor in the vicinity of island groups has

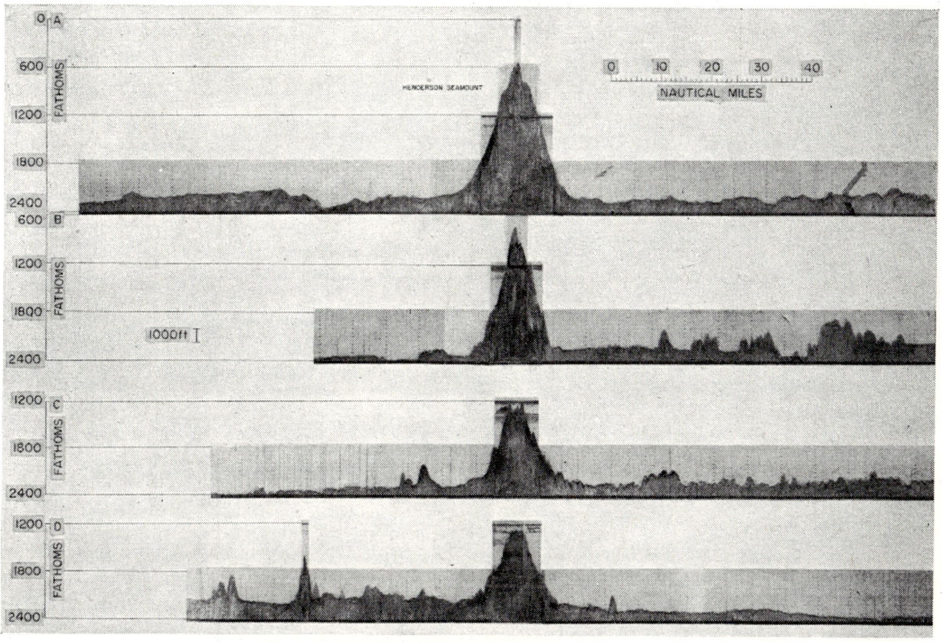

Fig. 29. Echograms of typical seamounts rising above abyssal hills in the Baja California seamount province. (After Menard, 1955.)

been relatively unknown until recently, however, and it has proved rewarding to study it. Most linear groups such as the Hawaiian, Marquesas, Samoan and Society Islands stand above a high ridge which may be caused in a small part by faulting but appears to be explainable solely as the result of piling up of lava flows between closely spaced volcanoes. Some linear groups, including the Austral Islands, are relatively isolated volcanoes rising directly from the deep-sea floor. The volcanoes of the latter groups are too widely spaced or have not been productive enough to pile up lavas between them and to build a ridge. Although they are linear on a small-scale map, the archipelagos in detail are spread in a band that may be one-fourth as wide as the length, if adjacent large seamounts are included with the islands. Consequently, they mark a zone of weakness in the crust, rather than a single major fault. The center of volcanic activity appears to migrate in a regular manner from one end of a zone of weakness to the other. An outstanding example of apparent shifting is the Hawaiian archipelago where former volcanoes at the northern end are now atolls, the central section is made up of inactive volcanoes, and the southern end has active volcanoes. The Austral Islands, in the drowned mid-Pacific mountains, have had a more complicated history with at least two periods of vulcanism. Sprinkled among the Austral Islands are five guyots at depths of

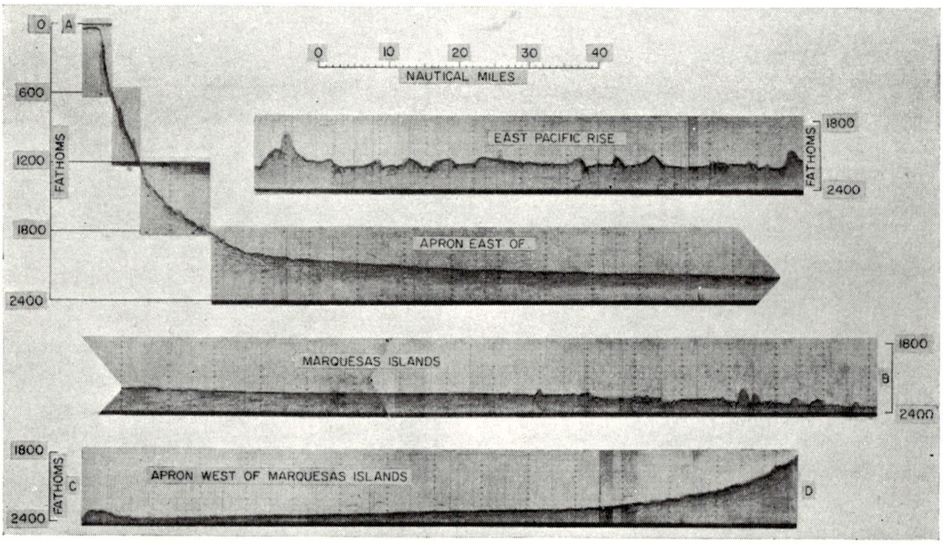

Fig. 30. Echograms of a typical archipelagic apron. The figure shows a well-developed archipelagic plain. Note the contrast of the archipelagic-plain topography with the blocky topography of the mid-oceanic ridge in the central east Pacific. Note abyssal hills in Profile B. (After Menard, 1956.)

1000 to 1500 m. Clearly, they are former active volcanoes which were islands but which became inactive and subsided long before the period of vulcanism in which the existing islands were formed.

The enormous load of a group of volcanic islands might be expected to depress the crust of the earth, and a moat or encircling depression around the southern Hawaiian Islands has been attributed to this cause (Dietz and Menard, 1953). The moat is only slightly deeper than average for the region but it, in turn, is encircled by a broad arch, which makes the moat more conspicuous (Fig. 24). Most archipelagos are not encircled by moats that are evident in the topography of the sea floor. However, abnormal thicknesses of several kilometers of volcanic rock lie under the sea floor in the vicinity of many archipelagos. Inasmuch as the sea floor is not elevated, the upper surface of the crust must be depressed in a moat that has been filled predominantly with lava flows. The moat filling is called an archipelagic apron. The surface of the moat-fill is usually smooth and, in most places, the existence of a smooth region surrounding an island group may be taken as evidence of an archipelagic apron (Fig. 30). However, the surface material is sediment and the smooth surface and extremely gentle slopes indicate that the mode of spreading of erosion products is by turbidity currents. The smooth surface of an archipelagic apron is known as an archipelagic plain.

The Bermuda Islands are surrounded by a thick ring of sedimentary and volcanic material. However, this thick ring extends much further than the smooth surface of the Bermuda archipelagic plain. It may be that a formerly smooth apron was deformed tectonically, forming abyssal hills. On the other hand, it may be that the sediments shed by Bermuda since its origin, which was at least prior to late Cretaceous time, have been insufficient to produce a smooth surface to cover the irregular surface of the volcanic blister and, thus, the archipelagic plain has not yet completely covered the archipelagic apron.

A startling recent discovery illustrates how little is known about the relief of the deep sea. The Hawaiian Islands are the focus of a very large number of sounding lines by naval vessels, and several reconnaissance studies have been made by oceanographic ships with recording echo-sounders. Nevertheless, a major element of the sea-floor topography completely escaped notice until a survey was made with the closely spaced sounding lines positioned with precise electronic equipment not previously available. Narrow asymmetrical mountains 2 km high and 100–200 km long have been found in abundance on the Hawaiian arch. They form a distinct pattern, trending northeastwards on the east side of the Hawaiian Islands and southwards on the west side. Similar mountains do not occur on the arch where it curves around the end of the Hawaiian structure.

k. Tectonic activity of the ocean-basin floor

The ocean-basin floor is generally aseismic. The number of actual volcanoes, or even volcanic islands, is small. For this reason, it is commonly generalized that the basin has been a stable area in geological time. Modern oceanography, however, has shown that the basin is covered with volcanoes and broken in many places by enormous faults, so that it cannot always have been stable. The question raised is whether the present is abnormally quiet tectonically or whether geological time is adequate to produce the volcanoes and faults in a relatively stable basin. The answer depends to a large degree on the time available. Demonstrable history of the ocean-basin floor, based on samples, is very brief. Cores penetrate only a small distance and could hardly be expected to reach very ancient sediments. Dredge hauls of seamounts might sample material of any age, because most submarine elevations are swept clear of sediment. To date, Cretaceous fossils are the oldest ones sampled. But fossils of any kind are rare in dredge hauls. The basalts dredged have not often been dated, and in many cases are undatable owing to their highly weathered condition. The potassium–argon method is capable of dating fresh basalts but the dredge can break off only weathered material unless the dredger has extraordinary luck. If present sampling is considered to be adequate, the ocean-basin floor is no older than Cretaceous. However, further sampling may indicate greater age and it seems more logical to assume that the basin is very old.

As to faulting, it can only be concluded that the ocean-basin floor was formerly more seismically active. The only alternative is to assume that the oceanic type of crust can be faulted without earthquakes and this is not entirely acceptable because large earthquakes occur in a few places within the oceanic crust.

When viewing the physiographic provinces of the North Atlantic or the South Atlantic, one sees a strong parallelism not only between the continental margins and the mid-oceanic ridge, but also between all the other province boundaries of the ocean basin. It has been pointed out that, in the Pacific, and

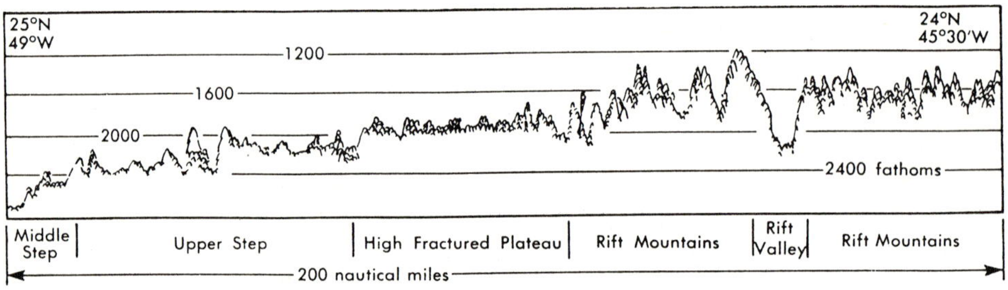

Fig. 31. Tracing of a Precision Depth Recorder record showing crest and western flank of Mid-Atlantic Ridge. (After Heezen, Tharp and Ewing.)

to a certain extent in the Atlantic, the minor relief of the continental margins parallels the continental slope. This is certainly true in those few areas where surveys are sufficiently detailed to permit a study of the minor relief of the abyssal hills and the lower steps of the mid-oceanic ridge. Here again, the minor relief features are parallel both to the continental margins on the one hand and the crest of the mid-oceanic ridge on the other. To a remarkable degree, the topography of the sea floor resembles a piece of expanded-metal grill-work in which the major topographic lines are all roughly parallel.

C. Mid-Oceanic Ridge (see Chapter 16)

The greatest topographic feature of the sea floor is an essentially continuous median elevation extending through the Atlantic, Indian, Antarctic and South Pacific Oceans for a total distance of over 30,000 miles. The relief of the elevation above the adjacent ocean-basin floor is 1–3 km, and the width in most places is more than 1000 km. This median elevation, known as the Mid-Oceanic Ridge, is essentially a broad, fractured swell occupying the center third of the ocean. The portion of this feature lying in the Atlantic Ocean—the Mid-Atlantic Ridge—can be divided into distinctive physiographic provinces, identifiable on most trans-Atlantic profiles (Fig. 31) (see Chapter 16).

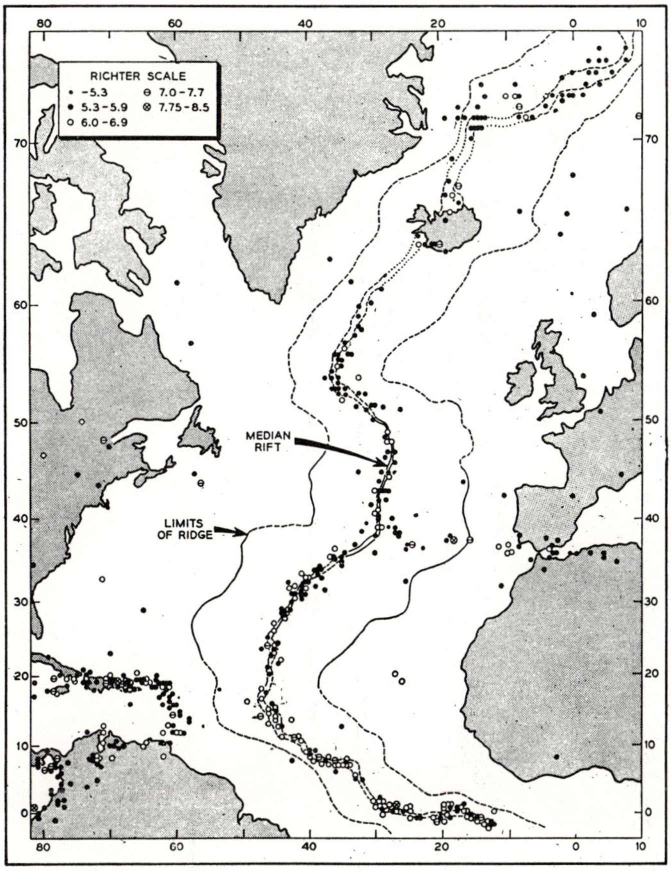

Fig. 32. Earthquake epicenters in the North Atlantic and the mid-oceanic rift. (After Elmendorf and Heezen, 1957.)

a. Crest provinces

The rift mountains and high fractured plateau, which constitute this category, form a strip 50–200 miles wide. In some places, particularly the North Atlantic, the rift valley is bounded by the inward-facing scarps of the rift mountains. The floor of the rift valley lies 500–1500 fm below the adjacent peaks of the rift mountains. The rift mountains drop abruptly to the high fractured plateau, which lies in depths of 1600–1800 fm on either side of the rift mountains. An earthquake belt follows the rift valley (Figs. 32 and 33). Heat-flow measurements

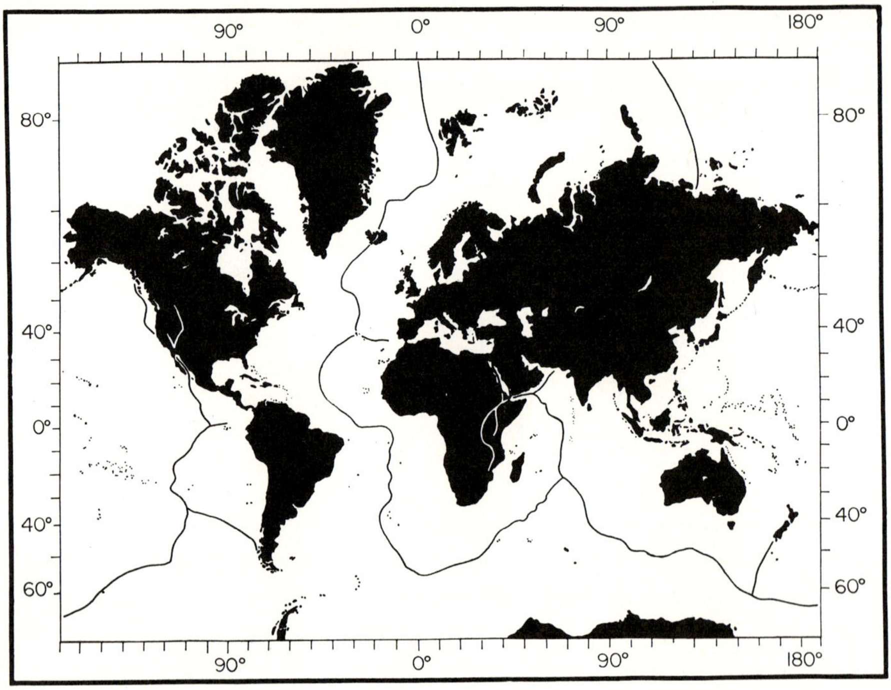

Fig. 33. Location of the mid-oceanic rift. (After Heezen *et al.*, 1959.)

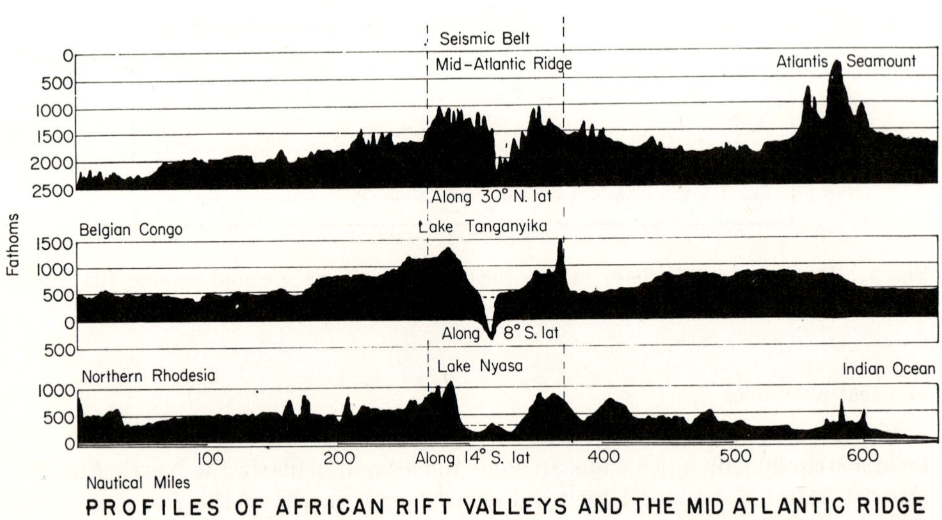

Fig. 34. Profiles of African Rift Valleys and the Mid-Atlantic Ridge. Note the virtual identity of topographic features shown in these three profiles across the seismic belt of the mid-oceanic ridge system. (After Heezen, 1958.)

in the crest provinces in the Eastern Pacific and in some places in the Atlantic give values several times greater than have been obtained in normal ocean-basin floor or continental areas (Von Herzen, 1959). A large positive magnetic anomaly is often but not always associated with the rift valley and some seismic-refraction measurements suggest a crust intermediate in composition between the oceanic crust and the mantle. The crest provinces of the mid-oceanic ridge can be traced directly into the rift valleys, rift mountains and high plateaus of Africa (Figs. 34 and 35) and of the western United States. The faulted topography characteristic of the crest provinces is well developed on an unnamed ridge trending NW from southern Chile. However, much of the Pacific portion of the system is smoother than the remainder of the mid-oceanic ridge.

b. Flank provinces

The flank provinces of the Mid-Atlantic Ridge can usually be divided into somewhat arbitrary steps or ramps, each bounded by prominent scarps. Parts of the flank provinces, particularly the upper step south of the Azores, are characterized by smooth-floored intermontane valleys; but, in general, the relief of the entire Mid-Atlantic Ridge is highly irregular and characterized by lack of any smooth areas (Fig. 35). Thus, these intermontane valleys stand out very distinctly. Photographs, cores and dredgings indicate that the crest of the Mid-Atlantic Ridge north of the Azores is being denuded of its sediments. These sediments, eroded from the crest provinces and deposited on either side, probably account for the smoothing of the relief of the flanks and the filling of the intermontane basins. Rocks dredged from the rift valley in the North Atlantic consist of fresh basalts and fresh-to-weathered serpentines and gabbros. Several large boulders of basalt dredged from the rift valley have recently been dated by the potassium–argon method and found to be less than 10,000,000 years old (G. Erikson, in press).

The serpentines, gabbros and peridotites dredged from the rift valley, and those which protrude above sea-level at St. Paul's Rocks, would seem to support Hess's (1955) theory that the ridge has been domed up by serpentinization of the upper mantle in a long linear belt lying under the Mid-Oceanic Ridge. On the other hand, the seismic velocity of 7.2–7.4 km/sec observed in the mid-oceanic ridges is remarkably similar to the seismic velocity found in the low-velocity channel lying 120 km below the continents and some 50 km beneath the ocean basins (Gutenberg, 1959).

The mid-oceanic ridge is conspicuously centered in the Atlantic and Indian Oceans, where it is called the Mid-Atlantic and Mid-Indian Ocean Ridges. In the South Pacific the position relative to the margin of the basins is less obvious but it can be determined by an arbitrary method. The technique used is to accept the 1000-m isobath as the margin of the ocean basin and to draw a family of lines approximately parallel and equidistant from a margin by swinging arcs on a globe (Fig. 30). Intersections of arcs with equal radii are connected to established median lines. Consequently, any major convexity of margin of an

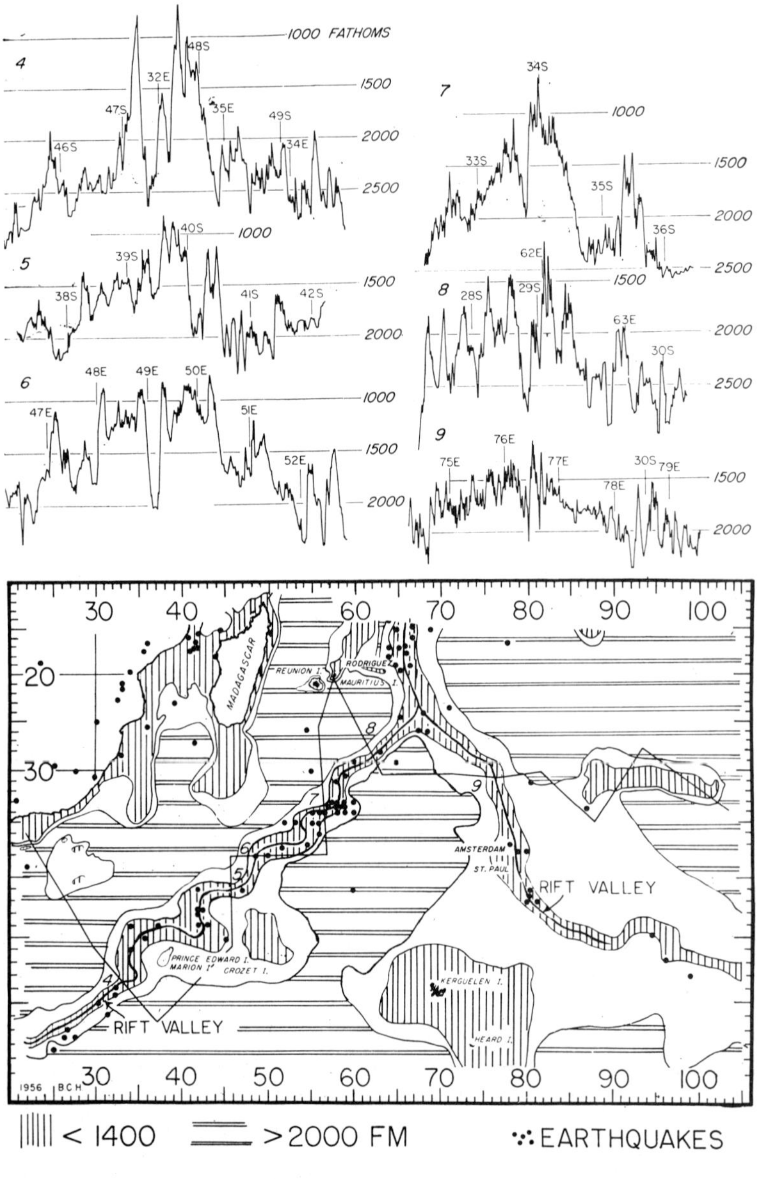

Fig. 35. Mid-oceanic ridge and rift valley in the southwest Indian Ocean. (After Ewing and Heezen, 1960.)

ocean basin has a median line in addition to the one for the whole basin. The great curve bounded by New Guinea, the Philippines and Japan is an example of such a convexity in the Pacific. This method of geometrical construction shows that the crest of the "East Pacific" Rise (part of which is also known as the Easter Island Ridge) lies near the median line between New Zealand and Antarc-

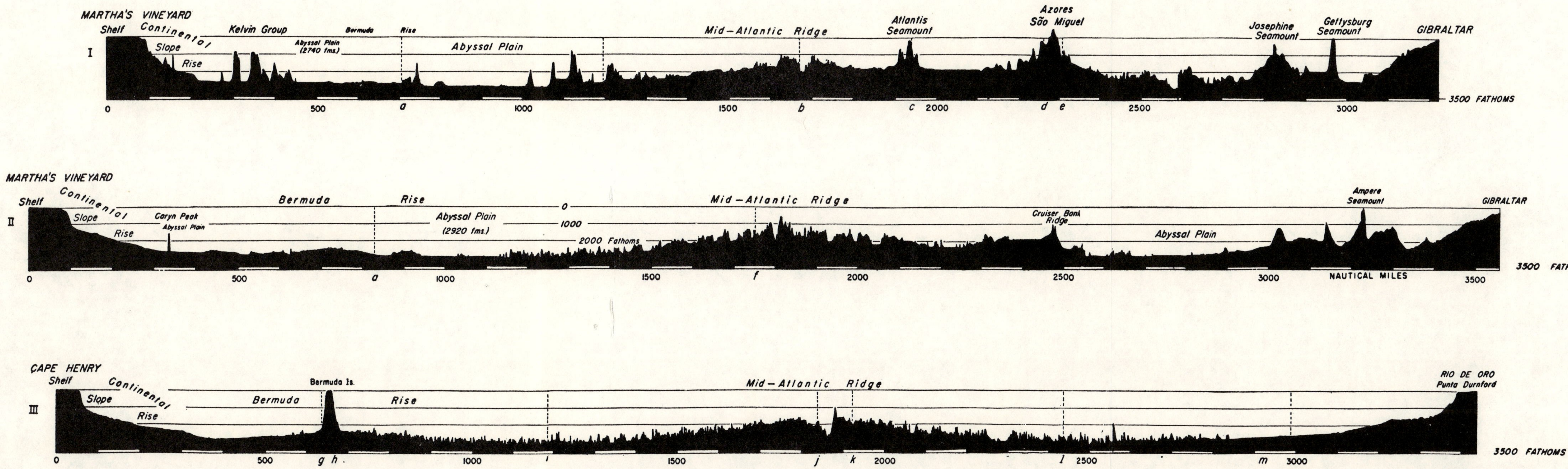

Fig. 36. Three profiles across the North Atlantic. (After Heezen *et al.*, 1959.)

tica; but at about 55° south latitude, the crest bends sharply to the east. Apparent fault scarps suggest that the bend is an offset caused by faulting or else a discontinuity along the margin of a large crustal block. It is only in the Central and North Pacific that an obvious correlation between the geometric median line

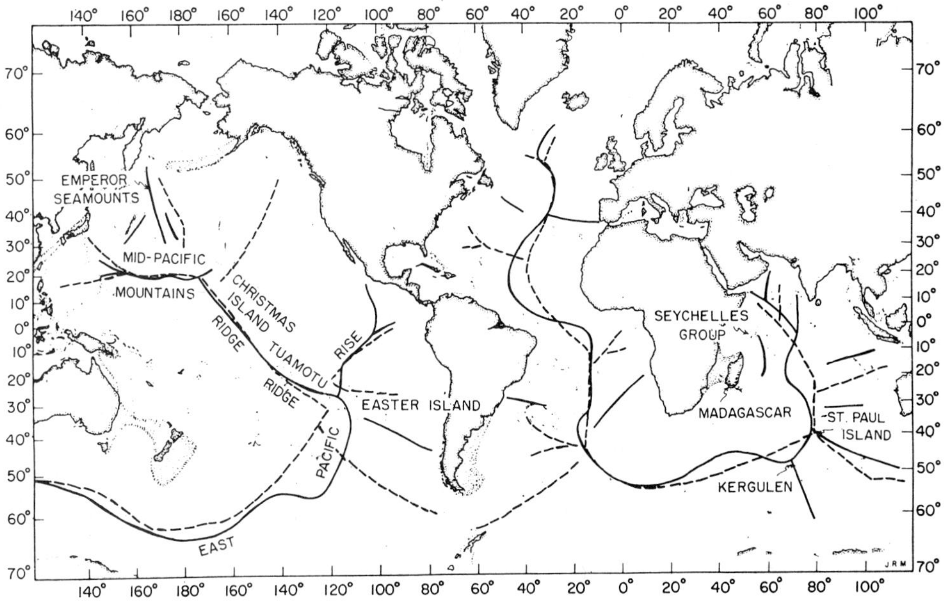

Fig. 37. Position of the crest of the mid-oceanic ridge in ocean basins compared with the center line of the ocean basins as defined by an arbitrary method discussed in the text. (After Menard, 1959.)

and a broad median elevation is not found. However, there is a correlation with another type of submarine topography—long, narrow, steep-sided ridges capped with atolls and seamounts. The median line of the whole basin trends northwest from Easter Island to the Tuamotu Islands and the Line Islands to Johnston Island. Tuamotu Ridge has an area of 200,000 km^2 shoaler than 3 km, and about 100 atolls and seamounts rise above the ridge. Some of the seamounts are guyots or former islands, now submerged. The Line Islands rise from the Christmas Island Ridge, which also has an extensive area, perhaps 100,000 mi^2 above 3 km in depth, and many seamounts and guyots between the islands. From Johnston Island, the median line has three spurs. The northeastern spur is not related to any conspicuous topography. The northern spur lies somewhat east but parallel to the Emperor Seamounts, which are a line of large guyots rising from a ridge with about 1 km relief. The western spur corresponds to the Marquis–Necker Rise, which, in its eastern half, the Mid-Pacific Mountains, has an area of 60,000 km^2 shoaler than 3 km. The Mid-Pacific Mountains include about 50 seamounts, mostly guyots, distributed broadly over the ridge. The wholly submerged mountains in the Tuamotu archipelago bear a close resemblance in every detail of topography except that the ancient islands of the mid-Pacific were not capped by reef corals which grew upward as the islands subsided. The mid-oceanic ridge in most places in the ocean has an average shape of a broad, gentle, more or less fractured arch. But in the Central and North Pacific, the aseismic median elevations are narrow and steep-sided. Do these different types of topography represent different stages of development? If they do, the broad, fractured arches may be an early stage, and the narrow ridges may be a later one. This is suggested by the fact that the broad fractured arches are the locus of a continuous narrow band of earthquakes and active, or recently extinct, volcanoes. In addition, the heat flow from the earth is high and variable under the broad fractured arch, suggesting that very powerful deforming forces are active. Central and North Pacific narrow ridges show no such signs of crustal instability. There are no earthquakes, active volcanoes, or abnormal values of heat flow. On the contrary, deep guyots and atolls bear witness to long-continued quiescence and subsidence.

c. Mid-oceanic plateaus

In the vicinity of the Azores Islands in the North Atlantic and near the Crozet Islands in the Indian Ocean, the mid-oceanic ridge passes along one flank of a large mid-oceanic plateau. The Azores Plateau has an area of 52,000 mi^2 and surrounds the Azores Islands. The depth of this plateau is generally less than 1000 fm. The Azores Islands are oriented S to SE–N to NW along a topographic trend striking off toward the Straits of Gibraltar. This topographic connection between the Azores Plateau and the Southern Iberian Peninsula has been called the Azores–Gibraltar Ridge, which Heezen *et al.* (1959) consider as a poorly developed mid-oceanic ridge, of the same general class as the Mid-Atlantic Ridge.

The Azores Plateau itself merges with the rift mountains of the Mid-Atlantic Ridge. The sea-floor topographic trends of the eastern part of the plateau are parallel to the known tectonic and volcanic trends of the Mid-Atlantic Ridge. The Azores Plateau or bulge is generally considered to be a highly fractured tectonic uplift in which vulcanism has played a very small part (Cloos, 1939).

The topography of the Azores Plateau is somewhat less irregular than that of the Mid-Atlantic Ridge. Although prominent topographic trends cross the area, isolated steep-sided volcanoes occur and long linear scarps are common. The minor relief on the surface of the plateau is relatively gentle. So little work has been done in the area of the Crozet Plateau that it is impossible to delineate the sea-floor features. One echogram taken midway between Prince Edward Island and the Crozet Islands reveals that the surface of the plateau is remarkably smooth, in fact much smoother than the Azores Plateau. The origin of this smooth topography is difficult to ascertain. However, it seems likely that an original irregular relief must underlie the smooth surface, which was smoothed by redistribution of sediments by strong bottom currents. The surface of the

Azores Plateau is characterized by a series of ridges trending NW–SE at approximately right angles to the trends in the Mid-Atlantic Ridge whose spacing is entirely different from the ridges forming the Mid-Atlantic Ridge. Instead of being close together with only a narrow, steep-walled valley separating two ridges, the ridges on the Azores Plateau are separated by broad valleys. There is a strong possibility that other mid-oceanic plateaus exist, particularly in the vicinity of St. Paul Island in the Indian Ocean; however, so little of the mid-oceanic ridge has been explored, even in a reconnaissance fashion, that it is impossible at this time to make an accurate prediction.

3. Summary of Submarine Topographic Forms

The ocean floor consists of three major morphological divisions: (1) continental margin, (2) ocean-basin floor, and (3) mid-oceanic ridge. The continental margin is formed by three categories of provinces, representing (1) the submerged continental platform, (2) the steep edge of the continental block, and (3) the raised or depressed edge of the ocean floor. The topographic detail of the continental margin is predominantly smooth except for the submarine canyons and minor irregularities of the upper continental rise. A close correspondence of topography and distribution of recent sediments is apparent. For example, deep-sea sands are found in the canyons and on the fans of the lower continental rise.

The continental slope appears to be a thinly veneered or bare outcrop of Tertiary and Mesozoic sediments. The upper continental rise can in general be considered a filled marginal trench and the lower continental rise is directly analogous to the outer ridge observed off marginal trenches. The initial form of the typical Atlantic continental margin must have been remarkably similar to the form of the present Pacific marginal trenches.

The ocean-basin floor lies between the continental margin and the mid-oceanic ridge in the Atlantic and Indian Oceans and comprises large parts of the North Pacific. The ocean-basin floor consists of the deeper abyssal floor and the elevated oceanic rises. On the abyssal floor adjacent to the continental margin are found the flattest surfaces of the Earth. These Abyssal plains were built by turbidity-current deposits. The unburied abyssal floor is represented by abyssal hills.

The mid-oceanic ridge is a broad fractured arch whose axis follows the median line of the ocean and covers the center third of the ocean. The crest provinces include: (1) a rift valley 15–30 mi wide and 500–1500 fm deep; (2) rift mountains which form the sides of the rift valley; and (3) the high fractured plateau, a rugged plateau bordering the rift mountains. These characteristics are found universally in the Atlantic and the Indian Oceans but the form of the mid-oceanic ridge is much more subdued in the eastern Pacific and it is not yet certain whether the rift valley system can be traced to the east Pacific.

The flank provinces consist on each side of steps generally separated by large scarps. This, again, has been confirmed in the North Atlantic and South Atlantic. Data are still insufficient in the Indian Ocean to indicate whether the large scarps observed can be correlated in successive profiles, and the more subdued topography of the east Pacific has not yet been studied in sufficient detail to tell whether a subdued system of steps occurs in this area. A seismic belt accurately follows the crest province throughout the Atlantic and Indian Oceans and can be traced along the crests of the Pacific portion of the mid-oceanic ridge.

References

Bourcart, J., 1949. *Géographie du fond des mers*. Payot, Paris, 307.

Cloos, H., 1939. Zur Tektonik der Azoren. *Geol. Rundschau*, **30**, 506–510.

Dietz, R. S., 1952. Geomorphic evolution of continental terrace (continental shelf and slope). *Bull. Amer. Assoc. Petrol. Geol.*, **36**, 1802–1820.

Dietz, R. S. and H. W. Menard, 1951. Origin of abrupt change in slope at continental shelf margin. *Bull. Amer. Assoc. Petrol. Geol.*, **35**, 1994–2016.

Dietz, R. S. and H. W. Menard, 1953. Hawaiian Swell, Arch, and Deep, and subsidence of the Hawaiian Islands. *J. Geol.*, **61**, 99–113.

Dietz, R. S. and G. Shumway, 1961. Geomorphology of the Arctic Basin. *Bull. Geol. Soc. Amer.*, **72**, 1319–1330.

Elmendorf, C. H. and B. C. Heezen, 1957. Oceanographic information for engineering submarine cable systems. *Bell Syst. Tech. J.*, **36**, 1047–1093.

Ericson, D. B., M. Ewing and B. C. Heezen, 1951. Deep-sea sands and submarine canyons. *Bull. Geol. Soc. Amer.*, **62**, 961–965.

Ericson, D. B., M. Ewing, B. C. Heezen and G. Wollin, 1955. Sediment deposition in the deep Atlantic, 205–220 in Poldervaart, A. (Ed.), "Crust of the Earth". *Geol. Soc. Amer. Spec. Paper* 62, 762 pp.

Ewing, J. I. and W. M. Ewing, 1959. Seismic refraction measurements in the Atlantic Ocean basins, Mediterranean Sea, on the Mid-Atlantic Ridge and in the Norwegian Sea. *Bull. Geol. Soc. Amer.*, **70**, 291–318.

Ewing, M., D. B. Ericson and B. C. Heezen, 1958. Sediments and topography of the Gulf of Mexico. In Weeks, L. (Ed.), *Habitat of Oil*. Amer. Assoc. Petrol. Geol., Tulsa, 995–1058.

Ewing, M. and B. C. Heezen, 1955. Puerto Rico Trench topographic and geophysical data. *Geol. Soc. Amer. Spec. Paper* 62, 255–268.

Ewing, M. and B. C. Heezen, 1956. Oceanographic research programs of the Lamont Geological Observatory. *Geol. Rev.*, **46**, 508–535.

Ewing, M., B. C. Heezen, D. B. Ericson, J. Northrop and J. Dorman, 1953. Exploration of the Northwest Atlantic Mid-Ocean Canyon. *Bull. Geol. Soc. Amer.*, **64**, 865–868.

Fisher, R. L., 1954. On the sounding of trenches. *Deep-Sea Res.*, **2**, 48–58.

Gibson, W. M., 1958. Gulf of Alaska trough parallels Aleutian Trench. *Bull. Geol. Soc. Amer.*, **69**, 611–614.

Gorsline, D. S. and K. O. Emery, 1959. Turbidity-current deposits in San Pedro and Santa Monica Basins off Southern California. *Bull. Geol. Soc. Amer.*, **70**, 279–290.

Gutenberg, B., 1959. *Physics of the Earth's Interior*. Academic Press, New York, 240 pp.

Hamilton, E. L., 1956. Sunken islands of the Mid-Pacific Mountains. *Geol. Soc. Amer. Mem.*, **64**, 97 pp.

Heezen, B. C. and M. Ewing, 1952. Turbidity currents and submarine slumps, and the Grand Banks earthquake. *Amer. J. Sci.*, **250**, 849–873.

Heezen, B. C. and M. Ewing, 1955. Orleansville earthquake and turbidity currents. *Bull. Amer. Assoc. Petrol. Geol.*, **39**, 2505–2514.

Heezen, B. C. and M. Ewing, 1960. Continuity of Mid-Oceanic Ridge and Rift Valley in the Southwestern Indian Ocean confirmed. *Science*, **131**, 1677–1679.

Heezen, B. C., D. B. Ericson and M. Ewing, 1954. Further evidence for a turbidity current following the 1929 Grand Banks earthquake. *Deep-Sea Res.*, **1**, 193–202.

Heezen, B. C., M. Ewing and D. B. Ericson, 1951. Submarine topography in the North Atlantic. *Bull. Geol. Soc. Amer.*, **62**, 1407–1409.

Heezen, B. C., M. Ewing, D. B. Ericson and C. R. Bentley, in press. The flat-topped Atlantis, Cruiser and Great Meteor seamounts.

Heezen, B. C., M. Ewing, R. J. Menzies and N. Granelli, 1957. Extending the limits of the Congo submarine canyon. *Bull. Geol. Soc. Amer.*, **68**, 1743–1744.

Heezen, B. C., M. Tharp and M. Ewing, 1959. The floors of the oceans, I. The North Atlantic. *Geol. Soc. Amer. Spec. Paper* 65, 122 pp.

Hess, H. H., 1955. The Oceanic Crust. *J. Mar. Res.*, **14**, 423–439.

Hill, M. N., 1960. A median valley of the Mid-Atlantic Ridge. *Deep-Sea Res.*, **6**, 193–205.

Kay, M., 1951. North American geosynclines. *Geol. Soc. Amer. Mem.*, **48**, 143.

Koczy, F. F., 1956. Echo soundings. *Reps. Swedish Deep-Sea Exped.*, IV, *Bottom investigations*, Facs. II, 99–158.

Luskin, B. and H. G. Israel, 1956. Precision depth recorder MK-V. *Lamont Geol. Obs.—Time Facs. Corp.*, Tech. Rep. 12, 31 pp.

Luskin, B., B. C. Heezen, M. Ewing and M. Landisman, 1954. Precision measurement of ocean depth. *Deep-Sea Res.*, **1**, 131–140.

Maurer, H. and T. Stocks, 1933. Die Echolotungen des *Meteor. Wiss. Ergebn. Deut. Atlant. Exped.* '*Meteor*', *1925–1927*, B.2, pls. iv–xxxii.

Menard, H. W., 1955. Deep sea channels, topography and sedimentation. *Bull. Amer. Assoc. Petrol. Geol.*, **39**, 236–255.

Menard, H. W., 1955a. Deformation of the northeastern Pacific Basin and the west coast of North America. *Bull. Geol. Soc. Amer.*, **66**, 1149–1198.

Menard, H. W., 1956. Archipelagic aprons. *Bull. Amer. Assoc. Petrol. Geol*, **40**, 2195–2210.

Menard, H. W., 1959. Geology of the Pacific sea floor. *Experientia*, XV/6, 205–213.

Menard, H. W., 1959a. Minor lineations in Pacific Basin. *Bull. Geol. Soc. Amer.*, **70**, 1491–1495.

Menard, H. W. and R. S. Dietz, 1951. Submarine geology of the Gulf of Alaska. *Bull. Geol. Soc. Amer.*, **62**, 1263–1285.

Menard, H. W. and R. S. Dietz, 1952. Mendocino Submarine Escarpment. *J. Geol.*, **60**, 266–278.

Menard, H. W. and R. L. Fisher, 1958. Clipperton fracture zone in the northeastern equatorial Pacific. *J. Geol.*, **66**, 239–253.

Menard, H. W. and V. Vacquier, 1958. Magnetic survey of part of the deep-sea floor off the coast of California. Office of Naval Research, Research Reviews, June, 1–5.

Moore, D. G., 1960. Acoustic-reflection studies of the continental shelf and slope off Southern California. *Bull. Geol. Soc. Amer.*, **71**, 1121–1136.

Northrop, J. and B. C. Heezen, 1951. An outcrop of Eocene sediment on the continental slope. *J. Geol.*, **59**, 369–399.

Shand, S. J., 1949. Rocks of the Mid-Atlantic Ridge. *J. Geol.*, **57**, 89–92.

Stocks, T. and G. Wüst, 1935. Die Tiefenverhaltnisse des offenen Atlantischen Ozeans. *Wiss. Ergebn. Deut. Atlant. Exped.* '*Meteor*', *1925–1927*, B.3, Teil 1, Lief 1, 31.

Stride, A. H., 1959. A linear pattern on the sea floor and its interpretation. *J. Mar. Biol. Assoc. U.K.*, **38**, 313.

Stride, A. H., 1960. Recognition of folds and faults on rock surfaces beneath the Sea. *Nature*, **185**, 837.

Sverdrup, H. U., M. W. Johnson and R. H. Fleming, 1942. *The Oceans; Their physics, chemistry and general biology*. Prentice-Hall, New York, 1087 pp.

Udintsev, G. B., 1954. Topography of the Kurile-Kamchatka Trench. *Trans. Inst. Oceanol.*, **12**, 16–61.

Udintsev, G. B., 1957. Relief of the Okhotsk sea bottom (Translation). *Trudy Inst. Okeanol. Akad. Nauk S.S.S.R.* **22**, 76 pp.

Vacquier, V., A. D. Raff and R. E. Warren, 1961. Horizontal displacements in the floor of the Northeastern Pacific Ocean. *Bull. Geol. Soc. Amer.*, **72**, 1251–1258.

Von Herzen, R., 1959. Heat-flow values from the south-eastern Pacific. *Nature*, **183**, 882–883.

Wiseman, J. D. H. and C. D. Ovey, 1954. Proposed names of features on the deep-sea floor, 1. The Pacific Ocean. *Deep-Sea Res.*, **2**, 93–106.

Wiseman, J. D. H. and C. D. Ovey, 1955. Proposed names of features on the deep-sea floor, 2. General principles governing the allocation of names. *Deep-Sea Res.*, **2**, 261–263.

Wüst, G., 1940. Zur Nomenklatur der Grossformen der Ozeanboden. *Assoc. Oceanog. Un. Geod. Geophys. Internat.* Publ. Sci., **8**, 12–124.

Zeigler, J., W. D. Athearn and H. Small, 1957. Profiles across the Peru–Chile trench. *Deep-Sea Res.*, **4**, 238–249.

13. CONTINENTAL SHELF AND SLOPE (CONTINENTAL MARGIN)

A. Guilcher

1. Definitions

The continental shelf is usually defined as the platform extending off the continents down to 100 fm, or 200 m; and the continental slope, according to the International Committee on the Nomenclature of Ocean Bottom Features (Wiseman and Ovey, 1953), is "the declivity from the outer edge of the continental shelf into great depths". If the foot of the slope is conventionally taken at 2000 m, the shelf and the slope cover respectively 7.6 and 8.5% of the ocean floor. Shelf and slope are parts of the continental margin, and they will be described and discussed together in this section, since they combine to form prominences off the continents and the problems involved are the same.

In fact, the definitions of these features cannot be very accurate for several reasons. First, the outer edge of the shelf does not always lie at the same depth. Shepard (1948) has shown that the average depth is 72 fm, so that the conventional definition is too comprehensive in most areas. But some continental shelves, especially in high latitudes, extend to depths much greater than 100 fm. The most typical case is the shelf around Antarctica, which commonly extends down to 300 fm or even more, especially in the southern Indian Ocean (Jivago and Lissitzin, 1957 and 1958). Off Norway, the edge between 64° and 67° 20′N is found from 280 to 410 m (H. Holtedahl, 1955, p. 45), and abnormally large depths are also found along parts of Greenland and in the Barents Sea (Frebold, 1950, p. 12). On the other hand, the shelf edge in the Beaufort Sea and the East Chukchi Sea, which were apparently not covered by ice-caps during the Pleistocene, is only 35 fm deep (Carsola, 1954). In other regions, two or several steps occur after a first edge is reached, and when the steps are wide it is difficult to decide if the first break of slope or the second one should be taken as the boundary of the shelf. The term *marginal plateau* was introduced by Heezen, Tharp and Ewing (1959) for such large steps at depths greater than 100 fm and less than 1200 fm (e.g. the Blake Plateau off Georgia, and the marginal plateau off southern Argentina).

Bourcart states in several papers (e.g. 1954) that no true edge exists in many areas between the shelf and the slope, the former merging into the latter in a convex curve. According to his investigations in the Mediterranean, this is the general situation off southern France. A very smooth, ill-defined, convex edge is also found along Porcupine Bank off Ireland (Hill, 1956). However, it must be noticed that such a feature remains quite exceptional, and the edge of the shelf, if not always very sharp, is generally easily distinguished on the echograms.

The shelves surrounding very large islands which lie in the deep sea, such as Madagascar, are true continental shelves, because these islands are small continents.

[*MS received June, 1960*].

But some shelves are not tied to continents, or to large islands, although they do not lie very far from them and cannot be considered as true oceanic features. The Bahama Banks fall into this category. They are only 4 to 6 fm deep and are surrounded by steep escarpments and deep bodies of water (Newell and Rigby, 1957). Other shallow, flat banks not connected with the mainland are found in the Caribbean Sea between Honduras and Jamaica. Rockall Bank, Faroe Bank, Bill Baileys Bank, Lousy Bank off Ireland and Scotland (Robinson, 1952; Hill, 1956) and the Flemish Cap near Newfoundland may also be quoted. They may be referred to as *continental borderlands*, a term introduced by Shepard and Macdonald in 1938 for irregular shelves, such as those off Southern California, which include deep basins. But the distinction from oceanic banks is not always easy. The structure, either basaltic or consisting of continental rocks, cannot be used as a general criterion, since the banks off North-west Europe are probably made of basalt, which outcrops widely in Northern Ireland: and Ireland has obviously a continental structure.

Another difficulty is the definition of the foot of the continental slope, which in many areas gradually merges into the deep-sea floor. The range in depth, between 750 and more than 2000 fm in the North Atlantic, is considerably larger than at the edge of the shelf. Before the abyssal plains are reached, a gently sloping continental rise is often encountered (Heezen, Tharp and Ewing, 1959).

2. Topography of the Continental Shelf

Minor ridges related to recent marine sedimentation need not be described here, since they are considered in Chapter 12 of this volume. Apart from these, other irregularities are of common occurrence, so that perfectly flat plains are fairly rare features on the shelf. Most of the irregularities belong to types well represented in the continental topography, especially in areas adjacent to the shelf.

Submarine valleys are probably the most interesting element. Although they have been obscured by sedimentation in many areas, they are clearly seen in others, for example off the coast of Brittany (Fig. 1), where strong tidal scour contributes to keeping them open: but the tidal currents have not been able to carve them into the solid rock, since the latter consists of granite or of hard Palaeozoic sediments. Beautiful patterns of branching valleys have been found by Molengraaf on the Sunda Shelf between Borneo, Sumatra and Java (see Kuenen, 1950, fig. 203), and others have been discovered more recently by French hydrographic engineers off Guinea, where the narrow African shelf widens considerably. The submarine Hudson channel (not canyon) on the east American continental shelf falls into the same category (Veatch and Smith, 1939). In all the areas mentioned above, the submarine valleys have evident connections with the subaerial valleys inland. They can be followed down to depths of 25 to 50 fm.

In other parts of the shelves, discontinuous elements of submarine valleys

Fig. 1. Drowned valleys on the shelf off Penmarc'h Point, South-west Brittany. Data from French chart. Lines of dots: valleys. Contours in metres below lowest spring tides. Probable submarine fault scarp at Penmarc'h. Higher part of shelf very rocky, with many granitic stacks resembling those found on dry land; lower part smoothed by sedimentation (mostly sand and broken shells).

have been revealed by soundings, down to depths of more than 100 fm sometimes: for example, the well-known Hurd Deep (94 fm) in the central part of the English Channel, the Ushant trough (105 fm) at the opening of the same, and several elongated and complex depressions in the North Sea (Devil's Holes, Silver Pits; Fig. 2, and others) (Lewis, 1935). They may occasionally be U-shaped troughs with a comparatively wide bottom and steep sides, but the most frequent transverse profile is a more or less V-shaped one.

Another type of depression consists of troughs, hundreds of fathoms deep, that is, much deeper than the former. They are of common occurrence in glaciated areas, e.g. Labrador, Alaska, Norway, Antarctica, etc. A large number of them run approximately at right angles to the general direction of the shelves, but others are roughly parallel, as O. Holtedahl (1940), H. Holtedahl (1955, 1958), and Jivago and Lissitzin (1957 and 1958) pointed out. In East Antarctica, the main longitudinal depression may reach down to 600 fm in some areas; it runs for hundreds of miles along the continent, and other

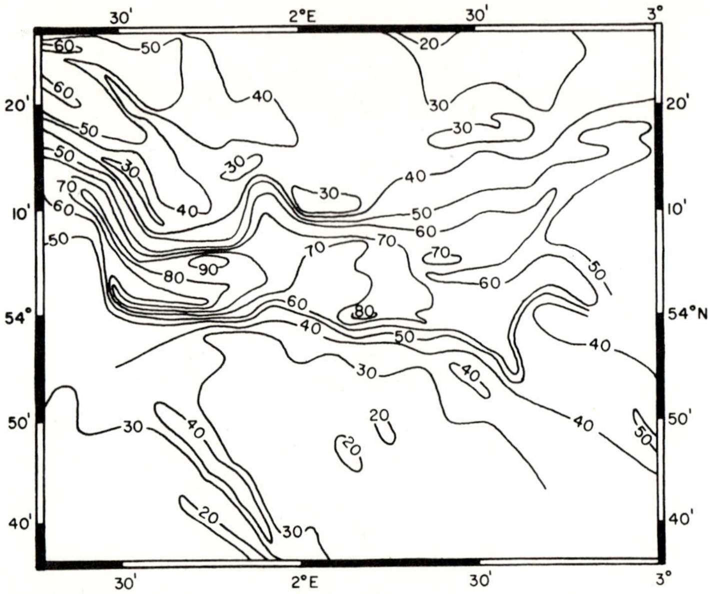

Fig. 2. Outer Silver Pit and Markhams Hole, North Sea. Contours in metres. (After Berthois and Brenot, 1957.)

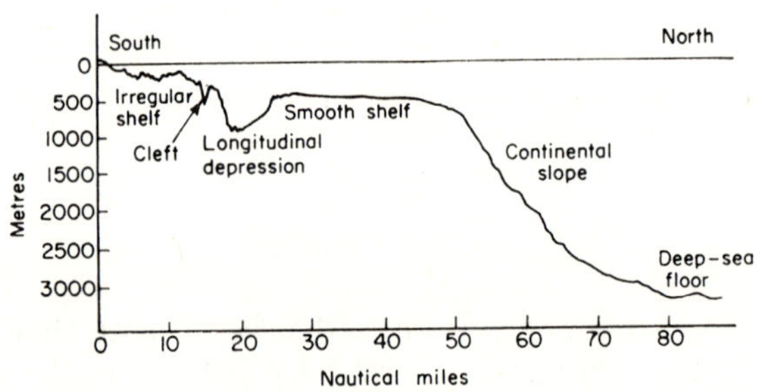

Fig. 3. Transverse profile across the shelf and slope of Antarctica in Sabrina Coast area (120°E). (After Jivago and Lissitzin, 1957.)

shallower and narrower channels of similar trend have been found between the mainland and it (Fig. 3). In northern Europe, a deep marginal channel is found in the Skagerrak close to the Norwegian coast: the deepest part lies to the east at the entrance of the Kattegat, and here reaches more than 350 fm. Farther north, off Norway, marginal depressions deeper than the outer shelf are also frequent, especially off More and Romsdal provinces (Fig. 4). Off Labrador, a longitudinal channel runs parallel to the coast for more than 250 nautical miles, with depths of 300–400 fm, inside a shallower outer shelf; another smaller channel is found to the north of Belle Isle. Although the depressions in the Scottish shelf are shallower than those in other glaciated areas, the Minch and the channel between the Outer Hebrides and St. Kilda (Ting, 1937) lie in a similar position.

Outside the glaciated areas, along the southern California coast, the continental shelf proper is restricted to a narrow coastal strip, 2 to 25 km wide. Outside this shelf, however, a continental borderland may be defined as a basin and range province, including 14 basins the bottom depths of which range from 627 to 2571 m, and the sill depths from 475 to 1902 m (Revelle and

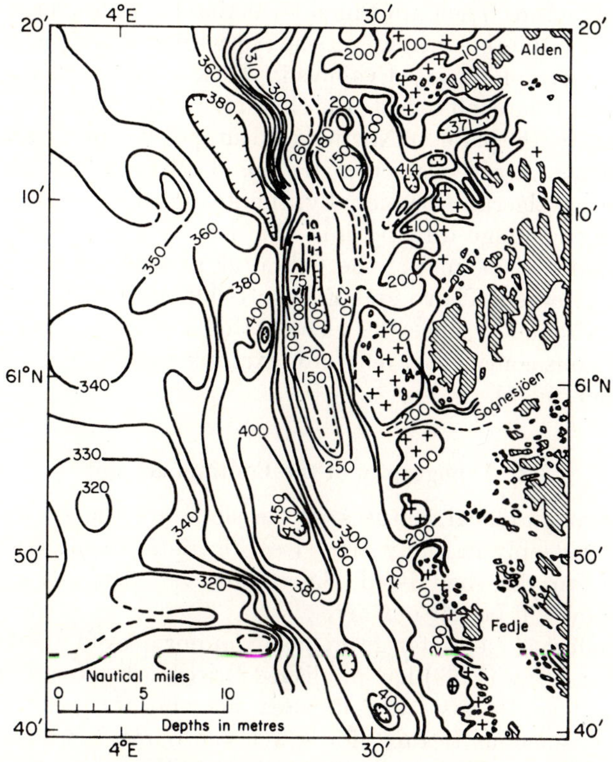

Fig. 4. Continental shelf near Bergen, Norway, with longitudinal depressions. Transverse depression at entrance to fjords. (After H. Holtedahl, 1955.)

Shepard, in Trask, 1939; Shepard and Emery, 1941; Emery, 1954, 1960). The basins are often steep-sided troughs. The intervening ranges, which run parallel to the coast in a NW–SE direction, include islands with fringing submerged shelves. Finally, a typical continental slope is found after the outermost ridge is crossed.

There are various elevations on the continental shelves, of the same types as those rising from the adjacent mainland. The hummocky topography in the Gulf of Maine strongly resembles the glaciated landscape in North-east America.

In East Antarctica, Jivago and Lissitzin (1958) describe series of sharp, cuesta-like ridges in the Davis Sea and off Queen Maud's Land, whereas they report groups of smooth mounds from the surroundings of Drygalski Island. On the west coast of Brittany, long ridges of submarine rocks rise from a sea-bottom smoothed by sedimentation. These ridges are extensions of the flattened crests of Ordovician sandstone which run across the Crozon Peninsula (Guilcher, 1958, p. 206; see also Fig. 1).

Such resemblances between the shelf and the continental morphology suggest that both belong to the same structural units, and were shaped by the same erosional factors. The Californian borderland is a submerged continuation of California, and the topography on the Norwegian and Scottish shelves calls to mind the pattern of glaciated valleys in Norway and Scotland. On the other hand, the width of the shelf reflects the character of the continental topography: wide platforms face the plains of North-west Europe, Argentina and eastern North America, whereas narrow platforms are found off mountainous areas such as West America or Algeria. Occasionally, there is no shelf at all, and mountains are known which plunge directly into the deep sea as at Nice in South-east France (Bourcart, 1959, map); the Gibraltar area is another area where no shelf is found (Heezen, Tharp and Ewing, 1959). But it must be noticed that this is quite infrequent, since the elements of a platform occur generally in regions of vigorous continental relief, for example in most parts of the Algerian coast (Rosfelder, 1955).

3. Topography of the Continental Slope

The main topographical features of the slope are the submarine canyons, which dissect it deeply in many areas (see Fig. 7 for example). Some of these even encroach on the shelf itself. Since, however, canyons are described in Chapter 20, they will be left out here.

The continental slope is far from being uniform even if canyons are not considered. Thus Day (1959) has published British echo-sounding profiles from the slope in the south-western Celtic Sea, showing its great irregularity. In this area, a conspicuous terrace, for which the name Meriadzek Terrace has been proposed, breaks the slope at 47° 30′N, 8° 30′W; it lies at a depth of 1100–1200 fm. Farther to the north-west, the slope consists, on the contrary, of the high and steep Pendragon Scarp, while it becomes much smoother and more gentle in the Porcupine Seabight, which forms a wide re-entrant between Pendragon Scarp and Porcupine Bank. French echo-soundings by Berthois confirm these findings (Guilcher, 1958a; Berthois and Brenot, 1962) and show other variations in profiles, hitherto unknown, between the Celtic Sea and the inner part of the Bay of Biscay.

In the Gulf of Mexico, Ewing, Ericson and Heezen (1958) point to a particular profile of the continental slope over several areas. Off West Florida, West Louisiana and Texas, and the Yucatan or Campeche Peninsula, "the continental slope is made up of two segments: a wide upper rugged portion with a gentle average seaward slope, and a precipitous lower scarp. . . . The Campeche and West Florida escarpments are approximately 1000 fathoms high, whereas the Sigsbee scarp south of Texas is 400–600 fathoms high." The foot of these scarps is extremely sharp and well defined. The West Florida scarp was mapped in detail and described by Jordan (1951), and by Jordan and Stewart (1959): it runs fairly straight to the north of 27°N, while in the south it is complicated by many small embayments. The southern part of the Campeche escarpment was investigated by Creager (1958), who found that the scarp ends in two parallel walls bordering a comparatively narrow trough. The slope off West Louisiana was mapped by Gealy (1955). On the other hand, no steep scarp is

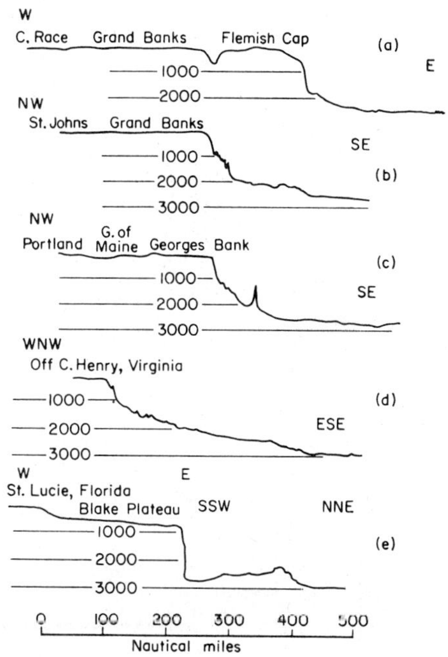

Fig. 5. Five transverse profiles across the shelf and slope of eastern North America. Depths in fathoms, lengths in nautical miles. (After Heezen, Tharp and Ewing, 1959.)

found off the Mississippi delta, but only a "relatively smooth and fairly continuous slope", which merges very gradually into the abyssal plain of the central part of the Gulf. Another impressive scarp is the Blake Escarpment, which begins at about 600 fm off the Blake Plateau in front of Georgia and Florida. Here the gradient exceeds 1 in 2 in some profiles (Heezen, Tharp and Ewing, 1959). Off the east coast of the United States, the transverse sections show many benches in the slope which are remarkably persistent in the Cape Hatteras region, and have been correlated from profile to profile (*ibid.*).

The Morocco continental margin is another area where accurate echo-soundings have been carried out in recent years by the French Navy as far south as Agadir, down to 1600–2200 fm (Brémond, 1953; Lacombe, 1955; Grousson, 1957; Gougenheim, 1959). The northern part is shown in Fig. 6. No true submarine canyon exists between 32° 40′ and 35°N, and the slope is straight in plan for the larger part of its course. Here also a conspicuous steepening has been revealed on several transverse profiles in the deeper part of the slope, particularly off Azemmour, but not on all profiles. In some areas, the foot of the scarp occurs at 1200–1300 fm, with a continental rise farther down; whereas in others the steep lower scarp extends down to 2200 fm and perhaps

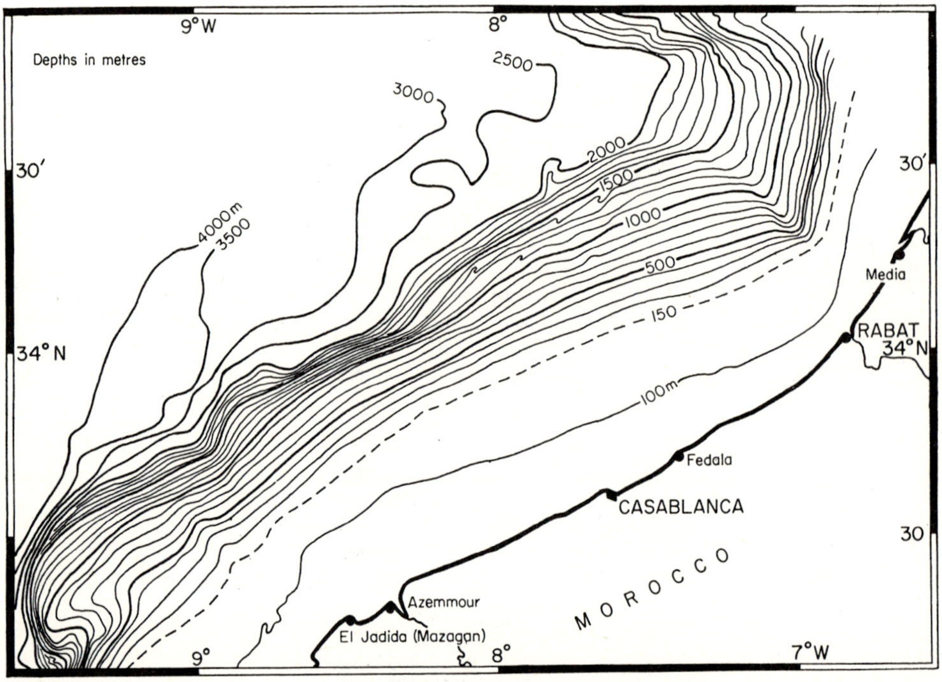

Fig. 6. Continental shelf and slope off North Morocco. Depths in metres. Wide re-entrant off the Rharb plain. (After French surveys, Gougenheim, 1959.)

more. A wide re-entrant has been discovered off the flat Rharb plain, which lies between the Meseta and the Rif chain: the sea floor shows a much more gentle slope in this flat-bottomed depression, which is bordered by two parallel sides like the trough surveyed by Creager in the Bay of Campeche (see above). Off the Rif chain, the topography of the slope is much more irregular than it is in front of the Meseta as far south as 33°N. From Cape Cantin to Agadir it is even more complicated, with many canyon-like depressions and a broad, convex spur off Cape Tafelneh (31° 15′N).

The existence of nine transverse profiles across the whole continental slope of the Norwegian Sea, from West Norway to North Spitzbergen, has been published by H. Holtedahl (1955, pp. 59–61). Most of these profiles show longitudinal ridges or seamounts, occurring principally at 700 fm. Only two profiles, off Lofoten Islands and off Tromsö, include no important break of slope below the continental shelf. The most irregular profile lies off the southern tip of Spitzbergen. Other areas including seamounts on the slope are found off San Francisco, California (Fig. 7), off Nantucket Island, New England and elsewhere.

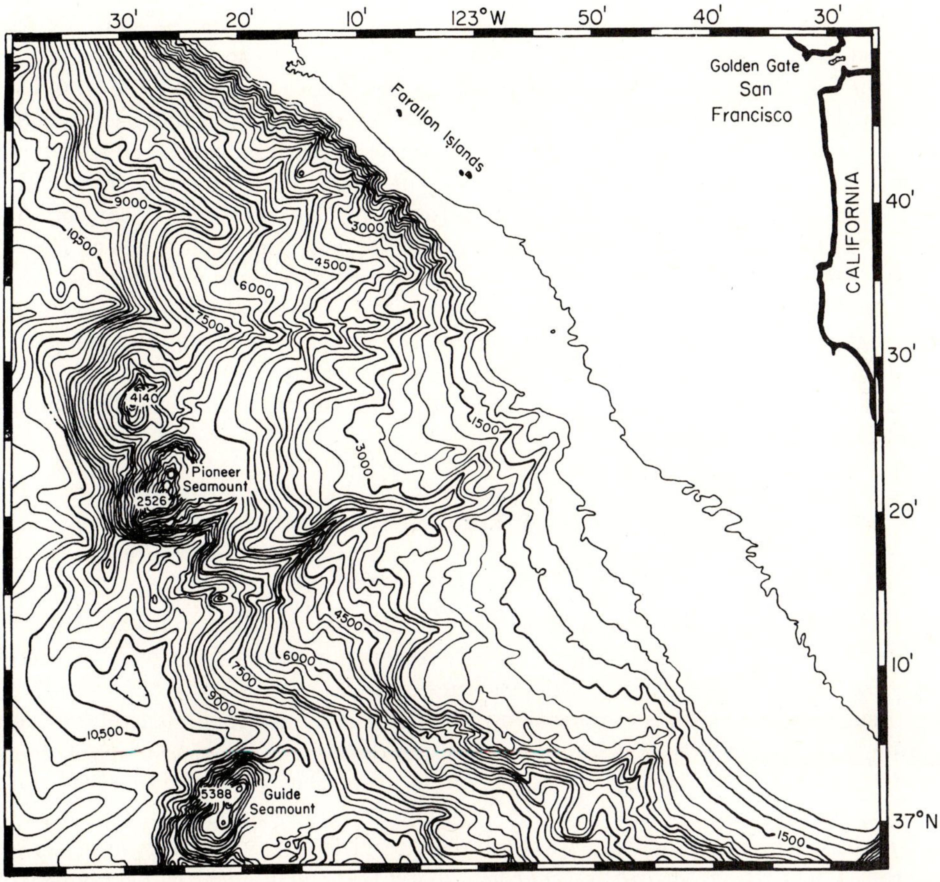

Fig. 7. Continental shelf and slope off San Francisco, California, with canyons and large seamounts on the lower slope. Contour interval 300 ft. (After Shepard and Emery, 1941.)

The Russian expeditions to East Antarctica (Jivago and Lissitzin, 1957–1958) have discovered that the slope off this continent displays a wide variety of types. In front of the Shackleton ice shelf (96°–100°E), and off the Princess Ragnhild Coast (25°–30°E), the slope consists of tilted plains down to 800 fm or more. Off Lützow–Holm Bay (38°E), and to the south of the Kerguelen–Gaussberg Ridge, it is often made up of tilted plains with gentle undulations. Steps or benches of moderate slope alternate with steep and short scarps in the lower section of the slope (1400–2000 fm) off the Princess Ragnhild Coast and elsewhere. In the Enderby Land region (45°–60°E), and near the Riisser–Larsen Peninsula (34°E), the slope breaks at variable depths into massive blocks of diverse forms, and intervening depressions with sharp and angular contours: this type is found in relatively small areas. In the western part of the region investigated are groups of rounded, small hills, 60 to 80 m high, sometimes covering the sloping sides of large submarine ridges. Similar ridges, but not bearing rounded hills, occur along the continental slope in some of its steepest parts, especially above 1100 fm, to the north of the Davis Sea and off the Princess Ragnhild Coast: they can rise to 500 fm or more above the troughs, and three or four ridges are often crossed before the deep-sea floor is reached. Near the Lars Christensen Coast (70°E), smaller longitudinal ridges, 80 to 100 m high, were discovered on the lower part of the slope; in some areas they combine with larger ridges. The last feature found in this wide region consists of transverse spurs with intervening embayments near the Kemp Coast (58°–60°E): the difference in height between spurs and depressions is 500–700 fm, and the bottom in the latter is very rugged and divided into small blocks. The spurs seem to be submarine extensions of continental mountains running at right angles to the general direction of the coast.

Such a diversity suggests that the continental slope, as well as the shelf, must have various origins.

4. Structure and Origins of the Continental Margins

A. *The Erosional and Constructional Classic Theory*

A classic theory of the origin of the continental margin has been exposed by D. W. Johnson in his basic work (1919). According to his conception, which was also supported by others, the continental shelf consists usually of two parts: (1) the inner one, which is an abrasion platform which has been levelled by wave action down to the wave base, i.e. the limit depth at which the waves can erode the solid rock: thus, this part is made up of the regional rocks; (2) the outer part and the continental slope, which are constructional features the material of which comes from wave abrasion and from river sedimentary discharge. The French term *talus continental* for the continental slope refers to this conception.

Such an explanation can no longer be entertained unless it is considerably modified, especially in its erosional part.

It now seems certain that the wave base is much shallower than it was admitted before. Bradley (1958) remarks that, during the nineteen-twenties and thirties, most American authors believed that the wave base lay between 600 and 100 ft below sea-level, whereas post-war authors think that it lies between 60 ft and a few feet only. The same opinion seems to prevail in Europe (e.g. in Kuenen, 1950, pp. 227–228; the examples of the contrary opinion which are quoted by Kuenen come mostly from Johnson). Berthois (1949) has found that, along the shores of Brittany, the rounding of pebbles is extremely slow as soon as the low-tide level is reached. Guilcher, Adrian and Blanquart (1959) have also shown that, in the same region, the roundness of pebbles is generally poor at low-spring-tide level, whereas it is very high on the upper parts of the same sections of the coast. This does not mean that the wave base is reached at low-tide level, since sand and gravel are found to be stirred farther out to sea by waves, but that the figures quoted above from American post-war authors seem to the writer to be correct as an approximation. The wave base is generally shallower in temperate and cold seas, where large seaweeds efficiently break the energy of the surf at low tide, than it is in warm seas where no kelp beds are found (McGill, 1958, see map). However, the contrary may sometimes be true; for instance in the Subantarctic Islands, which are beaten by the exceptionally heavy swell of the Southern Ocean. As a rule, wave action cannot be responsible for wide and deep platforms if the sea-level and the land remain stable.

During Pleistocene times, however, the sea-level underwent large variations as a consequence of glaciations. This can account for a wider range of wave action, extending down to several dozens of fathoms. On the other hand, a slow subsidence of land can also allow waves to erode wider platforms; and both factors may combine. Such has been the case off Southern California (Emery, 1958; Bradley, 1958), where submarine terraces have been recognized along the mainland, around islands and on bank tops: the larger part of them is too deep to be affected now to an appreciable extent by wave erosion. These terraces are distributed into sets which are related to different low sea-levels; moreover, they have suffered a differential warping (see below). Other submarine terraces are reported from the Persian Gulf off the Qatar Peninsula at 3–9 fm, 11–17 fm, 17–28 fm and 30–40 fm by Houbolt (1957), who assumes that they represent stages in the post-glacial rise of sea-level (however, this hypothesis makes one wonder whether the lapse of time has been sufficient). Others are found off western Australia (Carrigy and Fairbridge, 1954). On the West Florida shelf, the Florida Keys line is continued under water by a ridge 65 miles long, down to 90–100 fm in depth: this would be a Pleistocene barrier spit, possibly consolidated in beach-rock which may have enabled it to persist (Jordan and Stewart, 1959).

In addition, it must be pointed out that wave abrasion in solid rock is not the only force in action on the shores and continental shelves. In the intertidal zone, weathering plays an essential part in cliff recession and platform lowering: deep granular distintegration in granites, under tropical humid climates, exposes the rotten rocks, which can then be more easily removed; alternations of wetting and drying, action of salt in the spray zone, borings of marine organisms and especially of microscopic algae (Cyanophyceae) are very powerful in limestones, where solution is probably also in progress, although there has been much discussion about the processes of limestone solution on sea-shores. Again, with volcanic rocks, weathering is often most efficient and leads to spectacular features; in cold regions, frost splitting in the ice-foot bordering the cliffs carves benches into them (Tricart, 1957; Emery, 1946; Emery and

Foster, 1956; Fairbridge, 1952; Wentworth, 1938–1939; Guilcher and Pont, 1957; Guilcher, 1958; Nansen, 1922; etc.). In the writer's opinion, the result of these numerous processes is probably more efficient than wave attack on solid rock; and below low sea-level, where most of them disappear, tidal currents are able, in some areas, to erode quickly weak rocks, such as the Tertiary strata outcropping in estuaries of the Netherlands, provided that gravels acting as grinders are carried forward by them.

On the other hand, many submarine features on shelves are to be ascribed to former subaerial erosion or deposition. The submarine valleys (not the deep canyons) were almost certainly eroded during one or several lowerings of sea-level in Pleistocene times, when the river beds largely extended seawards. Similarly, the submarine ridges in solid rock, for example off West Brittany, must have been shaped by subaerial erosion. The hummocky landscapes, such as those reported from the Gulf of Maine, are drowned glacial deposits (drumlins, eskers or moraines). This is also true for the Baltic Sea, where a beautiful submarine esker 21 miles long, 150 yd wide and 5 to 10 fm high, occurring between Öland Island and Sweden, has been mapped in detail (Dannstedt, 1947). Pratje (1951) has explained stony strips in the northern part of the North Sea as terminal moraines belonging to the Pommern, Frankfurt and Warta stages; and, farther south in the same area, the Dogger Bank is probably another moraine (Stride, 1959). Many deep depressions extending across the glaciated shelves of Norway and North America off mountainous areas are obviously glacial troughs; but the longitudinal depressions in these shelves may have another origin and will be discussed below. The recent history is especially intricate in the heavily glaciated areas, since they have been isostatically uplifted after the melting of the Wisconsinian glaciers as much as 500 m in some places in Scandinavia. As a result, the parts of the continent adjacent to the sea have firstly suffered ice action, which is of continental nature; then they have been drowned by the post-glacial transgression; and finally they have been uplifted, and became dry land. This shows how much the continental shelf is intimately connected with the continent itself in such regions. Even outside the glaciated areas, the sea-level during the Pleistocene interglacial periods stood at higher levels than now (see, for example, Zeuner, 1950), so that the limit between dry land and sea floor also suffered fluctuations in these regions, as a response to glaciations. Innumerable "raised" beaches in South Britain, Normandy, Brittany, Portugal and Morocco (Guilcher, 1958), as well as those reported from Florida and Georgia (Stearns MacNeil, 1950) and elsewhere, point to these fluctuations.

The deep, discontinuous depressions in the English Channel and the North Sea (Fig. 2) raise difficult problems. Since some of them are as deep as 100 or 120 fm, they are not as easily explained by fluviatile erosion during the eustatic lowerings of sea-level as are the shallower submerged valleys; nor do they seem to be explained as being overdeepened by glaciers, because no glacier extended into the English Channel during the Pleistocene, and, in the North Sea, the ice had apparently no capacity for any important overdeepening since this area was flat and not mountainous, as are Norway and Scotland. Tidal scour does not seem to be quite satisfactory in explaining the Hurd Deep, even if it is entirely cut into Mesozoic and not Palaeozoic strata (as it appears on King's map, 1954), because it does not coincide exactly with the area where the tidal currents are strongest off the Cotentin peninsula. The Ushant trough might perhaps result from faulting, but more data are needed. In the North Sea, the deeps were perhaps carved by streams flowing under pressure beneath an ice-cap, a process which can account for their unusual depth and their complex pattern. An alternative suggestion is that they have been lowered by subsidence since they have been cut by subaerial rivers. The influence of subsidence on shelves is examined below.

Erosion by the various processes which have just been reviewed does not explain the nature of the continental margins. It may have occurred in rocks of the same structure as those of the adjacent continent, and, then, the continental shelf might be related to a progressive bevelling of the continent by erosional processes alone. But it may be also that erosion has affected only a superficial veneer of the continental margins, so that the shelf, as a whole, would result either from tectonic movements or from sedimentation. The true nature of the continental margins (and borderlands) is a problem which may be approached by seismic prospecting (see Chapter 1), or by drilling or by consideration of the structural geology of the nearby continent and of data from dredgings and corings on the shelf.

B. The Constructional, Subsiding Type (Fig. 10a)

Detailed information is not yet available about the structure of the larger part of the continental margins of the world. The areas where investigations are most accurate are eastern North America, the Bahama platform, the northern part of the Gulf of Mexico, North-west Europe (Celtic Sea, English Channel, North Sea), Southern California, and South France and Corsica. These shelves fall into a constructional, subsiding type, except for the last and the last but one.

In eastern North America, it is now ascertained that the submerged continental margin belongs to the same structural unit as the Atlantic Coastal Plain, which consists of Mesozoic and Cenozoic sediments covering the downwarped Appalachian basement. In the coastal plain, drillings for oil research in the Cape Hatteras area (Swain, 1947; Kuenen, 1950a) "show a prism of Tertiary to Lower Cretaceous strata resting on a weathered surface of granite. Each stratum thickens regularly seaward", and the basement, which is found at about 800 m below sea-level at 70 km to the WNW of Cape Hatteras, is encountered at 3200 m below the cape. In the coastal plain of New Jersey, the older the deposits are, the steeper the seaward dip is, the values averaging (Johnson and Richards, 1952) 11 ft per mile for the Cohansey formation (Miocene), 30 ft per mile for the Upper Cretaceous, 40 ft per mile for the top of the Raritan formation (base of Upper Cretaceous), and 65 to 102 ft per mile for the buried surface of the basement (see also Anderson, 1951; Spangler and

Peterson, 1950). In sections parallel to the coast, the depth of the basement is not regular: it decreases greatly on the "Carolina ridge" (south of North Carolina), and increases again farther south; however, the seaward thickening of the strata is a general feature of the coastal plain. In the submerged part, the structure is the same, since Cretaceous, Eocene, Miocene and Pliocene have been found in the canyons encroaching on the shelf edge in the Georges Bank region (Stetson, 1949). The benches breaking the general declivity of the

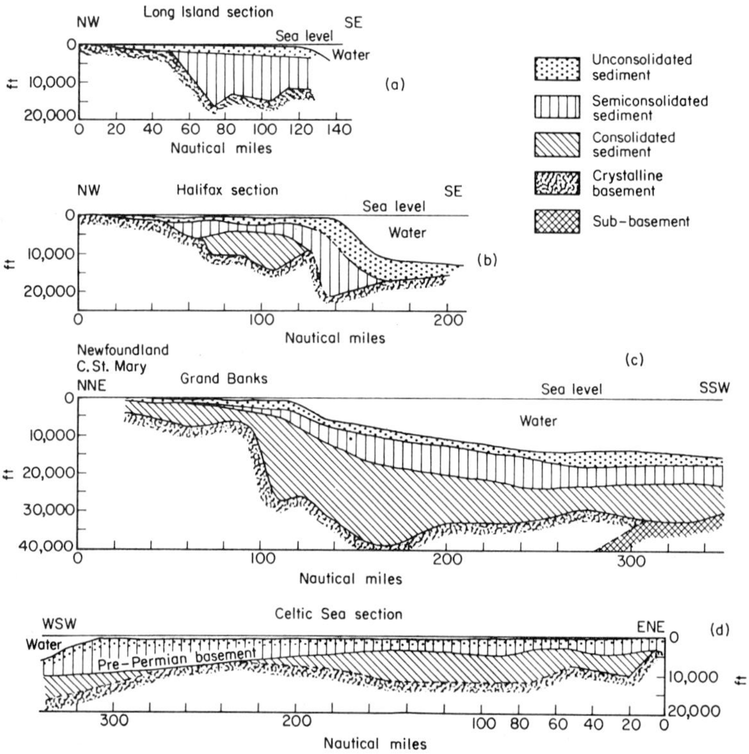

Fig. 8. Probable geological structure of the shelf and slope in north-eastern North America and in North-west Europe (Celtic Sea), according to geophysical investigations. (After Ewing, Bentley, Officer, Press, Steenland and Worzel, 1950–1954–1956, for America; after Day, Hill, Laughton and Swallow, 1956, for Europe.)

continental slope are likely to be structural: they have been correlated, between Cape Hatteras and Cape May, with the formations encountered in the Hatteras drilling, and the same interpretation is proposed for other benches along the Blake Escarpment from which sediments of Recent to Upper Cretaceous were obtained (Heezen, Tharp and Ewing, 1959, pp. 46–48). On the other hand, careful and numerous geophysical investigations (Ewing *et al.*, 1937 to 1956; Hersey *et al.*, 1959) have revealed the topography of the basement under the sedimentary strata along a number of transverse sections between Florida and Newfoundland Banks. Between Jacksonville, Florida and Cape Henry, Virginia, the basement beneath the shelf lies at depths ranging from 2800 to 20,000 ft. Off Chesapeake Bay, the slope of the basement is relatively simple down to 12,000 ft below sea-level. Off New York (Fig. 8a), the thickest part of the sedimentary basin lies under the middle shelf, where the contact with the basement reaches 16,000 ft; the contact rises again at the edge of the shelf where it is found at about 12,000 ft, a feature which points to a maximum subsidence located under the shelf and not extending far into the Ocean. In the Gulf of Maine, the basement is shallow (1000 to 2000 ft below sea-level or even less), and the sedimentary rocks seem to be Triassic in age; farther offshore, under Georges Bank, the basement slope lowers first very quickly, and then more slowly and irregularly beneath the continental slope. Off Nova Scotia (Fig. 8b), the basement slope is irregular and dips very steeply beneath the continental slope; a Triassic trough seems to exist off Halifax, and the banks around Sable Island consist only partly of glacial deposits, under which probably lie Cenozoic and Mesozoic beds. Between Misaine Bank and Saint Pierre Bank, the seismic prospection under the wide U-shaped Laurentian Trough, at the opening of Cabot Strait, indicates a considerable thickening of consolidated sediments upon the basement, so that the depression is likely to be initially a tectonic graben; but an important over-deepening by ice has probably occurred during the Pleistocene, explaining the present surface forms, as Shepard suggested in 1931. In a north–south section off Avalon Peninsula, Newfoundland (Fig. 8c), the maximum depth of the basement is found outside the continental slope, and the thickness of the sediments above it reaches about 40,000 ft; but further to the south the contact rises to 30,000 ft, as it does off New York. An interesting feature is the relatively great thickness of the consolidated sediments under the Grand Banks (about 6000 ft), which seems to indicate a Triassic, or possibly Palaeozoic, age for a very large part of the banks. Whatever the age, the Grand Banks as a whole are certainly not made up of glacial sediments, and came into existence well before Pleistocene times.

Summarizing these results, we may conclude that subsidence has been general along eastern North America, the continental margin of which is a huge sedimentary basin. Irregularities occur in this basin, since usually, especially in the northern part, the maximum subsidence has probably migrated from place to place.

In the Great Bahama Bank, a drilling has been made at Andros Island down to 14,587 ft: it reached Lower Cretaceous rocks at 12,480 ft after it cut through exclusively calcareous formations. The Quaternary limestones may be as much as 400 ft thick under Andros Island. The Eocene is found at 2200 ft and the Upper Cretaceous at 8760 ft. It is believed that the Bahama Banks result from marine deposition on a basement having the same general contours as the banks have today, and that this basement has subsided slowly, the platform remaining shallow as a consequence of sedimentation, whereas the latter had little effect in the deep channels between the banks (Illing, 1954; Newell and

Rigby, 1957). During the Pleistocene, the eustatic movements of sea-level have played a part in the geomorphological evolution of the banks, and the submergence after the melting of glaciers accounts for the drowning of a karst surface over them. Nevertheless, however conspicuous the effects of eustatism may be on the surface, the thickness of the strata they have affected is almost negligible when compared with what had been deposited during the long-continued subsidence before the eustatic lowering. The Bahama Banks afford an excellent example of the respective importance of Pleistocene events and former phenomena in the present features of a constructional, subsiding platform.

Another well-known constructional shelf is the so-called Gulf Coast geosyncline in and off South Louisiana. All geologists do not agree on the use of the term "geosyncline" when applied to such a structure (Aubouin, 1959), but the feature in itself is outside discussion. In this area R. J. Russell and R. D. Russell (in Trask, 1939) state "deltaic sedimentation by the Mississippi and its ancestral streams has been an essentially continuous process since at least the beginning of the Tertiary". Subsidence has been going on as an isostatic response to the deposition of a huge pile of sediments since the Cretaceous, and the aggregate thickness of post-Oligocene deposits, established from records of several thousand wells, is about 62,000 ft in southern Louisiana (Crouch, 1959). The axis of the trough migrated southward during the Tertiary, but not appreciably since early Pleistocene times. The Pleistocene terraces are distinctly tilted seawards and uplifted landwards, a hinge line limiting the zone of isostatic downwarp and the zone of uplift (Fisk, 1939); at the present day, subsidence is maximum in the active sub-delta, where natural levees are already sinking below sea-level in some areas (Welder, 1959). In older post-glacial sub-deltas the drowning is more pronounced, as it is, for example, to the north of the present mouths. Off the Mississippi delta, the fact that the continental slope is much more gentle than on either side (see above) is explained by Ewing, Ericson and Heezen (1959) by the building of a huge submarine cone by the river during the Pleistocene, which should have buried the steep scarps existing elsewhere in the lower parts of the slope around the Gulf of Mexico. In Ewing's interpretation, these scarps should thus be relatively old, and no important recent faulting has occurred in this area where seismicity is low; while, on the contrary, Greenman and LeBlanc (1956) have suggested that the Gulf was downfaulted to its present depth in late Pleistocene or early Recent time. On the other hand, the change in form of the West Florida scarp at 27°N is explained by Jordan and Stewart (1959) as a result of a change from clastic underlying rocks and a thick blanket of unconsolidated sediments in the north to carbonate rocks with only a veneer of sediments at most in the south.

In North-west Europe, Dangeard (1928), W. B. R. King (1948, 1954), Bourcart and Marie (1951), Day (1958), Curry (1960), and Boillot (1961) have collected a great deal of geological data concerning the English Channel and the Celtic Sea, and knowledge has also been largely improved by seismic prospecting in the same areas (Bullard and Gaskell, 1941; Hill and King,

1953; Day, Hill, Laughton and Swallow, 1956; Allan, 1961). The English Channel is an epicontinental sea in which sediments have been deposited since New Red Sandstone (Permian and Triassic) times. As in eastern North America, the shelf in the Channel, and the surrounding plains in South England and in the French Bassin de Paris, belong to the same sedimentary unit. In the emerged or coastal parts, several distinct depressions in the Armorican basement, more than 10,000 ft deep, have been found in Hampshire (Kent, 1949), in Champagne and in Lorraine (Cholley, 1960). In the Channel proper, the deposition has also been more or less thick according to place and period: for example, it seems that the Tertiary deposits occur in separate small basins. Trias and Cretaceous cover large areas. Although the sedimentation has been somewhat discontinuous, some 3000 ft of post-Carboniferous strata seem to exist across the western part of the Channel between Brittany and Cornwall; in the Celtic Sea (Fig. 8d), they merge into a post-Carboniferous synclinorium which becomes deeper and deeper to the SSW as far as the edge of the continental shelf, where the Palaeozoic floor is found at about 10,000 ft below sea-level. Formerly, a Palaeozoic geosyncline extended to the continental slope at least. Mesozoic and Cenozoic samples have been dredged from three places on the continental slope in the Celtic Sea (Bourcart and Marie, 1951; Day, 1959), which thus resembles the continental slope off eastern North America (if the samples were *in situ*).

The North Sea is another subsiding basin, separated from the former by a buried ridge running from Charnwood Forest and East Anglia to Brabant and the Ardenne shield. It includes North-east England, the Netherlands and North-west Germany. According to Dutch geologists quoted by Umbgrove (1945), the post-Carboniferous sedimentation would reach 7500 m and perhaps 9000 m in some parts; some 6000 m are assumed to occur under Lüneburg near Hamburg; the Pleistocene alone is more than 600 m thick in West Holland (Pannekoek, 1956). Recent seismic prospecting on the Dogger Bank (Stride, 1959) points to sediments more than 11,400 ft thick (3400 m) above the basement (thickness of supposed Pleistocene 1250 ft; Cenozoic 1850 ft; Mesozoic and possibly Pennsylvanian more than 8300 ft). A resemblance between the North Sea basin and the Gulf Coast geosyncline lies in the salt domes rising from the deepest parts to the surface, in Friesland, Germany and Louisiana.

Although the data concerning the continental shelves of western and north-western Australia are very sparse, Fairbridge has tried to gather the available material into a provisional synthesis (Fairbridge, 1953; Carrigy and Fairbridge, 1954), from which it results that large parts of these areas, where the shelf is 24 to 250 miles wide, seem to belong more or less to the type considered here (Fig. 9). The advance in geological knowledge on land in western Australia, clearly summarized and discussed by Teichert (1958), indicates that four large sedimentary basins (Bonaparte Gulf, Canning, Carnarvon and Perth), in which maximum thicknesses of sedimentary rocks reach from 17,750 to 40,000 ft, occur in this part of the continent, and are widely open to the sea. The Pre-Cambrian basement rises as blocks between them. The sequences are predominantly marine and range from Cambrian to Pleistocene. Post-Carboniferous rocks seem to outcrop only in the Carnarvon and Perth basins. It seems that corresponding features, namely submarine rises and depressions, exist on the shelf off the coast, so that it is believed that the geotectonic elements on the shelf and in the adjacent mainland are the same, except in the north-west where the North Kimberley block, between Bonaparte and Rowley depressions, divides seaward into two relatively narrow ridges between which the Browse depression opens. Unfortunately, no accurate information seems to

Fig. 9. Relationships between geotectonic units on continent and on shelf in North-west Australia. (Simplified from Fairbridge, 1953.) Maximum thicknesses in feet of sedimentary rocks in Canning Basin and in Bonaparte Gulf Basin, from Teichert (1958).

be available about the thickness of the sedimentary column on the shelves, and the gravimetric observations by Vening Meinesz are not very conclusive here, except for the Rottnest shelf off Perth basin where an anomaly of -140 mgals has been observed. A subsidence is supposed in depressions from bathymetric data, especially in the Bonaparte and Rowley depressions where the shelf is as deep as 300 fm—a fairly unusual figure. In this north-western region, it may be that sedimentation has not been able, everywhere at least, to keep pace with subsidence (Tertiary rocks are not reported from the Canning and Bonaparte basins); the downwarping, however, has been slow enough to allow coral reefs to grow as far as the edge of the shelf (Teichert and Fairbridge, 1948). Further to the south, off the Carnarvon and Perth basins, where the shelf is narrower, echo-soundings show faceted-like profiles which suggest another kind of subsidence. It may be provisionally assumed that the shelves off western Australia represent a transitional type between the constructional one and the type which will be discussed immediately below.

It is possible that the constructional subsiding type is very widely distributed in the world. It might occur, for example, in the Bay of Bengal (Morgan and McIntire, in Russell, 1959), off the Amazon (Pimienta, 1957), and in the Yellow Sea, but more information is needed for these shelves, and for others.

C. The Flexured Type

In a number of regions, it seems that the continental margin has been warped as it is in South Louisiana; but the difference from the type just described is either that no thick sedimentary column is apparently responsible for the subsidence, or that the shelf is very narrow as a result of a steep seaward slope in the basement.

The narrow continental margins on both sides of the West Mediterranean (Balearic Sea, or *Bassin algéro-provençal* in French terminology) afford a fine example of this type [Fig. 10 (b1)]. Many investigations by French geologists, on land as well as under the sea (Bourcart, 1953; Bourcart and Glangeaud, 1954; Bourcart, 1959), have provided a picture of the Cenozoic evolution of this area which may be summarized as follows. Evidence for a general emersion of the Balearic Sea in Oligocene times, during the main Alpine folding, derives from the fact that, on the coasts of South-east France and Algeria, Oligocene sediments include elements which could only have come from a continental area occupying the place of the Mediterranean, because their source is not found on the surrounding emerged lands. On the other hand, in the Nice region the sedimentary structure of the Flysch, which is hundreds of metres thick, points to a feeding by a northward flow, whereas the land is now to the north and the sea to the south (P. H. Kuenen, in Bourcart, 1959). In early Miocene times, the present orography was initiated: as the folding released, the slopes were inverted, and the collapse of the Balearic Sea continued to go on later. The numerous submarine canyons which have been surveyed by Bourcart and the French Navy off southern France and western Corsica fit fairly well with the conception of this evolution, and may be interpreted as former subaerial valleys, initiated when the slopes began to be inverted, and finally drowned and deeply downwarped under the sea. Thus, the continental margin off southern France consists of a large flexure which came into existence in Tertiary times as a consequence of Alpine foldings, and is probably still in progress in the Nice region, where Pleistocene beaches are tectonically much disturbed: this accounts for the absence of any continental shelf in this area. Off Algeria, the soundings are very much less accurate; however, provisional charts by Rosfelder (1955), in which the existing data are compiled, as well as Bourcart and Glangeaud's synthesis (1954), indicate a general marginal flexure of the Provençal type upon which are superposed many folds running in a north-east and south-west direction, that is, obliquely to the trend of the coast. The Algerian margin is thus both flexured and folded. Off North Tunisia, the shelf widens considerably and extends as far as Sicily, but north-east–south-west ridges (anticlines) are found again as far as 11°E, where the structural trend changes and becomes north-west–south-east after the transverse Zaghouan fault is crossed (Castany, in Bourcart, 1959).

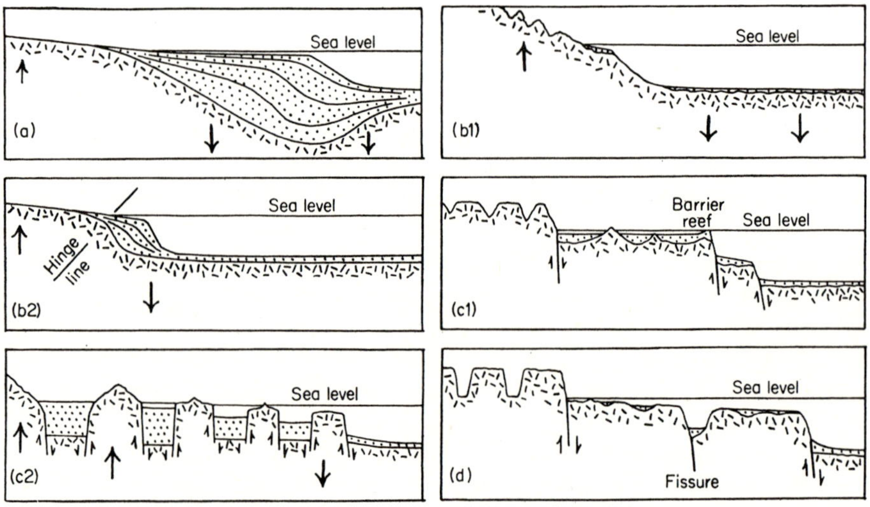

Fig. 10. Types of continental margins. (a) Constructional, subsiding type. Wide shelf, wide and thick sedimentary basin, e.g. eastern North America, North-west Europe. (b) 1. Flexured type. No significant sedimentation, e.g. Provence (at Nice, no shelf at all). 2. Flexured type with narrow shelf and sedimentary wedge. Intermediate between (a) and (b1). Probable examples, West and South Africa. (c) 1. Faulted type. Irregularities in basement filled up by sedimentation. Probable example, Queensland. 2. Block-faulted type. Sedimentation in downfaulted basins, erosion on ranges, e.g. continental borderland of Southern California. (d) Fissured type. Glacial overdeepening in transverse valleys, e.g. Norway, Labrador, East Antarctica.

According to Bourcart (1938, and other papers), the flexured type is very widespread throughout the world; he even thinks that this is the normal type of contact between land and sea. If the hinge line migrates outwards, it follows that a part of the sea-bed becomes dry land, and the contrary happens if the hinge line migrates inwards. Although Bourcart does not deny the existence of eustatic fluctuations of sea-level during the Quaternary, he tends to relate the emerged beaches to continental flexures rather than to eustatism. Umbgrove (1947) is in favour of many of Bourcart's ideas. Jessen (1943) has also come to similar conclusions in his book, in which all large coastal units in the world are considered. He points out that the marginal regions of the continents are often higher than the inner parts of them (India, Africa, Scandinavia, etc.), a fact which is, in Jessen's opinion, evidence for a marginal flexure. The lowering of the sea bottom by continental flexure may create a faceted submarine profile, with outer shelves more tilted than the inner ones: according to Bourcart, the continental margin off Morocco would be an example of this.

The recent echo-soundings off Morocco (Fig. 6) do not seem to support such a conclusion. Facets exist in the slope on some profiles, it is true, but not on all, and their distribution does not appear to be systematic. The most interesting feature off North Morocco is the wide trough off the Rharb plain: both the

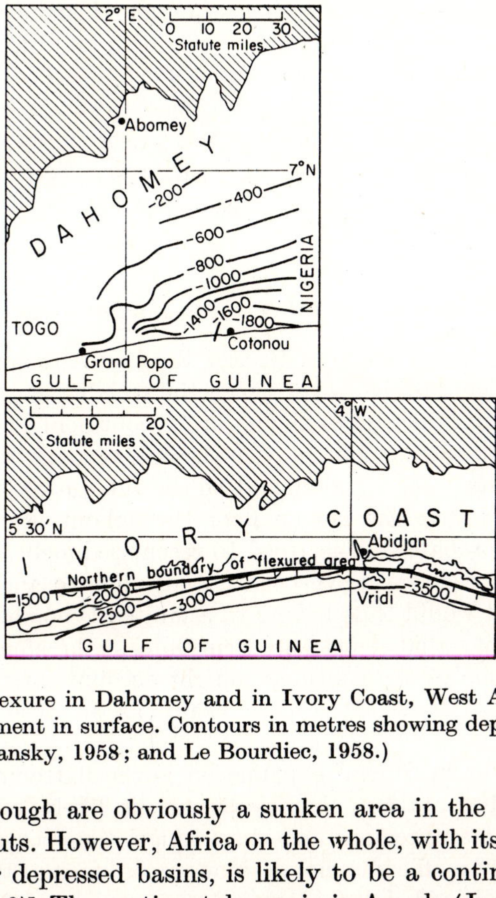

Fig. 11. Marginal flexure in Dahomey and in Ivory Coast, West Africa. Hatched: Pre-Cambrian basement in surface. Contours in metres showing depth of concealed basement. (After Slansky, 1958; and Le Bourdiec, 1958.)

Rharb and the trough are obviously a sunken area in the Rif foreland across which the coast cuts. However, Africa on the whole, with its bulging peripheral heights and inner depressed basins, is likely to be a continent with marginal flexures [Fig. 10 (b2)]. The continental margin in Angola (Jessen, 1936) seems to be a typical place of repeated warpings along a hinge line, which resulted in terraced planation surfaces on the land and facets (former shelves) on the continental slope. The Congo submarine canyon may thus be explained as an antecedent valley. In the Ivory Coast and Dahomey (Fig. 11), the Pre-Cambrian basement plunges to the south under narrow coastal basins covered by Mesozoic and Cenozoic sediments; between Abidjan and Vridi, Ivory Coast, a large concealed fault occurs beyond which the seaward dip steepens considerably, so that the basement is likely to lie as deep as 3500 m under the coastline (Le Bourdiec, 1958); at Cotonou, Dahomey, seismic prospecting indicates that the basement lies at about 1800 m below sea-level, and its slope steepens seaward here also (Slansky, 1958). These steep slopes in the downwarped rocks fit well with the narrowness of the shelf. There is no appearance of significant warping during Pleistocene times in Dahomey, but in the Ivory Coast the famous submarine canyon of Trou Sans Fond may be evidence for late subsidence. In South Africa too, the peripheral regions are strongly tilted outwards. The coast of Natal, for example, has been described as monoclinal: the sediments overlying the basement (Table Mountain Sandstone and Karroo formations), and the surface of the basement itself, have been flexured and dip to the sea (L. C. King, 1940). Here again, the shelf is narrow. Off Cape Province it becomes much wider on the Agulhas Bank, but several submarine channels may be followed to a depth of 400 fm to the east and to the north of Cape Town, a fact which points to an outward flexuring. Conversely, Tertiary marine strata stand, in places, 1200 ft above sea-level, suggesting an axis of tilting parallel to the coast (L. C. King, 1951, p. 190).

D. The Downfaulted or Block-Faulted Type

In a paper dealing with the Great Barrier Reef off Australia, Steers (1929) expressed the view that the shelf over which the reefs have grown off Queensland may be a downfaulted part of the adjacent continent [Fig. 10 (c1)]. The islands rising as hills in the lagoon between the outer barrier and the mainland are made of continental rocks, and should be the summits of the downfaulted block; and "the fact that the lagoon floor throughout is of comparatively uniform depth may probably be ascribed to recent sedimentation". The slope outside the outer barrier may have been initially due to another large fault. The Queensland shelf would thus belong to a faulted type. Jessen agrees with this opinion and thinks that deep-seated currents in the simatic substratum, flowing in the direction of the continent, might account for the faulting. The same mechanism may have acted periodically as a cause of the flexures examined above.

An interesting point in Steers' explanation is the flattening of the downfaulted shelf by sedimentation. Such a process is likely to have played a considerable part in many regions and may explain the occurrence of flat and moderately wide shelves off some mountainous areas, such as the Galician shelf off Cape Finisterre. Provided the block has sunk to an adequate depth, the filling up of initial topographic depressions between hills is able to create a smooth platform, apart from some stacks of solid rock rising from the general plain. It is much easier to understand in this way the formation of shelves in hard rocks such as those outcropping in Galicia and in Queensland than to imagine a planation by sea erosion, especially when tectonic movements and faulting are reported from the adjacent continent, as in Galicia.

A striking example of a block-faulted margin in the Red Sea on the Farsan

Bank near Lith, to the south of Jidda, has been described by Guilcher and Nestéroff (1955), who had an opportunity of examining this submarine area during the *Calypso* Expedition in 1951–1952 (Fig. 12). The lofty mountains of Asir are bordered to the south-west by a series of large fault scarps ending on the coastal plain of El Tihama. The submarine topography includes two steps perfectly distinct from one another: the higher one, near land, occurs at 30–45 fm, the lower one lies at 350–400 fm. The scarp between the two, which was sounded near Abulat Island, is precipitous and runs in the same direction as the general trend of the Red Sea graben. Abulat Island, on the higher platform,

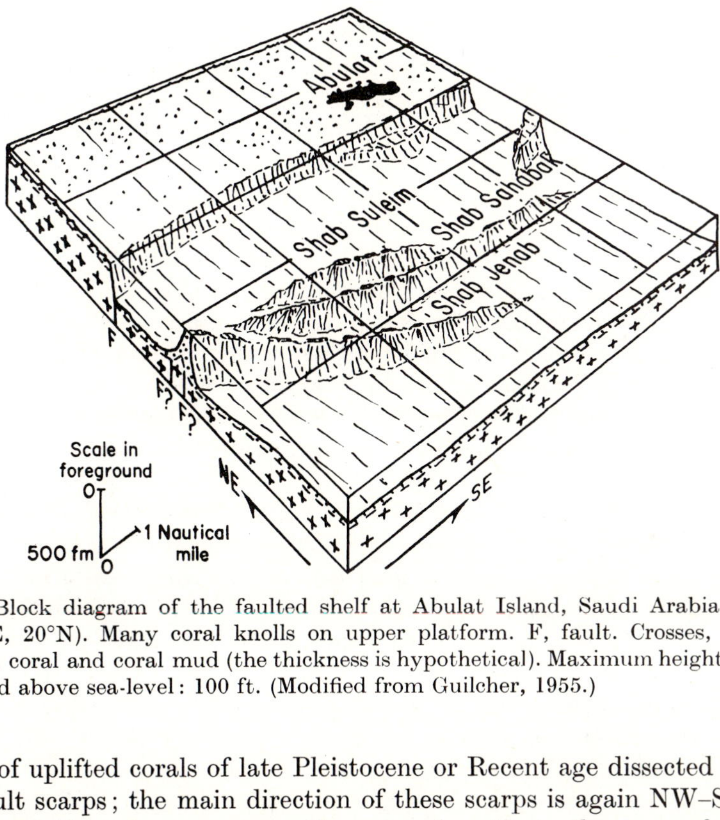

Fig. 12. Block diagram of the faulted shelf at Abulat Island, Saudi Arabia, Red Sea (40°E, 20°N). Many coral knolls on upper platform. F, fault. Crosses, basement. Dots, coral and coral mud (the thickness is hypothetical). Maximum height of Abulat Island above sea-level: 100 ft. (Modified from Guilcher, 1955.)

consists of uplifted corals of late Pleistocene or Recent age dissected by many small fault scarps; the main direction of these scarps is again NW–SE. From the lower platform long, narrow ridges rise abruptly to the sea-surface; these consist of coral reefs in their upper parts and are supposed to have initially been small faulted blocks rising moderately from the general level of the platform on which corals have grown during a comparatively slow subsidence. These high ridges are referred to as *shaban* (singular, *shab*) by the Arabs. Farther to the north, some submarine and coastal features off the Arabian coast, at the entrance to the Gulf of Aqaba, have been briefly described by Schick (1958). The Midian Islands, which are composed of a dozen islands and islets and innumerable reefs, stretch in a long line parallel to the coast, from which they are separated by a body of water more than 50 fm deep. A submarine scarp is found at the outer edge of the islands, which resemble the *shaban* near Lith except for their emersion. Tiran Island, at the opening of the straits, bears a beautiful series of uplifted terraces. Here we are dealing with one of the most faulted areas in the world.

Data are much more accurate for the continental borderland of Southern California [Fig. 10 (c2)]. According to Emery (in Bourcart, 1959, and 1960), this area, which includes deep, steep-sided troughs with intervening ridges, "may have been a geosyncline during the Jurassic period, but most of it was uplifted to form a broad land during the Cretaceous, Eocene and Oligocene. Erosional bevelling followed by a downwarp allowed the region to be covered by epicontinental seas during the Miocene. Near the end of the Miocene Epoch, the region became blockfaulted, so that, from the beginning of the Pliocene to the present, deep basins were separated by high areas that constituted islands and banks". The basins near the continent have been filled up by sediments, whereas in those lying farther offshore the sedimentation proceeded at a much slower rate. As seen above, Pleistocene submarine terraces occur on banks and around islands (Emery, 1958). After these terraces were cut (probably in Wisconsin times), they were warped, since a correlation diagram shows that each of them is deeper around offshore islands and banks than off the mainland, the differences in depth reaching about 225 ft. The average warp of the shelf edge is about 160 ft per 100 miles. The formation of the features of this continental borderland thus involves a number of factors : Mesozoic and Cenozoic sedimentation, subaerial planation, block-faulting, recent marine planation and deposition and regional warping. But the main features, namely the basins and ridges, are related to block-faulting.

The banks between the British Isles, Porcupine Bank and the Faroes, which may be considered as the continental borderland of North-west Europe, seem to belong structurally to the volcanic belt of the North Atlantic, extending from the Antrim Plateau to Greenland across Iceland. It is not possible to ascertain if they were initially tied together into a single platform and separated by further faulting, or if they were built as distinct units; consequently, their classification into a definite type remains doubtful. The seamounts on the slope, such as those shown in Fig. 7, are likely to be submarine volcanoes in a number of cases or salt domes in some areas.

E. The Fissured Type (Fig. 10d)

The fissured type consists of shelves in which a long and deep depression runs parallel to the direction of the coast, dividing the shelf into an inner and an outer part. Sometimes more than one depression of this kind is found. As it has been shown above (Figs. 3 and 4), this type is common in glaciated areas: Norway, Scotland, Labrador, South-east Alaska, Ellesmere Island, South-west Greenland, West Spitzbergen, East Antarctica (H. Holtedahl, 1958; *idem*, in Bourcart, 1959; Jivago and Lissitzin, 1957). It combines generally with deep troughs at right angles to the direction of the coast. The problem is whether these features can be explained only or essentially by glacial erosion.

The glacial, or, more generally, erosional, explanation for the longitudinal depressions meets with a serious difficulty in that the depressions are perpendicular to the normal flow of ice, or of rivers before and between the glaciations. When the shelves into which they are incised bore ice-caps, it is evident that the ice flowed through these depressions; but it is not likely to have deepened and enlarged them much because of their direction, whereas the overdeepening was much easier in the transverse troughs, running in the general trend of the ice flow. In other words, both the longitudinal and transverse depressions may be initially of structural origin; but the present forms probably owe much more to ice than to structure when they are transverse, and much more to structure than to ice when they are longitudinal.

The explanation put forward by O. and H. Holtedahl is that the longitudinal depressions result from rifting which occurred during a late Tertiary uplift of the continental margins. It agrees well with the evolution generally accepted for Norway and several other arctic lands from which marginal depressions are reported, but, as Kuenen and the writer separately pointed out, the coincidence of the fissured shelves with glaciated areas is curious, and one wonders whether there is any structural relation between the marginal fissures and the growing and melting of ice-caps. The same idea is expressed by Jivago and Lissitzin, who explain the facts observed in East Antarctica by a series of vertical isostatic movements resulting from modifications in ice load. The main marginal depression would be a hinge line allowing vertical "breathing" of the continent, alternately depressed and uplifted relative to the deep-sea floor, the continental slope and even the outer shelf. A rifting on a smaller scale, related to the same events, would account for smaller fissures.

Would it not be possible to combine the two explanations? A Tertiary uplift might have predisposed the shelves to be faulted; further, this predisposition might have led to marginal rift depressions in glaciated areas, owing to the isostatic breathing of the continental margin, whereas in other regions flexuring or step-faulting generally occurred instead of rifting. If this interpretation were right, it might be expected that the longitudinal rifts would not extend down to the continental slope, since the ice-caps did not overload this part of the continental margin. Here, however, a difficulty is found. The profiles across the continental slope of the Norwegian Sea, which have been published and discussed by H. Holtedahl (1955) and have already been referred to, show distinct rising blocks breaking the slope in steps. Thus, it appears as though the rifting affecting the shelf has also affected the slope down to a depth of 700 fm or more. In East Antarctica too, the description of the slope which has been summarized above suggests evidence for rifting or block-faulting on various scales in many areas lower than the level at which the Pleistocene Antarctic ice-caps are likely to have rested upon the bottom. But this difficulty may be solved if we consider that the peripheral belt of the area depressed under the weight of the ice-cap may have been somewhat bulged upwards during the glaciation, because of the outflow of magma in the subcrustal zone; and conversely, after the melting of the ice, the peripheral belt must have sunk to some extent when the central tract rose (Daly, 1934). Thus, isostatic movements contrary to those occurring in glaciated areas may account for block-faulting or rifting outside the main hinge line separating the glaciated belt from the peripheral area.

As to the origin of the shelf itself in Norway, in the diastrophic interpretation it "may be regarded as a *paleic* surface formed primarily during the Tertiary, and probably subject to block-faulting and to denudation by marine and subaerial agencies during periods of changing relative sea-levels" (H. Holtedahl, 1955, p. 197). In other words, it was a downfaulted part of the planation surface which bevelled the Scandinavian highlands with some further modifications. The same explanation may be suggested for other glaciated lands, in so far as a subaerial planation is likely to have affected them before they were uplifted. There has been much controversy about the origin of the Scandinavian shelf and the *strandflat* which occurs just below or above sea-level in the Norwegian coastal area. Other strandflats have been reported from Bear Island, Spitzbergen, Iceland and Greenland. (A review of the opinions concerning the possible origins may be found, with references, in Guilcher, 1958, p. 160.) As O. and H. Holtedahl suggested, the strandflat may be either a large downfaulted block or a coastal marine surface, partly developed by ice-foot corrosion, as Nansen believed. It may, on the other hand, be the result of the recession of glacial corries, as Dahl proposed, or an old subaerial peneplain (Ahlmann) or *Rumpftreppe* (Evers, 1941), partly reworked by wave action. Many authors admit a compound origin, with one prevailing factor. In the writer's opinion, faulting is likely to have played a major part; H. Holtedahl (1955, p. 49) has given strong evidence for an extreme slowness of wave abrasion along the coast of Norway; at greater depths in the same region, the sediments are poorly sorted and point to a glacial marine, not true marine, origin (see Chapter 24 in this volume); in East Antarctica, the shelf seems not to have been much smoothed by ice erosion, except for some rare areas.

The considerable depth of the shelf in many glaciated areas has been supposed to be due to overloading by ice-caps and incomplete postglacial isostatic adjustment. This implies that the whole shelf has been covered by non-floating ice as far as the outer edge, which is by no means sure, and even improbable in a number of regions.

5. Conclusion

So far as the present state of knowledge allows a conclusion, the continental margins appear to have a mainly diastrophic or constructional origin. Marine erosion in shallow and coastal waters, including a number of processes of which wave abrasion is probably less important, is responsible for some of the features, but its action has been of limited extent. Turbidity currents, slumping and sliding, which are discussed in Chapters 27 and 28 in this volume, must be most efficient in slope evolution. Subaerial agents, including glaciers, have carved valleys and troughs into the shelf. Tidal scour may have preserved these depressions from being filled up with sediments, and may have carved some others in weak rocks, but these factors do not explain the continental margin itself as a distinct unit. Warping, faulting and long-continued sedimentation are probably the essential forces which have built this outstanding feature off the continents.

References

Allan, T. D., 1961. A magnetic survey in the Western English Channel. *Q. J. Geol. Soc. London*, **117**, 157–170.

Anderson, J. L., 1951. Northeastern United States (petroleum resources). *Bull. Amer. Assoc. Petrol. Geol.*, **35**, 421–437.

Aubouin, J., 1959. Les aventures de la notion de géosynclinal. *Rev. Géog. Phys. Géol. Dynam.*, ser. 2, **2**, 135–188.

Berthois, L., 1949. L'érosion marine et la formation des galets. *C. R. Acad. Sci. Paris*, **229**, 841–843; *Bol. Soc. Geol. Portugal*, **8**, 56 pp.

Berthois, L. and R. Brenot, 1957. Topographie sous-marine de l'Outer Silver Pit et du Markhams Hole. *Rev. Trav. Inst. Pêch. Marit.*, **21**, 431–434.

Berthois, L. and R. Brenot, 1962. Bathymétrie du secteur atlantique du Banc Porcupine (Ouest de l'Irlande) au Cap Finisterre (Espagne). *Rev. Trav. Inst. Pêch. Marit.*, **26**, 219–246 (with set of 12 charts).

Boillot, G., 1961. La répartition des fonds sous-marins dans la Manche occidentale. *Cah. Biol. Mar. Roscoff*, **2**, 187–208.

Bourcart, J., 1938. La marge continentale. *Bull. Soc. Géol. France*, (5), **8**, 393–474.

Bourcart, J., 1949. *La géographie du fond des mers.* Payot, Paris, 303 pp.

Bourcart, J., 1952. *Les frontières de l'Océan.* Albin Michel, Paris, 319 pp.

Bourcart, J., 1953. Contribution à la connaissance du socle sous-marin de la France le long de la côte méditerranéenne. *19th Int. Geol. Cong., Alger, 1952.* **IV**, 4, 25–63.

Bourcart, J., 1954. Note sur les définitions des formes du terrain sous-marin. *Deep-Sea Res.*, **2**, 140–144.

Bourcart, J., ed., 1959. *Topographie et géologie des profondeurs océaniques.* 83e Colloque Int. C.N.R.S., Paris, 1958, 313 pp. (Chart of continental slope off SE France and Liguria.)

Bourcart, J. and L. Glangeaud, 1954. Morphotectonique de la marge continentale nord-africaine. *Bull. Soc. Géol. France*, (6), **4**, 751–772.

Bourcart, J. and P. Marie, 1951. Sur la nature du rebord continental à l'Ouest de la Manche. *C. R. Acad. Sci. Paris*, **232**, 2346–2348.

Bradley, W. C., 1958. Submarine abrasion and wave-cut platforms. *Bull. Geol. Soc. Amer.*, **69**, 967–974.

Brémond, E., 1953. Le talus continental marocain entre Mazagan et Fédala. *Bull. Comité d'Océanog. et d'Et. des Côtes*, **5**, 242–244.

Bullard, E. C. and T. F. Gaskell, 1941. Submarine seismic investigations. *Proc. Roy. Soc. London*, **A177**, 476–499.

Carrigy, M. A. and R. W. Fairbridge, 1954. Recent sedimentation, physiography and structure of the continental shelves of western Australia. *J. Roy. Soc. W. Austral.*, **38**, 65–95.

Carsola, A. J., 1954. Microrelief on Arctic sea-floor. *Bull. Amer. Assoc. Petrol. Geol.*, **38**, 1587–1601.

Cholley, A., 1960. Remarques sur la structure et l'évolution morphologique du Bassin Parisien. *Bull. Assoc. Géog. Français*, **288-289**, Jan.–Feb., 2–25.

Creager, J. S., 1958. A canyon-like feature in the Bay of Campeche. *Deep-Sea Res.*, **5**, 169–172.

Crouch, R. W., 1959. Inspissation of post-Oligocene sediments in Southern Louisiana. *Bull. Geol. Soc. Amer.*, **70**, 1283–1292.

Curry, E., 1960. Eocene limestones to the west of Jersey. *Geol. Mag.*, **97**, 289–298.

Daly, R. A., 1934. *The Changing World of the Ice Age*. Yale Univ. Press, New Haven, 271 pp.

Dangeard, L., 1928. Observations de géologie sous-marine et d'océanographie relatives à la Manche. *Ann. Inst. Océanog.*, **6**, 295 pp (thesis).

Dannstedt, G., 1947. Bottentopografien i södra Kalmarsund. *Geog. Ann.* (*Stockholm*), **29**, 1–19. (English summary.)

Day, A. A., 1958. The Pre-Tertiary geology of the western approaches to the English Channel. *Geol. Mag.*, **95**, 137–148.

Day, A. A., 1959. The continental margin between Brittany and Ireland. *Deep-Sea Res.*, **5**, 249–265.

Day, A. A., M. N. Hill, A. S. Laughton and J. S. Swallow, 1956. Seismic prospecting in the western approaches of the English Channel. *Q. J. Geol. Soc. London*, **112**, 15–44.

Emery, K. O., 1946. Marine solution basins. *J. Geol.*, **54**, 209–228.

Emery, K. O., 1954. Source of water in basins off Southern California. *J. Mar. Res.*, **13**, 1–21.

Emery, K. O., 1958. Shallow submerged marine terraces of Southern California. *Bull. Geol. Soc. Amer.*, **69**, 39–60.

Emery, K. O., 1960. *The Sea off Southern California*. Wiley, New York, 366 pp.

Emery, K. O. and H. L. Foster, 1956. Shoreline nips in tuff at Matsushima, Japan. *Amer. J. Sci.*, **254**, 380–385.

Evers, W., 1941. Grundzüge einer Oberflächengestaltung Südnorwegens. *Deut. Geog. Blätter, Bremen*, **44**, 159 pp.

Ewing, M. *et al.*, 1937 to 1956. Geophysical investigations in the emerged and submerged Atlantic coastal plain. *Bull. Geol. Soc. Amer.*, **48**, 733 *et seq.* (Virginia and Massachusetts); **50**, 257 *et seq.* (New Jersey); **51**, 1821 *et seq.* (New Jersey); **61**, 877 *et seq.* (New Jersey and Massachusetts); **62**, 1287 *et seq.* (Long Island); **65**, 299 *et seq.* (Grand Banks and Laurentian Channel); **65**, 653 *et seq.* (Nova Scotia and Georges Bank); **65**, 957 *et seq.* (Gulf of Maine); **67**, 1 *et seq.* (Grand Banks).

Ewing, M., D. B. Ericson and B. C. Heezen, 1958. Sediments and topography of the Gulf of Mexico. In Weeks, L. (Ed.), *Habitat of Oil*. Amer. Assoc. Petrol. Geol., Tulsa, 995–1053.

Fairbridge, R. W., 1952. Marine erosion. *7th Pacific Sci. Cong.*, **3**, 347–359.

Fairbridge, R. W., 1953. The Sahul Shelf, northern Australia: its structure and geological relationships. *J. Roy. Soc. W. Austral.*, **37**, 1–33.

Fisk, H. N., 1939. Depositional terrace slopes in Louisiana. *J. Geomorphol.*, **2**, 181–200.

Frebold, H., 1950. Geologie des Barentsschelves. *Deut. Akad. Wiss., Kl. Math. allg. Naturwiss.*, **5**, 151 pp.

Gealy, B. L., 1955. Topography of the continental slope in Northwest Gulf of Mexico. *Bull. Geol. Soc. Amer.*, **66**, 203–228 (chart).

Gougenheim, A., 1959. Levé bathymétrique de la côte du Maroc. *C.R. Acad. Sci. Paris*, **249**, 2599–2601 (chart).

Greenman, N. N. and R. J. LeBlanc, 1956. Recent marine sediments and environments of Northwest Gulf of Mexico. *Bull. Amer. Assoc. Petrol. Geol.*, **40**, 813–847.

Grousson, R., 1957. Bathymétrie de la côte atlantique du Maroc. *Bull. Comité d'Océanog. et d'Et. des Côtes*, **9**, 411–412.

Guilcher, A., 1958. *Coastal and Submarine Morphology*. Methuen and Wiley, London and New York, 274 pp.

Guilcher, A., 1958a. La pente continentale devant l'Europe occidentale, de l'Irlande à la Galice. *Norois* (*Poitiers*), **5**, 89–91.

Guilcher, A., B. Adrian and A. Blanquart, 1959. Les "queues de comète" de galets et de blocs derrière des roches isolées sur les côtes Nord-Ouest et Ouest de la Bretagne. *Norois (Poitiers)*, **6**, 125–145 (cf. pp. 142–144).

Guilcher, A., P. H. Kuenen and F. P. Shepard, 1957. Scientific considerations relating to the continental shelf. Paris, UNESCO memorandum, 1957, 13 pp. (mimeogr.).

Guilcher, A. and W. Nestéroff, 1955. Géomorphologie de l'extrémité nord du Banc Farsan (Mer Rouge). *Ann. Inst. Océanog.*, **30**, 8–100.

Guilcher, A. and P. Pont, 1957. Etude expérimentale de la corrosion littorale du calcaire. *Bull. Assoc. Géog. France*, **265–266**, March–April, 48–62.

Heezen, B. C., M. Tharp and M. Ewing, 1959. The floors of the Oceans. I: The North Atlantic. *Geol. Soc. Amer. Spec. Paper* 65, 122 pp. (chart).

Hersey, J. B., E. T. Bunce, R. F. Wyrick and F. T. Dietz, 1959. Geophysical investigations of the continental margin between Cape Henry, Virginia, and Jacksonville, Florida. *Bull. Geol. Soc. Amer.*, **70**, 437–466.

Hill, M. N., 1956. Notes on the bathymetric chart of the Northeast Atlantic. *Deep-Sea Res.*, **3**, 229–231 (chart).

Hill, M. N. and W. B. R. King, 1953. Seismic prospecting in the English Channel and its geological interpretation. *Q. J. Geol. Soc. London*, **109**, 1–19.

Holtedahl, H., 1955. On the Norwegian continental terrace, primarily outside Møre-Romsdal. *Univ. Bergen Årb., Naturvid. Rekke*, **14**, 1–209.

Holtedahl, H., 1958. Some remarks on geomorphology of continental shelves off Norway, Labrador, and Southeast Alaska. *J. Geol.*, **66**, 461–471.

Holtedahl, O., 1940. The submarine relief off the Norwegian coast. *Norsk Vidensk. Oslo*, Spec. Paper (charts).

Houbolt, J. J. H. C., 1957. Surface sediments of the Persian Gulf near the Qatar Peninsula. Thesis, Utrecht, 113 pp.

Illing, L. V., 1954. Bahaman calcareous sands. *Bull. Amer. Assoc. Petrol. Geol.*, **38**, 1–95.

Jessen, O., 1936. *Reisen und Forschungen in Angola*. Berlin, 397 pp.

Jessen, O., 1943. Die Randschwellen der Kontinente. *Petermann's Mitteil., Ergänzungsheft*, **241**, 205 pp.

Jivago, A. V. and A. P. Lissitzin, 1957–1958. Bottom topography and deposits in the southern Indian Ocean. *Izvest. Akad. Nauk S.S.S.R., Geog. Ser.*, 1957, **1**, 19–35; 1958, **2**, 9–21 and **3**, 22–36 (in Russian).

Johnson, D. W., 1919. *Shore Processes and Shoreline Development*. Wiley, New York, 584 pp.

Johnson, M. E. and H. G. Richards, 1952. Stratigraphy of the coastal plain of New Jersey. *Bull. Amer. Assoc. Petrol. Geol.*, **36**, 2150–2160.

Jordan, G. F., 1951. Continental slope off Appalachicola River, Florida. *Bull. Amer. Assoc. Petrol. Geol.*, **35**, 1978–1993 (chart).

Jordan, G. F. and H. B. Stewart, 1959. Continental slope off Southwest Florida. *Bull. Amer. Assoc. Petrol. Geol.*, **43**, 974–991 (chart).

Kent, P. E., 1949. A structure contour-map of the surface of the buried Pre-Permian rocks of England and Wales. *Proc. Geol. Assoc.*, **60**, 87–104.

King, L. C., 1940. The monoclinal coast of Natal, South Africa. *J. Geomorphol.*, **3**, 144–153.

King, L. C., 1951. *South African Scenery*. Oliver and Boyd, Edinburgh and London, 2nd ed., 379 pp.

King, W. B. R., 1948. The geology of the eastern part of the English Channel. *Q. J. Geol. Soc. London*, **104**, 327–338.

King, W. B. R., 1954. The geological history of the English Channel. *Q. J. Geol. Soc. London*, **110**, 77–101.

Kuenen, P. H., 1950. *Marine Geology*. Wiley and Chapman, New York and London, 568 pp.

Kuenen, P. H., 1950a. The formation of the continental terrace. *Brit. Assoc. Adv. Sci.*, **7** (25), 76–80.

Lacombe, H., 1955. Le talus continental marocain au large de Rabat. *Bull. Comité d'Océanog. et d'Et. des Côtes*, **7**, 201–206.

Le Bourdiec, P., 1958. Aspects de la morphogenèse plio-quaternaire en Basse Côte d'Ivoire. *Rev. Géomorphol. Dynam.*, **9**, 33–42.

Lewis, R. G., 1935. The orography of the North Sea bed. *Geog. J.*, **86**, 334–342.

McGill, J. T., 1958. Map of coastal landforms of the world. *Geog. Rev.*, **48**, 402–405 (map).

Nansen, F., 1922. The strandflat and isostasy. *Skrifter Videnskapsselskapet i Kristiania, 1, Math.-Naturvid. Klasse*, **2**, 313 pp.

Newell, N. D. and J. K. Rigby, 1957. Geological studies on the Great Bahama Bank. In *Regional aspects of carbonate deposition*, Tulsa, 15–72.

Pannekoek, A. J., 1956. *Geological history of the Netherlands.* Gov. Printing Off., The Hague, 147 pp.

Pimienta, J., 1957. La sédimentation à l'aval du graben de l'Amazone. *Bull. Soc. Géol. France*, (6), **7**, 729–736.

Pratje, O., 1951. Die Deutung der Steingründe in der Nordsee als Endmoränen. *Deut. Hydrog. Z.*, **4**, 106–114.

Robinson, A. H. W., 1952. The floor of the British seas. *Scot. Geog. Mag.*, **68**, 64–79.

Rosfelder, A., 1955. Carte provisoire au 1:500,000 de la marge continentale algérienne. *Serv. Carte Géol. Algérie*, n. s., Bull. no. 5, 57–106 (charts).

Russell, R. J., ed., 1959. Second Coastal Geog. Confer., Louisiana State Univ., Washington, 472 pp.

Schick, A. P., 1958. Tiran: the straits, the island and its terraces. *Israel Explor. J.*, **8**, 120–130 and 189–196.

Shepard, F. P., 1931. Saint Lawrence (Cabot Strait) submarine trough. *Bull. Geol. Soc. Amer.*, **42**, 853–864.

Shepard, F. P., 1948. *Submarine Geology.* Harper and Brothers, New York, 348 pp.

Shepard, F. P., 1959. *The Earth beneath the Sea.* Johns Hopkins Press, Baltimore, 275 pp.

Shepard, F. P. and K. O. Emery, 1941. Submarine topography off the California coast. *Geol. Soc. Amer. Spec. Paper* 31 (charts).

Slansky, M., 1958. Vue d'ensemble sur le bassin sédimentaire côtier du Dahomey-Togo. *Bull. Soc. Géol. France*, (6), **8**, 555–580.

Spangler, W. B. and J. J. Peterson, 1950. Geology of Atlantic coastal plain in New Jersey, Delaware, Maryland, Virginia, and North Carolina. *Bull. Amer. Assoc. Petrol. Geol.*, **34**, 1–132.

Stearns MacNeil, F., 1950. Pleistocene shorelines in Florida and Georgia. *U.S. Geol. Surv. Prof. Paper*, **221F**, 95–107.

Steers, J. A., 1929. The Queensland coast and the Great Barrier reefs. *Geog. J.*, **74**, 232–257, and 341–370 (see p. 365).

Stetson, H. C., 1949. The sediments and stratigraphy of the east coast continental margin: Georges Bank to Norfolk canyon. *Mass. Inst. Tech. Woods Hole Oceanog. Inst.*, 60 pp.

Stride, A. H., 1959. On the origin of the Dogger Bank. *Geol. Mag.*, **96**, 33–44.

Swain, F. M., 1947. Two recent wells in coastal plain of North Carolina. *Bull. Amer. Assoc. Petrol. Geol.*, **31**, 2054–2060.

Teichert, C., 1958. Australia and Gondwanaland. *Geol. Rundschau*, **47**, 562–590.

Teichert, C. and R. W. Fairbridge, 1948. Some coral reefs of the Sahul shelf. *Geog. Rev.*, **38**, 222–249.

Ting, S., 1937. The coastal configuration of western Scotland. *Geog. Ann.* (*Stockholm*), **19**, 62–83.

Trask, P. D., ed., 1939 and 1955. *Recent Marine Sediments: a Symposium.* London, 1939; 2nd ed., Tulsa, 1955, 736 pp.

Tricart, J., 1957. Aspects et problèmes géomorphologiques du littoral occidental de la Côte d'Ivoire. *Bull. Inst. Fr. Afrique Noire* (*Dakar*), **19**, ser. A, 1–20.

Umbgrove, J. H. F., 1945. Periodical events in the North Sea basin. *Geol. Mag.*, **82**, 237–244.

Umbgrove, J. H. F., 1947. *The Pulse of the Earth.* Nijhoff, The Hague, 358 pp.

Veatch, A. C. and P. A. Smith, 1939. Atlantic submarine valleys of the United States and the Congo submarine valley. *Geol. Soc. Amer. Spec. Paper* 7, 101 pp. (charts).

Welder, F. A., 1959. Processes of deltaic sedimentation in the Lower Mississippi River. *Coastal Studies Inst., Tech. Rep.*, **12**, Baton Rouge, 90 pp.

Wentworth, C. K., 1938–1939. Marine bench-forming processes. *J. Geomorphol.*, **1**, 6–32 and **2**, 3–25.

Wiseman, J. D. H. and C. D. Ovey, 1953. Definitions of features on the deep-sea floor. *Deep-Sea Res.*, **1**, 11–16.

Zeuner, F. E., 1950. *Dating the Past.* Methuen, London, 2nd ed., 474 pp.

14. ABYSSAL PLAINS[1]

BRUCE C. HEEZEN and A. S. LAUGHTON

1. Introduction

One of the most striking discoveries of the 1947 Mid-Atlantic-Ridge Expedition was that vast areas of the deep-ocean floor in the North Atlantic are occupied by flat, nearly level plains. In 1948 the Swedish Deep-Sea Expedition discovered a similar plain in the Indian Ocean south of the Bay of Bengal.[2] Subsequently, more abyssal plains were discovered throughout the world.

Although a rough idea of the extent of abyssal plains could be gained from a study of records made by standard echo-sounders, the large errors inherent in these instruments made a study of abyssal-plain gradients impossible. All irregularities observed in abyssal plains on standard echo-sounding records could be ascribed to errors and inaccuracies in the recording system; thus, it was not possible to determine exactly how smooth the abyssal plains were.

Soon after the discovery of abyssal plains, it became obvious that a much more accurate echo-sounder recorder was needed before an effective study of these features could be made. To fill the specific need, a Precision Depth Recorder (P.D.R.) was developed (Luskin, Heezen, Ewing and Landisman, 1954).

The ocean-basin floor is divided into two parts, the oceanic rises and the abyssal floor.[3] The abyssal floor, in turn, is divided into the abyssal plains, the abyssal hills and areas of gently undulating topography. An abyssal plain is an area of the ocean-basin floor in which the ocean bottom is flat and the slope of the bottom is less than 1 : 1000. An abyssal hill is a small, relatively sharply defined hill that rises in the ocean-basin floor to an elevation a few fathoms to a few hundred fathoms in height and is from a few hundred feet to a few miles in width. The term "abyssal-hills province" is applied to those areas of the abyssal floor in which nearly the entire area is occupied by hills, i.e. the province lies at approximately the depth of the adjacent abyssal plain but lacks a smooth floor. Isolated, sharply defined abyssal hills and groups of abyssal hills also protrude from the abyssal plains. The areas of undulating topography may either represent lower tectonic activity than the abyssal hills, or greater age and consequently greater thickness of pelagic sedimentation.

There are a few areas of the ocean-basin floor which approach abyssal-plain smoothness but are apparently not a part of the modern abyssal plains. These areas may represent relic abyssal plains. Large areas of the oceanic rises are extremely gentle and, although these areas contrast sharply with the rugged relief of the abyssal hills, they do not reach the featureless smoothness of abyssal plains.

[1] Lamont Geological Observatory (Columbia University) Contribution No. 591.

[2] In 1951 H.M.S. *Challenger* made a second crossing (Gaskell and Ashton, 1954).

[3] The major physiographic regions of the ocean are described by Heezen and Menard in Chapter 12.

[*MS received December, 1960*]

Characteristically, the abyssal plains lie at the base of the continental rise. The abyssal-hills provinces lie on the seaward edge of the abyssal plains. The local axis of maximum depth in an ocean basin is generally found within the abyssal-hills province or at the seaward edge of the abyssal plains. The seaward edge of the continental rise is generally marked by an abrupt change in slope. Here gradients change from values between 1 : 100 and 1 : 700, characteristic of the continental rise, to values of less than 1 : 1000. In some areas no distinct break is seen and the continental rise grades in an exponential form into the abyssal plain. In these cases the 1 : 1000 isopleth is taken as a boundary between the continental rise and the abyssal plain. Gradients of abyssal plains range from 1 : 1000 to 1 : 10,000.

Near the seaward edges of the continental rise, there is often a gradual merging of the abyssal cones and deep-sea fans with abyssal plains. In some areas deep-sea channels at the seaward ends of submarine canyons extend across the boundary onto the abyssal plains. Seamounts that rise in the abyssal plain do so abruptly in a manner suggesting that their bases have been buried.

Abyssal plains are also found in the marginal seas (western Mediterranean, Gulf of Mexico, Caribbean and doubtless in many others) and features with all the characteristics of abyssal plains are found in the floors of lakes, an excellent example being the Lake of Geneva (Dussart, 1957). The basins lying in the continental margin of Southern California are floored with abyssal plains, and deep-sea trench plains are found in the floors of the major ocean trenches.

An abyssal plain has been defined as "an area of the ocean-basin floor in which the bottom is flat . . .". What exactly is meant by "flat"? In the relatively well mapped Atlantic abyssal plains, variations in depth do not exceed one fathom per mile. Irregularities of less than one fathom are gentle, never sharp. Although abyssal hills and seamounts break through the surface of the plains as islands protrude above the surface of the sea, they do not otherwise disturb the nearly level surface of the abyssal plains. A typical P.D.R. record from the Cape Verde Abyssal Plain is shown in Fig. 1c.

Superficially, the smooth topography of the continental rise (see Chapter 12) is quite similar to the abyssal plains. However, when studied with the aid of P.D.R., the two are quite easily distinguished. Local areas on the continental rise are of abyssal-plain flatness and in many instances have gradients which would allow them to qualify under the definition of abyssal plains. However, these local perched plains are interrupted by areas of more irregular relief, often of only a few fathoms in amplitude (Fig. 1a) but distinctly different from the monotonous, featureless topography of the abyssal plains. The authors consider the term "abyssal plain" restricted to areas "of the ocean floor in which the bottom is flat and the slope to the bottom is less than 1 : 1000". It is clear that the smooth surface of the continental rise owes its origin to much the same processes that produced the abyssal plains. However, this fact should not obscure the marked morphological difference between these two features of depositional topography.

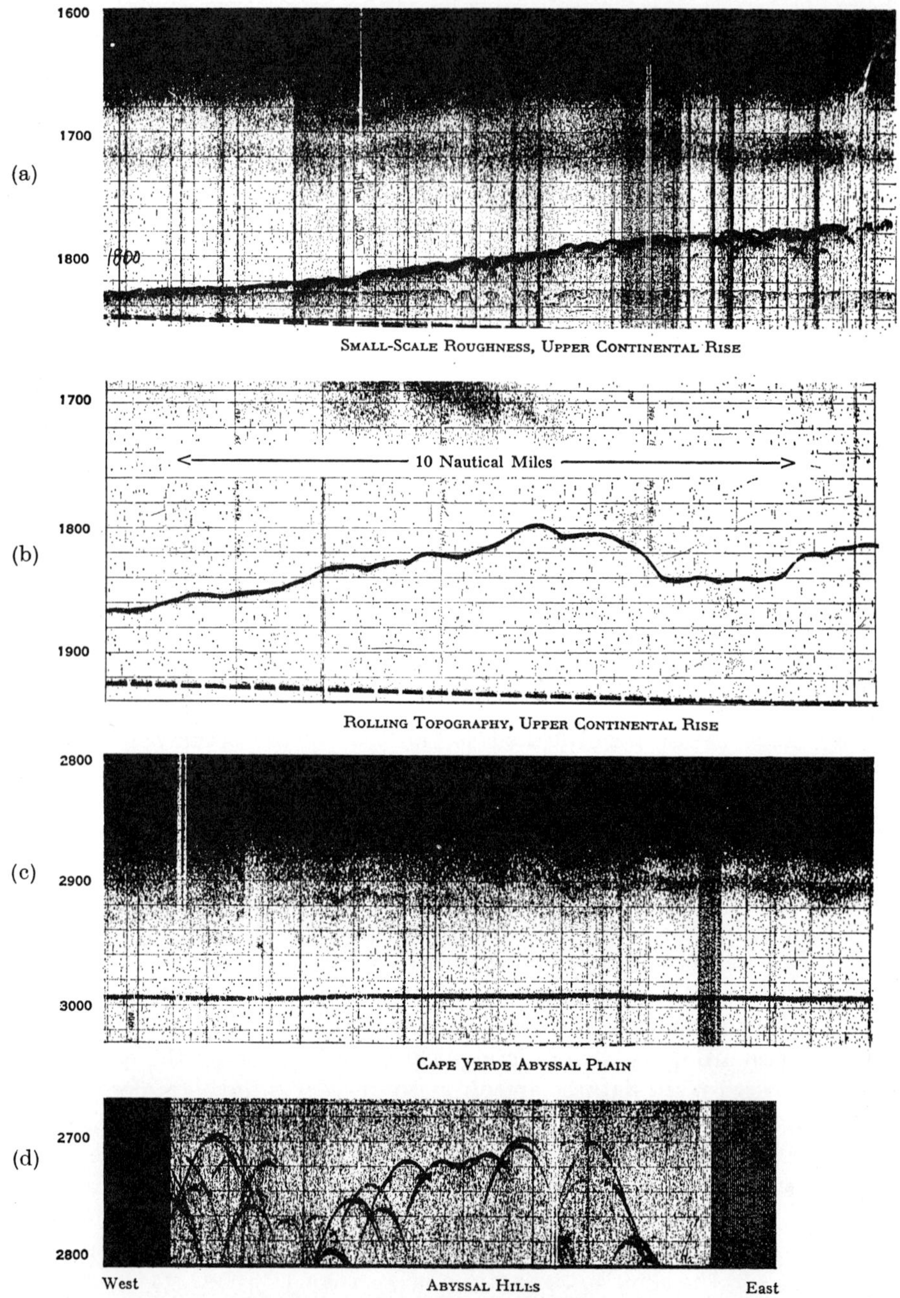

Fig. 1. Typical Precision Depth Recorder records from the Canary Basin. Note the marked contrast between the gentle topography of the continental rise, the extremely smooth Cape Verde Abyssal Plain, and the intensely rough abyssal hills. Depth in fathoms. (After Heezen *et al.*, 1959.)

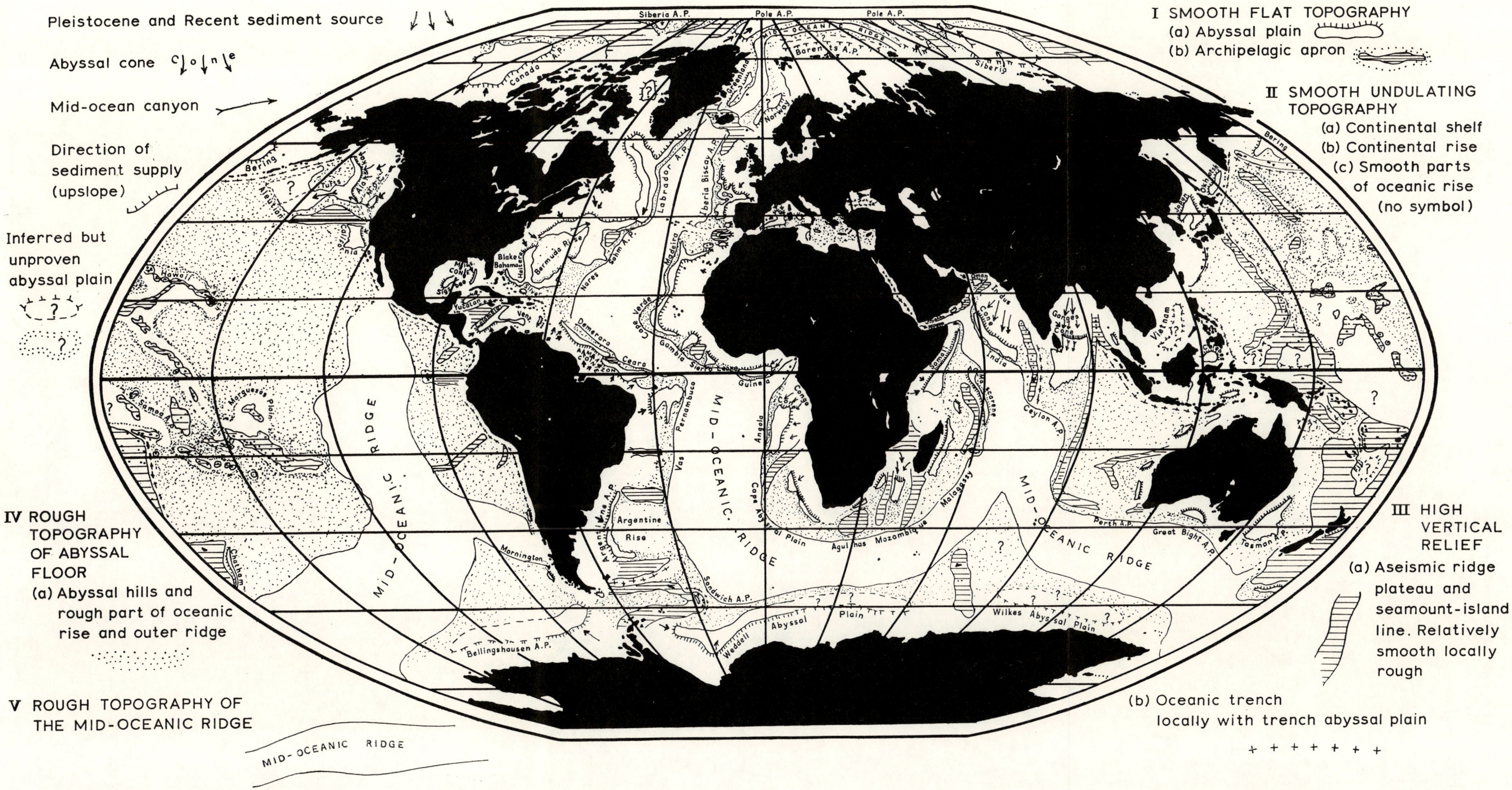

Fig. 2. Distribution of smooth and rough topography in the world ocean. (North Atlantic after Heezen, Tharp and Ewing, 1959 ; South Atlantic after Heezen and Tharp, 1961; Indian Ocean after Heezen and Tharp, in press; Pacific Archipelagic Aprons after Menard, 1956 ; Arctic after Dietz and Shumway, 1961.)

2. Regional Description

A. Western North Atlantic

The abyssal plains of the northwest Atlantic have been crossed dozens of times by ships employing precision depth recorders (see Fig. 3, Chapter 12, p. 235). Most of this work has been accomplished by the Lamont Geological Observatory's Research Vessel *Vema*.

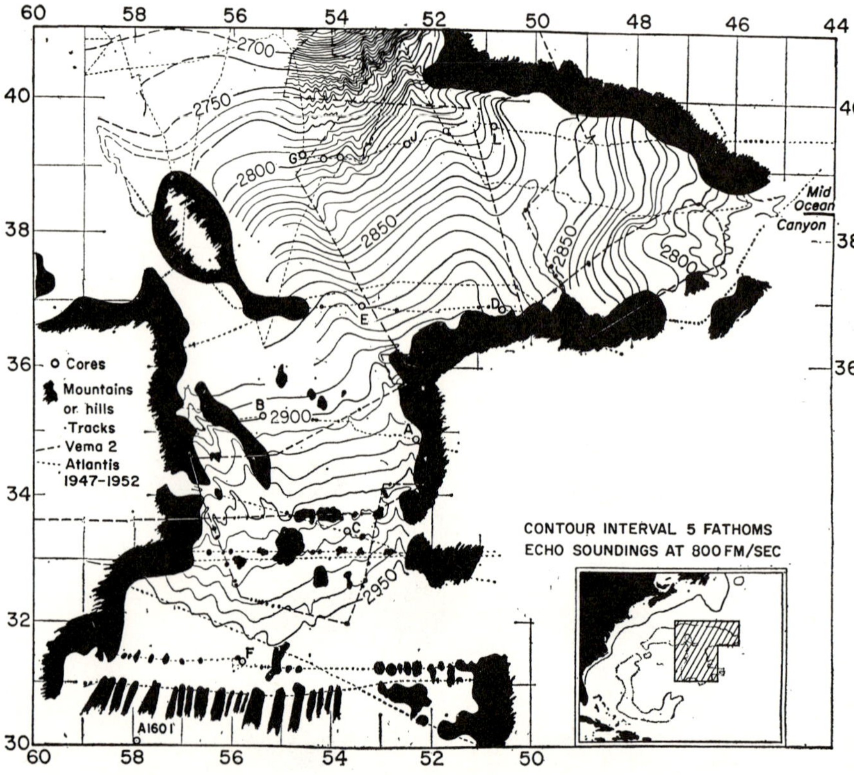

Fig. 3. Bathymetric chart of the Sohm Abyssal Plain. (After Heezen, Ewing and Ericson, 1955.)

a. Sohm Abyssal Plain

The best studied abyssal plain is the Sohm Abyssal Plain south of Newfoundland (Heezen *et al.*, 1955, 1959). The plain is T-shaped and generally about 200 mi wide (Fig. 3); the depth increases in each arm of the T toward the stem where the slope changes to south. The greatest depth (3181 fm corrected) is found at the south boundary of the plain, near 29°N (Fig. 4). In the northern part of the plain, small topographic irregularities are unknown; but toward the south, peaks 50 to 500 fm high increase in number until they finally replace the plain (Fig. 5). North of the Bermuda Rise the plain surrounds a line of huge

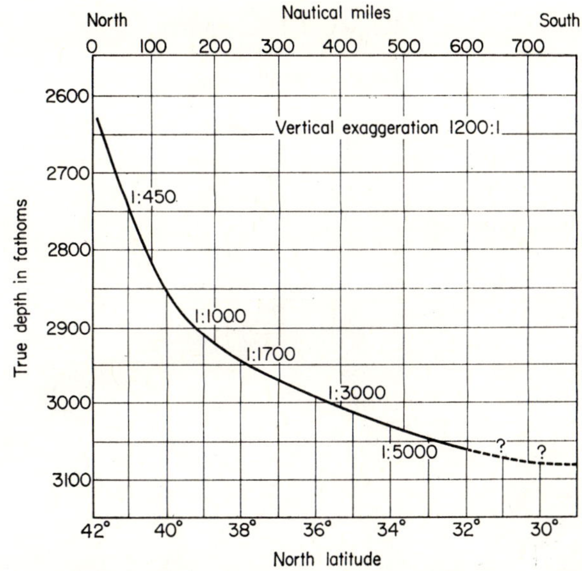

Fig. 4. North–south profile of the Sohm Abyssal Plain along long. 54°, from 32° to 42°N. (After Heezen, Ewing and Ericson, 1955.)

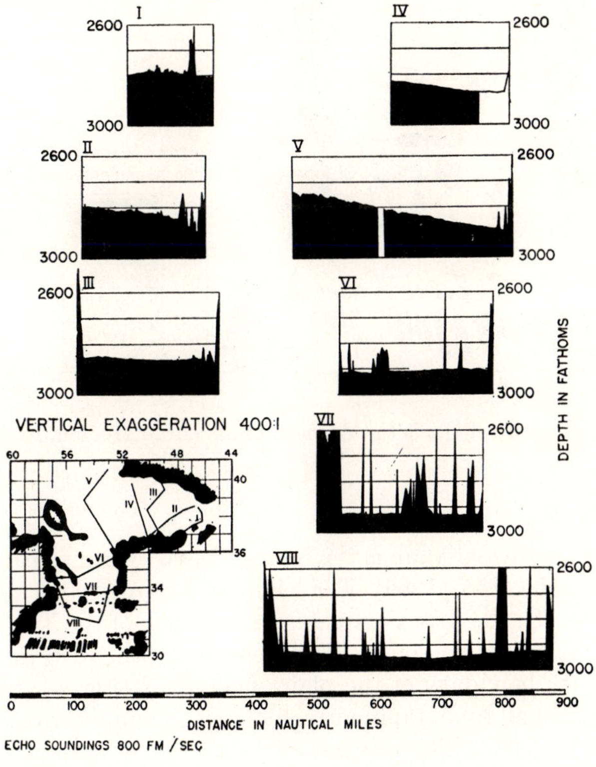

Fig. 5. Profiles across the Sohm Abyssal Plain. (After Heezen *et al.*, 1955.)

seamounts. However, even here no hills are found. In the strip extending between 30° and 37°N, the east and west boundaries of the plain are formed by scarps 200 to 800 fm high. The boundary of the north side of the Bermuda Rise is formed by a distinct shallowing but no scarp is seen. There is some justification for separating the northwest arm of the Sohm Abyssal Plain north of the Bermuda Rise from the rest of the Sohm Abyssal Plain, since a mid-ocean canyon runs from the northwest to the central portions of the plain. This northwest portion of the Sohm Abyssal Plain has sometimes been called the Nova Scotia Abyssal Plain. A sill might lie northeast of the northeast tip of the Bermuda Rise. Here the bottom gradient changes slightly and an extremely large magnetic anomaly is associated with the region. However, the plain narrows only slightly at this point and, since we cannot tell how many individual basins were filled to form any individual abyssal plain, we will consider that the Sohm Abyssal Plain now includes the whole area. South of Newfoundland the continental rise merges with the abyssal plain with no scarp or line of hills. Here the boundary is taken at the point where the gradient is gentler than 1 : 1000. This is not an entirely arbitrary definition since an appreciable change in gradient generally occurs near this point. South of the boundary the gradient gradually decreases to 1 : 3000 at 37°N, beyond which it decreases even more gradually and reaches 1 : 5000 at the southern edge of the plain. The Sohm Abyssal Plain covers 350,000 mi^2.

b. Hatteras, Blake–Bahama and Silver Abyssal Plains

The Hatteras and Sohm Abyssal Plains are separated either by abyssal hills or quite possibly by an irregular gap. The northwest margin of the Hatteras Abyssal Plain is abruptly formed by the lower continental rise hills. The eastern boundary is generally distinct and in places is marked by a 50–100-fm scarp. From 25–44°N the western margin of the plain gently laps the smooth Blake–Bahama Outer Ridge. Its southern margin laps the Antilles Outer Ridge. Only two or three small hills have been found within the Hatteras Plain.

The narrow Blake–Bahama Abyssal Plain which lies between the Blake–Bahama Outer Ridge and the Blake Plateau–Bahama Island region connects through Cat Gap at 24.5°N, 74.5°W with the Hatteras Abyssal Plain (see Fig. 14, Chapter 12, p. 247). Judging from width and regional slope this plain appears to have been fed largely from the Providence Channel.

The Silver Abyssal Plain which lies northeast of Silver Bank is bounded on the north by the Antilles Outer Ridge which separates it from the Nares and Hatteras Abyssal Plains. Only recently discovered, it appears to be fed from the southern Bahamas.

c. Nares Abyssal Plain

South and southeast of the Bermuda Rise lies the Nares Abyssal Plain, a 37,000 mi^2 abyssal plain that slopes eastward toward the Mid-Atlantic Ridge

(Fig. 2). It is connected on the west by a narrow gap to the Hatteras Plain. This gap, discovered by *Vema* and known as the Vema Gap, has much higher gradients than the abyssal plains lying both to the northeast and southwest. From 62° to 64°W the northern and southern boundaries of the Nares Abyssal Plain are abrupt and only three or four small abyssal hills have been observed in the plain. East of 64°W the gradient in the plain exceeds 1 : 2500 and the frequency of hills increases until, at 60°W, the plain consists of a series of fingers which extend into the abyssal hills. The Nares Abyssal Plain is the deepest of the broad abyssal plains of the North Atlantic Ocean Basin floor. It is also the farthest from land. If the material forming the Nares Abyssal Plain was transported along the sea floor, it would have had to flow through Vema Gap, since the outer ridge and marginal trench to the south and the outer ridge and marginal basin to the west prevent bottom-seeking currents from either the south or west from reaching the Nares Abyssal Plain. The idea that the sediment was transported on the bottom through Vema Gap is supported by the fact that the abyssal plain slopes eastward through the gap.

d. Demerara Abyssal Plain

A 1000-mile-long abyssal plain lies east of the Lesser Antilles and the Venezuelan and Guianan coasts of South America. The plain varies from 100 to 150 mi in width and includes an area of 130,000 mi^2. The Ceara Abyssal Plain lies along the same trend southeast of the Demerara Abyssal Plain. The two plains are separated by the 100-mile-wide Ceara Rise.

e. Ceara Abyssal Plain

The Ceara Abyssal Plain, which lies parallel to the coast of the Ceara Province, Brazil, covers an area of 45,000 mi^2. It does not seem to connect either with the Demerara Abyssal Plain to the northwest or with the Pernambuco Abyssal Plain to the southeast.

B. Eastern North Atlantic

The abyssal plains of the eastern North Atlantic differ from those of the western side in being smaller and more numerous. This is partly due to the greater complexity of large features in the eastern ocean basin, such as the Horseshoe group of seamounts west of Gibraltar, and partly to the lower supply of sediments from the adjacent continent (Fig. 6).

a. Porcupine Abyssal Plain

This plain occupies the deepest part of the shallow trough between the continental shelf SW of Ireland and the east flank of the Mid-Atlantic Ridge. It covers 10,000 mi^2 and is elongated in a NNW–SSE direction. Its slope is uncertain owing to lack of accurate soundings, but it appears to be connected to the western arm of the Biscay Abyssal Plain.

b. Biscay Abyssal Plain

The Bay of Biscay is occupied by a plain of 30,000 mi^2 extending both NW and SW, where there appear to be connections with the Porcupine Plain and with the Iberian Plain. The contours of the plain suggest that all the sediment is derived from the part of the continental shelf running NW from Cap Breton

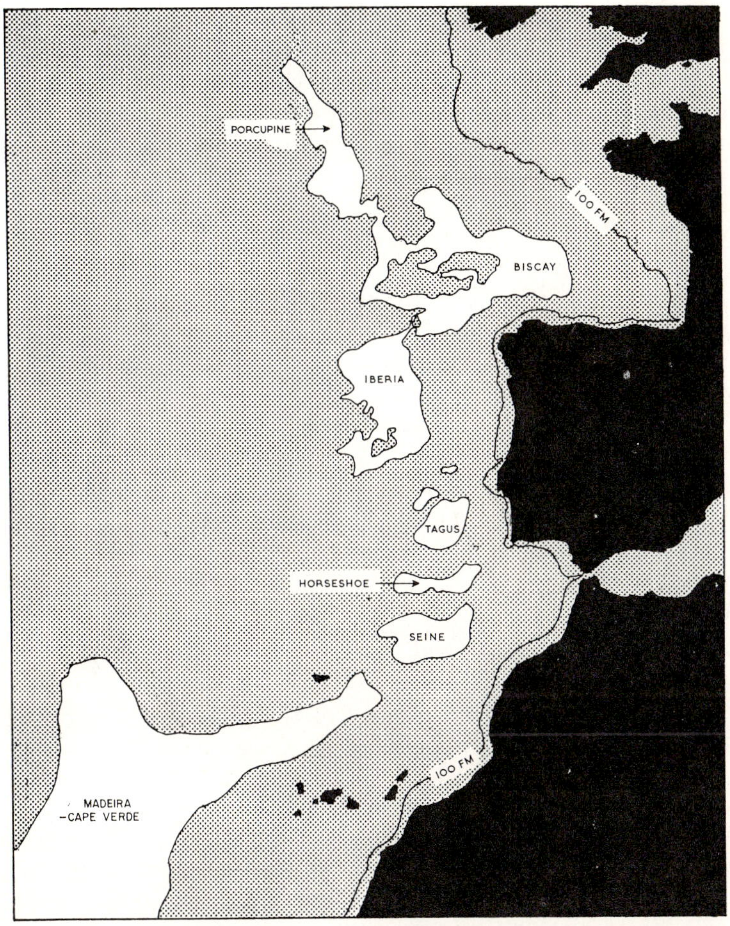

Fig. 6. Abyssal plains east of the Mid-Atlantic Ridge, as far south as 24°N. Land, black; non-abyssal-plain areas, stippled.

Canyon, bordering the coast of France, and the approaches to the English Channel. Very little sediment has come from the north coast of Spain where there are mountains close to the coast and the rivers are short. The deepest part of the plain is in the SW corner where the gradient is about 1 : 2500 and the depth is 2750 fm. In the extreme SW corner, channels in the plain converge on a system of narrow interplain channels leading through Theta Gap to the Iberia Plain (cf. Section 5, pages 342–345).

c. Iberia Abyssal Plain

The Iberia Plain is roughly rectangular, covering 30,000 mi^2, and near its western boundary is extremely flat, the gradient being 1 : 4000 from the north. The sediments are derived partly from the Biscay Plain via the Theta Gap interplain channels and partly from the SE from the large canyons off the coast of Portugal. The increase in the number of seamounts and small hills projecting from the plain toward the west suggests that the thickness of sediments steadily diminishes until the buried topography totally emerges. The depth of the plain at its northern end is 2850 fm, 100 fm deeper than the SW end of the Biscay Plain, and its deepest part is 2930 fm.

d. Tagus Abyssal Plain

West of southern Portugal lies an oval plain of 6000 mi^2 of depth varying from 2750 to 2800 fm sloping down to the SW. It is fed principally from the two large canyons SW of the River Tagus. To the north of this plain are two smaller and shallower plains (2680 and 2670 fm) which appear to be filled from locally derived sediments.

e. Horseshoe Abyssal Plain

A small abyssal plain of 5400 mi^2 lies in the center of a horseshoe-shaped ring of seamounts west of Gibraltar. There are insufficient soundings to establish the extent of this plain.

f. Seine Abyssal Plain

To the east of Seine Bank an oval plain covers 15,000 mi^2. While part of the sediment has come from the NE, most of it derives from the SW where there is a very smooth col 50 fm above the level of the plain and separating it from the Madeira Plain. This source appears to be shared with the Madeira Plain.

g. Madeira and Cape Verde Abyssal Plains

From Gibraltar to Cape Verde a vast abyssal plain of some 200,000 mi^2 parallels the wide continental rise of Northwest Africa. A northern finger extends to the west of the Madeira Rise. The lower continental rise and the Madeira Rise, both relatively smooth (Fig. 1), merge with the plain in a smooth transition and a distinct boundary is lacking. Northeast of the Canary Islands, continental rise gradients of greater than 1:1000 reach the abyssal hills and provide a natural boundary between the Cape Verde and Madeira Abyssal Plains.

The more southerly part differs in character from the northerly part. To the west of Madeira, the sediment is apparently derived from the hills and seamounts of the Azores–Gibraltar Ridge, and from the region between Madeira and the Canary Islands. The westward boundary of the plain here is determined by the abyssal hills abutting the chain of large seamounts comprising Atlantis, Plato, Cruiser and Great Meteor. The sediment supply has been sufficient to pond the sediments and create an abyssal plain in the area of maximum depth.

South of a line joining the Canary Islands and Greater Meteor Seamount, the abyssal hills west of the plain do not form a solid boundary but merely dissipate the energy of turbidity currents, which are then unable to reach the axis of maximum depth further west (Fig. 7).

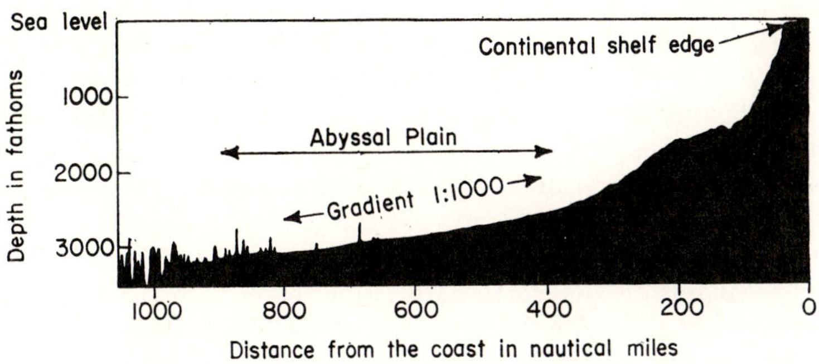

Fig. 7. Profile of Cape Verde Abyssal Plain. Vertical exaggeration, 100 : 1. There has been insufficient material transported westward by turbidity currents to cover the irregular topography 1000 mi from the coast. (After Laughton, 1959.)

h. Sierra Leone and Gambia Abyssal Plains

A 50,000-mi^2 abyssal plain lies between the continental rise off Sierra Leone and the Romanche Trench portion of the mid-oceanic ridge. This plain is separated from the Gambia Abyssal Plain to the north by the Sierra Leone Rise and appears to be separated from the Guinea Abyssal Plain to the east by an area of abyssal hills. The Gambia Abyssal Plain lies north of the Sierra Leone Rise and is separated from the Cape Verde Plain by abyssal hills.

C. South Atlantic

Abyssal plains are not as well developed in the South Atlantic as they are in the North Atlantic. An extensive abyssal plain of 80,000 mi^2, the Pernambuco Abyssal Plain, extends eastward of Pernambuco Province, Brazil; but the abyssal plains of the southern part of the Brazil Basin and the Argentine Basin are surprisingly small and intermittent. Whereas on either side of the North Atlantic, abyssal plains are found in all but a very few areas, in the South Atlantic there are many places where the continental rise merges with the abyssal hills without an intervening abyssal plain. The Argentine Abyssal Plain, which follows the base of the continental rise east of Argentina, is 30 to 150 mi in width and covers an area of 100,000 mi^2. It resembles in relative location the Hatteras and Nares Abyssal Plains of the North Atlantic but is somewhat narrower. Given a longer period of time and a source of detrital material, it can be inferred from the bathymetric configuration of the Argentine Basin that plains quite analogous to the Nares and Hatteras could eventually develop.

East of the Mid-Atlantic Ridge, in the Angola Basin, the large (140,000 mi^2) Angola Abyssal Plain spreads out from the seaward edge of the giant Congo Cone, which forms a distributory system of channels and fans leading from the Congo Canyon. The Angola Abyssal Plain is the largest of the abyssal plains of the South Atlantic which lie in temperate latitudes. A narrow abyssal plain lies along the base of the continental rise in the Cape Basin but its width probably does not exceed 100 mi. No detailed sounding surveys have been made in the Guinea Basin, but judging from the form of the gross bathymetric contours off the mouth of the Niger River, it is reasonable to suppose that a large abyssal cone lies seaward of the Niger Delta and that, seaward of this cone, a fairly extensive abyssal plain will be found.

A few minute abyssal plains have been found in the Scotia Sea. One of the largest abyssal plains discovered to date forms the floor of the Weddell Basin. For obvious reasons, the western extremities of the Weddell Basin have not been well explored, but the Weddell Abyssal Plain has been traced as far east as 20°E longitude, and in 20°W longitude it is over 200 mi wide from north to south.

D. Pacific Ocean

The flat zones of the Pacific Ocean can be divided into two distinct types, formed by somewhat different mechanisms. On the eastern boundaries in the north and extreme south, there are abyssal plains of the type described in the Atlantic. The other type includes large areas of the western and central Pacific which surround the chains of islands found there. These areas have been called archipelagic plains and are described in Section 4, page 336.

The flat areas of the Pacific, including the archipelagic plains, cover a greater area than those of the Atlantic, but the proportion of the ocean covered by abyssal plains is considerably less and reflects the lower influx of sediments per unit area owing to the higher ratio of area to circumference and the prevalence of sediment-trapping marginal trenches, which nearly surround the Pacific basin. The abyssal plains of the Pacific are associated with gaps in the circum-Pacific belt of trenches. The largest abyssal plains of the Pacific are found between the eastern end of the Aleutian Trench and the northern end of the Middle-America Trench. The other major area of abyssal plains in the Pacific lies off the northern shore of Antarctica. The only abyssal plain known at present in the northwestern Pacific is a small one south of Honshu.

a. Northeast Pacific Abyssal Plains

The most extensive abyssal plains of the Pacific are found between the eastern part of the Aleutian Trench and the Mendocino Escarpment, which

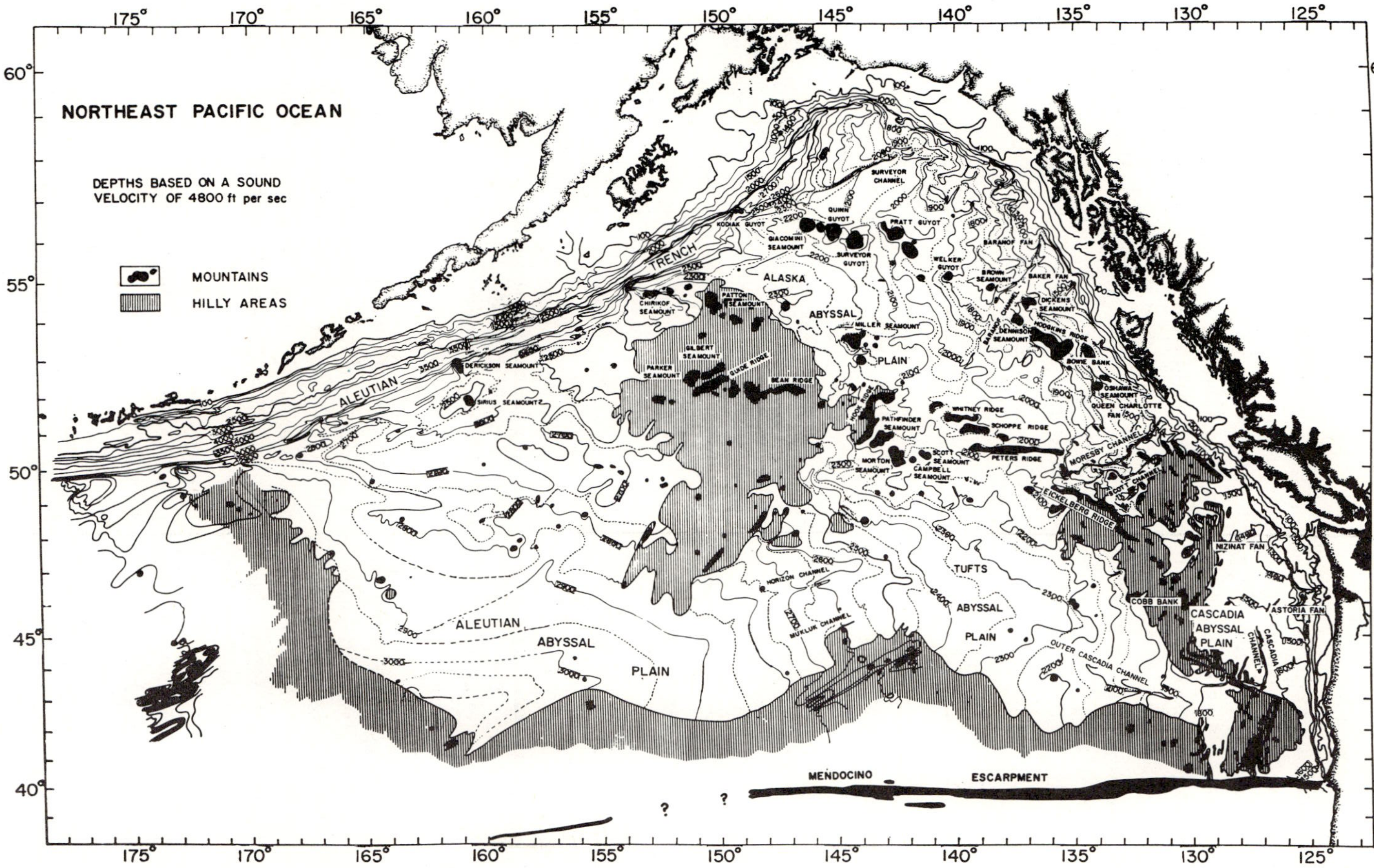

Fig. 8. Abyssal plains of the northeast Pacific Ocean. Limits of the abyssal plains are not yet well determined, and the titles simply indicate the general locations of the various abyssal plains. (After Hurley, 1960.)

runs east–west at 40°N latitude. As the result of a recent detailed study, Hurley (1960) has been able to divide the northeast Pacific into four separate plains (Fig. 8).

The Alaskan Plain lies at the base of an extensive continental rise which extends from the eastern end of the Aleutian Trench to the southern tip of the Queen Charlotte Islands (Menard and Dietz, 1951). This continental rise is interrupted by a series of seamounts. The Alaskan continental rise is crossed by several well mapped deep-sea channels, several of which can be shown to be the seaward extensions of continental slope submarine canyons. Centered along 150°W longitude is a low, irregular rise punctuated by many large seamounts. The Alaskan Abyssal Plain lies between the base of the Alaskan continental rise and the eastern edge of this oceanic-rise area. An east–west line of sea-

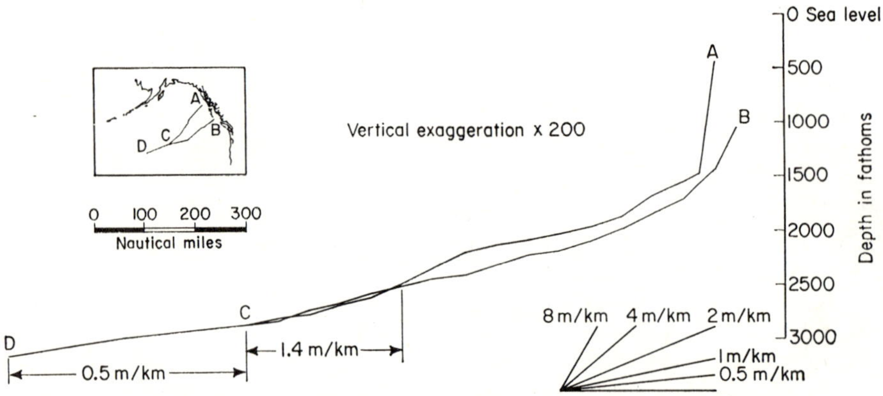

Fig. 9. Long profile of the northeast Pacific abyssal plains and continental rise. The Aleutian Abyssal Plain lies west of C. Immediately east of C are the steeper gradients of the abyssal gap which connects the Tufts and Alaska Abyssal Plains with the Aleutian Abyssal Plain. East of the steeper gradients of this gap are the abyssal-plain gradients of the Tufts and Alaska Abyssal Plain. Immediately to the west of A and B typical steep gradients of the continental slope drop to the lower gradients of the continental rise. (After Hurley, 1960.)

mounts and ridges at about 50°N separates the southern edge of the Alaskan Abyssal Plain from the Tufts Abyssal Plain, which lies to the south.

The Tufts Abyssal Plain is bounded on the south by the Mendocino Rise, on the east by the ridge and trough province, and on the northwest by the same oceanic rise which forms the western boundary of the Alaskan Abyssal Plain. The Alaskan Abyssal Plain is connected by several abyssal gaps to the Tufts Abyssal Plain. On the western boundary of the Tufts Abyssal Plain two distinct interplain channels, the Horizon Channel and the Mukluk Channel, pass through a broad abyssal gap connecting the Tufts Plain to the lower-lying Aleutian Abyssal Plain to the west (Fig. 9).

East of the ridge and trough province, which runs from Cape Mendocino to the southern tip of the Queen Charlotte Islands, lies the Cascadia Abyssal Plain.

The Cascadia Abyssal Plain is connected to the Tufts Abyssal Plain via the Cascadia Channel. The Cascadia Channel starts on the continental rise west of the Washington–Oregon boundary as a seaward extension of the Columbia River submarine canyon. The canyon runs south to the southern edge of the Cascadia Abyssal Plain, where it enters a tectonic trough through the ridge and trough province. When it emerges from this tectonic trough, it forms a large fan which gradually merges with the Tufts Abyssal Plain.

The actual area of the northeast Pacific occupied by abyssal plains is subject to some controversy, not only owing to inadequate data but also to the manner in which workers on that area have chosen to define abyssal plains. Hurley recognizes no continental rise and applies the term "abyssal plain" to all the relatively smooth sea floor which lies between the base of the continental slope and the abyssal hills. Less than a quarter of the area included by Hurley in the northeast Pacific abyssal plains classifies as abyssal plains according to the original definition. Unfortunately, those central portions of the Alaskan Abyssal Plain, the Tufts Abyssal Plain and the Aleutian Abyssal Plain which appear to be true abyssal plains having gradients less than 1 : 1000 are poorly surveyed; therefore, it is impossible to give accurate figures for the gradients or smoothness of any of the northeast Pacific abyssal plains. For this reason, the labels in Fig. 8 merely indicate the approximate positions of the various abyssal plains without indicating their boundaries. Those smooth portions of the northeast Pacific which have been investigated with the aid of a Precision Depth Recorder have proved not to be as smooth as the Atlantic and Indian Ocean abyssal plains. Minor irregularities of a few fathoms are common and the topography of most of the areas so far explored resembles more the continental rise than the abyssal plains of the Atlantic.

The area mapped as the "Aleutian Abyssal Plain" (Fig. 8) presents some very important problems. The slope indicated by Hurley on the basis of numerous Coast and Geodetic Survey soundings is from north to south, with the minimum soundings found along the crest of the outer ridge which bounds the Aleutian Trench on the south. The few Precision Depth Recorder echograms obtained in the northern part of the Aleutian Abyssal Plain certainly do not indicate as smooth a bottom as that found in what would normally be referred to as an abyssal plain. This is not surprising when one considers the possible source of sediments for the smoothing of the Aleutian Abyssal Plain. The northern portion of the Aleutian Abyssal Plain is cut off from turbidity currents originating on the Aleutian Arc by the deep Aleutian Trench. It is cut off from the Alaskan Panhandle by an oceanic rise. If the sediment which smoothed this area came downhill by turbidity currents or other bottom transportation mechanisms, it must have come before the Aleutian Trench was formed or before the oceanic rise to the east was uplifted. The northern part of the Aleutian Abyssal Plain must, therefore, be a relic plain and must no longer be receiving sediments of the type which probably smoothed the plain. The securing of long cores in this relic plain should stand high on any list of important deep-sea research projects.

b. California Abyssal Plain

The abyssal plain off California, bounded by the Mendocino Escarpment on the north and the Murray Escarpment to the south, covers an area of 130,000 mi^2. The large submarine canyons of central California have built many deep-sea fans and abyssal cones which dominate the continental rise. These deep-sea distributory channels lead toward the abyssal plain. The abyssal plain off California was originally described by Menard in 1955 and was simply referred to as the "Deep Plain". The main part of the California Abyssal Plain has a slope of about 1 : 1000 to the WSW and abuts a more irregular and deeper region approximately 450 mi from the coast. The plain is not a filling sedimentary basin but a region across which transport of sedimentary material by turbidity currents takes place. The slope may, therefore, be one of equilibrium between deposition and erosion by passing turbidity currents.

c. Bellinghausen Abyssal Plain

An abyssal plain of uncertain extent lies parallel to the continental rise in the Bellinghausen Sea area of the southeast Pacific. The area has only been crossed twice by a ship employing a P.D.R., hence we cannot give details as to the extent or nature of this abyssal plain. It would seem likely, however, that the plain slopes south to north and is related to the continental rise of Antarctica.

d. Mornington Abyssal Plain

A small plain of about 10,000 mi^2 lies west of Mornington Island, Chile, between about 48° and 52° S. A prominent canyon was found at about 49°S.

E. Indian Ocean

The northern Indian Ocean is divided into two triangular bights by the peninsula of India. To the east of India lies the Bay of Bengal, which widens from north to south. The Ganges and Brahmaputra Rivers empty into the northern apex of the bay. Soundings made southeast of Ceylon have revealed the existence of an abyssal plain at the base of the continental slope near Ceylon, occupying much of the northeast part of the Indian Ocean. In this plain mid-ocean canyons were discovered by the Swedish *Albatross* Expedition (Dietz, 1953). It is reasonable to suspect that this abyssal plain is related to the abyssal cone of the Ganges–Brahmaputra, which dominates the entire floor of the Bay of Bengal. The southern limit of this plain is poorly known but the plain reaches at least to 8° South latitude. The area of the Ceylon Abyssal Plain probably exceeds 200,000 mi^2.

West of the Indian peninsula, the Indus River empties into the Arabian Sea. A deep submarine canyon has been discovered off the mouth of the Indus and an abyssal cone and associated abyssal plain occupy much of the northeast

Arabian Sea. The India Abyssal Plain occupies an area 100,000 mi² of the eastern Arabian Sea. East of the Somali coast a large abyssal plain extends from 4°N to 7°S. The Mozambique Abyssal Plain sloping north to south between approximately 28° and 34°S lies between 37° and 40°E to the south of the Mozambique Channel.

East of Madagascar the Malagassy Abyssal Plain lies between approximately 12° and 25°S and extends over approximately 3° of longitude.

An extensive abyssal plain lies west of Australia. Its limits are ill defined but it appears to be centred between 20° and 30°S.

An extensive abyssal plain extends along the base of the continental rise in the Great Australian Bight. This plain probably exceeds 70,000 mi² in area.

F. Arctic Ocean

An extensive abyssal plain has been reported in the Canada Basin. It is probable that this abyssal plain, which may have an area exceeding 150,000 mi², is fed by the Mackenzie River. The Arctic Ocean is divided by three ridges or rises: (1) the Alpha Rise, (2) the Lomonosov Ridge and (3) the Arctic extension of the mid-oceanic ridge (Heezen and Ewing, 1961). Much of the area between each of these ridges is probably occupied by abyssal plains. The pioneering work of the U.S.S. *Nautilus* and U.S.S. *Skate* has shown that abyssal plains exist in the Canada Basin between the Alpha Rise and the Canadian–Alaskan–

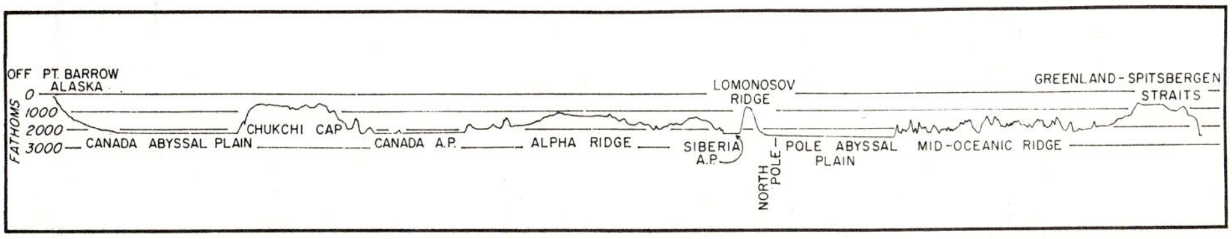

Fig. 10. Abyssal plains of the Arctic Basin. This profile, made by the submarine *Nautilus*, indicates the presence of three extensive abyssal plains in the Arctic Basin: the Canada Abyssal Plain, the Siberia Abyssal Plain and the Pole Abyssal Plain. (Modified after Dietz and Shumway, 1961.)

Siberian side (Fig. 10), between the Alpha Rise and the Lomonosov Ridge, and between the Lomonosov and the Arctic extension of the mid-oceanic ridge (Dietz and Shumway, 1961). It is probable that an abyssal plain also lies between the mid-oceanic ridge and the Barents Sea continental shelf, but it has not been mapped as yet. Recent soundings by H. Kutschale from Arlis Ice Island have revealed mid-ocean canyons in the Siberia Abyssal Plain.

a. The Norwegian Sea

An abyssal plain exists in the Norwegian Sea between the Norwegian continental rise and the mid-oceanic ridge. It is not known at present whether an abyssal plain lies between Greenland and the mid-oceanic ridge.

G. Abyssal Plains of Adjacent Seas

Although generally smaller in size than the great abyssal plains of the oceans, abyssal plains are also found in the adjacent seas.

a. Mediterranean Sea

The Mediterranean Sea is divided into three deep basins, the Western Mediterranean, the Tyrrhenian Sea and the Eastern Mediterranean. The floor of the Western Mediterranean is nearly completely filled by the Balearic

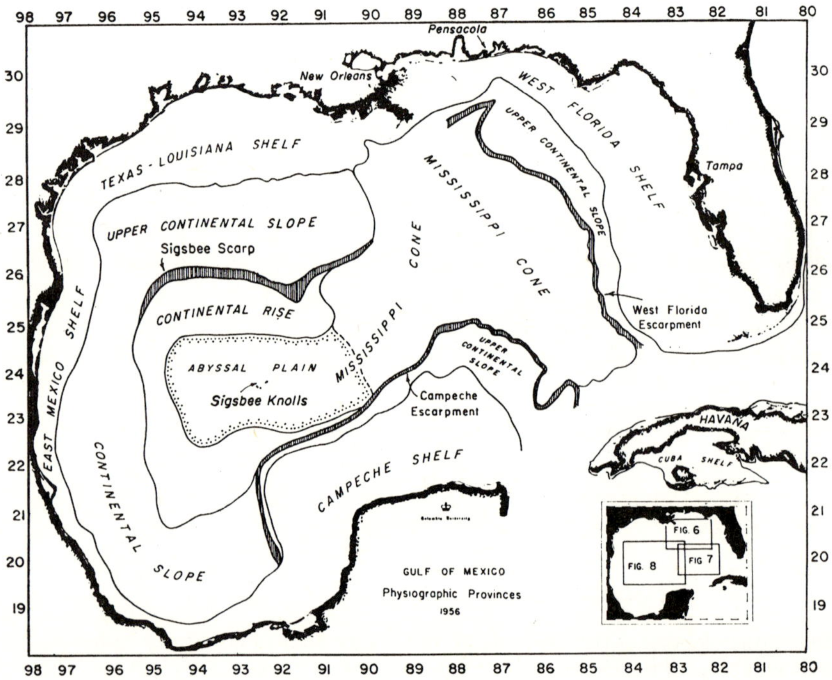

Fig. 11. Chart of the physiographic provinces of the Gulf of Mexico showing the location of the Sigsbee Abyssal Plain. The floor of the eastern Gulf of Mexico is dominated by the Mississippi Cone. Turbidity currents originating at the mouth of the Mississippi have built the large Mississippi Cone and the Sigsbee Abyssal Plain. (After Ewing, Ericson and Heezen, 1958.)

Abyssal Plain. The Balearic Abyssal Plain seems to be fed predominantly from the north. It apparently spreads out from the foot of the Rhône Abyssal Cone. However, the plain is probably also fed from the south, as indicated by the numerous submarine canyons of the Algerian continental slope and by the events associated with the Orleansville earthquake (Heezen and Ewing, 1955). The area of the Balearic Abyssal Plain probably exceeds 30,000 mi^2. A small abyssal plain is found in the center of the nearly circular Tyrrhenian Sea. Many small abyssal hills protrude from this abyssal plain. In the eastern

Mediterranean one would offhand expect to find a large abyssal plain lying seaward of the Nile. However, such a feature is not found. Although the region seaward of the Nile has been smoothed by sediments derived from the Nile delta, the smoothing has not proceeded to such a point that an abyssal plain has formed. An extremely small abyssal plain is found to the south of the Straits of Messina and is probably fed by material coming from the straits, judging from the events associated with the 1908 Messina earthquake (Heezen, 1957). It seems probable that many of the small basins making up the Aegean Sea have flat floors. However, little of this area has been investigated with precision sounders.

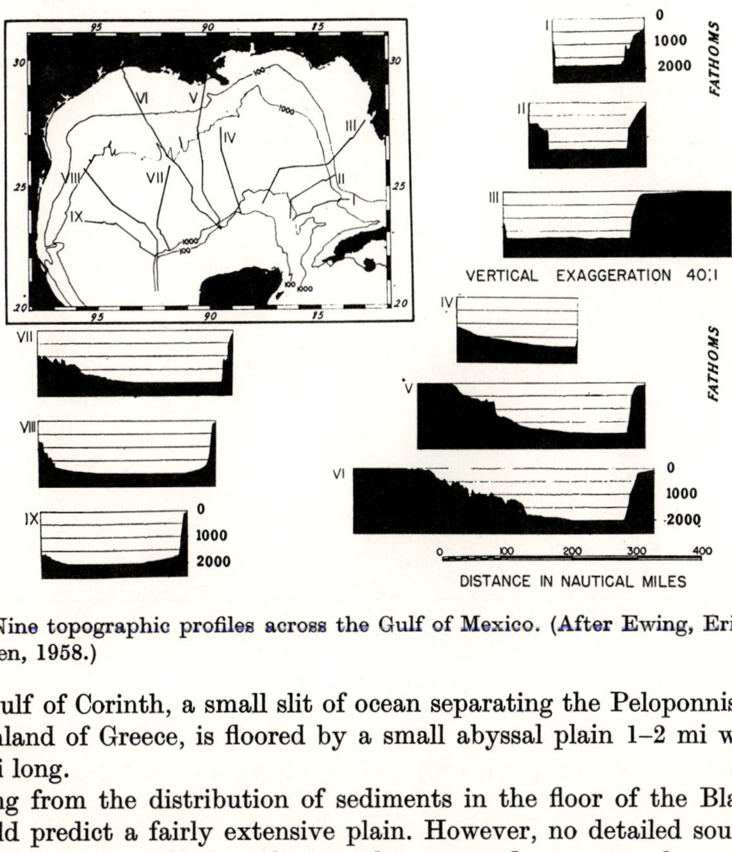

Fig. 12. Nine topographic profiles across the Gulf of Mexico. (After Ewing, Ericson and Heezen, 1958.)

The Gulf of Corinth, a small slit of ocean separating the Peloponnisos from the mainland of Greece, is floored by a small abyssal plain 1–2 mi wide and 10–12 mi long.

Judging from the distribution of sediments in the floor of the Black Sea, one would predict a fairly extensive plain. However, no detailed sounding is available for this area. Zenkovich (1958) has reported numerous abrupt changes in depth off the mouths of various rivers of the eastern Black Sea coast. Turbidity currents generated in this area would be expected to supply sufficient sediments to produce an abyssal plain seaward of the Caucasus.

b. Gulf of Mexico

The Sigsbee Abyssal Plain lies in the center of the Gulf of Mexico at the foot of the Mississippi Cone (Figs. 11 and 12). It is virtually level in the central part where gradients as low as 1 : 7000 have been measured (Fig. 13). The floor of the Gulf of Mexico is dominated by the Mississippi Cone, which has not only supplied sediments to the Sigsbee Abyssal Plain, but also to the smaller abyssal plain of the eastern gulf, which extends along the base of the West Florida Escarpment into the Campeche Tongue.

Sedimentation related to the building of individual distributory channels and fans of the Mississippi Cone has built this feature completely across the gulf floor to the base of the Campeche Escarpment, cutting the gulf floor into two

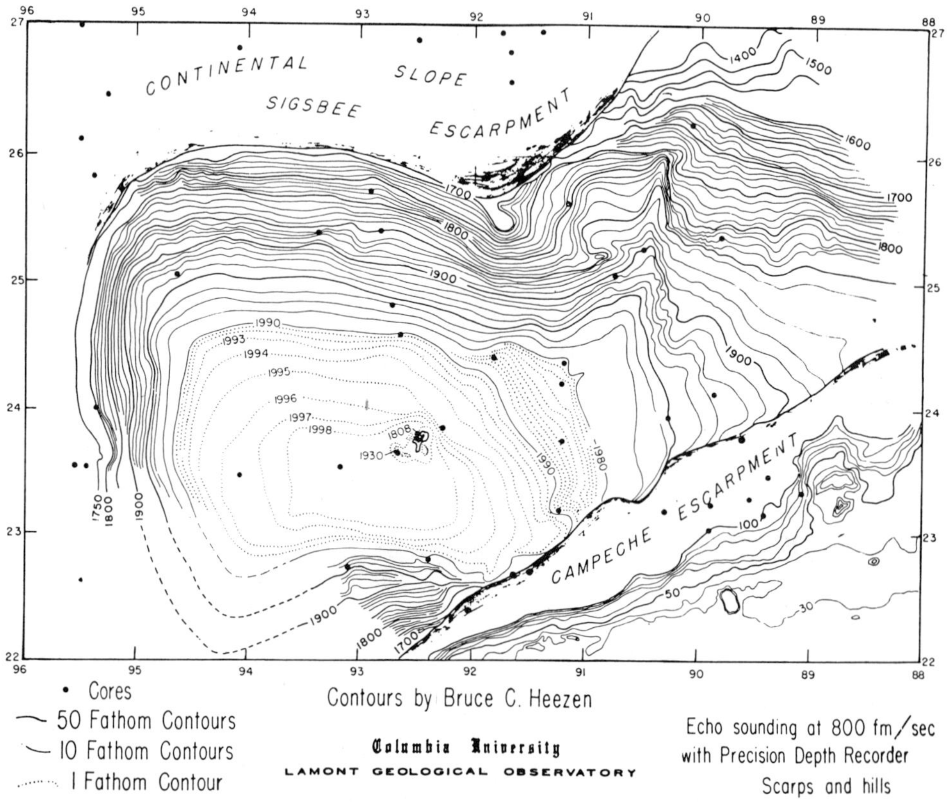

Fig. 13. Bathymetric chart of the Sigsbee Abyssal Plain. (After Ewing, Ericson and Heezen, 1958.)

halves. A small apron of sediment spreads out a few miles from the base of the Campeche Escarpment. The Sigsbee Abyssal Plain is probably fed in part by canyons leading from the east coast of Central Mexico and out of the Bay of Campeche (Ewing, Ericson and Heezen, 1958).

Over 10 m of fine silt was deposited on the Sigsbee Abyssal Plain during Wisconsin time (Ewing *et al.*, 1958). Rates of sedimentation exceeded 60 cm/1000 years during the Wisconsin but have averaged less than 8 cm/1000 years in Recent time. A number of small topographic prominences of about 6 fm elevation which occur in the plain (Ewing and Ewing, 1962) have been found by reflection studies to lie above deeper conical structures. The deeper structures have been interpreted as salt domes by Ewing, Worzel and Ewing (1962). Assuming that the plain was perfectly flat at the close of the Wisconsin, a differential movement of 1 m/1000 years is indicated. This distortion may indicate differential compaction or vertical migration of the sub-bottom structures.

c. Caribbean Sea

The southern Caribbean Sea is divided into the Colombia and the Venezuela Basins (Fig. 14). The Colombia Basin is floored by the well developed Colombia

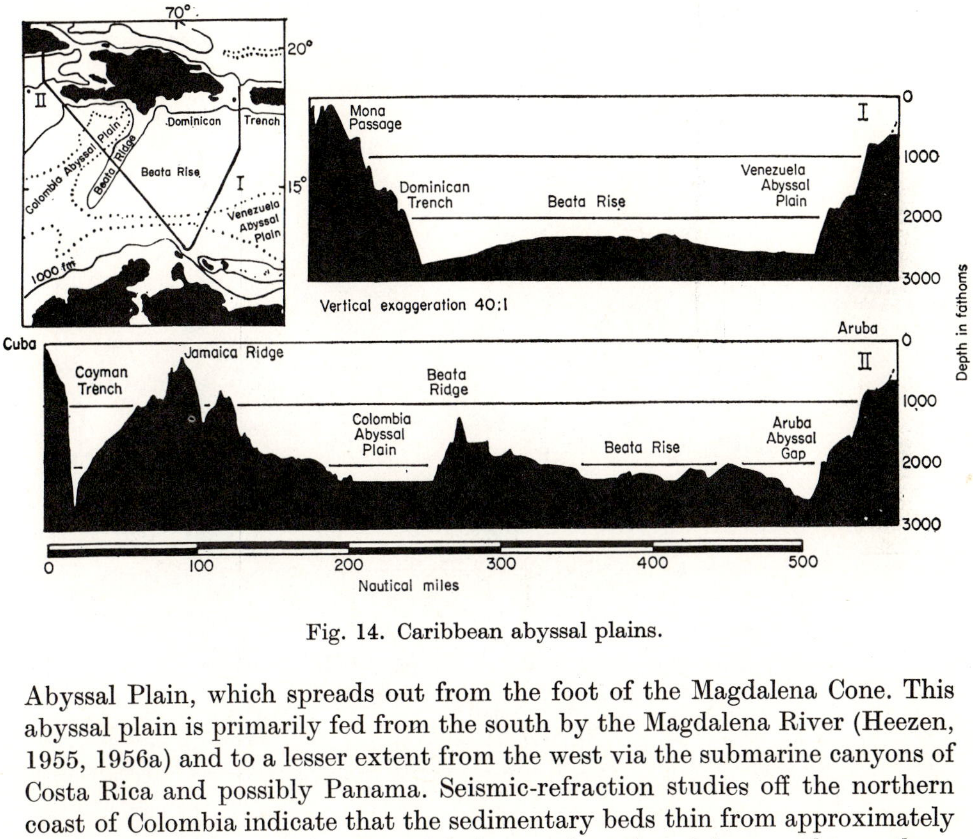

Fig. 14. Caribbean abyssal plains.

Abyssal Plain, which spreads out from the foot of the Magdalena Cone. This abyssal plain is primarily fed from the south by the Magdalena River (Heezen, 1955, 1956a) and to a lesser extent from the west via the submarine canyons of Costa Rica and possibly Panama. Seismic-refraction studies off the northern coast of Colombia indicate that the sedimentary beds thin from approximately 3 km at the base of the continental slope to less than 2 km on the northern part of the Colombia Abyssal Plain (Ewing, Antoine and Ewing, 1960). The Colombia Abyssal Plain is separated from the Venezuela Basin by the Beata Ridge, which extends in a southerly direction from Cape Beata on the island of Hispaniola. However, north of Aruba there is an abyssal gap between the southern end of the Beata Ridge and the continental slope of South America. The Aruba Gap provides a narrow connection between the Colombia Abyssal

Plain and the smaller Venezuela Abyssal Plain. The Venezuela Abyssal Plain is much less well developed than the Colombia Abyssal Plain, probably largely owing to its isolated position with respect to the continent, and the Venezuela Abyssal Plain may be largely fed from the Colombia Abyssal Plain via Aruba Gap. A very narrow abyssal plain exists along the northern side of the Venezuela Basin just south of the south coast of the Dominican Republic. This feature is known as the Dominican Abyssal Plain.

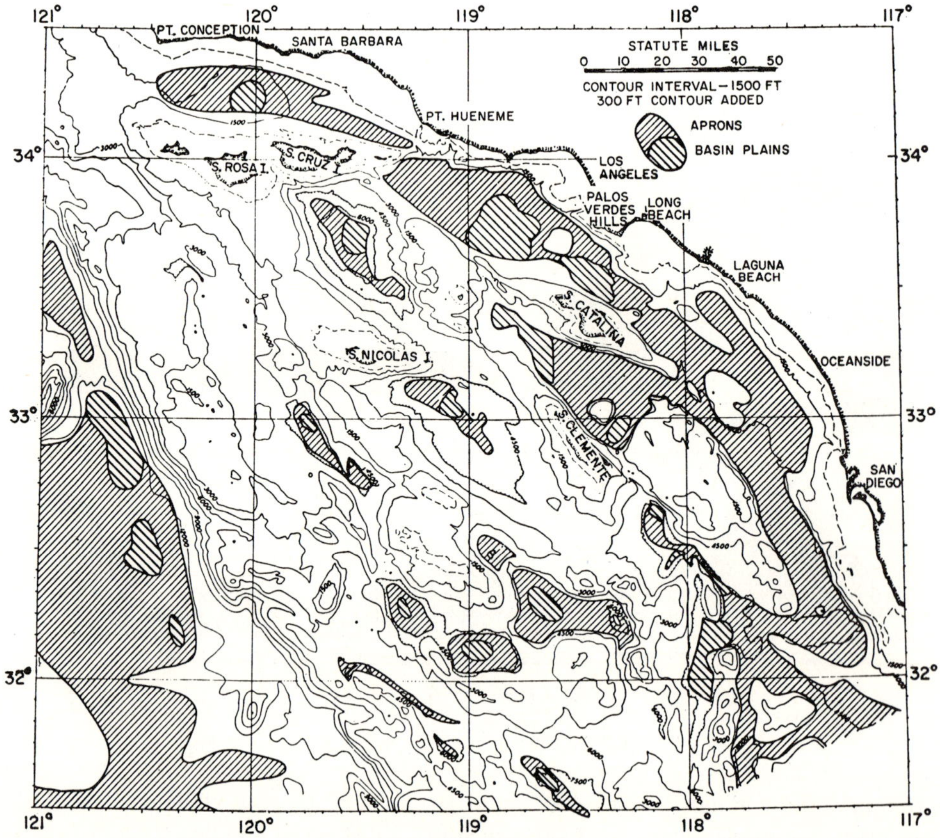

Fig. 15. Basin plains and aprons off Southern California. This summary chart is based on detailed basin maps. (After Emery, 1960.)

Between the Cayman Ridge and the south coast of Cuba lies the Yucatan Basin. In the center of the basin is found an abyssal plain probably exceeding 2500 mi^2 in area and lying in depths of 2450 to 2550 fm.

d. North Pacific marginal seas

The marginal seas which lie between the island arcs and the continent in the North Pacific area are floored by small abyssal plains. The largest is that of the Bering Sea (Udintsev *et al.*, 1959), in which the abyssal plain probably exceeds 100,000 mi^2 in area. The abyssal plain in the Okhotsk Sea probably does not exceed 30,000 mi^2 (Udintsev, 1957).

e. Southern California basins

Lying off the coast of Southern California is a series of thirteen isolated basins which lie in the continental margin, their long axes oriented parallel to the coastline (Fig. 15). The floor of each of these basins is occupied by a plain. The largest of these plains lie in the basins nearest the coastline (Fig. 16). Emery (1960) divides the basin floors into basin aprons and basin plains. The

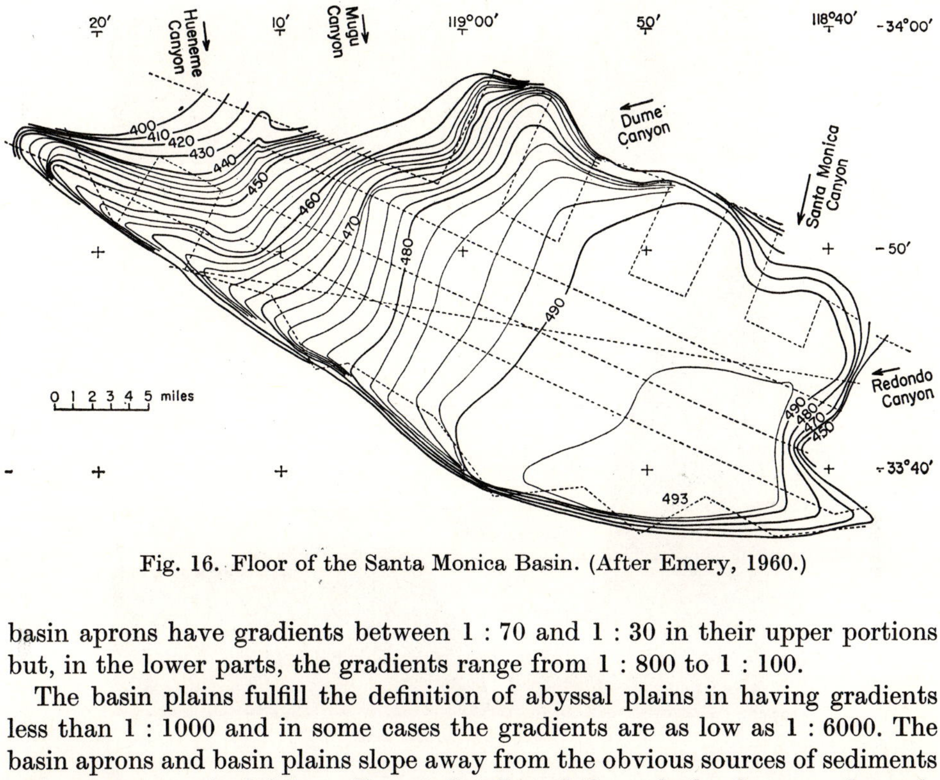

Fig. 16. Floor of the Santa Monica Basin. (After Emery, 1960.)

basin aprons have gradients between 1 : 70 and 1 : 30 in their upper portions but, in the lower parts, the gradients range from 1 : 800 to 1 : 100.

The basin plains fulfill the definition of abyssal plains in having gradients less than 1 : 1000 and in some cases the gradients are as low as 1 : 6000. The basin aprons and basin plains slope away from the obvious sources of sediments and are apparently fed by sediment distributed through the deep-sea channels associated with the many continental-slope submarine canyons. The distribution of sediments in this manner has been well demonstrated by the work of Gorsline and Emery (1959) in their study of the San Diego Basin.

3. Trench Abyssal Plains

A trench abyssal plain is an abyssal plain lying in the bottom of a deep-sea trench. Since the initial discovery of the Puerto Rico Trench Abyssal Plain in

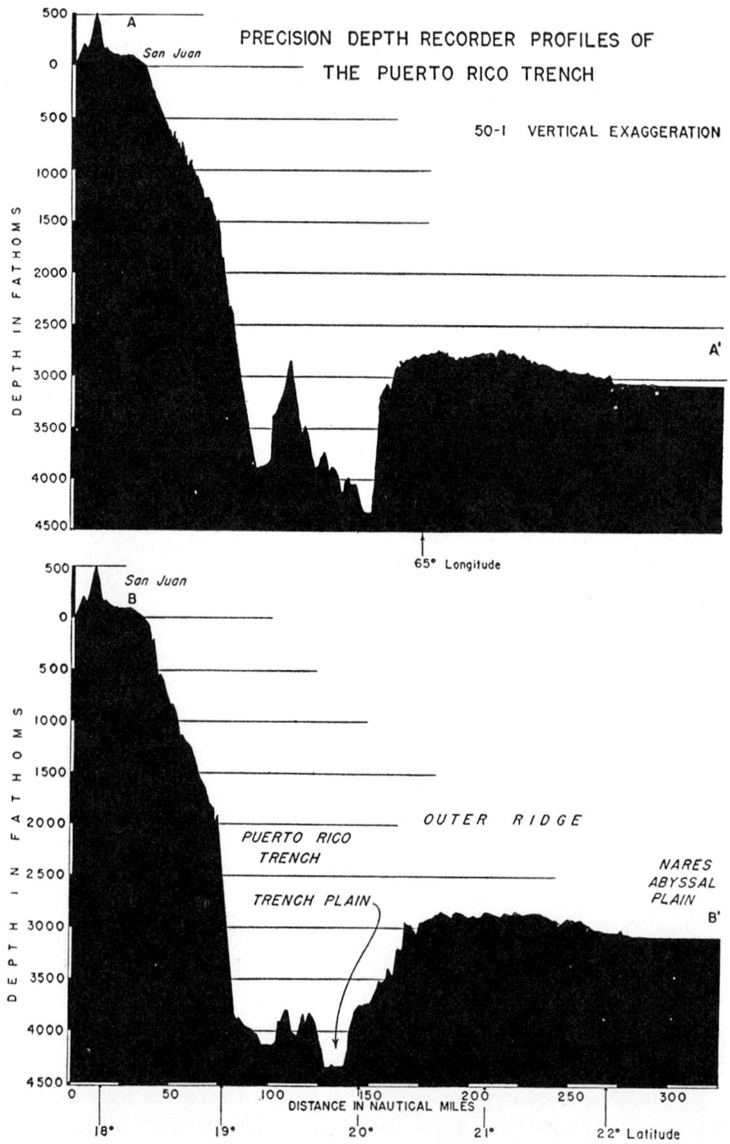

Fig. 17. Two profiles of the Puerto Rico Trench Abyssal Plain. (After Ewing and Heezen, 1955.)

1948, similar features have been reported in the Middle America Trench (Fisher, 1954), the Kurile–Kamchatka Trench (Udintsev, 1955) and the Peru–Chile Trench (Zeigler *et al.*, 1957). Two trench plains are known in the Puerto Rico Trench (Fig. 17). The smaller one occupies a basin south of a median ridge and the larger one lies along the deep axis of the trench (Fig. 18). They range in width from 1 or 2 mi to about 12 mi. Sediment cores taken in the trench plain contain beds of graded calcareous sand containing fragments of the calcareous alga *Halimeda* and shallow benthonic Foraminifera (Ericson, Ewing and Heezen, 1952). The depth of the deeper trench abyssal plain ranges only a few fathoms from 4585 fm.

The Puerto Rico Trench plain seems to be the widest and most extensively developed of all trench plains. Although most trenches are not well enough surveyed to reveal the maximum width of other trench plains, it is known that the Tonga Trench and the Kurile–Kamchatka Trench have relatively narrow trench plains, reaching maximum widths of 2 or 3 mi.

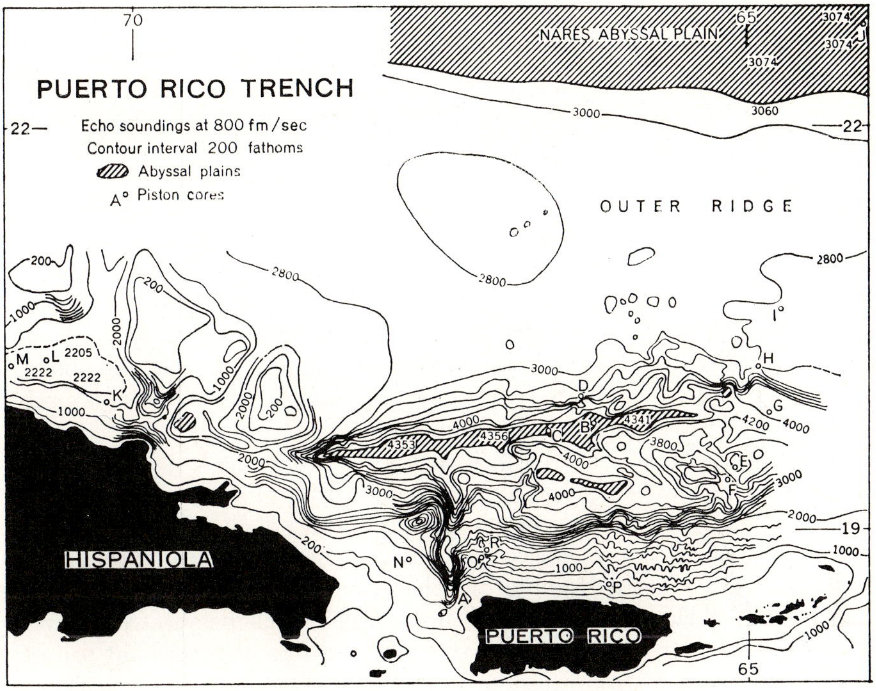

Fig. 18. Puerto Rico Trench Abyssal Plains. (After Ewing and Heezen, 1955.)

Although trench plains are common along the axis of the Peru–Chile Trench, these plains do not appear to be continuous over any considerable distance along the trench axis.

Enclosed basins in the continental margin which lie downhill from adjacent land-masses generally contain abyssal plains. Small abyssal plains are found in the floor of the Cariaco Trench, a 750-fm-deep depression in the continental shelf north of Venezuela (Heezen *et al.*, 1958). The flat floors of many fjords are probably similar in origin to the trench plains mentioned above. Indeed, much of the floor of the Lake of Geneva in Switzerland is of abyssal-plain character. This plain lies at the foot of a large cone spreading out from the mouth of the Rhône. These features, although similar in form and almost certainly identical in origin to the abyssal plains of the ocean-basin floor, lie in diverse tectonic and physiographic settings. They are only indicative of identical smoothing mechanisms whereby the bottoms of basins become flat.

4. Archipelagic Plains

Menard (1956) originally proposed the term "archipelagic apron" for the smooth sea floor around islands or island groups which "has the form of an apron" (cf. Chapter 12). Although most archipelagic aprons were found to be smooth, some were relatively rough and, therefore, Menard made the provision in his definition that an archipelagic apron "need not have a smooth surface". In view of this latter provision, it seemed necessary to propose the term "archipelagic plain" for those portions of an archipelagic apron which are smooth. An archipelagic plain is a smooth portion of the deep-sea floor which lies at the base of the pedestal of an island group or island chain. The characteristically low gradients and smoothness of archipelagic plains are extremely similar to abyssal plains. Menard has discovered dozens of archipelagic plains in the Pacific around the Hawaiian Islands, the Mid-Pacific Mountains, and the Marshall, Gilbert and Marquesas archipelagos. As Menard points out, aprons "often exhibit the features used elsewhere as criteria for partial burial of a hilly terrain, namely, steep-sided hills rising abruptly from smooth plains, apparent ponding of sediment behind ridges acting as dams and the absence of deeps below the level of the smooth surface. Consequently, it appears probable that the smooth aprons have buried an older terrain of hills."

The fact that there are archipelagic aprons which lack a smooth surface suggests that there was a rough-hewn apron or flange-like feature lying around the base of the island groups, which originally did not have a smooth surface, and, in many cases, this rough flange or apron may have been later smoothed by sedimentation. Menard points out that the total volume of material contained in that portion of an archipelagic apron which lies above the general level of the sea floor is many times the total volume of the island pedestal it surrounds. For this reason, it seems unlikely that the total volume of material in an archipelagic apron has accumulated through turbidity-current deposition of sediments derived from the pedestal. Likewise, it seems clear that the archipelagic plains are similar, if not identical in origin, to the abyssal plains. Both were probably produced by the burial of former relief by turbidity-current deposition.

Almost all the archipelagic plains of the world are found in the central Pacific. However, archipelagic plains are found in the Atlantic surrounding the pedestal of Bermuda and indications of similar features are found surrounding the oceanic islands of the South Atlantic and the Indian Ocean. If we accept the limits of the Pacific archipelagic aprons shown by Menard (1958), we must conclude that the area of the archipelagic aprons of the Pacific is nearly equal to the area of abyssal plains in the Atlantic. However, Menard does not state what percentage of the total area of the archipelagic aprons shown in his figures is smooth and thus analogous to abyssal plains, and what percentage is irregular.

5. Abyssal Gaps, Interplain Channels, Mid-Ocean Canyons and Deep-Sea Channels

In many areas of the flat deep-sea floor, especially on and associated with abyssal plains and the deep-sea fans found at the base of submarine canyons, are small depressions a few miles across and some tens of fathoms deep and varying in length from 10 mi to 1500 mi. They were first noted by Shepard and Emery (1941) leading from the base of submarine canyons off Southern California and have been found subsequently in most oceans of the world. Channels or canyons in the deep-sea floor have at least three modes of occurrence: (1) The extensions of submarine canyons which lead from the continental slope on across the continental rise to the abyssal plains. These are sometimes called "deep-sea channels". However, other investigators simply refer to them as submarine canyons. (2) Mid-ocean canyons, which are found in abyssal plains and have great length. They are usually unbranched and have not been traced to continental slope canyons. (3) Interplain channels. These are small channels or canyons which are associated with abyssal gaps. They begin in one abyssal plain and end in another and have no direct connection with submarine canyons of the continental slope.

A. Deep-Sea Channels Associated with Submarine Canyons and Deep-Sea Fans

Whatever the mechanism by which submarine canyons may have been cut into the edge of the continental shelf, it is well established that periodic turbidity currents flush sediment through them and build up depositional fans where the canyon reaches the lower gradients of the ocean basin. These deep-sea fans have been found at the bases of nearly all submarine canyons, notably the Hudson Canyon (Ericson, Ewing and Heezen, 1951), La Jolla Canyon (Menard and Ludwick, 1951), the Monterey Canyon (Dill, Dietz and Stewart, 1954), and many others off the California coast (Menard, 1955; Gorsline and Emery, 1959). A comparison of the volume of the deep-sea fans off central California with the available erosion products from the adjacent land areas has led Menard (1960) to date the fans as pre-Pleistocene. Recent surveys of the seaward extensions of other submarine canyons throughout the world has also revealed the existence of deep-sea fans or families of deep-sea fans, which are collectively referred to as abyssal cones (Heezen, Ewing, Menzies and Granelli, 1957).

Crossing these fans and abyssal cones are channels diverging from the mouth of the submarine canyon and gradually losing their identity once they have reached the abyssal plain. They are often characterized by leveed banks (Fig. 19) (Buffington, 1952; Dietz, 1958) and off the California coast there is a strong tendency for them to hook left, i.e. to the south, contrary to what one would expect if they were controlled by Coriolis forces. Menard has put forward an explanation of the left hook by considering the top surface of a turbidity current flowing in a canyon to be tilted and an asymmetry to be produced in the levees built up on either side so that the channel is continually forced to migrate to the left. He proposes a general rule for the orientation of leveed turbidity-current channels: "A turbidity current channel flanked by levees has a tendency to deviate from a straight line which varies directly with the amount of the deflecting force of the earth's rotation but is in the opposite direction" (Menard, 1955). However, not all turbidity currents from canyons can be confined to the channels on the fans, otherwise the fan may not be built up. Some flows would be too large to be contained by existing channels and

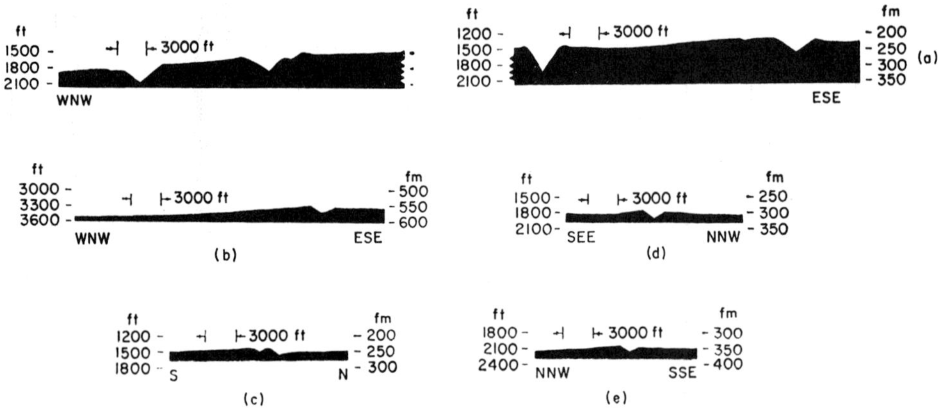

Fig. 19. Natural leveed submarine canyons off Southern California. (After Buffington, 1952.)

would spread laterally in the form of a sheet flow. The more rapid decay of the sheet-flow turbidity currents due to the two-dimensional spreading would result in the deposition of their sediment load near the base of the canyon and consequently in the formation of a deep-sea fan. For smaller flows, there might be sufficient topographic control to prevent two-dimensional spreading so that decay would be less rapid and well defined channels would be cut.

B. Mid-Ocean Canyons

In 1949, four canyon-like depressions were found in the abyssal floor east of Newfoundland (Figs. 20 and 21). A subsequent survey showed that these were part of a continuous mid-ocean canyon running for some 1500 mi from the seas around Greenland and feeding into the Sohm Abyssal Plain through a narrow gap in the South East Newfoundland Ridge (Ewing, Heezen, Ericson, Northrop and Dorman, 1953; Heezen *et al.*, 1959). Sections of the canyon typically have steep walls, sometimes even vertical, and flat floors. They vary in

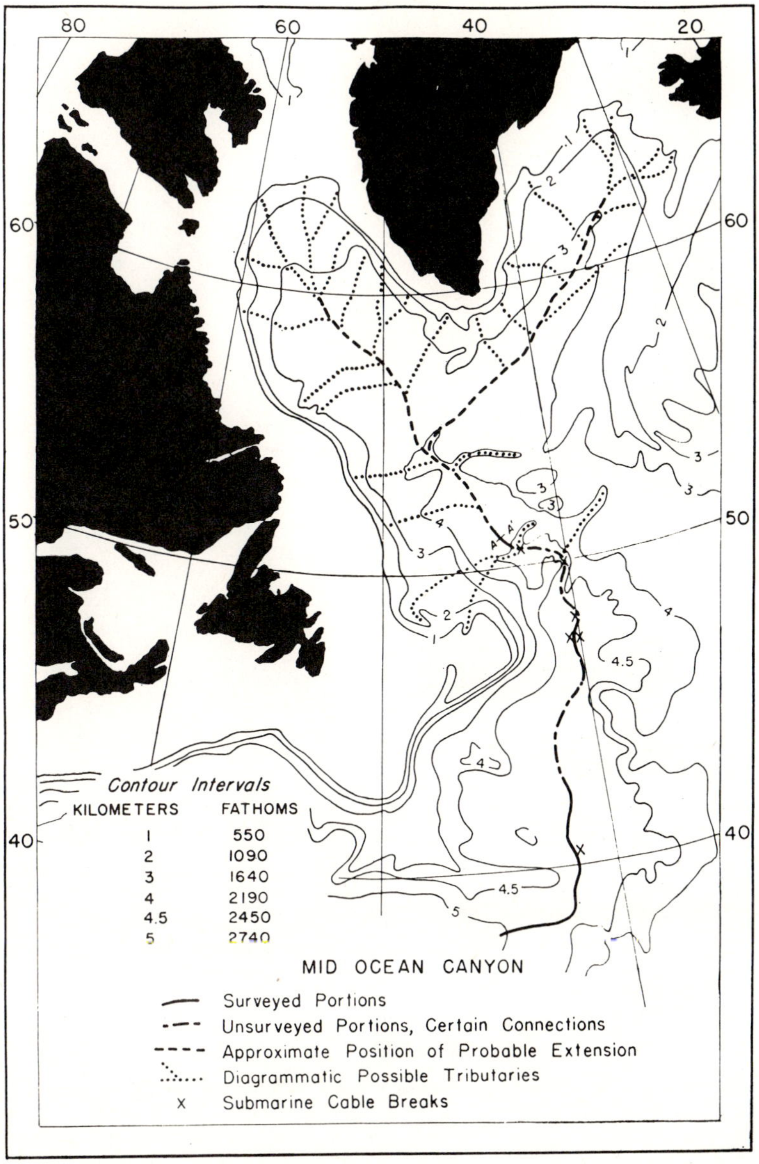

Fig. 20. Northwest Atlantic Mid-Ocean Canyon. (After Heezen *et al.*, 1959.)

width from 3 to 5 mi and in depth below the adjacent sea floor from 10 to 100 fm. There is a consistent asymmetry about the canyon, the west wall being 10–15 fm higher than the east wall (Fig. 22). The canyon has been mapped by more than 80 echo-sounding profiles. Surprisingly, the canyon does not follow the axis of greatest depth in the ocean basin but tends to the western side. The average slope of the sea floor at right angles to the axis of the canyon in most

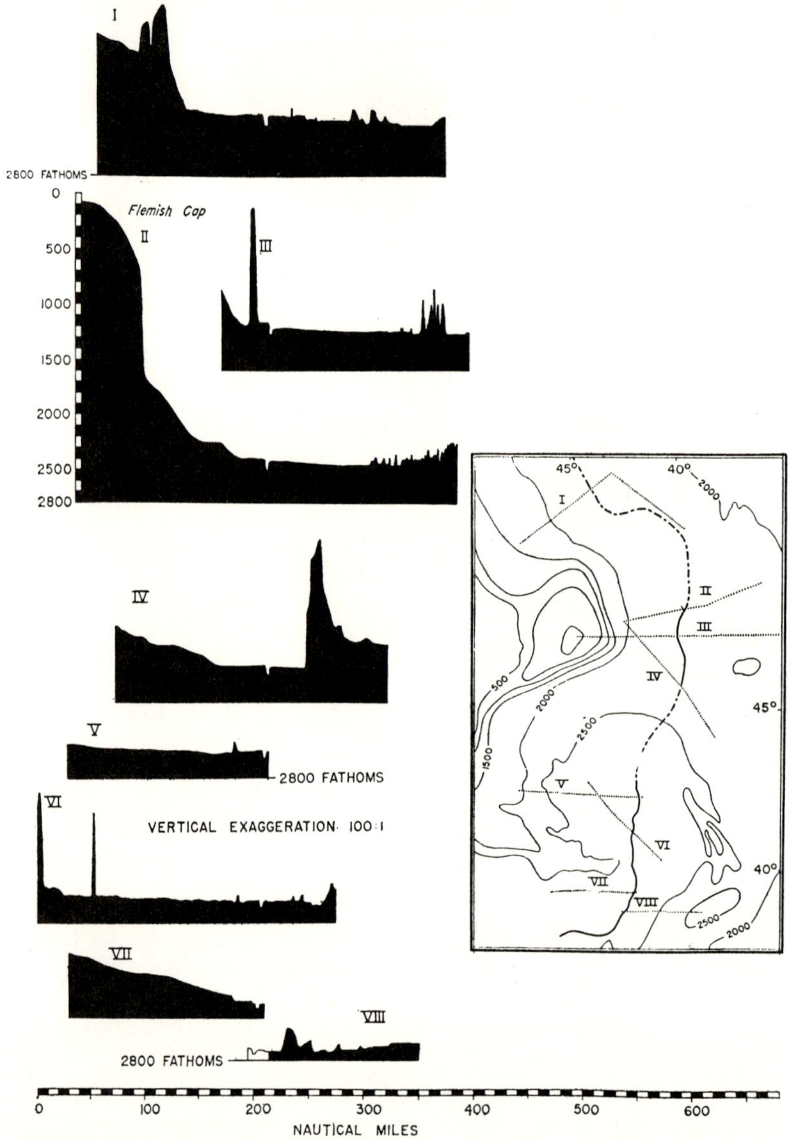

Fig. 21. Eight east–west profiles across the Northwest Atlantic Mid-Ocean Canyon. (After Heezen *et al.*, 1959.)

places exceeds the gradient of the canyon. The gradient west to east is about 1:1500 while the gradient of the canyon from north to south is about 1:2250. The canyon forms gentle meanders and, as its gradient decreases, it changes its cross-sectional shape from a narrow deep canyon to a broad shallow one.

A second mid-ocean canyon of similar form and shape has been traced for 350 mi in the northwest arm of the Sohm Abyssal Plain. It is possible that this

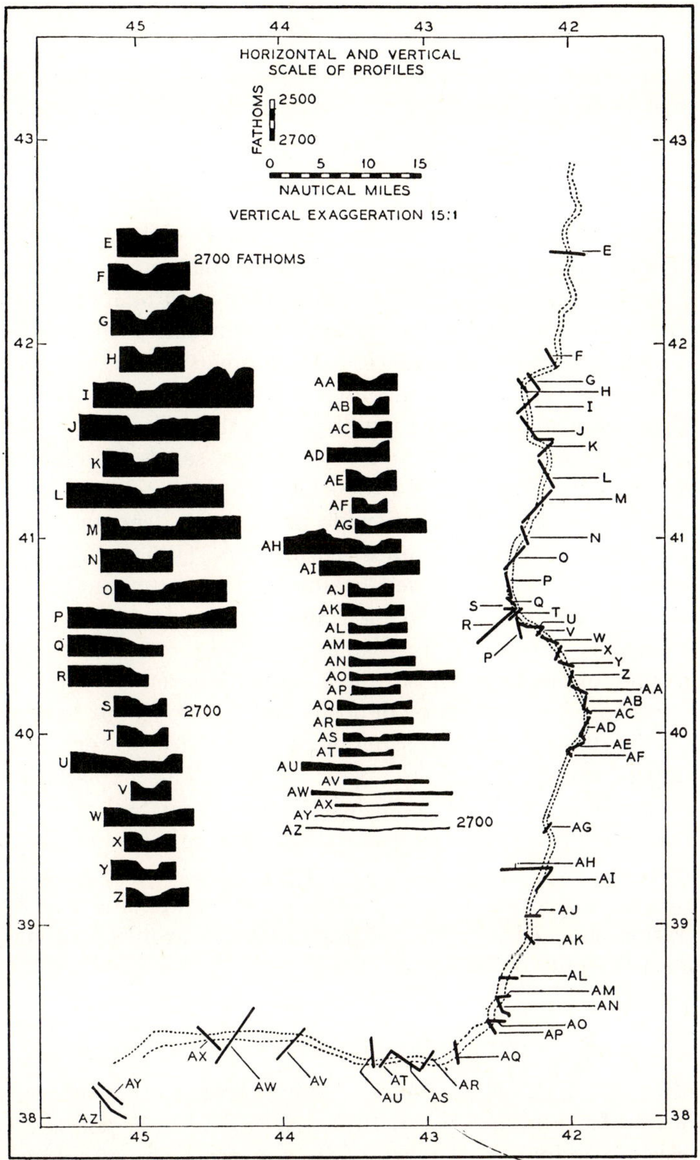

Fig. 22. Forty-eight profiles across the Northwest Atlantic Mid-Ocean Canyon. (After Heezen *et al.*, 1959.)

is associated with an abyssal gap through a ridge dividing the northwest arm of the Sohm Plain from the rest.

In the Indian Ocean, soundings taken by the Swedish Deep-Sea Expedition in the *Albatross* have shown the existence of trough-shaped features in an otherwise flat and featureless floor, which have been interpreted as grabens

(Koczy, 1954). However, these same features have been studied by Dietz (1953) and compared with the deep-sea channels off California to which they bear a close resemblance. If the Bengal Abyssal Plain is the result of turbidity currents, then it is plausible that these channels also have a similar origin. Further exploration of them will undoubtedly resolve this problem in the near future.

Other mid-ocean canyons have been found in other parts of the world, e.g. in the equatorial Atlantic (Heezen, Coughlin and Beckman, 1960) and in the Hikurangi Trench, and many more will no doubt be discovered.

The mechanism by which the mid-ocean canyons can be formed is still under discussion. Tectonic movements have already been discussed, but it is hard to see how such long, uniform and continuous channels can have been formed in this way when there is little evidence of vertical movements either side of the channels. Their association with abyssal plains, their similarity with the delta channels and interplain channels, and the deep-sea sands found in the canyon, argue a turbidity-current origin. However, it is difficult to envisage a turbidity current confined to a channel 2 mi wide and 50 fm deep travelling without decay for more than a thousand miles. Ewing *et al.* (1953) suggested a mechanism of successive filling of confined basins starting from the north, each one breaking through to the next by an interplain channel, and that these channels eventually joined to form one continuous one. However, there is little topographic evidence of these filled basins. Heezen *et al.* (1959) question whether the process is erosional or depositional. Dietz (1958) believes that the channels may be due to localized streams of clear dense water of high salinity and low temperature derived from Arctic regions. This explanation does not require the turbulence to keep particles in suspension and therefore permits lower current velocities, but fails to explain the mid-ocean canyons in the Indian Ocean and is difficult to apply in the equatorial Atlantic. The independence of canyon location and basement structure and the evidence of lateral migration shown by recent seismic reflection profiles support a depositional versus a tectonic or erosional origin.

C. Abyssal Gaps and Interplain Channels

If two adjacent but distinct abyssal plains have no through passage at or below the level of the higher plain, they are said to be separated by a sill, a ridge or a rise depending on the size of the feature involved. However, it has been found that several plains are interconnected by constricted passages which have been called variously "abyssal gaps" (Heezen *et al.*, 1959), and "interplain channels" (Laughton, 1960). However, both terms can be retained if "abyssal gap" is used to describe the larger features of a passage cutting across a ridge, and "interplain channel" more specifically to describe the continuously graded channels formed by turbidity currents travelling in an abyssal gap between two abyssal plains. Thus an abyssal gap may or may not contain interplain channels.

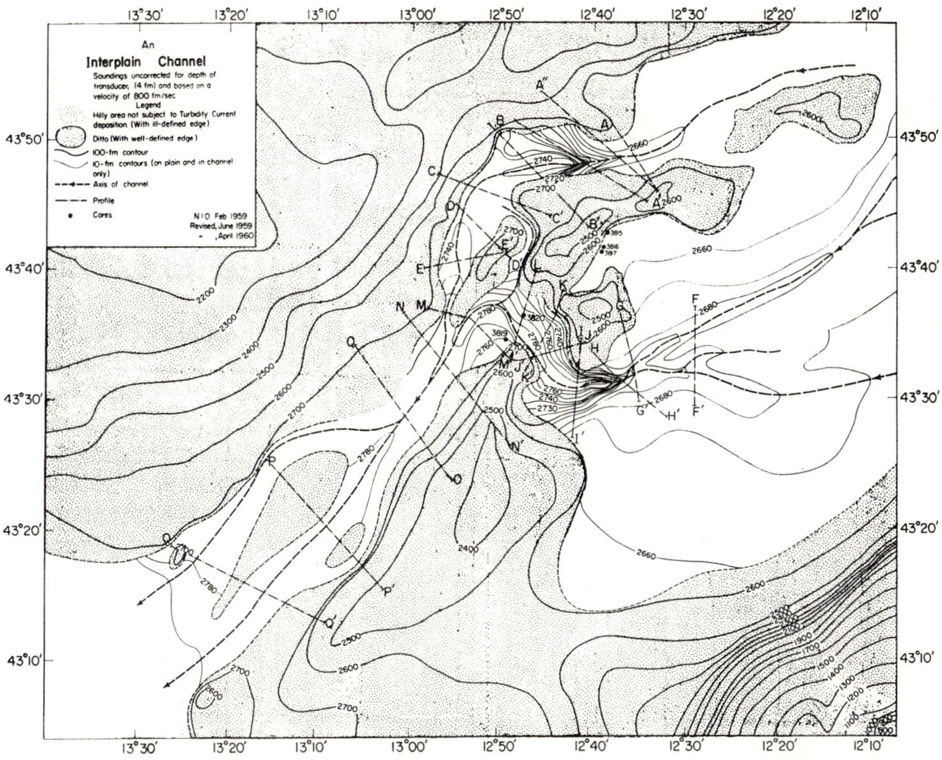

Fig. 23. Contour chart of an interplain channel system in Theta Gap. Depths in fathoms (uncorrected). (After Laughton, 1960.)

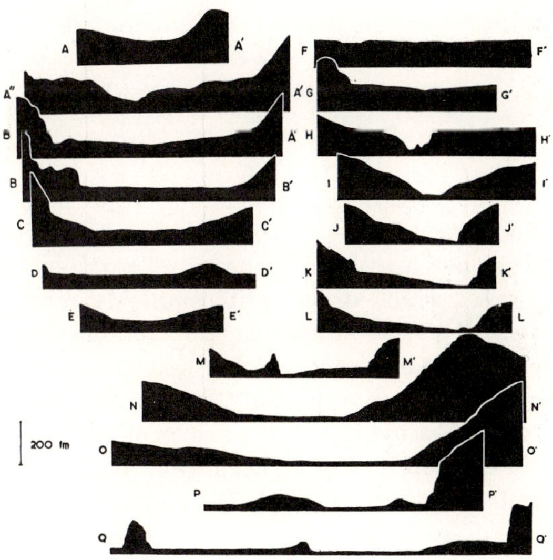

Fig. 24. Profiles of interplain channels in Theta Gap. Vertical exaggeration 8.4 : 1. Location of profiles indicated on Fig. 23. (After Laughton, 1960.)

The best explored abyssal gap (Figs. 23 and 24) and associated system of interplain channels is that connecting the southwest corner of the Biscay Abyssal Plain with the northern part of the Iberia Plain (Laughton, 1960). This has been called by Heezen *et al.* (1959) the Theta Gap. There is a difference in level between the two plains of about 100 fm which can supply enough kinetic energy that turbidity currents flowing through Theta Gap become actively erosive. On the Biscay Plain a dendritic system of feeding channels

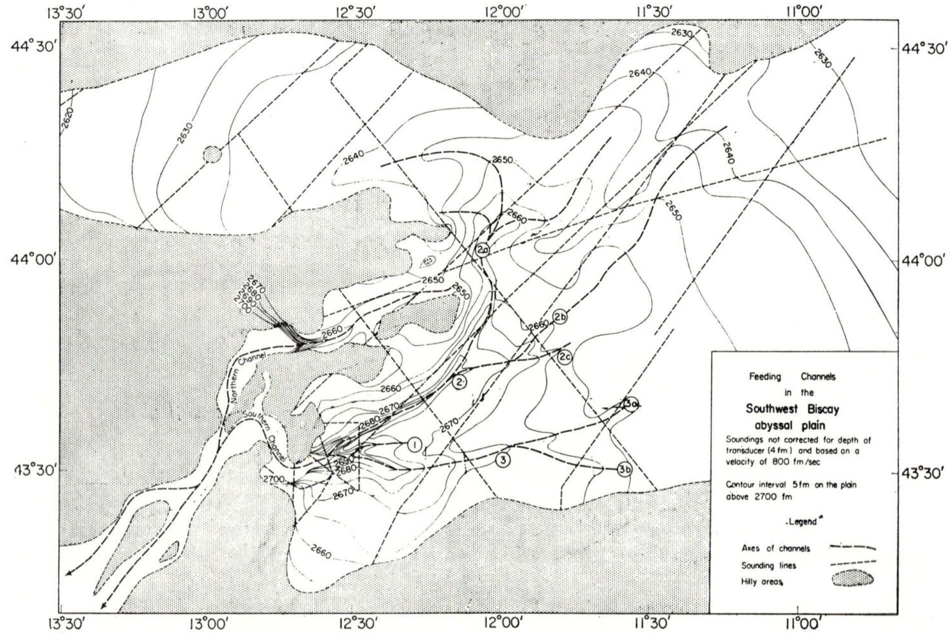

Fig. 25. Feeding channels in the southwestern part of the Biscay Abyssal Plain. (After Laughton, 1960.)

(Figs. 25 and 26) converge on an interplain channel which passes through the gap. In places these feeding channels are box-shaped, 2 or 3 mi wide and 10 or 20 fm deep: in other places they are V- or U-shaped. They join finally to form one broad V-shaped channel which drops quite suddenly more than 80 fm into the narrow, vertical-sided gorge of the interplain channel. Nearly all the change of level between the plains takes place within less than half a mile.

Thereafter the interplain channel winds through the abyssal gap exhibiting the characteristics of a meandering river, the deepest point swinging to the outside of a bend and the depth varying according to the width of the channel. It is joined half way by another branch of the channel system, divides around a "middle ground" hill and finally joins the Iberia Plain on a broad front. There is no sign of discrete channels on the Iberia Plain.

The feeding channels of the Biscay Plain extend some 80 mi from the head of the interplain channel, which is 50 mi long. The northern interplain channel has no feeding channels and appears to be an older feature, the approaches to which have since been silted up.

The pattern of these channels can only be explained by the horizontal flow of turbidity currents rejuvenated at the southwest corner of the Biscay Plain by an increase of gradient. Before the plains were so connected, one can imagine the Biscay Plain to have been filling with both pelagic and turbidity-current sediment faster than the northern part of the Iberia Plain where turbidity-current sediment was scarcer. When the level of the Biscay Plain reached the level of the sill, which is covered with pelagic sediment only, subsequent

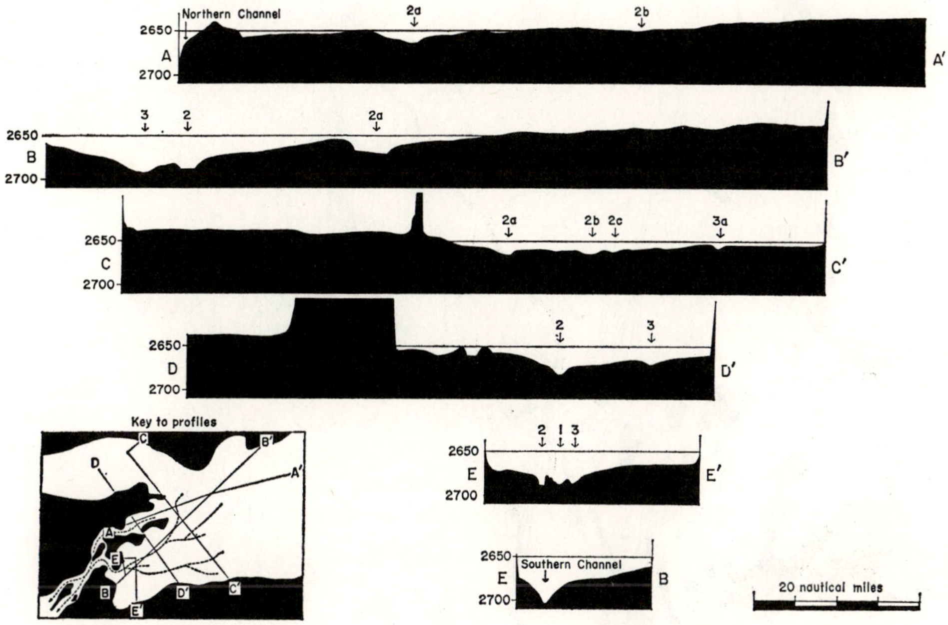

Fig. 26. Profiles of southwestern portion of the Biscay Abyssal Plain, showing feeding channels. Depth in fathoms (uncorrected). Vertical exaggeration, 100 : 1. (After Laughton, 1960.)

turbidity flows from the north would pour into the Iberia Plain and start to cut a channel back toward the north. The stage reached at present is that the cutting back has left the region of the gap itself and is in the Biscay Plain. Dredging, coring and photography of the head of this erosion have shown semi-consolidated pelagic sediments, vertical cliffs and exposed rocks.

A similar abyssal gap, the Vema Gap,[1] has been found between the Hatteras Abyssal Plain and the Nares Abyssal Plain. This is 20 mi wide, 70 mi long and the change of depth of the interplain channel is 170 fm. Associated with the

[1] For a new chart of Vema Gap and associated interplain channels see Chapter 12, Fig. 14, p. 247.

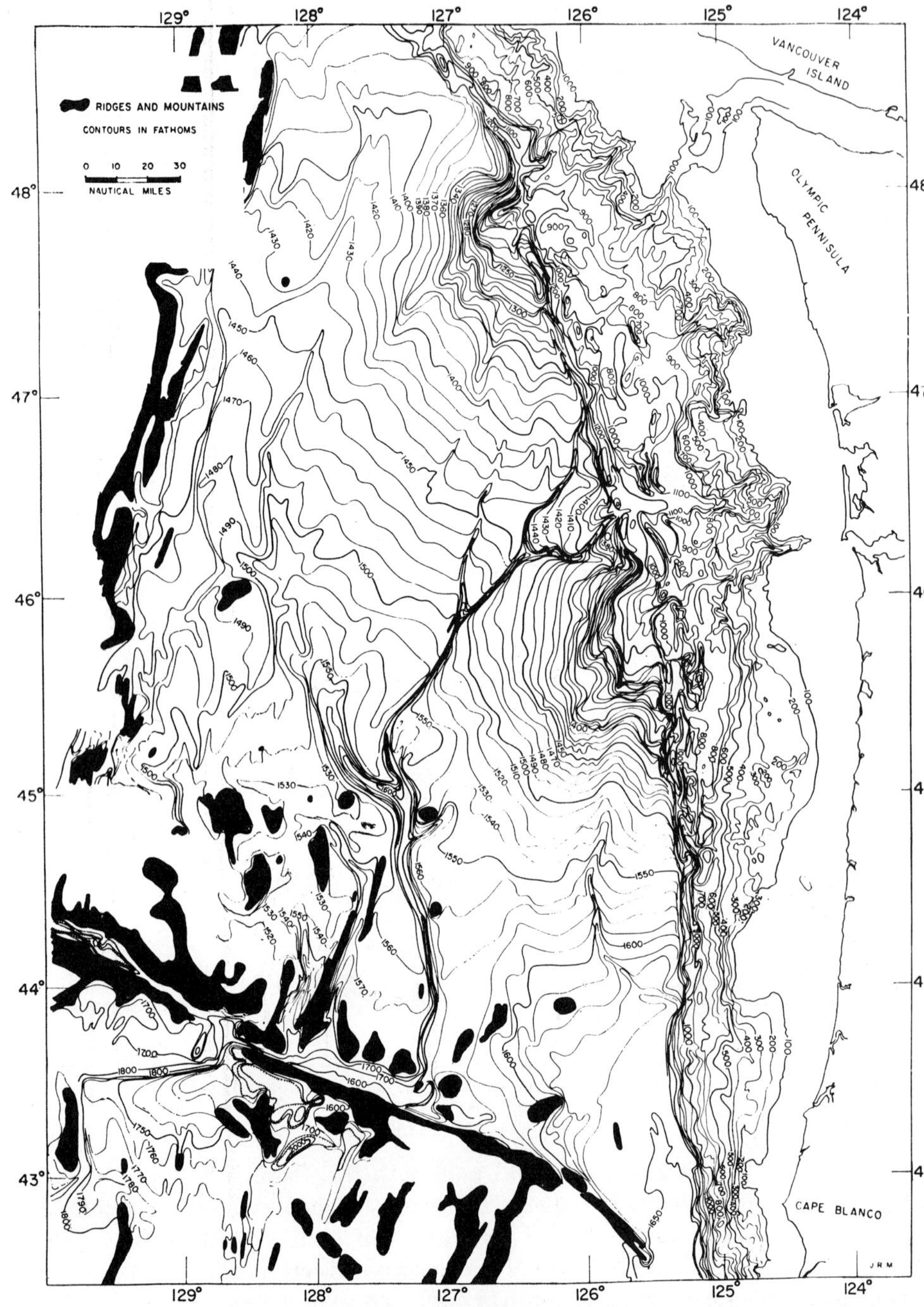

Fig. 27. Cascadia Canyon and Interplain Channel. Note how the tectonic trough between 43° and 44° North has acted as an abyssal gap. The Cascadia Submarine Canyon, which has apparently built the large cone shown in the northern part of the chart, passes through this abyssal gap and has supplied sediment to the abyssal plains lying to the west of the area shown on this map. (After Hurley, 1960.)

Vema Gap is a large magnetic anomaly suggesting a prominent buried ridge beneath the sill. In contrast, the anomalies over the Theta Gap are quite small and can be attributed to the visible topographic features of the area.

The Cascadia Channel (Fig. 27) mentioned previously is an example of a submarine canyon deep-sea channel system which directly connects through a tectonically controlled abyssal gap. The Cascadia Channel is both an extension of a continental slope canyon and an interplain channel. The tectonic trough which the Cascadia Channel follows through the ridge and valley province connects both to the north and to the south to a series of deeper trenches, leaving little doubt of the tectonic character of this feature.

An abyssal gap connects the Colombia and Venezuela plains in the Caribbean. Others will undoubtedly be discovered.

6. Sediments on the Abyssal Plains

The abyssal plains of the abyssal floor generally lie at such depths and distances from shore that the normal pelagic sediment covering them is deep-sea red clay. However, in the abyssal plains of the northeast Atlantic and Mediterranean, the normal pelagic sediment is *Globigerina* ooze and in the small abyssal plains of the continental margin basins lying off Southern California, the normal sediment is gray terrigenous clay. There is a sharp contrast in sediment type between the abyssal plains and adjacent, slightly elevated areas of nearly the same depth (Fig. 28).

On the rises only pelagic ooze is found. These sediments consist essentially of the shells of planktonic Foraminifera mixed in varying proportions with red clay or brown lutite. Because of the solution of calcium carbonate, the shells of Foraminifera decrease in number with increasing depth of water. Below about 5000 m, shells of Foraminifera are almost absent and the sediment consists essentially of brown lutite, which is the classical "red clay". This sediment accumulates through a constant rain of particles from above. The particles themselves are either wind-blown or transported by oceanic currents. Much of the calcium carbonate is precipitated originally in the upper layers of water in the form of the shells of planktonic plants and animals. In any case, the route of transportation of the particles is not along the ocean floor. Core A 160–1, at 5100 m depth on the Bermuda Rise, contains typical normal abyssal sediment, free from turbidity-current deposits. Throughout its length of 458 cm it is dark brown lutite. Shells of Foraminifera make up less than 1% by weight of the sediment. Micronodules of manganese oxide are common. Sand and silt layers are absent and there are no abrupt changes in lithology.

On the abyssal plains the normal pelagic sediments are interbedded with silt, sands, gravels and anomalous pelagic sediments normally found in shallower environments. Such detritus is absent from the rises and abyssal-hills area. For example, of 12 sediment-core stations, taken before 1954 in the Sohm Abyssal Plain, northwest of the Bermuda Rise, cores were obtained from only seven and at only three of these stations did the cores exceed 1 m in length. At five

DISTRIBUTION OF DEEP-SEA SANDS IN RELATION TO PHYSIOGRAPHIC PROVINCES

Fig. 28. Distribution of deep-sea sands in relation to physiographic provinces. Note that all cores from the abyssal plains contain sands or silt. (After Heezen *et al.*, 1959.)

Table I

Cores from the Abyssal Plain South of Newfoundland, 1948–1952

No.	N Lat.	W Long.	Depth, fm	Core lgth., m
(A) A152-131	34° 52′	52° 20′	2900	Nil
No core recovered, coring tube bent.				
(B) A152-132	34° 14′	55° 23′	2900	Nil
No core recovered, coring tube bent.				
(C) A153-141	33° 26′	53° 48′	2925	10.25
Red clay of normal deep-sea facies extends to 30 cm, below which is a series of alternating layers of red clay, gray clay, quartz silt and micaceous glauconitic sand.				
(D) A157-17	36° 56′	50° 40′	2905	0.25
Red clay was recovered in the top of the core, below which the pipe was empty; the lower portion of the core presumably ran out.				
(E) A157-18	36° 55′	53° 26′	2900	0.20
Red clay over silt below which the pipe was empty; presumably portion of the core ran out.				
(F) A160-16	31° 19′	55° 48′	2945	4.86
Alternating layers of red clay and gray clay with numerous thin layers of quartz silt.				
(G) A180-1	39° 07′	54° 32′	2840	1.00
Well-sorted graded silt extends from the surface to 130 cm, below which is a grayish-brown silty foraminiferal lutite except for a 5-cm layer of muddy sand at 320 cm below the top.				
(H) A180-2	39° 06′	54° 11′	2840	3.60
Well-sorted graded silt extends from the surface to 70 cm, below which is a foraminiferal lutite interbedded with two graded layers 5 and 9 cm thick. Both grade from well-bedded silty lutite at the top into very coarse subangular sand at the bottom.				
(I) A180-3	39° 08′	53° 54′	2840	Nil
Few grains of sand recovered, coring tube bent.				
(J) A180-4	39° 19′	52° 31′	2840	Nil
Few grains of sand recovered, coring tube bent.				
(K) A180-5	39° 30′	51° 48′	2840	Nil
Few grains of sand recovered, coring tube bent.				
(L) A180-7	39° 36′	50° 51′	2840	1.23
From the top to 30 cm brown foraminiferal lutite of abyssal facies. Below 30 cm gray lutite gradually grades downward into well-sorted silt between 95 and 100 cm. Below 100 cm dark gray foraminiferal lutite of abyssal facies.				

Letters A–L are indicated on Fig. 3.

stations only a few grains of sand were taken from a bent coring tube. At Stations D and E (Table I) only the upper layer of red clay, 10–20 cm thick, remained in the pipe after the lower portions of the core had run out between the leaves of the core retainer. The entire sample ran out of the pipe at Stations A, B, I, J and K. Therefore, the only cores of significant length obtained from the abyssal plain were F, G, H and I. All these cores contained sand and silt, so it is concluded that the reason for the lack of recovery on the other core stations is that the material was sand or silt which, unlike clay or ooze, could run out through the core retainer. In each case, a few grains of sand were imbedded in the grease around the core retainer. Cores G, H, I, J, K and L were taken for the specific purpose of testing the prediction of Heezen and Ewing (1952) that there would be in this area an uppermost layer of graded sand and silt. This prediction was based on the interpretation of the cable breaks following the Grand Banks earthquake as the work of turbidity currents triggered by slips, which in turn resulted from an earthquake shock. An upper layer of graded silt in G and H, and indications of a similar layer in I, J and K, confirmed this prediction.

In each core from an abyssal plain there is evidence of turbidity-current deposition, i.e. shallow-water Foraminifera, graded quartz sand, gray clays and bed of silt, all interbedded with lutite of abyssal facies.

Core A 153-141, at 5350 m depth on the Sohm Abyssal Plain, contains typical normal abyssal sediment of red clay facies to a depth of 30 cm. Below this is a series of alternating layers of red clay, gray clay and quartz silt to 400 cm. From 400 to 600 cm there is a micaceous glauconitic sand. At 460 cm the sand has a median diameter of 110 μ and is well sorted, having a quartile deviation of 0.75 φ. Between 600 and 718 cm there is a series of brown clay layers alternating with silt and sand. From 718 cm to the bottom of the core, there is again glauconitic sand, which has at 790 cm a median diameter of 103 μ and is also well sorted, having a quartile deviation of 0.85 φ.

The lower sand layer shows a gradual increase in particle size downward. Both sand layers contain a few Forminifera, among which are the shallow-water benthic forms *Elphidium incertum* var. *clavatum*, *Elphidiella arctica* and *Nonion labradoricum*.

This contrast in sediment type between the abyssal plains and the neighboring oceanic rises has been found in all oceans and seas containing abyssal plains. In several cases, more than 6 m of graded sand and silt have been found under a few centimeters of lutite in cores taken from abyssal plains. A further example is offered by the abyssal plain of the western equatorial Atlantic. Locher (1954) reports graded beds of terrigenous sand and silt in all cores taken from this plain. The shallow water origin of the material is strongly supported by Phleger (1954), who found shallow-water benthonic Foraminifera in the sands of four of these cores that he studied. Without exception, every abyssal plain from which data are available has yielded cores exhibiting the alternations in sedimentary type characteristic of turbidity-current deposits.

In Fig. 28 all deep cores taken in the North Atlantic are shown on a chart of physiographic provinces. The distribution of sands and silts correlates very well with the pattern of physiographic provinces. In the abyssal plain areas, all the cores contained sand or silt, whereas on the Bermuda Rise, sand or silt was definitely lacking, except on the archipelagic plain around the Bermuda Pedestal, where the sediments are obviously derived from Bermuda. On the continental rise, both sand-silt and purely clay-ooze cores are found, indicating that although canyons and channels running down the slope distribute the sands and silts over wide areas, broad, slightly elevated areas of the continental rise receive for a relatively long period of time only clays and pelagic oozes; whereas cores obtained from canyon floors, canyon banks, abyssal cones and deep-sea fans receive sands and silts intermittently and fairly regularly. A few of the sands and silts shown on Fig. 28 may be related to scour by bottom currents. Probably not all of them are due entirely to turbidity-current sorting (Heezen, 1959). The surface characteristics of the sediments on the abyssal plains, as revealed by preliminary studies of bottom photographs, do not seem to show any striking contrast with the sediments of the continental rise or of the oceanic rises. Both are covered with the small mounds, holes and tracks of benthic life (see Chapter 18).

7. Geophysical Characteristics

A. Seismic Reflection

Seismic-reflection measurements were the earliest geophysical investigations made in the abyssal-plain areas (Hersey and Ewing, 1949). It was early noted that seismic reflections taken from the abyssal plains were markedly different from those observed on the adjacent oceanic rises. The records obtained from the Bermuda Rise and the continental rise generally were relatively simple, showing one, two or three sub-bottom reflections, which, in general, could be correlated over relatively long distances. Records obtained on the abyssal plain in contrast showed a strong bottom reflection followed by a series of reflections so closely spaced that they formed a continuous reverberation (Fig. 29). At first the difference between the two types of records was not understood. The records from the rises, which showed a relatively few strong and persistent reflectors, were interpreted as evidence of persistent layering in the relatively thick sediments covering the rises. The lack of discrete correlatable sub-bottom reflections in the abyssal plain areas was interpreted as evidence of very thin sediment. Seismic-reflection records on the abyssal plains have been satisfactorily reinterpreted as multiple reflections from the thinly bedded sands and pelagic sediments underlying the abyssal plain. As examples, typical records from the Bermuda Rise and from the adjacent Sohm Abyssal Plain are illustrated in Fig. 29.

In the first seismic-reflection exploration of the deep sea, measurements were spaced every 2 to 10 miles, corresponding to a shot interval of 15 min to 1 hr. Reflecting horizons which would not correlate at this interval were considered either not significant or too complex to be investigated on a regional reconnaissance basis.

When in 1961 and 1962 a shot spacing of 2–3 min (roughly one shot each one-half mile) was first used this previous view was completely repudiated. Sub-bottom horizons completely uncorrelatable in earlier studies could be followed through complex sub-bottom topography. The contrast between oceanic rises and the abyssal plains mentioned above was substantiated. The new seismic-profiler records, however, add much new information on the total sediment thickness in the ocean floor. This work has been carried out under the

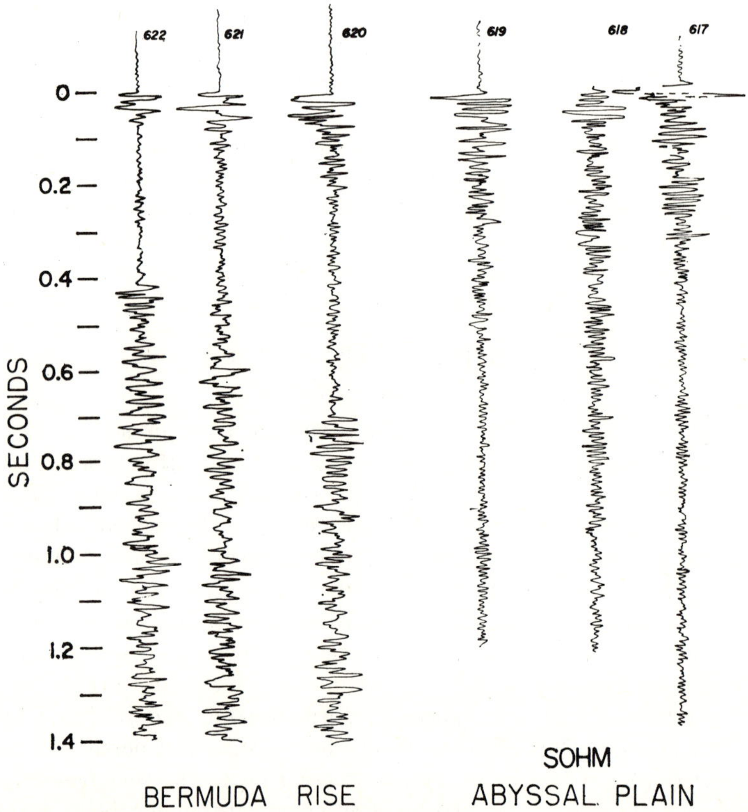

Fig. 29. Comparison of seismic-reflection records from the Bermuda Rise and the adjacent Sohm Abyssal Plain. Records from Bermuda Rise show strong discrete sub-bottom echoes, while those from the Sohm Abyssal Plain show a great many reflections which are extremely difficult to correlate.

supervision of Maurice Ewing and John Ewing of the Lamont Geological Observatory by scientists aboard the Research Vessel *Vema*.

The sediment thickness on the oceanic rises is surprisingly very similar to that beneath the abyssal plains, generally ranging from 1 to 2 km. The sediment thickness in the abyssal hills is generally too small to measure and the sediment thickness on the mid-oceanic ridge is generally extremely variable, being largely limited to intermontane basins and perhaps averaging one-tenth of the thickness found below the abyssal floor.

The sediment blanket beneath the abyssal plains covers a sub-bottom relief similar to the abyssal hills, strongly suggesting that the abyssal plain was created by the burial of pre-existing relief. The contrast in average sediment thickness between the abyssal hills and the mid-oceanic ridge on the one hand and the abyssal plain on the other is easy to understand in the above terms, but the similarity in sediment thickness between the abyssal plains and the abyssal rises presents a difficulty. The smoothing, it would seem, was accomplished in the course of the deposition of a small fraction of the total sediment thickness. Otherwise the sediment thickness beneath the plains should be measurably larger than beneath the oceanic rises if we assume that sedimentation began in both areas at the same time and has been of similar type in the two provinces. It is beyond the scope of the present paper to consider the validity of these assumptions. Suffice it to say that it may not be legitimate to assume that the

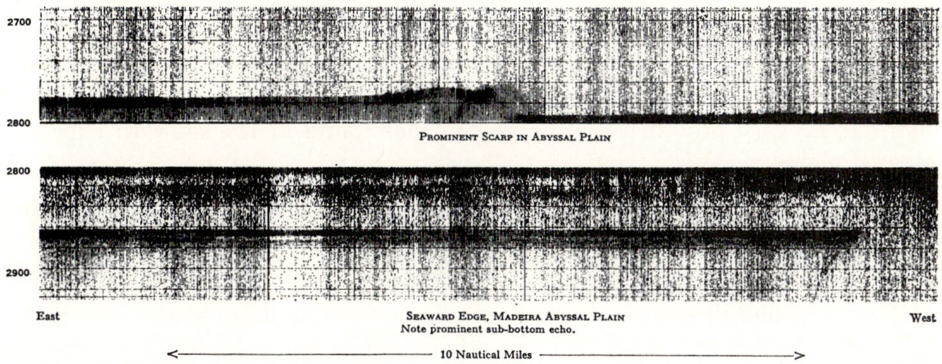

Fig. 30. Precision Depth Recorder records from the Madeira Abyssal Plain. Depth in fathoms. (After Heezen *et al.*, 1959.)

sedimentation began at the same time on the rises as on the plains. This problem is involved in the general problem of the geology of the ocean basins in relation to continental drift.

Sub-bottom reflections have been observed on Precision Depth Recorder records in the continental rise, on oceanic rises and many other environments of the deep-sea floor. However, the character of the sub-bottom reflections observed in the abyssal plains is generally quite different from those observed on the rises. Over most of the abyssal plains, sub-bottom reflections are either absent or, if present, persist as correlatable continuous horizons for relatively short distances. This is in sharp contrast to the oceanic rises, where sub-bottom horizons can be traced for dozens or hundreds of miles. However, a marked exception to this rule often occurs near the seaward edges of the abyssal plains, where sub-bottom reflections become very strong. Some of the deepest sub-bottom reflections traced by ordinary echo-sounders in deep water have been observed near the seaward edges of the Madeira and Nares Abyssal

Plains (Fig. 30). In such areas the sub-bottoms are prominent and reach to depths of 20 fm beneath the sea floor. The explanation of this typical distribution is similar to that offered for the more deeply penetrating seismic-reflection measurements; that is to say, in the portions of the plain close to the continental rise, turbidity currents have deposited thicker and coarser beds, which contrast markedly with the products of normal pelagic sedimentation. Therefore, many sharp reflecting horizons are present in these areas. However, due to the relatively limited area covered by the single turbidity-current deposit and the marked lateral variations in thickness in such a deposit, the resulting sub-bottom echos do not persist for great distances.

However, at the seaward edges of the abyssal plains, the sediment carried in by turbidity currents and that resulting from normal pelagic sedimentation are probably quite similar, both consisting of extremely fine clays. Sedimentation of clays by ponded turbidity currents may greatly increase the rate of sedimentation in these areas. In this region, certain notable events in pelagic sedimentation, such as vast ash falls or changes in the carbonate content of the sediments caused by relatively sudden changes in the circulation of the deep-sea water-masses, could form preservable major interfaces between sediments of different composition or different compaction and thus provide an explanation for the very prominent sub-bottom echos found at the seaward edges of abyssal plains.

B. Seismic Refraction

Many seismic-refraction measurements have been made in the "ocean-basin floor" of the western North Atlantic (Ewing and Ewing, 1959). These results can be divided into two categories, depending on whether the measurements were made (1) on the abyssal floor or (2) on an oceanic rise. Measurements on the abyssal floor of the western Atlantic revealed a simple pattern. Beneath 4–5 km of water lies 0.5 to 1 km of sediment and sedimentary rock with a compressional wave velocity of about 2 km/sec overlying a 4.5 km/sec layer 1 or 2 km thick. Beneath this lies 3–4 km of oceanic crustal rocks (approximately 6.5 km/sec) and the sub-M mantle rocks with a velocity of about 8.1 km/sec. This pattern has been observed by most workers in the abyssal floors of other oceans. In contrast, seismic-refraction measurements on the oceanic rises have revealed distinctly different and variable crustal layering. In the vicinity of Bermuda, Officer, Ewing and Wuenschel (1952) and Katz and Ewing (1956) failed to observe either the typical oceanic crustal velocities or the typical mantle velocities found in the abyssal floor. Instead, a thickened 4.5 km/sec layer overlay a 7.5 km/sec deepest layer. Although the oceanic rises have a distinctly different crustal structure from the abyssal plain no difference has been noted between observations made on the abyssal plain and those made in the adjacent abyssal hills. The sediment thickness in the abyssal hills is, however, markedly less than on the abyssal plain.

C. The Gravity Field and the Magnetic Field

On the basis of the relatively irregular gravity field shown by a series of closely spaced pendulum observations in the Bay of Biscay, Browne and Cooper (1950) concluded that there must be large topographic features buried beneath the Bay of Biscay Abyssal Plain. Subsequent studies of the Madeira Abyssal Plain with a continuous-recording Graf sea gravimeter gave further support to this conclusion (Worzel, 1959) (Fig. 31). In the region of the adjacent continental rise, the free-air gravity curve is by and large a perfect reflection of the topographic profile. However, on the abyssal plain the gravity field shows variations of the same amplitude as those observed over the adjacent

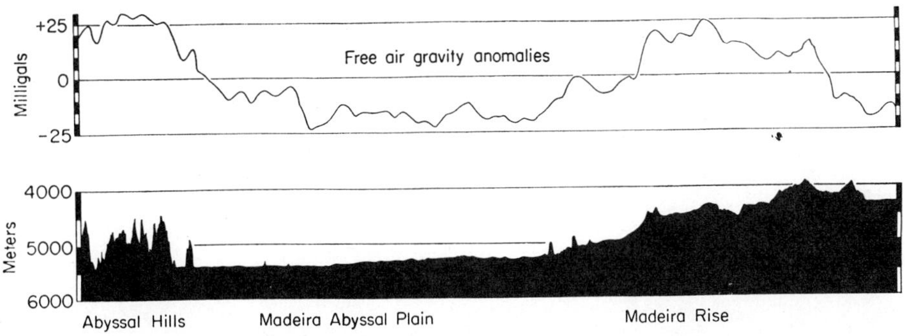

Fig. 31. Free-air gravity anomalies across the Madeira Abyssal Plain. (After Worzel, 1959.) Free-air gravity anomalies were observed with a Graf sea gravimeter from a surface ship. Note that the amplitude of the anomalies over the abyssal plain is not markedly different from that over the adjacent rises. This suggests that the basement structure is uniformly complex along the entire profile, and that the smoother topography of the abyssal plain is due to sediment blanketing the complex topography.

continental rise and abyssal hills. This clearly shows that the topography on the buried rock surface beneath the abyssal plain must be essentially similar to that of the continental rise.[1] The greater number of variable factors involved makes the interpretation of total magnetic intensity anomalies a much more difficult task than the interpretation of gravity anomalies. Profiles of total intensity of the Earth's magnetic field show the same irregularities of field over the abyssal plains as over the adjacent abyssal hills (Fig. 32).

D. Heat Flow

Heat flow measurements in the abyssal floor made by Bullard (1954), Bullard and Day (1961) and Von Herzen (1959) reveal values of approximately 1.2 μcal/cm²/sec (see Chapter 11 of this volume). Values of heat flow on the mid-oceanic ridges range to seven or eight times this value and heat flows measured in the

[1] Recent simultaneous observations of gravity and seismic-reflection profiles obtained during a round-the-world cruise have confirmed this correlation in several abyssal plains.

floor of oceanic trenches range from $\frac{1}{10}$ to $\frac{1}{2}$ this value. Sufficient measurements have not been made to determine whether there is any significant difference in heat flow between the abyssal plains and the abyssal hills, but except for effects of very recent turbidity currents, no such contrast would be expected.

8. Theories of the Origin of the Abyssal Plains

Any hypothesis or theory of the origin of abyssal plains must account for several critical characteristics. These are: (a) flatness of the plain; (b) continuity of gradient over large distances, and the exponential shape of the plain; (c) the terrigenous sediments and shallow-water detritus found on the

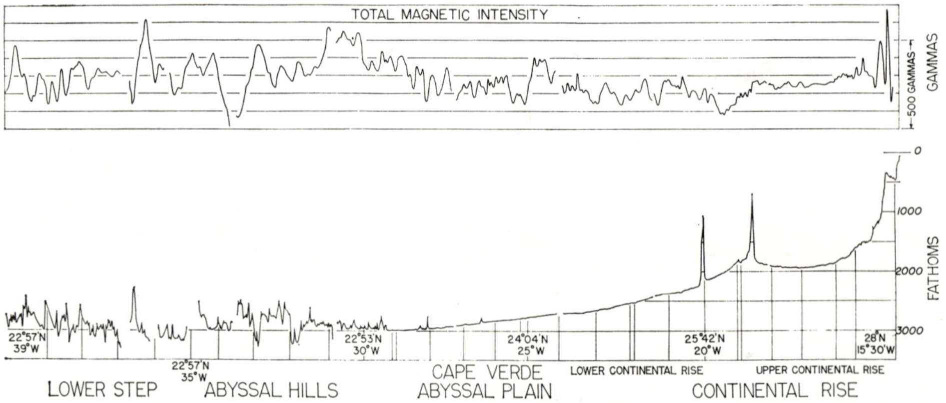

Fig. 32. Total magnetic intensity variations over the Madeira Abyssal Plain. Data obtained with a fluxgate magnetometer from the Research Vessel *Vema*. Note that the amplitude of the magnetic anomalies over the abyssal plain is not significantly different from the amplitude over the adjacent Mid-Atlantic Ridge and Madeira Rise. This apparently indicates that the structure lying beneath the abyssal plain is essentially similar to that lying beneath the adjacent features and that the only basic difference between these adjacent provinces is that the abyssal plain has been smoothed by turbidity-current sedimentation.

plain. Since the discovery and beginning of systematic investigation of abyssal plains in 1947, several explanations of their origin have been offered. Each of these is discussed in the following paragraphs, with special reference to the facts upon which it is based.

A. Atectonic Areas

The suggestion that abyssal plains represent atectonic areas in which the original crust of the Earth has been unaffected by diastrophism since the origin of ocean basins was made by Tolstoy and Ewing (1949). This hypothesis was offered in a tentative fashion largely on the basis of two facts: (1) "the plain was flat and nearly level"; (2) "the seismic reflections from the plain were of a type which was then interpreted as evidence of thin sediment". Thus, since the sediments were assumed to be very thin, the smoothing of the topography by the shifting of sediment seemed impossible, and even now no one would seriously consider the possibility that deep-ocean currents could effectively erode the crystalline rock to plain flatness. Thus, the theory was based solely on the lack of a leveling process, it being assumed that the smoothness of the topography was simply indicative of the lack of tectonic activity. The early seismic-reflection records from the abyssal plains were satisfactorily reinterpreted (Ericson, Ewing and Heezen, 1952) as reflections from a thinly bedded sequence in which alternate beds differ widely in lithology. Recent seismic-profiler records reveal abyssal-hill type topography buried beneath $\frac{1}{2}$–2 km of sediment. The fact remains that diastrophism has not been nearly as active in the abyssal floor as in the ridges and rises; for a mere $\frac{1}{2}$–2 km of sediment (Ewing, Sutton and Officer, 1954) has been sufficient to bury all minor topographic features in the abyssal plains. It is clear that a similar small thickness of sediment would be adequate to smooth many of the abyssal-hill areas. Tectonic activity in abyssal-hill areas has probably been no more intense than in the abyssal plains.

B. Mylonitic Plains

In direct contrast to the atectonic hypothesis, the suggestion has been made that abyssal plains represent areas of high and complex tectonic activity. According to this hypothesis, abyssal plains represent the nearly level surface of a great bed of mylonite, the rocks having been so completely sheared that nothing remains but a veritable sea of mylonite. This has been suggested apparently through a misunderstanding of the characteristics of abyssal plain deep-sea sands. Some of the first sands discovered in the deep sea were obtained from the Romanche Trench, an exceptionally deep and precipitous fracture zone which offsets the mid-oceanic ridge and rift in the equatorial Atlantic. These cores, obtained from depths of 3000–4000 fm, contained beds of sand. The sand grains were composed of minerals characteristic of basic and ultra-basic rocks (Mellis, 1958). The characteristics of the deep-sea sands occurring on abyssal plains contrast sharply with the sands of the Romanche Trench, and the only thing the two types of sand have in common is that they lie in a great depth of water. The sands of the Romanche Trench are mylonitic in character or, perhaps, are weathering products from the erosion of the Mid-Atlantic Ridge. However, they do not lie in an abyssal plain and are so vastly different from abyssal-plain sands as to have no real bearing on the origin of the normal abyssal-plain deep-sea sand.

Pettersson (1954), in an attempt to explain the occurrence on the equatorial Atlantic abyssal plains of graded deep-sea sands that contained "twigs, nuts and the bark of a dicotyledonous tree", offered the following explanation: "In the event that a large island harboring vegetation and with a fairly extensive shelf covered the Mid-Atlantic Ridge northwest of St. Paul's Rocks and became submerged during a catastrophe of seismic volcanic character a few hundred thousand years ago, material like that found in the *Albatross* cores might have become distributed over the adjacent sea bottom."

C. *Subaerial Erosion Surfaces*

The subaerial erosion hypothesis for abyssal plains was invoked to explain apparently widespread beaches and terraces found on the continental rise, continental slope, the Mid-Atlantic Ridge and on seamounts (Tolstoy, 1951). The abyssal plains were taken as the lowest of a great series of eustatic terraces. Evidence now available against subaerial erosion at great depths below present sea-level is very conclusive. One of the important arguments in favor of the subaerial erosion hypothesis was the existence of striking streamlike characteristics in the submarine canyons which reached the bottom of the ocean basins in 2500 to 3000 fm, 200–600 nautical miles from shore. The sediments in these canyons, however, demanded a submarine origin for the canyons. It was found that the canyon floors are covered by gravel and sand containing a cold shallow-water (glacial) fauna: Tertiary sediments crop out on the canyon walls. On the inter-canyon portions of the continental rise, normal, Recent and Wisconsin fine sediment (blue mud) was found. These findings give clear evidence that during the last glacial stage, some process of erosion and transportation was active in the canyons but not on intervening areas of the continental rise. This evidence clearly demands a submarine erosion process. Considering the great difficulties encountered by the subaerial erosion hypothesis even in connection with submarine canyons, the subaerial erosion of abyssal plains becomes completely impossible.

D. *Lava Plains*

The suggestion was made by Pettersson in 1953 that the abyssal plains were formed by vast outpourings of lava. This interpretation was based on two facts: (1) the existence of nearly level plains, and (2) the fact that the coring tube came up bent and empty after encounter with the abyssal plains. Damage to the coring tube was interpreted as being the result of an encounter with solid rock. The only plausible way to explain solid rock on the flat floor of the ocean basin was to appeal to lava flows. Pettersson believed that the apparent high thermal gradients in the deep-sea sediments might support the idea that widespread lava flows in the deep-sea basins were not uncommon. However, heat-flow values observed in the oceans are generally similar to those of non-volcanic regions of the continents. According to the experience of Heezen, Ewing and Ericson (1955), coring tubes are often bent and empty after encounters with beds of sand or gravel. Three serious difficulties of the lava plain hypothesis are: (1) the high fluidity of lava required to produce slopes as gentle as the typical 1:3000; (2) the high speed of flow required for lava to retain its fluidity while traversing distances of 500–1000 miles, particularly if required to start in abyssal depths instead of high on the continental slope; (3) in every plain where the slope has been determined, the slope is downward from an area adjacent to an obvious source of abundant sediments rather than from a source of vulcanism.

E. Areas of Long-Continued Sedimentation

Koczy (1954) suggests that the abyssal plains are merely "areas in which the sedimentation process has been in action for a long geological period". He apparently means to imply that the small-scale gravity movements—compaction, folding, slumping, etc.—would eventually produce a flat surface if given enough time. For sedimentation of the particle-by-particle type alone should cause no significant alterations of the topography, leaving the upper surface of the sediments parallel to the rock surfaces below. The theory of long-continued sedimentation would not explain the gradients or the type of sediments found on the plain. It is difficult to understand how small-scale gravity movements alone could produce such a flat plain. The study of sub-bottom reflections from known horizons on Precision Depth Recorder records has shown that even in very deep water the sediment is eroded from the crests of very small irregularities, even those a few hundred feet high, and that this erosion causes beds to thin over the rises and thicken in depressions. Thus, it is entirely conceivable that given a long period of time and a complete cessation of diastrophism, sediments slowly redistributed by currents or gravity dislocations could produce a smooth surface. But it seems unlikely that such a process could produce the exponential shape leading away from obvious sources of sediment, the continuous down-slope gradients, the shallow-water detritus and the other characteristics mentioned above.

F. Uniquely Fine Sediments Distributed by Bottom Currents

Browne and Cooper (1950) recorded a rough gravitational field over an abyssal plain west of the Bay of Biscay and inferred the existence of large topographic features in the buried rock surface. They concluded that on the abyssal plains "detrital mud is so finely divided that it can be carried in suspension for a very long time and gradually spread by bottom currents so as to fill in the hollows in the original topography". Browne (1954) abandons this hypothesis, for he states, "turbidity currents provide a natural explanation of how the ooze came to be contained in the trough of basement rocks and hence almost obliterated the surface indications of the underlying topography". The sands and silt occurring on all abyssal plains studied to date contradict the hypothesis that the flatness was produced by extremely fine sediments through the action of deep-ocean currents. However, as we stated in the previous paragraph, the action of deep-ocean currents can be important in producing smooth topography. Yet it seems improbable that this mechanism could have produced the characteristic shape of the abyssal plains and, of course, bottom currents would not be expected to produce beds of sand or silt exclusively of continental origin. One would expect that if a bottom current produced sand by winnowing, shells of pelagic organisms would play a large part. However, pelagic components are less abundant in turbidite deep-sea sands than in pelagic oozes.

G. Burial of Original Topography by Turbidity Currents

The hypothesis that abyssal plains were formed by deposits of turbidity currents, which have gradually covered the original topography, was made by Heezen, Ewing and Ericson (1951, 1955). This hypothesis was based on (1) the existence of flat, nearly level surfaces sloping away from areas adjacent to sediment sources; (2) the occurrence of graded sand layers, shallow-water benthonic Foraminifera, gray clays and silt layers in cores from the abyssal plains, all indications of turbidity-current transportation; (3) the partially buried appearance of seamounts protruding from the plain. Heezen and Ewing (1952) interpreted the submarine cable breaks which occurred for 13 h following the 1929 Grand Banks earthquake as the result of turbidity currents running down the continental rise and depositing a layer of graded sand on the abyssal plain to the south. In 1952 the prediction of an uppermost layer of graded sediment was confirmed by the results of five coring locations on the northern part of the Sohm Abyssal Plain south of Cabot Strait. Locher (1954), from a study of five cores taken from the abyssal plain of the western equatorial Atlantic, concluded that the deep-sea sands and shallow-water Foraminifera (Phleger, 1954) contained in these cores were transported from the South American continent to the plains by turbidity currents. He further suggested that the flatness of the plains was the result of this process.

Turbidity currents not only account for the existence of abyssal plains but provide explanations for many of the characteristic features found on the continental rise. The submarine canyons, deep-sea channels, deep-sea fans and great abyssal cones of the continental margin can best be explained through turbidity-current transportation, erosion and redeposition of sediment. Areas of the sea floor which turbidity currents can reach are in general smooth. Areas which turbidity currents cannot reach, because basins or trenches form sediment traps between the deep-sea floor and the continents, are areas of rough topography. The smooth topography of the continental rise and the abyssal plains does not necessarily slope away from the continent. In some areas the slope is parallel to the continental margin in places where the outward flow of turbidity currents is obstructed by mountains which act as dams. The basins with irregular bottoms are common in places which are shielded from turbidity currents by ridges or troughs. Similar basins have not been found in places which are not shielded from turbidity currents originating on the continental margin.

If one considers the highly hypothetical ocean basin which is just beginning to receive turbidity-current sediments, one might envisage a rough sea floor and a marginal trench bounding the continental slope. Turbidity currents would first fill up the trench and the rate of this filling would depend not only on the amount delivered to the submarine canyons and hence the frequency of turbidity currents, but it would also depend on the rate of subsidence of the trench. However, given a sufficiently long time, the trench would eventually be filled and the turbidity-current sediments would overflow the outer wall of the trench and flow out on the deep-sea floor. As long as slopes exceeded 1:1000, the growing smoothed area would resemble the surface of the continental rise with many deep-sea channels, submarine canyon systems, abyssal cones and deep-sea fans. However, as the basin floor begins to fill and the wedge of sediment building out from the shore reaches the greatest depths of the

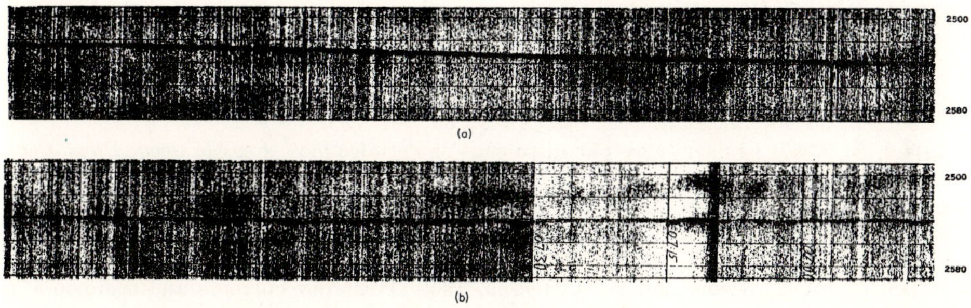

Fig. 33. Two Precision Depth Recorder records across the Biscay Abyssal Plain. (After Heezen *et al.*, 1959.)

ocean basin, a distinct change in the topography will begin to take place. As turbidity currents begin to build up against the seaward side of the oceanic basin, the depositional topography will change character from that resembling the topography of a continental rise to that resembling the topography of the modern abyssal plains. Gradually, as more sediment is supplied, the abyssal plain will widen and fill. If, after some time, tectonic warping uplifts the edge of the plain or a decrease in turbidity-current activity or a change in the type of turbidity-current sediment takes place, certain portions of the plain will become relic and no longer be receiving turbidity-current sediment.

These relic plains may be relic because they have been uplifted. Some will have been cut off from land by the subsidence of a new trench, as in the case of the northern "Aleutian Abyssal Plain". There are areas of the sea floor of very gentle relief which are distinctly elevated above the abyssal plains and no longer connected to them. Such an area lies in the center of the Argentine Basin. The very smooth topography in this area of nearly abyssal-plain flatness may mean that this area was at one time an abyssal plain which has later been uplifted and is now cut off from a supply of turbidity currents. The sediment (nearly 2 km) lying beneath this rise may have been partially contributed by turbidity currents before the elevation of the rise.

References

Bates, C. C., 1953. Rational theory of delta formation. *Bull. Amer. Assoc. Petrol. Geol.*, **37**, 2119–2162.

Buffington, E. C., 1952. Submarine natural levees. *J. Geol.*, **60**, 473–479.

Bullard, E. C., 1954. The flow of heat through the floor of the Atlantic Ocean. *Proc. Roy. Soc. London*, **A222**, 476–499.

Bullard, E. C. and A. Day, 1961. The flow of heat through the floor of the Atlantic Ocean. *Geophys. J.*, **4**, 282–292.

Browne, B. C., 1954. Gravity measurements and oceanic structure. *Proc. Roy. Soc. London*, **A222**, 398–400.

Browne, B. C. and R. I. B. Cooper, 1950. The British submarine gravity surveys of 1938 and 1946. *Phil. Trans. Roy. Soc. London*, **A242**, 243–310.

Dietz, R. S., 1953. Possible deep-sea turbidity current channels in the Indian Ocean. *Bull. Geol. Soc. Amer.*, **64**, 375–378.

Dietz, R. S., 1958. The channels on the deep sea floor. *New Scientist*, **4**, 946–949.

Dietz, R. S. and G. Shumway, 1961. Arctic Basin geomorphology. *Bull. Geol. Soc. Amer.*, **72**, 1319–1330.

Dill, R. F., R. S. Dietz and H. Stewart, 1954. Deep-sea channels and delta of the Monterey submarine canyon. *Bull. Geol. Soc. Amer.*, **65**, 191–195.

Dussart, B., 1957. Les sondages par ultra-sons et l'études lacs. *Rev. de Geog. Phys. Geol. Dynam.*, **1**, 99–102.

Emery, K. O., 1960. Basin plains and aprons off Southern California. *J. Geol.*, **68**, 464–479.

Ericson, D. B., M. Ewing and B. C. Heezen, 1951. Deep-sea sands and submarine canyons. *Bull. Geol. Soc. Amer.*, **62**, 961–965.

Ericson, D. B., M. Ewing and B. C. Heezen, 1952. Turbidity currents and sediments in the North Atlantic. *Bull. Amer. Assoc. Petrol. Geol.*, **36**, 489–512.

Ewing, J., J. Antoine and M. Ewing, 1960. Geophysical measurements in the Western Caribbean Sea and in the Gulf of Mexico. *J. Geophys. Res.*, **65**, 4087–4126.

Ewing, J. I. and W. M. Ewing, 1959. Seismic refraction measurements in the Atlantic Ocean basins, Mediterranean Sea, on the Mid-Atlantic Ridge and in the Norwegian Sea. *Bull. Geol. Soc. Amer.*, **70**, 291–318.

Ewing, J. I., J. Lamar Worzel and M. Ewing, 1962. Sediments and oceanic structural history of the Gulf of Mexico. *J. Geophys. Res.*, **67**, 2509–2527.

Ewing, M. and J. Ewing, 1962. Rate of salt-dome growth. *Bull. Amer. Assoc. Petrol. Geol.*, **46**, 708–709.

Ewing, M., D. B. Ericson and B. C. Heezen, 1958. Sediments and topography of the Gulf of Mexico (pp. 995–1058). In Weeks, L. (Ed.), *Habitat of Oil*. Amer. Assoc. Petrol. Geol., Tulsa, 1384 pp.

Ewing, M., B. C. Heezen, D. B. Ericson, J. Northrop and J. Dorman, 1953. Exploration of the northwest Atlantic mid-ocean canyon. *Bull. Geol. Soc. Amer.*, **64**, 865–868.

Ewing, M. and B. C. Heezen, 1955. Puerto Rico Trench topographic and geophysical data. *Geol. Soc. Amer. Spec. Paper* 62, 255–268.

Ewing, M., G. H. Sutton and C. B. Officer, 1954. Seismic refraction measurements in the Atlantic Ocean Basin, VI: Typical deep stations. *Bull. Seism. Soc. Amer.*, **44**, 21–38.

Fisher, R. L., 1954. On the sounding of trenches. *Deep-Sea Res.*, **1**, 48–58.

Gaskell, T. F. and W. Ashton, 1954. H.M.S. *Challenger*'s World Voyage 1950–1952, Part 2. *Intern. Hydrog. Rev.*, **31**, 101–111.

Gorsline, D. S. and K. O. Emery, 1959. Turbidity current deposits in San Pedro and Santa Monica basins off Southern California. *Bull. Geol. Soc. Amer.*, **70**, 279–290.

Heezen, B. C., 1955. Turbidity currents from the Magdalena River, Colombia (Abstract). *Bull. Geol. Soc. Amer.*, **66**, 12, Pt. 2, 1572.

Heezen, B. C., 1956. Origin of submarine canyons. *Sci. Amer.*, **195**, 36–41.

Heezen, B. C., 1956a. Corrientes de turbidez del Rio Magdalena. *Bol. Soc. Geog. Colombia*, **51** and **52**, 135–143.

Heezen, B. C., 1957. 1908 Messina Earthquake, Tsunami and Turbidity Current (Abstract). *Bull. Geol. Soc. Amer.*, **68**, 12, Part 2, 1743.

Heezen, B. C., 1959. Dynamic processes of abyssal sedimentation: erosion, transportation and redeposition on the deep-sea floor. *Geophys. J. Roy. Astr. Soc.*, **2**, 142–163.

Heezen, B. C., 1960. Some problems of Caribbean submarine geology. *Trans. 2nd Caribbean Geol. Conf.*, Puerto Rico, 12–16.

Heezen, B. C., Roberta Coughlin and W. C. Beckman, 1960. Equatorial Atlantic Mid-Ocean Canyon. *Bull. Geol. Soc. Amer.*, **71**, no. 12, 1886.

Heezen, B. C., D. B. Ericson and M. Ewing, 1954. Further evidence for a turbidity current following the 1929 Grand Banks earthquake. *Deep-Sea Res.*, **1**, 193–202.

Heezen, B. C. and M. Ewing, 1952. Turbidity currents and submarine slumps, and the 1929 Grand Banks earthquake. *Amer. J. Sci.*, **250**, 849–873.

Heezen, B. C. and M. Ewing, 1955. Orleansville earthquake and turbidity currents. *Bull. Amer. Assoc. Petrol. Geol.*, **39**, 2505–2514.

Heezen, B. C. and M. Ewing, 1961. The Mid-Oceanic Ridge and its extension through the Arctic Basin. *Geology of the Arctic* (Univ. of Toronto Press, Toronto), 622–642.

Heezen, B. C., M. Ewing and D. B. Ericson, 1951. Submarine topography in the North Atlantic. *Bull. Geol. Soc. Amer.*, **62**, 1407–1409.

Heezen, B. C., M. Ewing and D. B. Ericson, 1955. Reconnaissance survey of the abyssal plain south of Newfoundland. *Deep-Sea Res.*, **2**, 122–133.

Heezen, B. C., M. Ewing and R. J. Menzies, 1955. The influence of submarine turbidity currents on abyssal productivity. *Oikos*, **6**, 170–182.

Heezen, B. C., M. Ewing, R. J. Menzies and N. C. Granelli, 1957. Extending the limits of the Congo Submarine Canyon. *Bull. Geol. Soc. Amer.*, **68**, 1743–1744.

Heezen, B. C., R. J. Menzies, W. S. Broecker and M. Ewing, 1959. Date of the stagnation of the Cariaco Trench. *Intern. Oceanog. Cong. Preprints.* Amer. Assoc. Adv. Science, Washington, D.C., 99–100.

Heezen, B. C., M. Tharp and M. Ewing, 1959. The floors of the oceans, I: The North Atlantic. *Geol. Soc. Amer. Spec. Paper* 65, 126 pp.

Hersey, J. B. and M. Ewing, 1949. Seismic reflections from beneath the ocean floor. *Trans. Amer. Geophys. Un.*, **30**, 5–14.

Hurley, R. J., 1960. The geomorphology of abyssal plains in the northeast Pacific Ocean *Scripps Inst. Oceanog.*, Ref. 60–67, 105 pp. (unpublished manuscript).

Katz, S. and M. Ewing, 1956. Seismic-refraction measurements in the Atlantic Ocean, Part VII: Atlantic Ocean basin, west of Bermuda. *Bull. Geol. Soc. Amer.*, **67**, 475–510.

Koczy, F. F., 1954. A survey of deep-sea features taken during the Swedish Deep-Sea Expedition. *Deep-Sea Res.*, **1**, 176–184.

Laughton, A. S., 1959. The sea floor. *New Scientist*, **6**, 237–240.

Laughton, A. S., 1960. An interplain deep-sea channel system. *Deep-Sea Res.*, **7**, 75–88.

Locher, F. W., 1954. Ein Beitrag zum Problem der Tiefseesande im westlichen Teil des äquatorialen Atlantiks. *Heidelb. Beit. Min. u. Petrog.*, **4**, 135–150.

Luskin, B., B. C. Heezen, M. Ewing and M. Landisman, 1954. Precision measurement of ocean depth. *Deep-Sea Res.*, **1**, 131–140.

Mellis, O., 1958. Die Sedimentation in der Romanche-Tiefe (ein Beitrag zur Erklarung der Entstehung des Tiefseesands im Atlantischen Ozean). *Geol. Rundschau*, **47**, 1, 218–234.

Menard, H. W., 1955. Deep-sea channels, topography and sedimentation. *Bull. Amer. Assoc. Petrol. Geol.*, **39**, 236–255.

Menard, H. W., 1956. Archipelagic aprons. *Bull. Amer. Assoc. Petrol. Geol.*, **40**, 2195–2210.

Menard, H. W., 1958. Distribution et origine des zones plates abyssales. *Intern. Coll. C.N.R.S.* (*Nice*), **83**, 95–107.

Menard, H. W., 1960. Possible pre-Pleistocene deep-sea fans off Central California. *Bull. Geol. Soc. Amer.*, **71**, 1271–1278.

Menard, H. W. and R. S. Dietz, 1951. Submarine geology of the Gulf of Alaska. *Bull. Geol. Soc. Amer.*, **62**, 1263–1285.

Menard, H. W. and J. C. Ludwick, 1951. Applications of hydraulics to the study of marine turbidity currents. *Soc. Econ. Min. Pal., Spec. Pub.*, **2**, 1–13.

Officer, C. B., M. Ewing and P. C. Wuenschel, 1952. Seismic refraction measurements in the Atlantic Ocean. Part IV: Bermuda, Bermuda Rise, and Nares Basin. *Bull. Geol. Soc. Amer.*, **63**, 777–808.

Pettersson, H., 1954. *The Ocean Floor*. Yale Univ. Press, 179 pp.

Phleger, F. B., 1951. Displaced Foraminifera faunas. *Soc. Econ. Min. Pal., Spec. Pub.*, **2**, 66–75.

Phleger, F. B., 1954. North Atlantic Foraminifera. *Reps. Swed. Deep-Sea Exped.*, VII, 122 pp.

Shepard, F. P. and K. O. Emery, 1941. Submarine topography off the California coast canyons and tectonic interpretations. *Geol. Soc. Amer. Spec. Paper* 31, 171 pp.

Tolstoy, I., 1951. Submarine topography in the North Atlantic. *Bull. Geol. Soc. Amer.*, **62**, 441–450.

Tolstoy, I. and M. Ewing, 1949. North Atlantic hydrography and the Mid-Atlantic Ridges. *Bull. Geol. Soc. Amer.*, **60**, 1527–1540.

Udintsev, G. B., 1955. Topography of the Kurile–Kamchatka Trench. *Trudy Inst. Okeanol. Akad. Nauk S.S.S.R.*, **12**, 16–61. (In Russian.)

Udintsev, G. B., 1957. Relief of Okhotsk sea floor. *Trudy Inst. Okeanol. Akad. Nauk S.S.S.R.*, **22**, 3–76. (In Russian.)

Udintsev, G. B., J. G. Boitchenko and V. F. Kanaiev, 1959. The bottom relief of the Bering Sea. *Trudy Inst. Okeanol. Akad. Nauk S.S.S.R.*, **29**, 17–64. (In Russian.)

Von Herzen, R., 1959. Heat-flow values from the south-eastern Pacific. *Nature*, **183**, 882–883.

Worzel, J. L., 1959. Continuous gravity measurements on a surface ship with a Graf sea gravimeter. *J. Geophys. Res.*, **64**, 1299–1315.

Zenkovich, V. P., 1958. *Shores of Black and Asov Seas*. State Geographic Editions, Moscow, 371 pp. (In Russian.)

Zeigler, J., W. D. Athearn and H. Small, 1957. Profiles across the Peru–Chile Trench. *Deep-Sea Res.*, **4**, 238–249.

15. OCEANIC ISLANDS, SEAMOUNTS, GUYOTS AND ATOLLS

H. W. MENARD and H. S. LADD

1. Introduction

Oceanic islands, in general, are small in surface area, rise from the deep sea and lie far from continental shores. This definition excludes large continental islands such as Greenland, Iceland, the British Isles, Sicily, Madagascar, the islands of Indonesia, New Guinea and the "island continent" of New Zealand. It leaves in a borderline category such chains of islands as the Aleutians, Japan and other arcuate groups. Typical oceanic islands are volcanic. They may be isolated—at least as far as surface expression is concerned—or they may form part of a chain. Many of those lying in the warmer seas are encircled by reefs.

Seamounts, like oceanic islands, rise from the deep-sea floor and many of them lie far from continental shores. They, too, represent materials extruded from volcanic centers but not extruded in sufficient quantity to build their masses above sea-level. Again, like oceanic islands, they may be isolated or arranged in chains.

Guyots are flat-topped seamounts. They appear to have reached the surface of the sea at one stage in their development, at which time they were bevelled by wave action before sinking to their present levels below the sea. Atolls are reef-capped seamounts or guyots without exposed volcanic cores.

The distribution of islands and atolls has long been known but a better understanding of their distribution and their origin has been obtained during the past 10 years with the rising flood of echo-soundings. Many hundreds of thousands of kilometers of sound tracks have been recorded by oceanographic vessels operated by institutions in the United States, the Soviet Union, Sweden, Denmark and Great Britain.

In the case of guyots, the sounding data have been supplemented by information from dredging and bottom coring. The slopes of atolls have likewise been dredged and drill holes have been put down through their islets. Detailed studies of exposed reef surfaces have also been carried out.

Space does not permit full coverage of all recently acquired data but an attempt is made to summarize available information, particularly as regards its bearing on distribution, age and origin. In discussing the four major features, maps showing general distribution are given and typical examples are selected for brief description.

2. Oceanic Islands

Many of the oceanic islands that lie far from land are not as isolated as they appear on maps of the Earth's surface. Bathymetric charts show that they are high points on mid-oceanic ridges or rises that divide the Atlantic, Indian and South Pacific Oceans (Ewing and Heezen, 1956; Menard, 1959). In this category are the Azores, Ascension and St. Helena in the Atlantic, Prince

[*MS received June, 1960*]

Edward and St. Paul Islands in the Indian Ocean, and the Tuamotu Islands in the Pacific.

Oceanic islands of a second type occur in the Pacific. These are volcanic and are arranged in linear groups that are independent of the mid-ocean rise. The Hawaiian, Marquesan, Samoan and Society Islands are examples of this class. The submarine ridges from which they rise may be caused in part by faulting but the islands and their ridges may be explained solely as the result of piling up of lava flows between closely spaced vents. Other linear groups in the Pacific, such as the Tubuai (Austral) Islands, are made up of relatively isolated volcanoes rising directly from the deep-sea floor. The volcanoes in these chains appear to be too widely spaced or have not been productive enough to pile up lavas between them to build a ridge (Menard, 1959). Most oceanic islands lying in tropical latitudes are encircled by coral reefs. On some of these the volcanic foundation is partly or entirely covered by elevated shallow-water limestones.

Arcuate groups of islands are widely developed in the western Pacific. Two types are recognized: (1) the continental arcs, such as the Kamchatka–Kuril–Hokkaido Arc, the Philippine Arc and others lying between them, and (2) the "oceanic arcs" typified by the Marianas and Palau (Dietz, 1954). The first type is closely tied geologically to the continents and is not considered here. The "oceanic arcs", despite their name, are in a borderline category, because they lie outside the andesite line that encloses the Pacific Basin and they include rocks regarded as "continental". The arcs are believed to be tied to the tectogene, the postulated down-buckle in the Earth's crust into subcrustal material. Each arc is bordered by a deep trench on the convex side, the islands capping parts of a submarine ridge on the active or concave side. Saipan (Cloud *et al.*, 1956) and Guam (Tracey *et al.*, in press) are typical Mariana islands and both have been studied in detail. They are composed of volcanics (andesites, basalts) and are partially covered by limestones (mostly shallow water) that range in age from Eocene to Recent. Yap, an island forming part of a small arc between the Marianas and Palau, is underlain by schists and other rocks of unknown age overlain locally by Tertiary limestones (Cole, Todd and Johnson, 1960).

A typical example of an isolated volcanic island in the Atlantic is Ascension. It lies almost midway between Africa and South America, being 1500 km from the nearest point on the African coast. Ascension is roughly triangular in plan, measuring about 10 km on a side with an area of about 100 km^2 (Daly, 1925). It rises to a height of 858 m. It is composed entirely of volcanic rock (except for some diatomaceous material in one of the craters) and shows well preserved tuff cones, but there is no volcanic activity at the present time. The island tops a broad cone that rises from a depth of nearly 3000 m near the center of the Mid-Atlantic Ridge.

Ascension is only one of a number of isolated, reefless, volcanic islands that rise from the Atlantic Ridge. St. Helena, 1200 km to the southeast, is slightly larger but similar in essential features. Others include Tristan da Cunha and Gough Island.

In the Indian Ocean there are a number of isolated volcanic islands that are essentially similar to those of the Atlantic. Prince Edward and Marion Islands (known collectively as the Prince Edward Islands) lie southeast of the Cape of Good Hope. These islands rise from a deep-ocean ridge that is a continuation of the Mid-Atlantic Ridge around Africa into the Indian Ocean (Ewing and Heezen, 1956). Marion Island, the larger of the two, has an area of 70 mi^2 and an elevation of 4200 ft; Prince Edward has an area of 16 mi^2 and a height of 2370 ft. Like all volcanic islands in the cooler, reefless seas, these islands are strongly cliffed. They are composed of olivine basalt flows on which lie younger cones of scoria but neither island shows evidence of volcanic activity at the present time.

Two other examples of cliffed volcanic islands are Amsterdam and St. Paul, to the northeast of the Prince Edward Islands. These two also rise from a submarine ridge near the middle of the Indian Ocean (Menard, 1959, fig. 2).

In the Pacific, the Tuamotu (Paumotu) Islands rise from a mid-ocean ridge (Menard, 1959, fig. 2) but they are capped by limestone and have had a more eventful history. These low islands, in Dana's words, form "a vast island cemetery, where each atoll marks the site of a buried island" (Dana, 1849, p. 134).

The Hawaiian Islands are gently sloping, shield volcanoes spaced along the crest of a broad swell 2400 km in length, extending from northwest to southeast. The swell is nearly 800 km wide and is partly surrounded by a deep (Dietz and Menard, 1953), but these structures are independent of the mid-oceanic ridge. The islands have been built up largely by flows of olivine basalt. The largest island, carrying the only presently active volcanoes, lies at the southeast end of the chain. Early observers recognized that there appears to have been a progressive extinction of volcanic action from northwest to southeast (Dana, 1849). Today the northwestern end of the chain consists of coral reefs (Macdonald, 1956).

The age of the Hawaiian Islands is uncertain. Most geologists believe that they were built above the sea in Tertiary times and that the entire area has been deeply submerged (Stearns, 1945, 1946). Both conclusions have very recently been confirmed by dredging Miocene reef faunas on terraces deep on the flanks of Oahu. Cretaceous rocks have recently been dredged from the submerged Mid-Pacific Mountains less than 500 miles to the southwest (Hamilton, 1956) and there are biological data suggesting that the islands are older than is indicated by their meager fossil record (Ladd, 1960).

Other island groups in the central and west-central Pacific are spaced along lines that essentially parallel that of Hawaii. These include the Tubuai (Austral), Society, Cook, Samoa, Tuamotu and Line Islands (Macdonald, 1956). All of these, except the Tuamotus and the Line Islands, are primarily volcanic.

Most oceanic islands that lie within an area 30°N and 30°S are encircled by coral reefs. Water temperature in these low latitudes appears to be the chief control of reef growth. Reefs are best developed where the mean annual temperature is about 25°C; they are poorly developed or absent in areas where the temperature during the year falls below about 18°C for appreciable intervals (Wells, 1957). The importance of temperature in large part explains the absence of reefs around islands such as Ascension and St. Helena in the eastern Atlantic and the Galapagos Islands in the eastern Pacific. In the eastern parts of both oceans the upwelling of cold waters prevents reef development.

Many oceanic islands, for example Eua in the Tonga Group (Hoffmeister, 1932) and Oahu in Hawaii, are partly or completely encircled by *fringing* reefs

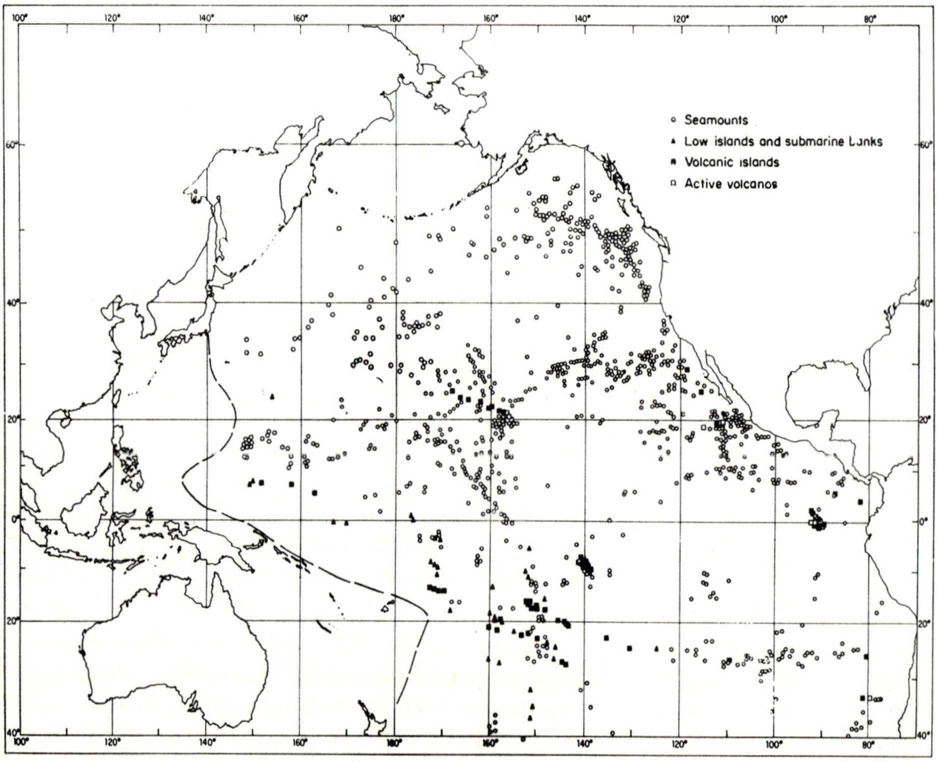

Fig. 1. Seamounts, low islands and submarine banks, volcanic islands and active volcanoes in the Pacific Basin. Broken line marks the structural boundary of the basin on the southwest (andesite line).

—platforms that may extend outward from the shore for distances of several hundred feet. Islands such as Ongea and many others in Fiji are protected by *barrier* reefs, structures comparable in size and constitution to fringing reefs but separated from the shore by a lagoon that may be several miles in width (Ladd and Hoffmeister, 1945). On some islands, such as Lakemba, Fiji, both fringing and barrier reefs are developed. Some reef-encircled islands are composed entirely of volcanic rock, others entirely of limestone; on some islands both types of rock are exposed, the limestones resting unconformably on the volcanic foundation.

3. Seamounts

A seamount may be defined as a more or less isolated elevation of the sea floor with a circular or elliptical plan, with at least 1 km (*ca.* 500 fm) of relief and comparatively steep slopes. This definition fits both flat-topped and pointed features. The former are a special class which will be discussed in more detail under the heading "guyots". Seamounts are indicated by soundings taken in the days of the *Challenger* Expedition, but it remained for Murray (1941) to make the first adequate description of a group of seamounts. Now an extensive literature exists and more than 1200 seamounts are known to be distributed in all the ocean basins (Figs. 1 and 2); roughly 200 have been surveyed

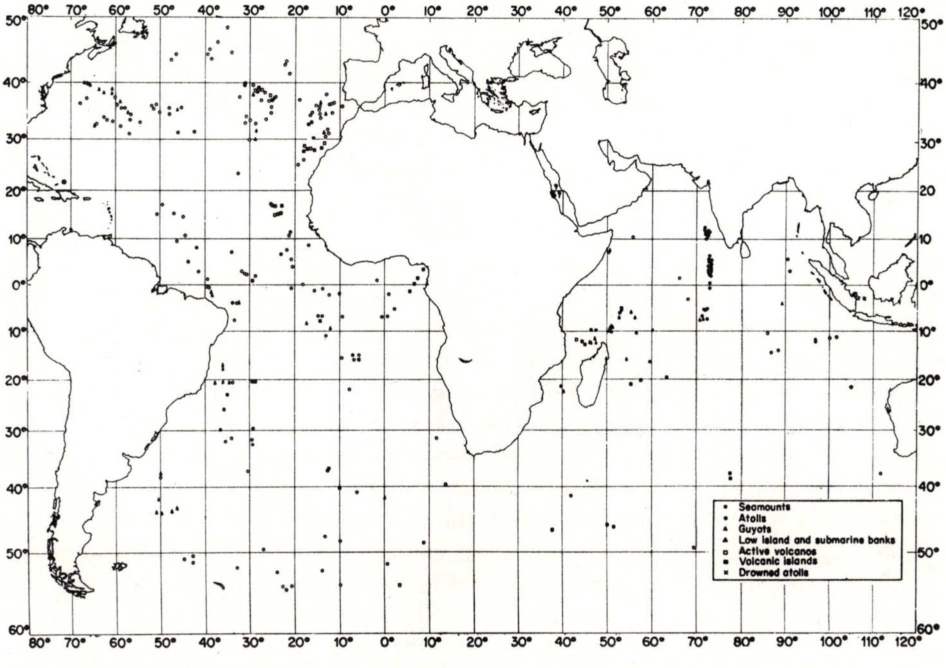

Fig. 2. Seamounts, atolls, volcanoes and related features in the world exclusive of the Pacific. Uncertainty of reported occurrences indicated by query.

in sufficient detail to establish the shape, and bedrock has been dredged from about 50. It seems clear that almost all, if not all, seamounts are submarine volcanoes because the bedrock is always basalt, and the shapes and slopes are characteristic of no other land-form. The only volcanic feature lacking in early surveys of seamounts was a crater at the top. This occasioned much speculation. However, the Precision Depth Recorder shows craters on many seamounts and it is highly probable that the earlier types of sounders simply concealed the crater in confused side echoes.

The total number of seamounts (including guyots) shown on the distribution charts is probably only a small fraction of the number that exists. The charts are particularly incomplete in the Atlantic and Indian Oceans where they are based on published soundings rather than on original and unpublished soundings as they are in the Pacific. Therefore, the charts should not be interpreted to indicate that seamounts definitely are more common in the Pacific than elsewhere. New seamounts are invariably discovered whenever an oceanographic ship crosses a previously unsounded region in the Pacific. This circumstance may be used as the basis for a prediction of the total number of seamounts that would be found if the whole Pacific were sounded along closely spaced lines (Menard, 1959). It is assumed that the seamounts of all sizes are cones with uniform slopes and that, in general, they are randomly distributed. No significant error is introduced by the former assumption, but the latter limits the size of the region that is studied because some topographic provinces have more seamounts than others. The Baja California Seamount province between Mexico and Hawaii was selected for study and all the seamounts were counted along 7400 km of sounding lines. About 10^3 seamounts with at least 1 km of relief appear to lie within the area of 1,400,000 km^2 or about one per 1400 km^2. Extrapolating to the whole Pacific Basin and considering various sea-floor topographic provinces, the total number would be roughly 10^4. A second method of estimating is to count all the seamounts in areas that are surveyed densely enough to find every seamount and to extrapolate to the remainder of the basin. The same number of 10^4 results. Extrapolating to the entire world ocean is more uncertain, but if the concentration per unit area is constant in any large region, the world total is about 2×10^4.

A typical large example is Gilbert Seamount in the Gulf of Alaska (Fig. 3). It consists of a roughly circular main cone with a diameter of 20 mi, an elevation of about 11,000 ft, and a smaller parasitic cone on one flank (Menard, 1955). The side slopes are slightly concave upward and reach a maximum of 22°. The maximum slopes are about the same on smaller seamounts but, in general, the slopes do not appear to be concave upward if the peaks are less than one mile high. Some seamounts are more elongate with a length as much as four times the width; in this characteristic, they resemble the shield volcanoes of the Hawaiian Islands, which tend to be elongate because they lie on linear rifts.

Most seamounts are clustered in lines or in elongate groups of roughly 10–100 and consequently, as groups, they closely resemble the volcanic archipelagoes of the Pacific Basin such as the Samoan and Hawaiian Islands. This is characteristic of the seamounts of the Pacific, but at least one prominent linear group, the Kelvin Seamounts (Heezen *et al.*, 1959), lies in the Atlantic. Most of the linear groups of seamounts of the northeastern Pacific have the same northwesterly trend as the linear island groups of the central and western Pacific. Likewise, the individual seamounts in a cluster are connected in various ways like archipelagoes. The Mid-Pacific Mountains (Hamilton, 1956) rise from a ridge with steep sides like the Tuamotu Archipelago. Other seamount clusters such as the outer ones in the Gulf of Alaska (Menard and Dietz, 1951) stand above low rises—resembling the Marquesas Islands in this respect. Some clusters of seamounts appear to rise from a relatively level, although irregular, sea floor and in this they are dissimilar from most archipelagoes. The difference probably is not significant with regard to origin, however; if the seamounts continued to grow, lava flows would eventually reach from the base of one volcano to the next and would build a connecting rise. In groups as well as individually, therefore, the seamounts resemble volcanic islands in all respects except size. The close similarity is further emphasized by the existence of intermediate types, namely, guyots, atolls and banks.

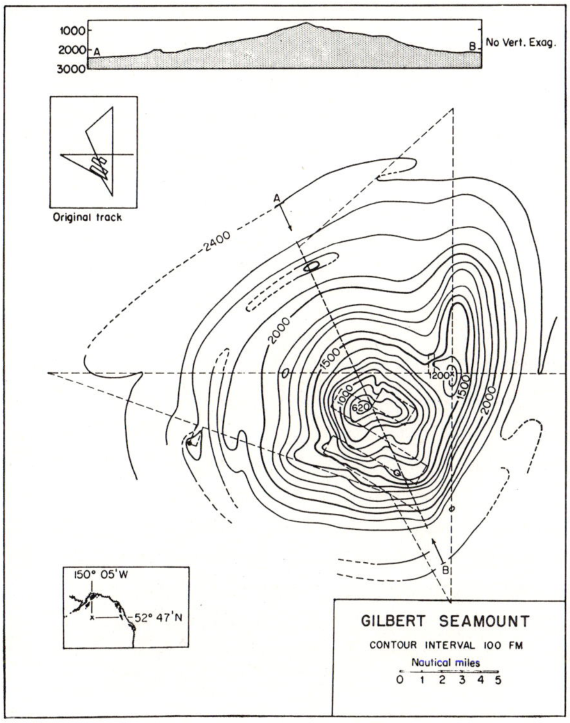

Fig. 3. Gilbert Seamount in the North Pacific.

4. Guyots

A guyot is a flat-topped seamount deeper than 100 fm. The discovery of these features was announced by Hess in 1946 and his pioneering efforts launched a continuing and very fruitful exploration of guyots. Hess personally sounded 20 guyots and found 140 probable guyots by examining unpublished soundings. The guyots described were limited to the northwestern Pacific, although Hess recognized that some of the seamounts described by Murray (1941) in the Gulf of Alaska were also flat-topped. From their topography alone, Hess concluded that the guyots were drowned ancient islands which had been truncated at sea-level. From the absence of coral capping he reasoned that they were Pre-Cambrian in age. The drowned island hypothesis has been confirmed repeatedly and can no longer be questioned. On the other hand, the supposed age definitely is incorrect. Confirmation of the origin of guyots developed quickly. In 1947, Carsola and Dietz (1952) dredged two guyots in the northeastern Pacific and found rounded basaltic cobbles which could only have been produced near sea-level. In 1949, the Mid-Pacific Expedition set out to explore guyots in the type locality discovered by Hess. There, in the Mid-Pacific Mountains, the flat tops were surveyed and rounded cobbles as well as an integrated reef fauna of Middle Cretaceous age were dredged from the guyots at a depth of about 1800 m (Hamilton, 1956). Very recently, Eocene reef debris has been found at a depth of about 1000 m on a guyot amid the atolls of the Tuamotu Archipelago, and similar debris of late Tertiary to Recent age has been dredged from guyots on the Nasca Ridge off Peru at a depth of about 400 m.

Guyots are not drowned atolls even though some are capped with reef debris. Profiles do not show the typical steep slopes of atolls but rather the more subdued slopes of submarine volcanoes (Fig. 9). In the Mid-Pacific Mountains, coarse rounded basaltic debris was dredged with coral from the tops of some guyots. At least in these places, the coral formed low banks which never developed into atolls. The reef-covered guyots of the Tuamotu Archipelago rise only a few hundred meters above a submarine plateau. Consequently, they cannot have a thick coral capping and any growth comparable to an atoll seems ruled out by gentle side slopes.

An intensive search has shown that the distribution of guyots is world-wide. The existence of guyots in the Gulf of Alaska (Fig. 4) was quickly confirmed by detailed surveys (Menard and Dietz, 1951). Additional guyots in the Emperor Seamounts in the northwestern Pacific were described by Dietz (1954), and new groups of guyots were recently discovered in the South Pacific. Description of guyots in the Atlantic has been hampered by classification of soundings but at least three groups exist (Heezen *et al.*, 1959). Two isolated examples show that there are guyots in the Indian Ocean (Wiseman and Hall, 1956) and it may be anticipated that more will be found. Guyots have not been reported in the Arctic Basin but the Lomonosov Ridge is truncated (R. S. Dietz, *in litt.*, 1959) and truncated seamounts would not be unexpected.

Despite the discovery of six new groups and many isolated guyots, the total number known differs insignificantly from the total of 160 reported by Hess (1946). A few of the probable guyots of Hess have proved to be seamounts with pointed tops; many others have not yet been confirmed by surveys. In contrast, all of the guyots reported since have been surveyed. Nevertheless, an increase of only 22 known guyots after 13 years of exploration suggests that the total number in the oceans probably does not exceed a few hundred and that at least half of them are concentrated in the central western Pacific.

Guyots range in depth from 200 m (by definition) to about 2500 m, but most are between 1000 and 2000 m. Even in small areas the peak depths normally are not concordant, suggesting that the guyots were islands at different times.

The Pratt–Welker group of guyots in the Gulf of Alaska, however, has relatively concordant peaks at a depth of 700–800 m; and the guyots among the Austral Islands are concordant at about 1500 m.

Individual guyots have the sizes and shapes that would be produced by planing off existing volcanic islands. The first random crossings of the "flat" tops indicated that the center was flat and that the margins sloped gently. This has not been confirmed by surveys. Instead the entire top slopes gently to a point in the center or a in places few pinnacles rise from near the center. The

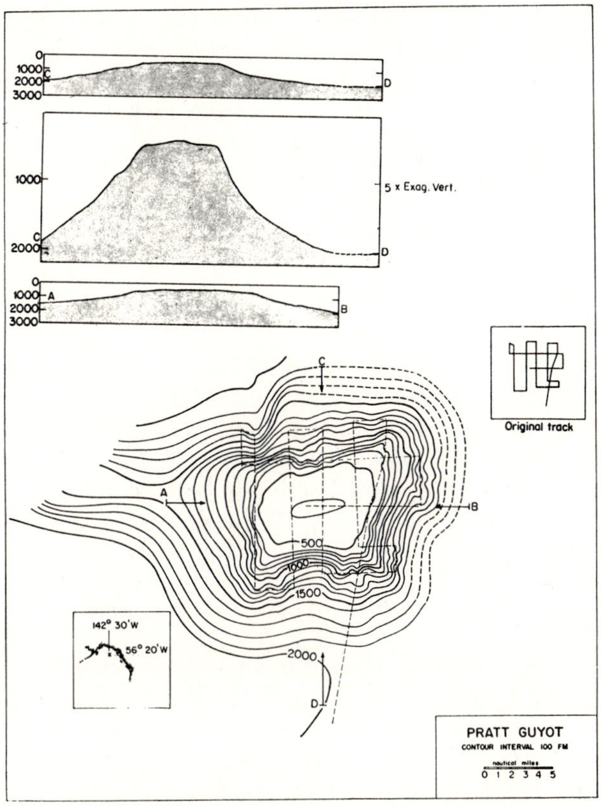

Fig. 4. Pratt Guyot in the North Pacific.

topography, in short, is exactly the same as that of an insular shelf except that the islands have been eroded away. Islands have terraces in their submarine slopes in some places, and random echogram crossings have suggested to various writers that similar terraces have been cut in the sides of guyots. Detailed surveys show that terraces that can be traced around a seamount or guyot are very rare at best. The broad sound cone of an echo-sounder normally causes a pinnacle on a steep slope to appear as a terrace. Consequently, many so-called "terraces" may be parasitic cones. In any event, any conclusions about changes in sea-level suggested by unsurveyed terraces appear unwarranted.

5. Atolls

An atoll is an annular reef, lying at or near the surface of the sea, which encloses a lagoon. Some atolls are found on the continental shelves but most of them are isolated structures that rise from the deep sea. Small rings, usually without islets, may be less than a mile in diameter, but in the western Pacific, where atolls attain their greatest development, many have a diameter of about 20 mi and bear numerous reef islets. Kwajalein in the Marshall Islands (the largest existing atoll) has an irregular outline and covers 840 mi^2.

The reefs of the atoll ring are flat, pavement-like areas, large parts of which may be exposed at times of low tide. They vary in width from narrow ribbons to broad bulging areas more than a mile across. Their continuity is interrupted, in most cases on the leeward side, by passes which may be as deep as the lagoon. The maximum depth of lagoons, in general, is proportional to their size. Atolls with a width of 20 mi have a maximum depth of about 300 ft. On the seaward side of many atolls a shallow terrace is developed, beyond which the slope is steep (35°); on the leeward side the seaward slope may be a vertical cliff for 200 ft or more; at greater depths, on all sides, the slopes flatten progressively until they become tangential with the floor of the ocean.

Most atolls lie in areas where unidirectional winds, such as the trades, prevail during most of the year and control water circulation. The reefs are complicated structures that are dynamically adjusted to their environment (Munk and Sargent, 1954). Detailed studies of atolls have been carried out at Funafuti in the Ellice Islands (Sollas and David, 1904) and at Bikini and nearby atolls in the Marshall Islands (Emery *et al.*, 1954). The generalized descriptions given below are based largely on work in these areas, with some additions from studies made elsewhere.

A. Reefs

On the windward side of an atoll the waves engendered by the prevailing wind bring a constant supply of food, nutrient salts and dissolved gases to the reef edge and the richest growth of reef-building organisms, particularly corals and algae, is found there. The seaward slope of the reef margin is cut by a series of regularly spaced grooves at right angles to the reef front which are separated by flattened spurs that appear to be developed largely by organic growth. Many of the grooves are continued through the reef margin as surge channels, ending on the reef flat as blow holes constricted by algal growth. The higher parts of the reef margin, particularly opposite islands, may be an algal ridge. This ridge and the channels that penetrate it form an effective baffle that robs the waves of destructive power and spreads the waters over the reef (Fig. 5).

The broad flats inside the seaward margin of the reef are called the reef pavement and in areas where the tide range is greater than average the pavement is widely exposed at low tide. Large parts of it consist of bare rock which may be veneered with fine sediment rich in smaller Foraminifera held together

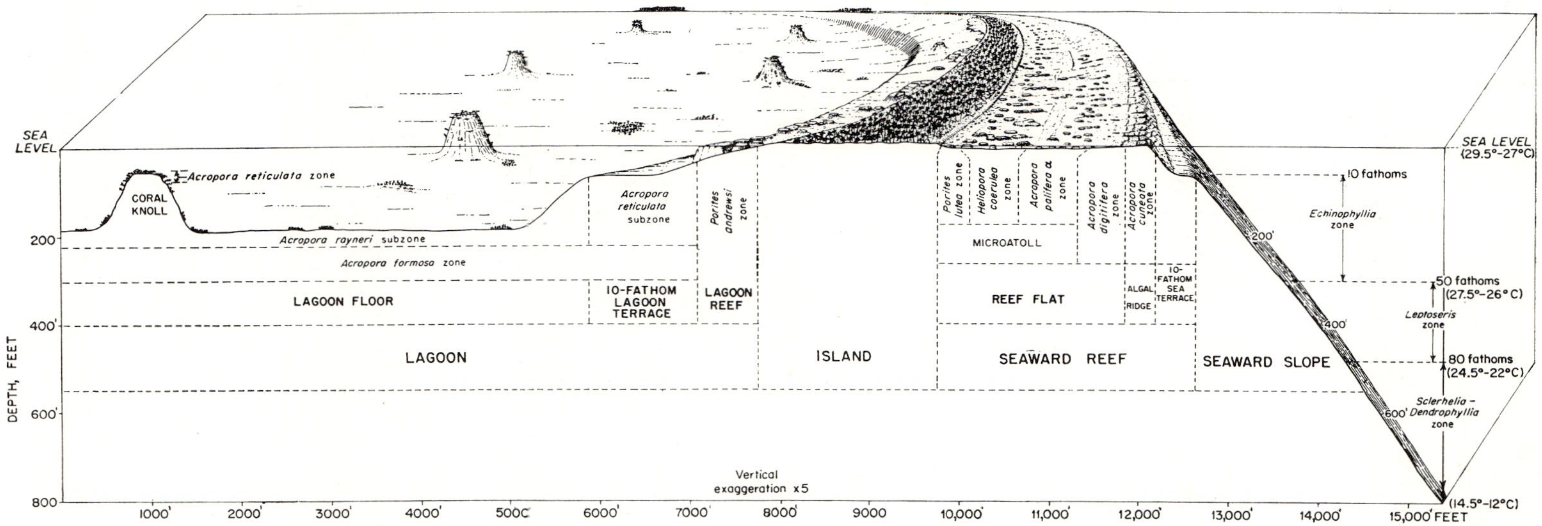

Fig. 5. Cross-section through the windward reef at Bikini Atoll in the Marshall Islands showing major zones, each characterized by distinctive coral growth. (After Wells, U.S. Geological Survey, 1954.)

by fibrous algae. The tests of the Foraminifera that live on the flat are eventually carried landward by waves to form the main constituent of the beaches that fringe all atoll islands. Many reefs show a definite zonation parallel to the reef front. Some of the zones are areas of rich coral growth. In the few places where the reef pavement has been drilled it is underlain by a solid plate of hard rock 10 ft or more in thickness, lying on unconsolidated reef sediments (Ladd and Schlanger, 1960).

The living organisms of the reef margin are damaged by wave attack and are the source of many of the sediments found upon the reef flat. Storm waves quarry large blocks from the reef edge and cast them upon the flat where they are later riddled by boring organisms and slowly broken up by wave action. The sand and silt formed by wave erosion are worked across the flat to the island beaches and may eventually be deposited in the lagoon.

On the leeward side of an atoll the reef edge lacks the algal ridge, grooves, buttresses and surge channels that characterize the windward reefs. Leeward reefs may be broader than those of the windward side but they are not distinctly zoned. Passes through the reef are more commonly developed to leeward than to windward. Most of these are shallow but in many atolls at least one pass is as deep as the deeper parts of the lagoon.

B. Islands

Most atolls bear a series of islands spaced at irregular intervals along the annular reef. The islands are formed on top of the reef platform. Most of them are small and all of them are low. They are composed of reef detritus piled up by currents, waves and winds. Except for layers of beach rock (consolidated conglomerate and sandstone) at intertidal levels, the material is unconsolidated. It consists of sand, gravel and rubble from pebble to boulder size, most of the coarser material being of coral origin. In the Marshall Islands about 15% of the island beaches show exposures of beach rock. In most areas the texture of the beach rock is similar to that of the material making up the existing beach and the strike and dip of the beds of rock are nearly or quite identical to those of the unconsolidated beach material, dipping away from shore at about 9°. Islands situated on reefs that are subjected periodically to heavy storms exhibit one or more ramparts built of blocks of reef rock and coral heads. These structures apparently can be built, moved, or destroyed by the waves of a single major storm. Inland from the beaches and ramparts, sand and gravel flats occur and large areas are covered with coconut palms, screw pines and thickets of bay cedar and other plants. Low sand dunes may occur inland from beaches.

C. Lagoons

Large lagoons with a fetch of 20 mi or more develop waves comparable to those of the open sea and lagoonal reefs on the down-wind side take on some of the characteristics of the seaward reefs on the up-wind side. The prevailing wind and waves bring a continuous flow of ocean water over the windward reefs to the lagoon. Much of the wind-driven surface water is returned by a slower and thicker bottom current, producing closed circulation. Von Arx measured these currents in the lagoons of Bikini and Rongelap where he also found evidence of a secondary circulation composed of two counter-rotating compartments (von Arx, 1954).

Some of the lagoons that have been accurately mapped by close soundings show a shallow terrace (Fig. 5) that separates the encircling reef from the deeper parts of the basin. Irregular depressions on the floors of such terraces may represent sinks that were formed by solution when sea-level was lowered during glacial times (Emery *et al.*, 1954).

Widely scattered soundings may suggest that the floors of lagoons are fairly flat but in lagoons that have been covered by dense soundings numerous isolated mounds of coral are revealed. These mounds were referred to long ago as coral knolls (Darwin, 1889). Some are small, measuring 10–15 ft across and 3–20 ft in height, but much larger ones are numerous and these may be several hundred feet to more than a mile in width and may rise to near low-tide level even from the deeper parts of the lagoon. The 18,000 soundings made in the 24-mi lagoon of Eniwetok Atoll in the Marshall Islands (Emery *et al.*, 1954) revealed more than 2000 coral knolls. Detailed maps of individual knolls show that the structures vary in outline as well as in size. Summits are broad, the sides having slopes of 45–60° with, in some cases, vertical slopes for as much as 10 fm of their depth (Fig. 5). Living corals cover up to 75% of the knoll crest. No coral knoll has ever been drilled but Nestéroff (1955) used T.N.T. to blow off the tops of several, finding concentric coral growth. There was no evidence to suggest that the knolls were residuals. In the Maldive Islands coral knolls shaped like miniature atolls are called "faros".

The coral knolls contribute sediment to the lagoon as do the marginal reefs. Additional sediment is derived from coral thickets on the lagoon floor and from large areas of the bottom that are covered by algal growth (*Halimeda*). Since very little sediment escapes from the lagoon its floor is slowly raised with the passage of time.

D. *Subsurface Geology*

Most atolls that rise from the deep sea have probably been developed on the tops of volcanoes that may or may not have been truncated by erosion prior to the initiation of reef growth. In the Pacific Basin (inside the circum-Pacific andesite line) the volcanic rock of the atoll foundation is probably all basalt. Rock of this sort has been dredged from the slopes of Bikini, Wotje and Ailuk Atolls in the Marshall Islands (Ladd and Schlanger, 1960) and has been recovered from a deep drill hole on Eniwetok (Fig. 6). Though the atolls of the Pacific Basin may all have basaltic foundations it is not safe to assume that they have all had similar histories of subsidence and reef growth. The drill holes of Bikini can be tied to those of Eniwetok and the volcanic foundations of these two atolls apparently subsided at similar rates. The same cannot be said, however, for Funafuti in the Ellice Islands some 1700 mi to the southeast. The known Quaternary section drilled there is almost twice as thick as that found in the Marshall Islands.

Practically nothing is known about the foundations of atolls lying outside the Pacific Basin. Deep drilling on Kita Daito Jima penetrated the Tertiary but failed to reach the bottom of the sedimentary section. No other atoll outside the basin has been drilled. Many of them, as, for example, the Argo Reefs

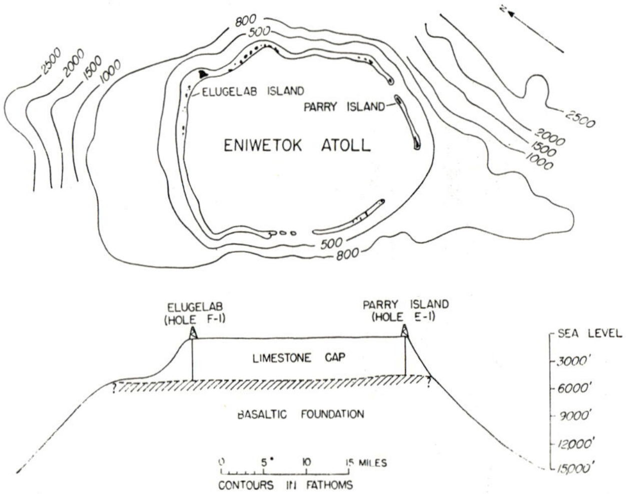

Fig. 6. Plan and structure section through Eniwetok Atoll in the Marshall Islands showing relations of limestone cap to volcanic foundation, as determined by drilling. (After Ladd and Schlanger, U.S. Geological Survey, 1955.)

and others in Fiji, lie in areas where there has been considerable elevation in post-Tertiary times. Some of the atolls are large and irregular. Even less is known about the foundations of atolls in the Indian Oceans.

E. Distribution

As shown in Figs. 2, 7 and 8, atolls are most abundant in the western Pacific, numerous in parts of the Indian Ocean and exceedingly rare elsewhere.

Structures comparable at least in general form to existing atolls have been described from rocks as old as Paleozoic (Stafford, 1959; Burnside, 1959). Many of these older structures are reservoirs of petroleum and are being intensively investigated.

F. Development and Age

Development of existing atolls began at least as far back as Eocene time. Most of those that rise from the deep ocean appear to be caps on ancient volcanoes that have subsided to permit the accumulation of shallow-water

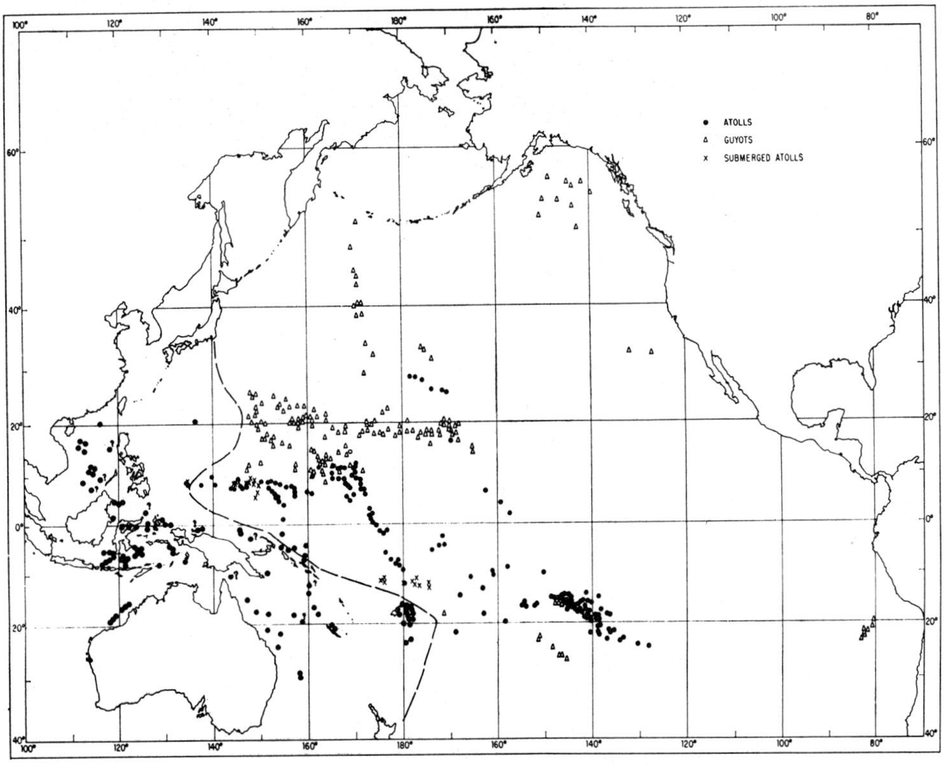

Fig. 7. Surface atolls, submerged atolls and guyots in the Pacific. Uncertainty of reported occurrences indicated by query. Andesite line shown by dashes. Atolls to west of line (after Cloud, 1958).

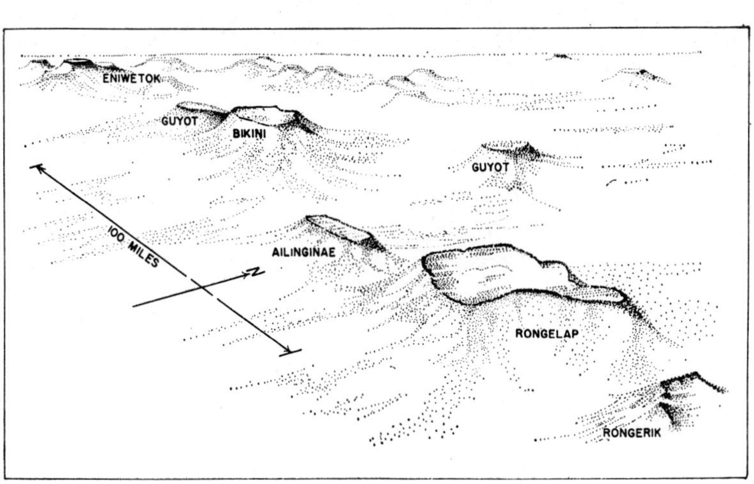

Fig. 8. Atolls and guyots in the Marshall Islands; perspective sketch as they might appear were the water removed. (After von Arx, U.S. Geological Survey, 1954.)

limestones nearly a mile in thickness. Upward growth appears to have been more important than lateral growth. Many atolls were developed on truncated volcanoes (guyots, see Fig. 9) and began their existence as atolls without

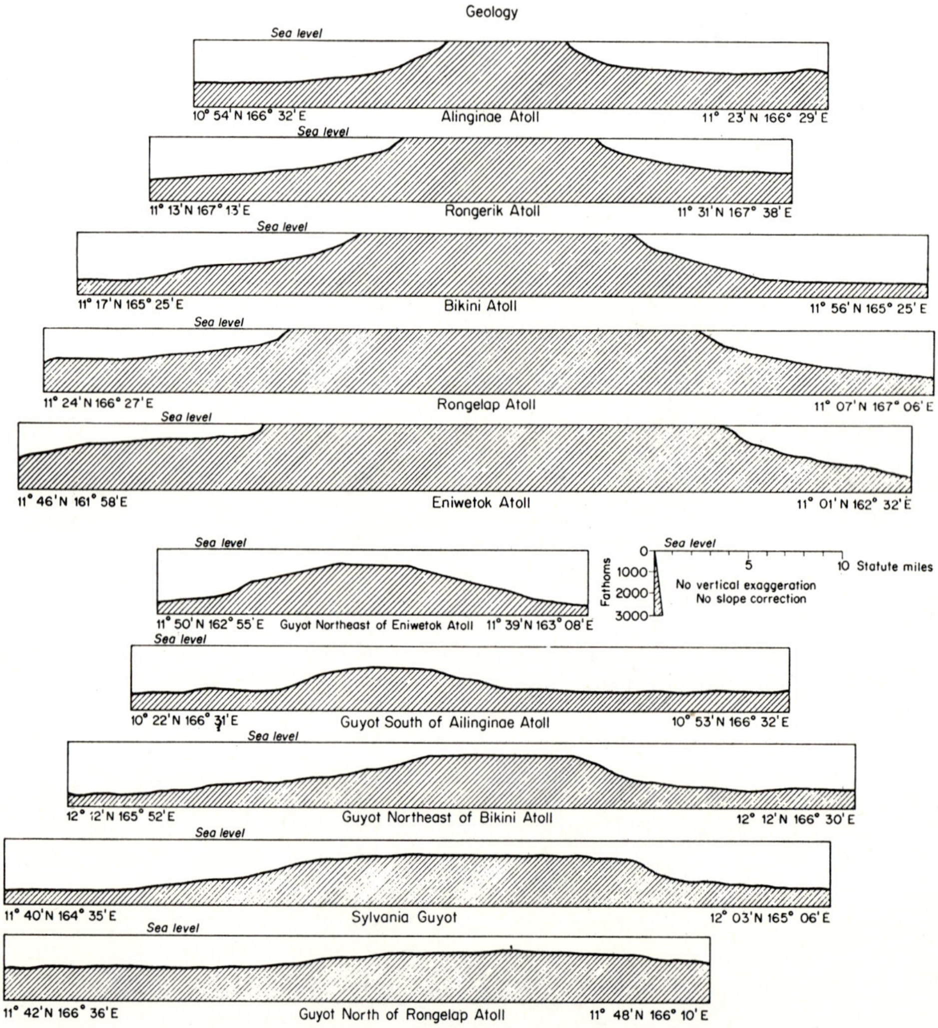

Fig. 9. Cross-sections of atolls and guyots; the peripheries of the guyots are rounded, those of the atolls comparatively sharp. (After Emery *et al.*, U.S. Geological Survey, 1954.)

passing through fringing reef and barrier reef stages. There is convincing evidence to indicate that their history has not been one of uniform and long-continued subsidence. Periodically the Pacific Basin atolls stood above the sea for appreciable intervals, some of them as high islands. During these emergent periods the tops of the atolls were exposed to atmospheric erosion. The aragonite of corals and other invertebrates was removed by solution or recrystallized to calcite. High island vegetation and land shells became established (Ladd, 1958; Schlanger, in press).

6. Submerged Atolls

Submerged atolls are shallow banks with a relatively continuous elevated rim and a deeper central area. The writers know of 22 in two groups in the Pacific Ocean and one isolated example in the Indian Ocean although many questionable occurrences suggest that they are much more common. One group of submerged atolls in the Pacific is interspersed with the atolls in the western Caroline Islands. Another group lies near the andesite line west of the Samoan Islands and was investigated during the *Capricorn* Expedition. Alexa Bank has a narrow elevated rim with a depth of about 20 m and a broad flat central "lagoon" at about 40–50 m except where a few pinnacles rise to near the rim depth. A ship-borne seismic-refraction station showed that the calcareous material exposed at the surface is more than 1 km thick (R. W. Raitt, *in litt.*, 1953) so that this is definitely a submerged atoll with a long history near sea-level. Diving scientists observed that the bank is covered with calcareous debris except for a few small patches of living lithothamnia and very sparse corals.

Why these shallow banks are not thriving atolls like those around them is unexplained.

7. Oceanic Volcanoes in Geological Time

Pre-Quaternary dated material is available from oceanic volcanoes at Bikini and Eniwetok Atolls in the Marshall Islands (Ladd *et al.*, 1953), the guyots of the Mid-Pacific Mountains (Hamilton, 1956), Erben Guyot of the north-eastern Pacific (Carsola and Dietz, 1952), the Tuamotu guyots and the Nasca Ridge guyots (W. Storrs Cole, *in litt.*, 1959), and in the Atlantic at Caryn Seamount, the Horseshoe seamount group, and the Atlantis–Great Meteor seamount group (Heezen *et al.*, 1959). Dredge hauls on all other seamount and guyot groups have obtained Quaternary organisms and undated basalt bed-rock. A geological history based on such a small sample requires the following assumptions:

1. The oldest fossils in a locality give the approximate date that atolls and guyots were volcanic islands.
2. Vulcanism in a group of oceanic volcanoes persists for a period that is geologically brief, so that the oldest date in a group is representative of a whole group.

The first "assumption" is a fact at Eniwetok atoll where the volcanic platform was drilled and at the Mid-Pacific Mountains where basalt and coral debris were dredged together from the top of a guyot. It seems a reasonable assumption for the Tuamotu and Nasca Ridge guyots because the dated material was deposited in shallow water and the side slopes do not indicate a thick reef capping. With regard to the other localities, the coating of pelagic material on a seamount is patchy and extremely thin and material of various ages may be readily available to the dredge or corer.

The second assumption seems reasonable considering the duration of vulcanism in a given province on land which is roughly 10–50 million years. There can be no doubt that vulcanism of oceanic-island groups persists or is intermittent during similar periods because of the mixture of deep guyots and islands or relatively shoal untruncated seamounts in some regions.

Sediment cores from ocean basins reveal widespread Cretaceous and Tertiary outcrops but nothing older. It is possible that older sediments have been buried by younger ones, however, and no general conclusions can be drawn from the cores alone. Atoll borings to basalt and seamount dredgings are more significant because the oldest sediments are available to the sampling device. The probability of sampling material of any given age, therefore, is proportional to the frequency of such material.

It may be important that no dredged seamounts, atolls or guyots have sediments older than Cretaceous on them. The number of samples is small but so is the number of atoll and guyot groups. All three large guyot groups in the North Atlantic have been sampled. Likewise, in the Pacific, four of the five large guyot groups and two large atoll groups have been sampled. In addition, perhaps a score of isolated seamounts and guyots has been dredged in widely separated parts of the ocean basins. Granting the stated assumptions, it appears probable that oceanic volcanoes large enough to form islands are almost all of Mesozoic or younger age. The myriad of small seamounts has not been adequately dredged and their age is unknown. However, either all oceanic volcanoes are post-Paleozoic or else the maximum size of the volcanoes increased drastically in Mesozoic times.

The number and distribution of oceanic islands in geological time may be of interest because of their bearing on faunal and floral migration. Little can be said about the Atlantic or Indian Ocean but the rough outline of faunal migration paths can be sketched for the Pacific. In pre-Mesozoic times islands may not have existed in the Pacific Basin and could, therefore, have no effect on biological migration. However, seamounts may have been present at depths as shallow as a few hundred meters and they might have provided steppingstones for suitably adapted biota.

In Mesozoic times a large group of active volcanoes existed in a roughly east–west band along the Marcus-Necker Rise. By Middle Cretaceous times many of the volcanoes in the eastern, or Mid-Pacific Mountains, region of the rise had grown up through water 4 km deep and formed large islands which were being planed down to sea-level by stream and wave erosion. Vulcanism and island formation continued for an indefinite time but the region gradually became submerged, and higher and higher submarine volcanoes were unable to reach the sea surface. Under the most favorable circumstances some 70 islands and shallow banks could have existed as closely spaced stepping-stones, reaching halfway across the equatorial Pacific.

From the fragmentary evidence available, it appears that late Mesozoic–early Cenozoic times saw a culmination of vulcanism in the basin. In the Eocene, volcanoes had grown up to form islands in the Marshall Islands and Tuamotu Archipelago, and islands probably still existed in the Mid-Pacific Mountains. By inference from the similarity of structural trends, the Gilbert Islands were probably also first formed at this time. With regard to biological migration, stepping-stones became much more common than before. The continuity of the stepping-stones depends on the age of the Line Islands. Inasmuch as the Tuamotu and Line Islands and the Mid-Pacific Mountains and Emperor Seamounts all lie along the median line of the Pacific Basin, it may be argued, by analogy with the Mid-Atlantic Ridge, that all the volcanoes of the groups were active at about the same time (Menard, 1958). If so, stepping-stones reached from northeast Asia to the southeastern Pacific. One of the puzzles of this time is the close juxtaposition in space and time of atolls and guyots in the Marshall and Tuamotu archipelagoes. Why did atolls develop from some banks and guyots from others in the same locality and of approximately the same age? In particular localities, such as at Bikini Atoll and the adjacent Sylvania Guyot, the atoll apparently is younger than the guyot because the atoll platform is much higher and is unterraced (Emery *et al.*, 1954). The guyot subsided below the level of reef growth before the atoll developed. However, the oldest fossils on the guyot are pelagic Foraminifera of earliest Eocene age (Hamilton and Rex, 1959), and it is quite improbable that the guyot was an island before Middle Cretaceous times when reef-forming organisms lived in the western central Pacific. The pattern of distribution (Fig. 7) suggests that some general environmental factor may have tended to restrict atoll development in the Pacific Basin to a narrow latitudinal band. It seems unlikely that temperature was restricting in the warm seas of the time. Current intensity may be a possibility. The mild climate of late Mesozoic–early Cenozoic time may have reduced the intensity of the wind-driven equatorial current system, and the reduced circulation may have decreased the chance that larval forms would be carried to a given island or bank. If the initial development of a reef is precarious, the reef stock may have to be replenished several times, and the influence of current velocity and eddy patterns would be correspondingly important.

Vulcanism continued through the Tertiary with the development first of Erben Guyot and the guyots of the Nasca Ridge, and later of the existing volcanic archipelagoes. The number of islands at any given time gradually increased because most islands became atolls and were preserved even though the volcanic platforms continued to subside. Also, as previously noted, there is considerable evidence suggesting that, periodically during the Quaternary, the present-day atolls stood above the sea (some as high islands) for appreciable intervals. The cause of these changes is not known. Correspondingly, the number of guyots formed by subsidence at any time decreased (Fig. 10).

The net rate of subsidence of atolls and guyots in geological time can be determined from the depth to datable horizons. For guyots, the depth to be considered is not the depth of the dredge haul but the break in slope at the edge

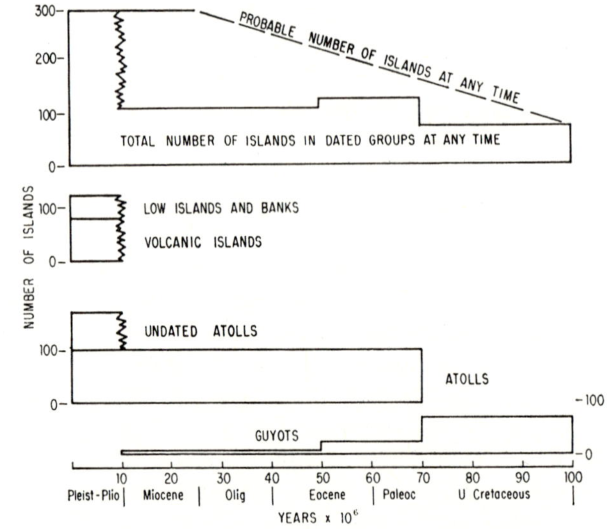

Fig. 10. Numbers and ages of islands, atolls and related features.

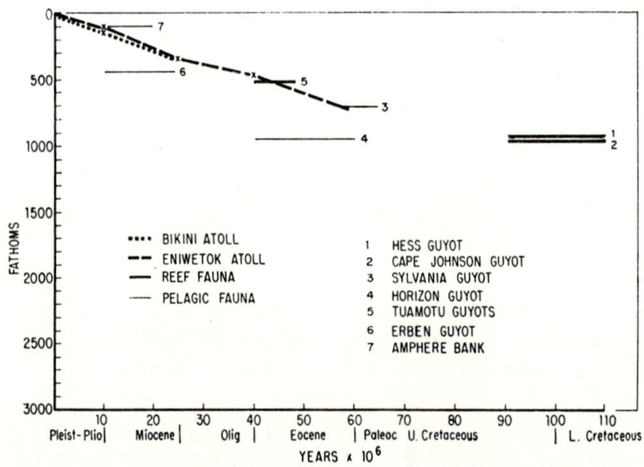

Fig. 11. Depths and ages of atolls and guyots.

of the flat top, because this represents sea-level when truncation began. Available dates suggest a constant rate of subsidence of 2 cm per 1000 years for the last 100 million years (Fig. 11). The relative subsidence of widely separated volcanic islands of various sizes at a constant net rate seems consistent with a rapid accumulation of juvenile water and a rise in sea-level relative to the sea floor, as proposed by Revelle (1955). Clearly more points are needed on the graph to resolve the important question of the cause of subsidence of oceanic volcanoes.

8. Importance of Islands and Seamounts

Oceanic islands have played important roles in the exploration of the world. In the Atlantic the Azores were an outpost for ships of all kinds and still function as a convenient stopping place for air traffic. In the Pacific the hundreds of islands and atolls now in existence were stepping stones in the migrations of man from Indonesia eastward and northward to Hawaii. Atolls with their thin soils and meager water supply are not ideal dwelling-places yet many of them support a population of several hundred each. In the days of sailing ships, many islands, including the Galapagos, were used to replenish supplies of water and other necessities. When steamships replaced sailing vessels, islands with good harbors were valuable as coaling stations and naval bases. The importance of islands and atolls in the Second World War can hardly be overemphasized and many are used as air stops today.

In the geological past Pacific islands and reefs were a controlling factor in the distribution of marine and terrestrial life over vast areas (Ladd, 1960). In the 19th century studies of the distribution of life on islands gave valuable support to the theory of evolution as developed by Darwin and Wallace. Islands of all sorts are still of value in surface navigation and as world-wide undersea navigation is now being developed, seamounts and guyots are of great value.

Less obvious, but perhaps of equal importance, is the part played by islands and atolls in oceanic circulation and their effect on the distribution of sea temperatures. The Marshall and Gilbert chains, for example—small and widely spaced as the units are—form a topographic barrier across the Equatorial Current system and give rise to large-scale eddies that affect the North Equatorial Currents, the Equatorial Counter Current and possibly the South Equatorial Current (Robinson, 1954).

References

Arx, W. S. von, 1954. Circulation systems of Bikini and Rongelap lagoons. *U.S. Geol. Surv. Prof. Paper*, **260B**, 265–273.

Burnside, R. J., 1959. Geology of part of the Horshoe Atoll in Borden and Howard Counties, Texas. *U.S. Geol. Surv. Prof. Paper*, **315B**, 21–34.

Carsola, A. J. and R. S. Dietz, 1952. Submarine geology of two flat-topped northeast Pacific seamounts. *Amer. J. Sci.*, **250**, 481–497.

Cloud, P. E., Jr., 1958. Nature and origin of atolls. *Proc. 8th Pacific Sci. Cong.*, **111-A**, 1009–1023.

Cloud, P. E., R. G. Schmidt and H. W. Burke, 1956. Geology of Saipan, Mariana Islands. *U.S. Geol. Surv. Prof. Paper*, **280A**, 1–126.

Cole, W. S., Ruth Todd and C. G. Johnson, 1960. Conflicting age determinations suggested by Foraminifera on Yap, Caroline Islands. *Bull. Amer. Paleont.*, **186**, 73–112.

Daly, R. A., 1925. The geology of Ascension Island. *Proc. Amer. Acad. Arts Sci.*, **60**, 1–80.

Dana, J. D., 1849. *Geol. U.S. Expl. Exped.* Wilkes, Philadelphia, **10**, 1–756.

Darwin, C., 1889. *The Structure and Distribution of Coral Reefs.* D. Appleton and Co., New York. 344 pp.

Dietz, R. S., 1954. Marine geology of northwestern Pacific: description of Japanese Bathymetric Chart 6901. *Bull. Geol. Soc. Amer.*, **65**, 1199–1224.

Dietz, R. S. and H. W. Menard, 1953. Hawaiian swell, deep, and arch, and subsidence of the Hawaiian Islands. *J. Geol.*, **61**, 99–113.

Emery, K. O., J. I. Tracey, Jr., and H. S. Ladd, 1954. Geology of Bikini and nearby atolls. *U.S. Geol. Surv. Prof. Paper*, **260A**, 1–265.

Ewing, M. and B. C. Heezen, 1956. Some problems of Antarctic submarine geology. *Geophys. Monog.*, **1**, 75–81.

Hamilton, E. L., 1956. Sunken islands of the Mid-Pacific Mountains. *Mem. Geol. Soc. Amer.*, **64**, 1–97.

Hamilton, E. L. and R. W. Rex, 1959. Lower Eocene phosphatized *Globigerina* ooze from Sylvania Guyot. *U.S. Geol. Surv. Prof. Paper*, **260W**, 784–798.

Heezen, B. C., M. Tharp and M. Ewing, 1959. The floors of the oceans, I. The North Atlantic. *Geol. Soc. Amer. Spec. Paper* 65, 1–121.

Hess, H. H., 1946. Drowned ancient islands of the Pacific basin. *Amer. J. Sci.*, **244**, 772–791.

Hoffmeister, J. E., 1932. Geology of Eua, Tonga. *Bull. Bishop Mus.*, **96**, 1–93.

Ladd, H. S., 1958. Fossil land shells from western Pacific atolls. *J. Paleont.*, **32**, 183–198.

Ladd, H. S., 1960. Origin of the Pacific island molluscan fauna. *Amer. J. Sci.*, **258A**, 137–150.

Ladd, H. S. and J. E. Hoffmeister, 1945. Geology of Lau, Fiji. *Bull. Bishop Mus.*, **181**, 1–399.

Ladd, H. S., E. Ingerson, R. C. Townsend, M. Russell and H. K. Stephenson, 1953. Drilling on Eniwetok Atoll, Marshall Islands. *Bull. Amer. Assoc. Petrol. Geol.*, **37**, 2257–2280.

Ladd, H. S. and S. O. Schlanger, 1955. Down to earth on Eniwetok. *Res. Rev.*, 13–16.

Ladd, H. S. and S. O. Schlanger, 1960. Drilling operations on Eniwetok Atoll. *U.S. Geol. Surv. Prof. Paper*, **260Y**, 863–905.

Macdonald, G. A., 1956. The structure of Hawaiian volcanoes. *Koninkl. Ned. Geol.-Mijn. Genootschap Verh. Geol. Ser.*, **16**, 274–295.

Menard, H. W., 1955. Deformation of the northeastern Pacific Basin and the west coast of North America. *Bull. Geol. Soc. Amer.*, **66**, 1149–1198.

Menard, H. W., 1958. Development of median elevations in ocean basins. *Bull. Geol. Soc. Amer.*, **69**, 1179–1186.

Menard, H. W., 1959. Geology of the Pacific sea floor. *Experientia*, **6**, 205–213.

Menard, H. W. and R. S. Dietz, 1951. Submarine geology of the Gulf of Alaska. *Bull. Geol. Soc. Amer.*, **62**, 1263–1285.

Munk, W. H. and M. C. Sargent, 1954. Adjustment of Bikini Atoll to ocean waves. *U.S. Geol. Surv. Prof. Paper*, **260C**, 275–280.

Murray, H. W., 1941. Submarine mountains in the Gulf of Alaska. *Bull. Geol. Soc. Amer.*, **52**, 333–362.

Nestéroff, W. D., 1955. Les récifs coralliens du Banc Farsan Nord (Mer Rouge). *Rés. Sci. Campagnes "Calypso"*, **1**, 1–53.

Revelle, R., 1955. On the history of the oceans. *J. Mar. Res.*, **14**, 446–461.

Robinson, M. K., 1954. Sea temperature in the Marshall Islands area. *U.S. Geol. Surv. Prof. Paper*, **260D**, 281–291.

Schlanger, S. O., in press. Subsurface geology of Eniwetok Atoll. *U.S. Geol. Surv. Prof. Paper*, **260BB**.

Sollas, W. J. and T. W. E. David, 1904. The Atoll of Funafuti. *Rep. Roy. Soc. London*, 1–428.

Stafford, P. T., 1959. Geology of part of the Horshoe Atoll in Scurry and Kent Counties, Texas. *U.S. Geol. Surv. Prof. Paper*, **315A**, 1–20.

Stearns, H. T., 1945. Late geologic history of the Pacific Basin. *Amer. J. Sci.*, **243**, 614–626.

Stearns, H. T., 1946. Geology of the Hawaiian Islands. *Bull. Hawaii Div. Hydrog.*, **8**, 1–106.

Tracey, J. I., Jr., S. O. Schlanger, J. T. Stark *et al.*, in press. Geology of Guam. *U.S. Geol. Surv. Prof. Paper*.

Wells, J. W., 1954. Recent corals of the Marshall Islands. *U.S. Geol. Surv. Prof. Paper*, **260I**, 385–486.

Wiseman, J. D. H. and G. P. D. Hall, 1956. Two recently discovered features on the floor of the Indian Ocean: Andrew Tablemount and David Seaknoll. *Deep-Sea Res.*, **3**, 262–265.

16. THE MID-OCEANIC RIDGE[1]

Bruce C. Heezen and Maurice Ewing

1. Introduction

The knowledge that there is a median ridge in the Atlantic gradually emerged from the observation of a few shoal soundings almost a century ago and became definite when the early oceanographic expeditions measured bottom-water temperatures in the eastern and western basins. Recent investigations of the morphology and seismicity of the ocean floor have revealed the wide extent and continuity of the mid-oceanic ridge, an active geotectonic feature of equal rank, in terms of areal extent and probable global significance, to the continents and the ocean-basin floor. After a ten-year study of the topography and geology of the North Atlantic, Heezen, Tharp and Ewing (1959) found that several major morphotectonic provinces, mapped in moderate detail in the North Atlantic, could be traced with certainty along a world-encircling belt, 40,000 miles long, passing through the South Atlantic, Indian, and South Pacific Oceans, the Norwegian Sea and the Arctic Basin. The mid-oceanic ridge, much longer and broader than the Alpine–Himalayan Mountain system, is morphologically identical to the East African rift valley system, which in fact forms a continental extension of the mid-oceanic ridge. The structure and origin of the mid-oceanic ridge is fundamental to the origin of oceans in general, as well as to the particular tectonics of the individual oceans.

2. North Atlantic

A typical trans-Atlantic profile between North America and Africa is shown in Fig. 1. The threefold division of the ocean into continental margin, ocean-basin floor and mid-oceanic ridge is well expressed in this profile. Each major physiographic division occupies approximately one-third of the area of each of the oceans which possesses a well-developed mid-oceanic ridge. The most thoroughly investigated part of the mid-oceanic ridge is that between 20°N and 50°N latitude in the Atlantic Ocean. It is in this area that the major morphological divisions of the ridge were first delineated. Cross-sections through this area are considered "type profiles" (Fig. 2). The lateral boundaries of the mid-oceanic ridge are defined by the first abrupt scarp, or abrupt gradient change, between the abyssal hills of the ocean-basin floor and the mid-oceanic ridge. The physiographic provinces of the mid-oceanic ridge can be divided into the crest provinces and the flank provinces

A. Crest Provinces

The most rugged topography of the Mid-Atlantic Ridge is found along the axis of the crest provinces. A steep cleft or rift valley is formed by the inward-facing scarps of the rift mountains which rise 500–2000 fm above the valley

[1] Lamont Geological Observatory Contribution No. 597.

[*MS received August, 1962*]

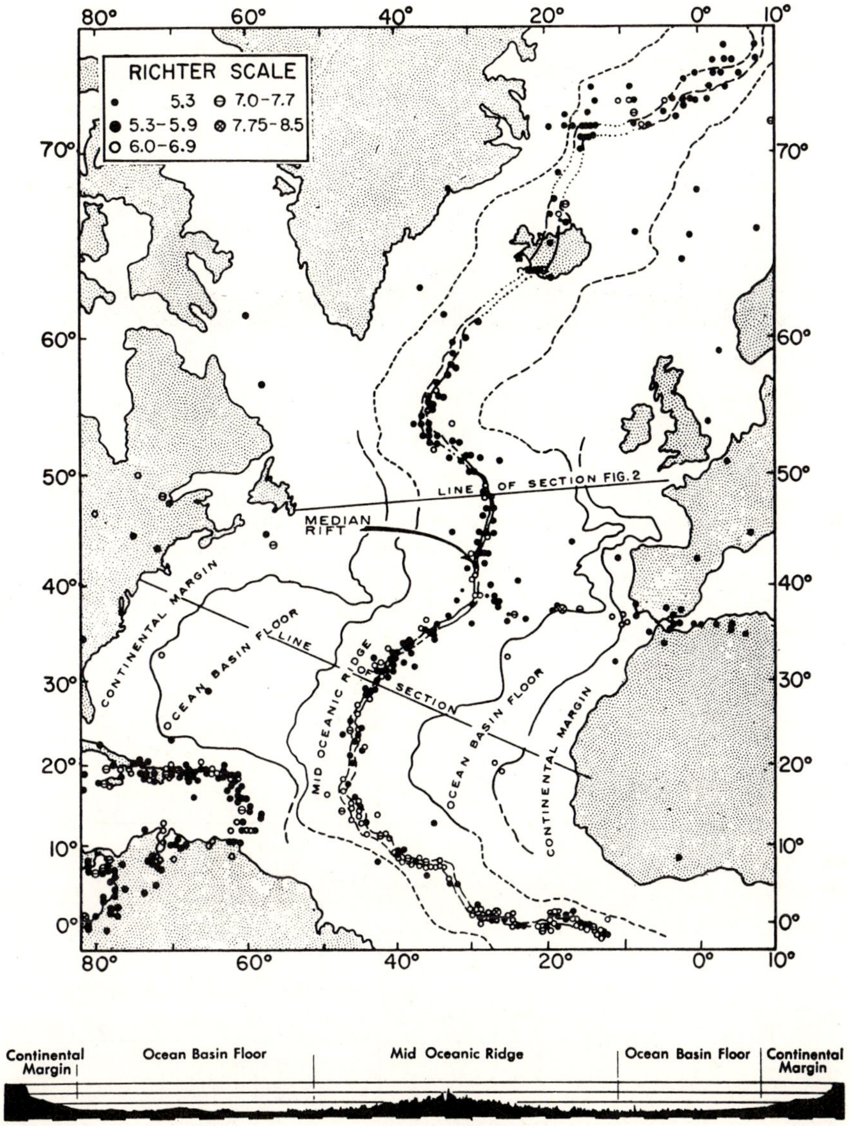

Fig. 1. Major physiographic divisions of the North Atlantic and earthquake epicenters 1905–56. This threefold pattern of continental margin, ocean-basin floor, and mid-oceanic ridge in the North Atlantic has been found in the South Atlantic, Indian, South Pacific and Arctic Oceans. Epicenters from Gutenberg and Richter (1954) and the epicenter cards of the United States Coast and Geodetic Survey. Accuracy of epicenter location varies from $\pm \frac{1}{2}°$ to $\pm 1\frac{1}{2}°$. The major morphological divisions and the Mid-Atlantic rift valley are shown on the profile, which extends from North America to Africa. (Vertical exaggeration of profile is 40:1.)

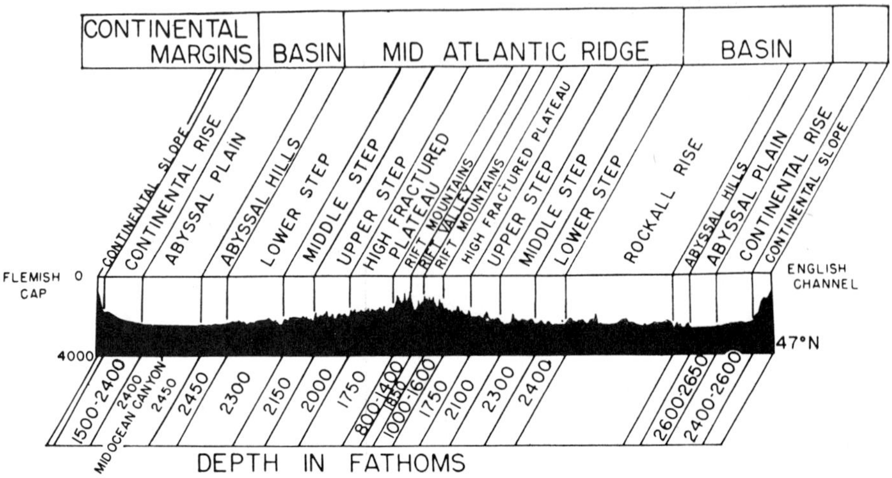

Fig. 2. Typical trans-Atlantic profile from Newfoundland to England. The depths indicated in the lower portion of the figure are characteristic of this profile only, as the depths of all provinces gradually fluctuate along the length of the ocean basin. The physiographic provinces have been traced throughout the Atlantic and are believed to extend, in a general way at least, throughout all oceans possessing a median-rifted ridge. (Vertical exaggeration 40:1.)

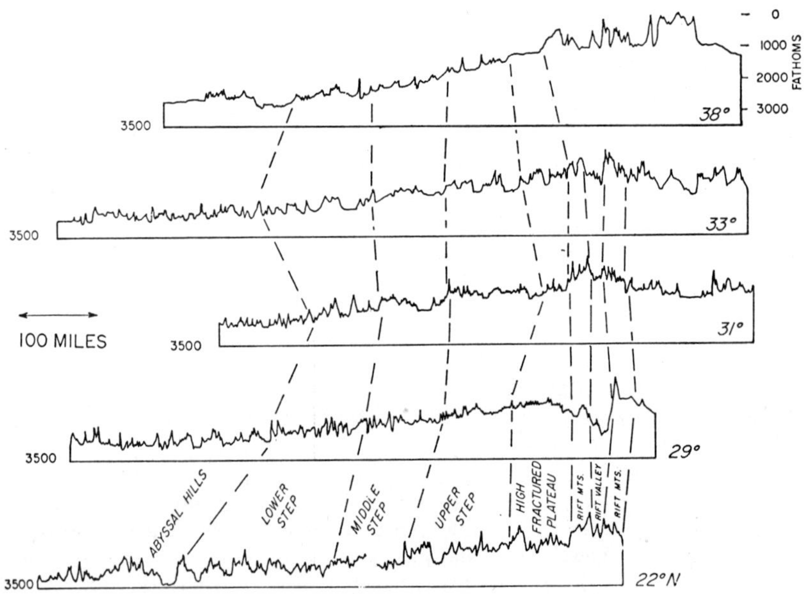

Fig. 3. On these profiles of the western North Atlantic the step provinces are shown as a series of tilted fault blocks bounded by scarps. The difficulty of tracing the steps from one profile to another is apparent. (Vertical exaggeration 40:1.)

floor. On either side the peaks of the rift mountains drop abruptly to depths of 1600–1800 fm of the high-fractured plateau. These features are illustrated in Fig. 2.

B. Flank Provinces

The flank provinces of the Mid-Atlantic Ridge have been arbitrarily divided into several steps parallel to the axis (Fig. 3). Outward-facing scarps of varying height and intermittent length form boundaries of these steps. Except for the local smooth-floored intermontane basins, peaks of 200–400 fm form the moderately rugged relief of the flank provinces. The smooth floors of the intermontane basins result from sediment filling, and, in general, a series of adjacent basin floors deepen progressively from the crest to the flank.

The continuity along the ridge axis of the step provinces has not been firmly established, perhaps because more detailed surveys are needed to evaluate the effects of fracture zones.

Fracture zones were originally discovered in the floor of the northeast Pacific. Studies of magnetic-anomaly patterns have disclosed large slip displacements along these zones. In the Atlantic similar zones offset the Mid-Atlantic Ridge. The magnitude of these displacements is judged by topographic discontinuities in the position of the axis of the ridge. These fracture zones are predominantly east–west in the Atlantic as they are in the Pacific, a fact of certain but unknown significance. Prominent fracture zones displace the Mid-Atlantic Ridge at 30°N, 15°N, 11°N, 1°N, 1°S and 2°S. There are indications of several additional fracture zones but data at present are insufficient for complete confirmation.

C. Seismicity

Even before the establishment of seismograph stations, a study of seaquakes (Rudolph, 1887) had indicated that the Mid-Atlantic Ridge is seismically active. As the world net of seismograph stations became more complete, the repeated observations of mid-oceanic earthquakes firmly established the seismicity of the Mid-Atlantic and Mid-Indian Ocean Ridges (Tams, 1931; Rothé, 1954). However, recent morphological investigations cited above reveal that although the epicenter belt in the North Atlantic is no more than 100 miles in width, the Mid-Atlantic Ridge is 1000–1200 miles in width. When we consider the probable inaccuracy of epicenter determination, it is clear that the actual width of the epicenter belt is probably even narrower (Fig. 1). The present accuracy of epicenter determinations in this area is about $\pm\frac{1}{2}°$ in the most favorable areas, but the earlier epicenters are probably never located better than $\pm 1°$. A careful study of each profile across the crest of the Mid-Atlantic Ridge reveals that the axis of the epicenter belt coincides with the axis of the rift valley. The plotted positions of most epicenters define a belt which includes the rift valley and whose width represents the probable error in epicenter determination. It therefore seems most probable that the

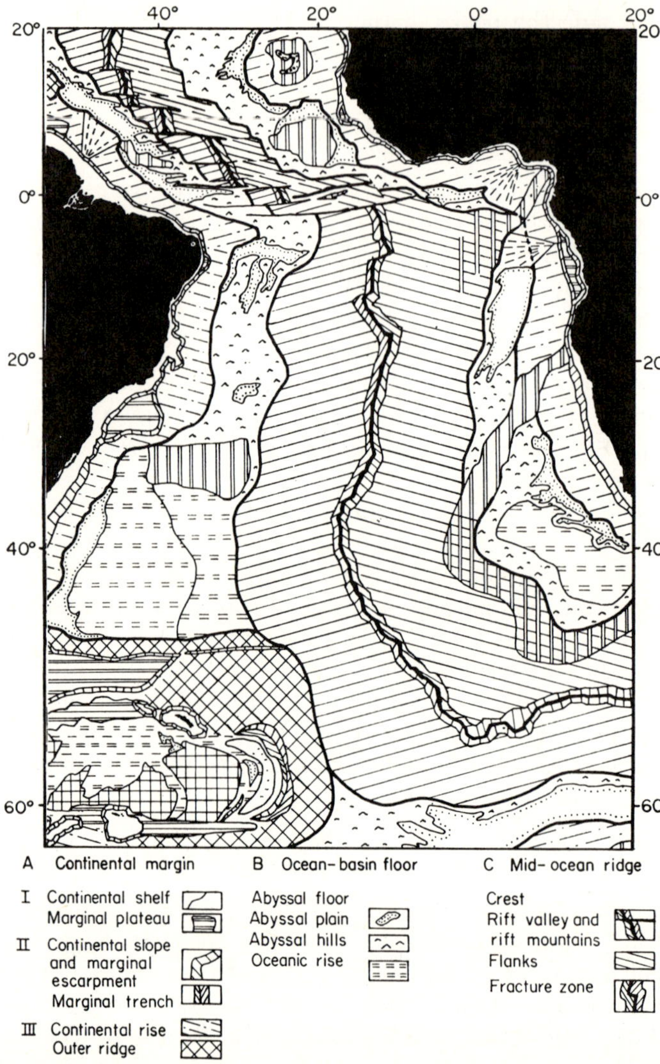

Fig. 4. A generalized province map of the South Atlantic showing the threefold division of the ocean into continental margin, ocean-basin floor and mid-oceanic ridge. In the equatorial Atlantic a series of left slip faults or fracture zones has offset the mid-oceanic ridge westward for over 2000 miles. Although fracture zones have been known since 1939 in the Pacific (Menard, 1955), this is the first time that they have been mapped in the Atlantic. (After Heezen and Tharp, 1961.)

epicenter belt actually coincides with the rift valley. The data available would, however, allow the conclusion that the rift mountains also lie within the seismic belt.

All epicenters of the mid-oceanic belt are of shallow-focus. This distinguishes the mid-oceanic belt from the circum-Pacific epicenter belt. The oceans are remarkably free of seismicity except along the epicenter belt. The continuity of the epicenter belt over its entire length of 40,000 miles has been established beyond doubt. Thus, the co-relation of the seismic belt and topography, in the moderately well surveyed portions of the North Atlantic, may be extended to areas less well sounded.

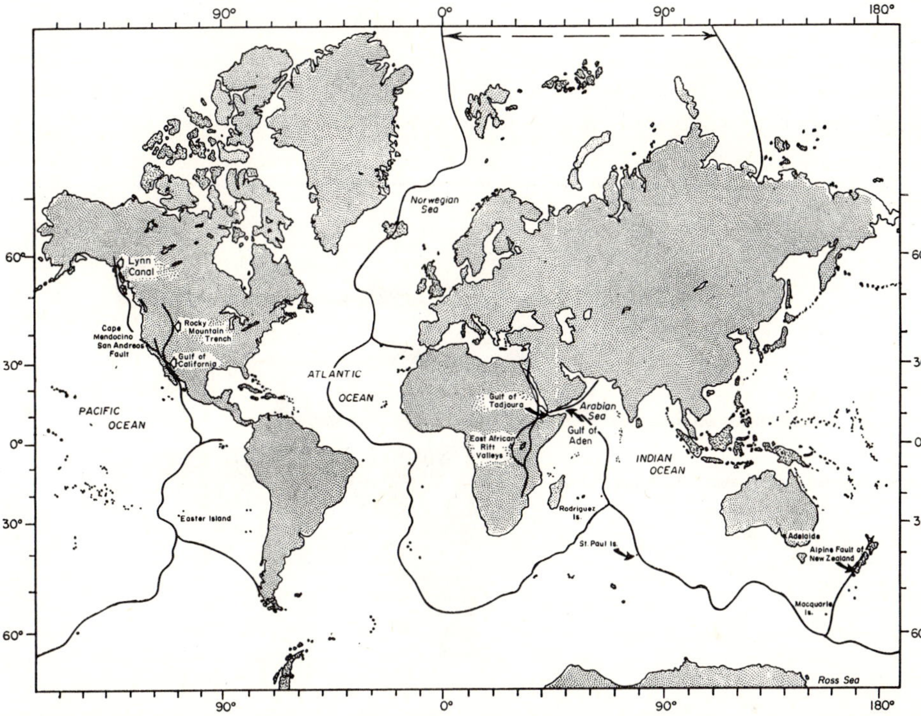

Fig. 5. World map of the mid-oceanic rift. This plot shows essentially the mid-oceanic epicenter belt. In every area in which the bathymetry is known, however crudely, the epicenter belt follows the crest of the mid-oceanic ridge. In every area where detailed submarine topographic studies have been carried out in the region of the epicenter belt a median rift valley has been discovered. Wherever the rift extends across the continental margins it joins either rift grabens, as in East Africa and Iceland, or great strike-slip faults, as in California and New Zealand.

3. South Atlantic

The major physiographic provinces, described above for the North Atlantic, have been recognized in 40 trans-Atlantic sounding lines across the South Atlantic (Fig. 4) (Heezen and Tharp, 1961). The pioneer sounding profiles of the *Meteor* Expedition across the South Atlantic (Stocks and Wüst, 1935) have been shown to be extremely dependable when compared with the more recent precision soundings of *Vema*, *Atlantis*, and *Crawford*. Epicenters, although few in number, at least partially owing to unfavorable location of seismographic stations, fall near the location of the rift valley.

The mid-oceanic ridge in the equatorial Atlantic in general is characterized by greater relief and has a more rugged and youthful appearance than in either the North Atlantic or South Atlantic. In the equatorial Atlantic the ridge is offset westward over 2000 miles by a series of left slip faults (Fig. 4). It is probable that fracture zones will be found when the South Atlantic is better surveyed.

Seamounts and seamount groups, which are found in all provinces of the oceans, are widely scattered throughout the Mid-Atlantic Ridge. Sometimes the seamounts are distributed in a random pattern but more often they occur along linear rows suggesting an association with fracture zones.

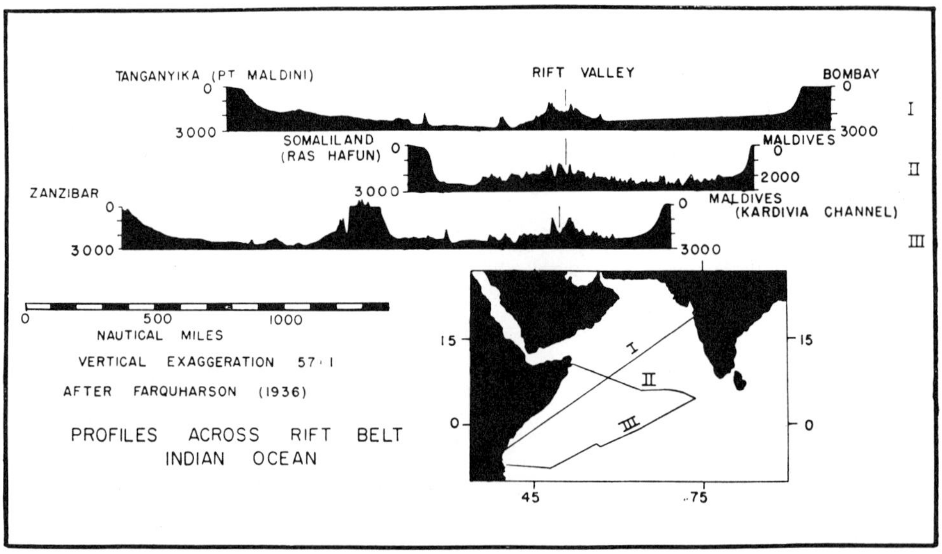

Fig. 6. Rift valley profiles, Indian Ocean. (After Farquharson, 1936. Soundings made by M.S. *Mahabis* during the John Murray Expedition.)

4. Indian Ocean

In the Indian Ocean there is a clearly delineated belt of epicenters which continues uninterrupted from the South Atlantic (Rothé, 1954) to the vicinity of Rodriguez Island, at which point the belt splits, one branch continuing towards Macquarie Island, the other entering the Gulf of Aden where the mid-oceanic ridge epicenter belt joins with the seismic belt associated with the East African rift valleys (Fig. 5). Sounding profiles reveal a median ridge which coincides with the epicenter belt. Soundings by *Mahabis* (Farquharson, 1936) reveal that the section of Mid-Indian Ocean Ridge from Rodriguez Island to the Gulf of Aden (known as the Carlsberg Ridge) is longitudinally split by a prominent central valley.

Recent cruises of *Vema* in previously unsounded areas have found a continuous rifted mid-oceanic ridge which coincides with the epicenter belt (Ewing and Heezen, 1960). Fig. 7 is the result of a preliminary study of the relation of the epicenter belt to the major topographic features of the Indian Ocean. An example of the very rough topography of the Mid-Indian Ocean Ridge is shown in the sections of Fig. 35, Chapter 12, p. 274.

The connection of the mid-oceanic seismic belt with the seismic belt of the African rift valleys, and the similarity of the morphologies of the mid-oceanic rift and the African rift valleys, give a clue to the structure and origin of the rifted mid-oceanic ridge. The 800-mile-wide mid-oceanic ridge of the Arabian Sea narrows to less than 100 miles where it enters the 200-mile-wide Gulf of Aden (Fig. 7). Its seismically active median rift trends west through the Gulf of Tadjoura in French Somaliland. As the Gulf of Aden shallows and narrows from east to west the rifted ridge maintains its median position, coinciding with the epicenter belt. The coastal escarpments of the Gulf of Aden are formed by rift faults, which converge toward the southwest and continue as the boundary scarps of the Ethiopian rift valleys (Fig. 5).

A. African Rift Valleys

In Fig. 34, Chapter 12, p. 272, a profile across the Mid-Atlantic Ridge is placed above two profiles across the East African rift valleys to illustrate their nearly identical form. This similarity, indeed virtual identity, of morphology and seismicity might still be considered superficial and accidental were it not for the fact that the three profiles lie on the same seismically active morphotectonic belt. The rifted mid-oceanic ridge seems to be closely similar in origin and structure to the African rift valleys.

Gregory (1921) gives evidence that the African rift valleys are tensional grabens. Wayland (1930) and others attempted to explain the features in terms of compression, but recent investigations have only added support for a tensional origin. McConnell (1951), and others who favor a dominantly strike-slip origin, attribute compressional, as well as tensional, features to minor complications along the strike-slip faults. Bullard (1936), who offered a compressional hypothesis to explain the negative gravity anomalies over the rifts, has now abandoned compression in favor of tension (Bullard, *in litt.*, 1959). Vening Meinesz (in Heiskanen and Vening Meinesz, 1958) has offered an explanation of the gravity anomalies in terms of simple extension. The compressional theory came mainly from those who found it difficult to comprehend large-scale extension on a shrinking earth; the shrinking, it was thought, being demanded by the Earth's folded mountain ranges.

5. Indian and South Pacific Oceans

The mid-oceanic ridge continues southeast from Rodriguez Island, passing near St. Paul Island, and then between Australia and Antarctica to the vicinity of Macquarie Island (Fig. 5). Minor seismically active ridges diverge in this region toward Tasmania, New Zealand and the Ross Sea. The grabens of

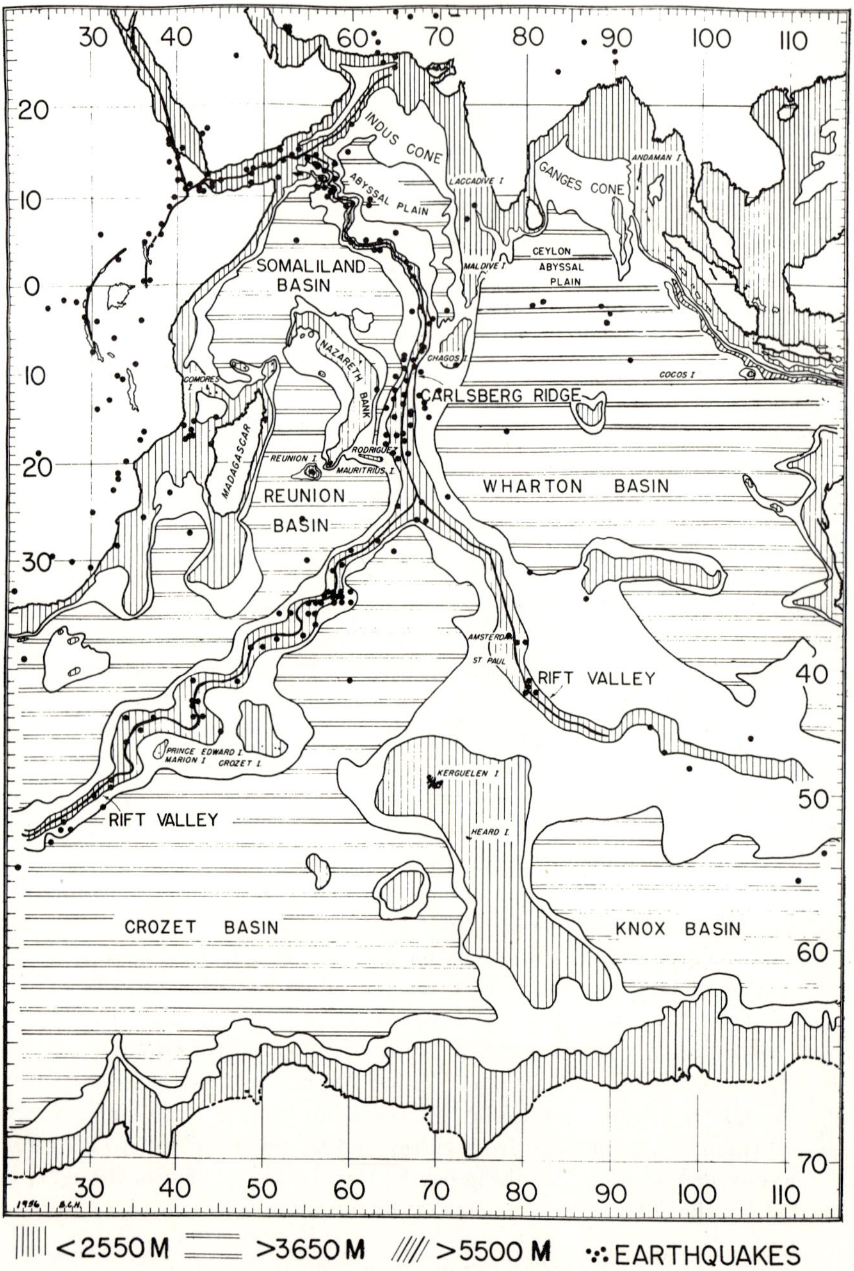

Fig. 7. A preliminary bathymetric chart of the Indian Ocean which shows the relation of the epicenter belt to the rifted Mid-Indian Ocean Ridge and to the major topographic features of the Indian Ocean. This chart, drawn in 1956, fails to show the recently discovered 90 East Ridge which extends from 10°N latitude to 40°S along 90°E longitude.

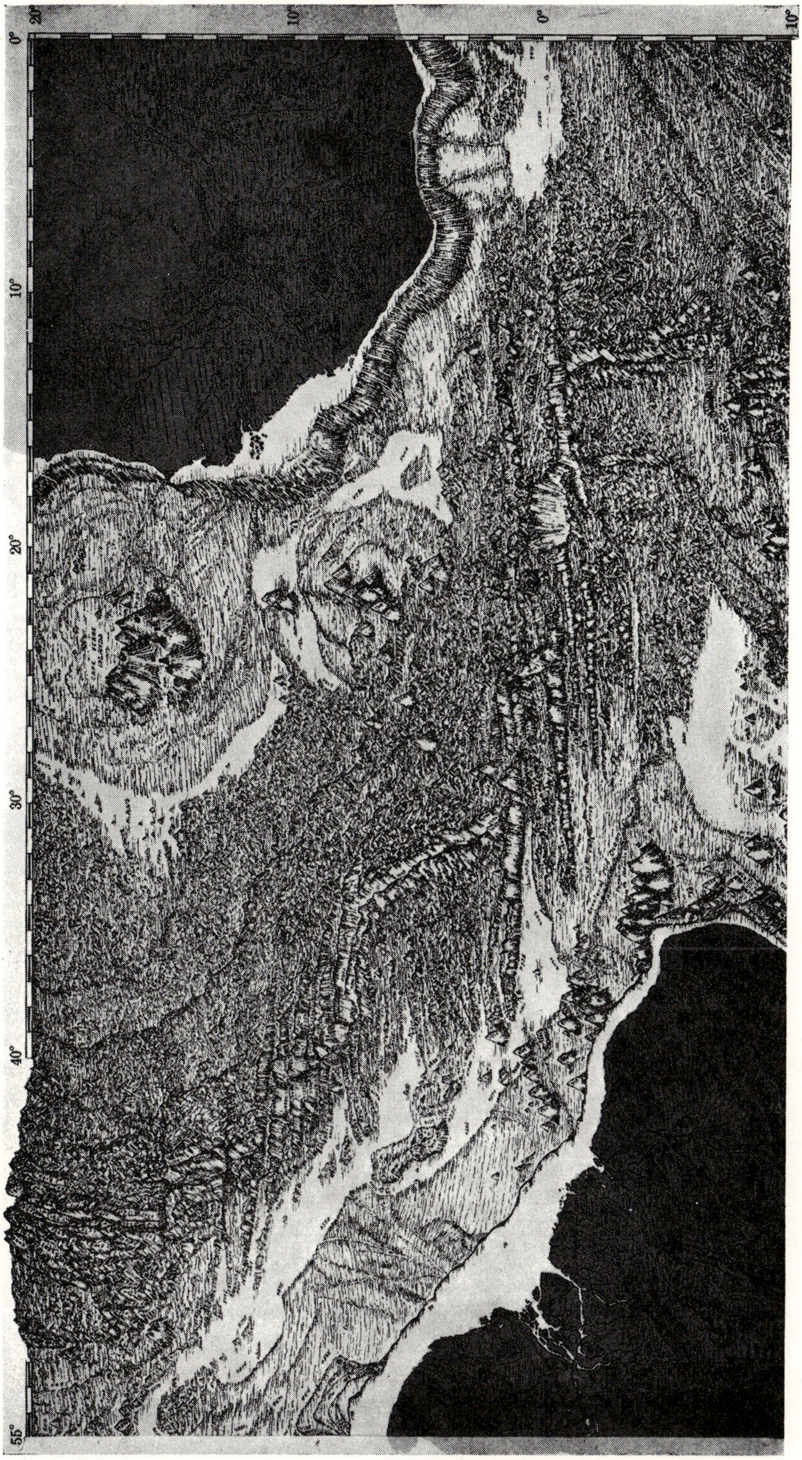

Fig. 8. Fracture zones offset the axis of the Mid-Oceanic Ridge in the Equatorial Atlantic. This is a portion of the Physiographic Diagram of the South Atlantic published by the Geological Society of America. (Copyright 1961 by Bruce C. Heezen and Marie Tharp. Reproduced by permission.)

Tasmania and those near Adelaide, Australia, may be related to the mid-oceanic ridge. The great Alpine Fault of New Zealand seems to be the landward extension of the Macquarie seismic ridge.

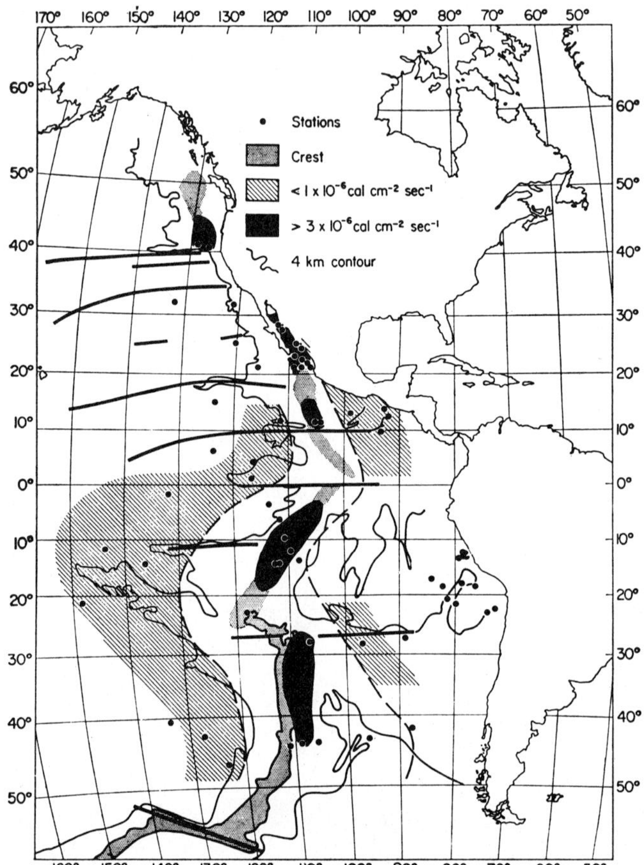

Fig. 9. Slip faults and heat flow in the mid-oceanic ridge of the east Pacific. The east–west trending black bars indicate the slip faults which cross the crest of the mid-oceanic ridge. The band of high heat flow associated with the crest of the ridge appears to pass through the western United States. (After Menard, 1960.) High heat flow has been observed along the crest of the mid-oceanic ridge in the Pacific, as well as at a few points in the North Atlantic. The slip faults in the southeast Pacific, although probable, are largely hypothetical. Insufficient data have been obtained to indicate whether slip faults have offset the crest of the mid-oceanic ridge in the equatorial or South Pacific.

From Macquarie Island to Easter Island soundings are sparse, but a mid-oceanic ridge can be delineated. The belt of epicenters continues without interruption. The scatter of plotted epicenters is somewhat greater here than elsewhere, because the accuracy of location is about 2° to 5°, but it coincides with the known position of the Easter Island Ridge and permits an inference of continuity of the ridge in this little known area.

Near Easter Island the earthquake belt splits, sending one branch southeast in the direction of southern Chile, and another northwards toward the Gulf of California. The existence of a ridge between Easter Island and Chile, predicted on the basis of the earthquake belt (Ewing and Heezen, 1956), has been confirmed by an I.G.Y. expedition of the Scripps Institution (Fisher, 1958).

The earthquake belt follows a well defined ridge north of Easter Island at least to the Equator. From the Equator to the mouth of the Gulf of California, the epicenter belt seems to follow a ridge similar to other mid-oceanic ridges, but Menard (1955), who has studied much unpublished and unavailable data, has come to the diametrically opposed view that the area is dominated by east–west-trending "fracture zones", and that no distinctive topographic feature is associated with the epicenter belt. This problem is possibly clarified by a later publication (Menard, 1960; and Fig. 9) which gives much information about the Pacific fracture zones and shows a continuation of the Easter Island Ridge ("East Pacific Rise") from Easter Island through the Gulf of California and western North America into the northeastern Pacific. Menard pointed out that the ridge is smoother in the southeast Pacific than in the Atlantic.

The offshore belt parallels the coast, at a distance of 100 miles from the continental slope, until it again extends into the continent along the linear Lynn Canal of southern Alaska. Menard found the topographic trend to be oriented parallel to the epicenter belt.

Apparently one end of the mid-oceanic ridge system is found in the neighborhood of the Gulf of Alaska.

6. Iceland and the Norwegian Sea

The belt of epicenters follows the crest of the Mid-Atlantic Ridge northward through the central graben of Iceland (Rutten and Wensink, 1960) into the Norwegian Sea, where it follows the median ridge system (Fig. 5). Soundings are rare in these northern waters, but those reported by Boyd (1948) give several clear indications of a central rift valley (Fig. 10). The line of epicenters clearly follows this rifted ridge through the Norwegian Sea (Fig. 11). Although some authors (Stocks, 1950) believe that the median ridge of the Norwegian Sea merges with the continental slope in the vicinity of Bear Island, the epicenter map suggests an alternative interpretation according to which the ridge bends to the north and parallels the western margin of the Barents continental slope west of Bear Island and Spitsbergen.

The recent work of Rutten and Wensink (1960) has further described the central graben of Iceland. It has long been recognized that essentially all earthquakes and all modern, and Recent, vulcanism in Iceland have been restricted to this fault-bounded continuation of the mid-oceanic rift. Bernauer (1943) described gaping fissures known by the local name of *gja* which occur in great numbers in the central graben, each aligned parallel to the main faults of the graben. From a study of the *gja* he estimates the post-glacial extension in Icelandic graben at 3.56 m/km/1000 years. This geology and structure have striking similarities with the East African rift valleys.

Dietrich (1959) and Ulrich (1960) report that in their survey of the Rekyjanes

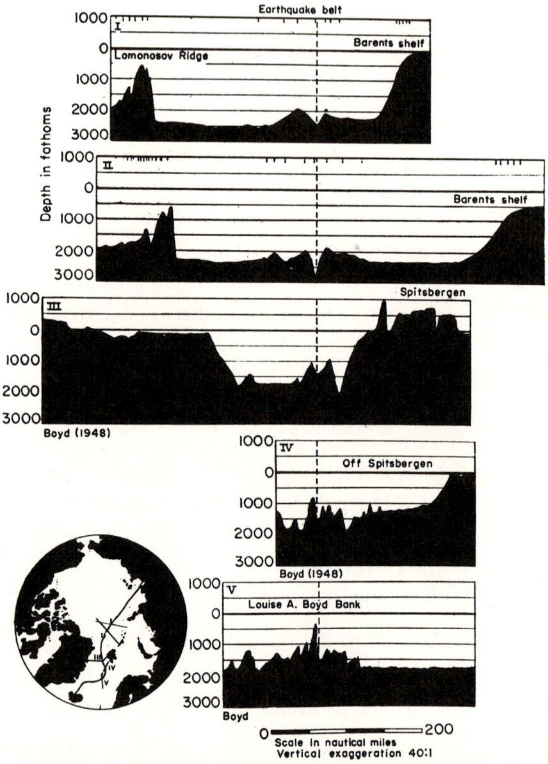

Fig. 10. Profiles across the Arctic Ocean and the Norwegian Sea. The two Arctic Ocean profiles are speculative interpretations of scattered soundings. The individual soundings are indicated as short vertical lines along the upper margin of each profile. Profiles IV and V are plotted from the early (1937–38) continuous echo-soundings published by Boyd (1948). The eastern half of profile III is based on continuous echo-soundings while the western half is plotted from discrete soundings (after Boyd, 1948). Note the strong similarity between profiles of this figure and profiles of Fig. 6.

Ridge south of Iceland the rift valley was typically absent north of 57° 30′N but was present on 5 crossings between 62°N and 62° 30′N.

In the Norwegian Sea the epicenter belt is continuous and is a reliable guide for following the structure in this complicated, narrow, and poorly-sounded region. The Mid-Atlantic Ridge, which in the south hugs the coast of Greenland, is found again in the north close to the continental margin off Spitsbergen. It is possible that this offset is due to a right-slip fault near Jan Mayen.

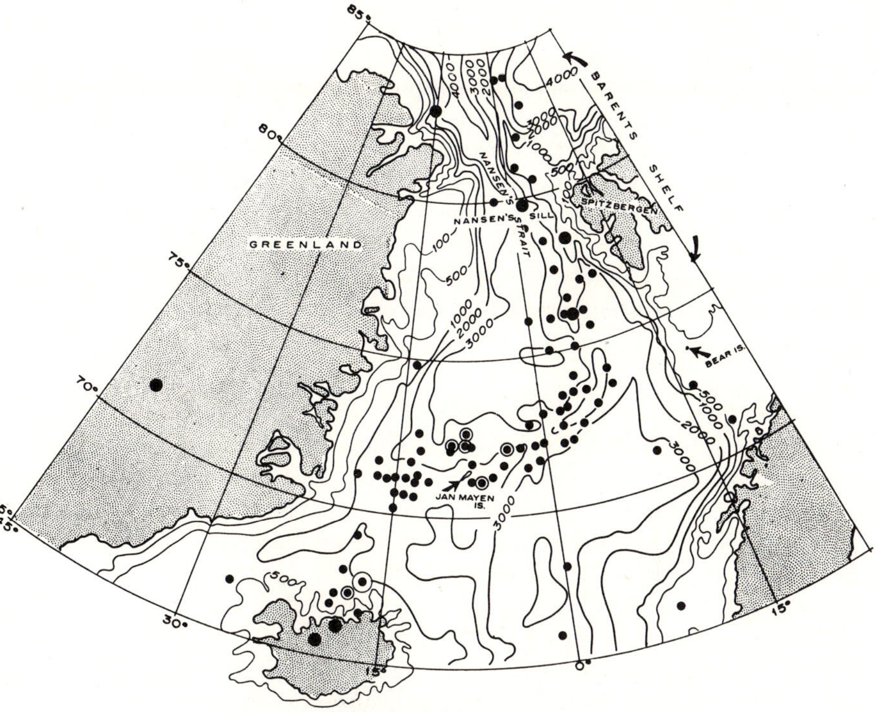

Fig. 11. Earthquake epicenters in the Norwegian Sea, 1905–56. Plotted from compilation of Gutenberg and Richter (1954) and the epicenter cards of the United States Coast and Geodetic Survey. (After Heezen and Ewing, 1961.)

7. Arctic Basin

Between Spitsbergen and Greenland, the Norwegian Sea basin narrows to less than 200 miles from shelf break to shelf break. This area has been referred to as the Nansen's Sill as it marks the sill or threshold for water-masses flowing between the Norwegian Sea and the Arctic Basin (Fig. 11; and Heezen and Ewing, 1961). That the depths across Nansen's Straits are less than those both to the north and to the south is apparent from existing soundings. However, the existence of an east–west ridge at Nansen's Sill is pure supposition. In considering the course of the mid-oceanic ridge through the length of the Gulf of Aden we saw that, as the Gulf narrowed, the depth of the basins flanking the median ridge shallowed accordingly. Nansen's Sill lies at the narrowest part of the ocean and hence, by analogy with the Gulf of Aden, the depth here should be the shallowest. Hope (1959), however, states that "recent Soviet exploration has in fact revealed a deep-water trough completely piercing Nansen's Sill". He also quotes Frolov and Pasecki (1958) as reporting that "on the 80th parallel the trough is 3000 m deep and 50 km wide between 1000 m isobaths. Its orientation is roughly northward". Hope points out that "this trough then lies in the seismic belt and it could mark the continuation of the Ewing–Heezen Rift north of Iceland and Jan Mayen".

Recent exploration by Soviet ships supports the present writers' interpretation that the mid-oceanic ridge follows the seismic belt north between Spitsbergen and Greenland. At the same time, the Soviet work refutes the traditional interpretation of an east–west trans-oceanic ridge at Nansen's Sill. The strongly north–south tectonic trends on Spitsbergen are in much closer agreement with a north–south submarine pattern in the Nansen Sill region than with the previously postulated east–west trend, which would have had to die out, or turn 90° between the shelf break and the shoreline, in order to join the Spitsbergen lineaments.

A. Arctic Basin Seismic Belt

The mid-oceanic seismic belt of the Atlantic Ocean and the Norwegian Sea continues across the Arctic Ocean (Tams, 1922; Heck, 1938; Raiko and Lindén, 1935; Hodgson, 1930). Unlike the Atlantic seismic belt, this Arctic continuation does not symmetrically divide the Arctic Ocean basin into two equal parts. Instead, the belt parallels the continental slope from Spitsbergen to Severnaya Zemlya (Fig. 12), maintaining an almost constant distance of 250 nautical miles seaward of the shelf break. Of thirty-two Arctic Ocean earthquakes shown on Fig. 12, twenty-four epicenters lie along this belt. Two occurred on the continental slope between Spitsbergen and Franz Joseph Land, one lies northwest of Banks Island, and three lie near the Bering Straits.

B. Arctic Continuation of the Mid-Oceanic Ridge

Although the extension of the mid-oceanic seismic belt through the Arctic Ocean has long been known, no clear evidence of a corresponding topographic ridge has been reported. As Hope (1959) points out, the discovery of the Lomonosov Ridge led some investigators to the conclusion that it was the continuation of the mid-oceanic ridge but, as we have already shown, the seismic belt lies parallel to the Lomonosov Ridge, about 250 miles distant, approximately bisecting the basin between the Lomonosov Ridge and the Barents Shelf continental slope. Clearly, the Lomonosov Ridge is not (at the present, at least) part of the mid-oceanic ridge system.

The area between the Lomonosov Ridge and the continental margin of the Barents and Kara shelves has usually been described as a single basin. We have, however, good reason to expect a rifted ridge along the mid-oceanic seismic belt. The ridge is shown in the contours of Fig. 12 in a new interpretation of the soundings largely obtained from *Fram* and *Sedov*. In contouring the soundings along the Arctic epicenter belt, Soviet investigators have generally shown broad submarine canyons running across the belt toward the continental slope off the Barents Shelf. The interpretation shown in this paper satisfies the soundings and also preserves the world-wide pattern of rift, ridge and seismicity which we have traced for over 40,000 miles. Post-war investigators have not added much

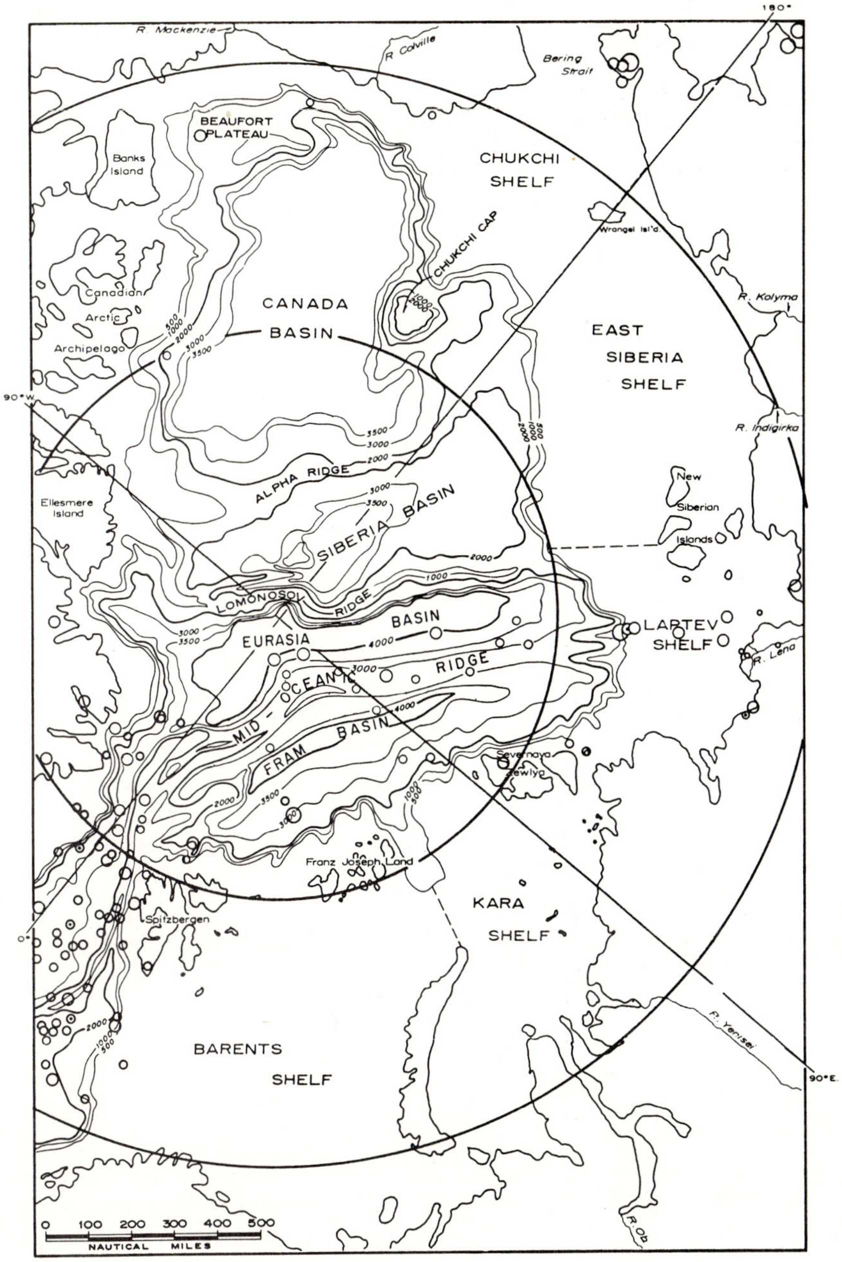

Fig. 12. Bathymetric sketch of the Arctic Ocean showing physiographic divisions and epicenters. The bathymetric sketch is modified from a chart prepared by E. R. Hope in 1958 which was largely based on Soviet data (to 1956). The probable original soundings were inferred by superimposing the tracks of drifting stations and aircraft landing sites on a recent Soviet bathymetric chart. The epicenters are from Linden (1959). (After Heezen and Ewing, 1961.)

to the older data, because drifting stations have not crossed this area and only a few aircraft landings have been made here. Since workers on *Fram* and *Sedov* were not able to take soundings on a daily schedule (the practice of the post-war drifting stations), soundings are widely spaced and hence only adequate for a speculative description of this rifted ridge.

Soundings west of Spitsbergen (Boyd, 1948) show a strong suggestion of a median-rifted ridge (Fig. 11). To the north, the recently discovered north–south cleft in Nansen's Sill (Hope, 1959) is probably a continuation of the rift. From 80°N to 85°N the rifted ridge appears to run north along the Greenwich meridian. The ridge is demanded by the available soundings which, however, are insufficient to confirm or deny its probable rifted character. From 85°N, 0°W the ridge turns east; between 30° and 45°W the soundings of both *Fram* and *Sedov* show a strong suggestion of the characteristic rifted ridge of the mid-oceanic belt (Fig. 10, profiles I and II).

Between 90°W and 120°W scattered soundings suggest a ridge but are insufficient to shed any light on the probable rift. As the seismic belt approaches the Siberian continental shelf, shallow soundings indicate an extension of shallow depths out along the epicenter belt. The Skado Trough, mentioned by Hakkel (Hope, 1959), may indeed actually be the end of the mid-oceanic rift as it cuts into the continental slope. As Hope points out, the Skado Trough seems too large to be described as a conventional erosional submarine canyon.

The Arctic seismic belt continues from the mid-oceanic ridge across the Laptev Shelf to the vicinity of the Lena delta where it crosses the shoreline and follows the Verkhoyansk Trough southward into the interior of Siberia. There is a suggestion that this seismic belt joins the Baikal-Rift-Valley seismic belt. The Verkhoyansk Trough, bounded by normal faults for 1000 km and filled with over 3.5 km of Jurassic, Cretaceous and Tertiary sediments, resembles the African and Lower California extensions of the mid-oceanic rift (Nalivkin, 1960).

At present the evidence for the extension of the mid-oceanic ridge through the Arctic Basin is strongly suggestive but not conclusive.

8. Crustal Structure

Thus, starting in the North Atlantic, we have followed a morphotectonic belt for over 40,000 miles along the axis of the Atlantic, Indian, South Pacific and Arctic Oceans. Extensions of the belt have been found continuing into the rift valleys of Africa, into the great strike-slip faults of New Zealand and southern California, Chilean canals, Gulf of Aden and Red Sea. The continuity of the epicenter belt is not open to question; the continuity of the rifted ridge is not proven beyond question due to sparsity of soundings in some areas, and to offsets at fracture zones. It is not certain whether the rift valley is a single valley or a belt of similar but *en echelon* features. The *en echelon* pattern of the East African rift valleys is probably a better generalization of the average situation than a single grand trough like the Red Sea.

Typical topographic profiles of the mid-oceanic ridge in the various oceans are shown in Fig. 13. Profile 7 of that figure represents the southeast Pacific crossing. It is conspicuous for the absence of a rift, the smoothness of the relief across the crest, and the continuous gentle gradient, unbroken by scarps.

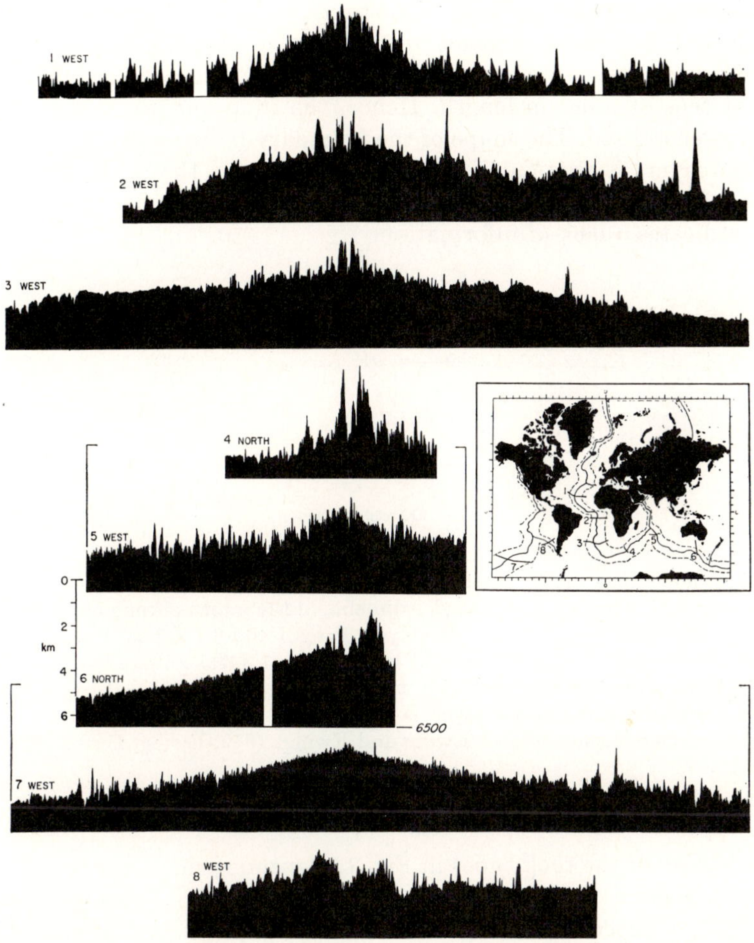

Fig. 13. Eight typical profiles of the mid-oceanic ridge in the North Atlantic, South Atlantic, Indian and South Pacific Oceans. The smooth relief across the crest of the mid-oceanic ridge in the South Pacific contrasts markedly with the rougher topography of the other oceans. Profile 2 was obtained aboard *Crawford* and Profile 8 was obtained aboard *Horizon*. The remaining profiles were obtained with a Precision Depth Recorder aboard R.V. *Vema*. (Vertical exaggeration of profiles is 100:1.)

Fig. 14 (Ewing, Ewing and Talwani, 1963) shows a ridge profile in the North Atlantic, between Dakar and Halifax, based on seismic-refraction (Ewing and Ewing, 1959) and reflection measurements.

The unconsolidated sediments, whose thickness was measured continuously along the traverse, form a thin and markedly discontinuous deposit. The pockety character of this deposit has been taken to indicate very free gravity transport of newly deposited sediments. It has been noted that the rough surface on which these sediments rest appears to form a continuous unit from the ridge crest to a point well into the basins.

The deeper interfaces depend upon seismic-refraction measurements at scattered points. Each such measurement gives average layer thickness over a segment tens of miles in length. Hence, the shapes of the deeper interfaces are highly generalized. The shape of the boundary between the 5.0 and the 7.3–7.5 km/sec layers is particularly uncertain, owing to the difficulty of making accurate refraction measurements in mountainous regions. The lack of relief on it really indicates a lack of information.

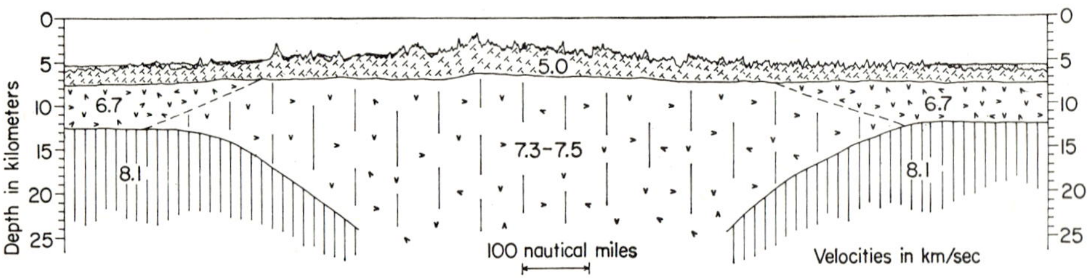

Fig. 14. A profile across the Mid-Atlantic Ridge based on seismic-refraction and reflection measurements. The unconsolidated sediments of the intermontane basins and abyssal plains rest on the rugged 5.0 km/sec layer. The 8.1 and 6.7 km/sec layers are standard upper mantle and "oceanic" crustal layers. The 7.3–7.5 km/sec is characteristic of mid-oceanic ridges throughout the world. Ewing, Ewing and Talwani have concluded that the 5.0 km/sec layer is "primitive" igneous, largely because of the intensely rough surface. (After Ewing and Ewing, 1959; Ewing, Ewing and Talwani, 1963.)

Identification of the rock types in Fig. 14 depends principally on the seismic-refraction velocities shown and on rocks dredged from a few places nearby. Shand (1949) described the dredge hauls of *Atlantis* Cruise 150 as olivine gabbro, serpentine, basalt and diabase. Some of them have been dated by the K–Ar method (Erickson, *in litt.*) at a few tens of million years. It is possible that the basalts consist largely of young volcanics, not identifiable with any of the seismic layers.

The 8.1 and 6.7 km/sec layers are standard upper mantle and "oceanic" crustal layers. The identity (and even the continuity) of the 5.0 km/sec layer are still debated, owing to the fact that many types of rock are characterized by this velocity; but Ewing, Ewing and Talwani (*loc. cit.*) have concluded that it is "primitive" igneous, largely on the basis of the rough upper surface.

The 7.3–7.5 km/sec layer is apparently characteristic of mid-oceanic ridges throughout the world (Raitt, 1956; Shor, 1959; Menard, 1960).

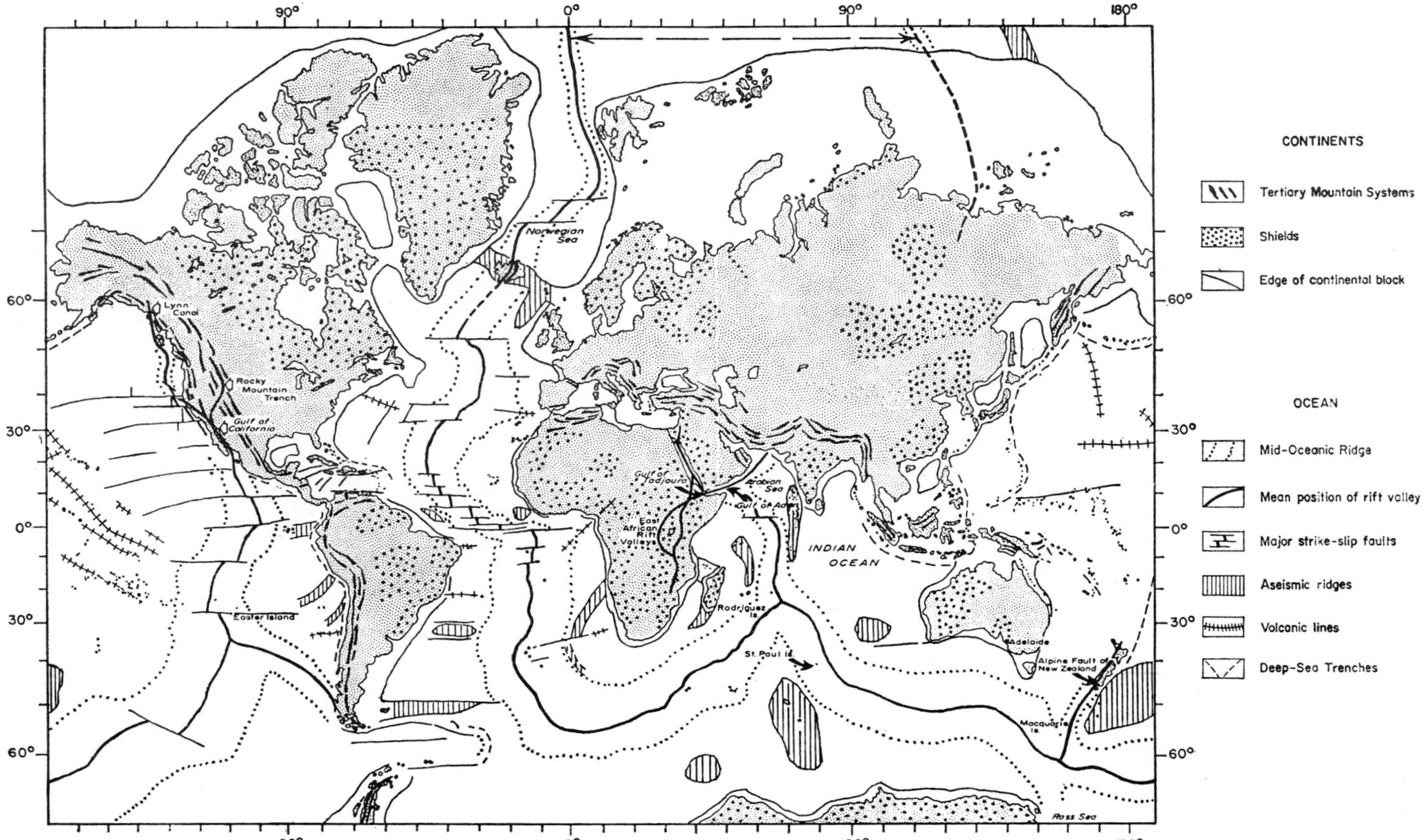

Fig. 15. Major tectonic features of the earth. The map of the mid-oceanic rift valley is essentially a plot of the mid-oceanic earthquakes. In every area in which the bathymetry is known, the epicenter belt follows the crest of the mid-oceanic ridge. In almost every area where detailed submarine topographic studies have been carried out in the region of the epicenter belt, the median rift valley has been discovered, with the exception of certain areas of the South Pacific. Wherever the rift valley extends across continental margins, it joins either rift, graben (as in East Africa or Iceland) or great strike-slip faults (as in California or New Zealand). The axis of the mid-oceanic ridge is displaced by fracture zones in the equatorial Atlantic and in the northeast Pacific. The oceans are so poorly surveyed that it cannot be known if additional fracture zones exist. A few fracture zones are highly speculative, particularly those in the South Pacific and the Norwegian Sea. Note that the mid-oceanic ridge is a much longer and more extensive feature than the Tertiary mountain systems of the continents. (After Heezen, 1962.)

9. Origin

The origin and structure (Fig. 15) of the mid-oceanic ridge have been the subject of much speculation by geologists and geophysicists. Some writers have suggested that the mid-oceanic ridge represents a pile of debris left on the sea floor when the continents drifted apart. Others have suggested that the ridge is composed of sediments which filled the cracks between drifting continents; but high heat flow, earthquake activity, and the results of the seismic-refraction and reflection exploration reveal a modern active feature with crustal characteristics not compatible with either of these hypotheses. Most current workers believe that the mid-oceanic ridge is the result of tension, and that material rising from the mantle beneath the crest of the mid-oceanic ridge is adding new rock to the floor of the rift valley. However, there are two schools of thought concerning the reason for this upflow of mantle rock. The one favored by most workers is the convection-current hypothesis, which attributes the fundamental features of the ocean floor to the effect of convection currents which rise beneath the crest of the mid-oceanic ridge and then flow laterally toward each continent. It is supposed that the drag of these currents on the underside of the continents causes the continents to be compressed and the oceans to be stretched. The second school attributes the stretching of the ocean floors to a general expansion of the interior of the earth.

Since the mid-oceanic ridge is the longest mountain range on the surface of the earth and covers an area almost equal to the area covered by the continents, its origin is of fundamental importance in considerations of the deformation of the earth's crust.

References

Bernauer, F., 1943. Junge Tektonik auf Island und ihre Ursachen. In O. Niemczyk, *Spalten auf Island*. Konrad Wittwer Verlag, Stuttgart, 14–64.

Boyd, L. A., 1948. The coast of Northeast Greenland. *Amer. Geog. Soc., Spec. Pub.* 30, 339 pp.

Bullard, E. C., 1936. Gravity measurements in East Africa. *Phil. Trans. Roy. Soc. London*, **A235**, 45 pp.

Dietrich, G., 1959. Zur Topographie und Morphologie des Meeresbodens im nördlichen Nord Atlantischen Ozean. *Deut. Hydrog. Z.*, Ergänzungsheft Reihe B, **3**, 26–34.

Ewing, J. I. and M. Ewing, 1959. Seismic-refraction measurements on the Atlantic Ocean basins, Mediterranean Sea, on the Mid-Atlantic Ridge and in the Norwegian Sea. *Bull. Geol. Soc. Amer.*, **70**, 291–318.

Ewing, M., J. Ewing and M. Talwani, 1963. Sediment distribution in the ocean—The Mid-Atlantic Ridge. *Bull. Geol. Soc. Amer.*, **74**.

Ewing, M. and B. C. Heezen, 1956. Some problems of Antarctic submarine geology, 75–81. In "Antarctica" in the I.G.Y. *Geophys. Monog.* No. 1, *Amer. Geophys. Un.*

Ewing, M. and B. C. Heezen, 1960. Continuity of the mid-oceanic ridge and rift valley in the southwestern Indian Ocean confirmed. *Science*, **131**, 1677–1679.

Farquharson, W. I., 1936. Soundings, the John Murray Expedition. *Sci. Reps.* 1, 2, *Brit. Mus.* (*Nat. Hist.*), 49 pp.

Fisher, R. L., 1958. Preliminary report on expedition *Downwind*, University of California, Scripps Institution of Oceanography I.G.Y. cruise to the Southeast Pacific. I.G.Y. World Data Center A, Nat. Acad. Sci., Washington 25, D.C., 58 pp.

Frolov, V. V. and V. M. Pasecki, 1958. Arctic Ocean research center. *Priroda*, 1958, no. 8, 56–62.

Gregory, J. W., 1921. *The Rift Valleys and Geology of East Africa*. Seely, Service and Co., London, 479 pp.

Gutenberg, B. and C. F. Richter, 1954. *Seismicity of the Earth*. Princeton University Press, 310 pp.

Heck, N. H., 1938. The role of earthquakes and the seismic method in submarine geology. *Proc. Amer. Phil. Soc.*, **79**, 97–108.

Heezen, Bruce C., 1962, The deep-sea floor, 235–288. In Runcorn, S. K. (Ed.), *Continental Drift*. International Geophysics Series 3, Academic Press, New York and London, 380 pp.

Heezen, Bruce C. and M. Ewing, 1961. The mid-oceanic ridge and its extension through the Arctic Basin, 622–642. In Raasch, G. O. (Ed.), Geology of the Arctic, *Proc. First Intern. Symp. Arctic Geol.* 1, Univ. Toronto Press, Toronto, 732 pp.

Heezen, B. C. and Marie Tharp, 1961. Physiographic diagram of the South Atlantic, the Caribbean, the Scotia Sea, and the eastern margin of the South Pacific Ocean. *Geol. Soc. Amer.*

Heezen, Bruce C., Marie Tharp and Maurice Ewing, 1959. The floors of the oceans. I. The North Atlantic. *Geol. Soc. Amer. Spec. Paper* 65, 122 pp.

Heiskanen, W. A. and F. A. Vening Meinesz, 1958. *The Earth and its Gravity Field*. McGraw-Hill, New York, 470 pp.

Hodgson, E. A., 1930. The seismicity of the Arctic. *Roy. Astr. Soc. Canada*, **24**, 201–210.

Hope, E. R., 1959. Geotectonics of the Arctic Ocean and the Great Arctic magnetic anomaly. *J. Geophys. Res.*, **64**, 407–427.

Linden, N. A., 1959. Seismic chart of the Arctic, 1959, seismological and glaciological researches during the International Geophysical Year; no. 2, *Acad. Sci., Moscow*, 7–17.

McConnell, R. B., 1951. Rift and shield structure in East Africa. *Rep. 18th Session Intern. Geol. Cong., London, 1948*, Pt. XIV, 199–207.

Menard, H. W., 1955. Deformation of the northeastern Pacific Basin and the west coast of North America. *Bull. Geol. Soc. Amer.*, **66**, 1149–1198.

Menard, H. W., 1960. The East Pacific Rise. *Science*, **132**, 1737–1746.

Nalivkin, D. V., 1960. *The Geology of the U.S.S.R.* (translated by S. J. Tomkeieff, translation editor, J. E. Richey). Pergamon Press, London, 170 pp.

Raiko, N. V. and N. A. Lindén, 1935. The earthquake of November 20, 1933, in Baffin Bay and the distribution of seismic foci in the Arctic. *Trans. Seism. Inst., Acad. Sci. U.S.S.R.*, no. 61, 1–8.

Raitt, R. W., 1956. Seismic-refraction studies of the Pacific Basin—I. *Bull. Geol. Soc. Amer.*, **67**, 1623–1640.

Rothé, J. P., 1954. La zone seismique Mediane Indo-Atlantique. *Phil. Trans. Roy. Soc. London*, **A222**, 387–397.

Rudolph, E., 1887, Über submarine Erdbeben und Eruptonen, *Gerl. Beit. Geophysik*, **1**, 133–373.

Rutten, M. G. and H. Wensink, 1960. Structure of Iceland. *Intern. Geol. Cong., Copenhagen*, **18**, 81–88.

Shand, S. J., 1949. Rocks of the Mid-Atlantic Ridge. *J. Geol.*, **57**, 89–92.

Shor, G. G., Jr., 1959. Reflection studies in the eastern equatorial Pacific. *Deep-Sea Res.*, **5**, 283–289.

Stocks, T., 1950. Die Tiefenverhältnisse des Europäischen Nordmeeres. *Deut. Hydrog. Z.*, **3**, 93–100.

Stocks, T. and G. Wüst, 1935. Die Tiefenverhältnisse des offenen Atlantischen Ozeans. *Wiss. Ergbn. Deut. Atlant. Exped. 'Meteor', 1925–27*, Bd. III, Lief. 1, 1–32.

Tams, E., 1922. Die Seismischen Verhältnisse des Europäischen Nordmeeres. *Centralbl. Min. Geol. u. Paleont. Jahrg.*, no. 13, 385–397.

Tams, E., 1931. Die Seismicity der Erde. *Handbuch der Experimentalphysik*, **25**, Pt. 2, 361–437.

Ulrich, Johannes, 1960. Zur Topographie des Reykjanes-Rückens, *Kieler Meeresforschungen*, 155–163.

Wayland, E. J., 1930. Rift valleys and Lake Victoria. *C. R. 15th Intern. Geol. Cong., South Africa*, **2**, 223–253.

17. TRENCHES

R. L. Fisher and H. H. Hess

Man's knowledge of the deepest parts of the oceans, the trenches that nearly ring the Pacific and border parts of the Atlantic and Indian Oceans, goes back about a hundred years. Early workers pointed out the arcuate shape of these features, and their common association with island arcs, mountain ranges and vulcanism. More recently, geophysical investigations of gravity, crustal structure, heat flow and earthquake activity indicate that trenches are surface expressions of large-scale processes acting deep within the earth.

1. Previous Work

By 1900, ships engaged in scientific reconnaissances or searching for cable routes had established that the greatest oceanic depths lie near land, seemingly closely related to the island groups of the western Pacific, to the Indonesian islands, the Antilles or the Andean mountain chain. Murray's bathymetric chart (Murray and Hjort, 1912) showed deeps in nearly all of the presently known trench areas. The only notable omissions are the Middle America Trench off western Mexico and Central America and the South Sandwich Trench in the South Atlantic Ocean. Subsequent bathymetric exploration has refined Murray's map, joining some separate deeps he plotted and named, but it has not revealed any trenches off Antarctica between New Zealand and South America, or between Mexico and southeastern Alaska. Recent work by *Vitiaz*, reported by Udintsev (1959), confirms the presence of a series of small deeps, shoaler than the usual trench, from western New Guinea to near Rotuma, northwest of the Fiji Islands. Another chain of trenches, southwest of or "interior" to the Solomon, Santa Cruz and New Hebrides groups, extends 3800 km from southeastern New Guinea to beyond the Loyalty Islands.

Prior to the 1920's, oceanic soundings were made by wire or line and lead, and, though each measurement consumed several hours, most soundings obtained a small sample of the bottom material. Modern echo-sounders record thousands of depths in a similar period; however, conclusions as to bottom materials and processes are commonly drawn by inference, from the shape and character of the trace recorded or from minor relief on the trench walls and bottom.

Vening Meinesz developed a method for measuring gravity at sea by pendulums installed in a submarine. In a series of cruises beginning in 1923, he demonstrated that negative isostatic anomalies characteristically are associated with trenches, and that nowhere else are such large negative values found (Vening Meinesz, 1948, for summary). Vening Meinesz' trench surveys were most detailed in the East Indies, but he, and later Ewing and Hess, made similar surveys in the Caribbean area (Ewing, 1937, 1938; Hess, 1938). Matuyama (1936) made gravity surveys in the Japan Trench. Since World War II

[*MS received May, 1961*]

scientists of the Lamont Geological Observatory, and others, have actively pursued marine gravity exploration in trench areas; notable are their surveys of the Puerto Rico (Talwani *et al.*, 1959), Peru–Chile (Ewing *et al.*, 1957), Tonga (Talwani *et al.*, 1961) and Middle America (Worzel and Ewing, 1952) trenches. Gajnanov (1955) reviews gravity observations in the Kuril–Kamchatka trench. All of these surveys confirm the existence of a strip of large negative isostatic anomalies centered slightly shoreward of the trench axis and, commonly, smaller positive anomalies closer to the island group or mainland and also seaward of the trench. In some localities, Timor and Barbados for example, the negative strip extends beyond the trench, continuing across the island or along a ridge that perhaps marks the position of a former deep (Hess, 1938). The shape and dimensions of the anomaly curve led Vening Meinesz (1930) to propose that in the region of the negative belt there is an elastic down-buckle of light crustal material into the substratum. Kuenen (1936) published the results of model experiments in which he duplicated this hypothetical structure that he called the "tectogene". Consideration of the probable stresses and elastic limits of crustal materials and the studies of Bijlaard (1936) as well as recent field studies of island arc structure caused Vening Meinesz to revise his hypothesis and to suggest that the anomalies reflect a plastic deformation of crustal material in the mobile belts, with thickening and down-buckling under compression (Vening Meinesz, 1954, 1955). He derived formulas (Vening Meinesz, 1955) that indicate an asymmetrical development of the two sides of a down-buckle at the ocean–continent boundary; such asymmetry is confirmed by the data of station and explosion seismology.

Compilations of earthquake epicenter data, such as those of Gutenberg and Richter (1954), establish that highest seismicity is associated with the circum-Pacific trenches and those of the Caribbean and South Antilles. They indicate that shallow shocks commonly occur near and slightly shoreward of the trench axis and intermediate depth-shocks (70–300 km) below and inside the island arc or mountain chain. When deep shocks (300–700 km) occur, their foci lie well inside the arcuate structure, below an interior basin or inland from the mountain range. From studies of the circum-Pacific earthquake sequences, Benioff (1949, 1954) has suggested that great reverse faults or fracture zones dip steeply shoreward along the island arcs of continental margins. He distinguishes two types of structure, the dual oceanic and the triple marginal fault. The former, typified by the Tonga and Kermadec earthquake sequences, has a lower component (70 to 550–700 km) that dips steeply away from the ocean basin. The marginal faults, perhaps most simply indicated by the Kuril–Kamchatka and Peru–Chile sequences, have an intermediate component (70–300 km) of moderate dip and a steep lower component extending to a depth of 550–700 km. For neither fault type is the observational precision great enough to indicate the dip of the shallow (0–70 km) component. Benioff believes the trench and uplifted shoreward block are surface manifestations of reverse motion on the faults or fracture zones.

Hodgson (1957) has questioned this postulated thrust-fault character of deep-focus earthquakes. From his analysis of specific examples he finds that the plane of movement is nearly vertical and that the displacement is of strike-slip character. His resolution, however, gives in each case two planes at right angles, one parallel, or nearly so, to the trench and one at right angles. McIntyre and Christie (1957) suggest that the set which is parallel is the preferred solution after analyzing Hodgson's earthquake data for the Tonga–Kermadec region. Hess and Maxwell (1953) had previously suggested strike-slip faults in this area at right angles to the trench and still hold this view.

Seismic refraction investigations at sea have provided information on crustal-layer structure down to the Mohorovičić discontinuity. Scripps Institution scientists have carried out such studies in the Tonga Trench (Raitt *et al.*, 1955), Middle America Trench (Fisher and Shor, 1956; Shor and Fisher, 1961), Peru–Chile Trench (Raitt and Shor, 1958) and Aleutian Trench (Shor, 1962). Lamont and Woods Hole personnel have investigated the Puerto Rico Trench (Officer *et al.*, 1957, 1959, for example) and the South Sandwich Trench (Ewing and Ewing, 1959). Russian scientists report seismic and gravimetric work across the Kuril–Kamchatka Trench (Galperin *et al.*, 1958), and Gaskell *et al.* (1958) report one short profile near the Challenger Deep in the Marianas Trench.

Measurements of the flow of heat through trench floors have been reported for the Middle America Trench (Bullard *et al.*, 1956) and the Peru–Chile Trench (von Herzen, 1959). Heat flow was considerably lower near the trench axes off Guatemala and Peru than was general in the deep-sea floor outside. No measurement was obtained in the deepest part of the Chilean segment, off Antofagasta, but a series of nearby measurements, normal to the trench axis, showed progressively lower values as the deep was approached. Though the low values observed (0.17–0.47 μcal/sec/cm^2) might be due to rapid sedimentation in trenches, these authors conclude that the abnormal values reflect real differences in processes occurring far beneath the sea floor. They suggest that convection currents in the mantle may bring heat to the surface in areas such as the East Pacific Rise, where high values of heat flow (up to 8.1 μcal/sec/cm^2) have been measured, and that the trenches may be regions where such currents are sinking (Bullard *et al.*, 1956).

Magnetic profiles have been published for the Tonga (Raitt *et al.*, 1955), Kuril–Kamchatka (Galperin *et al.*, 1958) and Aleutian (Keller *et al.*, 1954) trenches. From the absence of any consistent anomaly correlatable to the trench on their eight crossings, the latter authors state that the anomalies observed are produced by susceptibility contrast within the rocks of the sea floor rather than by the topography of their surface. They conclude that the structure which produced the susceptibility contrasts bears no relation to the trench. Mason (in Raitt *et al.*, 1955) reports an excess of magnetism on the island side of the Tonga Trench and a deficiency to the east that cannot be explained by reasonable basement configuration; he also considers this distribution may be due to susceptibility contrast rather than to structure.

Marshall (1911) pointed out that, in Melanesia, andesitic lavas occur on and to the southwest of a line drawn from New Zealand through Tonga, the Solomon Islands and New Guinea, and that this line closely parallels a series of oceanic trenches. Many geologists have taken this petrographic distinction, the "andesite line", extended northward along the deeps east of the Palau–Yap–Marianas–Japanese–Kuril islands as the chief criterion for distinguishing "continental" from Pacific basin areas. This raised the question of why, if trenches are continental border structures, two such topographic features, the Philippine and Ryuku trenches, lie so far outside the Pacific basin proper and two others, the New Britain and New Hebrides trenches, occur on the wrong side of the andesite line. Recent seismic work in the Caribbean, the deep Bering Sea and the Philippine Basin indicates that an oceanic crust lies between the island arcs and the continents. Hess (1955) suggested that the occurrence of andesite, rather than indicating the margin of the ocean basin proper, may provide an index of deformation; andesitic magmas might be produced where deformation is strong enough to force the basaltic or lower crustal layer downward sufficiently to cause partial fusion. It now seems more likely that andesite of island arcs as well as basalt of the ocean basins comes from deep within the mantle, the former occurring only in areas of strong shearing stress. Partial melting of a small percentage (5% to 10%) of a given volume of mantle will produce a less mafic liquid than basalt, but this magma would be retained in the rigid crystalline framework of the rock unless released by shearing. Partial melting of a large enough percentage of the same volume to destroy the rigidity of the framework could produce a separatable basaltic liquid under relatively passive conditions.

Another striking petrologic observation is the common occurrence of serpentinized peridotites in island groups bordering the East Indian, western Pacific and Antillean trenches (Hess, 1937, *et seq.*). Hess had proposed that these peridotites were intruded during the first great deformation of the present mountain belts, presumably during a down-buckle of the crust, and that their position approximated the axis of the down-fold. From various studies of the relations of these intrusions to datable rocks, Hess assigns a probable mid-Mesozoic age to the main deformation in the Japan–Formosa–Luzon–Borneo regions and a lesser age, perhaps late Cretaceous–Eocene, to those intrusions in the Bonin–Marianas–Palau and Mindanao–Celebes–New Guinea belts (Hess, 1948).

Various workers have noted that in both island arc-trench and continental margin-trench systems a row or double row of volcanoes, active at present or in the fairly recent geologic past, lie sub-parallel to the trench trend. The volcanoes lie 150–350 km landward from the trench axes, with the most common separation about 200 km. Such volcanoes, for example, comprise the islands of western Tonga, much of the Marianas, Bonin, Kuril and Aleutian groups, the peaks on the east shores of the southern Philippines and the Kamchatka peninsula, and the mountains bordering the west coast of Central America. The island volcanoes do not ordinarily rise from the deep-sea floor but from the crests or back slopes of geanticlinal ridges bordering the trenches. Gunn (1947) presented arguments and equations based on engineering theory to support his idea that the structures and spacings observed result from an elastic lithosphere (which overlies a weak magma and is subjected to strong horizontal compression) failing by shear near the continental boundary, with the landward segment overriding and depressing the seaward block to form the trench. Flexure of the lithosphere dies out to either side of the shear plane. He states that on the landward side total stresses on the bottom of the lithosphere are less than those in like position seaward of the trench. Hence, on the land side, highly compressed magmas could be injected into the crust and reach the surface to form the volcanic chain observed. More geophysical measurements of crustal structure, like those recently made in the Guatemala area treated by Gunn, and more data on the physical properties of crust and mantle are needed to test his hypothesis.

The most spectacular and direct observation of trench-bottom conditions was made in January, 1960, when the bathyscaph *Trieste*, piloted by Piccard and Walsh, reached the bottom of the Marianas Trench, near Guam, at 35,800 ft, the deepest point in the oceans (Walsh, 1960). They report a soft, gray, sediment bottom, and noted a small shrimp-like crustacean and one fish at this depth.

Let us turn to a more detailed discussion of trench topography and structure as revealed by recent bathymetric and geophysical explorations by the authors and others.

2. Topography of Trenches

Oceanic trenches are at once among the most appealing and frustrating subjects for bathymetric exploration. To the usual problems of ship positioning, frequency fluctuations and slope returns is added the difficulty of obtaining a recognizable echo from a sea bottom 8 to 10 km beneath the ship. Hence, it is not surprising that nearly every investigation of the very deep trench areas provides data significantly different from that collected by earlier, equally painstaking work in the same areas.

Methods and special problems of sounding in trenches have been discussed by several authors (Fisher, 1954; Kiilerich, 1959, for example). They published geometric reconstructions of profiles taken by echo-sounder or bomb-sounding. Such plots demonstrate that rarely does the first echo receive a return from the trench bottom, even when the ship is directly over the trench axis, unless a sounder with an extremely narrow sound beam width is used. Usually weak echoes from one or both trench walls precede the bottom return as the ship moves across the trench. Only by analyzing the entire echo train to resolve the multiple echoes can an equivalent bottom profile be drawn.

Even the very deep, V-shaped trenches commonly contain a narrow band of nearly flat bottom, $\frac{1}{2}$ to 3 km wide, usually attributed to sedimentary fill. This flat area characteristically gives a relatively strong echo; on a profile normal to the trench axis its lateral boundaries, the base of the trench walls, can be taken as the points where the strongly reflecting, nearly flat bottom apparently deepens perceptibly before fading out. Trench-bottom basins commonly extend a few kilometers or tens of kilometers along the axis. Sills separate the basins or ponds from one another, so that commonly the trench axis displays a series of terraces or steps differing in level by several tens or hundreds of meters (Ewing and Heezen, 1955; Fisher, 1961). Modern echo-sounders with high sensitivity, expanded recording scale and precise timing enable one to record accurately reflection-time differences equivalent to depth differences of less than 10 m, even at the greatest depths. Hence, by combining the assumption of sedimentary fill with a plot of multiple echoes from soundings taken at few-second intervals, a geologically reasonable and consistent topographic form can be deduced, even when a wide-beam sounder is employed. Isolated, momentary deep returns from an otherwise nearly flat trench bottom are probably spurious; usually they are due to a variation in electrical input frequency to the recorder drive, or to an especially strong side echo.

Figs. 1 and 2 are contoured plots of the deepest parts of the Philippine and Marianas trenches, respectively, based on soundings taken in 1959 (R.V. *Stranger*) employing a depth recorder with precise time base. The Philippine trench plot has been contoured both on first and on strong late arrivals to illustrate the above discussion. Late arrivals suggest that a basin deeper than 10,000 m, varying in depth by about 30 m along its mapped axis (reflection times of 12.93–12.97 sec), extends from near 10° 25′N to the bottom of the figure. It is indicated by first arrivals only near 10° 10′N and 10° 23′N. At about 10° 28′N, a sill separates this basin from a similar basin, of slightly shoaler depth, to the north. Reflection times by echo-sounder (vertical lettering) agree with bomb-sounding reflection times (slanted lettering) in the deep parts of the trench. However, generally more echoes can be distinguished on the bomb-records; greater energy from the explosion results in a detectable echo even from poorly reflecting surfaces like steep trench walls. Positions at the west ends of the *Stranger* lines were established by visual bearings and radar ranges on islands, and the ship stayed underway throughout the survey. Discrepancies between the trench axial position from this survey and from earlier surveys by *Cape Johnson* and *Galathea* are within the limits of navigational errors of the ships.

Challenger Deep has been investigated recently by *Challenger* (1951), *Vitiaz* (1957, 1958) and *Stranger* (1959), and by the bathyscaph *Trieste* (1960). The contours of Fig. 2 are based on *Stranger* soundings only. Again discrepancies in location of the greatest depths reported are probably due to uncertainties in fixing the ships' positions. It is likely that all the deep soundings were taken in the same flattish-floored basin. Part of the differences in the *Vitiaz* and *Stranger–Challenger* depths can be attributed to the velocity correction functions used by Del Grosso and Matthews respectively. Also, a variation of recorder-drive input frequency of only 1.7% would account for the range of values reported. *Trieste*'s depth measurement was made with a pressure gauge; it agrees well with the sonic sounding.

It is appropriate here to list the maximum depths reported for 20 of the trenches (Table I). Exploration to date has been sufficiently detailed so that

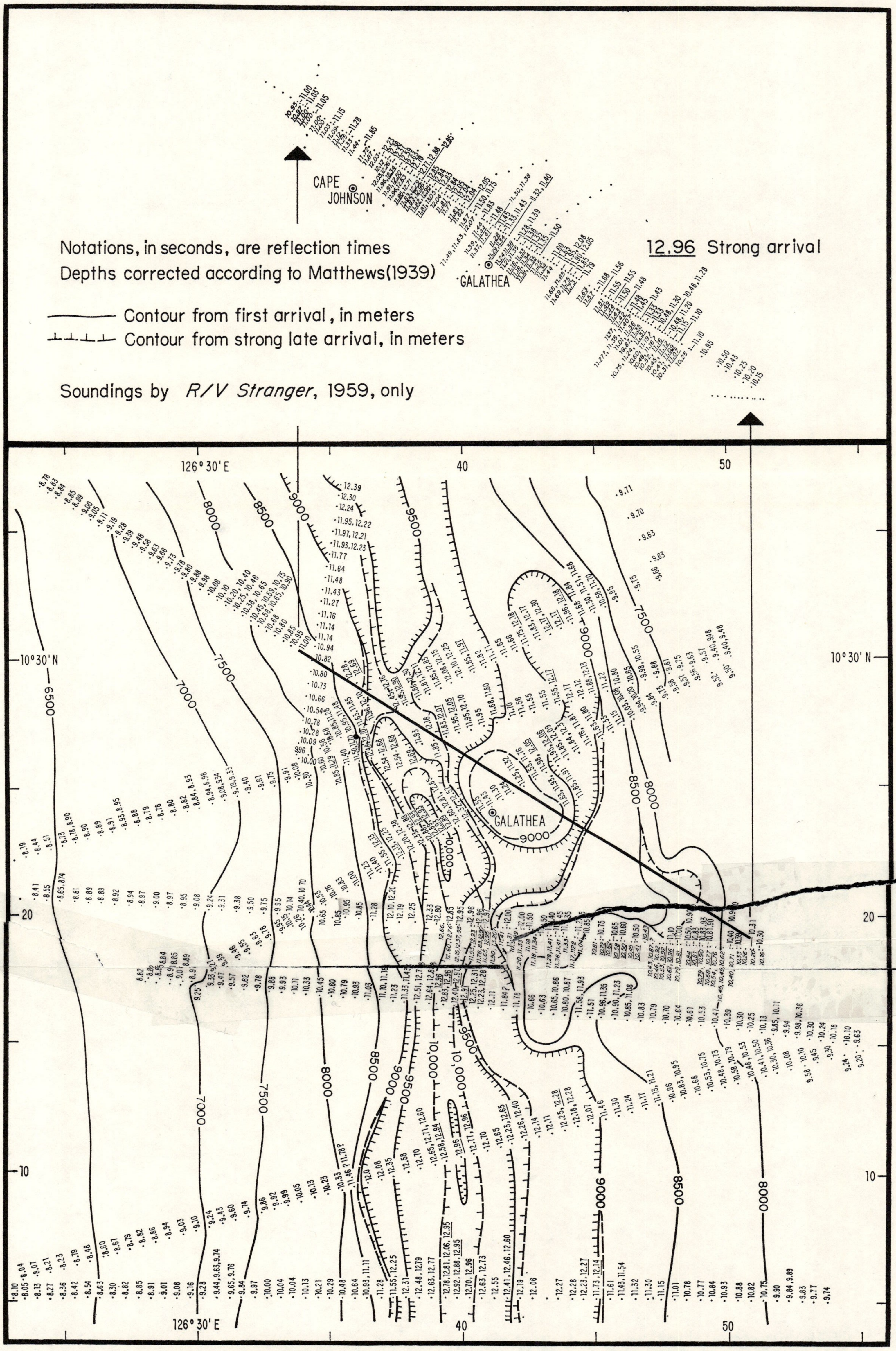

Fig. 1. Topography of the deepest basin in the Philippine Trench, in the vicinity of the Cape Johnson and Galathea Depths.

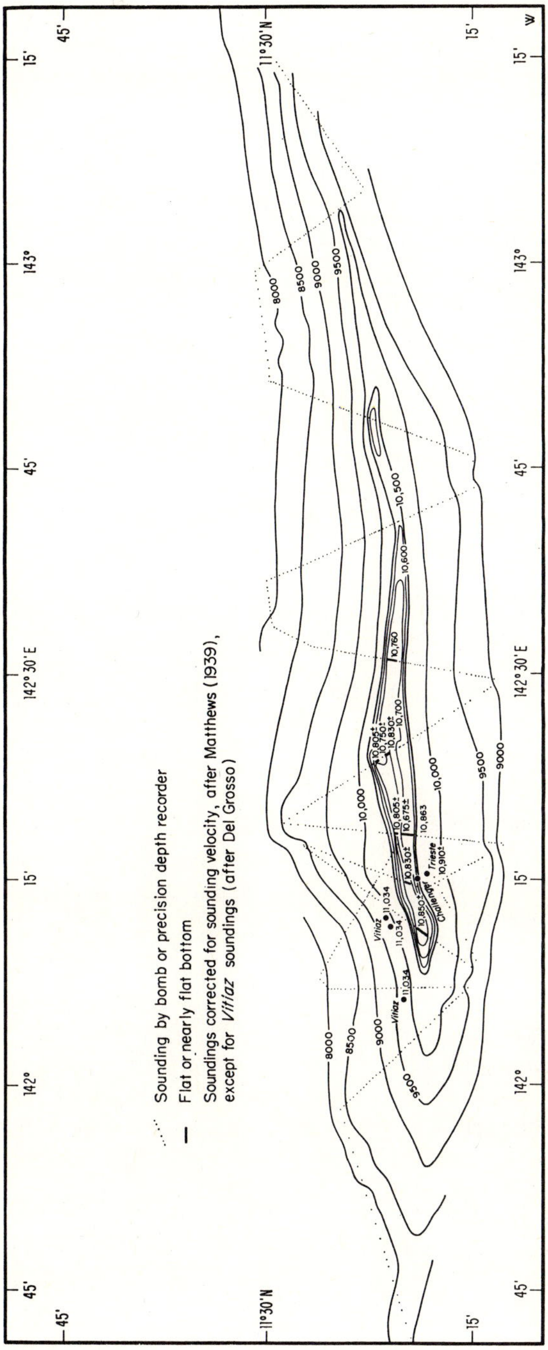

Fig. 2. Bathymetric exploration of the Challenger Deep, Marianas Trench, by S.I.O. R.V. *Stranger*, Naga Expedition, July, 1959.

TABLE I

Maximum Depths, in Meters, Reported for Major Trenches (in part after compilation of Wiseman and Ovey, 1955)

Marianas Trench (specifically Challenger Deep)
- 11,034 ± 50 (Hanson *et al.*, 1959)
- 10,915 ± 20 (this Chapter)[a]
- 10,915 ± (Lyman, *in litt.*, 1960)
- 10,863 ± 35 (Carruthers and Lawford, 1952)
- 10,850 ± 20 (this Chapter, Fig. 2)

Tonga
- 10,882 ± 50 (Hanson *et al.*, 1959)
- 10,800 ± 100 (Fisher and Revelle, 1954)

Kuril–Kamchatka
- 10,542 ± 100 (Udintsev, 1959)
- 9750 ± 100 (Delauze, *in litt.*, 1962)[b]

Philippine (vicinity of Cape Johnson Deep)
- 10,497 ± 100 (Hess and Buell, 1950)
- 10,265 ± 45 (Bruun, quoted in Wiseman and Ovey)
- 10,030 ± 10 (this Chapter)

Kermadec
- 10,047 ± (Udintsev, 1960)

Idzu-Bonin (includes "Ramapo Deep" of the Japan Trench)
- 9810 (Udintsev, 1959)
- 9695 (vicinity of Ramapo Depth, Nasu *et al.*, 1960)

Puerto Rico
- 9200 ± 20 (Lyman, 1954)

New Hebrides (North)
- 9165 ± 20 (this Chapter)[a]

North Solomons (Bougainville)
- 9103 ± (Udintsev, 1960)
- 8940 ± 20 (this Chapter)[a]

Yap (West Caroline)
- 8527 ± (Kanaev, 1960)

New Britain
- 8320 ± (Udintsev, 1960)
- 8245 ± 20 (this Chapter)[a]

South Solomons
- 8310 ± 20 (this Chapter)[a]

South Sandwich
- 8264 (Maurer and Stocks, 1933)

Peru–Chile
- 8055 ± 10 (Fisher, 1958)

Palau
- 8054 ± (Udintsev, 1960)
- 8050 ± 10 (this Chapter)[a]

(*continued*)

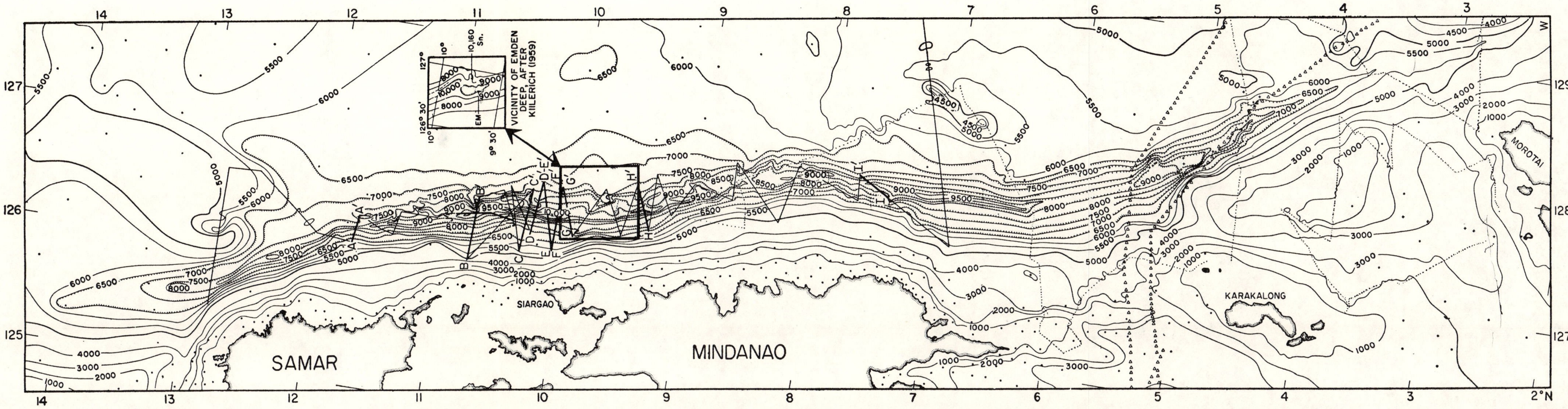

Fig. 3. Exploration of the central portion of the Philippine Trench by R.V. *Stranger*, Naga Expedition, August, 1959.

TABLE I—*continued*

Aleutian	
7679	(uncorrected, taken with nominal sounding velocity of 1500 m/sec)
Nansei Shoto (Ryuku)	
7507	
Java	
7450	(after van Riel, 1934)
New Hebrides (South)	
7070 ± 20	(this Chapter)[a]
Middle America	
6662 ± 10	

[a] These soundings were taken during Proa Expedition, April–June, 1962, aboard R.V. *Spencer F. Baird*. A Precision Depth Recorder was employed, and the ship's track crossed over (within the limits of celestial navigation) points from which maximum depths had been reported.

[b] This is the maximum sounding obtained in the vicinity of the Vitiaz Depth (Udintsev, 1959) by French and Japanese vessels in connection with dives of the bathyscaph *Archimède*, July, 1962.

future work will modify the ranking very little, except perhaps in the case of the Marianas–Tonga pair, where the spread in recent depth measurements is less than 250 m. As the preceding discussion and Figs. 1 and 2 indicate, more absolute values must await precise ship positioning and the development of powerful narrow-beam sounders with precise timing.

Recently prepared topographic charts, based largely or wholly on data from continuously recording sounders, include plots of the following trenches: Tonga (Raitt *et al.*, 1955), Middle America (Fisher, 1961), Peru–Chile (Fisher 1958), Philippine (Kiilerich, 1959; this Chapter), Kermadec (Brodie and Hatherton, 1958), Kuril–Kamchatka (Udintsev, 1955), New Britain–Bougainville–New Hebrides (this Chapter), Puerto Rico (Ewing and Heezen, 1955) and the western portion of the Aleutian Trench (Gibson and Nichols, 1953; Gates and Gibson, 1956).

The Philippine Trench (Fig. 3) is representative of the deep narrow trenches bordering island chains or small land areas of the western Pacific and associated with all the gravity and seismic characteristics discussed above. In plan nearly straight, its axis lies almost equally distant from the 4000-m contour at the base of the Philippine island chain and from the 6500-m contour marking the lip of the deep Philippine basin to the east. There is no ridge or swell paralleling this trench on the seaward side. The most marked change in trend occurs east of southern Mindanao, where the trench veers southeast but the negative-gravity-anomaly belt associated with it farther north continues southward to the west of Halmahera (Vening Meinesz *et al.*, 1934). A short southwest-trending ridge lies east of the trench in the vicinity of the change in trend. Northward the Philippine Trench ends against a broad area shoaler than 5000 m; to the south it dies out against the rise extending from Palau nearly to Halmahera. Although only 1250 km long at depths greater than 6000 m, it is deeper than 8500 m for more than 800 km of its length. Three basins deeper than 9500 m occur; the longest of these, near Siargao, is slightly greater than 10,000 m deep.

Where the Philippine Trench is best developed, at depths greater than 7500 m, the west or shoreward flank is steeper, with slopes of 10–16°, than the offshore flank, with general slopes of 6–8°. Shoaler than 7500 m, on the shoreward side, the slope is more gentle, generally 4–8° up to the nearshore 1000-m contour. On the east flank, above 7500 m, the bottom rises at a 1–4° slope to the deep Pacific sea floor. Locally, the general slope is steepened by pinnacles or ridges on the trench flank. The slopes listed here are similar to those reported for the Tonga Trench (Raitt *et al.*, 1955), Kuril–Kamchatka Trench (Udintsev, 1955) and the Japan (Idzu–Bonin) Trench (Nasu *et al.*, 1960). As the trench shoals to north and south, the slopes of the flanks become more nearly equal.

Bathymetric exploration of the upper slopes of the Philippine Trench has been insufficient to establish the presence or absence there of numerous submarine canyons and large re-entrants and depressions well described for the Kuril–Kamchatka Trench (Udintsev, 1955) and the Aleutian Trench (Gibson and Nichols, 1953; Gates and Gibson, 1956). These authors proposed a fault origin for most features of this kind, and related submarine fault systems to structural trends mapped on the adjacent islands. Udintsev, noting the change in character of trench bottom and island-trench nearshore topography from south to north along the Kurils, suggests that the most active trench development is taking place off Kamchatka and that the frequent vulcanism of that region is activated by transverse faults extending across the slope.

Cross-sections, traced from echograms (Fig. 4), of the deepest parts of the Philippine Trench clearly show a narrow flat bottom, 1–3 km wide, in this deep V-shaped trench. Projecting the lower trench walls to intersection, and allowing for steepening with depth, indicates a thickness of less than 300 m for the sedimentary fill in this basin, and thus a very small volume of sediment in this trench. This figure for fill is similar to that deduced from seismic-refraction observations in the Tonga Trench (Raitt *et al.*, 1955). The lower walls of this trench are irregular, with small knolls or pinnacles, particularly on the offshore side, that dam sediments or rocks slumping from further upslope. Bare rock, probably volcanic, may outcrop on such steep trench walls and pinnacles (Raitt *et al.*, 1955; Bezrukov, 1957). The small elevations shown on Sections I–I′ and J–J′ are spurs or toes of ridges constricting the basins and ponding sediments. Kiilerich (1959) notes that north of 10° 45′N, where a ridge constricts the trench, there is a valley or depression east of the trench axis and separated from it by a ridge up to 500 m high. Section A–A′ shows both the trench axis and this outer depression, which is interpreted as a fault trough.

Sediments of such very deep trenches have rarely been studied critically up to the present time, because of difficulties in obtaining adequate undisturbed samples and in knowing exactly where in the trench the samples were collected.

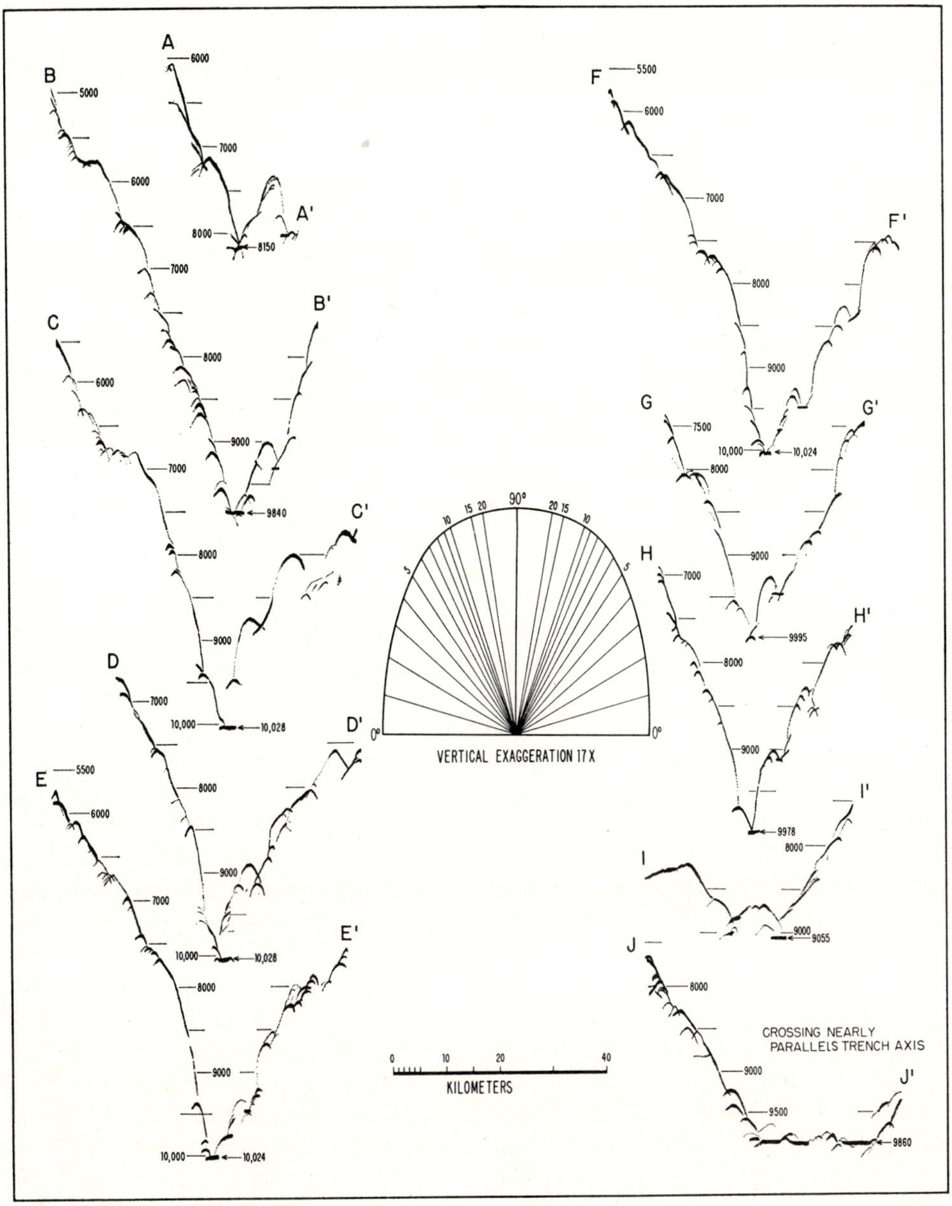

Fig. 4. Cross-sections of the Philippine Trench traced from R.V. *Stranger* Precision Depth Recorder tapes. Sounding scales corrected after Matthews (1939). (See Fig. 3 for location of the sections.) *Note:*—Since a wide-beam sounder was employed, many side echoes appear as early, late or faint returns. The upper surface, ordinarily accepted as the sounding trace, is actually the envelope of minimum reflection times.

In general, samples taken from depths shoaler than 4500 m on trench flanks are predominantly calcareous, composed of *Globigerina* oozes and silts; in tropical areas coralline fragments are common. The carbonate fraction is mixed with volcanic debris and pyroclastics. Further downslope the sediments are brown or red clays and silts together with ash and volcanic lapilli; organic remains commonly are siliceous. Surface deposits on the trench floors range from brown clays or silts to sand and fragments of volcanic glass and pumice; beneath the surface layers the sediments commonly are gray to black or dark brown. Coarse graded layers, sometimes several centimeters thick, result from intermittent turbidity-current deposition and from slumping of material down the steep walls. Occasionally such layers contain tests of shallow-water Foraminifera, attesting to turbidity-current origin and rapid burial. Bezrukov (1957) reported that large fragments or galls of older laminated clays litter portions of the western Pacific trench floors; some of the laminations in the fragments show distortion by sliding. He concluded that trench-floor sediments may show three types of stratification: (1) a micro-lamination of the solid clays, not yet explained; (2) an alternation, up to several millimeters or centimeters in thickness, of clays or silts with volcanic ash or sands and gravel, attributable to volcanic eruptions or to periodic stirring up, by earthquakes, of pyroclastic material on the trench slopes; (3) vertical differences, expressed in colors, of the oxidation states of iron and manganese in the silts and clays. This latter stratification, due to diagenetic processes, is controlled by rate of accumulation of the sediments and the incorporation of organic material into them.

The Middle America Trench (Fig. 5), shoalest of the 20 trenches listed in Table I, borders the west coast of Central America for more than 2600 km, from central Mexico to Costa Rica. In morphology and setting it strikingly resembles, on a smaller scale, the Peru–Chile Trench bordering the South American continent (Zeigler *et al.*, 1957; Fisher, 1958). Strongly arcuate over part of their extent, nearly straight in other sections, the axes of these trenches parallel closely the break in slope at the continental margin. Both are associated with gravity minima, shallow and intermediate depth earthquakes (deep-focus earthquakes also occur in the South American sequence) and, over part of their length, vulcanism. In both trenches a shallow northern section is separated from a deeper southern section by an offshore ridge intersecting the trench at a large angle. The ridges may be offshoots of the large fracture zones of the eastern Pacific (Menard and Fisher, 1958; Fisher and Norris, 1960). Rüegg (1960), however, considers that the ridge bisecting the Peru–Chile Trench represents very recent uplift of a former land-mass that foundered before or during the Nevadan movement, prior to the formation of the trench.

The Middle America Trench is intersected and approximately bounded on the north by the Clarion Fracture Zone that has been linked to the transverse volcanic province of Mexico (Menard, 1955). Topographically the northern termination is rather abrupt; farther to the northwest, the Gulf of California may mark a continuation of the same seismically active zone. To the south the

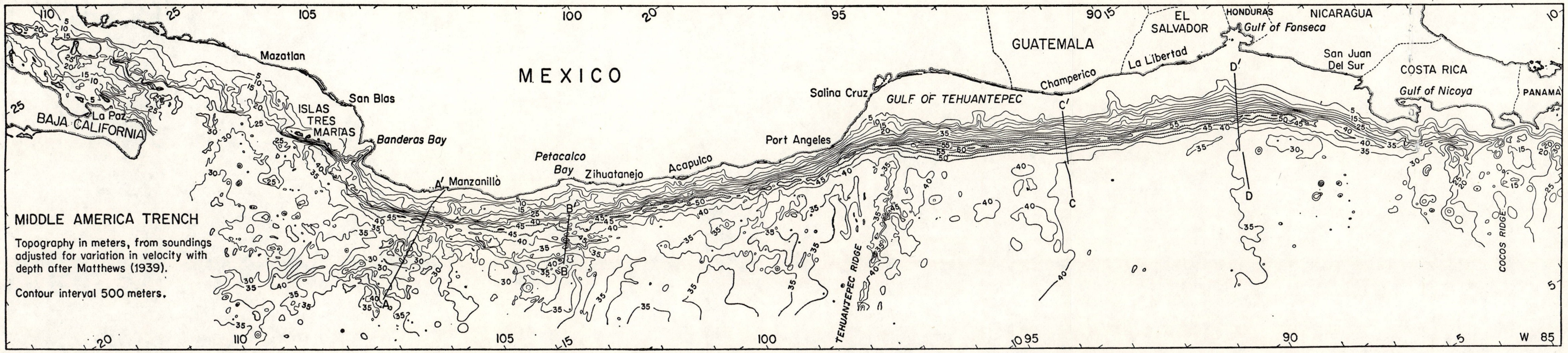

Fig. 5. Topography of the Middle America Trench, based chiefly on soundings from Scripps Institution investigations, 1952–1959.

trench dies out against the northeast-trending Cocos Ridge. The Tehuantepec Ridge intersects the trench near 15°N, 95°W (Menard and Fisher, 1958); at this boundary the depth of the trench, the topography outside the trench, the depth to the Mohorovičić discontinuity and the vulcanism related to the trench are changed completely (Fisher, 1961).

Northwest of the Gulf of Tehuantepec, depths along the trench axis vary abruptly over short distances. Small steps or terraces result from ponding of sediments; locally submarine hills constrict or even replace the usual trench floor (Fig. 6, A–A′, B–B′). However, northern cross-sections characteristically

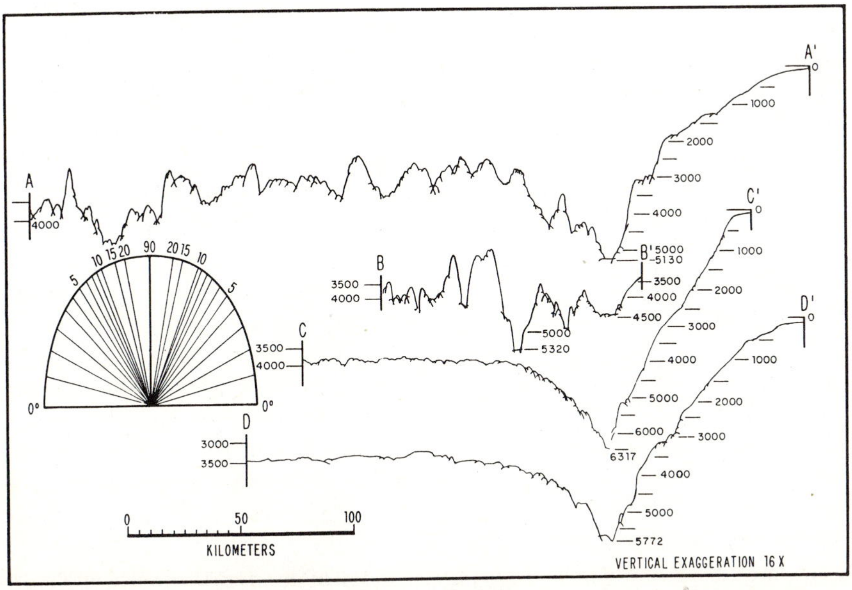

Fig. 6. Cross-sections of the Middle America Trench traced from continuous-sounding records. Sounding scales corrected after Matthews (1939). (See Fig. 5 for location of the sections.)

show the trench floor to be flat or gently basined; within such basins the bottom commonly rises 10–20 m toward the land and seaward flanks, with sharp breaks in slope at the basin edges. Seismic-refraction measurements (Shor and Fisher, 1961) and projection of flank slopes indicate $\frac{1}{2}$ to $1\frac{1}{2}$ km of sediments in several of the basins between Islas Tres Marías and Acapulco. In this segment of the trench, lying close to a coast composed dominantly of metamorphic and intrusive rocks, trench sediments are gray to greenish brown clays and silts. The fine sand layers occasionally present here in trench-floor cores are attributed to frequent slumping or turbidity-current transport down the steep flanks. Seismic-refraction measurements along the trench axis between Islas Tres Marías and Acapulco indicate that beneath the thick sediments and a probable basement (volcanic?) layer, the main crustal layer ($V_p = 6.5$–

6.9 km/sec) has a thickness about the same as at normal Pacific basin stations. Apparently, this portion of the Middle America Trench is relatively inactive at present; faulting and vulcanism seaward of the trench are more significant than the trench-forming process. If so, this trench may soon, in a geological sense, resemble the minor, strikingly flat-bottomed, sediment traps that border the west coast of central Baja California (Fig. 8f; Fisher and Revelle, 1955).

To the southeast of the intersection with the Tehuantepec Ridge, the axial depth of the trench increases abruptly, then shoals slowly toward the Cocos Ridge. This portion of the trench has the more usual V-shaped cross-section (Fig. 6, C–C′, D–D′) and a line of recently active volcanoes parallels its trend. However, the volume of trench-floor sediments is small. Coarse land-derived material is deposited on the broad shelf or upper slope; trench-floor deposits are fine-grained, with recognizable ash layers and higher proportions of volcanic glass. Seismic-refraction measurements in the deep part of the trench, off Guatemala (Fig. 8e) and El Salvador, indicate that here the main crustal layer is considerably thicker than it is seaward of the trench, although it thins somewhat to the outer portion of the continental shelf, where it is overlain by a thick section of sedimentary and basement (volcanic?) rock. Off Guatemala the crustal thickness and thickening at the axis of this minor, relatively shallow trench are nearly as great as that of the very active, deep Tonga Trench (Fig. 8c).

The linked trenches south of New Britain, southeast of the Solomon Islands and west of the New Hebrides (Fig. 7) represent a series concave toward the Pacific Ocean. This being the only present example of island arcs and trenches concave away from a continental area is the interesting example contrary to the general rule. The whole of this region is active seismically and evidence for Recent or sub-Recent vulcanism is common along the islands adjacent to the concave side of the trenches. Earthquake foci of moderate depth are common below the concave side of the arcs. The area is not well known and the bathymetry shown in Fig. 7 is based on the wartime data collected by the U.S.S. *Cape Johnson*, which was used to check such sounding lines and old charts as were available in the U.S. Hydrographic Office. The trench recently reported by *Vitiaz* (Udintsev, 1958) has been added.

One notable feature of the relatively small trenches in this area is that there is a distinct tendency to shoal or disappear off the middle of the arcs. In both the case of the Solomon Islands and the New Hebrides the trenches are well developed at either extremity but disappear off the centers of the arcs. A somewhat similar tendency may be noted in the case of the Marianas Trench.

The Cape Johnson–Vitiaz trench system is unique and has no known counterpart in other island arcs. If the ordinary trench off the convex side of an island arc represents the down-buckling of the crust, the Cape Johnson–Vitiaz Trench may represent some sort of strike-slip fault adjustment between adjacent blocks. This has been suggested by R. W. Fairbridge (*in litt.*, 1960).

The morphological details of the trenches paralleling the island arcs are

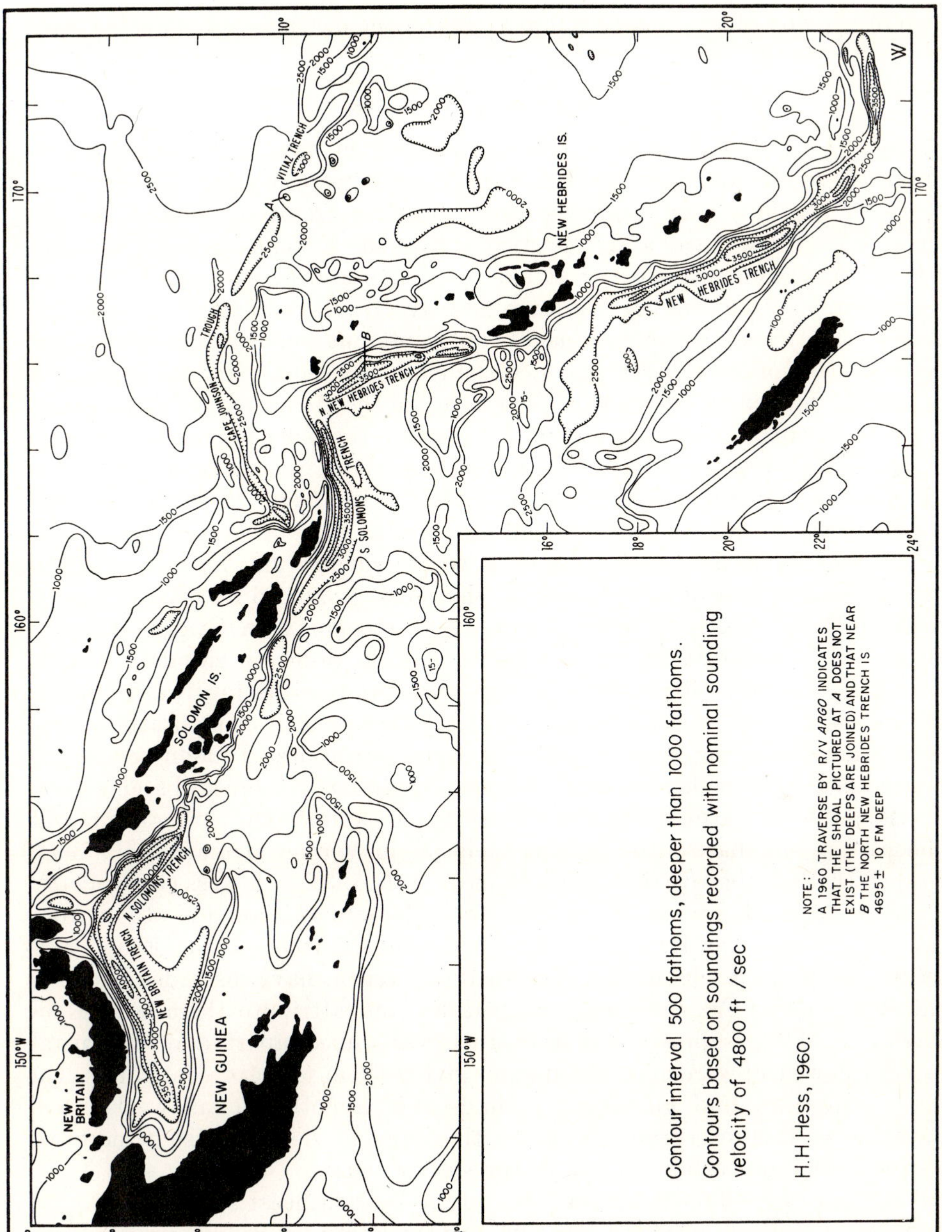

Fig. 7. Submarine topography of the New Britain–New Hebrides region, southwest Pacific. *A* near 10°S, 170°E; *B* in North New Hebrides Trench.

poorly known. In general the trenches are 1500 to 2000 fm deeper than the nearby sea floor. On the recent cruise of R.V. *Argo*, a depth of 4935 fm (corrected) was discovered in the North New Hebrides Trench, a depth considerably greater than thus far found in any of the other trenches of this area.

3. Structure of Trenches

Two major developments in this century which gave some insight into tectonic processes operating in the Earth's crust were Vening Meinesz' discovery in the late 1920's of very large negative gravity anomalies on the convex side of island arcs and in particular over the trenches, and the development of explosion seismometry at sea by Ewing and Worzel after the war. The prewar explanations for the large negative gravity anomalies were based on an oceanic crust not unlike that of continents but somewhat thinner. Thus a down-buckle of a sial layer provided the mass deficiency required to explain the gravity anomalies. Once the shallow depth of the Mohorovičić discontinuity under the oceans was established, modification of the older hypotheses became necessary. As a topic this has not yet been thoroughly reconsidered.

Laboratory experiments by Kuenen (1936) and by Griggs (1939) lent general support to the concepts then current. Both sets of experiments were predicated on a stronger crustal layer floating on a weaker substratum. Kuenen, starting with the assumption of compression in the crust, produced an elastic type of down-buckle to which he gave the name "tectogene". In Griggs' model a viscous liquid substratum was caused to flow in order to simulate mantle convection currents. Viscous drag on the crust produced a downward bulging of the crustal layer in a plastic manner which resulted in configurations over the downward-flowing limb resembling alpine structure. Though much has changed in the years subsequent to Griggs' experiments, there still seems to be a considerable degree of validity in the picture he painted. The earlier structural concepts are summarized by Vening Meinesz (in Vening Meinesz *et al.*, 1934, 1948), J. H. F. Umbgrove (in Vening Meinesz *et al.*, 1934) and Hess (1938, 1948).

More recently M. Ewing, Worzel, Shurbet, Heezen, Talwani and others from Lamont (Worzel and Ewing, 1954; Worzel and Shurbet, 1955; Ewing and Heezen, 1955), in discussing the Puerto Rico Trench, have discarded the idea that the trenches are compressional features and attribute them to tension (extension). They suppose that the crust was pulled apart and subcrustal material flowed upward into the fissure, giving them their present forms. On both geophysical and geological grounds this hypothesis has weaknesses. Nevertheless, it is interesting to discuss this point of view because it serves to focus attention on certain critical features of trenches.

The Puerto Rico Trench curves smoothly to the southeast and south around the Virgin Islands. Southward along its axis the Barbados Ridge arises, so that the trench changes into a rise flanked on either side by shallower troughs. The strongly negative gravity anomalies follow the axis from the trench to the axis of the Barbados Ridge, indicating the continuity of the structure and proving that the anomalies are the result of deeper structure, not the topography itself. A very similar situation is presented by the Java Trench and its central ridge. The Barbados Ridge was explained by Hess (1938) as the squeezing up of incompetent material from the axis of a down-buckle "like tooth paste out of a tube". Deep drilling for oil on Barbados in recent years gives a picture consistent with this interpretation so far as the sedimentary section is concerned. But stronger evidence is supplied by the seismic profile given by J. Ewing *et al.* (1957) showing the Mohorovičić discontinuity coming in from the Atlantic and bending downward approaching the Barbados Ridge (Fig. 8b). Similarly, Raitt, Fisher and Mason (1955) show the crust bending downward as one approaches the Tonga Trench from the Pacific (Fig. 8c). On the island side one might interpret the crust as behaving in much the same way except that the mantle velocity for P waves below it has dropped from 8.2 km/sec to 7.0–7.6 km/sec. This could be a result of serpentinization as discussed below. The Tonga Trench has large negative gravity anomalies (Talwani, Worzel and Ewing, 1961) but little sedimentary fill on its floor. On its eastern flank a guyot, Capricorn Seamount, with its upper surface tilted toward the trench axis has been described by Raitt *et al.* (1955). The bending downward of this upper topographic surface of the guyot may be assumed to parallel what the crustal layers have done beneath it and thus supports the seismic evidence mentioned above. Menard and Dietz (1951) report a guyot on the axis of the Aleutian Trench with its flat top about 900 fm deeper than guyots to the south of the trench.

The Puerto Rico Trench has been investigated much more intensively by geophysical methods than any other trench area. The geology of the islands to the south of it is now fairly well known (Mattson, 1960; Berryhill *et al.*, 1960; Donnelly, 1959; Helsley, 1960). More deductions may be drawn with regard to its structure than that of other trenches. One may hazard a guess that it first formed at about the beginning of Tertiary time, some 70 million years ago. Until late Eocene, active vulcanism and the intrusion of dioritic plutons were taking place but ceased at this time. Vulcanism, however, continued up to the present along the Lesser Antilles. Paralleling this, the sediments of the Barbados Ridge have been squeezed up from a trench axis but those in the Puerto Rico Trench apparently have not been. From this one may deduce that the Puerto Rico Trench is not actively going down at the present time, but that forces must be still acting upon it to keep it far out of isostatic equilibrium. Unfortunately, it is not a typical trench as would be the Tonga–Kermadec, Aleutian or Kuril Trenches. Large east–west strike-slip faults with a sinistral movement are known on the island of Puerto Rico, and N 60°E sinistral strike-slip faults paralleling the Anegada Trough have been mapped by Donnelly in the Virgin Islands. Such faults may have modified to some extent the Puerto Rico Trench, but probably as second-order effects not changing the gross structure drastically.

Talwani, Sutton and Worzel (1959) have computed very carefully mass distributions under the Puerto Rico Trench section and to the north and south

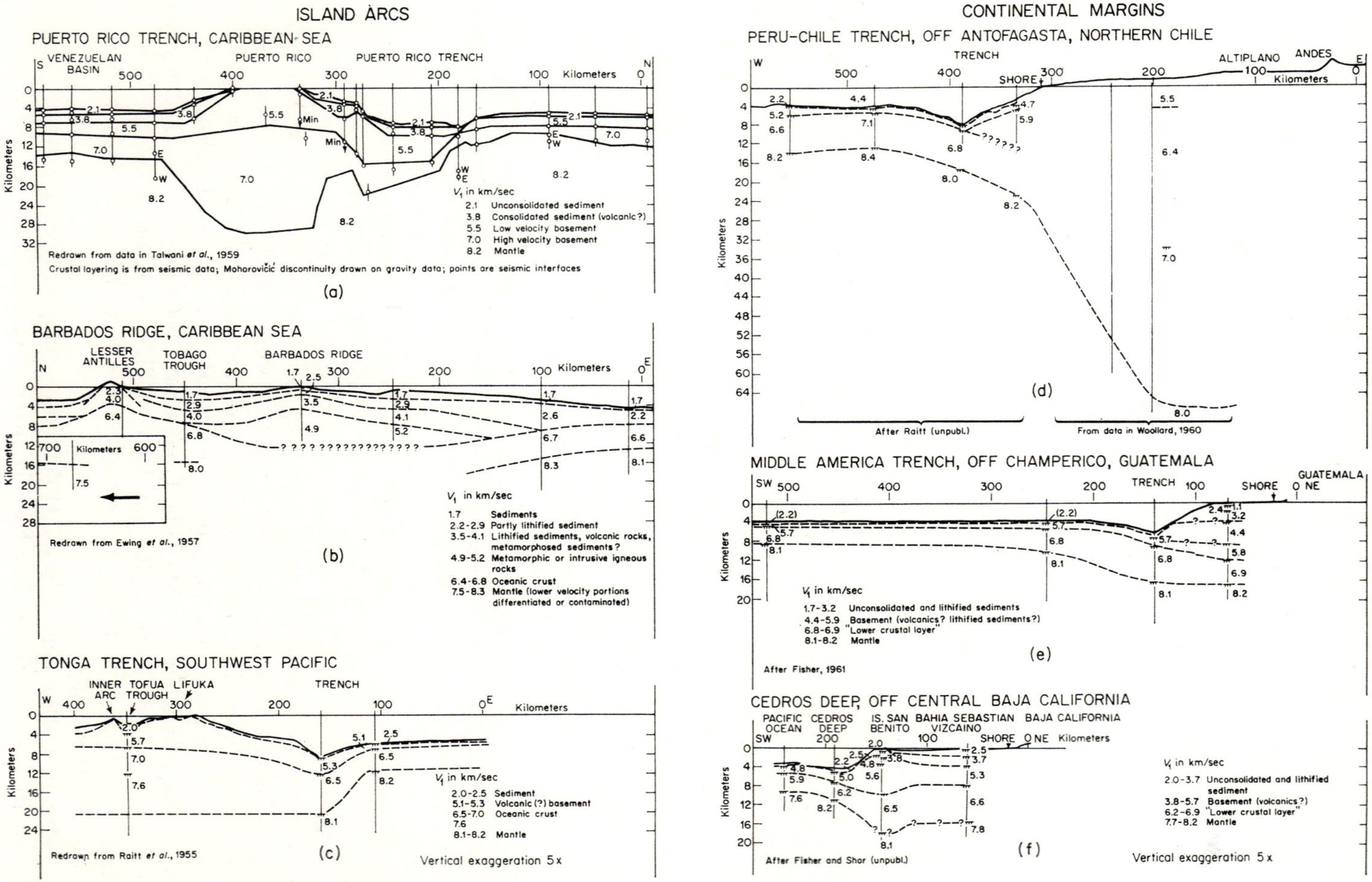

Fig. 8. Island arc and continental margin structure sections deduced from seismic-refraction data.

of it. These computations are based upon a fit between observed gravity, seismic velocities and indicated thicknesses of layers (Fig. 8a). Whereas more geophysical work might modify the minor detail in Talwani's section, there is enough information to accept his mass distribution as essentially established. The isostatic gravity anomalies over the trench (and the continuation along the Barbados Ridge) are still in the neighborhood of −180 mgal, indicating quite clearly that assumption of isostatic equilibrium here is incorrect; in other words, the required high-density antiroot under the trench to balance the mass deficiency represented by the water does not exist. The degree of departure from equilibrium may be estimated nicely from Talwani's data. Assuming that all the mass anomaly is above some arbitrary depth such as 40 km, one can compute the pressure at this depth for a crustal column in the middle of the trench and compare it to a standard oceanic column or to the columns on either side of it in Talwani's diagram. The result of such a calculation shows the pressure under the trench to be somewhat greater than 1000 kg/cm^2 less than a standard column. This is the critical factor in any attempt to explain the structure and dynamics of a trench.

Drawing an analogy to the behavior of the crust when loaded by ice accumulation gives a picture which can easily be appreciated. If a lobe of ice 120 km wide, 10 km thick at the center of the lobe, and several hundred kilometers long were resting on the crust in isostatic equilibrium and were instantaneously removed, then the resulting gravity anomalies and mass deficiency would be similar to the Puerto Rico Trench. Uplift would be rapid and equilibrium would be approached in at most a few tens of thousands of years. Thus, if one were to postulate formation of the trench by extension, any anomaly which developed would disappear rapidly; but an even more serious objection is that there is no reason why so large a departure from equilibrium should ever have come about. On a much smaller scale (10 km), one can appeal to the floating cork-block type of analogy. In this case various blocks are tilted, depressed or raised such that the positive and negative anomalies compensate for one another. The large negative one over the Puerto Rico Trench is not compensated by positive anomalies in the adjacent areas. An extension hypothesis of the origin of the trench cannot account for large departures from isostatic equilibrium.

Most geologists would agree to the probability that the Red Sea structure represents extension. Girdler (1958) analyzes the geophysical data for this area and comes to the same conclusion. The gravity anomalies are generally positive and small. Similarly, most of the free-air anomalies over the Bartlett Trough (Cayman Trench), generally considered to have formed by extension, are small, and the isostatic anomalies commonly strongly positive. The conclusion must be drawn that compressional forces are acting on the rock beneath the Puerto Rico Trench and other trenches sufficient to hold them down. The magnitude of the vertical component of this force is in the neighborhood of 1000 kg/cm^2.

It now seems necessary to construct a hypothesis which would adequately account for the mass distribution, the seismic velocities, and the forces acting to keep trenches far out of isostatic equilibrium. Such a hypothesis is necessarily highly speculative but it is at least better than no reasonable hypothesis —which is the present status with respect to this problem.

The forces necessary to produce the structure and departure from isostatic equilibrium will be attributed to mantle convection currents. The only other source which has been suggested, thermal contraction of the earth, seems the less likely of the two at this time. The supposed nature of the structure is shown in Fig. 9. It differs from previous proposals in that the mantle and "crust"

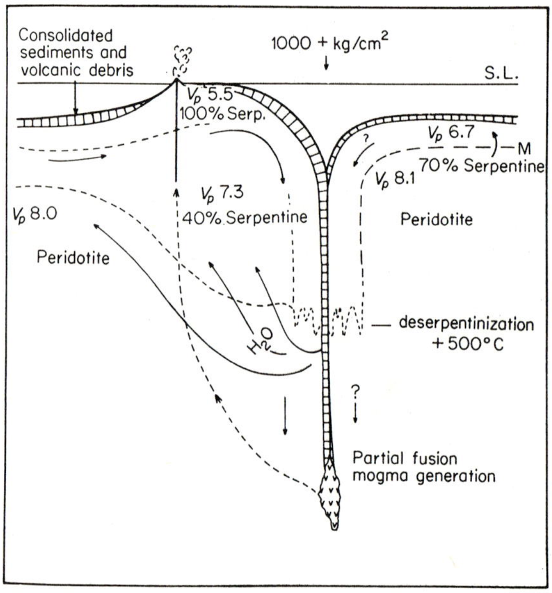

Fig. 9. Supposed structure, with typical seismic velocities (V_p in km/sec), for a hypothetical trench–island arc association.

are moving together as a unit rather than as in Griggs' (1939) experiment where the weaker mantle material was moving at a higher velocity and acting on the crust by viscous drag. It is postulated that the "crust" is serpentinized peridotite (70% serpentinized, $V_p = 6.7$ km/sec, density 2.86 g/cm³) rather than basalt. The crustal layer goes down with the descending limb of the convection cell until it reaches a temperature in the neighborhood of 500°C where a deserpentinizing reaction takes place, releasing water. Similarly, the original supracrustal volcanics and sediments descend until melting or partial melting occurs. Fluids, magma or water, rise, migrating toward the island arc or concave side of the structure. The reason for migration upwards on the concave rather than the convex side of the arc is given in Fig. 10. On this side there will be a tendency for open fractures to form at right angles to the direction of compression (at right angles to the trench axis). Strong deformation and plastic flow in the region where the mantle and crust are bending downward into the descending limb of the convection cell might prevent direct vertical rise of the fluids.

The above discussion treats largely with the curved trenches flanking island arcs. The question might be raised whether the linear trenches, such as those off Chile or middle America, have the same structure. One might reasonably

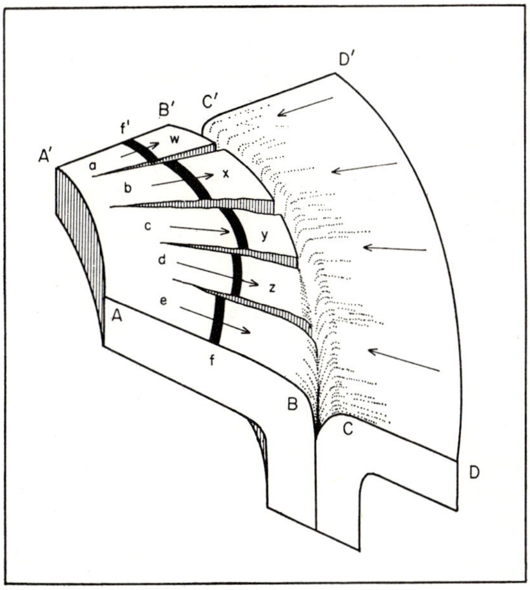

Fig. 10. Postulated mechanical situation where the crust moves into a curved trench. Movement of A′A to B′B would result in extension in the direction B′B, providing potential fissures to permit escape of magma to volcanoes at w, x, y and z. These volcanoes are in all cases on the concave side of the trench and about 180 km from the trench axis. Immediately beneath the trench, deformation is severe, and plastic flow might be expected to close the fissures. Movement on the convex side from D′D to C′C would cause compression along C′C so that open fissures would not form and hence volcanoes are absent. Individual sectors a, b, c, d and e might move differentially as portrayed by the displaced bed f′f. Strike-slip faults would bound the sectors and would tend to be sinistral toward A′B′ and dextral toward AB. *Note:*—This figure represents the deformation of a layer of arbitrary thickness, not necessarily the crust.

propose that these trenches are formed by the continental block overriding the oceanic area or the downward-flowing limb of a convection cell. Benioff (1949, 1955) postulated an overthrust mechanism and showed that the deep earthquake foci lie along a plane dipping under Chile at approximately 40°. This raises a series of unresolved problems which will be briefly considered below. Before proceeding to this subject, it should be noted that the Tonga–Kermadec Trench is also a linear feature but in other respects it is related to a province having all the characteristics of an island arc. Coulomb (1945) showed that the deep earthquake foci for island arcs also lie on a plane dipping under the concave side at about 45°.

Hodgson (1957) studied the first motion of large deep-focus earthquakes of the Tonga–Kermadec and South American regions. From these he deduced that the movement causing the earthquakes occurred as nearly horizontal displacement on nearly vertical planes. This type of analysis results in defining two planes at right angles to each other. Which is the real fault plane is indeterminate. In either case Benioff's postulated 40–45° thrust motion is ruled out.

McIntyre and Christie (1957) analyzed Hodgson's results for the Tonga–Kermadec region in an attempt to solve the problem of which was the real plane. They made fabric diagrams of the poles of the fault planes and concluded that the fabric which paralleled the main linear feature of the topography, the volcanic ridge and trench, was the more likely. In other words, the fault planes are parallel to the trench. Their reasoning is not compelling, and the present authors would prefer to choose the set of planes nearly at right angles to the trench. On the basis of topography, structural pattern, seismicity and volcanic activity, Hess and Maxwell (1953) postulated a sinistral strike-slip fault from the southeast end of the New Hebrides to the northern end of the Tonga Trench. Subsequently, Hodgson analyzed an earthquake lying on this line and found sinistral motion: one of his planes paralleled Hess and Maxwell's deduced fault, the other being perpendicular to it. Looking again at Hodgson's data, one finds that sinistral fault motions tend to occur to the north and dextral to the south in the Tonga–Kermadec region. The pattern thus resembles the displacements shown for the line f–f′ on Fig. 10. For the alternative hypothesis of McIntyre and Christie to have both dextral and sinistral faults paralleling the trench seems mechanically less likely and quite unexplained. On the other hand, many great strike-slip faults of the circum-Pacific do parallel the coast lines and trenches. To mention a few, there are the San Andreas, Kamchatka, 1960 Chilean earthquakes, Luzon to Mindanao in the Philippines, and the Alpine fault of South Island, New Zealand. These have shallow earthquake foci and are perhaps unrelated to the intermediate and deep-focus shocks on Benioff's 40° plane. In conclusion, it is evident that many of the problems relating seismicity to structure are unresolved.

References

Benioff, H., 1949. The fault origin of oceanic deeps. *Bull. Geol. Soc. Amer.*, **60**, 1837–1866.

Benioff, H., 1954. Orogenesis and deep crustal structure—additional evidence from seismology. *Bull. Geol. Soc. Amer.*, **65**, 385–400.

Benioff, H. 1955. Seismic evidence for crustal structure and tectonic activity. *Geol. Soc. Amer. Spec. Paper* 62, 61–73.

Berryhill, H. L., Jr., R. P. Briggs and L. Glover, III, 1960. Stratigraphy, sedimentation and structure of late Cretaceous rocks in eastern Puerto Rico—Preliminary report. *Bull. Amer. Assoc. Petrol. Geol.*, **44**, 137–155.

Bezrukov, P., 1957. On the sediments of the Idzu-Bonin, Marianas and Ryuku deep-sea oceanic trenches. *Doklady Akad. Nauk S.S.S.R.*, **114**, 387–390. [In Russian.]

Bijlaard, P. P., 1936. Théorie des déformations plastiques et locales par rapport aux anomalies négatives, aux fosses océaniennes, aux géosynclinaux, etc. *I.U.G.G. Rep., 6th Gen. Assembly, Assoc. Geod.*, Edinburgh.

Brodie, J. W. and T. Hatherton, 1958. The morphology of Kermadec and Hikurangi trenches. *Deep-Sea Res.*, **5**, 18–28.

Bullard, E. C., A. E. Maxwell and R. R. Revelle, 1956. Heat flow through the deep sea floor; in *Advances in Geophys.*, **3**. Academic Press, New York.

Carruthers, J. N. and A. L. Lawford, 1952. The deepest oceanic sounding. *Nature*, **169**, 601–603.

Coulomb, J., 1945. Séismes profonds et grandes anomalies de la pesanteur. *Ann. Géophys.*, **1**, 244–255.

Donnelly, T. W., 1959. The geology of St. Thomas and St. John, Virgin Islands. Ph.D. Thesis, Princeton University.

Ewing, J. and M. Ewing, 1959. Seismic refraction measurements in the Scotia Sea and South Sandwich Arc (abst.): *Intern. Oceanog. Cong. Preprints*. Amer. Assoc. Adv. Sci., Washington.

Ewing, J., C. B. Officer, H. R. Johnson and R. S. Edwards, 1957. Geophysical investigations in the eastern Caribbean: Trinidad Shelf, Tobago Trough, Barbados Ridge, Atlantic Ocean. *Bull. Geol. Soc. Amer.*, **68**, 897–912.

Ewing, M., 1937. Gravity measurements on the U.S.S. *Barracuda. Trans. Amer. Geophys. Un.*, **18**, 66–69.

Ewing, M., 1938. Marine gravimetric methods and surveys. *Proc. Amer. Phil. Soc.*, **79**, 47–70.

Ewing, M. and B. C. Heezen, 1955. Puerto Rico Trench topographic and geophysical data. *Geol. Soc. Amer. Spec. Paper* 62, 255–267.

Ewing, M., J. L. Worzel and G. L. Shurbet, 1957. Gravity observations at sea in U.S. submarines *Barracuda, Tusk, Conger, Argonaut* and *Medregal. Verhandl. Koninkl. Ned. Geol.-Mijn. Gen., Geol. Ser.*, **18**, 49–115.

Fisher, R. L., 1954. On the sounding of trenches. *Deep-Sea Res.*, **2**, 48–58.

Fisher, R. L., 1958. In *Preliminary Report on Expedition Downwind*. I.G.Y. General Report Ser., **2** (I.G.Y. World Data Center A, Washington).

Fisher, R. L., 1961. Middle America Trench: topography and structure. *Bull. Geol. Soc. Amer.*, **72**, 703–720.

Fisher, R. L. and R. M. Norris, 1960. Bathymetry and geology of Sala y Gomez, southeast Pacific. *Bull. Geol. Soc. Amer.*, **71**, 497–502.

Fisher, R. L. and R. Revelle, 1954. A deep sounding from the southern hemisphere. *Nature*, **174**, 469.

Fisher, R. L. and R. Revelle, 1955. The trenches of the Pacific. *Sci. Amer.*, **193**, 36–41.

Fisher, R. L. and G. G. Shor, Jr., 1956. Topography and structure of the Acapulco Trench (abst.). *Resúmenes Trab. Presentados, 20th Int. Geol. Cong.*, Mexico.

Gajnanov, A. G., 1955. Pendulum gravity measurements in the Sea of Okhotsk and the northwestern Pacific. *Akad. Nauk S.S.S.R. Inst. Okean., Trudy*, **12**, 145. [In Russian.]

Galperin, E. I., A. V. Goryachev and S. M. Zverev, 1958. The Pacific Geologo-Geophysical Expedition of the U.S.S.R. Academy of Sciences, 1957, XII (Seismology), no. 1. Akademia Nauk S.S.S.R., Moscow. [In Russian.]

Gaskell, T. F., M. N. Hill and J. C. Swallow, 1958. Seismic measurements made by H.M.S. *Challenger* in the Atlantic, Pacific and Indian Oceans and the Mediterranean Sea, 1950–1953. *Phil. Trans. Roy. Soc. London*, **A251**, 23–83.

Gates, O. and W. Gibson, 1956. Interpretation of the configuration of the Aleutian Ridge, Rat Islands—Semisopochnoi I. to west of Buldir I. *Bull. Geol. Soc. Amer.*, **67**, 127–146.

Gibson, W. and H. Nichols, 1953. Configuration of the Aleutian Ridge. *Bull. Geol. Soc. Amer.*, **64**, 1173–1188.

Girdler, R. W., 1958. The relationship of the Red Sea to the East African Rift system. *Q. J. Geol. Soc. London*, **114**, 79–102.

Griggs, D., 1939. A theory of mountain-building. *Amer. J. Sci.*, **237**, 611–650.

Gunn, R., 1947. Quantitative aspects of juxtaposed ocean deeps, mountain chains, and volcanic ranges. *Geophysics*, **12**, 238–255.

Gutenberg, B. and C. F. Richter, 1954. *Seismicity of the Earth.* Princeton University Press, Princeton.

Hanson, P. P., N. L. Zenkevich, U. V. Sergeev and G. B. Udintsev, 1959. Maximum depths of the Pacific Ocean. *Priroda*, no. 6, 84–88. [In Russian.]

Helsley, C. E., 1960. The geology of the British Virgin Islands. Ph.D. Thesis, Princeton University.

Hess, H. H., 1937. Island arcs, gravity anomalies and serpentinite intrusions. A contribution to the ophiolite problem. *17th Intern. Geol. Cong. Rep.*, **2**, 263–283.

Hess, H. H., 1938. Gravity anomalies and island arc structure with particular reference to the West Indies. *Proc. Amer. Phil. Soc.*, **79**, 71–96.

Hess, H. H., 1948. Major structural features of the western North Pacific, an interpretation of H. O. 5485, bathymetric chart, Korea to New Guinea. *Bull. Geol. Soc. Amer.*, **59**, 417–445.

Hess, H. H., 1955. The oceanic crust. *J. Mar. Res.*, **14**, 423–439.

Hess, H. H. and M. W. Buell, 1950. The greatest depth in the oceans. *Trans. Amer. Geophys. Un.*, **31**, 401–405.

Hess, H. H. and J. C. Maxwell, 1953. Major structural features of the Southwest Pacific: a preliminary interpretation of H. O. 5484, bathymetric chart, New Guinea to New Zealand. *7th Pacific Sci. Cong., N.Z.*, **2**, 14–17.

Hodgson, J. H., 1957. Nature of faulting in large earthquakes. *Bull. Geol. Soc. Amer.*, **68**, 611–643.

Kanaev, V. F., 1960. New data on the bottom relief of the western part of the Pacific Ocean. *Oceanological Researches*, **2**, 33–44. [In Russian.]

Keller, F., Jr., J. L. Mueschke and L. R. Alldredge, 1954. Aeromagnetic surveys in the Aleutian, Marshall and Bermuda Islands. *Trans. Amer. Geophys. Un.*, **35**, 558–572.

Kiilerich, A., 1959. Bathymetric features of the Philippine Trench. *Galathea Rep.*, **1**, 155–171.

Koczy, F. F., 1956. Echo soundings. *Reps. Swed. Deep-Sea Exped., 1947–48*, **4**, (3), 97–158.

Kuenen, Ph. H., 1936. The negative isostatic anomalies in the East Indies (with experiments). *Leidsche Geol. Neded.*, **8**, 169–214.

Lyman, J., 1954. The deepest sounding in the North Atlantic. *Proc. Roy. Soc. London*, **A222**, 334–336.

Marshall, P., 1911. Oceania; in *Handb. Reg. Geol.*, **7**, Heft. 2.

Matthews, D. J., 1939. Tables of the velocity of sound in pure water and sea water for use in echo-sounding and sound-ranging. *Brit. Admiralty Hydrog. Dep. Pub.*, H. D. 282, 2nd edn.

Mattson, P. H., 1960. Geology of the Mayagüez area, Puerto Rico. *Bull. Geol. Soc. Amer.*, **71**, 319–362.

Matuyama, M., 1936. Distribution of gravity over the Nippon Trench and related areas. *Proc. Imp. Acad. Tokyo*, **12**, 93–95.

Maurer, H. and T. Stocks, 1933. Die Echolotungen des *Meteor. Wiss. Ergebn. Deut. Atlant. Exped. 'Meteor', 1925–27*, **2**, 1–309.

McIntyre, D. B. and J. M. Christie, 1957. Nature of faulting in large earthquakes (discussion). *Bull. Geol. Soc. Amer.*, **68**, 645–652.

Menard, H. W., 1955. Deformation of the northeastern Pacific Basin and the west coast of North America. *Bull. Geol. Soc. Amer.*, **65**, 1149–1198.

Menard, H. W. and R. S. Dietz, 1951. Submarine geology of the Gulf of Alaska. *Bull. Geol. Soc. Amer.*, **62**, 1263–1285.

Menard, H. W. and R. L. Fisher, 1958. Clipperton Fracture Zone in the northeastern equatorial Pacific. *J. Geol.*, **66**, 239–253.

Murray, J. and J. Hjort, 1912. *The Depths of the Ocean.* Macmillan, London.

Nasu, N., A. Iijima and H. Kagami, 1960. Geological results in the Japanese Deep Sea Expedition in 1959. *Oceanog. Mag.*, **11**, 201–214.

Officer, C. B., J. I. Ewing, R. S. Edwards and H. R. Johnson, 1957. Geophysical investigations in the eastern Caribbean: Venezuelan Basin, Antilles Island Arc, and Puerto Rico Trench. *Bull. Geol. Soc. Amer.*, **68**, 359–378.

Officer, C. B., J. I. Ewing, J. F. Hennion, D. G. Harkrider and D. E. Miller, 1959. Geophysical investigations in the eastern Caribbean: summary of 1955 and 1956 cruises. In *Physics and Chemistry of the Earth*, **3**. Pergamon Press, London.

Raitt, R. W., R. L. Fisher and R. G. Mason, 1955. Tonga Trench. *Geol. Soc. Amer. Spec. Paper* 62, 237–254.

Raitt, R. W. and G. G. Shor, Jr., 1958. In *Preliminary Report on Expedition Downwind.* I.G.Y. General Report Ser., **2** (I.G.Y. World Data Center A, Washington).

Riel, P. M. van, 1934. The bottom configuration in relation to the flow of bottom water. *The 'Snellius' Exped.*, **2** (2), Chap. 2, Utrecht.

Rüegg, W., 1960. An intra-Pacific ridge, its continuation onto the Peruvian mainland, and its bearing on the hypothetical Pacific landmass. *Proc. XXI Intern. Geol. Cong., Copenhagen*, Part X (Submarine geology), 29–38.

Shor, G. G., Jr., 1962. Seismic refraction studies off the coast of Alaska: 1956–57. *Bull. Seism. Soc. Amer.*, **52**, 37–57.

Shor, G. G., Jr., and R. L. Fisher, 1961. Middle America Trench: seismic refraction measurements. *Bull. Geol. Soc. Amer.*, **72**, 721–730.

Talwani, M., G. H. Sutton and J. L. Worzel, 1959. A crustal section across the Puerto Rico Trench. *J. Geophys. Res.*, **64**, 1545–1555.

Talwani, M., J. L. Worzel and M. Ewing, 1961. Gravity anomalies and crustal section across the Tonga Trench. *J. Geophys. Res.*, **66**, 1265–1278.

Udintsev, G. B., 1955. Topography of the Kuril-Kamchatka Trench. *Trudy Inst. Okeanol.*, **12**, 16–61. [In Russian.]

Udintsev, G. B., 1958. Discovery of a deep-sea trough in the western part of the Pacific Ocean. *Priroda*, no. 7, 85–88. [In Russian.]

Udintsev, G. B., 1959. Relief of abyssal trenches in the Pacific Ocean (abst.). *Intern. Oceanog. Cong. Preprints.* Amer. Assoc. Adv. Sci., Washington.

Udintsev, G. B., 1960. Bottom relief of the western part of the Pacific Ocean. *Oceanological Researches*, **2**, 5–32. [In Russian.]

Van Huystee, Th., 1944. Echo soundings. *The 'Snellius' Exped.*, **2** (2), Chap. 3, Utrecht.

Vening Meinesz, F. A., 1930. Maritime gravity surveys in the Netherlands East Indies, tentative interpretation of the results. *Koninkl. Ned. Akad. Wetenshap., Proc.*, **33**, 566–577.

Vening Meinesz, F. A., 1947. Major tectonic phenomena and the hypothesis of convection currents in the earth. *Q. J. Geol. Soc. London*, **103**, 191–207.

Vening Meinesz, F. A., 1948. *Gravity expeditions at sea, 1923–1938*, **4**. Pub. Neth. Geod. Comm., Waltman, Delft.

Vening Meinesz, F. A., 1954. Indonesian Archipelago: a geophysical study. *Bull. Geol. Soc. Amer.*, **65**, 143–164.

Vening Meinesz, F. A., 1955. Plastic buckling of the earth's crust: the origin of geosynclines. *Geol. Soc. Amer. Spec. Paper* 62, 319–330.

Vening Meinesz, F. A., J. H. F. Umbgrove and Ph. H. Kuenen, 1934. *Gravity expeditions at sea, 1923–1932*, **2**. Pub. Neth. Geod. Comm., Waltman, Delft.

Von Herzen, R. C., 1959. Heat-flow values from the southeastern Pacific. *Nature*, **183**, 882–883.

Walsh, D., 1960. Our 7-mile dive to the bottom. *Life*, **48** (6), 113–121.

Wiseman, J. D. H. and C. D. Ovey, 1955. Proposed names of features on the deep-sea floor. *Deep-Sea Res.*, **2**, 93–106.

Woollard, G. P., 1960. Seismic crustal studies during the I.G.Y.: part II, continental program. *I.G.Y. Bull.*, no. 34, 1–5.

Worzel, J. L. and M. Ewing, 1952. Gravity measurements at sea, 1948 and 1949. *Trans. Amer. Geophys. Un.*, **33**, 453–460.

Worzel, J. L. and M. Ewing, 1954. Gravity anomalies and structure of the West Indies, part II. *Bull. Geol. Soc. Amer.*, **65**, 195–199.

Worzel, J. L. and G. L. Shurbet, 1955. Gravity interpretations from standard oceanic and continental crustal sections. *Geol. Soc. Amer. Spec. Paper* 62, 87–100.

Zeigler, J. M., W. D. Athearn and H. Small, 1957. Profiles across the Peru–Chile Trench. *Deep-Sea Res.*, **4**, 238–249.

18. MICROTOPOGRAPHY

A. S. Laughton

1. Introduction

The topographic forms of the sea floor that have so far been described are all large features delineated by echo-sounding survey. It will have been apparent, however, that there is a lower limit to the size of feature that can be resolved by echo-sounding because of the limitations of wave-length, beam angle and depth of water. Common wave-lengths are 3 cm (50 kc/s) in shallow water (i.e. less than 200 m) and 15 cm (10 kc/s) in deep water. Objects with dimensions less than several wave-lengths cannot be resolved. A more severe limitation is, however, imposed by the finite width of the beam which allows a point reflector to appear as a hyperbolic trace on the record as the ship passes over it, the width of the hyperbola being determined by the beam angle. Echo-sounders usually have an effective beam width of about 20°. In 200 m the beam, therefore, covers an area on the sea floor 70 m wide and, in 5000 m, an area nearly 1800 m wide. Although features somewhat smaller than these can often be resolved by studying the form of the echo, it is clear that the beam angle imposes a limitation on the lateral resolution. Vertical resolution, depending on time measurements, is generally much greater than this, being limited ultimately by the existence of side echoes, and often by the vertical movement of the ship on the surface.

In order to study features that cannot be resolved by echo-sounding, it is necessary to use either direct observation from a diving machine, such as a bathyscaph, remote-controlled underwater television or automatic underwater photography. The upper end of the size scale of features observed with these techniques is limited chiefly by the scattering and attenuation of light in the sea. In practice, in the clearest oceanic waters, the maximum range for the visibility of objects is about 50 m, whereas in coastal areas it may be as little as 20 cm. The lower limit depends on how close the observer is to the bottom and the resolution of the film.

The term microtopography is used here to describe those features of the sea floor that can be observed by visual means in the scale range of 50 m to 1 mm, which includes rock outcrops, sand waves and ripples, the benthic community and disturbances of the sediment by faunal activity. It is clear therefore, especially in deep water, that there is a gap in the size spectrum of observable features where they are too small to be detected by conventional echo-sounding but too large to be observed visually.

The techniques of bathyscaph observation, underwater television and photography are described in Volume 2, Chapter 23, and Volume 3, Chapter 19. On account of the limited number of observations of the sea floor made by the first two methods, most of the information discussed below has been derived from a study of underwater photographs obtained from published papers and from unpublished collections of photographs at the National Institute of

[*MS received August, 1960*]

Oceanography (England), the Lamont Geological Observatory (U.S.A.) and the Navy Electronics Laboratory (U.S.A.). Published photographs of the sea floor are scattered throughout the literature and only a limited number of references can be given. Whereas most papers deal with a regional study of the sea floor, some contain photographs from a variety of different physiographic regions (Schenck and Kendall, 1954; Owen, 1958; Heezen, Tharp and Ewing, 1959; Laughton, 1959).

The distribution of observations of the sea floors throughout the world is extremely uneven, some small areas having been studied in great detail and large areas having no observations. Furthermore, in publications, the photographs reproduced tend to be the most interesting ones rather than those most typical of a particular area, and, therefore, there is a bias towards showing a greater proportion of such things as boulders and outcrops.

In the survey of the microtopography described below, some 2000 photographs have been studied from as wide a variety of areas as possible (Fig. 1),

Fig. 1. World distribution of underwater photographs either published or inspected by the author.

and, as well as giving a general description, an attempt has been made to deduce some statistics about the distribution of bottom types. It must be realized, however, that the distribution of data is so poor that the conclusions drawn from these results can only be considered as very tentative.

Nothing will be said in this chapter about sea-floor observations in coastal zones where observations are easily made by shallow diving. The microtopography in these regions is well illustrated by the many photographs published in the popular and scientific literature.

2. Classification of Microtopographic Features

Three types of feature can be studied from random photographs of the sea floor and all pictures are analyzed under these three headings.

A. Surface geology. The nature of the bottom is listed under one or more of the classes; bedrock, boulders, cobbles, pebbles and granules, sand and mud, according to Table I, and rock types are identified where possible.

TABLE I

Classification of the Nature of the Bottom according to Size. (After Wentworth, 1922.)

Bedrock	*in situ*
Boulders	> 256 mm
Cobbles	64–256 mm
Pebbles and granules	2–64 mm
Sand	$\frac{1}{16}$–2 mm
Mud (silt and clay)	$< \frac{1}{16}$ mm

B. Indications of current activity. Ripple marks, scour pits, sediment cover and the orientation of fauna are related, if possible, to current strength and direction.

C. Benthic fauna. The fauna is identified and its ecology studied. Organic detrital material can sometimes be identified. Qualitative and quantitative studies are made of the disturbances of the sediment surface by benthic organisms.

A. Surface Geology

a. Bedrock outcrops

Whereas fractured rocks in a dredge may suggest the existence of bedrock, it can only be proved by direct observation or photographs. In searching for suitable outcrops to sample, experience of the correlation of photographs of outcrops with echo-sounding characteristics is invaluable. Most outcrops have been observed, as one would expect, on steep slopes or cliffs (Fig. 2), but comparatively recent submarine lava flows may solidify on a nearly level surface (Fig. 3). The majority of outcrops have proved to be basaltic and associated with seamounts of volcanic origin, although recent dredgings have indicated that some seamounts are capped by limestone (Fig. 4).

Bedrock not associated with topographic highs has been found in the Romanche Trench (Fig. 5) at 7500 m where erosion and slumping have been active, and on the cliffs of turbidity current cut channels in the abyssal plains (Fig. 6) (Laughton, 1960).

The identification of rock types from photographs is made difficult by a coating of manganese dioxide that accumulates on the rock surface, and by

Fig. 2. 41° 12′N, 15° 14′W. Depth 2939 m. Seamount west of Iberia plain. Area of picture $1\frac{1}{2}$ by 1 m. (Photo by N.I.O.)

A view into the face of a rock cliff. The bulbous appearance of the rock suggests either a water-cooled lava or a manganese crust. Many 10-armed commatulid crinoids can be seen, often attached to broken and manganese-covered coral fragments. An ophiuroid is visible in the top left corner.

Fig. 3. 35° 12′N, 15° 18′W. Depth 1317 m. Seamount northeast of Madeira. Area of picture 2 by 10 m. (Photo by L.G.O.)

Field of submarine pillow lavas with manganese-coated surface. Note the roundedness of the rocks and the lack of sediment cover on top indicating active currents.

Fig. 4. 42° 40′N, 11° 35′W. Depth 666 m. Top of Galicia Bank. Area of picture 1½ by 2½ m. (Photo by N.I.O.)

Bottom of limestone with marked linear fractures, strewn with dark erratic cobbles and boulders. The prominent fauna are commatulid crinoids with a pennatulid in the centre.

Fig. 5. 0° 10′S, 18° 21′W. Depth 7500 m. Romanche Trench. Area of picture 1 by 1 m. (Photo by Edgerton for N.G.S.)

Fractured rock-face showing coarse conglomerate-like structure of rock. The lack of fine sediments suggests strong currents. Four small Crustacea are swimming above the bottom. The dark patches are mud clots on the camera window.

Fig. 6. 43° 31.5′N, 12° 36′W. Depth 5200 m. Head of Biscay–Iberia interplain channel. (Photo by N.I.O.)

Semi-consolidated sedimentary rocks exposed on the cliffs of an eroded channel cut into the Biscay abyssal plain by turbidity currents. These may represent the degree of lithification of abyssal-plain sediments buried some 100 m deep.

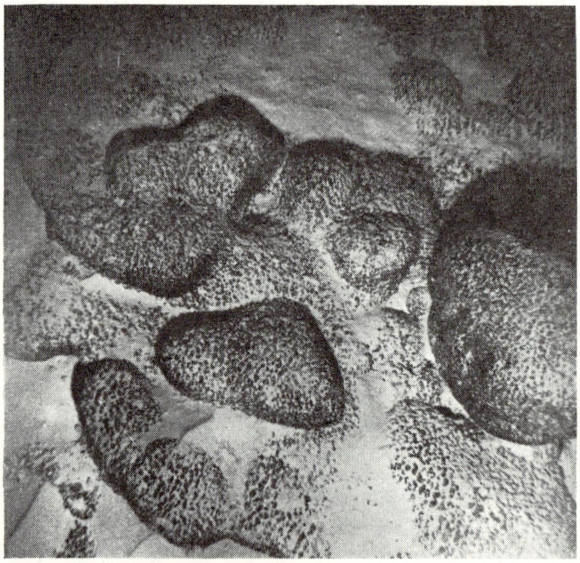

Fig. 7. 11° 51′N, 164° 55′E. Depth 1730 m. Sylvania Tablemount near Bikini. Area of picture $1\frac{1}{2}$ by $1\frac{1}{2}$ m. (U.S. Navy Electronics Laboratory photo by C. J. Shipek.)

Manganese-encrusted volcanic boulders and rippled *Globigerina* ooze. Similar boulders were dredged near by, showing these rocks to be olivine-basalt boulders coated by a thick crust of manganese.

layers of marine organisms which are often only made apparent by colour photographs. The manganese dioxide layer can vary in thickness from a few microns to several centimetres, first, discolouring the rock and, secondly, obscuring the surface texture with a botryoidal crust (Fig. 7). The extent of this is determined by the rate of manganese deposition and the length of exposure of the rock face, and may, therefore, be used as a guide to the relative ages of rocks within a given area.

b. Boulders and cobbles

Many photographs show areas strewn with boulders and cobbles. These may originate in three different ways. First they may be derived locally by the

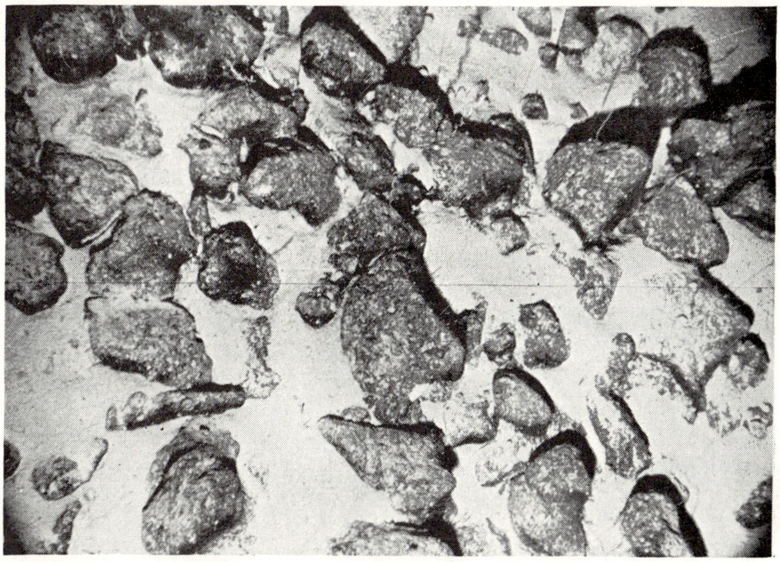

Fig. 8. 34° 53′N, 16° 32′W. Depth 1550 m. Near the peak of a seamount north of Madeira. Area of picture 3 by $2\frac{1}{2}$ m. (Photo by N.I.O.)

Boulders of volcanic basalt about 30 cm across partly buried by *Globigerina* ooze with a thin coating of manganese. The camera is looking into the side of the seamount.

breaking up of bedrock by some submarine weathering process: in the case of a cooling lava flow, the breaking up may take place very shortly after solidification as a result of internal stress set up by uneven cooling. Secondly, they may be transported to the area by some agency such as ice rafting or ejection from a volcano. Thirdly, they may have grown *in situ* by the accumulation of chemical or biochemical deposits to form, for example, manganese or phosphorite nodules.

Typical boulders produced by the fragmentation of solid rock are shown in Fig. 8 which are highly vesicular basalts on the sides of a seamount. Parallel

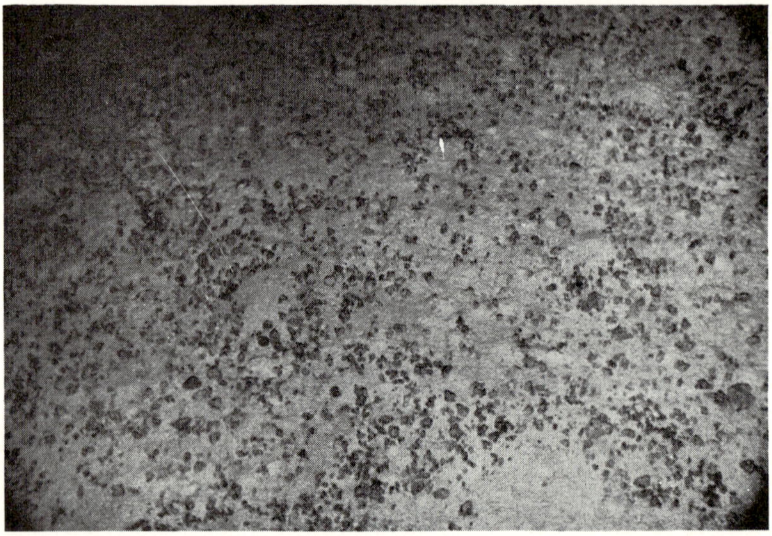

Fig. 9. 41° 20′N, 14° 29′W. Depth 5029 m. Swallow Bank on Iberia abyssal plain. Area of picture 3 by 4 m. (Photo by N.I.O.)

Small cobbles and pebbles of volcanic basalt near the top of a long, narrow ridge rising 400 m above the level of the plain. The absence of sediment cover suggests either recent volcanic activity or bottom currents.

Fig. 10. 46° 34′N, 27° 14′W. Depth 1480 m. Near the top of a peak on the Mid-Atlantic Ridge. Area of picture 3 by $4\frac{1}{2}$ m. (Photo by N.I.O.)

Broken rocks and boulders partially covered by a dark manganese coating, lying on a steep slope. White patches on boulders probably result from lack of manganese on what was the basal face before a recent disturbance. The complete absence of fine sediment and the angularity of the boulders also point to a recent disturbance, either by earthquake or volcano.

bands of vesicularity suggest successive flows of lava subsequently fractured and rounded by the erosive effect of sediment-laden currents. Small angular cobbles on a deep-sea ridge are shown in Fig. 9. On the Mid-Atlantic Ridge, recent earthquake and volcanic activity breaks up and dislodges rocks and can create screes of angular boulders and cobbles (Fig. 10). The variable covering of manganese on the boulders suggests that they have been recently moved and the pale uncovered basal surface exposed.

Manganese nodules, formed *in situ* on the sea floor, have frequently been photographed in the Pacific (Dietz, 1955; Menard and Shipek, 1958) (Fig. 11),

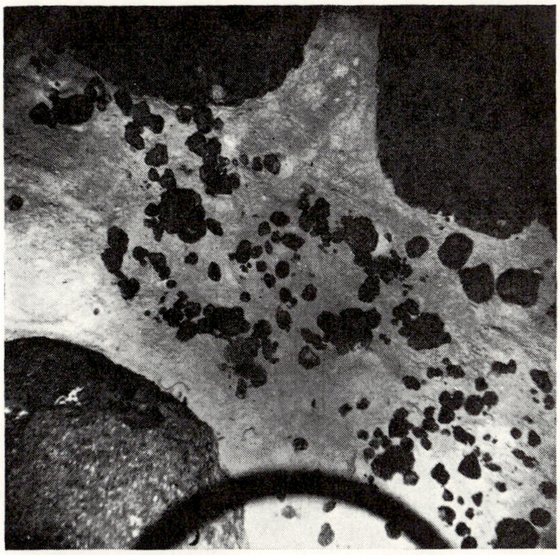

Fig. 11. 42° 50′S, 125° 32′W. Depth 4560 m. S.W. Pacific Basin. Area of picture 2½ by 2½ m. (U.S. Navy Electronics Laboratory photo by C. J. Shipek.)

Large and small manganese nodules or thickly encrusted rocks resting on a calcareous ooze. Note complete absence of sediment cover.

and more recently in the western Atlantic (Elmendorf and Heezen, 1957; Heezen, Tharp and Ewing, 1959) (Fig. 12), and in the Scotia Sea. In the Pacific they occur in sufficient quantities to constitute a potentially economic source for manganese, copper, cobalt and nickel. Sometimes they are extremely well rounded (in the Scotia Sea) and at other times they are moderately angular. They vary in size from microscopic up to half a metre or more in diameter and can be recognized in photographs from their uniform distribution over the bottom and from the fact that they are found in areas where clastic and erratic boulders are very rare. Nodules of phosphorite (Fig. 13) have been photographed, dredged and described by Emery and Dietz (1950).

In high latitudes, the proportion of erratic rocks caused by melting icebergs becomes appreciable and it is necessary to distinguish them from those locally derived. Although it may be possible to do this solely on the proportions of

Fig. 12. 29° 17′N, 57° 23′W. Depth 5618 m. Abyssal hills southeast of Bermuda Rise. Area of picture 2 by $2\frac{1}{2}$ m. (Photo by L.G.O.)

Round manganese nodules each capped with a thin covering of sediment. The scour pits around each nodule suggest currents which must have ceased before the sediment cap was laid down. Note the shark's tooth in lower right and meandering track.

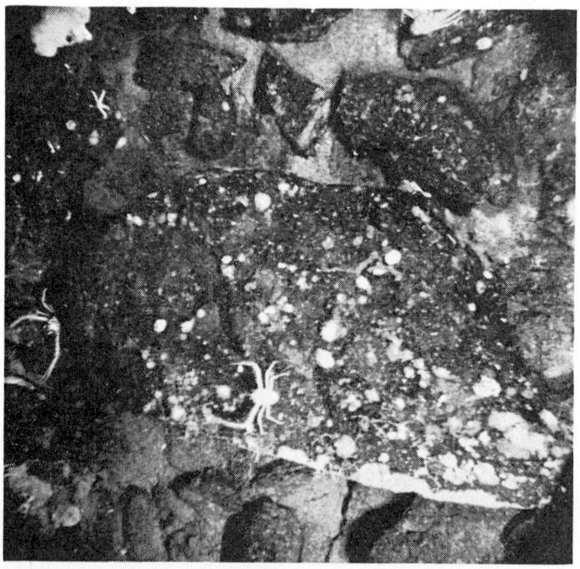

Fig. 13. 32° 34′N, 117° 47′W. Depth 256 m. Slopes of Thirty-Mile Bank. Area of picture $1\frac{1}{2}$ by $1\frac{1}{2}$ m. (U.S. Navy Electronics Laboratory photo by C. J. Shipek.)

Angular nodules of phosphorite half buried in sediment and richly coated with sponges, bryozoans and brachiopods. Note lobster-like crustaceans (*Galathea californiensis*) and starfish.

rock types, photography can sometimes help by observing how they lie on the bottom. Fig. 4 illustrates a case where dark erratic rocks, mainly igneous and metamorphic, are lying at random on top of a crumbling bedrock of limestone.

c. Pebbles and granules

Pebbles and granules generally occur in aggregates where they are conveniently described by the term gravel. In this size range it is much less easy to classify the different rock types since the normal resolution of the photographs is insufficient to examine individual particles. Each of the three processes which produce boulders and cobbles also produces finer gravel, and, where it occurs together with coarser material, it is reasonable to surmise a common origin (Fig. 10). But there are other areas where the largest particles present are no more than a few centimetres in diameter and where identification is extremely difficult unless samples are taken (Figs. 14 and 15).

Fig. 14. 47° 11′N, 27° 19′W. Depth 2798 m. East bank of median valley of Mid-Atlantic Ridge. Area of picture 4 by 5 m. (Photo by N.I.O.)

Shingle of very varied composition lying on a substratum of pteropod and *Globigerina* ooze. Samples of shingle recovered from the trigger weight of the camera include quartz basalts and limestones. This is probably ice-rafted from the north. Note the disturbance of the shingle by animal holes and two ophiuroids in the foreground.

One of the puzzles arising from underwater photographs is the formation of banks of what appears to be pebble-size gravel, but which have extremely sharp boundaries with what is evidently underlying *Globigerina* ooze (Fig. 16). The sharpness of the boundary precludes the possibility that the pebbles have settled from above, and the pebble size makes it unlikely that they are moved

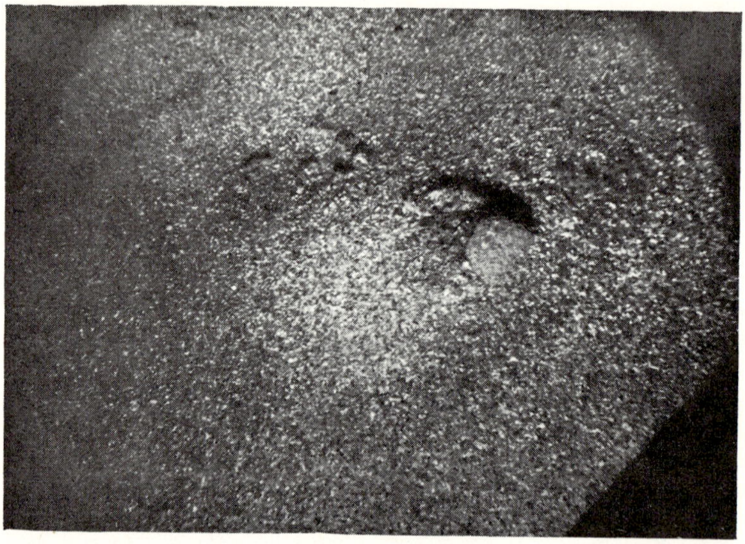

Fig. 15. 35° 07′N, 13° 04′W. Depth 2035 m. Ampère Bank. Area of picture $1\frac{1}{2}$ by $2\frac{1}{2}$ m. (Photo by L.G.O.)

Bank of well graded (1 cm diam.) but multicoloured shingle of unknown composition. In the centre right is a sponge.

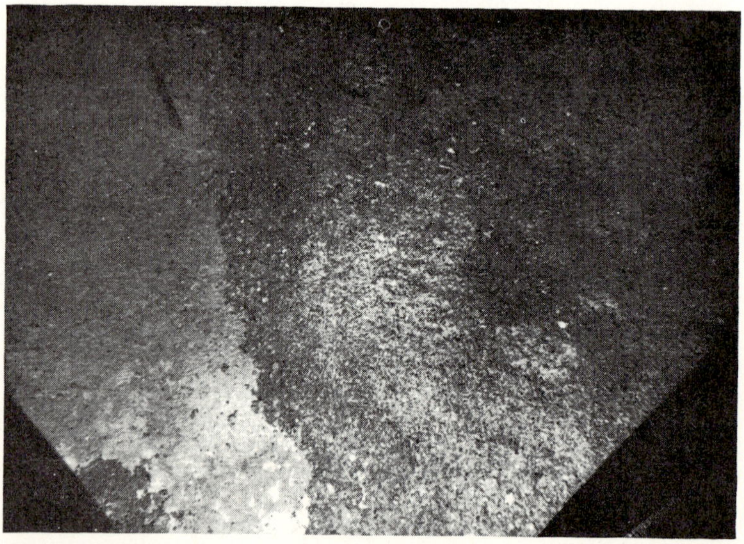

Fig. 16. 35° 10.5′N, 15° 28′W. Depth 905 m. Seamount northeast of Madeira. Area of picture $1\frac{1}{2}$ by $2\frac{1}{2}$ m. (Photo by L.G.O.)

Bank of shingle bordering an area of *Globigerina* ooze. The sharpness of the boundary and a detailed examination of the shingle suggest that the latter may be organic and spreading laterally over the ooze. Note the pennatulid at top left.

by currents along the bottom. There are two possible explanations, neither of which has any proof. It is possible that the particles, although pebble-size, are of organic origin, such as empty shells, and therefore have a much lower density than mineral particles. This would enable small currents to move them easily into banks (cf. current ridges described below). Secondly, the pebbles may in fact consist of a living colony of an abyssal mollusc which spreads laterally over the normal pelagic sediment. Only closely integrated photography and sampling can answer this question.

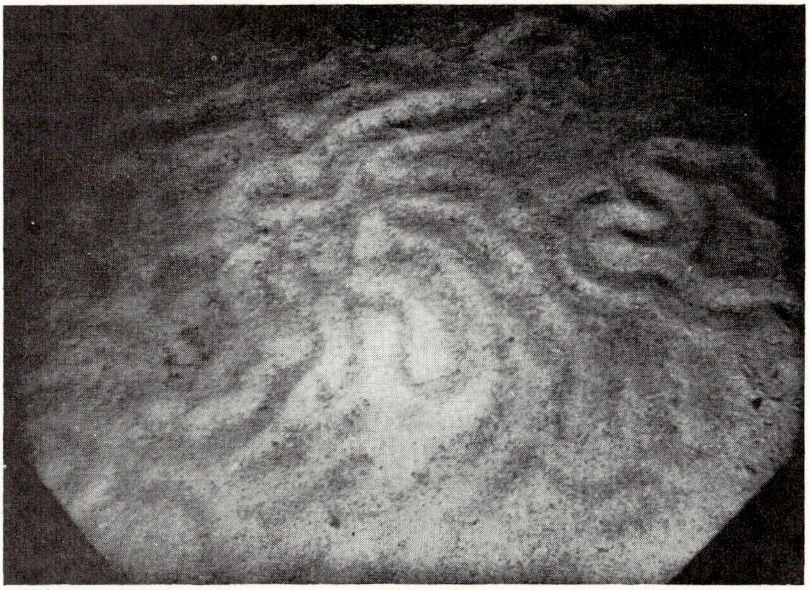

Fig. 17. 37° 20′N, 33° 25′W. Depth 838 m. Peak on Mid-Atlantic Ridge, southwest of Azores. Area of picture $1\frac{1}{2}$ by $2\frac{1}{2}$ m. (Photo by L.G.O.)

Sandy bottom with coarser dark particles heavily disturbed by the tracks of surface organisms.

d. Sand

The identification of sand on the sea floor from photographs depends chiefly on the effects on the sediment surface of disturbances by currents and faunal activity (Fig. 17). It is barely possible to distinguish it from oozes by its grain size except in the case of very coarse sands. The essential physical property that can be used to distinguish sand from ooze is its incoherence and, hence, its inability to maintain steep slopes. After disturbance it will take up its characteristic angle of repose. On this basis, a sandy sediment containing enough clay material to make it coherent will be classed as mud or ooze.

On the continental margins, the sands observed are primarily quartz sands derived from continental erosion and calcareous detritus. In the presence of comparatively strong currents they are well sorted and appear very frequently

Fig. 18. 41° 22′N, 14° 24′W. Depth 5340 m. Iberia abyssal plain. Area of picture 1½ by 1½ m. (Photo by N.I.O.)

Bottom of *Globigerina* ooze typical of deep-ocean area extensively reworked by benthic fauna making tracks, mounds and holes. At the top is either a worm (45 cm long) or a piece of rope.

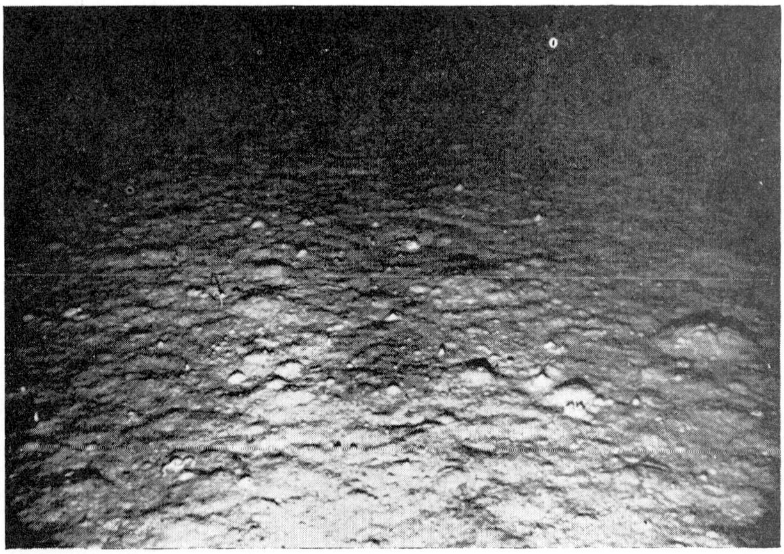

Fig. 19. 36° 36′N, 17° 13′W. Depth 4670 m. Ocean basin west of Josephine Bank. Area of picture 3 by 5 m. (Photo by N.I.O.)

Level bottom of *Globigerina* ooze extensively reworked by benthic fauna producing mounds. Note imprints of stars on mounds on the right. Note also worm tubes projecting above bottom.

with ripple marks. In the deep sea, the sands observed are primarily *Globigerina* sands, consisting of clean, well sorted tests of *Globigerina* and other shell debris. Having a lower mean density than quartz sands, they can be moved by slower currents. At most stations in the deep ocean where sand has been found, current ripple marks have also been found. These are discussed below.

e. Muds and oozes

A very high proportion of photographs of the sea floor contains some indication of mud or ooze. Gravitational forces will eventually concentrate the sediments into the deepest parts of the ocean basins and it is here that they are found in greatest quantities. The coherence of sediments containing an appreciable quantity of clay minerals usually prevents them from forming current ripples and, at the same time, enables the tracks and burrows of the benthic fauna to be preserved until they are eventually buried by subsequent sedimentation (Figs. 18 and 19).

B. Indications of Current Activity

a. Ripple marks

Apart from photographic observations of current-measuring instruments on the sea floor, there are many indications of currents that can be found from a study of small features on the bottom. The most obvious of these are the ripple marks that are found on areas of incoherent sands.

On the continental shelves and in the shallow seas around the continents it has long been known that ripple marks are readily formed where the current conditions and the bottom material are favourable. They have been observed in many geological formations and have been described and discussed at length in the literature on sedimentation (e.g. Twenhofel, 1950). Most naturally occurring ripple marks have been observed in the intertidal region or in shallow water where they can be seen from the water surface. But underwater photography and television have shown them to be very common over large areas of the continental shelves (Fig. 20). They vary in wave-length from a few centimetres up to the huge sand waves with a wave-length of a kilometre or more which have been described by Cartwright and Stride (1958). Often several systems with different wave-lengths are superimposed on one another (Fig. 21). At other times the ripple marks are confined to a thin layer of sand that appears to be moving as an isolated patch across a more consolidated bottom.

Ripple marks in the deep ocean were first described by Menard (1952) who found them at 1370 m on a seamount in the Pacific. Since then many deep photographs have shown them to exist on seamounts at depths as great as 3130 m (Figs. 25 and 26), and on the great oceanic ridges. Out of 860 photographs of sediment-covered areas on seamounts, 85 have shown indications of current action, by ripple marks, scouring, etc. None have been found, however, on the floor of the ocean basins, presumably on account of (a) insufficient current on an extensive flat bottom, and (b) coherence of the sediment.

Fig. 20. 48° 10′N, 07° 31′W. Depth 172 m. Near edge of continental shelf west of Ushant. Area of picture 2 by 3 m. (Photo by N.I.O.)

Short-crested ripple marks of wave-length 15 cm in quartz sands.

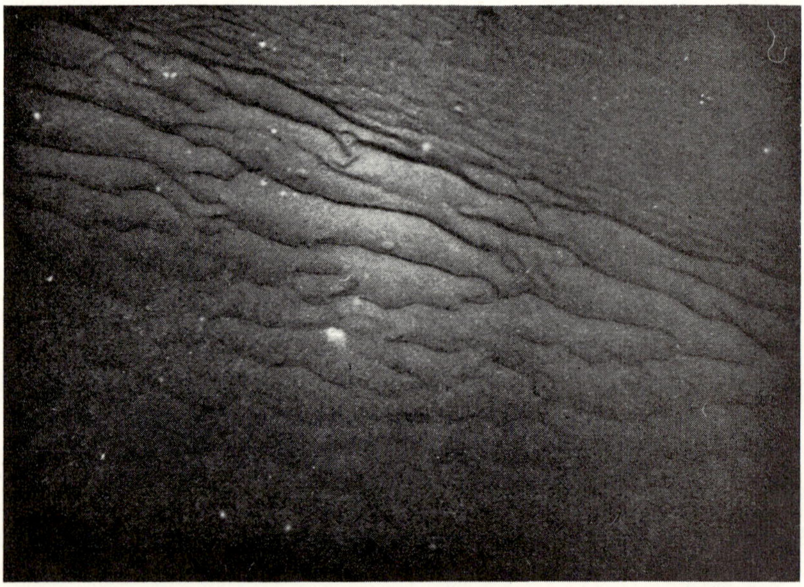

Fig. 21. 47° 42′N, 07° 34′W. Depth 475 m. Near the top of continental slope west of Ushant. Area of picture 2 by $2\frac{1}{2}$ m. (Photo by L.G.O.)

Small long-crested ripple marks superimposed on larger ones.

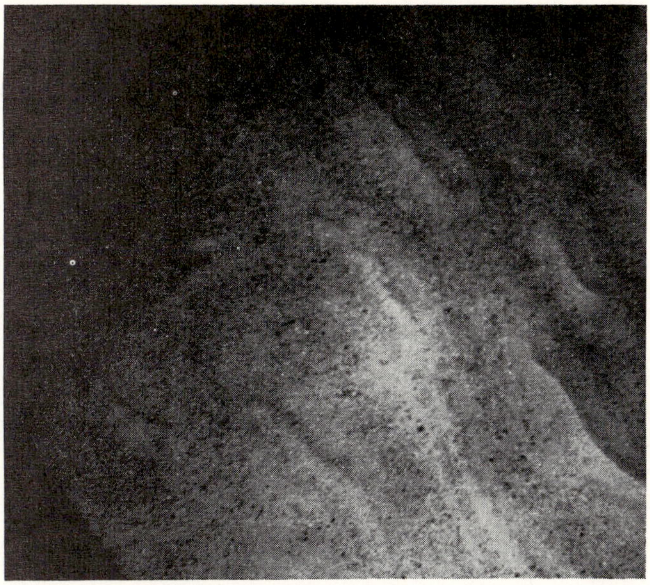

Fig. 22. 35° 12′N, 15° 18′W. Depth 1293 m. Seamount northeast of Madeira. Area of picture 1½ by 1½ m. (Photo by L.G.O.)

Symmetrical ripple marks in *Globigerina* sand (wave-length 25 cm).

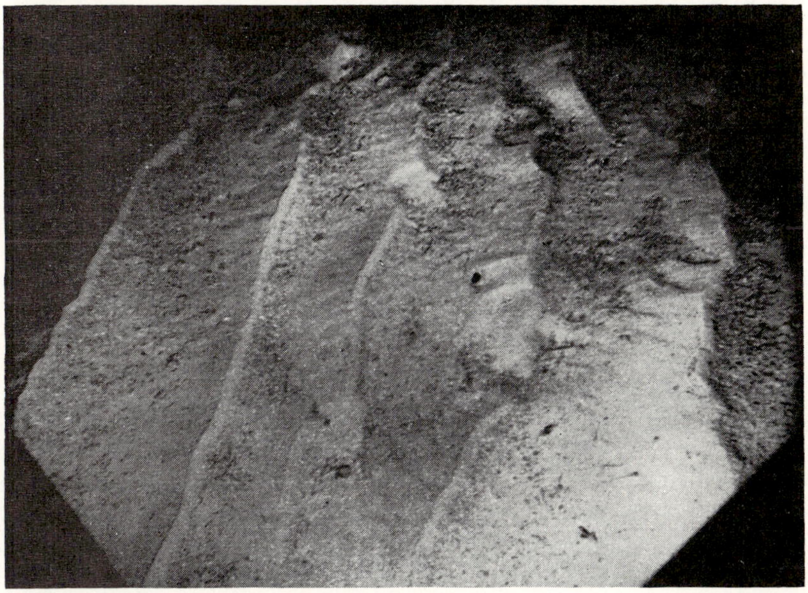

Fig. 23. 35° 12′ N, 15° 18′W. Depth 1317 m. Seamount northeast of Madeira. Area of picture 1½ by 2½ m. (Photo by L.G.O.)

Strongly asymmetrical ripple marks at same station as Fig. 22. Note very long and sharp crests, and coarse material in troughs (wave-length 30 cm).

Deep-sea ripple marks show as great a variety of forms as those in shallow water. Variations in wave-length, crest-length, amplitude and symmetry occur over very short distances and must result from local variations in current velocity, granular composition and in bed slope. Figs. 22, 23 and 24 were all taken at the same station within a hundred feet or so and show symmetrical, strongly asymmetrical and short-wave-length ripples, on a bottom that does not appear to be strongly sloping. One would be tempted from Fig. 22 to suppose the water movement to be oscillatory, whereas Fig. 23 suggests a steady current, or at least one in which there was a net transport in one direction. Little can be learnt, therefore, about the nature of the water movements from studying the photographs of ripple marks unless samples of the sediment are

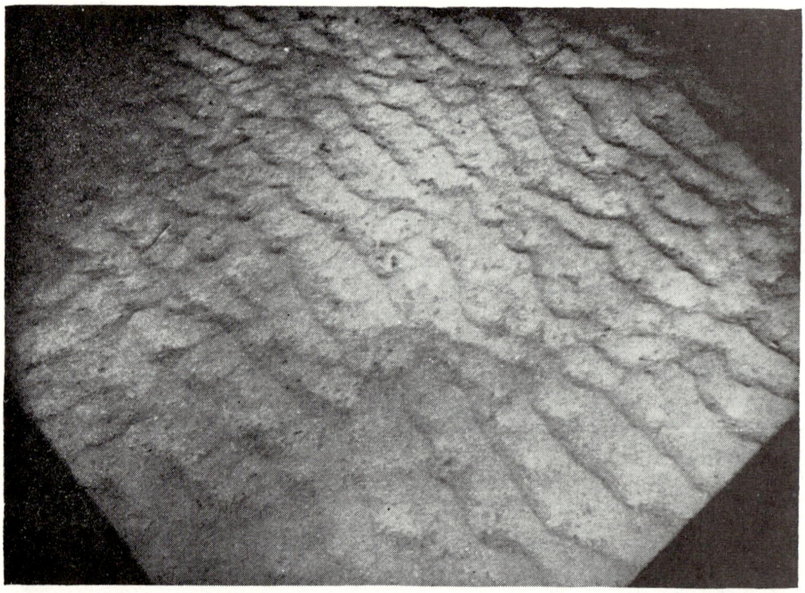

Fig. 24. 35° 12′N, 15° 18′W. Depth 1308 m. Seamount northeast of Madeira. Area of picture $1\frac{1}{2}$ by $2\frac{1}{2}$ m. (Photo by L.G.O.)
Short wave-length and fairly short-crested ripples at same station as Figs. 22 and 23 (wave-length 12 cm).

available for analysis and all details of the surrounding topography are known. At best one can only say that the existence of ripple marks indicates the presence of currents within a certain range of velocity, and, if one knows the orientation of crests, then the direction of movement can be deduced.

Recent experimental and theoretical studies of the formation of ripple marks by steady and oscillatory flow have been made by Bagnold (1946, 1947, 1955, 1956), Menard (1950), Anderson (1953), Manohar (1955) and Liu (1957). These have been carried out in laboratory flumes and tanks using a wide range of particle densities and sizes. In most cases of steady flow, the water depth has not been great enough to avoid the effects of waves on the free surface.

Fig. 25. 41° 12′N, 15° 14′W. Depth 3112 m. Seamount west of Iberia plain. Area of picture 4 by 5 m. (Photo by N.I.O.)

Globigerina sands showing a field of long-crested ripple marks parallel to the contours of a slope. Boulders and cobbles are partly buried by sediment.

Fig. 26. 41° 12′N, 15° 14′W. Depth 2917 m. Seamount west of Iberia plain. Area of picture $1\frac{1}{2}$ by $2\frac{1}{2}$ m. (Photo by N.I.O.)

Massive bedrock has channelled the current to produce strong ripple marks. Where the current has emerged and slowed down, organic debris containing such fragments as scaphopod shells and the spines of sea-urchins has been deposited. Several crinoids can be seen on the left.

Inman (1957) has studied the formation of ripples beneath beach waves using underwater swimmers.

In general the conclusion of all this work is that there is a threshold velocity for the formation of ripples depending on the particle characteristics and the regime of water flow over them, and that there is also a critical velocity above which the ripples are destroyed and the sediment is transported in sheet flow. According to a criterion derived by Manohar for the formation of ripple marks in oscillatory currents, based partly on theoretical and partly on experimental results, ripples will be initiated when the maximum water velocity during the oscillation exceeds $V_c = 57[D(\rho - 1)^2]^{0.2}$ cm/sec, where D and ρ are respectively the particle diameter and density in cgs units. They will be destroyed at approximately twice this velocity. For a particle density of 1.5 g/cm^3 (approximately the effective density of *Globigerina* sands), and a particle size of 0.2 mm, ripples will be formed for current velocities in the range 20 to 40 cm/sec. These figures are based on the maximum velocity of oscillatory currents, but it is reasonable to suppose that the same order of magnitude applies for steady currents since the initiation of ripples depends on getting the particles into movement along the sediment surface. The figure is also compatible with the criterion for the onset of ripple formation under steady flow derived by Liu (1957) based on the instability of a laminar boundary layer.

We have, therefore, to reconcile currents of the order of some 20 cm/sec deduced from the existence of ripple marks with observations of the deep-ocean circulation. Direct measurements of the deep-ocean currents made by Swallow (1961) show that velocities of up to 40 cm/sec exist in mid-ocean depths, and Swallow and Worthington (1961) have shown that velocities of at least 15 cm/sec can persist to within half a metre of the bottom. Superimposed on the mean velocities may be tidal oscillations of several cm/sec. Where an open-ocean current is obstructed by a seamount, the velocities may be locally increased by as much as a factor of two.

It is, therefore, plausible in the light of our present knowledge of deep currents to explain the formation of deep ripple marks in terms of steady currents. Variations in their symmetry may be ascribed to local disturbances in the flow of current and to tidal fluctuations, and it does not appear to be necessary to hypothesize short-period oscillatory motion of the water near the bottom.

b. Current ridges

Elongated piles of coarse, well-sorted material a metre or so in breadth have been observed at several stations (Fig. 27) superimposed on a bottom of finer and more coherent sediment. The linearity and parallelism of these ridges suggest that they are formed by current action. The granular material of the ridges is in the size range of granules and pebbles, and if they have been moved by the normal range of ocean currents, their density must be much lower than that of rock material. It is possible, therefore, that they are low-density shells and fragments of some marine fauna, perhaps pteropods, which would be easily concentrated by low-velocity currents.

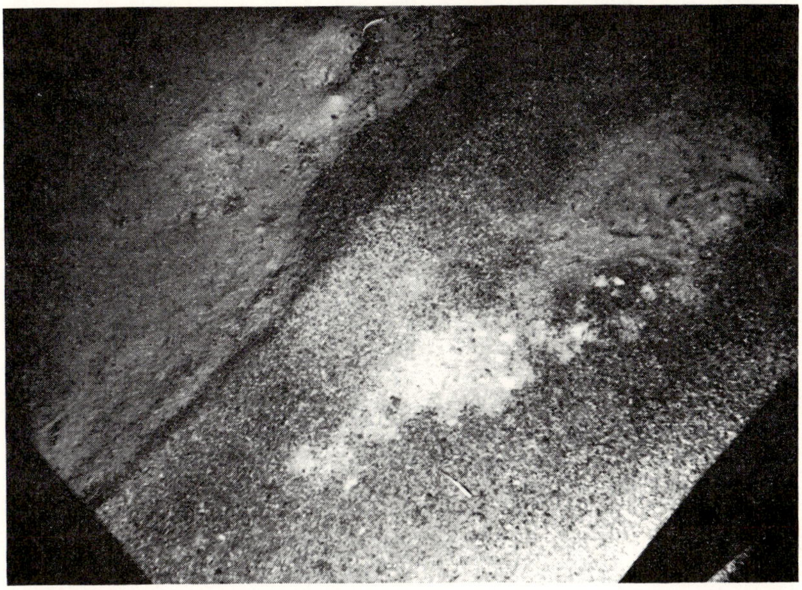

Fig. 27. 35° 07′N, 13° 04′W. Depth 2008 m. Ampère Bank. Area of picture $1\frac{1}{2}$ by $2\frac{1}{2}$ m. (Photo by L.G.O.)

Ridges of low-density shingle, thought to be organic in origin, lying on a level bed of *Globigerina* sand. The ridges are about 1 m between crests and are probably formed by currents perpendicular to the crests. Note the scaphopod-like shells top and bottom.

c. Scour pits and depositional drifts

Evidence of currents on the bottom is often found where a loose boulder or an attached benthic fauna, such as a sponge, creates scouring and drifting of the sediment. Sometimes this is observed in conjunction with ripple marks (Figs. 28 and 29), where it confirms the unidirectional nature of the current, but in other places no ripples are seen. Photographs of manganese nodules at a depth of 5834 m in the northwest Atlantic have shown scour pits on one side of the nodules (Fig. 12).

The existence of currents can sometimes be deduced from the lack of sediment cover on rocks, or from the way the sediment is lodged in cracks and crevices. Occasionally a rock face is seen where it has been abraded smooth by the passage of sediment-laden currents. This has been observed on seamounts where it is improbable that turbidity currents have been effective.

d. Orientation of benthic fauna

Certain filtration-feeding benthic fauna, such as gorgonians, crinoids, etc., have developed a planar shape that is suitable for the interception of maximum quantities of water, and they orient themselves perpendicular to the current. Some species, such as the common sea fan *Gorgonia*, are anchored rigidly to the bottom and are unable to change their orientation except by the slow process

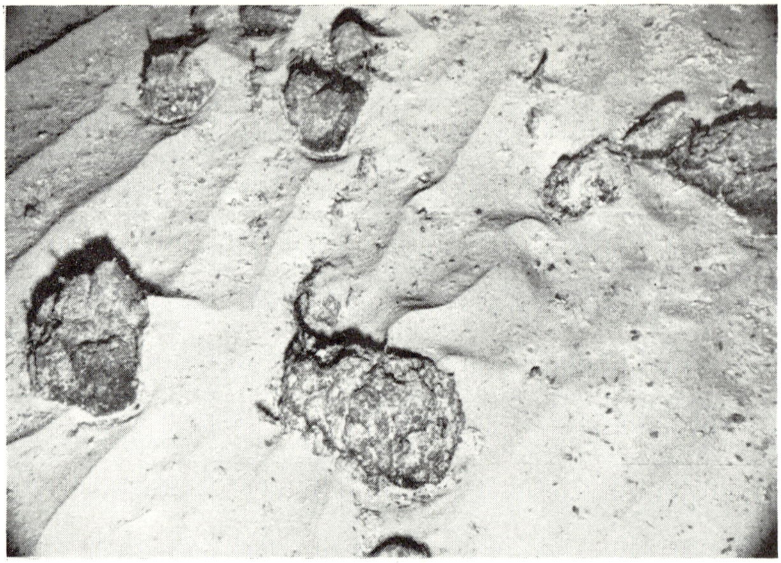

Fig. 28. 41° 12′N, 15° 14′W. Depth 3127 m. Seamount west of Iberia plain. Area of picture $1\frac{1}{2}$ by $2\frac{1}{2}$ m. (Photo by N.I.O.)

Calcareous sands with pronounced symmetrical ripple marks (wave-length 20 cm). Note the scour pits at the side of the half-buried boulders and the deposited sand drifts to the right of them, suggesting a predominant current from left to right.

Fig. 29. 35° 12′N, 15° 18′W. Depth 1302 m. Seamount northeast of Madeira. Area of picture $1\frac{1}{2}$ by $2\frac{1}{2}$ m. (Photo by L.G.O.)

Asymmetrical ripple marks with coarse material to the right (lee side) of ridges. Note the scour pits in front of and the deposited sand drifts behind the boulders, indicating predominant current from left to right.

of growth. The crinoids, however, are basically axially symmetric, but can move their arms so as to form a fan perpendicular to the current. Crinoids have been observed behaving in this way from the bathyscaph (Pérès and Picard, 1955) and a preferred orientation has been noted in fields of crinoids in underwater photographs. The orientation of many kinds of marine fauna has been studied by Bromhall in the waters near Hong Kong and he has noted the relationship between orientation and the funnelling of currents through a gap in the rocks.

A more obvious indication of current by marine fauna is the bending to the current of seaweed, sea-pens and other long and flexible organisms. By a study of both the slowly and quickly responding fauna appearing in photographs, it may be possible to obtain both the mean current direction and also the instantaneous current.

C. Benthic Fauna

The photographic method has been used to study the benthic fauna both qualitatively and quantitatively, especially in the littoral zone, since it is the most direct way of assessing the population and studying its environment and behaviour (Vevers, 1952; Posgay, 1958). Underwater television has also been used widely for this purpose (Barnes, 1955). However, much of the photographic information about the abyssal benthic fauna has been derived from photographs taken more for geological than zoological study. On account of the difficulties of deep-sea photography, it is important that the greatest use should be made of the collections of photographs from all aspects, and that they be made as widely available as possible.

It is not the intention to discuss here the general problems of the distribution of benthic fauna, but to mention only some of the principal fauna observed and their reaction with the sediment surface.

In the deep basins where there is sediment only, the most commonly observed fauna are the burrowing worms, sticking vertically through the sediment and sometimes showing fans of ciliated tentacles. Brittle stars (ophiuroids), sponges, scaphopods and holothurians are also frequently seen, or else their existence is deduced from their disturbance of the sediment surface. Mounds, burrows and tracks are almost universally observed wherever the sediment is coherent enough to preserve them.

Coming up out of the sediment basins onto the rocky slopes of seamounts, the benthic population changes and increases. Rock faces become obscured by matted bryozoa and sponges. Many species of coelenterates are common, including Hydrozoa, corals, gorgonians, pennatulids and sea anemones (Actiniaria). Echinoderm are often seen, especially sea-urchins and sand-dollars (echinoids), asteroids, ophiuroids and commatulid crinoids, the latter sometimes occurring in colonies covering large areas. Occasionally fish, deep-sea decapods and copepods and other nekton can be seen, but only occasionally can they be identified. Shallow waters on the tops of seamounts within the euphotic zone abound in benthos, with the same diversity of species as is found in the shallow coastal waters where so much work has already been done. However, from photographs alone it is extremely difficult to identify the fauna, and detailed work must depend on the collection of samples.

D. Disturbance of the Sediment–Water Interface by the Benthos

Most of the benthic fauna leave some marks on the soft sediments of the bottom where they move from place to place, when they search for food or when they seek seclusion or cover. The tracks, mounds and burrows that are produced are often preserved in the geological strata as the surface becomes buried by subsequent sedimentation. Furthermore, the sequence of layering of sediment in chronological order is disturbed over the range of burrowing activity, and the fine detail of stratification is often blurred. An extensive study of the fossilized traces of animal activity and their palaeobiological significance was made by Lessertisseur (1955) from the examination of sedimentary rocks. Many of the problems discussed by Lessertisseur can be illuminated by the study of contemporary animal traces on free sediment surfaces, and in some cases it is possible to observe the animals responsible for them.

Lessertisseur has classified these traces into exogene traces, comprising trails and tracks made by an animal on the surface (epifauna), and endogene traces, comprising the galleries and burrows of animals below the surface (infauna). This classification cannot always be rigorously adhered to since an exogene trace may suddenly be transformed into an endogene trace if a worm on the surface turns downward, and, furthermore, burrowing just beneath the surface may make an appreciable disturbance of the surface itself. When a burrow is formed, the sediment removed from below often accumulates on the surface as a mound, or else an empty burrow may subside and become filled with surface material leaving a depression.

The exogene traces can further be classified according to the nature of the operation which made them. Locomotion tracks are usually long and narrow, or in the form of a series of depressions if made by feet. Feeding tracks are characterized by spiral, radial or helminthoid patterns. (A helminthoid pattern consists of a long continuous track regularly folded back on itself with alternate left and right turns so that it efficiently covers a well defined area.) Resting marks are roughly equidimensional depressions. Similarly, the endogene traces can be divided into those of locomotion, feeding, dwelling and rest with somewhat similar characteristics.

Sea-floor photography can only observe the exogene traces (Emery, 1953) although there is sometimes evidence of further subsurface activity from the abrupt termination of tracks. The disturbances can be roughly grouped into four categories although they are often interrelated. Most prominent are the mounds which are found almost everywhere on the sea floor. They are roughly conical and vary in size from a few millimetres to 50 cm. Any suggestion that they represent a sediment covering over erratic rocks must be dismissed since dredging in the sediment basins has never produced such rocks. Some of the mounds appear to be due to a protective covering of sediment built over an animal in the sediment. Many of the mounds show the imprint of a five-cornered star on the top (Figs. 19 and 37), suggesting that they can be attributed to an asteroid or ophiuroid. Where there is a sharp change of colour of the sediment in the top few centimetres, the mounds have the colour of the lower layer, showing that the material has been brought up from underneath (Fig. 30). Another possible origin of the smaller mounds is that they have been formed from excreted material after digestion by mud-feeding animals such as holothurians. Studies of holothurians in the shallow waters around Bermuda

Fig. 30. 35° 11′N, 15° 20.5′W. Depth 2248 m. Seamount northeast of Madeira. Area of picture $1\frac{1}{2}$ by 4 m. (Photo by L.G.O.)

Level sediment covered by the workings of small bottom-burrowing organisms which have brought mounds of light calcareous ooze on top of the darker surface. Note the disturbed patches around the gorgonian (at the top) and the sponge (lower right).

(Crozier, 1918) showed that in 1.7 square miles, 500 to 1000 tons of sand passed annually through them, being equivalent to a layer of 1 cm thickness every 100 years. In deep water the population is very much less, but the sedimentation rate is low and excretions of this sort may contribute significantly to the sediment disturbance, and may, in fact, alter the chemical composition appreciably.

The second group consists of holes or depressions in the sediment. They are not so numerous in general as the mounds but in certain areas they are prominent. Occasionally the photographs show some organism in the hole but there is seldom sufficient detail for identification. An interesting feature of the holes

Fig. 31. 37° 55′N, 25° 58′W. Depth 1785 m. Mid-Atlantic Ridge west of San Miguel, Azores. Area of picture $1\frac{1}{2}$ by $2\frac{1}{2}$ m. (Photo by L.G.O.)

Line of burrows of unknown origin in calcareous clay. Note the tracks, typical of those made by scaphopod-like animals.

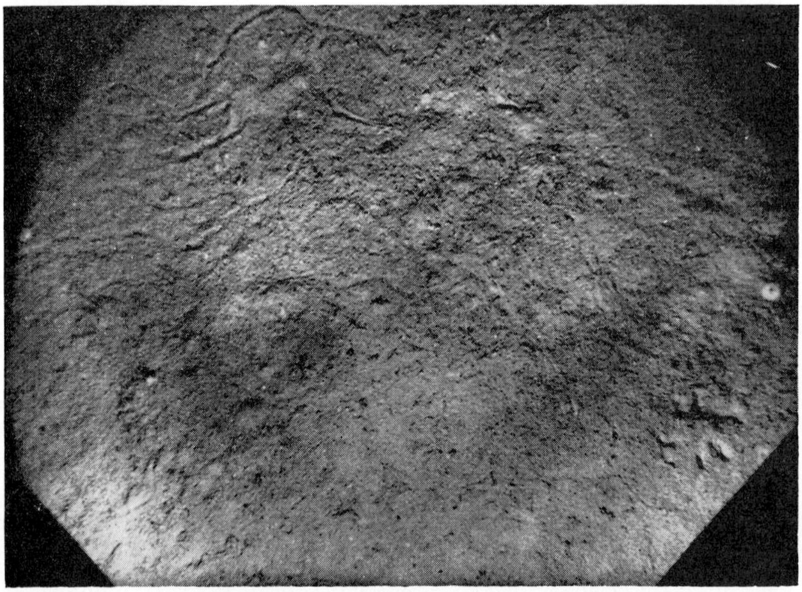

Fig. 32. 38° 20′N, 28° 01′W. Depth 1580 m. Mid-Atlantic Ridge near Pico, Azores. Area of picture $1\frac{1}{2}$ by $2\frac{1}{2}$ m. (Photo by L.G.O.)

Soft clay bottom showing ring of burrows (usually six) in lower right, and tracks of scaphopod-like animals, one of which can just be seen at the end of a track (top centre).

is that they are sometimes grouped in regular patterns such as straight lines or circles, six or eight holes comprising the group (Figs. 31 and 32).

Among the most spectacular features in the sediment of the sea floor are the tracks made by mobile fauna. The largest track so far observed is 15 cm wide showing a central ridge, parallel lines of small depressions on either side, and transverse markings on the ridge (Fig. 33). It seems likely that these tracks were made by a mud feeder such as a holothurian, the transverse marks being made by scooping the mud into suspension, and the depressions by its feet.

Fig. 33. 41° 18′N, 14° 23.5′W. Depth 5341 m. Iberia abyssal plain. Area of picture 1 by 1 m. (Photo by N.I.O.)

Broad tracks (15 cm wide) in *Globigerina* ooze made by a mud-feeding animal on the surface. Note the regular indentations (5 cm apart) either side of the tracks and the transverse markings (top left) suggesting a mud-scooping action resulting in a central ridge.

Smaller but more numerous tracks have been identified as having been made by smaller holothurians and sea-urchins (Fig. 34). Unidentified animals with long thin conical shells (resembling scaphopods), with two extended arms, have often been photographed at the end of thin and rather angular tracks (Fig. 35). Occasionally fish feeding near the bottom leave marks where the lower tip of the tail touches the sediment surface.

The greatest contribution to the disturbance of the sediments is due, however, to the galleries and dwelling holes of burrowing worms and other infauna which are visible to the camera only where they come to the surface. However, they are shown up well in cross-sections of cores in the region where the sediment colour changes. Galleries bored in the lower material become filled subsequently with the upper material and this constitutes a mixing process. The only evidence of these galleries in underwater photographs is the sudden

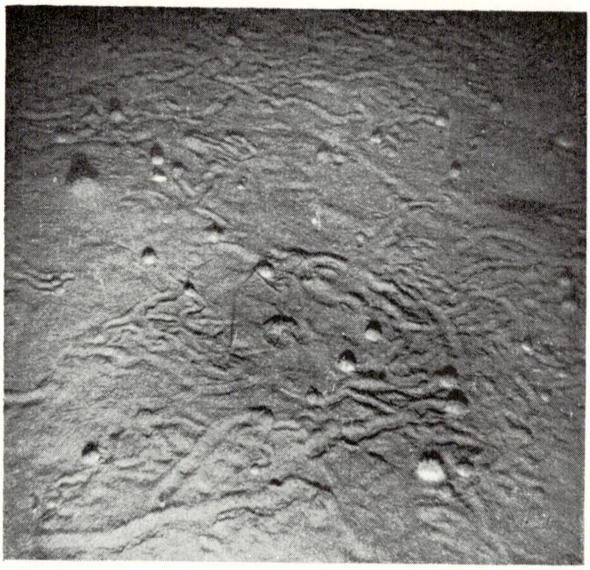

Fig. 34. 118° 38′W, 32° 22.5′N. Depth 549 m. Inshore continental slope off southern California. (U.S. Navy Electronics Laboratory photo by C. J. Shipek.)

A green mud bottom highly disturbed by benthonic activity showing a mound and depression of biological origin, plus numerous tracks formed by sea-urchins (echinoids).

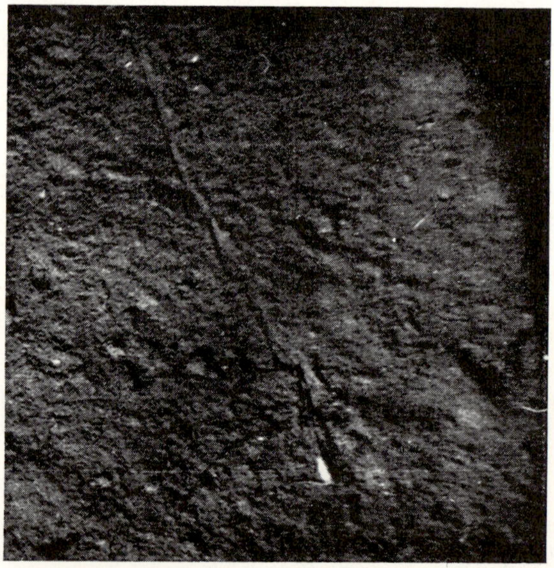

Fig. 35. 38° 20′N, 28° 01′W. Depth 1628 m. Mid-Atlantic Ridge near Pico, Azores. Area of picture approx. 1 m by 70 cm. (Photo by L.G.O.)

An example of one of many such tracks found on soft muddy bottoms being made by a scaphopod-like animal with two long arms. The track shows a characteristic angularity due to the area swept out when the rigid shell is pivoted.

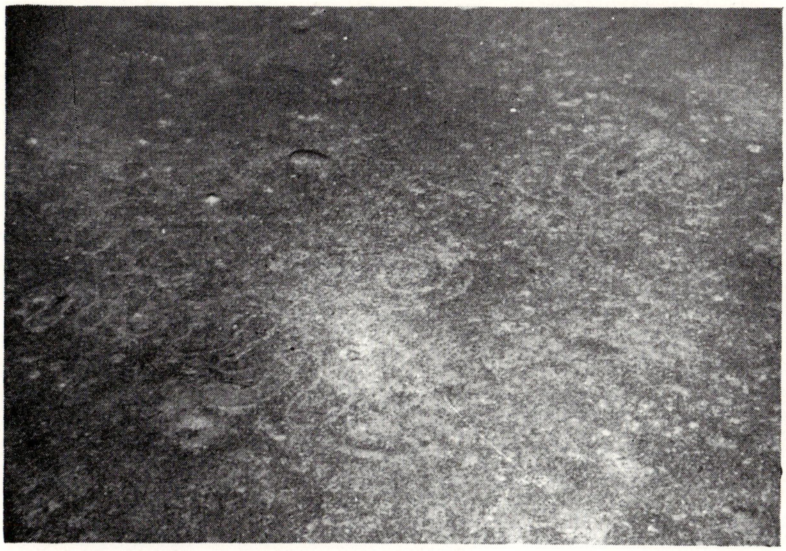

Fig. 36. 42° 53′N, 26° 45′W. Depth 3080 m. Foothills of the Mid-Atlantic Ridge. Area of picture 3 by 5 m. (Photo by N.I.O.)

Spiral and helminthoid raised tracks made by organisms burrowing just under the surface. Note the spiral ending in the centre where the organism has presumably burrowed straight down.

termination of tracks (Fig. 36), the presence of wormcasts, of worms themselves protruding from the bottom, and of radial feeding tracks where repeated excursions of tentacles from the burrow have made marks radiating from it (Fig. 37).

A quantitative approach to these disturbances can be made in two ways, by considering the surface distribution of disturbances and the effective diffusion downwards of surface material. To consider the surface distribution, nearly 500 photographs of sediment-covered areas were examined from 28 stations in the North Atlantic and in the Gulf of Mexico. The percentage area disturbed by mounds, holes and tracks was estimated for each station. The results, classified in terms of physiographic regions, are summarized in Table II.

TABLE II

Region	No. of pictures	No. of stations	Range of disturbance, % of area	Mean disturbance, % of area
Continental slope	47	5	2–20	13
Seamounts	152	8	1–70	20
Mid-Atlantic Ridge	157	7	1–30	7
Basins	128	8	1–20	8

It is evident that the higher percentage of disturbance on the continental slope and seamounts reflects the larger populations in shallower water. Particularly low percentage disturbance was found on the red-clay sediments in the deep basins.

The vertical diffusion of surface material can be estimated in the following way. Assume that within an area A, an area a is disturbed so that the surface material is uniformly mixed to a depth z. If m_0 is the mass of surface material per unit area before disturbance, then the mean concentration $\bar{c}$ of surface material at a depth z after n such disturbances is given by

$$\bar{c}/m_0 = \sum^{\text{at } z} a_n/z_n A. \quad (1)$$

Fig. 37. 41° 18′N, 14° 23.5′W. Depth 5341 m. Iberia abyssal plain. Area of picture 0.7 by 0.7 m. (Photo by N.I.O.)

A mound bearing the imprint of a star and tracks radiating from a point possibly made by repeated excursions of the tentacles of a buried animal in search of food.

The summation can be carried out over the four types of disturbance described above, each of which can be given a series of values of z_n and a_n estimated from the photographs of the bottom and from cross-sections of cores. Only surface disturbances with dimensions greater than 3 cm were used in the analysis and a constant term, used for the disturbance by internal galleries, was estimated from sections of cores. This term provides the major contribution to disturbance.

The results of this analysis for three deep-basin stations are shown in Fig. 38 at depth intervals of 0.1 ft (3 cm). It will be seen that the surface material is spread over a depth of 5 cm giving a mean concentration at that depth of about 0.05 g/cm^3 for a concentration of surface material of 1 g/cm^2.

In this analysis the problem has been treated as if the disturbance took place on one horizon only. In fact the sediment is slowly accumulating and the surface of activity is slowly rising. This means that the material of a narrow stratum will be diffused both upwards and downwards resulting in a general blurring of the stratification. Clearly many factors are responsible for the extent of disturbance by organisms on the bottom. The ecological conditions will

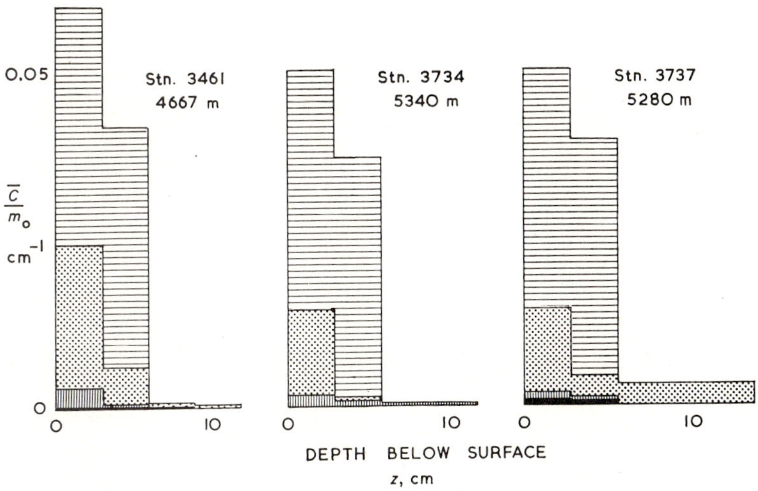

Fig. 38. Disturbance of the surface sediment by bottom-living animals at three stations in a deep-ocean basin, measured from bottom photographs and core sections.
Abscissa—Depth (z cm).
Ordinate—Mean concentration of surface material at depth z for unit mass per unit area on the surface ($\bar{c}/m_0$ cm^{-1}).
holes — mounds — tracks — galleries

determine first of all the faunal populations. The extent of the disturbance due to their activity will depend on the hardness and composition of the bottom, and on the rate of sedimentation in raising the surface of their habitation.

E. Description of Microtopography by Physiographic Regions

a. Continental shelves (Figs. 13, 20, 22, 34)

According to the classification of Heezen, Tharp and Ewing (1959) the continental margin can be divided into three categories, the first category consisting of the continental shelf, the second the epicontinental marginal seas and the last the marginal plateaus. These regions have been more closely studied photographically than any of the deeper ones since the bottom is more accessible. Early work on underwater photography by Ewing, Vine and

Worzel (1946), described also by Ewing, Woollard, Vine and Worzel (1946), showed some of the typical features of the continental shelf off the east coast of U.S.A., and these studies in the Atlantic have been furthered by Northrop (1951), Owen (1958) and Posgay (1958) on the western side, and Barnes (1955), Vevers (1952) and Whittard (1962) on the eastern side. In the north-western Pacific, Sasaki *et al.* (1957) have studied the shelves off Japan and Zenkevitch and Petelin (1956) those near the Kuril Islands.

The area off Southern California, which has been called the continental borderland, has been studied in great detail by Shepard and Emery (1941, 1946), Emery (1952), Emery and Terry (1956) and many other workers, and the results have been collected together and published in *The Sea off Southern California* by Emery (1960). This region is not a typical continental shelf, nor is it typically oceanic. It has islands, basins and trenches with maximum depths of a thousand fathoms, and is separated from both the land and the deep ocean by continental slopes; it is thus intermediate between shelf and ocean basin.

A summary of the characteristics of the microtopography of the different physiographic regions is given in Table III, which has been prepared from a study of all available photographic material both from published papers and private collections of underwater photographs.

TABLE III

Summary of Microtopographic Features of Different Physiographic Regions

	No. of photos examined	Percentage of photographs showing						Indications of currents (ripples, etc.)
		Mud	Sand	Gravel	Cobbles	Boulders	Bedrock	
Continental shelf	247	25	30	46	8	6	5	16
Continental slope and canyons	157	81	8	1	5	3	6	2
Ocean basins	266	100	4	5	6	5	4	0
Mid-ocean ridges	492	73	36	33	9	10	10	2
Seamounts	686	53	57	36	17	14	23	14

The continental shelves, excluding the littoral zones, are characterized by a lack of exposed rock. On the other hand, there is a large proportion of a coarse detrital material in the form of gravel, shells and sand. Ripple marks and sand waves are common where there is a sandy bottom, as one would expect from the tidal currents which surround the continents.

b. Continental slopes and submarine canyons (Fig. 21)

Less is known about the features of the continental slopes than of any other region. In places they are comparatively smooth and gentle, implying a sediment cover; many photographs have shown this to be so. On the other hand, the slopes are frequently cut by submarine canyons where one would expect to find exposed rock. Of this it is very difficult, with the present techniques of underwater photography, to take pictures. Northrop and Heezen (1951) found exposed Eocene rock south of Cape Cod, and McAllister (1957) took photographs of the vertical walls of the Scripps Canyon showing conglomerates. The granite walls of submarine canyons in Lower California have been described by Shepard (1959). A number of dives by bathyscaph have been made in or near canyons south of Toulon and west of Lisbon, most of which showed semiconsolidated muds, but a few revealed exposed rock (Pérès and Picard, 1955; Pérès, Picard and Ruivo, 1957; Picard and Dietz, 1957; Pérès, 1958).

c. Ocean basins (Figs. 6, 9, 11, 12, 18, 19, 33, 37)

The ocean basins include all those deep parts of the ocean bounded by the continental margin and the mid-oceanic ridges. For the purposes of the study of microtopography, seamounts are not included here and will be considered separately. The ocean basins can be divided into three categories; the continental rise leading gently up to the continental slope, the abyssal plains created by the horizontal distribution of sediments by turbidity currents and the gently undulating bottom where the roughness of the hard-rock topography is buried by great thicknesses of pelagic sediment.

However, the appearance of the bottom is very similar in these three categories and no systematic differences have been found. In areas of *Globigerina* ooze, there is very little to be seen other than a level sediment surface disturbed by the mounds, burrows and tracks described above. No photographs have yet shown an exposed sand surface on an abyssal plain due to recent turbidity currents. In the deeper parts of the Atlantic Ocean, where there is red clay, and in the Pacific, manganese nodules are often found (Dietz, 1955; Elmendorf and Heezen, 1957; Menard and Shipek, 1958; Shipek, 1960). In these regions the disturbance by benthic fauna is reduced.

Where deep-sea channels converge and cut down into the sediments in an abyssal gap, rocks of consolidated sediments have been exposed (Laughton, 1960). These are probably mudstones and sandstones derived from the lithification of deep-sea oozes.

d. Mid-oceanic ridges (Figs. 5, 10, 14, 17, 31, 32, 35, 36)

Best known of the mid-oceanic ridges is that in the Atlantic. It is an extremely rugged area with sharp peaks and deep valleys. Many of the peaks are very similar to typical seamounts showing patches of exposed bedrock, steep boulder-covered slopes and pockets of sediment. There is evidence in some areas of recent displacement of rocks on steep slopes resulting in screes where there is no sediment to be seen.

The central rift valley has been described by Heezen *et al.* (1959) and Hill (1960). Photographs taken on the floor of the valley by Cousteau have shown water-cooled lavas similar to pillow lavas. In the Romanche Trench, which cuts through the Mid-Atlantic Ridge, exposed rocks have been seen at 7500 m (Cousteau, 1958).

e. Seamounts (Figs. 2, 3, 4, 7, 8, 15, 16, 23, 24, 25, 26, 27, 28, 29, 30)

The features of microtopography on seamounts are very varied over short distances. Successive photographs in a camera station may show bedrock strewn with boulders followed by a current-rippled sand bed. There is a high proportion of visible rock and much coarse clastic material. Judging by the effect of disturbances of the sediment by animals and by currents, the finer sediments contain a high proportion of sand and are frequently found with ripple marks. It is probable that the clay portion is winnowed away by the currents leaving a comparatively well sorted sand.

Photographs of seamounts in the Atlantic have been published by Elmendorf and Heezen (1957), Heezen, Tharp and Ewing (1959), Laughton (1957, 1959), and Laughton, Hill and Allan (1960), and in the Pacific, apart from those in the seas off Southern California, by Menard (1952) and Carsola and Dietz (1952) and Emery, Tracy and Ladd (1954).

Acknowledgements

The following underwater photographs are published by permission of:

N.I.O. National Institute of Oceanography, Wormley, Surrey, England. Figs. 2, 4, 6, 8, 9, 10, 14, 18, 19, 20, 25, 26, 28, 33, 36, 37.

L.G.O. Lamont Geological Observatory, Palisades, New York, U.S.A. Figs. 3, 12, 15, 16, 17, 21, 22, 23, 24, 27, 29, 30, 31, 32, 35.

N.E.L. U.S. Navy Electronics Laboratory, San Diego, California, U.S.A. Figs. 7, 11, 13, 34.

N.G.S. National Geographical Society, Washington, D.C., U.S.A. Fig. 5.

References

Anderson, A. G., 1953. The characteristics of sediment waves formed by flow in open channels. *Proc. Third Midwest. Conf. Fluid Mech.*, 379–395.

Bagnold, R. A., 1946. Motion of waves in shallow water. Interaction between waves and sand bottoms. *Proc. Roy. Soc. London*, **A187**, 1–18.

Bagnold, R. A., 1947. Sand movement by waves: some small-scale experiments with sand of very low density. *J. Inst. Civ. Eng.*, **27**, 447–469.

Bagnold, R. A., 1955. Some flume experiments on large grains but little denser than the transporting fluid and their implications. *Proc. Inst. Civ. Eng.*, Pt. III, 174–205.

Bagnold, R. A., 1956. The flow of cohesionless grains in fluids. *Phil. Trans. Roy. Soc. London*, **A249**, 235–297.

Barnes, H., 1955. Underwater television and research in marine biology, bottom topography and geology. *Deut. Hydrog. Z.*, **8**, 213–236.

Carsola, A. J. and R. S. Dietz, 1952. Submarine geology of two flat-topped Northeast Pacific seamounts. *Amer. J. Sci.*, **250**, 481–497.

Cartwright, D. and A. H. Stride, 1958. Large sand waves near the edge of the continental shelf. *Nature*, **181**, 41.

Cousteau, J.-Y., 1958. Calypso explores an undersea canyon. *Nat. Geog. Mag.*, **113**, 373–396.

Crozier, W. J., 1918. The amount of bottom material ingested by holothurians (Stichopus). *J. Exp. Zool.*, **26**, 379–389.

Dietz, R. S., 1955. Manganese deposits on the Northeast Pacific sea floor. *Calif. J. Min.*, **51**, 209–220.

Elmendorf, C. H. and B. C. Heezen, 1957. Oceanographic information for engineering submarine cable systems. *Bell Syst. Tech. J.*, **36**, 1047–1093.

Emery, K. O., 1952. Submarine photography with the Benthograph. *Sci. Month.*, **75**, 3–11.

Emery, K. O., 1953. Some surface features of marine sediments made by animals. *J. Sediment. Petrol.*, **23**, 202–204.

Emery, K. O., 1960. *The Sea off Southern California*. John Wiley and Sons, New York–London.

Emery, K. O. and R. S. Dietz, 1950. Submarine phosphorite deposits off California and Mexico. *Calif. J. Min.*, **46**, 7–15.

Emery, K. O. and R. D. Terry, 1956. A submarine slope off Southern California. *J. Geol.*, **64**, 271–280.

Emery, K. O., J. I. Tracey and H. S. Ladd, 1954. Geology of Bikini and nearby atolls. *U.S. Geol. Surv. Prof. Paper*, **260-A**, 265 pp.

Ewing, M., A. C. Vine and J. L. Worzel, 1946. Photography of the ocean bottom. *J. Opt. Soc. Amer.*, **36**, 307–321.

Ewing, M., G. P. Woollard, A. C. Vine and J. L. Worzel, 1946. Recent results in submarine geophysics. *Bull. Geol. Soc. Amer.*, **57**, 909–934.

Heezen, B. C., M. Tharp and M. Ewing, 1959. The floors of the Oceans. I. The North Atlantic. *Geol. Soc. Amer. Spec. Paper* 65, 1–122.

Hill, M. N., 1960. A median valley of the Mid-Atlantic Ridge. *Deep-Sea Res.*, **6**, 193–205.

Inman, D. L., 1957. Wave generated ripples in nearshore sands. *Beach Erosion Bd., Corps of Engineers, Tech. Mem.* 100, 1–42.

Laughton, A. S., 1957. A new deep-sea underwater camera. *Deep-Sea Res.*, **4**, 120–125.

Laughton, A. S., 1959. Photography of the ocean floor. *Endeavour*, **18**, 178–185.

Laughton, A. S., 1960. An interplain deep-sea channel system. *Deep-Sea Res.*, **7**, 75–88.

Laughton, A. S., M. N. Hill and T. D. Allan, 1960. Geophysical investigations of a seamount 150 miles north of Madeira. *Deep-Sea Res.*, **7**, 117–141.

Lessertisseur, J., 1955. Trace fossiles d'activité animale et leur signification paléobiologique. *Mém. Soc. Géol. Fr.* **74**, 1–150.

Liu, H. K., 1957. Mechanics of sediment ripple formation. *Proc. Amer. Soc. Civ. Eng.*, **83**, 1197/1–23.

McAllister, R. F., 1957. Photography of submerged vertical structures. *Trans. Amer. Geophys. Un.*, **38**, 314–319.

Manohar, M., 1955. Mechanics of bottom sediment movement due to wave action. *Beach Erosion Bd., Corps. of Engineers, Tech. Mem.*, 75

Menard, H. W., 1950. Sediment movement in relation to current velocity. *J. Sediment. Petrol.*, **20**, 148–160.

Menard, H. W., 1952. Deep ripple marks in the sea. *J. Sediment. Petrol.*, **22**, 3–9.

Menard, H. W. and C. J. Shipek, 1958. Surface concentrations of manganese nodules. *Nature*, **182**, 1156–1158.

Northrop, J., 1951. Ocean bottom photographs of the neritic and bathyal environment south of Cape Cod, Massachusetts. *Bull. Geol. Soc. Amer.*, **62**, 1381–1384.

Northrop, J. and B. C. Heezen, 1951. An outcrop of Eocene sediment on the continental slope. *J. Geol.*, **59**, 396–399.

Owen, D. M., 1958. Photography underwater. *Oceanus*, **6**, 22–39.

Pérès, J. M., 1958. Trois plongées dans le canyon du Cap Sicié, effectuées avec le bathyscaphe F.N.R.S. III de la Marine Nationale. *Bull. Inst. Océanog. Monaco*, **1115**, 1–21.

Pérès, J. M. and J. Picard, 1955. Observations biologiques effectuées au large de Toulon avec le bathyscaphe F.N.R.S. III de la Marine Nationale. *Bull. Inst. Océanog. Monaco*, **1061**, 1–8.

Pérès, J. M., J. Picard and M. Ruivo, 1957. Résultats de la campagne de recherches du bathyscaphe F.N.R.S. III. *Bull. Inst. Océanog. Monaco*, **1092**, 1–29.

Picard, J. and R. S. Dietz, 1957. Oceanographic observations by the bathyscaphe Trieste (1953–1956). *Deep-Sea Res.*, **4**, 221–229.

Posgay, J. A., 1958. Photography of the sea-floor. *Oceanus*, **6**, 16–19.

Sasaki, T., S. Watanabe, G. Oshiba and N. Okami, 1957. Investigation of the sea floor of Tsugaru Straits by underwater camera. *Rec. Oceanog. Works in Japan*, **3**, 56–69.

Schenck, H. and H. Kendall, 1954. *Underwater Photography*. Cornell Maritime Press, Maryland.

Shepard, F. P., 1959. Granite-walled submarine canyons of Lower California. *Intern. Oceanog. Cong. Preprints*, New York, 661.

Shepard, F. P. and K. O. Emery, 1941. Submarine topography off the California Coast. Canyons and tectonic interpretation. *Geol. Soc. Amer. Spec. Paper* 31, 1–171.

Shepard, F. P. and K. O. Emery, 1946. Submarine photography off the California coast. *J. Geol.*, **54**, 306–321.

Shipek, C. J., 1960. Photographic study of some deep sea floor environments in the eastern Pacific. *Bull. Geol. Soc. Amer.*, **71**, 1067–1074.

Swallow, J. C., 1961. Ocean circulation. *Sci. Prog.*, **49**, 282–286.

Swallow, J. C. and L. V. Worthington, 1961. An observation of a deep counter current in the Western North Atlantic. *Deep-Sea Res.*, **8**, 1–19.

Twenhofel, W. H., 1950. *Principles of Sedimentation*. McGraw-Hill, New York.

Vevers, H. G., 1952. A photographic survey of certain areas of sea-floor near Plymouth. *J. Mar. Biol. Assoc. U.K.*, **31**, 215–221.

Wentworth, C. K., 1922. A scale of grade and class terms for clastic sediments. *J. Geol.*, **30**, 377–392.

Whittard, W. F. S., 1962. Geology of the Western Approaches of the English Channel: a progress report. *Proc. Roy. Soc. London*, **A265**, 395–406.

Zenkevitch, N. L. and V. P. Petelin, 1956. Photography of the sea bottom. *Priroda* No. 6, 95–99. [In Russian.]

19. UNDERWATER PHOTOGRAPHY

H. E. EDGERTON

1. Introduction

Next best to active personal observation or remote television viewing is the photographic camera record with its tremendous ability to memorize a scene for later and repeated study. Even when a person does descend into the sea to observe, he also greatly needs a camera: (1) to help him remember exactly what he saw; (2) to enable him to show the original subject or view to others; and (3) to record rapid events which he might not notice.

There are many technical difficulties to be solved when a vehicle is designed to transport a human observer to depths in the sea. The vessel must be capable of navigation in three dimensions under control of the crew. Safety is an important requirement. Support of auxiliary vessels is necessary. As a result, very few devices have been built for exploration in the sea because of the complications of construction and operation.

The camera itself can go into the sea without human accompaniment to record original information, which is then studied later by scientists. In this way, a relatively simple device, which can be considered expendable, goes into the deep inaccessible parts of the oceans with ease. During the wave of exploration of the oceans that is scheduled for the next decade, the camera, of both still and motion-picture type, will be exploited fully because of its ease of operation compared with other systems involving people or electrical connections.

It is fortunate that most of the sea is clear of light-diffusing or scattering material so that an optical camera can function. Only near the coast is the water full of sediment which makes it impossible to take photographs unless the dirty water in the optical path is displaced.

Since light from the sun is absorbed by water almost completely at shallow depths, the cameras require artificial light for photography. Flash seems the most advantageous since the light is only on when the photograph is being taken. The expendable flash bulb has been and still is greatly used for single exposures per camera lowering. The electronic-flash system is in very common use for camera systems involving more than one photograph.

Advantages of the electronic flash (strobe system) are: (1) many flashes can be obtained efficiently from a small battery; (2) the instant of flash can be electrically controlled; (3) the color is daylight in quality; and (4) the flash duration is short, producing clear photographs of rapidly moving objects.

Strobes with 20 W-sec input at 24 flashes per second have been used with motion picture cameras at depths of 3000 m. The camera–lamp assembly mounted on a rugged sled can be dragged across the bottom, obtaining a dynamic view of the scene that exists there. Single-flash still cameras with 100 W-sec of energy in the flash have been used in many ways to make both mid-water and bottom photos. The electronic-flash equipment used under

[*MS received January, 1960*]

water is composed of the same circuit elements as used in commercial equipment.

The pressure of the water on camera and lamp housings gives the designers of equipment some problems with strength of containers, optical windows, and electrical connections. At the maximum known depth of 36,000 ft the pressure is about 18,000 psi. Several authors (Ewing *et al.*, 1946; Edgerton and Hoadley, 1955; Thorndike, 1959; Shipek, 1957) give practical housing designs that have been found useful in oceanographic research.

From a strength-to-weight standpoint, the hollow sphere offers the most to a designer. However, the long cylinder is widely used although the wall thickness is about double that of the sphere for the same crushing strength. Various materials, such as aluminum and steel, are used. Tempered stainless steel, such as 17-4-PH, is excellent for long-time exposure to sea-water without rust. This steel can be machined before it is tempered. The tempering does not change the dimensions or cause surface scale.

The "O" ring has been almost universally adopted in all deep-sea housings in place of the gaskets of earlier designs. With the "O" ring, the seal consists of metal against metal with a rubber "O" ring filling the joint to seal against leakage at low pressure.

Windows of glass are best from optical considerations but, for various reasons, Plexiglas is often used. Glass is very strong in compression but very weak in tension, especially if there are any cracks or scratches. A window with a small scratch on the inner surface is, therefore, liable to develop cracks and it may even fracture. However, if the window inner surface is highly polished, the window will resist very high pressures and produce excellent photographs.

A standard lens behind a glass or Plexiglas window will show a restricted angular field which has serious chromatic aberrations at the edge of the field. Thorndike (1959) and A. Ivanoff, J. Baker and R. E. Hopkins have suggested several solutions to the aberration problem. A Hopkins lens of 35-mm focal length and *f* 11 aperture was used to take several of the examples that accompany this article. Hopkins and Edgerton, in 1961, described an *f* 4.5, 35 mm lens.

Stereophotography is made possible by the use of two cameras mounted together. The visual results are very revealing to the observer in giving details that may be missed in a single photograph. Color photographs also assist the observer greatly.

2. Camera Details

A. Data Chamber

A photographic record of pressure (depth), time, positional data (long. and lat.), date, vessel, lens, film, crew, etc. on the actual film with the photographs is a most desirable feature for a deep-sea camera. One advantage of such a marking system is the certainty that the data and the photographs are not separated. At sea and in processing, it is easy to get data and films marked

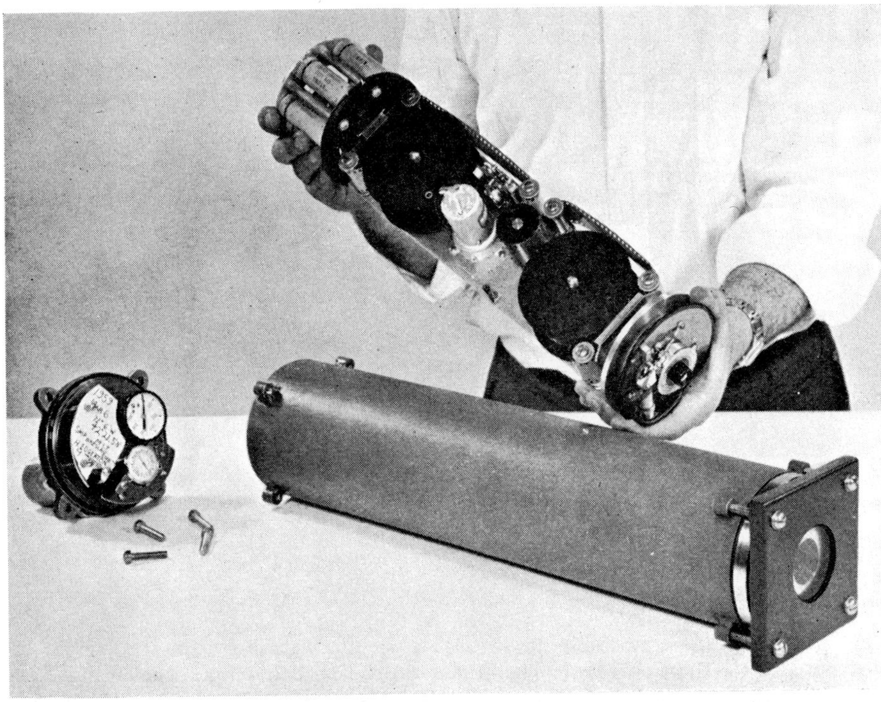

Fig. 1. Photograph of oceanographic camera interior showing "in-line" construction in a cylindrical steel housing. A "data" chamber at the rear of the camera gives a photographic record of pressure (depth), time (24-h clock), positional data, and notes.

wrongly or accidentally interchanged. One example of a data-chamber record is shown in Fig. 1.

B. Sonar

A sonar "pinger" on a camera, operated by internal batteries, has proved to be very useful for signalling information to the surface. The sonar can be used as a "go" or "no-go" device operated by a bottom switch. Also, it can be used to measure the camera-bottom surface by measuring the time interval between the direct ping and the reflected ping, since the velocity of sound in sea-water is known (about 5000 ft/sec). A 5-ft sonar-bottom distance corresponds to a 2-msec interval, which can be conveniently measured on a cathode-ray oscillograph or a standard graphical recorder such as is commonly used as a sonar receiver on oceanographic ships.

3. Cables and Winches for Cameras

Many of the practical problems of photography in the sea are caused by lowering methods. Most oceanographic ships are equipped with winches which do many duties besides camera lowering, such as the lowering of Nansen

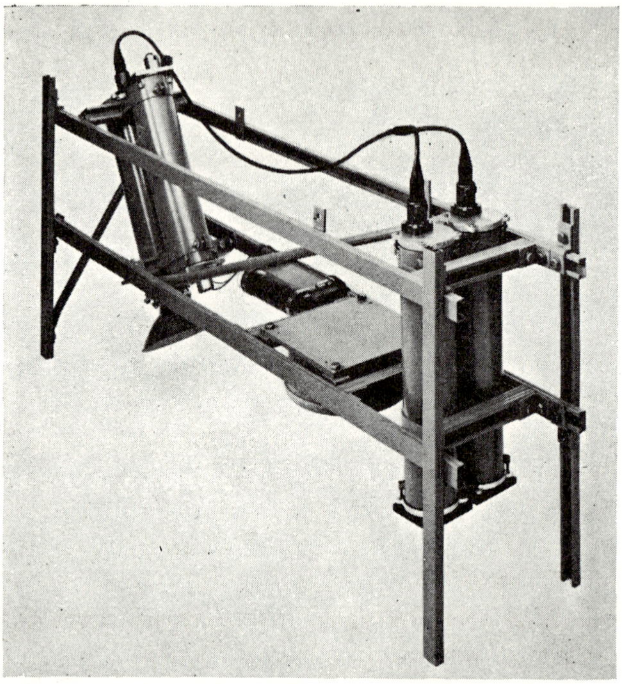

Fig. 2. A typical camera assembly for vertical stereophotography of the sea bottom. Two cameras at the right operate in synchronism with the flash lamp at the far left. Above the lamp is the sonar driver. The square flat object at the center of the framework is a 12-kc pulse sonar transducer for camera-bottom measurement.

Fig. 3. Example to show quality of special *f* 11, 35-mm Hopkins lens when used under water.

Fig. 4. Example to show data photo on same film as the underwater picture. Depth is shown as pressure in psi.

Fig. 5. One-half of a stereo pair of photographs taken from the R.V. *Chain* by the Woods Hole Oceanographic Institution at a depth of about 1100 m. The data section at the right shows pressure, time and other information. A compass in the field of view shows direction of the bottom.

Fig. 6. A bottom-photography equipment as used on the R.V. *Chain* by the Woods Hole Oceanographic Institution. Two cameras at the top produce a stereo pair of photographs. A 100 w.s. strobe is at the bottom. The "pinger" is located at the center.

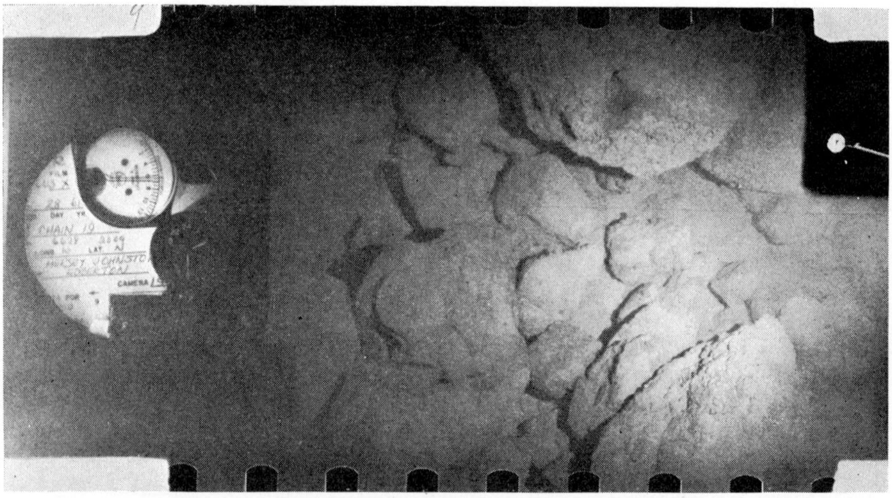

Fig. 7. A photograph of a rock outcrop on the North wall of the Puerto Rico Trench, taken with the equipment shown in Fig. 6.

bottles, corers and drags. The winch and cable design must meet the requirements of all these other duties.

There are times when a single lowering can accomplish several purposes. For example, the camera can be used as a bottom weight on a string of Nansen bottles. The sonar on the camera can be used to hold the camera at a required distance above the bottom and thus indicate to the crew on the deck the time to drop the weight messenger that operates the bottles. During the entire operation, the camera is recording both mid-water and bottom scenes.

References

Breslau, L. P. and H. E. Edgerton, 1958. The luminescence camera. Woods Hole Oceanog. Inst. Ref. 58-14; also in *J. Biol. Photog. Assoc.*, **26**, No. 2, May, 1958.

Breslau, L. P. and H. E. Edgerton, 1958a. Interruption camera. Woods Hole Oceanog. Inst. Ref. 58-27.

Cousteau, J.-Y., 1958. Calypso explores an undersea canyon. *Nat. Geog. Mag.*, **113**, 373–396.

Edgerton, H. E., 1955. Photographing the sea's dark underworld. *Nat. Geog. Mag.*, **107**, 523–537.

Edgerton, H. E., J.-Y. Cousteau, J. B. Brown and R. H. Hartman, 1957. Nylon rope for deep-sea instrumentation. *Deep-Sea Res.*, **4**, 287.

Edgerton, H. E. and L. D. Hoadley, 1955. Cameras and lights for underwater use. *J. Soc. Motion Pict. and Television Eng.*, **64**, 345–350.

Edgerton, H. E. and R. E. Hopkins, 1961. Lenses for underwater photography. *Deep-Sea Res.*, **8**, 312–317.

Ewing, M., A. Vine and J. L. Worzel, 1946. Photography of the ocean bottom. *J. Opt. Soc. Amer.*, **36**, 307–321. [Contains an extensive historical bibliography.]

Houot, G. S., 1958. Four years of diving to the bottom of the sea. *Nat. Geog. Mag.*, **113**, 715–731.

Houot, G. S., 1960. Deep diving by bathyscaphe off Japan. *Nat. Geog. Mag.*, **117**, 138–150.

Johnson, H. R., R. H. Backus, J. B. Hersey and D. M. Owen, 1956. Suspended echo-sounder and camera studies of mid-water sound scatterers. *Deep-Sea Res.*, **3**, 266.

Laughton, A. S., 1957. A new deep-sea underwater camera. *Deep-Sea Res.*, **4**, 120–125.

Laughton, A. S., 1959. Photography of the ocean floor. *Endeavour*, **18**, 178–185.

Owen, D. M., 1951. Deep-sea underwater photography and some recent stereoscopic applications. *Photogrammetric Eng.*, **17**, 13–19.

Shipek, C. J., 1957. NEL Type III deep-sea camera. *U.S. Navy Electronics Lab. Rep.* No. 768.

Shipek, C. J., 1960. A photographic study of some deep-sea floor environments in the eastern Pacific. *Bull. Geol. Soc. Amer.*, **71**, 1067–1074.

Thorndike, E. N., 1959. Deep-sea cameras of the Lamont Observatory. *Deep-Sea Res.*, **5**, 234–237.

20. SUBMARINE CANYONS

F. P. SHEPARD

1. Introduction

Submarine canyons have been discussed extensively in the literature during the past 30 years. Field investigations, however, have lagged and have been confined to a very few individuals and to a few localities. This is regrettable because the studies that have been made have indicated both that the canyons of the sea floor are of amazingly large proportions and that they play a very important role in sedimentary processes.

Most of the information available on submarine canyons comes as the result of coastal charting operations, particularly those of the U.S. Coast and Geodetic Survey. Detailed scientific studies have been largely centered around the canyons of southern California and Lower California, coming from the operations of Scripps Institution (Shepard and Emery, 1941, pp. 51–161; Shepard, 1949, 1951, 1961; Shepard and Einsele, 1962) and the Hancock Foundation (Emery, 1960, pp. 40–50). Several of the canyons off the eastern United States have been investigated by the Lamont Geological Observatory (Ericson *et al.*, 1951; Heezen, Tharp and Ewing, 1959) and by Woods Hole Oceanographic Institution (Stetson, 1936). French scientists have studied some of the Mediterranean canyons and the Japanese have collected some samples from the canyons off Japan. Otherwise the numerous submarine canyons found along the coasts of many parts of the world are known only from rather inadequate soundings.

Recently some decided advances have been made in canyon studies. These have come from aqualung explorations of the shallow canyon heads particularly off California. Also, Precision Depth Recorders used by the oceanographic laboratories have given much more detail concerning the configuration of a number of the submarine canyons.

2. Classification of Submarine Valleys

If we define canyons as deeply incised V-shaped valleys with steep walls, we find that many of the valleys of the sea floor to which the name "canyon" has been applied do not fit the definition. For example, the large valley off the Ganges Delta is more appropriately called a trough because of its broad flat floor; whereas the deep outer continuations of true submarine canyons, like those of Monterey Bay, resemble river channels with raised levees on the sides and therefore should be referred to as channels. Realization of the existence of decidedly different types of submarine valleys, which has come partly from the study of Precision Depth Recorder (P.D.R.) profiles, has now made it possible to subdivide submarine valleys and, hence, to discuss their origins separately rather than throwing them all into the same pot.

The types of submarine valleys which extend down the continental slopes can be divided into: (1) true submarine canyons which have winding courses, V-shaped profiles, steep walls with many rock outcrops, numerous entering tributaries, and virtually continuous outward-sloping floors; (2) trough-shaped,

[*MS received March, 1960*]

straight-walled valleys in tectonically active areas following structural trends rarely normal to the slope, containing basin depressions, and virtually lacking tributaries; (3) trough-shaped valleys found off deltas, lacking tributaries but differing from the second type in having virtually continuous outward-sloping floors; (4) short gullies found on the foreset slope off a few deltas; (5) the relatively shallow, flat-floored valleys with levees which extend beyond true submarine canyons; and (6) relatively straight, shallow valleys extending down submarine escarpments and having few tributaries.

The advantage of separating these types before considering their origin can be seen when we read such statements as "submarine canyons could not have been cut by subaerial erosion because they extend out to the deep ocean floor". One has but to ask what type of submarine valley is found extending to the ocean floor, before considering the merits of such a statement.

3. Canyon Descriptions

Not very many submarine canyons have been sufficiently surveyed for them to be described in any detail. The best known are along the California and Lower California coasts. Several of these have been sounded with P.D.R. recordings from their heads down to where they disappear on the outer fans. Furthermore, aqualungers have explored the canyon heads and brought back many pictures along with descriptions.

The most explored canyon is located off La Jolla, California, and has its two main branches, La Jolla and Scripps, coming into the coast on either side of Scripps Institution (Fig. 1). The numerous dives into the head of Scripps Canyon have revealed a series of narrow, rock-walled gorges with some overhanging precipices (Figs. 2 and 3). The inner ends of the canyons have what might be termed sand chutes which extend in almost to the shore. These chutes fill rapidly with sand carried to them along the shore by wave action and currents. Kelp and sea-grass accumulate below the zone of active sand deposition. Periodically they discharge much of their sand and kelp load and at such times the canyon floors deepen by amounts of as much as 5 fm (about 9 m) (Chamberlain, 1960). Two of the deepenings have been connected with earthquakes and two others with periods of large waves. Other deepenings have taken place without any apparent triggering effect. Chamberlain found that the sea-plant debris becomes compressed and this increases the instability of the inner sand deposits. According to Dill (1962), after the sand chutes have been evacuated, polished rock walls and the floors of these small tributaries are exposed, suggesting considerable wear from moving currents of sand. Traced outward, Scripps Canyon continues as a precipitous, rock-walled gorge as far as its juncture with La Jolla Canyon, a mile from the canyon head and at a depth of 150 fm (274 m).

La Jolla Canyon winds out to sea with relatively gentle walls except at the head where small cliffs are found in what appears to be Pleistocene or Holocene alluvium. Farther out the walls are of soft rock, but beyond axis depths of

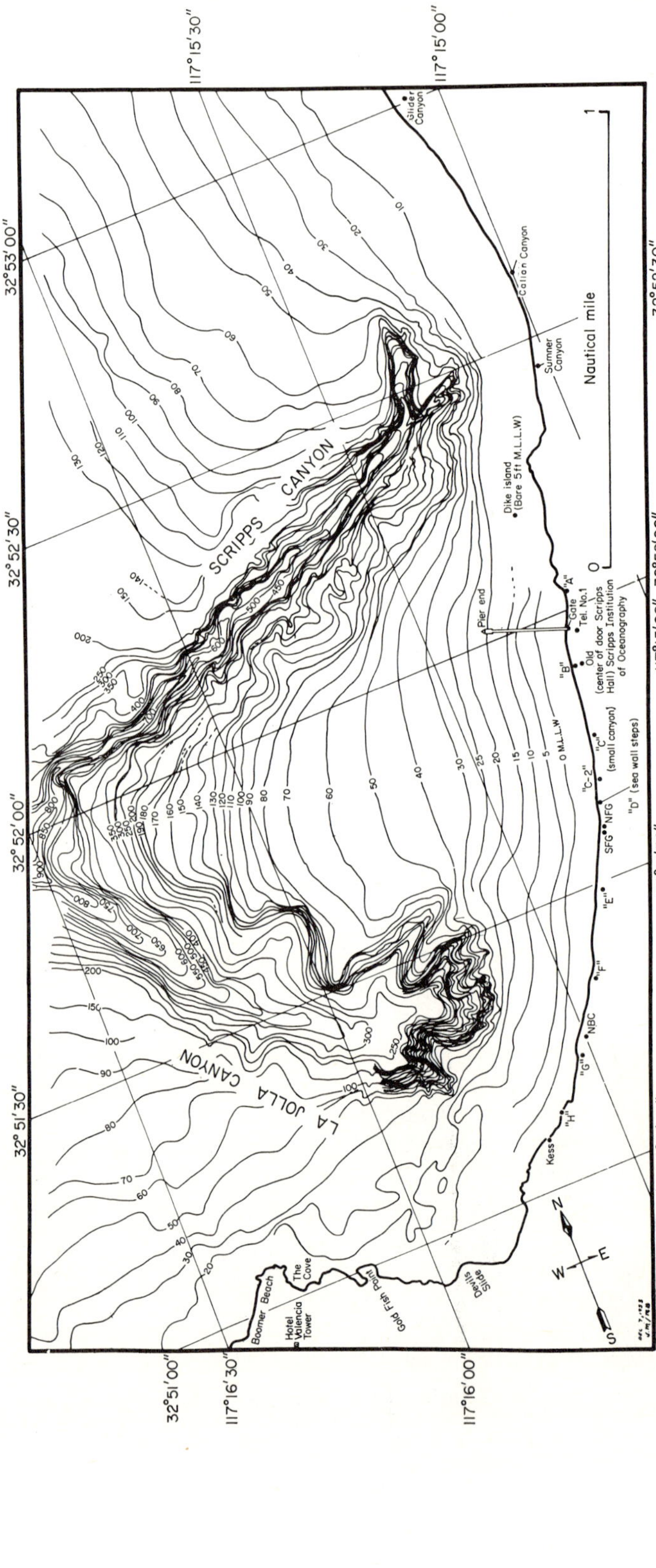

Fig. 1. Showing the head of La Jolla Canyon and its tributary, Scripps Canyon. Contour interval 10 ft out to 200 ft; 50 ft beyond.

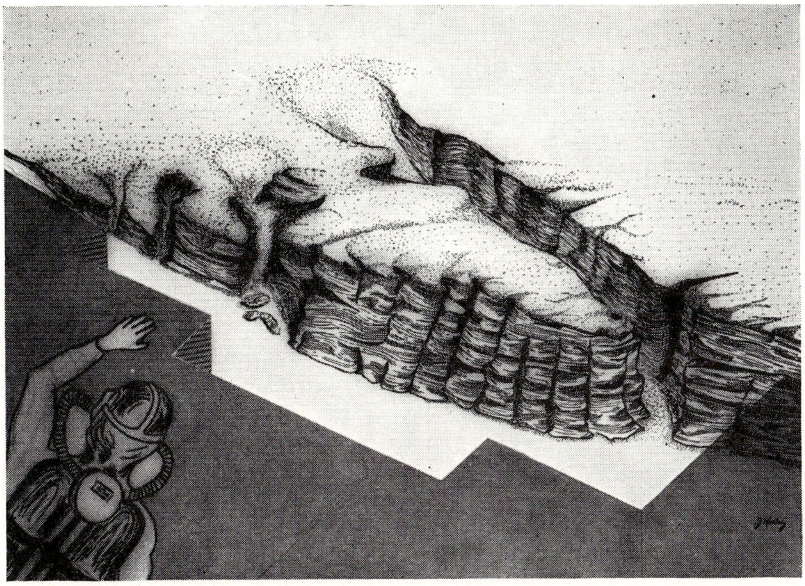

Fig. 2. Sketch of two of the branches at the head of Scripps Canyon, showing the vertical walls and the sand chutes. (Drawn by J. R. Moriarty from preliminary sketches by Earl Murray.)

Fig. 3. Photograph by R. F. Dill of the U.S. Navy Electronics Laboratory showing the overhanging walls, the terraces, and the encrusting organisms near the head of Scripps Canyon. (Official U.S. Navy photograph.)

300 fm (550 m), the rock apparently ceases and muddy walls with some sand are found. At depths of 350 fm (640 m) the walls appear to be related to ridges which can be termed natural levees. These levees traced still farther out have become definitely a part of the outer fan with a shallow channel in the mid-portion. All along the canyon and channel, sand occurs largely as layers between mud. The channel gets less and less deeply incised until it disappears

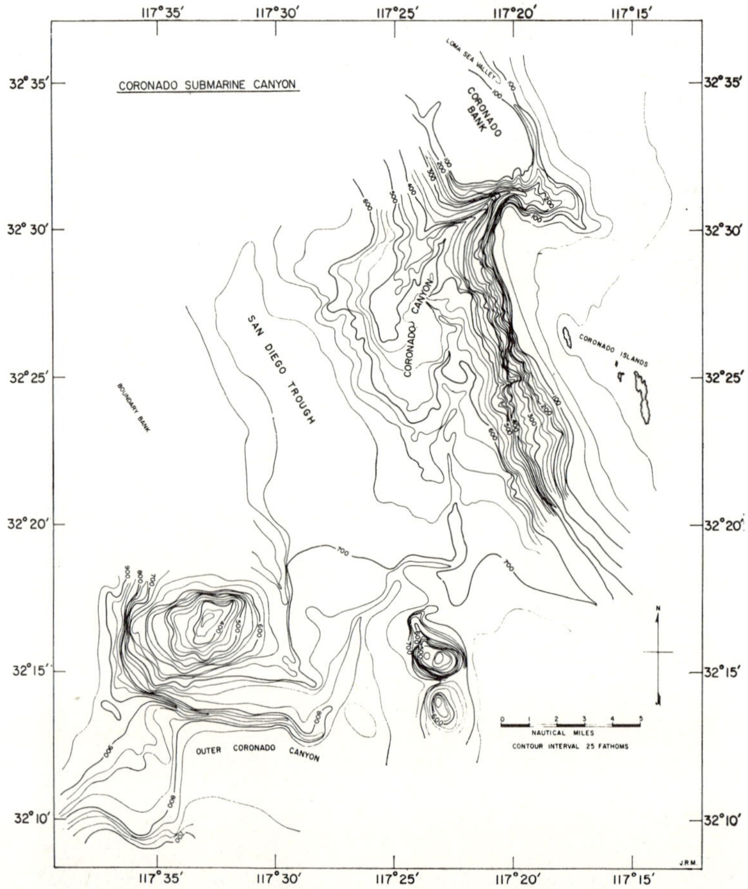

Fig. 4. Coronado Submarine Canyon and outer channel. Note the high levee (or dissected delta) on the outer curve below the inner gorge. Loma Sea Valley contains a divide draining north and south. (From Scripps Institution P.D.R. surveys.)

and the fan merges with the flat, but gently sloping, floor of San Diego Trough, where the depth is 600 fm (1100 m).

Coronado Canyon heads 8 miles off the Tia Juana River near the United States and Mexican border. The inner canyon has a rock gorge crossing what appears to be a hogback ridge (Fig. 4). Beyond the ridge, instead of directly entering San Diego Trough, the valley turns to the left and extends as a narrow trough for 11 miles along the base of the ridge leading down from the Mexican Coronados Islands. On the outside of the trough a great mass of sediment rises 100 fm (183 m) or more above the floor. This may well be a large natural levee although the alternative explanation given previously (Emery *et al.*, 1952) was that it was a delta which had undergone extensive slumping. Beyond, the channel crosses the lower end of San Diego Trough and shows definite levees on both sides. Coming to a rock ridge, Boundary Bank, the channel deepens considerably as its gradient increases in starting the descent toward San Clemente Trough. However, it terminates part way down the descent.

Fig. 5. Photograph by Conrad Limbaugh of sandfall observed in a tributary of San Lucas Canyon. Height of fall estimated as about 5 ft. The gently flowing sand above the fall was found in the lower end of a funnel-like narrow valley at a depth of about 15 to 25 fm.

San Lucas Canyon is located off the tip of Lower California. It heads in a small bay of a relatively large fan-filled valley coming down from the high mountains of the southern end of the peninsula. The main canyon head extends into the end of a pier at the small fishing village of San Lucas, as can be seen through the clear water from a plane or from the adjacent ridge. Along the tip end of the peninsula there are several tributaries, one being located off an intermittent tombolo which is occasionally topped by large waves from the open Pacific. At such times quantities of sand are carried across the tombolo and down into one of the tributaries. At the lower end of a funnel-shaped valley

Scripps Institution aqualungers found the sand moving along the axis until it fell over a cliff as a sandfall (Fig. 5). Tracing the main canyon outward (Fig. 6) it continues for 20 miles as a winding steep-walled V-shaped gorge and termi-

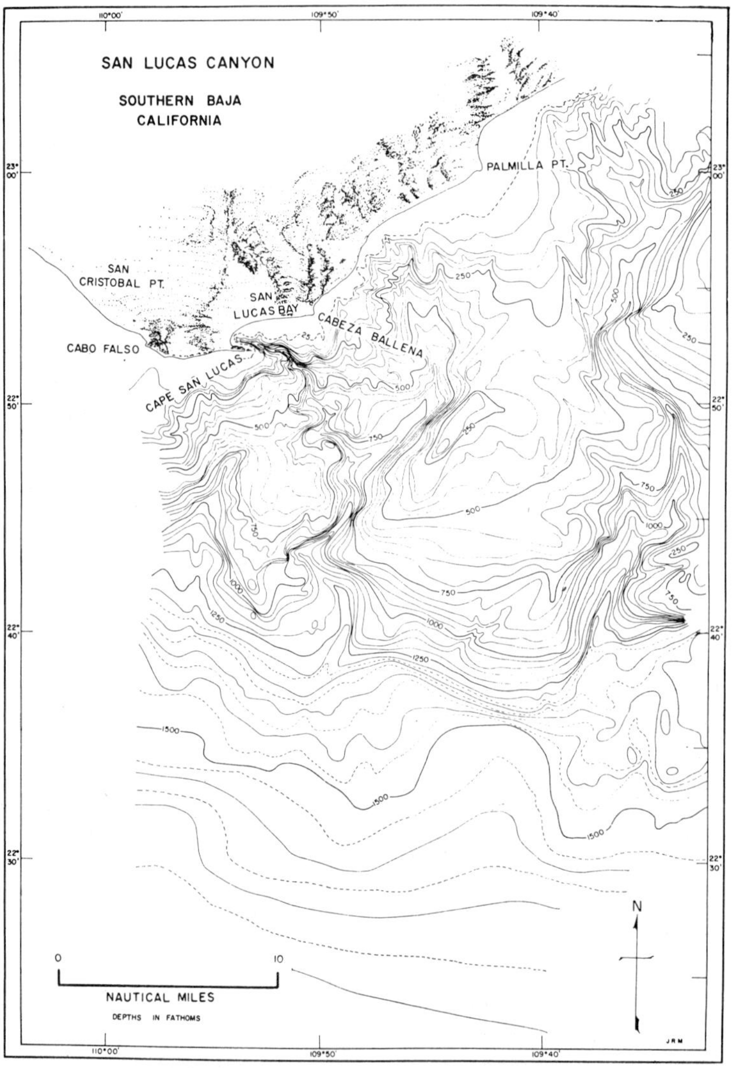

Fig. 6. The deeply dissected slopes south of Cape San Lucas. The two canyons are cut in granite and both terminate at about 1325 fm. Contour interval 50 fm, with 25 fm added at top and bottom. (From Scripps Institution P.D.R. surveys.)

nates in an irregular fan at a depth of 1325 fm (2420 m). The canyon walls are granite or another type of resistant rock. The canyon floor contains only sand, and the sand continues at least out onto the head of the fan. Unlike the end of

La Jolla Canyon there are no appreciable levees indicated in the sounding record.

Monterey Canyon and its tributary, Carmel Canyon, like the others considered previously, come in right to the shore (Fig. 7). The head of Monterey Canyon is located off the Salinas Valley and is definitely cut in unconsolidated materials. Also, as Woodford (1951) has emphasized, the gradient in the canyon far exceeds that of the Salinas Valley. The importance of turbidity currents in the canyon head is shown by cores with layers of relatively coarse

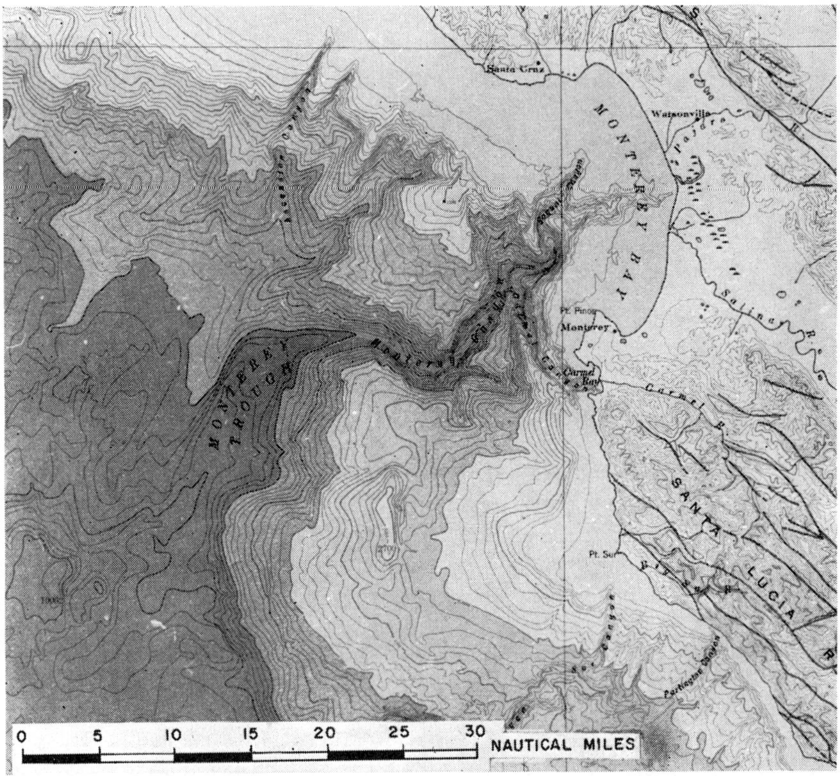

Fig. 7. Monterey Submarine Canyon and vicinity. Note the resemblance of the canyon axes to the river valleys on land. Contour interval 50 fm; shade changes at each 500 fm.

sand, similar to that of the adjacent beach, interbedded with muddy sediments. Out at a depth of about 100 fm (183 m) the walls become rocky and both sedimentary rock and granite have been dredged from the precipitous slopes at greater depths. The principal tributary, Carmel Canyon, is rock-walled throughout its length so far as has been determined. The rock in this branch is granite and the canyon has numerous entering tributaries. One tributary connects with a granite land canyon and the submarine continuation is similar to the land canyon but has a somewhat smaller gradient. Carmel Canyon joins Monterey Canyon at 1100 fm (2000 m). Beyond the junction the main canyon is somewhat wider with a relatively flat floor. This in turn connects with an outer fan having channels like those off the La Jolla Canyon, but these begin at much greater depths, around 2000 fm (3660 m) (Dill *et al.*, 1954). The connection is not as yet clearly demonstrated. Menard (1960) reported a gravel

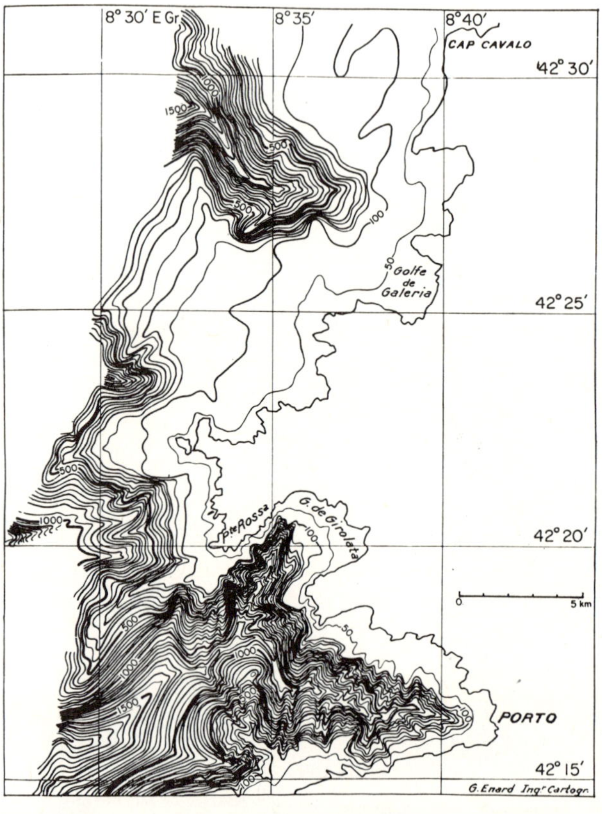

Fig. 8. Submarine canyons cut in the crystalline rocks along the west coast of Corsica. Note that the submarine contours conform to the shoreline. Contour interval 50 fm. (After Bourcart, 1959.)

core from the fan channel and the writer found gravel in the canyon at a depth of 835 fm (1530 m) (Shepard, 1951). The latter core showed the gravel at the bottom of a graded layer, indicative of turbidity-current deposits. In many ways the most impressive feature of Monterey and adjacent canyons is their dendritic and trellis drainage pattern (Fig. 7), showing structural trends that are similar to those on the land.

The Mediterranean canyons along the south coast of France and the west coast of Corsica are certainly the best known outside of those from western

America. The studies of Bourcart (1959) and of his associates have resulted in many interesting contour charts of these canyons (Figs. 8 and 9). Even before these canyons were surveyed by the French geologists, Kuenen (1953) had come to the conclusion that at least the Corsican and Riviera submarine canyons were the submerged continuations of river-eroded canyons. The newer surveys apparently confirm this opinion, at least so far as the shapes of the canyons are concerned. As off Cape San Lucas, Lower California, and Carmel,

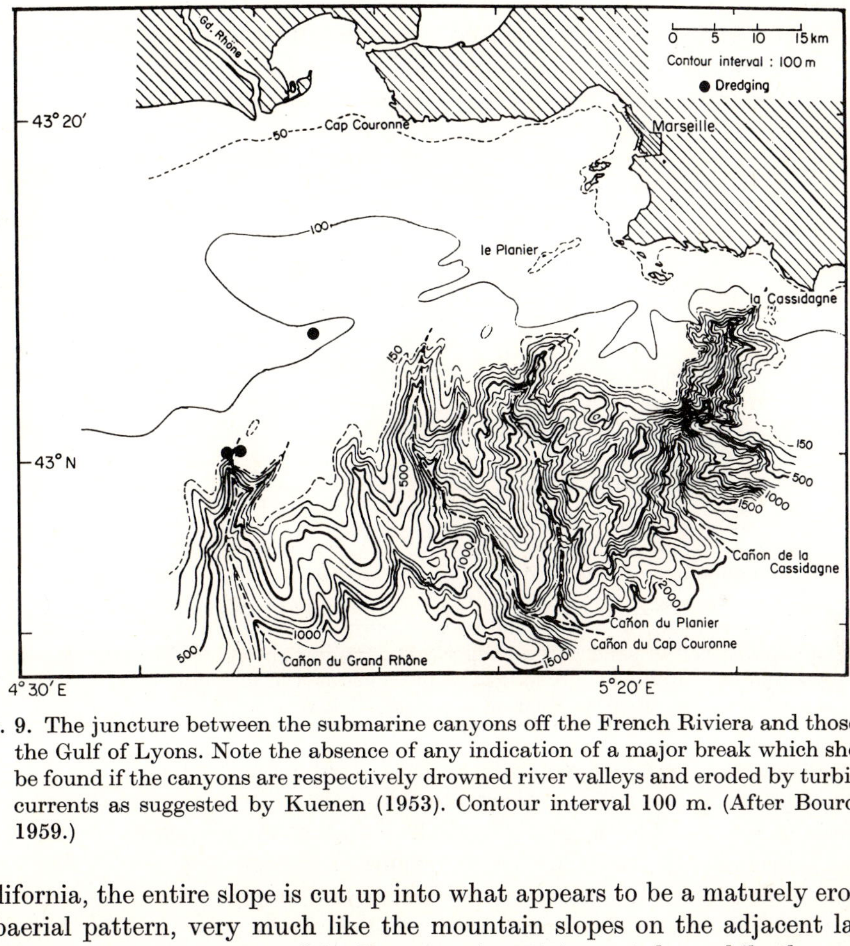

Fig. 9. The juncture between the submarine canyons off the French Riviera and those off the Gulf of Lyons. Note the absence of any indication of a major break which should be found if the canyons are respectively drowned river valleys and eroded by turbidity currents as suggested by Kuenen (1953). Contour interval 100 m. (After Bourcart, 1959.)

California, the entire slope is cut up into what appears to be a maturely eroded subaerial pattern, very much like the mountain slopes on the adjacent land. Canyon center cores reported by Bourcart (1959) (some taken while the writer was assisting in the operation) indicated the presence of sand layers interbedded between mud, as in many other submarine canyons. The wall rocks are evidently crystalline along the deeply indented coast of Corsica and both granite and metamorphosed Paleozoics were found by Bourcart along the canyon walls off the Maritime Alps. The rather remarkable conformity between the canyon heads and the west coast bays of Corsica (Fig. 8) is unique. Yet the submarine canyons are similar to most of the well surveyed canyons in other parts of the world.

The canyons west of Marseilles have lower gradients than those to the east but otherwise do not differ noticeably (Fig. 9). In both groups there are many tributaries and some of the branches extend parallel to the coast rather than directly down the slope. The Mediterranean canyons are said to terminate at depths of from about 1200 to 1400 fm (2200 to 2600 m).

Another canyon of rather special interest enters the long estuary at the mouth of the Congo River. This has been studied by Heezen and his associates (1959) and to them we are indebted for a contour chart which shows how the canyon terminates outward in a huge submarine fan creased by channels (Fig. 10). The canyon appears to contact the fan at a depth of 550 fm (1000 m).

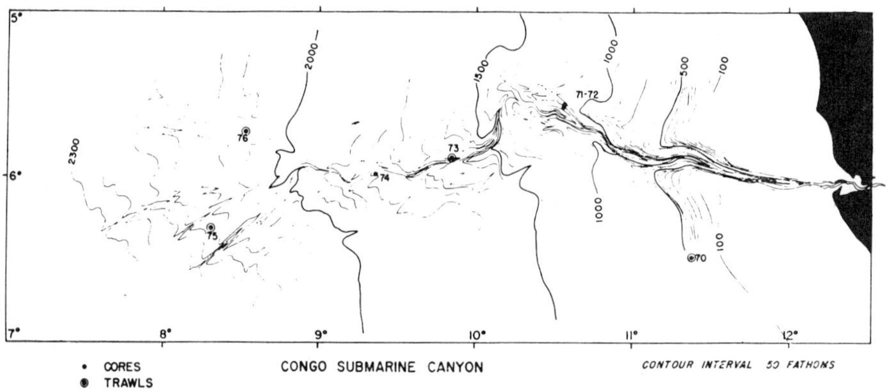

Fig. 10. Congo submarine canyon. Turbidity currents occur at the mouth of the Congo River at the rate of 50 each century. These have built up an enormous abyssal cone of coalescing natural levee systems. The sediment core from position 76 contains an uppermost bed of sand and a deeper bed containing wood and leaves. The trawl at position 76 contained many twigs and branches as well as an abundant bottom fauna. (Caption and figure from Heezen, Menzies and Ewing, 1959.)

Unfortunately there has been no dredging of the inner canyon walls so that we do not know whether it is entirely cut into soft sediments although the depth of about 400 fm (730 m) at the mouth of the estuary suggests the likelihood that this is not the case. The investigations of Heezen have shown that cables were broken after short intervals whenever laid across the canyon head. This suggests that slides and/or turbidity currents are very active in disposing of the excess sediment introduced by the Congo River. The turbidity currents can also account for the coarse sediments and the large amounts of plant debris reported by Heezen from the outer fan. In two respects the Congo Canyon differs from the others described previously. It has a very gentle gradient, about 1%, and it lacks large tributaries although small ones enter the walls in the inner portion.

Along the east coast of Ceylon there is a canyon which in many respects resembles those of west Corsica. At Trincomalee (the former British naval base) a branching rock-walled canyon comes into the bay and extends its arms so close to the shore that it appears to be a truly drowned valley (Fig. 11). One branch cuts directly across quartzite ridges and the walls of this gorge have been visited by aqualungers who reported to the writer that there are precipitous rocky slopes. The outer end of the canyon has not been surveyed but it definitely extends to 870 fm (1600 m) where it has walls rising 600 fm (1100 m)

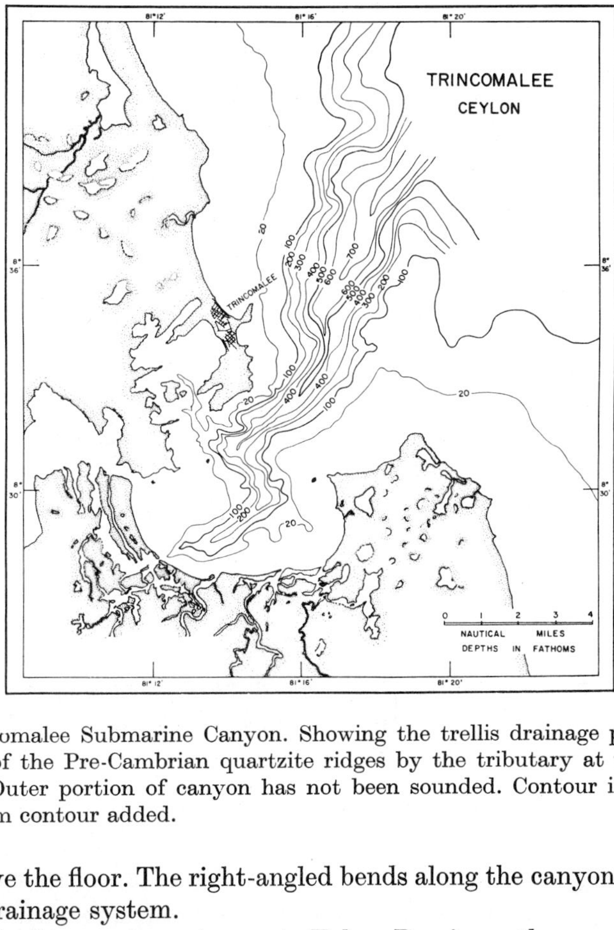

Fig. 11. Trincomalee Submarine Canyon. Showing the trellis drainage pattern and the crossing of the Pre-Cambrian quartzite ridges by the tributary at the head of the canyon. Outer portion of canyon has not been sounded. Contour interval 100 fm, with 20 fm contour added.

or more above the floor. The right-angled bends along the canyon are indicative of a trellis drainage system.

Coming into the curving entrance to Tokyo Bay is another canyon which has the steep rocky walls, the V-shaped floor, and the dendritic tributaries characteristic of land canyons (Shepard, 1948, fig. 74). It contrasts considerably with the fault-trough valleys found in the vicinity. Like many other submarine canyons its wall heights are measured in thousands of feet. It terminates seaward where it enters the Sagami Bay fault trough at 770 fm (1400 m). Some of the tributaries come into land embayments, but unlike the Corsican submarine canyons the main head terminates in broad shallow Tokyo Bay.

4. East Coast U.S. (New England Type) Canyons

Kuenen (1953) has called attention to some differences between the Corsican and Riviera-type canyons, which he called "ravines", and the east coast canyons, which he called "New England type". Although the continental slope off northeastern United States has been moderately well surveyed down to 1000 fm (1830 m), the outer portion of the slope valleys are still known only from sporadic lines which were not well controlled in position. Furthermore, the only well known contours of the canyons were those made by Veatch and Smith (1939), which have been widely criticized for their interpretative character.

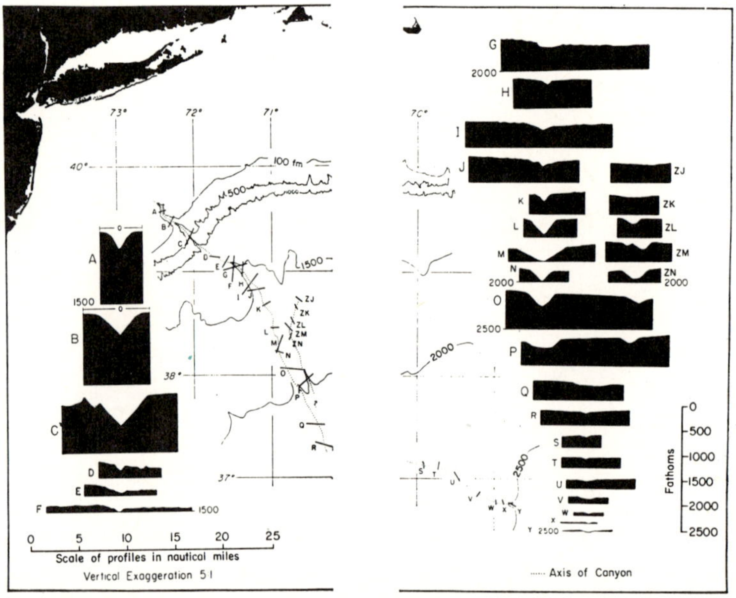

Fig. 12. Cross-sections of Hudson Submarine Canyon. Locations of the profiles are indicated on the map. (After Heezen, Tharp and Ewing, 1959.)

However, enough information is available from this survey and the sounding lines of Lamont Geological Observatory to allow some generalizations. They resemble those of the west coast of the United States and the Corsican canyons in their V-shape, and in having occasional rock outcrops on their walls. Their gradients are similar to those of the west-coast canyons (Shepard and Beard, 1938) although they appear to lack any rock-walled gorges at their heads with precipitous gradients. As Kuenen points out, the canyons differ also in extending directly down the dip of the submarine slope with only minor bends. They have far fewer major tributaries than found in the canyons previously described.

Profiles of Hudson Canyon (Fig. 12) made by Lamont scientists (Heezen *et al.*, 1959) indicate that it changes from a gorge more than 500 fm (900 m) deep to a channel only 50 fm (90 m) deep at the base of the continental slope, but again deepens to 300 fm (550 m) in the "lower gorge". This is similar to the changes in Coronado Canyon.

5. Trough-Shaped Valleys

The trough-shaped valleys of the sea floor have often been included with submarine canyons whereas they are strikingly different in character. There are two types of troughs which extend down the submarine slopes; one is found in areas of active tectonism and the other directly off deltas.

A. Tectonic Troughs

The shelves and slopes around the Aleutian Islands are notable for a series of sea-floor valleys which extend somewhat diagonally down the slopes reaching depths of several thousand fathoms (Gibson and Nichols, 1953). These valleys have a trough shape with a relatively flat floor; they have basin depressions and relatively straight sides with virtually no tributaries. They are evidently fault troughs. Another type of tectonic valley with a V-shape occurs along the southeast end of San Clemente Island (Shepard and Emery, 1941, fig. 11). There seems to be no reason to include these troughs among the erosion type of submarine valleys although some erosion, either marine or subaerial, may have occurred.

B. Delta Front Troughs

Recent surveys have shown that several of the valleys outside deltas are also quite distinct from submarine canyons. The valley beyond the Ganges Delta (Fig. 13) is now well surveyed to the edge of the shelf and several lines have been run on the slope and on the deep-ocean floor outside. The latter show that it apparently continues as a deep-sea channel for hundreds of miles seaward, perhaps as far south as Ceylon (Dietz, 1954). The valley, in crossing the wide shelf, has a low gradient (about 1%), a flat floor with relatively straight sides, and only very minor tributaries. The same type of valley is found off the Indus Delta and probably off the extreme west end of the Niger Delta. There is also a trough starting near the outer end of the shelf off the west side of an old lobe of the Mississippi Delta (Shepard, 1937). This also has approximately a 1% gradient and a flat floor. An inner filled portion of this trough has been discovered through geophysical prospecting. It continues landward, connecting with a buried channel of the Mississippi (Fisk and McFarlan, 1955). The Mississippi Trough extends seaward to a depth of 900 fm (1600 m), where it appears to terminate in a great fan (Ewing *et al.*, 1958, fig. 6). This trough is thought by the Lamont group to have been the means of transporting sediment out to the Sigsbee Deep during the Pleistocene low-sea-level stage. Based on this trough, many quotations have been made of Fisk and McFarlan's (1955) supposed 450-ft (134-m) sea-level stand, but neither the charts nor the geophysical measurements seem to give any reason for believing that there is any break at such a level. The writer feels that this is not evidence of a sea-level stand because the valley does not stop at this 450-ft (134-m) level, because turbidity currents are obviously involved at least in the present configuration of the Mississippi Trough, and because the delta area is subsiding.

6. Delta Foreset-Slope Gullies

Submarine canyons have sometimes been referred to as gullies or ravines. The fact that most submarine canyons have depths below their walls of several hundred meters suggests that neither term is very appropriate. Beyond the Mississippi Delta, however, there are some slope valleys which are small

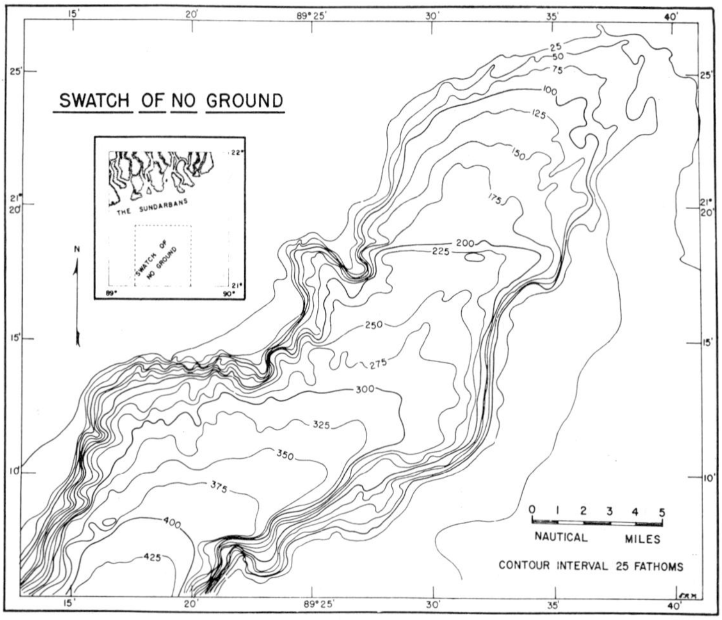

Fig. 13. The Ganges Submarine Trough. Contours are from a detailed survey by the Pakistan Navy. The change in gradient between 200 and 225 fm may be an error of position or alternatively due to a small fault on the floor of the trough. Note the absence of any appreciable tributaries along the sides of the trough.

enough for either ravines or gullies to be applicable names (Fig. 14) (Shepard, 1955). These stop on the slope at depths of approximately 33 fm (60 m). They are found off the portions of the Birdfoot Delta which have recently built across the shelf so that sediments are being added to the continental slope. The gullies are about 8 fm (15 m) deep on the average. They have small hills and depressions along their courses and mounds out beyond. Tributaries are almost entirely missing among these gullies. A few valleys are found in the foreset slope of the Fraser Delta (Mathews and Shepard, 1962). These also stop on the slope with many hills outside. The gullies in the Mediterranean off the Rhône Delta (van Straaten, 1959) differ only in having a suggestion of natural levees along

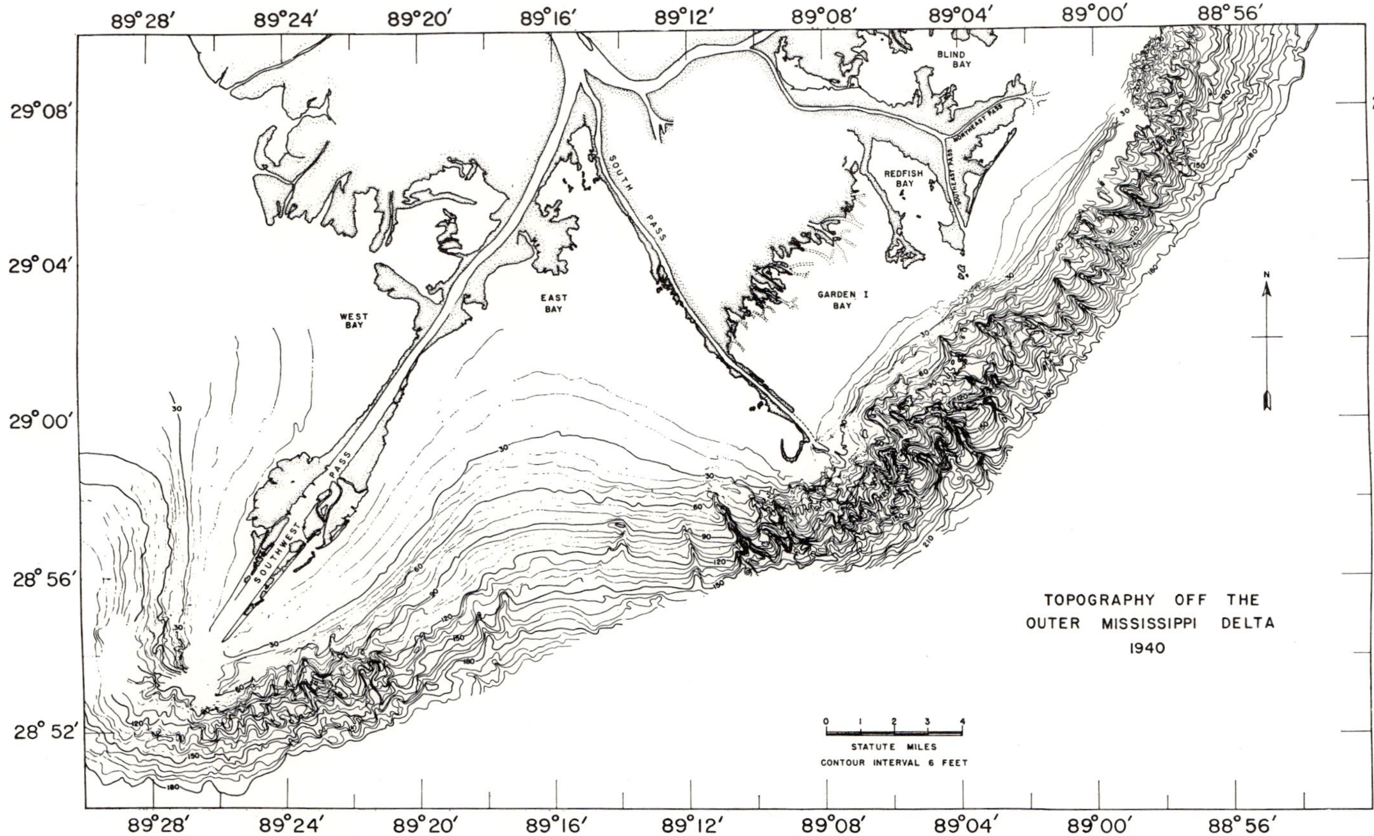

Fig. 14. Landslide ravines off the most advanced passes of the Mississippi Delta. Absence of turbidity currents is indicated by similarity of cores taken from valley floors and ridges. Note also that the valleys terminate outward at about 200 ft (33 fm).

their sides. The delta built by the Rhône into Lake Geneva has a valley which extends the entire length from the river mouth to the plain at the bottom of the lake. This ravine, unlike the marine delta-front gullies, has very large natural levees and has a sand floor, contrasting with mud on the levees. R. F. Dill has found currents here.

7. Shallow Straight-Walled Valleys of Submarine Escarpments

Along many steep submarine escarpments—many of them definitely fault scarps—there are incised valleys starting at various depths and some of them extending to the base of the slope. These have been described by Buffington (1951) and Emery and Terry (1956). These valleys differ from submarine canyons in lacking the dendritic tributaries and in having relatively straight sides. The depth of these valleys below the surrounding sea floor is mostly less than 100 fm. They do not penetrate into the shelf to any extent. They also lack the marginal levees of the fan valleys and are rather closely similar to the gullies on the front of deltas, although they have much steeper gradients and are somewhat more deeply incised. There seems to be little doubt that these straight-walled valleys in escarpments, as well as the gullies of the delta fronts (except in fresh-water lakes), are the product of submarine slumping. They have been explained in this way by almost everyone who has studied them.

8. Origin of Submarine Canyons

Many hypotheses have been offered to explain submarine canyons, although most of them seem now to have disappeared leaving only two principal contenders: turbidity-current erosion and subaerial erosion followed by downwarping during which the canyons have been preserved by submarine processes. A number of writers now consider that turbidity currents are the only possible solution, but the principal authors of the turbidity-current hypothesis (Daly, 1936; Kuenen, 1953) have expressed the opinion that at least some of the canyons were cut by rivers.

In considering the origin of the submarine valleys it seems useless to consider them as a unit having only one explanation. As has been indicated both by Kuenen (1953) and the present writer there are many different types of valleys and there is no reason for believing that they are all of the same origin. It is felt that little can be gained by just giving the arguments favoring one hypothesis along with the objections to the other. For many years Kuenen and the present writer have been discussing the hypotheses both in articles and in personal correspondence. The attempt here will be to point out the various arguments showing how new information bears on them but leaving the reader to judge the merits of the case.

A. Turbidity Currents as a Cause of the Formation of Canyons

When Daly (1936) first suggested turbidity currents as the agents which cut submarine canyons there was very little evidence to support his idea. It was known that the relatively heavy water from rivers which entered certain reservoirs sank below the surface developing an under-flow which sometimes followed the bottom as far as the dam. Later Kuenen and Migliorini (1950) showed in a large experimental tank that turbidity currents could transport sand and other coarse sediments along the sloping bottom without much loss of material. Next came the finding of sand layers on the sea floor at the foot of submarine canyons, first reported by Scripps Institution operations (Shepard, 1951) but better documented by Lamont Geological Observatory (Ericson *et al.*, 1951, 1961). These sands, along with occasional gravel deposits, indicated that rather appreciable currents must sweep along the submarine canyons in order to carry out such coarse material along the canyon lengths. It has gradually been found that there are huge fans at the base of many of the submarine canyons probably containing much greater quantities of material than would come from the mere excavation of the canyons on the slopes above (Menard, 1960).

To many investigators the most convincing argument favoring turbidity-current erosion has been the suggestion by Heezen and Ewing (1952) that the successive outward breaks of the Grand Banks cables in 1929 were the result of turbidity currents moving down the slope after the great earthquake of that year. Taking the times and distances between cable breaks they estimated that the currents at first were moving at a speed of 55 knots and then slowed gradually to 12 knots at the outer portion of the breaks. A re-examination of the data, however, seems to indicate that there is little basis for their estimate of such excessive speeds (Fig. 15). Heezen and Ewing appear to have overlooked the fact that at the time of the earthquake the cables broke instantaneously for a distance of about 100 nautical miles from the epicenter so that turbidity currents could also have started immediately at these considerable distances. With this in mind a more reasonable interpretation appears to be that the times between cable breaks indicate velocities of 15 knots without much, if any, change of speed from top to bottom of the slope.

If turbidity currents move at even a 15-knot speed, they could well be considered as an erosion agent since this is faster than most rivers flow. There is, however, an alternative explanation for the time intervals between breaks. Terzaghi (1956) has suggested that the advancing front of a wave of spontaneous liquefaction spread down this slope after the earthquake. Terzaghi considered that the character of the broken ends of the cables later recovered was suggestive of the irregular disturbances accompanying spontaneous liquefaction and that an advance of 15 knots is not out of line with other documented instances of spontaneous liquefaction. Such an explanation was well received by C. S. Lawton, General Plant Engineer of Western Union. It seems possible that the spontaneous liquefaction could have caused turbidity currents which might have acted in conjunction with the spontaneous liquefaction.

Although Heezen, Ericson and Ewing (1954) have argued that the sediments in the cable-break area support the idea of a tremendous turbidity current, it must be pointed out that their evidence is largely negative, consisting of not having recovered cores and finding bent core tubes in an area near the outer

(a)

(b)

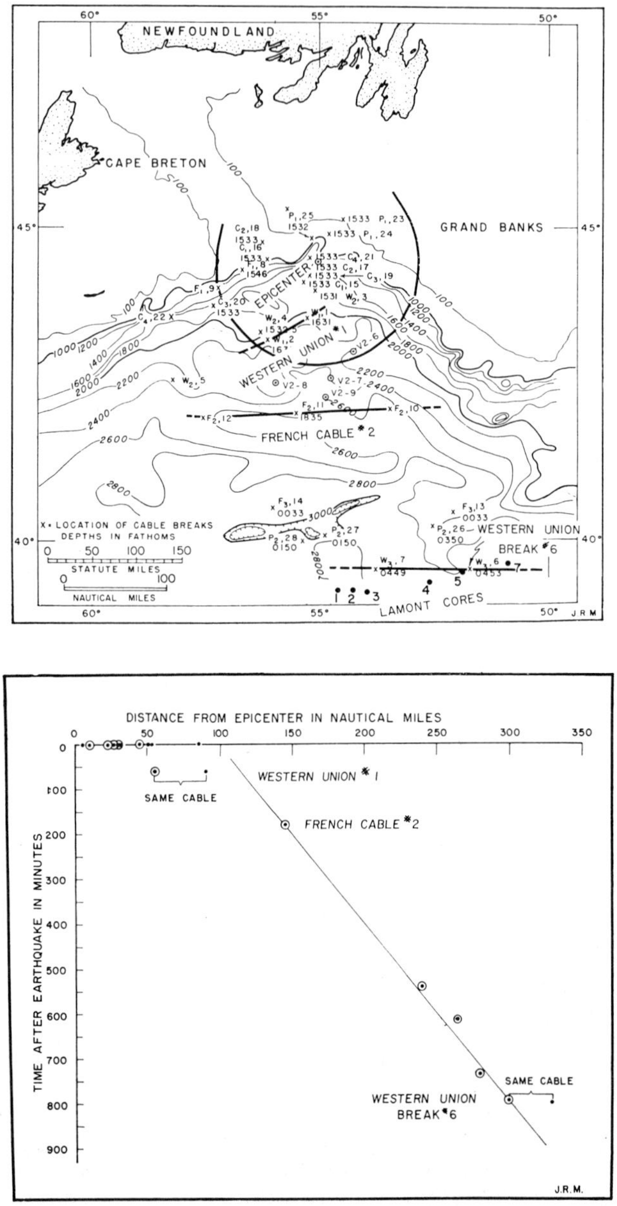

Fig. 15. (a) Showing the times of cable breaks after and during the Grand Banks earthquake of 1929. Note that instantaneous breaks occurred out to a distance of approximately 100 nautical miles from the epicenter. Positions of Lamont cores are indicated (1 to 7, V2–6 to V2–9). The contours are not well established because of the absence of recent accurate surveys.

(b) Indicating the relation of time of cable breaks to distance from epicenter in nautical miles. Note the line which gives the best fit for a velocity curve. This indicates a uniform velocity of 15 knots.

end of the cable breaks. In the experience of the writer similar findings have come from areas with a thick covering of fine sediment. Furthermore, an examination by Ericson of a group of four Lamont cores taken in the zone where Heezen and Ewing had estimated a speed of 55 knots showed nothing to indicate that there had been any recent turbidity current of appreciable size. Only one of the cores contained a layer at the surface with material coarser than a silty clay. The investigations of the cable companies in the area where the breaks occurred showed the sediment was a clayey material with a mixture of stones (Kindle, 1931). This is entirely what would be expected without turbidity currents since there are sufficient icebergs drifting past the area to introduce the scattered stones. It seems quite possible that the stones caused the bending of Heezen's coring tubes and prevented recovery of cores. In all fairness, however, it should be emphasized that the situation relative to sediments in the cable-break area is still relatively unknown and detailed exploration is required before more than tentative conclusions are justified.

The turbidity-current deposits found associated with various submarine canyons provide some evidence concerning current velocities. The vast majority of turbidity-current sediments collected to date, both in the canyons and on the fans beyond, are fine sands with a considerable amount of silt. Such material can be moved by currents of velocity considerably less than one knot. Currents of 15 knots or more should be capable of gouging out the rocks on the canyon floors and spreading coarse gravel debris, even boulders, over the fans. Little action of this sort has been found anywhere. The single cobble block reported by Lamont scientists (Ericson *et al.*, 1951) from the outer Hudson Channel could have been introduced by flotation by an iceberg during glacial stages. Gravel of small dimensions, however, is found in a number of places.

Another troublesome feature about turbidity currents as erosion agents is the finding in the axes of various submarine canyons, where turbidity currents are now operating, that the deposits in cores alternate between sands representative of turbidity currents and muds which are normal to the present depths. This alternation seems to indicate that the typical turbidity currents are not capable of eroding even the accumulation of muddy sediments formed after the occurrence of the previous turbidity current. As yet no case has been found where the turbidity-current deposit directly overlies a rock floor, as might be expected if the currents are eroding the rocky slopes where they are found.

B. Turbidity Currents as the Cause of Channels

Whether or not the turbidity currents can excavate the rock canyons there seems little doubt that they are responsible for the channels found in the fans on the outside of canyons. The depth of these channels is ordinarily only a matter of 20–50 fm (37–92 m), but several places have been found where the channels are well over 100 fm (183 m) deep at a considerable distance out along the fans. Such a place is indicated in Fig. 4 where the fan channel beyond Coronado Canyon shows an increased gradient in crossing a ridge leading down into San Clemente Basin. Similarly, where the Cascadia Channel off Oregon flows across a deep ridge, it appears to have cut a deep gorge (Hurley, 1959). The same was reported by Laughton (1959) for a channel crossing the fan off the Bay of Biscay. The size of some of these outer channels is so large and the depth below the natural levees so great that it seems quite possible that the currents which must fill them from time to time[1] to form the natural levees have relatively high velocities, provided there is any appreciable density difference with the overlying water. These channels are far larger than any stream channels on land and have higher gradients than most of the large rivers (Hurley, 1959).

The channel found in the Rhône Delta front in Lake Geneva must also be due to turbidity currents. This is indicated by the presence of coarser sediments in the channel than on the levees. However, here the channel is probably more non-depositional than erosional since the floor is somewhat higher than it is out beyond the natural levees. The gullies in the fronts of marine deltas are probably not due to turbidity currents because, so far as is known, they do not have coarser sediment than the adjacent slopes. An exception may exist on the Rhône–Mediterranean Delta since small natural levees exist here.

C. Turbidity Currents and Delta-Front Troughs

The troughs which extend across the shelf off the Ganges, Indus, and some other deltas are not easily explained. Most of them have gradients of only about 1%, far less than is typical of canyons. Little is known about their sediments except in the partly filled trough off the Mississippi. Apparently the latter has had turbidity currents flowing along it because there are indications of displaced faunas (Phleger, 1956) and there is considerable likelihood that sediments formerly were being transported into the Sigsbee Deep along its course (Ewing *et al.*, 1958; Curray, 1960). All of these troughs appear to have walls of unconsolidated sediment. It is possible that they have been maintained by turbidity currents during slow submergence as the delta grew upward on either side. The turbidity currents may be provided by the abundance of sediment brought in by the rivers. This could account for the location of the troughs off the Indus and Ganges near the present mouths.

9. Subaerial Erosion

When only a few submarine canyons were known most geologists had little hesitation in ascribing them to the drowning of river canyons. When, however, reports of the canyons began to come in from all parts of the world it was only natural that skepticism should grow despite the fact that the newly discovered canyons appeared to have many of the characteristics of river canyons. At first, it seemed to the writer that the situation could only be explained by a great sea-level lowering and in turn this seemed to indicate that the glaciers might

[1] More recent study suggests that these channels are being deepened off La Jolla and that the levees were formed before the deepening.

have been much larger during the Pleistocene than had commonly been considered to have been the case. As studies of the oceans continued, Pleistocene sea-level lowering was definitely excluded because the dredging from the oceanic seamounts refuted such a relatively recent lowering. On the other hand, these seamounts, and especially the flat-topped guyots, have yielded ample evidence of subsidence over a long period of time. Similarly, the continental margins in many places contain sediment thousands of feet thick which is now definitely far below the level at which it was deposited (see, for example, U.S. Geological Survey's *Tectonic Map of the United States*, 2nd edn.). Therefore, it should not be a matter of surprise if coastal canyons had been deeply submerged, at least in many places.

Proceeding from the premise that coastal canyons have probably been submerged, at least locally, we run into some obstacles in using this as an explanation for submarine canyons. First: why are the canyons world-wide? This would indeed be a problem, if we were certain that they are world-wide, but this we do not know. As has been explained previously there are numerous types of submarine valleys, only a few of which have been sufficiently surveyed so that they can be shown to possess true river-canyon characteristics. Thus the Corsican canyons, those of the French Riviera, Carmel Canyon, the head of La Jolla Canyon in California[1], the canyons at the tip of Cape San Lucas in Baja California, Tokyo Canyon, and Trincomalee Canyon of Ceylon can be considered as definitely within the fold. All but one of these is cut into hard rock. A number of others, particularly a group of canyons along the California coast can be considered as probable, but most of the reported but unexplored marine valleys could be due to such other causes as faulting, turbidity-current channels kept open during the deposition of deltaic masses, or even as scars left by giant landslides. Therefore, at the time of writing, there seems to be nothing particularly disturbing about the idea that some coasts have submerged, bringing river canyons deep below the sea.

The next obstacle offered to the idea of a remote submergence of river canyons is that the canyons should have been filled. The strength of this objection is, however, considerably weakened by the discoveries of landslides or of some types of sand flow or mud flow operating in the heads of many canyons. These slides undoubtedly set up turbidity currents carrying the sediment out of the canyon and depositing it as a fan on the outside. The writer knows of no reason why such periodic flushing of the canyons may not have operated for periods of millions of years. In fact the huge size of the fans outside the canyons indicates that this has been going on for extended periods (Menard, 1960). Therefore, this objection does not seem to be very valid.

Perhaps more troublesome is the question of why the submarine canyons, if due to ancient erosion, still have continuous outward slopes, particularly in such unstable areas as California. Why has there been no faulting or folding action which has cut off sections of the canyons? The head of Delgada Canyon

[1] Recent evidence reported by R. F. Dill shows that the head of this valley is retreating owing to submarine erosion.

along the northern California coast, which looks like a river-canyon type, has actually been cut off, although the cut took place at the present coast (Fig. 16). Delgada Canyon is along a northward continuation of the fault at Shelter Cove where a horizontal shift occurred at the time of the San Francisco earthquake; therefore it may be a part of the San Andreas fault system (Shepard and

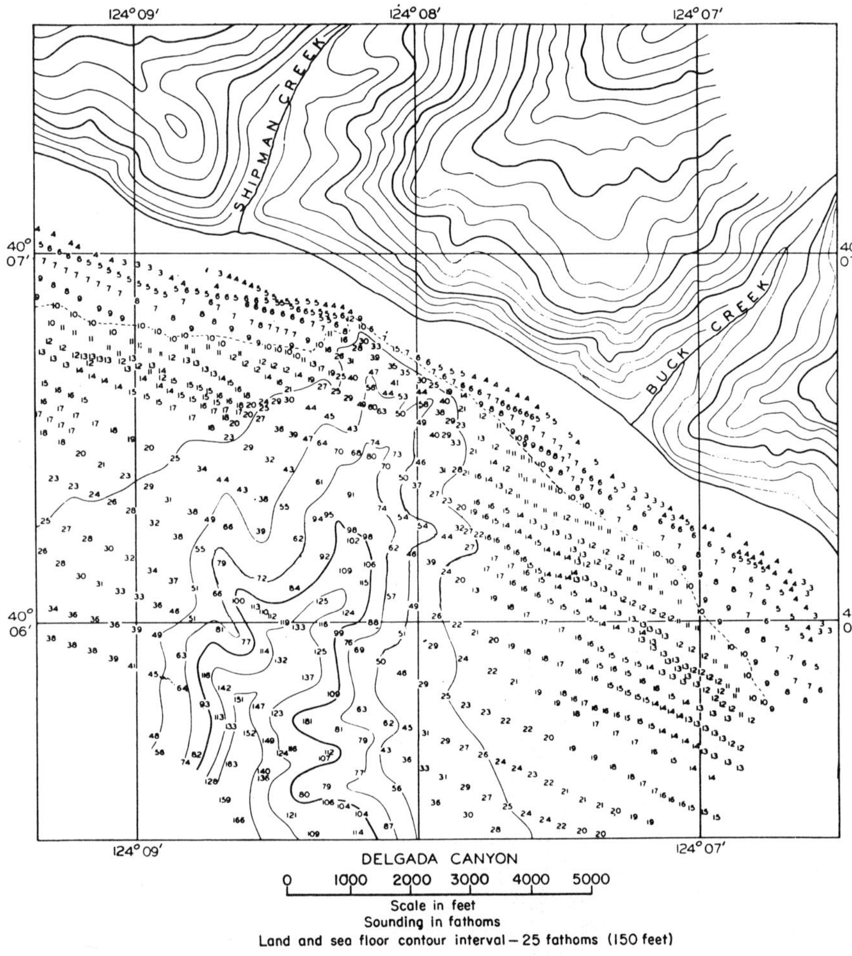

Fig. 16. Escarpment at the head of Delgada Canyon with a 45° slope which appears to have been produced by the cutting off of the canyon by movement along the San Andreas Fault. Note that the submarine canyon head does not coincide with the land canyon.

Emery, 1941, p. 39). The cutting of the canyon by such faulting is thoroughly understandable. The large right lateral movement along the San Andreas fault elsewhere would certainly result in developing the scarp at the canyon head and shifting the canyon along the coast out of line with the land canyons. It is rather surprising that similar features are lacking along the outer portions of such canyons as that of Monterey Bay (Fig. 7). This canyon can be traced seaward for 100 miles (including the outer trough) across a zone that must have undergone extensive faulting provided the canyon was submerged in the Miocene. This could be considered likely in view of the Miocene filling of the deeply submerged canyon under the Salinas Valley (Starke, 1956). As a possible reply to this objection one can suggest that there have been displacements from time to time on the sea floor but that the effects of landslides and turbidity currents have re-established the contacts and reconnected the severed canyon sections. Such action on the part of turbidity currents requires far less erosion than does the complete excavation of the canyons. It is also possible that the soundings are not sufficiently detailed as yet to reveal such breaks. Furthermore, if the continuity is an objection to the drowned valley hypothesis, it should also be an objection to canyons cut by turbidity currents unless such cutting has occurred in very recent times (see below).

The idea that the submarine canyons off the east coast of the United States were cut originally by rivers runs into some serious difficulties which have been discussed by Kuenen (1953). As he has emphasized, the upper walls of these canyons are cut through Tertiary sediment, some of it as young as Pliocene. Therefore, he argues, the canyons must have been cut since that time and there is no evidence of any uplift of the coastal margin, at least in such recent periods. The writer made an alternative suggestion that the canyons were cut before the Tertiary deposits were laid down and were kept open as Tertiary material accumulated on the shelf slope (Shepard, 1952). This is rejected by Kuenen on the ground that such a history would be recorded by finding nick points on the wall slopes of the canyons, with gentle slopes above indicative of the post-canyon cutting deposition. Kuenen's profiles do not show such nick points, but the results of using an acoustic probe on the sides of Hudson Canyon (Ewing *et al.*, 1960) show that, near the rim, the upper beds dip toward the canyon but have been covered by other beds that lie nearly horizontal and no nick point is shown along the slope. The writer is inclined to think that the absence of nick points is due to the accumulation of mud on the canyon slopes largely masking the outcrops and developing a slope that is temporarily stable but too steep for permanency. Such steep slopes are well known on the canyons of southern California where they have been seen by the aqualung divers. Dredging along these mud slopes has encountered occasional rock layers which protrude slightly. In any event the objection offered by Kuenen cannot be dismissed lightly and it may mean that the subaerial erosion hypothesis does not apply to the east coast canyons, as Kuenen believes. As we have seen, these canyons are rather different from the others and may require different explanations. On the other hand, it is hard to agree with Kuenen that the canyons off the French Riviera are drowned river valleys whereas the neighboring canyons off the Gulf of Lyons are due to turbidity currents. Such an explanation requires a large sea-floor displacement at the juncture, for which there is no evidence (Fig. 9).

10. Pleistocene Canyon Cutting by Turbidity Currents

Daly, Kuenen, and the Lamont scientists have all considered that the cutting of submarine canyons took place during the Pleistocene glacial epochs when the sea-level was lowered so as to allow the rivers to carry sediments directly to the edge of the shelf and hence supply materials for canyon cutting on the slope below. Such a lowering certainly must have helped develop turbidity currents but there are some serious objections to this idea. In the first place such erosion should affect only the slopes, but the majority of the known canyons penetrate deeply into the continental shelves inside. It seems obvious that if the canyons were cut by turbidity currents, these currents must have operated across the shelves and, in fact, this is well demonstrated by cable breaks across canyon heads in various places. These breaks may be due to landslides but these in turn can be expected to produce turbidity currents.

Further opposition to a Pleistocene cutting of the canyons comes from their vast size, particularly where they have walls of granite or other hard rocks. If we assume that these hard-rock canyons are submerged river valleys, we are still confronted by the vast size of most other canyons. The walls of the east coast canyons rise hundreds of fathoms and we know that some of them are cut in moderately well consolidated rocks of Cretaceous age. To think that the necessarily infrequent turbidity currents could have produced such vast cuts in the small fraction of a million years during which the sea uncovered the shelf is giving a role to turbidity currents that is certainly not indicated by their typical fine-grained sediments nor by their apparent failure to remove soft underlying sediments. The turbidity currents have been estimated to come down the southern California canyons about once in 100 years (Gorsline and Emery, 1959) and, so far as we know, these do not erode the canyons appreciably.

If Menard (1960) is correct about the size of the fans along the west coast, it would appear that the normal products of sand transportation along the shore to the canyon heads would have had to have been in operation for many millions of years to develop the fans, quite in addition to any erosion within the canyons themselves. The turbidity currents coming into the Los Angeles and Ventura basins were very likely directed along canyons which were, of course, far older than the glacial stages. Whether we believe in a turbidity-current origin or not, it is far easier to explain the canyons as older than Pleistocene. The investigations now in progress in Scripps Canyon (Dill, 1962) appear to lend considerable weight to erosion at the canyon heads by slumping action.

References

Bourcart, J., 1959. Morphologie du précontinent des Pyrénées à la sardaigne. In *La Topographie et la Géologie des Profondeurs Océaniques*. Centre Nat. Rec. Sci., Paris, 33–52.

Buffington, E. C., 1951. Gullied submarine slopes off southern California (abstr.). *Bull. Geol. Soc. Amer.*, **62**, 1497.

Chamberlain, T. K., 1960. Mechanics of mass sediment transport in Scripps Submarine Canyon, California. Ph.D. Thesis, Univ. of Calif., La Jolla, 200 pp.

Curray, J. R., 1960. Sediments and history of the Holocene transgression, continental shelf, northwest Gulf of Mexico. In Shepard *et al.*, *Recent Sediments, Northwestern Gulf of Mexico*. Amer. Assoc. Petrol. Geol., Tulsa, 221–226.

Daly, R. A., 1936. Origin of submarine canyons. *Amer. J. Sci.*, ser. 5, **31**, 401–420.

Dietz, R. S., 1954. Possible deep sea turbidity current channels in the Bay of Bengal. *Curr. Sci.*, **23**, 254–256.

Dill, R. F., 1962. Sedimentary and erosional features of submarine canyon heads. Abstr. in D. S. Gorsline (Ed.), *Proc. First National Coastal and Shallow Water Research Conf.*, National Science Foundation and Office of Naval Research, 531.

Dill, R. F., R. S. Dietz and H. Stewart, 1954. Deep-sea channels and delta of the Monterey submarine canyon. *Bull. Geol. Soc. Amer.*, **65**, 191–194.

Emery, K. O., 1960. *The Sea off Southern California*. John Wiley & Sons, New York, 366 pp.

Emery, K. O., W. S. Butcher, H. R. Gould and F. P. Shepard, 1952. Submarine geology off San Diego, California. *J. Geol.*, **60**, 511–548.

Emery, K. O. and R. D. Terry, 1956. A submarine slope of southern California. *J. Geol.*, **64**, 271–280.

Ericson, D. B., M. Ewing and B. C. Heezen, 1951. Deep-sea sands and submarine canyons. *Bull. Geol. Soc. Amer.*, **62**, 961–965.

Ericson, D. B., M. Ewing, G. Wollin and B. C. Heezen, 1961. Atlantic deep-sea sediment cores. *Bull. Geol. Soc. Amer.*, **72**, 193–286.

Ewing, J., B. Luskin, A. Robert and J. Hirshman, 1960. Sub-bottom reflection measurements on the continental shelf, Bermuda Banks, West Indies Arc, and in the West Atlantic basins. *J. Geophys. Res.*, **65**, 2849–2859.

Ewing, M., D. B. Ericson and B. C. Heezen, 1958. Sediments and topography of the Gulf of Mexico. In Weeks, L. (Ed.), *Habitat of Oil*, a symposium. Amer. Assoc. Petrol. Geol., Tulsa, 995–1053.

Fisk, H. N. and E. McFarlan, Jr., 1955. Late Quaternary deltaic deposits of the Mississippi River. In "The Crust of the Earth", *Geol. Soc. Amer. Spec. Paper* 62, 279–302.

Gibson, W. M. and H. Nichols, 1953. Configuration of the Aleutian Ridge, Rat Islands—Semisopochnoi I. to west of Buldir I. *Bull. Geol. Soc. Amer.*, **64**, 1173–1187.

Gorsline, D. S. and K. O. Emery, 1959. Turbidity current deposits in San Pedro and Santa Monica Basins off southern California. *Bull. Geol. Soc. Amer.*, **70**, 279–280.

Heezen, B. C., D. B. Ericson and M. Ewing, 1954. Further evidence for a turbidity current following the 1929 Grand Banks earthquake. *Deep-Sea Res.*, **1**, 193–202.

Heezen, B. C. and M. Ewing, 1952. Turbidity currents and submarine slumps and the 1929 Grand Banks earthquake. *Amer. J. Sci.*, **250**, 849–873.

Heezen, B. C., R. J. Menzies and M. Ewing, 1959. Influence of modern turbidity currents on abyssal productivity. *Abstr. 1959 Intern. Oceanog. Cong.*, Amer. Assoc. Adv. Sci., Washington, pp. 375–376.

Heezen, B. C., M. Tharp and M. Ewing, 1959. Floors of the Oceans. I. The North Atlantic. *Geol. Soc. Amer. Spec. Paper* 65, 122 pp.

Hurley, R. J., 1959. Geomorphology of abyssal plains in the northeast Pacific Ocean. Ph.D. Thesis, University of California, 105 pp.

Kindle, E. M., 1931. Sea-bottom samples from the Cabot Strait earthquake zone. *Bull. Geol. Soc. Amer.*, **42**, 557–574.

Kuenen, Ph. H., 1953. Origin and classification of submarine canyons. *Bull. Geol. Soc. Amer.*, **64**, 1295–1314.

Kuenen, Ph. H. and C. I. Migliorini, 1950. Turbidity currents as a cause of graded bedding. *J. Geol.*, **58**, 91–127.

Laughton, A. S., 1959. The exploration of an interplain deep-sea channel. *Abstr. 1959 Intern. Oceanog. Cong.*, Amer. Assoc. Adv. Sci., Washington, 36–38.

Mathews, W. H. and F. P. Shepard, 1962. Sedimentation of the Fraser River Delta, British Columbia. *Bull. Amer. Assoc. Petrol. Geol.*, **46**, 1416–1437.

Menard, H. W., 1960. Possible pre-Pleistocene deep-sea fans off central California. *Bull. Geol. Soc. Amer.*, **71**, 1271–1278.

Phleger, F. B., 1956. Foraminiferal faunas in cores offshore from the Mississippi Delta. *Paper Mar. Biol. Oceanog., Deep-Sea Res. Suppl.*, **3**, 45–57.

Shepard, F. P., 1937. Salt domes related to Mississippi submarine trough. *Bull. Geol. Soc. Amer.*, **48**, 1349–1362.

Shepard, F. P., 1948. *Submarine Geology*. Harper & Bros., New York, 338 pp.

Shepard, F. P., 1949. Terrestrial topography of submarine canyons revealed by diving. *Bull. Geol. Soc. Amer.*, **60**, 1597–1612.

Shepard, F. P., 1951. Mass movements in submarine canyon heads. *Trans. Amer. Geophys. Un.*, **32**, 405–418.

Shepard, F. P., 1952. Composite origin of submarine canyons. *J. Geol.*, **60**, 84–96.

Shepard, F. P., 1955. Delta-front valleys bordering the Mississippi distributaries. *Bull. Geol. Soc. Amer.*, **66**, 1489–1498.

Shepard, F. P., 1961. Submarine canyons of the Gulf of California. *Reps. 21st Intern. Geol. Cong.*, Part 26, 11–23.

Shepard, F. P. and C. N. Beard, 1938. Submarine canyons: distribution and longitudinal profiles. *Geog. Rev.*, **28**, 439–451.

Shepard, F. P. and Gerhard Einsele, 1962. Sedimentation in San Diego Trough and contributing submarine canyons. *Sedimentology*, **1**, 81–133.

Shepard, F. P. and K. O. Emery, 1941. Submarine topography off the California coast. *Geol. Soc. Amer. Spec. Paper* 31, 171 pp.

Starke, G. W., 1956. Genesis and geologic antiquity of the Monterey Submarine Canyon (abstr.). *Bull. Geol. Soc. Amer.*, **67**, 1783.

Stetson, H. C., 1936. Geology and paleontology of the Georges Bank canyons. *Bull. Geol. Soc. Amer.*, **47**, 339–366.

Terzaghi, Karl, 1956. Varieties of submarine slope failures. *Proc. 8th Texas Conf. Soil Mech. and Found. Eng.*, 41 pp.

van Straaten, L. M. J. U., 1959. Littoral and submarine morphology of the Rhone Delta. *2nd Coastal Geog. Conf.*, Louisiana State Univ., 233–264.

Veatch, A. C. and P. A. Smith, 1939. Atlantic submarine valleys of the United States and Congo submarine valley. *Geol. Soc. Amer. Spec. Paper* 7, 101 pp.

Woodford, A. O., 1951. Stream gradients and Monterey Sea Valley. *Bull. Geol. Soc. Amer.*, **62**, 799–852.

III. SEDIMENTATION

21. BEACH AND NEARSHORE PROCESSES

PART I. MECHANICS OF MARINE SEDIMENTATION

R. A. Bagnold

(For list of symbols and definitions see p. 550)

1. Definition, Origin and Relevant Physical Properties

A sea sediment, as distinct from native rock or from a coral accretion, may be defined as solid granular matter heavier than the sea-water, the separate pieces or "grains" of which, irrespective of their size, are or were at one time loose upon the sea bed or are in the process of falling through the sea.

Strictly, sedimentation denotes the process by which the grains of a sediment attain their ultimate resting places, or become "deposited". Since this process in general extends over a period of time during which the grains are being moved from place to place, sedimentation in this sense is inseparable from sediment transportation. But the term sedimentation is often used in the rather different geological sense to denote the study of the composition of existing sedimentary deposits in their ultimate static state, by mechanical, chemical and biological analysis, for purposes of classification. Here again, however, any such classification which excludes from consideration the dynamic process by which the composition has been brought about must remain to some extent arbitrary.

Grains of sea sediment may be classified as follows according to their origin.

a. Terrigenous.
   (i) Introduced by rivers and by polar glaciers,
   (ii) introduced by winds,
   (iii) eroded by the action of sea waves on shores,
   (iv) disintegrated as the debris of marine avalanches.
b. Cosmic. Introduced through the atmosphere as dust from extra-territorial sources.
c. Biogenous. Created as the debris of marine organisms.
d. Hydrogenous. Created as mineral nodules by crystallization from solution, e.g. manganese nodules.
e. Volcanic. Introduced as products of submarine vulcanism.

The size of sediment grains ranges from that of boulders, originated as a(i) and (iii) above, to microscopic particles. Both the size and density affect the dynamic process of transport in or by the sea-water. The effective density, ρ_s,

[*MS received October, 1960*]

is measured by the ratio of mass to overall volume, irrespective of the density of the constituent material. The effective density of marine organic debris, which frequently has hollow or spiny grains, and of clay crystals may be considerably less than the true density of the solid material composing them. Taking buoyancy into account, the dynamic effects of gravity vary as the excess density $(\rho_s - \rho)$. The true density of most solid mineral grains is approximately that of quartz (2.65 in c.g.s. units), which gives an excess density of 1.65 in water. However, if a marine organic grain with material density of 2.65 has an effective density ρ_s of half this amount (1.32), the effective excess density will be only 0.32, or only one-fifth that of the solid mineral grain.

Neglecting variations due to grain shape, the general criterion for dynamic similarity of grain behaviour under a combination of gravity and fluid forces may be derived very simply as follows. If the grain diameter is D, the net gravity force acting on it, i.e. the immersed weight, will vary as $(\rho_s - \rho)gD^3$; and if the fluid stress (fluid force per unit of effective grain area) is f, the fluid force on the grain varies as D^2. Whence the fluid stress per unit of immersed grain weight varies as $f/(\rho_s - \rho)gD$. This dimensionless quantity, which will be denoted by θ, is useful because it combines the effects of both grain size and grain density.

For instance, if f is the fluid shear stress exerted on a grain bed, the threshold of grain movement is definable in terms of θ. It is found that, under conditions of steady turbulent flow, the threshold value of θ approaches a universally constant value of approximately 0.06 as $(\rho_s - \rho)gD$ is increased.

Again, when a grain is falling freely, the fluid and gravity forces acting on it are equal and opposite; whence θ is unity. Expressing the fluid stress as $c_d \rho w^2$, where c_d is the drag coefficient, the fall velocity, w, is given at once by

$$w = \sqrt{\left[\frac{(\rho_s - \rho)gD}{\rho c_d}\right]}.$$

2. Transport during Fall through the Sea

According to its origin a grain of sea sediment may begin its existence as such at a great distance above the sea bed, or it may be already at or very close to the sea bed, e.g. if introduced as bed load by a river or as moraine material by a polar glacier, or if pre-existing as a marine bed organism. Ultimately, being by definition heavier than the sea-water, all sediment grains must find themselves so close to the sea bed as to be within a zone of water whose movement is affected by the presence of the bed boundary. But in the former case this state will be preceded by one in which the grains are falling steadily downwards through the sea at the fall velocity appropriate to their size and density.

During this fall the grains are suspended by the water, which exerts an upward force on them equal to their immersed weight; and the grains conform without resistance to any motion which the surrounding water may have. From a knowledge of the fall velocity the problem of predicting the drift of such grains away from the vertical, from a given point in the sea to that at which they enter the sea-bed zone, is straightforward in theory but difficult in practice. In common with the very similar problem of predicting the ground destination of fall-out from a nuclear explosion it requires a detailed knowledge of the speed and direction of the sea currents in every fluid layer.

The fall velocity of a quartz-density cobble 10 cm in diameter which might be rafted out to sea by an iceberg and loosed into deep water is about 150 cm/sec or 5 ft/sec. It would fall to a depth of 10,000 ft in less than an hour and would drift but little during its fall. On the other hand, since the fall velocity of fine grains decreases as the square of their size, very small particles may drift immense distances. The fall velocity of a solid quartz particle 10 μ in diameter

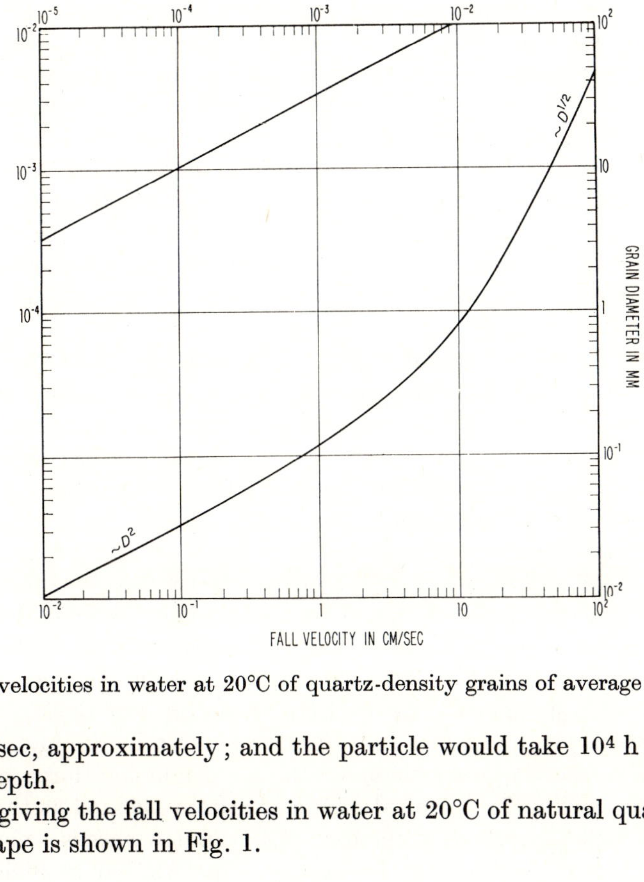

Fig. 1. Fall velocities in water at 20°C of quartz-density grains of average natural shape.

is 10^{-2} cm/sec, approximately; and the particle would take 10^4 h to reach the 10,000-ft depth.

A graph giving the fall velocities in water at 20°C of natural quartz grains of average shape is shown in Fig. 1.

3. Transport of Sediment over the Sea Bed

The transport of sediment over a gravity bed by the combined action of fluid flow and of a gravity component parallel to the bed surface is a phenomenon which occurs in nature under many different conditions. It appears in such varied guises that, until recently, no common characteristics have been looked for as clues to the little understood physical processes underlying the phenomenon as a whole.

The phenomenon covers sediment transport by wind and by water streams both in nature and in industrial ducts, and by sea currents with or without superimposed fluid oscillations due to surface waves. It also covers turbidity currents, the creep of super-saturated soils, etc., and the flow of slurries. It extends finally to the simple avalanching, under the action of the gravity component alone, of sand down the slip-face of a dune, and of material down the sides of a spoil heap.

While little serious attempt has been made in the past to investigate the underlying physics, particular aspects of the phenomenon have been subjects of interest to one or other branch of engineering technology, each for its own practical purposes, and have been studied in isolation as separate phenomena. Thus, though a number of separate quantitative theories have been suggested by which the measured rate of sediment transport may in particular fields be related to definable attributes of the sediment and of the fluid flow, the theories treat of different sets of conditions, assume different attributes to be relevant, and are expressed in different and often conflicting terms and symbols.

Moreover, since the majority of experiments made have been designed for practical rather than scientific ends, factual data are in most cases insufficient to test the general applicability of theoretical conclusions even within a particular field.

As a subject for broad general study, the phenomenon as a whole essentially involves the "flow" of dispersed granular solids within fluids, and therefore occupies a much neglected field between solid and fluid physics. So, it is unlikely that a proper understanding of the phenomenon can be attained by any approach based wholly on the concepts of either the one or the other discipline; or indeed that the physical processes involved can be described in any commonly conventional set of terms and symbols.

In this connection it will be as well to be clear at the outset as to the meanings to be attached to certain terms currently used in very different senses.

The term "friction factor" or "friction coefficient" has entirely different meanings in solid physics and in hydraulics. In solid physics, the limiting friction coefficient signifies the ratio of the shear force or stress, T, needed to be exerted over a plane of static contact between two bodies in order to maintain relative motion between them to the perpendicular force or stress, P, across that plane (Fig. 2a). The ratio T/P is usually defined as the tangent of the "friction angle" ϕ, which in the case of a granular mass is a measure of the mean angle of contact between the grains. Thus friction coefficient $= T/P = \tan \phi$.

In hydraulics, on the other hand, the friction factor or coefficient, c_d, relates the shear stress over a flow boundary not to any fluid stress normal to the boundary, but to the square of some representative flow velocity, u. Thus $c_d = \tau/\rho u^2$. The flow velocity of a fluid is independent of any normal stress it may exert on the flow boundary.

In the present context the term "friction coefficient" will be used exclusively in the former, solid sense to denote the ratio of shear stress to normal stress, as applied to the solid phase alone.

The term "concentration", N, applied to one substance dispersed within another is generally understood in physics to denote the ratio of the space occupied by one substance to the whole space (at a particular place if the dispersion is not uniform). It will here be used only in this sense. Hydraulic engineers currently use the same term to mean the ratio by weight of the time rates of sediment-to-water transport in a stream. This is a different quantity and may have a very different value.

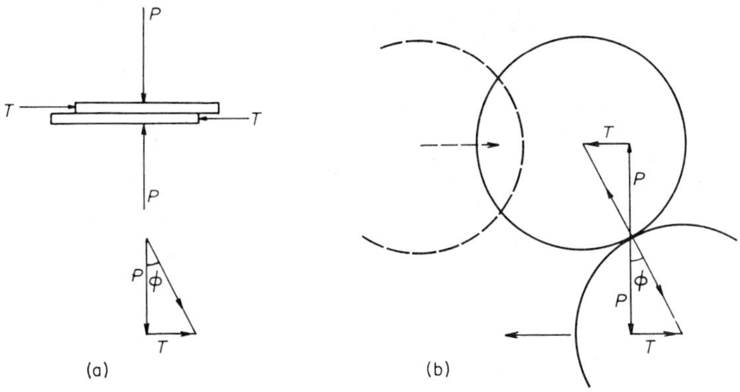

Fig. 2. (a) Continuous contact friction; (b) intermittent contact friction by diffusion of grain momentum.

Taking an aloof view of the phenomenon as a whole, it is at once apparent that all its aspects without exception share several common characteristics.

(i) The sediment is heavier than the fluid. Its grains are, therefore, pulled towards the bed by a gravity component normal to the bed surface. If a mass m of sediment of density ρ_s overlies unit area of any plane parallel to the bed surface which is inclined at an angle β to the horizontal, equilibrium demands a normal stress to be exerted across the plane, of value $[(\rho_s - \rho)/\rho_s]gm \cos \beta$ whether the sediment is moving or stationary. Here ρ is the density of the pervading fluid.

(ii) There must also exist in all cases a tangential stress $[(\rho_s - \rho)/\rho_s]gm \sin \beta$ due to the tangential component of gravity.

(iii) In all cases the movement of sediment over the bed involves the shearing of the pervading, inter-granular fluid and a shearing of the sediment grains over one another and over the bed surface.

(iv) The shearing of any layer of grains settled on the bed and in static contact with one another requires, as O. Reynolds showed, some degree of dispersion or dilatation of the grains from one another.

(v) In all cases this dispersion must necessarily be upwards normally to the bed surface and therefore against the normal component of gravity.

(vi) Hence, in all cases, continued steady movement of sediment over the bed requires that however much or little the moving grains are dispersed from one another, a normal dispersive stress $[(\rho_s - \rho)/\rho_s]gm \cos \beta$ must be maintained across every plane of shear in order that the immersed sediment weight may remain supported. This must also be true, on the average, in the case of oscillatory motion due to the passage of surface water waves.

Across the shear plane of the bed surface this stress must, over a representative period of time, have a mean value identical with that existing when the whole sediment load is settled stationary upon the bed.

The reality of this normal dispersive stress acting between the moving sediment grains and the bed boundary is logically indisputable, as are also the inference (a) that this stress must result from the dynamics of the shearing either of the fluid or of the sediment grains or of both, and (b) that the mechanism of sediment transport cannot be quantitatively understood until the dynamics of the combined shearing are understood.

It is clear from observational evidence of a qualitative kind that the normal stress which supports the moving sediment load must arise from two different dynamical mechanisms. For, though it is inescapable that fine sediment grains in low concentration are maintained suspended by dynamic fluid forces exerted by random eddies of turbulence, sediment is also maintained in a statistically dispersed state as "bed load" under conditions where fluid turbulence is either absent altogether or incapable of exerting the necessary forces on large grains.

Far too little is understood at present about the mechanism of fluid turbulence to allow of any theoretical prediction as to the load of suspended sediment which a given flow can carry. But enough light has been thrown on the bed-load mechanism by a study (Bagnold, 1954) of the collision dynamics of sheared grain dispersions within fluids to suggest that the most useful approach to the sediment-transport problem is likely to be from the direction of energetics rather than that of dynamics.

4. Transport Rate and Fluid "Power"

The general background of the new approach to sediment-transport problems outlined in this and the following sections has been discussed in a general way in a paper already published (Bagnold, 1956). The results are introduced here with special reference to marine conditions, together with summarized versions of further developments since 1956, a more ample discussion of which it is hoped will shortly be published as a U.S. Geological Survey Professional Paper in the series 282.

A. Work Done in Transporting a Bed Load

Repeated collisions between the solid grains, as they are sheared over one another and over the bed grains, must necessarily create components of grain momenta both normal and parallel to the planes of shear (Fig. 2b). The diffusion of the normal components from grain to grain constitutes a normal stress

P per unit area of shear plane which has the effect of making each grain layer repel its neighbours. Thus, if this stress were not resisted by an equal and opposite stress applied either between two parallel boundaries impenetrable by the grains or by the normal component of the gravity body-force pulling the grains towards the bed boundary, the grains would continue to disperse ever further through the fluid. When restrained from expansion by a normal gravity component, the grain dispersion may be likened to the Earth's atmosphere, whose dispersive pressure is in equilibrium with its weight. The stress P pertains, of course, only to the solid phase.

The diffusion of the components of grain collision momenta parallel to the planes of shear constitutes a shear stress T which is likewise transmitted from grain to grain and pertains only to the solid phase. It is additional to the shear stress τ maintained by the shearing of the pervading fluid. The total applied stress necessary to maintain the shearing of the combined substances may therefore be written:

$$\mathscr{T} = T + \tau \tag{1}$$

It was found experimentally that at high grain concentrations such as occur within the bed-load zone close to the bed surface the value of the solid-phase element T may exceed 100 times that of the fluid element τ, even when the densities of solid and fluid are the same. At such concentrations, therefore, the whole of the resistance to motion may be considered as exerted by the shearing of the solid phase. The fluid element did not begin to dominate until the concentration had been reduced to less than 9%.

Since the same grain collision gives rise simultaneously to impulsive components in both directions, the dynamic grain stresses, T and P, are proportional to one another, the proportionality depending on the mean angle ϕ' of collision. Therefore, the ratio T/P is to be considered as a friction coefficient, denotable by $\tan \phi'$, just as in the static case of continuous contact. Moreover, the mean angle ϕ' was found to be of the same order as the limiting angle given by the angle of repose of the granular material.

These results open a new approach to the sediment-transport problem. They render the conventional approach of the hydraulic engineer untenable. For the latter, being largely based on the kinematics of fluid flow, has been handicapped by the assumption that the sediment grains move frictionlessly, and without appreciable effect on the internal motion of the fluid.

We can now write down the condition of tangential stress equilibrium at the bed surface (suffix 0) independently of the degree of grain dispersion above as:

$$\mathscr{T}_F = \frac{\rho_s - \rho}{\rho_s} g m_b \cos \beta \,(\tan \phi - \tan \beta) + \tau_0, \tag{2}$$

where $\mathscr{T}_F$ is the stress externally applied via the agency of the fluid flow, m_b is the whole mass of the bed load and τ_0 is the residual fluid stress exerted directly on the bed surface and may be neglected when any considerable grain movement is taking place. If the bed inclination, β (measured downwards in the direction of flow), is small, the term $\tan \beta$ can also be neglected. But when β is increased until $\beta = \phi$ the stationary grain bed will move as an avalanche without any fluid stress being applied.

It should be noted that relation (2) expresses a state of shear equilibrium between an applied and a resisting thrust, the thrusts being equal in value but different in kind. For this reason the externally applied fluid thrust is distinguished for clarity by a special symbol $\mathscr{T}$—a capital script T. The whole applied thrust, including the tangential gravity component, if any, on the solid phase, would be $\mathscr{T}_F + [(\rho_s - \rho)/\rho_s] g m_b \sin \beta$.

The distinctions between the four different stresses essentially involved in two-phase equilibrium are shown schematically in Fig. 3.

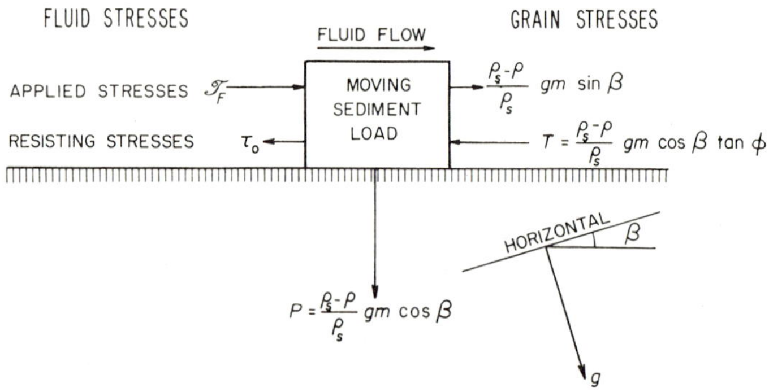

Fig. 3. Schematic diagram showing conditions of two-phase stress equilibrium at base of a bed load. A part of the applied fluid stress, $\mathscr{T}_F$, is transferred to the bed-load grains above the bed boundary, and is transmitted to the bed as a grain stress. The overall equilibrium condition is expressed by

$$\mathscr{T}_F + \frac{\rho_s - \rho}{\rho_s} g m \sin \beta = \tau_0 + \frac{\rho_s - \rho}{\rho_s} g m \cos \beta \tan \phi.$$

The implications of the general equilibrium relation (2) as applied to transport over a homogeneous grain bed have been discussed in a further paper (Bagnold, 1956). It was there shown that the residual fluid element τ_0 cannot increase above its value $\tau_t = \mathscr{T}_{Ft}$ at the threshold of grain motion over the bed when m_b is zero; and as $\mathscr{T}_F$ is increased τ_0 must, in fact, decrease progressively to a negligible value.

Any increase in $\mathscr{T}_F$ must cause erosion of the bed grains, with a resulting increase in the bed-load weight $[(\rho_s - \rho)/\rho_s] g m_b \cos \beta$ imposed on the new bed surface and holding it down, together with a corresponding increase in the solid-phase resisting thrust as given by the first term on the right of relation (2). Moreover, as the bed load increases, so does the concentration of grains in its bottom-most sheared layer, with the effect that the intensity of turbulence in the fluid between the grains is appreciably reduced. It is reasonable to conclude that when a certain transitional stage of bed movement is exceeded the fluid flow may be considered as unaffected by the existence of an underlying stationary boundary. The whole resistance to flow, in the two-dimensional case, is that exerted by the solid phase.

Unfortunately, no experimental methods have been devised for measuring either the load m_b of grains moving over a grain bed or the mean speed $\overline{U}_b$ of their travel. But the product $m_b\overline{U}_b$ is the rate j_b of mass transport of the load per unit width. This is measurable provided there is no additional load transported in suspension by fluid turbulence.

Since the tangential thrust stress required to maintain the bed load in motion is $[(\rho_s - \rho)/\rho_s]gm_b \cos\beta(\tan\phi - \tan\beta)$, multiplication by $\overline{U}_b$ gives the work done by the fluid in unit time, i.e. the fluid "power" expended per unit of bed area in transporting the bed load. Whence fluid power expended $=$ $[(\rho_s - \rho)/\rho_s]gj_b \cos\beta\,(\tan\phi - \tan\beta)$.

Calling the quantity $[(\rho_s - \rho)/\rho_s]gj_b \cos\beta$ the "dynamic transport rate" i_b, we have:

$$\text{Fluid power expended} = i_b\,(\tan\phi - \tan\beta). \tag{3}$$

This is the fluid energy that is converted directly into heat in unit time by inter-granular friction per unit of bed area in the process of transporting the bed load.

B. Available Fluid Power

Now let us regard the fluid as a transporting machine, and apply the principle of energy conservation. The power, i_b $(\tan\phi - \tan\beta)$, expended in transporting the bed load must be some proportion less than unity of the whole available fluid power.

The available fluid power per unit boundary area is the whole fluid energy dissipated into heat in unit time per unit of bed area by the effects of boundary resistance. Let this be denoted by ω. We can now write:

$$i_b = \omega\,\frac{\epsilon_b}{\tan\phi - \tan\beta}, \tag{4}$$

where ϵ_b is an efficiency factor which must be less than unity, and both i_b and ω are measurable in the same units, i.e. of power.

In the case of a steadily flowing water stream, and also in that of a train of stable surface waves propagated in shallow water, the value of ω can be found directly.

The power, ω, of a stream, per unit area of its bed boundary, is $\mathscr{T}_F\bar{u}$, where $\bar{u}$ is the steady mean velocity of flow. Since $\mathscr{T}_F$ is the decrement of flow energy per unit distance, x, travelled by the water, we can write:

$$\omega = \mathscr{T}_F\bar{u} = \rho gh\,\frac{dH}{dx}\,\bar{u}, \tag{5}$$

where h is the constant flow depth, H is the energy head, and dH/dx is the energy gradient, $\tan \beta$, due to the gravity slope.

The whole power transmitted forward by a train of surface waves of height H is given by $\frac{1}{8}\rho g H^2 Cn$, where n is the proportion of energy travelling forward with wave phase velocity C. Hence the power ω per unit of bed area is in the present context the decrement of the transmitted power per unit of distance travelled by the waves. Thus, $\omega = \frac{1}{8}\rho g\, d(H^2 Cn)/dx$, which for propagation over constant water depth and therefore at constant velocity, becomes

$$\omega = \tfrac{1}{4}\rho g H \frac{dH}{dx} Cn, \tag{6}$$

which is of precisely the same form as relation (5).

In arriving at (6) it is, of course, assumed that the decrement of the transmitted power is due entirely to energy losses by fluid drag at the bed surface. When the waves break, the much larger power decrement is likely to be due mainly to other internal energy losses not associated with bed drag. This needs further study, in particular of the variation in the amount of energy loss by bed drag with variation in the kind of break which takes place.

In the case of a ground wind, as in that of a sea-bed current, the fluid flow relative to the boundary is not maintained by a gravity fall but by an indeterminate combination of a pressure gradient, of the inertia of large fluid masses, and of a downward transmission of shear stress from above. In such cases the power ω may be inferred from the following general considerations applicable to all cases.

The power, ω, is essentially measureable as the product of the boundary stress, $\mathscr{T}_F$, arising directly or indirectly from the motion of the fluid, times its flow velocity, u', at some effective distance from the bed boundary. An examination of the experimental data on wind-blown sand (Bagnold, 1936) suggests the effective distance to be that of the "centre of pressure" at which fluid momentum is transferred to the bed-load grain dispersion.

In general, writing $u' = c_r u_* = c_r \sqrt{(\mathscr{T}_F/\rho)}$, where c_r is a numerical coefficient to be determined experimentally, we can define the available power as:

$$\omega = c_r \mathscr{T}_F^{3/2}/\sqrt{\rho} = c_r \rho u_*^3. \tag{7}$$

In this form, by assuming c_r to be a general constant under conditions of no suspended load, and by making certain quite general assumptions as to the theoretical value of $c_r \epsilon_b$, it was found possible (Bagnold, 1956) to predict the value of i_b to a good approximation from relation (4) both for water-driven sands in laboratory flumes and for wind-driven sands.

Moreover, the bed-load theory as given by Bagnold (1956) appears to account satisfactorily not only for the variation of i_b with grain size but also for the fact that beds of grains smaller than the order of 1 mm diameter become spontaneously rippled under the flow of water whereas beds of larger grains do not.

C. Work Done by the Fluid in Transporting a Suspended Load

The fluid power expended in transporting a sediment load in suspension by fluid turbulence can be found by reasoning similar to that leading to relation (3).

Suppose a mass m_s of sediment grains is maintained supported by fluid forces arising from the diffusion of upward eddy momentum components, and is transported at a mean speed $\bar{U}_s$. Under steady transport conditions the fluid must exert a normal stress $[(\rho_s-\rho)/\rho_s]gm_s \cos\beta$ in order to support the load. If the fall velocity of the grains is w relative to the fluid surrounding them, the fluid must in unit time do work equal to $[(\rho_s-\rho)/\rho_s]gm_s w \cos\beta$ in raising the grains at the velocity w to keep them statistically at a constant distance from the bed. Substituting $i_s/\bar{U}_s$ for $[(\rho_s-\rho)/\rho_s]gm_s \cos\beta$, the work done by the fluid in unit time, i.e. the power expended, will be $wi_s/\bar{U}_s$. But the existence of an energy gradient in the direction of flow, owing to a gravity slope $\tan\beta$, will reduce the work done by the fluid by an amount $i_s \sin\beta$.

Hence, the power expended by the fluid in transporting a suspended load at the dynamic rate i_s must be

$$i_s\left(\frac{w}{\bar{U}_s}-\tan\beta\right) \tag{8}$$

The power expended in transporting the bed load being $\epsilon_b\omega$, the remaining power, $(1-\epsilon_b)\omega$, is notionally available in the form of a rate of supply of kinetic turbulent energy to maintain the suspended load. Writing ϵ_s as the suspension efficiency, the total transport rate, in terms of $i=i_b+i_s$, is expressible by

$$i = i_b+i_s = K\omega$$

or

$$i = i_b+i_s = \omega\left\{\frac{\epsilon_b}{\tan\phi-\tan\beta}+\frac{\epsilon_s(1-\epsilon_b)}{(w/\bar{U}_s)-\tan\beta}\right\}, \tag{9}$$

where the dimensionless coefficient $K=i/\omega$ is the value of the group within brackets on the right.

It is not possible experimentally to make any separation of the measured values of i into its elements i_b and i_s. So the relative factual values of ϵ_b and ϵ_s remain unknown. Further, our understanding of the precise mechanism of fluid turbulence is insufficient to enable ϵ_s to be deduced by theoretical reasoning. However, on empirical evidence obtained from water-flume experiments, augmented recently by unpublished evidence from a number of natural rivers in the U.S.A., it appears, provided the gravity slope $\tan\beta$ is less than either $\tan\phi$ or $w/\bar{U}_s$, the overall coefficient K attains a maximum value which is a constant for the system concerned. In other words, it appears that a critical value of the power ω exists at which both ϵ_b and ϵ_s attain constant values, and at which, therefore, the total sediment transport rate becomes proportional to the power ω.

The evidence so far available suggests that the ultimately constant value of K is a function of the scale of the flow in terms of h/D, where h is the flow depth and D the mean grain diameter of the transported sediment; and, further, provided the gravity slope $\tan \beta$ is small compared to $w/\bar{U}_s$, K is independent of the absolute value of D. Over the large range of scale from miniature laboratory streams to the Mississippi river, K_{max} appears to have values of the order of $\frac{1}{2} \log_{10} (h/D)$. The reason for this apparent scale effect is not clear. The scale of the turbulent eddies may be involved.

D. Transport under Oscillating Water Motion due to Surface Waves

From the generality of the reasoning leading to (9), it seems likely that the mean rate, i, of to-and-fro transport of sediment over the sea bed by a purely oscillatory motion of the water close to the bed will also be found by experiment to attain proportionality to the power ω as given by (6).

Since the mean rate, $\bar{j}$, of the to-and-fro transport is equal to $\bar{m}\bar{U}_0$, where $\bar{m}$ is the mean sediment mass supported over unit bed area and $\bar{U}_0$ is the mean velocity of the sediment's to-and-fro motion, the immersed weight of the supported sediment will be $i/\bar{U}_0$, which by (9) can be written $K'\omega/\bar{U}_0$. Since the sediment is already supported by the wave action, no additional tangential stress should be needed to produce a transport of sediment at velocity U_θ by a bed current of velocity u_θ in any direction θ. Hence, assuming as a reasonable approximation that $U_\theta/\bar{U}_0 \sim u_\theta/u_0$, where u_0 is the orbital velocity measured at the same distance from the bed as u_θ, the sediment transport rate in the direction θ should be given by

$$i_\theta = K'\omega \frac{u_\theta}{u_0}. \qquad (10)$$

Before proceeding to discuss the property of surface waves in shallow water of creating a bed-drift current of their own, it is convenient here to point out a particular implication of relation (9) which appears to throw light on a number of puzzling phenomena including turbidity currents.

5. Auto-suspension of Sediment: Turbidity Currents, etc.

The theoretical possibility of this phenomenon was briefly alluded to by Bagnold (1956, p. 270). Since then factual evidence of its widespread existence has emerged in the course of research carried out under the auspices of the U.S. Geological Survey.

Just as a sediment bed will transport itself in unlimited quantity as an avalanche of bed load, without any energy supply from the fluid, when the gravity gradient, $\tan \beta$, becomes equal or exceeds the internal friction gradient,

$\tan\phi$, so we should expect a suspended load to transport itself to an unlimited extent when $\tan\beta$ exceeds the gradient $w/\bar{U}_s$.

The power expended by the fluid in supporting the suspended sediment load is $i_s w/\bar{U}_s$. The power expended by gravity directly on the sediment is $i_s \tan\beta$. But the latter power is transmitted from the sediment grains to the fluid surrounding them. Hence, if $\tan\beta$ exceeds $w/\bar{U}_s$, the suspended sediment, owing to its excess density, has the effect of pushing the fluid along instead of vice versa. A vessel floating down a river has no excess density so it cannot push the water along.

This concept appears first to have been suggested by Knapp (1938, p. 501), but its significance was overlooked and it was arrived at independently by Bagnold (1956, 1962). The reality of the effect, which is confined to conditions of turbulent fluid flow, should be manifested in the following ways:

(a) The transport rate, i_s, of such sediment grains as have fall velocities w less that $\bar{U}_s \tan\beta$ should be unlimited, except by availability of material, and, ultimately, at very high concentrations, by the suppression of fluid turbulence as the apparent viscosity is increased.

(b) Contrary to existing theories of turbulent suspension, all of which postulate a continuous increase in concentration downwards towards the bed boundary, sediment grains for which $w < \bar{U}_s \tan\beta$ should tend to distribute themselves uniformly throughout the fluid.

(c) Grains for which w just exceeds $\bar{U}_s \tan\beta$ should be found in streams in greatest concentration where $\bar{U}_s$ is greatest, i.e. in the upper fluid layers where the fluid velocity u is greatest.

All these manifestations have actually been observed. Extremely large sediment-transport rates have on occasion been measured in rivers. But, as far as the available data have been analysed, the mean sediment size is in every case that for which the fall velocity w is less than $\bar{U}_s \tan\beta$ ($\bar{U}_s$ being measured approximately as the mean flow velocity, $\bar{u}$, of the river). It is well known to hydraulic engineers that fine silt carried in suspension by water streams does not tend to concentrate towards the bed, but is distributed uniformly throughout the flow. The results of sediment sampling in rivers frequently disclose the anomaly that although the overall concentration of suspended solids increases downwards towards the bed the concentration of fine suspended grains diminishes downwards towards the bed instead of increasing as conventional theory requires.

The evidence for auto-suspension appears sufficiently strong to justify applying the reasoning which predicted it to the problem of turbidity currents along the sea bed.

Consider, from the viewpoint of energetics, the conditions necessary for the self-maintenance of a turbidity current of uniform thickness h flowing over a gravity bed inclined downwards at an angle β (Fig. 4). Let the concentration, N, of solid grains of density ρ_s and having a fall velocity w be supposed uniform and let the current velocity of both solids and liquids be $\bar{u}$.

The immersed load weight perpendicularly over unit bed area is $(\rho_s - \rho)gNh$;

and the fluid in this case contributes nothing to the applied power. The available power is therefore given by

$$\omega = (\rho_s - \rho) g N h \sin \beta \cdot \bar{u}, \tag{11}$$

which is the tangential gravity force on the solids times the velocity of their movement.

Of this power a part $(\rho_s - \rho) g N h w$ must be expended via the agency of the turbulence in the driven water in maintaining the load of solids at a constant height above the bed. A further part must be expended in maintaining the flow of the current water at the velocity $\bar{u}$. The power for this is $\tau_0 \bar{u}$, where τ_0 is

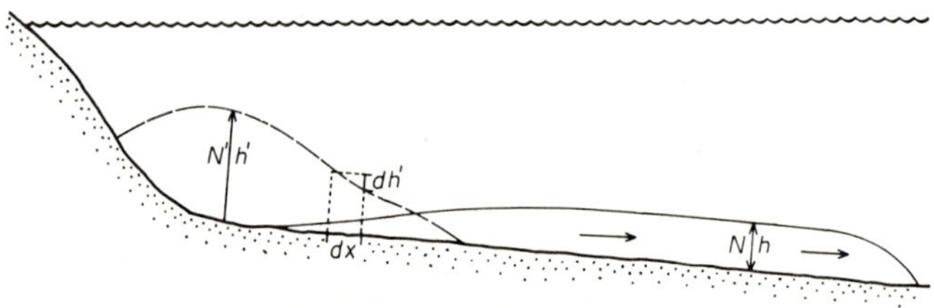

Fig. 4. Initiation and progress of a self-maintaining turbidity current.

the fluid resistance exerted at the flow boundary. Neglecting for a moment the extra resistance exerted at the upper flow boundary, we can regard the water current as a river. The stress τ_0 can then be expressed in terms of $\bar{u}$ and h by the semi-empirical relationship

$$\bar{u} = 5.75 \left(\frac{\tau_0}{\rho} \right)^{\frac{1}{2}} \log_{10} \frac{13.2h}{D_1},$$

which appears (Francis, 1957) to be a rational generalization consistent both with the Prandtl–Kármán theory and with well established practical flow formulae applicable to broad straight flow in open channels. This relationship gives

$$\tau_0 \bar{u} = \frac{\rho \bar{u}^3}{33} \log_{10}^{-2} \frac{13.2h}{D_1}, \tag{12}$$

where D_1 is the effective size of the boundary roughness.

The criterion that the turbidity current shall be self-maintaining is, therefore,

$$(\rho_s - \rho) g N h (\bar{u} \sin \beta - w) \not< \frac{\rho \bar{u}^3}{33} \log^{-2} \frac{13.2h}{D_1}$$

or

$$\frac{1}{\bar{u}^2} \frac{\rho_s - \rho}{\rho} g N h \log^2 \frac{13.2h}{D_1} \left(\sin \beta - \frac{w}{\bar{u}} \right) \not< 0.03, \tag{13}$$

where $\bar{u}$ must always exceed $w/\sin \beta$.

For beds of fine sand or silt which become rippled, the length D_1 is somewhere between the ripple length and the particle diameter. So a length of 1 cm would not be far wrong as a rough estimate.

To allow for the effect of the upper flow boundary, the number 0.03 on the right of (13) could be increased to say 0.045 or 0.06. But an assumption has to be made here that the concentration N remains sufficiently small for the dispersion of solids in the fluid to have no appreciable effect on its flow resistance. If N exceeds, say, 1% (by volume), the flow resistance, and, therefore, the number on the right, is likely to increase.

The criterion (13) predicts that a turbidity current cannot continue to flow indefinitely down a declined bed under still water unless it has initially attained a finite velocity $\bar{u}$ such that $\bar{u} \sin \beta$ exceeds the fall velocity w of its solid particles. It also predicts that for a given value of $(\sin \beta - w/\bar{u})$ the "scale factor", $Nh \log_{10}^2 13.2h/D_1$, must exceed a finite limiting value; for given values of $\sin \beta$ and w, the scale factor should be a minimum when $w/\bar{u} = 2 \sin \beta/3$.

Further it suggests that, if the initiating event, e.g. an avalanche, results in the value of the expression on the left exceeding the number on the right, the flow might become unstable and might begin to "snow-ball". The excess power might tend to increase as the current progresses by the entrainment of more sediment from the bed, until ultimately a power balance is set up by a resulting increase in $\bar{u}$.

The general idea seems sufficiently reasonable to provide a useful guide both to what is physically possible in nature, and also to the design of laboratory experiments whereby the phenomenon could be studied under controlled conditions.

For instance, suppose in nature a major avalanche down a steep continental slope were to result initially in a stationary ball of turbidity of height h' containing a mean concentration N' by volume of dispersed solids of quartz density; and suppose that the excess hydrostatic pressure at the base were to accelerate a current, of ultimate thickness $h = 500$ m and of concentration $N = 0.01$, to an ultimate speed $\bar{u}$ over an ocean floor sloping at 0.6° ($\sin \beta = 0.01$). Evaluation of (13) for fine sand ($w \ll \bar{u} \sin \beta$), taking the number on the right as 0.045, gives the speed, $\bar{u}$, attainable to be 25 m/sec. If the entrained sediment were such that $w = \frac{2}{3}\bar{u} \sin \beta$, $\bar{u}$ would be reduced by a factor $1/\sqrt{3}$ to 14.5 m/sec, for which $w = 10$ cm/sec (grains of 1 mm diam.).

The current might well, of course, drive still larger grains over the floor as un-suspended "bed load".

Acceleration of the current to these high speeds must, however, involve considerable initial pressure heads. The excess height, h', of the original turbidity must exceed $\bar{u}^2/2(\rho_s - \rho)gN'$. If $\bar{u} = 25$ m/sec, h' must exceed 1900 m for $N' = N = 0.01$, but N' might have a considerably larger temporary value during the initial stage, while large solids were still falling out.

The criterion (13) suggests that a self-maintaining turbidity current of the kind described should be easily reproducible on a laboratory scale on somewhat exaggerated slopes, using sediment which has an appreciable fall velocity. An important requirement here would be that the mean eddy velocities of the turbulence sufficiently exceed the fall velocity w, so that fully developed suspension is ensured. Thus, independently of (13), the current velocity $\bar{u}$ created by the initial disturbance might have to be 20 to 30 times w.

The experimental verification of (13) would be of considerable interest. In particular one would like to know the precise value to be given to the number on the right when an upper-fluid-flow boundary exists, and also the limiting concentration N at which this number begins to increase appreciably.

Auto-suspension may also occur when the fresh water of a sediment-laden river outflows over the heavier sea-water (sometimes for distances of several hundred kilometres). By existing theories suspension of sediment by turbulent flow cannot be maintained unless there exists a continuous increase in concentration downwards to a solid flow boundary. In this case there is no such boundary, and a downward increase in concentration would seem to be impossible. So the suspended sediment would be expected to fall out at the same rate as if the outflow of the overlying fresh water were non-turbulent.

But auto-suspension allows of the retention of sediment whose fall velocity w is less than $\bar{u} \tan \beta$ provided the flow remains turbulent. Here $\bar{u}$ is the flow velocity of the fresh water over the salt water and $\tan \beta$ is the residual flow gradient of the river. Accordingly, the turbidity should disappear somewhat abruptly, as it is observed to do when the Reynolds Number, $\bar{u}h/\nu$, of the flow falls below, say, 2500 and the turbulence dies out. For observed end-thicknesses of the fresh turbid layer of 15 cm, its velocity $\bar{u}$ of flow should be about 1.6 cm/sec. Assuming a residual surface water gradient of 10^{-6}, the critical fall velocity w would be 1.6×10^{-6} cm/sec, and particles 0.1 μ in size would persist at the water surface until the final cessation of turbulence.

Measurements of the depths of the fresh-water flow, of its velocity relative to the underlying sea-water, and of the sediment size and concentration at varied seaward distances from the river mouth would be of much interest.

Finally, auto-suspension may account for the existence of a lower limit to the size of shore sand. Though the beds of estuaries often consist of fine silts and muds the sands of open sea shores rarely contain grains less than 0.15 mm in size. Taking $\frac{1}{2}$ knot or 25 cm/sec as a fair average for the onshore and offshore currents due to waves, and 3° ($\tan \beta = 0.05$) as a fair average bed slope offshore of the breaker zone, all sediment grains whose fall velocity is less than $25 \times 0.05 = 1.25$ cm/sec should tend to become uniformly dispersed in the turbulent water of the breaker zone, and, therefore, to diffuse outwards into deeper water. From Fig. 1 the grain size corresponding to this fall velocity is in fact 0.15 mm.

6. Wave Drift

Wave theory considers water motion relative either to still water "elsewhere" or to axes stationary with regard to the mean position of all the water particles in the whole water body. It takes no account of general water drifts relative to the bed boundary, such as may be caused by wind, tides and other agencies. Water motions due specifically to the passage of waves must, therefore, be superimposed on any other motions that may exist.

It is usual in the experimental study of wave effects as such to prevent any drift of the water as a whole by confining the water within a closed channel. Under these simplified conditions classical theory predicts, and experiment confirms quantitatively, that in deep water a wave train of finite height causes a drift of surface water in the direction of wave travel, and a corresponding return drift beneath.

When in shallow water the wave oscillations extend down to the bed boundary, classical wave theory, which neglects all boundary drag, permits the return drift to extend down to the bed. Experiment (Caligny, 1878; Bagnold, 1947) has shown that on the contrary a forward drift occurs along the bed as well as at the surface, the return flow being confined to intermediate depths.

Longuet-Higgins (1953) has shown that classical theory can be extended to take account of bed drag by the introduction of a thin boundary layer within which the to-and-fro displacements close to the bed are assumed to dwindle to zero at the actual bed surface. This extended theory predicts both the fact and the magnitude of the forward bed drift, provided the boundary layer is supposed free of mobile sediment.

For a wave of full height H in a water depth h, the orbital velocity u_0 close to the bed is assumed to be given by

$$u_0 = \frac{\pi H}{T} \Big/ \sinh \frac{2\pi h}{L},$$

where T is the period and L the wavelength. If the depth to wavelength ratio, h/L, is small, this reduces to

$$u_0 = \frac{H}{2h} C,$$

where C is the wave phase velocity.

Taking u_0 as the orbital velocity just above the boundary layer, the predicted drift velocity, $\bar{u}$, is given by

$$\begin{aligned} \bar{u} &= \frac{5}{4} \frac{u_0^2}{C} \\ &= \frac{5}{4} \left(\frac{H}{2h}\right)^2 C \quad \text{when } h \ll L. \end{aligned} \tag{14}$$

It will be noticed that in the case of "long" waves ($h \ll L$), a case where the wave velocity C depends mainly on the depth h and not on the wavelength, neither the orbital velocity u_0 nor the drift velocity $\bar{u}$ is, according to the theory, dependent on the wavelength.

Longuet-Higgins' theory, as applied to a sediment-free bed and as tested over a wide range of experimental conditions in the laboratory, has been found (Russell and Osorio, 1958)[1] to be remarkably consistent with the facts except in one significant case. There can be little doubt also that, in general, the passage of a wave train creates a differential shear stress on the bed boundary in the direction of its propagation.

Unfortunately, Longuet-Higgins' theory, which involves the dynamical distortion of the boundary layer as the waves pass over it, is inapplicable when the boundary layer contains dispersed sediment. This is because the dynamical properties of the layer substance will be changed in an undefinable way.

Nevertheless, experiments in a wave tank (Bagnold, 1947) have demonstrated qualitatively that the passage of "long" waves over a sediment bed creates the same forward drift of water close to the bed, together with a progressive transport of sediment in the same direction. The drift velocity was found to be less, but the drift extended to a greater height.

It appeared, however, that contrary to the sediment-free case, both the drift velocity and the sediment-transport rate for waves of a given height H decreased to zero as the wavelength was reduced, although in all cases the waves would be classed as "long". The larger the wavelength the greater the sediment-transport rate was found to be, which seems consistent with experience in nature.

Thus in tentatively substituting $\overline{u}$ of relation (14) for u_θ in (10) to give,

$$\begin{aligned} i_{(\text{wave drift transport})} &= K''\omega u_0/C \\ &= K''\omega H/2h, \end{aligned} \tag{15}$$

it must be remembered that K'' is likely to be a function of the wavelength. This aspect deserves experimental study.

The exceptional experimental condition mentioned above suggests that a complication may be found when, as in the open sea, the constraints imposed in a narrow channel are removed. When the channel width in Russell's experiments exceeded a certain multiple of the water depth, the wave drift along the bed, which was otherwise uniform across the section, showed signs of instability, becoming greater on one side than on the other. This suggests that, in still wider channels, the wave drift may break up into circulations in a plane parallel to the bed. If so, a random scattering of sediment would tend to be superimposed upon a smaller forward drift.

It is possible that the same kind of lateral instability gives rise to the well known circulatory motion of the swash on a sand beach.

Another complication is likely to be introduced in the open by the effects of surface wind. Still confining attention to the simple conditions of a closed channel in which it is possible to make controlled experiments, a superimposed wind in the wave direction would be expected to affect matters in two ways, i.e.

(a) A direct wind drag on the water surface will, if sustained, create a

[1] This contains a useful summary by Longuet-Higgins of his theory, which he has extended to include the case of a turbulent boundary layer.

circulation in the vertical plane, forwards at the water surface and backwards below. The latter, backward drift below is likely to reduce or even reverse the forward wave drift, but in the absence of controlled experiments it is not known for sure whether the circulation due to wind effect is continuous or of a repeating cellular type.

(b) Under an imposed, surface-wind drag the water waves, being no longer wholly self-propagating, have a different character. Considerable turbulence develops from the water surface downwards, in addition to the turbulence due to bed drag. The effect of this on the Longuet-Higgins bed-water drift attributable to the waves cannot be predicted theoretically. Its experimental study, in common with that of many other wind effects, awaits the construction of suitable laboratory apparatus.

7. Threshold of Bed Disturbance by Fluid Action

A. Steady Unidirectional Flow

The threshold at which cohesionless bed grains are just disturbed and set in motion is usually assumed definable by the critical value of the mean distributed stress, τ_0, exerted by the fluid flow on the bed surface. The critical value, τ_t, is that at which gravity and frictional forces holding the least firmly bedded grains in place are just overcome by fluid forces.

On this view, following the work of Shields (1936), the natural threshold criterion is the ratio θ_t of the critical shear stress τ_t to the immersed weight, represented by $(\rho_s - \rho)gD$, of the topmost grain layer of the bed. θ_t thus has the nature of a generalized friction coefficient which takes both grain size and grain density into account. We have, therefore,

$$\theta_t = \tau_t/(\rho_s - \rho)gD. \tag{16}$$

Under precisely similar conditions of grain size and bed packing, of bed surface configuration, fluid turbulence and degree of steadiness of the flow, and of the mode of application of the applied fluid forces to the individual exposed grains, the threshold friction coefficient θ_t would be expected to be a constant number.

For fully turbulent flow over flat beds of natural grains exceeding about 1 mm in size, which do not spontaneously become rippled under continued flow over them, the value of θ_t has been found, in fact, to be very nearly constant, varying only between 0.04 and 0.06. In the case of beds of finer grains, the ripples of which have already become fully developed by a previous flow, θ_t is found to increase with decreasing grain size, from 0.04 to 0.25 or more, owing to the "form drag" exerted by the ripple features.

The variations of θ_t with grain size, calculated from Shields' experimental results, are shown in Fig. 5 by curve A for grains of quartz density, $\rho_s = 2.65$, and by curve B for lighter grains, $\rho_s = 1.5$. But Shields' results refer to initial movement over beds which had been artificially smoothed out. For reasons explained in Bagnold (1956), however, uniform grain movement over smooth beds of grains less than about 0.7 mm in size is associated with an inherent instability owing to which the bed surface becomes rippled under any continuing movement.

Though the primary ripple system thus created offers but little form drag, it becomes increasingly liable to a second instability as the grain size is further reduced. This second instability is associated with the interaction between the raised ripple features and the scouring effects of the vorticity the raised boundary form creates. It results in the formation of another ripple system which offers considerable form drag.

Though much more controlled experiment is necessary, it seems on existing evidence that the threshold value of θ_t, if measured after the secondary ripple

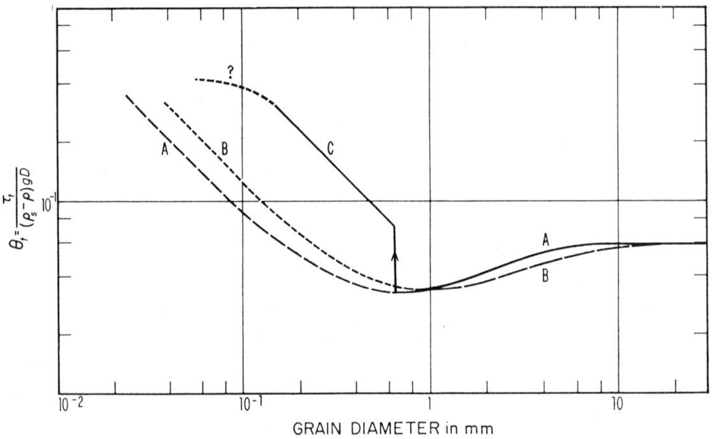

Fig. 5. Variations of the threshold friction coefficient θ_t with grain size. (After Shields, 1936.)

system has been allowed to develop fully under a previous flow, is approximately as shown by curve C. The factual evidence does not, however, extend to beds of grains less than 0.2 mm.

Owing to the sudden jump in the value of θ_t, due to the secondary instability, it is possible for two very different values to be obtained for the same experimental sand, and this has undoubtedly led to considerable misunderstandings in the past.

It is noteworthy that the sudden transformation of a primary ripple system into another system, which is clearly associated with the effects of local vorticity, has also been observed under the oscillating water motion due to surface waves.

It is also noteworthy that few if any experiments have been done on the behaviour of bed grains of controlled size less than 0.2 mm either under steady or oscillating conditions. Nor have the effects of varying either the distribution of grain sizes or the depth of the flow on the threshold by bed movement received the systematic study they deserve.

Under wholly laminar flow, θ_t appears to have the same high order of value as is found (curve C) for a grain size of 0.2 mm under turbulent flow.

Provided no other grains are moving in suspension at the time the threshold stress τ_t for a particular grain size D is considered, the value of τ_t may be calculated from the measured velocity gradient of the fluid above the bed by the Kármán–Prandtl relation which gives:

$$u_* = \sqrt{\left(\frac{\tau}{\rho}\right)} = \kappa \frac{u_2 - u_1}{\log_e (z_2/z_1)}. \tag{17}$$

If velocity measurements are made at distances such that $z_2 = 10z_1$, $\log_e (z_2/z_1) = 2.3$, and for sediment-free flow of a fully turbulent fluid, $\kappa = 0.4$, whence $u_* = 0.174\{u_{(10z)} - u_{(z)}\}$, where u_1 and u_2 are measured flow velocities at distances z_1 and z_2 from the bed, and κ is the Kármán constant, approximately 0.4 for a homogeneous fluid. When suspended sediment is present the fluid is not homogeneous, and the value of κ is found to be appreciably reduced. Stress determinations from measured velocity gradients are, therefore, unreliable.

In the case of a laboratory flume, τ_t can of course be measured directly as $\rho gh \tan \beta$.

B. Oscillatory Water Motion under Waves

There appears to be no theoretical or experimental justification for assuming that the effective threshold stress, τ_t, bears any similar relation to the orbital velocity near the bed, in the case of oscillatory water motion, as τ_t bears to the velocity gradient in the steady flow case. The Kármán–Prandtl relation in the latter case presupposes the establishment of a constant and persisting turbulent boundary layer, and such a layer cannot become established when the flow is continually changing in direction, unless the amplitude of the water oscillation is very large, as in tidal currents.

Experimental evidence (Bagnold, 1946) has shown that for semi-amplitudes of up to 30 cm the water oscillated over flat, unrippled, grain beds without forming any appreciable turbulent boundary layer. The distortion of the water close to the bed occurs as simple viscous shear.

It is clear from the evidence that the threshold of grain movement on an unrippled bed is determined not only by the orbital velocity but also by the relative value of the end-of-stroke acceleration which, for the same orbital velocity, increases with decreasing amplitude.

In Fig. 6 the experimental data from Bagnold (1946) are plotted as threshold orbital velocity, u_{ot}, against the semi-amplitude, r. It will be seen that it is impossible to extrapolate with any certainty to larger amplitudes at which u_{ot} might be expected to become constant and independent of r. Quantitative experiments will have to be made on a considerably larger scale.

There appears to be no similar information relative to conditions where the bed surface is already rippled; and it seems that, at present, we have but little reliable knowledge about the threshold of bed movement under full-scale sea conditions.

Further, while there can be little doubt that the threshold of bed disturbance is definable as a threshold stress, it is by no means certain that under all conditions the available power ω will be sufficient to maintain the transport of the disturbed grains, either in the steady or in the oscillating cases. If the value of ω were to be insufficient, it is reasonable to suppose the result would be a piling up of the excess disturbed grains to form raised features of some kind.

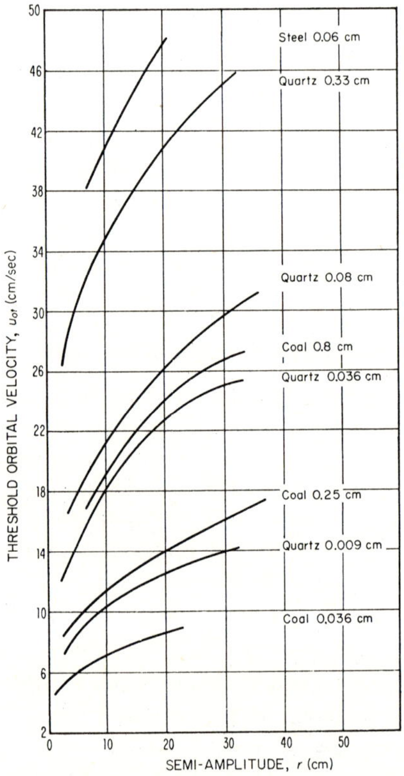

Fig. 6. Threshold of grain movement on an unrippled bed by oscillating water movement in terms of orbital velocity and amplitude of water displacement over the bed.

On this view it may well be that the threshold of bed movement will be found to be better defined in terms of the power ω_t in the dimensionless form $\omega_t/(\rho_s - \rho)gD[(\rho_s - \rho)gD/\rho]^{1/2}$ than in terms of the bed stress, τ_t, in the dimensionless form $\theta_t = \tau_t/(\rho_s - \rho)gD$.

This problem, like so many others in the field of sediment movement, calls for critical experiment designed specifically to throw light on the physics. It is unlikely to be solved by speculations based on the results of engineers' experiments designed for particular practical purposes.

PART II. LITTORAL PROCESSES

D. L. INMAN[1] and R. A. BAGNOLD

(For list of symbols and definitions, see p. 550)

8. Introduction

The application of the general principles of sedimentation to conditions close to a shore as propounded in Part I of this Chapter is complicated by the essential presence of an upward bed slope, and by the fact that this slope, extending to the water surface, introduces a barrier to the mass transport of water associated with surface waves and winds. Further complications are caused by tidal variations of water depth relative to any given point on the littoral slope. Not only does the upward littoral slope distort the oncoming waves until, becoming unstable, they begin to dissipate their energy by breaking, and thus cease to behave consistently with existing wave theory, but also it introduces special currents having components both longshore and offshore.

Littoral processes have been watched at work and their effects frequently described. The physical mechanics of the processes have been speculated upon, but no claim to a reliable understanding of a natural process can be accepted until it can be explained in terms of basic natural principles, and all of its effects can be predicted to the extent of successful reproduction at will under controlled conditions in laboratory models. It is our intention here to outline the degrees to which the natural processes controlling the form assumed by the littoral sea bed have been successfully reproduced in models, and to offer some tentative physical explanations.

9. Littoral Profile: Two-Dimensional Case

A. Definitions

Although the precise demarcations are blurred in nature by tidal variations of levels, it is convenient to divide the littoral boundary into three marine zones:

(1) *The offshore zone*, extending from the bed contour at which the depth admits disturbance of bed sediment by wave motion inwards to the contour at which the waves begin to break.
(2) *The surf zone*, in which the waves dissipate their energy by surface turbulence.
(3) *The swash zone*, comprising the beach proper, in which the residual wave motion consists of successive surges up and down the beach face.

The great majority of model experiments designed to gain an understanding of the natural processes, rather than to further engineering projects, have been restricted, on grounds of economy, to the ideal two-dimensional case. Waves of specified height and wavelength (or period), generated mechanically, are

[1] The support of the Office of Naval Research over many years has made this work possible.

made to travel in a narrow water tank over a bed of specified sediment material inclined upwards at an arbitrary initial slope to the water surface. The action is continued until progressive change of the bed by wave action has ceased.

The inherent limitations of such experiments made in narrow tanks are that the mean mass transport of water over the bed is zero everywhere, and that all motions both of water and of bed material are constrained to take place only in the plane of wave motion.

A further, human limitation, applicable to all model experiments in this field, is imposed by the time available for experiment. The rate of sediment displacement in the offshore zone decreases with increasing depth, till it becomes very slow. In order, therefore, to achieve a mature littoral profile in the time available, the initial littoral slope is usually made to extend steeply to the depth at which bed movement ceases. This limits the quantity of sediment amenable to movement, so that no progressive building out or recession of the profile is possible.

Nevertheless, in general, the mature model profiles reproduce the main characteristics of natural littoral profiles to a degree which is surprising in view of the great discrepancy in the sediment-scale ratio in terms of wave height to grain size.

It is surprising, too, that the well known and marked difference in profile form and slope that occurs between cobble and sand beaches in nature is reproduced in small models using materials of natural size. This fact is of sufficient interest to demand a physical explanation.

B. Energy Profile

The mean gradient of wave energy due to energy losses by bed friction, and by surface turbulence within the surf zone, is always onshore. But since the energy gradient due to bed friction is always in the direction of local flow relative to the bed, we have, superimposed, local fluctuations onshore and offshore due to the wave oscillations.

Consider any unit area of a littoral slope, locally inclined at an angle β to the horizontal. The energy loss, ΔE_1, over this unit area due to bed friction, involving appreciable sediment displacement during the time of onshore water displacement, may be assumed proportional to the mean force exerted by the water on the displaced sediment times the distance x that it is displaced. If the coefficient of intergranular friction is $\tan\phi$, and the displaced sediment mass is m_1 per unit area, we have

$$\Delta E_1 = a\,\frac{\rho_s-\rho}{\rho_s}\,g m_1 x_1 \cos\beta\,(\tan\phi+\tan\beta),$$

where, following the notation of Fig. 3, the $\tan\beta$ term represents the gravity component tangential to the slope, $(\rho_s-\rho)/\rho_s$ is the buoyancy factor, and a is a coefficient of proportionality.

Owing to losses of various kinds, including percolation into the beach, the energy of the returning water will be less. So the corresponding frictional energy loss, ΔE_2, over the same unit area will be less. For the return water displacement we have

$$\Delta E_2 = a \frac{\rho_s - \rho}{\rho_s} g m_2 x_2 \cos\beta\,(\tan\phi - \tan\beta).$$

But if the littoral profile is mature, as much sediment must be carried down the slopes as is carried up. Hence $m_2 x_2 = m_1 x_1$ and

$$\frac{\Delta E_1 - \Delta E_2}{\Delta E_1} = \frac{\delta E}{\Delta E_1} = \frac{2\tan\beta}{\tan\phi + \tan\beta},$$

thence

$$\tan\beta = \delta E \tan\phi / (\Delta E_1 + \Delta E_2),$$

or, writing $\Delta E_2 = c\,\Delta E_1$,

$$\tan\beta = \tan\phi \frac{1-c}{1+c}. \tag{18}$$

The friction coefficient $\tan\phi$ being a constant (ϕ being approximately the angle of repose of the material), the littoral angle β should be independent of the linear scale, but dependent on the ratio of the energy deficit $\Delta E_1 - \Delta E_2 = \delta E$ elsewhere to the loss ΔE_1 over the local area. If the deficit is large, so that c approaches its limiting value of zero,

$$\tan\beta \rightarrow \tan\phi$$

and the beach stands at the angle of repose of the material.

If there were no deficit, i.e. if the water tranquilly returned without loss of energy having flowed down a frictionless bed, the angle β would approach zero.

C. *Permeability and Sand Size*

The loss of energy between flow and return, which gives rise to the deficit δE, may be due wholly to bed friction, in which case it may be expected to decrease with decreasing grain size, for the grain size should determine not only the boundary roughness but the additional energy loss arising from underwater percolation tangentially through the bed under the large tangential pressure gradients associated with the travelling waves.

The permeability of the bed material may be expressed, independently of wave scale, as the water discharge perpendicularly through unit area of a bed of thickness equal to the perpendicularly applied head of water. The percolation discharge decreases very rapidly as the grain size of the material is decreased from that of pebbles to that of sand. This is because the flow within the pore spaces then changes from turbulent to laminar. Thus, while sands are relatively impermeable, pebbles and larger materials are very permeable and permit considerable energy losses within the bed. This probably explains why the same wave attack displaces a far greater mass of large-grain material than of small.

Permeability also affects the slope of a swash beach of coarse material in another way. A loss of swash energy occurs over the upper part of the beach by a mass loss of water by downward gravity percolation. As a consequence the value of c in relation (18) decreases progressively towards the beach crest where it may actually become zero, there being here no return flow at all over the beach surface. Hence, on model pebble beaches the angle β increases progressively upwards, and may approach the angle ϕ of repose, just as it is found to do in nature. It has been shown (Bagnold, 1940) that the increase in beach slope towards the beach crest can be prevented in the model by the insertion of an impermeable plate just below the beach surface.

Fig. 7a shows the profile of a typical model beach of small pebbles (diameter

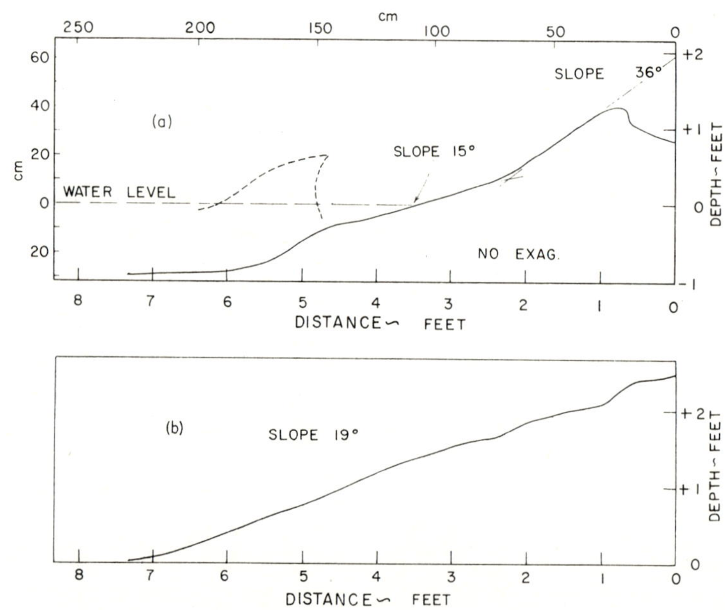

Fig. 7. Comparison of model beach profiles formed by waves during: (a) a stationary water level; and (b) a falling tide. The model beach consisted of pebbles with a mean diameter of 7 mm and a natural angle of repose of $33\frac{1}{2}°$; breaker height at the plunge point was approximately 29 cm. (After Bagnold, 1940.)

7 mm). The profile and its slope reproduce those formed in nature very well. When sand is introduced below the beach the well known abrupt discontinuity in angle between sand and pebble is also reproduced. For waves of a given height, the profiles are but little affected by changes in the wavelength of the attacking waves. The effect of a slowly falling tide on the same pebble beach is to produce a nearly plane beach face with a uniform slope of 19°, as shown in Fig. 7b. This draw-down of the beach face by falling tides is similar to that observed to occur in nature where the tide range is large compared with the wave height (Inman and Filloux, 1960).

The transition from pebble material to sand material is associated both in models and in nature with a considerable decrease in the general littoral slope. Since the height to which the swash rises on a beach face cannot be appreciably less than the wave height H, the width of the beach must be given approximately by $H \cot \beta$. For small slopes the beach width will increase very rapidly with decrease in β. Thus, whereas the width of the pebble or cobble beach rarely exceeds the minimum possible wavelength, that of sand beaches may exceed the wavelength of short, steep waves. Similarly, whereas the period of up-surge and down-surge on the pebble beach rarely exceeds the wave period either in models or in nature, the corresponding period on a sand beach is relatively far larger. Both of these factors, which are inter-related, appear to affect the manner in which the wave breaks, and in consequence the resulting littoral profile.

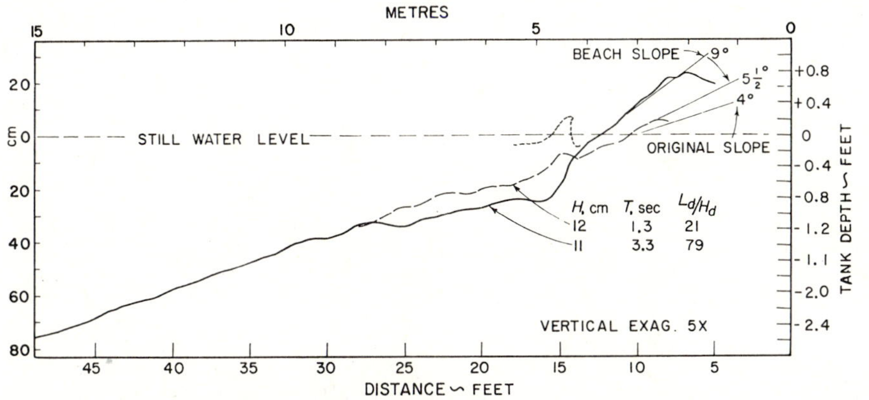

Fig. 8. Comparison of mature model beach profiles formed by long waves ($L_d/H_d = 79$), and by short waves ($L_d/H_d = 21$), of nearly constant height. The model beach consisted of coarse sand with a median diameter of 0.56 mm; in each case the beach was given an initial slope of 1 : 15 (4°) and then subjected to waves for a period of 10 h. Waves were generated in water 3.0 ft deep, but tabulated parameters are those computed for deep-water conditions. (After Corps of Engineers, 1947.)

D. Wave Steepness[1]

Fig. 8, reproduced from the Corps of Engineers (1947, pl. 4), shows the mature profiles of two coarse-sand model beaches. While the material (sand of 0.56 mm diameter), the wave height (11.3 cm), and the still water depth (91.5 cm) were the same in both cases, the wavelengths alone were different. The continuous line represents the profile formed by long waves with a wavelength to wave-height ratio of 79, while the broken line depicts the profile formed by shorter and steeper waves, for which the ratio length to height was 22. The change from long to steep waves had the effect of removing material

[1] Ratio of wave height to wavelength.

from the beach and depositing it offshore, with the formation of a pronounced bar at the point at which the waves started to break.

This reproduces qualitatively what is seen to happen in nature, but quantitatively it has so far not been found possible to reproduce the same profiles on differing model scales. One probable reason is that the smaller the model scale, the more do surface-tension effects modify the mechanics of energy dissipation within the upper water layers when the waves have broken. The dynamics of a breaking wave are not yet understood. For this reason, perhaps, only little experimental attention has been paid to the marked difference in the manner of breaking, as between long and steep waves breaking on flat and steep beaches. Also, little attempt has been made to reproduce the features of natural fine sand beaches in the laboratory. Here again, the very gentle slopes (Fig. 9)

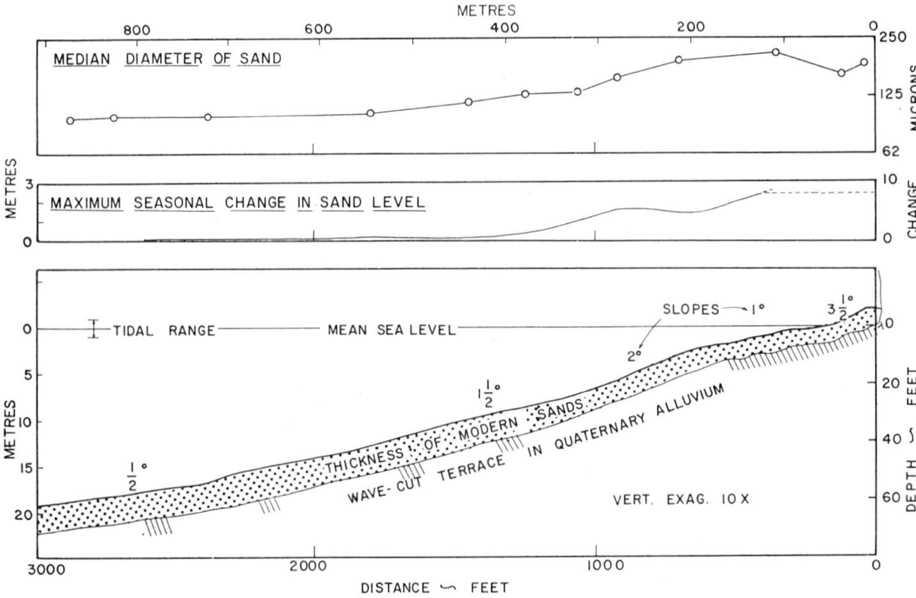

Fig. 9. Profile of a natural fine sand beach at La Jolla, California, showing the median sand size, the maximum seasonal change in the level of the sand associated with winter and summer waves, and the thickness of the modern littoral sand over the wave-cut terrace. Note the very gentle slopes. (After Inman, 1953.)

and the small size of sand together appear to give rise to a type of transverse sweeping motion which cannot be reproduced in narrow two-dimensional models. It seems likely that this sweeping motion, the cause of which is not understood, may give rise to the formation of beach cusps.

For reasons already suggested little attention has been paid to the littoral profile offshore of the plunge line. In many regions the sea bed remains sufficiently shallow to be affected by wave action for great distances offshore. Formation of ripples and active movement of sand on the sea bed has been observed by divers at depths exceeding 52 m (Inman, 1957). Everywhere outside of the surf zone, where the effect of wave action is moderate, the sandy bottom is rippled. In and near the surf zone the intense bottom shear produced by the higher orbital velocities destroys the ripples and the bottom is smooth. A seasonal fluctuation in the level of the sand, caused by differences between summer and winter waves, is clearly discernible at depths greater than 10 m, as shown in Fig. 9. The long low waves cause a net shoreward transport of sand so that the beach builds seawards and the offshore profile steepens. High steep waves erode the beach and transport material offshore, causing a decrease in the beach slope.

Model experiments (Bagnold, 1947) showed that waves of small wavelength to height ratio caused merely a to-and-fro motion of the horizontal sand bed offshore of the plunge line, with little or no net transport to or from the shore. Waves of the same height and longer wavelength caused general shoreward transport, accompanied by a general shoreward drift of the lowest water layer. However, while the shoreward bed-water drift continued up the sloping bed towards and as far as the plunge line, the corresponding drift of sand was appreciably reduced by the upward bed inclination. As a result, sand tended to accumulate at the foot of the slope and thus to reduce its steepness. Owing to the slowness of the movement, a steady-state profile was not reached in the time available. It is noteworthy, however, that no progressive building out of the beach face was detectable. All shoreward movement of sand stopped at the plunge line.

It has been shown (Bagnold, 1946; Inman, 1957) that the wavelength of sand ripples is approximately equal to the orbital diameter (horizontal particle displacement) of the waves generating the ripples, up to a limiting maximum ripple size beyond which the ripple cannot grow. The limiting maximum ripple wavelength is a function of sand size and varies from about 15 cm for very fine sand to 125 cm for coarse sand. Thus the long, low waves that produce an onshore transport of sand along ocean beaches have orbital diameters many times greater than the wavelength of the bottom ripples, and the sand moved by each wave may cross over several ripples. In models, on the other hand, the orbital diameter of the small waves does not commonly exceed the limiting ripple size, and the ripples have the same wavelength as the orbital diameters. For this condition Russell[1] has observed that long waves in models may cause a net offshore transport of sediment even though the bottom is horizontal. The rapid onshore acceleration accompanying the passage of a wave crest forms a vortex of suspended sand over each ripple crest, with relatively little onshore displacement of sand. The offshore motion accompanying the wave trough, although more gradual, results in a net offshore transport of sand already placed in suspension by the preceding onshore motion. Steep waves on the other hand, because there is less asymmetry between the crest and trough orbital velocities, merely cause a to-and-fro motion of sediment with no

[1] Personal communication from R. C. H. Russell, Hydraulics Research Station, Wallingford. England.

net transport. The dependence of sediment transport on the degree of asymmetry of orbital motion, when ripple size and orbital diameter are of similar magnitude, suggests that the ratio of ripple wavelength to orbital displacement is a measure of similitude that should be considered in models.

The shoreward transport of sand effected by long stable waves outside the plunge line is reduced or inhibited by an upward slope in the bed. This reduction in transport appears to be readily explainable by the reasoning leading to relation (18). A condition of tangential force equilibrium on the sand grains is approached. Further, if this equilibrium condition were to be disturbed by a reduction of the onshore water drift, due to a steepening of the waves, the result should be a net offshore transport of sand down the slope. This would explain the apparent anomaly; while offshore-flowing bottom currents have never been detected outside of the plunge line in two-dimensional models, offshore sand transport has been reported both in models and in nature.

In nature, high short waves also produce an offshore transport of sediment by increasing the turbulence and velocity of the currents in the nearshore circulation system, as discussed in a later section. Sand in suspension is transported through the breaker zone and into deeper water by rip currents. It is probable that a significant net seaward transport of sediment occurs by a general mixing between the waters of the surf zone, where the concentration of suspended sediment is high, and the offshore waters, where the concentration is relatively low. In addition it is likely that auto-suspension (Section 5) of fine sediments in the turbulent waters of the surf zone results in an outward diffusion of the finer grades of sediment.

E. Rise in Water Level

Waves transmit momentum in the direction of their travel, the momentum being relative, according to wave theories, to still water elsewhere. Also wave theories assume the momentum flux to continue on indefinitely. When this onward flux is prevented, however, the flux of momentum against the shore constitutes a shoreward force. Under steady-state conditions the shoreward force should be just balanced by a rise in water level on the beach. The rise should be proportional to the component of momentum flux normal to the shore: $C\rho q \cos \alpha$, where ρq is the momentum per unit crest length, C is the velocity of the breaking wave, and α is the angle between the breaking wave and the shore. However, in nature the grouping of high waves followed by low waves causes a pulsation of the water level in the surf zone, called surf beat (Munk, 1949a). Little attention has been paid to the degree of water-level rise which results under steady state of pulsating conditions, or to its gradient relative to water depth offshore. In a study of the second-order effects of the interaction between waves and currents Longuet-Higgins and Stewart (1962) have evaluated the additional flux of momentum due to the waves in terms of a "radiation stress". Assuming no loss of energy by friction or reflection, they show that the radiation stress associated with waves of steady amplitude should produce a draw-down or negative slope in the mean surface level as the depth diminishes. However, when the waves break, the water level rises and the slope becomes positive.

A rise of the water level nearshore is also produced by winds blowing onshore. Model experiments performed in the Geography Department at Cambridge University (King, 1959, p. 211) demonstrated that onshore winds caused offshore currents to flow along the bottom. Where combined with wave action, the wind-induced currents increased the rate of offshore sand transport of the waves in the surf zone, and decreased or reversed the onshore transport

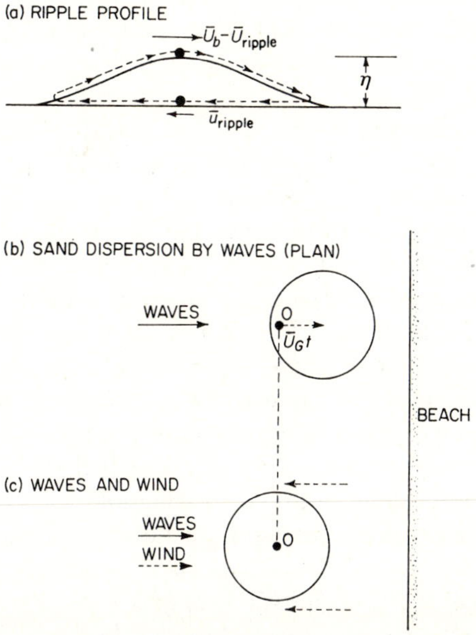

Fig. 10. (a) Path of a sand grain in a migrating ripple, observer moving forward with the ripple form; (b) dispersion pattern of marked sand introduced at "0" under the influence of waves; and (c) dispersion pattern under influence of waves and wind.

outside of the breaker point. Rise of the water level nearshore may be of considerable importance in the real three-dimensional case.

10. Migration of Sand Grains by Wave Action

The average speed, U_G, at which individual grains migrate from a given initial position on the sea bed is likely to be much less than the mean speed, $\bar{U}_b$, of mass transport. For owing to the presence and movement of bed ripples any individual grain is alternately exposed on the ripple surface and buried within it.

Consider the ripple cross-section sketched in Fig. 10a. To an observer moving with the ripple form, as the form slowly advances in the direction of sediment drift, marked grains would appear to circulate forwards over its surface at the bed transport velocity $\overline{U}_b$ and then backwards through the body of the ripple at the much slower speed of the latter's advance.

Assuming the ripple cross-section to be antisymmetrical about its mid-plane the mean thickness of the sediment composing it is $\eta/2$, where η is the whole height of the ripple from crest to trough. If D is the mean grain diameter there will be $\eta/2D$ layers of stationary grains within the ripple at any moment. If b is the number of eroded grain layers in motion at the speed $\overline{U}_b$, the average migration speed, $\overline{U}_G$, will be

$$\overline{U}_G = \overline{U}_b(2bD/\eta). \tag{19}$$

It is not easy to measure the number b of moving grain layers. Observed from above, the stationary bed grains become indistinct when $b=1$ and are invisible when b exceeds 2. Thus a rough estimate can be obtained from observation.

It is instructive to compare the average speed of migration as estimated from the above discussion and from the ideas of the preceding sections with the actual speed as measured on the sea bed under conditions of wave motion. Inman and Chamberlain (1959) re-deposited *in situ* a sample of about 1 kg of sea-bed sand after its grains had been irradiated, and traced the progress of the scatter during the succeeding 24 h. The site was outside the surf zone, 300 m offshore and opposite the Scripps Institution of Oceanography at La Jolla, in a water depth of 3 m. The average wave height was 38 cm and the average period 14 sec. The direction of wave travel was approximately onshore.

Putting the velocity of these long waves at $\sqrt{(gh)}=540$ cm/sec, relation (14) predicts a shoreward wave drift, $\bar{u}$, along the bed of 2.7 cm/sec (fixed bed), and on the very rough evidence given by Bagnold (1947) the velocity, $\overline{U}_b$, of sediment drift over a sediment bed should be about one third of this, namely 1 cm/sec.

The mean sediment size was 0.0125 cm, and the mean ripple height, η, was about $1\frac{1}{2}$ cm. If, as a reasonable estimate, we take $b=2$, the average migration velocity $\overline{U}_G$ should be 0.05 cm/sec, assuming that the motion is due wholly to bed load. If an appreciable proportion of transport is in suspension the migration velocity $\overline{U}_G$ would, of course, be larger.

Seven and one-half hours after the experimental re-deposition of the marked sample the pattern of the scatter was such that the greatest migration distance shoreward from the point of re-deposition was 30 m. This gives an average speed of 0.1 cm/sec in the onshore direction. The rapid onshore migration was in part due to transport of sand in suspension. During the stronger surges accompanying the occasional largest waves, a vortex of suspended sand extending approximately 15 cm above the bottom was observed. In such cases individual sand grains may be carried across several ripples before being deposited.

Irradiated grains were found by core sampling to be distributed to a depth below the surface approximately equal to the ripple height, which is just what would be expected from the above reasoning.

The approximate quantitative agreement in this single instance is obviously fortuitous, but the result suggests that a consistent understanding of what happens on the sea bed might best be got by combining further quantitative field measurement with laboratory experiments of extended scope designed to sort out the separate effects of wave, wind and boundary conditions. For instance, the field experiment of Inman and Chamberlain gave the seemingly anomalous result that the marked grains migrated, not only shorewards, as might be expected, but in all directions. In part this dispersion is associated with the onshore and offshore movement of sand, both in suspension and as bed load, by the orbital velocity of the waves. Also, a hint as to the reason for this may be given by the incipient horizontal circulation of bed water noticed by Russell and Osorio (1958) in their wave experiments. If such a circulation is real, the scatter due to wave drift alone would be expected after an elapsed time, t, to show the pattern sketched in Fig. 10b. Assuming no appreciable mass drift on the part of the sea-water as a whole, an onshore wind, as discussed previously, would be expected to superimpose an offshore drift along the bed, thus producing a modified pattern such as is sketched in Fig. 10c. However, no appreciable wind was observed during the field experiment.

It is possible that each of these effects can be studied separately in the laboratory, provided the equipment is designed with the specific objects in view, rather than in isolation.

Again, in the irradiated sand experiment the marked grains were detected by direct emission photography of samples of the bed surface taken on adhesive paper. This disclosed the location of the active grains on the paper, and their sizes should be measureable by microscopic observation; and since, by the reasoning leading to relation (19), bed surface samples should be representative of the sub-surface distribution also, it may be possible by this means to throw valuable light on the differential migration of sediment grains of different sizes —a subject on which no certain information is now available.

Following the experiment described above involving the dispersion of 1 kg of irradiated sand outside of the surf zone, a further series of experiments[1] using dyed sand was conducted on the swash zone of the beach face. In the first swash-zone experiment, 3 kg of dyed sand were placed in the wave uprush where the maximum depth of water in the uprush was about 20 cm. The second experiment involved 4 kg of sand and was conducted where the maximum uprush depth was about 15 cm.

The results of the two experiments showed that: (a) dispersion of sand grains in the uprush zone was extremely rapid; 4 kg of sand dispersed over 30 m^2 of beach in one minute; (b) individual sand particles had about 10 times the onshore–offshore trajectory as longshore trajectory; and (c) very small beach features, such as small cusps, drastically affected the dispersion of quantities of sand of the order of 4 kg.

In contrast to the area just outside the surf zone, where 1 kg of irradiated sand was sufficient to observe successfully sand movement for about 24 h, in

[1] As yet unpublished.

the surf zone or on the beach foreshore it appears that about 100 times that amount of irradiated material would be necessary to trace successfully the movement of sand particles for any distance along the beach.

11. Littoral Processes: Three-Dimensional Case

The transition from two-dimensional models to natural littoral processes involves the introduction of mass transport, momentum and like components in directions other than normal to the shore. It also involves the introduction of the effects of tides and of random or ill-definable variations in the character of waves, winds, currents and sediments. The complexity of these factors and the difficulties involved in their observation have resulted in inadequate descriptions of the motions of water and sediment in the field. No serious attempt has yet been made in the laboratory to sort out these motions according to their causes. Before proceeding to a working model for the littoral transport of sand, it is instructive to consider field and laboratory studies of currents in the surf zone as well as observations of sand transport rates.

A. Currents in the Surf Zone

Waves in deep water cause a volume transport, or "discharge", of water in the direction of their travel. This discharge increases as the waves pass into shallowing water near the shore. The beach, however, presents an almost impermeable obstacle which must reduce to zero the entire component of the discharge normal to the shore. In effect the rise in water level referred to previously constitutes a normal outward force which superimposes an outward discharge; this just balances the inward discharge by the waves. That such a state of equilibrium must exist on the average over a sufficient length of shore is indisputable, but it is doubtful how far the equilibrium is stable at any given point, i.e. to what extent the water transported shoreward at one point may return seaward elsewhere.

If, on the other hand, the direction of wave travel is not normal to the shore, the accompanying discharge has in any case a longshore component over and above that due to any local disequilibrium in the force components normal to the shore, and a definite longshore current must result. For some, as yet unexplained, reason this longshore current is found to be confined largely within the surf zone.

The waves discharge water into each unit length of the surf zone. So, assuming the resulting longshore current to be so confined, the velocity of the current would, were it not for the existence of outward-flowing rip currents at discrete places, go on increasing indefinitely with distance along the shore. On this view the existence of outward rip currents is inevitable.

The net onshore transport of water by wave action in the breaker zone, the lateral transport inside the breaker zone by longshore currents, the seaward return of the flow through the surf zone by rip currents, and the longshore movement in the expanding head of the rip current all constitute a nearshore circulation system (Fig. 11a). The pattern that results from this circulation commonly takes the form of an eddy or cell with vertical axis. The positions of the rip currents are dependent on the submarine topography and configuration of the coast, and the height and period of the waves. Periodicity or fluctuation of current velocity and direction is a characteristic of flow in the nearshore circulation system. This variability is primarily due to the grouping of high

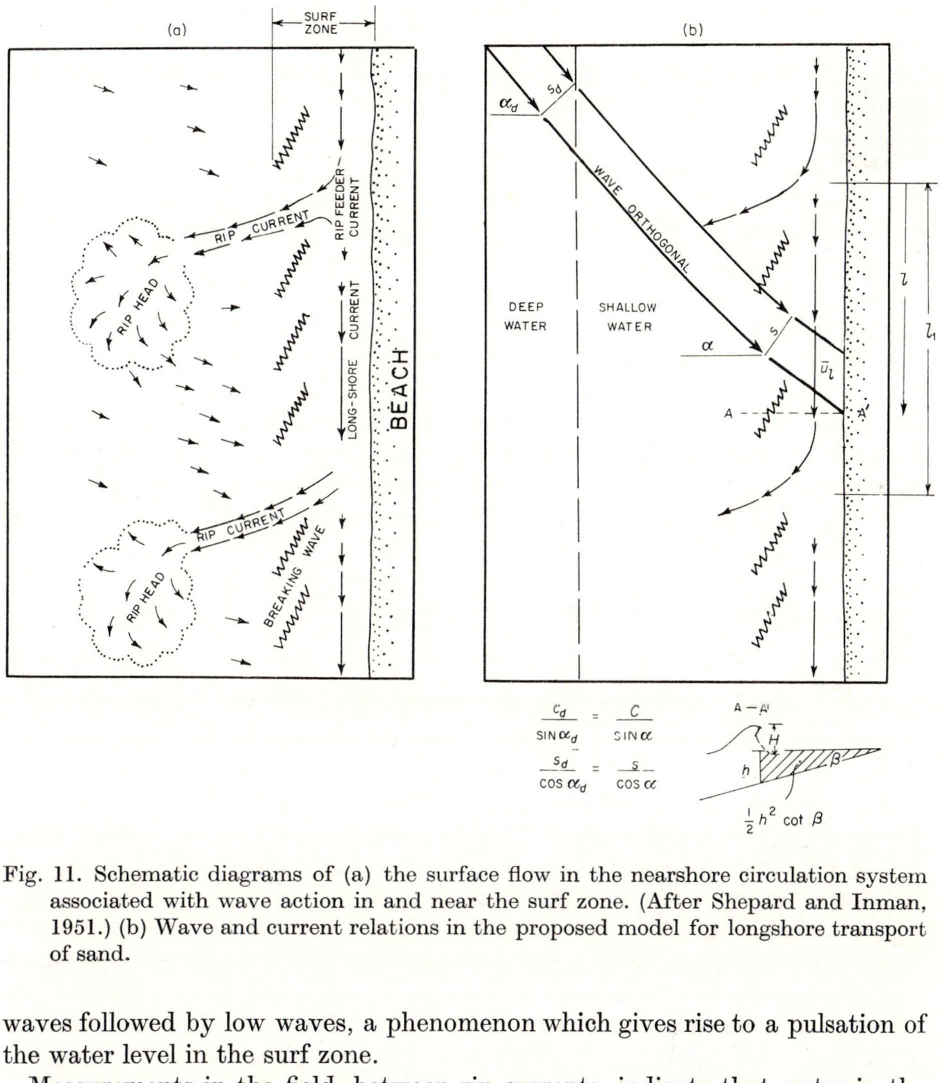

Fig. 11. Schematic diagrams of (a) the surface flow in the nearshore circulation system associated with wave action in and near the surf zone. (After Shepard and Inman, 1951.) (b) Wave and current relations in the proposed model for longshore transport of sand.

waves followed by low waves, a phenomenon which gives rise to a pulsation of the water level in the surf zone.

Measurements in the field, between rip currents, indicate that water in the surf zone has a slow net offshore flow near the bottom. This flow between rip currents does not extend through the breaker zone; rather, water outside of the surf zone moves towards the shore. Where beaches are not straight, or where offshore topography causes differential wave refraction, the longshore currents are dependent on the gradient of breaker height along the beach, as well as on the angle between the breaking wave and the beach. For such cases, the positions of the circulation cells are largely dependent upon the location of zones of wave convergence and divergence. Also, it has been observed that points, breakwaters, and piers all influence the circulation pattern and alter the direction of the currents flowing along the shore. In general, these obstructions determine the position of one side of the circulation cell.

Field measurements of longshore currents (Inman and Quinn, 1952) show that the velocity progressively increases from zero, immediately downstream from a rip current, to some maximum value just preceding its seaward deflection at the next rip current. Also, the spacing between rip currents decreases as the intensity of wave action increases, suggesting that the longshore flow has some limiting value above which it breaks seaward into rip currents. The maximum velocity of longshore currents appears not to exceed 2 to 3 knots. The observed spacing between rip currents ranges from about 30 to 1000 m. Measurements following a period of low waves along a 10-mile section of straight beach at Clatsop Spit, Oregon, gave a mean separation between rip currents of 400 m and a standard deviation of 145 m (Shepard and Inman, 1951).

Calculations indicate that rip currents along straight beaches discharge seaward between 2 and 10% of the total volume which is estimated to be transported shoreward by the waves between rip currents; whereas their seaward discharge is of the same order as the estimated longshore component of the shoreward transport. The implication is that the normal component of the shoreward wave transport has no real existence relative to the shoreline, because the general body of the water is being displaced outwards at an equal rate.

B. *Magnitude of the Longshore Current*

Putnam, Munk and Traylor (1949) derived a relation between the longshore current velocity and the characteristics of the oncoming waves in terms of the angle α of attack and of a beach drag coefficient c_d which defines the frictional resistance of the bed to the flow of the longshore current. Their relation, which was based on considerations of energy and momentum and tested by field and laboratory experiments, indicated that the beach drag coefficient c_d varied for medium sand from 0.04 in the laboratory to 0.008 in the field.

Later observations over a wider range of field conditions (Inman and Quinn, 1952) indicated that the relation of Putnam, Munk and Traylor gave values of c_d which varied unaccountably by a factor exceeding 10^4. These field observations show that rip currents, which were absent or only poorly developed in the laboratory, were an important form of return flow; also, longshore currents in the field did not obtain a steady state of flow, but were continually fluctuating or pulsating in response to the periodicity of the grouping of higher and lower waves impinging on the surf zone, whereas in the model they were constrained to the steady state. Again, since longshore currents are essentially pulsating, the remaining longshore component of momentum flux may be primarily expended in accelerating the water-mass along the shore from near rest to maximum current, which then turns to sea in the form of a rip current. It seems likely that the frictional dissipation in the rip current mostly takes place outside the surf zone.

The following alternative approach takes the principle of continuity of flow into consideration. Let q be the gross incoming wave discharge (volume of water) in unit time per unit length of wave crest entering the surf zone at an angle α to the beach normal (Fig. 11b). The normal component is supposed to be reduced to zero, as previously explained, by the superposition of an equal and opposite outward discharge. The tangential component along the beach is then $q \sin \alpha$ per unit length of wave crest, or $q \sin \alpha \cos \alpha$ per unit of beach length, l. Hence continuity requires that the total longshore discharge, Q_l, shall increase progressively along the beach.

Assuming that the longshore discharge Q_l is zero at a point immediately down-current from a previous outward-flowing rip current, its value through a beach section A–A′ distance l down-current from the previous rip current will be given by

$$Q_l = ql \sin \alpha \cos \alpha.$$

The average value of Q_l will be given when $l = \frac{1}{2}l_1$, where l_1 is the average separation between rip currents (Fig. 11b). Regarding the cross-section A–A′ of the longshore discharge as a triangular wedge whose base is the beach which is inclined at an angle β to the horizontal and whose outermost height is h at the plunge line, its area is $h^2/2 \tan \beta$. Whence the mean longshore current velocity, $\bar{u}_l$, over a long straight beach should be given by

$$\bar{u}_l = q \frac{l_1}{h^2} \tan \beta \sin \alpha \cos \alpha. \tag{20a}$$

The power required to accelerate the surf water to the velocity $\bar{u}_l$ and also to maintain the flow against both bed drag and the internal resistance to flow due to additional turbulence created by the breaking waves comes from the longshore component of the entering flux of wave momentum.

Both this applied power and that part of it which is used up in accelerating the surf water may be assumed uniform, per unit beach length, along the beach, but the power dissipated by internal turbulence would be expected to increase progressively with the distance l as the velocity of flow increases. So at some ultimate point, distant l_1, from a previous rip current, the acceleration would be expected to cease and the longshore discharge, therefore, to break outwards as another rip current because it can no longer be contained within the surf zone cross-section.

It remains to evaluate the mean longshore current velocity $\bar{u}_l$ in terms of the wave characteristics. For solitary waves (Munk, 1949a), the volume of water per unit crest length brought forward with a breaking wave is $4H^2/\sqrt{(3\gamma^3)}$

from which the average discharge, q, becomes $4H^2/\sqrt{(3\gamma^3)}T$, where $\gamma = H/h$ is the relative depth of water, H and h are respectively the wave height and depth of water, and T is the wave period. Substituting this value for the discharge gives

$$\bar{u}_l = 4\sqrt{\frac{\gamma}{3}}\frac{l_1}{T}\tan\beta\sin\alpha\cos\alpha. \tag{20b}$$

C. Longshore Transport of Sand (Field and Laboratory)

No satisfactory relationship has been developed between the rate of littoral transport and the waves and currents that cause it. The volume of littoral transport along oceanic coasts is estimated from observed rates of erosion or accretion, most commonly in the vicinity of coastal engineering structures such as groins or jetties. Such observations indicate that the transport rates vary from almost nothing to several million cubic metres per year, with average values commonly falling between 100,000 and 1,000,000 m^3 per year. In general, these may be conservative estimates since the volume of sand moved usually exceeds that indicated either by deposition or erosion.

Laboratory studies of littoral drift (Saville, 1949; Johnson, 1953; Savage, 1959) show that rates of littoral transport in models vary with wave energy, wave steepness, and the angle of wave approach. The maximum amount of total transport was observed to occur for wave steepnesses of about 0.025 and for angles of wave approach of about 30°. Also, the transport for waves of low steepness ratio took place largely in the wave-swash zone on the beach face, while the transport for steeper waves occurred principally on the longshore bar in the breaker zone. Marked changes in transport rate were observed as the beaches adjust from one set of wave conditions to a new set. Since natural sand beaches are in a continually transient state as the beach adjusts for changing conditions of waves and tides, the application of model to the prototype is difficult. However, the general comparisons between model and prototype appear to be consistent.

It was suggested (Scripps, 1947) that the work performed by waves in the nearshore zone might be a useful parameter for relating littoral transport rates of sand to wave action, and wave-work factors were computed for that purpose by wave period and direction. Caldwell (1956), observing the relationship between wave power and longshore transport rates for actual field conditions at Anaheim Bay, California, and South Lake Worth Inlet, Florida, obtained the following approximation for the relationship of the combined data from both localities:

$$S = 210[P\sin\alpha\cos\alpha]^{0.8} = 210P_l^{0.8},$$

where S is the longshore volume transport of sand in cubic yards per day, $P_l = P\sin\alpha\cos\alpha$ is the longshore component of wave-energy transmission (power) of the breaking waves in millions of ft-lb/day per ft of shoreline, and pound is a unit of force. The factor $\sin\alpha\cos\alpha$ converts P, the incident wave power per unit of wave crest, into the longshore component per unit of shoreline, where α is the angle the breaking wave makes with the beach. Savage (1959, fig. 4) gives a graph summarizing available data from field and laboratory experiments in terms of S and P_l. Inspection of this graph, excluding only the data for model experiments with sand of low density, indicates that the following relation, with a wave-power exponent of unity, fits the data equally well:

$$S = 125P_l,$$

where S and P are expressed in the same units as before.

These relations, which summarize the available facts, are empirical only, and are dimensionally incorrect. The ideas contained in Section 4 suggest that the transport rate of the sediment should be converted from a volume to an immersed weight basis by writing:

$$I_l = (\rho_s - \rho)ga'S = \text{constant} \times P_l,$$

where I_l is the longshore dynamic transport rate and a' is the correction for pore-space and may be taken as 0.6, approximately. The new constant is then dimensionless, has a value of 2.0×10^{-1} for natural beach sand, and is independent of the unit system used providing the units are consistent.

Since this relation is empirical, it still fails to suggest what other relevant factors may affect the value of the constant. A rational derivation is attempted in the following section.

D. Model for Longshore Transport of Sand

Let E be the mean energy per unit surface area of a wave which is near the breaking point and is approaching an infinitely long straight beach at an angle α (Fig. 11b). Then ECn is the rate of transport of energy (power) which is available for dissipation over a unit crest length of beach, where C is the wave phase velocity and Cn is the velocity at which energy is propagated forward. Assume that the rate of energy dissipation by friction on the beach bed is proportional to ECn, then the mean frictional force applied to the whole of the beach bed per unit of crest length is proportional to ECn/u_o, where u_o is the mean frictional velocity relative to the bed within the surf zone and is assumed to be proportional to the orbital velocity near the bottom just before the wave breaks. The mean force applied to the beach bed per unit of beach length then becomes proportional to $(ECn/u_o) \cos \alpha$. The immersed weight of sediment in motion should then be proportional to this applied force divided by the intergranular friction coefficient, $\tan \phi$. Once the sediment is in motion it becomes available for transport by any current, such as the longshore current $\bar{u}_l$ (see Section 4-D). Thus the total immersed weight of sediment transported in unit time past a section of beach A–A′ becomes

$$I_l = \frac{\rho_s - \rho}{\rho_s} gJ = K \frac{ECn}{u_o} \frac{\cos \alpha}{\tan \phi} \bar{u}_l, \tag{21}$$

where I_l is the dynamic transport rate, J is the sediment discharge in mass transported per unit time, $(\rho_s - \rho)g/\rho_s$ converts to immersed weight, and K is a factor of proportionality. In the above it should be noted that:

(1) Both sides of relation (21) have the nature and dimensions of power, or time rate of change of energy. The quantity J, usually measured as mass passing in unit time, can be expressed as transported mass per unit length in the direction of transport multiplied by a mean velocity of transport, here assumed proportional to $\bar{u}_l$. The dynamic transport rate $I_l = (\rho_s - \rho)gJ/\rho_s$ is, therefore, the immersed weight multiplied by the velocity of transport, and $I_l \tan\phi$ is the rate of work done in transporting the sediment.

(2) The quantity ECn is the power transmitted per unit width of wave. The power available to move sediments over unit bed area is the decrement of ECn per unit distance travelled by the wave provided it is assumed that the decrement is due entirely to bottom friction [see equation (6)]. If the whole of the power in the breaking wave is dissipated between the plunge line and the shore, then ECn would be the power available to move sediment over the entire width of the surf zone. In fact, however, a high proportion of the power dissipated in the surf zone is dissipated by means other than bottom friction, as discussed in Section 12, and thus the factor K is an index of the proportion of power dissipated in moving sediment.

If the position of rip currents is migratory and shows a random distribution with time or distance along the beach, then the most general expression for the mean longshore transport rate of sediment over a long straight beach will be obtained by substituting the value of the mean longshore current velocity, $\bar{u}_l$, as given in relation (20a):

$$I_l = K \frac{\tan\beta}{\tan\phi} \frac{ql_1}{h^2} \frac{ECn}{u_o} \sin\alpha \cos^2\alpha. \tag{22a}$$

The relation can be further simplified by assuming that the orbital velocity, u_o, equals $\frac{1}{2}(H/h)C$ and that the discharge of water, q, transported forward with the breaking wave, is given by the solitary wave relation leading to equation (20b):

$$I_l = \frac{8}{\sqrt{(3\gamma)}} \frac{\tan\beta}{\tan\phi} Kl_1 \frac{En}{T} \sin\alpha \cos^2\alpha. \tag{22b}$$

The empirical relation from the data of Caldwell and Savage (Section 11-C), in its dimensionally consistent form, can be derived from equation (22b) by substituting P_l for $ECn \sin\alpha \cos\alpha$. The relation then becomes

$$I_l = \frac{8}{\sqrt{(3\gamma)}} \frac{\tan\beta}{\tan\phi} K\left(\frac{l_1}{CT}\right) P_l \cos\alpha. \tag{22c}$$

It is possible that the dimensionless quantity in brackets may tend to a constant value for a given beach. If so, the spacing between rip currents, l_1, would increase with the wave period T as it is found to do in nature. Aside from the observation that l_1 decreases with increasing wave activity, little is known regarding its dependence on the characteristics of the beach.

In the above relations, γ, β and ϕ depend respectively upon the type of breaking wave, the slope of the beach, and the physical properties of the sand. Collectively the factor $8 \tan \beta/\sqrt{(3\gamma)} \tan \phi$ has a unitless value ranging from one-tenth to one for sandy beaches. The term $(l_1 En/T) \sin \alpha \cos^2 \alpha$ and its equivalent $(l_1/CT)P_l \cos \alpha$ are proportional to the total longshore component of power available for transporting sand, and other factors being equal, give a maximum in longshore power when the breaker angle α is about 35°. Larger and smaller angles should result in a decrease in the rate of sediment transport.

K is the ratio of the rate of the work done in transporting the sediment to the total power available, and can be considered as an efficiency coefficient. Approximate evaluation of K, using the laboratory data from Saville (1949, run 11) in equation (22b), gives values of about 3% if it is assumed that l_1 is equal to the whole length of the model beach. An estimation of the value of K for field conditions can be obtained by equating the coefficients of equation (22c) with the numerical constant 2.0×10^{-1} obtained for the empirical transport relation of the previous section. Taking measured field data from Inman and Quinn (1952, fig. 3), where $l_1 = 400$ m, gives K equal to 17%.

The theoretical relations expressed in equations (22a, b) refer to conditions well above the threshold of bed movement and, therefore, to a constant maximum value to which K is supposed to rise. At the threshold of sediment movement K is clearly equal to zero. If the field data leading to the empirical transport rates include threshold and sub-threshold wave conditions, then the calculated values of K will be lower than those for fully developed transport.

E. Application

An example of the possible application of the above model to the development of shoreline forms is given below and illustrated in Fig. 12. Imagine waves approaching a straight shoreline having a single symmetrical bulge or headland. The bulge could be caused by river deposition on an otherwise straight shoreline. In the absence of the bulge the rate of longshore transport of sand would be equal at all sections along the beach for any given wave and angle of approach. Assume the coast to be exposed to waves having a period of 5 sec, a deep-water-wave height $H_d = 61$ cm, and direction of approach $\alpha_d = 45°$. Under these conditions the transport rates will be approximately equal at beach sections A, C and E, and appreciably less at sections B and D.

The longshore transport rate given by equation (22b) can be expressed in terms of deep-water-wave characteristics when the bottom contours are straight and parallel by assuming that Snell's Law for ray optics applies: $C_d/C = \sin \alpha_d/\sin \alpha$; and that energy is conserved between wave orthogonals:

$E_d C_d n_d s_d = ECns$, where the subscript d refers to deep-water-wave conditions and s is the spacing between wave orthogonals (Fig. 11),

$$I_l = \frac{8}{\sqrt{(3\gamma)}} \frac{\tan \beta}{\tan \phi} K l_1 \frac{E_d n_d}{T} \sin \alpha_d \cos \alpha_d \cos \alpha. \tag{22d}$$

This equation implies that for a given deep-water-wave condition, the dynamic transport rate, I_l, should be proportional to $\sin \alpha_d \cos \alpha_d \cos \alpha = \frac{1}{2} \sin 2\alpha_d \cos \alpha$. For these wave conditions the sin–cos function has a value of

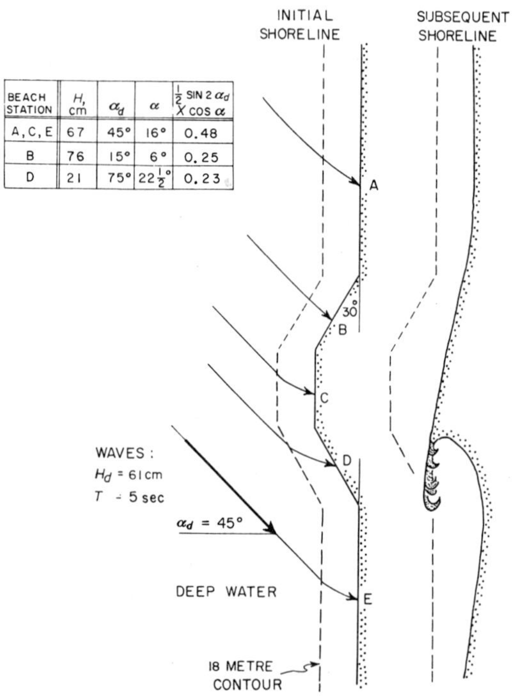

BEACH STATION	H, cm	α_d	α	$\frac{1}{2}$ SIN 2 α_d X COS α
A, C, E	67	45°	16°	0.48
B	76	15°	6°	0.25
D	21	75°	$22\frac{1}{2}$°	0.23

Fig. 12. Hypothetical wave and shoreline conditions, and the postulated subsequent modification of the shoreline. The waves in deep water approach the coast at an angle of 45° and have a height of $H_d = 61$ cm and a period of 5 sec.

0.48 at sections A, C, E and 0.25 and 0.23 respectively at sections B and D.[1] As a working approximation it would have been adequate to assume that $\cos \alpha$ is unity and the transport rate is proportional to $\frac{1}{2} \sin 2\alpha_d$. This function has values of $\frac{1}{2}$, $\frac{1}{4}$ and $\frac{1}{4}$ respectively for the three beach conditions.

The relationship implies that the transport rate is two times greater at beach sections A, C and E than at the inclined sections B and D. If this is the case,

[1] It is assumed that the bottom contours are straight and parallel off the beach sections in question, and that the wave refraction can be obtained from Snell's Law, assuming a wave velocity at the breaking point of $C = \sqrt{(2.28gH)}$, where $H = 0.30 H_d [L_d s_d / H_d s]^{1/3}$ (Munk, 1949a).

deposition will occur near B and cause the form of the bulge to migrate upstream, decreasing the angle of inclination, as shown in Fig. 12 (subsequent shoreline). An excess in deposition over erosion between C and D will favour the formation of a spit. Once formed, the spit will further decrease the transport in the vicinity of D by protecting it from wave action. This in turn will result in a downstream propagation of the disturbance by erosion. Certain waves and angles of approach will favour the formation of a spit, while others will merely destroy it. In general, if the bulge is large, waves approaching from a single direction as postulated in this example would be expected to form a spit. It seems quite probable that this process has been operative in the Gulf of California, in regions where waves are predominantly from one direction and large bulges of loose sand are rapidly deposited by flash floods along otherwise straight coastlines.

12. Discussion

A sound understanding of natural littoral processes awaits further experimentation and development of our knowledge, especially in the following areas: the mechanics of wave motion in and near the surf zone, the budget of water flowing in the nearshore circulation system, and the modes and mechanics of sand transport along the beach. In addition, it is necessary to gain a better understanding of the extent to which models are capable of reproducing nature.

The behaviour of breaking waves is not in agreement with existing wave theory, and it is not clear that waves in small models always follow the behaviour of their natural prototype. The distinctive differences both in the mode of breaking and in the after-effects of the wave once it has broken have not been satisfactorily reproduced in models. The persistence of the bore of white water once the wave has broken, common to natural beaches of fine sand (gentle slopes), appears to be absent from models; this may in part result from the attenuation of smaller waves by surface tension. Also, it appears doubtful that the effect of sand ripples on the rate of transport of sand is equivalent from model to prototype; ripples in the model frequently attain the relative size of what would be termed a bar in nature. Nevertheless, model profiles reproduce the main characteristics of natural coarse-sand beaches to a degree which is surprising, in view of the great discrepancy in the sediment scale ratio in terms of wave height to grain size.

While the two-dimensional model experiments appear in many respects to approach prototype conditions, it is apparent that there is yet a wide discrepancy between what has been made to occur in the three-dimensional model and the more complex littoral processes found along natural beaches. The currents in the model surf zone appear to lack many of the features, such as rip currents, that are common to nearshore circulation systems in nature. Special attention should be given to amplitude modulations of incident wave height in models so that the fluctuation in longshore current velocity characteristic of the prototype is produced. For example, a difference in phase between the maximum in force applied to the beach bed by breaking waves, ECn/u_o, and the maximum in longshore current velocity, $\bar{u}_l$, would result in an appreciable decrease in the longshore transport rate of sand as postulated by the relationship of equation (22). Also, in this regard, one would like to know what determines the spacing between rip currents, and why the longshore current is contained within the surf zone. Here again, the model appears to be in better agreement with the prototype in the case of very coarse sand beaches. Very coarse sand beaches, although less common in nature, appear to have poorly developed nearshore circulation systems when the waves break directly on the beach face. Further and more detailed observations of natural circulation are necessary before a rigorous testing of hypotheses is possible. Establishing a water budget for nearshore circulation along the various types of natural beaches which can then be tested and duplicated in the laboratory appears to be an essential step towards an understanding of the complex mechanism of the longshore transportation of sand.

In the present state of our knowledge, the most promising approach to the prediction of littoral sediment movement would seem to be from a broad consideration of how the incoming wave energy is transformed and dissipated. But progress here is limited by lack of sufficiently comprehensive data. This can best be remedied by laboratory experiment under controlled conditions. The data are required both as clues to, and as checks upon, general influences drawn from the basic principles of nature.

One of the most urgent needs is an experimental estimate of the proportion of wave energy which is dissipated in such ways that it is not available to transport sediment in the process of its dissipation. This unavailable energy falls under two heads: (a) that lost from the nearshore water system by radiation, as reflection, and as microseismic waves and noise; and (b) that converted into heat other than in the neighbourhood of the sediment, as in the decay of turbulence generated from above by the breaking wave rather than from below by boundary drag. Intuitively one might expect (b) to constitute the largest item in the energy budget. Thus a clearer understanding of the process by which a wave breaks would seem to be of prime importance.

Symbols

Symbol	Dimensions	Description
a, a', c	0	Coefficients of proportionality, $a'=\rho_s'/\rho_s$ is the correction for pore-space
b	0	Number of grain layers in motion
C	LT^{-1}	Phase velocity of surface wave
c_d	0	Drag coefficient, $f/\rho u^2$ or $\tau/\rho u^2$
c_r	0	Drag coefficient, $u'/u_* = 1/\sqrt{c_d}$
D, D_1	L	Grain diameter, roughness of bed boundary
d	0	Subscript denoting deep-water-wave properties
E	MT^{-2}	Mean energy per unit surface area of a wave, $\frac{1}{8}\rho g H_d^2$

Symbols—*continued*

Symbol	Dimensions	Description
$\Delta E_1, \Delta E_2, \delta E$	MT^{-2}	Energy loss by bed friction per unit area, $\Delta E_1 - \Delta E_2 = \delta E$
f	$ML^{-1}T^{-2}$	Fluid force on a grain, per unit of effective grain area
g	LT^{-2}	Acceleration of gravity
H	L	Whole height of a surface wave
H_1	L	Energy head of a stream
h	L	Water depth or flow depth
I, I_b, I_s	MLT^{-3}	Dynamic transport rates of sediment past a whole section of flow, i etc. times width of section
I_l	MLT^{-3}	Total longshore dynamic transport rate of sediment, $[(\rho_s - \rho)/\rho_s]gJ$
i	MT^{-3}	Dynamic transport rate of sediment past unit width of flow, total load $= i_b + i_s$
i_b, i_s	MT^{-3}	Bed load and suspended load contributions to i, $[(\rho_s - \rho)/\rho_s]g \cos \beta j_b$ or j_s
J, J_b, J_s	MT^{-1}	Dry mass of sediment passing in unit time across a whole section of flow, j etc. times width of section. J is also longshore mass transport rate of sediment; total dry mass passing beach section in unit time
j	$ML^{-1}T^{-1}$	Sediment transport rate, dry mass of sediment passing unit width in unit time $= m\bar{U} = j_b + j_s = m_b\bar{U}_b + m_s\bar{U}_s$
j_b, j_s	$ML^{-1}T^{-1}$	Bed load and suspended load contributions to j
K	0	Ratio i/ω
l, l_1	L	Distance along the shore; average separation between rip currents
m	ML^{-2}	Mass of total sediment load over unit bed area, $m = m_b + m_s$
m_b, m_s	ML^{-2}	Mass of bed, suspended load over unit bed area
N	0	Grain concentration by volume = grain-occupied space/whole space
n	0	Proportion of energy travelling forward with wave phase velocity C
P	$ML^{-1}T^{-2}$	Dispersive grain stress normal to shear planes (transmitted from grain to grain)
P_l	MLT^{-3}	Longshore component of wave power, per unit length of shore
p	$ML^{-1}T^{-2}$	Fluid pressure
Q_l	L^3T^{-1}	Longshore discharge of water; mean velocity times cross-sectional area
q	L^2T^{-1}	Volume discharge of water in unit time per unit width of flow
r	L	Semi-amplitude of wave displacement close to bed
S	L^3T^{-1}	Longshore bulk-volume transport rate of sand, volume is measured at rest and includes voids $S = J/\rho_s'$
s	L	Distance between wave orthogonals or rays (Fig. 11b)
T	$ML^{-1}T^{-2}$	Grain shear stress, transmitted from grain to grain
T	T	Wave period

Symbols—*continued*

Symbol	Dimensions	Description
$\mathcal{T}$	$ML^{-1}T^{-2}$	Applied tangential stress
$\mathcal{T}_F$	$ML^{-1}T^{-2}$	Applied tangential stress by the fluid flow only
U, U_b, U_s	LT^{-1}	Sediment transport velocity of total load, bed and suspended load
$\bar{U}_G$	LT^{-1}	Mean migration velocity of individual grains
U_0	LT^{-1}	Sediment transport velocity, in to-and-fro motion under waves
u, $\bar{u}$, $\bar{u}_l$	LT^{-1}	Fluid-flow velocity, mean, $\bar{u}_l$ is the longshore current velocity
u_o	LT^{-1}	Wave "orbital velocity" close to bed
u_θ	LT^{-1}	Fluid drift velocity in direction θ
u_*	LT^{-1}	"Drag velocity" $= \sqrt{(\tau/\rho)}$
w	LT^{-1}	Fall velocity of sediment grains
x	L	Distances in direction of flow or travel
z	L	Distances normal to bed surface
α	0	Angle between wave orthogonal (ray) and a normal to the beach; angle between wave crest and bottom contour (Fig. 11b)
β	0	Angle of inclination of bed surface to horizontal
γ	0	Relative wave height, H/h
ϵ_b, ϵ_s	0	Efficiency of bed-load, suspended load transport
η	L	Whole height (crest to trough) of a sand ripple
θ_t	0	The stress ratio, $\tau_t/(\rho_s-\rho)gD$
θ	0	Direction in horizontal plane
κ	0	von Kármán constant
ν	L^2T^{-1}	Kinematic viscosity
ρ	ML^{-3}	Density of fluid
ρ_s, ρ_s'	ML^{-3}	Effective density of sediment grains, dry bulk density of sand
τ	$ML^{-1}T^{-2}$	Fluid shear stress
τ_0	$ML^{-1}T^{-2}$	Fluid shear stress exerted at bed boundary
τ_t	$ML^{-1}T^{-2}$	Threshold value of τ_0
ϕ	0	Angle of solid friction, $\tan \phi = T/P$
ω	MT^{-3}	Available power attributable to unit boundary area. In the case of waves, ω is the decrement dP/dx of the transmitted power attributable to bed drag

References

Bagnold, R. A., 1936. The movement of desert sand. *Proc. Roy. Soc. London*, **A157**, 594–620.

Bagnold, R. A., 1940. Beach formation by waves; some model-experiments in a wave tank. *J. Inst. Civ. Eng.*, Paper 5237, 27–52.

Bagnold, R. A., 1946. Motion of waves in shallow water. Interaction between waves and sand bottoms. *Proc. Roy. Soc. London*, **A187**, 1–18.

Bagnold, R. A., 1947. Sand movement by waves: some small-scale experiments with sand of very low density. *J. Inst. Civ. Eng.*, Paper 5554, 447–469.

Bagnold, R. A., 1954. Experiments on a gravity-free dispersion of large solid spheres in a Newtonian fluid under shear. *Proc. Roy. Soc. London*, **A225**, 49–63.

Bagnold, R. A., 1956. The flow of cohesionless grains in fluids. *Phil. Trans. Roy. Soc. London*, **A249**, 964, 234–297.

Bagnold, R. A., 1962. Auto-suspension of transported sediment; turbidity currents. *Proc. Roy. Soc. London*, **A265**, 315–319.

Caldwell, J. M., 1956. Wave action and sand movement near Anaheim Bay, California. *Beach Erosion Bd., Corps of Engineers, Tech. Mem.* 68, 21 pp.

Caligny, A. de, 1878. Experiences sur les mouvements des molecules liquides des ondes courantes. *C. R. Acad. Sci. Paris*, **87**, 1089–1023.

Corps of Engineers, 1947. Laboratory study of equilibrium beach profile. *Bull. Beach Erosion Bd. (Washington)*, **1**, no. 1, 5–11.

Francis, J. R. D., 1957. *The Engineer*. London, 519 (5 April).

Inman, D. L., 1953. Areal and seasonal variations in beach and nearshore sediments at La Jolla, California. *Beach Erosion Bd., Corps of Engineers, Tech. Mem.* 39, 134 pp.

Inman, D. L., 1957. Wave generated ripples in nearshore sands. *Beach Erosion Bd., Corps of Engineers, Tech. Mem.* 100, 42 pp.

Inman D. L. and T. K. Chamberlain, 1959. Tracing beach sand movement with irradiated quartz. *J. Geophys. Res.* **64**, No. 1.

Inman, D. L. and J. Filloux, 1960. Beach cycles related to tide and local wind wave regime. *J. Geol.*, **8**, 225–231.

Inman, D. L. and W. H. Quinn, 1952. Currents in the surf zone. *Proc. Second Conf. Coastal Eng.*, Council on Wave Research, Univ. Calif., 24–36.

Johnson, J. W., 1953. Sand transport by littoral currents. *Proc. Fifth Hydraulics Conf.*, State Univ. of Iowa Studies in Engin., Bull. 34, 89–109.

King, C. A. M., 1959. *Beaches and Coasts*. Edward Arnold Ltd., London, 403 pp.

Knapp, R. T., 1938. Energy-balance in stream-flows carrying suspended load. *Trans. Amer. Geophys. Un.*, 19th Annual Meeting, Apr., 501–505.

Longuet-Higgins, M. S., 1953. Mass transport in water waves. *Phil. Trans. Roy. Soc. London*, **A245**, 535–581.

Longuet-Higgins, M. S. and R. W. Stewart, 1962. Radiation stress and mass transport in gravity waves, with application to surf beats. *J. Fluid Mech.* **13**, 481–504.

Munk, W. H., 1949. Surf beats. *Trans. Amer. Geophys. Un.*, **30**, 849–854.

Munk, W. H., 1949a. The solitary wave theory and its application to surf problems. *Ann. N.Y. Acad. Sci.*, **51**, 376–424.

Putnam, J. A., W. H. Munk and M. A. Traylor, 1949. The prediction of longshore currents. *Trans. Amer. Geophys. Un.*, **30**, 337–345.

Reynolds, O., 1885. Dilatancy of media composed of rigid particles in contact. *Phil. Mag. London*, **20**, 469.

Russell, R. C. H. and J. D. C. Osorio, 1958. An experimental investigation of drift profiles in a closed channel. *Proc. Sixth Conference Coastal Eng.*, Council on Wave Research, Univ. Calif., 171–183.

Savage, R. P., 1959. Laboratory study of the effect of groins on the rate of littoral transport: equipment development and initial tests. *Beach Erosion Bd., Corps of Engineers, Tech. Mem.* 114, 56 pp.

Saville, T., Jr., 1949. Preliminary report on model studies of sand transport along an infinitely long, straight beach. *Dept. Eng. Univ. Calif., Wave Investigations Tech. Report* HE-116-305, 11 pp.

Scripps, 1947. A statistical study of wave conditions at five open sea localities along the California coast. *Scripps Inst. Oceanog., Univ. Calif. Wave Report* No. 68 (mimeographed).

Shepard, F. P. and D. L. Inman, 1951. Nearshore circulation. *Proc. First Conf. Coastal Eng.*, Council on Wave Research, Univ. Calif., 50–59.

Shields, A., 1936. Anwendung der Ähnlichkeitsmechanik und der Turbulenzforschung auf die Geschiebebewegung. *Mitteil. Preuss. Versuchsanstalt für Wasserbau u. Schiffbau.* Heft 26.

22. SHALLOW-WATER CARBONATE SEDIMENTS

R. N. GINSBURG, R. MICHAEL LLOYD, K. W. STOCKMAN
and J. S. McCALLUM

1. Introduction

The purpose of this chapter is not to review the published literature on shallow-water carbonate sediments nor to consider many of the specific chemical and physical aspects of these sediments. Instead, this review will present the argument that carbonate sediments contain a faithful record of the hydrographic milieu and the activities of organisms at the time of sediment deposition. This record is frequently more complete and more subtle than that presented by sediments composed of terrigenous clastics, for in the latter only the depositional features of the sediments—their grain size, packing and sedimentary structures—reflect the environment of deposition. In the carbonate sediments, the particles themselves are produced in the environment of deposition. As a result, the variety and abundance of the constituent particles can be related to hydrographic conditions.

The material to be presented will be of most interest to geologists. Ecologists may also find useful ideas and data on the nature of the sea-floor sediments as a substratum and as a long-term record of the distribution and productivity of shelled marine organisms. For the oceanographer this chapter will suggest that variation in shallow-water sediments may reflect variation in gross water movements, a record that is difficult to develop from limited instantaneous observations.

The presentation is divided into two parts: the first treats the formation and distribution of particles derived directly from the skeletons of marine organisms (skeletal), and the second describes the nature and distribution of particles whose size, shape and consistency are a result of biochemical and/or physico-chemical processes (non-skeletal).

2. Types of Skeletal Particles

In many places on the shallow-sea floor, carbonate sediments are composed largely or even entirely of the whole and fragmented skeletons of marine plants and animals. Mollusk shells and fragments are the most obvious and familiar skeletal remains in sediments, but a wide variety of marine organisms also contribute—algae, corals, Foraminifera, echinoderms, worms, sponges, alcyonarians and many others. Skeletons of these organisms vary considerably. They differ in size, shape, specific gravity, microstructure, mineralogy and chemical composition. Many of the variations are sufficient to produce differences between the sediments derived from one or another type of skeleton; others can be magnified by selective organic and mechanical breakdown to yield different sediments. These variations in the nature of the skeletons determine to a large degree the grain size, mineralogy and chemical composition of skeletal sediments.

[*MS received December, 1960*]

The importance of the nature of the skeleton in determining the character of the sediments has long been recognized. Sorby (1879, p. 70) was, perhaps, the first to emphasize the need for considerations that transcend phyletic

Fig. 1. Representative sheaths and spicules.

A. Green and red algae that have fragile sheaths of calcium carbonate. Left to right: *Galaxaura* sp., *Udotea* sp., three specimens of *Penicillus* sp. and *Rhipocephalus* sp.

B. Negative print of a thin section of *Penicillus* sp. The large rings are cross-sections of the stem (see Fig. 1A), and the smaller ones are cross-sections of the filaments.

C. Electron micrograph of a shadowed carbon replica of the aragonite needles from the stem of *Penicillus* sp. The crystals are pseudo-hexagonal twins of aragonite. Crystals similar to these are reported from the sediments west of Andros (Lowenstam, 1955).

D. A specimen of the common Florida sea fan (alcyonarian), *Rhipidigorgia flabellum*, and its calcite spicules.

boundaries. His suggestions and the views of Lowenstam (1950, 1954) were the starting point for the following suggested grouping, which is based on the relative resistance to breakdown of the whole skeleton. Only the more common living members of each group are cited below, and no attempt is made to include all the sediment-producing organisms known. For data on the mineralogy and trace-element composition of these skeletons, see Chave (1954, table 1) and Revelle and Fairbridge (1957, table 8).

A. Sheaths and Spicules

This group includes all those species in which the mineral matter is so small or so loosely held by organic tissue that, upon the death of the organism, it disintegrates to fine particles. Most of these particles are of silt and finest sand size, but others are finer; all are subject to further size reduction.

Algae are the most common and abundant elements of this group (Fig. 1A). Many green algae (Codiaceae) have an external sheath of aragonite crystals (Figs. 1A, B). The crystals are so loosely bound together that when the plant dies and the organic matter decays, individual crystals or small aggregates of crystals are released (Fig. 1C). Some red algae (Fig. 1A) release equigranular particles of calcite when the organic matter decays.

Many animals have mineral spicules within their external tissue; sponges, alcyonarians (Fig. 1D), tunicates, and holothurians are perhaps the most common in Recent seas. When the animal dies and the organic tissue decays, these small spicules are freed to be incorporated into the sediment.

B. Segments

This group, first suggested by Sorby (1879, p. 70), includes all those plants and animals whose skeletons consist of segments held together in life by organic tissue. When the organism dies and the organic linkage decays, the segments separate to discrete particles most commonly of sand or gravel size.

Green and red algae are the most common members of this group (Fig. 2A). One green alga, *Halimeda* sp., is frequently the most abundant constituent in shallow-water sediments. Dasycladacean algae (Fig. 2A) are also included in this group. Although they are not particularly abundant in modern sediments, they have been major contributors to carbonate sediments in the geologic past. The articulated red algae (Fig. 2B) also produce segments, generally rod-shaped and somewhat more fragile than the heavily calcified *Halimeda*.

Echinoderms are the most familiar animal members of the segment group. Crinoids, asteroids and ophiuroids all yield sand-sized fragments upon their death (Fig. 2C).

C. Branches

Most plants and animals in this group have cylindrical branches such as those of some species of red algae, corals and bryozoans (Figs. 2D, F, G). Others have

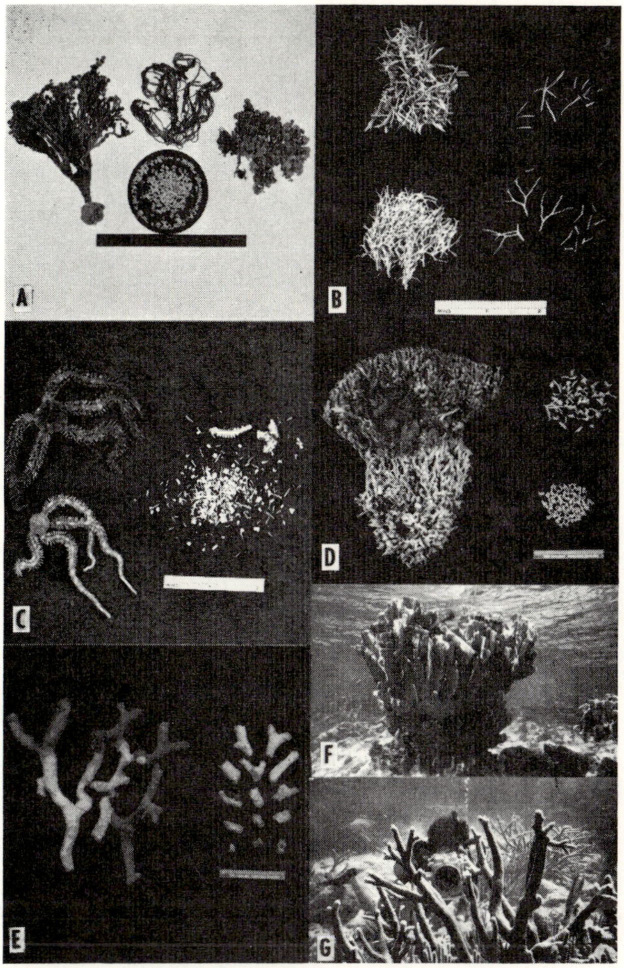

Fig. 2. Segments and branches.

A. Segmented green algae. Left, *Halimeda tridens*; middle, *Cymopolia* sp.; right, *Halimeda opuntia*.

B. Segmented red algae. Top, *Amphiroa* sp.; bottom, *Jania* sp.

C. Two ophiuroids and their segments.

D. The branching red alga *Goniolithon strictum* and its fragments.

E. The branching coral *Porites divaricata* and its fragments.

F. Underwater view of the bladed branches of the hydrocoral *Millepora*. (Photograph by Eugene Shinn.)

G. The branching coral *Acropora cervicornis*. (Photograph by Eugene Shinn.)

blade-shaped branches or projections, such as those of other species of corals, hydrocorallines, red algae and bryozoans (Fig. 2F).

Almost all of these branched forms are sufficiently well calcified that they do not disintegrate when the plant or animal dies. However, by reason of the

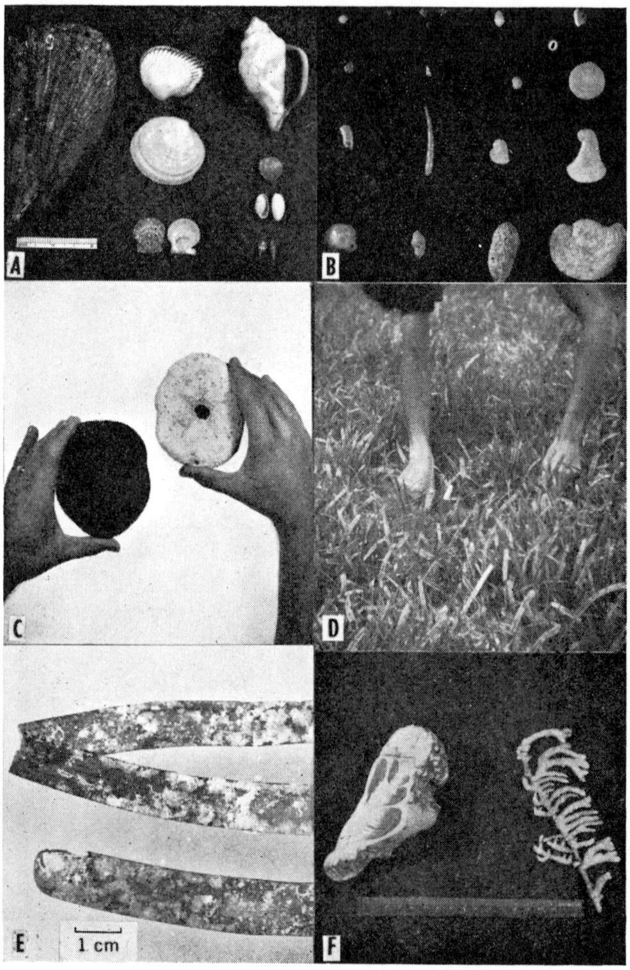

Fig. 3. Chambers and crusts.

A. Assorted molluscan chambers showing some of the variations in size, shape and shell thickness.

B. Assorted chambers of Foraminifera and ostracods showing some of the variations in size, shape and wall thickness.

C. The echinoid *Clypeaster roseaceus*. Left, a living specimen; right, an empty test collected from the bottom.

D. Submarine meadow of turtle grass, *Thalassia testudinum*, off the Florida Keys. Notice the white incrustations on the leaves (see Fig. 3E for detail).

E. Crusts of blades of turtle grass. The lobate areas are a red alga *Melobesia*(?). The branching network is perhaps an irregular foraminifer. The other irregularly shaped granular areas may be agglutinating Foraminifera.

F. Crust. Left, an incrustation of coralline algae on a gastrapod shell; right, the hydrocoral, *Millepora*, incrusting an alcyonarian.

small diameter or thickness of their branches relative to their length, or because of a delicate microstructure, they are quite fragile and are easily broken mechanically or by organisms. It is, therefore, not surprising to find that even where branching forms are the predominant sediment contributor, one rarely finds entire unbroken skeletons in the sediment. Instead, fragments are the rule. The size of the fragments depends primarily on the size and the strength of the branches and secondarily on the nature and intensity of organic and mechanical breakdown to which they are subject (Figs. 2D, E).

D. Chambers

All the skeletons that are hollow, or partly hollow, are included in this group —gastropods, Foraminifera, pelecypod valves, echinoderms, worms, brachiopods and some crustaceans. These chambers persist after the death of the animal, but there are wide variations in the resistance of different types of chambers to further breakdown. The strength of the chamber is determined by its absolute size, wall thickness, shape and microstructure. The persistence of thin-walled foraminiferal shells illustrates how size and specific gravity can prevent mechanical breakdown of the foraminiferal chamber. The effect of shape is clear when we consider the mechanical breakdown of a flat pelecypod valve and a gastropod shell of similar wall thickness. The pelecypod shell can be broken easily, but the curving gastropod shell will not break so readily. The echinoid test (Fig. 3C) is included in this class, but the spines which separate on death are classed as segments.

E. Crusts

This group includes all those plants and animals that encrust surfaces— species of algae, corals, annelid worms, Foraminifera, bryozoans, hydrocorallines and others. Most of these forms are well attached to their substratum and their breakdown is dependent mainly upon the breakdown of the substratum.

Where encrusting forms are on an organic substratum, such as grass (Figs. 3D, E), algae or alcyonarians, they will break down to finer particles when the host dies and decays. More often, however, encrusting forms occur on larger masses of hard mineral matter, usually corals, rocks or other resistant surfaces (Fig. 3F). These encrusting forms are quite persistent and are reduced in size only when the host is broken or disrupted.

F. Massive

Colonial corals, particularly those that form large hemispherical heads, are the predominant members of this group (Fig. 4A). They are quite resistant to breakdown by virtue of their size and the cellular nature of their microstructure. Some coralline algae that start as encrustations develop a massive form (Emery, Tracey and Ladd, 1954, p. 27) and, like the corals, are quite resistant to breakdown (*ibid.*, pls. 14-1, 16-1).

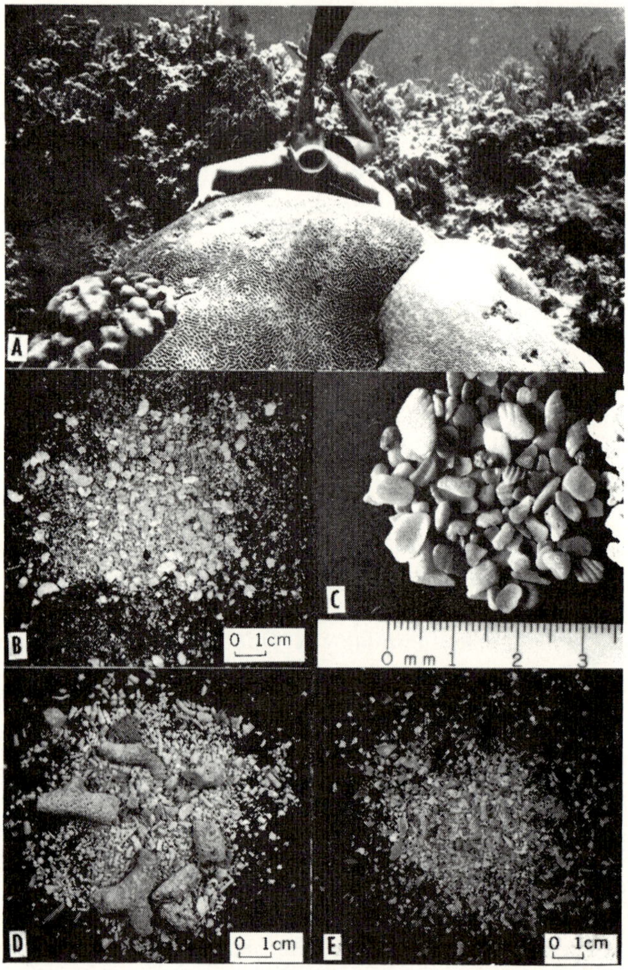

Fig. 4.

A. Underwater view of massive corals. Left, *Montastrea annularis*; center, *Diploria strigosa*(?); right, *D. labyrinthiformis*. (Florida reef tract photographs by Eugene Shinn.)

B. Specimen of *Halimeda* sand from Florida reef tract.

C. Fragments of rounded molluscan shells from north Miami Beach, east coast of Florida.

D. A skeletal sediment that consists of the fragments of branched forms, *Porites divaricata* and *Goniolithon strictum*. Rodriguez Key, Florida.

E. A skeletal beach sand from Nest Key, Florida Bay, that consists of thin-shelled mollusks and Foraminifera.

3. Skeletal Particles in Sediments

The grouping of organisms on the basis of skeletal form, and what might be termed the architecture of skeletons, can help us to evaluate the nature of the environment of deposition of sedimentary particles and to understand their subsequent history. To illustrate the application of this grouping to sediment particles, four sediment samples will be analyzed.

The sediment shown in Fig. 4B consists largely of the segments of *Halimeda opuntia.* From the coarse grain size alone, one might conclude that it represents an accumulation in relatively agitated waters. However, because these large segments are produced merely by the death and decay of the plant, agitation is not required.

The sediment of Fig. 4C consists of rounded fragments of mollusk shells. Because many of the mollusk skeletons (chambers) are quite strong and resistant to breakdown and erosion, considerable energy is required to produce the size and rounding observed here. It is, therefore, not surprising that this particular sample is a beach sand. Notice that the grain size and shape of its particles are almost indistinguishable from the specimen shown in Fig. 4B.

The sediment shown in Fig. 4D consists of fragments of the branching coral *Porites divaricata* and the branching red alga *Goniolithon strictum,* and fragments of *Halimeda.* Here again, the coarse grain size might suggest high agitation, but in fact it is the size and resistance to breakdown of these branched forms which determine the size of particles produced.

The sediment of Fig. 4E consists almost exclusively of whole and fragmented thin-shelled mollusks and Foraminifera. The size and specific gravity of these chamber forms control their deposition. The sample is from a beach on an island in the calm waters of Florida Bay. The particles are those that can be carried ashore in suspension and stranded as the water recedes.

The preceding examples are merely outlines of the way in which the variations in resistance to breakdown of the different types of skeletons can help to explain the origin and development of skeletal sediments. A more detailed and searching analysis should consider the strength of fragments of skeletons and the nature and intensity of the processes of breakdown.

4. Distribution of Skeletal Particles

The distribution of sediment-producing organisms on the sea floor is not haphazard but varies systematically with changes in physical, chemical and biological conditions. The specific factors that control the distributions are not well understood, but it is often possible to relate the variety and abundance of sessile organisms to the circulation of water. Circulation of water, or hydrography, as used here means the interaction of topography with water movements —tides, waves, wind-drift, ocean currents, and run-off from the land—that produces variations in temperature, salinity, nutrients, turbidity, and plankton.

Thorson (1957, p. 461) has emphasized the distinction between the *epifauna,* those animals that sit or crawl on rocks, stones, shells or vegetation, and the *infauna,* those animals that live within the bottom sediments. The relationships between the distribution of the infauna and water circulation may be established in the larval stage. The larvae settle on particular types of sediment because "the types of bottom in the open sea will serve as rough indicators for

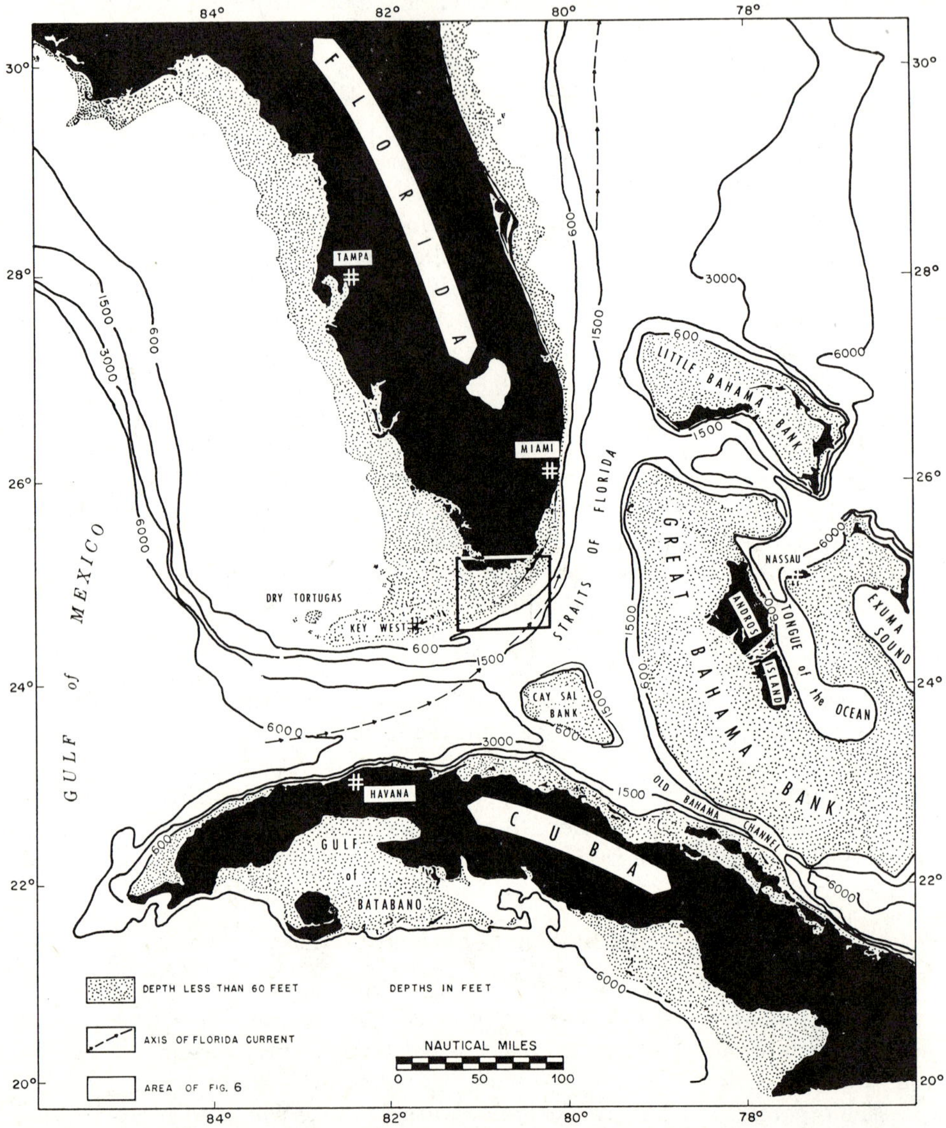

Fig. 5. Regional bathymetry and geography of the Florida–Bahama–Cuba area.

water movements, food conditions, and amplitudes of salinity and temperature in the surrounding milieu" (*op. cit.*, p. 486).

The observed relationships between infauna and water circulation may also be explained by the extinction or reduced growth of adult individuals by unfavorable water circulation acting directly through water properties, such as temperature, salinity, food substances, etc., or indirectly through the presence of predators or unfavorable bottom types (*op. cit.*, p. 471).

The distribution of the epifauna can also be related to water circulation, although to a lesser degree than the infauna. The distribution of the organic substrata for some elements of the epifauna may be controlled by the circulation of water; or some epifauna may require a combination of substrata and a specific water circulation. For further discussion of these problems and references to other works, the reader should consult Thorson (*op. cit.*) and Hedgpeth (1957).

Studies of skeletal sediments in several areas have shown that the variety and abundance of skeletal particles, like that of their parent organisms, can be related to variations in the circulation of water. These relationships help the marine geologist to interpret and predict the lateral and vertical changes in sediment type. In addition, skeletal sediments provide the ecologist with a long-term record of populations and their productivity. It can tell him something about the persistent fauna and flora and provide a base line for evaluating living populations.

The clearest examples of the relationships between variations in the circulation of water and skeletal sediments occur in areas where differences in water circulation are greatest.

A. Florida

The major environments of this region are shown in Figs. 5 and 6, and Table I summarizes their differences. The differences in physical environment produce quite distinct assemblages of sediment-producing organisms. In the

TABLE I

Major Environments of South Florida (from Ginsburg, 1956, p. 2396)

	Florida Bay	Reef tract
Bathymetry		
Depth range	0–10 ft	0–300 ft
Maximum local relief	6 ft	30 ft
Hydrography		
Circulation	Restricted, periodic tides only on margins	Open, semidiurnal tidal exchange with Florida Current
Variations in temperature and salinity		
Temperature	15°–40°C	15°–33°C
Salinity	10–40‰	32–38‰
Available plankton and nutrients	Low	Normal for tropical waters
Turbidity	Generally high	Periodically high only in lagoonal part

large restricted part of Florida Bay, the fauna is dominated by mollusks and Foraminifera (chambers), and the flora by the lightly calcified codiacean algae (sheaths). In contrast, the reef tract has a varied fauna that includes, in addition

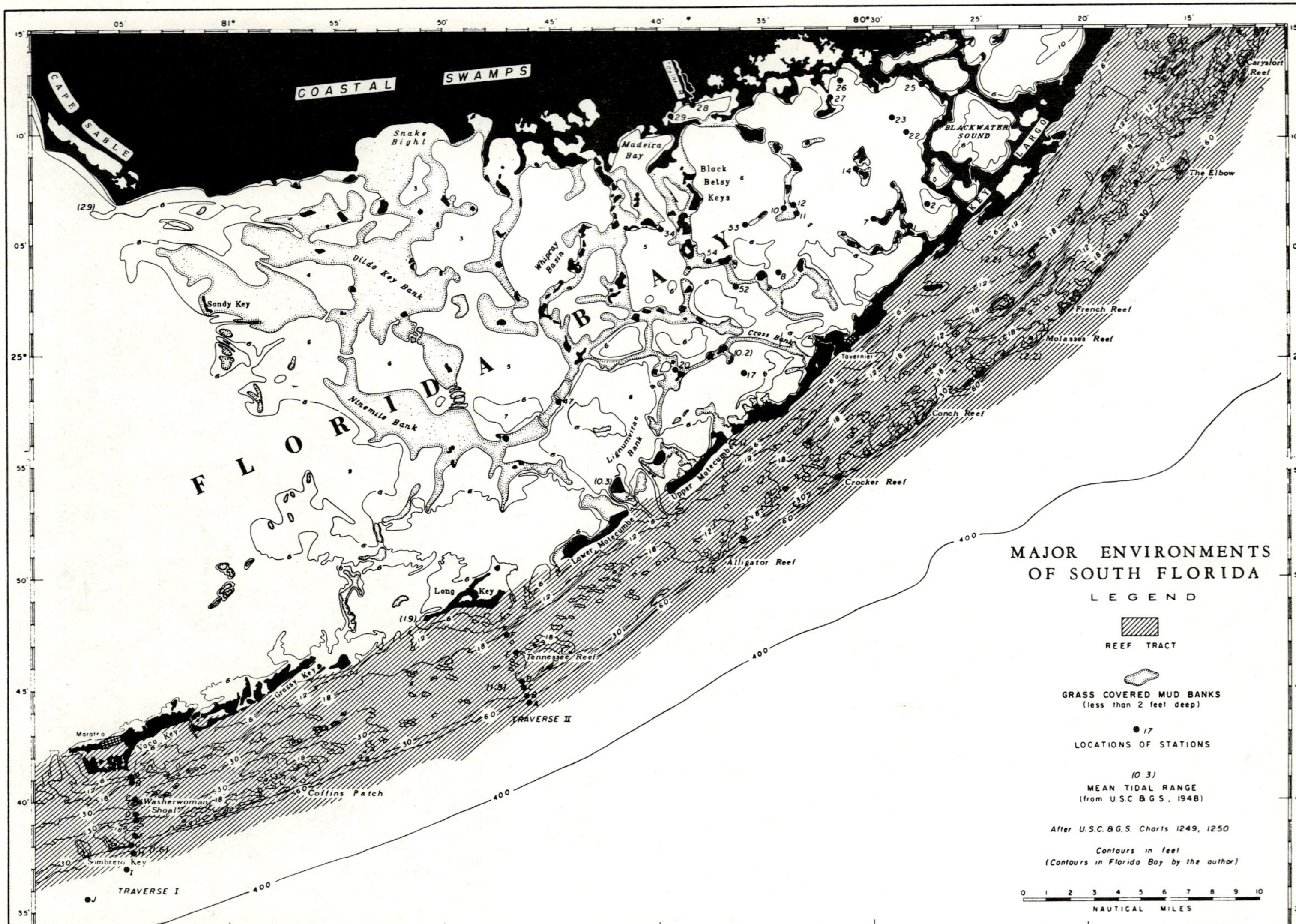
MAJOR ENVIRONMENTS OF SOUTH FLORIDA
LEGEND
REEF TRACT
GRASS COVERED MUD BANKS (less than 2 feet deep)
LOCATIONS OF STATIONS
(0.3) MEAN TIDAL RANGE (from U.S.C.&G.S., 1948)
After U.S.C.&G.S. Charts 1249, 1250
Contours in feet
(Contours in Florida Bay by the author)
NAUTICAL MILES
COASTAL SWAMPS
CAPE SABLE
FLORIDA BAY
Snake Bight
Madeira Bay
Black Betsy Keys
Whipray Basin
Dildo Key Bank
Sandy Key
Ninemile Bank
Lignumvitae Bank
Cross Bank
BLACKWATER SOUND
KEY LARGO
Carysfort Reef
The Elbow
French Reef
Molasses Reef
Conch Reef
Tavernier
Crocker Reef
Upper Matecumbe
Lower Matecumbe
Alligator Reef
Long Key
Tennessee Reef
Grassy Key
Vaca Key
Marathon
Washerwoman Shoal
Coffins Patch
Sombrero Key
TRAVERSE I
TRAVERSE II

to mollusks and Foraminifera, all the other skeletal types—corals, algae, bryozoans, worms, alcyonarians and others. The reef-tract flora includes the heavily calcified codiacean *Halimeda* (segments) and a variety of red algae (crusts, sheaths, segments). These differences in the organic populations are clearly reflected in the constituent composition of the sediments determined by microscopic analysis of the fraction greater than $\frac{1}{8}$ mm, as shown in Table II (Ginsburg, 1956, pp. 2424–2426). In addition to these very gross differences

TABLE II

Grain Size and Constituent Composition of Sediments from Florida Bay and the Reef Tract (from Ginsburg, 1956, p. 2399)

	Florida Bay (17 samples)		Reef tract (25 samples)	
	Average, %	Range, %	Average, %	Range, %
Grain size				
Weight percentage less than $\frac{1}{8}$ mm	49	10–85	17	0–68
Constituent composition of fraction greater than $\frac{1}{8}$ mm				
Algae	$\frac{1}{2}$	0–1	42	7–61
Mollusk	76	58–95	14	4–33
Coral	—	—	12	2–26
Foraminifera	11	1–32	9	3–32
Non-skeletal	3	0–3	12	3–24
Miscellaneous	$\frac{1}{2}$	0–4	9	2–23
Unknown	1	0–3	8	4–15
Ostracods	2	1–6	—	—
Quartz	6	0–20	—	—

on the phyletic level, differences on generic and specific levels occur in some groups (see, for example, Moore, 1957, pp. 730–731).

B. Great Bahama Bank

The eastern side of Andros Island, Bahamas, shown in Fig. 7, has a physical environment similar to that of the Florida reef tract (Newell, Rigby, Whiteman and Bradley, 1951). As in Florida, the fauna and flora are varied and abundant. The western side of Andros Island, on the other hand, is an area of restricted water circulation similar to Florida Bay (Smith, 1940, pp. 158–164), and, like Florida Bay, the fauna is dominated by mollusks and Foraminifera, and the flora by lightly calcified green algae (sheaths) (Newell *et al.*, 1959).

Analysis of the constituent composition of some samples from each of these areas by Thorp (1936, pp. 110–112) showed the same sort of contrast between the two environments as was found in Florida (Fig. 7).

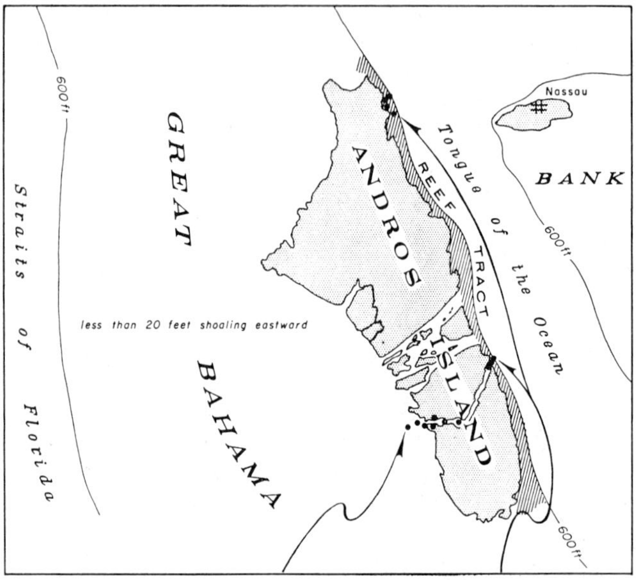

Fig. 7. Grain size and constituent composition of the sediments around Andros Island, Bahamas. (Data from Thorp, 1936.)

	West side—7 samples	Reef tract—11 samples
Weight % < $\frac{1}{8}$ mm	68	4
Constituent composition of fraction > $\frac{1}{8}$ mm		
Algae	5	30
Mollusks	14	15
Coral	0	12
Foraminifera	23	26
Pellets and aggregates	48	9

B. Gulf of Batabano

Daetwyler and Kidwell (1959, pp. 6–11) show a coincidence of high mollusk and Foraminifera content with fine grain size along the shoreline of a large bay on the southwest side of Cuba (Fig. 8). Although information on the circulation of water in the Gulf of Batabano is not available, it seems likely that the interior and shoreline parts of the bay have restricted circulation.

The predominance of mollusks and Foraminifera in the fine sediments of areas where the water circulation is restricted (lagoons) is quite common in areas of carbonate sediments and in areas of terrigenous sediments as well (Ladd, Hedgpeth and Post, 1957; Parker, 1959). The association is not unique to the lagoonal environment but can also be found at depth on open shelves (Shepard and Moore, 1955, p. 1502). However, distinct differences exist between

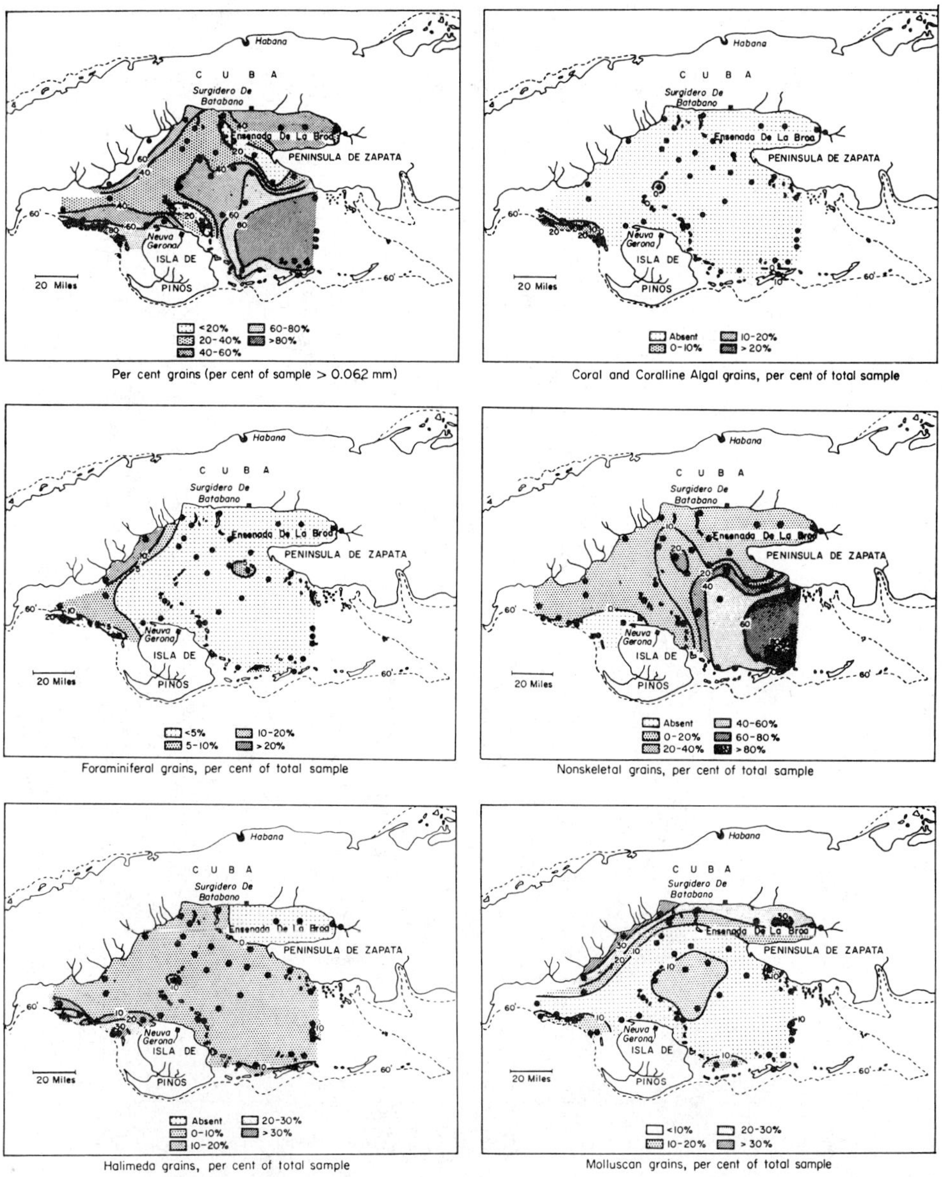

Fig. 8. Grain size and constituent composition in the Gulf of Batabano, Cuba. (After Daetwyler and Kidwell, 1959.)

the genera and species of mollusks and Foraminifera in the two types of occurrences.

The preceding examples are from regions where there are large and obvious differences in physical environment. In areas where the differences in physical conditions are smaller, we can also find differences in the flora and fauna and in the sediments derived from them. In these sediments, however, the differences are more a matter of the relative abundance of different organisms rather than their presence or absence.

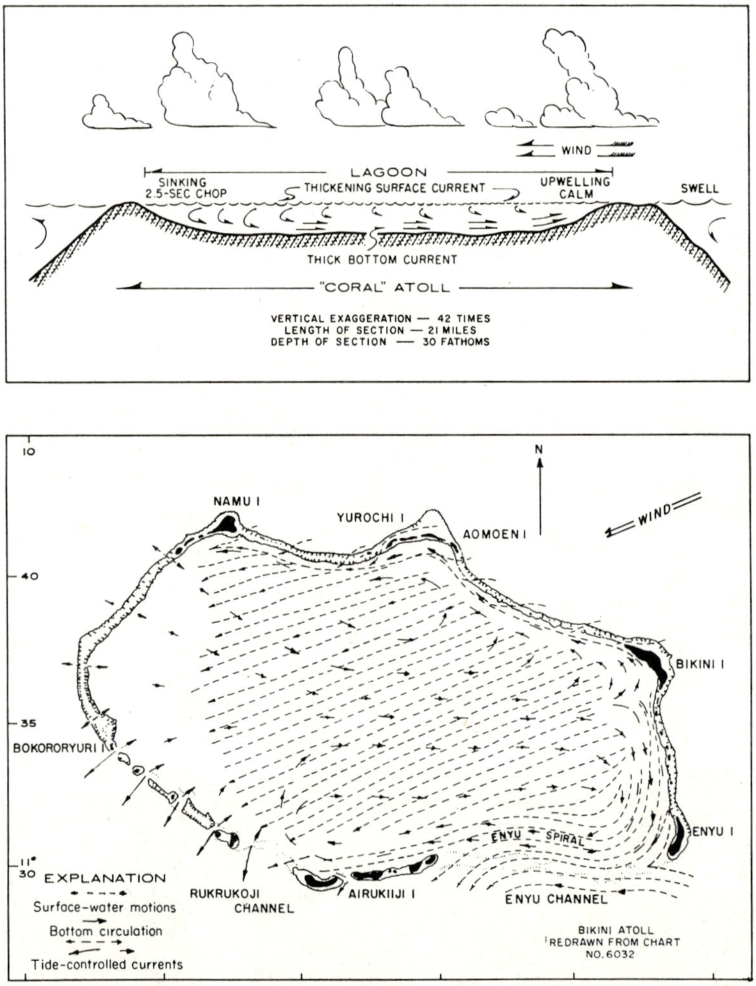

Fig. 9. Schematic water circulation, Bikini Atoll. (After von Arx, 1954.)

C. Bikini Atoll

The contrast between the physical environment of the coral–algal reefs and the lagoon at Bikini is obvious in Fig. 9 (von Arx, 1954). These differences in environment are faithfully recorded by the skeletal sediments (Fig. 10). *Halimeda*-rich lagoonal sediments from other Pacific atolls have been reported by Chapman (1901, p. 163), Cloud (1952, p. 27), and McKee, Chronic and Leopold (1959, pp. 507–509). Emery (1956, p. 1512) called attention to the small

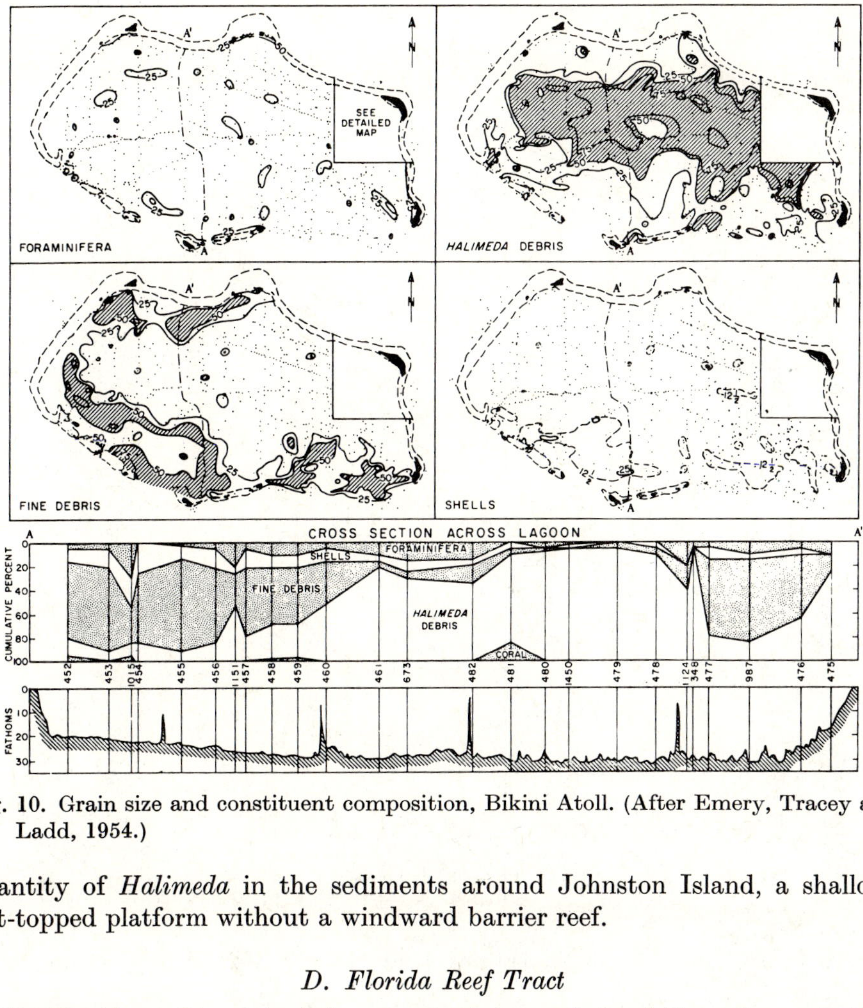

Fig. 10. Grain size and constituent composition, Bikini Atoll. (After Emery, Tracey and Ladd, 1954.)

quantity of *Halimeda* in the sediments around Johnston Island, a shallow, flat-topped platform without a windward barrier reef.

D. Florida Reef Tract

The Florida reef tract includes a protected area behind the outer reefs that is in many respects analogous to the atoll lagoons of the Pacific. It is much shallower, as seen in Figs. 6 and 11, and less removed from connection with the adjacent deeper water, but it does appear to have a similar physical environment. The sediments of the Florida back reef, like those of the atoll lagoons, are dominated by *Halimeda* (Fig. 11). The thin reef sands, on the other hand, are characterized by the presence of fragments of coral and coralline algae, the reef builders, and the fore-reef sediments by coralline algae and Foraminifera. These associations of particles coincide with the distributions of the zones of maximum abundance of their parent organisms. It should be noted that the composition of the reef sands is biased in favor of the fragile sediment-producing organisms because most of the reef builders, massive corals and encrusting algae, remain as cemented reef rock.

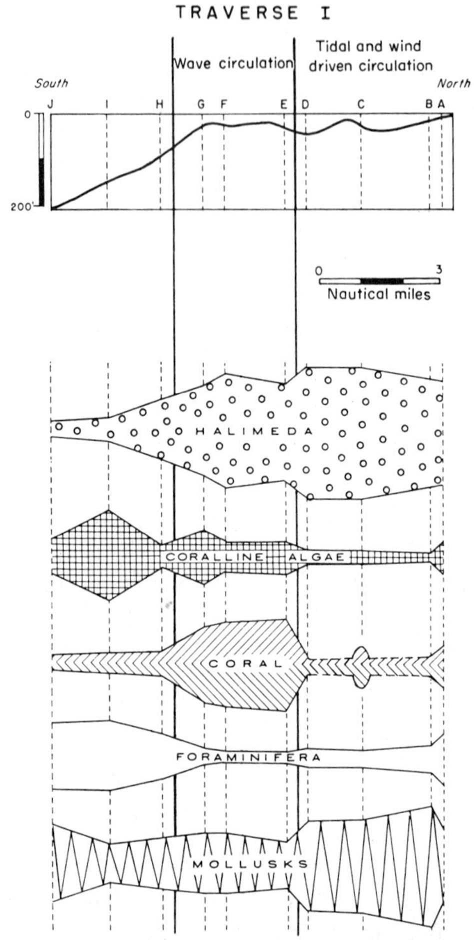

Fig. 11. Constituent composition, Traverse I, Florida reef tract. (After Ginsburg, 1956.) (See Fig. 6 for location.)

E. The Gulf of Batabano Reef Tract

The western part of this area, like the Florida reef tract, has a marginal zone high in coral and coralline algae succeeded landward by a zone high in *Halimeda* (Fig. 8). The rather abrupt decrease in the percentage of *Halimeda* northward probably reflects high temperature and fluctuating salinity or an increase in the amount of silt- and clay-sized particles (Fig. 8). Apparently sampling did not extend far enough south to encounter the coral and coralline algae zone along the shelf margin east of the Isle of Pines.

5. Discussion of the Distribution of Skeletal Particles

The last three examples from Bikini, the Florida reef tract and the Gulf of Batabano show consistent relationships between variation in the circulation of water and the relative abundance of sediment particles derived from coral, coralline algae and *Halimeda*. The seaward marginal zones are characterized by coral and coralline algae, and the lagoonal or back reef areas by *Halimeda*. The zone of *Halimeda* is best explained in terms of relative rates of production, for we know that living *Halimeda* is present in abundance in the marginal zone as well as in the lagoonal areas. In the marginal zones, however, the production of sediment by *Halimeda* is "diluted" by the considerable contribution of coral and coralline algae. Nevertheless, even the relative percentage of *Halimeda* in the marginal zone is at times as high as it is in the lagoonal zone (Fig. 11).

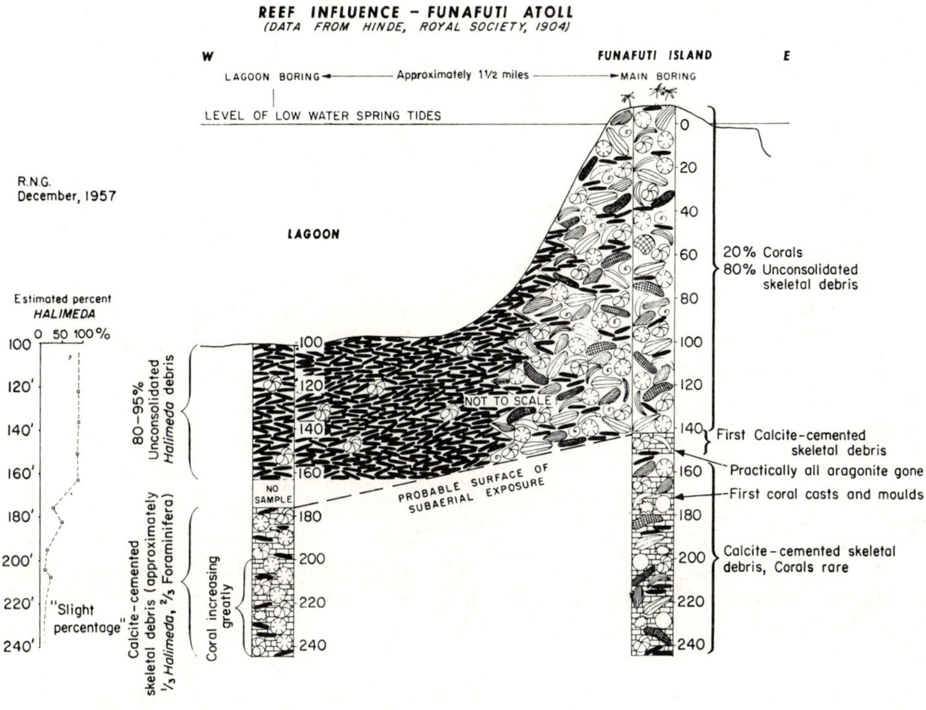

Fig. 12. Cross-section, Funafuti Atoll. (Data from Judd, 1904; and Hinde, 1904.)

The contribution of coral and coralline algae is, therefore, "excess" sediment production over the basic lagoonal percentage of *Halimeda*.

The examples above from Florida, Cuba and the Pacific suggest that *Halimeda* will predominate in the sediments of those areas where the intensity of water movements and circulation is intermediate between wave-agitated shelf margins and restricted bays. Fig. 12 illustrates how this relationship between abundant *Halimeda* and lagoonal or bank water circulation may help us to interpret the recent history of an atoll. In Fig. 12 the rather abrupt increase in the percentage of *Halimeda* in the lagoon sediments at a depth of 160 ft is believed to be a reflection of reef development. Without the reef barrier there could be no lagoon, no lagoonal water circulation and little *Halimeda*. When the reef developed and produced lagoonal circulation, *Halimeda* could flourish and produce the entire 60 ft of sediment.[1]

There appears to be little lateral transportation of sand-sized skeletal particles from one area to another, as indicated by the distinct distribution patterns for the different constituents. The lack of transportation of particles is in sharp contrast with the behavior of the particles in siliceous sediments, where transportation is the rule and particles often travel hundreds of miles along shorelines or down rivers. This difference between carbonate and siliceous sediments is the result of three factors:

(1) There is an absence of through-going water movements that can transport sediment over long distances in areas of carbonate sedimentation. Areas of carbonate sedimentation are characterized by much local relief (reefs, banks and islands) that impede the action of sand-transporting currents.

(2) Stabilization of the bottom sediments by organisms, chiefly plants, appears to be more intensive and extensive in the clear-water areas of carbonate sedimentation (Ginsburg and Lowenstam, 1958, pp. 312–314).

(3) The relative softness of the carbonate minerals will not permit traction transport without rapid size reduction.

There are notable exceptions to the general absence of transportation in carbonate sediments.[2] Houbolt (1957, p. 96) suggested that particles in the range of 125 μ to 250 μ can be transported in suspension for short distances. Two other examples of transportation of larger particles are shown in Fig. 4.

In the sediment shown in Fig. 4C, the resistance of the molluscan microstructure to mechanical erosion permitted the shell fragments to be brought into the beach and to withstand the intense grinding action there. In another sediment (Fig. 4E), the lightness of the skeletons and skeletal fragments allowed them to be carried in suspension to the beach. Both of these examples are from beaches, and similar selective transportation can be expected for dunes. However, beaches and dunes form but a small fraction of carbonate sediments; in the more abundant submarine sediments, we find little or no extensive transportation of sand-sized particles. Instead, these sediments are principally the products of *in situ* production. Silt- and clay-sized carbonate particles can, of course, be transported long distances in suspension.

The local source of skeletal particles makes them useful to the geologist and the ecologist. The geologist can use the relationships between water circulation and constituent-particle composition to organize his sampling, to interpret the lateral and vertical changes in the sediments, to evaluate transportation and to predict the general character of sediments in unexplored areas. For the ecologist, skeletal sediments provide a means of determining whether or not the living organisms on the surface are recent immigrants or long-time residents.

[1] The interpretation of a surface of subaerial exposure shown in Fig. 12 (Ladd *et al.*, 1948, p. 38) is supported by Ladd's recent discovery of a land snail in the main boring at 166–170 ft (Ladd, 1958, p. 187).

[2] See p. 582.

The amounts of different skeletal particles are also a record of relative rates of production, and they can provide a means of evaluating conclusions based on short-term observations of the life cycle and density of organisms on the sediment surface.

The sedimentary record of water circulation and organic production is not without bias. As we have seen, varying resistance of different skeletons to breakdown can distort the record toward the more resistant skeletons. Moreover, the sediment-churning of burrowing organisms can mix the records left during widely different conditions of deposition. Despite these qualifications, it is clear that the skeletal sedimentary record is a valuable source of information and has the potential to provide additional data that will help us to understand organic populations and sedimentation.

6. Non-skeletal Particles

Many of the particles and features of shallow-water carbonate sediments are not skeletal in the sense of the preceding section. Some, such as ooliths, are generally believed to be physicochemical precipitates; others, such as the hardened fecal pellets described by Illing (1954, pp. 24–25) and by Daetwyler and Kidwell (1959, p. 11) from the Gulf of Batabano, owe their shape and perhaps their hardness to organisms. However, because the criteria for the recognition of biochemical and physicochemical precipitation are not well established, the present discussion will treat them together as non-skeletal elements, following Illing (*op. cit.*). The non-skeletal category includes all those particles and features that are not skeletons or parts of skeletons.[1] The term is, unfortunately, ambiguous, since many non-skeletal particles contain large amounts of comminuted skeletal debris; but because these aggregates are hardened by biochemical and/or chemical precipitation, the particle as a whole is non-skeletal.

The two materials that are believed to be the products of non-skeletal precipitation are ooliths and the cement between and within pellets and other aggregates. A third material, the needle-shaped crystals of aragonite in the lime muds of western Great Bahama Bank, has long been considered a non-skeletal precipitate. Lowenstam (1955, p. 270) suggested that these crystals are produced by the decay of lightly calcified green and red algae (sheaths and segments). The following abbreviated discussion will treat the possible modes of origin for all three materials, their occurrence and abundance in the sediments, and their relationships to the circulation of water. More detailed descriptions and discussions of these non-skeletal materials will be found in the papers cited.

A. Ooliths

Ooliths are sand-sized sediment particles that consist of a core or nucleus surrounded by one or more shells of aragonite (Fig. 5A). The nucleus may be a

[1] Some workers would prefer to regard the extra-cellular carbonate of calcareous algae as a sort of non-skeletal precipitation produced as a result of algal photosynthesis. However, because this carbonate has a distinct structure that is recognizable and varies in different species, it is here considered as skeletal.

skeletal fragment, a fecal pellet, or a non-carbonate mineral grain. The shells of aragonite have only concentric structure.[1] From their optical behavior, they can be shown to consist of minute crystals of aragonite whose long axes are tangential to the shells (Sorby, 1879, p. 74; Illing, *op. cit.*, pp. 36–38).

Ooliths are generally believed to be the products of physicochemical precipitation. The main evidence for physicochemical precipitation is indirect but nonetheless rather strong:

(1) Marine ooliths resemble those produced by artificial precipitation of calcium carbonate in boilers (Cayeux, 1935) and water-purification plants.

(2) Areas of oolith formation are characterized by shifting bottom sediments and the lack of intense biological activity (Illing, *op. cit.*, p. 43; Newell, Purdy and Imbrie, 1960).

(3) The oxygen and carbon isotope ratio of some Bahamian ooliths is different from that of plants and animals living in the areas of oolith formation (Lowenstam and Epstein, 1957, pp. 371–372).

The organic origin for ooliths has received little support from most recent authors. Nestéroff (1956, p. 1049) has revived this interpretation on the basis of the rings of organic matter that occur in contemporary ooliths. Newell, Purdy and Imbrie (1960) have shown that some of this organic material is the remains of boring blue-green algae that penetrate almost all calcium carbonate in shallow-water sediments. However, these authors (*ibid.*) also call attention to algal growths on the surface of Bahamian ooliths and suggest that these growths may be responsible for some of the areas of unoriented cryptocrystalline aragonite that occur within the oolitic shells.

a. Occurrence and distribution of recent ooliths

Regardless of whether the precipitation is chemical, biochemical, or a combination of the two, the distribution of ooliths in contemporary marine sediments provides some idea of the water conditions necessary for their formation. Oolitic particles or those that have an external coating of concentric aragonite, regardless of thickness, occur in sediments from the Great Bahama Bank, from Florida, from the Gulf of Batabano in Cuba, and from the eastern Mediterranean off Egypt and Tunisia. The Bahamian occurrence appears to be by far the largest and most thoroughly studied. Illing (*op. cit.*, pp. 41–42) has described the somewhat confusing interpretations of early students of Bahamian ooliths. More recently, Black (1933, p. 462), Illing (*op. cit.*, pp. 35–43) and Newell, Purdy and Imbrie (1960) have all agreed that the oolitic sands of the surface are contemporary and that they are produced by physicochemical precipitation. The contemporary formation of ooliths has been verified by a radio-carbon date of less than 225 years for the exterior 10% of ooliths from the Cat Cay area on the northwestern margin of Great Bahama Bank (Olson and Broecker, 1959).

[1] The ooliths found in contemporary marine sediments do not have the radial structure that is characteristic of fossil ooliths.

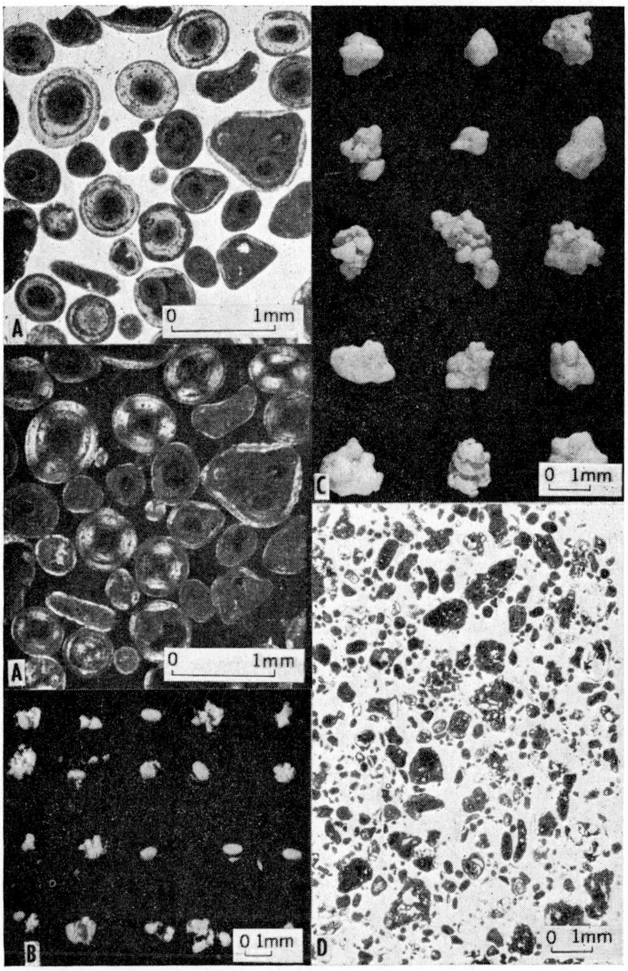

Fig. 13.

A. Ooliths from Great Bahama Bank; thin sections, ×25. Top, plane light; bottom, same area in polarized light. Simple and compound nuclei are shown.

B. Fecal pellets from sediments west of Andros Island, ×25. Each of the pellets was subjected to slight pressure from a needle. The different degrees of crushing reflect the variable hardness of the pellets.

C. Composite grains, "grapestone", from the sediments of the eastern Great Bahama Bank, ×15.

D. Thin section of the sample from which C was made, ×15, plane light.

Illing's pioneer study of the marginal sediments of the southeastern Bahamas suggested, as did Black's (*op. cit.*), that the slight warming of open ocean water as it flows up on the shallow Bahama Banks sets the stage for oolith formation. Illing believes that the increase in temperature produces the supersaturation of carbonate necessary for precipitation, and that the movement of the bottom sediments produces the specific oolith form (*op. cit.*, pp. 44–45). This conclusion is supported by the distribution of ooliths elsewhere in the Bahamas, in Florida and in the Mediterranean (Lucas, 1955, p. 53). In all of these occurrences the oolitic particles are found in shallow water less than 50 ft deep and within range of exchange with deeper-water areas nearby. All these occurrences are in the paths of strong currents, as evidenced by the rippled surface of the bottom sediments. Elevated salinity is not required for oolith formation.

B. Cements

Illing's classic study of some samples from the southeastern Bahamas described two types of cement: (1) cryptocrystalline aragonite that hardens and enlarges biological and physical aggregates of fine-grained material to form sand-sized particles, and then cements these and sand-sized skeletal particles together to form larger composite particles (*op. cit.*, pp. 24–31), and (2) fibrous aragonite that grows within the interstices of sand-sized sediments and within the empty chambers of shells and Foraminifera (*op. cit.*, p. 30).

The internal hardening of aggregates of fine sediment is most clearly demonstrable in fecal pellets. These ellipsoidal aggregates of silt- and clay-sized sediment particles are soft when produced by sediment-ingesting animals. The hardened pellets are similar in size and shape to soft ones (Fig. 13B) and, occasionally, faint sculpturing of the exterior can be observed. It is not possible to determine the precise nature of the aragonite that hardens these pellets because it is fine grained and thus indistinguishable in thin section from the detrital sediment of which the pellets consist. This "cementing" aragonite may be a purely physicochemical phenomenon occasioned by the trapping of water within the minute pores of the fecal aggregates, or, perhaps more likely, it may be of biochemical origin, a result of the local changes in water chemistry within the pellets from bacterial decay of organic matter.

Illing also proposed that the same fine-grained precipitate that hardens fecal pellets forms the cement in aggregates of sand-sized particles for which he suggested the term "grapestone lump", and for which Daetwyler and Kidwell used "composite grains" (*op. cit.*, p. 10). These composite grains consist of sand-sized skeletal and non-skeletal particles that are, according to Illing and Daetwyler and Kidwell, cemented around the points of contact by extremely fine-grained aragonite (Figs. 13C, D).

The fibrous aragonite, described by Illing, that occurs in the interstices of composite grains and in the voids within the shells of mollusks and Foraminifera and the pores of *Halimeda* is indistinguishable from the cement of some beachrock (Ginsburg, 1953, p. 88). Both consist of acicular aragonite that grows perpendicular to the particles within the voids. It is this habit, which resembles the growth of crystals in geodes, that suggests a precipitated origin. Illing believes that the submarine examples of fibrous aragonite were, like the cement of beachrock (Ginsburg, 1954), a chemical precipitate. Kaye (1959, pp. 73–76) has recently questioned this interpretation of the beachrock cement. Kaye (*ibid.*) follows Cloud (1952, p. 29) and others in attributing the growth of fibrous aragonite in beach sands to biochemical precipitation by microorganisms. It is certain that microorganisms are present within beach sands and within composite grains or lumps below sea-level. However, it seems likely that the abundance and rate of metabolism of these microorganisms are much less in these sand-sized sediments than in finer-grained sediments rich in organic matter. It therefore seems most likely that the fibrous aragonite is a physico-chemical precipitate.

We do not know precisely the amount of fibrous aragonite cement in the submarine sediments of the Bahama Banks. However, it seems probable that it occurs wherever there are composite grains (see Illing, *op. cit.*, pp. 78–79; Newell, Purdy and Imbrie, 1960) or numerous chambered forms.

C. Aragonite Needles

The acicular crystals of aragonite in the sediments west of Andros Island in the Bahamas have long been considered a non-skeletal precipitate. This conclusion was based on the apparent similarity of these euhedral crystals and those produced by artificial precipitation and on the unusual hydrographic conditions associated with the Bahamian occurrence. Smith (1940) and Cloud (1955) have shown that both the water conditions and the changes in the concentration of the dissolved calcium carbonate suggest that precipitation does occur in the area west of Andros Island. This conclusion implies that restriction of water movement with attendant increase of temperature and salinity favors precipitation.

Early workers suggested a microbiological origin for these aragonite needles (Drew, 1914; Bavendam, 1932). However, it appears that the conditions necessary for bacterial precipitation of calcium carbonate are only fulfilled in areas of unusually high organic matter (Black, 1933, p. 465). High concentrations of organic matter occur in the mangrove tidal swamps of Andros Island, and it has been supposed that aragonite precipitated there is swept westward into the sea by storms (*op. cit.*, p. 466).

Recently, the non-skeletal origin of aragonite needles has been seriously questioned by Lowenstam (1955) and by Lowenstam and Epstein (1957). Lowenstam (*op. cit.*, pp. 270–271) described acicular crystals of aragonite from several species of green algae that are similar in size and shape to those found in the sediments. He and one of the present authors have observed similar aragonite needles in sediments of the Florida reef tract, an area where there is no major hydrographic restriction that would suggest precipitation. Moreover, these marine occurrences are far removed from tidal swamps, where bacterial precipitation is believed to be most active.

The observed similarity between sediment and algal needles (Lowenstam, *op. cit.*; Fig. 1C) and the presence of these algae in the area indicate that some of the needles must be derived from the algae. Whether or not the algae can produce all the sediment needles depends on the rate of production and the area of supply as compared with the amount and age of the sediment needles.

At present, these data are not available. Should they be assembled, they might provide critical information on the relative contribution of algae.

Lowenstam and Epstein's study of the isotope ratios of sediment and algal needles from several areas on Great Bahama Bank provides additional evidence that the aragonite needles in the sediment are derived from fragile green algae (*op. cit.*). These authors found that the oxygen isotope ratios of the sediment needles fall within the range of values determined for the needles in green algae living in the area where the needle-bearing muds are found (*op. cit.*, p. 372).

7. Distribution of Non-skeletal Particles

All of the available information on the occurrence and distribution of non-skeletal particles has been obtained from studies in the Florida–Bahama–Cuba region. The studies of Illing, Daetwyler and Kidwell, Newell and his associates, Purdy, and the present authors have shown that non-skeletal particles are abundant—often even the dominant constituent of the sediments—and that their areas of maximum abundance show distinct relationships to the water circulation. Moreover, the distribution of the two types of particles appears to bear some relationship to smaller variations in the circulation of water.

Illing's samples from the southeastern Bahamas showed that the sediments of the marginal zone are skeletal and that those of the interior of the Bank are non-skeletal (Table III). More recently, an extensive study of the sediments on

TABLE III

Constituent Composition of the Surface Sediments on Great Bahama Bank (from Illing, 1954, p. 17)

Constituents	On the bank (7 samples)	On the edge of the bank (5 samples)
Non-skeletal		
Lumps	55%	5%
Grains	26%	8%
Aggregates	1%	—
Fecal pellets	6%	½%
Skeletal		
Mollusca	6%	18%
Foraminifera	4%	13%
Calcareous algae	1%	39%
Madreporaria	tr	12%
Other forms	1%	4%
Total skeletal grains	12%	86%

the northwestern Bank by Purdy has shown that non-skeletal sediments predominate on the Bank outside the area of lime mud (Fig. 14). Together, these studies indicate that non-skeletal sands are volumetrically the most abundant sediment on Great Bahama Bank, and that over the large interior they form 70% or more of the total sediment. Daetwyler and Kidwell have

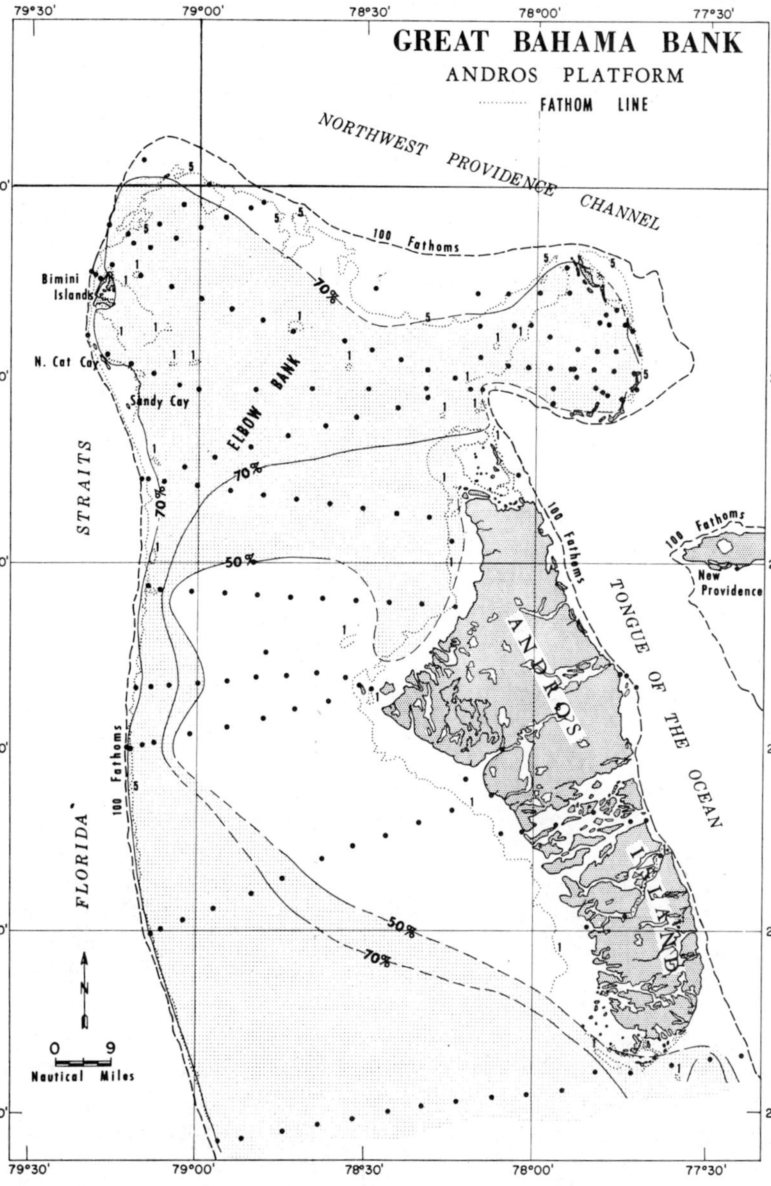

Fig. 14. Non-skeletal grains > $\frac{1}{8}$ mm in percentage of total sediment. (Data supplied by Edward Purdy.)

also mapped the percentage distribution of non-skeletal grains in the sediments of the Gulf of Batabano off the south coast of Cuba. Their results show the edge of what is probably a large area of non-skeletal sands (Fig. 8).

Both of these areas, Great Bahama Bank and the southeastern part of the

Gulf of Batabano, have a rather specific type of water circulation that might be described as "bank circulation". On these shallow banks or shoal areas, generally measured in tens of miles, there is frequent exchange of water with the adjacent oceans by tidal currents and wind drift. Salinities are seldom far above or below that of open ocean water, but temperatures are usually higher than those of surface ocean water.

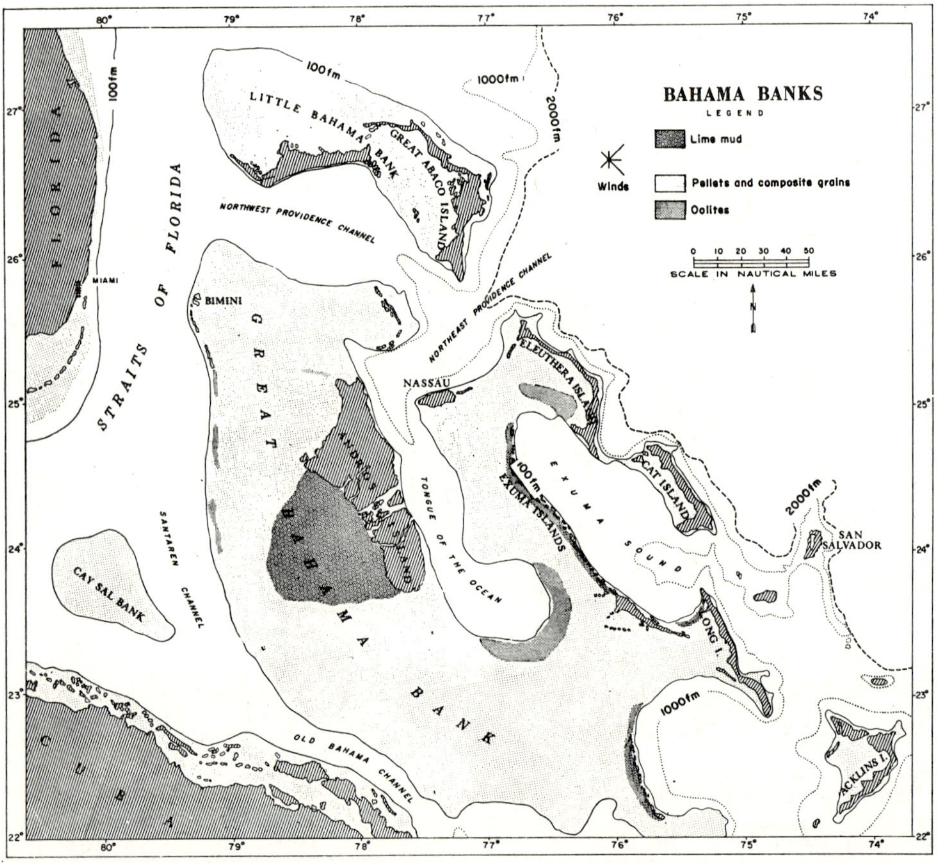

Fig. 15. Distribution of non-skeletal sands on Great Bahama Bank. (After Newell, 1955.)

Differences in the variety and abundance of non-skeletal particle types produce distribution patterns in non-skeletal sediments similar to those found in the skeletal sediments. This strongly suggests that the non-skeletal patterns are also related to variations in water circulation. At present the only example is the coincidence of oolith formation and strong currents (Fig. 15), but further study may disclose other more subtle relationships that will prove to be faithful records of the long-term pattern of water circulation.

References

Arx, W. S. von, 1954. Bikini and nearby atolls, II—Circulation systems of Bikini and Rongelap lagoons. *U.S. Geol. Surv. Prof. Paper*, **260-B**, 265–275.

Bavendam, W., 1932. Die microbiologische Kalkfallung in der tropischen See. *Archiv. Microbiol.*, **3**, 205–276.

Black, M., 1933. The precipitation of calcium carbonate on the Great Bahama Bank. *Geol. Mag.*, **70**, 455–466.

Cayeux, L., 1935. *Les Roches Sedimentaires de France, Roches Carbonatées.* Masson, Paris.

Chapman, F., 1901. Foraminifera from the lagoon at Funafuti. *J. Linn. Soc.*, **28**, **161–211**.

Chave, K. E., 1954. Aspects of the biogeochemistry of magnesium, I—Calcareous marine organisms. *J. Geol.*, **62**, 266–283.

Cloud, P. E., Jr., 1952. Preliminary report on the geology and marine environments of Onotoa Atoll, Gilbert Islands. *Atoll Res Bull.*, No. 12, National Research Council, **1–73**.

Cloud, P. E., Jr., 1955. Bahama Banks west of Andros Island [abstr.]. *Bull. Geol. Soc. Amer.*, **66**, 1542.

Daetwyler, C. C. and A. L. Kidwell, 1959. The Gulf of Batabano, a modern carbonate basin. *Fifth World Congress,* Sect. 1, Paper 1, 1–21.

Drew, G. H., 1914. On the precipitation of calcium carbonate in the sea by marine bacteria, and on the action of denitrifying bacteria in tropical and temperate seas. *Papers Tortugas Lab.*, **5**, Carnegie Inst., Washington, D.C., Pub. 182, 7–45.

Emery, K. O., 1956. Marine geology of Johnston Island and its surrounding shallows, central Pacific Ocean. *Bull. Geol. Soc. Amer.*, **67**, 1505–1520.

Emery, K. O., J. I. Tracey, Jr., and H. S. Ladd, 1954. Geology of Bikini and nearby atolls, I—Geology. *U.S. Geol. Surv. Prof. Paper*, **260-A**.

Ginsburg, R. N., 1953. Beachrock in south Florida. *J. Sediment. Petrol.*, **23**, 85–92.

Ginsburg, R. N., 1954. Early diagenesis of south Florida carbonate sediments [abstr]. *J. Sediment. Petrol.*, **24**, 138.

Ginsburg, R. N., 1956. Environmental relationships of grain size and constituent particles in some south Florida carbonate sediments. *Bull. Amer. Assoc. Petrol. Geol.*, **40**, 2384–2427.

Ginsburg, R. N. and H. A. Lowenstam, 1958. The influence of marine bottom communities on the depositional environment of sediments. *J. Geol.*, **66**, 310–318.

Hedgpeth, J. W., 1957. Concepts of marine ecology. *Geol. Soc. Amer. Mem. 67*, **1**, 29–52.

Hinde, G. J., 1904. Report on the materials from the borings at the Funafuti Atoll, Section XI. *The Atoll of Funafuti*, Roy. Soc. Lond., 186–362.

Houbolt, J. J. H. C., 1957. *Surface sediments of the Persian Gulf near the Qatar Peninsula.* Mouton & Co., Den Haag.

Illing, L. V., 1954. Bahamian calcareous sands. *Bull. Amer. Assoc. Petrol. Geol.*, **38**, 1–95.

Judd, J. W., 1904. General report on the materials sent from Funafuti and the methods of dealing with them, Section X. *The Atoll of Funafuti*, Roy. Soc. Lond., 167–185.

Kaye, C. A., 1959. Shoreline features and quaternary shoreline changes, Puerto Rico. *U.S. Geol. Surv. Prof. Paper*, **317-B**, 49–140.

Ladd, H. S., 1958. Fossil land shells from western Pacific atolls. *J. Paleont.*, **32**, 183–198.

Ladd, H. S., J. W. Hedgpeth and R. Post, 1957. Environments and facies of existing bays on the central Texas coast. *Geol. Soc. Amer. Mem. 67*, **2**, 599–639.

Ladd, H. S., J. I. Tracey, G. Lill, J. W. Wells and W. S. Cole, 1948. Drilling on Bikini Atoll, Marshall Islands, Report of the 18th Sess., *Intern. Geol. Cong.*, Pt. 8, 38–43.

Lowenstam, H. A., 1950. Niagaran reefs of the Great Lakes area. *J. Geol.*, **58**, 430–487.

Lowenstam, H. A., 1954. Systematic, paleoecologic, and evolutionary aspects of skeletal building materials. *Bull. Mus. Comp. Zool.*, **112**, 287–317.

Lowenstam, H. A., 1955. Aragonite needles secreted by algae and some sedimentary implications. *J. Sediment. Petrol.*, **25**, 270–272.

Lowenstam, H. A. and S. Epstein, 1957. On the origin of sedimentary aragonite needles of the Great Bahama Bank. *J. Geol.*, **65**, 364–375.

Lucas, G., 1955. Oolithes marines actuelles et Calcaires oolithiques récents sur le rivage africain de la Mediterranée Orientale (Egypte et Sud Tunisien). *Bull. Stat. Oceanog. Salammbo (Tunisie)*, No. 52, 19–38.

McKee, E. D., J. Chronic and Estella B. Leopold, 1959. Sedimentary belts in lagoon of Kapingamarangi Atoll. *Bull. Amer. Assoc. Petrol. Geol.*, **43**, 501–562.

Moore, W. E., 1957. Ecology of recent Foraminifera in northern Florida Keys. *Bull. Amer. Assoc. Petrol. Geol.*, **41**, 727–741.

Nestéroff, W. D., 1956. De l'origine des oolithes. *C. R. Acad. Sci. Paris.*, **242**, 1047–1049.

Newell, N. D., 1955. Bahamian platforms. In "The Crust of the Earth" (A Symposium). *Geol. Soc. Amer. Spec. Paper* 62, 303–315.

Newell, N. D., J. Imbrie, E. G. Purdy and D. L. Thurber, 1959. Organism communities and bottom facies, Great Bahama Bank. *Bull. Amer. Mus. Nat. Hist.*, **117**, 117–228.

Newell, N. D., E. G. Purdy and J. Imbrie, 1960. Bahamian oolitic sand. *J. Geol.*, **68**, 481–497.

Newell, N. D., J. K. Rigby, A. J. Whiteman and J. S. Bradley, 1951. Shoal water geology and environments, eastern Andros Island, Bahamas. *Bull. Amer. Mus. Nat. Hist.*, **97**, 1–29.

Olson, E. A. and W. S. Broecker, 1959. Lamont natural radiocarbon measurements. *Amer. J. Sci., Radiocarbon Suppl.*, 1.

Parker, R. H., 1959. Macro-invertebrate assemblages of central Texas coastal bays and Laguna Madre. *Bull. Amer. Assoc. Petrol. Geol.*, **43**, 2100–2166.

Revelle, R. and R. Fairbridge, 1957. Carbonates and carbon dioxide. *Geol. Soc. Amer. Mem. 67*, **1**, 239–295.

Shepard, F. and D. G. Moore, 1955. Central Texas coast sedimentation: Characteristics of sedimentary environment, recent history, and diagenesis. *Bull. Amer. Assoc. Petrol. Geol.*, **39**, 1463–1593.

Smith, C. L., 1940. The Great Bahama Bank. *J. Mar. Res.*, **3**, 147–189.

Sorby, H. C., 1879. The structure and origin of limestones, *Q. J. Geol. Soc. London*, **35**, 56–95.

Thorp, E. M., 1936. Calcareous shallow water marine deposits of Florida and the Bahamas. *Papers Tortugas Lab.*, **29**, Carnegie Inst., Washington, D.C., Pub. 452, 37–120.

Thorson, G., 1957. Bottom communities (Sublittoral or Shallow Shelf). *Geol. Soc. Amer. Mem. 67*, **1**, 461–534.

[2] Footnote to p. 572 added in proof. Recent studies of the sediments in the western Straits of Dover add much to our ideas of the extent of transportation in calcareous sediments by strong tidal currents. Boillot (1960, pp. 17–19) describes submarine dunes of skeletal debris (shell fragments, barnacle plates, and bryozoan debris) up to 20 m high off the French coast. This skeletal debris is produced by animals who live on the current-swept, pebbly bottoms surrounding the dunes. The distances that these particles are transported is relatively short judging from the study of a larger area, Boillot (1961, p. 203). Another study of the sediments of the embayment at Mont-Saint-Michel by Bourcart and Boillot (1960, p. 198) presents evidence of the transportation of skeletal debris up to several kilometers by waves and flood tidal currents.

Boillot, G., 1960, La Répartition des fonds sous-marins au large de Roscoff (Finistère). *Cah. Biol. Mar.*, **1**, 3–23.

Boillot, G., 1961, La Répartition des fonds sous-marins dans la Manche occidentale. *Cah. Biol. Mar.*, **2**, 187–208.

Bourcart, J. and G. Boillot, 1960. La Répartition des sediments dans la Baie du Mont-Saint-Michel. *Rev. Geog. Phy. et Geol. Dynam.* **3**, Fasc. 4, 189–199.

23. BASIN SEDIMENTATION AND DIAGENESIS

I. R. Kaplan and S. C. Rittenberg

1. Introduction

Large-scale depressions in the ocean are generally referred to as basins, troughs or trenches depending on their configuration. The basin is usually considered a more or less equidimensional structure, the trough a longitudinal broad depression with gently sloping sides, and the trench a narrow long structure with steep walls. All have bottom depths that are below the general elevation of the surrounding ocean bottom. Many such depressions are tectonically unstable and active subsidence accompanied by continuous deposition can result in the accumulation of great thicknesses of sediments.

Geologists have long recognized that the neritic environment is the most important for the formation of sedimentary deposits. Basins play a special role in this environment since they can act as traps for huge deposits of detrital material derived from terrigenous erosion. Because of their configuration many important biochemical and geochemical processes occur which lead to the release of nutrients, the accumulation of petroleum and the precipitation of insoluble metal compounds.

Since the changes that occur in basin sediments after deposition are duplicated in fjords, estuaries, lagoons and inland seas, it is our intention to draw attention to the similarities existing, and the term basin will generally be used to include all of these depressions.

A. Types of Basins, Water Circulation and Physiography

Marine basins have been only partially delineated. The major ones such as the Mediterranean, Red, Black and Baltic Seas and the Gulfs of Mexico and California are well recognized. Depressions on the continental shelves are, however, poorly charted and the sediments and diagenetic processes occurring in them are largely unknown. Perhaps the most thoroughly studied depressions of this type are the basins found off southern California. Future exploration may reveal other similar basins in continental shelves, particularly in tectonically active areas.

Physiography, water circulation, climate and latitude are major factors influencing the nature of basin sediments and the processes occurring within them; of these physiography is probably the most important. Many authors describing basins (Strøm, 1939; Fleming and Revelle, 1939; Kuenen, 1943; Emery, 1954; Rittenberg, Emery and Orr, 1955) have laid particular stress on the sill depth. This is the lowest point on the rim of the walls surrounding the depression and is the threshold depth for the free entrance and exit to and from the basin. Basins with shallow sills tend to have either stagnant waters or most certainly highly reducing sediments. Those with deep sills have restricted water circulation, but the degree of stagnation depends on other factors.

[*MS received September, 1960*]

Four major types of basins can be defined on the basis of water circulation. In submerged basins of the type found in southern California and the East Indies, the water and circulation characteristics are largely determined by sill depth. In more protected basins, such as Kaoe Bay (Kuenen, 1943), Cariaco Trench (Richards and Vaccaro, 1956), Norwegian (Strøm, 1939) and New Zealand fjords (Skerman, *in litt.*), Baffin Bay (Smith, Soule and Mosby, 1937) or inland seas such as the Black Sea (Arkhangel'sky and Strakhov, 1938) and Lake Maracaibo (Redfield, 1958), climatic conditions play the major role.

An arid climate such as in the Mediterranean and Red Sea areas causes evaporation and thus a sinking of heavy warm surface water and considerable vertical mixing. A humid area like the Norwegian fjord region supplies a large amount of low-density water which tends to overlay the heavier sea-water and prevent vertical mixing across the halocline, thus aiding stagnation. The basins close to the equator may belong to either type since both dry and wet seasons exist. Run-off and evaporation being equal, it appears that basins in higher latitudes are less prone to become stagnant than those in tropical waters (Strøm, 1939) because of overturn due to marked temperature differences in the summer and winter. Where climatic conditions allow formation of sea-ice a fourth basin type exists, for example Baffin Bay. The freezing process produces a cold brine denser than the surface water that sinks and renews the basin water.

Other forces influence water movement. Kuenen (1948) working on the Moluccan deep-sea basins, and Redfield (1958) from studies on Lake Maracaibo find evidence that both centrifugal and Coriolis forces have a dictatorial effect on water movement despite their small magnitude relative to other areas of the ocean. The centrifugal force may cause the accumulation of denser more saline water in the center of a basin with a flat bottom.

Little attention has been paid to the rate of renewal of water in basins. Kuenen (1943) concluded that 200–300 years were required to renew the water in the Moluccan basins, deriving his estimate from the horizontal velocities of the currents. Calculations based on rate of oxidation of organic matter and on heat flow from surface insolation gave a value of 111–160 years for the Cariaco Trench (Richards and Vaccaro, 1956). Studies on regeneration in the sediments and distribution of nutrients and oxygen in the water indicated that 2–20 years were required for water renewal in the basins off southern California (Rittenberg, Emery and Orr, 1955). Emery (1956) has since shown that deep standing internal waves having amplitudes of 130–200 m and causing bottom currents of 18 cm/sec are perhaps the principal cause of vertical turbulent mixing in these basins and are responsible for the rapid replacement of water. It is quite probable, therefore, that shorter times than were previously thought necessary are involved in water renewal.

B. Nutrient Distribution in Basin Water

The distribution of nutrients in basin waters depends mainly on sill depth and circulation type. Where the sill is within the euphotic zone, photosynthesis removes nutrients from the upper layer which are in part released below this zone by decomposition of organic matter. As a result, seasonal differences and gradients exist as in the open ocean. In non-stagnant and semi-stagnant basins having deep sills below the euphotic zone, the nutrients have a uniform concentration corresponding in general to that of the water outside the basin at sill depth. The effects of diagenetic processes in the sediments may be detectable in the bottom water if not masked by the rapid renewal of basin water.

The neritic basins off southern California, typical of the semi-stagnant type, show little change in nutrient distribution from sill depth to just above the sediment surface. The contents of oxygen and nutrient salts correspond to those of the surrounding ocean water at sill depth. Oxygen may be as low as 0.04 ml/l. but does not disappear completely. Hydrogen sulfide is never detected in the water, although the surface sediments of some basins, notably Santa Barbara, have a distinct odor of free sulfide.

Sulfate, phosphate and silicate may show slight to marked increases in concentration in a thin layer of water just above the sediment; nitrate concentration varies in both directions depending on the basin, although total fixed nitrogen probably increases in the bottom water of all the southern Californian basins as it does in the Black Sea (Danil'chenko and Chigirin, 1929). The changes at the bottom are due to biodiagenetic processes at the sediment–water interface and within the sediment column resulting in the release of ions to the water. Changes may also be due to ion-exchange processes and it cannot be taken for granted that an increase of ions in the overlying water invariably occurs. Koczy (1953), for instance, showed that in bottom waters of the Atlantic Ocean there is a marked decrease, not only of oxygen, but also of phosphate and silicate, which he explained by base exchange at the mud–water interface.

Stagnant basins are characterized by the presence of H_2S in the bottom waters. A number of such environments have been described, e.g. the Black Sea (Caspers, 1957), some Norwegian fjords (Strøm, 1939; Fleming and Revelle, 1939), Kaoe Bay (van Riel, 1943; Kuenen, 1943), Cariaco Trench (Richards and Vaccaro, 1956) and Lake Maracaibo (Redfield, 1958). Other areas of the world associated with seasonal upwelling (Callao Bay, Peru, and Walvis Bay, S.W. Africa) will have varying amounts of stagnation depending on the particular circumstances (Brongersma-Sanders, 1948).

It was shown by Russian workers (Danil'chenko and Chigirin, 1929; Chigirin, 1930) from studies on the Black Sea and by Strøm (1939) from Norwegian fjords that the increase of phosphate in waters parallels the increase of hydrogen sulfide. Strøm states that a maximum of 700 mg/m^3 P_2O_5 (5 mg A/l. P) was obtained, while Danil'chenko and Chigirin found a maximum of 632 mg/m^3 P_2O_5 (4.5 mg A/l. P). The same relation has been demonstrated by Richards and Vaccaro (1956) in the Cariaco Trench and by Redfield (1958) in Lake Maracaibo.

The data of Richards and Vaccaro fit the equation (by atoms)

$$PO_4 = 0.00426(AOU+SU)-0.095,$$

where AOU is the apparent oxygen utilization and SU is 4 times the sulfide. In other terms, for every 235 atoms of oxygen equivalents consumed, 1 atom of phosphorus is liberated. This value agrees reasonably well with what one would calculate on the basis of the mineralization of plankton of average composition. The assumption has been made (Redfield, 1958a) that phosphate release occurs as a result of decomposition of photosynthetic organisms in the water column, probably by bacterial action. However, using average values given by Caspers (1957, pp. 821, 823) for oxygen, H_2S and P_2O_5 with depth, it is possible to obtain a range of values for the oxygen equivalents to phosphorus ratio, and a ratio of approximately 900 is obtained from average data for stagnant fjords (Fleming and Revelle, 1939). It is suggested here, and elaborated on more fully in the section on diagenesis, that phosphorus release, in some areas at least, occurs most actively at the mud–water interface.

2. Physical Aspects of Sedimentation

A. Sediment Types

In discussing sediment types we will follow the suggestion of Emery (1952) that the sediments in the neritic environments, particularly the continental shelf, be given a generic classification according to the dominant cause of deposition. The principal sediment types thus classified are listed below with brief comments. Although the relative importance of each is largely a local phenomenon, the importance in present day basin sedimentation generally increases toward the last mentioned.

a. Eolian

These are wind-borne deposits of silt to fine sand that may be of particular importance in regions where sand dunes are well developed along a coast or where deserts reach the ocean. Visual observation of this means of deposition was made by Darwin (1887), and more recently Rex and Goldberg (1958) showed such an origin for certain deep-sea sediment components. Eolian sand grains would probably be well rounded and distinguishable from other sediments; Eolian silts would be difficult to recognize on this basis.

b. Volcanic

There are a limited number of locations where vulcanism occurs directly adjacent to neritic areas. Basins in the vicinity of such areas, namely Iceland, Alaska, New Zealand, Japan and Southeast Asia, could receive considerable volcanic contribution. Volcanic ash was described by Kuenen (1943) and Neeb (1943) from many of the Moluccan basins, and more recently by Worzel (1959) in the east Pacific.

Ash from submarine vulcanism may also contribute to basin sediments. Submarine seepage of hot springs may be responsible for chemical deposition of lime, silica or even manganese (Niino, 1959); these are more properly classified under authigenic sediments.

c. Rafted

These are of two types, those rafted by organic matter and those rafted by ice. Such deposits can generally be recognized by the occurrence of erratically distributed large pebbles or even boulders. Their origin can sometimes be determined—rounding and smoothness in gastrolytes, striations in ice stones. The importance of ice-rafted pebbles may be great in higher latitudes, especially where glaciers reach the sea.

d. Residual

Debris obtained from *in situ* weathering of rocks cropping out on the sea floor. These may be igneous or metamorphic basement rocks or ancient sediments exposed by erosion. Grippenberg (1939) describes outcrops of Silurian and Ordovician limestone in the Gulf of Bothnia, while Arkhangel'sky and Strakhov (1938) describe areas in the deep portions of the Black Sea where considerable areas of dark grey-black calcareous mud with organic contents of over 20% exist. These deposits belong to the Euxinian time, probably the Würm glacial period. In this case detrital deposition has probably been slow so that it has not covered the older deposits, while organic matter settling from the surface may have enriched them and a complex history of subaerial, subaqueous and submarine erosion may be involved. Topographic highs on the bottom of many basins may be due to outcrops of ancient beds covered by recent sediments.

e. Relict

These are ancient deposits formed under different environmental conditions than those now present, but which have remained uncovered by recent sediments. The most important are ancient sand dunes and beach ridges formed during the Pleistocene and now covered by as much as 100 m of water as a result of rise in sea-level. Uncovered glacial tills could conceivably be included in this category since ancient glaciers, extending into relatively isolated basins, may have deposited shallow-water sediments that are now covered by greater depths of water.

f. Authigenic

Sediments that are chemically or biologically deposited *in situ* make up this category. In stagnant basins or those with reducing conditions in the sediments, metal sulfides (e.g. pyrite) are important. In mildly reducing sediments, metal carbonates (e.g. siderite) can form, while under oxidizing conditions metal oxides will form. Carbonates, silicates and probably divalent cation sulfates will form in arid climate basins, principally lagoons, by evaporation. Complex silicates such as tourmaline, chlorite and glauconite along with phosphorite are known to exist in basins, often associated with organic matter.

g. Organic

Sediments of this class can be broken down into two categories. The first, of an organized nature and largely inorganic in composition, is derived from animal and plant exoskeletons such as calcareous foraminiferal tests, siliceous diatom frustules and coprolites from benthonic animals living on the bottom of well aerated basins. The contribution by siliceous tests and faecal pellets to a sediment can generally best be determined visually, since chemical analysis is misleading owing to the masking effect of detrital material. Carbonates can easily be determined chemically.

The second category comprises organic compounds, largely unorganized, of complex structure. A large number of analyses have been made to determine the absolute and relative contents of organic matter in sediments, but little is yet known of the actual constituents.

h. Recent detrital

This class undoubtedly represents the most important type of sediment being deposited in present day basins. Its origin is weathered and eroded terrigenous material with an overall composition of alumino-silicates. It is brought to the ocean by run-off, streams and rivers, and by wave erosion of cliffs. Such sediments are distributed in suspension by currents which allows for the settling out of the heavy and larger particles near shore and a concentration of finer particles in the deep water. As water movement in basins is generally at a minimum, the winnowing out effect of fine-sediment removal does not occur and fine particles of clay and silt can accumulate. Recent detrital sediments, as well as other types, may be redistributed by slumping or sliding down the wall of a steep-sided basin, or by turbidity currents which can disperse sediments over very large distances and introduce coarse-grained materials on an otherwise fine-grained bottom (Gorsline and Emery, 1959).

B. Structure

It is known that many ancient basin sediments have well preserved laminations (Bramlette, 1946; Baldwin, 1959). In present day basins this phenomenon is not as well represented. One of the best examples is that occurring in Malo Jezero (Seibold, 1958), a brackish-water lagoon connected to the Adriatic. Here annual cycles of carbonate and quartz deposits alternate, forming bed thicknesses of 0.25 mm; larger beds also occur at regular intervals. Grippenberg (1939) reports varves in Baltic sediments that were evidently formed in the late glacial period and contain little organic matter. Layering in fjords is not common, perhaps because rainfall and run-off are usually high and the sediments are coarser than normal basin deposits. Strøm (1939) found that of 32 fjords, two had irregular lamination and only one had regular varves. The upper sediment in the Cariaco Trench contains numerous thin, discontinuous horizontal laminae, distinguishable by lighter and darker shades of coloring as well as by slight differences in grain size of the sediments (Athearn, 1959; Heezen *et al.*, 1959). Seasonal variations, which might cause differences in the type and amount of sediment being deposited, may best explain these laminae.

Of the southern California basins, laminations have been observed only in the Santa Barbara Basin (Hülsemann and Emery, 1961), which is also the most reducing. The laminated beds are irregular, the thickest undisturbed layer observed so far being 7.5 cm. The individual light and dark layers have a thickness (wet) of 2 mm with the dark layers being slightly thicker than the light ones (dark 0.74, light 0.52; air dried). The thickest sequences were found exclusively in the green deposits, but did not appear to depend on the organic content.

An irregular stratification is known in many basins (Gorsline and Emery, 1959; Ewing, Ericson and Heezen, 1958). This is reflected in the presence of sand layers of varying thicknesses, compositions and grain sizes occurring between more regular and predominant clay and silt deposits (see Fig. 1). Such

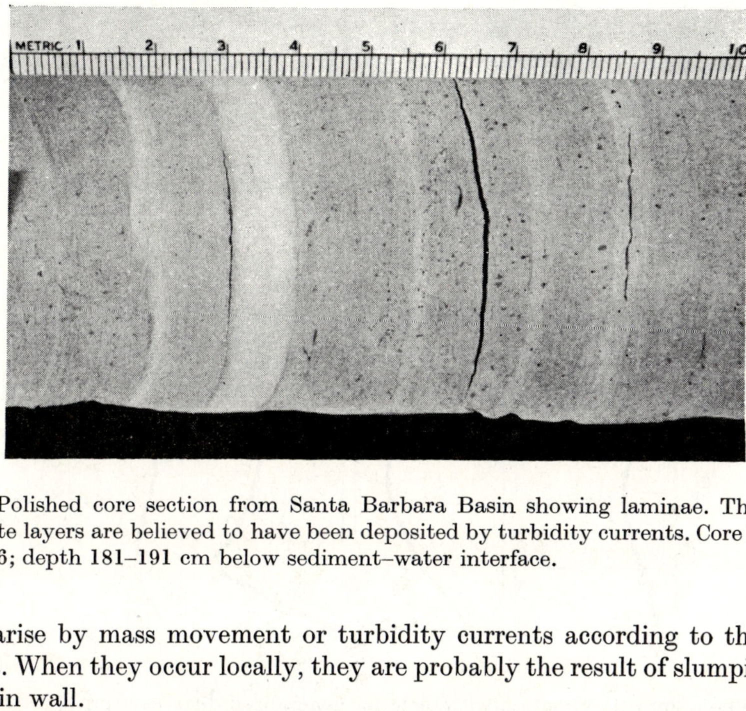

Fig. 1. Polished core section from Santa Barbara Basin showing laminae. The thicker white layers are believed to have been deposited by turbidity currents. Core No. AHF 5986; depth 181–191 cm below sediment–water interface.

layers arise by mass movement or turbidity currents according to the above authors. When they occur locally, they are probably the result of slumping from the basin wall.

Stratification was not observed in the Moluccan basins by Kuenen (1943) (although he did record it in Kaoe Bay). He explained this on the basis of a continual addition of coarser-grained material to the more slowly settling fine-grained detritus. He further considered that turbulence on the bottom was sufficient to stir up the finer material and always keep some of it in suspension. Another very important factor in preventing or disturbing stratification is the activity of benthonic animals (Emery, 1953, 1960). The burrowing activities of worms, brittle stars, holothurians and crinoids, amongst others, are very significant. Some species can function where only traces of oxygen exist, but, since animals are usually absent on stagnant bottoms, these locations are the most likely for the preservation of laminations.

C. Composition and Distribution

a. Grain size

This parameter shows a very clear tendency in most recent basins. In inland seas or land-locked basins the picture may be more complicated because a number of sources for the sediment may exist. The trend, however, is for a continual decrease in grain size with distance from the source. Thus, in land-locked basins such as the Black Sea, Baltic Sea or Lake Maracaibo, the deposits in the center are fine-grained silts and clays, those along the shores are coarse-grained sands or pebbles. This tendency was also observed by Kuenen (1943) in

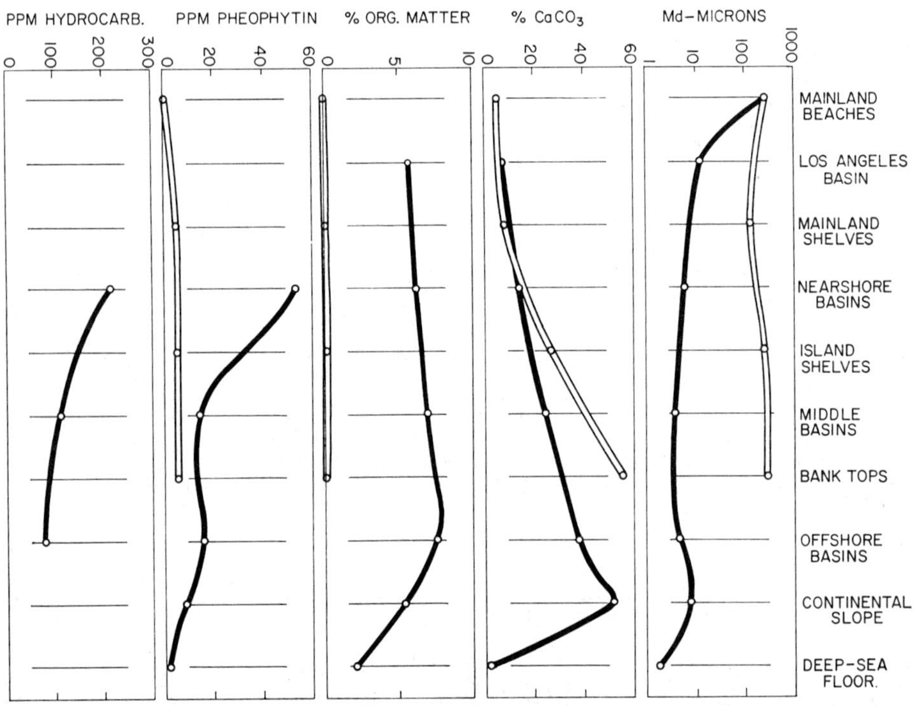

Fig. 2. Curves showing variation with facies or generalized distance from shore of several components of the sediments of southern California basins. (From Emery, 1960, fig. 203.)

the Moluccan basins, but is probably most dramatically illustrated in the series of basins extending seaward on the continental borderland off southern California.

Here it is possible to draw seven roughly parallel strips, which are also parallel to the coast, each strip containing a number of basins (Emery, 1960, p. 220). The Los Angeles Basin, now completely filled with sediment and wholly exposed, can be considered the first basin. Fig. 2 illustrates the decrease in grain size in going from the nearshore basins to the offshore basins. This distribution results because the primary mode of transport is by suspension and coarser detritus settles out faster. The increase in grain size of the sediments along the continental slope is partly due to the effects of currents removing finer material and partly due to the contribution of relatively coarse foraminiferal tests.

b. Calcium carbonate distribution

It is difficult to predict the distribution of carbonates in basins since their origin can be organic, relict, residual and even authigenic. In the Baltic Sea residual carbonates are present and hence their distribution is erratic. In areas with limestone deposits, reworked carbonates will be introduced by run-off, streams or wave abrasion. Thus, the Danube and other rivers carry $CaCO_3$ and $Ca(HCO_3)_2$ into the Black Sea (Caspers, 1957, p. 830) along with terrigenous material. The bulk of the calcium carbonate in Black Sea sediments is thought to be chemically precipitated, although some believe deposition by bacterial activity is also important.

Kuenen's (1943) data for the Moluccan basins show first a decrease of carbonate content with distance from shore and then a sharp increase at about 100 km. In the southern California basins there is a continuous increase with distance from shore (Fig. 2), the maximum being reached at the continental shelf break. The explanation is based on relative supply. Near shore, productivity may be lowered because of turbidity in the water preventing photosynthesis, but calcium carbonate is introduced with reworked material. The supply of vast amounts of terrigenous matter generally dilutes the carbonate when relative analyses are made. In the offshore basins, especially in areas of high productivity, calcium carbonate deposition is high and masking by detrital material will be minimal.

c. Organic matter

The organic matter is generally closely related to the grain size and often to the calcium carbonate content. In land-locked basins this relationship is due almost entirely to relative rates of settling and supply of detrital material, which in turn is related to the sources of supply and to the current circulation patterns. The organic content is usually greatest in the part of the basin where settling is the slowest and not necessarily in its deepest part (e.g. Lake Maracaibo).

In continental-shelf basins the pattern is different. Here there is first an increase in organic matter in going from the nearshore basins to the outlying basins and then a decrease in the basins furthest from shore. This pattern, shown in Fig. 2, can be explained on the basis of relative rates of supply of organic matter and terrigenous material. Away from the coast, detrital material decreases and hence the organic component becomes relatively more important. Still further offshore the depth of the basin bottom below the euphotic zone becomes so great that the settling organic debris is probably largely decomposed by aerobic bacterial action before it reaches the bottom.

Reducing conditions on the bottom preserve the organic matter since decomposition by anaerobic bacteria is generally not as efficient or as complete as decomposition by aerobes. Since clays have fewer pore spaces, fine-grained sediments will in general promote reducing conditions by preventing water circulation and so help preserve organic matter. The binding of organic material by clays may decrease its availability to bacteria.

The above discussion is concerned with the areal distribution of the components relative to the total dry weight of the sediment. It is possible to consider absolute distribution if the stratigraphy is well known. Thus, for the Black Sea, Strakhov (see Smirnow, 1958) found that by considering a unit area and a column of sediment equivalent to a constant time lapse (2500 years), the total amounts of insoluble residues, calcium carbonate and organic matter were greatest inshore and decreased toward the center.

The highest organic content reported for recent sediments is for a fjord, 23.4% (Strøm, 1939). In the recent Black Sea sediments the highest content of organic matter is about 7.7% (Caspers, 1957, p. 823), which is considerably lower than the 25% reported for the residual Euxinian deposits. In the Baltic Sea and Lake Maracaibo the maximum is about 8.5%. The highest content measured in the southern California basins is about 11.5%. These values are much greater than the 1–3% of the deep-sea sediments. All data available show a sharp decrease in the organic content with depth below the top 5–10 cm, indicating that diagenetic processes occur there most rapidly.

d. Composition of detrital material

The insoluble residue of the detrital material in basins is composed largely of silt (16μ–4μ) and clay (< 4μ). It has been found in southern California basins that the mineral composition of the silt is similar to that of the sand on the beaches and on the continental shelf. Here the dominant minerals are quartz and feldspar (ratio 2.2:1), the plagioclase to orthoclase ratio in the feldspar being 4.5:1 (Emery, 1960).

The clay minerals show an enrichment of illite and montmorillonite over kaolinite. The greater the distance from the shore the higher the illite–kaolinite ratio in the southern California basins (Emery, 1960). Montmorillonite appears to be an intermediate in the diagenetic alteration of land-derived kaolinite to illite.

e. Color

The color of basin sediments varies considerably depending on their state of oxidation and reduction. Highly oxidized sediments are generally brown; highly reduced sediments vary from gray to green to black. Strøm (1939) emphasized that muds covered by H_2S-containing waters are invariably black, but this is not always the case. Green muds occur in Walvis Bay (Copenhagen, 1934), Kaoe Bay (Kuenen, 1943) and in the Cariaco Trench (Athearn, 1959; Heezen *et al.*, 1959)—all stagnant environments. Lake Maracaibo, on the other hand, has black mud in the stagnant zone (Redfield, 1958) while the Black Sea appears to have predominantly gray sediments. The fine sediments in the southern California basins are generally an olive green color, although the Santa Monica basin, where sedimentation is most rapid, has dark gray-green clays.

No satisfactory explanation has yet been given for the color differences, and it is the author's opinion that many factors are involved. Those of significance may be the kinds and proportion of clay minerals, such as illite and chlorite, and the presence of highly pigmented organic matter, ferrous salts, large quantities of $CaCO_3$, hydrotroilite and pyrite. Of great importance may be the total amount of iron available, either in solution or particulate. A high iron content will bind the H_2S very rapidly to form black hydrotroilite. It would be expected that high iron would be related to areas with great run-off and perhaps a high content of volcanic rocks. We should, therefore, expect the color to be darker closer to the source. Estuaries, lagoons and marshes fed by streams usually contain black sediments.

f. Rate of deposition

The rate of sedimentation in basins is dependent on distance from sources, type and amount of material delivered into the basin and on the agents of transport. Many of the older data are not reliable because of uncertainties in the age determinations for the sediments. Other values were based on estimates of annual supply by rivers, run-off and wave erosions. Many of the earlier results were given in thickness accumulated per year; this has the added problem of neglecting compaction.

The most reliable work on relative rates so far published was carried out by Emery (1960) on the southern California basins. The results, Fig. 3, show a decrease in rate with distance from shore of all sediment components studied with the exception of calcium carbonate. These rates are given in $mg/cm^2/year$ for a column of sediment neglecting the turbidity-current deposits, and are based on ^{14}C-age-dating techniques.

3. Diagenesis in Sediments

A. Types of Organic Matter

The extensive work of Trask (1932, 1939) has contributed much to our knowledge of the organic content of various marine sediments. The composition of organic matter is different in different environments and also changes with depth of burial. Because of the complexities involved, no agreement has yet been reached as to what the average carbon content of the organic matter in sediment is. Emery and Rittenberg (1952) suggested that for southern California basins an average organic content of 1.7 times the organic carbon best fitted the observed data.

The carbon to nitrogen ratio also varies with sediment type, with depth of burial and with the analytical procedure. Trask obtained an average C/N of 8.0 while Emery and Rittenberg found a C/N of 13.0. The uncertainty of the measurements prompted Emery and Rittenberg to suggest the use of 10.0 as an average. Since nitrogen analysis can be simply and quickly carried out by the Kjeldahl method, this procedure is frequently used for estimation of total organic matter and the organic content is then taken as 17 times the nitrogen content of the sediment. Care must be taken in the interpretation of nitrogen values because the C/N ratio usually increases with depth of burial (nitrogen is preferentially lost by the sediment).

The studies on specific groups of organic substances in marine sediments are

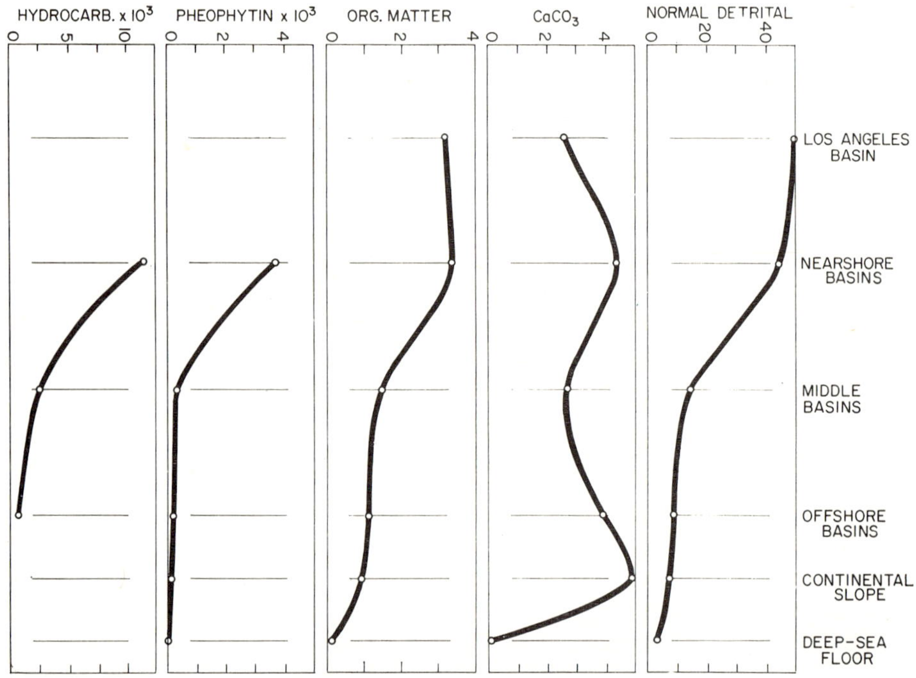

Fig. 3. Curves showing rates of accumulation of various components of southern California basin sediments. Note the similarities and differences between curves for absolute rates of deposition and for the percentage compositions given in Fig. 2. Values for organic matter are at the surface of the sediment. (From Emery, 1960, fig. 208.)

largely fragmentary although many of the major constituents present in the plankton of the overlying waters have been detected in recent marine deposits. The review of Vallentyne (1957) shows that more work has been done on the muds of lakes than on oceanic sediments and demonstrates the inadequacy of our state of knowledge. Only in the last two decades have adequate techniques become available for separating and identifying individual compounds rather than groups or families of organic substances.

Two objectives have recently led to the intensive study of organic substances. The first is to obtain information of use to paleontology and perhaps to give credence to the speculations on the origin of life. The second is to investigate diagenetic processes with the ultimate goal of gaining an insight into the enigma of petroleum genesis. As a result, information is now available on some specific groups of organic compounds which may help solve the outstanding problems mentioned above; four of these groups are discussed below.

a. Amino acids

The finding of proteins and amino acids in the shells of living animals by Abelson (1954) aroused interest in their application to paleobiochemistry. Definite changes in the stability of the proteins and polypeptides occur with age of burial. An aqueous environment and heat was found to be most detrimental to their preservation. Abelson (1959) found that shells of the clam *Mercenaria mercenaria*, an existing species but also present in the Miocene, showed a loss of protein with burial (a complete loss in the Miocene specimen), but a gain in free amino acids. No evidence is available from this study as to the stability of the individual amino acids with age.

Working on shelf sediments from the Gulf of Mexico, Erdman, Marlett and Hanson (1956) were able to show the presence of nine common amino acids, while Plunkett (1957) found eight in the core from the Cariaco Trench and somewhat fewer from the sediment outside the basin. Studies on the sediments from Santa Barbara Basin (Saltman, unpublished data) have yielded eleven amino acids. In all, the following have been isolated in marine sediments by the above mentioned workers; alanine, arginine, aspartic acid, glutamic acid, glycine, isoleucine, leucine, lysine, phenylalanine, proline, serine, threonine, tyrosine and valine. Since the adequacy of the techniques of separation and identification is questionable, interpretations based on the presence or absence of particular amino acids should be reserved for a later date.

b. Hydrocarbons

Although reports of "paraffins" in marine sediments had been made by Trask and Wu (1930), their actual isolation and identification was not effected until relatively recently by Smith (1952, 1954). Apart from steroids and carotenes, three groups of hydrocarbons are recognized: paraffin-naphthenes, aromatics and asphalts. They are extracted from the dried sediment by suitable solvent mixtures and separated on alumina columns. Identification is usually made by infrared spectroscopy. The molecular weights of the compounds are on the whole high. Smith (1952, 1954) found that average weight fell in the 250 to 300 range. From a large number of cores, Orr and Emery (1956) isolated only one compound that was liquid at room temperature. They found that the paraffin-naphthenes were colorless, the aromatics yellow and the asphalts black.

Smith (1954) found that the hydrocarbon content varied from 9 to nearly 12,000 parts per million of dried sample from a number of localities. Emery (1960) showed that basin sediments can have nearly 1600 ppm hydrocarbons, with asphalts the greatest fraction. Bonnett (1959) also found a higher content of asphaltic material than of the other hydrocarbons in some argillaceous sediments from Britain, but the difference was not as great as in the southern California basin sediments. Petroleum *per se* has not been recognized in recent marine sediments. Perhaps one reason for failure to isolate light-weight hydrocarbons is because of inadequate techniques. Removal of the low-boiling-point solvents used for extractions may also remove low-molecular-weight products, including gaseous hydrocarbons, which are known to be present (Emery and Hoggan, 1958).

The work with southern California basin sediments (Orr and Emery, 1956) reveals that the highest concentrations of hydrocarbons occur in the most reducing sediments. This suggests that aerobic degradation may be the only or certainly the major form of removal of these substances from the sediment. Except for the upper 5 to 10 cm, there appears to be little correlation in concentration with the depth of burial to about 500 cm; there is some indication, however, that the paraffin-naphthene fraction does decrease.

c. Lipids

These compounds are vital and major constituents of all living matter. Under certain circumstances they become the dominant substance in the organism, particularly in algae (Spoehr *et al.*, 1949). The most important type of lipid is the neutral fat or ester of fatty acids and glycerol. Such fats have apparently not been isolated in sediments, although Shabarova (1954) isolated salts of oleic acid from a Black Sea ooze, and Plunkett (1957) identified seven short chain mono-, di- and tricarboxylic acids from the Cariaco Trench. These fatty acids have been considered as source material for petroleum (Brooks, 1952). Little information is available on the lipid content of marine sediments, although complex lipids, including waxes, have been isolated from peat (Abelson, 1959). A wide range of fatty acids is present in petroleum.

d. Porphyrins

In recent years the porphyrins have assumed special importance in supporting the theories postulating the biological origin of petroleum. The vast majority, if not all, organisms contain porphyrins as components of blood, respiratory enzymes or chlorophylls. In the marine environment, it is generally accepted that the chlorophylls contributed by phytoplankton are the chief source of the porphyrins in the sediment. Chlorophyll, a tetrapyrrole with a central magnesium atom, has never been isolated in the sediments of marine basins (Orr, Emery and Grady, 1958) or lakes (Vallentyne, 1957). Instead, its first breakdown products, pheophytin (which is chlorophyll minus the magnesium), chlorophyllin (loss of phytol group) and pheophorbide (loss of magnesium as well as phytol group) are present. The diagenetic process leading to the chlorophyll transformation is not known, nor is it known whether this occurs in the water column soon after the death of the organism and before settling or after deposition has occurred. The conversion of chlorophyll-type porphyrin to the nickel or vanadyl type found in petroleum (Treibs, 1936) is still a mystery.

However, some interesting results were recently presented by Hodgson (1959) who was able to carry out the conversion in the laboratory by varying the temperature, pH and trace-metal content, using a lake gyttja.

In the southern California basins, the porphyrins and hydrocarbons have a similar distribution (Emery, 1960); their content is highest in the basins closest to shore, they are preserved best in a reducing environment and they diminish with depth most rapidly in the first few centimeters.

B. Bacteriological Processes

a. Aerobic and anaerobic metabolism

Bacteria are without doubt the most efficient scavengers on the ocean floor. In aerobic environments they compete for organic matter with fungi and protozoa, as well as metazoa, but under stagnant conditions they carry out their processes without challenge. It is of little value to enumerate all the known organisms, at the best such an effort would be totally inadequate. Reviews on the subject have been written by Waksman (1934), ZoBell (1946), Brisou (1955) and Wood (1958). As with many other aspects of oceanography, the studies have so far been largely empirical, and the field is open for the perusal of fundamental problems.

It is generally accepted that the phytoplankton and the zooplankton constitute the major source of organic matter to oceanic sediments. Close to shore, land plants and sessile algae may also be washed into the sediments. This matter represents organic molecules ranging from very simple structures to those which are stable and highly complex.

Lithified sediments, like shales, generally do not contain unaltered organic matter. The changes that have occurred can in part be due to the lithification process and in part to diagenetic changes which took place prior to lithification. Since the organic matter even in very recent unconsolidated sediments has largely lost its identity, it can be supposed that the diagenetic processes, having bacteria among their main agents, are the most significant.

Two groups of bacterial processes are involved in degradation, the aerobic in the presence of molecular oxygen and the anaerobic in its absence. During aerobic metabolism reduced substances of diverse structure are converted by stepwise processes to a few common intermediates which enter into terminal oxidation pathways, the tricarboxylic acid cycle being the most important, with their ultimate conversion to carbon dioxide. Since the energy release is greatest during the aerobic processes, the destruction of organic matter under these conditions is more complete and generally more rapid. Anaerobic metabolism must result in the formation of end products more reduced as well as more oxidized than the starting substrates, and the typical fermentation process causes the accumulation of simple organic molecules such as fatty acids. Thus the breakdown of organic matter is not usually as complete anaerobically, although in the special situations of sulfate reduction and denitrification, where the reduced product is inorganic, this need not be so.

Further, a smaller number of molecules are susceptible to attack under anaerobic conditions.

Complex carbohydrates, lipids and proteins, which may be either totally insoluble or only minutely so, are degraded. For example, agar, cellulose, long-chain fatty acids and gelatin are readily solubilized under both aerobic and anaerobic conditions. More stable substances, such as keratin and lignin, appear to require aerobic conditions and only select flora will attack them. The lignins appear to be the most stable of the common organic polymers (Vallentyne, 1957; ZoBell, 1946). This is seen from the fact that marine humus may contain from three to four times as much lignin as the phytoplankton (Waksman, 1933), and lignin accumulated in fresh water is considered as the source material for coal.

It must be understood that the marine environment, like the lacustrine environment and soils, has a tremendously adaptable and interdependent microbial flora, each preparing the environment for their associates. Thus, one particular set of organisms may be capable of hydrolyzing the bonds of a complex molecule, such as cellulose, and breaking it down to the individual sugars. Others may convert the sugars to carboxylic acid, which may in turn be ultimately oxidized to CO_2 or reduced to methane and hydrogen. In turn autotrophic bacteria may assimilate the end products of heterotrophic catabolism and either convert them back to complex organic matter or use them for their own energy-generating processes. The associations may extend to higher forms; thus vitamins essential to animals may be generated by bacterial metabolism.

b. Types and numbers

The types of bacteria present in marine sediments are essentially the same in terms of biochemical potential at least as those found in lake muds or soils. Very often the establishment of the presence of a particular organism is directly proportional to the effort undertaken in finding it.

It is recognized that the mud–water interface is the richest zone for microbial proliferation; it is here that a continual supply of nutrients, organic matter, oxygen, etc. is being provided and toxic end products removed. Bacterial counts range from 10^2 to 10^8/ml on an average for the bottom water and surface sediment as compared with 10^2 to 10^4 for the photosynthetic zone (ZoBell, 1959).

The bacterial count in the sediments drops rapidly in the first 10 to 20 cm from the surface and more slowly thereafter. The bacterial counts vary greatly according to the environment. In the North Pacific Ocean, 10^8/g of sediment have been reported (Kriss and Rukina, 1952); comparable populations have been detected by Rittenberg (1940) and by Emery and Rittenberg (1952) in sediments in the southern California basins. Actually, it is the sediment type rather than geographical location that determines numbers.

It is often assumed that stagnant waters and underlying sediments contain a restricted and impoverished microbial flora. However, Kriss (1959) reported that in the Black Sea bottom waters and sediments the bacterial counts were higher than in the North Pacific. Not only were anaerobes numerous, but nitrifying bacteria, thiobacteria and aerobic protein decomposers were also isolated in large numbers. According to Kriss, life in the bottom waters of the Black Sea consists primarily, if not exclusively, of bacteria, and the average bacterial biomass was 1.5 to 2 times that of the oxygenated zone. As a consequence of the large volume of the Black Sea containing hydrogen sulfide, the total biomass there is estimated as 18.5×10^6 tons as compared with 17.7×10^6 tons in the oxygenated and euphotic zones.

Care must be exercised in interpreting data pertaining to reported bacterial populations and types. All counts based on culture procedures have the limitation that only those types capable of development under the nutritional and physical environment imposed by the experimenter will be counted, and the diversity of bacteria is such that only an undeterminable and probably small fraction of a complex population is enumerated in any single procedure. Direct microscopic techniques not only involve technical difficulties, especially when dealing with sediments, but also fail to distinguish between living and dead organisms. More important, the detection or counting of a particular type by any technique does not establish its significance in the environment in which it is found. Many kinds of bacteria, for example nitrifiers or sulfate reducers, can be detected in appreciable numbers in environments in which they cannot be active, at least insofar as we understand their physiology and metabolism. Much more attention must be paid to the development of methods that will enable quantitative assessment of the *in situ* activity of the different groups of bacteria.

c. Nutrient regeneration

Perhaps the most vital role performed by bacteria on the ocean bottom is that of mineralization, the regeneration of nutrients previously fixed by other living systems. This occurs throughout the entire water column as well as in the unconsolidated sediments, but is most effective in, or just below, the photosynthetic zone (Riley, 1951) where the plankton content is highest and death is proportionally high, and at the mud–water interface where the dead organisms come to rest. The net result is a solution of the soluble inorganic ions by the deep-ocean waters which are then returned to the euphotic zone by upwelling. A survey of the nutrient budget for the world's oceans by Emery, Orr and Rittenberg (1955) indicates that far more nutrients were annually assimilated by plankton than were supplied by the major sources. The renewal times (the number of years that would be required for dissolved nutrients in rivers, plus rain in the case of nitrogen, to build up the nutrients dissolved in the ocean water to their present concentrations in the absence of withdrawal) are similar—12,000, 60,000 and 27,000 years for nitrogen, phosphorus and silicon, respectively. All are close to the renewal time for the water itself, 50,000 years (volume of ocean divided by annual volume of river flow). The brevity of these renewal times and the character of the geological record, which suggests that the total life in the ocean has been more or less constant over long periods, indicate that steady-state conditions exist.

Under steady-state conditions as much nitrogen, phosphorus and silicon must be deposited annually in sediments, or otherwise lost from the ocean, as is contributed to the ocean from non-marine sources. Total fixed nitrogen is contributed in far greater quantity than is lost to sediments; maintenance of balance can be attributed to denitrification. On the other hand, both phosphorus and silicon are brought to the ocean and deposited on its floor in approximate balance. Because the contribution of each is dominantly in the form of mineral grains, the percentages of phosphorus and silicon in the ocean-bottom sediments are not materially different from their percentages in stream-borne sediment.

Comparative studies of the early diagenetic processes in sediments of the Santa Barbara, Santa Monica and Santa Catalina Basins have been made (Emery and Rittenberg, 1952; Rittenberg, Emery and Orr, 1955). The first is highly reducing with free hydrogen sulfide in the interstitial water, the second is mildly reducing with no free H_2S and the third is oxidizing at the surface and reducing below a depth of about 200 cm. The variation of nitrogen compounds in these sediments is shown in Fig. 4. Although the cumulative content does not vary appreciably in the surface sediments of the three basins, the dominant species is different in each, as is the change with depth. The existing patterns are understandable in terms of environments and the biochemistry of the bacterial processes involved.

Ammonia, which results from deamination of amino acids (proteins) and other organic constituents, can be produced under both aerobic and anaerobic conditions by a large variety of organisms. Bacterial deamination continues down to a depth of over 400 cm (representing several thousand years since the time of deposition), as can be judged from the increase of ammonia with depth. The increase is most noticeable in the reducing basin since here losses occur only by diffusion upward into the overlying water.

In the presence of molecular oxygen, ammonia is oxidized first to nitrite and then to nitrate by the *Nitrosomonas* and *Nitrobacter* groups respectively. An oxidized environment should, therefore, contain nitrate and this is precisely what is found in the Santa Catalina Basin sediments. Here nitrate forms to considerable depths as it probably also does in certain sandy layers of the Santa Monica Basin sediments introduced by turbidity currents.

Nitrate, like ammonia, can be lost from the sediment by diffusion into the overlying water. In addition, the conversion of nitrate to molecular nitrogen (denitrification) can occur. Laboratory investigations have shown that denitrification can take place only in the absence of molecular oxygen. The recent field data of Richards and Benson (1959), who demonstrated by measurements of the $^{15}N/^{14}N$ ratios in molecular nitrogen from the bottom waters of Cariaco Trench and Oslo fjord that biological nitrogen is liberated in the anaerobic environment only, are in agreement. In Santa Catalina Basin nitrate formed in the upper zone of oxidized sediment can be denitrified in the reduced subsurface zone, thus allowing a greater net loss of nitrogen than in Santa Barbara Basin where the absence of an oxidized zone makes this cycle impossible.

The California basin waters contain small quantities of oxygen and this appears to be enough to allow the oxidation of any ammonia which may escape from the sediment. The result is that the basin waters have no detectable nitrite or ammonia but contain about 40 μg atoms/l. of nitrate (a little less for

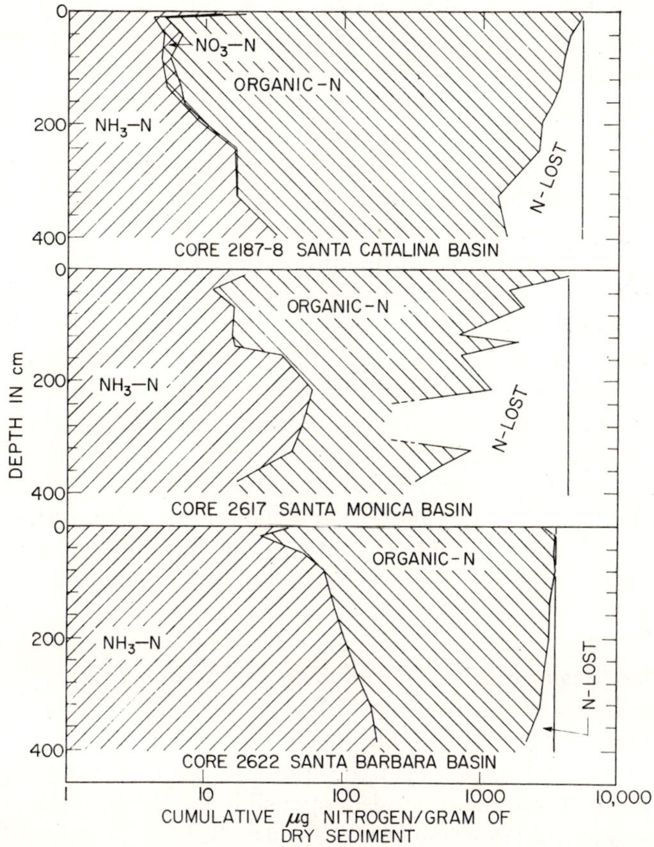

Fig. 4. Depth distribution of nitrogen compounds in sediments of three southern California basins. Note the use of a logarithmic scale of concentration. (From Rittenberg, Emery and Orr, 1955, fig. 4.)

the bottom waters of the reducing basins and more for the oxidizing basins), which is considerably lower than the 240 μg atoms/l. present in the surface interstitial water of Santa Catalina Basin.

In contrast to nitrogen, phosphorus is supplied to basin sediments largely as detrital phosphate minerals (e.g. apatite) and phosphate adsorbed on hydrolyzates and oxidates (e.g. clays and ferric hydroxide) in addition to organic phosphate. The conversion of organic phosphorus to phosphate occurs with bacterial decomposition of the organic matter; thus, a decrease in organic and an increase in inorganic phosphorus with depth in the sediment should be found if these forms could be distinguished analytically. The solubility of the phosphate ion is very susceptible to pH changes and the presence of divalent cations. At a low pH it exists as the acid $H_2PO_4^-$ whose salts are relatively soluble. At more alkaline pH, above 7.2, HPO_4^{2-} is dominant and may even dissociate to PO_4^{3-}, which forms insoluble phosphates with divalent cations. Thus the phosphate content of the interstitial water is not necessarily directly related to the state of decomposition of organic matter. In fact, the sediments in the southern California basins show concentrations of soluble phosphate up to fifty times the values reported for sea-water. The content varies from basin to basin and with depth in the sediment column.

Although phosphorus has only one known valence state in the sediments—pentavalent—its concentration in solution appears to be markedly affected by the oxidation-reduction potential. This is very probably only a secondary phenomenon; more directly the phosphate is proportional to the sulfide content of the sediment. Baas Becking and MacKay (1956) found that treatment of suspensions of metal phosphates in sea-water with H_2S released HPO_4^{2-} and $H_2PO_4^-$ causing a marked drop in pH. They also demonstrated that marine algae (e.g. *Enteromorpha*) are capable of releasing bound phosphate from the sediment as a result of liberating organic acids which have a marked solubilizing action. As mentioned earlier, the stagnant waters of the Black Sea, Lake Maracaibo and some Norwegian fjords contain abnormally high phosphate. Sugawara, Koyama and Kamata (1957) demonstrated the direct relationship between the H_2S content and the phosphate in the interstitial water of a series of freshwater and brackish lakes in Japan. The same relationship is found in the southern California basin sediments; phosphate is abundant throughout the Santa Barbara Basin where free H_2S is available at all depths so far cored, and increases in the Santa Catalina Basin where the reducing environment is reached. In the Santa Monica Basin, where no free H_2S is available, the phosphate content of the interstitial water is low at all depths.

The distribution of phosphate in the interstitial water can be explained as follows. Bacterial sulfate reduction is responsible for the formation of H_2S and CO_2 in the sediments. These molecules effectively bind Fe^{2+} as FeS and Ca^{2+} as $CaCO_3$, which are highly insoluble and precipitate preferentially to phosphates. The removal of the metal cations enables the phosphate ion to remain in solution. In the Santa Monica Basin the supply of detrital material rich in iron exceeds the sulfide production. There is, therefore, no competition for the calcium and iron binding the phosphate and it remains in the insoluble state.

Silicon, like phosphorus, is known to have only one valence state in the natural environment. It seems to be little affected by the oxidation state but is influenced by the pH. According to Krauskopf (1956) and White, Brannock and Murata (1956), working on the solubility of opal, silica solubility increases rapidly at a pH above 9, but is little affected below it. Of the various forms of silicon depositing on the ocean bottom, amorphous silica is probably the most soluble (120–140 ppm below pH 9) as well as being an important source continually provided by diatoms (Siever, 1957). The oceans on an average contain about 5 ppm and so are highly undersaturated. Deep waters, especially in the higher latitudes, contain considerably larger quantities. Interstitial waters from the basins in southern California (Rittenberg, Emery and Orr, 1955) showed a tenfold increase in silicon relative to the overlying basin waters. There is a steady increase with depth and in the three basins measured a maximum was reached at about 120–150 cm. It cannot be concluded at the present stage whether these depths constitute a real maximum until further and deeper measurements are made.

The actual mechanism influencing silica solution is not known. It is possible that a time factor is necessary. It is further possible that bacterial processes result in the breakdown of a protective film which encases the diatom frustule. Lewin (1959) found that removal of this film with nitric acid increased the solubility of the tests in dilute alkali by 50%.

Sulfur in the ocean is not usually considered in a discussion of nutrients since it is normally abundant (880 mg/kg as the sulfate ion) but nevertheless it is an essential constituent of all living matter. In the aerobic zones of the ocean the concentration of sulfate varies only slightly. In reducing basins, such as the Black Sea, the disappearance of sulfate, as well as the presence of other forms of sulfur (Danil'chenko and Chigirin, 1929), has been reported for the waters. The sediments of basins may present a complex mixture of sulfur compounds.

Studies on the sediments from the southern California basins show that apart from sulfate, free sulfide, acid volatile iron monosulfide (hydrotroilite), fixed sulfide (pyrite), elementary sulfur and organic sulfur can occur in varying concentrations. In the highly reducing sediments where sulfide formation is active, the sulfate concentration decreases rapidly in the first 50 or 60 cm and then more slowly, down to one-tenth or less of the concentration in sea-water at a depth of about 6 m. In the Santa Catalina Basin, where oxidizing layers overlie reducing ones, there is good evidence to suggest that a reoxidation of bottom sulfides has occurred to form sulfates at a higher concentration than initially present in the sea-water.

The concentration of elementary sulfur is directly proportional to the content of free and acid volatile sulfide. In the oxidizing basins, or those with small sulfide content, native sulfur is absent. Organic sulfur and pyrite appear to be proportional to the content of organic matter. Thus, in oxidizing or slightly reducing sediments having high organic contents and low free sulfide and hydrotroilite (e.g. Santa Catalina Basin), the content of pyrite is high. The mechanism of pyrite formation is as yet unknown, although most observations indicate that it probably has an authigenic origin.

d. Oxidation-reduction potentials and pH

The oxidation-reduction potential (E_h) of a marine sediment is influenced more by the activity of organisms than by any other factor. Microorganisms, because of their quantitatively greater importance, ultimately dictate the conditions. The larger organisms, particularly the worms, as well as some mollusks, sea-urchins and other animals which burrow or disturb the surface of the sediment by their movements (Hartman and Barnard, 1958), may have a marked influence in keeping the first few centimeters oxygenated.

The role of the microbes is to utilize the oxygen for the metabolism of the organic matter settling on the bottom and to release reducing substances. In a basin in which water is not constantly renewed, the extraction of oxygen plus the release of reduced substances lowers the E_h. From the work of Baas Becking, Kaplan and Moore (1960) it appears that the removal of oxygen alone does not lower the E_h. The most effective poising agent is hydrogen sulfide, or more correctly the sulfhydril ion (HS^-). Highly oxidized waters have an E_h of +500 mV, while stagnant hydrogen-sulfide-containing waters appear to have a minimum of about −200 mV (Strakhov, 1959). This corresponds to the SO_4^{2-}–HS^- couple. Basin sediments on the other hand generally show lower E_h values (Emery and Rittenberg, 1952; Rittenberg, Emery and Orr, 1955; Baas Becking, Kaplan and Moore, 1960); it is seldom that a potential exceeds +300 mV under maximal oxidizing conditions, and reducing values as low as −400 mV have been reached. This approaches the lower limit for the stability of water.

In sea-water the pH generally remains within the range 7.5 to 8.5; the water is buffered by the carbonate–bicarbonate system. Under conditions of extreme evaporation, addition of fresh water or high biological activity, a greater spread may result. The sediments of true marine basins also have a narrow pH range. There is usually no marked trend, although, in general, a slight increase of pH occurs with depth. There is often a change in pH between the sediment and the overlying water. The direction of this change is unpredictable and usually slight. In some basins there is a definite decrease in the pH of the sediment relative to the water, e.g. Milford Sound, New Zealand. This may be due to the release of CO_2 by biological activity and to a cation exchange process in which the alkali metals are preferentially adsorbed.

C. Subsurface Waters of Ancient Basins

Subsurface waters of ancient marine basins are thought to be remnants of sea-water trapped within the sediments at the time of deposition. It is, therefore, of some interest to examine the ionic composition of connate waters in order to gain a greater understanding of the processes causing diagenetic changes in the interstitial waters of freshly deposited marine sediments. The recent review by Chave (1960) draws adequate attention to the pitfalls involved in accepting all published data without critical evaluation. It may be added that because of the heterogeneous character of sediments, even analyses on freshly deposited material are often subject to insurmountable error.

From the many thousand published results, Chave (1960) draws some conclusions on the chemistry of the ancient waters. The predominant ion is chloride. There is always an increase of Cl^- with respect to sea-water and the chlorinity may range from 19 to 200‰. A decrease in chlorinity is interpreted by Chave as being due to contamination by meteoric water. The other principal anions, bromide and iodide, are also enriched. Sulfate is generally depleted (presumably by bacterial reduction), often to vanishingly small quantities, and sulfide may substitute for it. Enrichment in sulfate is thought to be due to the addition of hypersaline evaporite water. The cations also show a variation. Calcium is enriched, probably as a result of the solution of limestone and replacement by magnesium in the formation of dolomite. This process, and the adsorption of magnesium by clays during montmorillonite and illite formation, results in the depletion of magnesium. Potassium is also depleted, presumably by the processes of exchange in clay-mineral formation. Strontium and barium are enriched, probably through the release by fine sediments into sulfate-poor water.

Data on these ions in recent sediments in present-day basins are not as well established. This is partly due to lack of effort and partly to the inability of getting sufficiently long cores until relatively recently. The decrease of sulfate in reducing environments has been confirmed by many workers. The change in chloride is not so straightforward. Emery and Rittenberg (1952) and Rittenberg, Emery and Orr (1955) found no change with depth down to 200 cm; Bien (Chave, 1960) also apparently could not detect any significant changes. Shishkina (1959), however, studying the Black Sea sediments, found that a considerable decrease occurred with depth, amounting to 43% at about 900 cm in one core. Bromide was also found to decrease, but not to the extent of chloride. Potassium, sodium and magnesium were all found to diminish by Shishkina. In some cores, potassium almost disappeared at depths of 800–900 cm, although in others the changes to depths of 300–400 cm were not pronounced. Calcium generally showed an increase with depth.

These results, especially the analyses of anions in the interstitial waters of recent sediments, can be seen to differ from the measurements on connate waters. The results of the Black Sea measurements may not be truly representative of present-day conditions, having been influenced by post-glacial deposition in an environment of lower salinity. There is, however, a serious lack of data from present-day neritic basins.

4. Isotopic Studies Related to Biological Processes in Sediments

The recent interest in stable isotopes (Rankama, 1954) has proved of great value in elucidating many problems in nature, including the ocean. Surprisingly little work, however, apart from the oxygen isotope work, has so far appeared in the literature attempting to interpret mechanisms in oceanic sediments generally and basin sediments in particular.

The studies of Richards and Benson (1959) on denitrification were briefly mentioned earlier. By measuring the $^{15}N/^{14}N$ ratio in the dissolved molecular nitrogen of the Cariaco Trench, they were able to show that approximately 3% of the nitrogen was derived from denitrification processes occurring in the anaerobic zones only. This is probably the first *in situ* estimate that has been obtained. Much more work is required to establish a reliable figure that can be used on a world-wide basis in order to balance the nitrogen cycle. This technique should ultimately prove important in determining the amount of molecular nitrogen being fixed, both in the euphotic zone as well as at the mud–water interface.

Fractionation of hydrogen isotopes by bacteria has been shown in the laboratory. Using facultative organisms isolated from sediments of the Bahama Bank, Cloud *et al.* (1958) showed that there was an enrichment of the lighter isotope by a factor of 20 over its occurrence in sea-water in the metabolic hydrogen produced. Such drastic fractionations could have some influence on the H/D ratio of semi-isolated bodies of water, such as lagoons.

Studies on the isotopes of carbon are of particular interest in tracing breakdown and accumulation processes of organic matter. It has been confirmed by a number of workers that growing organisms are able to fractionate the carbon atoms by concentrating the lighter isotope in the protoplasm and the heavier one in the carbonate skeleton. Craig (1953) has shown that marine plants and animals have a definite range for the $^{12}C/^{13}C$ ratio of their total organic constituents. Certain extracts of phytoplankton, however, notably hydrocarbons and lipids, were found by Silverman and Epstein (1958) to be considerably enriched (15 to 20‰) in the lighter isotope with respect to the total protoplasm. The organic matter in basin sediments is also known to be enriched in the lighter isotope with respect to marine organisms (Emery, 1960; Silverman and Epstein, 1958). Chloroform–methanol extracts of the organic matter gave slightly higher ^{12}C concentrations than the average total material and closely coincided with the measurements made for petroleum. This can be used as an argument for the origin of oil directly from the hydrocarbon and lipid fractions of the organic matter. Carbon isotope work should be able to determine the importance of bacterial processes in the deposition of carbonate, since metabolic CO_2 is probably enriched in ^{12}C. This process has been proposed for the formation of the limestone associated with the salt-dome deposits of Texas and Louisiana (Feely and Kulp, 1957). Laboratory experiments have not yet substantiated these predictions.

There is a considerable literature on the isotopes of sulfur, which has most recently been reviewed by Ault and Kulp (1959). The investigations have been largely restricted to sulfur compounds in ancient sediments, ores and volcanic exhalations. Comparatively few results exist for recent marine sedimentary deposits. The investigations carried out on some sediments at Milford Sound by Kaplan, Rafter and Hulston (1960) can be used as an indication of the importance of such studies. The analyses on four samples are given in Table I. These represent surface sediments from stations A and G and the water overlying them. The results show that the isotope ratios for the sea-water sulfate fall very close to the average of 21.76 given by Ault and Kulp (1959). The slightly higher ratio for sample R240/1 may be due to the influence of intermixing with fresh water, since the area in which this restricted basin lies has an

TABLE I

Isotope Fractionation in Estuarine Mud and Water of Milford Sound, New Zealand

Station no.	Lab. no.	Sample description	SO_4^{2-} %	SO_4^{2-} $^{32}S/^{34}S$	Acid volatile S^{2-} %	Acid volatile S^{2-} $^{32}S/^{34}S$	Acid volatile S^{2-} $\delta^{34}S$ enrichment (‰)	Combustible S^{2-} %	Combustible S^{2-} $^{32}S/^{34}S$	Combustible S^{2-} $\delta^{34}S$ enrichment (‰)
H	R240/1	Bottom sea-water SO_4^{2-} only form of sulfur	0.234	21.80	—	—	—	—	—	—
A	R240/2	Bottom sea-water SO_4^{2-} only form of sulfur	0.268	21.75	—	—	—	—	—	—
A	R240/3	Surface mud $\sum S = 0.218\%$ (dry wt) $E_h = -100$ mV, pH = 7.20, $^{32}S/^{34}S = 22.47$	0.280	21.86	0.080	22.92	−46.2	0.041	22.78	−40.4
G	R240/5	Surface mud $\sum S = 0.219\%$ (dry wt) $E_h = +15$ mV, pH = 7.10, $^{32}S/^{34}S = 22.39$	0.262	21.99	0.061	22.83	−36.8	0.069	22.86	−38.0

annual rainfall of 270 in. and a very high run-off rate into the fjord. The influence of fresh water may also be reflected in the somewhat lower sulfate content for this sample.

The sulfate in the sediments shows a consistently higher ^{32}S content than that in the sea-water. This cannot be explained on the basis of sulfate reduction, since in that case there would be an enrichment of ^{34}S. Here again, as in other natural environments described by Kaplan, Rafter and Hulston (1960), it would appear that a reoxidation of sulfide is occurring with the formation of a sulfate enriched in ^{32}S which dilutes the ^{34}S-rich sulfate. This process could only occur at the surface, where a sufficiently high oxygen tension is present. On this assumption, an increased depth down a sediment profile should produce sulfate correspondingly increasing in ^{34}S content. This suggestion of sulfate formation by hydrogen sulfide oxidation is also reflected in the higher oxidation–reduction potential as well as in the higher isotope ratio in sample R240/5. The presence of *Thiobacillus* detected in surface muds supports these conclusions.

The isotope ratios for both the acid volatile and combustible sulfides (probably of the pyrite type as well as perhaps some other base-metal sulfides) show an expected enrichment in ^{32}S. The highest enrichment is 46.2‰ for the acid volatile sulfide of sample R240/3 with respect to the sulfate in this sample. This enrichment falls close to the value of 46‰ found by Feely and Kulp (1957) for samples of sulfides and sulfates from the salt domes of Louisiana and Texas. Since bacterial sulfate reduction in this fjord is still very active, and no other reasonable explanation can be offered for the formation of such high concentrations of sulfides, we must conclude that fractionations of 46‰ (or somewhat higher if based on the average $^{32}S/^{34}S$ ratio of SO_4^{2-} in sea-water) can be attained by such biological systems, under suitable conditions.

5. Special Problems

Basin sediments are thought to be particularly suited for the formation and accumulation of certain deposits commonly not present in other marine environments. These include petroleum, base-metal sulfides and uranium. The exact mechanism for their formation is unknown, and so far the evidence comes mainly, if not in some cases entirely, from ancient marine sediments; a short discussion of these topics is believed warranted in order to draw attention to possible lines of study.

A. Petroleum

The origin of oil has been a much disputed question for the past 100 years. Although a voluminous literature exists, no definitive theory has yet been proposed satisfying all the questions arising from the complex properties of petroleum and the special conditions needed to create them. At the present time the majority of workers consider that most deposits have a marine origin. Although a freshwater environment could likewise serve as a starting point, such an environment is quantitatively unimportant compared with the marine environment. It is generally agreed that the plankton serves as the source material.

The evidence for the marine environment is that petroleum is found in strata which are either entirely or predominantly marine. The evidence for the biological origin stems from the direction of optical rotation and more important from the presence of porphyrins (desoxophylloerythrin) in the petroleum. Since these substances are sensitive to high temperatures and since petroleum has been found in relatively young beds of the late Pliocene it has been concluded that oil can form at low temperatures ($<150°C$) and pressures. Complications do arise, since the highly paraffinic oils do not appear to contain porphyrins. This has generally been interpreted as being the result of their destruction by heat and pressure due to deep burial. However, paraffinic oils have been found at shallow depth in stratigraphic traps which do not appear to show deep burial.

This leads to the question of the nature of the organic substances from which petroleum was derived. As mentioned in an earlier section, crude oil as such does not exist in present-day basin sediments, although hydrocarbons have been separated from the surface muds of both marine basins and freshwater lakes. It would therefore appear that formation and accumulation begin soon after deposition and perhaps even during sedimentation of the dying plankton. The two theories suggested are (a) the accumulation of oil from residual undecomposed compounds such as lipids and fatty acids, (b) the production of hydrocarbons by diagenetic processes. The first proposal may account for the low density asphaltic petroleums containing porphyrins, the second for the high density paraffinic type. Evidence from recent basin sediments can be used to support either hypothesis.

Petroleum is generally found in porous rocks or in strata containing fissures, faults and voids, mainly in sands or limestones. Such sediments are known to be poor in organic matter. Shales, on the other hand, have high contents of organic matter but are poor reservoirs for petroleum. It is concluded that petroleum generally does not remain where it originates, but is transported either laterally or vertically as a result of compaction. This introduces the factors of selection and transformation through transportation. The recent mechanisms proposed by Baker (1959) and Meinschein (1959) for the accumulation of substances with the general characteristics of petroleum suggest that the transporting agent is of vital importance. It may be that this is the reason for the apparent absence of petroleum in continental-type sediments.

The production of organic matter in modern basins is usually high enough to account for the accumulation of hydrocarbons in large quantities. From a comparison of the Los Angeles Basin with the still submerged Santa Monica Basin, Emery (1960) concluded that for every barrel of petroleum formed, 19,000 barrels of equivalent organic matter had to be produced by plankton and 1200 barrels deposited on the basin floor; the present annual extraction of petroleum in the Los Angeles Basin is 100 million barrels per year. Sufficiently long cores are still lacking in present marine basins to determine whether petroleum is currently being formed in them.

B. Metal Sulfides

Iron sulfides are known constituents of reducing basin sediments. The two types most commonly met are the black hydrated iron monosulfide, hydrotroilite, and pyrite. The first is highly unstable and decomposes during drying or heating to iron oxides and elementary sulfur. Sulfides of other heavy metals have not yet been described as occurring in modern basin sediments, but this is probably due to their relative scarcity compared with the detrital material and they have probably been overlooked. There appears to be no reason why copper, manganese, zinc and other cations which form highly insoluble sulfides should not coexist with iron sulfide. Many such metals are present in living systems (e.g. copper in the blood of Crustacea) and are liberated on the death of the organism.

The geological literature contains many descriptions of sedimentary pyrite and marcasite deposits. Coals sometimes have as much as 10% sulfur in the form of pyrite. Shales are particularly rich in pyrite, as for example the Wabana District of Newfoundland which contains oolitic pyrite beds of 1 cm to about 30 cm in thickness. Sandstones and limestones also contain pyrite as found in the Upper Cretaceous of Wyoming. The large deposits of Nairne, South Australia, have been described as sedimentary by Skinner (1958) and LaGanza (1959). Copper, lead and zinc sulfides are also found in sedimentary rocks but their origin is not well established. The Kupferschiefer Basin in the Permian of

Germany, with shallow-water marine fossils, contains black shales rich in chalcocite, bornite and chalcopyrite as well as sulfides of iron, zinc and lead, together with small quantities of silver, nickel, cobalt, vanadium and molybdenum minerals. The Tri-State District in the Mississippi Valley is exceptionally rich in zinc and lead sulfides, which occur in limestones, dolomites, cherts or calcareous shales of the Paleozoic (Lindgren, 1928).

Whether these non-ferrous sulfide ores are epigenetic or syngenetic is still a lively question. The main difficulty lies in finding the source of the metals. The tendency has been to describe sulfide ores, even in sedimentary strata separated by large distances from igneous intrusions, as being of a hydrothermal or metasomatic origin. Stanton (1955) found that in the lower Paleozoic of the Bathurst region of New South Wales, Australia, sulfide ores occurred adjacent to limestone and volcanic rocks. He interpreted this as indicating a release of cations by volcanic exhalations then carried to the ocean in solution and deposited in lagoons adjacent to actively growing reefs, probably containing reducing sediments with a high sulfide content.

The exact mechanism for pyrite formation in marine sediments is not yet known, although it is commonly assumed to arise from hydrotroilite. This concept is strengthened by the similarity in the isotopic ratios of the hydrotroilite and pyrite sulfur in the Milford Sound sediments (Table I). Pyrite is found widely dispersed throughout reducing as well as oxidizing sediments. In the surface oxidizing layers of the Santa Catalina Basin, pyrite constitutes the most abundant sulfur compound, whereas in the Santa Monica Basin, which has a high hydrotroilite content, pyrite is low. It thus appears that pyrite can form in mildly oxidizing environments and at a comparatively rapid rate.

Since sulfate reduction is undoubtedly involved as the first step in pyrite genesis, it might appear paradoxical that pyrite is abundantly formed in mildly oxidizing sediments which should exclude sulfide production. The common occurrence of pyrite within foraminiferal and diatom tests suggests that reducing micro-environments adequate for sulfate reduction can occur in a milieu that is predominately mildly oxidizing.

Of great interest is the direct relationship between the pyrite content and organic content of the sediment, which was also observed by Harmsen *et al.* (1954) for marshes and soils in the low-lying coastal area of the Netherlands. Because of the tendency for pyrite to concentrate in sediments with high organic matter it is tempting to postulate a reaction occurring between the hydrotroilite first formed and the protoplasmic matter, resulting in the deposition of the disulfide.

With respect to the formation of other sulfides, Miller (1950) has shown that sulfate-reducing bacteria can tolerate relatively high contents of base-metal cations and Baas Becking and Moore (1961) have recently demonstrated the formation of galena, sphalerite, covellite and digenite in bacterial cultures. Present day neritic marine basins adjacent to areas of high volcanic activity or submarine seepage such as Alaska, Japan or the South Pacific Island arcs, may, therefore, prove useful places to initiate studies on the formation of such minerals.

C. Uranium and Transition Metals

Uranium, cobalt, molybdenum, nickel and vanadium are known to be enriched in bituminous marine shales, and vanadium and nickel are present in some crude oils. There is no doubt that these trace metals are authigenic; the mechanism of their deposition is, however, still ambiguous. Three processes can be suggested, but their relative importance must await further studies.

The first is the active concentration of the trace metals by living organisms. It is well known that the transition metals are important constituents of a number of biological substances. They are able to form chelate compounds which perform the important function of solubilizing particulate matter. Goldberg (1957) reviewed the literature and found ample evidence for their selective concentration by plankton, sessile algae, sponges and other organisms. Vanadium appears to be concentrated the most, by a factor of $>280,000$ over its concentration in sea-water.

Scavenging by clays, colloids and settling organic matter is another process that has been suggested (Krauskopf, 1955) for the concentration of trace metals. Krauskopf found that hydroxides of manganese and iron were the most efficient in the adsorption of ions from sea-water.

The third method, and according to Krauskopf (1955) the least important, is the chemical precipitation of oxides, carbonates or sulfides. In areas of the ocean where upwelling results in high productivity and later in stagnation in bottom waters, or in restricted stagnant basins discussed in this chapter, the E_h may drop sufficiently to transform the soluble complex ions into the lower valence insoluble oxides. The theoretical calculations by Garrels (1955) show that the reducing conditions produced by sulfate reduction create an environment in which the oxides of vanadium III and IV, uranium IV and sulfides of iron, copper, zinc and lead are stable. The isotopic studies by Jensen (1958) on the metal sulfides associated with the uranium deposits of the Colorado Plateau (freshwater deposits) give extra weight to the hypothesis that bacterial sulfate reduction may be directly involved in the deposition of uraninite and montroseite.

It is well known that primary uranium occurs in ancient marine sediments, principally those rich in phosphate and poor in lime, and the carbonaceous or black bituminous shales. The most favorable environments for its deposition seem to have been areas near margins of platforms during the Paleozoic. It appears that the uranium may substitute for calcium in the crystal lattice of the carbonate-fluorapatite, which is the predominant phosphate mineral in phosphatic sediments. In the organic shales there appears to be a direct relationship with the organic matter (McKelvey and Nelson, 1950). Thin layers particularly rich in organic matter can attain uranium concentrations of 0.7%. In general, the uranium content of these shales varies from a few thousandths to 0.02% (Klepper and Wyant, 1957). The shale beds are often several feet thick and cover areas of tens or hundreds of thousands of square miles. The uranium is preferentially concentrated in the fine-grained sediments, especially where the carbonate content is low, and, from the nature of the deposits, in areas that are tectonically relatively stable allowing for a slow accumulation. The source can sometimes be traced to overlying volcanic ash beds that have been leached (Becraft, 1958).

Because of the recent interest in nuclear energy, a considerable literature exists on ancient marine basin sediments enriched in syngenetic uranium. No reports have yet been published on the occurrence of uranium in recent neritic basin sediments. As in the case of the metal sulfides, it is recommended that lagoons or other reducing environments adjacent to volcanic areas or to localities particularly rich in granitic rocks, that can act as sources, be studied for enrichment of uranium as well as vanadium and the other trace metals mentioned above.

D. Phosphates

Phosphate deposits may be classified into six geological types: apatite deposits of igneous origin, residual deposits, phosphatized rocks, river-pebble deposits, guano and marine phosphorites (Curtis, 1957). There are two types of marine phosphates present in the geological column: (a) the phosphatic nodules and (b) the stratified phosphates associated with shales or limestones; both of these are authigenic. The former type is present along the continental shelves of many continents, and is especially high in regions where upwelling is dominant and sedimentation of detrital material low, e.g. the coasts of Peru and Chile, the southwest coast of Africa and the coast of North America (Dietz, Emery and Shepard, 1942). The second type has not yet been described in recent marine sediments.

The phosphorites associated with shales are widespread, often cover large areal distances and tend to accumulate uranium. The Phosphoria Formation of the Permian of Western United States of America is perhaps the best known deposit of this type. It is thought that deposition occurred in the miogeosyncline facies of the orthogeosyncline. The phosphate is often enriched in clastic or limestone members of a dominantly shale formation. For example, the Tennessee blue-rock phosphate occurs in the Hardin Sandstone member of the Chattanooga Shale.

Many authors have suggested origins for the phosphorites present in shales (Mansfield, 1931; McKelvey *et al.*, 1953); these explanations are based largely on field observations without giving much attention to the conditions prevailing in present-day environments. Emery (1960) has attempted to explain phosphatic nodule formation on continental shelves as the result of precipitation from supersaturated solution probably in the form of colloids. He concludes that upwelling bottom waters rich in phosphate may undergo changes in environmental conditions which cause the tricalcium phosphate to exceed its solubility product.

We wish to draw attention to a further possibility which appears to have been generally overlooked and could profitably be investigated. The close association of ancient marine phosphates with black shales indicates a reducing environment, as now exists in marine basin sediments and occasionally in waters, as described in this chapter. Under these conditions phosphate is released by diagenetic processes and may reach concentrations one hundred times as great as in the overlying surface waters. At the same time the pH of the basin waters is decreased (Rittenberg, Emery and Orr, 1955; Baas Becking and MacKay, 1956). The mixing of such water with oxygenated water at the sill or near the rim of reducing basins with a resulting increase in pH should cause phosphate to precipitate. As a consequence, areas in the open ocean surrounding such basins would be enriched in insoluble phosphate. The shelves surrounding the Cariaco Trench or the coastal regions of Norway where fjords are abundant could be present-day environments where phosphate deposition is occurring. Another possible mechanism could involve large-scale turbidity currents which might introduce sufficient oxygenated water and divalent ions into a stagnant basin to cause a precipitation of phosphate. In such an event phosphate would be enriched in the surface layers of the coarser sediments. This may account for the presence of phosphate in ancient basin sediments.

E. Carbonate Deposition

The deposition of carbonates on the ocean bottom has been a problem of primary interest to geologists for many years. The source and mechanism of precipitation of amorphous or microcrystalline carbonates is still not entirely solved to the satisfaction of all workers. One of the earliest explanations was given by Murray and Irvine (1889) who suggested an interaction between ammonium carbonate liberated by microbial decomposition of proteinaceous matter and the calcium sulfate in sea-water, producing calcium carbonate and ammonium sulfate. Since then there have been numerous advocates of the hypothesis of deposition of calcite by bacterial activity (Lalou, 1957). Much evidence has been based on the experimental work of Drew (1913) who suggested that denitrification was the primary cause of calcium carbonate deposition. Lipman (1924) critically repeated the work of Drew and concluded that the artificial conditions used in culturing the bacteria were so different from those actually existing in the ocean as to make a direct comparison untenable. The most recent support for bacterial precipitation comes from the work of Lalou (1957, 1957a) who set up mixed cultures in aquarium tanks using reducing mud covered with sea-water enriched with 0.5 to 1.0% glucose.

The studies with bacterial cultures have many drawbacks, the major one being the high content of nutrients and calcium added to the culture medium. The processes of denitrification to which has been attributed the liberation of ammonia are probably incorrect. Denitrification which results in the reduction of nitrate to ammonia or nitrogen can occur only in the anaerobic environment. In a highly anaerobic environment the nitrate content is low, and it is doubtful how much ammonia actually forms compared to the molecular nitrogen liberated. Deamination of the proteins in the mud is probably more important.

Bacterial sulfate reduction has been proposed as a direct method of formation of calcium carbonate through the substitution of the sulfate by a sulfide and then a carbonate. Lalou (1957, 1957a) found that the initial stage of decomposition in the reducing sediments was the formation of organic acids together with the liberation of CO_2, which caused a marked decrease in pH. This solubilized calcium and caused a twofold increase in its concentration in the sea-water. The next step was the initiation of sulfate reduction with an increase in H_2S, until the sulfate disappeared. The pH at this stage rose to that at the start. Up to this point little or no carbonate precipitated; only when the containers were placed in the light was there an observed deposition. In the presence of light there was growth of the photosynthetic purple bacteria and a rise in pH. The conclusion that must be reached from this is that CO_2 was removed by photosynthetic fixation displacing the equilibrium to the side of carbonate. There is no indication from this work whether carbonate is actually removed from seawater or whether it is just a recycling of the carbonate in the mud.

The necessity for photosynthesis suggests that those processes would occur only at the relatively shallow depths of the euphotic zone. Rapid escape of CO_2, an increase in temperature and a high pH of the water would all help this process. Shallow lagoons associated with atolls or fringing reefs would suggest themselves as such environments; most of these, however, are highly oxygenated. Stagnant basins, particularly in temperate zones, could produce this phenomenon. In such a case autotrophic bacteria living at the oxidizing–reducing interface could remove the CO_2. Such a process may be responsible for the reworking of carbonates in the Black Sea and other deep anaerobic basins.

The problem outlined above deserves careful attention. There is need for studies *in situ* rather than artificial culturing as has been done in the past. Since bacterial metabolism liberates the light isotope of carbon as CO_2, any carbonate precipitated by this method should be enriched in ^{12}C. On the other hand precipitation of carbonate by the biological adsorption of CO_2 will leave the calcite enriched in ^{13}C. This may be one method of obtaining more information from the sediments, as well as from controlled laboratory experiments.

Neritic marine basins, including fjords, lagoons and inland semi-isolated seas cover only a small fraction of the total oceanic area. Their significance, however, appears to be out of proportion to their extent. The epicontinental seas that were so abundant during the Paleozoic have left ample evidence of their importance in the deposition of neritic sediments. In particular, basins located within continental shelves have played a special role in the formation, preservation and accumulation of mineral resources. The two basic reasons for this role are (a) the slow rate of deposition preventing a dilution of authigenic minerals by terrigenous detritus, and (b) the isolation of the basin water from the overlying water because of restrictions imposed by sill depth.

The consequence of the above two factors is to permit relatively high contents of organic matter to accumulate in the sediment. Because of a restricted exchange between basin water and surrounding water, dissolved oxygen is removed by biological activity, mainly at the mud–water interface. In exchange, hydrogen sulfide is liberated through bacterial breakdown of organic matter and by sulfate reduction. In the extreme case, this results in the formation of oxygen-free, sulfide-containing water, that is, a stagnant environment.

Bacterial diagenetic processes, in particular the formation of highly reducing conditions through the liberation of sulfides, have many interesting consequences. The organic matter, which would normally be effectively destroyed in an oxidizing environment, is only partially degraded and permitted to accumulate, in as yet an unknown manner, to huge reservoirs of petroleum. Trace metals often associated with clays and living organisms are fixed in the sediments. Nutrients are regenerated within the sediment column and at the mud–water interface and released to the overlying water, ultimately to re-enter the productive cycle. The reduced environment appears to be particularly efficient for the regeneration of phosphate and plays a special role in maintaining nitrogen balance in the oceans through denitrification.

Although more is known about basins than many other marine environments, large gaps are still present. The preceding discussion has attempted to draw attention to our present state of knowledge and to indicate where further studies can be profitably undertaken.

References

Abelson, P. H., 1954. Annual Report of the Director of the Geophysical Laboratory, 1953–1954. *Carnegie Inst. Wash. Year Book*, **53**, 97–101.

Abelson, P. H., 1959. Geochemistry of organic substances. In *Researches in Geochemistry*. John Wiley and Sons, Inc., New York, 79–103.

Arkhangel'sky, A. D. and N. Strakhov, 1938. *Geological Structure and Developmental History of the Black Sea*. Academiia Nauk S.S.S.R., Moscow–Leningrad. (Cited by Caspers, 1957, p. 828.)

Athearn, W. D., 1959. Sediments of the Cariaco Trench. *Intern. Oceanog. Cong. Preprints*, 594.

Ault, W. U. and J. L. Kulp, 1959. Isotopic geochemistry of sulfur. *Geochim. et Cosmochim. Acta*, **16**, 201–235.

Baas Becking, L. G. M., I. R. Kaplan and D. Moore, 1960. Limits of the natural environment in terms of pH and oxidation-reduction potentials. *J. Geol.*, **68**, 243–284.

Baas Becking, L. G. M. and M. MacKay, 1956. Biological processes in the estuarine environment. V. The influence of *Enteromorpha* on its environment. *Koninkl. Ned. Akad. Wetenschap., Proc.*, Ser. B., **59**, 109–123.

Baas Becking, L. G. M. and D. Moore, 1961. Biogenic sulfides. *Econ. Geol.*, **56**, 359–272.

Baker, E. G., 1959. Origin and migration of oil. *Science*, **129**, 871–874.

Baldwin, E. J., 1959. Pliocene turbidity current deposits in Ventura Basin, California. Univ. Southern California, Unpubl. Master's Thesis, 1–66.

Becraft, G. E., 1958. Uranium in carbonaceous rocks in the Townsend and Helena Valleys, Montana. *Bull. U.S. Geol. Surv.*, **1046G**, 149–164.

Bonnett, B., 1959. Soluble organic matter in some argillaceous sediments in Great Britain. Preprints General Petroleum Geochemistry Symposium, Fifth World Petroleum Congress, Fordham University, New York, 7–8.

Bramlette, M. N., 1946. The Monterey formation of California and the origin of its siliceous rocks. *U.S. Geol. Surv. Prof. Paper*, **212**, 1–57.

Brisou, J., 1955. *La Microbiologie du Milieu Marin*. Editions Medicales Flammarion, Paris, 1–271.

Brongersma-Sanders, M., 1948. The importance of upwelling water to vertebrate paleontology and oil geology. *Koninkl. Ned. Akad. Wetenschap., Proc., afd. Natuurk.*, sec. 2, **45**, No. 4, 1–112.

Brooks, B. T., 1952. Evidence of catalytic action in petroleum formation. *Ind. & Eng. Chem.*, **44**, 2570–2577.

Caspers, H., 1957. Black Sea and the Sea of Azov. In "Treatise on Marine Ecology and Paleoecology". *Geol. Soc. Amer. Mem. 67*, **1**, 803–890.

Chave, K. E., 1960. Evidence on history of sea water from chemistry of deeper subsurface waters of ancient basins. *Bull. Amer. Assoc. Petrol. Geol.*, **44**, 357–370.

Chigirin, N. I., 1930. Phosphorus in the water of the Black Sea. *Trudy Sevastopol Biol. Sta.*, **2**, 143–163. (Cited by Caspers, 1957, p. 822.)

Cloud, P. E., I. Friedman, F. D. Sisler and V. H. Dibeler, 1958. Microbiological fractionation of the hydrogen isotopes. *Science*, **127**, 1394–1395.

Copenhagen, W. J., 1934. Occurrence of sulphides in certain areas of the sea bottom on the South African coast. *Un. S. Africa Fish. Mar. Biol. Surv.*, Rep. 11, investig. rep. 3, 1–11.

Craig, H., 1953. The geochemistry of the stable carbon isotopes. *Geochim. et Cosmochim. Acta*, **3**, 53–92.

Curtis, D., 1957. Selected annotated bibliography of the geology of uranium-bearing phosphorites in the United States. *Bull. U.S. Geol. Surv.*, **1059B**, 29–58.

Danil'chenko, P. T. and N. I. Chigirin, 1929. Unterlagen zur Chemie des Schwarzen Meeres. 2. Zur Frage des Stoffkreislaufes in Schwarzen Meer. *Zap. Krim. obsch. estest.*, **11**. (Cited by Caspers, 1957, p. 823.)

Darwin, C., 1887. *Journal of Researches into the Natural History and Geology of the Countries Visited During the Voyage of H.M.S. Beagle Round the World.* New ed., D., Appleton and Co., New York, 1–519.

Dietz, R. S., K. O. Emery and F. P. Shepard, 1942. Phosphorite deposits on the sea floor off southern California. *Bull. Geol. Soc. Amer.*, **53**, 815–848.

Drew, G. H., 1913. On the precipitation of calcium carbonate in the sea by marine bacteria, and on the action of denitrifying bacteria in tropical and temperate seas. *J. Mar. Biol. Assoc. U.K.*, **9**, 479–524.

Emery, K. O., 1952. Continental shelf sediments of southern California. *Bull. Geol. Soc. Amer.*, **63**, 1105–1108.

Emery, K. O., 1953. Some surface features of marine sediments made by animals. *J. Sediment. Petrol.*, **23**, 202–204.

Emery, K. O., 1954. Source of water in basins off southern California. *J. Mar. Res.*, **13**, 1–21.

Emery, K. O., 1956. Deep standing internal waves in California basins. *Limnol. Oceanog.*, **1**, 35–41.

Emery, K. O., 1960. *The Sea off Southern California, A Modern Habitat of Petroleum.* John Wiley and Sons, Inc., New York, 1–366.

Emery, K. O. and D. Hoggan, 1958. Gases in marine sediments. *Bull. Amer. Assoc. Petrol. Geol.*, **42**, 2174–2188.

Emery, K. O. and S. C. Rittenberg, 1952. Early diagenesis of California basin sediments in relation to origin of oil. *Bull. Amer. Assoc. Petrol. Geol.*, **36**, 735–806.

Erdman, J. G., E. M. Marlett and W. E. Hanson, 1956. Survival of amino acids in marine sediments. *Science*, **124**, 1026.

Ewing, W. M., D. B. Ericson and B. C. Heezen, 1958. Sediments and topography of the Gulf of Mexico. In Weeks, L. (Ed.), *Habitat of Oil.* Amer. Assoc. Petrol. Geol., Tulsa, 995–1053.

Feely, H. W. and J. L. Kulp, 1957. Origin of Gulf Coast salt-dome sulphur deposits. *Bull. Amer. Assoc. Petrol. Geol.*, **41**, 1802–1853.

Fleming, R. H. and R. Revelle, 1939. Physical processes in the ocean. In *Recent Marine Sediments.* Amer. Assoc. Petrol. Geol., Tulsa, 48–141.

Garrels, R. M., 1955. Some thermodynamic relations among the uranium oxides and their relation to the oxidation states of the uranium ores of the Colorado Plateau. *Amer. Mineralogist*, **40**, 1004–1021.

Goldberg, E. D., 1957. Biogeochemistry of trace metals. In "Treatise on Marine Ecology and Paleoecology". *Geol. Soc. Amer. Mem. 67*, **1**, 345–358.

Gorsline, D. S. and K. O. Emery, 1959. Turbidity current deposits in San Pedro and Santa Monica Basins off southern California. *Bull. Geol. Soc. Amer.*, **70**, 279–290.

Grippenberg, S., 1939. Sediments of the Baltic Sea. In *Recent Marine Sediments*. Amer. Assoc. Petrol. Geol., Tulsa, 298–321.

Gusow, W. C., 1959. Knowledge on the origin of oil. Preprints General Petroleum Geochemistry Symposium, Fifth World Petroleum Congress, Fordham University, New York, 97–100.

Harmsen, G. W., A. Quispel and D. Otzen, 1954. Observations on the formation and oxidation of pyrite in the soil. *Plant and Soil*, **5**, 324–348.

Hartman, O. and J. L. Barnard, 1958. The benthic fauna of the deep basins off southern California. *Allan Hancock Pacific Expeds.*, **22**, 1–67.

Heezen, B. C., R. J. Menzies, W. S. Broecker and W. M. Ewing, 1959. Stagnation of the Cariaco Trench. *Intern. Oceanog. Cong. Preprints*, 99–102.

Hodgson, G. W., 1959. Petroleum pigments from Recent lake sediments. Preprints General Petroleum Geochemistry Symposium, Fifth World Petroleum Congress, Fordham University, New York, 13–19.

Hülsemann, J. and K. O. Emery, 1961. Stratification in recent sediments of Santa Barbara Basin controlled by organisms and water character. *J. Geol.*, **69**, 279–290.

Jensen, M. L., 1958. Sulfur isotopes and the origin of sandstone-type uranium deposits. *Econ. Geol.*, **53**, 598–616.

Kaplan, I. R., T. A. Rafter and J. R. Hulston, 1960. Sulphur isotopic variations in nature. 8. Its application to some biogeochemical problems. *N.Z. J. Sci.*, **3**, 338–361.

Klepper, M. R. and D. G. Wyant, 1957. Notes on the geology of uranium. *Bull. U.S. Geol. Surv.*, **1046F**, 87–148.

Koczy, F. F., 1953. Chemistry and hydrology of the water layers next to the ocean floor. *Statens Naturvetenskapliga Forskningsråds Årsbok*. Sjatte Årgangen, 1951–1952, 97–102.

Krauskopf, K. B., 1955. Sedimentary deposits of rare metals. *Econ. Geol. Anniv. Vol.*, 411–463.

Krauskopf, K. B., 1956. Dissolution and precipitation of silica at low temperatures. *Geochim. et Cosmochim. Acta*, **10**, 1–26.

Kriss, A. E., 1959. Microbiology and the chief problems in the Black Sea. *Deep-Sea Res.*, **5**, 193–200.

Kriss, A. E. and E. Rukina, 1952. Microorganisms in the deposits of the bottom regions of the oceans. *Izvest. Akad. Nauk S.S.S.R., Biol. Ser.*, **6**, 67–79.

Kuenen, P. H., 1943. Bottom samples. *The 'Snellius' Expedition*, **5**, pt. 3, 1–46.

Kuenen, P. H., 1948. Influence of the earth's rotation on ventilation currents of the Moluccan deep-sea basins. *Koninkl. Ned. Akad. Wetenschap., Proc.*, **51**, 417–426.

LaGanza, R. F., 1959. Pyrite investigations at Nairne, South Australia. *Econ. Geol.*, **54**, 895–902.

Lalou, C., 1957. Étude experimentale de la production de carbonates par les bacteries des vases de la Baie de Villefranche-sur-mer. *Ann. Inst. Océanog. Monaco*, **33**, 201–267.

Lalou, C., 1957a. Studies on bacterial precipitation of carbonates in sea water. *J. Sediment. Petrol.*, **27**, 190–195.

Lewin, J. C., 1959. Dissolution of silica from diatom walls. *Intern. Oceanog. Cong. Preprints*, 950.

Lindgren, W., 1928. *Mineral Deposits*. McGraw-Hill Book Co., Inc., New York, 1–1049.

Lipman, C. B., 1924. A critical and experimental study of Drew's bacterial hypothesis on $CaCO_3$ precipitation in the sea. *Dep. Mar. Biol. Carnegie Inst.*, **19**, 179–191.

McKelvey, V. E. and J. M. Nelson, 1950. Characteristics of marine uranium-bearing sedimentary rocks. *Econ. Geol.*, **45**, 35–53.

McKelvey, V. E., R. W. Swanson and R. P. Sheldon, 1953. The Permian phosphorite deposits of western United States. *C. R. 19th Intern. Geol. Cong., Algiers*, Sec. **11**, 46–64.

Mansfield, G. R., 1931. Some problems of the Rocky Mountain phosphate field. *Econ. Geol.*, **26**, 353–374.

Meinschein, W. G., 1959. Origin of petroleum. *Bull. Amer. Assoc. Petrol. Geol.*, **42**, 821–841.

Miller, L. P., 1950. Formation of metal sulphides through the activities of sulphate reducing bacteria. *Contr. Boyce Thompson Inst.*, **16**, 85–89.

Murray, J. and R. Irvine, 1889. On coral reefs and other carbonate of lime formations in modern seas. *Proc. Roy. Soc. Edinburgh*, **17**, 79–109.

Neeb, G. A., 1943. Bottom samples. *The 'Snellius' Expedition*, **5**, pt. 3, 55–265.

Niino, H., 1959. Manganese nodules from shallow water off Japan. *Intern. Oceanog. Cong. Preprints*, 646–647.

Orr, W. L. and K. O. Emery, 1956. Composition of organic matter in marine sediments: Preliminary data on hydrocarbon distribution in basins off southern California. *Bull. Geol. Soc. Amer.*, **67**, 1247–1258.

Orr, W. L., K. O. Emery and J. R. Grady, 1958. Preservation of chlorophyll derivatives in sediments off southern California. *Bull. Amer. Assoc. Petrol. Geol.*, **42**, 925–962.

Plunkett, M. A., 1957. The qualitative determination of some organic compounds in marine sediments. *Deep-Sea Res.*, **5**, 259–262.

Rankama, K., 1954. *Isotope Geology*. Pergamon Press Ltd., London, 1–535.

Redfield, A. C., 1958. Preludes to the entrapment of organic matter in the sediments of Lake Maracaibo. In Weeks, L. (Ed.), *Habitat of Oil*. Amer. Assoc. Petrol. Geol., Tulsa, 968–981.

Redfield, A. C., 1958a. The biological control of chemical factors in the environment. *Amer. Scientist*, **46**, 205–222.

Rex, R. W. and E. D. Goldberg, 1958. Quartz contents of pelagic sediments of the Pacific Ocean. *Tellus*, **10**, 153–159.

Richards, F. A. and B. B. Benson, 1959. Nitrogen argon and nitrogen isotope ratios in anaerobic marine environments. *Intern. Oceanog. Cong. Preprints*, 896–897.

Richards, F. A. and R. F. Vaccaro, 1956. The Cariaco Trench, an anaerobic basin in the Caribbean Sea. *Deep-Sea Res.*, **3**, 214–228.

Riel, P. M. van, 1943. The bottom water. Introductory remarks and oxygen content. *The 'Snellius' Expedition*, **2**, pt. 5, 1–77.

Riley, G. A., 1951. Oxygen, phosphate and nitrate in the Atlantic Ocean. *Bull. Bingham Oceanog. Coll.*, **13**, 1–126.

Rittenberg, S. C., 1940. Bacteriological analysis of some long cores of marine sediments. *J. Mar. Res.*, **3**, 191–201.

Rittenberg, S. C., K. O. Emery and W. L. Orr, 1955. Regeneration of nutrients in sediments of marine basins. *Deep-Sea Res.*, **3**, 23–45.

Seibold, E., 1958. Jahreslagen in Sedimenten der mittleren Adria. *Geol. Rundschau*, **47**, 100–117.

Shabarova, N. T., 1954. The organic matter of marine sediments. *Uspekhi Souremennoi Biologii*, **37**, 203–208.

Shishkina, O. V., 1959. Metamorphism of chemical substances in the interstitial waters of the Black Sea. In *To the Understanding of the Diagenesis of Sediments*. Akad. Nauk S.S.S.R., Moscow, 29–50.

Siever, R., 1957. The silica budget in the sedimentary cycle. *Amer. Mineralogist*, **42**, 821–842.

Silverman, S. R. and S. Epstein, 1958. Carbon isotopic compositions of petroleums and other sedimentary organic materials. *Bull. Amer. Assoc. Petrol. Geol.*, **42**, 998–1012.

Skinner, B. J., 1958. The geology and metamorphism of the Nairne pyritic formation, a sedimentary sulfide deposit in South Australia. *Econ. Geol.*, **53**, 546–562.

Smirnow, L. P., 1958. Black Sea Basin. In Weeks, L. (Ed.), *The Habitat of Oil*. Amer. Assoc. Petrol. Geol., Tulsa, 982–994.

Smith, E. H., F. M. Soule and O. Mosby, 1937. Marion and General Green Expeditions to Davis Straight and Labrador Sea under the Direction of the U.S. Coast Guard 1928–1931–1933–1934–1935, Washington, 1–259.

Smith, P. V., 1952. The occurrence of hydrocarbons in recent sediments from the Gulf of Mexico. *Science*, **116**, 437–439.

Smith, P. V., 1954. Studies on the origin of petroleum: Occurrence of hydrocarbons in recent sediments. *Bull. Amer. Assoc. Petrol. Geol.*, **38**, 377–404.

Spoehr, H. A., J. H. C. Smith, H. H. Milne and G. J. Hardin, 1949. Fatty acid antibacterials from plants. *Pub. Carnegie Inst. Wash.*, 586, 1–67.

Stanton, R. L., 1955. Lower Paleozoic mineralization near Bathurst, New South Wales. *Econ. Geol.*, **50**, 681–714.

Strakhov. N. M., 1959. Forms of iron in the sediments of the Black Sea and their meaning to the theory of diagenesis. In *To the Understanding of the Diagenesis of Sediments.* Akad. Nauk S.S.S.R., Moscow, 92–119.

Sugawara, K., T. Koyama and E. Kamata, 1957. Recovery of precipitated phosphate from lake muds related to sulfate reduction. *J. Earth Sci., Nagoya Univ.*, **5**, 60–67.

Strøm, K. M., 1939. Land-locked waters and the deposition of black muds. In *Recent Marine Sediments*. Amer. Assoc. Petrol. Geol., Tulsa, 356–372.

Trask, P. D., 1932. *Origin and Environment of Source Sediments of Petroleum.* Gulf Publishing Co., Houston, 1–323.

Trask, P. D., 1939. Organic content of recent marine sediments. In *Recent Marine Sediments.* Amer. Assoc. Petrol. Geol., Tulsa, 428–453.

Trask, P. D. and C. C. Wu, 1930. Does petroleum form in sediments at time of deposition? *Bull. Amer. Assoc. Petrol. Geol.*, **14**, 1451–1463.

Treibs, A., 1936. Chlorophyll and hemin derivatives in organic mineral substances. *Angew. Chem.*, **49**, 682–686.

Vallentyne, J. R., 1957. The molecular nature of organic matter in lakes and oceans, with lesser references to sewage and terrestrial soils. *J. Fish. Res. Bd., Canada*, **14**, 33–82.

Waksman, S. A., 1933. On the distribution of organic matter in the sea bottom and the chemical nature and origin of marine humus. *Soil Sci.*, **36**, 125–147.

Waksman, S. A., 1934. The role of bacteria in the cycle of life in the sea. *Sci. Monthly*, **38**, 35–49.

White, D. E., W. W. Brannock and K. J. Murata, 1956. Silica in hot-spring waters. *Geochim. et Cosmochim. Acta*, **10**, 27–59.

Wood, E. J. F., 1958. The significance of marine microbiology. *Bacteriol. Revs.*, **22**, 1–19.

Worzel, J. L., 1959. Extensive deep sea sub-bottom reflections identified as white ash. *Proc. Nat. Acad. Sci.*, **45**, 349–355.

ZoBell, C. E., 1946. *Marine Microbiology*. Chronica Botanica Co., Waltham, Mass., 1–240.

ZoBell, C. E., 1959. Introduction to marine microbiology. Contributions to marine microbiology. *N.Z. Oceanog. Inst. Mem.* No. 3, 7–20.

24. ESTUARIES, DELTAS, SHELF, SLOPE

A. Guilcher

1. Estuaries, Tidal Marshes and Flats, and Deltas

A. Definitions

Estuaries and tidal marshes are river mouths and coastal areas where tides and tidal currents play an essential part in sedimentation (Latin *aestus* means tide), whereas in deltas the deposition results usually from a large excess of load in the river stream at the place where the river meets the sea. True marshes may exist, however, in some regions where the tidal range is low, for example at Esbjerg, Denmark, or at Newport Bay, California, where it does not exceed 5 or 6 ft at spring tides (Jacobsen, 1956; Stevenson and Emery, 1958), or even 2 ft at Porto Vecchio in Corsica (Blanc and Picard, 1954), but the marshes tend to be very poorly developed when the range falls below 3 ft. Marshes and tidal flats are usually found on both sides of estuaries (Chesapeake Bay, Delaware Bay, Gironde, Loire, Thames, Scheldt, Weser, Elbe, etc.), behind shingle- or sand-spits (Norfolk in England, New England, Guianas, Guinea, etc.), or in sheltered parts of bays (Bay of Fundy, Baie de l'Aiguillon and Marais Poitevin in France). A special type consists of very wide tidal flats extending in front of large reclaimed marshes: these flats are referred to as *Watten* (Dutch: *Wadden*) in Dutch, German and Danish Friesland (Fig. 1); other watts are those lying in the Wash in England, and in Mont Saint Michel Bay and Saint Brieuc Bay in Brittany. On some watts, the distance between high and low spring water marks may be as large as 10 nautical miles or even more (Johnson, 1925, pp. 504–588; Steers, 1946, chap. 14 and *passim*; Van Straaten, 1954; Gripp, 1956; Russell, 1959, including examples from tropical and subtropical countries; Phlipponneau, 1955; Verger, 1954).

Deltas, which are generally piles of sediments at the mouths of rivers, are more especially found in seas where the tidal range is insignificant, e.g. in the Mediterranean (Rhône, Ebro, Po, Nice, etc.), in the Baltic (Weichsel), and in the Gulf of Mexico (Mississippi), because these conditions are unfavourable to a wide dispersal of sediment carried to the sea. They can, however, sometimes coexist with large tidal ranges, for example in the Far East, provided that the sedimentary discharge is very high, and true tidal marshes can occur in deltas when the tidal range, or wind tides (surges), allow them to come into existence (Mississippi). Most deltas have several distributaries, and the Lena River is said to have no less than forty-five mouths; some, such as the Tiber, however, have only one, but deserve their name because they protrude into the sea. The classification of the mouth of the Senegal River is more controversial, since this river has but one mouth and does not bulge the shoreline; the sediments have only filled a former bay and continue to spread in the coastal region (see below). Another transition type is the submarine delta which partially fills the outer part of an estuary, as at the mouth of the River Loire (Guilcher, 1958, fig. 13, p. 111). Submarine deltas are even found with no river at all: these are called

[*MS received April, 1960*]

tidal deltas and occur at inlets between barrier islands, as off North Carolina and off Friesland (Fig. 1). They project inwards and outwards, and originate from the slackening of the flood and ebb currents on both sides of the pass. Thus, the concept of a delta is complex and comprehensive; protrusion forms are diverse (Gulliver, 1898–1899); caution is necessary in definition.

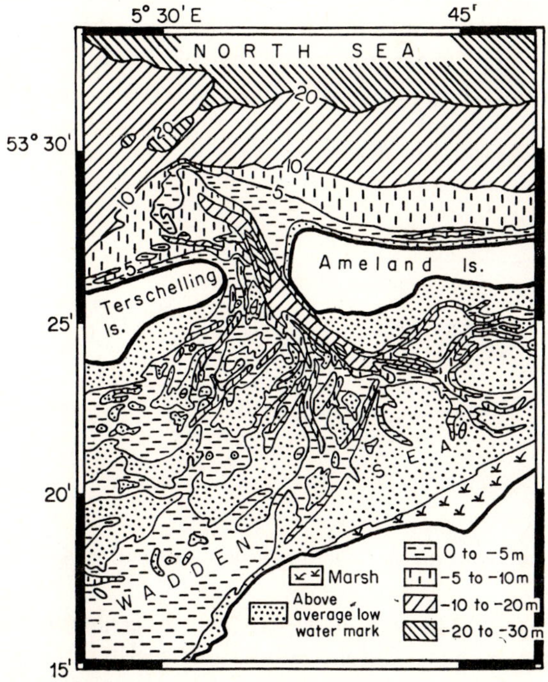

Fig. 1. Tidal delta and part of Wadden Sea in West Friesland. Inlet over-deepened by tidal scour. (From Waddensymposium, 1950, p. 448.)

B. Description of Tidal Marshes

A typical tidal marsh in temperate countries (Fig. 2) includes two main parts. The lower part, which is exposed at low tide and is entirely under water at high ordinary tide, is generally free of vegetation. It is referred to as the *slikke* in international terminology. Sometimes, it bears a *Zostera* cover, or mussel banks, or green algae, e.g. in the Netherlands and in Ireland (Van Straaten, 1954; Guilcher and King, 1961). The higher part is the *schorre* (salt marsh in England and in America), which is entirely covered by sea-water only at high spring tides and always bears halophytes. On the schorre proper, or "high marsh", different species of plants are found according to altitude: the most widespread one in Europe is *Puccinellia maritima*; *Aster tripolium* and *Obione portulacoides* are also numerous; with further increase in height, *Juncus maritimus* is often of common occurrence (see Steers, 1946, for more detail).

Between the high marsh and the slikke a transition area sometimes occurs—the "low marsh" (*haute slikke* in French papers)—with a sparse cover of *Salicornia*. But the *Salicornia* zone often disappears and is replaced by a low cliff, vertical or overhanging, which abruptly bounds the schorre from the slikke. In many marshes of the English Channel and the southern part of the North Sea, *Spartina townsendi* has rapidly spread in recent times. Corresponding zones, with somewhat different species, exist on both coasts of North America (Stevenson and Emery, 1958; Johnson, 1925). These halophytes are able to endure large changes in salinity, according to the time of the tide, the range of the tide and the season (rain, river discharge, etc.), but each zone has its particular ecology. On the other hand, peat can grow in the inner parts of the marshes, which usually remain outside the reach of the salt or brackish water.

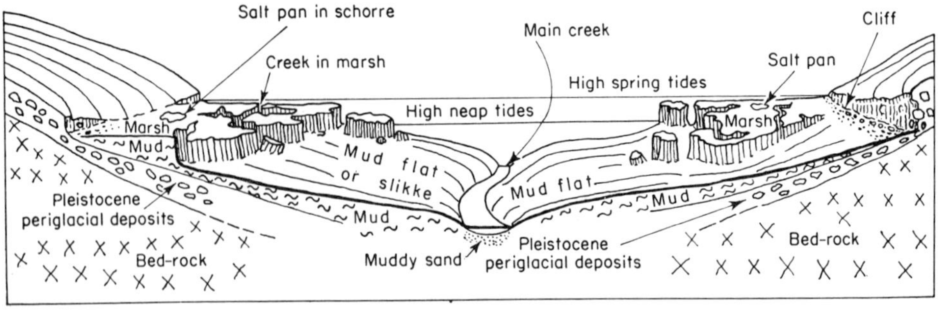

Fig. 2. Diagram showing parts and section of a small estuary in Brittany.

The marshes are fed by large tidal creeks (Dutch: *geulen*) meandering into the slikke, and smaller creeks (Dutch: *prielen*) dissecting the schorre. Straight sections are rare, and very pronounced meanders are much more common than in ordinary rivers, but their general evolution is the same, consisting of cut-offs, captures, etc. Interconnections between creeks are frequent. The ebb current is more important in creek evolution than the flood current, since the former flows during a longer time than the latter. But on many slikkes in watts or estuaries, the channels show an inter-digitated pattern: some of them are flood channels which gradually shoal as they enter the watt, whereas others are only followed by the ebb current: these are deepest in their inner parts and gradually merge into shallow banks at their outer reaches. Such ebb- and flood-channels in the Scheldt estuary have been described by Van Veen (in Waddensymposium, 1950), and on several tidal flats around Great Britain by Robinson (1956), and elsewhere. The surface of the high marsh between the creeks includes ponds which are devoid of plants and are known as salt pans, in which the water becomes highly saline at neap tides by evaporation, a fact which may account for the absence of vegetation. Salt pans are often remnants of silted-up creeks, although some of them may have other origins.

In the intertropical belt, grass is generally replaced by mangrove in tidal

Fig. 3. Meandering tidal creeks in mangroves, north-west Madagascar. (Photo. A. Guilcher.)

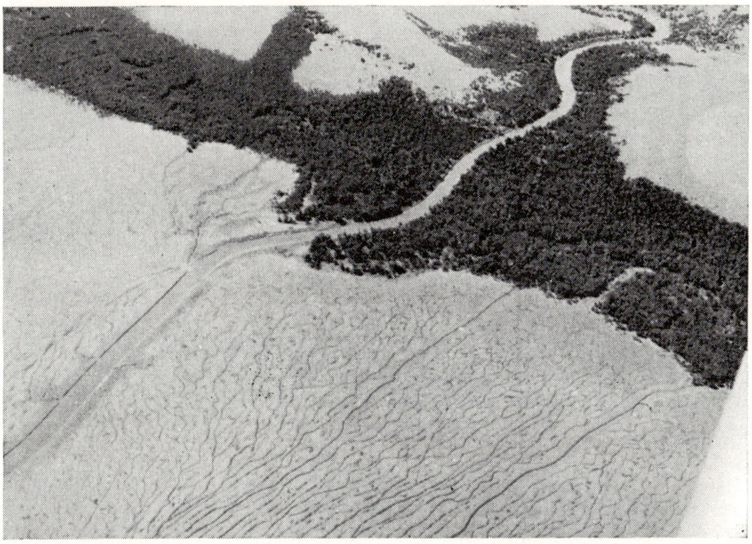

Fig. 4. Part of the Betsiboka estuary, Madagascar. Mud flats in foreground; mangrove in middle distance; bare zone in background, submerged only at high spring tides. (Photo. A. Guilcher.)

marshes (see for example Davis, 1940; Schnell, 1952, pp. 197–218; Lindeman, 1953). On Atlantic coasts, the mangrove trees consist mainly of *Rhizophora* and *Avicennia*, and also *Laguncularia* and *Conocarpus*. As a rule, *Rhizophora* lives on the outer flats, but this distribution may be inverted in young mangroves, and *Avicennia* may be found seaward of *Rhizophora*, for example in Guinea, because it grows more quickly. Similar mangrove successions have been described as evidence of coastal changes in Malaya (Carter, 1959). It must be pointed out that mangroves are not found only in muddy marshes, but occasionally also on rocks or pebbles, and that grassy areas, resembling the temperate high marshes, are found not infrequently among, or instead of, trees, for example in Dahomey where mangroves are rare. On the other hand, between the mangrove zone and the continental vegetation, the high-spring-tide zone often remains completely free of vegetation, for example in Madagascar. This feature may be due to the long dry season and to the high temperature, which

Fig. 5. Mangroves with alternating sand ridges (cheniers) in north-west Madagascar. (Photo. A. Guilcher.)

lead to complete dessication of the surface and crystallization of salt during neap tides so that no plant is able to grow there. The pattern of the tidal creeks in mangroves is the same as in tidal flats and marshes in temperate countries.

C. Estuarine Mud

In estuaries, tidal marshes and flats, the sediments consist principally of mud (*vase* in French; *Schlick* in German), a term which is intentionally imprecise, since the sand–silt–clay ratio is extremely variable in the mud (in American nomenclature, the medium grain size in clay is smaller than 3.9 μ; in silt it lies between 3.9 μ and 62.5 μ; in sand it is coarser than 62.5 μ). A silt or clay fraction is present in all muds, so that the mud always hardens when it dries, whereas sand remains loose. But in some intertidal muds the medium grain size may be as coarse as 250 μ, while in others it may be as small as 2.5 μ (Berthois, 1955). As a rule, the grain size is coarser in the tidal creeks, where the sand fraction may be very large, and it decreases on tidal flats and salt marshes, where much higher clay contents occur. Examples of this distribution are found in the Wadden Sea, West Friesland (Fig. 6), off Cuxhaven, East Friesland (Gellert, 1952), and in North Friesland (Hansen, 1951). True sand may be seen on parts of the outer watts bordering the North Sea and the English Channel, and also on tidal flats behind sand-spits in Kerry, Ireland (Guilcher and King, 1961). This is connected with processes of deposition which are considered below. As a consequence of the wide range of the grain size in muds, the sorting

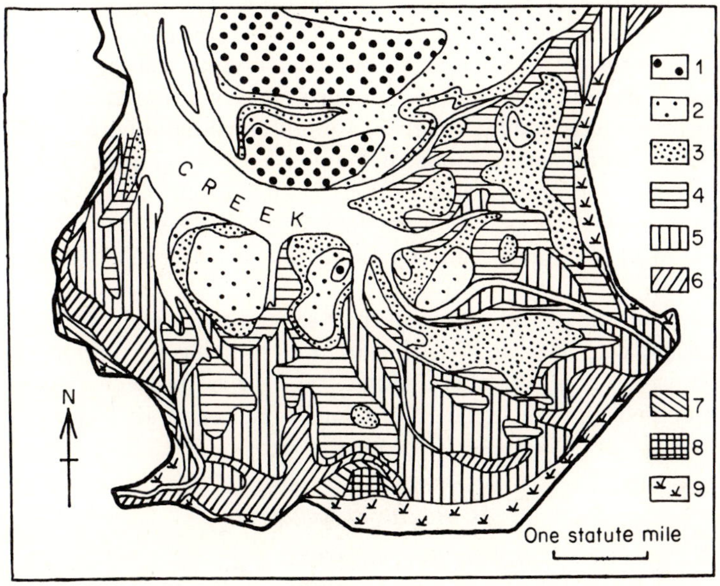

Fig. 6. Lutite (< 2 μ) contents of sediments in Lauwerszee, Wadden Sea, West Friesland: general increase towards the coasts and the watersheds. (From Van Straaten and Kuenen, 1957.)

Legend—1: 0–1½%; 2: 1½–3%; 3: 3–5%; 4: 5–8%; 5: 8–12%; 6: 12–17%; 7: 17–25%; 8: > 25%; 9: marshes.

is always much poorer than on beaches and in submarine sands on the shelf where strong currents act upon them. In deltas, the sorting is also poor, and the grain size is even more variable than in estuaries: the sediments can include large pebbles in Mediterranean deltas, such as the Var delta in south-east France, while in the Mississippi delta the coarser fraction consists of fine sand. On the other hand, the particles are not the same in all parts of the deltas, since they depend on the processes of deposition (see below).

Estuarine and deltaic sediments also contain an organic fraction, consisting of algal particles, small pieces of roots, remnants of planktonic animals, and so on. The organic fraction sometimes constitutes a large volume, and it leads, either in fresh or in salt water, to the formation of suspended flakes in which the mineral particles are enclosed in fibrous networks. Foam also includes small mineral grains, giving them more buoyancy and allowing their transportation to the higher parts of the marsh (Waksman, 1933; Trask, 1939–1955; Bourcart and Francis-Bœuf, 1942; Francis-Bœuf, 1947; Rajcevic, 1957). Nevertheless, the organic fraction is always considerably smaller in estuarine muds than the mineral fraction, which usually exceeds 90% by weight (Berthois, 1955). The exceptions are peats (which are not truly muds) and *tangues* deposited on the watts of north-east Brittany (Bourcart and Charlier, 1959), in which the lime content is unusually high (more than 60% sometimes) and consists of comminuted fragments of marine shells and spicules of sponges. These tangues are comparatively coarse, since the clay is almost completely lacking, and fine sand predominates with some silt. Fine calcareous sediments deposited in shallow water in warm seas, and euxinic or poorly ventilated basins, where the organic matter content is much higher, are not considered here.

Iron is found in estuarine muds, in percentages ranging from 2.1 to 7.3, in examples given by Bourcart and Francis-Bœuf (1942, p. 21). Along with the organic content, it acts as a binding factor, and bacteria, which live in very large amounts at the surface of the tidal flats, seem essential in its fixation (Harder, 1919).

The water content of the soft mud may exceed the weight of the dried sediment, and those who have walked on tidal mud flats have experienced how deep one can sometimes sink into them. However, mud is plastic, and the footmarks remain impressed into its surface for a long time, even after many tides have passed over it.

D. Source of Sediments

The source of the mineral fraction in deltas is obviously the catchment area of the river, or a part of it. Van Andel (1955) has thus shown, by means of heavy-mineral analyses, that "at least four-fifths of the Rhône delta sediments have been derived from the Alps; a comparatively small contribution is made by the Massif Central". This is not surprising, because the tributaries coming from the Alps are by far the most important ones. According to the same author, the sand in the Lower Rhine is supplied by reworked Pleistocene deposits, which came partly from Switzerland and partly from the Rhenish Schiefergebirge (Van Andel, 1950, p. 114).

Concerning estuaries and tidal marshes, the problem of the source of the sediments is more complex. A mass of material in suspension is carried to and fro with the rising and falling tide, but whence this *bouchon vaseux* (Glangeaud, 1938) comes is disputed, and various views are held. Many Dutch authors, especially Crommelin (1940, 1947), have come to the conclusion that all fractions of the sediments of the Wadden Sea in West Friesland are derived from the North Sea, since the heavy minerals are the same as those found in the fine particles on the bottom of the North Sea, and different from those carried in suspension in the rivers Ems, Weser and Elbe. Van Straaten and Kuenen (1957) support this conclusion with some reservations: the main source is the North Sea; the River Ijssel may have supplied some mud, but this contribution was "only of subordinate importance". The evidence which has been produced is convincing, and this result seems to be well established.

Rajcevic (1957) also thinks that 75% of the mud in the Seine estuary is of marine origin; however, evidence for this is scanty and doubtful, and, in the writer's opinion, the Seine estuary needs more attention before a definite conclusion can be reached. It might be supposed that a large supply of marine particles is made easily available by the fact that, in the salt wedge which has been proved to exist in many estuaries in the world, the resulting current goes upstream at and near the bottom, whereas the resulting outflow to the sea occurs near the surface. Such a distribution of the currents is encountered, for example, in Chesapeake Bay (Pritchard, 1952). Nevertheless, Burt (1955) assumes that, in Chesapeake Bay, the settling particles are mostly coming from the rivers ending in the estuary. Spectrophotometer readings show that the maximum of suspended materials in the estuary occurs in spring, a fact which can be explained neither by the production of organic matter, which is as large in summer as it is in spring, nor by the tidal currents, which have no annual cycle, nor by the wind stress; on the contrary, the peak in river discharge occurs in spring, and this is in accord with observations. It may be suggested that the bottom current carries along particles of terrestrial origin, which first travelled downstream in the surface current.

A continental origin of the estuarine sediments has also been well established by careful and numerous observations over a period of several years carried out by Berthois in the Loire estuary (several papers summarized in Guilcher, 1956–1957; see especially Berthois, 1956). Multiple measurements at the entrance of the estuary to the north of Saint Gildas Point have shown that turbidity is always higher at all depths during ebb than during flood tides, so that a net inflow of marine suspended material is impossible. The continental suspended material can hardly be carried outside at neap tides and in periods of low river discharge, but it is partly driven to the sea during the floods, or at spring tides in summer, when the viscosity of the water is low. On the other hand, the mineral fraction in the *tangues* of the Breton tidal flats has been supplied by Pleistocene loess or head existing in nearby cliffs, and even from the foreshore beneath the *tangue*; only the calcareous fraction (see above) is derived from marine shells. Burrowing organisms (worms, lamellibranchs, amphipods, etc.) largely contribute to converting the loess into *tangue* (Milon, 1935). In the small estuaries along the coasts of Brittany, thermal differential, X-ray and grain-size analyses and mineralogy support the conclusion that the mud is derived from periglacial Pleistocene deposits (rubble drift) covering the slopes of the estuaries (Fig. 2), which are washed by waves at high spring tides. The deposits are cliffed: the coarser fraction settles at the foot of the cliffs and the finer fraction is brought into suspension in water and feeds the mud flats and tidal marshes (L. and C. Berthois, 1954–1955; Berthois, 1955; Guilcher and Berthois, 1957). In Derrymore marsh, Tralee Bay, Kerry, mud is also derived mostly from the washing of continental deposits outcropping in the immediate vicinity (Guilcher and King, 1961); and, in Newport Bay, California, mud is supplied by slumping of shales and siltstones outcropping in the surrounding cliffs (Stevenson and Emery, 1958, p. 51).

In tropical countries, where observations are not so numerous, it may be assumed in a number of cases that estuarine mud is also derived from the continent. A typical example may be cited at Majunga, Madagascar, where the Betsiboka estuary has been filled up at a tremendous rate by red lateritic muds, as a consequence of deforestation in the catchment area of the river (LaFond, 1957). The lagoons stretching along the coast of the Gulf of Benin in Ivory Coast, Ghana, Togo and Dahomey, are likewise fed by rivers during the annual floods, and a large part of the suspended mud spreads into the sea through the outlets (Guilcher, in Russell, 1959).

Thus, the mud deposited in estuaries and tidal marshes seems to be generally ascribed to a terrestrial supply, according to the evidence given by accurate investigations. The case of the Wadden Sea in Friesland appears to be a notable exception.

E. Processes of Deposition and Erosion

Present-day sediments in estuaries and deltas are the ultimate result of deposition processes which were in progress during the so-called Flandrian transgression, corresponding to the melting of Pleistocene ice-caps. Practically all the river mouths in the world rest upon a sedimentary column filling an old valley which was carved during Pleistocene low sea-levels. The few exceptions consist of small hanging valleys in which estuaries are lacking. Examples for Great Britain may be found in Codrington (1898) and for Brittany in Guilcher (1948, pp. 157, 320 and 584–585). In the United States, the borings in Chesapeake Bay (Hack, 1957) have shown that the bottom of the old valley system lies at 200 ft below present sea-level near Annapolis. The gradients of the profiles of some tributaries steepen near the main river; this is an inheritance of the Pleistocene deepening. The oldest deposits of the fill consist of coarser sediments (sand and gravel) than the present estuarine muds. This occurs widely in, for example, the Mississippi valley and in many European estuaries and is probably related to overloaded and braided-channel streams at the end of the Pleistocene. Another typical area in this respect is the Texas and southwestern Louisiana coast, for which data are very accurate (LeBlanc and Hodgson, in Russell, 1959; Shepard and Moore, 1955; Gould and McFarlan, 1959; and others).

The eustatic rise of sea-level is not alone responsible for the recent aggradation in estuaries and deltas. In several large coastal plains, a long-continued subsidence must also be taken into account. Two typical examples are the Lower Mississippi Plain and the Low Countries along the North Sea. In both areas, the Pleistocene is some 2000 ft thick under present sea-level in its thickest sections (see for instance Akers and Holck, 1957; Pannekoek, 1956; and Chapter 13 in this book). In the coastal marshes and estuaries bordering the North Sea from Zeeland to the Elbe mouth, the combination of subsidence and eustatic shifts of sea-level has resulted in most complicated stratigraphy. The marine deposits related to the last two interglacial periods, which are referred to as the Holsteinian (or Needian) and the Eemian deposits (Woldstedt, 1950), are depressed under the marshes, while they stand as subaerial terraces in coastal areas where no subsidence occurs. During the last 10,000 years—the Holocene—the following strata were laid down in order (Fig. 7): a layer of peat (the lower peat), deposited before the post-glacial sea-level was high enough to reach these countries; the older tidal flat deposits and marine clays, deposited behind barrier beaches; the upper peat, which may tentatively be related to a fall of sea-level by a few metres after the post-glacial climatic optimum; and young

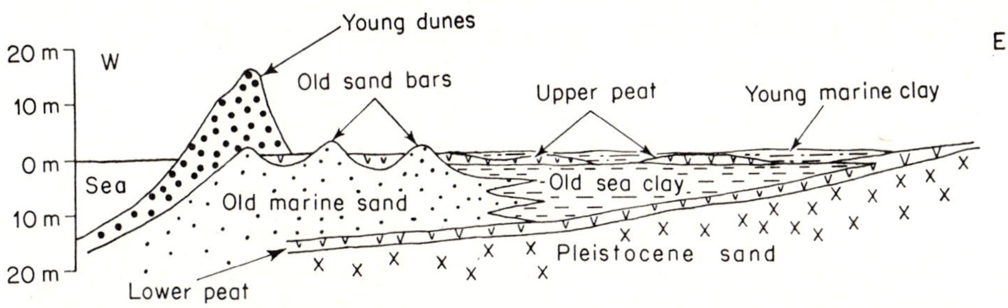

Fig. 7. Diagrammatic section through the Holocene deposits of Holland. (Simplified from Pannekoek, 1956.)

marine clays and sands, resulting from the fact that the small lowering of sea-level was finally surpassed by the subsidence, which led to invasions of the peat bogs by the sea. Such alternations of marine and terrestrial deposits, with peat layers in the sub-surface, have morphological consequences in the coastal marshes (see below).

If we consider now the processes acting today, we must describe separately the deltas, the estuaries and the related marshes.

a. Deltas

Deltas (see, for example, Russell and Russell, in Trask, 1955) consist of a complex of distributary channels, natural levees bordering the channels, and shallow lakes or swamps lying between them at a slightly lower level. The coarser elements are concentrated in the channels and on the crests of natural levees; the finer ones are found in inter-levee lakes or swamps, where clay can settle to the bottom in quiet waters. Organic sedimentation is also important in this environment. As the lakes are slowly filled up, they gradually enlarge by wave erosion along their banks, and their contours become more and more regular and rounded. During floods, crevassing can occur in the levees, so that the quiet sedimentation in the lakes is disturbed and coarser layers consisting of sand or silt are deposited upon the clays. Gullying and coarse lenses in clay beds are thus common. The counterpart of these deposits in the emerged part of the delta is a subaqueous offshore deposition, including topset beds which form a shallow submarine plain, foreset beds extending outwards and dipping downwards in inclined and parallel strata, and bottomset beds on the sea-floor beyond the submarine delta front (Fig. 8). As a rule, the foresets are coarser than the topsets, and the bottomsets are the finest ones; and, as the delta is prograded, the foresets gradually overlap the older bottomsets. Actually, this is an idealized picture which may be found in small deltas but rather infrequently in large ones, where it is disturbed in various ways. On the east margin of the Mississippi delta (Fig. 17), which has been thoroughly investigated by Shepard (1956), the foreset beds, or "pro-delta slope", are definitely finer than the bottomset and topset beds, with a percentage of clay as high as 55%, and their dip scarcely exceeds 1 degree, whereas it reaches commonly 10 to 25 degrees in smaller deltas. In the submarine part of the Rhône delta (Van

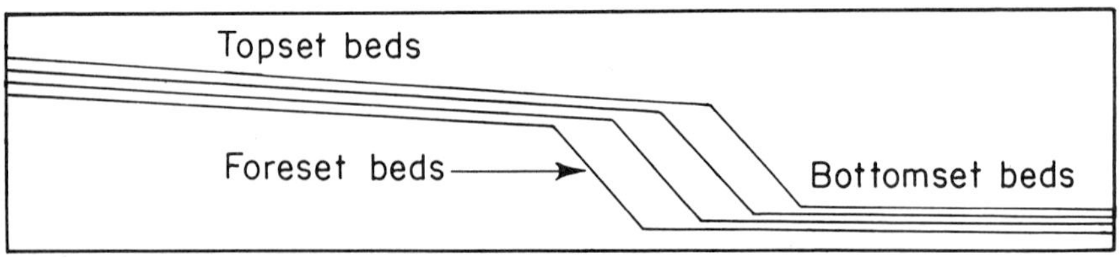

Fig. 8. Schematic development of the subaqueous parts of a small delta.

Straaten, in Russell, 1959), the recent foreset beds are much steeper and coarser. In both areas, submarine gullies in the delta front complicate the picture.

A major phenomenon in delta evolution is the shifting course of the river into successive distributaries, which may lead to the building of successive sub-deltas. As the delta protrudes more and more into the sea, the average slope and the capacity decrease, while much shorter routes can be found across the swamps. An occasional crevasse initiates a new course, and a new sub-delta comes into existence, till it in turn is abandoned, and so on. But the inactive parts are quickly eroded by sea waves, since they are no longer fed with sediment by the river, and the balance between sea and river forces then favours erosion. The Mississippi delta is classic in this respect: within the last 5000 years the river has built five large sub-deltas (Fig. 9). The oldest ones (Cypremort and Teche) are the most decayed; the intermediate ones (Lafourche and Saint Bernard) have subsided below sea-level in their outer parts and barrier islands have been built by waves in front of them with the sand fraction. These islands are rapidly migrating landwards, but the old river courses (the so-called *bayous*) and levees remain as yet conspicuous. Even the present sub-delta (Balize) is not fully active in all its parts, and some of its distributaries tend to become tidal creeks (Welder, 1959). Other good examples of shifting courses are found in the Medjerda delta, Tunisia (Pimienta, 1959), which is also undergoing subsidence, and

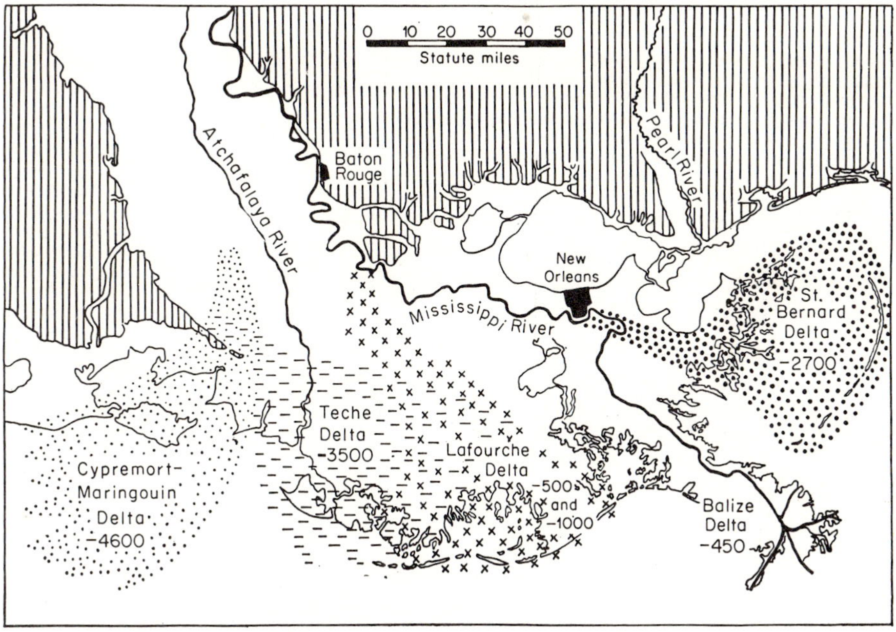

Fig. 9. Mississippi River Holocene sub-deltas. (From Louisiana Coastal Studies Institute, Russell, 1959, p. 370.) Hatchings: Pleistocene terraces. Figures indicate beginning of Mississippi River occupation before present time.

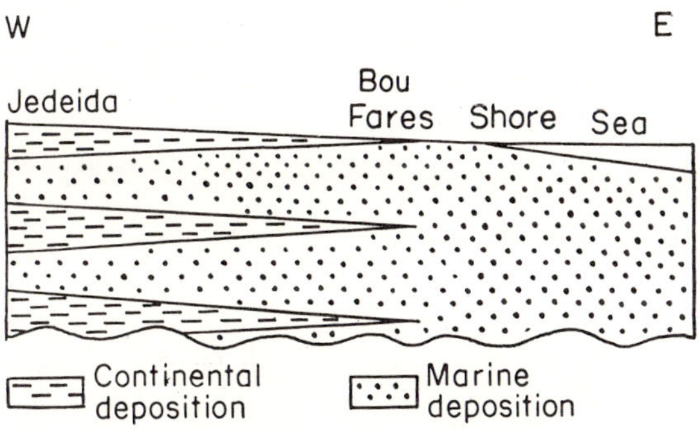

Fig. 10. Scheme of Utiquian cyclothems in the Medjerda River delta, Tunisia. (From Pimienta, 1959.)

in the Rhône delta, which has been carefully investigated in recent years (Duboul-Razavet, 1956; Kruit, 1955; Van Straaten, in Russell, 1959). According to Pimienta, a Pleistocene sub-delta of the Medjerda River is found immediately to the south of Tunis, and the modern delta includes a number of *cyclothems* in which marine and continental sediments alternate (Fig. 10). This type of stratification is common, and points to alternating advances and retreats of the delta front resulting from changes in balance between supply of sediments, subsidence, shiftings in sea-level, strength of waves and long-shore drift. The Rhône delta does not divide into true sub-deltas, but large changes in discharge in its various distributaries have led to recent accretion at the eastern mouth and erosion further to the west, where the submarine foreset beds have been truncated and partially reworked into beaches.

When the amount of sediment distributed over the delta sea-front by waves is very small, as at the present mouths of the Mississippi, beach ridges may completely fail to occur, but submarine bars appear and can emerge at the mouths of the distributaries: they are known as *theys* in the Rhône delta (Duboul-Razavet, 1956) and cause frequent bifurcations at the passes of the Mississippi (Welder, 1959). Combinations of natural levees along distributaries

Fig. 11. Pass a Loutre, Mississippi delta, showing growing levees. (Photo. A. Guilcher.)

and recurved courses of these channels under the influence of long-shore drift which lead them to follow a course parallel to the coast are described as "perilittoral levees" by Mrs. Duboul-Razavet and Pimienta. More purely marine features reworking the coarse fraction of deltaic deposits are the *cheniers* or successive beach ridges of south-western Louisiana (Byrne *et al.*, 1959). They form beautiful birdfoot complexes in the Rhône delta, where they have been mapped by C. Kruit. Every kind of transition exists between digitated deltas such as the Mississippi and stunted deltas such as the Guadalaviar mouth near Valencia in Spain (see Gulliver's classification, 1898–1899).

Climatic conditions diversify the processes in deltaic sedimentation. One of the most curious climatic types is the arctic or periglacial delta, principally represented in Siberia (Lena, Yana, Indigirka, Kolyma), and existing also in Canada (Mackenzie). The deposits are deeply frozen during the greater part of the year, but in summer catastrophic floods rush from the south, where the break-up of the ice occurs distinctly earlier than in the delta: the mean discharge of the Lena River in June is said to be as large as 2,275,000 ft^3/sec; in other words it nearly reaches the absolute maximum discharge reported from the Congo River. This huge mass of water, carrying ice-rafts and trees, is often not able to find its way in the delta through the existing distributaries, since these are still frozen, so that it commonly overtops the banks and creates new channels every year. The resulting network is very confused. Concerning the grain size of the sediments, Carsola (1954) gives information for the Mackenzie River, which appears to carry to the open sea only fine silts and clays, either in suspension or through ice-rafts (Fig. 22). It would be interesting to know the

Fig. 12. North Mahavavy delta, Madagascar, showing a braided-channel pattern. (Photo. A. Guilcher.)

part played by wave action on the delta front: it may be supposed that it would be of very small extent, because the sea is frozen at the time when waves could overcome the river forces.

Another type includes intertropical deltas in which disastrous floods occur owing to the great rains which fall over large areas during hurricanes or typhoons. The North Mahavavy delta in the north-west of Madagascar belongs to this type (Fig. 13). Each big hurricane results in a new course. The channels are braided and basins exist as distinct units. The sediments in channels are of two types: coarse sand deposited during floods and fine sand, silt and some clay deposited during low discharge. Micaceous silts of low density are found on levees. The clay content is a maximum in the basins, although coarse quartzose elements are present. On the seaward side, between meandering tidal creeks extensive mangroves grow on mud which includes an important coarse fraction of quartzose sand. Waves can rework the fluviatile sand into beaches or bars, when the river discharge is low during the dry season.

Another deltaic environment is found at the mouth of tropical rivers flowing into the sea after a long and low course across arid or semi-arid countries, e.g. the Senegal River (Tricart, 1955, 1956). At the time of high river discharge (August to November), numerous basins adjoining the lower valley are flooded and fed with silt and clay. Then the discharge becomes extremely low or even practically nil for several months, so that, owing to the exceedingly low gradient of the Senegal River, sea-water enters deeply into the distributaries with the tide, and is driven into the basins when the floods of the next rainy season arrive. Other sources of salt for basins lying close to the sea are spray from the

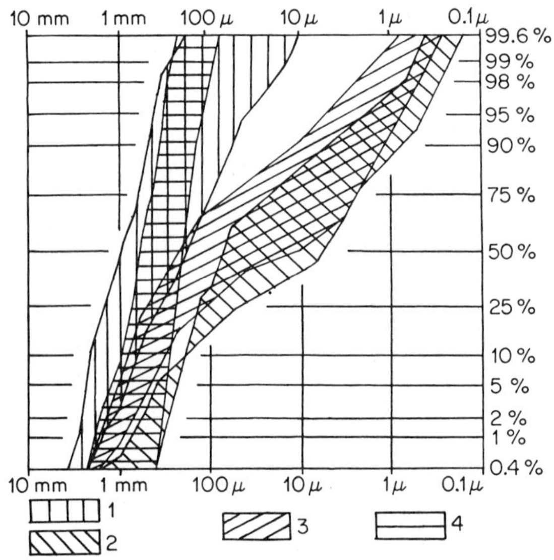

Fig. 13. Enveloping cumulative curves of grain-size distribution in four environments in the North Mahavavy delta, Madagascar. (From data in Berthois and Guilcher, 1956.)
Legend—1: coarse sediments in channels (7 samples); 2: basins (2 samples); 3: coastal mangroves (3 samples); 4: beaches on delta sea-front (6 samples).

heavy surf beating upon the outer beach and percolation of sea-water through the coastal dunes at high tide. As a consequence of this, the water in the basins is saline. Since a large proportion of the basins are dry at the end of the dry season or even before, the salt crystallizes and salt mixed with clay, silt and sand is transported by the wind and deposited in more sheltered areas. Owing to the climatic environment, the inter-levee basins are thus *sebkhas*, or saline bahadas, and the rôle of salt in deltaic sedimentation is much more important than it would be under a humid climate. Rather similar processes have been reported from the Laguna Madre flats, coastal south Texas, by Fisk (in Russell, 1959), Price (1958) and others.

b. Estuaries and marshes

In these the processes and rates of deposition on schorres or high marshes have been studied and measured in Europe (Wales, Denmark, England, Brittany, Poitou, Normandy) by means of sand layers scattered in patches between the plants on the muddy or silty marshes. Borings show the thickness of mud which has settled upon the sand (Richards, 1934; Nielsen, 1935; Steers, 1938–1948, 1946; Guilcher and Berthois, 1957; Verger, 1956; Elhai, 1959). The rate of retreat of the low cliffs at the outer edge of the schorres has also been measured by Guilcher and Berthois.

The rate of accretion is extremely variable, ranging from 0.5 mm in five years to 8 mm per month. It depends on the marsh, and, in any individual marsh, on the height and on the distance to big creeks acting as feeders, and also on the type of vegetation. The relative importance of these factors differs according to the marshes. In Brittany and Poitou, it was found that the height is generally the controlling factor if the low marsh (or high slikke) is not considered: the maximum accretion takes place on the lowest parts of the schorre. In north Norfolk, England, the distance to the creeks is important in determining the accretion rate, particularly if dense *Obione* colonies grow along the creeks. *Obione* is also of particular importance in Normandy, since the thickest deposition occurs in it. Irregularities in deposition exist from year to year, but the sand patches are very rarely removed.

At the same time, the edges of the schorres are retreating at the places where they are cliffed. The rate of retreat is also highly variable, since the recession is caused by blocks of hardened mud falling down from time to time onto the mud flats. Photographs repeatedly taken in the same areas show that these blocks are progressively disintegrated and their mud returns once more to suspension in the water, which in turn feeds the marsh surface and allows it to continue to grow. This may be defined as a true cyclic evolution of the mud.

On low marshes or high slikkes, the same method has been used in *Salicornia*. The rate of deposition has proved to be higher than on schorres in many cases; but deposition alternates with erosion, since sand patches have often disappeared, sometimes after they had first been covered by a rapidly growing layer of mud. The conclusion is that deposition is more irregular than higher up on the marsh, probably because *Salicornia* is an annual plant the vegetal cover of which no longer exists in winter. Moreover, the cover is always more sparse than on schorres, and as the plants are stirred by the wind they gather some mud and can erode the flats.

The fact that deposition generally exceeds erosion in marshes has been explained by Van Straaten and Kuenen (1957) as follows. The velocity of the tidal current entering the marsh decreases as the current goes farther into the marsh. In such a current, there is a time lag between the moment at which it is no longer able to hold its sedimentary material in suspension and the moment at which this material reaches the bottom. This "settling-lag effect" increases the distance for transportation to the inner reaches of the marsh. On the other hand,

Fig. 14. Part of Le Conquet estuary, Brittany, on September 5, 1951. (Photo. A. Guilcher.)

Fig. 15. Same part of Le Conquet estuary on September 9, 1955, showing erosion of high marsh edges and outliers since 1951. (Photo. A. Guilcher.)

the minimum velocity required to erode a sediment after it has been deposited is higher than the maximum velocity at which the same particles can settle: this "scour lag" favours an excess of deposition. Likewise, the decrease in depth from creeks to watersheds makes easy the deposition of suspended particles reaching the watersheds at the end of the flood, whereas the particles reaching the outer parts of the channels at the end of the ebb have more difficulty in settling in these deeper waters. This accounts for the deposition of fine sediments, especially on watersheds and inner marshes (Fig. 3).

Berthois's studies (1953–1958) in the Loire estuary have also added much to the knowledge of the processes. Deposition is more active in warm water than in cold water (from which we can infer that its rate should be high in mangroves) and in salt water than in fresh water; it occurs principally at slack tide, and not during the ebb as Francis-Bœuf (1947) suggested. It reaches a peak in a lense of immobile water which has been proved to persist for several hours near the bottom in a part of the Loire estuary, and also in smaller Breton rivers, especially during the ebb. The silting up of the lateral marshes is favoured by a

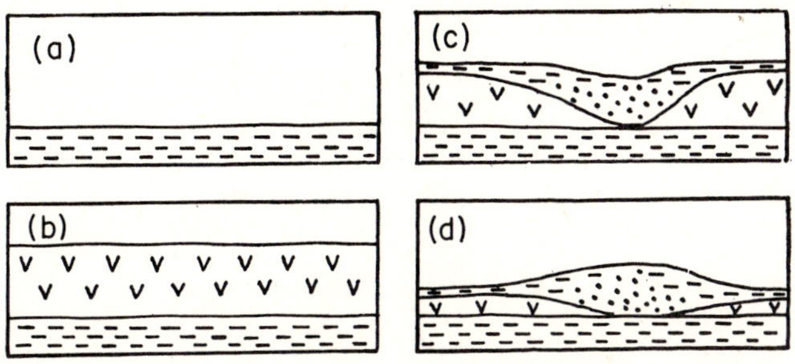

Fig. 16. Inversion of relief by shrinking of peat in Flanders. (From Tavernier, 1947.) (a) Deposition of old marine clay; (b) formation of peat; (c) peat eroded by creeks; deposition of young marine clay and sand; (d) shrinking of peat and inversion of relief.

transverse pulsation of the zone of maximum turbidity, which occurs during the flood. Glangeaud (1938) has also explained the median mud-banks in estuaries as a consequence of vertical eddies on both sides of the channel, with descending and converging currents which feed the banks. Marshes lying to the lee of sand dunes are often very sandy, owing to the winnowing of dunes by wind, as in Kerry, Ireland (Guilcher and King, 1961).

When marshes include peat lenses, a drying-up of the marshes causes the peat to shrink to a small fraction of its original volume. In Flanders, the Netherlands and elsewhere, the shrinkage of the peat leads to an inversion of relief (Fig. 9): when the sea invaded the peat bogs, creeks were carved into the peat and were partially or completely filled with sand and clay, while clay settled on the flats where the peat had not been eroded. After the marshes have been reclaimed, the peat shrinks beneath the clay which now occurs in depressions, whereas the former creeks stand as winding ridges at a higher level (Tavernier, 1947; Pannekoek, 1956). Peat also shrinks when it is compressed by a sand bar retreating landwards and encroaching on the marsh (Johnson, 1925). Whether the marsh is peaty or clayey, multiple sand barriers, or cheniers, alternating with long, narrow swamps, are a common feature and point to successive stages in progradation: such areas have been studied in the Guianas (Van der Eyk, 1957, with an excellent map; Vann, in Russell, 1959; and others), in Texas and Louisiana (Price, 1955; LeBlanc and Hodgson, in Russell, 1959; Byrne, LeRoy and Riley, 1959; Gould and McFarlan, 1959; and others), in Dahomey (Guilcher, in Russell, 1959), etc.

Laminations, or the presence of thin successive layers, are expected to occur in estuarine and deltaic deposits. Actually, the situation is not always the same, since it depends on processes of deposition and on ecological conditions (Van Straaten, 1959; Häntzschel, in Trask, 1939). Salt marshes (schorres) show laminations which are often graded "as a result of decreasing competency of the currents during each stage of inundation of the marshes". On banks of tidal creeks in flats, other laminations are due to variations in competency of tidal currents, and show irregularities related to local erosion. Laminations are also found in some kinds of deltaic deposit, but in other tidal environments they disappear owing to burrowing animals, especially in tropical mangroves where innumerable crabs, mud-skippers (gobies), etc. hide in the mud. At the places where such disturbances occur, it does not seem possible to study the deposition by the method of artificially scattered sand patches.

2. Continental Shelf

The simplest conceivable distribution of sediments over the continental shelf includes well sorted sands near the shore, followed by silts and clays more and more poorly sorted on the median and outer parts of the shelf as the depth increases and the waves become less efficient in stirring the sediments on the bottom. Such a distribution would fit with a well graded slope of the shelf, consisting of a smooth curve, concave upward, which may be defined as the profile of equilibrium of the shelf. Before accurate sedimentological investigations were carried out on the shelf, this feature was expected to be of general occurrence, and, according to H. C. Stetson (in Trask, 1939, 1949), it is actually found in sections off the east coast of the United States.

However, Stetson himself admits that "the fully graded profile . . . would be the exception. . . and represents the theoretical form that should eventually be arrived at, but seldom is". The profile shows irregularities in several New England traverses, and examples are given in the same area by Stetson in which the sediments near the outer edge of the shelf "are coarser than some of those found in shallower water near shore". As early as 1939, Twenhofel insisted on the same idea, which was also supported by findings of pebbles in the Celtic Sea and in the Bay of Biscay at depths ranging from 30 to 55 fm, where the waves are no longer able to stir them (Furnestin, 1937).

Shortly after World War II, Shepard gave in his treatise (1948) the results of an extensive examination of detailed charts covering the principal shelves around the world. He came to the same conclusion already reached for the continental borderland off Southern California by Revelle and Shepard (in Trask, 1939), namely, that "there is no progressive decrease in average grain size with either increasing depth or distance from shore": the general distribution is irregular, and it is not at all unusual to find coarse sediments on the outer shelf.

During the last fifteen years, a great deal of new data has been gathered concerning shelf sediments, especially in European and North American waters but also in hitherto little-known areas such as the platform off Venezuela and

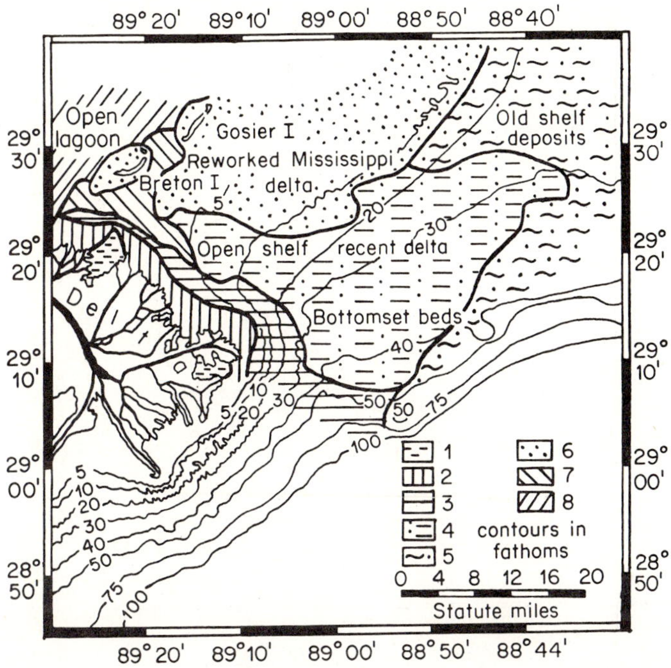

Fig. 17. East Mississippi delta sediment environments. (Slightly simplified from Shepard, 1956, fig. 5.)

Legend—1: interdistributary bays; 2: delta-front platform; 3: pro-delta slope; 4: open shelf recent delta; 5: old shelf deposits; 6: reworked Mississippi delta deposits; 7: open inlet deposits; 8: open lagoon deposits.

Western Guiana where accurate investigations were initiated in 1952 under Kuenen's leadership. The results agree entirely with the provisional conclusions drawn from chart examination. An increase in grain size on the outer shelf is now well established for a large number of areas, and this feature is an inheritance from Pleistocene conditions, as Stetson (in Trask, 1939) and Daly (1934, p. 206) had already suggested for the New England shelf.

One of the most striking examples is afforded by the eastern margin of the Mississippi delta (Fig. 17). Although the Mississippi River at present carries to the sea a tremendous sedimentary discharge, averaging perhaps two million tons per day, the present-day sediments do not settle farther east than some forty statute miles off the eastern passes. This part of the shelf, 20 to 40 fm deep, is occupied by old sediments, much coarser than the present bottomset shelf deposits nearer shore, since sand on the average constitutes almost 60% of these older sediments compared with only 2% in the latter. They include macro-organisms representing an older environment characteristic of a shallow or even estuarine water, in which a ^{14}C determination on some of the shells showed an age of about 7000 years. They seem thus to have been related

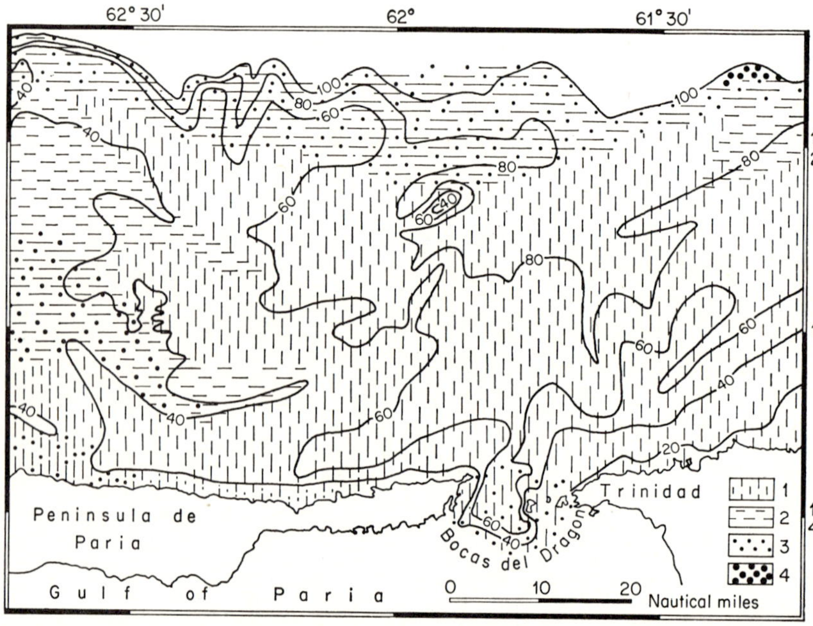

Fig. 18. Distribution of terrigenous sediments on the shelf off Peninsula de Paria, Venezuela, and to the north of Trinidad Island, showing an increase of coarse particles near the edge of the shelf. (Simplified from Koldewijn, 1958.)

Legend—1: percentage of pelites (particles smaller than 50 μ) exceeding 30; 2: percentage of pelites between 30 and 5; 3: percentage of sand particles between 5 and 30; 4: percentage of sand particles exceeding 30. Contours in fathoms.

to a lower sea-level at the end of the Pleistocene or the beginning of the Holocene. Off West Louisiana and Texas, however, Greenman and LeBlanc (1956) assume that Recent clastic deposits have accumulated on the shelf (although they do not give figures for the thickness of accumulation on the shelf, but only on the slope). The causes would be the discharge from the Mississippi and Texas Rivers and the prevailing east-to-west current. But "Recent" is defined by these authors as including the last 20,000 years, that is, a large part of the Wisconsin or Würm glacial stage (see Woldstedt, 1958), so that it cannot be said that these sediments have been deposited in Holocene conditions.

On the shelf off Trinidad, Venezuela and Western Guiana (Van Andel and Postma, 1954; Koldewijn, 1958; Nota, 1958), the distribution is much the same as to the east of the Mississippi delta (Fig. 18). Pelitic sediments, including more than 30% of terrigenous particles finer than 50 μ and less than 5% of sand, are found in the median part of the shelf off the Orinoco River and near shore to the north of Trinidad and of the Peninsula of Paria, while sediments including 5% to more than 30% of sand occur on the outer part of the shelf, down to 100 fm at least. It is believed that the finer sediments are now being supplied by the Orinoco River and transported to the west by the North Equatorial Current, whereas offshore sands have been deposited during one or several low glacial sea-levels; since that time, fine particles coming from the Orinoco River have been added to the sands on the outer shelf, but they do not conceal the older deposits. The study of the heavy minerals off Trinidad and Tobago demonstrates that the coarser fraction on the outer shelf cannot have been supplied at the present time, because the assortment of minerals is different from that existing on Trinidad and Tobago beaches: it has come either from local sources on the shelf or from the Orinoco, which flowed into the sea during the lowerings of sea-level farther north than now.

In the Mediterranean off south-eastern France (Blanc, 1958b; Nestéroff, 1958), the distribution resembles more, it is true, the classic pattern described in old textbooks (Fig. 19). Terrigenous muds supplied by the continent occur on an average deeper than 30 fm down to great depths, while, closer to the shore, the sediments consist of terrigenous or shell sands, often covered with dense meadows of sea-grasses, *Cymodoceae* or *Posidoniae*, which trap the particles in their rhizomes. But it must be noted that, even in this area, the classic distribution is disturbed by the occurrence of a large number of small pebbles, widely scattered among the muds and muddy sands of the shelf and consisting of quartz, sandstones, phyllites, variolites, etc. They have been dredged mostly from gently sloping grounds beginning at 40 fm and extending down to 60 fm, but others have been discovered (Bourcart) at much greater depths. In the Bays of Marseilles and Cassis, they have encrustations which are associated with thanatocoenoses, usually associated with shallow waters. Although their age is not well known, they might represent fluviatile or coastal deposits belonging to the Pliocene or Lower Pleistocene. Be that as it may, they are neither of present nor of recent age, and they show that present-day sedimentation on the shelf has not been thick enough to bury them deeply.

Obviously this remark also applies to other coarse sediments quoted above: since old deposits still lie at the surface near the edge of the shelf, it must be concluded that the rate of sedimentation is very low or nil there today. On the other hand the deposition has been very considerable in many inner parts of the shelf. In the Gulf of Paria, Venezuela, Van Andel and Postma conclude tentatively that it ranges from 12 cm to 100 cm per century for the last 700 years, except at and near the inlets of the bay where scouring by currents has kept the bottom free from deposition. Off Trinidad, figures ranging from 1.2 cm to 27 cm per century have been found by radiocarbon analyses for the inner shelf (Koldewijn, 1958). On the other hand, the recent Mississippi sub-delta has been built on the shelf within the last 450 years (Fig. 9), and continues to be prograded off South Pass at the rate of 200 ft a year.

Observations in Europe also point to very low amounts of recent marine deposition in many regions of the shelf lying far offshore, especially in the Celtic Sea and the Bay of Biscay (Berthois, 1955a, 1959; Berthois and Le

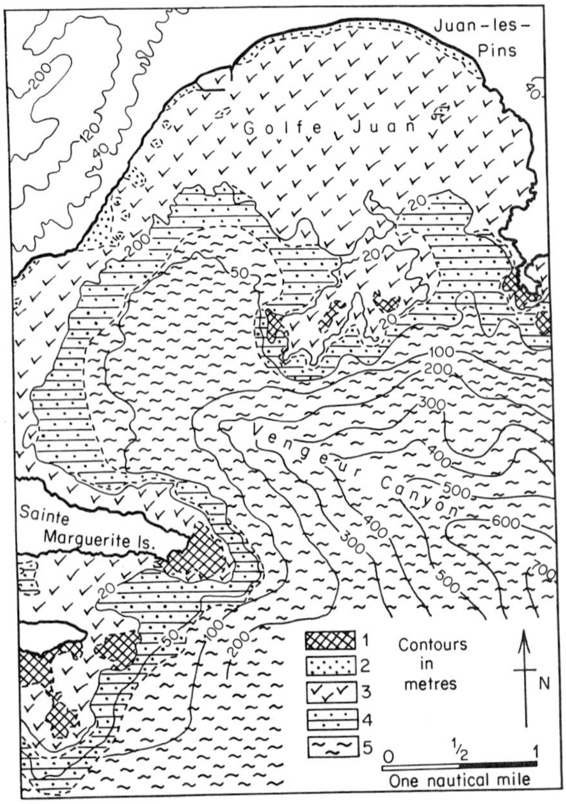

Fig. 19. Sediments in Golfe Juan, south-east France, showing a typical distribution over the continental margin in the north-west Mediterranean. (From Nestéroff, 1958.)

Legend—1: rock; 2: terrigenous sand; 3: *Cymodoceae* and *Posidoniae* meadows, trapping sediments; 4: skeletal sand; 5: mud.

Calvez, 1959). In the latter, the sediments to the west of a line joining the mouth of the Gironde to Penmarc'h Point are sands; their grain-size distribution resembles closely the beach and dune sands of the Landes of Gascony, although they extend to 100 fm depth. They have been partly reworked by sluggish currents, and this reworked type has been reproduced in laboratory experiments from samples collected in the dunes of the Landes. Moreover, many sand grains on the shelf of Biscay are wind-worn. Berthois concludes that these submarine sands are the continuation of the dunes of Gascony, which extended over the shelf as far as south-west Brittany before the post-glacial transgression and have not been covered by other sediments after they were drowned. The same suggestion has been put forward by Stetson (in Trask, 1939) for rounded and frosted sand grains found off the coast of southern New England, which are believed to be remnants of dunes formed when the sea retreated from the continental shelf. Another conclusion concerns the so-called "Grande Vasière", a broad strip of mud which runs, according to charts, across the shelf of Biscay: in fact, this mud layer is so thin that it is dispersed when corings are attempted in it and only sand remains in the corer. In the Celtic Sea, large sand ripples, more than 30 ft high and several hundreds of yards apart, have been known for a long time. Berthois has shown that they are probably inactive, since no current can account for them at the present time. They probably came into existence during a Pleistocene low sea-level, when currents were much more efficient in shallower waters. Today, weak currents are only able to produce much smaller ripples which are superimposed on the large ones. Here again the deposition is now insignificant.

In the North Sea, the distribution of grain size in sediments is most intricate, as may be seen on a very accurate map published by Jarke (1956). As early as 1936, Baak showed that these submarine sediments have intimate petrological connections with the emerged lands which they lie in front of, namely, that English–Scottish, Dutch and Scandinavian groups exist as distinct units which have not been mixed together. According to Cailleux (1942, p. 117), the English and Scandinavian groups often include a very high content of wind-worn grains in their southern extension, probably because they have undergone severe periglacial conditions during the last glaciation (Wisconsin, or Weichsel). More recently, Berthois (1957) has found that some of the North Sea sands have been reworked by currents and are now well-sorted, while others remain ill-sorted and are probably glacial or fluvio-glacial sediments in which sea action is hardly perceptible. Völpel (1959) has also studied them, and observes that those which are well-sorted lie generally at 14–17 fm and are lacking at depths below 22 fm: they occur in strips coinciding in position and direction with strong nearly alternating tidal currents, in which the ellipses are very narrow. The same differences in sorting are also reported from the bottom of the Irish Sea (Berthois, 1957). It may be said that in the North and Irish Seas, a winnowing of sands by present marine action may occur in shallow depths, but it does not apply anywhere near to the entire bottom; the influence of Pleistocene sub-aerial or sub-glacial action remains the deciding factor.

Tidal currents are sometimes so strong that they are able to keep large shelf areas free of sediments. The English Channel, especially around the Cotentin Peninsula and off north Brittany, is well known in this respect. Dangeard (1928) and Pratje (1950) have emphasized the fact that "hard grounds", on which the solid rock outcrops, are common in these areas. Such localities correspond to places where the current velocity increases. Pratje's calculations led him to the conclusion that 100% of the rocks or gravel grounds are found in areas in which the maximum velocity exceeds 3½ knots. Along the English coast, tidal ranges are lower and sediments are finer grained; nevertheless, hard grounds exist off headlands, such as the Lizard and Saint Alban's Head, where currents are stronger. Similarly, in the Sicilian Channel, which belongs entirely to the shelf, currents are fairly strong owing to the narrowing of the Mediterranean, and hard grounds are reported from several places by Blanc (1958). Tidal scour is certainly responsible for the deep clefts, or *zeegaten*, between the Frisian Islands (more than 22 fm below mean low tide between Texel and the mainland, and between Borkum and Rottum in the West Ems: see Van Straaten, 1954, chart of the Wadden Sea; and Fig. 1).

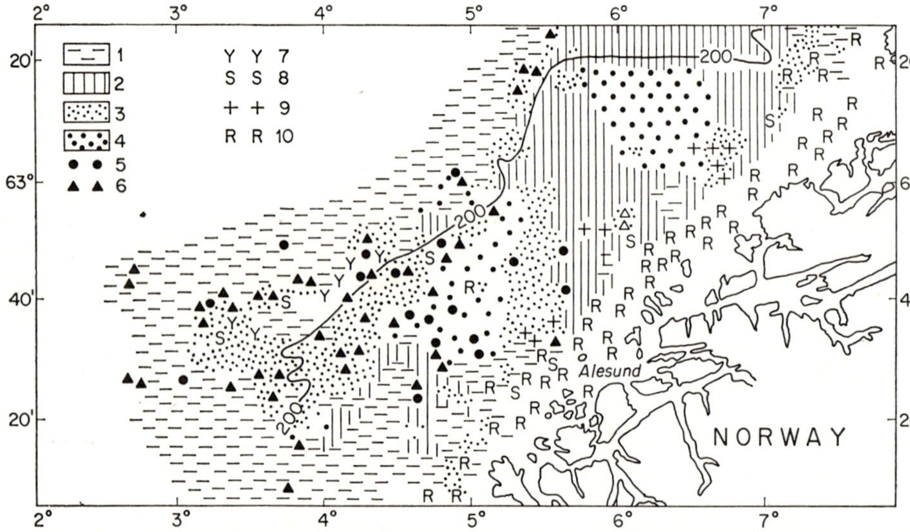

Fig. 20. Distribution of sediments over the continental shelf and slope off Møre and Romsdal, West Norway. (From Holtedahl and Berthois, in Berthois, 1957.)

Legend—1: mixed clay and sand; 2: fine sand; 3: sand; 4: coarse sand and gravel; 5: pebbles; 6: angular cobbles; 7: corals; 8: broken shells; 9: numerous stones; 10: rocky bottom. Depth contour to 200 m. Depressions on shelf not shown.

However, geologists must be very cautious in generalizations, for sediments can settle in unexpected areas. Between the islands of the Molène archipelago, off West Brittany, hydraulic "dunes" of different shapes, consisting of gravel and coarse sand, exist in many places below low water at spring tides, in spite of the fact that the maximum velocity of the tidal currents exceeds 5 knots. This results from numerous skerries which divide the current and create small shelters where deposition is possible. It can derive from the difference in velocity between the ebb and the flood; this causes deposition at one side of the rocks, whereas at the opposite side the bottom remains rocky (Guilcher, in Bourcart, 1959). In such areas of very rugged submarine topography, the local environment plays an essential part in sedimentation: the characteristics of the deposits may be completely different at equal depths and 50 yd apart as a result of local shelters. Conversely, rocky grounds exist in shelf areas where current velocities do not seem to be especially high, for example on outer shelves off Southern California (Revelle and Shepard, in Trask, 1939), and in East Antarctica, where they often correspond to places from which the inland ice has retreated only recently (Lissitzin and Jivago, 1958).

An interesting type of deposit on shelves consists of the sediments situated in front of present or former glaciated areas in America, North Europe and Antarctica. Those lying off Møre and Romsdal, Norway, have been studied by Holtedahl (1955) and Berthois (1957) (Figs. 20 and 21). The samples include stones, sand, silt and clay in variable percentages. "The medians show wide

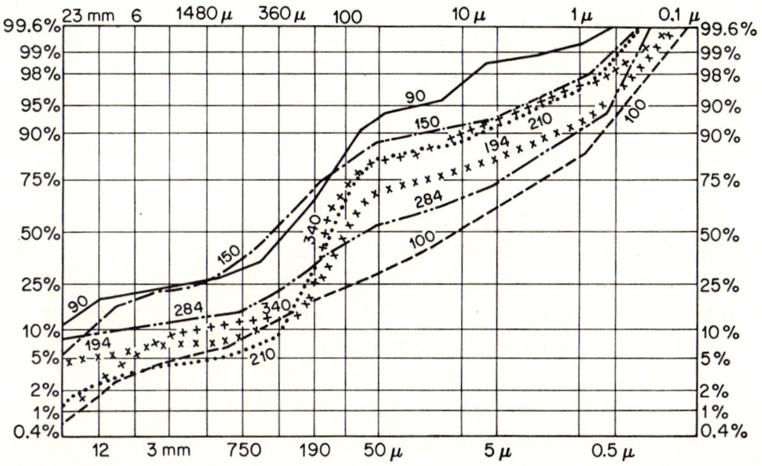

Fig. 21. Cumulative curves of particle diameters of glacial marine sediments of continental shelf and slope off Møre and Romsdal, West Norway. Figures on curves indicate the depths of samples in fathoms. (From Berthois, 1957.)

variations at the same depths, both sand and gravel being found in the depressions as well as on the shallow banks." Only pebbles are dredged in some places. Their roundness is not at all typical of a former coastal environment: on the contrary, they closely resemble glacial pebbles, and the bulk of the sediments has the character of glacial tills. Some samples consisting of sand and lying at 100 fm are fairly well sorted and might represent an old shoreline, but generally the sorting is extremely poor and the size distribution curves show such individual characteristics that it is difficult to classify them. The nature of the pebbles covers a wide range of crystalline and metamorphic rocks; limestones, sandstones and flints are sometimes present. These data point to a Pleistocene glacial origin and a transportation to the shelf by ice, either through glaciers extending over it or through icebergs dropping their material when they melted. Transportation by drifting icebergs is most probable for flints, which may have come from Denmark where Cretaceous rocks exist. Such deposits are known as glacial marine sediments. Off Brittany, the pebbles dredged on the shelf may also come in part from the melting of Pleistocene ice-rafts, as suggested long ago by French and British geologists; others may be drowned strand lines.

The Russian expeditions to East Antarctica during the International Geophysical Year have added to our knowledge of the glacial marine sediments from the shelves of the Southern Continent. These had, however, been reported before, especially by the scientists of the German South Polar Expedition at the beginning of this century. Here we are concerned with *present* glacial deposition. Lissitzin and Jivago (1958) have clearly defined this environment, which is, at the present day, unique in the world for such a wide area. Run-off is practically absent on land (whereas it exists even in Greenland); no chemical weathering occurs, so that the material is affected only by mechanical disintegration; wave abrasion is hindered by the pack and the ice-foot; thermal conditions prevent calcium carbonate precipitation; the ice-cap and diatoms are essential in the supply of mineral and biogenic particles; the rocky bottom, on which the sediments settle, is often very uneven and rugged (see Chapter 13); icebergs sometimes drift for years before melting completely, so that their material becomes widely scattered. Nevertheless, the greater part of the iceberg deposits falls to the bottom in front of areas in which ice calving is more important and the rocky content in ice is higher. Between 50° and 70° E, where land ice supply is comparatively poor, the iceberg deposits are restricted to a narrower area than elsewhere. Several mineralogical provinces have been distinguished on the shelf, corresponding probably to differences in the structure of the adjacent parts of the continent, which does not consist only of a Pre-Cambrian shield, since Mesozoic and Tertiary stones have been collected on the shelf. The sorting is typically poor and stones are numerous near the continent, although fine-grained muds occur in some depressions. The fine fraction includes, along with terrigenous particles, an important skeletal content consisting of diatoms and siliceous spicules of sponges. Carbonates are practically absent. The rate of deposition seems to be high in some areas, since a core 14.5 m long failed to reach the basement of the glacial marine sediments. It may, however, have encountered Pleistocene deposits.

The continental shelf in the Beaufort Sea and the East Chukchi Sea, off north Alaska and north-west Canada, represents another environment in very cold water (Carsola, 1954; and Fig. 22). Poorly sorted sediments consisting of mixtures of cobbles, pebbles, sand and mud are deposited on the shelf, but only sea-ice or river-ice rafts and not icebergs contribute to this sedimentation, since no ice-cap exists on the adjacent continent. The situation appears to have been the same during the Pleistocene. The dirty ice does not drift very far from land, so that these "glacial" marine sediments, which would be more adequately described as ice-borne marine sediments, extend to a much narrower strip than in areas where they derive from land-ice.

The recent results summarized above lead to a general observation. It appears that the sedimentation on continental shelves results largely from Pleistocene processes, which were much more efficient than those acting now, except, perhaps, for the Arctic Ocean and Antarctica, which may provide a picture of the conditions existing formerly on other shelves, and for some restricted areas where large overloaded rivers flow into the sea (e.g. the Hwang Ho and the Yangtze in the western parts of the Gulf of Pohai and the East China Sea: see Niino and Emery, 1961). Even in coastal regions which were not directly affected by glaciations, the terrestrial supply is likely to have been much larger during the cold periods of the Pleistocene, when the shelf, under periglacial or "pluvial" climates, was widely exposed as dry land and the sedimentary discharge was higher. The deposition proceeds now at a much lower rate, especially on the outer parts of the shelves which are no longer within the reach

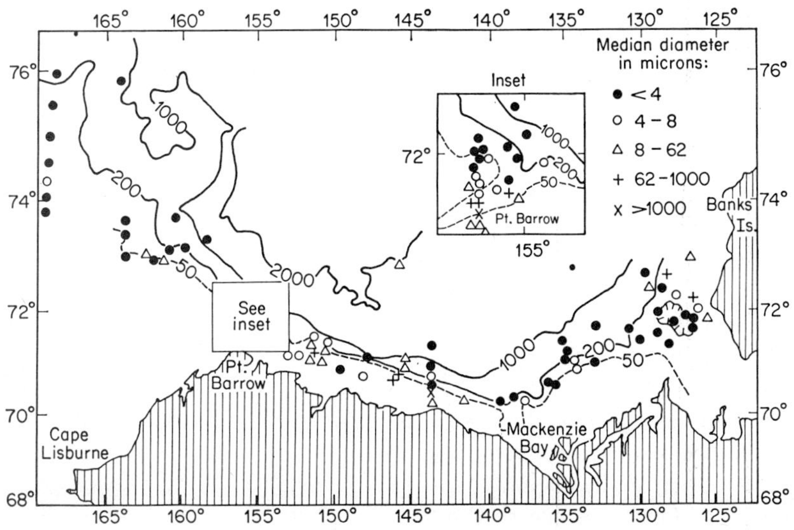

Fig. 22. Median diameter, expressed in microns, of sediments on shelf and slope in Beaufort and north-east Chukchi Seas. Contours in fathoms. (From Carsola, 1954, fig. 4.)

of big sedimentary supplies owing to the retreat of shorelines after the postglacial transgression. The reduction in grain size from outer to inner shelves may also be due to a general reduction in river competence, and thus have a climatic significance.

The low rates in terrigenous sedimentation may have prevailed for a much longer time off arid or semi-arid regions, where river discharge, although it was increased during pluvial Pleistocene periods, was not able to compete with more humid climates. This is probably the case for the continental shelf of Western Australia (Carrigy and Fairbridge, 1954), and the Persian Gulf (Emery, 1956; Houbolt, 1957), where the hot dry climate results in slow accumulation, relative abundance of organogenic calcareous deposition, and dominance of silt and clay in the non-calcareous fraction, partly contributed by wind supply.

This does not mean that organogenic deposition is always insignificant on the shelves outside the warm areas. Sediments including more than 75% of calcium carbonate are sometimes found on shelves at fairly high latitudes, for example, off Brittany, where more than 80% or even 95% of skeletal remains may be found in some places, especially in underwater "dunes" piled up by tidal currents (Berthois, 1955a, p. 543; Guilcher, in Bourcart, 1959; Hinschberger and Guilcher, 1962).

3. Continental Slope

The continental slope is not so well investigated as the shelf. At the present status of knowledge, the sediments on this part of the ocean floor seem to cover a wide range of types as far as grain size is concerned. According to Shepard (1959, p. 129), whose opinion is based largely on chart notations, the world distribution on the continental slope would be approximately as follows: mud, 60%; sand, 25%; rock and gravel, 10%; shells or ooze, 5%. On the average, mud is more common than on shelves, but it may be seen from the figures above that coarse sediments are not infrequent.

Many instances are known in which the sediments on the slope have strong connections with those lying higher up the shelf. This may be particularly true for slopes lying off glaciated areas in Norway and East Antarctica (Holtedahl, 1955; Lissitzin and Jivago, 1958; also Fig. 20). Off Norway, mixtures of sand, silt and clay, sometimes including pebbles, are reported down to 520 fm. As a rule, no stratification is to be seen in the clays. Grades finer than sand are not so abundant above 270 fm, although they are present as admixtures. The process of deposition by Pleistocene icebergs has been the same as on the shelf. No sedimentation has occurred down to a depth of about 500 fm after the deposition of the glacial marine sediments.

In the western Mediterranean, the mud which covers the lower part of the shelf continues farther down to the abyssal plain (Fig. 19), more or less mixed with sands (Bourcart, 1954; Bourcart and Ottmann, 1957). The calcium carbonate content is relatively high, consisting of organogenic particles. Rocky ledges are frequent on the walls of the submarine canyons, which may be visualized from the descriptions of those who have visited them on board the first French bathyscaph (Houot, 1957). These walls stand out as imposing, winding and stepped escarpments with a general coating of mud, resembling furniture and stairs covered with thick dust in an old castle neglected for a very long time. The slightest impact results in avalanches and big clouds of mud.

On both sides of the North Atlantic also, rock outcrops alternate with clay and sand on the slope. Off the eastern United States (Elmendorf and Heezen, 1957), the slope is often "covered with low-density grey clay in which the coring rig completely buries itself. Gravel and sand form the floors of some continental slope canyons, while others are deeply covered with low density mud. In many areas ancient, partly consolidated clays crop out on canyon walls." In the Bay of Biscay (Berthois, 1955a; Berthois and Le Calvez, 1959), dredgings and corings on the slope have proved that the coarse sediments occurring on the outer shelf exist also at lower levels. Sand, gravel and pebbles have been dredged between 250 and 410 fm: the pebbles range from 23 to 60 mm in length and include a variety of crystalline, metamorphic and sedimentary rocks. But another sample, collected in a Kullenberg piston corer at 480 fm, consists of a much finer, plastic sediment including a microfauna, which points to a gradual warming of the environment from the base to the top. The core is 2.67 m long, and it may be assumed that the rate of deposition has been relatively high here since the end of the last glaciation. At another locality on the Biscay shelf, a Mesozoic rock has been obtained. The Lamont Geological Observatory took photographs from the upper part of the slope off Saint Nazaire, France, at 260 and 285 fm, showing different kinds of ripples on a sedimentary bottom (Heezen, Tharp and Ewing, 1959, pl. 11). More data are needed for an accurate picture of this area, which appears to be complicated but promising.

In the Beaufort and East Chukchi Seas (Carsola, 1954), clays have been deposited on the slope, while poorly sorted sediments including pebbles have been dredged on the shelf (Fig. 22). The decrease in grain size from the shelf to the slope is due to the fact that the ice-rafts carrying coarse sediments outwards do not reach the slope, as seen above, so that they do not disturb the "normal" sedimentation.

The north-west Gulf of Mexico (Greenman and LeBlanc, 1956) is another area where the sediments are distinctly finer grained on the slope (clay) than on the shelf (sand, silt, shells, with poor correlation from core to core). But the outstanding feature on the Gulf slope is found further to the east: it consists of a huge cone gently sloping into great depths from the Mississippi delta (Ewing, Ericson and Heezen, 1958: see also Chapter 13). This cone is believed to have been built by the Mississippi River, whereas the slope on both sides of the cone seems to have remained largely free from deposition. The building of the cone must belong essentially to the Pleistocene, since warm water planktonic Foraminifera are found only in an upper layer 2–3 ft thick; the lower layer, consisting of a grey silty clay with occasional beds of sand and silt, generally exceeds in thickness the length of the cores, which are 6 to 10 m long. The abrupt change in sedimentation has been dated by radiocarbon as being about 11,000 years B.P. We meet here again the idea of a fundamental modification and drastic decrease in deposition at the end of the Ice Age which has been put forward concerning the continental shelf.

As to the processes of deposition on the slope, little will be said in this section, because they involve a discussion of turbidity currents, submarine landslides and other mechanisms which are considered elsewhere (see Chapters 27 and 28). Here it will be merely pointed out that, if turbidity currents and/or landslides are acting up to the present time, as shown by spectacular breaks of submarine cables, they were probably much more efficient during the Pleistocene, and may largely account for changes in deposition on the slope such as those which have just been mentioned in the Gulf of Mexico. They may also explain why solid rock is so often found in dredgings on the slope.

References

Akers, W. H. and A. J. J. Holck, 1957. Pleistocene beds near the edge of the continental shelf, Southeastern Louisiana. *Bull. Geol. Soc. Amer.*, **68**, 983–992.

Baak, J. A., 1936. Regional petrology of the Southern North Sea. Thesis, Leiden, Wageningen, 128 pp.

Berthois, L., 1953 to 1958. (Studies on processes in sedimentation in the Loire estuary.) *C.R. Acad. Sci. Paris*, **237**, 465–467; **239**, 820–822; **243**, 1343–1345; **245**, 1739–1741; **246**, 141–142; **247**, 947–950.

Berthois, L., 1955. Les variations de la composition granulométrique des vases. *C.R. Acad. Sci. Paris*, **241**, 231–232.

Berthois, L., 1955a. Contribution à l'étude de la sédimentation dans le Golfe de Gascogne. *Rev. Trav. Inst. Pêch. Marit.*, **19**, 501–579.

Berthois, L., 1956. Turbidité des eaux à l'entrée de l'estuaire de la Loire. *C.R. Acad. Sci. Paris*, **243**, 2113–2115.

Berthois, L., 1957. Recherches sur les sédiments de la Mer du Nord et de la Mer d'Irlande. *Rev. Trav. Inst. Pêch. Marit.*, **21**, 485–554.

Berthois, L., 1959. Remarques sur les rides sous-marines. *Rev. Trav. Inst. Pêch. Marit.*, **23**, 225–239.

Berthois, L. and C. Berthois, 1954–1955. Étude de la sédimentation dans l'estuaire de la Rance. *Bull. Lab. Marit. Dinard*, **40**, 4–15; **41**, 3–18.

Berthois, L. and C. Berthois, 1955. Étude lithologique des sédiments dragués par le "Président Théodore Tissier" en Rade de Brest et en Manche (campagne 1949). *Rev Trav. Inst. Pêch. Marit.*, **19**, 467–499.

Berthois, L. and A. Guilcher, 1956. La plaine d'Ambilobé (Madagascar), étude morphologique et sédimentologique. *Rev. Géomorphol. Dynam.*, **7**, 33–52.

Berthois, L. and Y. Le Calvez, 1959. Deuxième contribution à l'étude de la sédimentation dans le Golfe de Gascogne. *Rev. Trav. Inst. Pêch. Marit.*, **23**, 323–377.

Blanc, J. J., 1958. Sédimentologie sous-marine du détroit siculo-tunisien. *Ann. Inst. Océanog.*, **34**, 91–126.

Blanc, J. J., 1958a. Recherches géologiques et sédimentologiques en Méditerranée nord-orientale. *Ann. Inst. Océanog.*, **34**, 157–211.

Blanc, J. J., 1958b. Recherches de sédimentologie littorale et sous-marine en Provence occidentale. Thesis, Paris, 140 pp.

Blanc, J. J. and J. Picard, 1954. Les schorres de l'estuaire du Stabbiaco (Corse). *Rev. Géomorphol. Dynam.*, **5**, 2–24.

Boillot, G. 1960–1961. La répartition des fonds sous-marins au large de Roscoff (Finistère) et dans la Manche occidentale. *Cah. Biol. Mar. Roscoff*, **1**, 3–23; **2**, 187–208

Bourcart, J., 1954. Les vases de la Méditerranée et leur mécanisme de dépôt. *Deep-Sea Res.*, **1**, 126–130.

Bourcart, J., 1957. Géologie sous-marine de la baie de Villefranche. *Ann. Inst. Océanog.*, **33**, 137–200.

Bourcart, J., ed., 1959. La topographie et la géologie des profondeurs océaniques. 83e Colloque Intern. C.N.R.S., Paris, 1958, 313 pp.

Bourcart, J., 1959. Le plateau continental et les sédiments de la Méditerranée occidentale. *C.R. Acad. Sci. Paris*, **249**, 1380–1382, 1699–1700 and 1783–1784.

Bourcart, J. and R. H. Charlier, 1959. The tangue, a "non-conforming" sediment. *Bull. Geol. Soc. Amer.*, **70**, 565–568.

Bourcart, J. and C. Francis-Bœuf, 1942. *La Vase*. Hermann, Paris, 67 pp.

Bourcart, J. and F. Ottmann, 1957. Recherches de géologie marine dans la région du Cap Corse. *Rev. Géog. Phys. Géol. Dynam.*, Ser. 2, **1**, 66–78.

Burt, W. V., 1955. Interpretation of spectrophotometer readings on Chesapeake Bay waters. *J. Mar. Res.*, **14**, 33–62.

Byrne, J. V., D. O. Le Roy and C. M. Riley, 1959. The Chenier Plain and its stratigraphy, Southwestern Louisiana. *Trans. Gulf Coast Assoc. Geol. Soc.*, **9**, 237–260.

Cailleux, A., 1942. Les actions éoliennes périglaciaires en Europe. *Mém. Soc. Géol. Fr.*, **46**, 176 pp.

Carrigy, M. A. and R. W. Fairbridge, 1954. Recent sedimentation, physiography and structure of the continental shelves of Western Australia. *J. Roy. Soc. Western Austral.*, **38**, 65–95.

Carsola, A. J., 1954. Recent marine sediments from Alaskan and North-West Canadian Arctic. *Bull. Amer. Assoc. Petrol. Geol.*, **38**, 1552–1586.

Carter, J., 1959. Mangrove succession and coastal changes in South-West Malaya. *Trans. Inst. Brit. Geog.*, **26**, 79–88.

Codrington, T., 1898. On some submerged rock valleys in South Wales, Devon and Cornwall. *Q. J. Geol. Soc. London*, **54**, 251–278.

Crommelin, R. D., 1940. De herkomst van het zand van de Waddenzee. *Tijd. Koninkl. Ned. Aardr. Gen.*, **57**, 347–361.

Crommelin, R. D., 1947. Pétrologie des fractions fines des sédiments marins aux Pays-Bas. Géol. Terr. Réc. Ouest Europe (Sess. Extraord. Soc. Belges Géol.), Brussels, 114–124.

Daly, R. A., 1934. *The Changing World of the Ice Age.* Yale Univ. Press, 271 pp.

Dangeard, L., 1928. Observations de géologie sous-marine et d'océanographie relatives à la Manche. *Ann. Inst. Océanog.*, **6** (1), 295 pp.

Davis, J. H., Jr., 1940. The ecology and geologic rôle of mangroves in Florida. *Papers Tortugas Lab.*, **32**, *Carnegie Inst. Wash. Publ.*, 517, 303–412.

Duboul-Razavet, C., 1956. Contribution à l'étude géologique et sédimentologique du delta du Rhône. *Mém. Soc. Géol. Fr.*, **76**, 234 pp.

Elhai, H., 1959. Un havre de la côte ouest du Cotentin. *Bull. Assoc. Géog. Fr.*, **284–285**, May–June, 2–19.

Elmendorf, C. H. and B. C. Heezen, 1957. Oceanographic information for engineering submarine cable systems. *Bell Syst. Tech. J.*, **36**, 1047–1093.

Emery, K. O., 1956. Sediments and water of Persian Gulf. *Bull. Amer. Assoc. Petrol. Geol.*, **40**, 2354–2383.

Ewing, M., D. B. Ericson and B. C. Heezen, 1958. Sediments and topography of the Gulf of Mexico. In Weeks, L. (Ed.), *Habitat of Oil.* Amer. Assoc. Petrol. Geol., Tulsa, 995–1053.

Fairbridge, R. W., 1947. Coarse sediments on the edge of the continental shelf. *Amer. J. Sci.*, **247**, 146–153.

Francis-Bœuf, C., 1947. Recherches sur le milieu fluvio-marin et les dépôts d'estuaire. Thesis, Paris, 193 pp.

Furnestin, J., 1937. Les dragages géologiques de la croisière du "Président Théodore Tissier" (mai-juillet 1935). *Rev. Trav. Off. Pêch. Marit.*, **10**, 223–258.

Gellert, J. F., 1952. Das Aussenelbwatt zwischen Cuxhaven-Duhnen und Scharhörn. *Petermann's Mitt.*, **96**, 103–109 (map).

Glangeaud, L., 1938. Transport et sédimentation dans l'estuaire et à l'embouchure de la Gironde. *Bull. Soc. Géol. Fr.*, Ser. 5, **8**, 599–631 (see also *Rev. Géog. Phys. Géol. Dynam.*, **11**, 323–370).

Gould, H. R. and E. McFarlan, Jr., 1959. Geologic history of the Chenier Plain, Southwestern Louisiana. *Trans. Gulf Coast Assoc. Geol. Soc.*, **9**, 261–270.

Greenman, N. N. and R. J. LeBlanc, 1956. Recent marine sediments and environments of Northwest Gulf of Mexico. *Bull. Amer. Assoc. Petrol. Geol.*, **40**, 813–847.

Gripp, K., 1956. Das Watt: Begriff, Begrenzung und fossile Vorkommen. *Senckenbergiana Lethœa*, **37**, 149–181.

Guilcher, A., 1948. Le relief de la Bretagne méridionale. Thesis, Paris, La Roche sur Yon, 682 pp.

Guilcher, A., 1956–1957. Les travaux de Berthois sur l'estuaire de la Loire. *Norois (Poitiers)*, **3**, 87–90; **4**, 106–107 (summary with refs.).

Guilcher, A., 1958. *Coastal and Submarine Morphology.* Methuen and Wiley, London and New York, 274 pp.

Guilcher, A. and L. Berthois, 1957. Cinq années d'observations sédimentologiques dans quatre estuaires-témoins de l'Ouest de la Bretagne. *Rev. Géomorphol. Dynam.*, **8**, 67–86.

Guilcher, A. and C. A. M. King, 1961. Spits, tombolos and tidal marshes in Connemara and West Kerry, Ireland. *Proc. Roy. Irish Acad.*, **61B**, 17, 283–338.

Gulliver, F. P., 1898–1899. Shoreline topography. *Proc. Amer. Acad. Arts Sci.*, **34**, 151–258.

Hack, J. T., 1957. Submerged river system of Chesapeake Bay. *Bull. Geol. Soc. Amer.*, **68**, 817–830.

Hansen, K., 1951. Preliminary report on the sediments of the Danish Wadden Sea. *Med. Dansk Geol. For.*, **12**, 1–26 (see also *ibid.*, **13**, 1956, 112–117).

Harder, E. C., 1919. Iron-depositing bacteria and their geologic relations. *U.S. Geol. Surv. Prof. Paper*, **113**, 89 pp.

Heezen, B. C., M. Tharp and M. Ewing, 1959. The floors of the Oceans. I: The North Atlantic. *Geol. Soc. Amer. Spec. Paper* 65, 122 pp.

Hinschberger, F. and A. Guilcher, 1962. Les bancs de sable sous-marine des abords d'Ouessant et de la Chaussée de Sein (Finistère). *C.R. Acad. Sci. Paris*, **254**, 4065–4067.

Holtedahl, H., 1955. On the Norwegian continental terrace, primarily outside Møre-Romsdal: its geomorphology and sediments. *Univ. Bergen Årb., Naturvik, Rekke*, **14**, 209 pp.

Houbolt, J. J. H. C., 1957. Surface sediments of the Persian Gulf near the Qatar Peninsula. Thesis, Utrecht, The Hague, 113 pp.

Houot, G., 1957. Promenades dans les canyons sous-marins. *Rev. Géog. Phys. Géol. Dynam.*, Ser. 2, **1**, 56–57.

Jacobsen, N. K., 1956. Jordbunsundersøgelser i Tøndermarsken. *Geog. Tidssk.* (*Copenhagen*), **55**, 106–146. (English summary.)

Jarke, J., 1956. Eine neue Bodenkarte der südlichen Nordsee. *Deut. Hydrog. Z.*, **9**, 1–9.

Johnson, D. W., 1925. *The New England–Acadian Shoreline.* Wiley, New York, 608 pp.

Koldewijn, B. W., 1958. Sediments of the Paria–Trinidad shelf. Thesis, Amsterdam, 109 pp.

Kruit, C., 1955. Sediments of the Rhône delta. Thesis, Groningen, The Hague, 141 pp.

Kuenen, P. H., 1950. *Marine Geology.* Wiley, New York, 568 pp.

LaFond, L. R., 1957. Aperçu sur la sédimentologie de l'estuaire de la Betsiboka (Madagascar). *Rev. Inst. Fr. Pétrol.*, **12**, 425–431.

Lindeman, J. C., 1953. The vegetation of the coastal region of Suriname. Thesis, Utrecht, 135 pp.

Lissitzin, A. P. and A. V. Jivago, 1958. Bottom deposits in the Southern Indian Ocean. *Izvest. Akad. Nauk S.S.S.R., Geog. ser.*, **3**, 22–36 (in Russian).

Milon, Y., 1935. Observations sur l'origine et la formation des tangues et vases littorales. *C.R. Somm. Soc. Géol. Minéral. Bretagne* (*Rennes*), **5**, Nov.–Dec., 8–9.

Nestéroff, W., 1958. Recherches sur les sédiments marins actuels de la région d'Antibes. Thesis, Paris, 347 pp.

Nielsen, N., 1935. Eine Methode zur exakten Sedimentationsmessung: Studien über die Marschbildung auf der Halbinsel Skalling. *Kgl. Danske Videnskab. Selskabs., Biol. Medd.*, **4**, 98 pp.

Niino, H. and K. O. Emery, 1961. Sediments of shallow portions of East China Sea and South China Sea. *Bull. Geol. Soc. Amer.*, **72**, 731–762.

Nota, D. J. G., 1958. Sediments of the Western Guiana shelf. *Med. Landbouwhogeschool Wageningen*, **58** (2), 98 pp.

Pannekoek, A. J., 1956. *Geological History of the Netherlands.* Government Printing Off., The Hague, 147 pp.

Phlipponneau, M., 1955. La baie du Mont Saint Michel, étude de morphologie littorale, *Mém. Soc. Géol. Minéral. Bretagne*, 215 pp. (thesis, Paris).

Pimienta, J., 1959. Le cycle pliocène-actuel dans les bassins paraliques de Tunis. *Mém. Soc. Géol. Fr.*, **85**, 176 pp.

Pratje, O., 1950. Die Bodenbedeckung des Englischen Kanals und die maximalen Gezeitenstromgeschwindigkeiten. *Deut. Hydrog. Z.*, **3**, 201–205.

Price, W. A., 1955. Environment and formation of the chenier plain. *Quaternaria* (*Rome*), **2**, 75–86.

Price, W. A., 1958. Sedimentology and Quaternary geomorphology of South Texas. *Trans. Gulf Coast Assoc. Geol. Soc.*, **8**, 41–75.

Pritchard, D. W., 1952. Salinity distribution and circulation in the Chesapeake Bay estuarine system. *J. Mar. Res.*, **11**, 106–123.

Rajcevic, B. M., 1957. Étude des conditions de sédimentation dans l'estuaire de la Seine. *Ann. Inst. Techn. Bâtiment Trav. Pub.* (*Paris*), **117**, 745–775.

Richards, F. J., 1934. The salt marshes of the Dovey estuary. IV: The rates of vertical accretion, horizontal extension, and scarp erosion. *Ann. Botany*, **48**, 225–259.

Robinson, A. H. W., 1956. The submarine morphology of certain port approach channel systems, *J. Inst. Navigation*, **9**, 20–46.

Russell, R. J., editor, 1959. *Second coastal geography conference, Louisiana State University*. Washington, 472 pp.

Schnell, R., 1952. Les groupements et les unités géobotaniques de la région guinéenne. *Mém. Inst. Fr. Afr. Noire*, **18**, 41–236 (mangroves, pp. 197–218).

Schott, C., 1950. Die Westküste Schleswig-Holsteins. *Schr. Geog. Inst. Univ. Kiel*, **13**, 34 pp.

Shepard, F. P., 1948. *Submarine Geology*. Harper and Brothers, New York, 348 pp.

Shepard, F. P., 1956. Marginal sediments of Mississippi delta. *Bull. Amer. Assoc. Petrol. Geol.*, **40**, 2537–2623.

Shepard, F. P., 1959. *The Earth beneath the Sea*. Johns Hopkins Press, Baltimore, 275 pp.

Shepard, F. P. and D. G. Moore, 1955. Central Texas coast sedimentation. *Bull. Amer. Assoc. Petrol. Geol.*, **39**, 1463–1593.

Steers, J. A., 1938–1948. The rate of sedimentation on salt marshes on Scolt Head Island. *Geol. Mag.*, **75**, 26–39; **85**, 163–166.

Steers, J. A., 1946. *The Coastline of England and Wales*. Cambridge Univ. Press, 644 pp.

Stetson, H. C., 1949. *The sediments and stratigraphy of the East coast continental margin: Georges Bank to Norfolk Canyon*. Mass. Inst. Tech., Woods Hole Oceanog. Inst., 60 pp.

Stevenson, R. E. and K. O. Emery, 1958. Marshlands at Newport Bay, California. *Allan Hancock Foundation Pub., Occas. Papers*, **20**, 109 pp.

Tavernier, R., 1947. L'évolution de la plaine maritime belge. *Bull. Soc. Belge Géol.*, **56**, 332–343.

Trask, P. D., editor, 1955. *Recent Marine Sediments*, a symposium. London, 1939 (2nd edit.), Tulsa, 736 pp.

Tricart, J., 1955. Aspects sédimentologiques du delta du Sénégal. *Geol. Rundschau*, **43**, 384–397.

Tricart, J., 1956. Le delta du Sénégal, type zonal de delta. *Com. Trav. Hist., Bull. Sect. Géog.*, **69**, 289–314.

Twenhofel, W. H., 1950. *Principles of Sedimentation*. McGraw-Hill, New York, 673 pp.

Van Andel, Tj. H., 1950. Provenance, transport and deposition of Rhine sediments. Thesis, Wageningen, 129 pp.

Van Andel, Tj. H., 1955. Sediments of the Rhône delta. II: sources and deposition of heavy minerals. *Verhandl. Koninkl. Ned. Geol.-Mijn. Gen.*, **15**, 516–543.

Van Andel, Tj. H. and H. Postma, 1954. Recent sediments of the Gulf of Paria. *Verhandl. Koninkl. Ned. Akad. Wetenschap., Afd. Natuurk.*, **20**, 245 pp.

Van der Eyk, J., 1957. Reconnaissance soil survey in Northern Surinam. Thesis, Wageningen, 99 pp.

Van Straaten, L. M. J. U., 1954. Composition and structure of recent marine sediments in the Netherlands. *Leidse Geol. Med.*, **19**, 1–110.

Van Straaten, L. M. J. U., 1959. Minor structures of some recent littoral and neritic sediments. *Geol. Mijnb.*, n.s., **21**, 197–216.

Van Straaten, L. M. J. U. and P. H. Kuenen, 1957. Accumulation of fine grained sediments in the Dutch Wadden Sea. *Geol. Mijnb.*, n.s., **19**, 329–354.

Verger, F., 1954. La morphologie de l'Anse de l'Aiguillon. *Bull. Assoc. Géog. Fr.*, **246**, 157–165.

Verger, F., 1956. Quelques remarques sur la formation et le relief de schorres. *Bull. Assoc. Géog. Fr.*, **259–260**, 146–156.

Völpel, F., 1959. Studie über das Verhalten weitwandernder Flachseesande in der südlichen Nordsee. *Deut. Hydrog. Z.*, **12**, 64–76.

Waddensymposium, 1950. *Tijd. Koninkl., Ned. Aardr. Gen.*, 148 pp.

Waksman, S. A., 1933. On the distribution of organic matter in the sea bottom and the chemical nature and origin of marine humus. *Soil Sci.*, **36**, 125–147.

Welder, F. A., 1959. Processes of deltaic sedimentation in the Lower Mississippi River. *Coastal Stud. Inst., Baton Rouge, Tech. Rep.* 12, 90 pp.

Woldstedt, P., 1950. *Norddeutschland und angrenzende Gebiete im Eiszeitalter*. Koehler, Stuttgart, 464 pp.

25. PELAGIC SEDIMENTS[1]

G. Arrhenius

1. Concept of Pelagic Sedimentation

The term pelagic sediment is often rather loosely defined. It is generally applied to marine sediments in which the fraction derived from the continents indicates deposition from a dilute mineral suspension distributed throughout deep-ocean water. It appears logical to base a precise definition of pelagic sediments on some limiting property of this suspension, such as concentration or rate of removal. Further, the property chosen should, if possible, be reflected in the ensuing deposit, so that the criterion in question can be applied to ancient sediments.

Extensive measurements of the concentration of particulate matter in sea-water have been carried out by Jerlov (1953); however, these measurements reflect the sum of both the terrigenous mineral sol and particles of organic (biotic) origin. Aluminosilicates form a major part of the inorganic mineral suspension; aluminum is useful as an indicator of these, since this element forms 7 to 9% of the total inorganic component,[2] and can be quantitatively determined at concentration levels down to 3×10^{-10} (Sackett and Arrhenius, 1962). Measurements of the amount of particulate aluminum in North Pacific deep water indicate an average concentration of 23 μg/l. of mineral suspensoid, or 10 mg in a vertical sea-water column with a 1 cm^2 cross-section at oceanic depth. The mass of mineral particles larger than 0.5 μ constitutes 60%, or less, of the total. From the concentration of the suspensoid and the rate of fallout of terrigenous minerals on the ocean floor, an average passage time (Barth, 1952) of less than 100 years is obtained for the fraction of particles larger than 0.5 μ. For the finer particles the average passage time is longer, such as more than 200 (but considerably less than 600) years.

A mechanism which possibly contributes significantly to the removal of coarse suspensoid is aggregation in the gut of filter-feeding animals; this phenomenon has been observed by Rex and Goldberg (1958). Gravitative settling of single grains could account for the deposition of most particles larger than a few microns, but for smaller grains this mechanism is inadequate since the settling time required by Stokes' law[3] is several orders of magnitude larger than the passage time actually observed ($< 10^2$ years).[4]

[1] Much of the information presented in this chapter is the result of research partly carried out under Contract No. AT (11–1)–34 with the U.S. tomic Energy Commission, partly sponsored by the Petroleum Research Fund of the American Chemical Society (ACS PRF Unsolicited Award 875–C6). The generous support from these agencies is gratefully acknowledged.

[2] Data from the Pacific Ocean (Sackett and Arrhenius, 1962); determinations from the English Channel give similar values (Armstrong, 1958).

[3] 7800 years for a 0.1 μ spherical particle of density 2.6 through 2900 m of the oceanic-water column at 10°C.

[4] The passage time is defined as the time at which the mass of particles originally present in the water column has been reduced to $1/e$ of the original value. This occurs when the

[*MS received February, 1961*]

Coagulation of unprotected colloids at high electrolyte concentration has been demonstrated by von Smoluchovski (1917) to be a second-order reaction with a rate constant of the order 2×10^{-2}. If in the coarse ($> 0.5\ \mu$) part of the oceanic mineral suspensoid the average particle diameter is taken to be 1 μ, in agreement with the particle-size distribution found in the sediment,[1] the number of particles per ml is 7×10^3. In 3×10^9 sec (100 years, which is the maximum passage time given above for this fraction), coagulation by such a reaction would reduce the original particle concentration by a factor of 4×10^{11}, i.e. all particles in 5×10^7 ml of sea-water, or a 100 cm^2 oceanic-water column, would be aggregated together.

If a similar reasoning is applied to the fine fraction of the suspensoid (0.01–0.5 μ), in which an average particle diameter of 0.05 μ is assumed, the initial particle content of 3×10^7 per ml should be reduced to 6×10^{-15} of the original value in 10^{10} sec (300 years) which, as indicated above, is a probable passage time for this fraction. This aggregation would comprise all particles in this size range contained in a 400-cm^2 water column.

It is obvious that the rate of coagulation by such a reaction is far more rapid than the rate of removal actually observed in the open ocean. Rates of the order required by von Smoluchovski's theory prevail, however, in concentrated suspensions such as in shallow seas and off river mouths (see e.g. Gripenberg, 1934). It is possible that one of the organic components of sea-water, present in concentrations several orders of magnitude higher than that of the mineral suspensoid, might decrease its rate of coagulation. Whether this is correct or not, the low concentration of the suspensoid in the open ocean, its abnormally low rate of coagulation, and, therefore, its long passage time, permitting wide areal distribution, are observed properties which contrast with the properties of the more concentrated, rapidly flocculating, and, therefore, locally varying hydrosol observed in some coastal areas. Pelagic sediments may consequently be defined on the basis of a maximum value for the rate of deposition of the terrigenous component. This value seems to fall in the range of millimeters per thousand years. Within the basins accumulating pelagic sediments, the terrigenous deposition rate appears to vary not much more than one order of magnitude (5×10^{-5} to 5×10^{-4} cm/year), whereas values much higher and varying by several orders of magnitude are characteristic of the sediments fringing the continents. An attempt to outline the area covered by pelagic sediments, as defined above, is made in Fig. 1.

The rapidly accumulating sediments on the continental slope are unstable, and when the structures fail, coherent masses of sediment slide or slump. When water infiltrates the sliding masses, the concentrated suspensions slide as

---

particles have settled $(1-1/e)$ of the height of the water column, or $(1-1/2.72) \times 4600 = 2900$ m on the average.

[1] The particle-size frequency distribution of the sediment is not identical with the corresponding distribution in the suspensoid for reasons discussed in Sackett and Arrhenius (1962), but this effect is too small to be considered in the order-of-magnitude computation above.

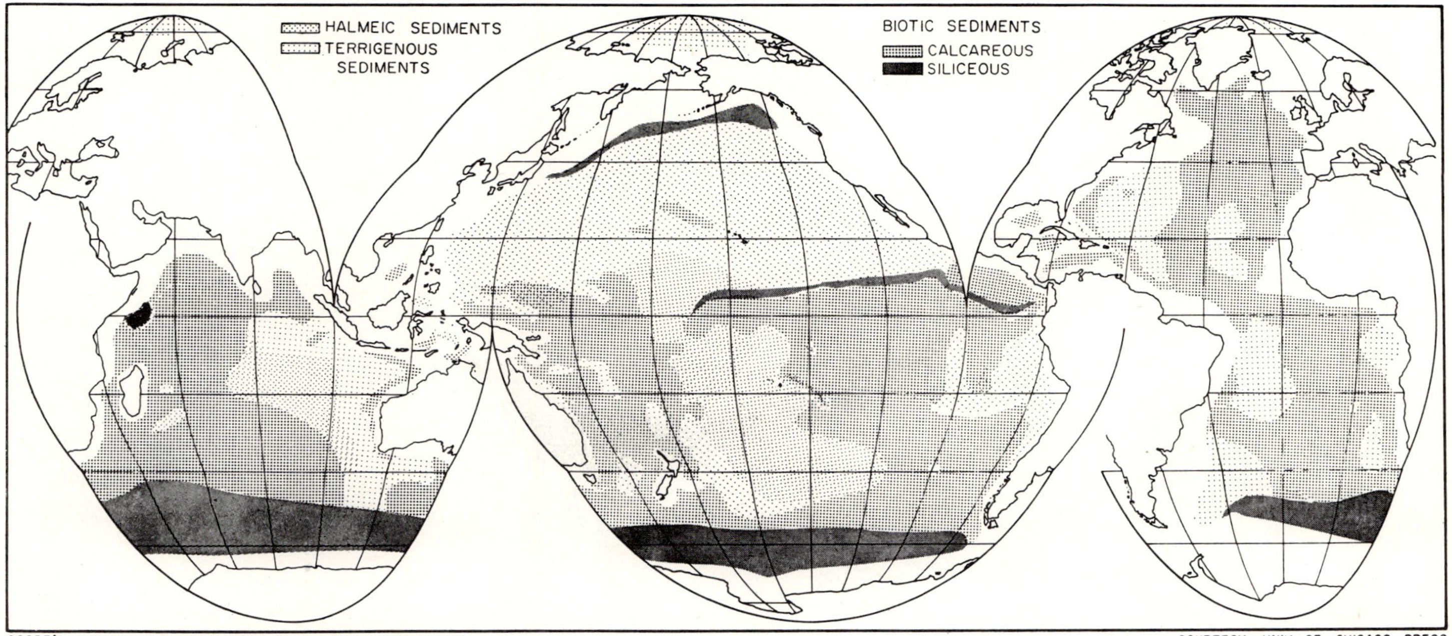

Fig. 1. Present areal distribution of pelagic sediments. The term halmeic has been used to indicate those sediments where the major component is formed from solution in sea-water. Sediments dominantly made up of minerals transported from the continents (regardless of origin of exchangeable cations) have been designated as terrigenous.

The boundaries shown in the map are approximate in most areas owing to lack of precise information on the critical properties of the sediment. The generalization due to the small scale has necessitated the omission of features with small areal extension. Local occurrences of pyroclastic and halmeic deposits, the latter mostly phosphorite and manganese oxide rocks, and of biotic sediments such as coral reefs and pteropod ooze are consequently not shown. Features with intricate limits such as the fast accumulating sediments funneled into pelagic areas along submarine canyons, as in some parts of the Atlantic Ocean, are omitted for the same reason.

The map attempts to represent present conditions, characterized by relatively narrow zones of rapidly depositing sediments around the continents, compared to the conditions during Pleistocene epochs with a maximum rate of sediment transport.

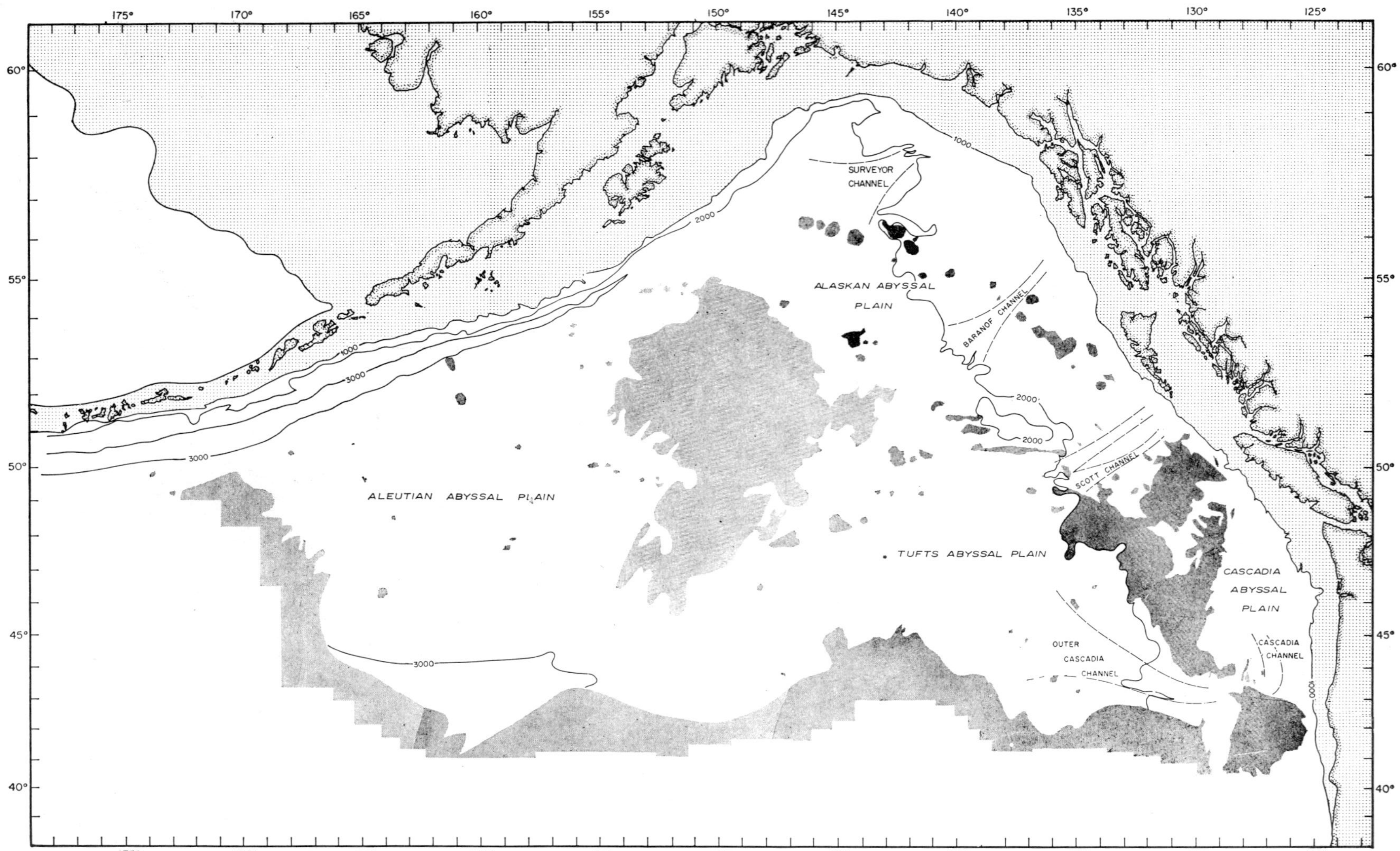

Fig. 2. Abyssal plains and feeding channels in the North Pacific. Depths in meters. Shaded areas indicate hilly topography. (Modified from Hurley, 1960.)

high-speed turbidity currents, invading the unprotected areas of the deep-ocean floor adjacent to the shelf slope, and smoothing the original topography to almost level abyssal plains (Kuenen, 1950; Heezen *et al.*, 1959; Menard, 1959; Hurley, 1960). Rises or trenches protect areas separated from the continental shelf from invasion by turbidity currents; in the absence of topographic barriers such as in the Gulf of Alaska and in large parts of the Atlantic Ocean, detrital sediments are, or were once, spread over extensive areas of the deep ocean by this mechanism (Fig. 2). On the other hand, some pelagic deposits accumulate close to the coast in areas where river discharge is low, where topographic protection is provided and where currents prevent fine-grained clastics from accumulating, such as on banks and rises, or where the clastic erosion products are efficiently funneled into deep catchment basins by submarine canyons (Shepard, 1948; Kuenen, 1950; Emery, 1960; Inman and Chamberlain, 1960). Iron and manganese oxide rocks, phosphorite and glauconite deposits, coral reefs, pteropod and foraminiferal oozes are thus frequently found in local areas close to the continents.

Although few measurements exist which permit quantitative estimates of the rate of deposition of the terrigenous component, i.e. the parameter suggested as a basis for division of marine sediments into pelagic and rapidly accumulating ones, the large differences in accumulation rate between these two sediment types often permit their recognition on the basis of a number of easily observed features. One of these is evidence of reworking of the sediment by organisms. Benthic animals appear to be distributed over all areas of the ocean floor where free oxygen is available, even at the greatest depths. Studies of the mixing of sediments across unconformities demonstrate that although single worm burrows might occasionally penetrate as deep as 20–30 cm, the mean mixing depth, above which 50% of the extraneous material is located, is of the order of 4–5 cm. In pelagic sediments the time required for burial of such a layer under another equally thick one varies between 10^3 and 10^5 years. The longer time is typical of areas with a low rate of deposition of organic remains, where a correspondingly low population-density of benthic animals is sustained. The total amount of reworking of a given stratum before ultimate burial might, therefore, not vary as much as the total rate of deposition within the area of pelagic sedimentation. When adjacent strata have different colors or shades, the mixing process causes a typical mottled appearance (Fig. 3). In pelagic sediments without a color stratification, the mud-eating animals leave less conspicuous traces, but their presence is indicated by fecal pellets, annelid jaws, and other fossil remains including chemical reduction structures. Non-pelagic deep-sea sediments, on the other hand, are deposited so rapidly that the sparse benthic population does not have enough time to disturb the strata as extensively as in pelagic deposits with similar population densities, and the original stratification is preserved, often in minute detail. Examples are the thin laminations often present in graded beds, deposited by turbidity currents, and laminae of volcanic ash.

Another effect of the low rate of detrital deposition, characteristic of pelagic

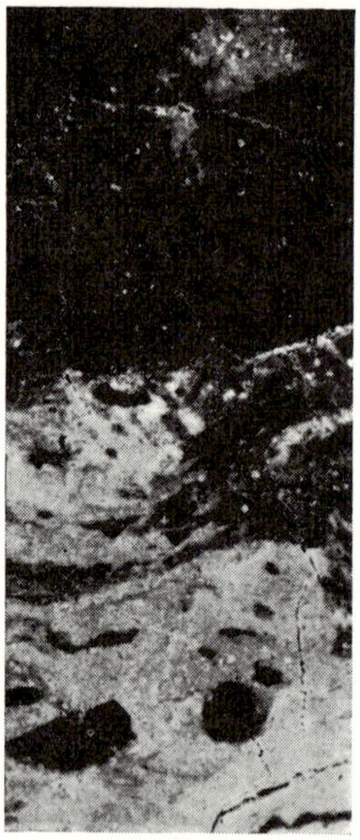

CH 40
0–8 cm

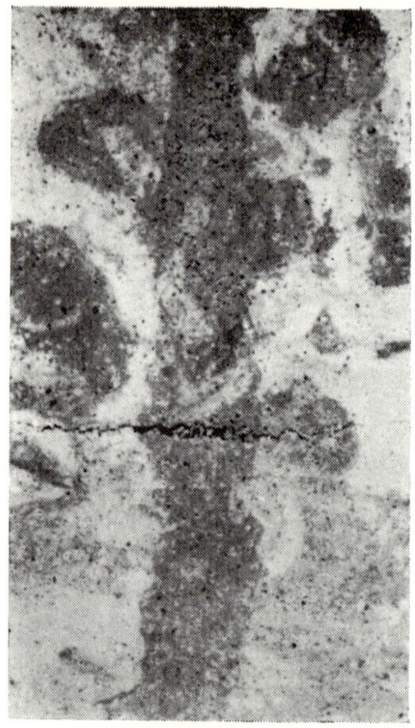

CH 30
16–24 cm

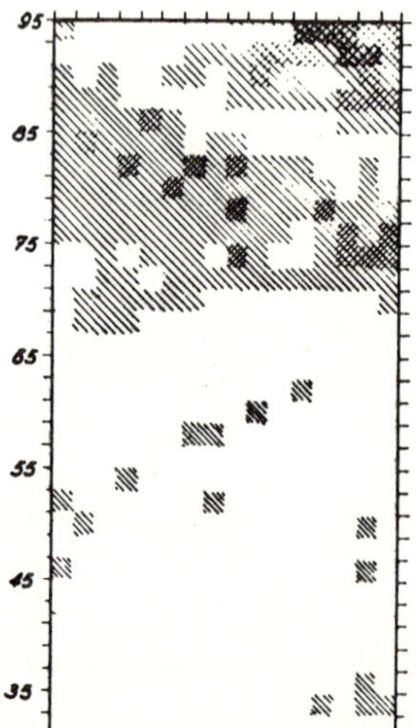

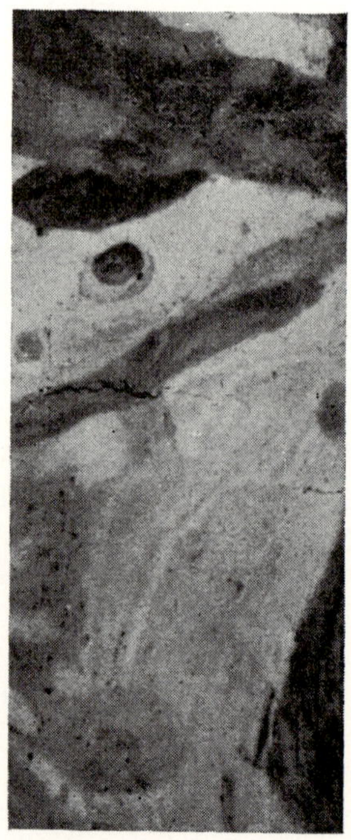

CH 30
40–48 cm

Fig. 3. Mixing by animal burrowing in pelagic sediments. The photograph CH 30 (16–24 cm) contains an unusually long vertical worm burrow extending through the section (8 cm). CH 40 (0–8 cm) demonstrates by tone contrast the redistribution of sediment over the boundary between layers of clay and calcareous ooze.

The section CH 30 (40–48 cm) contains an unconformity between Middle Tertiary and Quaternary sediment. The evaluation of a radioautograph of this section (adjacent to the photograph) distinguishes between the highly alpha-active Quaternary sediment and the inactive Tertiary, and demonstrates the extent and distribution of mixing of material from the two strata.

The total number of alpha tracks generated in the surface of the section CH 30 (40–48 cm) during six weeks was counted, and the average activity in each 4 mm^2 surface unit is shown in the radioautograph evaluation in the following intensity shades, graded in units of 10^{-4} alpha particles per cm^2 sec:

No surface tone	0–2.0
Line hatched	2.1–5.0
Cross hatched	5.1–9.3
Double cross hatched	9.4–16.6

The linear scale in millimeters (relative to an arbitrary origin) is indicated at the edge of the diagram. (From Picciotto and Arrhenius, unpublished.)

---

sediments, is the high percentage of authigenic minerals, extra-terrestrial material, and fossil remains of planktonic organisms. Further, with a similar rate of diagenesis in pelagic and neritic deposits, considerably more rapid increase in shear strength with depth in the sediment is found in the former than in the latter (Fig. 4). Bramlette (1961) has pointed out that the redox

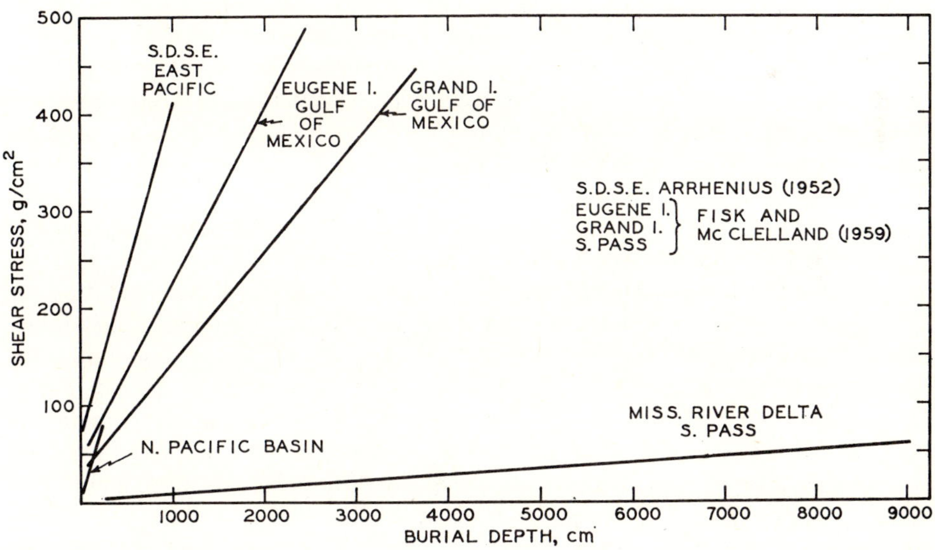

Fig. 4. Induration of sediment as a function of burial depth in pelagic sediments (Swedish Deep-Sea Expedition, East Pacific, N. Pacific basin) and in rapidly accumulating clay sediments (max. rate: Miss. delta). (From Moore, 1960.)

state of the sediment is mainly determined by the duration of contact with the oxidizing bottom water. This property consequently reflects the total rate of deposition, and Bramlette suggests a high degree of oxidation as one criterion for pelagic sediments.

Any single one of the criteria for pelagic versus high-rate sedimentation might be insufficient when applied alone. In cases where several indications occur jointly, conclusions regarding the order of magnitude of the sedimentation rate can apparently be safely drawn. The reliability of such conclusions is illustrated by the fact that estimates of the rate of deposition of different types of pelagic sediments, made before the advent of nuclear age-determination methods, have in general proved correct within surprisingly narrow limits. Even if the number of absolute age determinations will remain relatively limited, the suggested definition of pelagic sediments on the basis of a maximal value for the rate of deposition of the terrigenous component thus appears rational and practical.

2. Composition

Attempts to classify pelagic sediments have been based either on appearance and composition, or on the ultimate origin of the components. A rigorous application of the latter scheme has been attempted by Grabau (1904) and Schott (1935), and recently in a modified form by Goldberg (1954), further applied by Arrhenius (1959). However, at this time not enough is known of the ultimate origin and mode of accretion of some of the minerals making up the sediment (cf. Pettijohn, 1949, p. 184). To indicate the general distribution of different types of pelagic sediments it is consequently necessary to limit genetic classifications to major groups, except in the case of biotic sediments which can be identified morphologically. The origin of the major inorganic component in most areas has not been established with certainty; the term "red clay" has come into general use, although a red hue is seldom dominant and clay minerals sometimes are not the major mineral group.

Although introducing additional technical terminology is deprecable, the discussion of pelagic sediments on a genetic basis warrants some modification of existing terms. It appears feasible to distinguish minerals which crystallized in sea-water from those which formed in magmas, in hydrothermal solution, or by weathering under acidic conditions. This distinction is important, inasmuch as only the first group can be used to interpret the physicochemical state of the ocean in the past. These minerals are here described as *halmeic* (from ἅλμη; sea-water). Conversely, the properties of the third group of components frequently furnish information on the processes acting on the lithosphere and on the transport of this group of minerals into the pelagic environment. The minerals derived from the exposed surface of the lithosphere are here designated as terrigenous. Minerals and mineraloids deposited into the ocean by volcanic eruption are described as pyroclastic. Solids secreted by living organisms are referred to as *biotic*.

In the map showing the distribution of pelagic sediments (Fig. 1), this broad genetic classification is attempted, based in some instances on assumptions which, admittedly, are as yet unproven. However, the account given below of the composition of such sediments is based solely on the observed properties of the constituent minerals, whose possible sources of origin are discussed subsequently. In this way the basic observations have been separated from the interpretations.

A. Elements and Oxides

Iron and manganese oxide mineral aggregates constitute one of the major types of rock encountered on the ocean floor; according to Menard (unpublished) about 10% of the pelagic area of the Pacific is covered by such nodules. Measurements from the northeast Pacific by Skornyakova (1960) and Mero (1960a) give similar averages and indicate a considerable local variability in concentration (Fig. 5).

The nodules consist of intimately intergrown crystallites of different minerals; among those identified, besides detrital minerals and organic matter, are opal, goethite, rutile, anatase, barite, nontronite, and at least three manganese oxide minerals of major importance, described by Buser and Grütter (1956), Grütter and Buser (1957, 1959). One of these minerals, identical with a synthetic phase described in the chemical literature as δ-MnO_2, forms aggregates of randomly oriented sheet units as small as 50–100 Å. The other two minerals possess a double layer structure similar to that of lithiophorite and to synthetic phases, in chemical nomenclature called manganites.[1] These consist of ordered sheets of MnO_2, alternating with disordered layers of metal ions, co-ordinated with water, hydroxyl and probably also other anions. The most prevalent cations in the disordered layer are Mn^{2+} and Fe^{3+}. It is assumed that Na, Ca, Sr, Cu, Cd, Co, Ni, and Mo also substitute in this part of the structure. Two species of this general type have been observed, one with a basal spacing of 10 Å, the other with 7 Å. Buser has interpreted the 10 Å spacing as due to the existence of two discrete layers, OH and H_2O, whereas in the collapsed structure these groups are assumed to form a single layer, but owing to the disorder, the structure has not yet been precisely described. Synthesis experiments demonstrate that the three manganese oxide minerals represent increasing degrees of oxidation at formation in the order 10 Å "manganite," 7 Å "manganite", and δ-MnO_2. Buser has pointed out the potential use of this information in interpreting the conditions of formation of pelagic sediments.

Recent determinations of the bulk composition of manganese nodules have been made by Goldberg (1954), by Riley and Sinhaseni (1958), and by Mero (1960, and unpublished), and specific elements have been reported on by a number of authors. Mero's data, compiled in Table I, demonstrate the wide

[1] Not identical with the mineral manganite, γ-MnOOH, which is monoclinic–pseudo-rhombic and isotypical with the diaspore group (Strunz, 1957).

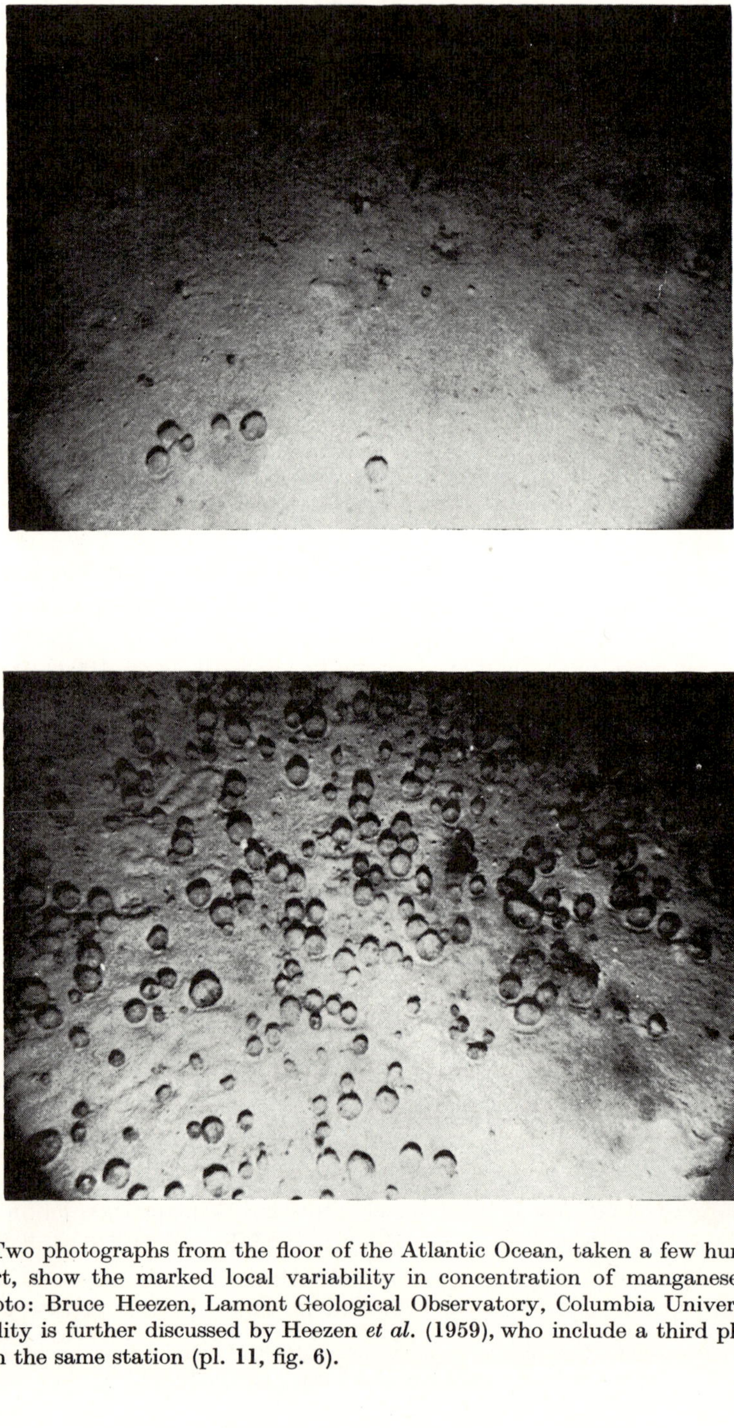

Fig. 5. Two photographs from the floor of the Atlantic Ocean, taken a few hundred feet apart, show the marked local variability in concentration of manganese nodules. (Photo: Bruce Heezen, Lamont Geological Observatory, Columbia University.) The locality is further discussed by Heezen *et al.* (1959), who include a third photograph from the same station (pl. 11, fig. 6).

TABLE I

Bulk Chemical Composition of Manganese Nodules from the Pacific and Atlantic Oceans as Determined by X-Ray Spectrography[a]

Element	Pacific Ocean			Atlantic Ocean		
	Maximum	Minimum	Average	Maximum	Minimum	Average
B	0.06	0.007	0.029	0.05	0.009	0.03
Na	4.7	1.5	2.6	3.5	1.4	2.3
Mg	2.4	1.0	1.7	2.4	1.4	1.7
Al	6.9	0.8	2.9	5.8	1.4	3.1
Si	20.1	1.3	9.4	19.6	2.8	11.0
K	3.1	0.3	0.8	0.8	0.6	0.7
Ca	4.4	0.8	1.9	3.4	1.5	2.7
Sc	0.003	0.001	0.001	0.003	0.002	0.002
Ti	1.7	0.11	0.67	1.3	0.3	0.8
V	0.11	0.021	0.054	0.11	0.02	0.07
Cr	0.007	0.001	0.001	0.003	0.001	0.002
Mn	50.1	8.2	24.2	21.5	12.0	16.3
Fe	26.6	2.4	14.0	25.9	9.1	17.5
Co	2.3	0.014	0.35	0.68	0.06	0.31
Ni	2.0	0.16	0.99	0.54	0.31	0.42
Cu	1.6	0.028	0.53	0.41	0.05	0.20
Zn	0.08	0.04	0.047	—	—	—
Ga	0.003	0.0002	0.001	—	—	—
Sr	0.16	0.024	0.081	0.14	0.04	0.09
Y	0.045	0.033	0.016	0.024	0.008	0.018
Zr	0.12	0.009	0.063	0.064	0.044	0.054
Mo	0.15	0.01	0.052	0.056	0.013	0.035
Ag	0.0006	—	0.0003	—	—	—
Ba	0.64	0.08	0.18	0.36	0.10	0.17
La	0.024	0.009	0.016	—	—	—
Yb	0.0066	0.0013	0.0031	0.007	0.002	0.004
Pb	0.36	0.02	0.09	0.14	0.08	0.10
Ignition loss at 500°C	39.0	15.5	25.8	30.0	17.5	23.8

[a] Concentrations are given in weight per cent. (Data from Mero, unpublished.)

variability in the concentrations of cobalt, nickel, copper, iron and molybdenum relative to manganese in the nodules. Arrhenius and Korkisch (1959) have pointed out that eddy diffusion is an insufficient transport mechanism for a number of the heavy metals found in manganese nodules, including the rare earth elements, thorium and uranium. They consequently assume that the vertical-transport mechanism described on page 677 below controls the transfer of many of the heavy metals from sea-water to sediment. High concentrations of a specific element at the ocean floor would under these conditions be expected in areas with a high rate of biological extraction of this element in the surface layer. Mero (*op. cit.*) has established a regional regularity in the elemental composition of the nodules (Fig. 6), and assumes that the regional variations

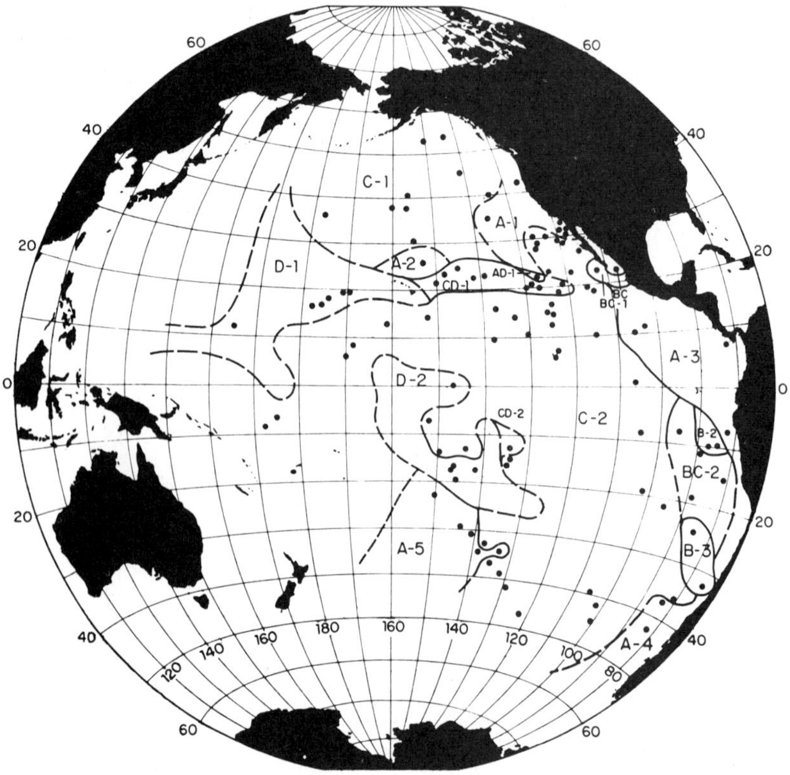

Fig. 6. Regional variation in chemical composition of manganese nodules. Regions marked A are characterized by ratios Mn/Fe less than 1; B-areas conversely have exceptionally high ratios Mn/Fe (ranging from 12 to 50). In areas marked C the nickel and copper content of the nodules is unusually high. D denotes regions with large amounts of cobalt (0.7–2.1%) in the nodules. (Mero, unpublished, modified from Mero, 1960a.)

in copper and nickel content of the nodules are controlled by the process referred to above.

Arrhenius and Korkisch (1959) have attempted to separate from each other the different minerals constituting the nodules, in order to establish the details of their structure and the localization of the heavy metal ions. The results demonstrate (Table II) that copper and nickel are concentrated in the manganese oxide phases. Cobalt, part of the nickel and most of the chromium are distributed between these and the acid-soluble group of the non-manganese minerals, dominated by goethite and disordered FeOOH. With increasing cobalt concentration in the rock, more of it substitutes in the manganese oxide minerals, which may achieve concentrations of well above 1% of cobalt, nickel, and copper. A variable fraction (mostly $\frac{1}{3}$ to $\frac{2}{3}$) of the soluble iron prevails in the form of goethite and in some instances as nontronite (the acid-soluble fraction of sample no. 2, Dwd HD 72, in Table II contains one-third nontronite). The remainder of the soluble iron is located in the manganese oxide minerals substituting for divalent manganese (Buser, *op. cit.*). Conversely, between a tenth and one per cent of manganese is always found in the separation group which contains the goethite, probably substituting as Mn^{3+} for Fe^{3+} in this

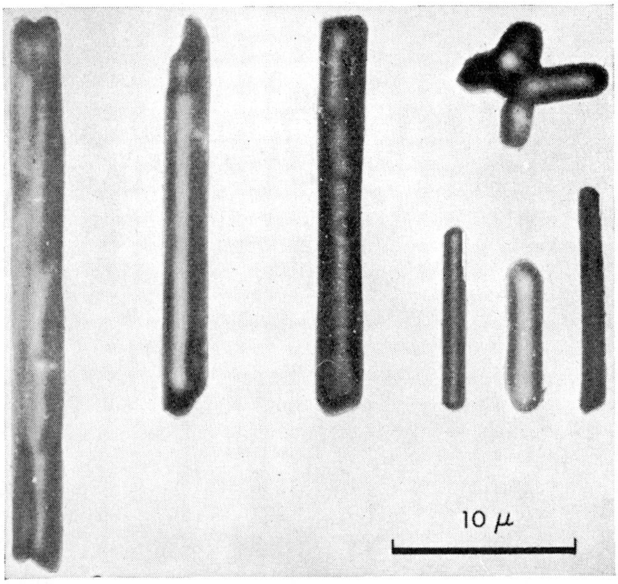

Fig. 7. Rutile needles and multiple twin, Cap. 49 BG, Quaternary, North Equatorial Pacific.

mineral. This group also tends to concentrate the major fractions of molybdenum, lead, titanium, scandium, the rare earth elements, and thorium. It is as yet uncertain which of the minerals constituting the group is responsible for the accumulation of the first four of these elements; it appears that thorium and the rare earth elements largely prevail as phosphates.

The elements barium and strontium in manganese nodules appear to be partitioned between the manganese oxide structures and the barite-celestite solid solutions (see also Section 2-B). These latter crystals account for the occasional high barium content of the acid-soluble and insoluble fractions in Table II. Titanium in the insoluble residue (Table II) is present as small euhedral crystals of rutile (Fig. 7) and anatase, the rutile frequently twinned on 011. The relatively high niobium content of the insoluble residue is probably

TABLE

Distribution of Elements between Minerals Soluble in 1M Hydroxylamine
Residue of

Sample number, region and position	Fraction	Percent of total mass	Cu	Ni	Co[a]	Cr	Fe	Mn
AA2	reducible	43.6	2000	1100	33	7	4400	major
high Mn	acid sol.	47.6	200	300	80	20	4400	8000
28° 23′N, 126° 57′W	residue	8.8	45	30	0	60	15,000	800
Dwd HD 72	reducible	70.0	800	2200	3300	7	30,000	major
high Mn	acid sol.	13.8	210	1100	9900	28	240,000	8000
25° 31′S, 85° 14′W	residue	16.0	20	7	35	0	100,000	45
AA 4711	reducible	84.4	3000	> 1000	250	4	10,000	major
high Mn, Ni, Cu	acid sol.	4.6	500	> 1000	700	50	90,000	4500
07° 48′S, 94° 06′W	residue	11.0	90	40	0	17	17,000	300
Alb 13	reducible	82.5	10,000	9900	1100	14	80,000	major
high Ni, Cu	acid sol.	3.3	200	200	40	50	60,000	2500
09° 57′N, 137° 47′W	residue	14.2	25	25	0	30	6000	40
AA 4721	reducible	54.5	7000	7700	440	3	130,000	major
high Ni, Cu	acid sol.	22.8	450	1400	660	70	120,000	2500
08° 08′S, 104° 11′W	residue	22.5	90	20	0	12	18,000	200
Dwd BD 4	reducible	84.25	1000	3100	9300	6	30,000	major
high Co	acid sol.	11.4	250	150	180	60	140,000	3500
13° 20′S, 146° 30′W	residue	4.3	20	25	0	5	3,000	p. tr.
Nth Hol 10	reducible		5900	1000	900	83		major
low Mn	acid sol.		1000	1700	1000	81		8000
40° 14′N, 15° 52′W	residue $\rho > 3.0$		250	25	8	62		700
	residue $\rho < 3.0$		140	25	9	58		500

[a] 0 = below sensitivity limit.

The reducible fraction contains the manganese oxide minerals. The acid-soluble fraction consists mainly of goethite and, if present, the cations from hydrous aluminosilicates (the high iron content in the acid-soluble fraction of Dwd HD72 is partly caused by a nontronite with $n\gamma = 1.598$; $Fe^{3+} = 15\%$). Further, such apatite crystals in which substitution with rare-earth ions is only partial are dissolved to a small extent in the weakly acid hydroxylamine hydrochloride and extensively in the hydrochloric acid, which accounts for the high concentration of rare earths, thorium, and possibly for the molybdenum and lead found in the acid-soluble fraction. When considerable barite is present, as in AA 4711, the corrosion of the crystallites yields appreciable amounts of barium, but the major fraction of the mineral is left in the residue.

II

Hydrochloride (Reducible); in 1M HCL (Acid-Soluble) and in the Insoluble Manganese Nodules

Mo	Pb[b]	Ti[c]	Sc	La	Y	Yb	Zr[c]	Th[c]	Ba	Sr	Nb
220	p. tr.	242	0	220	17	2	70	0	3000	450	p. tr.
150	p. tr.	1400	8	150	55	6	114	8.5	3000	45	p. tr.
2	0	4000	15	2	15	1	390	37.5	1000	100	0
250	60	785	4	200	130	9	61.4	0	500	200	p. tr.
700	150	7600	0	300	250	20	36.3	58.4	1000	50	30
11	0	3330	0	0	0	0	343	0	140	3	40
250	p. tr.	2860	0	90	60	4	162	16.6	3000	300	p. tr.
1000	65	5800	60	350	450	45	112	0	2000	60	16
0	0	2500	18	0	10	1	1670	8.3	20,000	300	20
300	35	182	0	250	120	12	20	17	500	350	p. tr.
800	20	15,800	110	0	15	2	236	58.3	3000	200	40
4	0	4000	12	0	13	1	1430	57.1	5000	250	110
130	p. tr.	555	0	250	150	14	125	0	900	1000	p. tr
400	60	6200	30	170	200	18	84.2	5.1	600	60	10
1	0	3570	0	0		1	1570	0	2000	50	10
300	40	562	5	280	130	13	120	57.1	3000	350	p. tr.
1000	300	15,600	16	450	400	20	363	27.2	500	65	50
15	0	122,900	20	0	0	0	4290	114.0	80	0	700
0	0	250	0	0	0	0	820	3.6	400	2600	0
26	300	3820	p. tr.	500	240	20	828	60.0	19,000	400	20
0	100	10,000	14	0	50	10	930	0.7	3200	300	20
0	110	1560	11	0	53	8	831	0.8	1000	330	21

[b] p. tr. = possible traces. [c] Colorimetric determination.

The residue also contains the terrigenous minerals, which are embedded in the nodules in varying amounts (mainly quartz, feldspar [of which the calcic plagioclase is partly dissolved by HCl], pyroxene, hornblende, mica residue and spinels), further authigenic opal, authigenic and terrigenous rutile and anatase (carrying the niobium), and finally the more complete authigenic pseudomorphs of rare-earth zirconium–thorium phosphates after apatite. The residue of Nth Hol 10 was separated in two density fractions by centrifugation in a Clerici solution with $\rho = 3.0$.

Concentrations are given in parts per million. (Spectrographic analyses by A. Chodos and E. Godijn, California Institute of Technology; colorimetric measurements of Zr, Ti, and Th by J. Korkisch, University of Vienna.)

due to substitution in rutile. The origin of the titanium oxide minerals is still uncertain; it appears probable that the rutile is terrigenous but the anatase might develop *in situ* (cf. Correns, 1937, 1954; Teodorovich, 1958).

The ferromanganese nodules range in size from a few microns, suspended in the sediment or coating other minerals, to intergrowths forming slabs several meters wide. The nodules have alternating growth zones of high and low

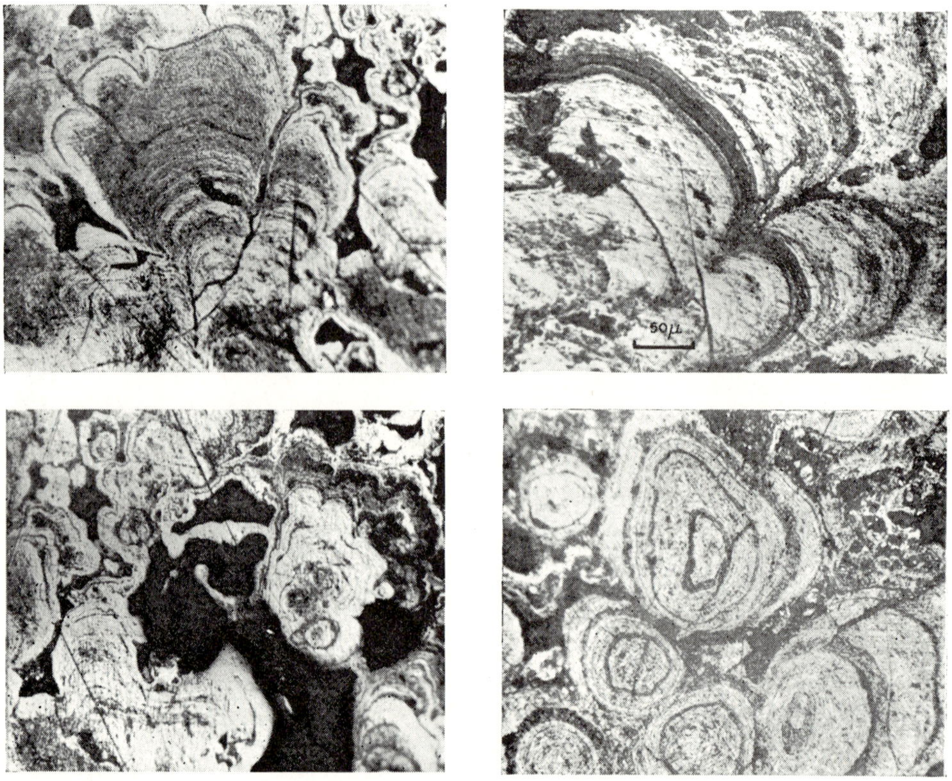

Fig. 8. Zonal growth in manganese nodule (16810, South Pacific). Reflected light. White = manganese oxide minerals; grey = goethite; black = mounting medium (polyvinyl resin) filling voids.

goethite content (Figs. 8 and 9). Detailed information on the distribution of macroscopic concretions over the present sediment surface has been obtained through deep-sea bottom photography (Fig. 5) such as by Owen, Shipek, Menard and Dietz in Dietz (1955), Menard and Shipek (1958), Heezen *et al.* (1959), Shipek (1960), and Zenkevitch (1959), and by sampling of the sediment surface (Skornyakova, *op. cit.*). Large nodules and crusts appear to accrete on topographic highs, or in other areas with a low total rate of deposition, where the growing nuclei are not buried by other sedimentary components. The rate of accretion was in the range of 10^{-5} to 10^{-6} cm year^{-1} in a nodule investigated by von Buttlar and Houtermans (1950) applying the results of Goldberg and

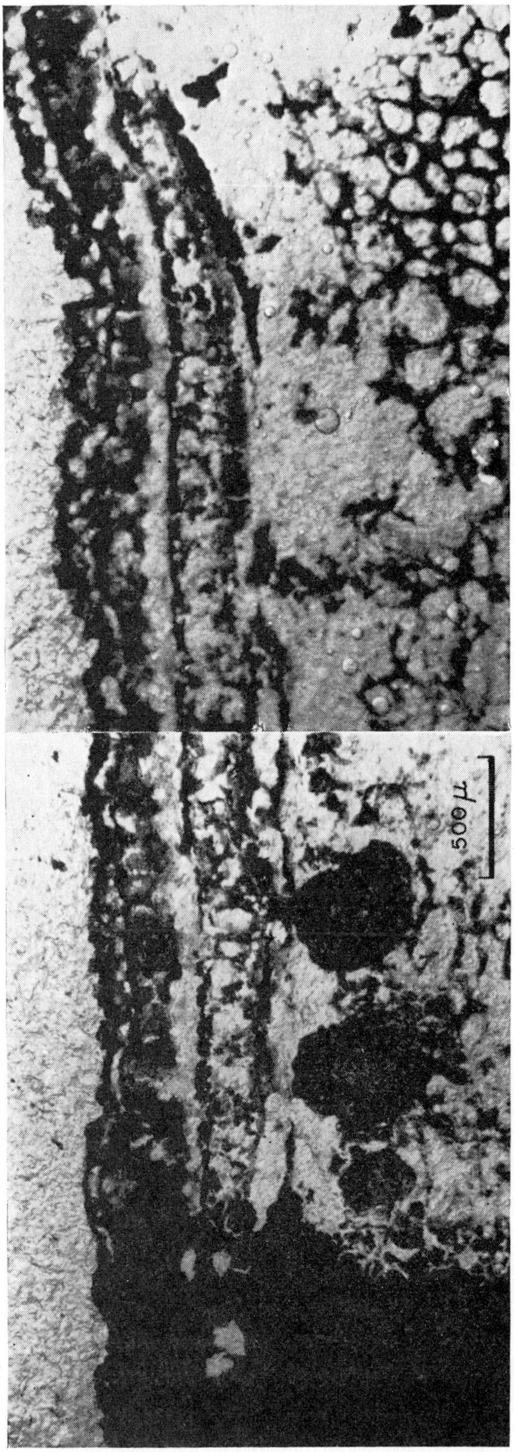

Fig. 9. Goethite network in manganese nodule from North Pacific (PAS 19121). Section cut normal to surface of nodule and manganese oxide minerals removed from right part of section by reduction with hydroxylamine, showing residual goethite texture. Untreated part of section, with remaining manganese oxide minerals, in opaque area to the left. Transmitted light.

Picciotto (1955) (Goldberg and Arrhenius, 1958, p. 198). Similar values are obtained for manganese micronodules in pelagic clays and zeolitites by integrating the manganese content in a column of known interval of time.

Several hypotheses have been advanced regarding the ultimate source of the manganese. Murray, in contrast to Renard (Murray and Renard, 1884) and lately Pettersson (1945, 1955, 1959), suggested that the manganese is derived from pyroclastics decomposing on the ocean floor; so far, no decomposition residue correspondingly deficient in manganese has been observed. Further, by order of magnitude, the manganese found on the ocean floor is accounted for by the amount of this element known to be continually lost from the continents (Kuenen, 1950, p. 390; Goldberg and Arrhenius, *op. cit.*; Wedepohl, 1960). Consequently many authors assume that the manganese on the ocean floor, other than the relatively small part which can be accounted for by decomposition *in situ* of basaltic pyroclastics, originates from dissolution of this element from continental rocks and from volcanic exhalates. Recent geochemical balance computations by Wedepohl (*op. cit.*) indicate that the volcanic exhalates are quantitatively important sources of manganese, iron, and other heavy metals with high halide vapor pressures. World-wide or large regional changes in the absolute rate of deposition of manganese in pelagic sediments could accordingly be due to variations in the rate of weathering on the continents, or in volcanic activity. Local or regional differences in the concentration of manganese in pelagic sediments, such as between the North and South Pacific at the present time or between Atlantic and Pacific sediments, can be accounted for by differences in dilution of the sediments with terrigenous material. Strata with a markedly increased manganese concentration, frequently found in pelagic sediments (see, for example, Arrhenius, 1952, pls. 2.51, 2.56; Revelle *et al.*, 1955, fig. 7; Pettersson, 1959), appear to correspond to periods of a lowered rate of total deposition, resulting in a decreased dilution of the halmeic oxide minerals with terrigenous silicates.

Whatever the ultimate source and mode of transportation of manganese and associated elements, several processes have been suggested to explain the mode of subsequent accretion of the manganese oxide minerals. One group involves various inorganic reactions (a review of these is given in Goldberg and Arrhenius, 1958); another group assumes organic (bacterial) mechanisms (Dieulafait, 1883; Butkevich, 1928; Dorff, 1935; Kalinenko, 1949; Ljunggren, 1953; Graham, 1959; Kriss, 1959). Goldberg and Arrhenius suggest specifically that manganese is removed from the bottom water by catalytic oxidation of manganous ion by colloidal ferric hydroxide at the sediment–water interface. In support of the biotic transfer Graham has demonstrated the presence of organic matter in the nodules. Although investigations in process (Galen Jones, unpublished) have demonstrated that the nodules contain bacteria capable of reducing manganese, it is difficult at the present time to evaluate the biotic hypothesis against an inorganic one.

Under the oxidizing conditions on the ocean floor, the only elemental mineral observed, besides carbon from burning forests and grasslands, is

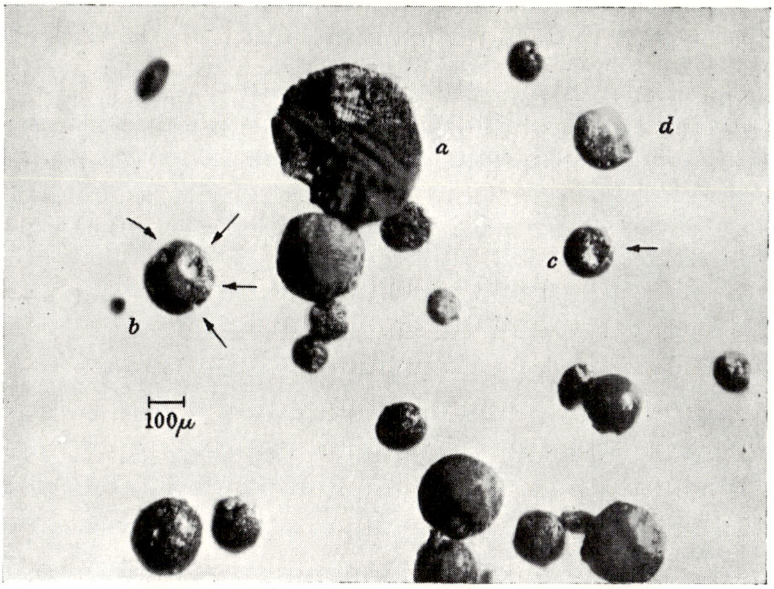

(a)

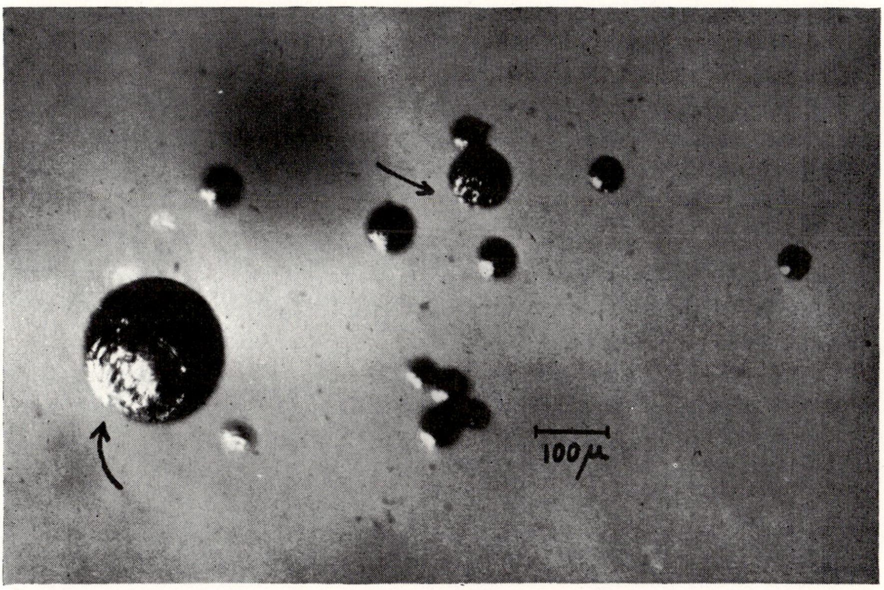

(b)

Fig. 10. Cosmic spherules from deep-sea sediments: (a) silicate spherules; (b) magnetite-coated nickel–iron spherules. (From Hunter and Parkin, 1960.)

nickel iron in the form of droplets, formed by ablation of iron meteorites during their passage through the atmosphere (Fig. 10b). The total amount of such spherules has been estimated at a few thousand tons per year over the surface of the earth (Pettersson and Fredriksson, 1958), which is only a fraction of a per cent of the total accretion rate of cosmic matter estimated from satellite measurements of micrometeorite impacts (LaGow and Alexander, 1960). Associated with the partly oxidized metal spherules are chondrules (Fig. 10a) consisting of olivine and pyroxene,[1] which are crystalline and thus could hardly

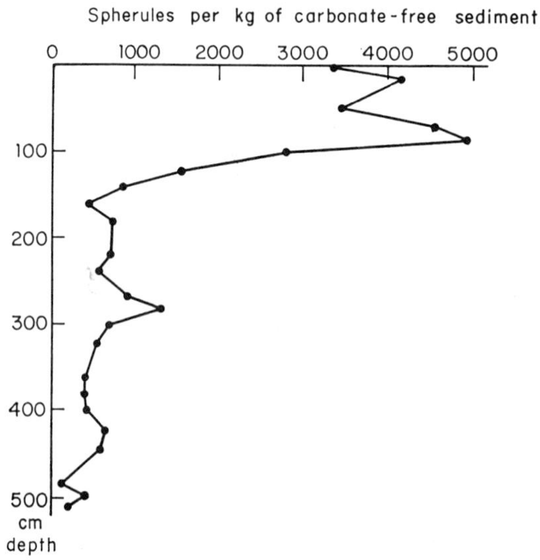

Fig. 11. Frequency of cosmic iron spherules as a function of depth in core 90, Swed. Deep-Sea Exped., from the West Pacific. (After Pettersson and Fredriksson, 1958.) Paleontological investigations by Bramlette and Riedel (quoted by Olausson, 1961) indicate an unconformity between Quaternary and Lower Miocene at about 56 cm depth.

have formed from molten drops in the atmosphere; these are probably original cosmic particles or fragments of chondritic meteorites breaking apart at entry into the atmosphere. Studies of the cosmic components of pelagic sediments were originally carried out by Murray and Renard (*op. cit.*). Recent extensive investigations by Hans Pettersson have focused on the variation in time of the accretion of cosmic material. The results of Pettersson and his co-workers (Pettersson and Fredriksson, 1958; Fredriksson, 1958; Castaing and Fredriksson, 1958; Laevastu and Mellis, 1955; Pettersson, 1959a; Fredriksson, 1959) indicate marked, world-wide changes in the rate of accretion of metallic spherules during Cenozoic times (Fig. 11). Hunter and Parkin (1960) have also

[1] The presence of this latter mineral is not evident from the work of Hunter and Parkin (1960) but has been established in similar chondrules by Murray and Renard (1891) and Bramlette (unpublished).

studied the metallic spherules and have investigated the nature of the silicate chondrules. The size distributions both of the nickel–iron spherules and of the olivine–pyroxene chondrules found on the deep ocean floor (Fig. 12) are in remarkable disagreement with the tentative size distribution of micrometeorites derived by impact counting from the satellites 1958 Alpha and 1959 Eta (LaGow and Alexander, *op. cit.*). An exponential increase in number of particles with decreasing size below 10–20 μ is indicated by the satellite observations, although cosmic particles from the ocean floor show a marked decrease in

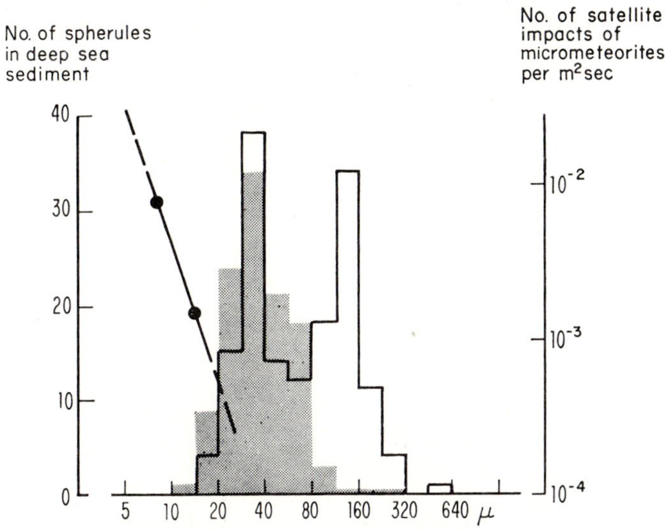

Fig. 12. Size distribution of cosmic spherules from pelagic sediments, and from micrometeorites in outer space. Shaded histogram = iron spherules; line-bounded histogram = silicate spherules from pelagic sediments (data from Hunter and Parkin, 1960); filled circles = impact observations from satellites 1958 Alpha and 1959 Eta (computed from data in LaGow and Alexander, 1960, assuming an average particle density of 3.0).

frequency below 25 μ and a comparatively large number of big (90 to 300 μ) silicate chondrules. This corroborates the idea that the cosmic material found on the ocean floor represents debris of meteorites, as suggested by Fredriksson (1959), rather than the original cosmic dust.

Large numbers of magnetic spherules of unknown origin and composition, ranging in size from less than a micron to several microns, are frequently observed in sediments (see, for example, Crozier, 1960). Analyses of such bodies indicate a heterogeneous origin. Many of the magnetic spherules previously assumed to have been derived from outer space have been shown by Fredriksson (1961) to consist of volcanic glass with inclusions of magnetite and metallic iron. Others, in size-range of 0.5–5 μ, appear to consist of goethite and might have been accreted by marine bacteria. It therefore appears necessary to define individual cosmic spherules not only on the basis of shape and magnetic properties but also on chemical composition. Considering the difficulties in

Fig. 13. Authigenic celestobarite from Pacific pelagic sediment showing frequent elongated prismatic habit and barrel-shaped growth form (B). Varying degree of corrosion by aqueous solution at separation is shown by A–H. I and J were protected against corrosion by use of excess sulfate ion at separation.

preservation, separation, and diagnosis of spherules in sediments, polar ice is a promising source for recovery and quantitative study of Pleistocene cosmic accretion (Thiel and Schmidt, 1961).

A material of geophysical significance found on the deep-ocean floor is maghemite (R. Mason, unpublished), which is a diagenetic alteration of magnetite (Hägg, 1935; B. Mason, 1943). The presence of maghemite increases the magnetic susceptibility of the solid. The martite reported by Mellis (1952, 1959) might possibly also be a pseudomorph of maghemite after magnetite.

B. Sulfates

The high concentration of sulfate ion in sea-water, occasionally increased in the interstitial water of the sediment by oxidation of proteinaceous matter, probably limits the solubility of strontium, barium, radium, and lead (Arrhenius, 1959). Radioactive solid solutions of celestite ($SrSO_4$) and anglesite ($PbSO_4$) in barite ($BaSO_4$) thus constitute geochemically important mineral species on the ocean floor (Arrhenius, Bramlette and Picciotto, 1957; Arrhenius, 1959) (Fig. 13). A sample consisting of a large number of celestobarite crystals, obtained from equatorial North Pacific sediments, showed the average amount of substitution to be 5.4 mole per cent celestite and 0.05 mole per cent anglesite.

Comparatively high concentrations of barium, strontium, and lead are found in some marine planktonic organisms, which also contain considerable quantities of other heavy-metal ions (Table III). This suggests that biological extraction from surface sea-water and subsequent sinking is an important mechanism in accreting these elements to the sediment. Among the organisms notable in this respect are some species of Foraminifera, pteropods and heteropods. The acantharid Radiolaria (Schewiakoff, 1926) are particularly efficient in extracting strontium; the celestite ($SrSO_4$) skeletons of these protozoans contain in addition 0.4 mole per cent of barite in solid solution. The skeletons consist of dart-like spines of radially arranged celestite microcrystals with Y oriented in the radial direction of the spine. At decomposition of the protoplast, the spines become detached and presumably settle at a high rate owing to their shape and high density. In contact with sea-water they are, however, rapidly corroded and dissolve entirely before burial in the sediment occurs.

Thus, marine organisms provide a conveyor mechanism for certain elements from the surface of the ocean to the deep water or the bottom, where the dissolved elements are released by dissolution. Phosphorus, silicon and nitrogen have been found in higher concentrations in intermediate and deep water than in surface water (ref. in Sverdrup *et al.*, 1946). Still higher gradients are found from the near-bottom water and the interstitial water of the sediment for silicon and phosphorus (Koczy, 1950, fig. 2.3; and Table VI of this chapter), for radium (Koczy, 1958), for barium (Chow and Goldberg, 1960), and for nitrate (3 to 15 times excess over bottom-water concentration in interstitial water in South Pacific sediments, Arrhenius and Rotschi, 1953, fig. 29). Part of the ions released into the bottom water or interstitial water are returned to the sediment

TABLE IIIa

Trace-Element Composition of Some Marine Planktonic Organisms. Concentrations in parts per million of ash. The blank consists of spectrally pure quartz treated identically as the samples. Spectrochemical analyses by A. Chodos and E. Godijn. (From Arrhenius, Bradshaw and Kharkar, unpublished.)

Sample no.	Description	Ba	Sr	Mg	Cu	Ag	Zn	Pb	Ti	V	Cr	Mn	Fe	Ni	B
16 256	Foraminifera (listed in Table IIIb)	700	400	1000	25	3	0	10	15	p. tr.	0	8	300	9	0
17 121	Spines of acantharid Radiolaria	5400	major	–	750	0	600	105	300	0	90	0	300	57	0
16 257	Pteropods and heteropods listed in Table IIIc														
16 257–1	Tissue remaining after short treatment with dilute HCl	60	0	–	140	0	660	0	180	0	100	0	–	0	960
16 257–2	Fraction removed by dilute HCl (mainly shells)	31	480	–	15	0	0	0	0	16	0	0	–	0	0
16 257–3	Blank	0	0	–	0	0	0	0	0	0	0	0	–	0	0

p. tr. = possible trace 0 = below sensitivity limit – = no determination

TABLE IIIb

Species Distribution in Composite Sample (16 256) of Planktonic Foraminifera Used for Analysis in Table IIIa. The sample was collected at Tethys Expedition Station 28, 26° 13.9′N, 141° 34.5′W, 9 Aug., 1960, and Station 29, 27° 15.5′N, 137° 58.0′W, 11 Aug., 1960. (Arrhenius, Bradshaw and Kharkar, unpublished.)

Hastigerina pelagica (d'Orbigny)	66.4%
Globigerinoides conglobatus (Brady)	19.5
G. sacculifer (Brady)	6.4
Orbulina universa (d'Orbigny)	3.0
Globerigerinoides ruber (d'Orbigny)	2.6
Candeina nitida (d'Orbigny)	1.5
Globoquadrina conglomerata (Schwager)	0.3
Globorotalia tumida (Brady)	0.2
G. cultrata (d'Orbigny) = *menardii* (d'Orbigny)	0.1
Total number of specimens	2077
Dry weight	37.5 mg

TABLE IIIc

Relative Abundance of Pteropods and Heteropods in Composite Sample (16 257) Used for Analysis in Table IIIa. The sample was separated from plankton collected at the same stations as Sample 16 256 (Table IIIb)

	Relative abundance
Pteropods	
Styliola subula	3
Hyalocylix striata	1
Creseis acicula	2
Creseis virgula conica	1
Limacina bulimoides	2
L. lesueuri	3
L. inflata	1
L. trochiformis	1
Heteropods	
Atlanta peroni	2
A. fusca	1
A. turriculata	1
Shells, dissolved by adding dilute HCl dropwise until effervescence stopped, weighed as $CaCl_2$, anhydr.	685 mg
Ash of soft tissue, oxidized in perchloric acid	10.8 mg

by crystallization of less soluble solids; barium, strontium, and lead separate into harmotome-type zeolites, manganese oxide minerals, possibly including psilomelane, and further crystalline solid solutions of celestite, barite, and anglesite in proportions mentioned above. At slight solution of the celestobarite in distilled water, the crystal faces develop a typical pitted appearance (Fig. 13, A–H). The lack of these etching features in the crystals as found in the sediment (Fig. 13, I–J) indicates that dissolution of celestobarite is not taking place on the ocean floor.

Besides the extensive cation substitution in the celestobarite crystals, replacement of SO_4 by BF_4 and possibly CrO_4 is indicated by the presence of 1000 ppm of boron and 1400 ppm of chromium in this mineral. These and the cation substitutions are of potential interest as indicators of the physico-chemical conditions in the sea-water and in the interstitial solution.

The exceedingly slow crystal growth on the ocean floor probably produces a close approach to thermodynamic equilibrium between the liquid and the solid solution. If in the relation

$$D = \frac{(B_c/A_c)}{(B_1/A_1)},$$

B_c and A_c denote the concentrations of the substituting species and the substituted main species respectively in the crystal, and B_1 and A_1 the corresponding concentrations in the liquid, the partition coefficient, D, indicates the enrichment (if > 1) or depletion of the substituting foreign ion in the crystal structure. If B_c and A_c are known from analysis of the actual crystals, and D from controlled experiments, the ionic ratio, B/A, in the bottom water or interstitial solution from which the crystals formed can be derived. For the cation substitution couple, Sr^{2+}/Ba^{2+}, the ionic ratio in the halmeic crystals is 0.057, and D has been determined to be 0.030 ± 0.004 (Gordon, Reimer and Burtt, 1954). The solution in equilibrium with these crystals should consequently have an ionic ratio, Sr^{2+}/Ba^{2+}, of 1.9. If 114×10^{-3} mmole/l. (the average strontium concentration in sea-water) is accepted as a minimum concentration value in the interstitial solution of the sediment, then a minimum concentration of 60×10^{-3} mmole/l. of barium is needed to maintain the ratio indicated by the crystal composition. Such a barium concentration would be 136 times higher than that observed in deep water. A part of this apparent discrepancy might be due to lowering of the strontium concentration in the interstitial water by the zeolite and oxide species observed in co-existence with the barite; however, a considerably higher barium concentration in the interstitial water than in the deep water is suggested by these data. Chow and Goldberg (*op. cit.*) have interpreted the deep-water concentrations observed by them as close to saturation with the sedimentary barite. However, considerably higher saturation concentrations should be expected since both the cation and the anion substitution in the barite structure contribute to a markedly increased solubility of the crystalline solid solution above that of pure barite. In pure water the solubility of celestite is 587 μmole/l. at 25°C and 1 bar[1] as compared with 9.5 μmole/l. for barite. Under the assumption of a linear increase in solubility with substitution, the observed 5.4 mole per cent of celestite in the barite structure would result in a solubility of 40.7 μmole/l., or 4.3 times the solubility of barite. As is seen in Table IV, the other substitutions observed further increase the solubility of the crystal.

TABLE IV

Solubility and Substitution Relations

	Solubility, μmole/l. (25°C, 1 atm)	Substitution in pelagic celestobarite (mole per cent)
$BaSO_4$	9.5	94.3
$SrSO_4$	587	5.4
$PbSO_4$	140	0.046
$BaCrO_4$	13	0.12
$Ba(BF_4)_2$	no data	0.11

Gypsum has been observed in several instances in pelagic sediments with a high content of organic matter. Whether these crystals were formed on the ocean floor or during storage of the samples, as a result of sulfide oxidation, is not known.

C. Phosphates

The phosphate concentration in surface sea-water is limited by photosynthetic organisms. These settle toward the bottom or are consumed by higher members of the food chain, and a fraction of the phosphorus is ultimately used by vertebrates for building an apatite skeleton. A continuous transport of phosphorus from the surface layer of the ocean to the bottom is thus maintained. The solubility of the apatite series is not accurately known (Sillén, 1959, p. 566); but it is a well-established fact that skeletal apatite is slowly dissolving on the deep-ocean floor (Koczy, 1952; Arrhenius, 1959), and that no precipitation occurs there of the phosphate formed by hydrolysis of organic matter. It is thus obvious that the solubility of apatite at the temperature and hydrogen ion concentration of present deep-ocean water is higher than the phosphate concentration observed in bottom water and in interstitial water of deep-sea sediments (5.0 μmole/l., Arrhenius and Rotschi, 1953; 5.5 mole/l., Bruejewicz and Zaytseva, 1958). Consequently, all the organic phosphorus and a large fraction of the skeletal apatite reaching the deep-ocean floor are returned to solution in the deep water and ultimately brought back to the upper layer of the ocean. The recycling of phosphate in the ocean due to temperature and pH control of the solubility leads to partial removal of phosphate from the deep-ocean floor and to permanent accumulation of skeletal apatite and of authigenic

[1] Owing to the presence of sulfate ion the solubility is somewhat lower in sea-water (approx. 500 μmole/l. at 25°C and 1 bar) than in pure water (Müller and Puchelt, 1961).

phosphate minerals (mainly francolite, $Ca_5[F|PO_4CO_3]_3$) in shallow low-latitude areas where saturation is reached. Upwelling of phosphate-rich deep water produces exceptionally high concentrations of phosphate minerals in such areas (Kazakov, 1950). The ensuing increase in organic productivity in the euphotic zone leads to a high rate of accumulation of organic remains on the bottom, and a high rate of crystallization of phosphates is maintained by decomposition of the organic phosphorus compounds (McKelvey *et al.*, 1953). Bruejewicz and Zaytseva (1958) measured concentrations of dissolved phosphorus as high as 27 μmole/l. in Pacific sediments of this type.

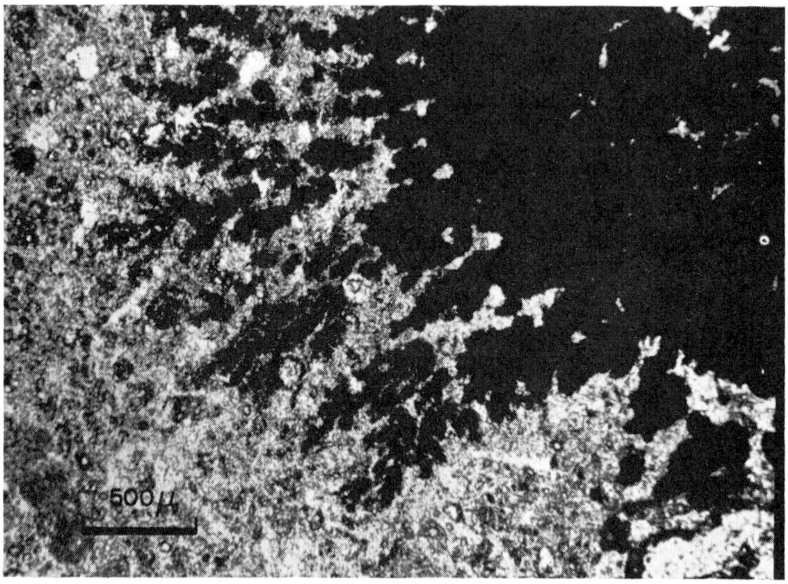

Fig. 14. Dendritic intergrowth of manganese oxide in marine phosphorite from oxidizing environment (Cape Johnson Guyot; 17° 10′N, 177° 10′W). Transmitted light.

The solubility relations of francolite are complicated by variable substitution of calcium with zirconium and rare earth ions, which drastically reduce the solubility; zirconium phosphate concentrations up to 2800 ppm Zr have been observed in marine inorganic apatite (Arrhenius and Korkisch, unpublished).

The phosphorite rock accumulating in areas of high organic productivity consists of a microcrystalline matrix of francolite with phosphatized tests of Foraminifera (originally consisting of calcite), skeletal apatite of marine vertebrates, thin flakes of opaline silica derived from diatom frustules, films and grains of glauconite, interspersed dark-colored organic matter and terrigenous minerals. Other widespread deposits of marine phosphorite occur, without association with exceptionally high organic productivity, in shallow areas of the tropical ocean, where calcareous deposits are exposed to relatively warm sea-water, such as on seamounts and on drowned coral reefs; some of these are now elevated above sea-level and commercially exploited. Complete pseudomorphous replacement of calcite and aragonite fossils is attained in some cases (Rex, in Hamilton, 1956, p. 35). In contrast to the phosphorite from areas of high productivity, intergrowth with manganese and ferric oxide minerals is common in this type of phosphorite (Fig. 14), but glauconite and organic

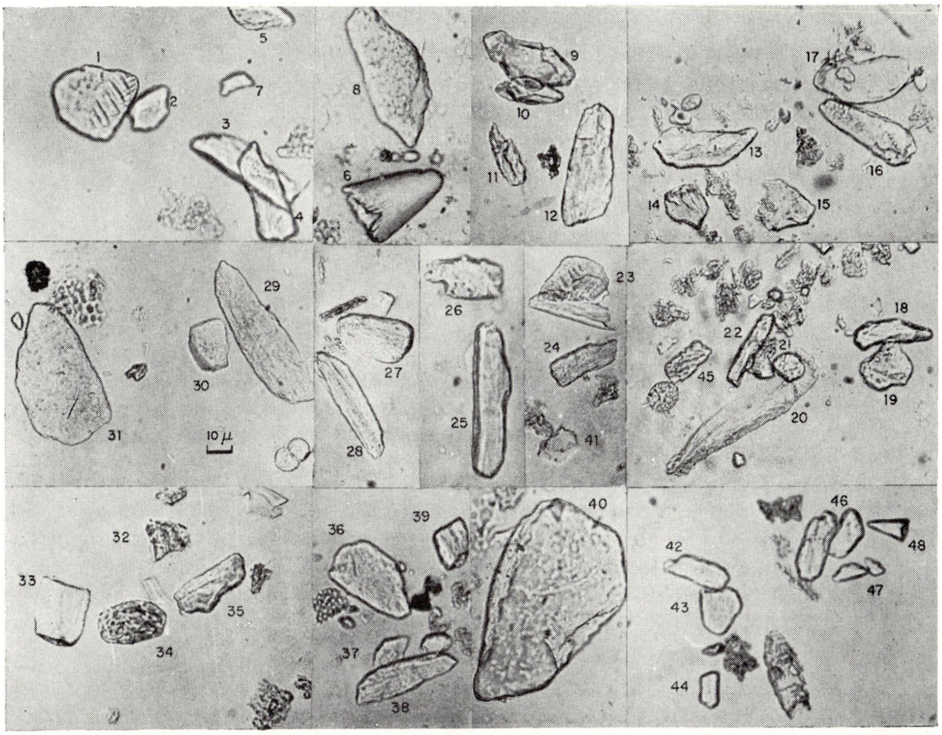

(a)

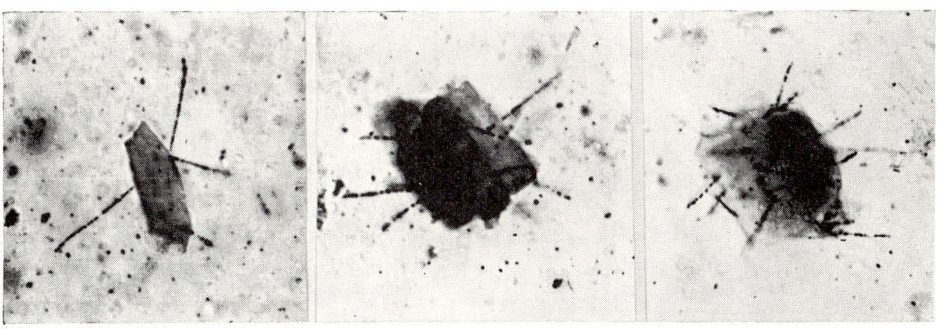

(b)

Fig. 15. Fish-bone debris: (a) Diaphane mounting medium, phase contrast. Fish-bone fragments labeled 6 and 48 are teeth. (b) Nuclear emulsion, showing tracks of alpha particles emitted from fish debris. [(b) from Picciotto and Arrhenius, unpublished.]

matter are absent. The oxidizing conditions of deposition are further indicated by the low uranium content of this type of marine francolite. Uranium probably prevails in the hexavalent state (ionic radius 0.80 Å) under the normal oxidizing conditions in sea-water, presumably as a carbonate complex (Starik and Kolyadin, 1957), and does not easily substitute for the 20% larger calcium ion in the francolite structure. However, where uranium is locally reduced to the tetravalent state, its increased ionic radius (0.97 Å) is close enough to that of calcium (0.99 Å) to substitute extensively in the crystal structure (Altschuler, Clarke and Young, 1958). Consequently, the concentrations of uranium in the seamount- and reef-type francolite, formed under oxidizing conditions, amount to only one twelfth to one hundredth of the concentrations found in francolite when it is deposited under reducing conditions in areas of high organic productivity (Arrhenius and Korkisch, 1959) (Table V).

TABLE V

Substitution of Uranium in Marine Authigenic (Halmeic) Apatite and in Microcrystalline Fish-Bone Debris

Those apatite deposits which crystallized under reducing conditions, indicated by the presence of organic compounds, glauconite and other ferrous minerals, are high in uranium, presumably substituting as U^{4+}. Conversely, those formed under oxidizing conditions, indicated by lack of organic matter and ferrous minerals and by the coexistence with manganese oxide minerals, have a low uranium content, probably due to the difference in size between Ca^{2+} and U^{6+}, and to the excess charge of the latter ion.

Similar relations appear to govern the sorption of uranium on microcrystalline bone apatite in oxidizing and reducing environments as shown in the lower part of the table. (Data from Arrhenius and Korkisch, 1959.)

Sample no.	Mineral type	Environment	U, ppm	Description of sample
16 182	Halmeic francolite	Oxidizing	8.5	Pseudomorph after calcite (late Eocene Foraminifera). Guyot 19171, Central Pacific
16 176	,, ,,	Oxidizing	0.7	Pseudomorph after calcite (Foraminifera) Sylvania Seamount, Marshall Is.
16 175	,, ,,	Oxidizing	2.3	Microcrystalline, massive. Cape Johnson Seamount
16 174	,, ,,	Reducing	102	Microcrystalline, massive, with organic matter and glauconite. Coronado Bank
16 803	,, ,,	Reducing	68	Strongly phosphatized surface layer of bark from Pleistocene tree trunk, exposed at sediment surface off Tehuantepec (also described in Goldberg and Parker, 1960)

TABLE V—*continued*

Sample no.	Mineral type	Environment	U, ppm	Description of sample
16 195–8.2	Fish-bone apatite	Oxidizing	1.5	From low organic productivity area, 17–21°N in East Pacific; brown clay sediment with numerous manganese nodules
16 193–10+ +16 194–6	,, ,, ,,	Oxidizing but less than in 16 195–8.2	8.5	From siliceous-calcareous sediments below Equatorial Current system. Ferrous iron preserved in zeolites, 1–3 mmole NH_4^+/l. of interstitial solution
16 181–1	,, ,, ,,	Weakly reducing	16.3	Modelo Formation, Calif. (Tertiary). Comparatively rapidly accumulating epicontinental sediment. Organic matter present at the level of 2–4% carbon in the sediment; ferrous iron preserved or generated, no sulfides

Goldberg and Parker (1960) have proposed that the low oxygen content of the bottom water in areas with a high rate of deposition of organic matter is a causal factor in the precipitation of francolite. The widespread occurrence of this mineral, deposited under oxidizing conditions as well as under reducing ones, would seem to point instead to a reaction which is insensitive to the redox state of the system.

The high concentrations of phosphorus, often attaining values of 1–2% P_2O_5 in slowly accumulating clay sediments and zeolitites on the deep-ocean floor, were previously attributed to precipitation of iron phosphate (Goldschmidt, 1954); it now seems quite evident that the phosphate prevails in the form of skeletal debris of fish (Arrhenius, Bramlette and Picciotto, 1957). Apatite is soluble at the deep-ocean floor, so that delicate structures such as scales and minute bone fragments are preserved only in the most recent parts of rapidly accumulating calcareous deep-sea sediments.

In deep-sea sediments, including manganese nodules, which accumulate slowly, only the most resistant structures with a large crystallite size, such as teeth of fish and earbones of whales, remain undissolved. The dissolution of skeletal apatite is counteracted by the incorporation into the crystal structure of large quantities of the rare earth elements thorium and zirconium, adsorbed from solution in sea-water. The ensuing highly insoluble phosphates eventually remain as microscopic crystals after the disappearance of the original skeletal debris in the most slowly accumulating types of deep-sea sediments, such as manganese nodules.

D. *Carbonates*

The abundant planktonic and benthic organisms which build skeletal structures of calcite or aragonite provide a mechanism for deposition of these minerals over the ocean floor. Quantitatively most important in this respect are Foraminifera, which are Protozoa with calcareous, arenaceous, or chitinous tests (Fig. 16), and coccolithophorids, which are unicellular algae with minute calcite platelets—coccoliths—that have been secreted from the cell

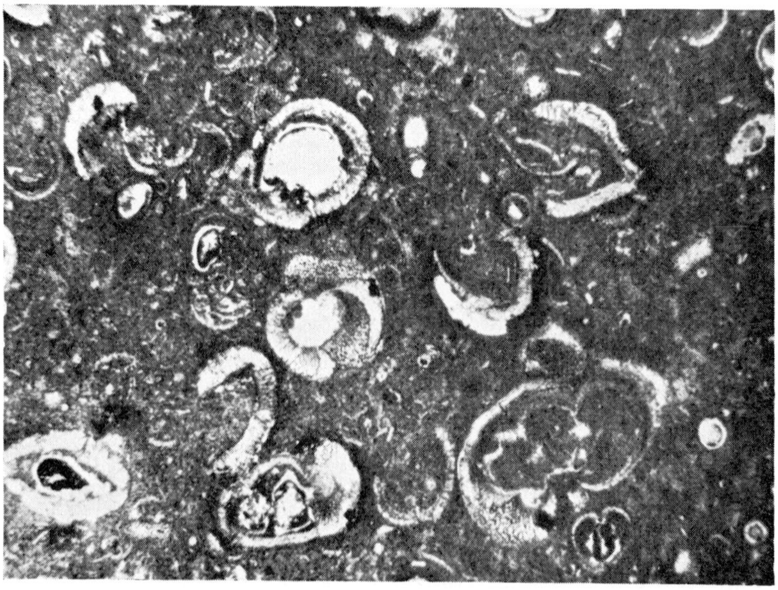

Fig. 16. Calcareous ooze with tests of Foraminifera. Thin section, polarized light.

wall (Fig. 17). Most Foraminifera contributing to the pelagic sediments are planktonic, and all of the planktonic species form calcite tests. Benthonic Foraminifera constitute less than 10% of the fossil assemblages in pelagic deep-sea sediments but are more abundant in shallow-water pelagic sediments such as coral reefs. The variation in species distribution as a function of ambient water temperature has successfully been used, initially by Schott and later by a number of workers (ref. in Phleger, 1960), to trace variations in the surface-water temperature of the ocean in the past. Recently, similar information has been obtained by mass spectrometric determination of the temperature-dependent ratio O^{18}/O^{16} in the carbonate of Foraminifera in deep-sea pelagic sediments (Emiliani *et al.*, see ref. in Emiliani, 1955 and 1956; Ericson and Wollin, 1956) and in the skeletal structures of a number of reef organisms (Lowenstam and Epstein, 1957). The usefulness of pelagic coccolithophorids for this purpose has recently been indicated by Craig (unpublished). A comprehensive survey of marine paleo-temperature work is presented by Emiliani and Flint elsewhere in this volume (Chapter 34).

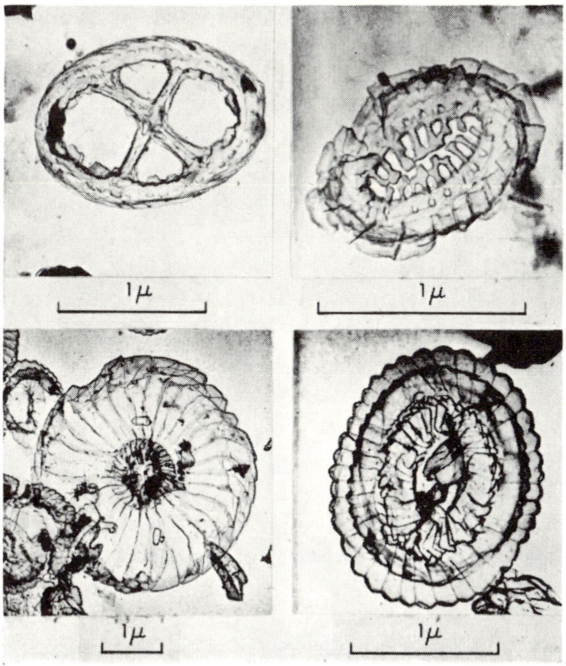

(a)

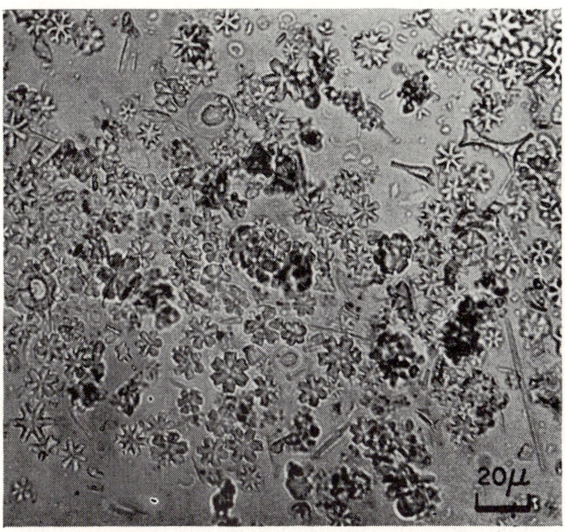

(b)

Fig. 17. (a) Coccoliths (electron micrographs of carbon replicas) (from J. Markali, Oslo, unpublished); (b) Coccolith and *Discoaster* ooze (Aquitanian), Equatorial Pacific. (From Bramlette, 1961.)

Owing to relatively rapid and easily distinguished evolutionary changes of the coccolithophorids, it has been possible to use their fossil remains for indication of geological age by highly precise biostratigraphic correlations (Bramlette and Riedel, 1954). The mass of coccoliths in the carbonate fraction of deep-sea sediments sometimes exceeds that of foraminiferal tests; such coccolith oozes were particularly abundant in the Cretaceous and the Tertiary. Coccolith ooze also constitutes the glacial maximum productivity stage below the Pacific Equatorial Divergence (Fig. 36).

In sediments older than the Middle Mesozoic, identifying coccoliths is usually difficult, owing to gradual recrystallization of the calcite; in fact, many fine-grained limestones in older geological formations may be recrystallized coccolith oozes (Bramlette, 1958). Although the calcareous planktonic Foraminifera did not evolve until Mesozoic time, calcium from the continents might thus have been deposited as carbonate over the deep-ocean floor by planktonic organisms early in the history of the Earth.

Aragonite shells are secreted by some groups of planktonic organisms, of which pteropods are the quantitatively most important. Because of the relatively high rate of dissolution of aragonite in sea-water at low temperatures and high pressures, pteropod ooze is accumulating only in comparatively shallow areas.

Extensive substitution in the marine calcium–magnesium carbonate structures is essentially limited to strontium in aragonite. A concentration of other trace elements such as lead, copper, barium, strontium, titanium, and iron is to be found in the protoplasts of some species of Foraminifera (Table III) and in pteropods, whose organic shell structures interlayered with aragonite provide a possible site for heavy-metal ions concentrated from sea-water (Curl and Nicholls, unpublished). Thus, a number of trace elements can be related to calcareous fossils or their dissolution residue (Correns, 1937; Graf, 1960). Several of these organism groups might consequently play an important role in the vertical-transport mechanism mentioned in Section 2-B of this chapter.

Pelagic sediments, as defined in the present work, include shallow-water pelagic carbonate sediments which constitute bioherms of various types and particularly coral reefs. This group of sediments is described elsewhere in this volume (Chapter 22).

Calcite and aragonite are highly soluble, so the original distribution of these minerals over the ocean floor is extensively modified by post-depositional dissolution favored by low temperatures, high hydrostatic pressures, rapid removal of the solution, and high partial pressures of CO_2, generated by oxidation of organic matter at the sediment surface. The dissolution process in a given stratum is slowed as new sediment accumulates above it, producing a diffusion barrier of increasing thickness, separating the carbonate from the solvent. The ultimate rate of accumulation of carbonate at a given depth and circulation is consequently determined by the rate of deposition of carbonate minerals and by the total rate of accumulation of sediment. The regional distribution of calcium carbonate in ocean sediments illustrates the influence of these parameters. From the distribution of carbonate concentrations with depth in areas with different productivities, the combined influence of hydrostatic pressure and temperature and likewise the influence of productivity are quite evident (Figs. 18 and 19). Further, at a higher total rate of deposition, providing rapid pro-

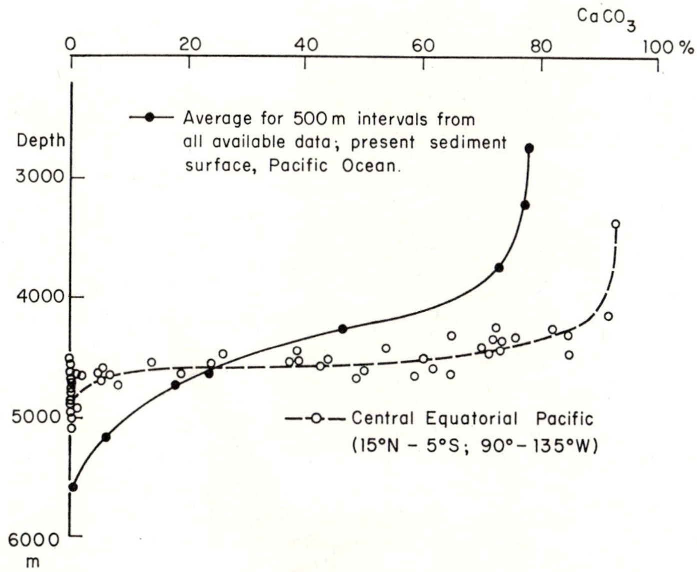

Fig. 18. Concentration of calcium carbonate at the present sediment surface as a function of depth in the Pacific Ocean. The average curve shows a pronounced effect of depth on the rate of solution of carbonate. In the central equatorial Pacific, characterized by a wide regional variability in biological productivity, this latter factor dominates the carbonate distribution. (From Bramlette, 1961.)

tection of the carbonate, the disappearance of the carbonate occurs at greater water depths (Fig. 20). Finally, the influence of the removal of the dissolved calcium ion is demonstrated by the carbonate distribution in the South Atlantic basins; at equal depths the dissolution of calcite is far more extensive in the Brazil Basin, which is flushed by Antarctic deep water, than in the neighboring Congo Basin, which has a slower renewal of the bottom water (Wattenberg, 1933, 1937; Correns, 1939).

At shallow and intermediate depths where the rate of dissolution of calcium carbonate is low, calcite and aragonite are sometimes dolomitized. Outcrops of dolomitized limestone, such as on the Nasca Ridge in the South Pacific (Bramlette, unpublished), are coated with manganese–iron oxide minerals, indicating that the dolomitization was active in the past, and that it has since been replaced by oxide deposition. A similar sequence is observed in many oceanic limestones and phosphorite deposits, which also display a coating of manganese–iron oxide minerals (Fig. 14). Growth of sparse euhedral dolomite crystals in

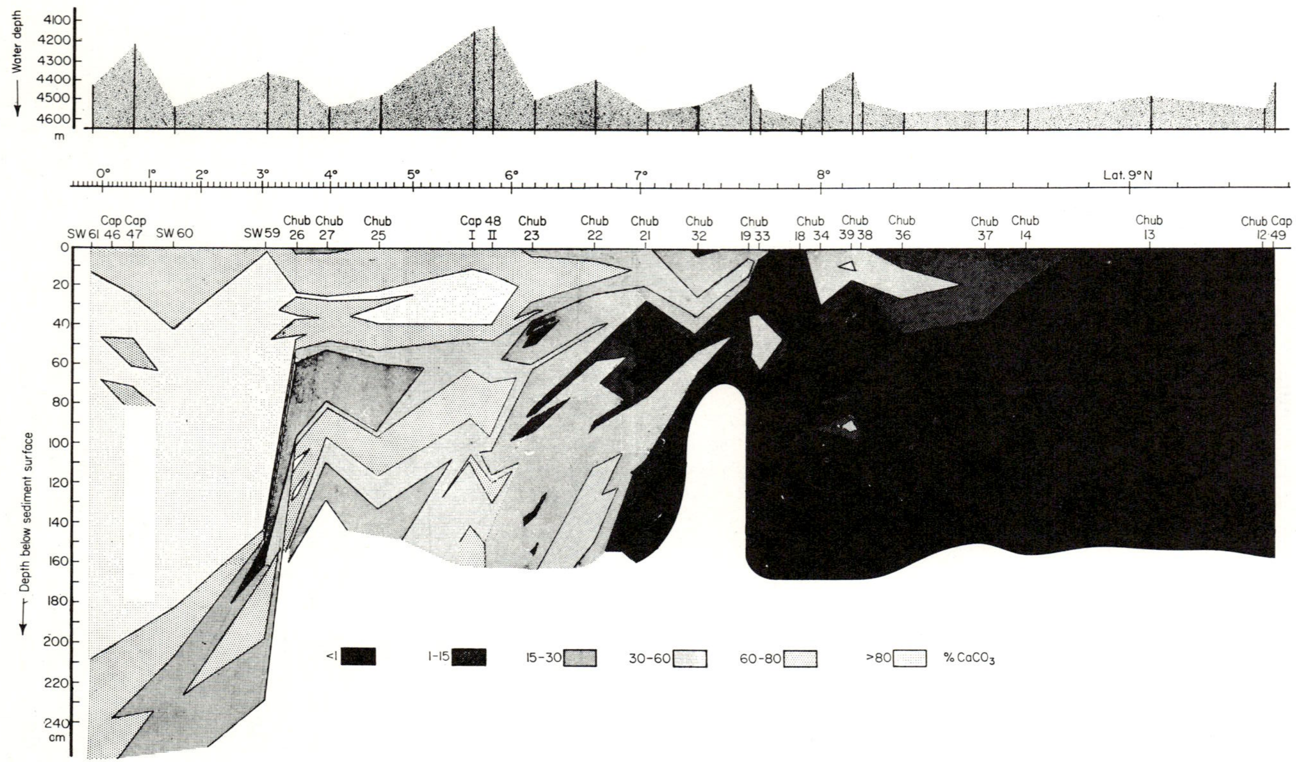

Fig. 19. Distribution of calcium carbonate in a meridional section through the central equatorial Pacific, showing the variation of carbonate content with time (depth of sediment) and with latitude (productivity; falling off from equator). For a general discussion of these features, see Section 3 of this chapter, and Fig. 36, which is based on the section shown above, and its deeper parts. In addition to the productivity effects in time and space, the superimposed effect of water depth (hydrostatic pressure) is manifest as deviations from the regular change with latitude, correlating with the deviations from average depth shown in the upper graph. Diagenetic solution is presumably responsible for the general recession toward lower latitude of the carbonate compensation surface ($CaCO_3 = 0$) with increasing depth in the sediment. The rate of such solution is low but has a large relative effect on the low original carbonate content of the sediment near the compensation surface.

Recent deep-sea sediments have been observed in several instances, both in the Pacific (Bramlette, unpublished) and in the Atlantic (Correns, 1937; Radczewski, 1937). Extensive dolomitization is observed to occur at the interface of basaltic intrusions, such as the one encountered in the experimental Mohole off Guadalupe Island.

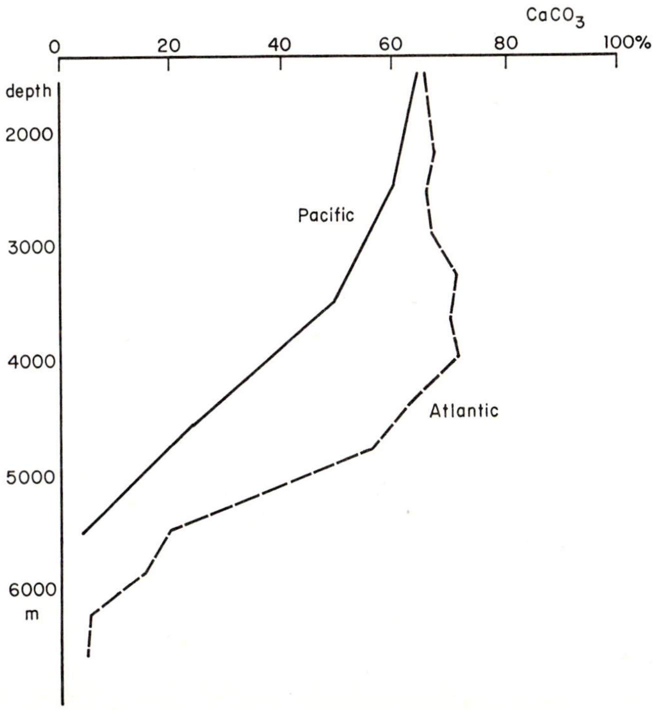

Fig. 20. Average carbonate content as a function of water depth in the Atlantic and Pacific, demonstrating the effect of total rate of deposition on the rate of solution of calcium carbonate at the ocean floor. Under similar depth conditions the ultimate carbonate content of the sediment is, on the average, higher in the Atlantic than in the Pacific. This is due to more rapid burial of the sediment in the Atlantic Ocean, which, per surface area, receives considerably more solid and dissolved material from the continents than does the Pacific. (From Revelle, 1944.)

Much detrital carbonate appears to be carried by wind into the pelagic environment. Calcite and dolomite are often the dominant components of eolian dust, which occasionally also contains small amounts of breunnerite (Radczewski, 1937). In Mediterranean sediments, calcite is the dominant eolian material (Norin, 1958). In deep pelagic sediments most of the fine-grained eolian calcite is dissolved, but in areas with a heavy eolian fallout and a high preservation of accumulating carbonates, eolian calcite and dolomite might be quantitatively important.

E. Silicates

Skeletal structures of opaline silica are secreted by a number of marine organisms, including the planktonic diatoms, radiolarians and silicoflagellates, and some benthic sponges. Much of this opal is dissolved or peptized soon after the death of the organisms; however, the remaining fraction of relatively robust skeletons often forms a significant portion of the sediment, especially below areas of high productivity such as the Subarctic Convergence, the Equatorial Divergence, and the divergences along the west coasts of the

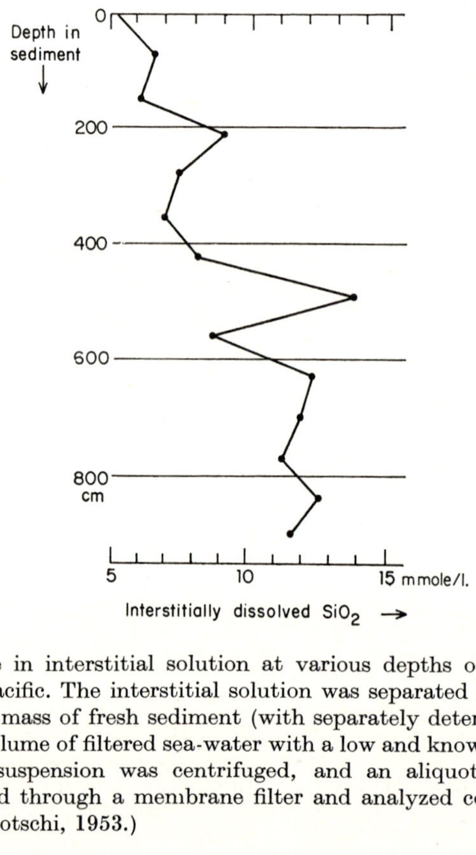

Fig. 21. Monosilicate in interstitial solution at various depths of the core Capricorn 38 BP, South Pacific. The interstitial solution was separated from the sediment by adding a known mass of fresh sediment (with separately determined water content) to a measured volume of filtered sea-water with a low and known silica content. After dispersion, the suspension was centrifuged, and an aliquot of the supernatant liquid was passed through a membrane filter and analyzed colorimetrically. (From Arrhenius and Rotschi, 1953.)

continents (cf. Fig. 1). The instability of silica in the interstitial water of the sediment causes continuous dissolution of the siliceous fossils after deposition; the silicoflagellates disappear first, followed by diatoms, then radiolarians, and finally even the robust sponge spicules. Part of the dissolved silica reacts to form authigenic aluminosilicates observed as overgrowths in partly dissolved siliceous skeletons (see below). In some cases reprecipitation of the silica as opal takes place, resulting in thin flakes of this mineral along the bedding planes in phosphorite, or in laminae of chert, observed to occur at depth in pelagic sediments in a few instances. Another part of the dissolved silica is returned to the bottom water by diffusion. Thus, a marked vertical gradient is maintained in the concentration of silica in the interstitial solution of the sediment, and at the water interface (Table VI and Fig. 21).

TABLE VI

Concentration of Dissolved Monosilicate in the Water immediately above the Sediment Surface and in the Interstitial Solution of the Sediment in the South Pacific. (Arrhenius and Rotschi, 1953.)

Station	Position Lat S° ° ′	Position Long W° ° ′	Depth, m	SiO_2, µmole/l. Bottom water	SiO_2, µmole/l. Interstitial water	Sediment type
Cap 17 B	19 56	172 40	5730	122	103	Sandy volcanic mud
24 B	19 29	173 44	2780	124	7900	Siliceous volcanic mud
30 B	17 28	160 59	4710	116	12000	Dark brown zeolitite
33 B	12 46	143 33	4380	129	13130	,, ,, ,,
34 B	11 20	142 26	4500	132	9400	Calcareous brown clay
40 B	14 47	112 08	3020	132	—	Foraminiferal ooze
41 B	13 33	113 21	2940	—	12.4	,, ,,
42 B	07 19	118 40	4230	—	1110	Calcareous brown clay
44 B	03 42	121 03	4170	141	—	Foraminiferal ooze

Wind-transported diatom frustules have been observed in quantity in eolian dust derived from arid areas with exposed diatomaceous deposits (Hustedt, 1921; Radczewski, 1937). Such transport appears to be the most probable explanation of the observation by Kolbe (1955) of freshwater diatoms in equatorial Atlantic sediments. The same author has reported the presence in these sediments of silicified grass cells, presumably suspended in the atmosphere by fires, an explanation which is further indicated by the similar distribution of charcoal fragments (Radczewski, *op. cit.*).

Quartz was recognized as a significant component in deep-sea sediments by Murray and Renard (1884), who drew attention to the importance of wind transport of this mineral. Radczewski (1937, 1939) investigated quantitatively the distribution of eolian dust and found that eolian quartz in pelagic sediments of the equatorial Atlantic frequently exceeded 30% of the total sediment. He further established the regional variations in concentration of eolian quartz and demonstrated a decrease in median size and concentration of the eolian component with increasing distance from the continent. Furthermore, Radczewski found windborne quartz in sediments of glacial and interglacial age as well as in the Recent. Subrounded and pitted quartz has been further reported by Norin (1958) as, together with calcite, the quantitatively most important non-volcanic mineral in the eolian component of Mediterranean sediments.

Revelle (1944) observed notable amounts of silt-size quartz in Pacific pelagic sediments. The present author found a narrow size frequency distribution of quartz centred at 3–10 μ in north equatorial Pacific clay sediments, and suggested eolian transport as an explanation. Rex and Goldberg (1958), in subsequent extensive studies of Pacific sediments, demonstrated a regional regularity in the quartz distribution (Fig. 22), which they interpreted as an

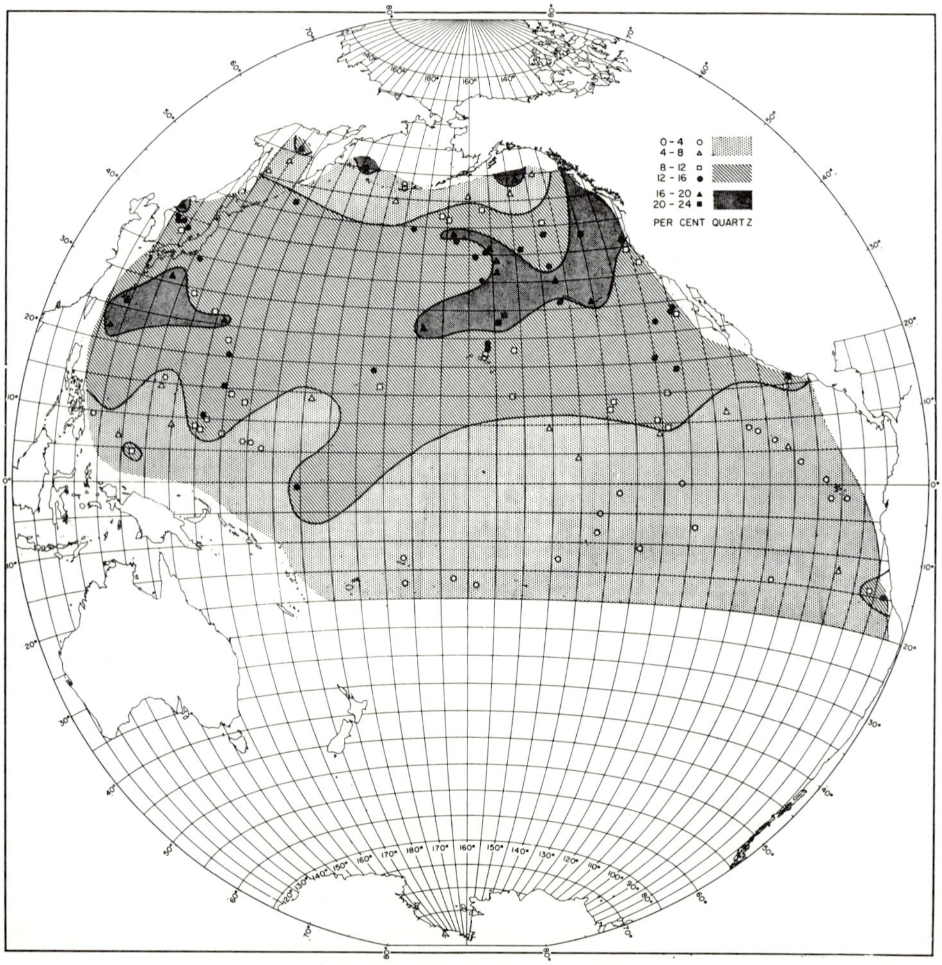

Fig. 22. Areal distribution of quartz in the Pacific. (After Rex and Goldberg, 1958.)

indication of fallout of dust from the high altitude jet streams. The importance of tropospheric transport is indicated by the numerous observations of extensive dust-falls at sea in the equatorial Atlantic (Radczewski, *op. cit.*; Seefried, 1913) and Indian Oceans, and in the South Pacific (Marshall and Kidson, 1929), and further by the correlation between the wind pattern from arid areas and the occurrence of dry atmospheric haze at sea (Fig. 23). Rex and Goldberg (*op. cit.*)

report low concentrations of quartz in the Tertiary pelagic sediments of the North Pacific, as compared to the unconformably overlying Pleistocene, indicating a smaller arid source area or weaker transporting wind currents during Tertiary time. Another indication of low intensity of the Tertiary circulation is the low organic productivity in the Equatorial Divergence during the

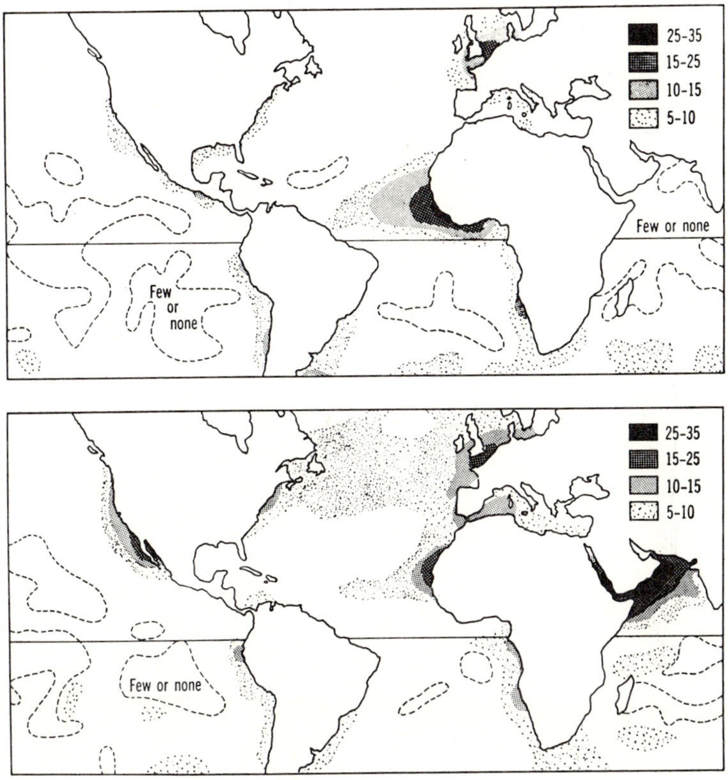

Fig. 23. Tropospheric transport of dust, as indicated by average frequency of haze (dry aerosol) during the northern winter (upper map) and summer (lower map). Frequencies are given in per cent of total number of observations. (From Arrhenius, 1959.)

Pliocene, followed by conditions suggesting a sequence of epochs with markedly increased intensities of zonal winds during the Pleistocene (Fig. 24).

Eolian dust contains little material in the size range below one or a few microns (Fig. 25), though pelagic sediments often display a secondary maximum in this range. Most of this fine-grained pelagic quartz was probably transported together with the aluminosilicate suspensoid in sea-water from the continents to the areas of deposition.

Mixed assemblages of mainly plagioclase feldspar with a grain-size distribution similar to that of quartz are ubiquitous in pelagic sediments. The optical properties indicate that the dominant species range between labradorite and oligoclase in most Pacific sediments; commonly several species are present together. Large grains of orthoclase, microcline and sanidine, indicating occurrences of acid igneous rocks, are not uncommon within the predominantly basaltic province of the Pacific basin (Revelle, 1944, pp. 10, 18–39). A detailed

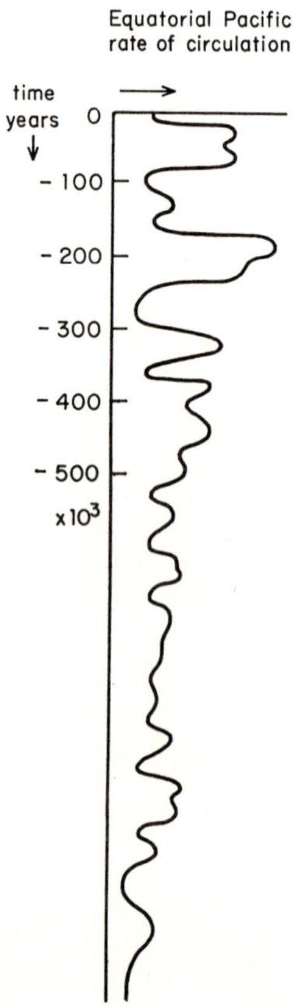

Fig. 24. Relative trade-wind intensity as a function of geological time. The relative intensity is estimated from the ensuing productivity changes at the Equatorial Divergence. (From Arrhenius, 1952.)

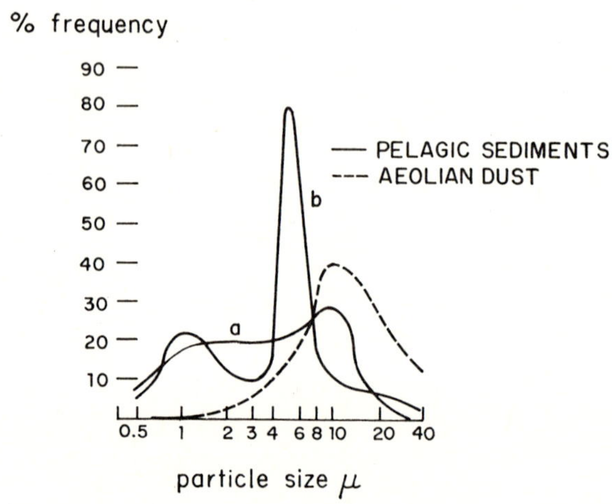

Fig. 25. Particle-size distribution in pelagic sediments from the North Pacific and in eolian dust from Illinois. (Data from Rex, 1958; Udden, 1914.)

investigation by Peterson and Goldberg (1962) shows that such feldspar species prevail over a large part of the East Pacific Rise. For the Atlantic, Radczewski (1937) and Leinz (1937) showed that acidic rocks must have been the main source of feldspar in the Cape Verde and the Guinea Basins, inasmuch as the majority of grains had a composition An < 30. As is true of quartz, most of the fine-grained feldspar is probably wind-transported from continental clastics.

Authigenic orthoclase has been observed in a few instances; Mellis (1952, 1959) described a sample from the west Atlantic in which orthoclase had replaced the original plagioclase in altered andesite boulders. This sample indicates highly alkaline conditions during crystallization, as it is known from controlled synthesis experiments that orthoclase grows only above a pH of about 9 at low temperatures; directly below this limit, zeolites form the stable phases.

Rex (1958) attempted to establish the regional variability in the total amount of feldspar in reference to the known amount of quartz in the Pacific. The great variability of X-ray diffraction from given crystallographic planes in the feldspar series, as well as interference from other minerals in the range of inter-

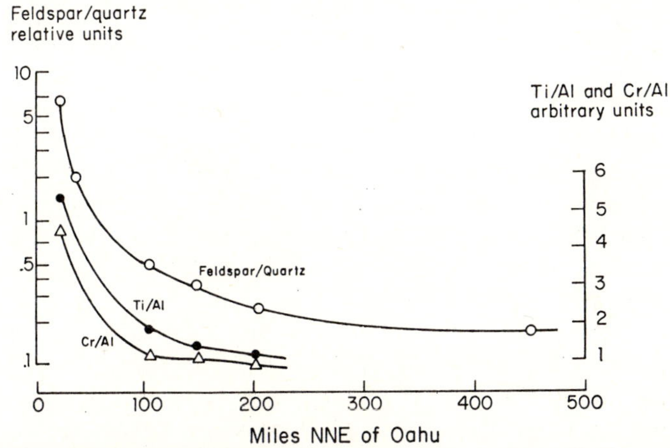

Fig. 26. Relative ratio feldspar/quartz in the North Pacific off Hawaii, showing the rapidly vanishing influence of the basaltic province. This is also reflected by the decrease in the relative concentration of titanium and chromium, mainly occurring in pyroxene crystallites. (From Rex, 1958.)

planar spacings used, limits the interpretation of these data. Nevertheless, some general relationships appear well established, such as an increase in the relative amount of feldspar around the Hawaiian Islands (Fig. 26) indicating the contribution from the basaltic province.

Olivine is unstable on the deep-ocean floor and hence is not observed in quantities in pelagic sediments. However, in association with cosmic nickel–iron spherules, olivine and pyroxene chondrules have been observed which strongly resemble those in chondritic meteorites, and are probably of extra-terrestrial origin (see further page 674 and Fig. 10b).

Pyroxene minerals are ubiquitous and are useful indicators of the origin of pyroclastics. The absence of hypersthene–enstatite among the frequently occurring pyroxene grains in South Pacific sediments indicates an origin within the province of oceanic basalts, rather than transport from the circum-Pacific andesite regions. Mellis (1959) has found examples of uralitization of pyroxene.

The conditions determining the rate of decomposition and devitrification of volcanic glass in marine sediments are still obscure. While some minute glass shards in Mesozoic sediments are unaltered, some Quaternary deposits of ash and pumice have been entirely altered to montmorillonoid minerals or to phillipsite. The occurrence of unaltered glass in calcareous deposits indicates that a high calcium ion concentration might preserve the glass. Layers and laminae of volcanic glass are potentially useful as stratigraphic markers because of the wide dispersal and the short duration of the supply of volcanic ash. Successful attempts to correlate strata over large distances in marine sediments have been made by Bramlette and Bradley (1942), Mellis (1954) and Nayudu (1962a). A correlation has also been suggested between ash layers observed (Worzel, 1959) off the Middle American coast in the Pacific and in the Gulf of Mexico (Ewing *et al.*, 1959). Further observations of these deposits suggest, however, that they are due to eruptions from different centers and probably at different times.

Kaolinite, which commonly occurs as a weathering product of basaltic glass under acid weathering conditions (Sigvaldason, 1959), has not been found in marine decomposition products of pyroclastics. This agrees with the experimental observations of the influence of hydrogen ion concentration on the stability of low temperature silicates, referred to below.

Palagonitization of basaltic glass is frequently observed on the ocean floor; Correns (1939) and Nayudu (1962) indicate that this alteration takes place mainly at the interaction between hot lava and sea-water or interstitial water, rather than during the continued exposure after cooling.

The zeolite phillipsite was early recognized as quantitatively important in slowly accumulating pelagic sediments (Murray and Renard, 1884). Recently, it has been found that other members of the phillipsite–harmotome series are widespread, although less conspicuous because of their microcrystalline habit, the low intensity of their first order diffraction and their diffractive interference with feldspar. Zeolite concentrations higher than 50% are not infrequent, and zeolitite consequently is a common pelagic sediment type. Controlled synthesis of the mineral indicates phillipsite and harmotome to be stable from a lower pH limit between 7 and 8 to an upper between 9 and 10 in a potassium–hydrogen cation system. Above this limit feldspar is stable at low temperatures, and directly below the limit $1M$ mica forms from solution. These relations might be useful in estimating the hydrogen ion concentration in ancient marine environments.

It is not entirely clear how much of the silicon and aluminum of the zeolite framework, and the interstitial cations, is directly inherited from the igneous silicates and glass decomposing on the ocean floor, and what proportion is derived from ions which have long been in solution and circulation in the ocean. The amount of aluminum passing through the ocean in a dissolved state is of the same order of magnitude as the rate of accumulation of zeolites (Sackett and

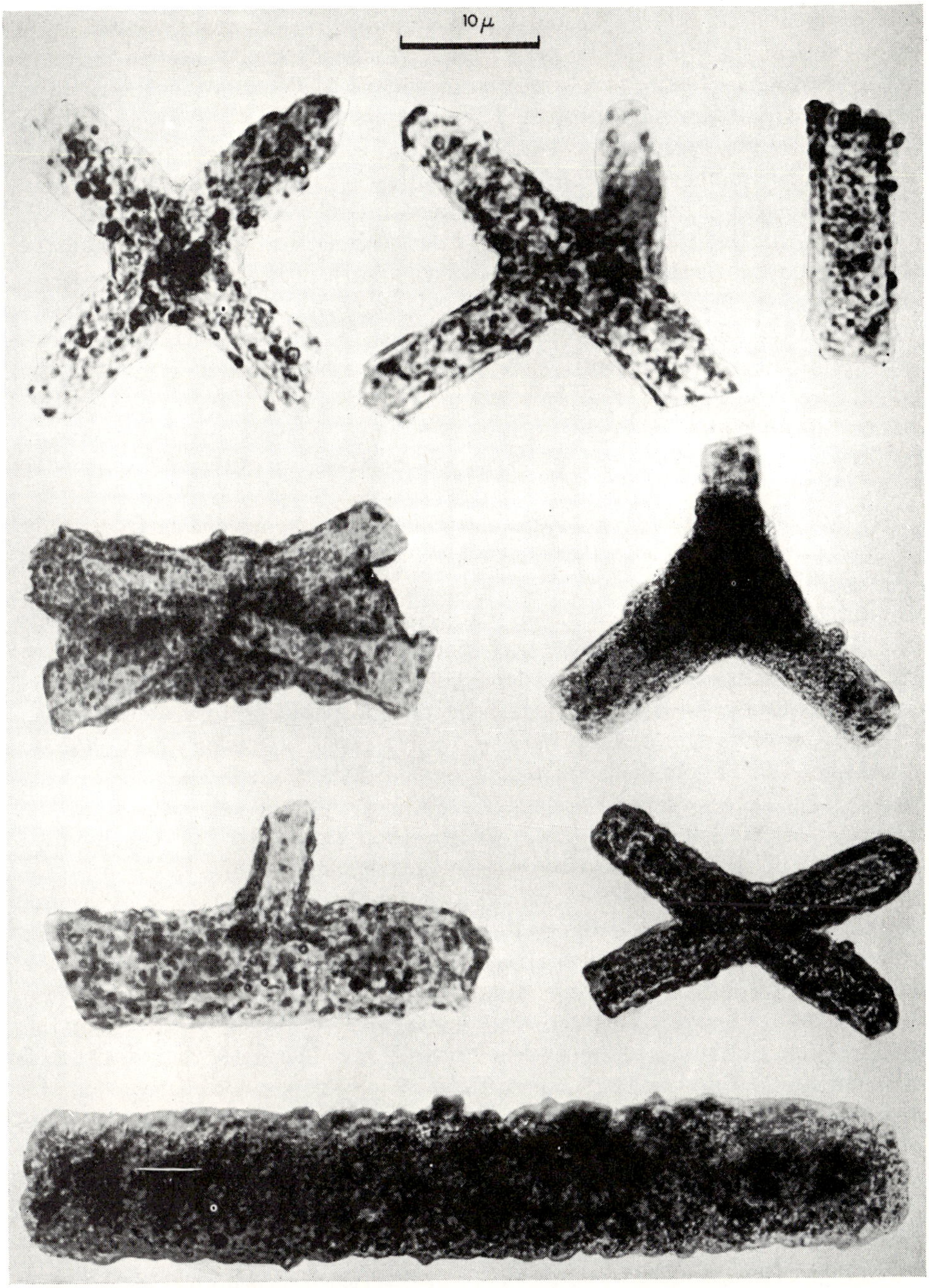

Fig. 27. Twinned zeolites of the phillipsite (K^+ Na^+; Ca^{2+})–harmotome (Ba^{2+}) series. The crystals typically display extensive intergrowth with other phases, primarily nontronite, which often form as much as one half the volume.

Arrhenius, 1962). The synthesis experiments mentioned above also indicate that precipitation of aluminum species from solution in sea-water would form zeolites rather than phyllosilicates. Finally, microcrystalline zeolite is found within dissolving skeletons of siliceous organisms in sediments without significant quantities of igneous silicates. This suggests that the high concentrations of dissolved silica from biotic sources at the sediment–water interface (cf. Table VI) induce the formation of the zeolites in these cases. On the other hand, phillipsite is frequently observed in intimate physical association with pyroclastics or their remnants on the deep-ocean floor, suggesting that the solution from which the zeolite crystallized is derived directly from the pyroclastic aluminosilicates (Murray and Renard, *op. cit.*).

A third mode of formation of zeolites, known to operate on the deep-ocean floor, is the interaction between sea-water or interstitial water and hot basaltic lava during its intrusion. This phenomenon, together with extensive palagonitization of the glass and serpentinization of olivine, is beautifully illustrated in the basalt found at the bottom of the experimental Mohole drilled at Guadalupe Island in 1961.

Phyllosilicates frequently form a major part of the non-biotic fraction in pelagic sediments. Some are clearly terrigenous, with kaolinite the major representative. In solutions containing alkali or alkaline earth ions, kaolinite is a stable phase at only a relatively high concentration of hydrogen ion; it is thus precluded as a mineral forming in the ocean. Oinuma *et al.* (1959; see also Kobayashi *et al.*, 1960) have demonstrated higher concentrations of this mineral on the continental shelf than in the pelagic areas of the Western Pacific. Further, the investigations by Correns (1937) and Yeroshchev-Shak (1961a) indicate considerably higher kaolinite content in Atlantic sediments than in the Pacific. This agrees with the greater influx of terrigenous material in the Atlantic, indicated independently by deposition rates and by the chemical composition of the sediment (Goldberg and Arrhenius, 1958).

Other phyllosilicate species are obviously halmeic, while a third group, comprising much of the poorly crystallized material with a grain size less than a few microns, is of uncertain origin. Highly ordered polymorphs like biotite and chlorite prevail in relatively large grain sizes (3–30 μ) in areas receiving considerable terrigenous material from metamorphic rocks. However, in the finer fraction of these sediments the crystallinity decreases, and disordered $1M$ mica species[1] dominate. Some of these may have come directly from the continents; their coexistence with coarser, ordered polymorphs seems to indicate that the disordered mica species are partly diagenetic alteration products. Rimšaite (1957) has reported progressive alteration of different mica species.

On the basis of the high boron content and the bonding of the boron in the grain-size fraction containing the $1Md$ micas, Arrhenius (1954 and 1954a) suggested that these micas have crystallized from solution in the ocean with boron

[1] The term illite has been used as a group name for these minerals, together with mixed layer structures (Grim, Dietz and Bradley, 1949). Yoder (1957) has pointed out the need for a modified terminology.

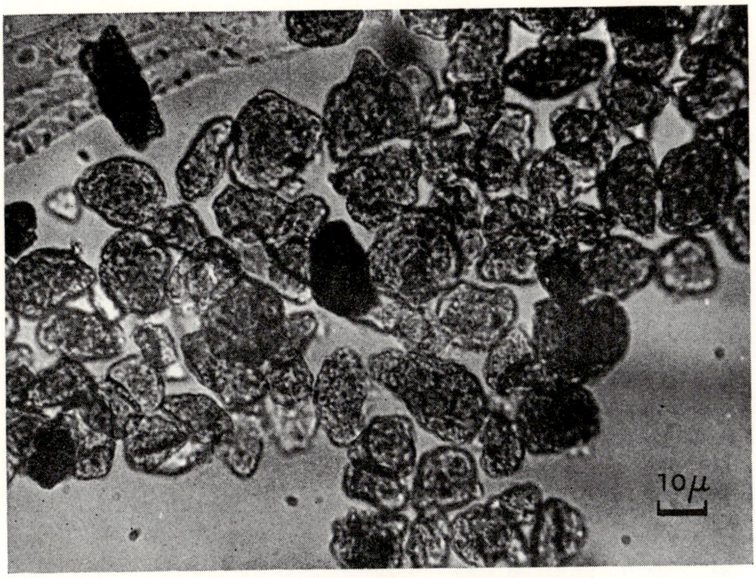

(a)

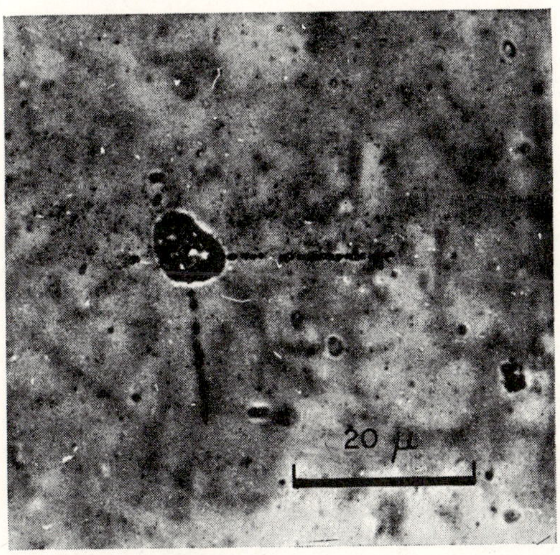

(b)

Fig. 28. (a) Aggregates of microcrystalline harmotome and nontronite. The aggregates contain Ba^{2+} and Fe^{2+} as principal cations, and often considerable amounts of thorium (in one instance 2200 ppm). The density varies owing to substitution of lighter cations for barium; the highest density measured is 3.2. The aggregates have a high magnetic susceptibility. The microcrystallites rarely exceed a few microns in length. (b) Radioautograph of harmotome–nontronite aggregate showing emerging alpha particle tracks. [(b) from Picciotto and Arrhenius, unpublished.]

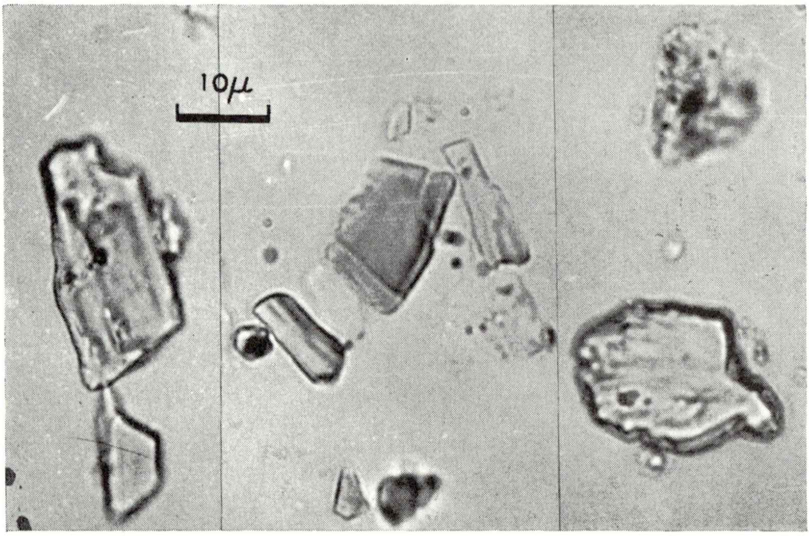

Fig. 29. Chlorite. North Pacific abyssal plain. Polarized light; canada balsam.

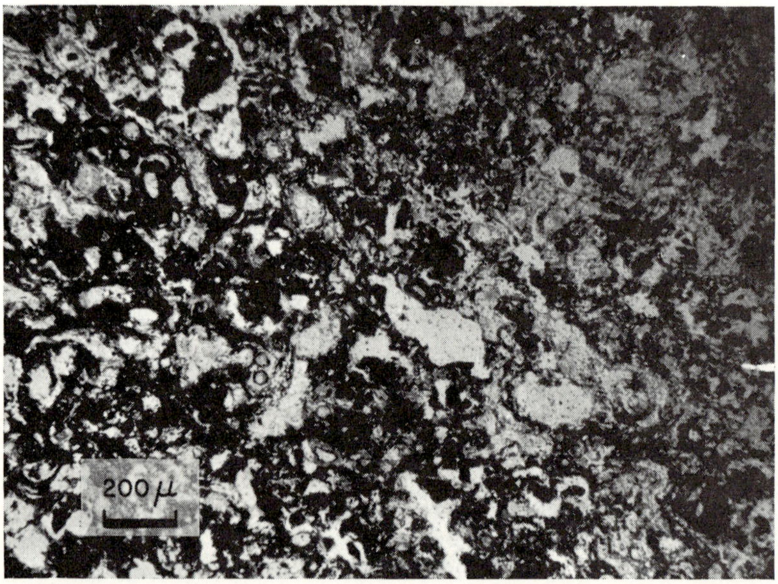

Fig. 30. Nontronite (grey) intergrown with manganese oxide minerals (black) in nodule, (white = voids). Noil Tobe, Timor (Danau Formation).

in coordination with oxygen or fluorine substituting in the tetrahedral silicate sheets. However, Harder (1959) has recently demonstrated that borate ion is readily and irreversibly fixed from solution by natural 1*Md* mica. Further, measurements of the rate of passage of dissolved aluminum through the ocean (Sackett and Arrhenius, *op. cit.*) show that only a small fraction of the total aluminosilicate deposition can be accounted for in this way (cf. Sillén, 1961).

Montmorillonoid minerals of nontronite type occur in significant quantities around centers of volcanic activity, like the Cape Verde and the Hawaiian Islands, and in general in areas with a high proportion of pyroclastic sediments (Table VII). These minerals are usually assumed to have formed *in situ* by decomposition of the volcanic glass. The highest relative crystallinity is found

TABLE VII

Concentration and Relative Crystallinity of Montmorillonite Species in Marine Sediments of Varying Geological Age.

A purified sample of the API clay mineral standard 25a, Upton, Wyoming, was arbitrarily selected to represent pure and perfectly crystalline montmorillonite. Concentrations were measured against synthetic boehmite as internal standard.

Sample	Geological age	Montmorillonite Concentration, %	Montmorillonite Crystallinity, %
Bentonite, Gotska Sandön, Sweden	Ordovician[a]	33	65
Bentonite, Mossen, Kinnekulle, Sweden	Ordovician[a]	41	49
"Red Clay", Noil Tobe, Timor	Jurassic[b]	36	53
McBride Chalk, Alabama	Tertiary	16	57
Pelagic Clay, S. Pacific, Dwd 61	Quaternary	26	43
Pelagic Clay, S. Pacific, Cap 32B	Quaternary	16	40
Pelagic Clay, S. Pacific, Cap 31B	Quaternary	15	38

[a] Thorslund (1948, 1958); Waern *et al.* (1948); Hagemann and Spjeldnaes (1955).
[b] Molengraaff (1920).

in sediments of Paleozoic and Mesozoic age (Table VII), whereas the Cenozoic–Recent marine montmorillonoid is poorly crystallized and has a limited expansibility. Pelagic montmorillonoids and 1*M* micas may show high disorder because they form as metastable structures, which is often the case at low temperature synthesis; the well ordered zeolites, on the other hand, are stable phases at the composition and in the pH range of sea-water (molar ratio of large cations to magnesium ion = 9, pH approx. 7.3–8.5). This suggestion is supported by laboratory studies of the stability relations of the minerals in question. The higher crystallinity of the ancient montmorillonites may be the result of a diagenetic process active after the removal of the sediment from the original alkaline environment.

F. Organic Matter

At the low rate of deposition characteristic of pelagic sediments, and under the oxidizing conditions usually prevailing at the deep-ocean floor, most of the

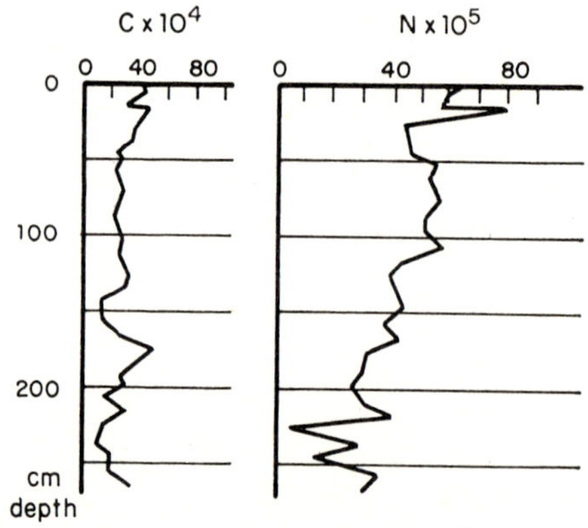

Fig. 31. Carbon and nitrogen concentrations as a function of depth in a pelagic clay sequence (Sw 55, 11° 20′N, 127° 37′W; 4916 m). (From Arrhenius, 1952.)

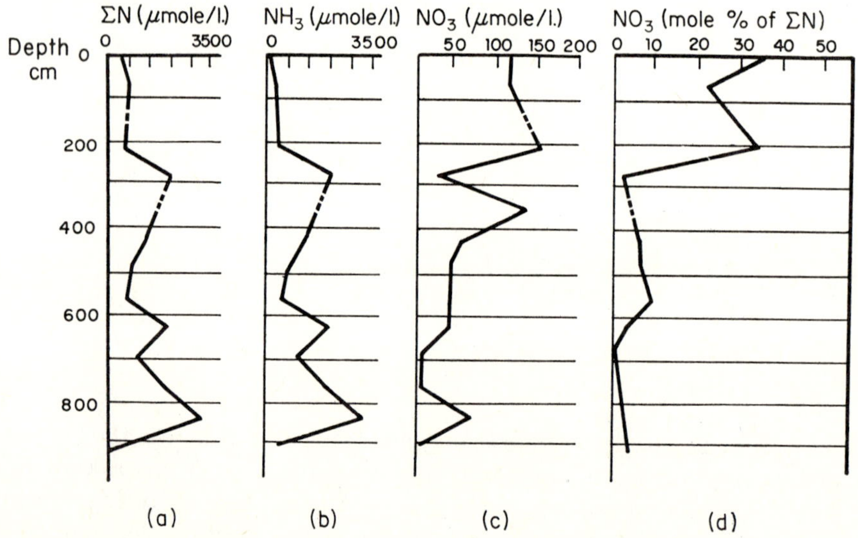

Fig. 32. Distribution of nitrogen between ammonia and nitrate in interstitial solution of core Capricorn 38 BP (South Pacific, 14° 16′S, 119° 11′W; 3400 m). (From Arrhenius and Rotschi, 1953.)

organic matter is oxidized soon after deposition. The oxidation probably is accomplished mainly by bacteria and mud-eating animals concentrated in the surface layer of the sediment (cf. Section 5, and ZoBell, 1946, p. 91). In pelagic sequences far from any terrestrial source of organic matter, the concentration of organic carbon and nitrogen first decreases rapidly with depth in the sediment and then slowly as the remainder contains an increasingly larger proportion of refractory compounds (Fig. 31) (Mironov and Bordovsky, 1959). Correspondingly, less resistant structures, such as scales of fish and the chitinous beaks of cephalopods, are found only in the surface layer of the sediment. The decomposition of proteins and aminosugar in the sediment produces ammonia and nitrate, which distribute themselves between the interstitial solution and sorption sites on the minerals. Typical values range between 0.1 and 4 mmole of NH_4^+ per kg of solid, and similar amounts per liter of interstitial solution

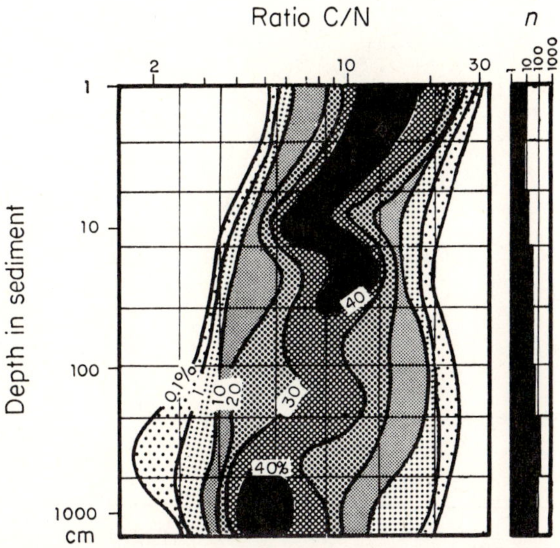

Fig. 33. Frequency distribution of the carbon/nitrogen ratio as a function of depth in Pacific north equatorial clay sediments. The right-hand graph shows the number of samples measured within each of the depth classes limited by horizontal lines. (From Arrhenius, 1952.)

(Fig. 32b) (Bruejewicz and Zaytseva, 1959; Shiskina, 1958). The oxidation state of the interstitially dissolved nitrogen compounds changes rapidly with depth in the sediment, so that at the surface a considerable portion is nitrate whereas at depth ammonium ion predominates in solution (Fig. 32d). In the progressive decomposition of pelagic organic matter, carbon is lost more rapidly than nitrogen and consequently the ratio of carbon to nitrogen in the sediment decreases with depth below the sediment surface, i.e. with geological time (Fig. 33). The combined influence of origin and diagenesis of the organic matter on its composition is illustrated in Fig. 34.

Most of the refractory solid organic matter consists of the original and decomposed organic molecules interstratified with, and protected by, the apatite

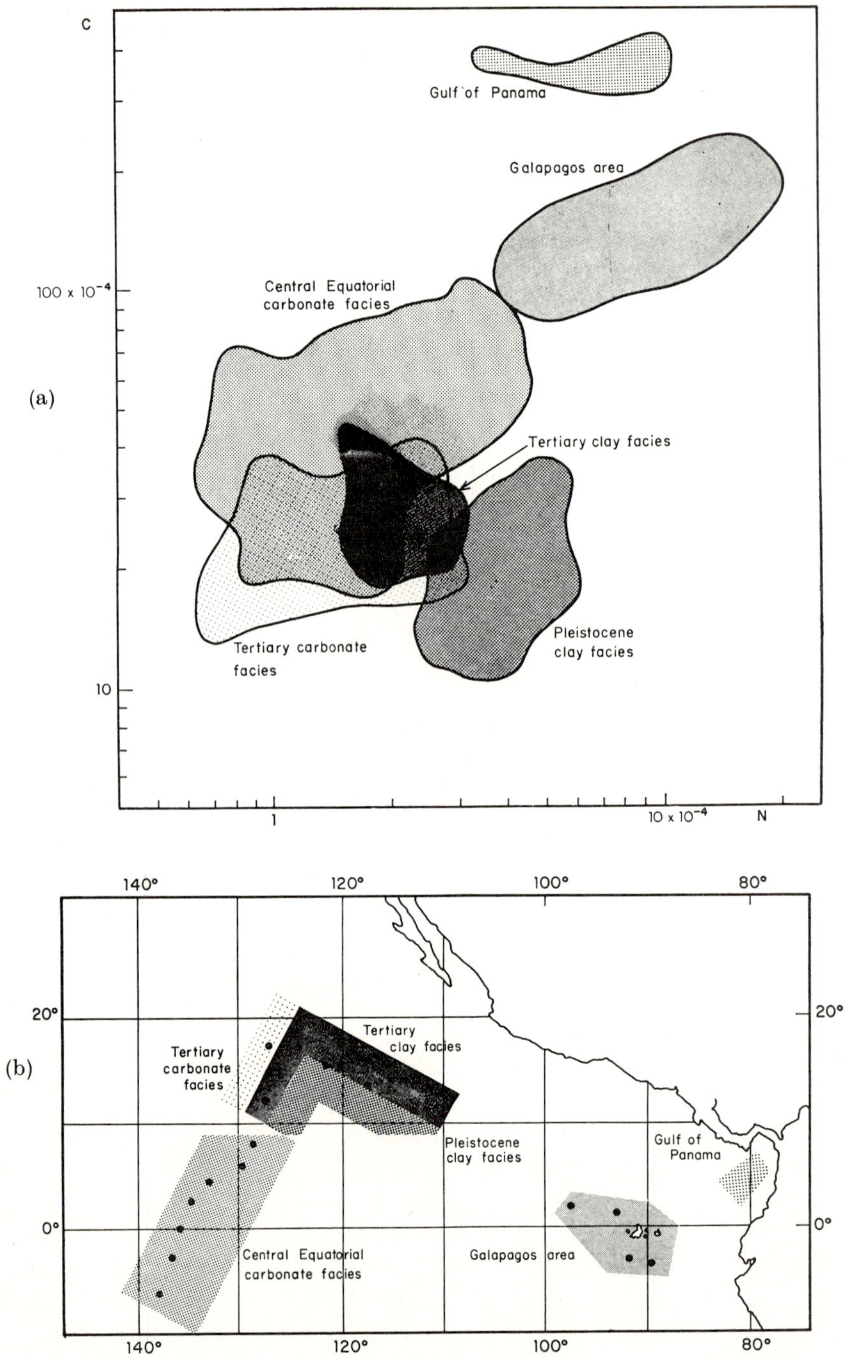

Fig. 34. Influence of origin and diagenesis on the composition of organic matter (a) in sediment cores with location shown in (b). The distribution is based on approximately 2000 analyses. The envelopes in (a) enclose the central 90% of the total number of samples from the respective areas. (After Arrhenius, 1952.)

crystallites in the ubiquitous fish-bone debris (Fig. 15a) (Arrhenius, Bramlette and Picciotto, 1957), and by the calcite crystallites in foraminiferal tests (Correns, 1937). In the former case higher fatty acids form much of the decomposition products precipitated by salt formation with zinc adsorbed from sea-water. Copper, nickel, lead, and silver, also adsorbed from sea-water, are fixed by the proteinaceous residue in the bone debris, whereas the large amounts of rare-earth elements zirconium and thorium from the same source replace calcium in the apatite crystallites. The slow progressive decomposition of the organic matter in the fish debris is further indicated by the gradual disappearance of the deep amber color typical of fragments in Recent strata (Bramlette,

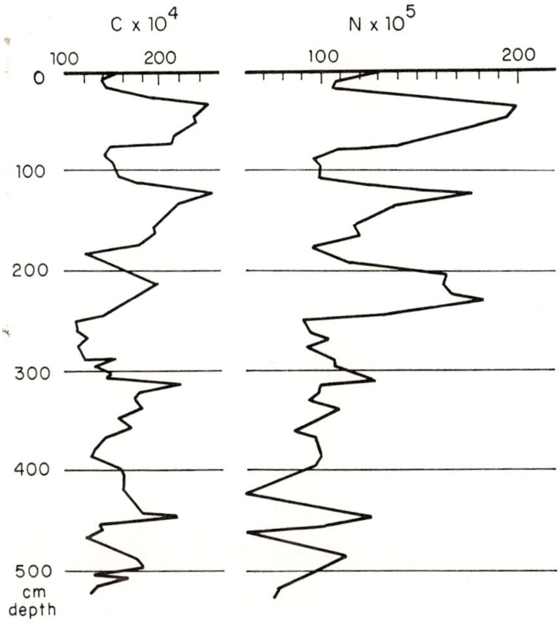

Fig. 35. Variation of carbon and nitrogen concentrations with depth in core Sw 39 from the Galapagos area. The stratigraphic subdivision of this sequence is shown in Fig. 37, and the absolute age of the stratigraphic units is indicated in Fig. 38. (From Arrhenius, 1952.)

unpublished). Complete fading of the color has occurred in pelagic fish debris of Tertiary age.

Variations in geological time of the rate of accumulation of organic matter, large enough to overshadow the diagenetic effects, and the effects of varying dilution with calcium carbonate, have been found only below areas of exceptionally high productivity such as at the eastern end of the equatorial current system in the Pacific. In this area the ensuing sedimentary strata, representing times of low and high organic productivity at the surface of the ocean, differ by a factor of two in their content of carbon and nitrogen (Fig. 35). A similar stratification might be expected below the Peru Current and the Kuroshio.

Because of low density, particles of organic matter from land might spread extensively over adjacent oceanic areas, a minor fraction being windborne (cf. Fig. 21) and the majority transported into the ocean by rivers. Some pelagic sediments, in areas near land with heavy tropical vegetation and protected by marginal trenches (such as on the Pacific side of Middle America), may contain several per cent of terrestrial organic detritus.

The decomposing organic matter on the ocean floor rarely forms well defined organic minerals. Weddellite (calcium oxalate dihydrate) and earlandite (calcium citrate tetrahydrate) have been described by Bannister and Hey (1936) from Antarctic sediment samples, together with authigenic gypsum. Small amounts of strontium and barium substitute for calcium in the organic crystals. Since later attempts to find these minerals in freshly collected Antarctic sediments have failed, and as gypsum is often observed to form in drying bottom samples from calcium carbonate and decomposing proteins, it is possible that the weddellite and earlandite formed during storage of the samples.

3. Productivity Control of Pelagic Sedimentation

One may select approximately level areas of pelagic deposition where the pressure factor and (at least during Pleistocene–Recent times) the temperature factor influencing the post-depositional dissolution of calcite is small compared to the production factor and can be separately estimated (Fig. 19). This facilitates investigating the combined effects of organic production and total rate of deposition on the concentration and rate of accumulation of calcium carbonate. Such studies have been carried out in an approximately level area in the east equatorial Pacific where the Equatorial Divergence maintains a high organic productivity.

The meridional distribution of the rate factors involved, the causative circulation mechanism, and the ensuing sedimentary record are shown graphically in Fig. 36. The top diagram shows a north–south section through the surface layer of the Pacific Ocean at approximately 130°W. The temperature distribution is indicated by the isotherms for 15, 17, 20, 25 and 27°C. The cold intermediate water is represented by a dotted area, and the vertical circulation is indicated by arrows.[1] Divergence of the trade-wind drift currents produces upwelling along the equator of the nutrient-rich intermediate water, resulting in a high rate of production of phytoplankton, sustaining an increased productivity also of higher members of the food chain. At higher tropical latitudes the lack of a mechanism enriching the euphotic layer keeps the productivity low.

The rate of production of skeletal calcium carbonate from coccolithophorids and Foraminifera is illustrated in the middle diagram of Fig. 36, where the full line curve marked "Pleistocene minimum rate of production (interglacial)" corresponds to present-time conditions. The inferred rate of dissolution of calcium carbonate before final burial of the fossils is indicated by the dashed

[1] The heavier shading marks the extension of the Cromwell Current (Knauss, 1960).

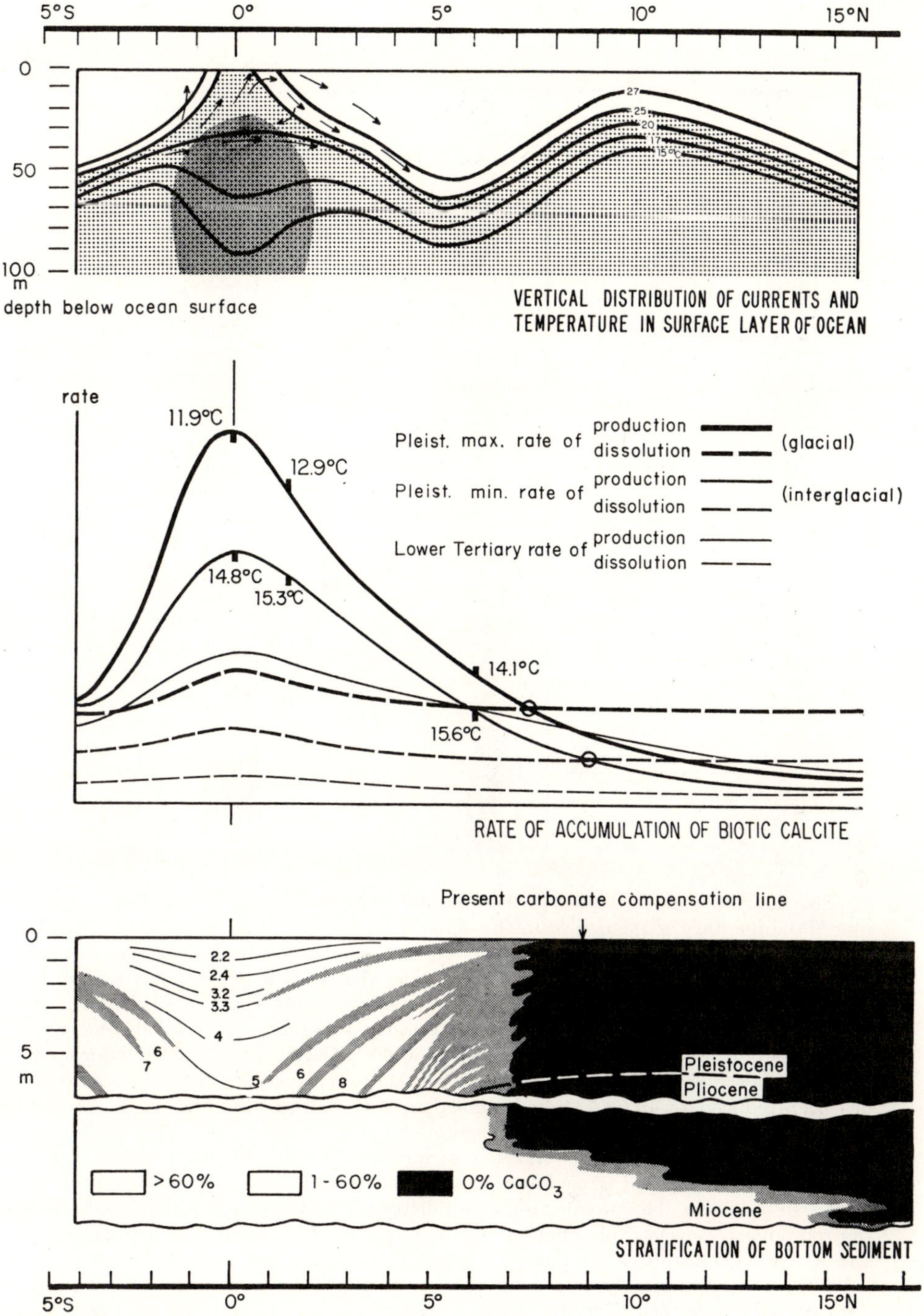

Fig. 36. The equatorial sedimentation mechanism. For explanation, see text.

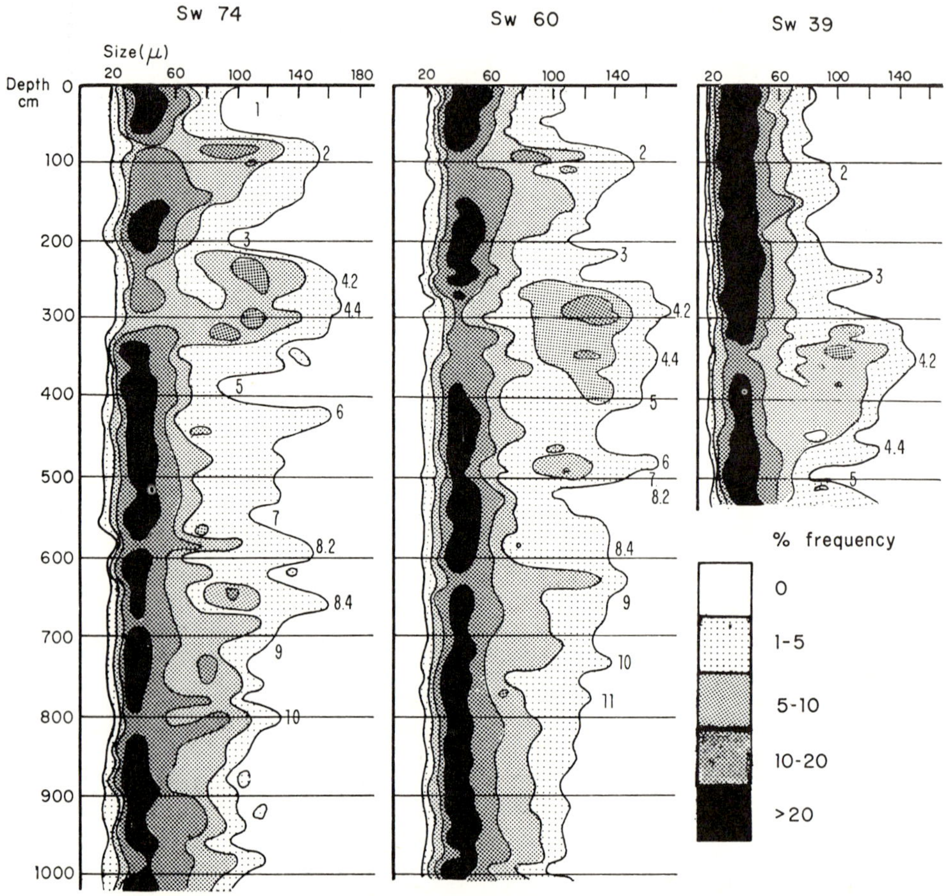

Fig. 37. Size distribution of frustules of the diatom *Coscinodiscus nodulifer* in three widely spaced sedimentary sequences from the equatorial Pacific (Sw 39, long. 93°W; Sw 60, long. 135°W; and Sw 74, long. 150°W). The flat cylindrical frustule of this diatom consists of two halves, the upper one fitting over the lower like a top over a box. At cell division, each daughter cell retains half a frustule as a top and secretes a new bottom (smaller) half. Continuous reproduction by division leads to a distribution of frustule sizes from a few hundred microns to approximately 13 μ. In the majority of cases the frustules are shed and sexual reproduction takes place when the frustule has been reduced to 40 μ. A naked auxospore is then formed, which grows out to maximum size before a new frustule is secreted.

During times of maximum circulation, as indicated by the rate of accumulation of calcareous, siliceous, and phosphatic fossils (cf. Fig. 38), a secondary maximum at large size occurs in the frustule frequency distribution, indicating a modification of the reproductive cycle. This phenomenon is employed to confirm the correlation of strata over wide distances in equatorial facies. The distance between Sw 39 and Sw 74 amounts to approximately one earth radian. Stratigraphic stages are indicated by numbers in the graphs. (For the age-and-productivity relations of these stages, see Figs. 38 and 35. Also, compare D. Ericson, Chapter 31 of this volume, and Ericson *et al.*, 1955, 1956.)

curve "Pleistocene minimum rate of dissolution (interglacial)", which presupposes an increase under the equatorial productivity maximum due to the effect of carbon dioxide produced by the respiration of bottom-living animals that feed on the rain of organic detritus.

The resultant rate of accumulation of calcium carbonate corresponds to the ordinate difference between the two curves and constitutes the parameter actually measured. At the intersection of the two curves near 9°N the carbonate accumulation accordingly drops to zero. Siliceous clay sediments cover the bottom of the ocean north of the present loci of this intersection, the carbonate compensation line.

Emiliani (1955) has used the calcium carbonate tests of both benthonic and pelagic Foraminifera for paleotemperature measurements in this section. The planktonic results for the present sediment surface are indicated at the present-time production curve in the middle diagram of Fig. 36. The benthonic tests indicate a temperature of about 1°C, in good agreement with the directly measured temperature of the bottom water. The planktonic tests display a gradient, with temperature rising from the equator northward (14.8°C at the equator, 15.3°C at 2°N and 15.6°C at 7°N). This trend agrees with the observed temperature distribution in the surface layer shown in the top graph.[1]

The lower diagram in Fig. 36 shows a generalized cross-section through the bottom deposits. The ordinate for the present sediment surface is taken as 0. The lower detached part of the graph, showing the stratification of the Middle Tertiary sediments, lacks detail because of the paucity of observations. The vertical scale is in this latter case relative, and reference to the zero level is not available.

The distribution of carbonate at the present sediment surface is a function of the productivity distribution in the surface layer of the ocean with the superimposed effect of dissolution. From high values below the Equatorial Divergence, the concentration drops toward higher latitudes, and the present northern carbonate compensation line is found near latitude 9°N.

The profile further demonstrates the marked stratification of the calcareous deposits. Below the surface layer, which is characterized by relatively low carbonate concentrations, lies a stratum (2) subdivided into units 2.2, 2.3 and 2.4, high in carbonate and coinciding with the last glaciation in higher latitudes. This stratum is preceded by another low carbonate unit (3) with a small maximum, 3.2, between the minima 3.1 and 3.3.[2] Below this is a high-carbonate stratum (4) indicating extraordinarily high rates of accumulation of biogenous carbonate and silica. Altogether about nine major carbonate maxima occur in the Pleistocene part of the sequence. The underlying Pliocene sediments are

[1] For no known reason, the magnitudes of temperature derived from the oxygen isotope ratio fall considerably below the temperature range in the top 50-m layer of the ocean, where the meridional gradient occurs, and which is the main habitat of the Foraminifera.

[2] Not all the details of the stratification are included in this generalized figure, but they can be seen in Figs. 38 and 39.

characterized by less variability with time of the carbonate content. The Pliocene–Pleistocene transition is indicated by a drop in the temperature of the bottom water, shown by the oxygen isotope distribution in benthonic Foraminifera (Emiliani and Edwards, 1953), and by a marked change in the fossil

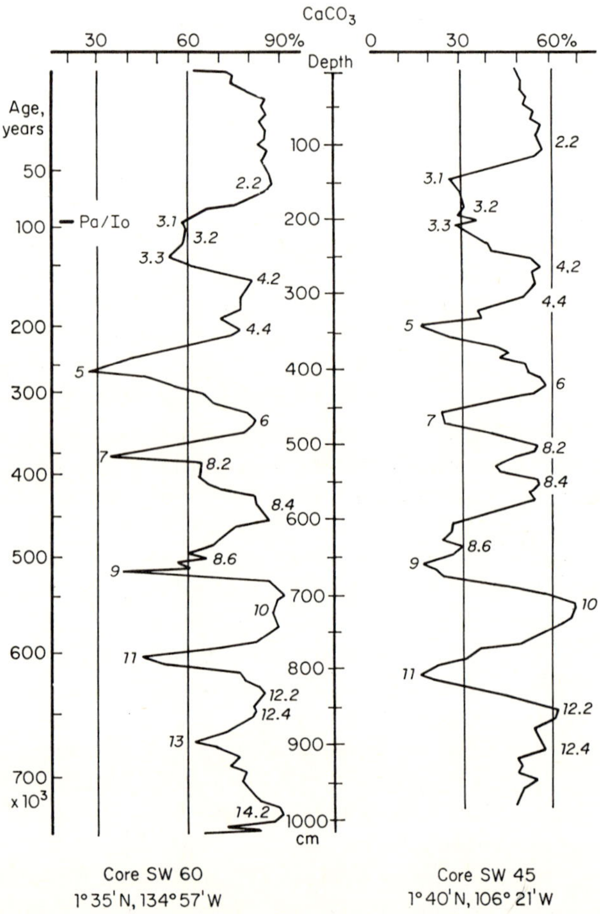

Fig. 38. Age relations of strata in Pacific equatorial carbonate facies as shown by two cores, 3600 km apart. The 95,000-year level in core 60 is determined by the protactinium/ionium ratio at this level, as compared to the sediment surface. The extension of the time scale beyond this age is based on extrapolation assuming a constant rate of deposition of inorganic components but allowing for the wide variability in accumulation of biotic compounds. Strata interpreted as isochronous on the basis of stratigraphic correlation are marked by numbers in the graph (cf. Fig. 37).

assemblages of coccolithophorids (Bramlette, unpublished). In this profile, the transition is marked by a dashed line crossing over the carbonate compensation surface.

From measurements in stratum 4 the full line curve marked "Pleistocene maximum rate of production (glacial)" in the middle diagram, and the dashed curve marked "Pleistocene maximum rate of dissolution (glacial)" have been constructed. The production appears to have increased greatly, especially in the Equatorial Divergence, and the dissolution seems to have increased somewhat, owing to respiration of benthonic animals, although not enough to counterbalance the greatly increased rate of deposition of calcareous skeletal remains. Thus, the rate of accumulation of calcium carbonate reaches a maximum during this time. The greatest increase occurs at the equator, where values of 5 to 10 g of calcium carbonate per cm^2 in a thousand years are found.

Although the amount of calcium carbonate relative to the inorganic components of the sediment is an analytically convenient measure of past variations

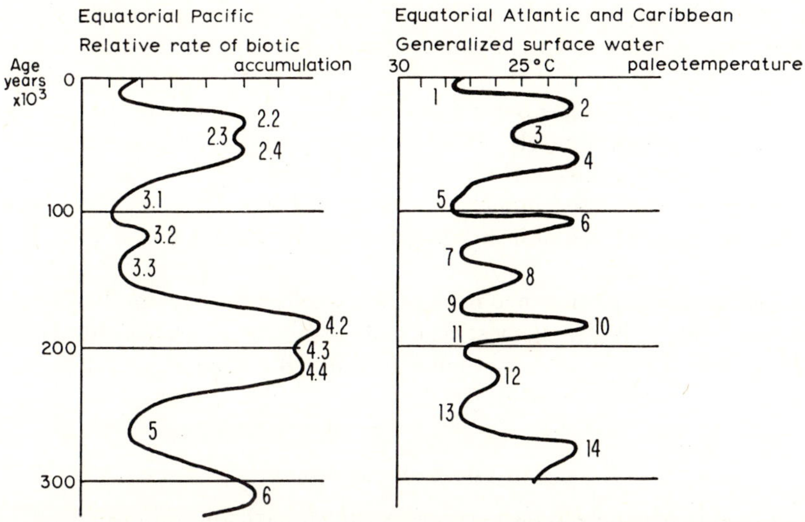

Fig. 39. Age relations and stratigraphic designation of climatically controlled strata in the equatorial Pacific and Atlantic Oceans. (Data from Arrhenius, 1952; Emiliani, 1955; and Emiliani, *in litt.*, 1961.)

in organic productivity, it is not the only one available. The variations described above are accompanied by still more marked changes in the relative amounts of phosphate from fish debris and of opaline silica from diatoms. The diatoms also display characteristic changes in their reproductive cycle (Fig. 37).

As is seen from the planktonic paleotemperature data at the curve "Pleistocene maximum rate of production (glacial)" in the middle graph of Fig. 36 (11.9°C at the equator, 12.9°C at 2°N and 14.1°C at 7°N), the surface temperature gradient away from the equator displayed by the oxygen isotope ratio in the foraminiferal tests is even steeper during the time of circulation maximum (Riss [Illinoian] Glacial Age) than under the present conditions of interglacial type. This further confirms that an increased rate of upwelling in the Equatorial Divergence is the direct cause of the increased productivity. A displacement toward the Equator of the high pressure centers now found between latitudes

20 and 35° could explain this intensified vertical circulation during glacial times. The stratification of biotic components in the Equatorial Pacific under these conditions provides a record of the general climatic evolution of the earth; new drilling techniques now being introduced (Bascom, 1961) could extend the time record beyond the few million years now accessible.

Sediments from Eocene to Pliocene age, outcropping or covered by relatively thin Quaternary strata, have been sampled on topographic highs within the area in question (Arrhenius, 1952, 2.57; Riedel, Chapter 33, this volume). This sedimentation information from the early Cenozoic has been summarized in the lower disconnected part of the bottom graph of Fig. 36, and in the curves marked "Lower Tertiary" in the upper graph of the figure. The northern carbonate compensation line in Lower Tertiary was much farther north than at present, probably north of latitude 45°. However, it was gradually displaced southward, passed latitude 16°N in Miocene time, and approximately reached its present location in Upper Miocene or Pliocene times. Oxygen isotope measurements of benthonic Foraminifera from these strata indicate bottom-water temperatures of approximately 10°C as compared to 1–2° throughout the Pleistocene (Emiliani, *op. cit.*; Emiliani and Edwards, *op. cit.*). Apparently the large meridional extent of carbonate sediments in Lower Tertiary times was due to a decreased rate of dissolution rather than to an increased rate of production.

The meridional drop in concentration of fossils away from the equatorial zone appears also in the Tertiary strata, indicating the operation of the Equatorial Divergence already in early Cenozoic times. Its accurate location in relation to the present equator is not yet known.

4. Physical Stratification

The detailed record on the ocean floor of the climatically induced oscillations in the Equatorial Current system makes it possible to correlate the sedimentary strata of this facies over large distances. Such a correlation is indicated in Fig. 38, which shows two sequences, 3200 km apart, or one-third of the earth's quadrant. The identification of the individual strata is further corroborated by the sequence of properties other than carbonate content, such as the distribution of diatoms shown in Fig. 37.

Attempts have been made to determine the absolute ages of the Pacific equatorial Pleistocene strata by the radiocarbon activity of the carbonate (Arrhenius, Kjellberg and Libby, 1951), and recently by the protactinium/ionium method (Sackett, 1960). Because the short half-life of ^{14}C (5800 years) limits direct age determinations by this method to the Late Glacial and Postglacial strata (stages 1 and 2.2), the age measurement was used to determine the Late Pleistocene and Recent rate of deposition of inorganic constituents; and the ages of older strata were extrapolated under the assumption of a constant average rate of deposition of inorganic compounds throughout the Pleistocene within the area selected. The error inherent in this assumption is probably not much larger than the experimental error of the absolute age determination. This statement is based on two facts. First, accumulation measurements have demonstrated that within the area in question and during the Pleistocene time-intervals studied, the rate varies only little, and apparently randomly from one station to another (Arrhenius, 1952, fig. 1.1.6.1). This approximate spatial constancy demonstrates that the depositional area is far enough removed from the source area of the settling particles for the suspension, from which sedimentation takes place, to have assumed a uniform character. The time that the particles have spent in suspension must consequently be long compared to the half-life of the suspension, which is controlled by coagulation or other precipitation mechanisms. If these reactions, as is likely, are of second or higher order, even large variation in the rate of influx of suspensoids in the ocean will have a negligible effect on the terminal precipitation rate (Arrhenius, 1954). Only small variations with time of the rate of accumulation of inorganic components would consequently be expected in an area where the conditions mentioned above are met, and as long as the global ratio of chemical to mechanical weathering is not markedly changed.

Second, the Tertiary clay sediments in the Pacific, as compared with the Quaternary, clearly indicate less dilution of the halmeic and biotic components (such as manganese oxide minerals and skeletal phosphate) with terrigenous material, and a variation of this dilution with time (Fig. 40). The lack of such marked variations in the Quaternary part of the sequences confirms the assumption of a relatively steady inorganic accumulation rate during this time, particularly when averaged over at least one climatic cycle. This conclusion pertains to only one specific region and is not valid in some other areas investigated (cf. section from East Pacific Rise in left part of fig. 1.1.5, Arrhenius, 1952; Broecker *et al.*, 1958). Local deviations from the average conditions, attributed to topographic control, are discussed below (Section 5).

The time scale obtained for the Upper and Middle Pleistocene events is indicated in Fig. 38, together with the recent protactinium/ionium date. This absolute age determination confirms the extrapolated age within the limits of experimental error (95,000 ± 13,000 years for stage 3.1 by protactinium/ionium, as compared to 90,000 ± 6000 years by radiocarbon extrapolation).

The Lower Pleistocene and Upper Pliocene strata, which are conformably represented in a few investigated sequences, are beyond the reach of present nuclear age determination methods. In this stratigraphic range two events characterize the Tertiary–Quaternary boundary: a distinct evolutionary change in the coccolithophorids (Bramlette, unpublished) and a marked drop in the surface temperature of the ocean, as indicated by the carbonate paleotemperature (Emiliani, 1955). These events have been recorded in two sequences, one near the equator at an extrapolated age level of 1.4 million years (core Sw 62), the other near the north equatorial carbonate compensation surface (Fig. 36), at 1.7 million years (core Sw 58). Because the first area is smoother, its value is more probable; the second area might contain redistributed sediment from a nearby seamount which causes a rate of accumulation of inorganic matter higher than average and thus an overestimate for the extrapolated age. These conditions illustrate the desirability of stratigraphic control in sequences where absolute age determinations are attempted.

Methods similar to those applied in the equatorial Pacific have been used by Emiliani (1955) and Broecker *et al.* (1959) to determine the absolute age relations of the Upper Pleistocene climatic stages in the Atlantic Ocean.

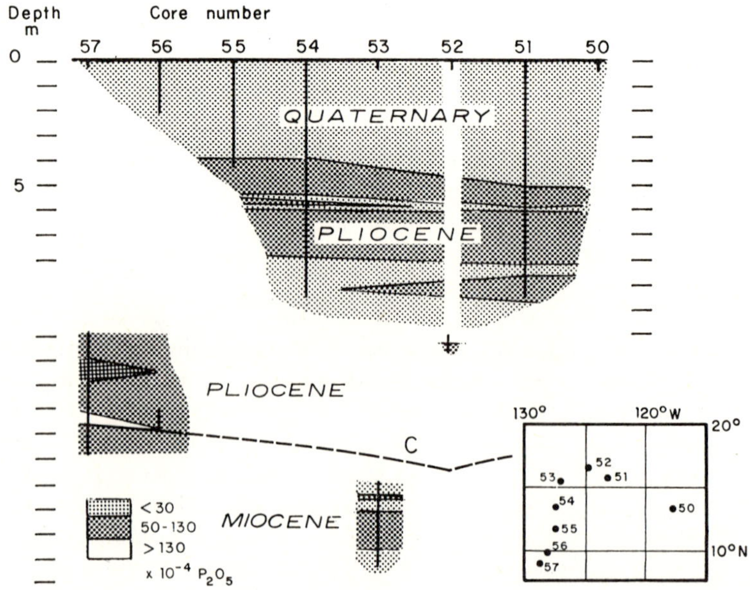

Fig. 40. Stratigraphy of the present north equatorial clay facies in the Pacific on the basis of phosphate content of the sediment. The phosphate is due to skeletal debris of pelagic fish, and the concentration appears to be inversely proportional to the rate of accumulation of terrigenous components. Concentrations are given in g/g.

Vertical lines correspond to the sediment cores indicated by numbers in the graph and in the map insert. The lower detached parts of the figure show older sediments now near the surface below unconformities of various age.

C indicates the heterochronous calcium carbonate compensation surface, separating calcareous from non-calcareous deposits. This boundary moved south past approximately latitude 15°N in the Upper Miocene and appears to have arrived near its present position (cf. Fig. 36) in Pliocene times. (After Arrhenius, 1952.)

Emiliani assumes a low variability in time for both the inorganic and biotic sedimentary components. The stratigraphic subdivision used by Emiliani is based on changes in ocean surface temperature, whereas the stratification in the equatorial Pacific, as mentioned above, is presumably caused by productivity changes, induced by variations in trade-wind intensity. Since certainly both equatorial Pacific circulation and equatorial Atlantic surface temperature were controlled by Pleistocene climatic changes, one would expect agreement in phase but not necessarily in amplitude between the Pacific and Atlantic records.

Fig. 39 shows data for both oceans based on protactinium/ionium dates at approximately the 100,000-year level in both cases, and on extrapolation of the age below this level. A higher resolution is obtained in the Atlantic sequences owing to the higher total rate of deposition, which protects the record more efficiently from the smoothing effects of animal burrowing and post-depositional dissolution. The higher Atlantic resolution indicates a bipartition of the Pacific productivity maximum 3.2 and confirms the bipartition of the Pacific maxima 2 and 4, only vaguely indicated in most Pacific sequences. A cumulative error from extrapolation might account for the difference in apparent age between the Pacific maximum 6 and the Atlantic 14. The suggested interoceanic correlation is summarized in Table VIII. The climatic implications of the stratification are discussed in detail elsewhere (Emiliani, 1955; Arrhenius, 1959a; Emiliani and Flint, Chapter 34 of this volume).

TABLE VIII

Suggested Correlation between Productivity Stratification in Equatorial Pacific Sediments (Arrhenius, 1952) and Paleotemperature Stratification in Atlantic and Caribbean Sediments (Emiliani, 1955)

Stages and substages Equatorial Pacific	Equatorial Atlantic and Caribbean
1	1
2.2	2
2.3	3
2.4	4
3.1	5
3.2	6
3.3	7+8+9
4.2	10
4.3	11
4.4	12
5	13
6	14

The marked effects on circulation and organic production, and ultimately on sediment deposition, caused by the vanishing of the Coriolis force at the equator, open a promising possibility for determining its position in the past. Any displacement of the geographic equator should be accompanied by a corresponding displacement of the thickness maximum of the strata (cf. Fig. 36) and of the maximum in concentration of biotic sedimentary components, resulting from the high productivity in the Equatorial Divergence. If the position of the equator at a given time were established at two points, less than 180° and preferably 90° apart, the position of the poles would also be determined. The estimated possible resolution of this method is of the order of a few degrees of meridional displacement. The sediment cores now existing at the equator go back to only the Middle Pleistocene, about half a million years. During this time the equator has remained constant within the resolution of the present sampling grid, which is about four degrees of latitude. If 500-m cores could be drilled in the equatorial sediment bulge, Cretaceous or older sediments might be reached, corresponding to postulated positions of the North Pole in the North Pacific (Runcorn, 1956, 1959). Determination of polar wandering by this method would be free from the uncertainty imposed by possible continental drift.

Biostratigraphic means available for subdividing into stages and correlating Upper Cenozoic formation units in their calcareous–siliceous facies are lacking in the adjacent clay and zeolitic facies. The stratigraphic units so far resolved in the North Equatorial Pacific clay facies consist of a Pleistocene–Recent bed of buff clay, comparatively low in halmeic minerals, conformably overlying a series of dark brown clay members (Pliocene), high in manganese–iron oxide minerals and other halmeic phases, and in fish bone apatite (Fig. 40). Older clay sediments (Middle Tertiary) occasionally sampled in outcrops on topographic highs (cf. Section 5) show a petrological composition similar to the Pliocene ones. Where the buff top formation is found without interruption of the sequence by unconformities, the thickness amounts to a few meters, but its lower boundary cannot be defined precisely. Accepting for the north equatorial Pleistocene–Recent clay facies a rate of inorganic accumulation around 2 mm/1000 years, indicated by protactinium/ionium and radiocarbon dates (see above) and ionium/thorium measurements (Goldberg and Koide, 1958), it appears that the buff clay formation comprises the last one or two million years. The stratigraphy suggests a relatively low rate of transportation of terrigenous matter into the area during Tertiary times through Middle Pliocene, periodically varying transition conditions in the Upper Pliocene, and a general increase in terrigenous contribution during Pleistocene and Recent.

5. Topographic and Tectonic Control of Sedimentation

Deformation of the ocean floor into hills, mounts, and fault scarps has undoubtedly occurred many different times. Much of the present small-scale topographic relief in the Pacific was probably created by extensive laccolithic injection of basaltic lava, which updomed the overlying sediments (Arrhenius, 1952, 2.57.0, 3.4.1). This process is assumed to have created the low-relief, hilly topography typical of vast areas in the Pacific Ocean basins (cf. Figs. 41 and 43). Lava erupting at oceanic depths—unlike conditions on land or in shallow water—does not encounter marked discontinuity in the confining pressure at the sediment surface. Lateral flows are consequently not particularly probable at this surface, but are likely to occur at petrological discontinuities at depth in the sediment. Submarine lava flows at great water depth are expected to spread extensively owing to low viscosity resulting from the flow of interstitial water or sea-water into the lava at high water pressure. The discontinuity in sound velocity a few hundred meters below the sediment over wide areas in the east

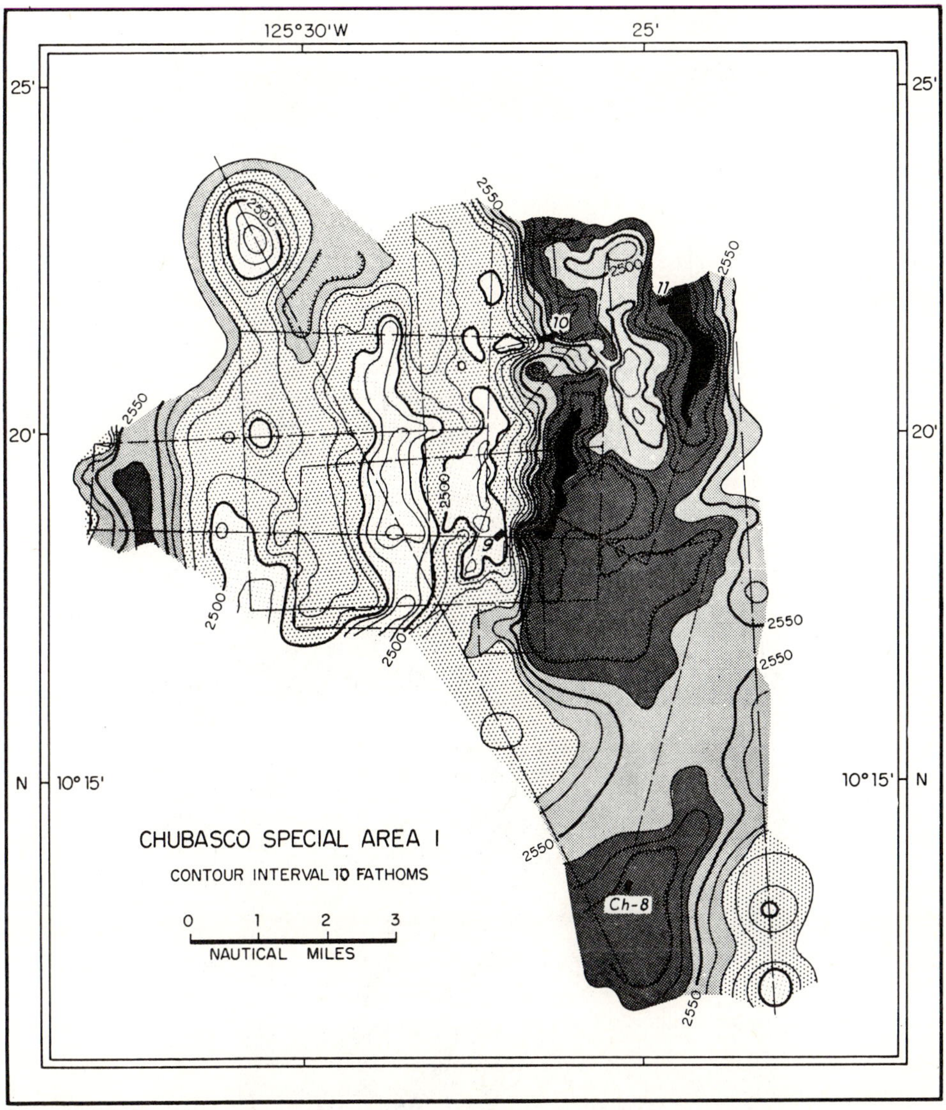

Fig. 41. Topography of area around stations Chub 8, 9, 10 and 11 in North Equatorial clay facies. (Dashed lines indicate the sounding track.) At stations 9 and 10 on the topographic high the Tertiary clay formation is covered only by about 40 cm of Quaternary. The cores 8 and 11, however, which are 160 and 175 cm long respectively, do not penetrate through the Quaternary formation in the valley bottom. The similarity in stratification between cores 8 and 11 suggests that the greater thickness of the Quaternary formation in the valley is not caused by slumping. The abnormal thinness of the Quaternary strata on the slope and crest of the hill consequently appears to be due either to a persistently lower rate of accumulation or to a period of non-deposition represented by the Tertiary–Quaternary unconformity. (From Bramlette and Arrhenius, unpublished; topographic survey by R. L. Fisher.)

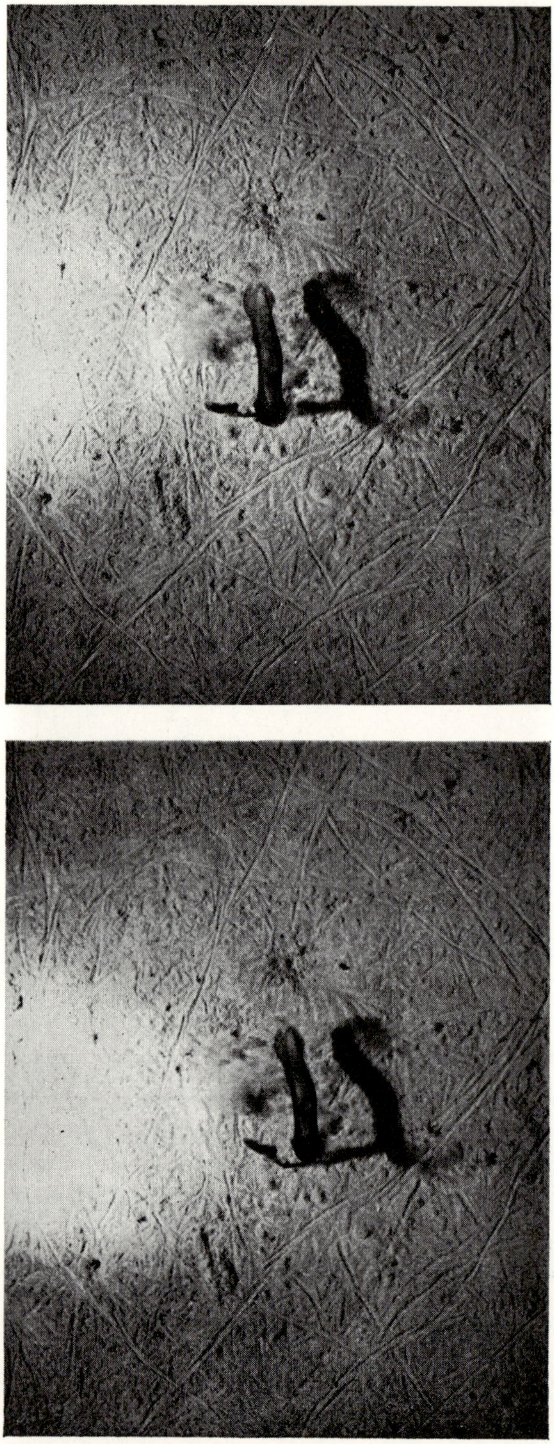

Fig. 42. Sediment suspension cloud stirred up or expelled by a holothurian. (Also visible in the picture: a sea-pen, tracks and burrows of benthic organisms.) 31° 22′N, 117° 29′W; 1930 m (SW of Ensenada, Continental Borderland). (Stereophoto: D. Krause and D. Corrigan.)

Pacific might well be due to such a stratigraphically determined preferred level of intrusion.

Probably because of the uplift, sedimentation appears in many cases to have ceased on the topographic highs for varying lengths of time, except for halmeic minerals accreting by precipitation. In some cases erosion seems to be or has been active on the crests and slopes of the hills, while accumulation continues around them (Fig. 41). The older deposits on the topographic highs, outcropping or covered by relatively thin layers of younger sediments, appear to range in age between Oligocene and Pliocene (Arrhenius, 1952; Riedel, Chap. 33 of this volume). If it is correct to postulate control of the topography by volcanic intrusions, it would then seem that such events have occurred frequently during the Cenozoic era.

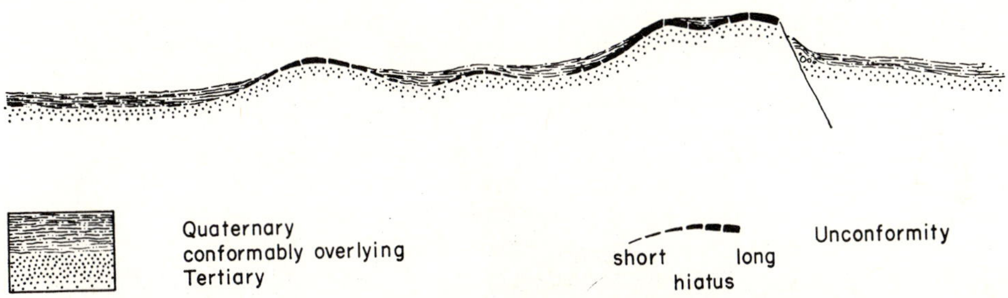

Fig. 43. Hypothetical schematic profile illustrating topographic control of pelagic sedimentation in low, hilly topography. The order of magnitude size of the hills is a few kilometers horizontally and up to a few hundred meters vertically. The thickness scale of the Quaternary formation is exaggerated; the maximum thickness found in pelagic clay facies is a few meters.

Scouring even by weak currents might prevent accumulation of fine-grained clay sediment from taking place on hill-tops and slopes and may account for lateral displacement of resuspended sediment, but is unlikely to be responsible for the resuspension which appears to affect Pleistocene and older sediments with comparatively high shearing strengths. The dispersion of Tertiary Radiolaria through the Pleistocene sediments adjacent to the eroded hills (Riedel, Chapter 33 of this volume) indicates a continuous process of resuspension and transport downhill of the deposit from the topographic highs. This is probably caused by the churning action of benthic animals, some of which have been observed forming a suspension cloud (Fig. 42), which may be displaced by any horizontal movement of the near-bottom water. Repetition of this process seems to result in slow abrasion of the elevated areas and movement of the resuspended sediments downhill, until a level is reached where the streamlines are parallel to the sediment surface. The accumulating sediment, mostly new material and including some redeposited sediment, appears gradually to invade the elevated areas of non-deposition, as schematically indicated in Fig. 43. The upper Cenozoic clay sediments thus form a blanket with frequent holes and thin spots over hills and seamounts. The smoothing of the topographic irregularities, accompanied by gradual closure of the holes, is apparently more rapid in calcareous sediments on account of their higher accumulation rate.

References

Altschuler, Z. S., R. S. Clarke and E. J. Young, 1958. Geochemistry of uranium in apatite and phosphorite. *U.S. Geol. Surv. Prof. Paper*, **314-D**.

Armstrong, F. A. J., 1958. Inorganic suspended matter in sea water. *J. Mar. Res.*, **17**, 23.

Arrhenius, G., 1952. Sediment cores from the East Pacific. *Reps. Swed. Deep-Sea Exped.*, **5**, Göteborg.

Arrhenius, G., 1954. Origin and accumulation of aluminosilicates in the ocean. *Tellus*, **6**, 215.

Arrhenius, G., 1954a. Significance of carbonate stratification in pelagic deposits. *Bull. Geol. Soc. Amer.*, **65**, 228.

Arrhenius, G., 1959. Sedimentation on the ocean floor. *Researches in Geochemistry*. John Wiley, New York.

Arrhenius, G., 1959a. Climatic records on the ocean floor. *Rossby Memorial Volume*, edit. B. Bolin, Rockefeller Inst. Press.

Arrhenius, G., 1961. Geological record on the ocean floor. *Oceanography*. Amer. Assoc. Adv. Sci., Publ. 67. Wash. D.C.

Arrhenius, G., M. Bramlette and E. Picciotto, 1957. Localization of radioactive and stable heavy nuclides in ocean sediments. *Nature*, **180**, 85–86.

Arrhenius, G., G. Kjellberg and W. F. Libby, 1951. Age determination of Pacific chalk ooze by radiocarbon and titanium content. *Tellus*, **3**, 222.

Arrhenius, G. and H. Korkisch, 1959. Uranium and thorium in marine minerals. *Intern. Oceanog. Cong. Preprints*, 497. Edit. M. Sears. Amer. Assoc. Adv. Sci., Wash. D.C.

Arrhenius, G. and H. Rotschi, 1953. SIO Expedition Capricorn. Unpubl. Rep. Scripps Institution of Oceanography.

Bannister, F. A. and M. H. Hey, 1936. Report on some crystalline components of the Weddell Sea deposits. *Discovery Reps.*, **12**, 60.

Barth, T. F. W., 1952. *Theoretical Petrology*. John Wiley, New York, 369 pp.

Bascom, W., 1961. *A Hole in the Bottom of the Sea*. Doubleday, New York.

Bramlette, M. N., 1958. Significance of coccolithophorids in calcium carbonate deposition. *Bull. Geol. Soc. Amer.*, **69**, 121–126.

Bramlette, M. N., 1961. Pelagic sediments. *Oceanography*. Amer. Assoc. Adv. Sci., Publ. 67 Wash. D.C.

Bramlette, M. N. and W. Bradley, 1942. Geology and biology of North Atlantic deep-sea cores between Newfoundland and Ireland. Lithology and geologic interpretations. *U.S. Geol. Surv. Prof. Paper*, **196**.

Bramlette, M. N. and W. R. Riedel, 1954. Stratigraphic value of discoaster and some other microfossils related to Recent coccolithophores. *J. Paleont.*, **28**, 385.

Broecker, W. S., K. K. Turekian and B. C. Heezen, 1958. The relation of deep-sea sedimentation rates to variations in climate. *Amer. J. Sci.*, **256**, 503–517.

Broecker, W. S., A. Walton, B. C. Heezen and K. Turekian, 1959. Sedimentation rates in the deep ocean. *Intern. Oceanog. Cong. Preprints*, 87. Edit. M. Sears. Amer. Assoc. Adv. Sci., Wash. D.C.

Bruejewicz, S. V. and E. D. Zaytseva, 1958. The Bering Sea sediments. *Trudy Inst. Okeanol. Akad. Nauk S.S.S.R.*, **26**, 8–108.

Buser, W., 1959. The nature of the iron and manganese compounds in manganese nodules. *Intern. Oceanog. Cong. Preprints*, 962. Edit. M. Sears. Amer. Assoc. Adv. Sci., Wash. D.C.

Buser, W. and A. Grütter, 1956. Untersuchungen an Mangansodimenten. *Schweiz. Mineralog. Petrograph. Mitt.*, **36**, 49–62.

Butkevich, E. S. 1928. The formation of marine iron and manganese deposits and the role of microorganisms in the latter. *Verh. Wiss. Meeresinst. Moscow*, **3**, 63.

Buttlar, H. von and F. G. Houtermans, 1950. Photographische Bestimmung der Aktivitätsverteilung in einer Manganknolle der Tiefsee. *Naturwissenschaften*, **37**, 400–401.

Castaing, R. and K. Fredriksson, 1958. Analyses of cosmic spherules with an x-ray microanalyzer. *Geochim. et Cosmochim. Acta*, **14**, 114.

Chow, T. J. and E. D. Goldberg, 1960. On the marine geochemistry of barium. *Geochim. et Cosmochim. Acta*, **20**, 192.

Correns, C. W., 1937. Die Sedimente des äquatorialen Atlantischen Ozeans. *Wiss. Ergebn. Deut. Atlant. Exped. 'Meteor', 1925–1927*, **3**, 1.

Correns, C. W., 1939. Die Sedimentgesteine. Part 2. In Barth, T. F. W., C. W. Correns and P. Eskola, *Die Entstehung der Gesteine*. Springer, Berlin.

Correns, C. W., 1954. Titan in Tiefseesedimenten. *Deep-Sea Res.*, **1**, 78–85.

Crozier, 1960. Black magnetic spherules in sediments. *J. Geophys. Res.*, **65**, 2971.

Dietz, R. S., 1955. Manganese deposits on the northeast Pacific sea floor. *Calif. J. Mines Geol.*, **51**, 209–220.

Dietz, R. S., K. O. Emery and F. P. Shepard, 1942. Phosphorite deposits on the sea floor off southern California. *Bull. Geol. Soc. Amer.*, **53**, 815–848.

Dieulafait, L., 1883. Le manganese dans les eaux de mers actuelles et dans certains de leur depôts. *C. R. Acad. Sci. Paris*, **96**, 718.

Dorff, F., 1935. *Biologie des Eisen- und Mangankreislaufes*. Springer, Berlin.

El Wakeel, S. K. and J. P. Riley, 1961. Chemical and mineralogical studies of deep-sea sediments. *Geochim. et Cosmochim. Acta*, **25**, 110.

Emery, K. O., 1960. *Geology of the Sea Floor off California*. John Wiley, New York.

Emiliani, C., 1955. Pleistocene temperatures. *J. Geol.*, **63**, 538.

Emiliani, C., 1956. Oligocene and Miocene temperatures of the equatorial and subtropical Atlantic Ocean. *J. Geol.*, **64**, No. 3, 281.

Emiliani, C. and G. Edwards, 1953. Tertiary ocean bottom temperatures. *Nature*, **171**, 887.

Ericson, D. B., M. Ewing, B. Heezen and G. Wollin, 1955. Sediment deposition in deep Atlantic. *Geol. Soc. Amer. Spec. Paper* 62, 205–220.

Ericson, D. B. and G. Wollin, 1956. Correlation of six cores from the equatorial Atlantic and Caribbean. *Deep-Sea Res.*, **3**, 104–125.

Ericson, D. B. and G. Wollin, 1956a. Micropaleontological and isotopic determinations of Pleistocene climates. *Micropaleontology*, **2**, 257.

Ewing, M., B. C. Heezen and D. Ericson, 1959. Significance of the Worzel deep-sea ash. *Proc. Nat. Acad. Sci.*, **45**, 355.

Fredriksson, K., 1958. A note on investigations of cosmic spherules and other small meteoritic particles. *Astronomical Notes*. Univ. of Göteborg, Sweden, p. 21.

Fredriksson, K., 1959. On the origin, age, and distribution of cosmic spherules. *Intern. Oceanog. Cong. Preprints*, 456. Edit. M. Sears. Amer. Assoc. Adv. Sci., Wash. D.C.

Fredriksson, K., 1961. Origin of black spherules from Pacific islands, deep-sea sediments and Antarctic ice. *Symposium Abst. 10th Pacific Sci. Cong., Honolulu.*

Goldberg, E. D., 1954. Marine geochemistry 1. Chemical scavengers of the sea. *J. Geol.*, **62**, 249–265.

Goldberg, E. D. and G. Arrhenius, 1958. Chemistry of Pacific pelagic sediments. *Geochim. et Cosmochim. Acta*, **13**, 153–212.

Goldberg, E. D. and M. Koide, 1958. Ionium-thorium chronology in deep-sea sediments of the Pacific. *Science*, **128**, 1003.

Goldberg, E. D. and R. H. Parker, 1960. Phosphatized wood from the Pacific sea floor. *Bull. Geol. Soc. Amer.*, **71**, 631.

Goldberg, E. D., C. Patterson and T. Chow, 1958. Ionium-thorium and lead isotope ratios as indicators of oceanic water masses. *2nd U.N. Intern. Conf. Peaceful Uses of Atomic Energy*, 15/P/1980.

Goldberg, E. D. and E. Picciotto, 1955. Thorium determinations in manganese nodules. *Science*, **121**, 613–614.

Goldschmidt, V. M., 1954. *Geochemistry*. Clarendon Press, Oxford.

Gorbunova, Z. N., 1960. Fine grained minerals in the sediments of the Indian Ocean. *Doklady Akad. Nauk S.S.S.R.*, **134**, 935.

Gorbunova, Z. N., 1960a. Composition of clay minerals in different horizons of the bottom deposits of the Indian Ocean. *Doklady Akad. Nauk S.S.S.R.*, **134**, 1201.

Gordon, L., C. C. Reimer and B. P. Burtt, 1954. Coprecipitation from homogeneous solution. *Anal. Chem.*, **26**, 842.

Grabau, A. W., 1904. On the classification of sedimentary rocks. *Amer. Geol.*, **33**, 228.

Graf, C. L., 1960. Geochemistry of carbonate sediments and sedimentary carbonate rocks. Part III; minor element distribution. *Ill. State Geol. Surv.*, Urbana, Ill. Circular 301.

Graham, J. W., 1959. Metabolically induced precipitation of trace elements from sea water. *Science*, **129**, 1428.

Grim, R. E., R. S. Dietz and W. F. Bradley, 1949. Clay mineral composition of some sediments from the Pacific Ocean off the California coast and the Gulf of California. *Bull. Geol. Soc. Amer.*, **60**, 1785–1808.

Gripenberg, S., 1934. A study of the sediments of the North Baltic and adjoining seas. *Fennia*, **60**, No. 3, Helsingfors.

Grütter, A. and W. Buser, 1957. Untersuchungen an Mangansedimenten. *Chimia*, **11**, 132–133.

Hagemann, F. and Spjeldnaes, N., 1955. The Middle Ordovician of the Oslo region, Norway. 6. Notes on bentonite (K-bentonites) from the Oslo-Asker District. *Norsk Geol. Tidsskr.*, **35**, 29–52, Oslo.

Hägg, G., 1935. The crystal structure of magnetic ferric oxide. *Z. physik. Chem.*, **29**, 95.

Hamilton, E. L., 1956. Sunken islands of the mid-Pacific mountains. *Geol. Soc. Amer. Mem.*, **64**.

Harder, H., 1959. Beitrag zur Geochemie des Bors. *Nachr. Akad. Wiss. Göttingen*, **6**, 123.

Heezen, B. C., M. Tharp and M. Ewing, 1959. The floors of the oceans. *Geol. Soc. Amer. Spec. Paper* 65.

Hunter, W. and D. W. Parkin, 1960. Cosmic dust in Recent deep-sea sediments. *Proc. Roy. Soc. London*, **A255**, 382–397.

Hurley, R. J., 1960. Unpublished manuscript, SIO ref. 60–67.

Hustedt, F., 1921. Untersuchungen über die Natur der Harmattantrübe. *Deut. Ubersee. Meteorol. Ber.*, **23**.

Inman, D. and T. Chamberlain, 1960. Littoral sand budget along the southern California coast. *21st Intern. Geol. Cong. Copenhagen*, 245.

Jerlov, N. G., 1953. Particle distribution in the ocean. *Reps. Swed. Deep-Sea Exped.*, **3** (Fasc. 2), 73–97.

Kalinenko, V. O., 1949. The origin of Fe–Mn concretions. *Mikrobiologiya*, **18**, 528.

Kato, K., 1955. Accumulation of organic matter in marine bottom and its oceanographic environment in the sea to the southeast of Hokkaido. *Bull. Fac. Fish., Hokkaido Univ.*, **6**, 125–151.

Kazakov, A. V., 1950. Geotectonics and the formation of phosphorite deposits. *Bull. Acad. Sci. U.S.S.R., Ser. Geol.*, **5**, 42–68.

Knauss, J. A., 1960. Measurements of the Cromwell Current. *Deep-Sea Res.*, **6**, 265–286.

Kobayashi, K., K. Oinuma and T. Sudo, 1960. Clay mineralogical study on Recent marine sediments. *Oceanog. Mag.*, **11**, 215 (Tokyo).

Koczy, F., 1950. Die bodennahen Wasserschichten der Tiefsee. *Naturwissenschaften*, **37**, 360.

Koczy, F. F., 1952. On the properties of the water layers close to the ocean floor. *Assoc. Oceanog. Phys. Procès-Verbaux.*, No. 5, **146**.

Koczy, F., 1958. Natural radium as a tracer in the ocean. *Proc. 2nd U.N. Intern. Conf. Peaceful Uses of Atomic Energy, Geneva*, **18**, 351–357.

Kolbe, R. W., 1955. Diatoms from equatorial Atlantic cores. *Reps. Swed. Deep-Sea Exped.*, **7**, 3.

Kriss, A. Y., 1959. *Marine Microbiology*. Moskva (in Russian).

Kuenen, Ph. H., 1950. *Marine Geology*. John Wiley and Sons, New York.

Laevastu, T. and O. Mellis, 1955. Extraterrestrial material in deep-sea deposits. *Trans. Amer. Geophys. Un.*, **36**, 385.

LaGow, H. E. and W. M. Alexander, 1960. Satellite measurements of cosmic dust. *IGY Bull. U.S. Nat. Acad. Sci.*, **12**.

Leinz, V., 1937. Die Mineralfazies der Sedimente des Guinea-Beckens. *Wiss. Ergebn. Deut. Atlant. Exped. 'Meteor', 1925–1927*, **3**, 245.

Ljunggren, P., 1953. Some data concerning the formation of manganiferous and ferriferous bog ores. *Geol. Fören. Stockholm Förhandl.*, **75**, 277.

Lowenstam, H. A. and S. Epstein, 1954. Paleotemperatures of the Post-Aptian Cretaceous. *J. Geol.*, **62**, 207.

Lowenstam, H. A. and S. Epstein, 1957. On the origin of sedimentary aragonite needles of the Great Bahama Bank. *J. Geol.*, **65**, 364.

Marshall, P. and E. Kidson, 1929. Dust storms of October, 1928. *N.Z. J. Sci. Tech.*, **10**, 291.

McKelvey, V. E., R. W. Swanson and R. P. Sheldon, 1953. The Permian phosphorite deposits of western United States. *C. R., XXIX, Cong. Géol. Intern., Alger 1952*, **11**, 45.

Mason, B., 1943. Mineralogical aspects of the system $FeO–Fe_2O_3$, $MnO–Mn_2O_3$. *Geol. Fören. Stockholm, Förhandl.*, **65**, 95.

Mellis, O., 1952. Replacement of plagioclase by orthoclase in deep-sea deposits. *Nature*, **169**, 624.

Mellis, O., 1954. Volcanic ash horizons in deep-sea sediments from the Eastern Mediterranean. *Deep-Sea Res.*, **2**, 89.

Mellis, O., 1959. Gesteinsfragmente im roten Ton des Atlantischen Ozeans. *Meddel. Oceanog. Inst. Göteborg*, Ser. B., **8**, No. 6, 1–17.

Menard, H. W., 1959. La topographie et la géologie des profondeurs océaniques. *Coll. Intern. Centre Nat. Rech. Sci.*, **83**, 95.

Menard, H. W. and C. J. Shipek, 1958. Surface concentrations of manganese nodules. *Nature*, **182**, 1156–1158.

Mero, J. L., 1960. Mineral resources on the ocean floor. *Sci. Amer.*, **203**, No. 12.

Mero, J. L., 1960a. Mineral resources on the ocean floor. *Mining Cong. J.*, **46**, 48–53.

Mironov, S. I. and O. K. Bordovsky, 1959. Organic matter in bottom sediments of the Bering Sea. *Intern. Oceanog. Cong. Preprints*, 970–974. Edit. M. Sears. Amer. Assoc. Adv. Sci., Wash. D.C.

Molengraaff, G., 1920. Manganknollen in Mesozoische Diepzeeafzettingen van Nederlandsche-Timor. *Versl. Wisen. Nat. Afd., Koninkl. Akad. Wetenschap. Amsterdam*, D. 1.29, 677.

Moore, D. G., 1960. Submarine slumps. *J. Sediment. Petrol.*, **31**, 343.

Müller, G. and H. Puchelt, 1961. Die Bildung von Coelestin aus Meerwasser. *Naturwissenschaften*, **48**, 301.

Murray, J. and A. F. Renard, 1884. On the nomenclature, origin and distribution of deep-sea deposits. *Proc. Roy. Soc. Edinburgh*, **12**, 495.

Murray, J. and A. F. Renard, 1891. Deep-sea deposits. *Rep. 'Challenger' Exped.*, London.

Nayudu, Y. R., 1962. A new hypothesis for origin of guyots and seamount terraces. *Amer. Geophys. Un. Monog.*, 6.

Nayudu, Y. R., 1962a. Volcanic ash deposits in the Gulf of Alaska and problems of correlation of deep-sea ash deposits. *J. Sediment. Petrol.*, in press.

Niino, H., 1959. On the distribution of organic carbon in the deep-sea sediments of the Japan Sea. *Intern. Oceanog. Cong. Reprints*, 974-6. Edit. M. Sears. Amer. Assoc. Adv. Sci., Wash. D.C.

Norin, E., 1958. The sediments of the central Tyrrhenian Sea. *Reps. Swed. Deep-Sea Exped.*, **8**, 1.

Oinuma, K., K. Kobayashi and T. Sudo, 1959. Clay mineral composition of some Recent marine sediments. *J. Sediment. Petrol.*, **29**, 56–63.

Olausson, E., 1961. Remarks on Tertiary Sequences of two cores from the Pacific. *Bull. Geol. Inst. Univ. Uppsala*, **40**, 299.

Ostroumov, E. A. and L. S. Fomina, 1960. On the forms of sulfur compounds in the bottom deposits of the Marianas trench. *Doklady Akad. Nauk S.S.S.R.*, **126**, no. 1–6.

Peterson, M. N. and E. D. Goldberg, 1962. Feldspar distribution in South Pacific pelagic sediments. *J. Geophys. Res.*, **67**, 3477.

Pettersson, H., 1945. Iron and manganese on the ocean floor; a contribution to submarine geology. *Göteborgs Kungl. Vet. o. Vitterh. Samh. Handl.*; 6. följden, ser. B, **2**, no. 8, 1–43.

Pettersson, H., 1945a. Oceanbottnens problem. *Ymer.*, 161–178.

Pettersson, H., 1955. Manganese nodules and oceanic radium. *Deep-Sea Res.*, **3**, Suppl. 335–345.

Pettersson, H., 1959. Manganese and nickel on the ocean floor. *Geochim. et Cosmochim. Acta*, **17**, 209–213.

Pettersson, H., 1959a. Frequency of meteorite falls throughout the ages. *Nature*, **183**, 1114.

Pettersson, H. and K. Fredriksson, 1958. Magnetic spherules in deep-sea deposits. *Pacific Sci.*, **12**, 71–81.

Pettijohn, F. J., 1949. *Sedimentary Rocks*. Harper, New York.

Phleger, F. B., 1960. *Ecology and Distribution of Recent Foraminifera*. Johns Hopkins Press, Baltimore.

Radczewski, O. E., 1937. Die Mineralfazies der Sedimente des Kapverden-Beckens. *Wiss. Ergebn. Deut. Atlant. Exped. 'Meteor', 1925–1927*, **3**, 262.

Radczewski, O. E., 1939. Eolian deposits in marine sediments. *Recent Marine Sediment Symposium* (P. D. Trask, ed.). Amer. Assoc. Petrol. Geol., Tulsa, 496–502.

Revelle, R. R., 1944. Marine bottom samples collected in the Pacific Ocean by the Carnegie on its seventh cruise. *Carnegie Inst. Wash. Pubs.*, 556, Washington, D.C.

Revelle, R. R., M. N. Bramlette, G. Arrhenius and E. D. Goldberg, 1955. Pelagic sediments of the Pacific. *Geol. Soc. Amer. Spec. Paper* 62, 221–236.

Rex, R., 1958. Quartz in sediments of the Central and North Pacific Basins. Unpub. thesis, University of California.

Rex, R. W., and E. D. Goldberg, 1958. Quartz contents of pelagic sediments of the Pacific Ocean. *Tellus*, **10**, 153–159.

Riley, J. P. and P. Sinhaseni, 1958. Chemical composition of three manganese nodules from the Pacific Ocean. *J. Mar. Res.*, **17**, 466.

Rimšaite, J., 1957. Über die Eigenschaften der Glimmer in den Sanden und Sandsteinen. *Beitr. Min. Petrol.*, **6**, 1.

Runcorn, S. K., 1956. Paleomagnetism, polar wandering and continental drift. *Geol. Mijnb.*, **18**, 253–256.

Runcorn, S. K., 1959. Rock magnetism. *Science*, **129**, 1002–1012.

Sackett, W. M., 1960. Protactinium-231 content of ocean water and sediments. *Science*, **132**, 1761.

Sackett, W. M. and G. Arrhenius, 1962. Distribution of aluminum species in natural waters. *Geochim. et Cosmochim. Acta*, **26**, 955.

Schewiakoff, W., 1926. Fauna e flora del golfo di Napoli. Acantharia, *Pub. Staz. Zool. Napoli*, Mono. **37**.

Schott, W., 1935. In G. Schott, *Geographie des Indischen und Stillen Ozeans*. Boysen, Hamburg, 110–111.

Seefried, V., 1913. Untersuchungen über die Natur der Harmattantrübe. *Mitt. Deut. Schutzgeb.*, Bd. H.I. **26**.

Shepard, F. P., 1948. *Submarine Geology*. Harper, New York.

Shipek, C. J., 1960. Photographic study of some deep-sea floor environments in the eastern Pacific. *Bull. Geol. Soc. Amer.*, **71**, 1067–1074.

Shiskina, O. V., 1959. On the salt composition of the marine interstitial waters. *Intern. Oceanog. Cong. Preprint Abst.* Edit. M. Sears. Amer. Assoc. Adv. Sci., Wash. D.C., 977.

Sigvaldason, G. E., 1959. Mineralogische Untersuchungen über Gesteinszersetzung durch postvulkanische Aktivität in Island. *Beitr. Min. Petrol.*, **6**, 405.

Sillén, L. G., 1961. Chemistry of sea water. *Oceanography*. Amer. Assoc. Adv. Sci., Publ. 67. Wash. D.C.

Skornyakova, N. S., 1960. Manganese concretions in sediments of the north-eastern Pacific Ocean. *Doklady Akad. Nauk S.S.S.R.*, **130**, 653–656.

Smoluchovski, M. von, 1917. Versuch einer Mathematischen Theorie der Koagulationskinetik Kolloider Lösungen. *Z. physik. Chem.*, **92**, 129.

Starik, I. Ye. and L. B. Kolyadin, 1957. The conditions of occurrences of uranium in sea water. *Geokhimiya*, no. 3, 213.

Starik, I. Ye. and A. P. Zharkov, 1961. Rate of sediment accumulation in the Indian Ocean by C^{14}. *Doklady Akad. Nauk S.S.S.R.*, **136**, 203.

Strunz, H., 1957. *Mineralogische Tabellen*. Akad. Verlagsges., Leipzig.

Sverdrup, H. U., M. W. Johnson and R. H. Fleming, 1946. *The Oceans: Their physics, chemistry and general biology*. Prentice-Hall, New York.

Teodorovich, G., 1958. *Autigennye mineraly osadochnykh porod*. U.S.S.R. Acad. Sci. Press (translation into English; Consultant's Bureau, N.Y., 1961).

Thiel, E. and R. A. Schmidt, 1961. Spherules from the Antarctic Ice Cap. *J. Geophys. Res.*, **66**, 307.

Thorslund, P., 1948. Silurisk bentonit från Gotland. *Geol. Fören. Stockholm Förhandl.*, **70**, 245–246.

Thorslund, P., 1958. Djupborrningen på Gotska Sandön. *Geol. Fören. Stockholm Förhandl.*, **80**, 190.

Udden, J. A., 1914. Mechanical composition of clastic sediments. *Bull. Geol. Soc. Amer.*, **25**, 655.

Waern, B., P. Thorslund and Q. Henningsmoen, 1948. Deep boring through Ordovician and Silurian strata at Kinnekulle, Västergötland. *Bull. Geol. Inst. Uppsala*, **32**, 433–474.

Wattenberg, H., 1933. Kalziumkarbonat- und Kohlensäuregehalt des Meerwassers. *Wiss. Ergebn. Deut. Atlant. Exped. 'Meteor', 1925–1927*, **8**.

Wattenberg, H., 1937. Die Bedeutung anorganischer Faktoren bei der Ablagerung von Kalziumkarbonat im Meere. *Geol. Meere u. Binnengewäss*, **1**, 237.

Wedepohl, K. H., 1960. Spurenanalytische Untersuchungen an Tiefseetonen aus dem Atlantik. *Geochim. et Cosmochim. Acta*, **18**, 200.

Worzel, J. L., 1959. Extensive deep sea sub-bottom reflections identified as white ash. *Proc. Nat. Acad. Sci.*, **45**, 349.

Yeroshchev-Shak, V. A., 1961. Kaolinite in the sediments of the Atlantic Ocean. *Doklady Akad. Nauk S.S.S.R.*, **137**, 695–697.

Yeroshchev-Shak, V. A., 1961a. Illite in the sediments of the Atlantic Ocean. *Doklady Akad. Nauk S.S.S.R.*, **137**, 941–953.

Yoder, H. S., Jr., 1957. Experimental studies on micas: a synthesis. *Proc. 6th Nat. Conf. on Clay and Clay Minerals*, no. 1284, 42–60.

Zen, E-An, 1959. Mineralogy and petrology of some marine bottom sediment samples off the coast of Peru and Chile. *J. Sediment. Petrol.*, **29**, 513–539.

Zenkevitch, N. L., 1959. Methods and some results of photographing the floor of the Pacific Ocean. *Intern. Oceanog. Cong. Preprints*, 493. Edit. M. Sears. Amer. Assoc. Adv. Sci., Wash. D.C.

ZoBell, C., 1946. *Marine Microbiology*. Chronica Botanica Co., Waltham, Mass.

26. CLAY-MINERAL DISTRIBUTIONS IN THE PACIFIC OCEAN

J. J. Griffin and E. D. Goldberg

1. Introduction

The clay minerals are often the predominant components of the calcium-carbonate-free phases of marine sediments, yet only very limited studies have been made on their areal and depth distributions. Many significant problems in marine geochemistry await resolution as to their compositions, origins or modes of formation, such as the use of clay minerals in geochronology, source areas for detrital components in marine sediments, and changes with time of the reactants and products within the major sedimentary cycle.

Grim (1953) has summarized much of the earlier work and concludes that many of the reported analyses suffer on the basis of incomplete or inadequate data. Determinations made prior to 1944–45 did not involve the glycerine or ethylene glycol solvation technique (MacEwan, 1944; Bradley, 1945) for the analysis of montmorillonite. Electron microscopic and diffraction techniques in combination are today definitive for the identification of halloysite (Bates, Hildebrand and Swineford, 1950; Studer, *in litt.* 1960). Finally, information on mixed layer clay minerals and the variation in types of illite, chlorite and montmorillonite were unknown to the pioneer workers. Thus the efforts of Revelle (1944) on the *Carnegie* samples and Correns (1937) on the *Meteor* material from the Atlantic are difficult to interpret and collate with modern analyses.

All of the analyses in the recent literature are restricted to samples recovered close to land and with a small areal distribution. Zen (1957) has studied the clay minerals in the sediments off the coast of Peru and Chile; Grim, Dietz and Bradley (1949), the Gulf of California and the Pacific Ocean off the California coast; Pinsac and Murray (1960), and Murray and Harrison (1956), the Gulf of Mexico (Sigsbee Deep and environs); Murray and Sayyab (1955), the Atlantic coast; and Suzuki and Kitazaki (1954) and Oinuma and Kobayashi (1957), the coastal Japanese area. There are a host of other investigations on even more restricted areas including seas, estuarine environments, bays, etc.

In order to obtain general background information for marine sedimentation problems on an oceanic basis, clay-mineral analyses were made on materials covering the entire Pacific Ocean. Although the density of the samples is relatively low, the clay-mineral suites were found not to change markedly from one location to another, and it is felt that the coverage is probably adequate for a general distributional pattern. Many of the samples have been subjected to other analyses: quartz contents (Rex and Goldberg, 1958); chemical composition (Goldberg and Arrhenius, 1958); and rates of sedimentation (Goldberg and Koide, 1962). Most of the materials were collected by expeditions of the Scripps Institution of Oceanography.

[*MS received December, 1960*]

2. The Clay Minerals

The clay minerals were identified by their X-ray basal reflections. Initially, the samples were dispersed in distilled water and the soluble salts removed through a filter cone by repeated washings. The $< 2\ \mu$ size fraction was collected by centrifugation and the clays were pipetted on glass microscope slides and air dried, and these oriented samples were then subjected to X-ray diffraction analysis.

A large number of the samples were preserved continuously under refrigeration after removal from the sea floor and were in a moist condition. In certain cases no moist samples were available and samples dried at room temperature were used. To evaluate the effect of using such samples, controlled experiments were conducted upon the moist materials. One portion was dried at 110°C overnight and then rehydrated and the other left in its original wet condition. The diffraction patterns of these two samples were identical within the limits of error of the X-ray technique.

A. Montmorillonites

A diffraction peak that shifts to 17 Å after ethylene glycol solvation was assigned to the montmorillonite group (Bradley, 1945). This mineral could be a true montmorillonite (low tetrahedral substitution) or an illite and/or chlorite which has its inter-layer cations removed and now sorbs two layers of ethylene glycol. This material would have a high tetrahedral charge. In some cases, and maybe in all, it is possible to have a mixture of the two types either as discrete mineral species or as intergrowths (Greene-Kelly, 1955; Jonas and Thomas, 1960). With glycol–glycerine sorption as the sole criterion, it is often difficult to tell what phases are present (Johns and Tettenhorst, 1959).

B. Illite

Diffraction maxima at 10, 5 and 3.3 Å which do not shift after ethylene glycol solvation are characteristic of the illite group of clay minerals. In the present investigation no attempt was made to distinguish 2*M* micas from the 1*M* and 1*Md* types.

C. Chlorite–Kaolinite

Diffraction maxima at 7 and 3.5 Å were attributed to the chlorite–kaolinite groups. A distinct maximum at 14 or 4.7 Å was used as evidence of chlorite. The presence of kaolinite in those samples which contained chlorite could not be determined with any degree of accuracy. Therefore in such cases the minerals are reported as "chlorite + kaolinite".

D. Halloysite

The electron-microscope pictures of the typical tubular morphology, coupled with an electron diffraction pattern, uniquely determined halloysite.

In order to show trends in the clay-mineral distribution, the ratio of the integrated peak areas of the minerals was used. This tends to compensate for fluctuations in orientation and to a certain degree the effect of variations in the crystallinity. Compensation for the effects of isomorphous substitution in both the illite and chlorite groups of minerals is difficult to make. Most probably this effect will be slight for clay minerals from the marine environment where the major cation concentrations in the water are nearly constant.

To give equal weight to illite when comparing it to montmorillonite (i.e. the M/I ratio), the 10 Å peak area was multiplied by 4. This compensates for the difference in the form factor of the two minerals (Bradley, 1945a; Johns, Grim and Bradley, 1954).

3. Surface Distribution of Clay Minerals

The clay-mineral distribution in the upper few centimeters of cores taken from the Pacific is shown in Figs. 1, 2, 3 and 4. A most dramatic observation is the systematic distributional trends in the North Pacific, whereas there appears to be no obvious pattern in the South Pacific. The equatorial region acts as a natural boundary between these two provinces of clay-mineral distribution. The following generalizations are derived from our experimental data, partly shown in the plots.

A. North Pacific

1. Abundant illite was found in all samples, whereas montmorillonite, chlorite and kaolinite were generally present and their abundances are a function of location.

2. Montmorillonite was generally more abundant in nearshore sediments. The montmorillonite/illite ratio (M/I) is >1 along the coast of North America and Asia; however, the ratio decreases in high latitudes (Fig. 1). Above 50°N the coastal sediment ratio generally has a value <1. Between the coastal area and the mid-ocean areas of the eastern Pacific there is a band of sediment which has M/I ratios varying between 0.5 and 1.0. This band is not defined off the Asiatic coast. In the mid-oceanic areas, the ratio is generally quite low and has values lying between 0 and 0.5.

3. Chlorite abundance, based on the presence of the 14 and 4.7 Å peaks, increased with increasing latitude in nearshore sediments (Fig. 2). However, the relatively high abundances in coastal sediments off Northern Canada, Alaska and the Aleutian Islands decrease with increasing distance from land.

4. Kaolinite abundance, which was based on the presence of the 7.1 and 3.5 Å peaks and the absence of the 14 and 4.7 Å peaks, was confined to nearshore areas on both the east and west coasts. However, with increasing distance from land, it is difficult to evaluate the kaolinite abundances because the 14 and 4.7 Å chlorite peaks begin to appear in the diffraction tracings.

5. The 7.1/10 Å peak-area ratios (chlorite + kaolinite)/(illite) show a correspondence to the M/I ratio (Fig. 3). The ratio is usually <1 along the eastern

coastal area up to 50°N at which point it becomes >1. The $(C+K)/I$ ratio decreases from values around 1 along the coast of Canada, Alaska and the Aleutian Islands to 0.5–1.0 in Mid-Pacific waters. The 3.5/3.3 Å peak-area ratio (chlorite + kaolinite)/(illite) and the 7.1/10 Å were quite similar; however, both ratios were complicated by kaolinite and the former ratio was further complicated by quartz (Fig. 4). In order to eliminate these interferences the 4.7/5.0 Å ratios were determined (Fig. 2). This plot showed a gradual increase in the ratio with increasing latitude in coastal waters. There is also a decrease in the ratio from the coastal waters off Northern Canada, Alaska and the Aleutians where the ratio is approximately 1.0 to mid-ocean where the ratio is about 0.5. The similarity between the 7.1/10, 3.5/3.3 and 4.5/5.0 Å peak-area ratios indicates that chlorite is the varying parameter, not kaolinite. However, the influence of increasing amounts of illite in mid-ocean areas cannot be overlooked and may exert an influence in decreasing the chlorite/illite ratio. Isomorphous substitution is thought to have little effect since the observed trend is noted in all three peaks, 7.1, 4.7 and 3.5 Å. If this effect was present it would show in one of these peak ratios unless this substitution was extremely complex.

B. South Pacific

1. Montmorillonite was the most abundant clay mineral in those samples where clay minerals were present. The M/I ratio was always >1 and no obvious variation in the ratio with either latitude or longitude was observed (Fig. 1).

2. Illite was generally present but in much smaller quantities than in the North Pacific.

3. Kaolinite was present but it was difficult to evaluate its abundance because the 7.1 Å peak of kaolinite coincided with one of the strong peaks of the zeolite phillipsite. The 4.7 Å peak of chlorite was found in the South Pacific in only a few samples (Fig. 2).

4. The 3.5/3.3 Å ratio was variable but with high values clustered near land areas.

5. In the area between 100–140°W and from the equator to 50°S, which includes the East Pacific Rise, the highly ferruginous and calcareous cores have no detectable clay minerals in the calcium-carbonate-free residue. However, the residue further treated to remove iron oxide (Mehra and Jackson, 1960) contained trace amounts of montmorillonite, illite and chlorite.

The factors regulating these surface-clay-mineral assemblages can best be approached by some general considerations of marine sedimentation. It is evident from an inspection of maps of the Pacific area that the North Pacific borderlands have more rivers emptying into the ocean and that these rivers have larger drainage basins than do the South Pacific borderlands. Furthermore, the North Pacific Ocean has a larger ratio of land to water area. Quantitative estimates of river-water influx to oceanic areas, based on river lengths, drainage areas, annual rainfalls of land, etc., certainly would enhance our understanding; yet, because the necessary data are either lacking or uncertain in accuracy,

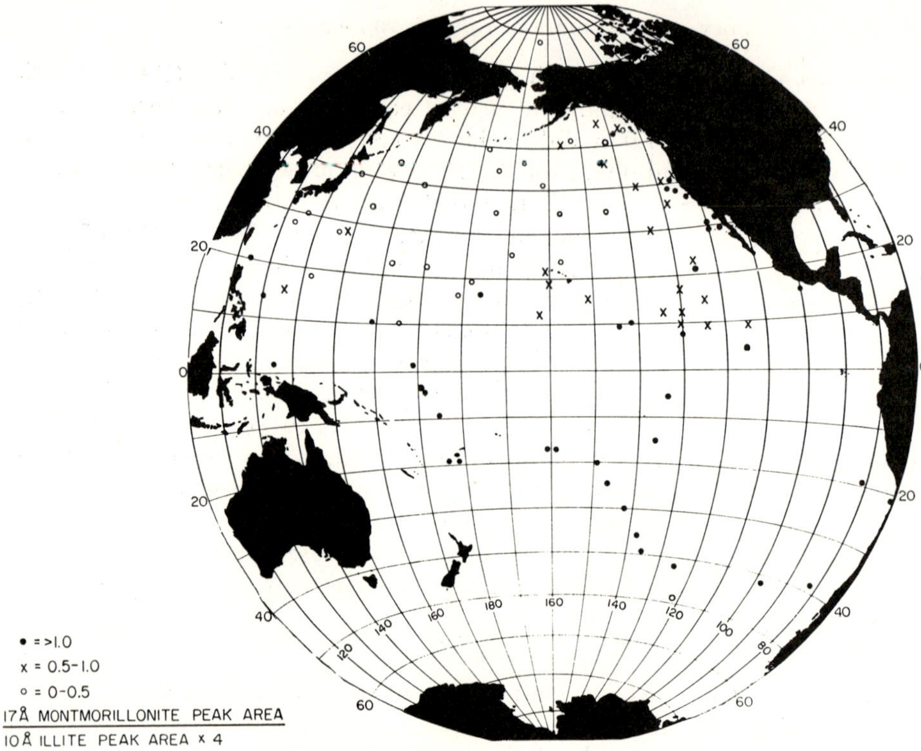

Fig. 1. The 17 Å montmorillonite/10 Å illite ratio.

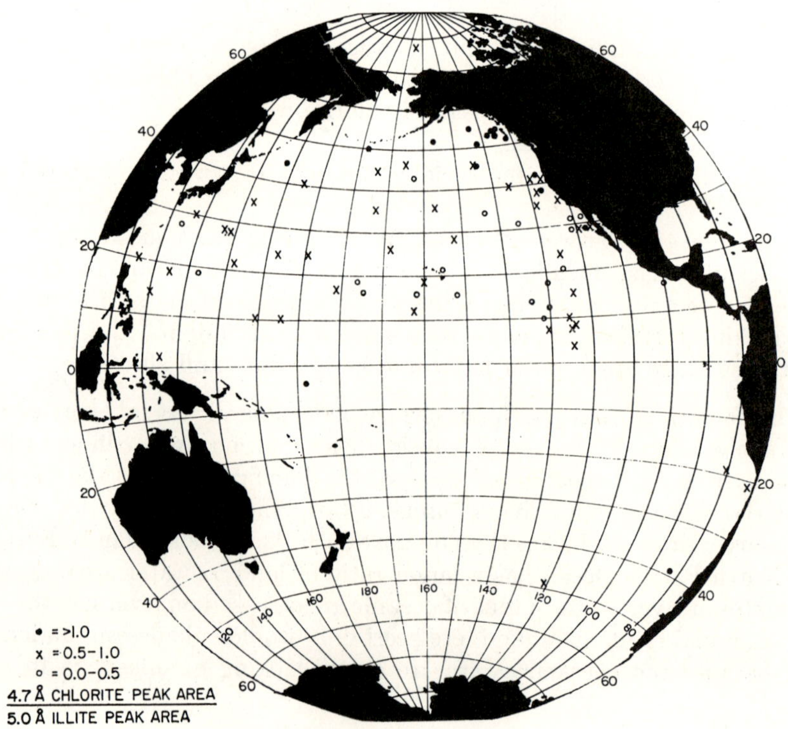

Fig. 2. The 4.7 Å chlorite/5.0 Å illite ratio.

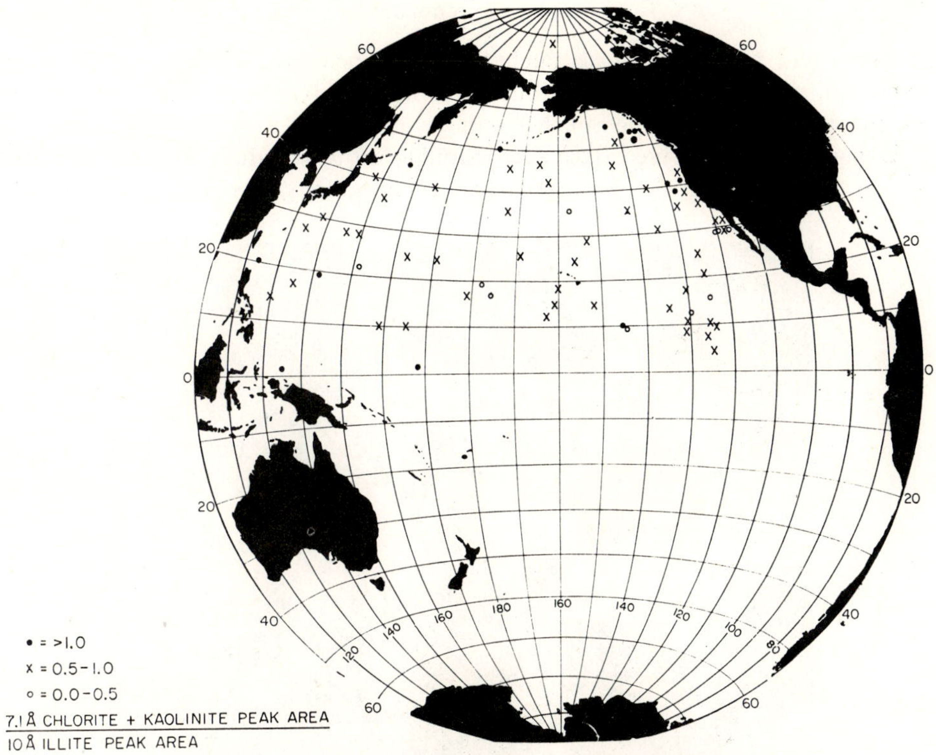

Fig. 3. The 7.1 Å chlorite + kaolinite/10 Å illite ratio.

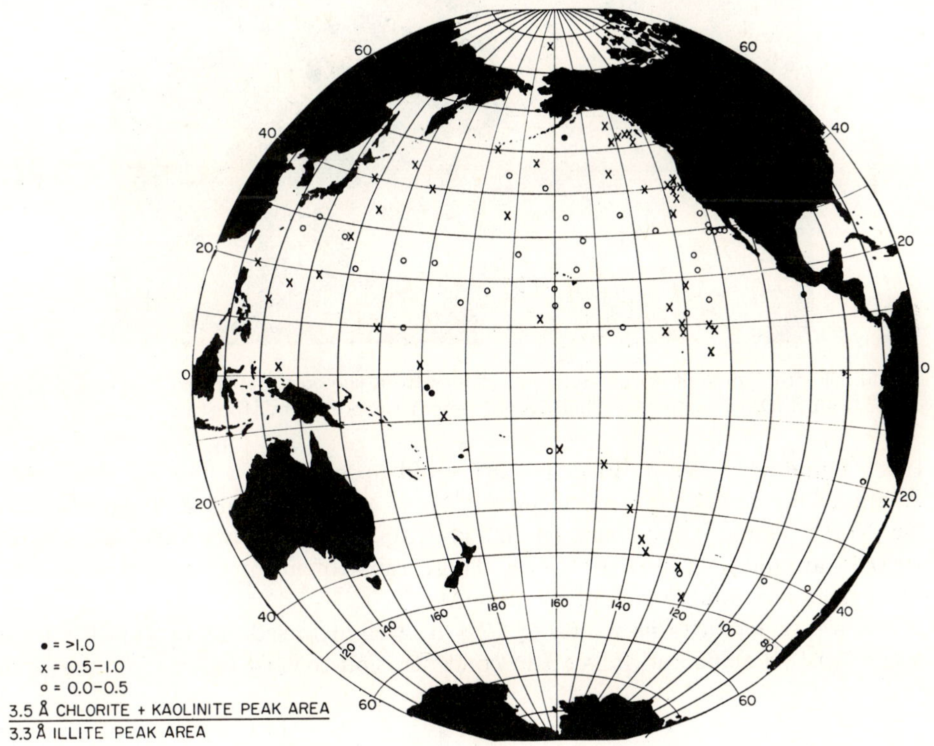

Fig. 4. The 3.5 Å chlorite + kaolinite/3.3 Å illite ratio.

recourse will not be made to such considerations. Nonetheless, one *a priori* expects a greater relative contribution of land-derived material to the North Pacific as compared to the South Pacific.

Experimentally derived data give strength to this deduction. Goldberg and Koide (1962) using the ionium/thorium chronology (Goldberg and Koide,

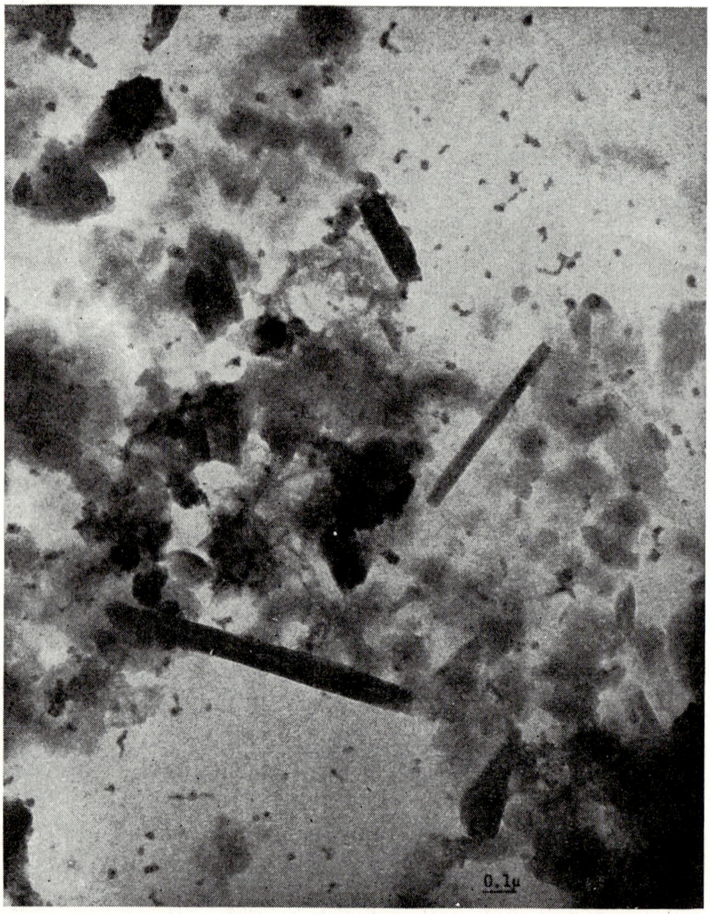

Fig. 5. Electron micrograph of halloysite tubes from the core Capricorn BG 9, 19° 01′S, 177° 46′E. (By courtesy of Shell Development Corporation, Houston, Texas.)

1958) found a rather uniform rate of build-up of the sea floor in the South Pacific of 0.4 mm/10^3 years, whereas values of the rate of sedimentation in the North Pacific vary between 0.5 and 10 mm/10^3 years with an overall average around 1–2 mm/10^3 years. These rates of accumulation are for the non-calcareous components of the deposits.

Detrital quartz attains values of 25% by weight in sediments from the mid-ocean North Pacific and falls off in concentration toward polar and equatorial latitudes (Rex and Goldberg, 1958). The South Pacific is essentially devoid of quartz with abundances rarely exceeding 6% in mid-oceanic areas. These authors attribute the latitudinal dependence to an eolian path for the quartz particles from the arid regions of the continents which have their greatest areal extents in the mid-latitudes of the North Pacific.

Hurley *et al.* (1959) point out that the potassium–argon ages of illites taken from the surface sediments of the North Pacific are of the order of hundreds of millions of years old, clearly suggesting a continental source for these clay minerals.

Finally, the more common occurrences of authigenic minerals, the zeolite phillipsite and the ferromanganese minerals, in sediments from the South Pacific (Goldberg and Arrhenius, 1958) emphasize the significance of *in situ* mineral formation here. The finding of the large area, the East Pacific Rise, almost free of any detrital minerals is added evidence for a lack of land-derived materials in these sediments.

Thus, an entry into the understanding of clay-mineral distributions can be initiated with considerations of source areas in the North Pacific and diagenetic processes in the South Pacific. The complexity of the North Pacific clay-mineral assemblages suggests a number of source areas. On the other hand, the rather homogeneous distribution of clay minerals in the South Pacific, combined with the uniformity in chemical and physical properties of near-bottom waters, is not in conflict with a primarily authigenic origin of these materials.

The formation of montmorillonite from volcanic ash is well documented in the literature (Ross and Hendricks, 1945) and it is quite probable that such an alteration occurs in this area. Furthermore, a few samples examined by electron microscopy and diffraction showed halloysite tubes (Fig. 5), which are found as weathering products of volcanic debris on land.

The source areas for the clay minerals in the North Pacific may be derived by studying their concentration gradients (Fig. 6). Increasing chlorite abundances tend toward the coasts of Canada, Alaska and the Aleutian Islands. Increasing montmorillonite concentrations approach the coastal regions of the east and west Pacific. The implication of source areas for these two minerals from such plots is clear. On the other hand, illite concentrates in mid-ocean areas. This pattern, similar to the quartz band but more diffuse, may well result from an eolian origin from the arid regions of the world for illite. However, because of the much smaller grain size of the clay minerals ($< 2\ \mu$) than the quartz (3–10 μ), the illite would be subject to a greater dispersion by both atmospheric and oceanic currents.

The 17 Å material found in coastal sediments of the North Pacific may well contain contributions of expandable illites and chlorites. In continental environments where proton attack, oxidation and hydration reactions occur, K^+ and Mg^{2+} can be removed from the inter-layer structures of the illites and chlorites respectively. These stripped clays react like montmorillonite and expand to 17 Å under solvation with ethylene glycol. Because these stripped illites and chlorites expand to 17 Å in glycol, they have been identified as montmorillonite whereas they are distinctly different from true montmorillonite. A true montmorillonite has its excess negative charge located in the octahedral layer and little, if any, negative charge in the tetrahedral layer. The micas, on the other hand, will have a high net negative charge in the tetrahedral layer. Furthermore, the net charge on the montmorillonite will be much smaller than that on the micas.

So it would be expected that the stripped illite and chlorite would upon entry into a proper environment revert back to better crystalline mica and

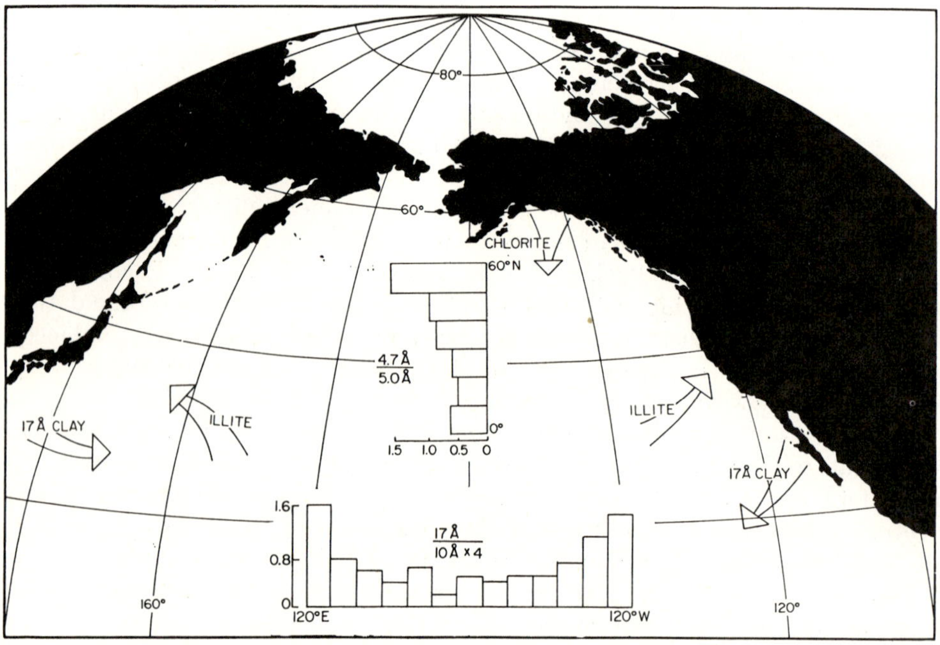

Fig. 6. Concentration gradients of clay minerals in the North Pacific.

chlorite by sorption and "fixation" of K^+ and Mg^{2+}. The phenomenon of K^+ "fixation" is well established and a voluminous literature on the subject has been developed by soil scientists (Volk, 1938; Barshad, 1954). Recently, Weaver (1958) used the phenomenon to differentiate montmorillonite from stripped micas. He showed that the montmorillonite would not "fix" K^+ and then collapse, whereas stripped micas would readily "fix" K^+ and collapse.

An examination of the 17 Å peak from diffraction tracings taken from the surface sediment in the east to west traverse across the North Pacific lends support to the hypothesis that the 17 Å material is sorbing K^+ and Mg^{2+} and collapsing. In nearshore sediments the 17 Å peak is sharp. However, further from land, the peak broadens and loses intensity because of random interlayer collapse. In the mid-oceanic areas the peaks disappear (Fig. 7).

At first glance this striking correlation of the 17 Å peak broadening and decreasing in intensity with concomitant increase in intensity and resolution of the illite and chlorite peaks would lead one to believe that the 17 Å material was transforming to a well crystallized illite and chlorite. Undoubtedly this is one of the factors that is influencing the clay-mineral distribution pattern in the

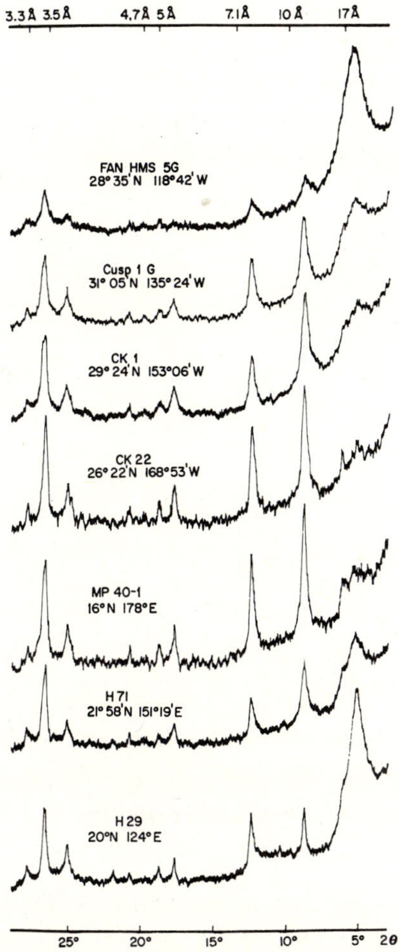

Fig. 7. X-ray diffraction tracings of the ethylene-glycol-treated ($< 2\mu$) clay fraction in the surface sediments in a traverse across the North Pacific Ocean.

North Pacific. However, a closer look at the data will show that the sediment along the coast of Canada, Alaska and the Aleutian Islands is high in chlorite and illite and deficient in montmorillonite. Transport of these sediments to mid-oceanic areas would also tend to give the observed increase of illite and chlorite with increasing distance from land. Furthermore, the fall-out of air-borne material from continental areas as postulated by Rex and Goldberg (1958) no doubt becomes important in the sediments of the mid-oceanic areas.

One permutation upon the source area pattern is the introduction of Tertiary sediments to the surface deposits. This phenomenon is well known in certain parts of the Pacific (Riedel, 1957) and may account for anomalously high montmorillonite concentrations in certain North Pacific areas. The following section develops evidence that at least some of the depositional parameters in the North Pacific during the Tertiary were quite similar to present-day parameters in the South Pacific.

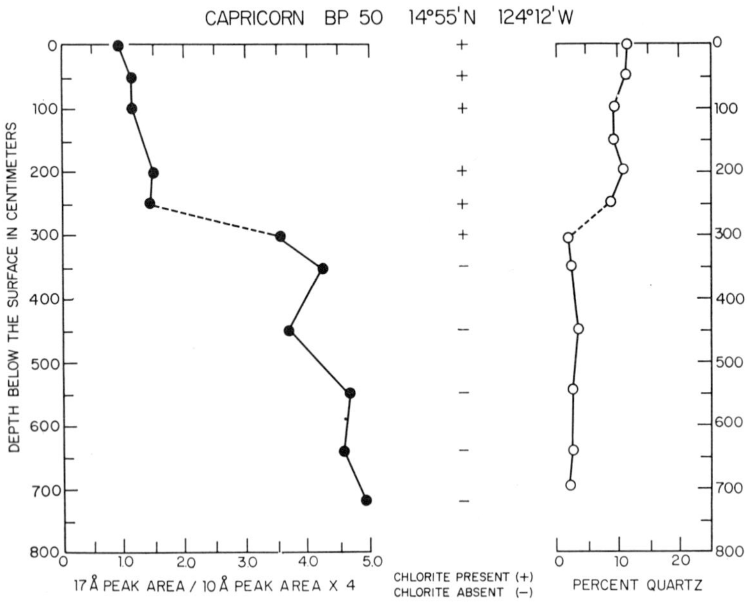

Fig. 8. The quartz, M/I and chlorite values in Capricorn 50 BP as a function of depth.

4. Vertical Distribution

A. North Pacific

The distribution of clay minerals in the North Pacific falls into two categories of variation versus depth profiles:

(1) There is a continuous decrease in the M/I ratio with depth in a few cores taken very close to shore. A concomitant increase in the 4.7/5.0 Å ratio and a decrease in the 7.1/10 Å ratio was observed. The 3.5/3.3 Å ratio remained constant with depth.

(2) The M/I ratio shows abrupt changes with depth where Tertiary sediments are encountered. In Capricorn 50 BP, the typical M/I ratio discontinuity, due to sediment passing downwards from the Recent to the Tertiary, is accompanied by large discontinuities in the 4.7/5.0 and 3.5/3.3 Å ratios. Below the discontinuity the 10, 5.0, 4.7, 3.5 and 3.3 Å peaks were nearly or completely absent (Fig. 8), whereas the 17 Å peak increased in intensity.

The first group includes those sediments deposited very rapidly which apparently do not equilibrate with sea-water before burial. However, the majority of cores which penetrate only Recent sediments show no variation in the vertical distribution of clay minerals.

B. South Pacific

The few cores examined from the South Pacific showed a uniform vertical distribution. The M/I ratio was uniform through the depth of sampling. Like the Tertiary sediments in the North Pacific, the 5.0, 4.7, 3.5 and 3.3 Å peaks were missing or too diffuse so that their ratios were not amenable to interpretation. The 7.1/10 Å ratio could not be used because of the interference of phillipsite.

It is interesting to note that a core taken from an area in the East Pacific Rise contained no appreciable clay minerals either at the surface or at depth. Samples taken through a length of 9 m did not reveal any indication of clay minerals in diffraction patterns for either the untreated sample or those free of carbonate. Only after removal of iron oxide did any evidence of clay minerals become apparent and then only in trace amounts.

The homogeneous distribution of clay minerals with depth in any given core emphasizes the stability of clay minerals in the marine environment at least over periods of hundreds of thousands of years. Certainly, the burrowing action of marine organisms may have homogenized the sediments so that subtle changes in composition are smoothed out or are not detectable. Further, it is evident that the physical-chemical environment parameters have not fluctuated with time and that the supply of sediments from the source areas have remained constant within the Recent.

The older Tertiary sediments from the North Pacific are distinctly similar to the present-day South Pacific ones. The chemical composition (Goldberg and Arrhenius, 1958), the quartz contents (Rex and Goldberg, 1958) and the clay minerals make it inviting to suggest that the North Pacific during Tertiary times was similar as a depositional environment to the South Pacific today, i.e. an area that receives small amounts of detrital minerals and large amounts of pyroclastics. Ross (1955) reported that during the Tertiary Period, the western U.S. received thousands of cubic miles of pyroclastics. No doubt large quantities of this material were ejected into the atmosphere and transported by winds for great distances. If the wind patterns were directed to the west, the North Pacific would receive large quantities of fall-out. With time this fall-out would lose its volcanic identity by reacting with sea-water and altering to montmorillonite or other more stable minerals.

5. Summary

The clay-mineral assemblages taken from the surfaces of the marine deposits of the Pacific Ocean can be interpreted on the basis of both terrestrial sources and authigenic formation. The mid-North Pacific shows marked concentrations of illite, normally associated with high quartz contents. Both of these components are presumed to originate from the arid regions of the earth and are primarily brought to the oceanic environment by eolian paths. The North Pacific coastal deposits show an enrichment in a "montmorillonite type" mineral which shows evidence of being composed mostly of expandable stripped illite or chlorite. Such clays probably have as their source the coastal continental areas and are mainly transported to the oceans by water paths. Finally, there is a latitudinal decrease in chlorite going from high latitude regions southward, apparently resulting from a chlorite source in the high latitudinal land areas.

The South Pacific surface samples show little terrestrial influence with montmorillonite being the predominant clay mineral. Apparently this clay arises from the alteration of volcanic materials. In strong association with it is often the zeolite phillipsite. These South Pacific sediments are slowly accumulating (0.4 mm/10^3 year on a carbonate-free basis) as compared with their northern mid-ocean counterparts (1–2 mm/10^3 years) which receive large amounts of terrestrial materials.

The vertical distribution of clay minerals in Recent sediments of the North Pacific is essentially uniform, although a few near-coastal cores show a continuous decrease in montmorillonite relative to illite with depth. The Tertiary deposits of the mid-North Pacific are enriched in montmorillonite and phillipsite and resemble both mineralogically and chemically those of the present-day South Pacific. The South Pacific cores show no recognizable changes in clay-mineral composition with depth in the sedimentary column.

References

Barshad, I., 1954. Cation exchange in micaceous minerals. II Replaceability of ammonia and potassium from vermiculite, biotite and montmorillonite. *Soil Sci.*, **78**, 57–76.

Bates, T. F., F. A. Hildebrand and A. Swineford, 1950. Morphology and structure of endellite and halloysite. *Amer. Mineralogist*, **35**, 463–484.

Bradley, W. F., 1945. Molecular dissociation between montmorillonite and some polyfunctional groups. *J. Amer. Chem. Soc.*, **67**, 975–981.

Bradley, W. F., 1945a. Diagnostic criteria for clay minerals. *Amer. Mineralogist*, **30**, 704–784.

Correns, C. W., 1937. Die Sedimente des äquatorialen Atlantischen Ozeans. *Wiss. Ergebn. Deut. Atlant. Exped. 'Meteor', 1925–1927*, **3**, Pt. 3.

Goldberg, E. D. and G. Arrhenius, 1958. Chemistry of Pacific pelagic sediments. *Geochim. et Cosmochim. Acta*, **13**, 153–212.

Goldberg, E. D. and M. Koide, 1958. Ionium–thorium chronology in deep sea sediments of the Pacific. *Science*, **128**, 1003.

Goldberg, E. D. and M. Koide, 1962. Geochronological studies of deep-sea sediments by the ionium/thorium method. *Geochim. et Cosmochim. Acta*, **26**, 417–450.

Greene-Kelly, R., 1955. Dehydration of the montmorillonite minerals. *Mineral. Mag.*, **30**, 604–615.

Grim, R. E., 1953. *Clay Mineralogy*. McGraw-Hill, New York, 384 pp.

Grim, R. E., R. S. Dietz and W. F. Bradley, 1949. Clay mineral composition of some sediments from the Pacific Ocean off the California coast and the Gulf of California. *Bull. Geol. Soc. Amer.*, **60**, 1785–1808.

Hurley, P. M., S. R. Hart, W. H. Pinson and H. W. Fairbairn, 1959. Authigenic versus detrital illite in sediments. *Bull. Geol. Soc. Amer.*, **70**, 1622 (Abstract).

Jonas, E. C. and G. L. Thomas, 1960. Hydration properties of potassium-deficient clay micas. *Clays and Clay Minerals, Proc. Eighth Nat. Conf.*, 183–192.

Johns, W. D., R. E. Grim and W. F. Bradley, 1954. Quantitative estimates of clay minerals by diffraction methods. *J. Sediment. Petrol.*, **24**, 242–251.

Johns, W. D. and R. T. Tettenhorst, 1959. Differences in the montmorillonite solvating ability of polar liquids. *Amer. Mineralogist*, **44**, 894–896.

MacEwan, D. M. C., 1944. Identification of the montmorillonite group of minerals by X-rays. *Nature*, **154**, 577–578.

Mehra, O. P. and M. L. Jackson, 1960. Iron oxide removal from soils and clays by a dithionite-citrate system buffered with sodium bicarbonate. *Clays and Clay Minerals*, NAS-NRC, **5**, 317–327.

Murray, H. H. and J. L. Harrison, 1956. Clay mineral composition of recent sediments from Sigsbee Deep. *J. Sediment. Petrol.*, **26**, 363–368.

Murray, H. H. and A. S. Sayyab, 1955. Clay mineral studies of some Recent marine sediments off the North Carolina coast. *Clay and Clay Minerals, NAS-NRC Pub.* 395, 430–441.

Oinuma, I. and K. Kobayashi, 1957. Mineralogical study of marine bottom sediment. Pt. 1. *Jap. Hydrog. Reps.*, **54**, 8–38.

Pinsac, A. P. and H. H. Murray, 1960. Regional clay mineral patterns in the Gulf of Mexico. *Clays and Clay Minerals, Proc. Seventh Nat. Conf.*, 162–177.

Revelle, R. R., 1944. Marine bottom samples collected in the Pacific Ocean by the *Carnegie* on its seventh cruise. *Carnegie Inst. Wash. Pub.*, 556, 1–180.

Rex, R. W. and E. D. Goldberg, 1958. Quartz contents of pelagic sediments of the Pacific Ocean. *Tellus*, **10**, 153–159.

Riedel, W. R., 1957. Radiolaria: A preliminary stratigraphy. *Reps. Swed. Deep-Sea Exped.*, **6**, 59–96.

Ross, C. S., 1955. Provenience of pyroclastic materials. *Bull. Geol. Soc. Amer.*, **66**, 427–434.

Ross, C. S. and S. B. Hendricks, 1945. Minerals of the montmorillonite group. *U.S. Geol. Surv. Prof. Paper*, **205B**, 1–77.

Suzuki, K. and U. Kitazaki, 1954. Mineralogical studies on the red clays from the western Pacific Ocean, east of the Bonin Islands. Studies on the clay mineral fractions of the recent marine sediments. *Jap. J. Geol. Geog.*, **24**, 171–180.

Volk, G., 1938. Nature of potassium fixation in soils. *Soil Sci.*, **45**, 263–276.

Weaver, C. E., 1958. The effects and geological significance of potassium "fixation" by expandable clay minerals derived from muscovite, biotite, chlorite and volcanic materials. *Amer. Mineralogist*, **43**, 839–861.

Zen, E.-An, 1957. Preliminary report on the mineralogy and petrology of some marine bottom samples off the coast of Peru and Chile. *Amer. Mineralogist*, **42**, 889–903.

27. TURBIDITY CURRENTS[1]

Bruce C. Heezen

1. Early Views

The idea that dense flows of sediment–water mixtures could be of importance in creating subaquatic morphological features, and in transporting sediment, was first suggested by F. A. Forel in 1885. Forel, who had studied the morphology of the bottom of Lake Léman, discovered that there was a large sublacustrine canyon which ran from the vicinity of the mouth of the upper Rhône River to the flatter floor of the lake. He deduced that repeated flows of watery sediment, emanating from the river, had created the canyon, partially by erosion, but largely by deposition along the lateral margins of the bottom-seeking current.

This idea was resurrected and championed again by R. A. Daly, whose classic paper published in 1936 encouraged a number of investigations which ultimately led to the verification of turbidity currents as an important submarine and sublacustrine phenomenon. Daly proposed that, during the glacial stages, storm waves breaking on beaches created mixtures of sediment and water which flowed beneath the sea, eroding submarine canyons into the continental slopes.

At the time Daly made this suggestion there was no experimental evidence for turbidity currents, and although considerable detail was known concerning the morphology of the submarine canyons (largely owing to excellent surveys of the United States Coast and Geodetic Survey) little else was known concerning the canyons. It was, however, known, through the work of Stetson and others, that rock cropped out in the canyons. The canyons seemed definitely to have been eroded into the continental slope.

Kuenen, stimulated by Daly's paper, conducted tank experiments in which he was able to model turbidity currents. Kuenen demonstrated that mixtures of clay, silt and sand introduced at one end of a tank would indeed flow down the sloping tank floor beneath the clear water and deposit a layer of sediment. However, he was unable to demonstrate erosion by such currents.

Support also came from the work done in reservoirs. Grover and Howard (1938) had observed turbid underflows in Lake Mead Reservoir and had discovered that the bottom of the lake immediately behind the dam was being flattened by the deposits of these bottom-seeking sediment flows. However, the density contrast between these weak turbidity currents and the lake water was small, and the velocity of the currents did not exceed a small fraction of a mile per hour.

It was generally thought that although turbidity currents may be of local importance in the ocean, they probably do not account for submarine canyons,

[1] Lamont Geological Observatory Contribution No. 599.

[*MS received May, 1961*]

for it is difficult to believe that halting flows of muddy water could cut rock-walled, giant canyons into the continental slope.

Stetson and Smith (1937) believed that the density contrast between turbidity currents and the superadjacent water was so small that a marine turbidity flow would leave the bottom and flow as a turbid interflow when it reached a depth where the density of the water exceeded that of the turbidity current. However, later experiments by Kuenen, in which he used higher density mixtures, yielded much greater velocities and suggested that erosion could be accomplished by turbidity currents.

Although high density turbidity currents had been effectively demonstrated in Kuenen's later experiments, scepticism and lack of confidence in scale factors prevented a general acceptance of the process when projected to full scale. The concept of low-density turbidity currents which leave the bottom to form interflows was also adopted by Menard and Ludwick (1951). Although they concluded that submarine landslides "are the only likely sources of dense turbidity currents thus far suggested", they had serious reservations, and felt that such turbidity currents resulting from slides "may be tenuous rather than dense". They also concluded that "turbidity currents resulting from suspended sediment in rivers must be rare". Concerning the generation of turbidity currents along the shore they concluded:

"If they exist, marine turbidity currents produced by wave agitations, or any other stirring agent, probably do not have high effective density....

"No means exist for holding the current in place until the suspended sediment becomes very concentrated; therefore, it is unlikely that very dense turbidity currents can be generated even by violent wave and current agitation....

"...turbidity currents may flow on the ocean surface or within the ocean due to marine density stratification.... Surface flows and interflows may be important geologically because they can spread sediment over wide areas."

For this reason they conclude that "marine turbidity current deposits may be difficult to identify".

The concept thus projected was not appealing and most oceanographers and geologists could not believe that turbidity currents were an important process in oceans, and many completely rejected the concept.

2. Full-Scale Experiment

The case for modern high-density, high-velocity turbidity currents was finally proved by a study involving the explanation of both the breakage of submarine cables and the origin of deep-sea sands.

A. Grand Banks Turbidity Current

In 1952 Heezen and Ewing concluded that the "...events associated with the Grand Banks earthquake of 1929 November 18 may be considered as a

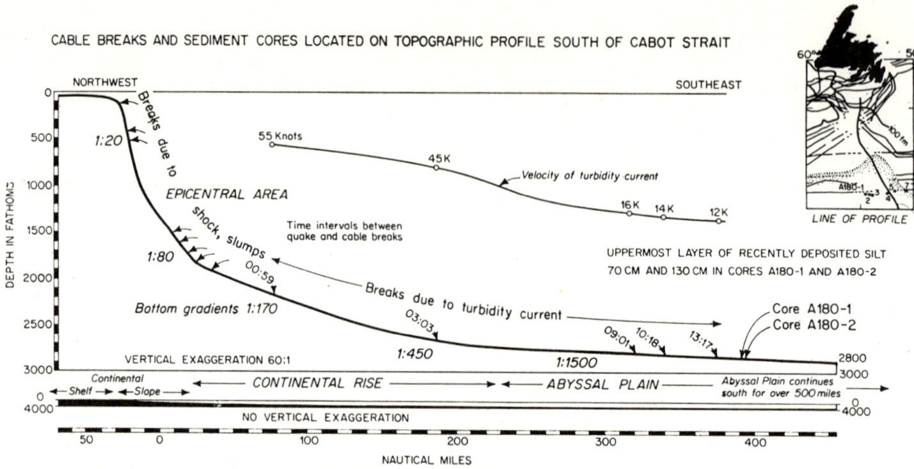

Fig. 1. Cable breaks and sediment cores located on the topographic profile south of the Cabot Strait, with superimposed graph of velocity of the Grand Banks turbidity current. (After Heezen and Ewing, 1952.)

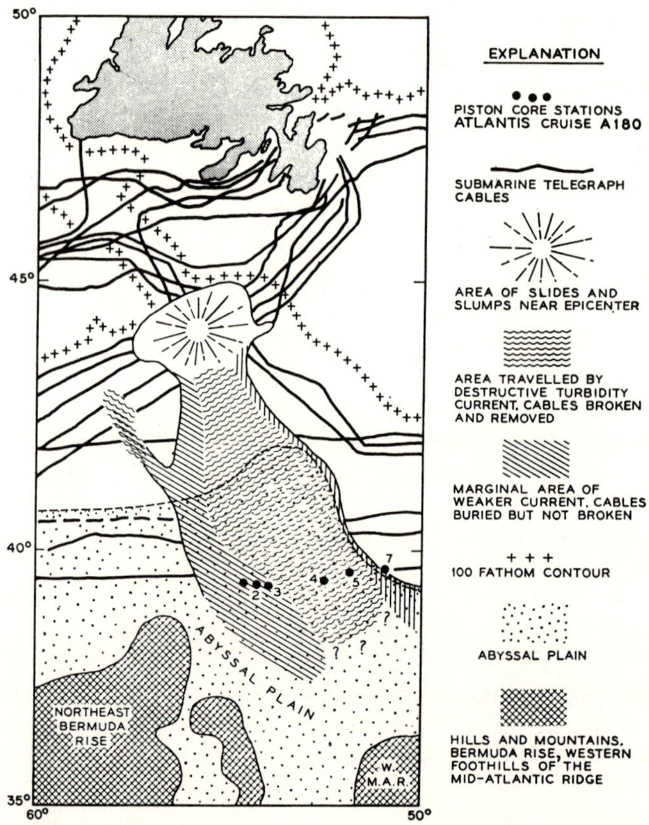

Fig. 2. Map of the 1929 Grand Banks turbidity current. (After Heezen, Ericson and Ewing, 1954.)

full-scale experiment in erosion, transportation, and deposition of marine sediments by a turbidity current in which the submarine telegraph cables served to measure its progress and to give tangible evidence of its force".

Following this earthquake, which shook the continental slope south of Newfoundland, all the submarine telegraph cables lying downslope (south) of the epicentral area were broken in sequence from north to south (Fig. 1). According to Heezen and Ewing's explanation each successive cable was broken by a turbidity current which had originated as a series of slumps in the epicentral

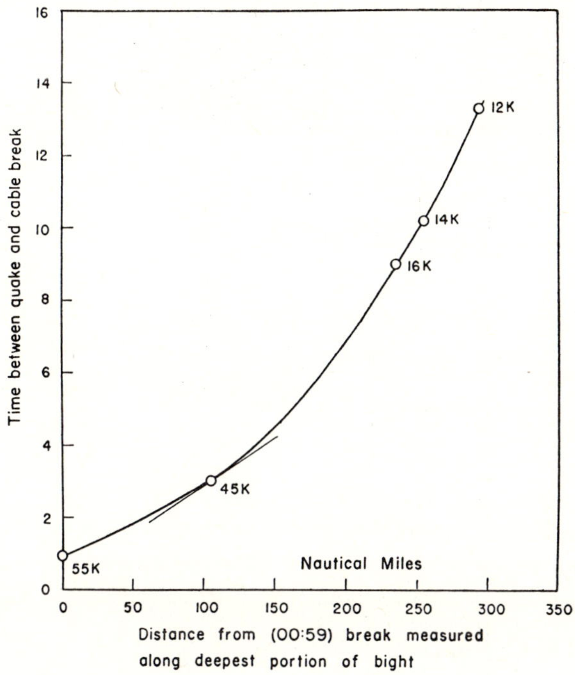

Fig. 3. Time–distance graph of the 1929 Grand Banks turbidity current. (After Heezen and Ewing, 1952.)

area. The current travelled down and across the continental slope, continental rise and ocean basin floor, and, continuing far out onto the abyssal plain, reached a limit well over 450 miles from its source area on the continental slope (Fig. 2).

On the continental slope where the bottom gradient is 1:10 to 1:30 the velocity of the current exceeded 50 knots; and out on the abyssal plain where the gradient is less than 1:1500 its velocity, although diminished, still exceeded 12 knots (Fig. 3). Subsequent exploration of the area with the piston corer revealed a one-meter thick, uppermost bed of graded silt containing shallow-water microfossils (Heezen, Ericson and Ewing, 1954) overlaying abyssal lutite

(Figs. 4 and 5). The thickness of the bed was nearly identical with that calculated by Kuenen (1952) on the basis of the slopes and observed velocities.

This bold challenge to the concept of the motionless, pristine abyss was immediately attacked. Shepard (1954) based his objections largely on the claim

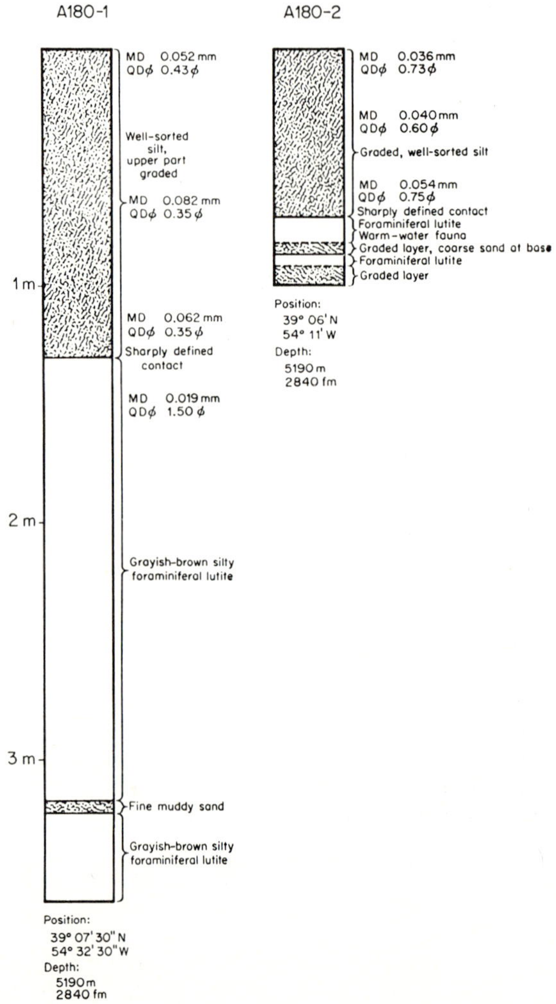

Fig. 4. Logs of cores 180-1 and 180-2 from the Sohm Abyssal Plain south of the cable break area. Note the upper layer of graded silt in both cores, deposited by the 1929 Grand Banks turbidity current. (After Heezen, Ericson and Ewing, 1954.)

that the velocities of 10–50 knots were larger than could be expected, but offered no more serious objections. Kullenberg (1954), misled by some erroneous soundings, claimed that Heezen and Ewing's explanation required the turbidity current to run uphill and over a range of hills, an ability which, he correctly

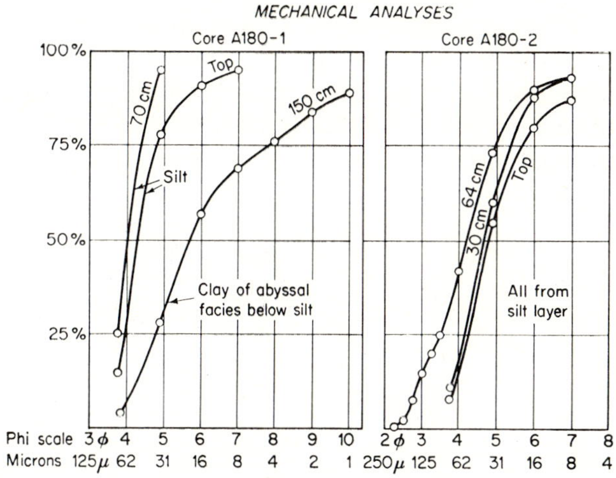

Fig. 5. A mechanical analysis of cores 180-1 and 180-2. Note the well defined particle grading in both cores. Also note the sharply contrasting particle size distribution in the abyssal clay facies at 150 cm in core 180-1. (After Heezen, Ericson and Ewing, 1954.)

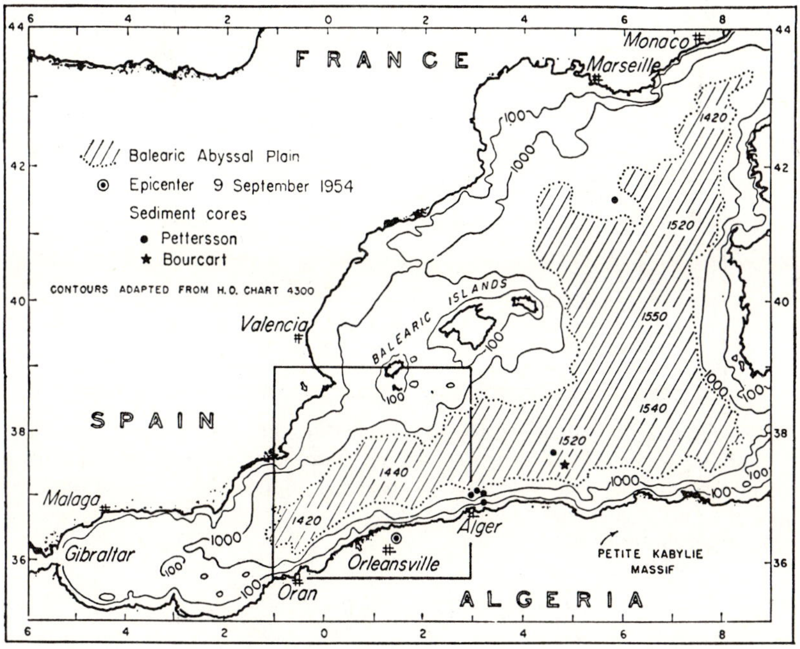

Fig. 6. Generalized bathymetric map of the western Mediterranean Sea, showing the area in which submarine cables were broken on the Balearic Abyssal Plain, following the 1954 Orleansville earthquake. Depths in fathoms. (After Heezen and Ewing, 1955.)

pointed out, could not be attributed to turbidity currents. However, subsequent exploration confirmed Heezen and Ewing's belief that the alleged mountain range was the result of poor contouring of bad soundings. In fact, a survey by Heezen, Ewing and Ericson (1954) revealed nothing but a smooth abyssal plain in the indicated position of Kullenberg's mountains.

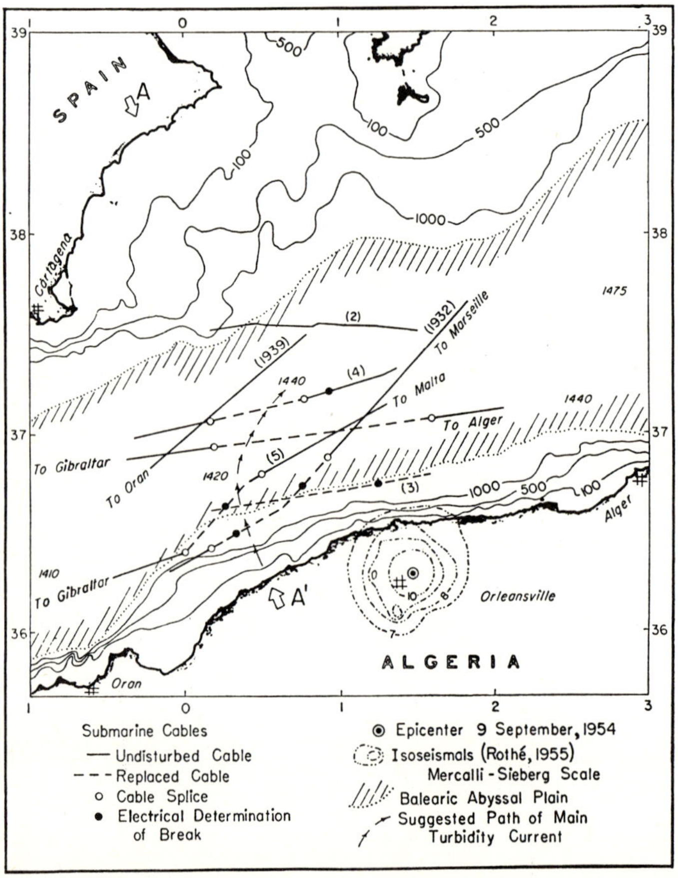

Fig. 7. Map showing locations of Orleansville earthquake, path of main turbidity current and submarine cables. Depths in fathoms. Note that the northernmost cables were not interrupted by the turbidity current. It is probable that several coalescing turbidity currents joined to produce the pattern of breaks indicated on this map. (After Heezen and Ewing, 1955.)

B. Orleansville Turbidity Currents

A series of cable failures on the floor of the western Mediterranean (Figs. 6, 7 and 8), essentially similar in distribution and order to the Grand Banks sequence, followed the 1954 Orleansville earthquake (Heezen and Ewing, 1955). These events further support the interpretation of the Grand Banks turbidity current and provide additional evidence of the mode of emplacement of the Mediterranean deep-sea sands which Bourcart (1953) had previously interpreted as turbidity-current deposits.

Bourcart and Glangeaud (1956, 1958), however, objected to the interpretation claiming that turbidity currents had broken only the cables on the steep slopes. To explain the later breaks out on the nearly level ocean floor they suggested that undetected or at least unlocated aftershocks had caused the breaks.

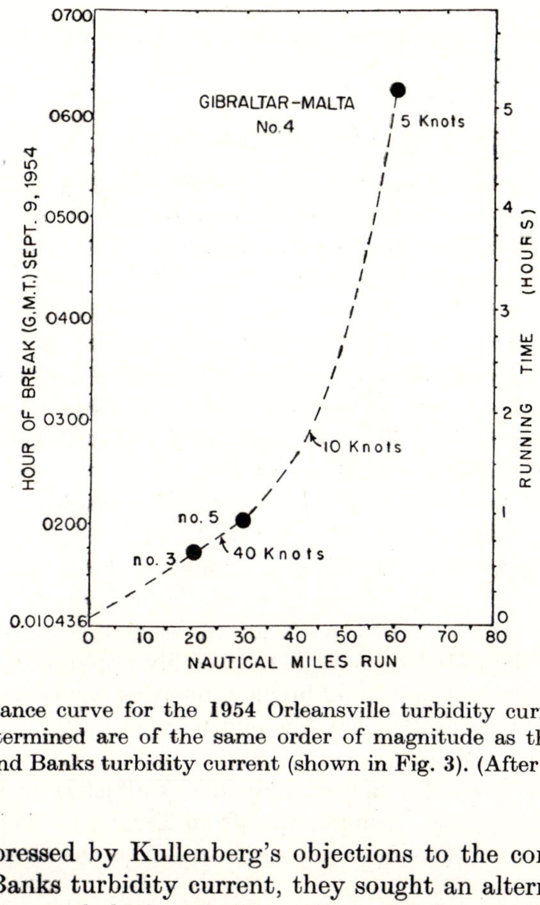

Fig. 8. Time–distance curve for the 1954 Orleansville turbidity current. Note that the velocities determined are of the same order of magnitude as those determined for the 1929 Grand Banks turbidity current (shown in Fig. 3). (After Heezen and Ewing, 1955.)

Apparently impressed by Kullenberg's objections to the computed velocities for the Grand Banks turbidity current, they sought an alternative interpretation not requiring such high velocities. Also misled by bad soundings, they denied the existence of an abyssal plain in the area. Precision soundings, however, have confirmed the existence of the Balearic Abyssal Plain.

3. Alternative Explanations

A number of novel explanations had been offered in an attempt to provide an alternate explanation of the events following the Grand Banks earthquake.

Shepard (1954) now admits that turbidity currents transport sediments, but denies that they have the ability to erode. Unable to accept the high velocities indicated by the Grand Banks sequence of cable breaks, Shepard (1954) suggested that an unrelated series of local and independent submarine landslides had broken the cables. He wrote: "Is it not possible that slides developed more slowly on the gentle outer slopes and built up pressure gradually against the cables until they broke? . . . otherwise the sequence could be explained as at least in part a *coincidence* and in part due to the slower development of slides at an increasing distance from the epicenter."

Jones (1954) also believed the breaks were ". . . much more likely to have been caused by slumping than by the thin stream of material carried in a turbidity current (flowing at) an *incredible* velocity".

A. Spontaneous Liquefaction

Encouraged by the objections to the turbidity-current hypotheses raised by Shepard and others, but not satisfied with their alternate explanations, Terzaghi (1957) offered a new explanation for the cable breakage which followed the 1929 Grand Banks earthquake. He supposed that the sediment in the epicentral area and over the whole area of cable breakage to the south existed in a metastable state. He supposed that the earthquake shock caused spontaneous liquefaction over the epicentral area and that the apparent velocity of the turbidity current was instead: "the velocity of the advance of the boundary of the liquified portion of the sediments underlying the ocean floor and not the velocity at which the slide material moved over the floor".

The only fact he attempts to explain is the velocity indicated by the sequence of cable breaks. He is concerned neither with the problem of how the sediment came to be in the supposed metastable state, nor with why the sequence of breaks was downslope and not in all directions from the epicenter. Nor was he concerned with the emplacement of the shallow-water sediments in deep water or the grading of the deposited beds. It is not readily apparent why the passage of such a wave of liquefaction should break submarine cables with at least 10% slack. One might think that the cable would be required to sink thousands of feet into the sediment before breakage would occur. Although it cannot be denied that the mechanism he suggests may locally affect sediments, the weight of evidence seems definitely to point away from Terzaghi's theory of propagation of spontaneous liquefaction as an explanation of the Grand Banks cable breaks.

B. Seismo-volcanic Catastrophe

One of the most severe critics of the turbidity-current concept is Professor Hans Pettersson (1954), the well-known leader of the *Albatross* Expedition. To explain the occurrence on the equatorial Atlantic abyssal plains of graded, deep-sea sands which contained "twigs, nuts, and the bark of dicotyledonous trees", he offered the following alternative to turbidity-current deposition: "In the event that a large island, harbouring vegetation, and with a fairly extensive shelf covered the Mid-Atlantic Ridge northwest of St. Paul's Rocks and became submerged during a catastrophe of seismic-volcanic character a few hundred thousand years ago, material like that found in the *Albatross* cores might have become distributed over the adjacent sea bottom."

C. Creep

Lombard (1956), referring to the transportation of ancient sediments, and Koczy and Burri (1958), referring to the present deep-sea environment, favor as an alternative to turbidity currents a slow creep which might require hundreds or thousands of years to traverse a distance that a turbidity current would cross in only a few hours. That slumps, slides and other slower and more massive slope failures occur in the ocean is clear from the evidence of topography, of sediments and of sedimentary structures. But the results of such slow movements are quite different from the graded, well-sorted deposits laid down by turbidity currents travelling at express train speeds over the ocean floor. Koczy (1954) notes that: "Near the shelf the surface is often undulating because of the presence of the sediment sliding down the continental slope. This movement can be assumed to be very slow."

This slow creep, though certainly of importance over wide areas, cannot be considered as a substitute for turbidity currents, but rather as an additional and independent process (Kuenen, 1956).

D. Tsunami

Bucher (1940) formerly thought that tsunami waves set up by the Grand Banks earthquake had caused the later sequence of cable breaks. Indeed there seems to be a relationship between tsunami and cable breaks; however, they are probably not cause and effect but both result from the motion of the slump and ensuing turbidity current. Heezen and Ewing (1952) concur with Gutenberg's (1939) view that the "principal cause of each tsunami is submarine slumping". However, the turbidity currents themselves probably are also capable of generating tsunami. That tsunami could not be the direct result of earthquakes was proved by the fact that not all large submarine earthquakes produce tsunami. In order to prove that turbidity currents and slumps alone could produce a tsunami, we clearly need to find a tsunami and a related series of cable failures not preceded by an earthquake; or to disprove the concept, to discover a turbidity current lacking an associated tsunami.

4. Physiographic Evidence

Following World War II continuously recording echo-sounders were first extensively used in deep water. One of the first discoveries was that, while most of the deep-sea floor is covered by hills and mountains, large areas in the deepest portions of the ocean are covered by abyssal plains (Heezen, Ewing and Ericson, 1954). The origin of these vast flat areas posed a real problem until the concept that turbidity currents were filling the oceanic depressions offered a ready solution (Heezen, Ewing and Ericson, 1951). Pettersson and some of his followers, including Koczy, remained unconvinced, and to this day prefer to explain the abyssal plains as the result of great outpourings of lava (Koczy and Burri, 1958).

During the exploration of the abyssal plains, long narrow steep-sided channels were discovered which cut into the abyssal plains. These features, called mid-ocean canyons (Ewing *et al.*, 1953) or deep-sea channels (Dietz, 1953; Menard, 1955), were indeed an enigma to those who believed that the plains are the result of the outpourings of lava. Koczy and Burri (1958), in an attempt to explain the occurrence of features normally ascribed to erosion and deposition in an area where these authors deny the existence of such processes, offered the suggestion that mid-ocean canyons were, in fact, fault-bounded rift valleys, and that their natural levees were instead the sides of tilted fault rocks. This explanation says nothing of the graded sand beds found in the canyon floor, the graded silt beds found on the levees, or of any of the other streamlike characteristics of the Northwest Atlantic Mid-Ocean Canyon (Ewing *et al.*, 1953).

A. Submarine Natural Levees

Submarine natural levees have been observed along the continental slope canyons of southern California (Buffington, 1952), as well as along the mid-ocean canyons of the Atlantic (Ewing *et al.*, 1953), the Indian Ocean (Dietz, 1953), and the deep Pacific (Menard, 1955). The largest deep-sea natural levees observed to date are those of the Congo Submarine Canyon (Heezen *et al.*, 1957), which in a depth of 2250 fm are 100 fm high and over 20 miles wide. It is safe to conclude that the submarine natural levees are of nearly the same origin as the subaerial ones. When turbidity currents overflow the tops of the canyon walls they suddenly deposit a considerable portion of their load as they abruptly slow down and disperse.

Since the natural levees of the Northwest Atlantic Mid-Ocean Canyon are nearly identical in form with those bounding the deeper parts of the Congo Submarine Canyon, it seems unlikely that they are the result of a radically different process, as proposed by Koczy and Burri (1958).

5. Sediments

A. Deep-Sea Sands

Ericson *et al.* (1952, 1961) presented an account of their study of turbidity-current sediments deposited in the abyssal plains of the North Atlantic.

The lithological contrast between pelagic sediments and sediments of turbidity-current origin is generally very striking. The turbidity-current beds consist of silt and sand and sometimes even gravel. They are characterized by extremely sharply defined bases, and with particle size grading from coarse at the bottom to fine at the top. Since the sediments grade generally into the normally fine-grained pelagic sediments, the tops of the beds are generally much less distinct than the bottoms of the beds. Frequently the blurring of the top of a turbidity-current bed is due to churning by sediment-burrowing organisms. Apparently most burrowers confine themselves to the upper 10 or 20 cm. Thus, the lower parts of graded beds, greater than 10 or 20 cm in thickness, are rarely disturbed by burrowing.

The shallower-water origin of at least a part of the coarser material making up turbidity-current layers is frequently proven by the presence of a certain species of Foraminifera and particles of calcareous algae or other plant debris.

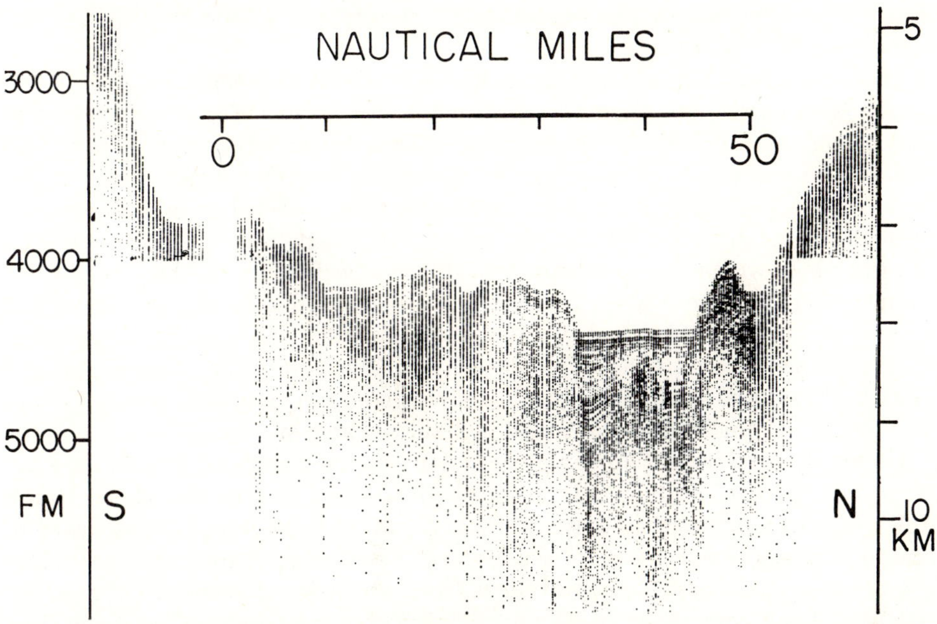

Fig. 9. Seismic reflection profile across the Puerto Rico Trench. This profile, obtained by John Ewing indicates over 1 km. of sediments below the trench abyssal plain. Beds of calcium carbonate-rich graded sediment have been found intercalated in abyssal clays beneath the trench abyssal plain. These sediments containing fragments of calcareous algae were transported by turbidity currents from south to north into the trench.

In some layers there occurs a concentration of coarse vegetal debris several centimeters thick. These concentrations are usually associated with silts or fine sands, and individual particles may be several millimeters in diameter. In other layers, organic material is present but only as dark gray or nearly black pigment in the upper lutite fraction.

Frequently turbidity-current beds deposited in the greater depths of the ocean have a much higher lime content than the normal abyssal lutite. This is particularly true of the sharply defined layers of graded lutite often found in cores of the abyssal plains of the North Atlantic. In these, the coarse fraction is either absent or is represented by a mere film of silt at the base. The lime content is from two to three times as great as the interbedded abyssal red clay.

A peculiarity common to all these turbidity-current layers, whether composed of sand or very fine sediment, is their discordance in physical, chemical and biological make-up from the environment in which they occur. They are truly displaced sediments. Usually some of the material, and sometimes most of it, is of shallow-water origin. However, in most cases, good sorting, thin bedding and grading set deep-sea turbitites distinctly apart from typical shallow-water marine deposits.

Another highly important characteristic of these sediments is their distribution with respect to ocean-bottom topography. Graded layers, and layers of gray calcareous lutite, have been found in submarine canyons, on the continental rise and on the gently sloping abyssal plains and in the abyssal plains of the oceanic trenches. They have never been found on isolated rises. Some of the most dramatic evidence comes from two cores taken from the narrow trench abyssal plain which lies at about 4580 fm at the bottom of the Puerto Rico Trench (Fig. 9). These cores consisted of abyssal red clay at their base; but each contained a graded bed of calcareous sand, silt and clay which included fragments of the alga *Halimeda*. Since living *Halimeda* requires sunlight this detritus must have originated near the shore. The graded layers of calcareous sand also contained shells of pteropods and shallow-water species of Foraminifera.

However, cores taken on the ridge north of the trench, in depths between 5000 and 6000 m contain only red clay of abyssal facies. Elsewhere, depth differences of no more than 100 m have been found to be enough to separate an almost continuous succession of graded beds from uniform foraminiferal lutite.

Over 200 cores have been obtained from the abyssal plains of the North Atlantic. Each core taken in an abyssal plain contained either sands or silts, many with shallow-water Foraminifera (Figs. 10 and 11). Most cores from the flanks of the Mid-Atlantic Ridge, the abyssal hills and the oceanic rises contained only pelagic deposits, although a few at the bases of islands or from the crest zone of the ridge contained coarser material (Ericson *et al.*, 1961; Heezen, Tharp and Ewing, 1959). Deep-sea sands and even gravels have been found associated with fans or cones at the mouths of submarine canyon systems (Fig. 12).

A concept which might be described by the term "elevator tectonics" (or perhaps "lift faulting") has often been employed in the explanation of displaced sediments interbedded in pelagic sequences (Cushman, 1941; Landes, 1952). The latest proponents of this concept are Greenman and LeBlanc (1956) who invoke a 10,000-ft subsidence of the entire central portion of the Gulf of Mexico at the close of the Wisconsin (11,000 years ago) to account for the emplacement of gray silts now found on the floor of the Sigsbee Abyssal Plain. These gray silts have been explained by Ewing, Ericson and Heezen (1958) as the results of turbidity-current deposition of sediment contributed by the Pleistocene Mississippi River. Cores from low knolls rising above the abyssal

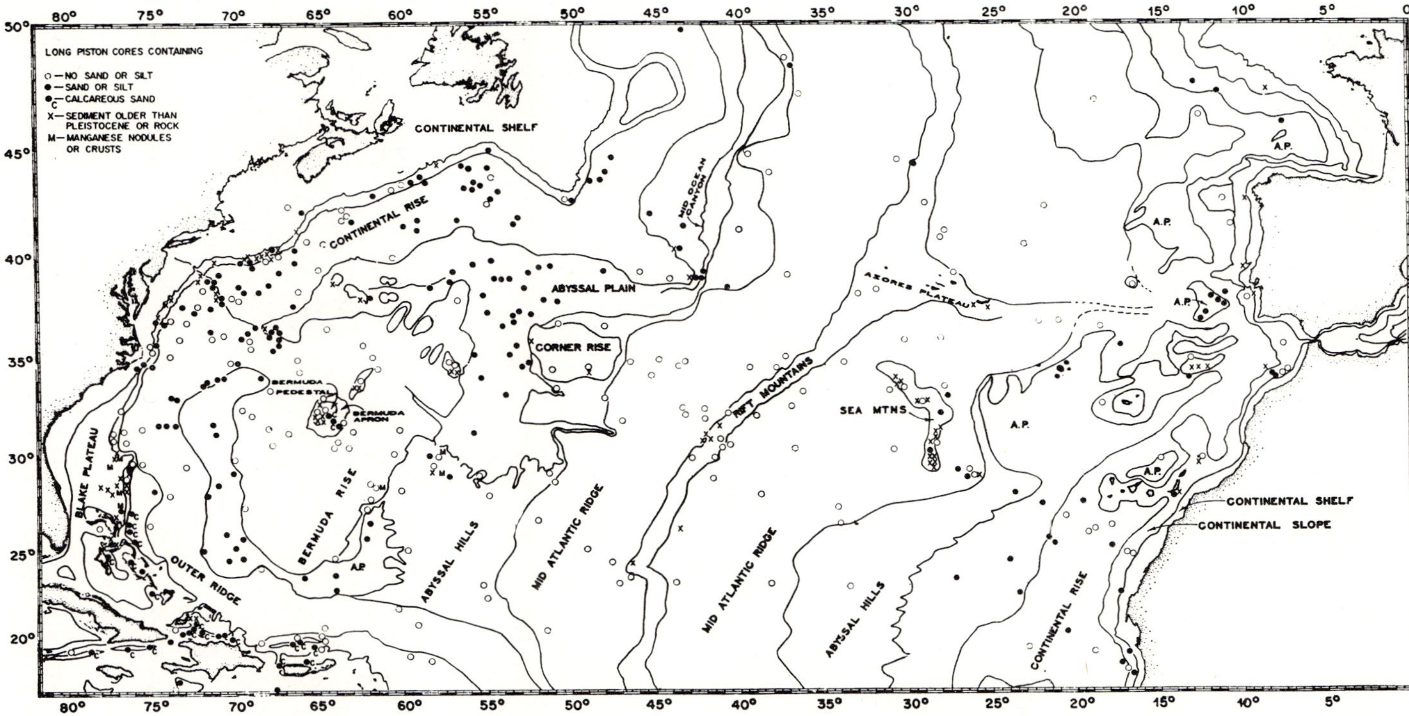

Fig. 10. Distribution of deep-sea sands in relation to physiographic provinces. Note that no cores obtained from the abyssal plains are devoid of silts or sand layers. Note the distribution of calcareous sand near the Bahamas. Note also that the sediments from the Bermuda Rise and the Mid-Atlantic Ridge consist entirely of pelagic sediments, except in certain rare areas where winnowing of sediments near the crest of the Mid-Atlantic Ridge, or on the flanks of seamounts, has altered this pattern. (After Heezen, Tharp and Ewing, 1959.)

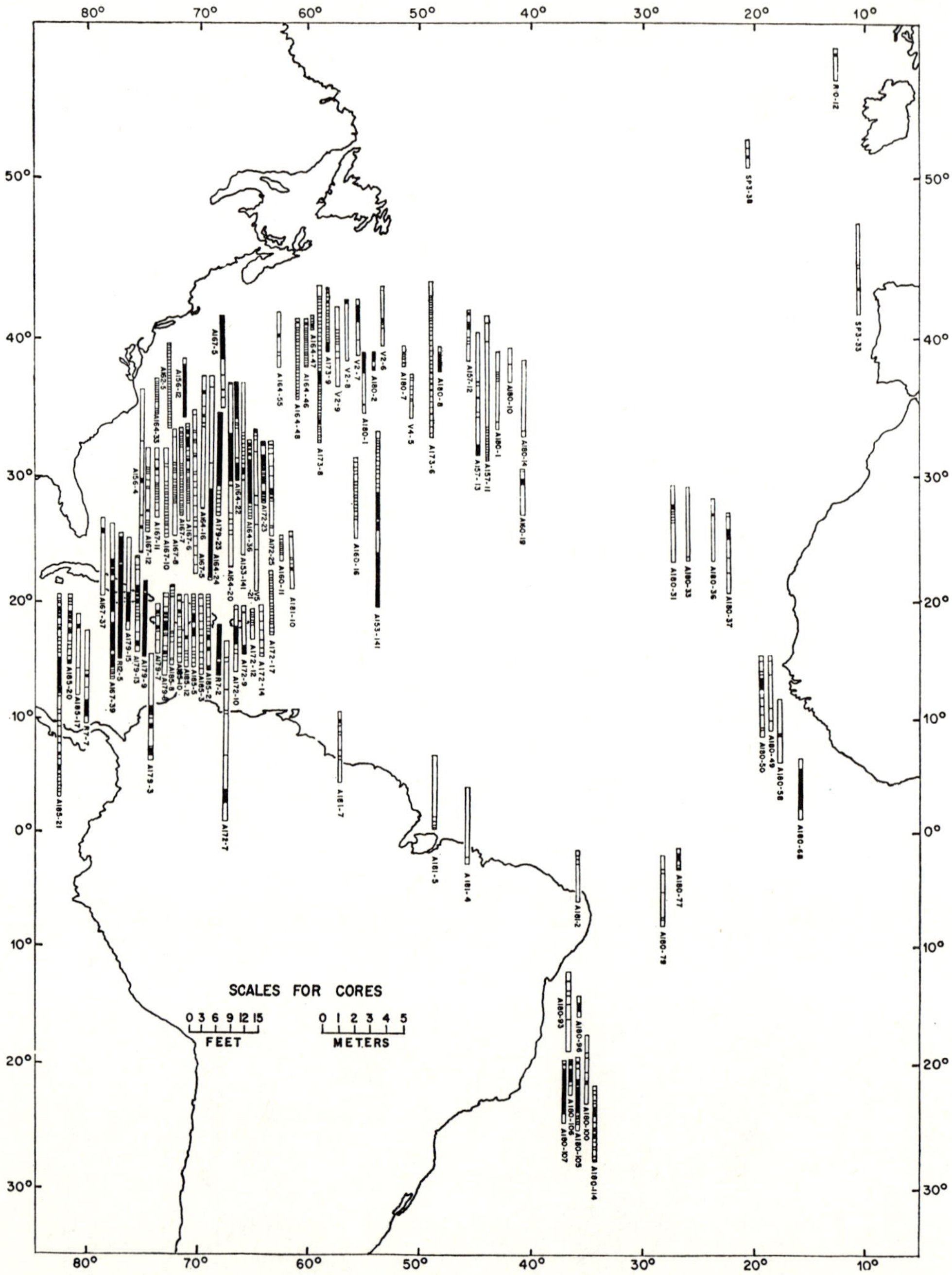

Fig. 11. Distribution of sediment in cores containing sand and silt layers. The top of the column is approximately the location of the core station. A black zone denotes sand and/or silt. A line denotes a thin sand or silt layer. An open zone denotes lutite. (After Ericson, Ewing, Wollin and Heezen, 1961.)

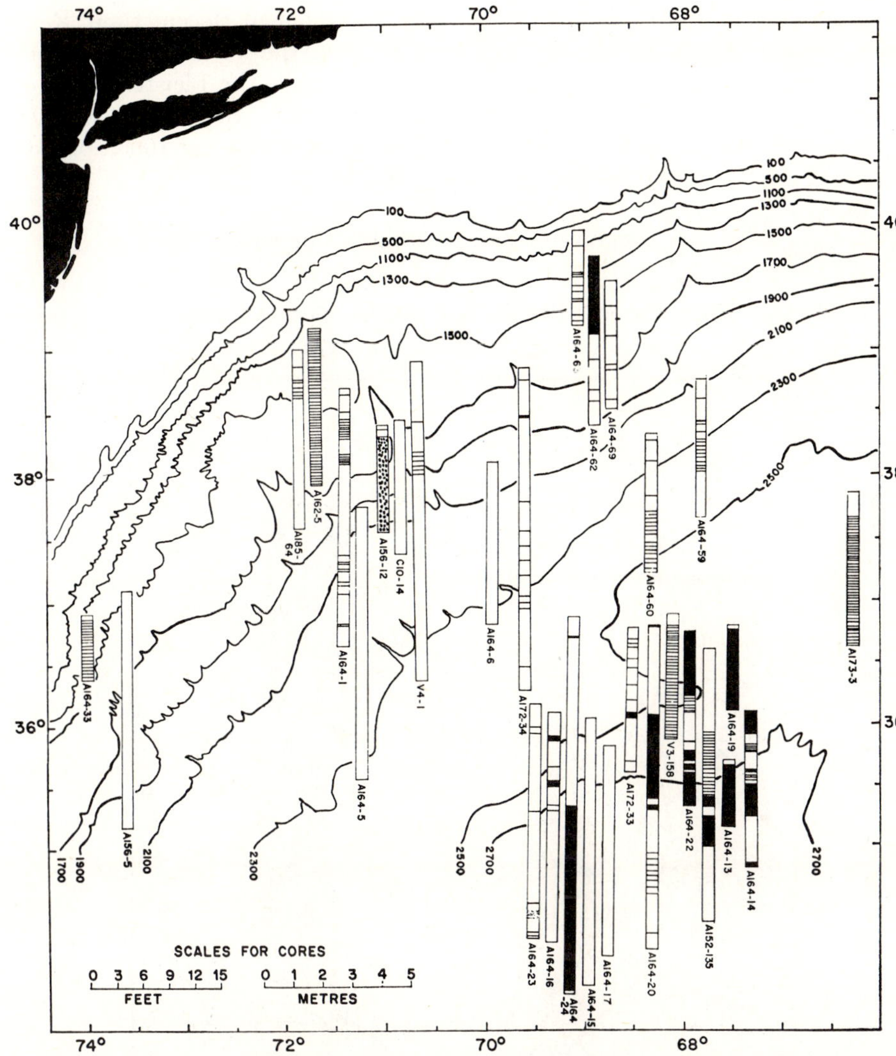

Fig. 12. Distribution of sediment cores in the vicinity of the Hudson Submarine Canyon. Contours in fathoms. The top of the column is the approximate location of core stations. A black zone indicates quartz sand and/or silt. A line indicates a thin quartz sand or silt layer, an open zone indicates lutite. The dotted zone in Core A 156-12 denotes gravel. Note that Cores A 164-5 and A 164-6, which lie on either side of the Hudson Canyon on the continental rise, are entirely free of sand or silt, whereas most of the cores near the mouth of the canyon contain appreciable percentages of sand or silt. Cores A 164-63, -62 and -69 are associated with Hydrographer Canyon. (After Ericson, Ewing, Wollin and Heezen, 1961.)

plain contained only slow, uniformly deposited, pelagic sediments. According to Greenman and LeBlanc's explanation, the knoll core stations would have stood more than 150 m above sea-level until commencement of Recent time. Radiocarbon measurements on the pelagic deposits from the knolls indicated post-glacial rates of deposition of 3 cm per 1000 years, while the Pleistocene rate in the same cores ranged from 10 to 20 cm per 1000 years. In contrast, the sediments cored from the abyssal plain showed post-glacial rates of 10 cm per 1000 years and glacial rates of 60 cm per 1000 years.

6. Requirements of Turbidity Currents

Turbidity currents first of all require sediment, just as rivers require water; without it they do not exist. Turbidity currents require a trigger mechanism;

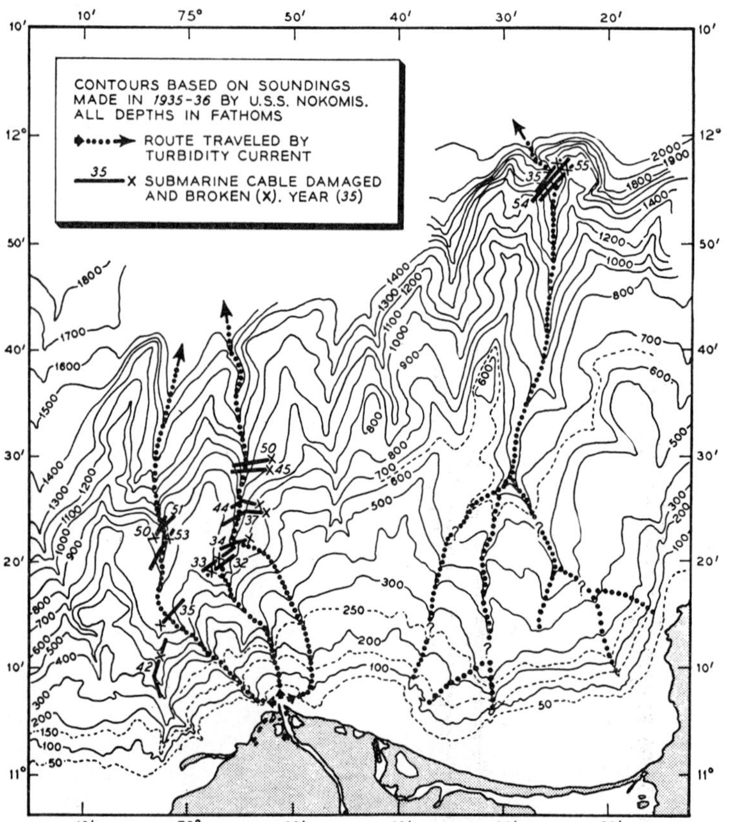

Fig. 13. Cable breaks in the submarine canyons at the mouth of the Magdalena River, Colombia, South America. All cable breaks indicated on the diagram were in a single submarine cable. Most of the cable failures shown occurred following extensive slumps of the Magdalena river bar. Preceding the 1935 and 1945 breaks, 450 and 250 m respectively of the western and eastern river-mouth control jetties respectively were carried away, along with the river bar. (After Heezen, 1956.)

this can be supplied by earthquakes, hurricanes impinging on the shore, high bedload discharge of rivers; and, lacking a trigger mechanism, turbidity currents probably can occur after a long-continued deposition simply by slope failure resulting from gradual over-steepening of a depositional slope. Finally, turbidity currents require a slope. The optimum conditions for the generation of a turbidity current probably would be a large body of fine, well-sorted sand triggered by an earthquake on a steep slope.

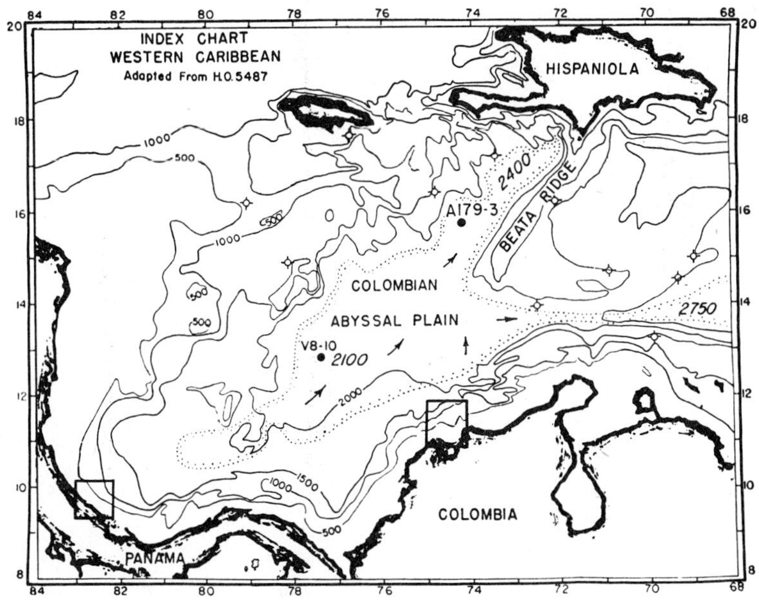

Fig. 14. Sediment cores from the western Caribbean. Sediment cores indicated by open circles contain only pelagic sediment, without any evidence of turbidity-current deposition. Sediment cores A179-4 and V8-10 indicated by solid dots both contain beds high in vegetal debris. The log of core A179-4 is shown in Fig. 15. The core V8-10 contained two beds almost entirely composed of twigs and leaves. Turbidity currents are frequently initiated in the area off the Magdalena and Sixaola Rivers. These areas are indicated by black rectangles. (After Heezen, Ewing and Menzies, 1955.)

Heezen (1956), studying cable failures from various parts of the world, has found evidence for several different trigger mechanisms. Besides earthquakes, hurricanes have, in at least one instance, triggered turbidity flow off Cape Hatteras. One of the most interesting types of trigger mechanism is that related to the high bedload transport at the mouths of rivers. Potentially, such a source should have a greater geologic importance since both the trigger effect and the source of sediments are supplied by the river. In such an area repeated flows from a single source could build up a submarine abyssal cone, and at the same time repeated currents down the same path on the continental slope could erode a canyon. Significant periodic changes in depth have been observed at the heads of submarine canyons by Shepard (1948), Zenkovich (1958), Heezen (1956) and others.

It was found that cable failures in the submarine canyon of the Magdalena (Heezen, 1956) occurred after prominent changes had taken place in the depth and shape of the river mouth (Fig. 13). These changes in turn had occurred only during the highest flood stages of the river. A similar pattern was also observed off the Congo. Both rivers have a common characteristic in their lack

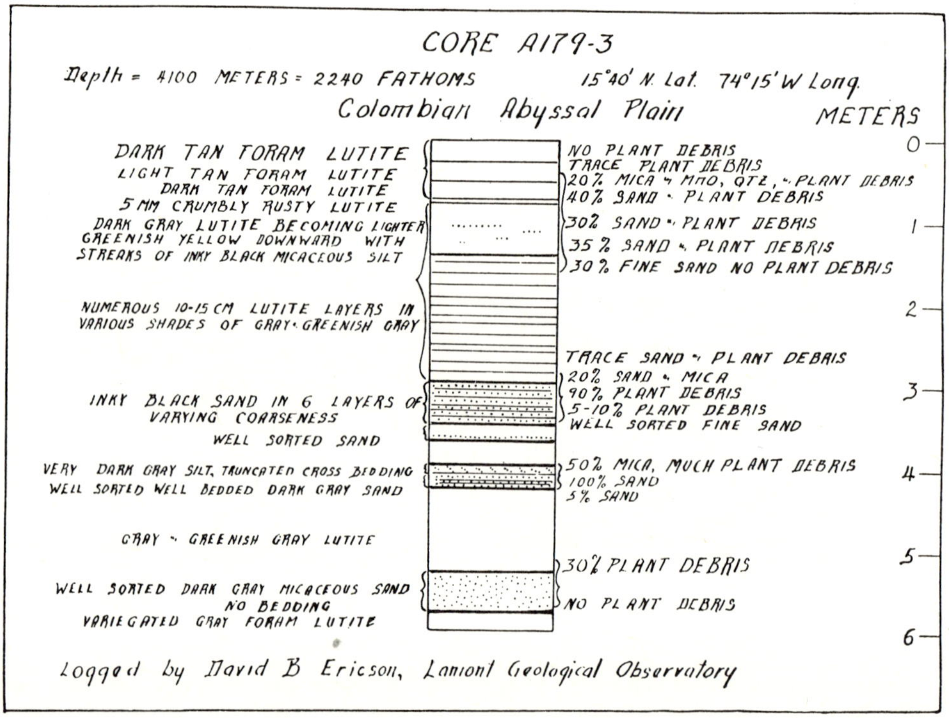

Fig. 15. Log of core A179-3 from the Colombia Abyssal Plain. Sediment in this core presumably came from the mouth of the Magdalena River through the action of turbidity currents, which carried this sediment nearly all the way across the Colombia Basin. Note that many of the beds contain up to 90% of plant debris. Other cores in this general area were found to be extremely rich in plant debris. (After Heezen, 1956a.)

of a modern subaerial delta. The delta sediments in both cases are to be found in the alluvial cones at the mouths of their submarine canyons; and, indeed, exploration in both areas has disclosed graded beds containing a large proportion of terrestrial organic remains as well as tree leaves and wood (Figs. 14 and 15). A dramatic fact associated with the 1935 cable failure off the Magdalena was recorded when grass, still green, was brought up wrapped around the damaged cable which lay in 750 fm, 12 miles seaward from the river mouth.

7. Tectonic Significance of Turbidity Currents

A. Modern Geosynclines

Turbidity currents are gradually filling the modern deep-sea trenches. Turbidity-current smoothed trench (abyssal) plains have now been reported from every trench which has been sounded in detail. Trenches bound most of the circumference of the Pacific Ocean, but in the Atlantic such marginal topographic trenches are exceedingly rare. However, wherever seismic-refraction studies have been made at the base of the continental slope, a deep trough of lower velocity material ascribed to unconsolidated sediments and sedimentary rock has been found paralleling the base of the continental slope (Fig. 16). These former trenches were probably filled largely by turbidity-current sediments. They constitute in all probability the modern geosynclines which quietly await the process of orogeny which, through deformation and uplift, may eventually turn them into folded mountains.

The effect of turbidity currents over a long period of geologic time is to fill in the hollows of the deep-sea floor by building vast abyssal plains, and to erode canyons into the continental slopes. Of course, if the rate of subsidence of the marginal trenches exceeds the rate of deposition, broad basin-floor abyssal plains would not form since all turbidity-current deposits would be trapped in the trenches. The effect of turbidity currents in the deep sea produces a topography essentially similar to that produced by periodic downpours in desert mountains. The coalescing alluvial fans resemble the continental rise, and the playas resemble the abyssal plains.

Unfortunately, the turbidity-current concept has been occasionally overworked even in regard to recent deposits. By invoking turbidity-current action Rigby and Burckle (1958) attempted to explain the emplacement of thin beds (millimeters thick) of freshwater diatoms found in cores of mid-equatorial Atlantic eupelagic sediment. This explanation is no less ridiculous than the theory proposed by Malaise (1957) who believes them to be deposits laid down in ponds during a recent and complete emergence of the Mid-Atlantic Ridge. The Congo River at present carries freshwater diatoms hundreds of miles offshore. Kolbe (1957, 1957a) points out that they are also blown hundreds of miles to sea by the wind. To emplace diatoms on the Mid-Atlantic Ridge as suggested by Rigby and Burckle (1958) would require turbidity currents to climb over 3000 m in 500 miles, travelling all the way over an intensely rugged bottom, an ability not attributable to turbidity currents. Rigby and Burckle do not doubt that surface currents, or perhaps the wind, can transport diatoms as far as the core stations on the Mid-Atlantic Ridge, but they do not believe, for some unspecified reason, that ocean currents could deposit and concentrate beds of freshwater diatoms.

The geological record contained in the ancient sediments is sufficiently complex for other misapplications undoubtedly to occur.

Turbidity-current deposits have been described from the modern basins of sedimentation off southern California by Gorsline and Emery (1959). These

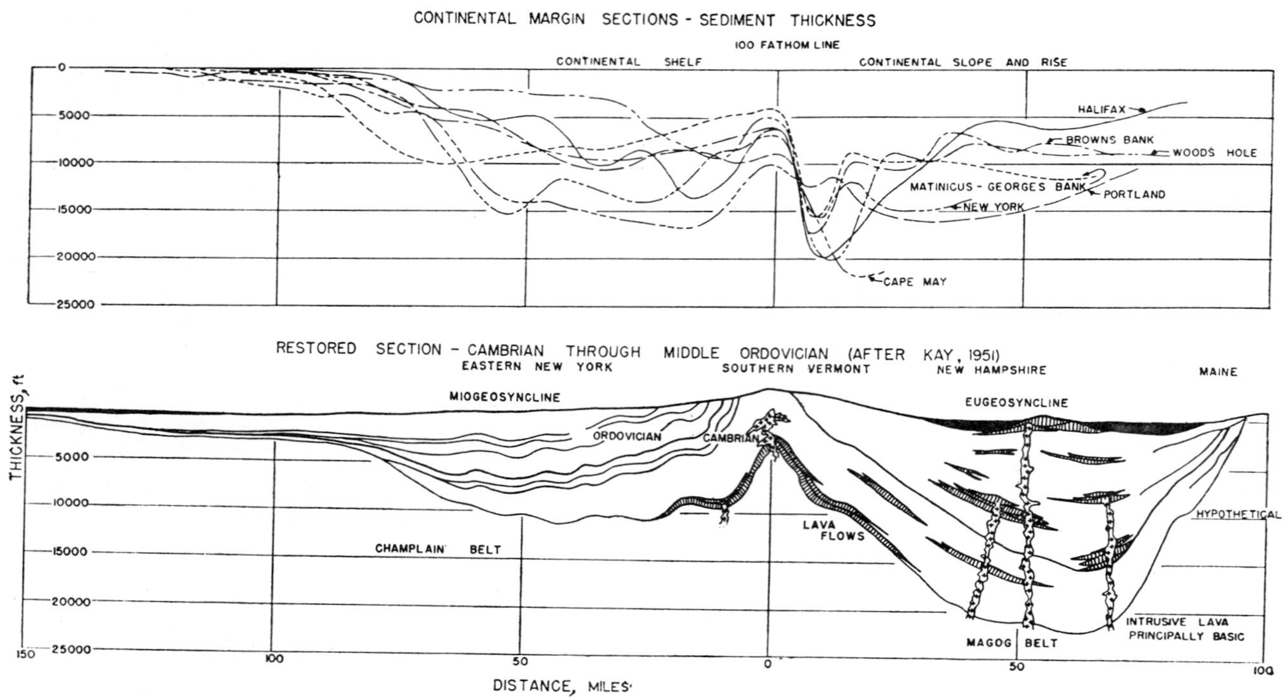

Fig. 16. Turbidity currents and modern geosynclines. The profiles in the upper part of the figure represent the base of the sedimentary column in profiles off the eastern coast of North America, from Halifax to Cape May. Note that about ten miles east of the 100-fm line there is a major thickening of the sediment on all profiles. It is believed that this deep sedimentary basin, lying below the upper continental rise, is being largely filled by turbidity-current deposition. It is also presumed that the present continental margin off eastern United States is a modern analogy of the ancient Paleozoic geosyncline (Kay, 1951) as indicated in the restored section. (After Drake, Ewing and Sutton, 1959.)

modern turbidity-current deposits have been compared by them to those described by Natland and Kuenen (1951) and others from the filled sedimentary basins of southern California. Considerable attention has been given to these deposits since the analogous deposits in the filled basins are productive of petroleum.

Menard (1955) has employed the turbidity-current hypothesis to explain the smooth topography, and the canyons, and the deep-sea channels of the North Pacific. Locher (1954) and Phleger (1951), who studied the lithology and fauna of the deep-sea sands from the equatorial Atlantic, became convinced of their turbidity-current origin. Bourcart (1953) and Duplaix and Calleux (1956) described deep-sea sands from the abyssal plains of the Mediterranean which they identified as the deposits of turbidity currents. The concept has found an

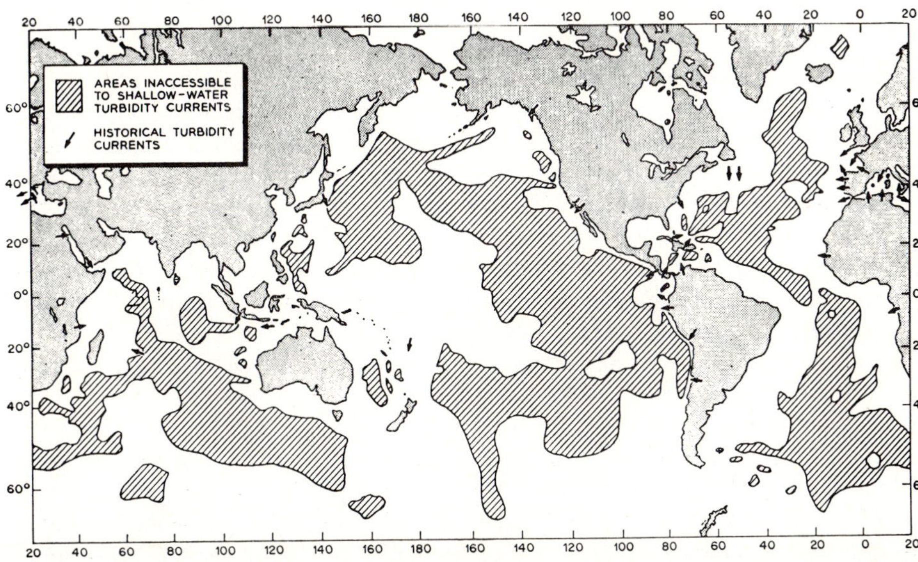

Fig. 17. Historical turbidity currents (since 1880). Each arrow denotes one or more turbidity currents. With the exception of two off southern California, all are based on submarine cable breakage. (After Elmendorf and Heezen, 1957.)

increasingly greater application in paleogeographical reconstruction and a host of recent papers have appeared which we will not attempt to review here. It will suffice to say that Kuenen, his associates, and his students, as well as many other independent workers, have had considerable success in identifying ancient turbidity-current deposits throughout the world.

There is now little doubt that turbidity currents play an important part in the deep sea (Fig. 17). Their velocities have been determined by sequences of cable breaks. The distribution of sands and silts in the deep-sea basins has conclusively shown that the abyssal plains, archipelagic plains and smooth continental rise areas owe their existence to the smoothing activity of turbidity currents. The deep-sea sands, long an enigma of geology, are adequately explained by this process. It is clear that turbidity currents occur both as broad, sweeping flows, such as the Grand Banks turbidity current, and as narrowly channelized flows which create characteristic physiographic features, such as natural leveed channels, deep-sea fans and deep-sea cones. All investigators are now convinced of the existence of turbidity currents, and their relatively frequent occurrence in the deep sea.

However, there is one point which is still quite controversial and that is the capacity of turbidity currents to erode. It is held by several workers that turbidity currents cannot erode into solid rock. Indeed, if turbidity currents are to be considered the primary agent for the erosion of submarine canyons, it would seem that they must have appreciable capacity to erode. In Chapter 12 on Submarine Topography and in Chapter 14 on the Abyssal Plains, physiographic evidence for turbidity currents has been more extensively summarized.

8. Biological Significance of Turbidity Currents

The occurrence of carbonaceous deposits in the deep-sea alluvial cones led to speculation on the effect of this material on the nutrition of the abyssal fauna (Heezen, Ewing and Menzies, 1955, 1957). It was reasoned that turbidity flows would have two radically contrasting effects on the abyssal fauna. Rapid deposition of a thick bed of sediment over the abyssal floor must have a tremendous destructive effect, smothering by burial much of the indigenous fauna. In contrast, the large windfalls of organic debris brought to the deep sea by turbidity currents must support a greater productivity than the normal, slow, pelagic deposition.

Repeated turbidity currents from an organic-rich source build an organic-rich submarine deposit containing both the transported organic debris and the trapped and entombed abyssal fauna.

Another important effect of turbidity currents on petroleum occurrence is independent of the bottom fauna. Orr, Emery and Grady (1958) found that while the content of chlorophyll derivatives in sediment immediately below turbidity-current-deposited sands was anomalously high, that of the sands themselves was anomalously low. The turbidity current which brought in the sands was apparently derived from a source low in organic matter and thus in this case only the blanketing effect of the turbidity-current deposit is seen.

Since the geologist's key to the past is the present, the implications of submarine turbidity currents as regards paleographic reconstructions are tremendous. No longer can scattered shallow-water fossils be considered evidence of shallow-water deposition, nor does evidence of strong currents indicate shallow depths. Even ripple marks, long considered an absolute proof of shallow-water origin, have been found as deep as 2000 fm.

9. Other Processes of Deep-Sea Erosion and Transportation

A. Abyssal Ripples

When photographs of the deep-sea floor were first taken in appreciable numbers the frequent discovery of ripple marks and scour marks was simply astounding. In 1948 a well-known physical oceanographer, upon seeing a picture of ripple marks at 410 fm, was so amazed that he seriously suggested to the writer that the ripples were fossil and had been formed before a local subsidence of the sea floor. Ripples were found in many localities inaccessible to downslope movements. It is improbable that turbidity currents are related in any general way to the processes forming the ripples seen in bottom photographs. Actually, the ripples should not have come as a surprise, since early studies of sediments taken from submarine peaks and in deep straits in depths ranging down to 2000 fm had revealed winnowing of the finer particles

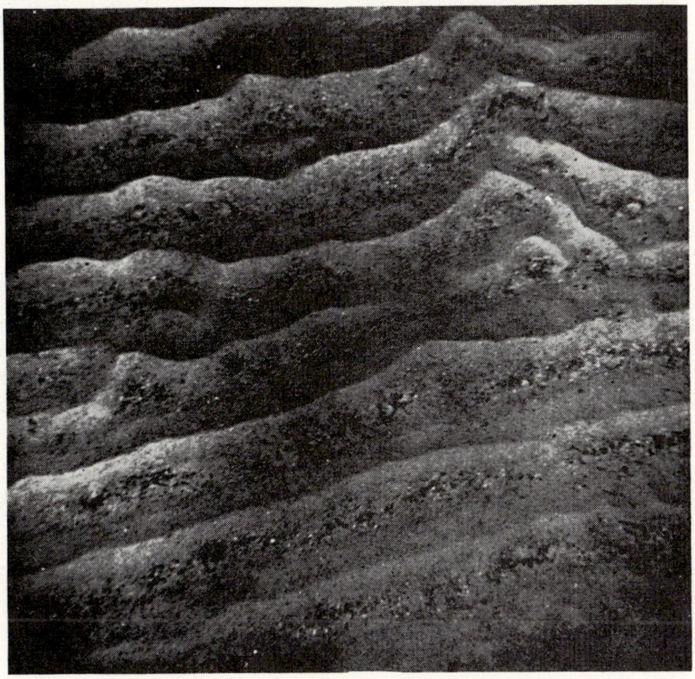

Fig. 18. Ripple marks on the flank of Atlantis Seamount in 410 fm. Photograph made on *Atlantis* Cruise 152, August, 1948. Area of photograph approximately 6 m^2.

(Murray and Renard, 1891). In fact, Agassiz (1888) early recognized that current scour could be effective to depths down to at least 650 fm. He concluded that "... the bottom of the Gulf Stream along the Blake Plateau is swept clean of slime and ooze, and is nearly barren of animal life".

Weber, in 1900, later noted coarse sand and rock bottom to depths of 1500 m in the passages between the Indonesian Islands. But concerning these findings Grabau (1913) concluded that: "It is probable that we have here a prevention of sedimentation by the removal of the particles before they reach the bottom, rather than any effect of eroding work of the currents at such depth."

Ripple marks (Fig. 18) and scour marks (Fig. 19) have been found on every seamount photographed to date (Menard, 1952; Heezen, Tharp and Ewing, 1959). Ripple marks have been observed on the crest of the Mid-Atlantic Ridge down to 2000 fm (Elmendorf and Heezen, 1957). They have also been observed on the continental slopes to similar depths. The deepest photographs of the North Atlantic abyssal floor had indicated probable scour marks in 3200 fm depth (Dietz, 1952; Elmendorf and Heezen, 1957). The deepest Atlantic photographs recently taken revealed rock bottom at 4000 fm in the Romanche

Fig. 19. Photograph from western rift mountains of the Mid-Atlantic Ridge in 1410 fm depth (48° 38′N, 28° 48′W). Note the patches of recent sediment lying in the depressions between outcropping rocks.

Trench Fracture Zone (Equatorial Atlantic) (Cousteau, 1958). These sharper topographic features are being scoured by ocean currents which remove the finer sediment and produce the ripple marks. The direction of these currents has not been determined, although a collection of oriented photographs might go far to solving the problem. It must be supposed that the sediment eroded from these peaks by the currents is deposited in their lee. A study of these deposits may produce evidence of the direction of the deep currents and the nature of their variations in late Tertiary, Pleistocene and Recent times (Figs. 20 and 21).

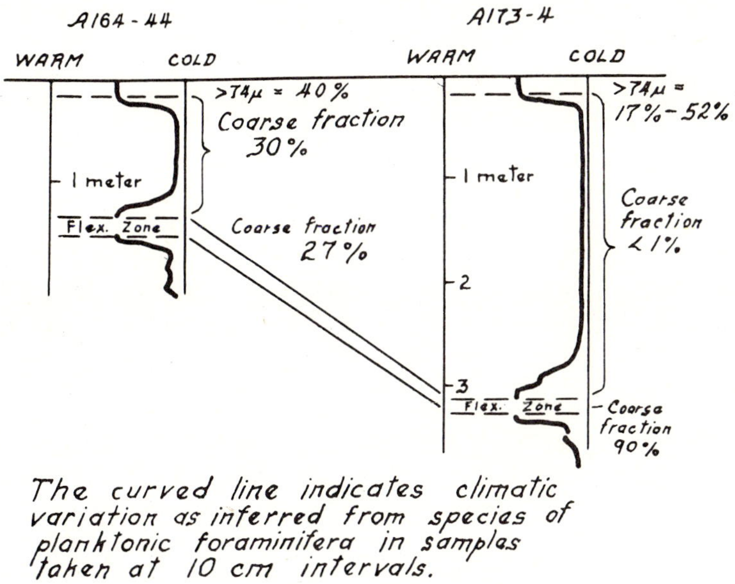

Fig. 20. Log of two cores on the flank of Muir Seamount. Note the coarser, thinner layer representing the last cold period in core A164-44, as compared to the finer thicker section in core A173-4. The thicknesses of the recent warm period and the previous warm period, which correlates with the *Globorotalia menardii flexuosa* zone, are near the same thickness in both cores. (After Ericson and Heezen, 1959.)

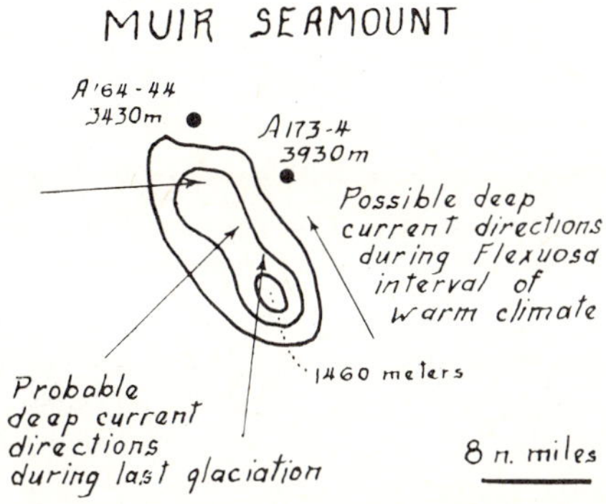

Fig. 21. Location of cores A164-44 and 173-4 on the flank of the Muir Seamount, showing the interpretation of the data presented in Fig. 20. It is presumed that during the present warm period and during the previous period of warm climate, currents flowed northwest, parallel to the seamount. However, during the last glaciation, currents are presumed to have moved from the southwest to northeast, winnowing material from the top and the flanks of the seamount, and depositing the finer material in the lee of the seamount on the northeast flank. This would account for the thick bed of fine sediments representing the last glaciation in core A173-4 and the thin, winnowed layer in core A164-44. (After Ericson and Heezen, 1959.)

Wüst (1957) concluded that relatively strong bottom currents swept along the continental slope off eastern South America. His calculations indicated a 5–20 cm/sec western boundary current in the South Atlantic. The same range of velocities was recently measured in the North Atlantic western boundary current through the use of neutral floats (Swallow and Worthington, 1957).

Ocean-bottom photographs of rock outcrops and ripple marks taken on the continental slope show the eroding capability of this current (Northrop and Heezen, 1951; Heezen *et al.*, 1959). Dietrich (1957) and Cooper (1955) have described boluses of cold Norwegian sea-water which flow through the Denmark Strait and descend to the depths of the Atlantic at velocities of 10 to 30 cm/sec. Jarke (1958) has shown the influence of bottom currents on the distribution of sediments on the Iceland–Faeroe Ridge. Strong currents have been calculated from the fine systematic observations made by J. B. Tait (1956) in the Shetland Channel.

It has been suggested that the boluses of cold water flowing across the northern margins of the Atlantic may be related in some way to the turbidity currents which built or eroded the Northwest Atlantic Mid-Ocean Canyon (Ewing *et al.*, 1953).

Except where shallow-water benthonic organisms are found incorporated in the deposit, it may be difficult to distinguish in every case the effects of bottom scour from the deposits of turbidity currents.

More must be learned of the causes and further effect of these strong bottom currents which have denuded the crest of a large part of the Mid-Atlantic Ridge to depths of 2000 fm (Heezen *et al.*, 1959). These currents wear so gradually that trans-Atlantic submarine cables crossing the crest of the ridge often last 25 years or more before their armour wires are finally worn through. Thus, this incessant motion must never become too violent or the cables would snap more frequently.

Swallow (1957) has recently made measurements of deep-sea currents in the western Atlantic with neutral buoyant floats. He has found that velocities up to 43 cm/sec can be observed in depths as great as 4000 m and that these currents vary in direction and velocity over very short periods of time. Such currents are of such magnitude that the occurrence of ripple marks, scour marks and denuded surfaces in the deep sea is no longer a mystery, and certainly no longer conflicts with the concepts of modern physical oceanography.

B. Homogenized History

Evidence of deep-sea erosion, transportation and deposition by bottom currents comes from the fine studies of the minute, fragile tests of silicious Radiolaria and diatoms found in mid-Pacific abyssal red clays. The long, red-clay cores of the *Albatross* Expedition were chosen for detailed study since, based on estimates of sedimentation rates, they were thought to contain the undisturbed record of most of the late Tertiary. At the time Riedel (1957) began his investigation of the Radiolaria, he "expected that the assemblages in the tops of the cores could be used to determine the areal distribution of various species at the present day, while samples from lower levels of the cores would show, in an orderly manner, the development of present-day species through at least Pleistocene and Recent time".

However, the results of his study forced him to conclude "that the situation is actually more complicated than was expected. Late Tertiary Radiolaria were found to be reworked into Recent sediments, and evidence of disconformities were found in the sediment sequences. Re-examination of some of the samples from the *Challenger* Expedition showed that Tertiary sediments outcrop on the deep-sea floor. Erosion, and reworking of older Radiolaria into younger sediments, obscure phylogenetic developmental sequences in the long sediment cores. . . ."

The "epic poem" contained in these layers of sediment had, it seemed, been frequently reduced to a font of jumbled type as cryptic to read as a typesetter's pie.

Kolbe (1957, 1957a), who found pre-Quaternary diatoms in several Pacific and Indian Ocean cores, concluded that their presence is "unimportant" and "accidental, although no plausible explanation of the occurrence can be given". Riedel, however, "speculated that a more or less continuous agency (? current) was transporting Radiolaria on or near the sea floor" and that "the transporting agency has acted either continuously, or at such short intervals as to obscure intermittance".

Recent seismic-reflection studies indicate that the thickness of sediments beneath the abyssal floor is much greater than that beneath the adjoining abyssal hills and Mid-Atlantic Ridge. This suggests that ocean currents may gradually concentrate fine sediments in the deep ocean basins.

In the light of Riedel's conclusions, it is not surprising that difficulty was encountered in the stratigraphic interpretation of the chemical analyses of these cores (Berrit and Rotschi, 1956) and that, in general, the ionium and radium methods of dating of deep-sea sediments have been more often a failure than a success (Kroll, 1955; Volchok and Kulp, 1957). Probably the most important result of the later studies has been the conclusion that conditions of uniform, undisturbed, unreworked sedimentation required by these dating methods only rarely prevail on the deep-sea floor.

C. Sub-bottom Echoes

The character of the echo returned to the echo-sounder from the deep-sea floor can be examined in great detail by use of the high-resolution Precision Depth Recorder.

The echo may be sharp and discrete, diffuse and prolonged, single, double or multiple. Over distances of a few hundred yards, the bottom may appear absolutely smooth or completely disrupted on a minute scale.

Sub-bottom echoes are frequently observed in deep-sea echograms. In some cases the sediment horizon responsible for the sub-bottom echo has been identified by coring or drilling. Worzel (1959) has correlated a prominent sub-bottom echo in the east tropical Pacific with a 2–4-in. thick bed of volcanic ash. Heezen *et al.* (1958) have correlated a prominent sub-bottom echo with the base of the Recent anaerobic layer in the Cariaco Trench. Gorsline and Emery (1959) have mapped turbidity-current beds in the San Pedro Trough. In areas of particle by particle deposition where one or two prominent 5–20-fm deep sub-bottom echoes can be traced over a wide area they typically show deepening in the small depressions, thinning over gentle rises, and are totally absent on steeper inclines, indicating that, owing to gentle scour, sedimentation has been greater in the depressions (Fig. 22).

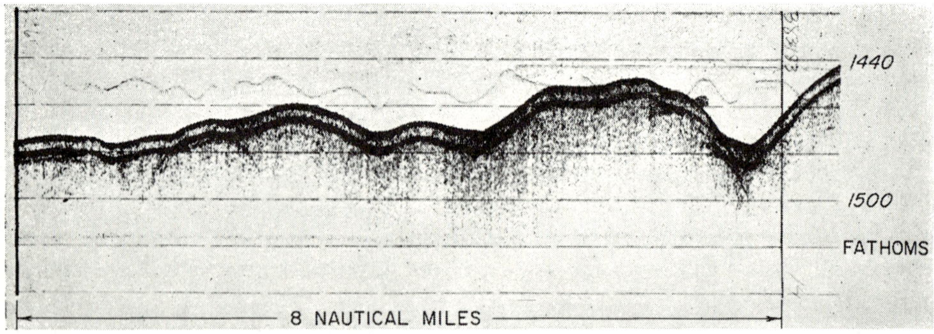

Fig. 22. Precision Depth Recorder echogram made west of Panama by R.V. *Vema*. The strong, sub-bottom echo, 5–10 fm below the sea floor, is reflected from the Worzel Ash. Note the slight, distinct variations in thickness of the sediment above the ash. These variations are presumably due to the combined action of bottom scour and minor gravitational movements.

The gentle but continuous wafting suggested by Riedel in explanation of the distribution of Radiolaria would seem to be a satisfactory explanation of these configurations. In other areas the simple sub-bottom echo, while still detectable, became intensely disturbed, possibly suggesting that the bed is intermittent or has been folded.

The occurrence of a series of closely spaced reflectors on an otherwise smooth bottom (Heezen *et al.*, 1959, pl. 5) may indicate deep-sea dunes (2400 fm) produced by strong bottom scour. Echogram profiles taken in depths of 300 fm beneath the Gulf Stream were interpreted in a similar manner.

Photographs of manganese nodules (Dietz, 1952; Elmendorf and Heezen, 1957) taken in great depths of the Atlantic Ocean floor, and more recently on the Easter Island Ridge of the southeast Pacific (Revelle, 1958), tell us that sedimentation in these areas is virtually at a standstill. Features in the sediment around the nodules resemble scour marks. Thus the gentle scour and reworking of the abyssal red clay is supported both by the study of cores and by examination of the bottom photographs.

10. Conclusion

We can now view the vast abyss with somewhat less mystery and more real understanding. The picture of an unremitting snowfall of sediment which softly casts a thin veil over the rugged, grand and imposing relief of the deep-sea must be tempered with occasional, sometimes frequent, turbidity flows, which, like snow avalanches in the Alps, sweep down and smother the relief of the lower slopes. Just as winds striking the mountains drive the snow from the peaks, so the ocean currents sweep the sediment from the higher and bolder submarine mountains.

Just as wind and water shape the land, so do the submarine currents of water and mud, by erosion, by solution and by scour shape the ocean floor. To be sure, erosion dominates on the land, as does deposition in the deep sea.

References

Agassiz, A., 1888. *Three Cruises of the Blake*, **1**. Harpers, New York, 220 pp.

Andrée, K., 1920. *Geologie des Merresbodens*. Gebrüder Borntraeger, Leipzig, 689 pp.

Arrhenius, G., 1952. Sediment cores from the East Pacific. *Reps. Swed. Deep-Sea Exped.* **5** (1, 2, 3 and 4).

Bally, A., 1957. Turbidity currents—a list of selected references. *J. Alberta Soc. Petrol. Geol.*, **5**, 89–98.

Bell, H. S., 1942. Density currents as agents for transporting sediments. *J. Geol.*, **50**, 512–547.

Benest, H., 1899. Submarine gullies, river outlets, and fresh-water escapes beneath the sea level. *Geog. J.*, **14**, 394–413.

Berrit, G. R. and H. Rotschi, 1956. Chemical analyses of cores from the central and west equatorial Pacific. *Reps. Swed. Deep-Sea Exped.*, **6** (2), 51–58.

Bourcart, J., 1953. Sables "néritiques" à 2750 m de profondeur au large de Bougie (Algérie). *C.R. Acad. Sci. Paris*, **236**, 738–740.

Bourcart, J. and L. Glangeaud, 1956. Perturbations sous-marines et courants de turbidité du Seisme d'Orleansville. *C.R. Acad. Sci. Paris*, **242**, 1504–1506.

Bourcart, J. and L. Glangeaud, 1958. Perturbations sous-marines et courants de turbidité resultant du tremblement de terre d'Orleansville. *Bull. d'Information C.O.E.C.* (*Paris*), **10** (10), 642–656.

Bowden, K. F., 1954. The direct measurement of subsurface currents in the oceans. *Deep-Sea Res.*, **2**, 33–47.

Bramlette, M. N., 1958. Significance of coccolithophorids in calcium-carbonate deposition. *Bull. Geol. Soc. Amer.*, **69**, 121–126.

Bramlette, M. N. and W. H. Bradley, 1942. Geology and biology of North Atlantic deep-sea cores. *U.S. Geol. Surv. Prof. Paper*, **196**, 1–32.

Broecker, W. S., M. Ewing and B. C. Heezen, 1960. Evidence of an abrupt change in climate close to 11,000 years ago. *Amer. J. Sci.*, **258**, 429–448.

Broecker, W. S., K. K. Turekian and B. C. Heezen, 1958. The relation of deep-sea sedimentation rates to variations in climate. *Amer. J. Sci.*, **256**, 503–517.

Bucher, W. H., 1940. Submarine valleys and related geologic problems of the North Atlantic. *Bull. Geol. Soc. Amer.*, **51**, 489–512.

Buffington, E. C., 1952. Submarine natural levees. *J. Geol.*, **60**, 473–479.

Carson, R., 1951. *The Sea around Us*. Oxford Univ. Press, New York, 230 pp.

Clarke, F. W., 1924. The data of geochemistry, 5th ed. *U.S. Geol. Surv. Bull.*, **770**, 841 pp.

Cooper, L. H. N., 1955. Deep-water movements in the North Atlantic as a link between climatic changes around Iceland and biological productivity of the English Channel and Celtic Sea. *J. Mar. Res.*, **14**, 347–362.

Cousteau, J.-Y., 1958. *Calypso* explores an undersea canyon. *Nat. Geog. Mag.*, **113**, 373–396.

Cushman, J. A., 1941. A study of the Foraminifera contained in cores from the Bartlett Deep. *Amer. J. Sci.*, **239**, 128–147.

Daly, R. A., 1936. Origin of submarine canyons. *Amer. J. Sci.*, **31**, 401–420.

Defant, A., 1950. On the origin of internal tide waves in the open sea. *J. Mar. Res.*, **9**, 111–119.

Dietrich, G., 1957. *Allgemeine Meereskunde.* Gebrüder Borntraeger, Berlin, 492 pp.

Dietrich, G., 1957a. Ozeanographische Probleme der deutschen Forschungsfahrten im Internationalen Geophysikalischen Jahr 1957–58. *Deut. Hydrog. Z.*, **10**, 41–61.

Dietz, R. S., 1952. Geomorphic evolution of continental terrace. *Bull. Amer. Assoc. Petrol. Geol.*, **36**, 1802–1819.

Dietz, R. S., 1953. Possible deep-sea turbidity-current channels in the Indian Ocean. *Bull. Geol. Soc. Amer.*, **64**, 375–378.

Drake, C. L., M. Ewing and G. H. Sutton, 1959. Continental margins and geosynclines: the east coast of North America, north of Cape Hatteras. In *Physics and Chemistry of the Earth*, **3**, 110–198. Pergamon Press, London.

Duplaix, S. and A. Calleux, 1956. Sur quelques sables des fonds de 1850 à 4270 m de la Méditerranée. *C.R. Acad. Sci. Paris*, **244**, 104–106.

Elmendorf, C. H. and B. C. Heezen, 1957. Oceanographic information for engineering submarine cable systems. *Bell Syst. Tech. J.*, **26**, 1047–1093.

Emery, K. O. and S. C. Rittenberg, 1952. Early diagenesis of California basin sediments in relation to origin of oil. *Bull. Amer. Assoc. Petrol. Geol.*, **36**, 735–806.

Emiliani, C. and G. Edwards, 1953. Tertiary ocean-bottom temperatures. *Nature*, **171**, 887–888.

Ericson, D. B., M. Ewing and B. C. Heezen, 1951. Deep-sea sands and submarine canyons. *Bull. Geol. Soc. Amer.*, **62**, 961–965.

Ericson, D. B., M. Ewing and B. C. Heezen, 1952. Turbidity currents and sediments in the North Atlantic. *Bull. Amer. Assoc. Petrol. Geol.*, **36**, 489–512.

Ericson, D. B., M. Ewing, G. Wollin and B. C. Heezen, 1961. Atlantic deep-sea sediment cores. *Bull. Geol. Soc. Amer.*, **72**, 193–286.

Ericson, D. B. and B. C. Heezen, 1959. Distribution of fine sediment as an indication of deep current directions. *Intern. Oceanog. Cong. Preprints*, 454–456.

Ewing, J. I. and M. Ewing, 1959. Seismic refraction measurements in the Atlantic Ocean basins, Mediterranean Sea, on the Mid-Atlantic Ridge and in the Norwegian Sea. *Bull. Geol. Soc. Amer.*, **70**, 291–318.

Ewing, M., D. B. Ericson and B. C. Heezen, 1958. Sediments and topography of the Gulf of Mexico. In Weeks, L. (Ed.), *Habitat of Oil.* Amer. Assoc. Petrol. Geol., Tulsa, 995–1058.

Ewing, M., B. C. Heezen, D. B. Ericson, J. Northrop and J. Dorman, 1953. Exploration of the Northwest Atlantic Mid-Ocean Canyon. *Bull. Geol. Soc. Amer.*, **64**, 865–868.

Forel, F. A., 1885. Les ravins sous-lacoustre des fleuves glaciaires. *C.R. Acad. Sci. Paris*, **101**, 725–728.

Gaskell, T. F., 1958. Drilling deep holes into the earth. *Proc. Roy. Inst. G.B.*, **37**, 189–209.

Goldberg, E. D. and G. Arrhenius, 1958. Chemistry of Pacific pelagic sediments. *Geochim. et Cosmochim. Acta*, **13**, 153–212.

Gorsline, D. S. and K. O. Emery, 1959. Turbidity current deposits in San Pedro and Santa Monica basins off southern California. *Bull. Geol. Soc. Amer.*, **70**, 279–290.

Gould, H. R., 1951. Some quantitative aspects of Lake Mead turbidity currents. *Spec. Pub. Soc. Econ. Paleont. Mineral.*, **2**, 34–52.

Grabau, A. W., 1913. *Principles of Stratigraphy.* Seiler, New York, 1185 pp.

Greenman, N. N. and R. J. LeBlanc, 1956. Recent marine sediments and environments of northwest Gulf of Mexico. *Bull. Amer. Assoc. Petrol. Geol.*, **40**, 813–847.

Grover, N. C. and C. S. Howard, 1938. The passage of turbid water through Lake Mead. *Trans. Amer. Soc. Civ. Eng.*, **103**, 720–790.

Gutenberg, B., 1939. Tsunamis and earthquakes. *Bull. Seism. Soc. Amer.*, **29**, 517–526.

Heezen, B. C., 1956. Origin of submarine canyons. *Sci. Amer.*, **195**, 36–41.

Heezen, B. C., 1956a. Corrientes de turbidez del Rio Magdalena. *Bol. Soc. Geog. Colombia*, **51** and **52**, 135–143.

Heezen, B. C., 1959. Dynamic processes of abyssal sedimentation: erosion, transportation and redeposition on the deep-sea floor. *Geophys. J. Roy. Astr. Soc.*, **2**, 142–163.

Heezen, B. C., D. B. Ericson and M. Ewing, 1954. Further evidence for a turbidity current following the 1929 Grand Banks earthquake. *Deep-Sea Res.*, **1**, 193–202.

Heezen, B. C. and M. Ewing, 1952. Turbidity currents and submarine slumps and the Grand Banks earthquake. *Amer. J. Sci.*, **250**, 849–873.

Heezen, B. C. and M. Ewing, 1955. Orleansville earthquake and turbidity currents. *Bull. Amer. Assoc. Petrol. Geol.*, **39**, 2505–2514.

Heezen, B. C., M. Ewing and D. B. Ericson, 1951. Submarine topography in the North Atlantic. *Bull. Geol. Soc. Amer.*, **62**, 1407–1409.

Heezen, B. C., M. Ewing and D. B. Ericson, 1954. Reconnaissance survey of the abyssal plain south of Newfoundland. *Deep-Sea Res.*, **2**, 122–133.

Heezen, B. C., M. Ewing and R. J. Menzies, 1955. The influence of submarine turbidity currents on abyssal productivity. *Oikos*, **6**, 170–182.

Heezen, B. C., M. Ewing, R. J. Menzies and W. S. Broecker, 1958. Date of stagnation of the Cariaco Trench. *Bull. Geol. Soc. Amer.*, **69**, 1579.

Heezen, B. C., M. Ewing, R. J. Menzies and N. C. Granelli, 1957. Extension of the Congo Submarine Canyon. *Bull. Geol. Soc. Amer.*, **68**, 1743.

Heezen, B. C., M. Tharp and M. Ewing, 1959. The floors of the oceans: I, The North Atlantic. *Geol. Soc. Amer. Spec. Paper* 65, 126 pp.

Hill, M. N., 1957. Recent geophysical exploration of the ocean floor. In Ahrens, L. H. *et al.* (edit.). *Physics and Chemistry of the Earth*, 129–163. Pergamon Press, London.

Houot, G. S., 1958. Four years of diving to the bottom of the sea. *Nat. Geog. Mag.*, **113**, 715–731.

Jarke, J., 1958. Sedimente und Mikrofaunen im Bereich der Grenzschwelle zweier ozeanischer Räume, dargestellt an einem Schnitt über den Island-Färöer-Rücken (nordatlantischer Ozean Rosengarten-europäisches Nordmeer). *Geol. Rundschau*, **47**, Heft 1, 234–249.

Johnson, D., 1939. *Origin of Submarine Canyons*. Columbia Univ. Press, New York, 216 pp.

Jones, O. T., 1954. The characteristics of some Lower Palaeozoic marine sediments. *Proc. Roy. Soc. London*, **A222**, 332-372.

Kay, M., 1951. North American geosynclines. *Geol. Soc. Amer. Mem.*, **48**, 143 pp.

Koczy, F. F., 1954. A survey of deep-sea features taken during the Swedish Deep-Sea Expedition. *Deep-Sea Res.*, **1**, 170–184.

Koczy, F. F. and M. Burri, 1958. Essai d'interpretation de quelques formes du terrain sous-marin. *Deep-Sea Res.*, **5**, 7–17.

Kolbe, R. W., 1957. Diatoms from equatorial Indian Ocean cores. *Reps. Swed. Deep-Sea Exped.* **9** (1), 1–50.

Kolbe, R. W., 1957a. Fresh-water diatoms from Atlantic deep-sea sediments. *Science*, **126**, 1053–1056.

Kroll, V. S., 1955. The distribution of radium in deep-sea cores. *Reps. Swed. Deep-Sea Exped.*, **10** (1), 1–32.

Kuenen, Ph. H., 1937. Experiments in connection with Daly's hypotheses on the formations of submarine canyons. *Leidsche Geol. Med.*, **8**, 327–335.

Kuenen, Ph. H., 1950. *Marine Geology*. Wiley, New York, 568 pp.

Kuenen, Ph. H., 1950a. Turbidity currents of high density. *Rep. 18th Intern. Geol. Cong., London, 1948*, Part VIII, 44–52.

Kuenen, Ph. H., 1952. Estimated size of the Grand Banks turbidity current. *Amer. J. Sci.*, **250**, 874–887.

Kuenen, Ph. H., 1956. The difference between sliding and turbidity flow. *Deep-Sea Res.*, **3**, 134–139.

Kuenen, Ph. H. and C. I. Migliorini, 1950. Turbidity currents as a cause of graded bedding. *J. Geol.*, **58**, 1–127.

Kullenberg, B., 1954. Remarks on the Grand Banks turbidity current. *Deep-Sea Res.*, **1**, 208–210.

Landes, K. K., 1952. Our shrinking globe. *Bull. Geol. Soc. Amer.*, **63**, 225–240.

Locher, F. W., 1954. Ein Beitrag sum Problem der Tiefseesande im westlichen Teil des aquatorialen Atlantiks. *Heldelberger Beitr. Mineral. Petrog.*, **4**, 135–150.

Lombard, A., 1956. *Géologie sédimentaire; les séries marines*. Masson, Paris, 722 pp.

Malaise, R., 1957. Oceanic bottom investigations and their bearings on geology. *Geol. Fören. Stockholm Förhandl.*, **79**, 195–224.

Maury, M. F., 1855. *Physical Geography of the Sea*. Harpers, New York, 389 pp.

Menard, H. W., 1952. Deep ripple marks in the sea. *J. Sediment. Petrol.*, **22**, 3–9.

Menard, H. W., 1955. Deep-sea channels, topography, and sedimentation. *Bull. Amer. Assoc. Petrol. Geol.*, **39**, 236–255.

Menard, H. W., 1956. Archipelagic aprons. *Bull. Amer. Assoc. Petrol. Geol.*, **40**, 2195–2210.

Menard, H. W. and J. C. Ludwick, 1951. Applications of hydraulics to the study of marine turbidity currents. *Spec. Pub. Soc. Econ. Paleont. Mineral.*, **2**, 2–13.

Menzies, R. J. and J. Imbrie, 1958. On the antiquity of the abyssal fauna. *Oikos*, **9**, 192–210.

Milne, J., 1897. Suboceanic changes. *Geog. J.*, **10**, 129–146, 259–289.

Murray, J. and J. Hjort, 1912. *Depths of the Ocean*. John Murray, London, 821 pp.

Murray, J. and A. F. Renard, 1891. *Report on the Scientific Results of the Voyage of H.M.S. 'Challenger', 1873–76*. Neill, Edinburgh, 525 pp.

Natland, M. L. and Ph. H. Kuenen, 1951. Sedimentary history of the Ventura Basin, California, and the action of turbidity currents. *Spec. Pub. Soc. Econ. Paleont. Mineral.*, **2**, 78–107.

Northrop, J. and B. C. Heezen, 1951. An outcrop of Eocene sediment on the continental slope. *J. Geol.*, **59**, 396–399.

Orr, W. L., K. O. Emery and J. R. Grady, 1958. Preservation of chlorophyll derivatives in sediments off southern California. *Bull. Amer. Assoc. Petrol. Geol.*, **42**, 925–962.

Pettersson, H., 1954. *The Ocean Floor*. Yale Univ. Press, 179 pp.

Pettersson, H., 1955. Manganese nodules and oceanic radium. *Deep-Sea Res.*, **3**, suppl., 335–345.

Phleger, F. P., 1951. Displaced Foraminifera faunas. *Spec. Pub. Soc. Econ. Paleont. Mineral.*, **2**, 66–75.

Raitt, R. W., 1956. Seismic-refraction studies of the Pacific Ocean basin, Part I, Crustal thickness of the equatorial Pacific. *Bull. Geol. Soc. Amer.*, **67**, 1623–1640.

Revelle, R., 1958. The Downwind Expedition to the South-east Pacific. *Trans. Amer. Geophys. Un.*, **39**, 429–528.

Revelle, R., M. N. Bramlette, G. Arrhenius and E. D. Goldberg, 1955. Pelagic sediments of the Pacific. *Geol. Soc. Amer. Spec. Paper* 62, 221–236.

Riedel, W. R., 1957. Radiolaria, a preliminary stratigraphy. *Reps. Swed. Deep-Sea Exped.* **6** (3), 59–96.

Rigby, J. K. and L. H. Burckle, 1958. Turbidity currents and displaced fresh-water diatoms. *Science*, **127**, 1504.

Roher, E., 1954. Staublawinen–ja oder nein? *Die Alpen*, **3**, 62–69.

Rubin, M. and H. E. Suess, 1955. U.S. Geol. Survey radiocarbon dates II. *Science*, **121**, 481–488.

Shepard, F. P., 1948. *Submarine Geology*. Harpers, New York, 348 pp.

Shepard, F. P., 1954. High velocity turbidity currents, a discussion. *Proc. Roy. Soc. London*, **A222**, 323–326.

Spencer, J. W., 1903. Submarine valleys off the American coast and in the North Atlantic. *Bull. Geol. Soc. Amer.*, **14**, 207–226.

Stetson, H. C. and J. F. Smith, 1937. Behavior of suspension currents and mud slides on the continental slope. *Amer. J. Sci.*, **35**, 1–13.

Stoneley, R., 1957. On turbidity currents. *Koninkl. Ned. Geol.-Mijn. Gen.*, **18**, 279–285.

Swallow, J. C., 1955. A neutral-buoy float for measuring deep current. *Deep-Sea Res.*, **3**, 74–81.

Swallow, J. C., 1957. Some further deep current measurements using neutrally-buoyant floats. *Deep-Sea Res.*, **4**, 93–104.

Swallow, J. C. and L. V. Worthington, 1957. Measurements of deep currents in the western North Atlantic. *Nature*, **179**, 1183.

Tait, J. B., 1956. Recent oceanographical investigations in the Fåroe–Shetland Channel. *Proc. Roy. Soc. Edinburgh*, **64**, 239–289.

Terzaghi, K., 1957. Varieties of submarine slope failures. *Norw. Geotech. Inst. Pub.*, **25**, 1–15.

Volchok, H. L. and J. L. Kulp, 1957. The ionium method of age determination. *Geochim. et Cosmochim. Acta*, **11**, 219–246.

Weber, M., 1900. Die niederlandische SIBOGA Expedition zur Untersuchung der marinen Fauna und Flora des Indischen Archipels. *Petermann's Geog. Mitt.*, **46**, 187.

Worzel, J. L., 1959. Extensive deep-sea sub-bottom reflections identified as white ash. *Proc. Nat. Acad. Sci.*, **45**, 349–355.

Wüst, C., 1957. Quantitative Untersuchungen zur Statik und Dynamik des Atlantischen Ozeans; Stromgeschwindigkeiten und Strommengen in den Tiefen des Atlantischen Ozeans. *Wiss. Ergebn. Deut. Atlant. Exped. 'Meteor', 1925–27*, **6** (2), 420 pp.

Zenkovich, V. P., 1958. *Shores of Black and Azov Seas*. State Geographic Editions, Moscow, 371 pp.

28. ORGANIC TRANSPORTATION OF MARINE SEDIMENTS

K. O. Emery

1. Introduction

The chief agents of erosion of the ocean floor are inorganic in nature; however, organic agents locally may be more important. Organisms make use of both mechanical and chemical methods of erosion. Simple mechanical erosion is illustrated by the biting off of coral tips by fishes in search of the polyps within, and by the breaking away of projecting rocks by attached kelp lifted by storm waves. Erosion of rock by pholads, snails, chitons, limpets, echinoids and worms (Barrows, 1917; Emery, 1960, pp. 15–19) is also mostly mechanical in nature, but is aided to an unknown extent by biochemical solution of the rocks. Sponges and blue-green algae (Ranson, 1955; Revelle and Fairbridge, 1957, pp. 279, 280) appear to do their work mostly by biochemical methods. Biochemical activity alone, or almost alone, is responsible for the development and growth of solution basins and the leveling of reef flats (Revelle and Emery, 1957).

Organisms transport and deposit some sediments which are produced by inorganic agents as well as those which they themselves eroded. The organic transporting agents play roles which make difficult an assignment of order of importance. For example, worms probably transport a greater quantity of sediment than do all other organisms combined, but the movement is only a few centimeters; in contrast, driftwood and birds carry small quantities for thousands of kilometers. Each of the major kinds of transporting organisms will be considered in turn, but in a more or less arbitrary order.

Many oceanographers appear to know of several instances of organic transportation of stones, but few realize the extent of this transport or the number of organic agents involved; accordingly, a relatively large bibliography is included for its possible value in other studies.

2. Kelp

The larger seaweeds, particularly the genera of larger brown algae, are known as kelps. Largest of these is *Macrocystis*, which reaches lengths of 200 m, and *Pelagophycus* and *Nereocystis*—about 30 m. All of these plants live in waters mostly between 5 and 20 m deep, exceptionally 30 m. Several other genera (*Egregia* and *Pterygophora*) occur in shallower water, up to the level of low tide. Gas-filled floats keep most of the plant near the water surface where the blades spread out in a broad canopy to catch the sunlight. At the base is a holdfast, or rootlike mass, composed of branching structures attached to the bottom. Usually attachment is to bedrock or to boulders and cobbles, but many holdfasts also enclose much shelly fine gravel and sand and some are imbedded only in sand (Thompson, 1959). The writer has also seen two holdfasts of *Macrocystis* attached to the tops of abalone (*Haliotis*) shells.

Large waves from storms often exert a greater lifting force on the plant and its floats than the attachment can stand. Unless the stipe, or stem, parts

[*MS received March, 1960*]

first, the holdfast can break off and lift part of the bottom and the plant be set adrift. If one plant drifts against another, the combined pull of both may break the second one loose, finally tearing open a wide swath through a kelp bed. The anchor, buoyed by the plant, may be floated away from the kelp bed (Fig. 1), occasionally touching or dragging on the bottom to form markings which in geological strata sometimes have mistakenly been identified as plant

Fig. 1. Two kelp holdfasts (*Macrocystis pyrifera*) afloat in 25 m depth off Coronado Islands, Mexico. Largest is about 40 cm long and both consist of probable Middle Miocene sandstone and conglomerate. Note that intermittent touching of the bottom can produce holes and other markings on the sand. (U.S. Navy photograph by R. F. Dill.)

impressions. Most of the rocks are drifted inshore, as shown by the fact that beaches near kelp beds usually contain many rocks having remnants of holdfasts and encrustations of calcareous red algae, bryozoans, rock oysters, and other organisms which live below the low tide level (Emery and Tschudy, 1941), Single rocks rafted to beaches by kelp weigh as much as 10 kg, and one holdfast was found to contain an estimated 25,000 pebbles and granules coarser than one millimeter in diameter.

Some plants also float seaward where masses of the larger ones have been observed several hundred kilometers from shore in waters several kilometers deep. In fact, free-floating kelp was taken as one of the signs of nearness to the rugged California coast by pilots of the Manila galleons en route to Acapulco; thereupon, the galleons were turned to a southeasterly course to reach Acapulco (Schurz, 1959, p. 239). An unknown percentage of these freely drifting kelps still tow their rocky anchors through the water. An example reported by Menard

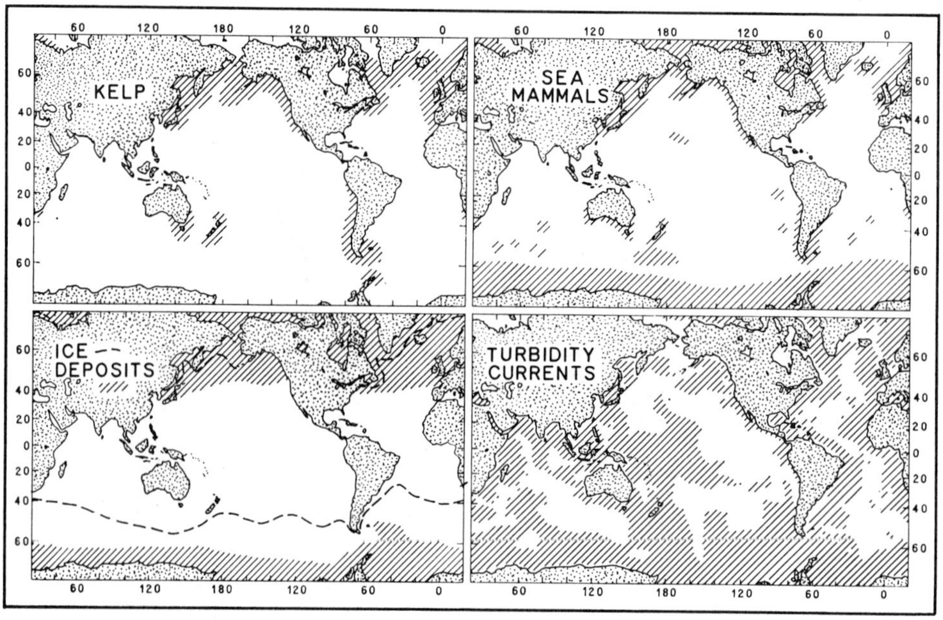

Fig. 2. Areas of greatest importance to chief transporting agents of coarse sediments.

Kelp: cross-hatching shows areas likely to be reached by kelp rafting from coastal zones where kelp lives. (After McGill, 1958.)

Sea mammals: cross-hatching shows areas where transportation by sea-lions and most other sea mammals may occur. (After Scheffer, 1958.)

Ice: dashed line indicates equatorward limit of present-day floating ice; cross-hatching shows areas where bottom sediment may be classed as glacial-marine origin. (From Dietrich and Kalle, 1957; Hough, 1956; Menard, 1953; and others.)

Turbidity currents: cross-hatching shows areas accessible to turbidity currents, i.e. omits mid-oceanic ridges, continental shelves. (Modified from Elmendorf and Heezen, 1957.)

(1953) was a raft of *Nereocystis* about 500 km offshore in the Gulf of Alaska suspending a subround pebble about 8 cm in diameter. At least two rocks having remnants of holdfasts were dredged from the bottom at depths of about 215 m off southern California, but many other rocks have been similarly rafted and their holdfasts decomposed or eaten by benthic animals. Rounded and partly pholad-bored pebbles and cobbles of various rock types dredged from about 3800 m off western Mexico have thus been attributed to kelp rafting

(Shumway, 1953). In addition to rafting rocks, kelp also transports shells, foraminiferal tests and other organic remains far from the environment in which they had lived. Since representatives of nearly every phylum are known to live in or on holdfasts, their transportation to deep water can raise problems in ecology and paleoecology.

Smaller algae such as sea lettuce (*Ulva*) and some brown algae (*Lessoniopsis* and *Laminaria*) have no floats but provide a sail-like surface which permits somewhat stronger than normal waves and currents to move attached pebbles and shells. Examples of transportation of 2-cm pebbles and shells several kilometers were reported by Brown (1911), the movement leaving trails along the bottom.

The kelps at present have an antipodal distribution (Fig. 2), being confined to cool waters of high latitudes or areas of strong upwelling (McGill, 1958). However, in order to achieve this distribution they must have lived to some extent at low latitudes during the colder glacial epochs of the Pleistocene. It is likely that even then the kelps were more abundant at high latitudes than at low ones. Because the plants can drift freely only a limited time owing to losses of floats by decomposition and eating by fishes, kelp-rafted rocks and organic remains are probably confined to an area within a few hundred kilometers from shore: farther in cold water at high latitudes than in warm water nearer the equator (Fig. 2). Rafting by kelp has probably also occurred during the geological past because brown algae are well known from the Miocene (Gardner, 1923; Emery, 1960, pp. 162–166) and lived much earlier, even in Silurian and Devonian times (Krausel and Weyland, 1930).

Locally, sand and shells are entangled and transported a short distance (usually less than 100 m) in the balls of eel grass, *Phylospadix*, and of *Posidonia* (Aleem, 1955) and of other plants (Olson, 1957) which are rolled onto beaches from depths of a few meters, sometimes in enormous numbers. According to Guilcher (1958, p. 32), *Posidonia* may cause a rise in the sea floor near such beaches amounting to one meter per century by trapping of sediment particles in the balls. Examples of algal balls, possibly of this origin, with enclosed sand are reported from rocks as old as Silurian (Croneis and Grubbs, 1939).

3. Driftwood

Trees which grow in rocky soil commonly enclose masses of earth, pebbles, cobbles and even boulders in and between their roots. These trees are sometimes undermined by streams and are washed out to sea or they fall directly into the sea by cliff erosion. Once afloat they have been known to be carried thousands of kilometers by ocean currents (Brooks, 1875; Rouch, 1954; Barber, Dadswell and Ingle, 1959). One floating tree is reported to have held a boulder three or four meters in diameter (Emery, 1955). Eventually, the wood loses its buoyancy by attacks of wood-boring animals and by soaking up water so that it sinks to the bottom carrying its cargo of rocks. Other logs drift ashore onto the mainland or islands. Especially on the islands of the central Pacific Ocean the rocks so carried were highly prized by the natives, frequently being designated as belonging to the chiefs. Occasional loose exotic stones found on the beaches of coral islands by Chamisso (1821) and by many later travellers probably arrived via driftwood which afterward decomposed. Although the quantity of rocks transported by driftwood is small, both the size of individual rocks and the distances are great.

Distance from shore is much less limited for rafting by driftwood than by kelp, so one might suppose that rocks carried by driftwood might be distributed thinly but widely throughout the oceans. There may be some relationship to major rivers and forested regions. Probably most of the rocks so carried are angular and many are weathered.

Erratic rocks are common in ancient sedimentary strata, particularly in coals, of Britain, Europe, Australia and the United States (Emery, 1955). Most likely these rocks were held by tree roots and rafted to the sites of their burial in coastal swamps while others were carried to the ocean beyond.

4. Sea Mammals

Gastroliths have been found in most of the smaller sea mammals—sea-lions, sea-elephants, fur seals, harbor seals and walrus (Hamilton, 1934; Scheffer, 1950; Fleming, 1951, 1953; Brooks, 1954; Loughrey, 1959; see other references in Emery, 1941). Most gastroliths consist of stones which are slightly to well rounded. Occasional mollusk shells, usually beach worn, are also present; Andrée (1920, p. 393) reported more than 500 shells of *Mya truncata* in the stomach of a single walrus. Individual stones have been recorded with diameters up to 10 cm and weights to 2 kg, but single stomachs have contained more than a hundred individual stones. The largest aggregate weight reported is 37 kg in a sea-lion (Mohr, 1952), but weights of more than 5 kg are unusual. There is no agreement as to why stones are picked up and carried as gastroliths; the animals may use the stones as an aid in triturating their food, as ballast, to destroy nematode worms, or perhaps they merely pick them up in play. In any event stones have been found in males and females, in old and young, even in milk-feeding pups, and in all seasons. Where stones are lacking, other objects may be picked up. For example, a harbor seal in an open tank at the Steinhart Aquarium in San Francisco became a floating bank; according to E. S. Herald, Curator (*in litt.*), "it had swallowed $7.48 in coins, ranging from French centimes to U.S. 50¢ pieces. Corrosion of copper pennies indicated that some of the coins had been in the stomach for a considerable length of time. Eventually, of course, this resulted in the demise of the harbor seal."

More is known of gastroliths in stomachs of the California sea-lion (*Zalophus californianus*) than of other sea mammals, owing to collections made by A. L. Kelly (Scripps Institution of Oceanography) and Emery along the mainland and islands of southern California and Mexico. Stones were found in 73 of 170 stomachs (Fig. 3). Where land geology is well known and the local rocks are of diagnostic types, it is possible to determine whether or not the sea-lions picked

TABLE I

Source of Gastroliths in some California Sea-Lions. Based on Autopsies by Arthur L. Kelly and K. O. Emery

Area where sea-lion was killed	Number of sea-lions having stones	Source of stones same as at collection site
North Coronado Island, Mexico[a]	11	8
South Coronado Island, Mexico[a]	5	5
San Nicolas Island, California[b]	3	2
La Jolla, California[c]	5	4

[a] North Coronado Island has only Miocene red sandstone, whereas South Coronado Island, only 3 km distant, has only Miocene yellow sandstone and a conglomerate composed of metamorphic and igneous rocks.

[b] San Nicolas Island is entirely Eocene shale and sandstone.

[c] La Jolla gravel beaches are mostly of a Mesozoic metavolcanic rock reworked from Eocene conglomerates in sea cliffs.

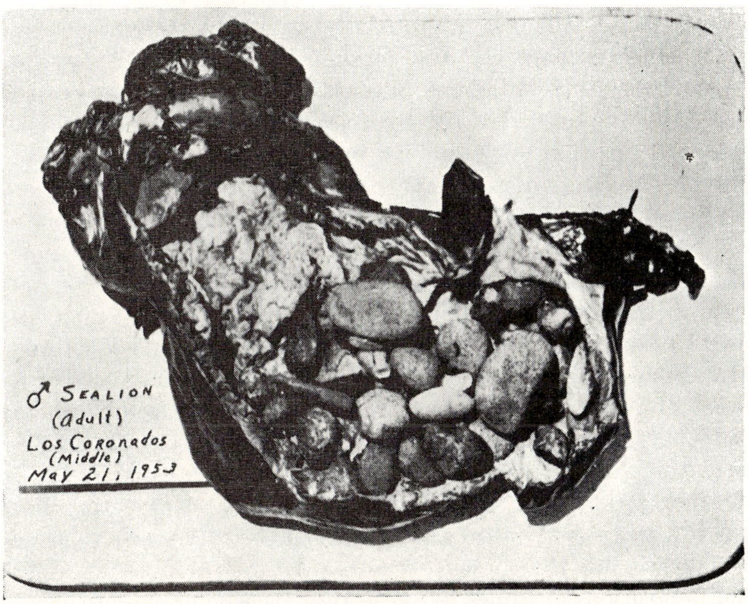

Fig. 3. Dissected stomach of sea-lion (*Zalophus californianus*) taken at Coronado Islands, Mexico, by A. L. Kelly. Largest stone is about 5 cm long. All are well rounded and are of Middle Miocene sandstone similar to that of which the islands are composed.

up the stones locally. Where identifiable the stones are mostly of local types, as shown by Table I, indicating that the sea-lion probably does not carry stones very far or very long. This conclusion is verified by observations that in zoo tanks stones are picked up after meals and regurgitated a few hours later, and by the fact that of 22 fur seals collected in the open sea by Austin and Wilke

(1950), none contained stones. Support is also given by the fact that the stones are only slightly polished, far less than typical gastroliths of dinosaurs (Wieland, 1907). On the other hand, Fleming (1951, 1953) reported finding basalt pebbles, presumably sea-lion gastroliths, on Snares Island off New Zealand about 250 km from the closest shallow water or land source of basalt.

Data are scanty, but porpoises also are known to carry stomach stones. According to Kenneth S. Norris (Marineland of the Pacific, Palos Verdes, California, *in litt.*) three or four freshly caught porpoises, *Tursiops gilli*, were found to contain shells and gravel to about 1 cm in diameter. One stomach was so tightly packed with gravel as to be a possible contributing cause of death. During captivity porpoises are observed frequently to pick up and play with stones in the tank, sometimes swallowing them. Stomach stones have not been reported in whales, not even in the sperm whale, which feeds on benthos by plowing its lower jaw through the bottom at water depths as great as 1135 m, in the process sometimes becoming entangled in telegraph cables (Heezen, 1957).

An interesting method of transportation is provided by sea-otters, which use stones as tools to break up and detach prey (Kenyon, 1959; Limbaugh, 1961). Sea-otters may thus be the only mammals in the natural state other than man to use tools. The animals commonly float on their backs carrying a stone as large as 20 cm in diameter on their stomachs. Mussels and other shells are rapped against the stone until broken open. The otter carries its stone during dives, keeping the same one for periods that are unknown but in at least one instance more than eight hours (R. A. Boolootian, University of California at Los Angeles, *in litt.*). During its foraging the otter may transport a stone as far as 20 km along the shore.

Another curious form of transportation is as enteroliths, concretionary bodies which grow in the kidneys, gall bladder, lungs, and other organs. They are best known in land mammals, but also occur in fishes and lobsters (Hutton, 1942, 1945). Some enteroliths of land mammals reach diameters of 8 cm and most consist of several phosphate minerals; others are composed of various organic salts. Some are probably passed from the system during life by sea mammals as they are by land mammals.

The sea mammals which most commonly carry stones are those which inhabit shallow water and spend a large part of their lives ashore or on floe ice. Their distribution, like that of kelp, is antipodal (Scheffer, 1958), and probably most of the depositions of stones by these animals is on or close to shore (Fig. 2). Bones of sea mammals are fairly common in strata as old as Miocene, and at least one instance is known of gastroliths associated with fossil seal and walrus bones—on an island off Massachusetts (Woodworth and Wigglesworth, 1934, pp. 19–20).

5. Fishes

Many fishes of coral islands are known to bite off coral tips in search of polyps and worms and to graze upon algae growing upon dead coral and other rocks (Wood-Jones, 1912, pp. 164, 664; Emery, 1956, and included references; Gohar and Latif, 1959, and included references). Shells, coral, and calcareous algae are usually present among the stomach contents of parrot fish (*Scarus*), surgeon fish (*Acanthurus*), trigger fish (*Melichthys*), and puffer (*Arothron*) from Pacific islands and the Red Sea. When frightened, the fishes defecate clouds of the material comminuted to sand and silt grain size. The great role of fishes in producing fine-grained sediments in lagoons of coral reefs has scarcely been recognized. According to Bardack (1961) such coral-browsing fishes at Bermuda ingest about one ton of calcareous material per hectare per year.

Shallow-water fishes of non-calcareous areas also transport stones. According to Conrad Limbaugh (Scripps Institution of Oceanography, *in litt.*) cabezone (*Scorpaenichthys*) carries stones in its stomach in shallow water (less than 10 m). The stones range up to 10 cm in diameter and weigh as much as a kilogram. Possibly these stones are transported seaward to the 100-m depth limit of this fish. Halibut, a bottom-living fish generally in deep water, was once noted by Thompson (1919) to contain a stone weighing about one kilogram, but its presence was ascribed "to reckless eating, not foresight". According to J. E. Fitch (California State Fisheries Laboratory, Terminal Island, *in litt.*) many other bottom-living fishes also carry gastroliths. Among them are flounders, sculpins and rock cods. Others, which live just above the bottom, sheepshead, ling cod and ocean whitefish, also sometimes carry them. R. V. Vaughan (San Diego, *in litt.*) stated that about 60% of the first arrivals in southern California of the pelagic albacore tuna (*Thunnus alalonga*) contain stones. Most are granitic or basaltic types, rounded and polished, about one cm in diameter. Aggregate weights range up to about 100 g. Later in the season gastroliths appear to be rare; in fact, they have not been recorded among the thousands of tuna stomachs examined by personnel of the California State Fisheries Laboratory. Some bottom-living freshwater fishes have the same habit, and, according to R. J. Miller of Long Beach State College (*in litt.*), 8.2% of 146 specimens of *Cottus beldingi* from Lake Tahoe, California, contained stones.

In addition to carrying stones as gastroliths, some fishes carry them for other purposes. Skates and rays, for example, often cover themselves with sand, presumably for concealment; when they swim away some of the sand is carried for tens of meters. According to Limbaugh, pile perch (*Damalichthys vacca*) at Guadalupe Island, off Mexico, carry flat pebbles in their mouths, "presumably as an aid for crushing shellfish". He also reported that jawfishes in the Bahama Islands line their burrows with pebbles, and that during excavation of burrows under rocks, fishes, lobsters, crabs, shrimps, and octopi carry small stones for short distances. Octopi in fact are known to carry pebbles for tens of meters and have been reported to drop stones into clams and oysters to prevent closure (Lane, 1957, pp. 69–72).

Probably the bulk of stone transportation by fishes occurs in coral reef areas, but some exists in other shallow-water coastal regions. Doubtless, fishes have had similar habits during most of post-Ordovician times, though no observations are available of fossil fish gastroliths. A distant relative, the Mesozoic reptile, *Plesiosaurus*, however, is known to have carried gastroliths (Shimer, 1924,

p. 357), but according to W. E. Schevill (Woods Hole Oceanographic Institution, Massachusetts, *in litt.*) in at least several instances these are only poorly polished. Some modern reptiles do carry stones. An 8-m cayman in the Philippine Islands was found to contain 70 kg of stones to about 12 cm diameter (de la Gironiere, 1853), and 545 out of 681 specimens of the Nile crocodile examined by Cott (1961) contained stones averaging in total about 1% of body weight.

6. Birds

It is well known that chickens must have available a supply of grit or small pebbles to serve as grinding tools against seeds in their gizzards. Many other seed-eating land birds pick up sand and grit for the same purpose. The largest land bird, the extinct moa of New Zealand, carried gizzard stones as much as 2 cm in diameter, as noted by the writer during the excavation of a moa skeleton, and according to Hamilton (1892) stones associated with a single skeleton may weigh in aggregate nearly 3 kg. The instinct or need for grinders is so well developed that in the absence of sand or pebbles swans at Tule Lake and nearby game refuges in California pick up so much of the estimated 52-ton annual fall-out of lead shot that many sicken and die of lead poisoning (Rosen and Bankowski, 1960).

Transportation of pebbles by shore-based sea birds is less well known but definitely occurs. Penguins, living far from known sources, commonly carry stones of a wide variety of rock types (Murphy, 1936, pp. 353, 364, 377, 397, 451). As much as 300 g of pebbles having diameters up to about 1 cm have been found in young penguin chicks (J. M. Sieburth, Virginia Polytechnic Institute, *in litt.*). Adults of the large emperor penguin have been reported (Sclater, 1888) to contain pebbles of 3.3 cm diameter with an aggregate weight of 5 kg. Since the stomachs are not muscular, the stones cannot serve as trituration aids, but possibly are used for ballast. They are commonly regurgitated near the nests and picked up again when the penguin returns to the water. Similar pebbles have been dredged by M. Ewing of Lamont Geological Observatory (Fellows, 1958) from deep water near Antarctica where they may have fallen from floating ice on which nests had been built. Many other sea birds from the Antarctic, such as shearwaters, storm petrels, diving petrels, steamer ducks, oyster catchers, sheath bills, cape pigeons and ducks are reported by Murphy (1936, pp. 666, 672, 770, 786, 962, 968, 980, 981, 1006, 1056, 1081) to contain pebbles, shells and sand in their stomachs. At Cedros Island, Mexico, Kelly found three sub-round 1.5-cm pebbles of a local kind of schist in a cormorant, but two other cormorants taken at La Jolla, California, contained none. At least one instance is recorded of fossil gizzard stones—in phosphate deposits of Kita–Daito–Zima (Yabe and Sugiyama, 1934).

Some sea birds, especially gulls, are well known for their shell-opening tactics, by dropping clams, mussels and other mollusks from a height of 10 to 20 m onto sandy or rocky shores. An entertaining description of the technique is given by

Oldham (1930), who also mentioned the application of the same technique to the opening of hermit crabs, turtles, and even walnuts by various birds. According to Fitch, it is sometimes a little hazardous to walk on the causeway at Mission Bay, near San Diego, or along the shore at Pebble Beach, California, when gulls are preparing clam dinners. Accumulations of broken shells formed in this manner are not uncommon (Andrée, 1926, pp. 719, 730; Teichert and Serventy, 1947). Many broken tubes of vermetid mollusks noted by the writer atop a small island near Guaymas, Mexico, were probably so deposited. A curious indirect role is played by woodpeckers through their habit of placing pebbles in holes drilled in tree trunks when nuts are not available (Ritter, 1938); the possible later floating away of the trees results in transportation of the pebbles.

An occurrence of a new species of mollusk in a guano deposit, as described by Smith (1952), resulted from the swallowing of the mollusk by a fish which was later caught and eaten by a cormorant which deposited the mollusk by regurgitation or excretion in its nesting area.

Living animals and plants may also be deposited unintentionally by birds as contamination on dirty feet (Ladd, 1960), or stuck to feathers (Wiens, 1962). Murphy (1936, p. 422) even mentioned the presence of barnacles growing on the feet of a penguin presumably long at sea. B. C. Cotton described a Black Duck from Australia on whose foot a 30-g mussel had closed; the bird had been seen flying with its foot hanging and was shot. The same species of mussel withstood six months' exposure in mud after a river flood; another specimen survived 231 days out of water followed by shipment to England (Cotton, *in litt.*). In other instances birds have been trapped by large clams (Jones, 1959).

Some sea birds move quantities of stones or mud in building their nests. Penguins use pebbles and have been observed to walk a kilometer to get a single nest pebble, often by robbing an untended nest (Murphy, 1936, pp. 373, 393, 394, 412, 447, 449). According to Limbaugh, noddy terns at Clipperton Island have carried hundreds of tons of coral rubble and shell for courtship and nesting materials onto the rocky surface around the lagoon. Albatrosses, pelicans and boobys mostly use mud for their nests, carrying it from more or less distant shore areas to places where such marine sediments would not naturally occur.

During migrations or other travels, birds probably transport sand, shells and pebbles as far as thousands of kilometers, eventually depositing them by regurgitation, excretion or death. The quantities of sediment so deposited, however, are so small that they may escape notice. In at least one instance Sayles (1931, p. 443) suggested a bird-borne source of quartz grains in the soils of Bermuda; however, more recent work indicates that the quartz sand more probably came from underlying igneous rocks.

7. Invertebrates

Worms, holothurians, mollusks, crustaceans and other invertebrates contribute in varying degrees to sediment transport. Most important are the various kinds of worms which live in the bottom sediments. Since most are deposit feeders, they annually pass large quantities of sediment through their digestive tracts, in the process making burrows and dumps of fecal pellets and destroying the bedding of sediments. Such features are common in ancient as well as in recent marine sediments (Schwartz, 1932; Shrock, 1935; Seilacher, 1953; Schäfer, 1956).

As shown by Darwin (1883, pp. 159, 305), earthworms on average ground of England bring up subsoil at a rate corresponding to a layer 0.5 cm thick each year; or in other terms, the average of 6 earthworms (6 g) per square meter moves about 2500 g of soil each year. The work of marine worms has not yet been well measured, but biomasses commonly amount to more than 100 g/m^2 in bottom depths of less than 2000 m, suggesting that worms may even be more important transporters of sediment on the sea floor than on land. It should be noted, however, that most of this transport is vertical, rather than lateral, and for only a few centimeters; thus, its main function is that of mixing sediment. Outstanding is the work of the echiuroid worm, *Echiurus echiurus*, which lives in U-shaped burrows about 30 cm deep in water depths of 10 to 30 m in western Sweden (Gislén, 1940). Specimens measure about 10 cm in length and number from 20 to 335 per square meter of bottom. A medium-sized specimen may excrete daily a total of 540 pellets, each about 2 mm long. Similar intense activity is probably characteristic of a larger echiuroid worm, *Listriolobus pelodes*, which lives in such abundance in muds of the shelf off Santa Barbara, California, as to constitute biomasses of 1100 grams per square meter (Barnard and Hartman, 1959).

During their passage between the hard tooth-like grinder stones in the worm's gizzard, sand grains become somewhat rounded and worn, especially the softer grains. Other modification of grain size of sediment occurs through the sorting activities of some of the sand-reef building polychaetous worms, notably those of the genera *Phragmatopoma* and *Pectinaria*. Species of *Phragmatopoma* are known to select sand grains with such precision that the species can be identified on the basis of grain size of sand composing their tubes. The form endemic to Peru selects grains of the smallest size, the one in Ecuador takes medium sand, and the ones in southern California and the West Indies use the coarsest sand (Hartman, 1944). These species are not known from any other parts of the world. Some terebellid worms select foraminiferal tests of particular kinds or sponge spicules of a certain length and thickness; where these species are abundant, they can account for deposits of considerable extent.

In shallow-water areas of coarse sand in the tropics holothurians probably play a greater role than worms as a geological agent. Studies by Crozier (1918) at Bermuda indicate that *Stichopus mobii* ingests 6 to 7 kg of sand per square meter each year. During the passage of this sand through the intestinal tract it becomes immersed in digestive fluids having a pH as low as 5.0. The resulting solution of calcareous grains corresponds to about 400 g/year, according to experiments by Mayor (1924) at Samoa; this amounts to approximately one per cent of all the sand which is ingested. Analyses by Emery (1962) of sand in and near five specimens of *Holothuria atra* from a reef flat at Guam showed that the finer grains are dissolved first, causing the excreted residual sand to have a coarser median diameter than the total which was ingested. Unlike worms, holothurians move laterally as far as 50 m per day, producing more lateral than vertical transportation of sand.

Many other invertebrates also act as sediment transporters, but those whose role is best known live in the intertidal region, and even there they are only of minor importance as compared to inorganic and other organic agents. Accordingly, they will not be further considered here.

8. Comparison with Inorganic Transporting Agents

Sand and stones are transported by inorganic agents over, on and under the water surface to eventual deposition on the ocean floor. According to Rex and Goldberg (1958) high-altitude jet streams may be responsible for the presence of fine-grained quartz in sediments at mid-latitudes. Other winds carry seaward so much desert silt and sand from the Sahara Desert as to discolor sails of ships (Darwin, 1887, p. 5) and to make an appreciable contribution to the ocean-floor sediment in the east central Atlantic (Radczewski, 1939). The writer has noted similar contributions of wind-blown sediment in the Red Sea, Persian Gulf and Yellow Sea. Darwin mentioned that the dust settling on to the Atlantic consists largely of diatoms, 65 of 67 species being of freshwater origin. Such may well be the origin of the thin layers of freshwater diatoms recently found in cores in the east central Atlantic (Kolbe, 1957), rather than the extreme eustatic or diastrophic movements proposed by Landis (1959) to convert the sea floor into sites of freshwater lakes.

Layers of volcanic ash are widespread in oceanic sediments, being well known in cores from the North Atlantic (Bramlette and Bradley, 1942), Mediterranean (Mellis, 1955) and Pacific (Menard, 1953; Worzel, 1959). The ash owes its presence, of course, both to vulcanism and wind. Coarser ejecta from volcanoes must make its way to distant sites by floating, an easy feat owing to the high porosity of pumice. Probably the chief contributing volcanoes are those which have grown upward from the sea floor to near sea-level. During the past decade eruptions from such volcanoes have been reported off Japan (Niino *et al.*, 1953), off Mexico (Richards, 1958), near Guam and in the Azores. Large quantities of pumice were produced in the nearly simultaneous eruptions off Japan and Mexico, and rounded pieces floated as far as 12,000 km from the Mexican source. The writer has noted much pumice stranded on beaches of the Mediterranean Sea probably from Greek or Sicilian volcanoes. As pointed out by Ladd (1960), floating pumice may transport for long distances mollusks, corals and other attached invertebrates. The same high porosity permits pieces of dried coral to float alone for long distances (Kornicker and Squires, 1962).

Non-volcanic sediments denser than water can also float on calm bodies of water owing to surface tension. Patches of floating sand are common on tide pools (Emery, 1945) and have also been seen on a quiet bay of Japan (Kotaka and Hayasaka, 1954) and on tidal rivers of Australia (Wolf, 1960), but they probably do not occur on the open sea where waves roughen the surface. Flotation of stones as large as 2.5 cm in diameter has been observed (Nordenskiold, 1900) in a Patagonian bay, probably because the local shale was impregnated with petroleum. The drifting ashore of buoyant tar from offshore seeps and from ships is commonplace; however, after some oxidation the tar becomes denser than sea-water and sinks to the bottom. Pieces have been taken from cores in basins off southern California at depths corresponding to deposition about 10,000 years ago and they are dead with respect to radio-carbon dating.

Widespread at high latitudes are deposits of sediment rafted by ice in the form of floes and bergs. These sediments range from clay to boulder size, with the boulders being most easily recognized as of ice origin by their mixed exotic lithologies, angularity, occasional striations, sometimes adhering rock flour and associated poorly sorted sediment (Carsola, 1954; Lisitsin, 1958). Bottom photographs in the Arctic Sea also show many small pits in the mud surface, presumably made by falling ice-rafted pebbles (Hunkins, Ewing, Heezen and Menzies, 1960). Sediments are frozen into the bottom of ice floes in shallow water or are deposited by melt-water streams on the tops of floes and on glaciers which later become water-borne. During the travels of the floating ice the sediment is released mostly by melting of the ice raft. Because of their larger masses, icebergs drift into lower latitudes than floes before melting away; but only floes occur in the North Pacific and eastern North Atlantic Oceans. At present, bergs and floes reach farther equatorward in the Northern Hemisphere than in the Southern, according to charts of Dietrich and Kalle (1957, p. 172). However, glacial marine sediments in the North Pacific (Menard, 1953; Kuno, Fisher and Nasu, 1956) and in the North Atlantic (Peach, 1912; Bramlette and Bradley, 1942) occur far south of the present limits of floating ice (Fig. 2). These deposits are doubtlessly of Pleistocene age. In the Southern Hemisphere the limits of glacial marine sediments recognized by Hough (1956), Lisitsin (1960), Needham (1962), and others are more restricted than is present-day floating ice. The difference between the limits of floating ice and limits of ice-rafted bottom sediment in the Southern and the Northern Hemispheres is probably a result of the relative lack of access of ice to rock on Antarctica (leading to small loads of sediment) as compared to the great ease of access on the continents of the Northern Hemisphere. In any event, ice-rafted sediment is widespread at high latitudes both at the sediment surface and beneath thin layers of post-glacial marine sediment.

Coarse-grained sediments are also transported beneath the ocean in the form of turbidity currents. Field studies have shown widespread deposits of sand on the floors of the Atlantic (Ericson, Ewing and Heezen, 1952) and the Pacific Oceans (Menard, 1955). Associated with the sand, but more restricted to the base of steep slopes, are accumulations of gravel. Pebbles and cobbles have been obtained even from the deep floor of the Philippine Trench by Bruun (1951). Detailed studies of gravels and sands from the basins off southern California

(Gorsline and Emery, 1959) clearly show that the coarser materials there are concentrated at the mouths of submarine canyons. Presumably the sediments were transported by turbidity currents flowing down submarine canyons or down steep slopes where they were initiated by slumping. After deposition of the coarser parts of their loads, the turbidity currents can carry fine sands and silts for additional hundreds of kilometers. Where ridges cross their paths the turbidity currents are blocked, but the areas of their effective deposition are well indicated by smooth, gently sloping topography even at great depths (Elmendorf and Heezen, 1957). Unlike ice-rafted sediments, the turbidity-current deposits can occur at all latitudes, but are largely in strips adjacent to continental areas (Fig. 2).

Possibly the bulk of sea-floor sediment is transported in suspension from shore areas. Transportation of sediment coarser than sand probably does not occur by suspension. The grain size appears to decrease outward to a point beyond which further decrease owing to limitation of transportation is masked by changes caused by diagenetic alteration of clay minerals.

9. Criteria for Recognition of Rafting Agent

The criteria to be used for possible recognition of the agent which rafted a given deposit of sand or gravel in recent, and especially in ancient, sediments must necessarily be somewhat elastic. However, the following are suggested for possible use:

(*a*) Kelp: within 500 km of shore, between 20° and 60° of latitude, often rounded, pholad-bored and encrusted by shallow-water organisms, in clusters.

(*b*) Driftwood: anywhere in ocean, angular, weathered, isolated.

(*c*) Sea mammals: on beach or in shallow water at high latitudes and possibly in deep water below margin of floe ice, less than 10 cm in diameter, rounded, slightly polished, in clusters.

(*d*) Fishes: less than 2 cm in diameter, many of coral, much fine-grained material if calcareous.

(*e*) Birds: less than 1 cm in diameter, varied lithology, in clusters.

(*f*) Wind: sand or finer, angular shards of silt to well-rounded sands, in layers.

(*g*) Ice: at high latitudes, clay to boulders, angular, unweathered, some striated, in layers.

(*h*) Turbidity currents: on flat areas of sea floor, silt to gravel, often graded upward from coarse to fine, gravel usually rounded, in layers.

References

Aleem, A. A., 1955. Structure and evolution of the sea grass communities *Posidonia* and *Cymodocea* in the southeastern Mediterranean. *Essays in the Natural Sciences in Honor of Captain Allan Hancock*. Univ. Southern California Press, 279–298.

Andrée, K., 1920. *Geologie des Meeresbodens*. Gebrüder Borntraeger, Leipzig, 1–689.

Austin, O. L. and F. Wilke, 1950. Japanese fur sealing. Fish and Wildlife Service. *Spec. Sci. Rep., Wildlife*, No. 6, 1–91.

Barber, H. N., H. E. Dadswell and H. D. Ingle, 1959. Transport of driftwood from South America to Tasmania and Macquarie Island. *Nature*, **184**, 203–204.

Bardack, J. E., 1961. On the transport of calcareous fragments by reef fishes. *Science*, **133**, 98–99.

Barnard, J. L. and Olga Hartman, 1959. The sea bottom off Santa Barbara, California, Biomass and community structure. *Pacific Naturalist*, **1** (6), 2–15.

Barrows, A. L., 1917. The geologic significance of fossil rock-boring animals. *Bull. Geol. Soc. Amer.*, **1**, 965–972.

Bramlette, M. N. and W. H. Bradley, 1942. Geology and biology of North Atlantic deep-sea cores between Newfoundland and Ireland. *U.S. Geol. Surv. Prof. Paper*, **196**, 1–34.

Brooks, C. W., 1875. Report of Japanese vessels wrecked in the North Pacific Ocean from the earliest records to the present time. *Proc. Calif. Acad. Sci.*, **6**, 50–66.

Brooks, J. W., 1954. A contribution to the life history and ecology of the Pacific walrus. *Alaska Cooperative Wildlife Res. Unit., Spec. Rep.*, **1**, 1–103.

Brown, A. P., 1911. The formation of ripple-marks, tracks, and trails. *Proc. Phil. Acad. Natural Sci.*, **63**, 536–547.

Bruun, A. F., 1951. The Philippine Trench and its bottom fauna. *Nature*, **168**, 692–693.

Carsola, A. J., 1954. Recent marine sediments from Alaska and northwest Canadian Arctic. *Bull. Amer. Assoc. Petrol. Geol.*, **38**, 1552–1586.

Chamisso, A., von, 1821. *Bemerkungen und Ansichten auf einer Entdeckungs Reise unternommen in den Jahren 1815 bis 1818 auf Kosten Sr. Erlaucht des Herrn Reichekanzlers Grafen Romanzoff auf dem Schiffe Rurick unter dem Befehl des Lieutenants der Russisch-kaiserlichen Marine Otto von Kotzebue*, **3**, Weimar.

Cott, H. B., 1961. Scientific results of an inquiry into the ecology and economic status of the Nile crocodile (*Crocodilus niloticus*) in Uganda and Northern Rhodesia. *Trans. Zool. Soc. London*, **29** (4), 211–356.

Croneis, C. and D. M. Grubbs, 1939. Silurian sea balls. *J. Geol.*, **47**, 598–612.

Crozier, W. J., 1918. The amount of bottom material ingested by holothurians (*Stichopus*). *J. Exper. Zool.*, **26**, 378–389.

Darwin, C., 1883. *The Formation of Vegetable Mould through the Action of Worms with Observations on their Habits.* D. Appleton & Co., New York, 1–326.

Darwin, C., 1887. *Journal of Researches into the Natural History and Geology of the Countries Visited during the Voyage of H.M.S. Beagle Round the World* (new edition). D. Appleton & Co., New York, 1–519.

de la Gironiere, P. P., 1853. *Twenty Years in the Philippines.* Vizetelly and Co., London, 1–372.

Dietrich, G. and K. Kalle, 1957. *Allgemeine Meereskunde, Eine Einführung in die Ozeanographie.* Gebrüder Borntraeger, Berlin, 1–492.

Elmendorf, C. H. and B. C. Heezen, 1957. Oceanographic information for engineering submarine cable systems. *Bell Syst. Tech. J.*, **36**, 1047–1093.

Emery, K. O., 1941. Transportation of rock particles by sea-mammals. *J. Sediment. Petrol.*, **11**, 92–93.

Emery, K. O., 1945. Transportation of marine beach sand by flotation. *J. Sediment. Petrol.*, **15**, 84–87.

Emery, K. O., 1955. Transportation of rocks by driftwood. *J. Sediment. Petrol.*, **25**, 51–57.

Emery, K. O., 1956. Marine geology of Johnston Island and its surrounding shallows, central Pacific Ocean. *Bull. Geol. Soc. Amer.*, **67**, 1505–1520.

Emery, K. O., 1960. *The Sea off Southern California, A Modern Habitat of Petroleum.* John Wiley & Sons, Inc., New York, 1–366.

Emery, K. O., 1962. Marine geology of Guam. *U.S. Geol. Survey Prof. Paper*, **403-B**, 1–76.

Emery, K. O. and R. H. Tschudy, 1941. Transportation of rock by kelp. *Bull. Geol. Soc. Amer.*, **52**, 855–862.

Ericson, D. B., M. Ewing and B. C. Heezen, 1952. Turbidity currents and sediments in North Atlantic. *Bull. Amer. Assoc. Petrol. Geol.*, **36**, 489–511.

Fellows, L., 1958. Mystery pebbles found in sea. *New York Times*, 30 March, 1958, 66.

Fleming, C. A., 1951. Sea lions as geological agents. *J. Sediment. Petrol.*, **21**, 22–25.

Fleming, C. A., 1953. The geology of the Snares Islands, Part I: General geology. *Cape Expedition Series*, Bull. No. 13, New Zealand, 9–27.

Gardner, N. L., 1923. Two new fossil algae from the Miocene. *Proc. Phil. Acad. Natural Sci.*, **75**, 361–363.

Gislén, T., 1940. Investigations on the ecology of Echiurus. *Lunds Univ. Årzsk. N.F. Avd.* 2, **36** (10), 1–39.

Gohar, H. A. F. and A. F. A. Latif, 1959. Morphological studies on the gut of some scarid and labrid fishes. *Pub. Mar. Biol. Station, Ghardaqa, Red Sea, Egypt*, **10**, 145–190.

Gorsline, D. S. and K. O. Emery, 1959. Turbidity-current deposits in San Pedro and Santa Monica basins off southern California. *Bull. Geol. Soc. Amer.*, **70**, 279–290.

Guilcher, A., 1958. *Coastal and Submarine Morphology.* John Wiley & Sons, Inc., New York, 1–274.

Hamilton, A., 1892. Notes on Moa gizzard-stones. *Trans. and Proc. N.Z. Inst. for 1891*, **24**, 172–175.

Hamilton, J. E., 1934. The southern sea lion *Otaria byronia* (de Blainville). *Discovery Reps.*, **8**, 269–318.

Hartman, Olga, 1944. Polychaetous annelids, Part VI. *Allan Hancock Pacific Exped.*, **10**, 311–384.

Heezen, B. C., 1957. Whales entangled in deep sea cables. *Deep-Sea Res.*, **4**, 105–115.

Hough, J. L., 1956. Sediment distribution in the southern oceans around Antarctica. *J. Sediment. Petrol.*, **26**, 301–306.

Hunkins, K. L., M. Ewing, B. C. Heezen and R. J. Menzies, 1960. Biological and geological observations on the first photographs of the Arctic Ocean deep-sea floor. *Limnol. Oceanog.*, **5**, 154–161.

Hutton, C. O., 1942. The chemical composition and optical properties of the mineral substance composing an enterolith. *N.Z. J. Sci. Tech.*, **23**, 9B–12B.

Hutton, C. O., 1945. The nature of an enterolith. *N.Z. J. Sci. Tech.*, **26**, 304–307.

Jones, S., 1959. An unusual instance of a bird getting trapped by a clam. *J. Mar. Biol. Assoc. India*, **1**, 97.

Kenyon, K. W., 1959. The sea otter. *Smithsonian Inst. Ann. Rep. Pub.*, 4354, 399–408.

Kolbe, R. W., 1957. Fresh-water diatoms from Atlantic deep-sea sediments. *Science*, **126**, 1053–1056.

Kornicker, L. S. and D. F. Squires, 1962. Floating corals: a possible source of erroneous distribution data. *Limnol. Oceanog.*, **7**, 447–452.

Kotaka, T. and S. Hayasaka, 1954. Floating transportation. On one method of transportation in an embayment (preliminary report). *J. Geol. Soc. Japan*, **60**, 113–117.

Krausel, R. and H. Weyland, 1930. Die Flora des deutschen Underdevons. *Preuss. Geol. Landesanstalt*, **131**, 7–12.

Kuno, H., R. L. Fisher and N. Nasu, 1956. Rock fragments and pebbles dredged near Jimmu Seamount, northwestern Pacific. *Deep-Sea Res.*, **3**, 126–133.

Ladd, H. S., 1960. Origin of the Pacific island molluscan fauna. *Amer. J. Sci.*, **258A**, 137–150.

Landis, K. K., 1959. Illogical geology. *Geotimes*, **3** (6), 19.

Lane, F. W., 1957. *Kingdom of the Octopus.* Jarrolds, London, 1–287.

Limbaugh, Conrad, 1961. Observations on the California sea otter. *J. Mammalogy*, **42**, 271–273.

Lisitsin, A. P., 1958. Types of marine sediments connected with the activity of ice. *Doklady Akad. Nauk S.S.S.R.*, **118**, 373–376. (Translation into English by Consultants Bureau, Inc.)

Lisitsin, A. P., 1960. Sedimentation in southern parts of the Indian and Pacific Oceans. *Intern. Geol. Cong. 21st Sess., Reps. of Soviet Geologists, Problem 10, Marine Geology, Acad. Sci. USSR., Moscow*, 69–87.

Loughrey, A. G., 1959. Preliminary investigation of the Atlantic walrus *Odobenus rosmarus rosmarus* (Linnaeus). *Wildlife Management Bull.* (*Ottawa*), ser. 1, no. 14, 1–123.

McGill, J. T., 1958. Map of coastal landforms of the world. *Geog. Rev.*, **68**, 402–405.

Mayor, A. G., 1924. Causes which produce stable conditions in the depth of the floors of Pacific fringing reef-flats. *Pub. Carnegie Inst. Wash.* 340, 27–36.

Mellis, O., 1955. Volcanic ash-horizons in deep-sea sediments from the Eastern Mediterranean. *Deep-Sea Res.*, **2**, 89–92.

Menard, H. W., 1953. Pleistocene and Recent sediment from the floor of the northeastern Pacific Ocean. *Bull. Geol. Soc. Amer.*, **64**, 1279–1294.

Menard, H. W., 1955. Deep-sea channels, topography, and sedimentation. *Bull. Amer. Assoc. Petrol. Geol.*, **39**, 236–255.

Mohr, E., 1952. Die Robben der Europäischen Gewässer. *Mon. Wildsäugetiere* (*Frankfurt*), **12**, 1–283.

Murphy, R. C., 1936. *Oceanic Birds of South America.* Amer. Mus. Nat. Hist., N.Y., 2 vols., 1–1245.

Needham, H. D., 1962. Ice-rafted rocks from the Atlantic Ocean off the coast of the Cape of Good Hope. *Deep-Sea Res.*, **9**, 475–476.

Niino, H. *et al.*, 1953. Report on the submarine eruption of Myojin-Sho. *J. Tokyo Univ. Fisheries*, **40** (2), 1–43.

Nordenskiold, E., 1900. Floating stones. *Nature*, **61**, 278.

Oldham, C. 1930. The shell-smashing habits of gulls. *Nature*, **6**, 239–243.

Olson, F. C. W., 1957. A seaball from Escambia County, Florida. *Q. J. Fla. Acad. Sci.*, **20**, 93–94.

Peach, B. N., 1912. Report on rock specimens dredged by the *Michael Sars* in 1910, by H.M.S. *Tritan* in 1882, and by H.M.S. *Knight Errant* in 1880. *Proc. Roy. Soc. Edinburgh*, **32**, 262–291.

Radczewski, O. E., 1939. Eolian deposits in marine sediments. In *Recent Marine Sediments.* 496–502. Amer. Assoc. Petrol. Geol., Tulsa.

Ranson, G., 1955. Observations sur l'agent essentiel de la dissolution du calcaire dans les régions exondées des îles coralliennes de l'Archipel des Tuamotu. Conclusions sur le processus de la dissolution du calcaire. *C.R. Acad. Sci. Paris*, **240**, 1007–1009.

Revelle, R. and K. O. Emery, 1957. Chemical erosion of beach rock and exposed reef rock. *U.S. Geol. Surv. Prof. Paper*, **260-T**, 699-709.

Revelle, R. and R. W. Fairbridge, 1957. Carbonates and carbon dioxide. *Mem. Geol. Soc. Amer.*, **67**, 239–296.

Rex, R. W. and E. D. Goldberg, 1958. Quartz contents of pelagic sediments of the Pacific Ocean. *Tellus*, **10**, 153–159.

Richards, A. F., 1958. Transpacific distribution of floating pumice from Isla San Benedicto, Mexico. *Deep-Sea Res.*, **5**, 29–35.

Ritter, W. E., 1938. *The California Woodpecker and I.* Calif. Univ. Press, Berkeley, 1–340.

Rosen, M. N. and R. A. Bankowski, 1960. A diagnostic technic and treatment for lead poisoning in swans. *Calif. Fish and Game*, **56**, 81–90.

Rouch, J., 1954. Les îles flottantes. *Arch. Met. Geophys. Bioklim.*, ser. A, **7**, 528–532.

Sayles, R. W., 1931. Bermuda during the Ice Age. *Proc. Amer. Acad. Arts Sci.*, **66**, 381–467.

Schäfer, W., 1956. Wirkungen der Benthos-Organismen auf den jungen Schichtverband. *Senckenbergiana Lethaea*, **37**, 183–263.

Scheffer, V. B., 1950. The food of the Alaska fur seal. *U.S. Dept. Interior, Fish and Wildlife Service*, Wildlife Leaflet 329, 1–16.

Scheffer, V. B., 1958. *Seals, Sea Lions and Walruses.* Stanford Univ. Press, Stanford, California, 1–179.

Schurz, W. L., 1959. *The Manila Galleon.* E. P. Dutton & Co., New York, 453 pp.

Schwartz, A., 1932. Der tierische Einfluss auf die Meeres-sedimente (besonders auf die Beziehungen zwischen Frachtung, Ablagerung und Zusammensetzung von Wattensedimenten). *Senckenbergiana.* **14**, 118–172.

Sclater, P. L., 1888. Notes on the Emperor Penguin (*Aptenodytes forsteri*). *Ibis*, **6**, 325–334.

Seilacher, A., 1953. Die geologische Bedeutung fossiler Lebenspuren. *Z. Deut. Geol. Ges.*, **105**, 214–227.

Shimer, H. W., 1924. *An Introduction to the Study of Fossils.* Macmillan Co., New York, 1–450.

Shrock, R. R., 1935. Probable worm castings in the Salem Limestone of Indiana. *Proc. Indiana Acad. Sci.*, **44**, 174–175.

Shumway, G. A., 1953. Rafted pebbles from the deep ocean off Baja California. *J. Sediment. Petrol.*, **23**, 24–33.

Smith, A. G., 1952. Shells from the bird guano of southeast Farallon Island, California, with description of a new species of *Liotia*. *Proc. Calif. Acad. Sci.*, 4th ser., **27**, 383–387.

Teichert, C. and D. L. Serventy, 1947. Deposits of shells transported by birds. *Amer. J. Sci.*, **245**, 322–328.

Thompson, W. C., 1959. Attachment of the giant kelp, *Macrocystis pyrifera*, in fine sediment and its biological and geological significance. *Intern. Oceanog. Cong. Preprints*, New York, 586–587.

Thompson, W. F., 1919. Halibut eats large rock. *Calif. Fish and Game*, **5**, 157–158.

Wieland, G. R., 1907. Gastroliths. *Science*, n. s., **25**, 66–67.

Wiens, H. J., 1962. *Atoll Environment and Ecology.* Yale Univ. Press, New Haven, 1–532.

Wolf, K. H., 1960. Transportation of sand grains by flotation. *Edmonton Geol. Soc. Bull.*, **2** (3).

Wood-Jones, F., 1912. *Coral and Atolls.* Lovell Reeve & Co., Ltd., London, 1–392.

Woodworth, J. B. and E. Wigglesworth, 1934. Geography and geology of the region including Cape Cod, the Elizabeth Islands, Nantucket, Marthas Vineyard, No Mans Land and Block Island. *Mus. Comp. Zool., Harvard Univ., Mem.*, 52, 1–322.

Worzel, J. L., 1959. Extensive deep sea sub-bottom reflections identified as white ash. *Proc. Nat. Acad. Sci.*, **45**, 349–355.

Yabe, H. and T. Sugiyama, 1934. Subfossil gizzard stones probably of birds found in the phosphate deposit of Kita-Daito-Zima. *Proc. Imp. Acad. Tokyo*, **10**, 361–364.

29. PHYSICAL PROPERTIES OF MARINE SEDIMENTS[1]

J. E. NAFE and C. L. DRAKE

1. Introduction

The unconsolidated sediments that blanket the ocean floor are of widely varying thickness, but seismic observations indicate that 200 to 400 m in the Pacific (Revelle *et al.*, 1955) and 1 km in the Atlantic (Ewing and Ewing, 1959) are fairly typical values for deep water. At present direct observation of these sediments is limited to such samples as may be recovered by dredging or coring operations, for drilling has been carried out only in the shallow waters of the coastal shelves. Knowledge of the physical properties of the great bulk of the sediments deeper than the few tens of feet reached by coring equipment is thus necessarily derived from geophysical observations.

The recoverable sediments of the topmost part of the sedimentary column may be studied in detail. Physical properties, such as elastic wave velocities, attenuation, density and thermal conductivity, are directly measurable. The bulk properties depend ultimately on the values of other observables. These include water content, particle density, size and shape of particles, age and chemical composition. Measurements may be static or dynamic. They may be carried out at elevated pressure, they may be made in various frequency ranges, or they may sometimes be made *in situ*. Measured properties on small samples are related only in some average sense to the results of large-scale measurements.

It will be noted that, whether on the scale of small samples or on the scale of seismic refraction experiments, acoustic measurements play a major role in any study of physical properties of marine sediments. They provide direct information on compressional-wave velocity, shear-wave velocity, impedance, attenuation and Poisson's ratio. If density is known independently, they supply in addition dynamic values of compressibility, shear modulus and other elastic constants. Combined with appropriate travel-time analysis they give the variation with depth of elastic-wave velocities and, consequently, the depth variation of any derived quantities. In certain cases, such as dispersion of elastic waves in shallow water, some information on the depth variation of density is obtained. Approximate depth variation of many physical properties may be derived from velocity measurements by combining them with empirical relationships connecting velocity with density or velocity with porosity. These relationships are based on experimental observation, and inferences from them are capable of being partially checked by comparison of gravity with seismic interpretations. The existence of such empirical connections is a consequence of the limited range of substances ordinarily occurring in sedimentary deposits and is to be regarded as a fact of geology rather than the operation of physical law.

[1] Lamont Geological Observatory (Columbia University) Contribution No. 598.

[*MS received July, 1960*]

2. Relationships among Observables

The physical properties to be dealt with in this chapter are listed in Table I, together with the symbols and units employed. Observables that specify particle statistics or sediment type have been omitted from the table but some will be discussed later in connection with results. Also omitted are observables that are not properties of the sediments alone. The group velocity of a dispersed wave in shallow water, for example, depends to some extent upon the density and elastic properties but it is especially a characteristic of a particular depth variation of these quantities. Heat flow is not a property of the sediments at all except to the extent that sources within the sediments might contribute to the total flow.

Equations relating physical properties that make possible the computation or estimation of one quantity from measured values of others are given in Table II. Some are generally valid both for sediments and sedimentary rocks;

TABLE I

Property	Symbol	Unit
Compressional velocity	V_p or V	km/sec
Shear velocity	V_s	km/sec
Compressibility	C	$(\text{dynes/cm}^2)^{-1}$
Incompressibility	k	dynes/cm^2
Rigidity	μ	dynes/cm^2
Young's modulus	E	dynes/cm^2
Poisson's ratio	σ	Dimensionless
Attenuation	γ	$(\text{length})^{-1}$ or dB/length
Characteristic impedance	$Z=\rho V$	g/cm^2 sec
Density	ρ	g/cm^3
Porosity	ϕ	Dimensionless
Thermal conductivity	K	cal/cm °C sec

others are valid only in limiting cases. Those equations that are strictly valid only for isotropic elastic media may, when applied to porous media, be considered to define dynamic values of such quantities as k, μ, σ and E.

A marine sediment is a particle aggregate with the interstices completely water-filled. A number of theoretical discussions of these porous or granular media have been reported. The reader is referred to papers of Ament (1953), Biot (1956, 1956a), Zwikker and Kosten (1949), Gassman (1951) and Paterson (1956) for information on their elastic behavior. Parasnis (1960) has computed the thermal conductivity to be expected under certain special assumptions with regard to grain shape. Porous media are in general neither isotropic nor perfectly elastic. Even when isotropic on a scale large compared with grain sizes, two dilatational waves and one rotational wave, each with a different velocity, can occur. Both attenuation and the velocities have been shown to increase with increasing frequency. Sutton *et al.* (1957) have shown, however, that over the frequency range of from a few tens of cycles per second to one megacycle the velocity change is not likely to exceed 1% for sediments usually encountered. Birch and Bancroft (1938) noted no significant difference between velocities measured on rocks at high frequency and corresponding seismic velocities. In effect, then, particles and fluid may usually be considered to move together. If compressional velocities are observed that are higher than predicted by the Wood equation (Table II, equation 8), the increment will be attributed to an increase of quantity $k+\frac{4}{3}\mu$ caused by compaction and cementation.

TABLE II

Equation	Applicability	Reference
1. $V_p = \sqrt{\dfrac{k+(4/3)\mu}{\rho}}$	isotropic elastic solid	Bullen (1947)
2. $V_s = \sqrt{(\mu/\rho)}$	isotropic elastic solid	Bullen (1947)
3. $\left(\dfrac{V_p}{V_s}\right)^2 = \dfrac{2(1-\sigma)}{(1-2\sigma)}$	isotropic elastic solid	Bullen (1947)
4. $E = 3k(1-2\sigma)$	isotropic elastic solid	Bullen (1947)
5. $Q = \pi/\gamma cT$	plane waves $\begin{cases} c = \text{phase velocity} \\ T = \text{period} \end{cases}$	Birch *et al.* (1942)
6. $\rho = \sum_i f_i \rho_i$ [a]	general	Shumway (1960)
7. $C = \sum_i f_i C_i$ [a]	emulsion or suspension	Shumway (1960)
8. $V^2 = 1/\rho C$ [b]	emulsion or suspension (Wood's equation)	Wood (1941)
9. $1/V = \sum_i f_i/V_i$	transmission perpendicular to a layer sequence (time-average equation)	Wyllie *et al.* (1956)

[a] f_i = volume fraction of ith constituent ($f_{\text{water}} = \phi$).
[b] ρ and C from equations (6) and (7) respectively.

Despite the exceedingly complex nature of granular media and the multiplicity of processes that affect the behavior of the aggregate as a whole, it is clear from Fig. 1 that some broad general trends may be discerned in the experimental results. The Wood equation (equation 8) is an approximate lower limit to observed compressional velocities for a given porosity. As Wyllie, Gregory and Gardner (1956) have pointed out the time average equation (equation 9) is a fair representation of the main trend of observations for consolidated sediments. The spread of observations is sufficiently limited that rough predictions of porosity and density may be made if velocity is known. Thus, if the compressional velocity is 4 km/sec, it is unlikely that porosity will be outside the range 0.15 to 0.25 no matter what other attributes the sediment may have. From equation 6 (Table II) density is a linear function of porosity provided the individual grains of the sediment are all of one kind. Nafe and Drake (1957) have shown that nearly the same straight line would fit the observations for sandstones, limestones, shales and ocean sediment cores. A porosity range of 0.15 to 0.25 means that density is in all likelihood between 2.3 and 2.5 g/cm³. Thus the similarity of particle densities in common sediment-forming substances leads to great simplification in empirical description of physical properties.

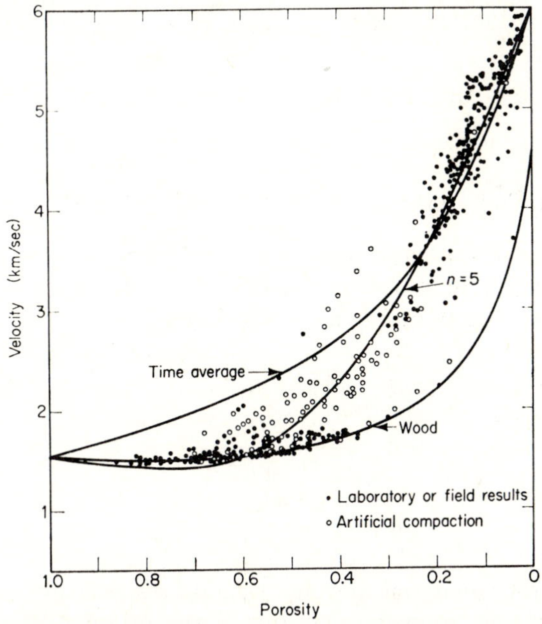

Fig. 1. Observations of compressional velocity as a function of porosity. The data are compared with the Wood equation, the time-average equation and an empirical equation of Nafe and Drake. Solid points at low velocity represent measurements on ocean sediments.

Before proceeding to a discussion of methods and the summary of results it is necessary to comment on the small number of actual measurements that have been made. Of reported observations, there are disproportionally many from easily accessible areas, particularly from shallow water. The only *in situ* measurements other than seismic have been made in depths that permit operations by divers. Precautions against disturbing the sediment samples to be studied are more effective in the case of shallow-water sediments. To include in the discussion of physical properties sufficient information to make possible predictions and estimates of properties not directly observed, some data from bore holes and from artificial compaction experiments are presented. A distinction between observations on actual marine sediments and others of possibly similar character will be maintained throughout the chapter.

3. Methods of Measurement

Among the methods described below some are direct measurements and some yield information on physical properties only through a comparison of experiments and theory involving observables other than the ones sought. Thus, velocity may be measured by direct timing or it may be determined indirectly by comparison of measured with computed curves of phase velocity as a function of period. It is not intended to provide a comprehensive summary of available methods but rather to indicate some that are typical and that are currently in use. Methods of static determination of elastic constants are not included on the grounds that they are adequately treated elsewhere. Moreover, results are frequently incompatible with observed velocities, which depend on dynamic values. Particular notice is given to methods having special advantages for use at sea.

A. Density, Porosity and Average Grain Density

These three quantities have been measured (Sutton *et al.*, 1957; Hamilton *et al.*, 1956) directly for sediment samples by weighing a sample of known volume and reweighing after drying at 100° to 115°C.

Then

$$\rho = W/V$$

$$\rho_g = \frac{D\rho_w}{W(\rho_w/\rho) - W + D}$$

$$\phi = \left(\frac{W-D}{W}\right)\frac{\rho}{\rho_w},$$

where W and D represent wet and dry weights respectively, ρ_w the density of water, ρ_g the average grain density and V the original volume. Alternative expressions may be written by replacing W by $V\rho$.

Sutton *et al.* (1957) have described a method for measuring bulk density that is particularly useful aboard ship where accurate weighing is not possible. The density of a known volume of sediment from a freshly obtained core was determined by suspending the sample beneath a modified spirit hydrometer in distilled water. The sample was obtained by inserting a small cylinder into the core, at the same time holding a piston fixed at the core surface to prevent compaction of the sample. After density measurement, samples were preserved in sealed containers so that dry and wet weights could also be determined in the laboratory for determination of ρ_g and an independent measurement of ρ.

Hamilton *et al.* (1956) have described underwater density sample collection by a diver. A metal tube, after insertion into the sediment, was sealed at the top by a plastic disk. Another disk was placed at the bottom of the tube by burrowing down along the side. As the methods of Hamilton *et al.* and Sutton *et al.* suggest, it is important to avoid any compaction in the process of sample collection. An excellent example of the distortion in length that can be produced by the act of sampling is shown by Ericson and Wollin (1956). They compare the lengths of corresponding intervals of cores taken with and without a piston. Nearly a 50% reduction in length occurs in the upper part of the core taken with no piston.

B. Compressional Wave Velocities and Attenuation

The methods of seismic refraction have been described elsewhere in this book and consequently will not be repeated here. It is sufficient to remark that compressional velocities and thicknesses of principal layers, and sometimes shear velocities, are obtained by this method. Since frequencies are typically of the order of 100 c/s (wavelengths of the order of 20 m) and path lengths within the sediment may be many kilometers, the resulting velocities are average values over large volumes of material.

Velocity measurements on sediments *in situ* were reported by Hamilton *et al.* (1956). This was accomplished by insertion of two probes each with a barium titanate transducer into the sea floor by a diver. The probes were 1 ft apart and penetrated to a depth of 6 in. Velocities were then measured by direct timing of 100 kc/s pulses.

Sutton *et al.* (1957) measured sound velocities in freshly obtained cores by direct timing of an acoustic pulse across the diameter of the core. A velocity standard was used to obtain the origin time of the transmitted pulse.

Velocity measurements by direct timing, though not on marine sediments, have also been described by Hughes (1957), Paterson (1956) and Wyllie *et al.* (1956). Laughton (1957) and Kershaw (*in litt.*) have employed direct timing methods to study the effect of compaction on physical properties of marine sediments.

Shumway (1956, 1960) measured both compressional velocity and attenuation by exciting axial modes of oscillation in sediment samples contained in thin-walled plastic cylinders. The values of velocity reported by Shumway were obtained by comparison of resonance frequencies for a given mode of a sediment-filled container with that from a water-filled container. Attenuation was measured by determining the Q of the system from the resonance widths at half power points. Toulis (1956) has derived equations relating Q and the resonance frequency to the attenuation and sound velocity respectively for this system. He has also shown how to introduce corrections for finite thicknesses of the cylindrical container.

C. Shear-Wave Velocities

Although there is no doubt that shear waves occur in marine sediments, there are many practical difficulties to overcome in measurements and few reliable observations have been made. The coefficient of rigidity is normally so small that measurements by direct timing on recovered samples have been accomplished only under elevated pressure. Laughton (1957) measured shear-wave

TABLE

Sample or station	Location	Water depth, fm	Method	V_p, km/sec	V_s, km/sec	μ 10^{-11} dynes/cm^2	κ 10^{-11} dynes/cm^2
(*A*) *Marine Sediment*							
Gray clay	Abyssal Plain	2570	Core	1.57			
Tan and gray clay	"	2570	"	1.51			
Cream calcarenite	Seamount	940	"	1.70			
Red arenaceous clay	Continental slope	820	"	1.61			
Fine sand	Shallow water	< 15	*in situ*	1.68		(0.06)	(0.472)
Very fine sand	" "	"	"	1.66		(0.05)	(0.458)
Silty very fine sand	" "	"	"	1.55			
Medium silt	" "	"	"	1.47		(0.00)	(0.366)
Clayey fine silt	" "	"	"	1.46		(0.00)	(0.332)
Coarse silt	" "	< 30	Resonance	1.514			
Medium sand	" "	"	"	1.785			
" "	" "	"	"	1.75			
Fine sand	" "	"	"	1.550			
" "				1.773			
Silty coarse clay	Continental slope	560	"	1.491			
Clayey med. silt	Cont. border land	100	"	1.506			
Glob. ooze		2577	Compaction	1.62			
" "		"	"	2.68	1.20	(0.31)	(1.13)
" "		"	"	2.89	1.42	(0.45)	(1.15)
" "		"	"	3.06	1.57	(0.56)	(1.38)
Terrig. mud		1140	"	1.51			
" "		"	"	1.83			
" "		"	"	2.46			
A179-8	Abyssal Plain	2830	Sub-bottom refl.	1.7–2.0			
A179-8	" "	2830	Crit. angle	1.88			
A179-9	" "	2730	Sub-bottom refl.	1.8–2.3			
A179-9	" "	2730	Crit. angle	1.81			
G 8	N. of Tobago Trough	960	Refraction	1.68			
Br. zeolitic clay		2440	Needle probe				
Clayey calc. ooze		1625	" "				
Buff calc. ooze		1680	" "				
Clayey calc. ooze		1440	" "				
Calc. silicious ooze		2080	" "				
(*B*) *Non-Marine Sediments*							
Organic silt	Hudson River	2	Refraction	0.538			
Ignimbrite	" "	—	Direct timing	2.08	0.80		
Pierre shale		—	Refraction	2.28	0.85		
" "		—	Refraction	2.36	0.86		

[a] $\phi = -\log_2$ (grain diameter in millimeters).

[b] Over 50 to 450 c/s range = $0.12f$ for vertically traveling compressional wave in dB per thousand feet and over 20 to 125 c/s = $1.0f$ for horizontally traveling shear waves in dB per thousand feet.

III

σ	γ, dB/ft	f, c/s	ρV 10^{-5} g/cm^2 sec	ρ, g/cm^3	ϕ	κ 10^{-3} cal/°C cm sec	Med. grain size ϕ units[a]	CO_3, %	Ref.
				1.62	0.61		7.2	23.8	Sutton *et al.* (1957)
				1.52	0.68		8.6	25.0	,,
				1.62	0.61		4.1	94.8	,,
				1.77	0.55		7.3	68.4	,,
(0.44)			(3.25)	1.93	0.462		2.5		Hamilton *et al.* (1956)
(0.45)			(3.18)	1.92	0.477		3.0		,,
			(2.60)	1.68	0.631		4.5		,,
(0.50)			(2.49)	1.69	0.609		5.0		,,
(0.50)			(2.34)	1.60	0.656		5.5		,,
	2.202	23,214		1.62	0.604		4.2		Shumway (1960)
	3.811	30,965							,,
	5.72	31,406		1.94	0.432		2.00		,,
	3.65	23,335		1.99	0.362		1.75		,,
	6.27	23,943		1.80	0.526		2.68		,,
	4.43	31,239		1.89	0.479		2.62		,,
	0.435	20,500		1.32	0.812		8.05		,,
	0.401	22,768							,,
	1.315	30,225							,,
	0.569	23,068		1.594	0.651		5.60		,,
	1.364	30,655							,,
				1.59	0.65			54	Laughton (1957)
(0.37)				2.14	0.32				,,
(0.34)				2.22	0.28				,,
(0.32)				2.26	0.26				,,
				1.54	0.69			21	,,
				2.13	0.34				,,
				2.41	0.17				,,
									Katz & Ewing (1956)
									,,
									,,
									,,
									Ewing *et al.* (1956)
				1.19	(0.88)	1.63			Von Herzen & Maxwell (1959)
				1.50	(0.71)	2.33			,,
				1.45	(0.73)	2.35			,,
				1.39	(0.77)	2.14			,,
				1.50	(0.70)	2.07			,,
				1.35	0.87				Drake
(0.41)				1.95					Evison (1956)
(0.42)	[b]								McDonal *et al* (1958)
(0.42)	[b]								,,

velocities in a sample of *Globigerina* ooze compressed between porous disks. Shear waves were observed at pressures above 500 kb/cm^2 on the initial compression. On decompression shear waves could be identified at pressures as low as 64 kg/cm^2. Indirect evidence from seismic refraction measurements for the existence of shear waves in marine sediments has been reported by Nafe and Drake (1957). An excellent measure of an average sedimentary shear-wave velocity is to be found in the dispersion of oceanic Love waves and higher mode Rayleigh waves. The observed periods and low values of group velocity near the group velocity minimum cannot be accounted for without assigning a rigidity to the sedimentary column. According to Oliver and Dorman (Chapter 8) an average shear velocity between 0.25 and 0.60 km/sec is required.

D. Dispersion Analysis

Surface waves traveling in a layered half space are, in general, dispersed with phase and group velocity determined by the depth variation of elastic wave velocities and density. Normal mode propagation in shallow water, and Rayleigh and Love waves in the crust and upper mantle of the earth are familar examples of such dispersed wave trains. The dispersion may be computed from an assumed structure. By comparison of observed with computed phase and group velocities the validity of the assumed velocity–density depth variation may be tested. A comprehensive treatment of layered media is to be found in Ewing, Jardetsky and Press (1957).

There are numerous examples of dispersion studies relating to shallow-water sediments. Pekeris (1948) has computed dispersion for a number of two- and three-layer cases, each layer being assumed liquid and of constant density and velocity. Press and Ewing (1950) and Tolstoy (1954) have treated the case of a liquid over a semi-infinite elastic solid. Tolstoy (1960) and Sâto (1959) have considered cases of continuously varying material properties.

The most detailed dispersion studies involving properties of sediments apply to shallow water. Some work yields information on deep-water sediments. Oliver *et al.* used the fluid thickness indicated by Rayleigh-wave dispersion across ocean basins to estimate average sediment thickness. Katz and Ewing (1956) identified a constant-frequency arrival occurring on deep-water refraction records at ranges of the order of 60 km as a wave propagating for part of its path in a thin layer just below the water–sediment interface with a velocity lower than that in the water. Much more information on deep-water sediments may be expected in the immediate future as machine computations of dispersion for a wide variety of assumed structures become available for comparison with observations.

E. Thermal Conductivity

Thermal conductivity has been measured both by steady-state and transient methods. Ratcliffe (1960) described apparatus for steady-state measurements for a few ocean sediments. He used two disk-shaped samples of like dimensions with a disk-shaped heat source between them. The sediment samples were confined in ebonite rings. The ends of the sediment disks opposite the heat source were cooled by circulating water. The apparatus was subjected to a vertical load to ensure good thermal contact and the whole apparatus was surrounded by glass wool insulation to reduce loss of heat at the edges. From the dimensions, power input and temperatures at the hot and cold end plates, the thermal conductivity was determined. Zierfuss and van der Vliet (1956) employed a steady-state method to measure thermal conductivity of sedimentary rocks. They also used cylindrical samples but arranged in such a way that all of the heat passing through the sample passed also through a crown glass standard. Thus the only measurements required were of temperature gradient in sample and standard. The ratio of the thermal conductivities is then equal to the inverse ratio of the temperature gradients. Zierfuss and van der Vliet depended for shielding on a heated glass tube having temperatures at all points very near to those at corresponding points of the sample and standard. In this way edge losses were minimized.

Von Herzen and Maxwell (1959) measured thermal conductivity on freshly obtained cores by insertion of a small needle probe containing both a heating element and a thermistor. The needle constitutes basically a line source and, after a sufficiently long time, the measured temperature varies linearly with the logarithm of the time. The slope of the line is $q/4\pi K$, where q is the heat input per unit length of cylinder and K is the thermal conductivity. For validity of the linear relationships the time must be large compared with the square of the probe radius divided by the diffusivity (diffusivity $= K\rho \times$ thermal capacity). The probe used by Von Herzen and Maxwell was constructed from a hypodermic needle 6.4 cm long and 0.086 cm in diameter. The power input was of the order of 2 W and the time required for measurement was 10 min or less. Measurements compared favorably with steady-state measurements carried out on selected samples. For application of measured values to heat-flow determination, corrections for change of temperature and pressure between place of measurement and the sea bottom are required. These corrections are discussed by Bullard *et al.* (1956). They amount to a 5 to 10% reduction on account of the temperature decrease and something less than a 3% increase to correct for pressure.

4. Summary of Results

A. Density–Porosity

In Fig. 2 are plotted measurements of density as a function of porosity for water-saturated sediments and sedimentary rocks. For a two-component mixture of fluid and particles of a single kind, the relationship is necessarily linear. It is remarkable that when observations are plotted without regard to sediment type the observed points come so close to fitting a single straight line. In the figure are plotted some 300 points representing measurements on limestones, dolomites, sandstones, shales, clays, sands, gravels and ocean sediments.

Measurements have been taken from Birch, Schairer and Spicer (1942), Hamilton *et al.* (1956), Shumway (1960), Sutton *et al.* (1957), Kershaw (*in litt.*) and others. The range of porosity values is 0 to 0.25 for limestones, 0 to 0.4 for sandstones and 0 to 0.5 for shales. Clays, sands and gravels cover the middle range from $\phi = 0.2$ to 0.6. Ocean sediments range from $\phi = 0.4$ nearly to 0.9. Deviations from the straight line drawn on the figure are greatest in the case of limestones and dolomites. Of all the points plotted, two-thirds lie almost on the line and the remaining one-third lies above the line. A few large deviations occur in the middle range of porosities for clays and gravels but even most of these lie close to the mean line.

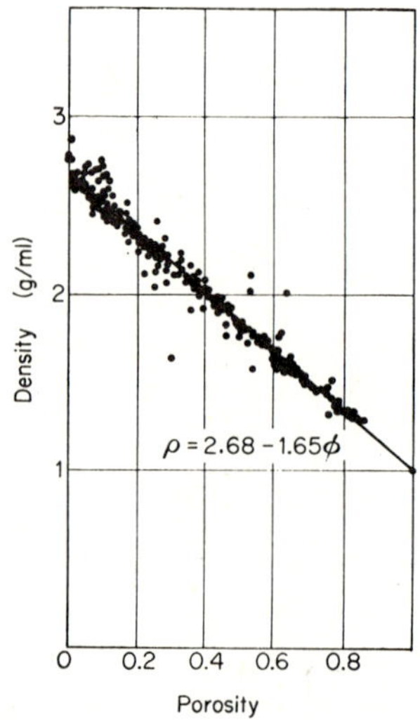

Fig. 2. Porosity–density data for sediments and sedimentary rocks. Measurements on ocean sediments range from porosities of about 0.40 to 0.85.

B. Compressional Velocity–Porosity

Observed compressional velocities are plotted against porosity in Figs. 1 and 3. It is apparent that they fall into three groups: the sedimentary rocks at high velocity, the ocean sediments at low velocity, and artificially compacted sediments intermediate between the others. The apparent separation into groups is probably a consequence of easy accessibility of samples in the high- and low-velocity ranges. Wells are cased at the top; coring devices have limited penetration. In Fig. 1 observations are compared with the Wood equation, the time average equation, and with a representative of a one-parameter family of equations proposed by Nafe and Drake (1957). Plotted points for ocean sediments have been obtained from Sutton (1957), Hamilton (1956), Shumway (1960), Laughton (1957) and Kershaw (*in litt.*) and those for sedimentary rocks from a variety of sources. In Fig. 3 observations are distinguished according to sediment type. It will be noted that most of the observations fall within the $n=4$ to $n=6$ range defined by the equations of Nafe and Drake. Since these equations roughly take into account the effect of grain-to-grain contacts, they do not fit the distribution of observations on uncompacted samples. On the other hand, they do outline the main trend of points and in particular fit well to trajectories on the V–ϕ graph of artificially compacted sediments. It is clear

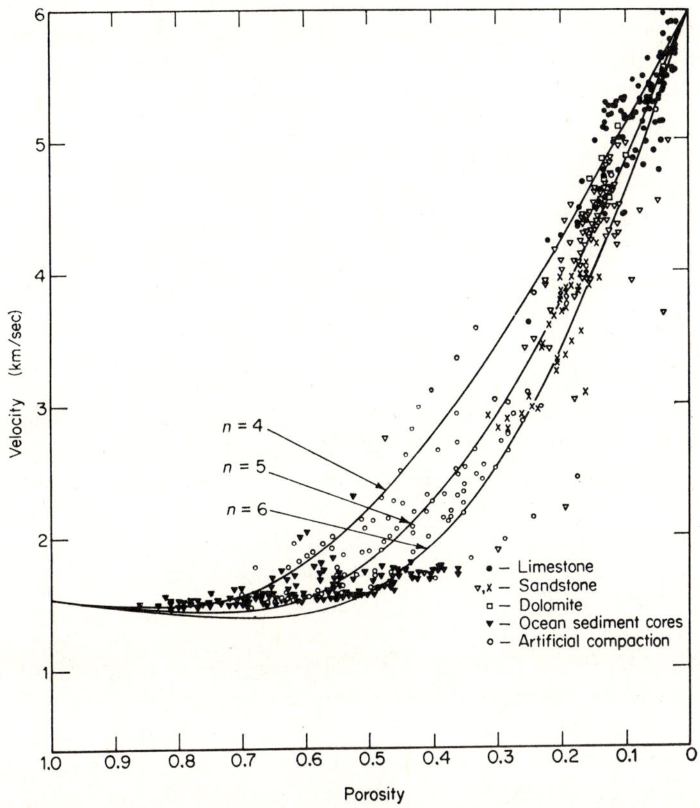

Fig. 3. Compressional velocity–porosity data for sediments and sedimentary rocks compared with empirical equations of Nafe and Drake.

from Figs. 1 and 3 that many samples can be found for which the observed points lie close to the Wood equation even at low porosity. Observations suggest that initial distribution on the V–ϕ plot is governed by the Wood equation but that subsequent compaction and cementation have the effect of displacing the points upward along the direction of the curves shown in Fig. 3.

The occurrence of measured velocities less than water velocity is now well documented. Katz and Ewing (1956) found evidence of a low-velocity layer at the top of the sediments on twelve seismic refraction stations in the Atlantic

Ocean basin west of Bermuda. Similar conclusions were reported by Officer (1955) on the basis of wide-angle reflection studies. Hamilton (1956) and Sutton *et al.* (1957) have observed velocities less than water velocity on recovered sediment samples. Such low velocities are not to be confused with much lower values that occur in gassy sediments. The effect is caused by initial loading of the fluid by particles of higher density in the high-porosity range. Within the experimental error the reduction of velocity is just that predicted by the Wood equation. Lowest measured values are about 1.45 km/sec.

Porosity is clearly a major factor in determining compressional velocity as Figs. 1 and 3 indicate. The scatter about the main trend may be partially understood by examination of the effects of other variables. Sutton *et al.* (1957) carried out a detailed study of a group of cores from a variety of environments in the Atlantic Ocean. They measured median grain size, porosity, carbonate content, sorting coefficient, density, average grain density and salt content with the aim of determining separately the influence of each variable on sound velocity. Using a multiple regression analysis involving as variables, porosity, carbonate content and median grain size, they found the velocity to increase with decreasing porosity and increasing grain diameter and carbonate content. Shumway (1960) in a study of a group of samples from the Pacific and Arctic Oceans found a positive connection between velocity and the porosity and between the grain size and porosity. Unlike Shumway's samples those described by Sutton *et al.* showed no strong correlation between porosity and grain size. Thus the effects of grain size and porosity on velocity could be clearly separated.

In the high-velocity region shown in Fig. 3 it should be noted that limestones lie for the most part above 4.5 km/sec and sandstones between 3.5 and 5 km/sec. The choice of 6 km/sec for the upper limit of the plotted curves is arbitrary. As the density data will show, some observations on limestones and dolomites occur at velocities as high as 6.5 to 7 km/sec. The dependence of measured velocities on temperature and pressure has been discussed by Shumway (1956) and Sutton *et al.* (1957). The change with temperature results almost entirely from the change of velocity of the water fraction. Change with temperature for a deep-water sample raised to the surface is approximately the same, though opposite in sign, as the change with pressure. Since bottom- and surface-water velocities differ by about 1% the *in situ* velocity should differ from that measured on deck by about the same amount.

C. Compressional Velocity–Density

Velocity–density data for sediments and sedimentary rocks are shown in Fig. 4. Included in the figure are results for most of the samples plotted on the velocity–porosity figures. Independent measurements of density were made for most of the ocean sediments occurring in the low-velocity range and for most of the values at very high velocity. Points representing artificially compacted sediments have been derived from measurements of volume of expressed water together with a measurement of initial density. The curve drawn through the points is intended only as a fair indication of the main trend and one that may serve to provide estimates of density from velocity in the absence of any other information. Cracking of samples or the presence of small amounts of gas in samples of ocean sediments will have the effect of reducing velocity for a given density, and anhydrites have exceptionally high densities for a given velocity. In drawing the curve of Fig. 4, therefore, relatively less weight was given to observations for which velocity reduction or density increase from such causes is a possibility.

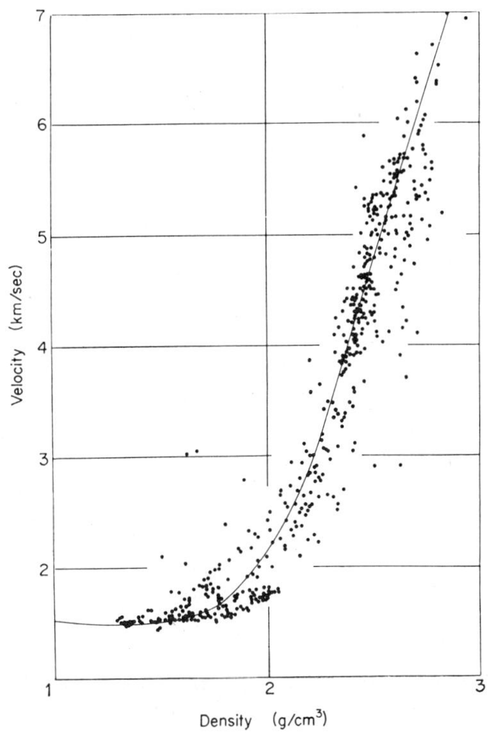

Fig. 4. Compressional velocity as a function of density. Most observations on ocean sediments lie below 2 km/sec in velocity.

All points plotted in Fig. 4 represent observations on water-saturated materials. Those published values for which the degree of saturation was not indicated have been omitted.

It is interesting to note that when data for igneous and metamorphic rocks are plotted together with those for sedimentary rocks the velocity–density curve may be continued upward to higher velocities with no noticeable offset, though the curve has a slight decrease in slope.

D. Shear-Wave Velocities and Poisson's Ratio

The information that exists on shear-wave velocities in marine sediments consists mainly of direct measurements by Laughton (1957) on artificially compacted sediments, the occurrence reported by Nafe and Drake (1957) of refracted arrivals consistent with the supposition that they propagated over part of their path as shear waves, and the evidence of Oliver and Dorman (Chapter 8) that higher mode Rayleigh- and Love-wave dispersion is affected by sediments having a shear-wave velocity of 0.25 to 0.60 km/sec. Hamilton has computed values of rigidity by attributing the discrepancy between observed compressional velocities and those predicted by the Wood equation as arising entirely from rigidity. Actually, an increment to the incompressibility may also be a contributing factor so that Hamilton's values are in fact upper limits to the rigidity. The available information is summarized in Fig. 5 on a graph of Poisson's ratio plotted against porosity. Poisson's ratio is related to shear velocity

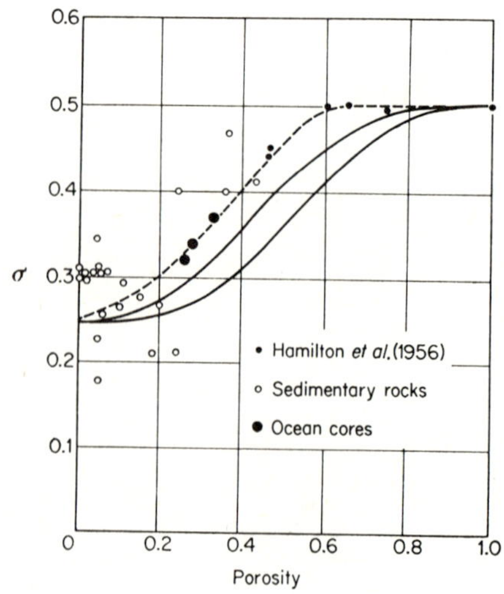

Fig. 5. Poisson's ratio plotted against porosity. Hamilton's values are upper limits to the ratio. The solid points are those reported by Laughton (1957) for artificially compacted *Globigerina* ooze. The region outlined by solid curves represents an estimate by Nafe and Drake (1957) of the range of values for ocean sediments.

through equation 5 (Table II). The solid curves were given by Nafe and Drake (1957). They represent an estimated range based on an approximate theory of the compaction process. The three solid points were derived by Laughton (1957) from his measurements of compressional and shear velocities in compacted *Globigerina* ooze. Points obtained by Hamilton *et al.* (1956) are indicated on the figure and are upper limits. The remaining points are measurements on sedimentary rocks by Evison (1956), Hughes and Jones (1951), Birch *et al.* (1942) and others. The range of average sedimentary shear velocities required by Oliver and Dorman to fit their dispersion data is an indication that estimates of σ by Nafe and Drake were too low for a given porosity. It is also likely that Laughton's points are higher than would be found for sediments *in situ* where cementation can occur. Thus the dashed curve may represent an approximate upper limit except at low porosities. Seismic refraction data normally yield values of σ higher than 0.25. Limestones at low porosity tend to group around 0.30. Evison (1956) has measured a value of σ for concrete of 0.20 at low porosity. Thus considerable scatter may be expected.

The degree of water saturation is especially important in determining σ, for unsaturated sedimentary rocks actually show a decrease of σ with increasing porosity rather than an increase as do saturated rocks.

E. Thermal Conductivity–Porosity

Observations of thermal conductivity for ocean sediments and sedimentary rocks are plotted against porosity in Fig. 6. They are compared with computa-

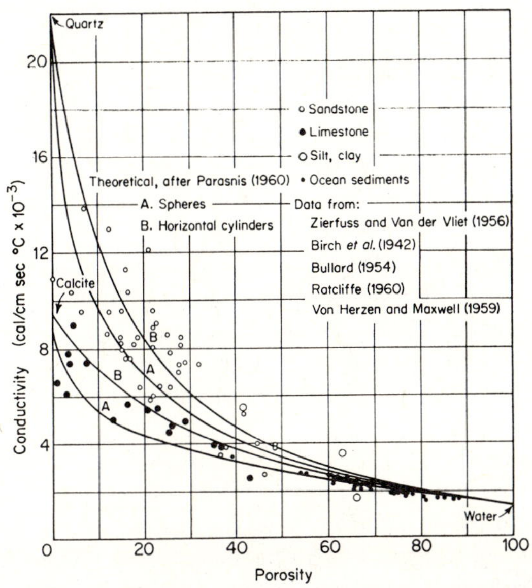

Fig. 6. Thermal conductivity–porosity data for ocean sediments, sandstones and limestones compared with theoretical curves adapted from Parasnis (1960).

tions of Parasnis (1960) for thermal conductivity of mixtures of particles and fluid for two convenient choices of particle shape. The results are applied in one case to the particles of quartz and in the other to calcite. While shape is an important factor there appears also a clear separation of materials according to composition as is shown by the data for sandstones and limestones.

Conductivities of ocean sediments are principally affected by the water content, as has been pointed out by Ratcliffe (1960) and Bullard *et al.* (1956).

F. Selected Values of Other Measured Physical Properties

Table III provides a selection of values of physical properties for ocean sediments from a number of different marine environments measured by several different methods. These values are not necessarily typical. The table has been constructed to give at least one value of each quantity listed, to illustrate measurements at high and low porosity, to compare shallow-water and deep-water environments to show changes with frequency and to provide some indication of results that may be expected for different sediment types. Numbers in brackets are not directly measured but are derived through one or other of the formulae in Table II. Although most of the observations were made on marine sediments a few entries for non-marine sediments have been added at the bottom of the table to illustrate unusual situations, such as inclusions of gas, or to provide estimates of probable values where actual measurements on marine sediments are few or lacking altogether. It is particularly important to assemble information on shear-wave velocities.

The only values of impedance listed are those of Hamilton *et al.* (1956). These are not measured but are derived by taking the product ρV. It is probably better to find impedance by taking this product than to derive it from a measured reflection coefficient, for measured reflection coefficients are almost invariably complicated by the occurrence of multiple reflections with a consequent increase in the apparent value of the impedance. The error so introduced is much more serious below 1000 c/s than it is above that frequency.

Velocities derived from critical angle reflections and sub-bottom reflections (Katz and Ewing, 1956) doubtless apply not to sediments close to the water–sediment interface but to those at some distance below. On the same refraction profiles leading to the tabulated numbers there was clear-cut evidence for the existence of a layer close to the interface with a velocity less than water velocity.

The frequency dependence of the attenuation coefficient γ is of particular interest. Shumway (1960a) has reported γ to be proportional to $f^{1.79}$ for 65 samples with a standard deviation of 0.98 for the exponent. Measurements were made in the frequency range 20 to 40 kc/s and attenuations apply to compressional waves. At much lower frequencies, though not for ocean sediments, McDonal *et al.* (1958) have observed an approximately linear dependence of attenuation on frequency, the attenuation in dB per thousand feet being $1.05f$ for horizontally traveling shear waves from 20 to 152 c/s and $0.12f$ for vertically traveling compressional waves from 50 to 450 c/s. Their measurements were made in bores in the Pierre Shale. Since the exponent may depend on the degree of water saturation (Born, 1941) this example is introduced mainly to illustrate the fact that attenuation for shear waves may differ widely from that for compressional waves.

It will be noted that in not one single instance have measurements been made of all the quantities of interest. In fact, it is frequently the case that in publication of results no clear indication is given of the degree of water saturation, though all evidence points to degree of saturation and water fraction as the most significant variables in determining physical properties. All quantities listed for ocean sediments apply to cases of 100% saturation.

G. Depth Dependence of Compressional Velocity

Evidence for depth variation of compressional velocity in marine sediments is derived from reflection and refraction measurements. Arrivals reflected at a sub-bottom interface or refracted in the upper part of the sedimentary column commonly show the effects of gradients ranging from 0.5 to 2 sec^{-1}. Hill (1952) observed gradients of the order of 2 sec^{-1} in the eastern Atlantic. Officer (1955) reported reflections at a location northeast of Bermuda that indicated an average gradient of about 1.0 sec^{-1}. Katz and Ewing (1956) reported gradients of about

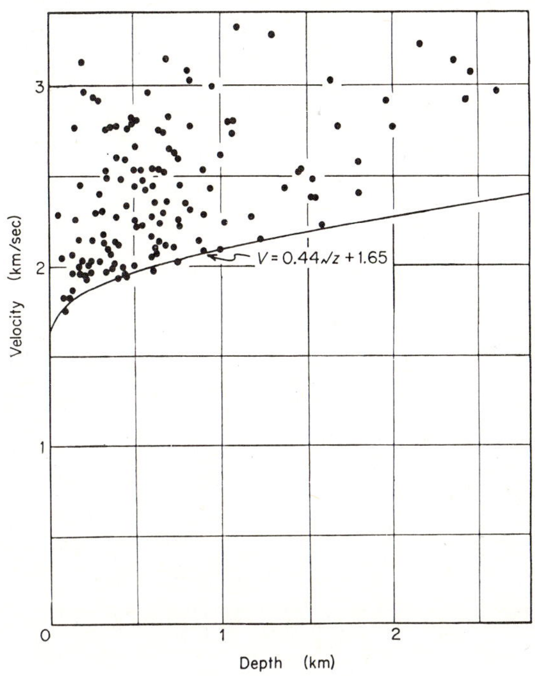

Fig. 7. Compressional velocity as a function of depth of burial. Data were obtained from seismic refraction measurements. The solid curve was derived by Laughton (1954) from the artificial compaction of *Globigerina* ooze. It is a rough lower limit to the observations.

0.6 sec^{-1} at a number of stations in the western Atlantic. Evidence from both sub-bottom reflections and from refracted arrivals was summarized by Nafe and Drake (1957). In Fig. 7 are plotted observations of velocity against thickness of sediment overlying the refracting horizon for a number of stations in the western Atlantic. Points of the figure are not distinguished as to water depth. The only discernible pattern is shown by comparison of the measurements with velocities found by Laughton (1954) for artificially compacted *Globigerina* ooze. Laughton's velocity–depth curve, based on an assumed parabolic gradient, is an effective lower limit to the points measured by the refraction method. This is to be expected, for processes of cementation occurring over an extended time interval should have the effect of increasing the velocity.

In Fig. 8 points representing observed velocities are distinguished according to locality. The observations from deep-water stations appear to have velocities

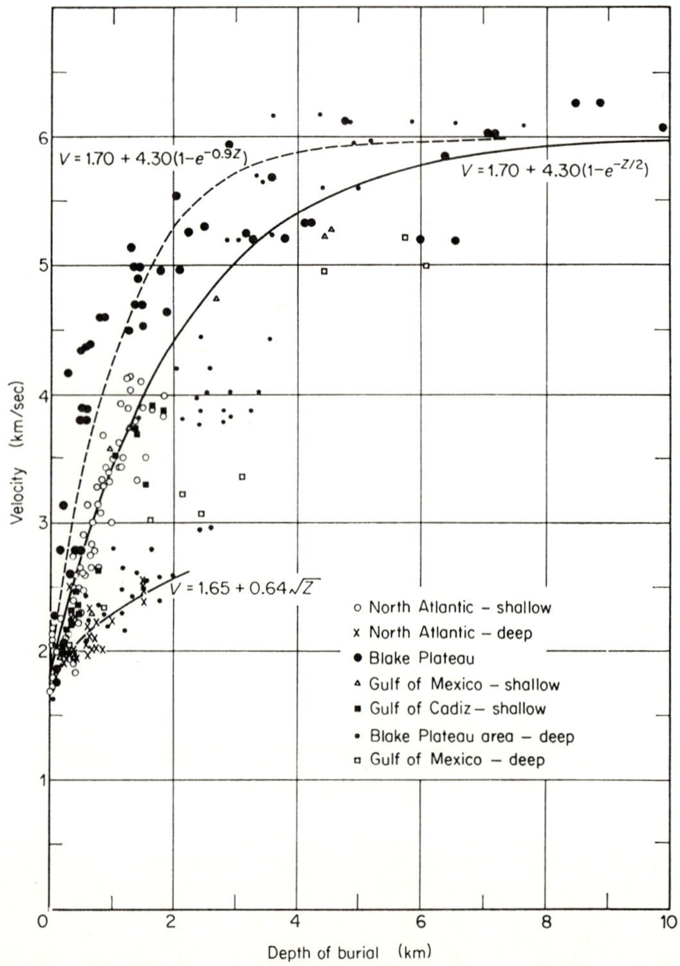

Fig. 8. Compressional velocity–depth of burial data derived from seismic refraction measurements. The indicated depth of burial does not include water depth. The slowest rate of increase of velocity with depth occurs in the case of deep-water sediments (water depth > 2000 fm).

distinctly lower than those from shallow-water stations with the same depth of burial. It should be pointed out that most shallow-water observations in Fig. 8 are from the continental shelf of northeastern America. Shallow-water environments with velocity–depth curves similar to those for deep-water sediments are known to exist. The equations given in the figure are not significant in themselves. They are simply convenient representations of the main trends of the data. Velocities observed on the Blake Plateau, east of Florida, are higher than those at corresponding depths of burial elsewhere in the shallow waters of the North Atlantic. This is probably a consequence of the high carbonate content of the sediments of the Blake Plateau and is consistent with the observation that, for a given porosity, sediments of high carbonate content have higher velocities than those of low carbonate content. Fig. 8 suggests, as do the velocity–porosity data, that the zero porosity limit for the velocity of commonly occurring sediments is near 6 km/sec. Additional evidence for the existence of this approximate velocity limit may be seen in the trend of deep-water points at depths of burial greater than about 1.5 km. The trend to much higher velocities below 1.5 km may be a consequence of loading of the sediments by the water column when water-filled interstices are isolated from each other and do not have free access to the sea.

Nafe and Drake (1957) have suggested that in deep water the principal contribution to the increase of velocity with depth is that of compaction alone. Many other processes can occur in a shallow-water environment such as circulation of water, deposition of solid material in the interstices between particle grains, and erosional unloading. All would tend to increase the compressional velocity. Thus, in a shallow-water environment, velocity might well increase at a greater rate with depth than in deep water. It is unlikely that the rate of increase could be less.

5. Conclusions

Consideration of available observations together with data on sedimentary rocks suggests that a known value of one quantity may limit the range within which values of other quantities are likely to be found. Estimation of such limits may be useful in determining the consistency of results of measurements of different kinds. For example, densities to be used in gravity interpretation have been assigned in accordance with seismic measurements of velocity. However useful, estimation is not measurement, and much more detailed information is needed on marine sediments. The close connection between density and porosity seems well established. Very little is known about shear velocity or about attenuation at low frequency. Compressional velocities are easily measured for near-surface samples but direct measurements of velocity as a function either of density or porosity at points well within the compacting sedimentary column are needed. Artificial compaction experiments suggest that the trend of observations will be roughly that of the $n=4$ to $n=6$ curves of Fig. 3.

The value of a physical quantity is in a sense a function of the use for which that quantity is required. A static determination of Young's modulus, as Evison (1956) has pointed out, may differ widely from the Young's modulus obtained dynamically through substituting into equation (4) constants found by measuring compressional and shear velocities. Indeed, the dynamic value of Young's modulus found by timing compressional and shear velocities over 100 ft of sediment *in situ* may differ widely from that obtained by measurements on a hand sample in the laboratory. If interpretation of seismic data is to provide information on sedimentary structure, then the physical constants most needed are dynamic values and values determined by relatively large-scale measurement.

References

Ament, W. S., 1953. Sound propagation in gross mixtures. *J. Acoust. Soc. Amer.*, **25**, 638–641.

Biot, M. A., 1956. Theory of propagation of elastic waves in a fluid saturated porous solid; I, Low frequency range. *J. Acoust. Soc. Amer.*, **28**, 168–178.

Biot, M. A., 1956a. Theory of propagation of elastic waves in a fluid saturated porous solid; II, Higher frequency range. *J. Acoust. Soc. Amer.*, **28**, 179–191.

Birch, F. and D. Bancroft, 1938. Elastic and internal friction in a long column of granite. *Bull. Seism. Soc. Amer.*, **28**, 24–31.

Birch, F. and D. Bancroft, 1938a. The effect of pressure on the rigidity of rocks. *J. Geol.*, **46**, 59–141.

Birch, F., V. F. Schairer and H..C. Spicer. 1942. Handbook of Physical Constants. *Geol. Soc. Amer. Spec. Paper* 36.

Born, W. T., 1941. The attenuation constant of earth materials. *Geophysics*, **6**, 132–148.

Bullard, E. C., 1954. The flow of heat through the floor of the Atlantic Ocean. *Proc. Roy. Soc. London*, **A222**, 408–429.

Bullard, E. C., A. E. Maxwell and R. Revelle, 1956. *Advances in Geophys.*, **3**, 153–181. Academic Press Inc., New York.

Bullen, K. E., 1947. *Introduction to the Theory of Seismology*. Cambridge University Press, London.

Ericson, D. B. and G. Wollin, 1956. Correlation of six cores from the equatorial Atlantic and the Caribbean. *Deep-Sea Res.*, **3**, 103–125.

Evison, F. F., 1956. Seismic determination of Young's modulus and Poisson's ratio for rocks in situ. *Geotechnique*, Sept., 1956, 119–123.

Ewing, J. I. and W. M. Ewing, 1959. Seismic refraction measurements in the Atlantic Ocean Basins, in the Mediterranean Sea, on the Mid-Atlantic Ridge, and in the Norwegian Sea. *Bull. Geol. Soc. Amer.*, **70**, 291–318.

Ewing, J. I., C. B. Officer, H. R. Johnson and R. S. Edwards, 1957. Geophysical investigations in the Eastern Caribbean: Trinidad Shelf, Tobago Trough, Barbados Ridge, Atlantic Ocean. *Bull. Geol. Soc. Amer.*, **68**, 897–912.

Ewing, W. M., W. S. Jardetsky and F. Press, 1957. *Elastic Waves in Layered Media*. McGraw-Hill Book Co., Inc., New York.

Gassman, F., 1951. Elastic waves through a packing of spheres. *Geophysics*, **16**, 673.

Hamilton, E. L., 1956. Low sound velocities in high porosity sediments. *J. Acoust. Soc. Amer.*, **28**, 16–19.

Hamilton, E. L., G. Shumway, H. W. Menard and C. J. Shipek, 1956. Acoustic and other physical properties of shallow water sediments off San Diego. *J. Acoust. Soc. Amer.*, **28**, 1–15.

Hill, M. N., 1952. Seismic refraction shooting in an area of the Eastern Atlantic. *Phil. Trans., Roy. Soc. London*, **A224**, 56–69.

Hughes, D. S., 1957. Variation of elastic wave velocities in basic igneous rocks with pressure and temperature. *Geophysics*, **22**, 277–284.

Hughes, D. S. and H. J. Jones, 1951. Elastic wave velocities in sedimentary rocks. *Trans. Amer. Geophys. Un.*, **32**, 173–178.

Katz, S. and W. M. Ewing, 1956. Seismic refraction measurements in the Atlantic Ocean, Part VII, Atlantic Ocean basin, West of Bermuda. *Bull. Geol. Soc. Amer.*, **67**, 475–570.

Laughton, A. S., 1954. Laboratory measurements of seismic velocities in ocean sediments. *Proc. Roy. Soc. London*, **A222**, 336–341.

Laughton, A. S., 1957. Sound propagation in compacted ocean sediments. *Geophysics*, **22**, 233–260.

McDonal, F. J., F. A. Angona, R. L. Mills, R. L. Sengbush, R. G. Van Nostrand and J. E. White, 1958. Attenuation of shear and compressional waves in pierre shale. *Geophysics*, **23**, 421–439.

Nafe, J. E. and C. L. Drake. 1957. Variation with depth in shallow and deep water marine sediments of porosity, density and the velocities of compressional and shear waves. *Geophysics*, **22**, 523–552.

Officer, C. B., 1955. A deep sea reflection profile. *Geophysics*, **20**, 270–282.

Oliver, J., W. M. Ewing and F. Press, 1955. Crustal structure and surface wave dispersion: IV, Atlantic and Pacific Ocean basins. *Bull. Geol. Soc. Amer.*, **66**, 913–946.

Parasnis, D. S., 1960. The compaction of sediments and its bearing on some geophysical problems. *Geophys. J.*, **3**, 1–28.

Paterson, N. R., 1956. Seismic wave propagation in porous granular media. *Geophysics*, **21**, 691–714.

Pekeris, C. L., 1948. Theory of propagation of explosive sound in shallow water. *Geol. Soc. Amer. Mem.*, **27**.

Press, F. and W. M. Ewing, 1950. Propagation of explosive sound in a liquid layer overlying a semi-infinite elastic solid. *Geophysics*, **15**, 426–446.

Ratcliffe, E. H., 1960. The thermal conductivities of ocean sediments. *J. Geophys. Res.*, **65**, 1535–1541.

Revelle, R., M. Bramlette, G. Arrhenius and E. D. Goldberg, 1955. Pelagic sediments of the Pacific. *Geol. Soc. Amer. Spec. Paper* 62, 221–236.

Sâto, Y., 1959. Numerical integration of the equation of motion for surface waves in a medium with arbitrary variation of material constants. *Bull. Seism. Soc. Amer.*, **49**, 57–77.

Shumway, G., 1956. A resonant chamber method for sound velocity and attenuation measurements in sediments. *Geophysics*, **21**, 305–319.

Shumway, G., 1960. Sound speed and absorption studies of marine sediments by a resonance method, Part I. *Geophysics*, **25**, 451–467.

Shumway, G., 1960a. Sound speed and absorption studies of marine sediments by a resonance method, Part II. *Geophysics*, **25**, 659–682.

Sutton, G., H. Berckhemer and J. E. Nafe, 1957. Physical analysis of deep sea sediments. *Geophysics*, **22**, 779–812.

Tolstoy, I., 1954. Dispersive properties of a fluid layer overlying a semi-infinite elastic solid. *Bull. Seism. Soc. Amer.*, **44**, 493–512.

Tolstoy, I., 1960. Guided waves in a fluid with continuously variable velocity overlying an elastic solid: Theory and experiment. *J. Acoust. Soc. Amer.*, **31**, 81–87.

Toulis, W. J., 1956. Theory of a resonance chamber to measure the acoustic properties of sediments. *Geophysics*, **21**, 299–304.

Van Nostrand, R. G. and J. E. White, 1953. Attenuation of shear and compressional waves in pierre shale. *Geophysics*, **23**, 421–439.

Von Herzen, R. and A. E. Maxwell, 1959. The measurement of thermal conductivity of deep sea sediments by a needle probe method. *J. Geophys. Res.*, **64**, 1557–1563.

Wood, A. B., 1941. *A Textbook of Sound*. Macmillan, New York.

Wyllie, M. R. J., H. R. Gregory and L. W. Gardner, 1956. Elastic waves in heterogeneous and porous media. *Geophysics*, **21**, 41–70.

Zierfuss, H. and G. Van der Vliet, 1956. Laboratory measurements of heat conductivity of sedimentary rocks. *Bull. Amer. Assoc. Petrol. Geol.*, **40**, 2475–2488.

Zwikker, C. and C. W. Kosten, 1949. *Sound Absorbing Materials*. Elsevier Publishing Company, Inc., New York.

30. AGE DETERMINATION IN SEDIMENTS BY NATURAL RADIOACTIVITY

F. F. Koczy

1. Introduction

The exact method of age determination in geology is based on the decay of radioactive elements. The simple mathematical expression for this process is given by the differential equation stating that the number of atoms of a radioactive isotope disintegrating per time unit is linearly proportional to the existing number of atoms and independent of exterior parameters:

$$dN/dt = -\lambda N, \tag{1}$$

where N is the number of atoms present and λ the decay constant. By integration we obtain the following equation:

$$N = N_0 e^{-\lambda(t-t_0)}, \tag{2}$$

where N_0 is the number of atoms at the time t_0, and N the number of atoms at the time t.

The radio-nuclides used for the purpose of age determination belong to three categories, as follows: (1) The three natural radioactive series: uranium, protactinium and thorium; (2) the natural radioactive elements: potassium-40 and rubidium-87; and (3) cosmic-ray-produced radio-nuclides. The elements belonging to the two latter categories generally decay by one step into a stable isotope, but the three natural radioactive series cover all the heavy elements down to lead.

The time span $(t-t_0)$, of particular interest in deep-sea research, is the time elapsed since a given sediment layer settled on the ocean floor. Other ages important to know include the time when the sediment entered the ocean, and, of further significance, the age of the formation of rock material found in the sediment on the ocean floor.

In order to be able to determine an age, it is obvious that two parameters must be known: one connected with the time of formation of the investigated material, the other with the present time.representing a reference point. In the most simple cases, these parameters are N_0, the number of atoms present t years ago, and N_1, the number of atoms present now. The first of these parameters has to be determined in some way other than by direct measurement, whereas the latter is accessible to precise measurements. One of the problems of age determination is connected with this indirect evaluation, which is usually dependent on certain assumptions that cannot always be established as valid.

A further problem is encountered through the fact that, in most cases, we cannot determine the number of atoms without relating it to a volume or a mass. Only concentrations of radioactive elements in the sediment, therefore, can be

[*MS received April, 1961*]

measured. Variations in the concentration of these elements are caused by the decay as well as the differences in the rate of sedimentation. This difficulty is serious when the addition of the radioactive elements to sediments is governed by a process independent of sedimentation.

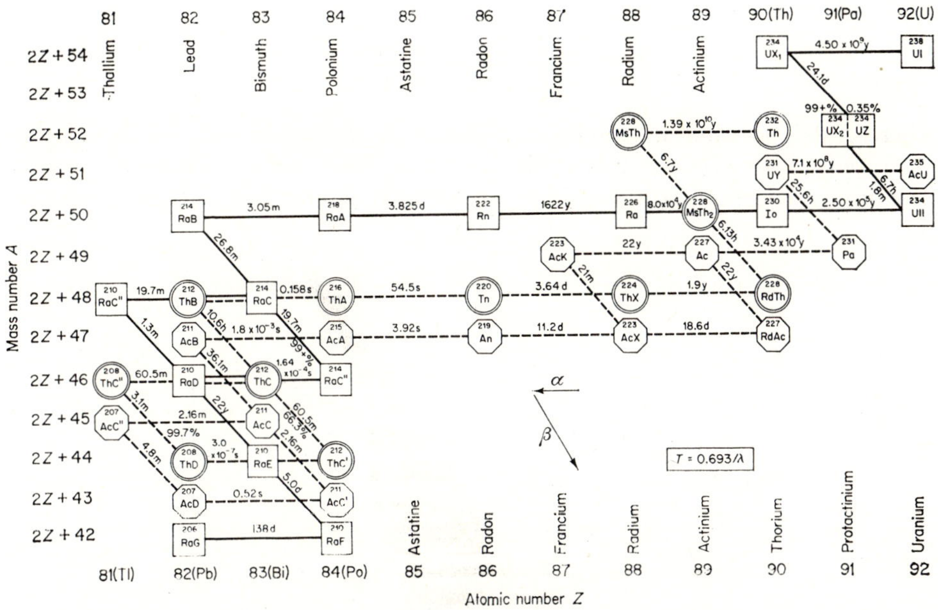

Fig. 1. Decay series of ^{238}U, ^{235}U and ^{232}Th, showing half lives and branching ratios of the isotopes. The old symbols shown in the circles, squares and octagons in the figure are giving way to the modern terminology shown on the top and bottom; i.e. ^{227}Th, not RdAc. (Data from Hollander, Perlman and Seaborg, 1953.)

2. Theoretical Considerations

In principle, three basic methods of age determination in sediments can be delineated. Some variations of them are noted, but each of these can be reduced to one of three basic types. In all cases, it must be assumed that the separation of the radio-nuclides used for the determination of age has been achieved to a known degree by a geochemical process in nature.

A. Theoretical Method of Age Determination, Type 1

In this method, the equilibrium between two radioactive nuclides is used to obtain an estimate of an age. The parameters known, and sometimes available for measurements, are the concentration of both radio-nuclides and their decay constants. The equation is the following:

$$N^2 = N_0{}^1 \frac{\lambda_1}{\lambda_2 - \lambda_1} (e^{-\lambda_1 t} - e^{-\lambda_2 t}). \tag{3}$$

(*a*) If λ_1 is much smaller than λ_2, $N_0{}^1 \rightarrow N^1$ and the equation can be approximated as follows:

$$N^2 = N^1 \frac{\lambda_1}{\lambda_2} (1 - e^{-\lambda_2 t}), \tag{4}$$

which gives the formula for age as

$$t = \frac{1}{\lambda_2} \ln \left(1 - \frac{N^2 \lambda_2}{N^1 \lambda_1}\right), \tag{5}$$

where $N_0{}^1$ is the initial amount of the decaying nuclide and N^2 the nuclide produced. This method is applicable when one of the long-lived nuclides, such as uranium or thorium, is incorporated without daughter products in a mineral.

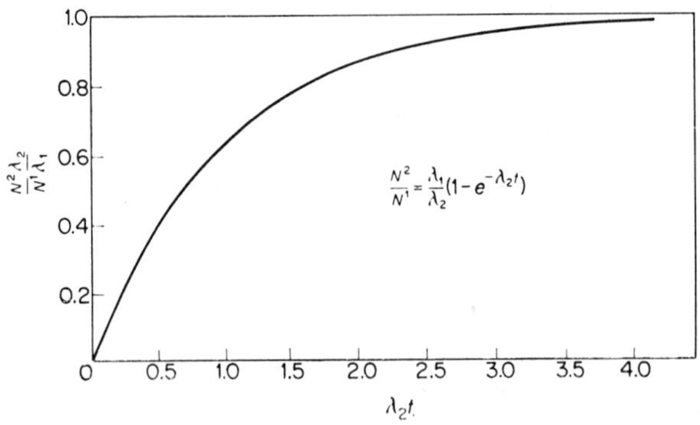

Fig. 2. Age determination with the aid of the ratio of two radioactive elements from the same series where the decay constant of the daughter product is large compared with the decay constant of the parent element. The ordinate gives the ratio of activities of the two radio-elements. The abscissa gives the time scale in non-dimensional units as multiples of mean lifetime.

The produced nuclides with half life large enough to be used for the purposes of marine geology are, in that case, ionium (^{230}Th), protactinium (^{231}Pa) and radium (^{228}Ra, ^{226}Ra).

(*b*) In the case where λ_1 is much larger than λ_2 the equilibrium equation is approximately

$$N^2 = N_0{}^1 (1 - e^{-\lambda_1 t}), \tag{6}$$

but as $N_0{}^1 = N^1 e^{\lambda_1 t}$, we obtain as a formula for the age

$$t = \frac{1}{\lambda_1} \ln \left(1 + \frac{N^2}{N^1}\right), \tag{7}$$

where the same notation is used as above.

This equation is used when a radio-nuclide produces a stable isotope, such as potassium-40 giving argon-40, and rubidium-87 producing strontium-87. The

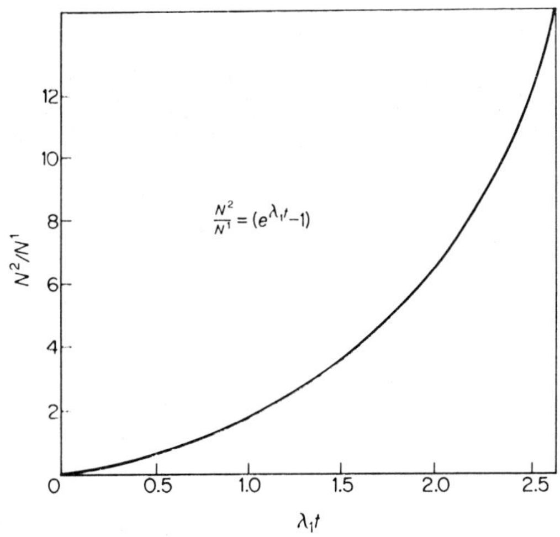

Fig. 3. Age determination using the ratio of two elements from radioactive series for the case when the decay constant of the parent element is much larger than the decay constant of the daughter element. The ordinate gives the ratio of the two radio-isotopes. The abscissa gives the time scale in non-dimensional units as multiples of mean lifetime.

same equation is applicable when the production of lead is used in age determination by the use of the natural radioactive series.

B. Theoretical Method of Age Determination, Type 2

When two isotopes of the same element are radioactive and each forms a new radioactive element, the ratio of the radio-isotopes produced can be used for age determination. A condition is that the ratio of the two isotopes of the parent element is constant in nature. This is nearly always true of elements with a high atomic weight. The decay equation of each of the isotopes is

$$N_t^2 = \frac{\lambda_1}{\lambda_2} N^1(1-e^{-\lambda_2 t}) \quad \text{and} \quad N_t^4 = \frac{\lambda_3}{\lambda_4} N^3(1-e^{-\lambda_4 t}). \tag{8}$$

By taking the ratio of these two, we obtain

$$\frac{N_t^2}{N_t^4} = \frac{\lambda_1 \lambda_4}{\lambda_2 \lambda_3} \frac{N^1}{N^3} \frac{(1-e^{-\lambda_2 t})}{(1-e^{-\lambda_4 t})}. \tag{9}$$

When the ratio of the parent isotopes is constant, when concentrations are given in units of per cent-equivalent, and when the precipitation is faster than 1000 years, we obtain an equation for the age:

$$t = \frac{1}{\lambda_2 - \lambda_4} \ln \left[\frac{\lambda_2}{\lambda_4} \cdot \frac{\overline{N_t^4}}{\overline{N_t^2}}\right]. \tag{10}$$

In this equation, N_0 is eliminated and we are able to use this method for absolute age determination without making any theoretical assumption. The uranium isotopes, 235 and 238, produce protactinium-231 and thorium-230. The ratio of the latter permits the determination of age when the two elements have been separated from their mother elements.

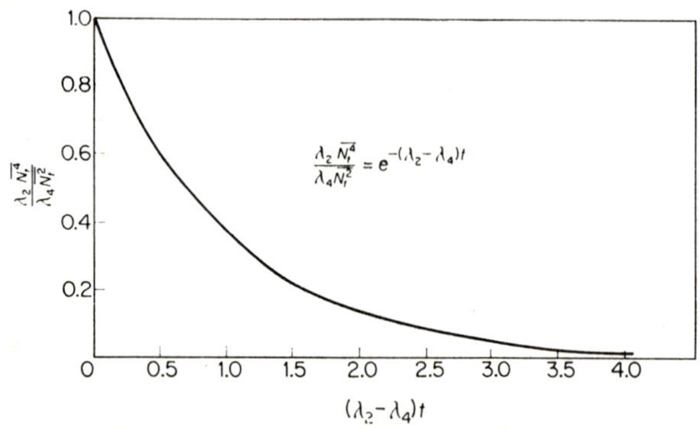

Fig. 4. Age determination by the ratio of two radio-nuclides that are produced by two radio-isotopes of the same element. The ordinate gives the ratio of activities of the two radio-nuclides and the abscissa gives the time scale in non-dimensional units as multiples of mean lifetime.

C. Theoretical Method of Age Determination, Type 3

In this group, we combine all age determinations that employ an assumption in order to estimate the original N_0. The most common assumption is that of a constant ratio of the radioactive isotope to a stable isotope of the same element. This includes the assumption of a constant rate of production of the radio-isotope. The equation is then given by

$$\frac{N}{I} = \frac{N_0}{I} e^{-\lambda t}. \tag{11}$$

The radio-isotope is assumed to be replenished at a constant rate, and a constant ratio of the radio-nuclide to the stable isotope is obtained when the nuclide is added to a large reservoir in which the stable isotope is present. The degree of geochemical separation of the isotopes must be known or we must be able to determine it. When a part of this element is separated from the reservoir, it decays and the change of the ratio gives the age:

$$t = \frac{1}{\lambda}\left[\ln\frac{N_0}{I} - \ln\frac{N}{I}\right] = \frac{1}{\lambda}\ln\frac{N_0}{N}. \tag{12}$$

This method is used with all isotopes produced by cosmic radiation, such as ^{14}C, ^{10}Be, ^{7}Be and ^{3}H. It has been shown (Lal and Peters, 1962) that the production of these isotopes is constant over some million years.

An improvement of this method, as proposed by Lal (*in litt.*), is obtained when the ratio of two cosmic-produced isotopes, such as ^{26}Al to ^{10}Be, can be employed.

For the natural radio-isotopes, the ratio of ionium to thorium has also been used. The use of this ratio is possible when it is assured that only thorium, which is precipitated from sea-water together with ionium, is measured.

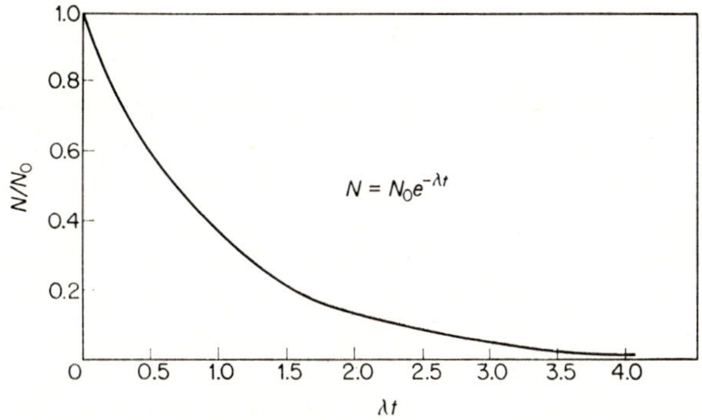

Fig. 5. Decay of radioactive substance. The ordinate gives the ratio of the original amount of the radio-nuclide to the amount left after a certain time. The abscissa gives the non-dimensional units indicating the time as multiples of mean lifetime.

3. Geochemical Considerations

The basic condition for all age determinations is the elimination or addition of radio-nuclides to a closed system which, after the completion of the processes, does not exchange with the surroundings. Age determination in marine sediments is based on the same principle. Because the sediments are a mixture of many components of different origins and different histories during the transport to their final burying ground on the ocean floor, many difficulties arise. In the following sections, some of the most important processes that have a bearing on age determination are discussed.

A. Thorium

As early as 1908, Joly had found a high radium content in sediment samples he had received from the *Challenger* Expedition. Hans Pettersson gave the right explanation in 1937 by suggesting that ionium (^{230}Th) is precipitated in the ocean and added to the sediment. His hypothesis was confirmed by Koczy *et al.* (1957) and later reconfirmed by Sackett, Potratz and Goldberg (1958). The yield of separation of ^{230}Th from sea-water was more than 98%. It is believed that it is as high as 99.8%. If we assume that the uranium content of ocean water is 3×10^{-9} g U/ml and that the produced ionium is evenly spread over the ocean floor, about 1.9×10^{-13} g ^{230}Th is precipitated per square centimeter per year on the ocean floor. The first assumption, based on several measurements by Rona *et al.* (1956), Stewart and Bentley (1954) and Hecht *et al.* (1956), gives values of 2.5 to 3.3 by 10^{-9} g U/ml. The second assumption is a crude approximation and valid only when a constant depth of the ocean is assumed.

The ionium precipitation is constant as long as the uranium content in ocean water is constant and the pH value of ocean water is about the same as at present. But it is unlikely that the uranium concentration is constant in the ocean, since the amount of uranium transported to the ocean is rather high (8×10^{-9} g/cm^2 per year) as compared with the amount of the uranium contained in a water column of 1 cm^2 base (12.6×10^{-4} g). (See later discussion of uranium.)

The mechanism involved in the precipitation of thorium is not very clear. Pettersson (1937) assumed co-precipitation with iron. Holland and Kulp (1954) assumed adsorption on clay. Arrhenius *et al.* (1957) also mentioned incorporation into bone structures. Thorium-232 is a thorium isotope found in greatest abundance in sea-water. It has not yet been possible to obtain exact values, but only an upper limit of the concentration. According to the measurements by Koczy *et al.* (1957), this is 2×10^{-11} g Th/ml. The thorium content of fresh water is so small that it has not yet been possible to determine it. It is believed, therefore, that the thorium content found in sediments must be transported while incorporated in the minerals which settle on the ocean floor, as the following calculation will substantiate. The thorium content in the deep-sea sediment is about 5 to 10×10^{-6} g Th/g. If we assume a rate of sedimentation of about 1 g sediment per square meter per year, about 5 to 10×10^{-6} g Th must be transported per square meter per year to the sea. This would make necessary the conclusion that the thorium content in fresh water is one order of magnitude higher than the upper limit estimated. Further thorium measurements in river water are necessary in order to substantiate this conclusion.

B. Uranium

Selective enrichment of uranium is known to occur by the formation of corals and oölites (Barnes, Lang and Potratz, 1956; Tatsumoto and Goldberg, 1959). In these calcareous organic remains the growth of thorium-230 and protactinium-231 from the parent elements uranium-235 and -238 and the age of the formation of the carbonate can be determined by the formulas given above. Uranium is also enriched in many of the sediments formed under reducing conditions.

The uranium content of deep-sea sediment is generally lower than in sediments found on the continent or in the rocks where it originated. On the other hand, black shales and sediments in land-locked water show a rather high uranium content (Strøm, 1948; Koczy, Tomic and Hecht, 1957). It may, therefore, be concluded that uranium is leached from the inorganic material formed in the deep-sea sediment and enriched in some shelf sediments. This process makes the uranium content in sea-water dependent on the distribution and the character of shelf sediments. Variation by a factor of 2 may occur during a period of 200,000 years. It should be evident that the variation of sea level during the fluctuations of the Ice Age had a pronounced effect on the uranium content of sea-water. During maximum glaciation, the area favorable for uranium precipitation would be considerably smaller than during the interglacial stages. In addition, the uranium in the layers where it was concentrated would be oxidized and could consequently be transported in solution to the ocean.

C. Radium

Radium is found in excess over ionium in sea-water (Koczy *et al.*, 1957). Further ionium determinations are necessary, however, in order to substantiate this statement, since it is based on the determination of the thorium isotopes in North Sea water and in one sample of Pacific deep-sea water (Sackett, Potratz and Goldberg, 1958). The explanation for this high radium content in sea-water was given by Koczy (1954). He assumed it is supplied from the ocean floor. In favor of this explanation is the deficit of radium in the surface layer of sediments (Picciotto and Wilgain, 1954; Picciotto, 1960). An attempt was made to determine the rate of diffusion of radium in sediments by Koczy and Bourret (1958). They were able to establish the equation for the distribution of the radium content in a sediment layer under the assumption that the rate of sediment precipitation, the rate of ionium precipitation, and the rate of radium diffusion are constant:

$$c(x) = c_0 e^{-\surd(\lambda/D)} + \frac{\mu T}{(\mu/v)^2 D - \lambda} \left[e^{-x\surd(\lambda/D)} - e^{-\mu x/v}\right], \tag{13}$$

where $c(x)$ is the radium concentration at depth x, λ is the half life, D is the effective diffusion coefficient, μ is the decay constant of ionium, v is the rate of total sedimentation, and T is the ionium content. This formula was applied to the radium distribution of a core measured by Piggot and Urry (1942) and remeasured by Volchok and Kulp (1957). The upper part of the curve allows us to determine the diffusion rate of radium, the lower part of the curve gives the rate of sedimentation. In this particular core, the maximum rate of diffusion was found to be of the order of 10^{-9} cm^2/sec. Thereby, it was assumed that about 10^{-10} g radium are given off to sea-water from the sediment per square meter per year. The low value of the radium diffusion indicates that radium is rather strongly adsorbed on clay. The ratio of the radium in solution in the interstitial water to the radium adsorbed must be about 1 to 10,000. Consequently, the transport of radium from sediments to sea-water must be rather dependent on the adsorption rate. The rate of radium release from the ocean floor was determined from the radium distribution close to the water–sediment interface and also from the total amount of radium that has to be sustained in sea-water. The calculation made for six different stations during the Swedish Deep Sea Expedition gave a value of the same order of magnitude, i.e. between 100 and 180×10^{-12} g radium/m^2 per year. Further studies of the geochemical processes near the sediment–sea-water interface will have to be carried out to obtain better understanding of the radium release.

Radium is also built into carbonate shells (Koczy and Titze, 1958). The radium content varies between 0.014 to 0.136×10^{-12} g/g. The highest radium content was found in a fresh plankton sample mainly composed of diatoms that contained 0.275×10^{-12} g/g. It is, therefore, evident that the radium content of sea-water is depleted in the surface layer by incorporation in organisms and may be released by solution of the organisms when they are slowly sinking in the ocean. Further investigation of the effect of this process on radium distribution has to be made.

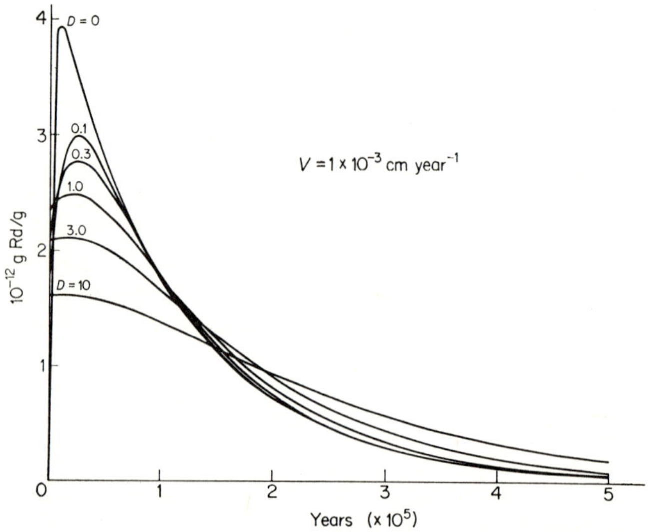

Fig. 6. The distribution of radium with the depth in a sediment layer. The settling rate of the sediment is assumed to be constant at 1×10^{-3} cm per year. It is further assumed that a constant amount of radium diffuses out from the sediment into the ocean water and the rate of ionium precipitation is constant during the time and corresponds to the total ionium produced in the water column of average depth containing 3×10^{-6} grams uranium per liter. The different curves indicate the distribution assuming different rates of diffusion from $D=0$ to $D=10$ cm^2 per year. The depth in the sediment is given as the time when the respective layer of sediment was settling.

D. Carbon-14 and Beryllium

The composition of sea-water with respect to carbon-14 has been rather intensely studied and a summary is given by W. Broecker in Volume 2 of this work (Chapter 4). Carbon-14 enters into carbonate shells formed in the ocean in a certain ratio to carbon-12 which is determined by isotopic fractionation.

Beryllium-10, which is added to the ocean from the atmosphere, follows the geochemistry of beryllium in ocean water. No specific survey has been carried out, but it can certainly be assumed that the main part of beryllium contained in deep sediments has been transported in mineral resistates and only a small amount has been in ionic solution in sea-water. Although it has been employed, the ratio of ^{10}Be to ^{9}Be is a very uncertain isotope ratio to be used for age determination.

4. Practical Methods of Age Determination

The few methods available that allow us to determine the time period elapsed since a component of a sediment layer reached the ocean floor are not generally applicable. Some of them are limited to a single component, which may be the carbonate or the clay; some do not work in shallow-water sediment; and unproved assumptions are often made for others. The theoretical equations applied and the geochemical processes involved are discussed above. In the following section, the methods in practice are described.

A. Cosmic-Ray-Produced Radio-nuclides

a. Carbon-14

This method was first used by Arrhenius, Kjellberg and Libby (1951) for a Pacific core taken during the *Albatross* Expedition. Further determinations on deep-sea samples were carried out by Rubin and Suess (1955), Rubin (1956), Ericson, Broecker, Kulp and Wollin (1956), and Broecker, Ewing and Heezen (1960). It is assumed that the $^{14}C/^{12}C$ ratio in the carbonates is constant at the time of the carbonate formation, and the change of the ratio is used to determine the age according to theoretical method Type 3 given on page 820. The method is generally limited to about 30,000 years and may be extended, by using extremely carefully developed techniques, to 65,000 years (Haring, De Vries and De Vries, 1958).

b. Other nuclides

Potential methods are given by the use of beryllium-10 (Peters, 1957; Arnold, 1956), silicon-32 (Lal, Goldberg and Koide, 1959) and aluminum-26.

c. Aluminum-26/beryllium-10 ratio

By using two cosmic-ray-produced radio-nuclides, a mixture of Types 2 and 3 is obtained and would allow absolute age determination with the only assumption that both nuclides are produced in constant ratio. Lal (*in litt.*) proposed the use of aluminum-26 in connection with beryllium-10. The difficulties involved in the use of this method are great and would entail the development of very low background counters. It would allow age determination from 500,000 to about 10,000,000 years.

B. Natural Radio-nuclides

a. Ionium/uranium or protactinium/uranium ratios

When uranium is incorporated in a mineral without its following products, the increase of the daughter elements ionium (^{230}Th) and protactinium (^{231}Pa)

can be used for age determination. First it must be established that the separation of uranium from its following product is complete or at least the degree of separation is constant and known. This is established for corals (Barnes, Lang and Potratz, 1956) and for oölites (Tatsumoto and Goldberg, 1959). The formula (7) applied is given above under Type 1 (b) (p. 818). The upper limit is about four half lives, i.e. 300,000 years by using ionium (^{230}Th) and 100,000 using protactinium (^{231}Pa). The use of the ratio of protactinium to ionium makes the determination of uranium unnecessary.

b. Protactinium/ionium ratio

The age of one level in a Pacific core was determined by Sackett (1960), and the ages of several levels in cores from the Caribbean and the Atlantic are reported by Rosholt *et al.* (1961). This method is based on the assumptions that (1) protactinium and thorium are precipitated to the same extent in sea-water, i.e. with a yield of more than 98%; (2) the isotope ratio or uranium-235 to uranium-238 is constant in sea-water; and (3) the uranium in the sediment is in equilibrium with its following product. In order to correct for the uranium content. the equation (10) of Type 2 is changed to

$$t = 8.66 \ln \left[2.33 \frac{\overline{^{230}\mathrm{Th}} - \bar{u}}{\overline{^{231}\mathrm{Pa}} - \bar{\bar{u}}} \right] \times 10^4 \text{ yr}, \tag{14}$$

where the bars indicate uranium equivalents. The limit of age determination is about 175,000 years. Of the three given assumptions, the first two seem to be valid as deduced from the chemistry of protactinium and thorium and the results of isotope measurement by Senftle *et al.* (1957). The assumption that uranium is in equilibrium with the daughter product may not be true as outlined in geochemical considerations, but the error introduced is rather small. This method gives the age of the material to which the protactinium and ionium are attached during precipitation. The determined age is not necessarily the same for other components of the sediment because of the reworking of a part of the sediment.

c. Ionium/thorium ratio

Picciotto and Wilgain (1954) proposed the use of the ratio of the two thorium isotopes with atomic weights 230 and 232. The method was used by Baranov and Kuzmina (1958) and by Goldberg and Koide (1958). The validity of this method is doubtful. It assumes that the total thorium-232 content extracted from the sediment is precipitated from sea-water. A further assumption is the constant ratio of uranium to thorium in ocean water, which seems unlikely. The development of suitable extraction methods yielding only the adsorbed thorium-232 may improve the method. The limit of age determination would be in the order of 400,000 years.

d. Ionium or protactinium

This method is based on the following assumptions: (1) The uranium content of sea-water is constant, and (2) the precipitation of thorium isotopes is constant. Piggot and Urry (1939) used for age determination the hypothesis proposed by Pettersson (1937) that thorium is precipitated in ocean water. They assumed that the concentration of uranium-unsupported ionium is constant in each type of sediment.

The difficulties in connection with these assumptions are many. A variation of the concentration of a radioactive element in the sediment can be caused by decay or changes of rate of sedimentation when we assume ionium is precipitated at a constant rate or by changes in the nature of the sediment. Furthermore, diffusion of radium was not considered. It was shown by Picciotto and Wilgain in Picciotto (1960) that this is sometimes the case in old sediments but never in recent sediments, where a diffusion of radium is always found. The difficulty arising is evident, as variations in the concentration of ionium in the sediment may be caused by decay or by changes in the rate of sedimentation, as outlined above.

If the assumptions of Piggott and Urry are accepted as valid, a change in the character of the sediment must be accompanied by a change in the concentration of ionium. Since very few deep-sea sediments are homogeneous in their composition, it is impossible to know the original ionium concentration by only knowing the type of sediment. Small variations in concentrations of ionium, then, may be attributed either to changes in the type of sediment or to the decay of ionium. If Pettersson's assumption is adopted, variations in the concentrations of ionium may be caused by the decay of ionium, thereby making the determination of age possible, or by a change in the rate of sedimentation. In the latter case, age cannot be determined, since the change in the rate of sedimentation is not known.

Both investigators encountered an additional difficulty in the fact that, at that time, no attempt had been made to determine the thorium-230 directly and only radium was measured. It was thereby assumed that radium was in equilibrium with ^{230}Th. If this assumption is correct, at least for the period earlier than 10,000 years ago when radium had reached secular equilibrium with ionium, the following considerations are applicable.

The ionium or the radium content is obtained as a concentration in the sediment, c. It can be assumed that for the time instant when the sediment settles, ionium and the sediment settle independently, having the setting rates i^0 and s^0. The concentration of ionium is then

$$c^0 = i^0/s^0 \text{ g/g}. \tag{15}$$

After t years the concentration of ionium becomes

$$c(t) = c^0(t)e^{-\lambda t}. \tag{16}$$

The age is then given by

$$t = -1/\lambda\,[\ln c(t) + \ln s(t) - \ln i^0(t)]. \tag{17}$$

It is obvious that, since λ is known and $c(t)$ can be measured, assumptions must be made about $s(t)$ and $i^0(t)$ before this equation is applicable. These are the assumptions discussed above.

(1) $$i^0/s(t) = c^0(t) = \text{constant} \quad \text{(Piggott–Urry assumption).}$$

The concentration of ionium is constant during the past for the moment when the sediment settles. The resulting radium distribution with depth is then smooth and must be logarithmic. Three cores have demonstrated this to be true, but they are exceptions.

(2) $$i^0(t) = \text{constant} \quad \text{(Pettersson assumption).}$$

This assumption alone does not allow age determinations and an additional assumption has to be made. If it is assumed that s^0 is constant, the same case as (1) results and $c^0(t)$ would be constant.

In order to explain the irregularities of the concentration with depth, radium diffusion and selective adsorption on some layers were assumed to blur the picture. Picciotto (1960) showed that this explanation could apply in some instances, but extremely high radium maxima in old layers also have a high ionium content unsupported by uranium. The only possible conclusions are that the rate of ionium precipitation and the rate of sedimentation are variable with time, or that ionium is mobile. The latter is rather unlikely because of geochemical reasons. Kröll (1953) designed a method giving the average rate of sedimentation, but not the age of different levels (see later discussion).

5. Rate of Sedimentation

The rate of sedimentation is estimated by dating of the sediment layers. In general, the rate is given in length per time, which is rather inexact, as has been discussed by Arrhenius (1947) and Koczy (1949). The appropriate unit, expressed as mass/time/area, is independent of compaction of sediment and water content. The rate of sedimentation is easily obtained for a given period of time from the difference in the ages of the upper and the lower boundary and the amount of sediment in between. But other methods may be used. If it is known that an element settles at a constant rate in the ocean and is added to the independently settling sediment, the concentration of the element gives the rate of sedimentation. One necessary condition is exact knowledge of the constant rate of precipitation of this element, and another is that the element is not contained in the sediment at the moment it settles. Thorium-230 and protactinium-231 are hydrated in ocean water and are consequently unstable. Both are produced by the rather soluble uranium isotopes. The amount produced seems to be much higher than the amount soluble in sea-water. Thorium-230 and protactinium-231 are precipitated at a constant rate as long as the uranium content in sea-water is constant. This situation is unique, and in no other known instance is there an independent and rather constant rate of precipitation of an element in the ocean.

If it is assumed that the uranium content of ocean water is 3×10^{-9} g/ml and the decay constant is 1.52×10^{-10} yr^{-1},

$$\frac{230}{238} \times 1.52 \times 10^{-10} \times 3 \times 10^{-9} \text{ g } I_0 = 4.45 \times 10^{-9} \text{ g } I_0$$

are produced per year per milliliter. When this amount is precipitated in a 4000-m deep ocean, 1.78×10^{-9} g of ionium settle per square meter per year. The ionium content of the surface sample then directly gives the rate of sedimentation:

$$R(\text{g/m}^2 \text{ yr}) = \frac{1.78 \times 10^{-9}}{\text{concentration of ionium}}. \tag{18}$$

The ionium content must be corrected for the ionium in equilibrium with the uranium in the original sediment. It is not necessarily true that the original ionium content of the sediment corresponds to the uranium content, but it is the only possible assumption at the moment. It may be, as mentioned before, that uranium is leached from the sediment, leaving a higher amount of ionium than the amount in equilibrium with the final uranium content. It is believed that this correction for slowly settling deep-sea sediments is very small and may be of the order of 1%, assuming that 25% of the original uranium content has been lost during the residence time of the sediment in the ocean.

Urry (1950) assumed a constant concentration of ionium in the settling sediment. With this assumption, the ionium content must decrease monotonously with depth in the sediment. The decrease would allow us to determine age and hence the rate of sedimentation would be obtained. His assumption has been disproved, since only a certain amount of ionium is produced per time unit in ocean water. Accounting for a constant rate of ionium precipitation as proposed by Pettersson (1937) and in order to obtain sedimentation rates and ages, Kröll (1953) reasoned that the total ionium unsupported by uranium in the sediment must be in equilibrium with the total uranium in sea-water. If it is further assumed that the rate of sedimentation is constant, the distribution of ionium in the sediment should show a logarithmic decrease with depth in the sediment. All deviations from this theoretical curve must then be connected with changes in the rate of sedimentation. Kröll could not use the above-mentioned rate of ionium precipitation, as he did not know the correct value of the uranium content of sea-water. He assumed that, at a level where the ionium content was twice as large as the equilibrium amount with uranium, the average rate of sedimentation and of ionium precipitation for the whole core could be applied. In that way, he was able to construct a normalized curve for radium distribution. By this method, he found that the uranium content of sea-water had to be larger than it was assumed to be at the time he carried out the investigation.

Another method, giving only the average rate of sedimentation, is obtained when it is assumed that each of two equally long subsequent parts of a sediment core covers an equally long period of time and that the ionium precipitation during each of these periods is the same. Then the time period covered by each section is given by

$$t = \frac{1}{\lambda} \ln \frac{A_1}{A_2}, \tag{19}$$

where A_1 and A_2 are the total ionium contents in the two subsequent parts of the core.

References

Arnold, J. R., 1956. Beryllium-10 produced by cosmic rays. *Science*, **124**, 584–585.

Arrhenius, G., 1947. Den glaciala lerans varvighet en studie over uppsalatraktens varviga margel. *Sveriges Geologiska Undersökning*, Ser. C, **486**, 1–74.

Arrhenius, G., M. Bramlette and E. Picciotto, 1957. Localization of radioactive and stable heavy nuclides in ocean sediments. *Nature*, **180**, 85–86.

Arrhenius, G., G. Kjellberg and W. F. Libby, 1951. Age determination of Pacific chalk ooze by radiocarbon and titanium content. *Tellus*, **3**, 222–229.

Baranov, V. I. and L. A. Kuzmina, 1958. The rate of silt deposition in the Indian Ocean. *Geochemistry* (English translation of Geokhimiya), No. 2, 131–140.

Barnes, J. W., E. J. Lang and H. A. Potratz, 1956. Ratio of ionium to uranium in coral limestone. *Science*, **124**, 175–176.

Broecker, W. S., W. M. Ewing and B. C. Heezen, 1960. Evidence for an abrupt change in climate close to 11,000 years ago. *Amer. J. Sci.*, **258**, 429–448.

Ericson, D. B., W. S. Broecker, J. L. Kulp and G. Wollin, 1956. Late-Pleistocene climates and deep-sea sediments. *Science*, **124**, 385–389.

Goldberg, E. D. and M. Koide, 1958. Ionium-thorium chronology in deep-sea sediments of the Pacific. *Science*, **128**, 1003.

Haring, A., A. E. De Vries and H. De Vries, 1958. Radiocarbon dating up to 70,000 years by isotopic enrichment. *Science*, **128**, 472–473.

Hecht, F., J. Korkisch, R. Patzak and A. Thiard, 1956. Bestimmung Kleinster Uranmengen in Gesteinen und natürlichen Wässern. *Mikrochim. Acta* (*Wien*), **7/8**, 1283–1309.

Holland, H. D. and J. L. Kulp, 1954. The transport and deposition of uranium, ionium and radium in rivers, oceans and ocean sediments. *Geochim. et Cosmochim. Acta*, **5**, 197–213.

Hollander, J. M., I. Perlman and G. T. Seaborg, 1953. Table of isotopes. *Revs. Mod. Phys.*, **25**, 469–651.

Joly, J. J., 1908. On the radium content of deep-sea sediments. *Phil. Mag.*, **16** (6), 190–197.

Koczy, F. F., 1949. Thorium in sea water and marine sediments. *Geol. Fören. Stockholm Förhandl.*, **71**, 238–242.

Koczy, F. F., 1954. Geochemical balance in the hydrosphere. *Nuclear Geology* (ed. H. Faul). John Wiley and Sons, Inc., New York, 120–127.

Koczy, F. F. and R. Bourret, 1958. Radioactive nuclides in ocean water and sediments. Progress Report, The Marine Laboratory, Univ. of Miami.

Koczy, F. F., E. Picciotto, G. Poulaert and S. Wilgain, 1957. Mesure des isotopes du thorium dans l'eau de mer. *Geochim. et Cosmochim. Acta*, **11**, 103–129.

Koczy, F. F. and H. Titze, 1958. Radium content of carbonate shells. *J. Mar. Res.*, **17**, 302–311.

Koczy, F. F., E. Tomic and F. Hecht, 1957. Zur Geochemie des Urans im Ostseebecken. *Geochim. et Cosmochim. Acta*, **11**, 86–102.

Kröll, V., 1953. Vertical distribution of radium in deep-sea sediments. *Nature*, **171**, 742.

Lal, D., E. D. Goldberg and M. Koide, 1959. Cosmic-ray-produced Si^{32} in nature. *Phys. Rev. Letters*, **3** (8), 15 Oct., 1959.

Lal, D. and B. Peters, 1962. *Progress in Elementary Particle and Cosmic Ray Physics*, Vol. VI, Chapter I. North Holland Publishing Co., Amsterdam.

Peters, B., 1957. Über die Anwendbarkeit der Be-10 Methode für Messung komischer Strahlungintensität und der Ablagerungsgeschwindigkeit von Tiefseesedimenten von einigen millionen Jahren. *Z. Physik*, **148**, 93–111.

Pettersson, H., 1937. Der Verhältnis Thorium zu Uran in den Gesteinen und im Meer. *Anz. Akad. Wiss. Wien, Math-nat. Kl.*, 127–128.

Picciotto, E., 1960. Géochimie des éléments radioactifs dans l'océan et chronologie des sediments océaniques. *Ciel et Terre*, No. 3–4, March–April, 1960.

Picciotto, E. and S. Wilgain, 1954. Thorium determination in deep-sea sediments. *Nature*, **173**, 623–633.

Piggot, C. S. and W. D. Urry, 1939. The radium content of an ocean bottom core. *J. Wash. Acad. Sci.*, **29**, 405–415.

Piggot, C. S. and W. D. Urry, 1942. Time relations in ocean sediments. *Bull. Geol. Soc. Amer.*, **53**, 1187–1210.

Rona, E., L. O. Gilpatrick and L. M. Jeffrey, 1956. Uranium determination in sea water. *Trans. Amer. Geophys. Un.*, **37**, 697–701.

Rosholt, J. N., C. Emiliani, J. Geiss, F. F. Koczy and P. J. Wangersky, 1961. Absolute dating of deep-sea sediments by the Pa^{231}/Th^{230} method. *J. Geol.*, **69**, 262–285.

Rubin, M., 1956. U.S. Geological Survey radiocarbon dates III. *Science*, **123**, 442–448.

Rubin, M. and H. E. Suess, 1955. U.S. Geological Survey radiocarbon dates II. *Science*, **121**, 481–488.

Sackett, W. M., 1960. The protactinium-231 content of ocean water and sediments. *Science*, **132**, 1761–1762.

Sackett, W. M., H. Potratz and E. D. Goldberg, 1958. Thorium content of ocean water. *Science*, **128**, 204–205.

Senftle, F. E., L. R. Stieff, F. Cuttitta and P. K. Kuroda, 1957. Comparison of the isotopic abundance of U^{235} and U^{238} and radium activity ratios in Colorado Plateau uranium ores. *Geochim. et Cosmochim. Acta*, **11**, 189–193.

Stewart, D. C. and W. C. Bentley, 1954. Analysis of uranium in sea water. *Science*, **120**, 50–52.

Strøm, K., 1948. A concentration of uranium in black muds. *Nature*, **162**, 922.

Tatsumoto, M. and E. D. Goldberg, 1959. Some aspects of the marine geochemistry of uranium. *Geochim. et Cosmochim. Acta*, **17**, 201–208.

Urry, W. D., 1950. Radioactivity in ocean sediments, VII: Rate of deposition of deep-sea sediments. *J. Mar. Res.*, **7**, 618–634.

Volchok, H. L. and J. E. Kulp, 1957. The ionium method of age determination. *Geochim. et Cosmochim. Acta*, **11**, 219–246.

31. CROSS-CORRELATION OF DEEP-SEA SEDIMENT CORES AND DETERMINATION OF RELATIVE RATES OF SEDIMENTATION BY MICROPALEONTOLOGICAL TECHNIQUES[1]

D. B. Ericson

1. Introduction

The new era in the study of deep-sea sediments which opened with the invention of the piston corer by Kullenberg (1947) has taught us many things. We now know that uninterrupted slow particle-by-particle accumulation should never be taken for granted. On the contrary, experience at Lamont with hundreds of cores has shown that when a core is taken without regard to bottom topography it is more likely than not to contain evidence of slumping or catastrophic deposition by turbidity currents. Often the evidence for such interruptions of orderly sediment accumulation is clear enough, as for example when the normal sequence of foraminiferal lutite is broken by a layer of graded sand containing parts of organisms of shallow-water origin. At other times, and particularly when a part of the section has been removed by slumping, there is no hint from the gross lithology of the core that the section is incomplete. This is not surprising in view of the lithological uniformity of the Pleistocene section and the mixing or blurring effect of burrowing by mud-feeders. How then is the student of the deep-sea record of Pleistocene climatic changes to distinguish between complete and incomplete records? The only satisfactory way of avoiding confusion is to study suites of cores and to persist until layer-by-layer correlations have been established between several cores.

As a by-product of the cross-correlation of cores from the North Atlantic, evidence has emerged that rates of sediment accumulation may vary greatly from place to place apparently quite without the intervention of turbidity currents. The plausible and virtually proven explanation is that deep oceanic circulation, contrary to older ideas, is a fairly effective agent in the horizontal transportation of fine sediment. This opens up the possibility of determining the general direction of deep circulation wherever there are cores adjacent to a seamount whose top has been swept by deep currents. An even more interesting possibility is that of detecting changes in direction and velocity of deep circulation during the Pleistocene and in connection with climatic changes.

2. Methods of Correlation

The stratigraphy of Pleistocene deep-sea sediments is peculiar in its small scale not only in the vertical dimension but also in the time dimension. For example, the deep-sea stratigrapher must establish the time equivalence of zones which are often no more than a few decimeters thick. Because of the short time inter-

[1] The laboratory work upon which this report is based was supported by research grants G4174 and G6540 from the National Science Foundation.

Lamont Geological Observatory Contribution No. 589.

[*MS received June, 1960*]

vals involved most of the faunal zones cannot be defined by extinctions or by first appearances of new species. Instead, the stratigrapher must learn to deal with a series of oscillations between dominance by low latitude species and dominance by high latitude species. The distinction between zones of low and high latitude species in a single core is usually clear enough, but to establish the time equivalence of various zones in two or more cores from stations separated by many kilometers is quite a different matter. For this purpose the mere matching of species is hardly good enough. The need is rather for some variable or variables of the foraminiferal assemblages which can be measured with a fair degree of precision and which can be expressed numerically.

The foraminiferal number of Schott (1937), defined by the number of tests of planktonic Foraminifera in a gram of dry sediment, can be measured with sufficient precision, but curves of variation of the foraminiferal number are not very satisfactory in cross-correlation. Unfortunately, the foraminiferal number is often strongly, sometimes overwhelmingly, influenced by the supply of terrigenous and fine calcareous sediment, which may vary enormously from place to place. The percentage of coarse fraction ($> 74\ \mu$) determined by wet-sieving the samples taken for foraminiferal analyses is essentially the same as the foraminiferal number since in most deep-sea foraminiferal lutites the coarse fraction is composed almost exclusively of the tests of planktonic Foraminifera. It has been used a good deal at Lamont as an invaluable aid in distinguishing different processes and environments of deposition, but it has not been found to be particularly reliable in making cross-correlations.

The frequency-to-weight ratio (Ericson and Wollin, 1956) is obtained by counting the tests of a particular species in a weighted sample of coarse fraction. It is therefore, in effect, a measure of the abundance of the particular species with respect to the total number of tests of planktonic Foraminifera in the sample. In this way the disturbing influence of variation in supply of fine terrigenous material is eliminated. It has the further advantage that no time is wasted in counting all of the eurythermal species which are important only as a background against which the good climatic indicators can be contrasted. As an index of climatic change in individual cores the frequency-to-weight ratio is valuable, and it is sometimes useful in cross-correlation. Its weakness lies in the well-known fact that the species compositions of the thanatocoenoses of deep-sea sediments are strongly influenced by differential solution. For this reason curves of variation in the frequency-to-weight ratios of various species do not provide a very satisfactory way of making firm correlations between cores from stations differing considerably in depth and, therefore, subject to markedly different degrees of solution of the tests.

The most satisfactory method of establishing correlatable faunal zones depends upon variations in the coiling directions of certain species of planktonic Foraminifera. As yet no physical or chemical difference between the dextral and sinistral tests has been found. It is, therefore, very probable that the two kinds of a given species are equally subject to solution, and consequently coiling ratios should be independent of solution. The species which have proven to be most useful are *Globorotalia truncatulinoides*, *G. hirsuta* and *Globigerina pachyderma*. These species construct fairly robust tests which sometimes occur where the tests of many other species have been completely destroyed. In designating the mirror image forms as right or left the gastropod convention has been followed, that is the test is oriented with the dorsal or aboral side up. If when so viewed the chambers have been added in a clockwise sense or *cum sole*, the test is called dextral, and vice versa. In counting the tests we have found it convenient to use two tally counters. When the species is abundant we continue to count until one hundred dextral or sinistral tests have been counted. This takes from

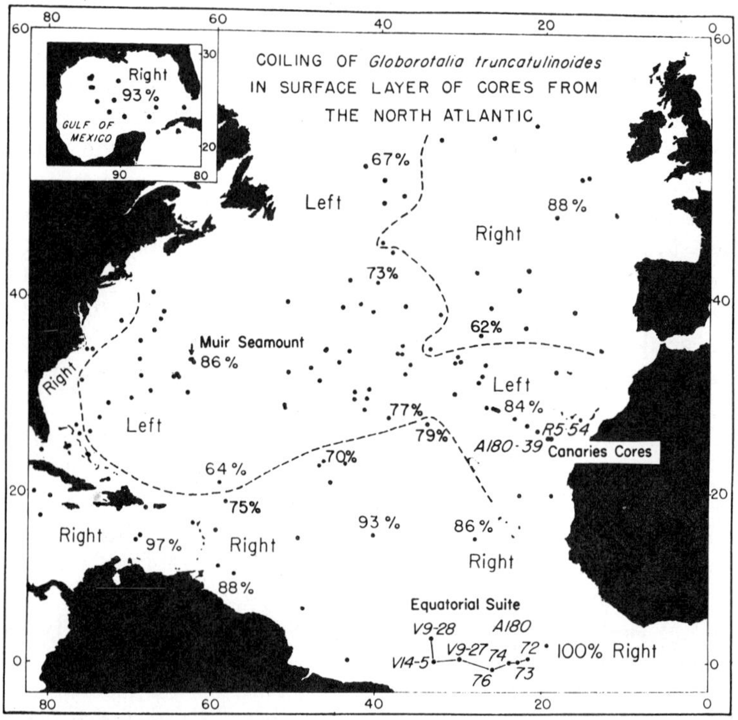

Fig. 1. Chart of coiling of *Globorotalia truncatulinoides* in the surface layer of sediment in the North Atlantic.

one to two minutes. The coiling ratio may be expressed either as a percentage of the dextral tests in the total count or as a percentage of tests of the dominant coiling direction in the total count.

Curves of variation in coiling direction have been used successfully in the correlation of oil wells by Bolli (1950, 1951), Vašíček (1953) and Nagappa (1957). However, the method can be used only where planktonic species are common. Although chiral symmetry is frequent among benthic species, dextral and sinistral tests are normally about equally abundant. Apparently the geometry of the tests plays a critical role only in the planktonic way of life.

A well defined interdependence between coiling direction and temperature tolerance has been found in *Globigerina pachyderma* (Ericson, 1959). On the other hand, no interrelationship between coiling direction and surface water temperature is discernible in the geographical distributions of dextral and sinistral populations of *Globorotalia truncatulinoides* (Fig. 1) and *G. hirsuta*.

3. Time Equivalence of the Faunal Zones

If these Pleistocene faunal zones are to serve as indicators of variation in relative rates of sediment accumulation from place to place, their time-equivalence must be more rigorously assessed than is usually necessary in stratigraphical geology where a few tens of thousands of years one way or the

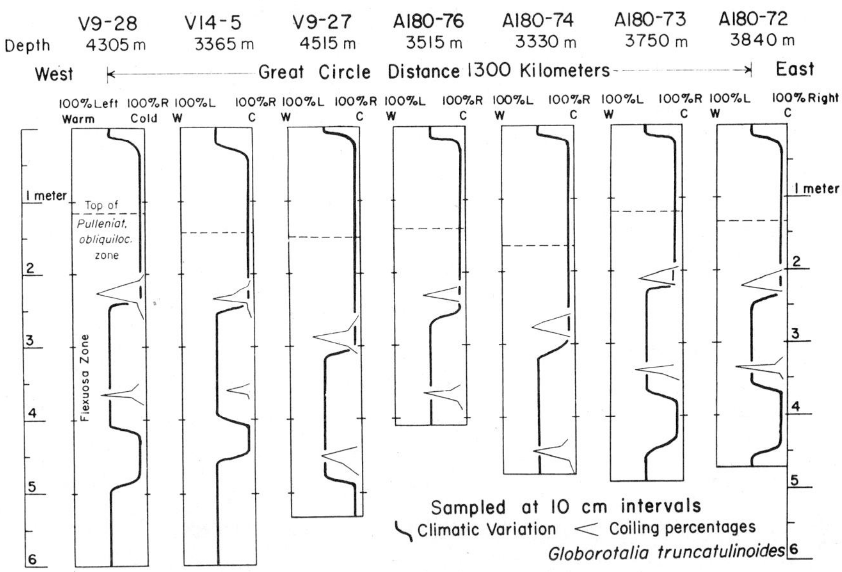

Fig. 2. The equatorial suite of cores.

other are negligible in comparison with the gross time intervals involved. The radiocarbon method of dating is of little use here because of its limited range. It is, in fact, directly applicable only to the horizon defined by the rather abrupt appearance of *Globorotalia menardii* at the base of the uppermost post-glacial zone in cores from the equatorial and mid-latitude Atlantic. Radiocarbon assays of samples from above and below this horizon indicate that it is essentially isochronous within the suite of cores sampled (Ericson *et al.*, 1956).

Perhaps the strongest argument for close time equivalence of the faunal zones over wide areas is the present distribution of the various planktonic species in the North Atlantic. In the east–west dimension their present distribution is truly ocean-wide. It appears reasonable to suppose, therefore, that the faunal zones of the suite of cores shown in Fig. 2 are essentially isochronous even though the total length of the profile is 1300 km. When the distances between cores are much smaller, as for example the distance between two cores from the vicinity of a seamount, essentially complete time equivalence of faunal zones is a reasonable assumption.

Fig. 1 shows the general distribution of dextral and sinistral populations of *Globorotalia truncatulinoides* in the North Atlantic. Here again the pattern of distribution is on an ocean-wide scale. Study of widely scattered cores has shown that this pattern has persisted for at least several thousand years, a time interval amply sufficient to have permitted thorough mixing of the populations. Evidently there is some environmental condition which tends to maintain the present distributional pattern. Coiling dominance may, therefore, be regarded as characteristic of the waters of certain regions much in the same way that a particular salinity, temperature or oxygen content may characterize a water-mass. Presumably changes in coiling direction with depth in the cores reflect wide geographical shifts in environmental conditions during the late Pleistocene. On theoretical grounds it seems improbable that the areal distribution of controlling conditions in the open Atlantic could ever have been on any other than a very broad scale. This implies that zones determined by changes in coiling dominance in cores ought to be very nearly isochronous over distances of at least hundreds of kilometers, if not thousands.

4. Some Examples

The general locations of the suites of cores discussed in the following section are shown in Fig. 1. The geographical positions and depths of the individual coring stations are given in the following table.

TABLE I

Position and Depths of Coring Stations

Station no.	Latitude	Longitude	Depth in meters
A164-44	33° 57′N	62° 39′W	3430
A173-4	33° 52′N	62° 32′W	3930
A180-39	25° 50′N	19° 18′W	3470
A180-72	00° 36′N	21° 47′W	3840
A180-73	00° 10′N	23° 00′W	3750
A180-74	00° 03′S	24° 10′W	3330
A180-76	00° 46′S	26° 02′W	3515
V3-126	23° 45′N	92° 28′W	3485
V3-127	23° 38′N	92° 40′W	3540
V3-128	23° 46′N	92° 29′W	3495
V9-27	00° 23′N	29° 52′W	4515
V9-28	02° 54′N	33° 10′W	4305
V14-5	00° 15′N	32° 51′W	3365
R5-54	25° 52′N	19° 03′W	3295

A. The Equatorial Suite

Fig. 1 shows the positions of the cores, and Fig. 2 shows the faunal zones in the cores in graphic form. The temperature or climatic curve (heavy line) is based upon the abundance of *Globorotalia menardii* and other temperature-sensitive species. The subspecies *G. menardii flexuosa* (Koch) is especially abundant in the *flexuosa* zone. Thus the heavy line represents primarily fluctuations in faunal composition. In drawing the graphs these fluctuations have been translated into climatic terms. Here it should be emphasized that the variations in faunal composition are very strikingly apparent; their detection does not depend upon careful counts or complicated statistical methods. For example, the contrast between washed samples of Foraminifera from the uppermost zone of "cold" climate and samples from the *flexuosa* zone is such that the difference can be seen easily without taking the samples out of the vials in which they are stored. The dashed line labeled "Top of *Pulleniat. obliquiloc.* zone" in the diagram of core V9-28, Fig. 2, marks the level in the cores above which the planktonic species *Pulleniatina obliquiloculata* is rare (1–5 tests) or absent. Below this level the species is very abundant (> 100 tests in each sample). It is again abundant in the uppermost post-glacial layer of these same cores.

The light line representing coiling percentages of *Globorotalia truncatulinoides* has been drawn only where there is strong deviation to the left. Elsewhere in the cores coiling is consistently between 90 and 100% to the right. Apparently twice during the past 100,000 years a boundary between provinces of right and left coiling dominance has swept back and forth across the equatorial region. If the boundary line was essentially parallel to the core profile, it must have crossed the positions of the various core stations at about the same time. If, on the other hand, it moved from east to west and the rate of movement was slow, it must have passed successively over the stations at appreciably different times. But in that case the zones of left coiling in core A180-72 would be thicker than the corresponding zones in core V9-28, because station A180-72 would then have lain within the left coiling province an appreciably longer time than station V9-28. Since no difference in thickness is discernible, we conclude that movement of the boundary across all the stations took place virtually "instantaneously" or occupied a time interval which was so short as to be inappreciable in terms of sediment accumulation at these stations. From this we infer that the coiling zones in these cores are for practical purposes isochronous.

Clearly evident in this suite of cores is the small variation in rate of sediment accumulation from station to station. The almost exact equivalence of rates at the extreme stations, A180-72 and V9-28, is particularly striking. This, together with the fact that three other cores show closely similar rates, makes us believe that, in these cores, we have examples of orderly accumulation by vertical settling of particles with little or no interference from horizontal transportation. Extrapolation of radiocarbon dates in various cores (Ericson and Wollin, 1956) and one direct dating by the ionium method (Urry, 1948) show that the interval of warm climate represented by the *flexuosa* zone came to an end between 65,000 and 70,000 years ago. If we assume a compromise date of 67,000 years B.P. for the top of the *flexuosa* zone in the five cores, A180-72, A180-73, A180-76, V14-5 and V9-28, we obtain an average rate of sediment accumulation of 3.3 cm in one thousand years.

B. Cores from the Sigsbee Knolls

These were taken on or near the tops of some low rises at about the center of the abyssal plain of the Sigsbee Deep in the Gulf of Mexico. The faunal zones in

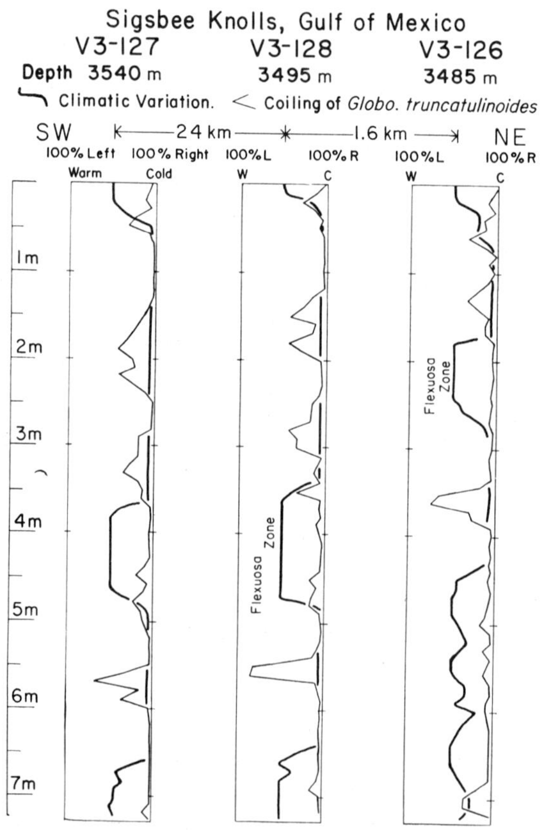

Fig. 3. Cores from the Sigsbee Knolls, Gulf of Mexico.

the cores are shown graphically in Fig. 3. In this instance time equivalence of the faunal zones is virtually certain in view of the following considerations. At present the deep and open Gulf is homogeneous with respect to coiling of *Globorotalia truncatulinoides* and the various species of planktonic Foraminifera. It is highly improbable that a faunal boundary could have maintained a position between such closely spaced stations, about 25 km, in so homogeneous a body of water for any appreciable time during the late Pleistocene.

Good correlation between cores V3-127 and V3-128 is evident from the graphs of Fig. 3. This amounts to conclusive evidence that these two cores contain continuous records of late Pleistocene events. Core V3-126 is in striking contrast in spite of the fact that it came from a point only 1.6 km distant from station V3-128. Comparison of the graphic logs shows that the section which lies between 1.70 and 3.50 m in core V3-128 is missing from V3-126. No layer of coarse winnowed sediment occurs at the level of the stratigraphical hiatus in V3-126, as would be expected if the hiatus were due to deep-current scour. Furthermore, current scour of sufficient energy to prevent deposition of about 1.8 m of sediment ought to have made some impression on the corresponding section in the adjacent core, V3-128. Erosion by a turbidity current cannot be seriously considered; not only are the topographical positions of the core stations such as to put them above the influence of the turbidity currents which flowed over the surrounding abyssal plain, but station V3-126 as the highest of the three ought to have been the least accessible to turbidity currents. The most plausible explanation is that slumping was responsible for removal of the missing section.

As is often the case the gross lithology of V3-126 gives no evidence of a depositional hiatus at 1.8 m. The lesson to be learned from this example is clear. A single core, however free of disturbance it may appear to be, should not be relied upon to yield a continuous record of Pleistocene events, and no regional stratigraphical sequence should be regarded as established until a layer by layer cross-correlation has been found between at least two cores.

C. The Canaries Cores

Fig. 1 shows the positions of the coring stations. Curves of variation in coiling percentages of *Globorotalia truncatulinoides* are shown in Fig. 4. Time equivalence of the zones is reasonably certain in view of the short distance, 28 km, between the coring stations. Completeness of the sections is assured by the close correspondence between the two curves. It is obvious that there has been much disparity in rates of sediment accumulation at the two stations, that at station A180-39 having been continuously about 80% faster. Samples from core A180-39 are consistently deficient in coarse fraction ($> 74\ \mu$), that is in the tests of planktonic Foraminifera. Presumably the tests have been settling at about the same rate at both stations, but the deeper station, A180-39, has continuously received an excess of fine material. Preferential accumulation of this fine sediment at the deeper station suggests that the route of travel was determined by bottom topography, but the continuity of accession of fine material makes it appear improbable that transportation was due to turbidity currents, which are characteristically spasmodic in effect. More probably the energy responsible for horizontal transportation of the fine material was supplied by deep oceanic circulation.

Comparison of the coiling curves shows that the ratio of the rates of accumulation in the two cores has remained very nearly constant. Since both cores pass through the *flexuosa* zone and into sediment which was probably deposited during the last interglacial, we conclude that deep circulation in this part of the North Atlantic has remained essentially constant in both direction and velocity during the late Pleistocene.

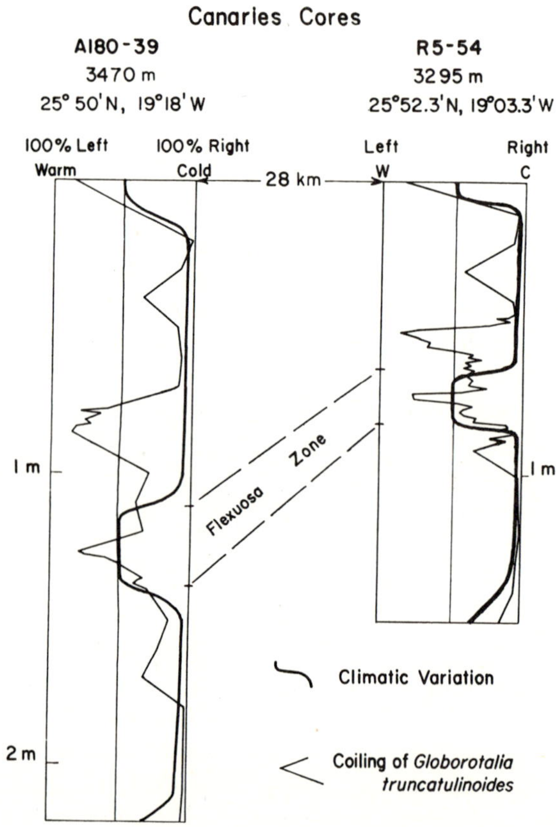

Fig. 4. The Canaries cores.

D. *The Muir Seamount Cores*

The location of the Muir Seamount is shown in Fig. 1. The positions of the cores with respect to the seamount, the coiling curves of *Globorotalia truncatulinoides* and the positions of the *flexuosa* zone in the cores are shown in Fig. 5. Here again, as in the case of the Canaries cores, we find a marked contrast in rates of sediment accumulation within a short distance, 15 km. Sediment samples from the top of the mount are almost devoid of fine fraction. Presumably this is due to winnowing by a deep current of sufficient velocity to prevent fine material from coming to rest on the top of the mount. Percentages of fine fraction ($< 74\ \mu$) in the cores are shown in Fig. 5. In the zone between the top of the *flexuosa* zone and the base of the post-glacial zone in core A164-44, the percentage of fine fraction averages 65%. This is about normal for foraminiferal lutites in the North Atlantic. Apparently horizontal transportation of sediment has had a negligible effect at station A164-44. In sharp contrast is the corresponding zone in core A173-4 from the northeast flank of the mount. Here the percentage of fine fraction in most samples exceeds 99%. Evidently this station received much fine material from the top of the mount during the latter part of the last ice age. From this we infer that the general direction of deep circulation

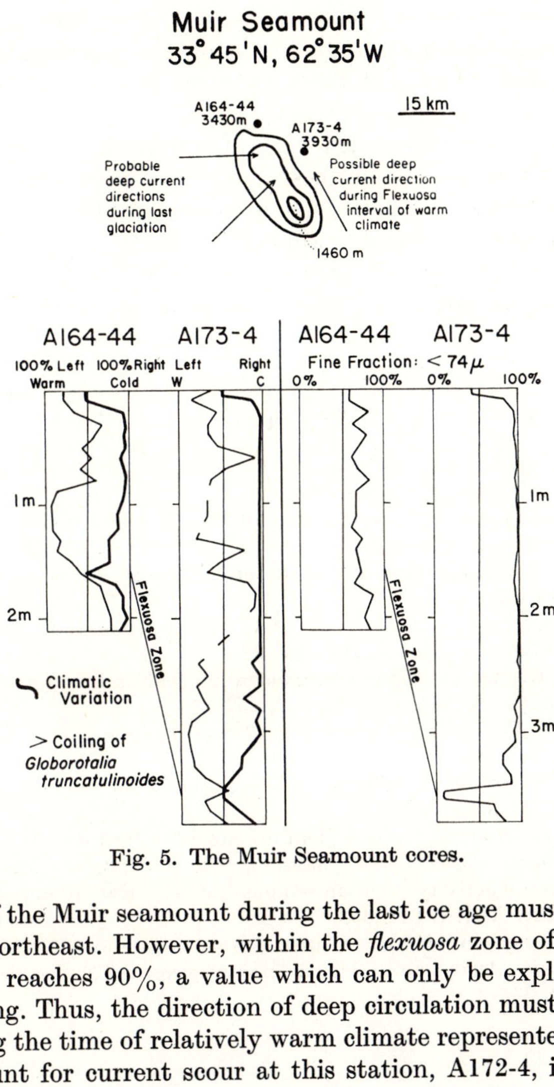

Fig. 5. The Muir Seamount cores.

in the region of the Muir seamount during the last ice age must have been from southwest to northeast. However, within the *flexuosa* zone of core A173-4 the coarse fraction reaches 90%, a value which can only be explained by current scour winnowing. Thus, the direction of deep circulation must have been quite different during the time of relatively warm climate represented by the *flexuosa* zone. To account for current scour at this station, A172-4, it is necessary to suppose that the direction of deep circulation was parallel to the long axis of the seamount. The fact that the percentage of coarse fraction in the *flexuosa* zone of core A164-44 is normal for the depth of water suggests that the direction of deep circulation was in general from southwest to northeast. With circulation in the reverse sense station A164-44 would probably have been exposed to current scour judging from its position with respect to the seamount.

It is significant that the uppermost post-glacial layers in both cores contain fairly large percentages of coarse fraction. While this may be partly due to greater productivity of large species of planktonic Foraminifera such as *Globorotalia menardii* under present climatic conditions, the 52% of coarse fraction in the top layer of core A173-4 appears to be too great to be accounted for in this way. More probably station A173-4 is presently subject to weak current scour. This suggests that the direction and velocity of deep circulation in the vicinity of the Muir seamount is now just about the same as it was during the period of warm climate represented by the *flexuosa* zone.

With these few data no closer definition of current directions is possible. The important thing is that we have in these two cores compelling evidence of significant changes in deep circulation which are correlatable with climatic changes of the late Pleistocene. Presumably application of this method of analysis to a series of cores suitably spaced around a seamount could yield interesting evidence regarding variations in deep circulation in the Atlantic during the Pleistocene epoch.

References

Bolli, H., 1950. The direction of coiling in the evolution of some Globorotaliidae. *Contrib. Cushman Found. Foram. Res.*, **1**, 82–89.

Bolli, H., 1951. Notes on the direction of coiling of Rotalid Foraminifera. *Contrib. Cushman Found. Foram. Res.*, **2**, 139–143.

Ericson, D. B., 1959. Coiling direction of *Globorotalia pachyderma* as a climatic index. *Science*, **130**, 219–220.

Ericson, D. B., W. S. Broecker, J. L. Kulp and G. Wollin, 1956. Late Pleistocene climates and deep-sea sediments. *Science*, **124**, 385–386.

Ericson, D. B. and G. Wollin, 1956. Micropaleontological and isotopic determination of Pleistocene climates. *Micropaleontology*, **2**, 257–270.

Kullenberg, B., 1947. The piston core sampler. *Svenska Hydrog.-Biol. Komm. Skr.*, s. 3, **1** (2), 1–46.

Nagappa, Y., 1957. Direction of coiling in *Globorotalia* as an aid in correlation. *Micropaleontology*, **3**, 393–398.

Schott, W., 1937. Die Foraminiferen in dem äquatorialen Teil des Atlantischen Ozeans. *Wiss. Ergebn. Deut. Atlant. Exped. 'Meteor', 1925–1927*, **3**, 43–131.

Urry, W. D., 1948. Radioactivity of ocean sediments: VII—Rate of deposition of deep-sea sediments. *J. Mar. Res.*, **7**, 618–633.

Vašíček, M., 1953. Changes in the ratio of sinistral and dextral individuals of the foraminifer *Globorotalia scitula* (Brady) and their use in stratigraphy. *Sborník Ústředního Ústravu Geologického*, **20**, 345–420.

History

32. THE ORIGIN OF LIFE

STANLEY L. MILLER

1. Early Considerations of the Problem

The problem of the origin of life has occupied man's mind since history began. The problem was not as troublesome for the ancients as it is for us today, since it was quite clear to them that life arose by spontaneous generation. Thus when the Egyptians saw snakes arise from the muds of the Nile and when the Greeks saw rats arise from piles of garbage on exposure to warmth, they interpreted this to mean that such animals arose spontaneously.

That organisms could arise spontaneously, as well as by sexual and asexual reproduction, was not seriously questioned until the middle of the 17th century. In 1688 the Tuscan physician, Francesco Redi, showed that worms did not develop in meat if the flask was covered with muslin so that flies could not lay their eggs on the meat.

The demonstration that micro-organisms, which had been discovered in 1676, did not develop spontaneously was much more difficult. In 1765 Lazzaro Spallanzani showed that micro-organisms did not appear in various nutrient broths if the vessels were sealed and boiled. Objections were raised that the heating had destroyed the "vital force" in the broth and air. This "vital force" was postulated as necessary for life to develop. By readmitting air it was possible to show that the broth could still support the growth of micro-organisms. But Spallanzani could not demonstrate that the air in the sealed flask had not been altered, and the doctrine of spontaneous generation persisted widely.

This problem was finally solved by Pasteur in 1862. He used a flask with broth, but instead of sealing the flask he drew out a long S-shaped tube with its end open to the air. The air was free to pass in and out of the flask, but the particles of dust, bacteria and molds in the air were caught on the sides of the long S-shaped tube. When the broth in the flask was boiled and allowed to cool, no micro-organisms developed. When the S-shaped tube was broken off at the neck of the flask, micro-organisms developed. These experiments were extended by Pasteur and by Tyndall to answer all the numerous objections that were raised. Thus, after two centuries of experimentation and many more centuries of belief, the doctrine of spontaneous generation was disproved. (For an extension of this history, see Oparin, 1938.)

Shortly before 1858, Darwin and Wallace had published, simultaneously and independently, the theory of evolution by natural selection. This theory could account for the evolution from the simplest single-celled organism to the most complex plants and animals, including man. Therefore, the problem of the origin of life no longer involved how each species developed, but only how the first living organism arose.

To answer this question it has been proposed that life was created by a supernatural event. This is not a scientific theory, since by its very nature it is

[*MS received December, 1960*]

not subject to experimental investigation. S. A. Arrhenius offered a second theory that life developed on the Earth as a result of a spore or other stable form of life coming to the Earth in a meteorite from outer space or by the pressure of sunlight driving the spores to the Earth. One form of this theory assumes that life had no origin, but like matter, has always existed. The presence of certain long-lived radioactive elements shows that the elements were formed about 5×10^9 years ago. If the elements have not always existed, it is difficult to understand how life could have always existed. Another form of this theory assumes that life was formed on another planet and traveled to the Earth. This hypothesis does not answer the question of how life arose on the other planet. In addition, it is doubtful that any known form of life could survive very long in outer space and fall through the atmosphere without being destroyed. Therefore, while this theory has not been disproved, it is held to be highly improbable. If micro-organisms are carried on interplanetary rockets, space is likely to become contaminated, and this would make it impossible to disprove this theory conclusively. For this reason, space vehicles are being sterilized.

A third hypothesis holds that the first living organism arose from inorganic matter by a very improbable event. This organism, in order to grow in an inorganic environment, would have to synthesize all of its cellular components from carbon dioxide, water and other inorganic nutrients. Our present knowledge of biochemistry shows that even the simplest bacteria are extremely complex and that the probability of the spontaneous generation of a cell from inorganic compounds in a single event is much too small to have occurred in the approximately 5×10^9 years since the Earth's formation.

The fourth proposal is that life arose spontaneously in the oceans of the primitive Earth, which contained large quantities of organic compounds similar to those which occur in living organisms. This theory was outlined mainly by Oparin (1938) and it forms the basis of most of the present ideas on the origin of life. Oparin argued that if large amounts of organic compounds were in the oceans of the primitive Earth, these organic compounds would react to form structures of greater and greater complexity, until a structure would form which we would call "living". In other words, the synthesis of the first living organism would involve many non-biological steps, and none of these steps would be highly improbable. Oparin further proposed that the atmosphere was reducing in character and that organic compounds might be synthesized under these conditions. This hypothesis implied that the first organisms would be heterotrophic, meaning that they would obtain their basic constituents from the environment instead of synthesizing them from carbon dioxide and water.

In spite of the argument by Oparin, numerous attempts were made to synthesize organic compounds under the oxidizing conditions now present on the Earth (Rabinowitch, 1945). Various sources of energy acting on carbon dioxide and water failed to give reduced carbon compounds except when contaminating reducing agents were present. The one exception to this was the synthesis of formic acid and formaldehyde in very small yield (10^{-7} molecule of H_2CO per ion pair) by the use of 40×10^6 eV helium ions from a 60-in. cyclotron (Garrison *et al.*, 1951). While the simplest organic compounds were indeed synthesized, the yields were so small that this experiment can best be interpreted to mean that it would not have been possible to synthesize organic compounds non-biologically as long as the Earth had oxidizing conditions. This experiment is important in that it induced a re-examination of Oparin's hypothesis of the reducing atmosphere (Urey, 1952, 1952a).

2. The Primitive Atmosphere

The geological record gives us little (and contradictory) evidence as to the nature of the atmosphere when the Earth was formed. Opinions as to the nature of the early atmosphere range from regarding it as strongly reducing to strongly oxidizing and include all types of atmospheres of intermediate degrees of oxidation.

The discussion here is based on the assumption that the conditions on the primitive Earth were favorable for the production of the organic compounds which make up life as we know it. There are many sets of conditions under which organic compounds could have been produced. All these conditions are more or less reducing. However, before accepting a set of conditions for the primitive Earth, one must show that reactions known to take place will not rapidly change the atmosphere to another type. The proposed set of conditions must also be consistent with the known laws for the escape of hydrogen.

Cosmic dust clouds, from which the Earth is believed to have been formed, contain a great excess of hydrogen. The planets Jupiter, Saturn, Uranus and Neptune are known to have atmospheres of methane and ammonia. There has not been sufficient time for hydrogen to have escaped from these planets owing to their lower temperatures and higher gravitational fields. It is reasonable to expect that the Earth and the other minor planets also started out with reducing atmospheres and that these atmospheres became oxidizing due to the escape of hydrogen.

The meteorites may be the closest approximation we have to the solid material from which the Earth was formed. They are observed to be highly reducing, with the iron mostly as metallic iron with some ferrous sulfide, the carbon as elemental carbon or iron carbide, and the phosphorus as phosphides. The carbonaceous chondrites contain several per cent of organic carbon.

The overall chemical change has been the oxidation of the reducing atmosphere to the present oxidizing atmosphere. This is caused by the loss of hydrogen which results in the production of nitrogen, nitrate, sulfate, free oxygen and ferric iron. As we shall discuss, many complex organic compounds would have been formed during this overall change, thereby presenting a favorable environment for the formation of life. It is extremely doubtful that life could begin unless large quantities of organic compounds were present on the Earth. Therefore, the first living organism must have been formed by the time significant quantities of oxygen appeared in the atmosphere, for otherwise the organic compounds formed during the reducing conditions would have been oxidized rather rapidly by the molecular oxygen (Miller, 1957a). Whether the surface carbon of the present Earth was all part of the initial atmosphere or whether it has been escaping from the Earth's interior in a somewhat reduced condition is not important to the overall picture to be presented.

A. The Escape of Hydrogen

We have learned in recent years that the temperature of the high atmosphere is 2000°K or more, and there is no reason to suppose that the same temperature was not present in the past. One might expect that a reducing atmosphere would be cooler than an oxidizing atmosphere because methane and ammonia can emit infrared radiation while nitrogen and oxygen cannot. However, Curtis and Goody (1956) have shown that carbon dioxide is ineffective in emitting infrared radiation in the high atmosphere. This is due to the low efficiency of energy transfer from the translational and rotational to the vibrational degrees of freedom, and it seems likely that this would apply to methane as well.

The loss of hydrogen from the Earth is now believed to be limited by the diffusion of H_2 to the high atmosphere, since almost all the water is frozen out before it reaches the high atmosphere. Urey (1959) has discussed this problem and finds that the loss is entirely due to these effects and not to the Jean's escape formula.

The present rate of escape is 10^7 atoms of hydrogen cm^{-2} sec^{-1}, and it is proportional to the concentration of molecular hydrogen in the atmosphere, which is now 10^{-6} atm at the Earth's surface. This rate would result in an escape of hydrogen equivalent to 20 g of water/cm^2 in the last 4.5×10^9 years. This rate is not sufficient to account for the oxygen in the atmosphere (230 g/cm^2).

In addition, we must account for the oxidation of the carbon, ammonia and ferrous iron to their present states of oxidation. The oxidation of the 3000 g/cm^2 of surface carbon on the Earth from the 0 to +4 valence states (i.e. from C or H_2CO to CO_2) would require the loss of 1000 g/cm^2 of hydrogen. At the present rate of escape this would require 2.5×10^{12} years. If this escape was to be accomplished in 2.5×10^9 years (i.e. from 4.5 to 2.0×10^9 years ago) the pressure of hydrogen at the surface of the Earth would be 0.7×10^{-3} atm. In order also to oxidize the nitrogen, sulfur and iron, even larger losses and a higher pressure of hydrogen are needed. We shall use a figure of 1.5×10^{-3} atm for the hydrogen pressure in the primitive atmosphere.

These calculations are much oversimplified since methane and other volatile hydrogen compounds would be decomposed in the high atmosphere, and, therefore, a higher concentration of hydrogen might exist in the high atmosphere than indicated by surface partial pressures. However, the results of the calculation would be qualitatively the same for hydrogen pressures different from the chosen value by an order of magnitude.

B. The Equilibria of Carbon Compounds

The partial pressure of carbon dioxide in the atmosphere is kept low by two buffer systems. The first system, which is rapid, is the absorption of carbon dioxide in the sea to form CO_3^{2-}, HCO_3^- and H_2CO_3; the second, which is slow, is the reaction of carbon dioxide with silicates, e.g.

$$CaSiO_3 + CO_2 = CaCO_3 + SiO_2 \qquad K_{25^\circ} = 10^{-8}.$$

This reaction, which is the basic cause of the low CO_2 content in the atmosphere, is known as the Urey equilibrium. The equilibrium partial pressure of CO_2 is 10^{-8} atm, while the observed partial pressure at sea-level of 3.3×10^{-4} atm is somewhat higher.

The equilibrium constant at 25° in the presence of liquid water for the reaction

$$CO_2 + 4H_2 = CH_4 + 2H_2O_{(1)}$$

is 8×10^{22}. Assuming equilibrium was attained, and using $P_{CO_2} = 10^{-8}$ atm and $P_{H_2} = 1.5 \times 10^{-3}$ atm, we find that $P_{CH_4} = 4 \times 10^3$ atm. It is evident then that the partial pressure of CO_2 would be less than 10^{-8} atm and limestone ($CaCO_3$) would not form if equilibrium conditions are assumed.

Complete thermodynamic equilibrium could not exist in a reducing atmosphere because of the dependence of the equilibrium proportions of compounds on pressure and hence on altitude. It is more likely that the steady-state concentrations of CO_2 and CH_4 would not be determined by the equilibrium at sea-level, but rather by the equilibrium at higher altitude where the ultraviolet light would provide the necessary activation energy to bring about rapid equilibrium. Under these conditions water would be a gas, and the equilibrium constant is 10^{20}, so

$$K_{25^\circ} = 10^{20} = P_{CH_4} P_{H_2}{}^2 = (X_{CH_4} X_{H_2O}{}^2 / X_{CO_2} X_{H_2}{}^4) p^{-2},$$

where the X's are the mole fractions and P is the total pressure. If the surface partial pressures were $P_{CH_4} = 1$, $P_{CO_2} = 3.3 \times 10^{-4}$ (the present value), $P_{H_2} = 1.5 \times 10^{-3}$, the X's would be equal to these partial pressures. We shall use $X_{H_2O} = 10^{-6}$, which is the present value for H_2O above the tropopause. Equilibrium will be established under these conditions when $P = 2.5 \times 10^{-9}$ atm, which is the present atmospheric pressure at about 180 km. It is reasonable to assume that equilibrium was established at some high altitude, and, therefore, that carbon dioxide and hydrogen could both be present at small partial pressures and methane present at a moderate partial pressure in a reducing atmosphere where the pressure of hydrogen is 1.5×10^{-3} atm.

Carbon monoxide should not have been an important constituent of the atmosphere as can be seen from the following reactions,

$$CO_2 + H_2 = CO + H_2O\ (1), \quad K_{25^\circ} = 3.2 \times 10^{-4}$$

$$P_{CO}/P_{CO_2} = 3.2 \times 10^{-4} P_{H_2}$$

Using $P_{H_2} = 1.5 \times 10^{-3}$, we have the ratio $P_{CO}/P_{CO_2} = 5 \times 10^{-7}$, which is independent of pressure. Furthermore, carbon monoxide is a relatively reactive compound, and should any significant quantities appear in the atmosphere, it would react in various ways to give organic compounds, carbon dioxide and hydrogen, and formate. The hydration of carbon monoxide to formate is a base-catalyzed reaction and proceeds at a significant rate, and the other reactions would also be relatively rapid.

Rubey (1955) and Abelson (1956) have argued that the surface carbon and nitrogen came from out-gassing the interior of the Earth instead of coming from the remainder of the cosmic dust cloud from which the Earth was formed. The carbon from the out-gassing may be in the form of CO_2 and CO or CH_4, and hydrogen may be present. While this may have been a significant process on the primitive Earth, it does not require that the atmosphere was composed of CO_2 and CO. The above thermodynamic considerations still apply. The carbon dioxide would dissolve in the ocean to form HCO_3^- and CO_3^{2-}, and $CaCO_3$ would be deposited, and the CO would be unstable as demonstrated above. The state of oxidation of gases from volcanoes is difficult to determine, since they come in contact with the oxygen of the air and with oxides of iron, which in turn may owe their oxidation state directly or indirectly to the oxidizing atmosphere.

The buffer systems of the ocean and the calcium silicate–calcium carbonate equilibrium are of sufficient capacity to keep the partial pressure of the carbon dioxide in the atmosphere at a low value, so that the principal species of carbon in the atmosphere would be methane even though the fraction of surface carbon in the oxidation state of carbon dioxide was continuously increasing. This would be true until the pressure of H_2 fell below about 10^{-6} atm. At this point the small partial pressure of carbon dioxide would have made it the principal carbon species in the atmosphere. It is likely that shortly after this, significant quantities of molecular oxygen would appear in the atmosphere.

C. The Equilibrium of Nitrogen Compounds

The equilibrium concentrations of ammonia can be discussed by considering the reaction

$$\tfrac{1}{2}N_2 + \tfrac{3}{2}H_2 = NH_3, \quad K_{25^\circ} = 7.6 \times 10^2$$

Using $P_{H_2} = 1.5 \times 10^{-3}$, we have

$$P_{NH_3}/P_{N_2}^{1/2} = 0.04.$$

The nitrogen pressure could not be the present value and was probably much lower.

Ammonia is very soluble in water and this would displace the above reaction toward the right, giving

$$\tfrac{1}{2}N_2 + \tfrac{3}{2}H_2 + H^+ = NH_4^+$$

$$[NH_4^+]/P_{N_2}^{1/2}P_{H_2}^{3/2} = 8.0 \times 10^{13}[H^+],$$

which is valid for pH's less than 9. At pH = 8 and $P_{H_2} = 1.5 \times 10^{-3}$, we have

$$[NH_4^+]/P_{N_2}^{1/2} = 47,$$

which shows that most of the ammonia would be in the ocean instead of in the atmosphere. The ammonia in the ocean would be largely decomposed when the pressure of hydrogen fell below 10^{-5} atm, assuming that the pH of the ocean was the present value of 8. A higher pH would make the ammonia less stable and conversely for a lower pH.

The presence of large quantities of ammonia in the ocean does not necessarily imply that the oceans had a pH greater than at present. The NH_4^+ would weather acidic rocks, thereby lowering the pH. It is even possible that the oceans under reducing conditions were more acidic by one or two units of pH.

It is unlikely that there exist any effective inorganic syntheses of nitrogenous organic compounds of biological interest which could use nitrogen in a higher oxidation state than ammonia, unless strong reducing agents were present. It seems almost necessary that ammonia was present in oceans of the primitive Earth but not necessarily in the atmosphere. This point alone is a strong argument for the presence of reducing conditions on the primitive Earth.

All the oxides of nitrogen would have been unstable and should have been rare. Hydrogen sulfide would have been present in the atmosphere only as a trace constituent because it would precipitate as ferrous and other sulfides. Sulfur would be reduced to hydrogen sulfide by the reaction,

$$H_2 + S = H_2S, \quad K_{25^\circ} = 6 \times 10^4.$$

It is evident that the calculations do not have a quantitative validity because of many uncertainties, such as temperature, processes by which equilibrium could be approached, the atmospheric level where such processes were effective, the partial pressure of hydrogen required to provide the necessary rate of escape, etc.

It should also be noted that these thermodynamic calculations do not mean that compounds which are highly unstable in a reducing atmosphere—such as oxygen, oxides of nitrogen, oxides of sulfur, etc.—were entirely absent, but that they were present in no more than a few parts per million. In the present atmosphere, hydrogen, ozone, methane and nitrous oxide are highly unstable thermodynamically, but they are present only in a few parts per million.

The above discussion does not prove that the primitive atmosphere contained low partial pressures of hydrogen and ammonia and moderate partial pressures of methane and nitrogen. However, this atmosphere is stable and can account for the loss of hydrogen from the Earth, which is not the case for atmospheres containing large quantities of carbon dioxide and carbon monoxide. The reducing atmosphere also provides favorable conditions for the origin of life.

D. Sources of Energy

At the present time the direct or indirect source of free energy for all living organisms is the sunlight utilized by photosynthetic organisms. But before the evolution of photosynthesis other sources of free energy must have been used. It is of interest to consider the sources of such free energy as well as the origin of the appropriate chemical compounds prior to the existence of what should be called living organisms, and before the evolution of photosynthesis.

TABLE I

Present Sources of Energy Averaged over the Earth

Source	cal cm^{-2} yr^{-1}
Total radiation from sun	260,000
Ultraviolet light	
$\lambda < 2500$ Å	570.0
$\lambda < 2000$ Å	85.0
$\lambda < 1500$ Å	3.5[a]
Electric discharges	4.0[b]
Cosmic rays	0.0015
Radioactivity (to 1.0 km depth)	0.8[c]
Volcanoes	0.13[d]

[a] Includes the 1.9 cal cm^{-2} yr^{-1} from the Lyman α at 1216 Å (Rense, 1953).

[b] Includes 0.9 cal cm^{-2} yr^{-1} from lightning and about 3 cal cm^{-2} yr^{-1} due to corona discharges from pointed objects (Schonland, 1953).

[c] The value 4×10^9 years ago was 2.8 cal cm^{-2} yr^{-1} (Bullard, 1954).

[d] Calculated assuming the emission of 1 km^3 of lava ($C_p = 0.25$ cal/g, $p = 3.0$ g/cm^3) per year at 1000°C.

Table I gives a summary of the sources of energy in the terrestrial surface regions. It is evident that sunlight is the principal source of energy, but only a small fraction of this is of wavelengths below 2000 Å which can be absorbed by CH_4, H_2O, NH_3, CO_2, etc. If more complex molecules were formed, the absorption could move to the 2500 Å region or longer wavelengths where substantial energy is available. With the appearance of porphyrins and other pigments, absorption in the visible becomes possible.

Although probable, it is not certain that the large amount of energy from ultraviolet light would make the principal contribution to the synthesis of organic compounds. Most of the photochemical reactions at these low wavelengths would take place in the upper atmosphere. The compounds so formed would absorb at longer wavelengths and, therefore, might not get into the oceans before they were decomposed by the ultraviolet light. The question is whether the rate of decomposition in the atmosphere is greater or less than the rate of transport to the oceans.

Next in importance as a source of energy are electric discharges which, as lightning and corona discharges from pointed objects, occur closer to the Earth's surface; hence more efficient transfer to the oceans would have occurred.

Cosmic-ray energy is negligible at present, and there is no reason to assume it was greater in the past. The radioactive disintegration of uranium, thorium and potassium was more important 4.5×10^9 years ago, but still the energy was largely expended on the interior of solid materials of the rocks, and only a very small fraction of the total energy was available in the oceans and atmosphere. Volcanic energy is not only small but its availability is very low. A continuous source of energy is needed. It contributes little to the evolutionary process to have a lava flow in one part of the Earth at one time and to have another flow on the opposite side of the Earth years later. For a brief time heat is available at the surface of the lava, but the surface cools and heat flows slowly from the interior for years, giving a slightly warmer surface. Only a very small contribution to the evolutionary process could be contributed by these energy sources.

3. The Synthesis of Organic Compounds

A. Electric Discharges

While ultraviolet light is a greater source of energy than electric discharges, the greatest progress in the synthesis of organic compounds under primitive conditions has been made with electric discharges. The apparatus used by Miller in these experiments was a closed system of glass, except for tungsten electrodes. The water is boiled in a 500-ml flask which mixes the water vapor and gases in a 5-l. flask where the spark is located. The products of the discharge are condensed and flow through the U-tube back into the small flask. The first report (Miller, 1953) showed that when methane, ammonia, water and hydrogen were subjected to a high-frequency spark, milligram quantities of glycine, alanine and α-amino-n-butyric acid were produced. A more complete analysis (Miller, 1955, 1957) of the products gave the results shown in Table II. The compounds in the table account for 15% of the carbon as methane, with the yield of glycine alone being 2.1%. Indirect evidence indicated that polyhydroxyl

TABLE II

Yields in moles ($\times 10^5$) from Sparking a Mixture of CH_4, NH_3, H_2O and H_2. 710 mg of Carbon was added as CH_4.

Glycine	63	Succinic acid	4
Glycolic acid	56	Aspartic acid	0.4
Sarcosine	5	Glutamic acid	0.6
Alanine	34	Iminodiacetic acid	5.5
Lactic acid	31	Iminoaceticpropionic acid	1.5
N-Methylalanine	1	Formic acid	233
α-Amino-n-butyric acid	5	Acetic acid	15
α-Aminoisobutyric acid	0.1	Propionic acid	13
α-Hydroxybutyric acid	5	Urea	2.0
β-Alanine	15	*N*-Methyl urea	1.5

compounds (possibly sugars) were synthesized. These compounds were probably formed from condensations of the formaldehyde that was produced by the electric discharge. The alanine was demonstrated to be racemic, as would be expected in a system which contained no asymmetric reagents. It was shown that the syntheses were not due to bacterial contamination by performing blank runs in the absence of a spark, which gave no amino acids, and by autoclaving the whole apparatus prior to sparking, which gave the same yields as when the autoclaving was omitted. The addition of ferrous ammonium sulfate did not change the results, and the substitution of N_2 for the NH_3 changed only the relative yields of the compounds produced.

This experiment has been repeated and confirmed by Abelson (1956a), Pavlovskaya and Passynskii (1957) and by Heyns, Walter and Meyer (1957). Abelson worked with various mixtures of H_2, CH_4, CO, CO_2, NH_3, N_2, H_2O and O_2. As long as the conditions were reducing—either H_2, CH_4, CO or NH_3 present in excess—amino acids were synthesized. The products were the same and the yields as large in many mixtures as in the methane–ammonia–water case. If the conditions were oxidizing, no amino acids were synthesized. These experiments have confirmed that reducing atmospheres are required for the formation of organic compounds in appreciable quantities. A number of the mixtures of gases used by Abelson are highly unstable. The primitive atmosphere could not have been composed of such mixtures, in view of the discussion in Section 2-B.

Heyns, Walter and Meyer also performed experiments with different mixtures of gases with results similar to Abelson's. These workers also used CH_4, NH_3, H_2O and H_2S. They obtained ammonium thiocyanate, thiourea and thioacetamide in addition to compounds formed when H_2S was absent. All of these experiments indicate that reducing conditions are required for the production of appreciable amounts of these carbon compounds.

The mechanism of synthesis of the amino acids is of interest if we are to extrapolate the results in these simple systems to the primitive Earth. Two alternative proposals were made for the synthesis of the amino and hydroxy acids in the spark discharge system. (1) Aldehydes and hydrogen cyanide react in the aqueous phase of the system to give amino and hydroxy nitriles, which are hydrolyzed to amino and hydroxy acids. This mechanism is essentially a Strecker synthesis. (2) The amino and hydroxy acids are synthesized in the gas phase from the ions and radicals produced in the electric discharge.

It was shown that most, if not all, of the amino acids were synthesized according to the first hypothesis, since the rate of production of aldehydes and hydrogen cyanide by the spark and the rate of hydrolysis of the amino nitriles were sufficient to account for the total yield of amino acids (Miller, 1957).

This mechanism accounts for the fact that most of the amino acids were α-amino acids, the ones which occur in proteins. The β-alanine was not formed by this mechanism, but probably by the addition of ammonia to acrylonitrile (or acrylamide or acrylic acid), followed by hydrolysis to β-alanine. Similarly, the addition of hydrogen cyanide to acrylonitrile would give the succinic acid on hydrolysis.

The experiments on the mechanism of the electric discharge synthesis of amino acids indicate that a special set of conditions or type of electric discharge is not required to obtain amino acids. Any process or combination of processes that yield both aldehydes and hydrogen cyanide would have contributed to the amount of α-amino acids in the oceans of the primitive Earth. Therefore, it is not fundamental whether the aldehydes and hydrogen cyanide came from ultraviolet light or from electric discharges, since both processes would contribute. It may have been that electric discharges were the principal source of hydrogen cyanide and that ultraviolet light was the principal source of aldehyde, so that the two processes complemented each other.

B. Ultraviolet Light

It is clear from Table I that the greatest source of energy would be ultraviolet light. The effective wavelengths would be $CH_4 < 1450$ Å, $H_2O < 1850$ Å, $NH_3 < 2250$ Å, $CO < 1545$ Å, $CO_2 < 1690$ Å, $N_2 < 1100$ Å and $H_2 < 900$ Å. It is more difficult to work with ultraviolet light than with electric discharges because of the short wavelengths involved.

The action of the 1849 Å Hg line on a mixture of methane, ammonia, water and hydrogen produced only a very small yield of amino acids (Miller, 1957a). Only NH_3 and H_2O absorb at this wavelength, but apparently the radical reactions formed active carbon intermediates. The limiting factor seemed to be the synthesis of hydrogen cyanide.

Groth (1957) and Groth and von Weyssenhoff (1957) found no amino acids produced by the 1849 Å line with a mixture of methane, ammonia and water, but amines and amino acids were formed when the 1470 Å and 1295 Å lines of argon were used. The 1849 Å line produced amines and amino acids with a mixture of ethane, ammonia and water. The mechanism of this synthesis was not determined. Terenin (1957) has also obtained amino acids by the action of the krypton lines on methane, ammonia and water.

We can expect that a considerable amount of ultraviolet light of wavelengths greater than 2000 Å would be absorbed in the oceans, even though there would be considerable absorption of this radiation by the small quantities of organic compounds in the atmosphere. Few experiments have been performed which simulate these conditions.

Ellenbogen (1958) has worked with a suspension of ferrous sulfide in aqueous ammonium chloride through which methane was bubbled. The action of ultraviolet light from a high-pressure mercury lamp gave small quantities of a substance with peptide frequencies in the infrared. Paper chromatography of a hydrolysate of this substance gave a number of spots with ninhydrin, of which phenylalanine, methionine and valine were tentatively identified. Unfortunately, this experiment is difficult to reproduce.

Bahadur (1954) has reported the synthesis of amino acids by the action of sunlight on concentrated formaldehyde solutions containing ferric chloride and nitrate or ammonia. Pavlovskaya and Passynskii (1958) have found a number of amino acids by the action of ultraviolet light on a 2.5% formaldehyde solution containing ammonium chloride. These high concentrations of formaldehyde would not have occurred on the primitive Earth. It would be interesting to see if similar results could be obtained with 10^{-4}–10^{-5} M formaldehyde. This type of experiment deserves further investigation.

C. Radioactivity and Cosmic Rays

Because of the small amount of energy available, it is highly unlikely that high-energy radiation could have been very important in the synthesis of organic compounds on the primitive Earth. However, a good deal of work has been done using this type of energy, and some of it has been interpreted as bearing on the problem of the origin of life.

Dose and Rajewsky (1957) produced amines and amino acids by the action of X-rays on various mixtures of CH_4, CO_2, NH_3, N_2, H_2O and H_2. A small yield of amino acids was obtained by the action of 2 MeV Van de Graaff electrons on a mixture of CH_4, NH_3 and H_2O (Miller, unpublished experiments).

The formation of formic acid and formaldehyde from carbon dioxide and water by 40 MeV helium ions was mentioned previously. These experiments were extended by using aqueous formic acid (Garrison, Morrison, Haymond and Hamilton, 1952). The yield per ion pair was only 6×10^{-4} for formaldehyde and 0.03 for oxalic acid. Higher yields of oxalic acid were obtained from $Ca(HCO_3)_2$ and $(NH_4)HCO_3$ by Hasselstrom and Henry (1956). The helium ion irradiation of aqueous acetic acid solution gave succinic and tricarbolic acid along with some malonic, malic and citric acids (Garrison *et al.*, 1953).

The irradiation of 0.1 and 0.25% aqueous ammonium acetate by 2 MeV Van de Graaff electrons gave glycine, aspartic acid and a third compound, probably diaminosuccinic acid (Hasselstrom, Henry and Murr, 1957). The yields were very small. Massive doses of gamma rays from cobalt-60 on solid ammonium carbonate yielded formic acid and very small quantities of glycine and possibly some alanine (Paschke, Chang and Young, 1957).

The concentrations of carbon compounds and dose rates in the above experiments were very much larger than could be expected on the primitive Earth, and the products and yields may depend markedly on these factors, as well as the effect of radical scavengers such as HS^-, and Fe^{2+} and Fe^{3+}. It is difficult to exclude high-energy radiations entirely, but if one is to make any interpretations from laboratory work, the experiments should be performed at much lower dose rates and concentrations of carbon sources.

D. Thermal Energy

The older theories of the formation of the Earth involved a molten Earth during its formation and early stages. These theories have been largely abandoned, since the available evidence indicates that the solar system was formed from a cold cloud of cosmic dust. The mechanisms for heating the Earth are the gravitational energy released during the condensation of the dust to form the Earth and the energy released from the decay of the radioactive elements. It is not known whether the Earth was molten at any period during its formation, but it is clear that the crust of the Earth would not remain molten for any length of time.

Studies on the concentration of some elements in the crust of the Earth indicate that the temperature was less than 150°C during this lengthy fractionation, and that it was probably near ordinary terrestrial temperatures (Urey, 1953).

Hydrogen cyanide, acetylene and other compounds are synthesized by heating methane and ammonia or nitrogen to temperatures of 400 to 1000°C for short periods of time. The short contact times of the atmospheric gases with lava flows might be useful in synthesizing some of these compounds, although the efficiency would probably be low. Long contact times with the lava would give an equilibrium mixture of C, CO, CO_2, CH_4, H_2 and H_2O.

Fox *et al.* (1956) have maintained that organic compounds were synthesized on the Earth by heat. When heated to 150°, malic acid and urea were converted to aspartic acid and ureidosuccinic acid, and the aspartic acid decarboxylated to α- and β-alanine. The difficulty with these experiments is the source of the malic acid and urea, a question not discussed by Fox. He has also synthesized some peptides by the well known reaction (Katchalski, 1951) of heating amino acids to 150–180°C. The reaction will proceed at lower temperatures using polyphosphoric acid (Vegotsky and Fox, 1959) as a catalyst and dehydrating agent, but it is not likely that polyphosphoric acid would have been present on the primitive Earth. Salts of polyphosphoric acid are possible but the reaction has not been shown to work with these salts. These polymerizations were carried out with pure amino acids. It is questionable whether this polymerization would work in the presence of large quantities of salts and other impurities.

There is a difficulty connected with heating amino acids and other organic compounds to high temperatures. Abelson (1957) has shown that alanine, one of the more stable amino acids, decarboxylates to methylamine and carbon dioxide. The mean life of alanine is 10^{11} years at 25°C but only 30 years at 150°C. Therefore, any extensive heating of amino acids will result in their destruction, and the same is true for most organic compounds. In the light of this and since the surface of the primitive Earth was probably cool, it is difficult to see how the processes advocated by Fox could have been important in the synthesis of organic compounds.

E. Organic Phosphates, Porphyrins

Gulick (1955) has pointed out that the synthesis of organic phosphates is a difficult problem because phosphate precipitates as calcium and other phosphates under present Earth conditions in the oceans. He proposes that the presence of hypophosphites, which are more soluble, would account for higher concentrations of phosphorus compounds in early times when the atmosphere was reducing. Thermodynamic calculations show that *all* lower oxidation states of phosphorus are unstable in the presence of the pressures of hydrogen assumed in this paper. For example, the equilibrium constant for the reaction

$$H_2PO_4^- \text{ (aq)} + 2H_2 = H_2PO_2^- \text{ (aq)} + 2H_2O_{(1)}$$

is 2.5×10^{-27} at 25°C. It is possible that stronger reducing agents than hydrogen (such as formaldehyde or hydrogen cyanide) reduced the phosphate, or that some process other than reduction solubilized the calcium phosphate. This problem deserves careful attention.

The synthesis of porphyrins is considered by many authors to be a necessary step for the origin of life. Porphyrins are not necessary for living processes if the organism obtains its energy requirements from fermentation of sugars or other energy-yielding organic reactions. According to the heterotrophic theory of the origin of life, the first organisms would derive their energy requirements from fermentations. The metabolism of sulfate, iron, N_2, hydrogen and oxygen appears to require porphyrins as well as photosynthesis. Therefore, porphyrins would probably have to be synthesized before free energy could be derived from these compounds. While porphyrins may have been present in the environment before life arose, this is apparently not a necessity. Porphyrins may have arisen as a result of biochemical experimentation of very primitive organisms.

F. Intermediate Stages in Chemical Evolution

The major problems remaining for an understanding of the origin of life are: (1) the synthesis of peptides, (2) the synthesis of purines and pyrimidines, (3) a mechanism by which "high energy" phosphate or other types of bonds could be synthesized continuously, (4) the synthesis of nucleotides and polynucleotides, (5) the synthesis of polypeptides with catalytic activity (enzymes), and (6) the development of polynucleotides and the associated enzymes which are capable of self-duplication.

The free energy change for the reaction

$$\text{amino acid} + \text{amino acid} = \text{dipeptide} + H_2O$$

is about +3 to +5 kcal ($K = 10^{-3}$). Since the concentrations of amino acids in the oceans would be low, the percentage of amino acid in the form of dipeptide would be extremely small. The equilibrium constant for a peptide of n amino acids is about K^n. The K^n must be corrected for the fact that the addition of an amino acid to a long peptide requires less energy than to a small peptide, but this difference still leaves the corrected K^n extremely small. This reaction can be pushed to the right by removing water or by heating as Fox has done, but, as pointed out earlier, these methods would probably not be effective on the primitive Earth. Adsorption on mineral surfaces might aid in the synthesis of peptides, but this would require the peptides to be firmly bound to the mineral.

If the peptide is desorbed from the mineral then the mineral is simply a catalyst, and a catalyst cannot change the equilibrium constant of a reaction.

In biological systems energy is "pumped" into the amino acids by adenosine triphosphate. It seems clear that some method of "pumping energy" into the amino acids is necessary in order to form peptides.

There has been no synthesis of the purines and pyrimidines starting from the constituents of the primitive atmosphere. However, adenine and possibly guanine and uracil have been synthesized by allowing ammonium cyanide to polymerize in aqueous solution (Oró, 1960; Oró and Kimball, 1961). Glycine, alanine and aspartic acid are also formed by the polymerization of ammonium cyanide (Oró and Kamat, 1961). Hydrogen cyanide is readily produced from methane and ammonia (or nitrogen) by electric discharges and some other sources of energy (Miller, 1955, 1957). However, these experiments were performed in concentrated ammonium cyanide solutions (1.5 M) which could not have been present on the primitive Earth. It has not been shown that such reactions will take place in more dilute solutions.

If the purine and pyrimidine bases and the sugar (ribose, deoxyribose or other suitable sugar) are present then the problem of making a nucleotide or adding phosphates to make a nucleotide is similar to making a peptide bond. Energy must be pumped into the system. The bases and sugar are less stable than amino acids, so drastic treatments would not be effective here.

If the nucleotides (in the di- or tri-phosphate form) can be made, then it appears that their polymerization to polynucleotides should not be too difficult. Ochoa has found an enzyme (Steiner and Beers, 1961) which will polymerize ribonucleotide diphosphates and Kornberg (1960) has found an enzyme which will polymerize the deoxyribonucleotide tri-phosphates. It is possible that certain mineral surfaces might be able to accomplish this polymerization, or enzymes might have developed by this time which could catalyze this reaction.

The synthesis of polypeptides with catalytic activity is an extremely difficult one. Granick (1953) and Calvin (1956) have proposed that this synthesis proceeded in steps. First a metal ion catalyzed a reaction, followed by the formation of a chelate complex to give the metal greater activity. The next step was the addition of a protein to give greater activity, followed by the "evolution" of this protein to improve on the catalytic activity and/or specificity. In order for a molecule to evolve, there must be reproduction, mutation and selection. This process of molecular evolution could go on in an organism that could reproduce, but one would think that the enzymes for a very primitive organism would need some degree of specificity. If such molecular evolution took place outside a living organism, then this hypothesis does not answer how such primitive enzymes could have been reproduced so that there could be mutation and selection. The problem might better be approached by looking for some process for synthesizing enzymes that is more direct and does not require a large degree of chance and evolution.

4. The Origin of Life

In order to consider the problem of the origin of life, it is necessary to define life. There appears to be no general agreement on how to define life, and it has been asserted that life is incapable of definition. While it may be difficult to find a single definition that will cover all aspects of life, we will define life as follows: A living organism is an entity that is both (a) capable of making a reasonably accurate reproduction of itself, with the copy being able to accomplish the same task; and (b) is subject to a low rate of alterations (mutations) that are transmitted to its progeny.

Using this definition, one would say that the development of polynucleotides and associated enzymes capable of self-duplication would be the origin of life. The enzymes, polynucleotides and other factors required for this process might require a membrane to hold them together, or they might be held together on mineral surfaces.

Kornberg's enzyme appears to be able to duplicate the DNA used as a primer for the reaction (Kornberg, 1960). This enzyme is synthesized by bacteria, but if this system could be synthesized outside an organism and held together under conditions that might have been present on the primitive Earth, this would probably be considered the synthesis of life.

Oparin does not view the first organism as a polynucleotide capable of self-duplication, but rather as a coacervate colloid which accumulates proteins and other compounds from the environment, grows in size and then splits into two or more fragments which repeat the process. The coacervate would presumably develop the ability to split into fragments which were similar in composition and structure, and eventually a genetic apparatus would be incorporated which could make very accurate duplicates.

These two hypotheses for the steps in the formation of the first living organism differ mainly in whether the duplication first involved the relatively accurate duplication of nucleic acids followed by the development of cytoplasm duplication, or in the reverse order. Other sequences could easily be enumerated, but it seems clear that any significant hypothesis on this process requires a much greater knowledge of the preliminary steps.

5. Evolution of Early Organisms

While little can be said about the development of the first living organisms, reasonable hypotheses can be made for the early evolution of the most primitive organisms. The theory that the primitive oceans contained large quantities of organic compounds implies that the first organisms were heterotrophic. Heterotrophic organisms do not synthesize their basic constituents such as amino acids, nucleotides, carbohydrates, vitamins, etc., but obtain them from the environment. Autotrophic organisms synthesize all their cell constituents from CO_2, H_2O and other inorganic materials. Heterotrophic organisms are simpler than autotrophic organisms in that they contain fewer enzymes and specialized structures to carry out their metabolism. Hence heterotrophic organisms would have formed first.

A mechanism by which heterotrophic organisms could acquire various synthetic abilities was proposed by Horowitz (1945). It has been found that the presence of an enzyme in an organism is often dependent on a single gene. This is known as the *one gene–one enzyme hypothesis.* Suppose that the synthesis of A involves the steps

$$D \xrightarrow{c} C \xrightarrow{b} B \xrightarrow{a} A,$$

where a, b and c are the enzymes and A, B and C are compounds which the organism cannot synthesize. If the necessary compound A becomes exhausted from the environment, then the organism must synthesize A if it is to survive. But it is extremely unlikely that there would be three simultaneous mutations to give the enzymes a, b and c. However, a single mutation to give enzyme a would not be unlikely. If compounds D, C and B were in the environment when A was exhausted, the organism with enzyme a could survive while the other would die out. Similarly when compound B was exhausted, enzyme b would arise by a single mutation, and organisms without this enzyme would die out. By continuing this process the various steps of a biosynthetic process could be developed, the last enzyme in the sequence being developed first, and the first enzyme last.

It is necessary for all living organisms to have a source of energy to drive the biochemical reactions that synthesize the various structures of the organism. According to the heterotrophic hypothesis the first organisms would obtain their free energy from fermentation reactions. For example, the lactic acid bacteria obtain their free energy from the reaction

$$C_6H_{12}O_6 \text{ (glucose)} = 2CH_3CH(OH)COOH \text{ (lactic acid)}.$$

Yeasts also ferment glucose, but they produce ethyl alcohol and CO_2 instead of lactic acid, and there are bacteria which carry out many other types of fermentation reactions. It is likely that organisms obtained their free energy from fermentation reactions until the supply of fermentable compounds was exhausted. At that point it would have been necessary for the development of photosynthetic organisms which could obtain their free energy from light.

The first type of photosynthesis was probably similar to that of the sulfur bacteria and blue-green algae which carry out the reactions

$$2H_2S + CO_2 \xrightarrow{\text{light}} 2S + (H_2CO) + H_2O$$

$$H_2S + 2CO_2 + 2H_2O \xrightarrow{\text{light}} H_2SO_4 + 2(H_2CO),$$

where (H_2CO) means carbon on the oxidation level of formaldehyde (carbohydrates). It is much easier to split H_2S than to split H_2O, and so it would seem likely that organisms would develop photosynthesis with sulfur first. When the H_2S and S were exhausted, it would become necessary to split water and evolve O_2, the hydrogen being used for the reduction of CO_2.

When the methane and ammonia of the primitive atmosphere had been converted to carbon dioxide and nitrogen by photochemical decomposition in the upper atmosphere, water would be decomposed to oxygen and hydrogen. The hydrogen would escape, leaving the oxygen in the atmosphere and thereby resulting in oxidizing conditions on the Earth. It is likely, however, that most of the oxygen in the atmosphere was produced by photosynthesis from plants instead of by the photochemical splitting of water in the upper atmosphere.

The evolution of multicellular organisms probably occurred after the development of photosynthesis. The evolution of primitive multicellular organisms to more complex types and the development of sexual reproduction can be understood on the basis of the theory of evolution.

6. Life on Other Planets

Life as we know it requires the presence of water for its chemical processes. We know enough about the chemistry of other systems, such as those of silicon, ammonia and hydrogen fluoride, to realize that no highly complex system of chemical reactions similar to that which we call "living" would be possible in such media. Also, much living matter exists and grows actively on the Earth in the absence of oxygen, so that oxygen is *not* necessary for life, as is stated so often. Moreover, the protecting layer of ozone in the Earth's atmosphere is not necessary for life since ultraviolet light does not penetrate deeply into natural waters, and also because many carbon compounds capable of absorbing the ultraviolet would be present in a reducing atmosphere.

It is possible for life to exist on the Earth and to grow actively at temperatures ranging from 0°C, or perhaps a little lower, to about 70°C. It seems likely that if hot springs were not so temporary, many plants and possibly animals would evolve which could live in such temperatures. Plants are able to produce and accumulate substances which lower the freezing point of water and hence can live at temperatures below 0°C. At much lower temperatures the rates of reaction would probably be too slow to proceed in reasonable periods of time. At temperatures much above 120°C reaction velocities would probably be so great that the nicely balanced reactions characteristic of living things would be impossible. In addition, it is doubtful if the organic polymers necessary for living organisms would be stable much above 120°C, even allowing for the amazing stability of the enzymes of thermophilic bacteria and algae.

Only Mars, Earth and Venus conform to the general requirements so far as temperatures are concerned. Mars is known to be very cold and Venus to be too hot. Recent observations of the black-body emission of radio waves from Venus indicate surface temperatures of 290 to 350°C (Mayer *et al.*, 1958). Infrared bands of water have been detected recently by observing Venus from a balloon at high altitude. The presence of water on Venus would make life possible, but if the radio-wave temperature is in fact the temperature on the surface, then life could only exist on dust particles in the atmosphere.

Mars is known to be very cold with surface temperatures of +30 to −60°C during the day. The colors of Mars have been observed for many years by many people. The planet exhibits seasonal changes: green or bluish in the spring and brown and reddish in the autumn. Sinton (1959) has observed absorptions at 3.43 μ, 3.56 μ and 3.67 μ in the reflected light of Mars. The 3.43 μ absorption corresponds to the C—H stretching frequency of most organic compounds at this wavelength. The changing colors of Mars and the 3.5 μ absorption are the best evidence, however poor, for the existence of life on the planet. One thing that can be stated with confidence is that if life exists there, then liquid water must have been present on the planet in the past, since it is difficult to believe that life could have evolved in its absence. If this was so, water must have escaped from the planet, as very little water remains there now, and no liquid water has been observed. Hence, oxygen atoms must escape from the planet. This is possible if the high atmosphere has a temperature of 2000°K, and this is not impossible in view of the high temperatures in the high atmosphere of the Earth.

Surely one of the most marvellous feats of 20th-century science would be the firm proof that life exists on another planet. All the projected space flights and the high costs of such development would be fully justified if they were able to establish the existence of life on either Mars or Venus. In that case, the thesis that life develops spontaneously when the conditions are favorable would be far more firmly established, and our whole view of the problem of the origin of life would be confirmed.

References

Abelson, P. H., 1956. Paleobiochemistry: Inorganic synthesis of amino acids. *Carnegie Inst. Wash. Year Book*, **55**, 171.

Abelson, P. H., 1956a. Amino acids formed in "primitive atmospheres". *Science*, **124**, 935.

Abelson, P. H., 1957. Some aspects of paleobiochemistry. *Ann. N.Y. Acad. Sci.*, **69**, 276.

Bahadur, K., 1954. Photosynthesis of amino acids from paraformaldehyde and potassium nitrate. *Nature*, **173**, 1141. Also in Oparin *et al.* (Eds.), *Proc. 1st Intern. Symp. Origin of Life on the Earth*. Pergamon Press, London, 1959, p. 140.

Bernal, J. D., 1949. The physical basis of life. *Proc. Phys. Soc. London*, **62A**, 537; **62B**, 597.

Bullard, E., 1954. The interior of the earth. In G. P. Kuiper (Ed.), *The Earth as a Planet*, Chicago Univ. Press, Chicago, Ill., 110.

Calvin, M., 1956. Chemical evolution and the origin of life. *Amer. Sci.*, **44**, 248.

Curtis, A. R. and R. M. Goody, 1956. Thermal radiation in the upper atmosphere. *Proc. Roy. Soc. London*, **A236**, 193.

Dose, K. and B. Rajewsky, 1957. Strahlenchemische Bildung von Aminen und Aminocarbonsäuren. *Biochim. Biophys. Acta*, **25**, 225.

Ellenbogen, E., 1958. Photochemical synthesis of amino acids. *Abstr. Amer. Chem. Soc. Meeting, Chicago, 1958*, p. 47C; and personal communications.

Fox, S. W. *et al.*, 1956. On biochemical origins and optical activity. *Science*, **124**, 923; *Ann. N.Y. Acad. Sci.*, **69**, 328 (1957); *J. Chem. Ed.* **34**, 472 (1957).

Garrison, W. M., D. C. Morrison, J. G. Hamilton, A. A. Benson and M. Calvin, 1951. Reduction of carbon dioxide in aqueous solutions by ionizing radiation. *Science*, **114**, 416.

Garrison, W. M., D. C. Morrison, H. R. Haymond and J. G. Hamilton, 1952. High-energy helium-ion irradiation of formic acid in aqueous solution. *J. Amer. Chem. Soc.*, **74**, 4216.

Garrison, W. M., H. R. Haymond, D. C. Morrison, B. M. Weeks and J. Gile, Melchert, 1953. High energy helium ion irradiation of aqueous acetic acid solutions. *J. Amer. Chem. Soc.*, **75**, 2459.

Granick, S., 1953. Inventions in iron metabolism. *Amer. Naturalist*, **87**, 65.

Groth, W., 1957. Photochemische Bildung von Aminosäuren und anderen organischen Verbindungen aus Mischungen von H_2O, NH_3 und den einfachsten Kohlenwasserstoffen. *Angew. Chemie*, **69**, 681.

Groth, W. and H. von Weyssenhoff, 1957. Photochemische Bildung von Aminosäuren aus Mischungen einfacher Gase. *Naturwissenschaften*, **44**, 510.

Gulick A., 1955. Phosphorus and the origin of life. *Amer. Sci.*, **43**, 479; *Ann. N.Y. Acad. Sci.*, **69**, 309 (1957).

Hasselstrom, T. and M. C. Henry, 1956. New synthesis of oxalic acid. *Science*, **123**, 1038.

Hasselstrom, T., M. C. Henry and B. Murr, 1957. Synthesis of amino acids by beta radiation. *Science*, **125**, 350.

Heyns, K., W. Walter and E. Meyer, 1957. Modelluntersuchungen zur Bildung Organischer Verbindungen in Atmosphären einfacher Gase durch elektrische Entladungen. *Naturwissenschaften*, **44**, 385.

Horowitz, N. H., 1945. On the evolution of biochemical syntheses. *Proc. Nat. Acad. Sci.*, **31**, 153.

Katchalski, E., 1951. Poly-α-amino acids. *Advances in Protein Chem.*, **6**, 123.

Kornberg, A., 1960. Biologic synthesis of deoxyribonucleic acid. *Science*, **131**, 1503.

Mayer, G. H., T. P. McCullough and R. M. Sloanaker, 1958. Observations of Venus at 3.15 cm wave length. *Astrophys. J.*, **127**, 1.

Miller, S. L., 1953. A production of amino acids under possible primitive earth conditions. *Science*, **117**, 528.

Miller, S. L., 1955. Production of organic compounds under possible primitive earth conditions. *J. Amer. Chem. Soc.*, **77**, 2351.

Miller, S. L., 1957. The mechanism of synthesis of amino acids by electric discharge. *Biochim. Biophys. Acta*, **23**, 480.

Miller, S. L., 1957a. The formation of organic compounds on the primitive earth. *Ann. N.Y. Acad. Sci.*, **69**, 260. Also in Oparin *et al.* (Eds.), *Proc. 1st Intern. Symp. Origin of Life on the Earth.* Pergamon Press, 1959, p. 123.

Oparin, A. I., 1938, 1957. *The Origin of Life.* Macmillan, New York, N.Y. Third edition. Academic Press, New York, N.Y.

Oparin, A. I., A. E. Braunshtein, A. G. Pasynskii and T. E. Pavlovskaya (Eds.), 1959. *Proc. 1st Intern. Symp. Origin of Life on the Earth.* Pergamon Press, 1959, p. 151.

Oró, J., 1960. Synthesis of adenine from ammonium cyanide. *Biochem. Biophys. Res. Commun.*, **2**, 407.

Oró, J. and S. S. Kamat, 1961. Amino acid synthesis from hydrogen cyanide under possible primitive earth conditions. *Nature,* **190**, 442.

Oró, J. and A. P. Kimball, 1961. Synthesis of purines under possible primitive earth conditions, I. Synthesis of adenine. *Arch. Biochem. Biophys.*, **94**, 217.

Paschke, R., R. Chang and D. Young, 1957. Probable role of gamma irradiation in origin of life. *Science*, **125**, 881.

Pavlovskaya, T. E. and A. G. Passynskii, 1957. The original formation of amino acids under the action of ultraviolet rays and electric discharges. In Oparin *et al.* (Eds.), *Proc. 1st Intern. Symp. Origin of Life on the Earth.* Pergamon Press, 1959, p. 151.

Pavlovskaya, T. E. and A. G. Passynskii, 1958. Amino acid formation when exposing formaldehyde and ammonium salt solutions to ultraviolet irradiation. *Abstr. 4th Intern. Cong. Biochem.*, 12.

Rabinowitch, E. I., 1945. *Photosynthesis*, Vol. 1, p. 81. Interscience, New York.

Rense, W. A., 1953. Intensity of Lyman-alpha line in the solar system. *Phys. Rev.*, **91**, 299.

Rubey, W. W., 1955. Development of the hydrosphere and atmosphere, with special reference to probable composition of the early atmosphere. *Geol. Soc. Amer. Spec. Paper* 62, 631.

Schonland, B., 1953. *Atmospheric Electricity*. Methuen, London, 42, 63.

Sinton, W. M., 1959. Further evidence of vegetation on Mars. *Science*, **130**, 1234; *Astrophys. J.*, **126**, 231.

Steiner, R. F. and R. F. Beers, Jr., 1961. *Polynucleotides*. Elsevier, Amsterdam.

Terenin, A. N., 1957. In Oparin *et al.* (Eds.), *Proc. 1st Intern. Symp. Origin of Life on the Earth*. Pergamon Press, 1959, p. 136.

Urey, H. C., 1952. On the early chemical history of the earth and the origin of life. *Proc. Nat. Acad. Sci.*, **38**, 351.

Urey, H. C., 1952a. *The Planets*. Yale Univ. Press, New Haven, 149.

Urey, H. C., 1953. On the concentration of certain elements at the earth's surface. *Proc. Roy. Soc. London*, **A219**, 281.

Urey, H. C., 1959. The atmospheres of the planets. In S. Flügge (Ed.), *Handbuch der Physik*, **52**, 363. Springer-Verlag, Berlin.

Vegotsky, A. and S. W. Fox, 1959. Pyropolymerization of amino acids to proteinoids with phosphoric acid or polyphosphoric acid. *Federation Proc.*, **18**, 343.

33. THE PRESERVED RECORD: PALEONTOLOGY OF PELAGIC SEDIMENTS

W. R. Riedel

1. Introduction

The paleontology of the deep and relatively permanent ocean basins represents the last frontier of a discipline which, in the course of its development over the past hundred years, has been largely concerned with fossil remains deposited in epicontinental seas, on continental shelves, and in terrestrial environments. Study of fossil assemblages deposited in shallow marine waters has yielded a great deal of information about the history of the oceans, but this field is excluded from consideration here because its most significant results have recently been summarized in the *Treatise of Marine Ecology and Paleoecology* (edited by Hedgpeth and Ladd, 1957). Differences which exist between deep-sea and shallow-water paleontology depend principally on the different types of organisms constituting the bulk of the paleontologic records of the two kinds of environment, and the relative stability and uniformity of the deep-sea environment. More superficial differences in methods of study result from the relative inaccessibility of ancient deep-sea sediments.

As the terms "pelagic" and "planktonic" tend to be confused by some geologists and paleontologists, it may be well to explain their usage in this chapter. "Pelagic" is here used to mean "of the open sea", as in the classical usage of the word. Thus, sediments deposited near shores or in narrow bodies of water cannot be "pelagic". "Planktonic" is applied to organisms to denote incapability of strong directed movement—their distribution is to a great degree dependent upon water currents. Under certain circumstances, sediments containing a high proportion of planktonic microfossils can accumulate in a non-pelagic environment.

2. Fossils of Deep-Ocean Basins versus Those of Shallow Waters

Sediments deposited in shallow marine waters present a wide range of biogenous components, from coquinas and reef deposits formed almost exclusively of the remains of benthic macro-organisms to fine-grained calcareous and siliceous sediments consisting largely of skeletons of planktonic micro-organisms. Sequences of such sediments at any one locality are not generally uniform through periods of many millions of years, and at any one time the areal extent of a uniform sedimentary unit seldom exceeds a few hundred square miles. Changes in the physical and chemical characteristics of shallow and nearshore waters occur relatively frequently, affect relatively small areas, and are reflected by sharp and frequent changes in the specific and general composition of the associated fossil assemblages. Abundance, rarity or absence of any fossil group in a shallow-water sedimentary rock is a function mainly of the contemporary presence or absence of that group of organisms at the locality, dilution of the fossils by inorganic sedimentary components, and processes during or after lithification which may have affected the skeletal material physically or

[*MS received August, 1960*]

chemically. In most sequences of sedimentary rocks of shallow-water origin, planktonic microfossils are far outweighed in mass by the skeletons of benthic organisms. These latter, owing principally to their usually larger size, have been the subject of many investigations and form the basis for most interpretations of paleoecology and age of sedimentary sequences. Finally, sediments deposited in shallow water frequently have their character influenced profoundly by water movements (shallow currents, wave action, etc.) which frequently change intensity and direction; these movements rework, sort, and abrade the associated fossils.

In sediments of the deep-sea basins, on the other hand, skeletal remains of benthic organisms practically never constitute a high proportion of the contained fossils. The fossil assemblage here is composed principally of the remains of a few groups of micro-planktonic organisms. The little direct evidence at present available indicates that in many areas the type of sediment accumulating and the contained fossil assemblage have remained constant in most of their characters over millions of years: and if a change of sediment type and fossil content does occur, it is most frequently occasioned by a change in the calcium carbonate compensation depth and concomitant increase or decrease of the calcareous components as a whole. Another principal cause of changes in the sediment at any locality is migration of the boundaries of the main surface water masses of the open oceans: such changes apparently occur much less frequently than their analogues in shallow and nearshore waters. The areal extents of most of the uniform pelagic sedimentary units being deposited at the present day are measurable in thousands of square miles, and there are reasons to believe that this has been the case for as long as the ocean basins have existed. All of the main groups of planktonic organisms contributing skeletal remains to deep-sea sediments are ubiquitous, only their relative proportions differing from one locality to another. Their frequency in sediments is controlled largely by the extent of their dissolution during descent to the sea floor and during their burial to a depth of a few centimeters below the sediment surface: dilution by varying rates of supply of inorganic sedimentary components plays a relatively minor role. Once deposited, deep-sea sediments and their associated fossils are not so liable to disturbance by strong current action as are shallow-water deposits, though even at great depths there are physical agencies capable of re-distributing sediments, as indicated in the section on "Reworking", page 881.

3. Objectives of Deep-Sea Paleontology

Following the early observation that patterns of distribution of present-day organisms are strongly dependent upon certain measurable physical and chemical parameters, it was inevitable that natural historians should attempt to deduce past environmental conditions through investigation of fossil assemblages. A knowledge of past oceanic conditions is fundamental to many aspects of oceanography, and, therefore, one of the most important objectives of deep-sea paleontology is the elucidation of the history of the oceans on the basis of biogenous components of sediments. The several steps toward this goal are:

(1) Mapping the distribution and abundance of the skeletons of pelagic organisms in Recent deep-sea sediments, and determination of the fidelity with which these thanatocoenoses reflect the biocoenoses of the overlying water. The more fundamental investigation of the relation of species of organisms to present-day physical and chemical oceanographic parameters is the province of the biogeographer.

(2) Determination of ages of various levels in sediment cores, or at least the establishment of successive isochronous levels in the cores.

(3) Comparison of isochronous fossil assemblages from many localities over large areas, and interpretation (in terms of oceanic conditions) of observed patterns in the distribution of species and assemblages.

4. Shelled Pelagic Micro-organisms and Their Distribution

The principal groups of organisms contributing skeletal material to pelagic sediments are:

1. Plants (restricted in life to the photic zone)
   a. with calcareous skeletons
      Coccolithophorids
   b. with siliceous skeletons
      Diatoms
      Silicoflagellates

2. Animals
   a. with calcareous skeletons
      Foraminifera
      Pteropods
      Fish (phosphatic remains)
   b. with siliceous skeletons
      Radiolaria
      Sponges

A. Coccolithophorids

These organisms consist in general of a spherical or ellipsoidal protoplast of the order of 5–50 μ in diameter surrounded by a layer of joined or separated circular, elliptical or angular calcareous plates (coccoliths, rhabdoliths, pentaliths, etc., Fig. 1). After death, the spherical tests (coccospheres, rhabdospheres, braarudospheres, etc.) usually disintegrate into their component elements, though they are occasionally preserved intact in sediments. Hentschel (1936) and

Hasle (1959) have reported that the numbers of coccolithophorids in the upper 100–200 m of tropical to subantarctic waters are of the order of thousands/ml. Bramlette (1958) estimated that in typical Quaternary calcareous ooze from the tropical Pacific, coccolithophorids constitute 10–15% of the total calcium carbonate of the sediment, and that in many early Tertiary calcareous oozes and chalks, coccoliths and related forms account for about 50% of the total calcium carbonate. In recent years it has come to be generally recognized that practically none of the calcium carbonate in deep-sea sediments is of inorganic origin—electron microscope examination has revealed that the fine-grained calcium carbonate previously loosely described as "impalpable" is formed principally of fragments of coccolithophorids. "Coccolithophorids" in this context is used to embrace the true coccolithophores, known from plankton

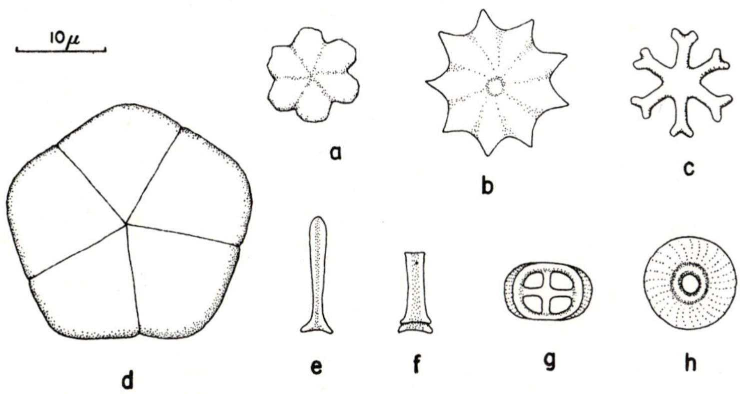

Fig. 1. Coccolithophorids and related forms. a–c, discoasters; d, pentalith of *Braarudosphaera*; e, f, rhabdoliths; g, h, coccoliths.

samples, and other small calcareous bodies which are inferred to have been formed by rather similar organisms. Quantitatively the most important of these microfossils of uncertain affinities are the discoasters (Fig. 1a–c)—usually flat, stellate bodies of approximately the same size as coccoliths, formed by a group of organisms which are believed to have become extinct at about the end of Tertiary time.

Although information on the distribution of species is difficult to obtain because of the difficulty of routine species identifications, Hentschel (1936) was able to recognize different associations of coccolithophorids characteristic of different water masses of the Atlantic. Preliminary results of Hasle (1959) indicate that some of the species in Pacific plankton can be distinguished as warm-water forms and others as cold-water forms, and that different species live at different depths, at least within the upper 150 m in the equatorial Pacific. Much more work on Recent plankton is required, however, before it will be possible to make ecological interpretations of the abundant members of this group in sediment cores.

B. Diatoms

Members of this ubiquitous group have siliceous frustules of a variety of shapes (Fig. 2), most of them ornamented with characteristic areolae or other surface patterns. The forms most obvious in sediment preparations are discoidal, elliptical, and, much less commonly, triangular or quadrangular in the usual plan view, generally 10–100 μ in diameter, but fragments of elongate rod- or thread-like forms up to several hundred microns in length are common at some localities. Some species form frustules so delicate as to dissolve soon after the death of the organism, in many cases apparently during their descent from the photic zone to the sea floor. Highly diatomaceous sediments are especially characteristic of high latitudes, and occur also under some areas of coastal upwelling along the western coasts of Africa and America. In tropical regions, radiolarians generally exceed diatoms in the mass of silica contributed to the

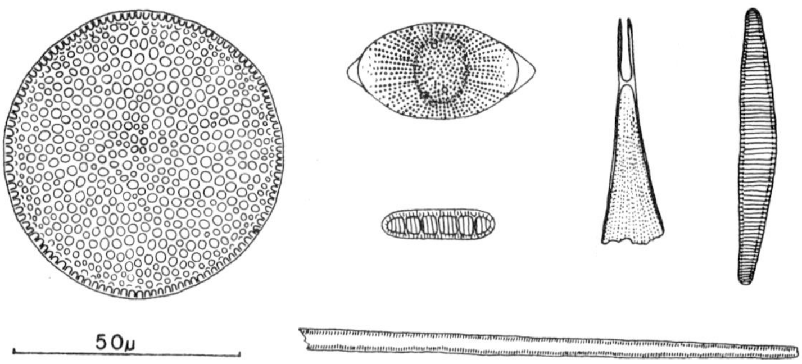

Fig. 2. Some Quaternary planktonic diatoms.

sediments, but under conditions especially favorable to the preservation of delicate siliceous skeletons (relatively shallow depth and high rate of sediment accumulation) the contribution of biogenous silica by the diatoms approaches or exceeds that of the radiolarians. Well defined diatom oozes in some areas contain as much as 30–40% biogenous silica, but it must not be assumed that such a high content of siliceous skeletons is characteristic of all of the areas marked as diatom ooze on charts of Recent sediments: the sediments of a part of the North Pacific usually mapped as diatom ooze are now estimated to contain only up to about 10% biogenous silica (Riedel, 1959; N. S. Skornyakova, *in litt.*). The floral assemblages of most diatomaceous sediments comprise a large number of species: the *Ethmodicus* oozes reported from some tropical and temperate localities of the open oceans represent a special case in which fragments of only one genus of large diatom dominate the assemblage (Kolbe, 1957; Jousé, Petelin and Udintsev, 1959).

Patrick (1948) comprehensively reviewed earlier work elucidating the factors affecting distribution of living diatoms, and it is possible here to mention only a few of the biogeographic studies which have broadened our understanding of the relation between diatom species and assemblages and the main water masses of the oceans. Cleve (e.g. 1900) was one of the first of a series of European workers to show that different diatom assemblages are associated with the different water masses of the North Atlantic. Hendey (1937) distinguished cold-water and warm-water floras in the near-surface waters of the Southern Ocean and reported a relation between physical and chemical conditions of the environment and the morphology of diatom frustules; and Hustedt (1958) investigated some of the South Atlantic and Antarctic diatoms in greater detail. Semina (1958) and Karohji (1959), investigating plankton diatoms from the northwest Pacific, traced a pronounced difference between the boreal assemblage of the colder northern water and the warm-water assemblage of the Kuroshio. In his study of the sediments of the tropical parts of the Pacific,

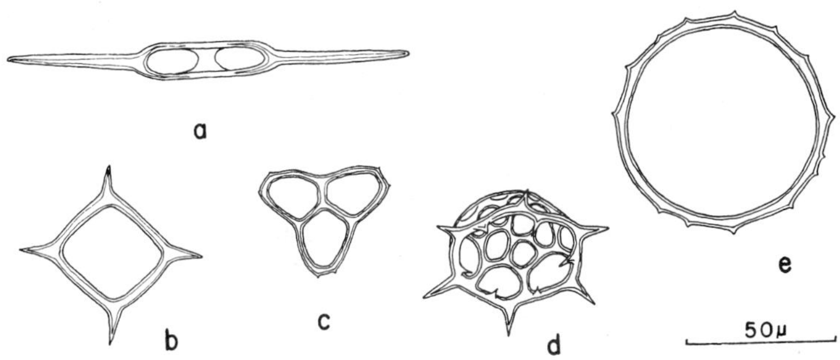

Fig. 3. Silicoflagellates from pelagic sediments. a–c, Eocene; d, e, Miocene.

Atlantic and Indian Oceans, Kolbe (1957) found the warm-water assemblage to be similar in the three oceans, and different from the assemblages characteristic of high latitudes.

C. Silicoflagellates

The living organism comprises a small, flagellated cell with an associated siliceous skeleton of hollow bars which form a spiny ring, variable in form and in most species simply reticulated (Fig. 3). Silicoflagellates occur in all parts of the oceans, and although their skeletons are found in almost all sediments containing diatoms, they never constitute a significant proportion of the mass of pelagic sediments. As a result probably of the small size of these organisms (generally 10–50 µ), little is known of their biogeography (Gemeinhardt, 1930).

D. Foraminifera

The chambered tests of these protozoans, generally about 50–1000 µ in size, are the major component of the calcareous microfossils in Quaternary pelagic sediments. Most of the tests accumulating on the deep-sea floor are of planktonic species (Fig. 4) which inhabit principally the upper part of the water column, to a depth of 100–300 m. Tests of benthonic species, some of them calcareous and others formed of agglutinated foreign grains, are widespread but at present form a quantitatively minor constituent. However, planktonic Foraminifera have not always formed a major constituent of sediments—they first became abundant during the Cretaceous. The sparse benthonic foraminiferal assemblages in pelagic sediments are relatively uniform in species-composition, compared to assemblages of planktonic forms (Phleger, Parker and Peirson, 1953).

Planktonic foraminiferal species at present inhabiting the world's oceans number only about 30–50, a surprisingly small total when compared to the hundreds of described Recent species in such groups as the diatoms and Radiolaria: the possibility should be borne in mind, however, that the large number of species described in some groups may be to some extent a result of over-splitting, not truly reflecting the number of non-interbreeding units. Murray

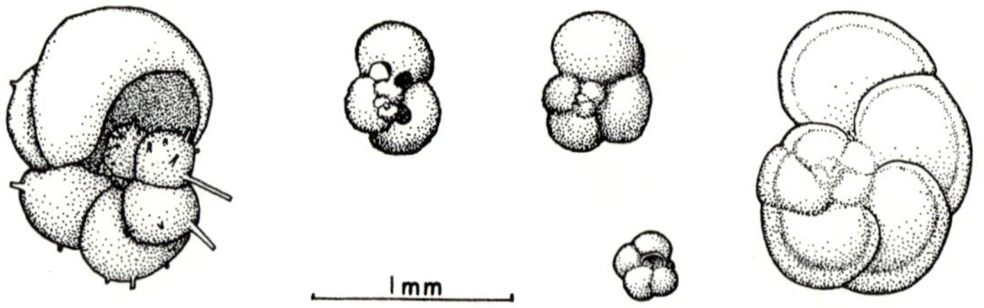

Fig. 4. Some Recent planktonic Foraminifera.

(1897) noticed that the Foraminifera obtained in plankton tows in high latitudes are different from those of temperate and tropical waters, and a few investigators have since made more detailed biogeographic studies on this group. Schott (1935) mapped the distributions of living planktonic Foraminifera in the equatorial part of the Atlantic Ocean and noted that at least some of the species showed distribution patterns apparently related to surface water masses as defined by temperature and chemical characteristics. Bé (1959), investigating the Foraminifera in the upper 200–350 m of water in the North Atlantic over an area bounded by 35–44°N, 55–65°E, found the biocoenoses to differ in the three water masses represented (the slope-water in the north, the Gulf Stream and the Central Atlantic water of the Sargasso Sea). Most of the nineteen species occurring in this area were found throughout the region, but groups of species separated as cold-tolerant and warm-tolerant differ in relative abundance from one water mass to another. Bradshaw (1959) found it possible to group the twenty-seven species of planktonic Foraminifera which he found in the North and Equatorial Pacific Ocean into four faunas (subarctic, transitional, central and equatorial-west-central) with limits corresponding in a general way with those of the surface water masses. Many of the species occur in two or more of his faunal regions, but zones of relative abundance are more restricted. In Chapter 31 of this book Ericson discusses the interrelation between water temperature and coiling direction of the planktonic foraminifer *Globorotalia truncatulinoides*.

E. Pteropods

Some members of this group of pelagic molluscs have shells which are generally spirally coiled, conical or compressed-conical, up to several millimeters in length (Fig. 5). Biogeographic studies of this group have been made by Meisenheimer (1905), Tesch (1946, 1948) and Hida (1957). Characteristically tropical, subtropical, boreal and ubiquitous species are known. The majority of species live within the upper few hundred meters of the water column, but there are some bathypelagic species living predominantly in depths greater

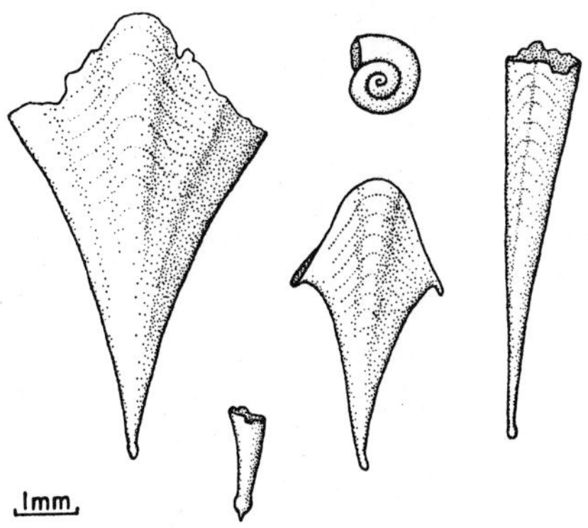

Fig. 5. Pteropods from a Recent sediment from the Tuamotu Archipelago.

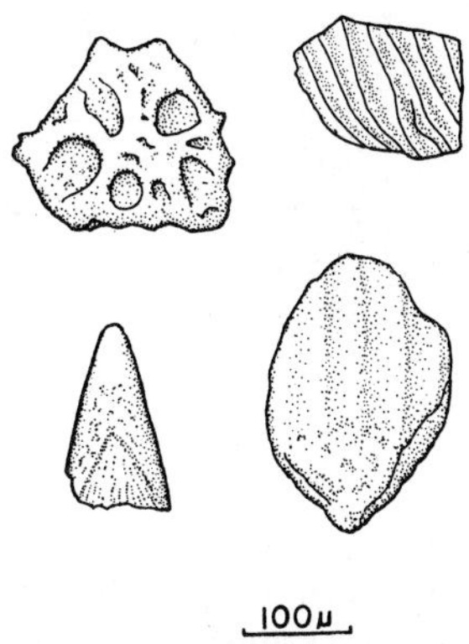

Fig. 6. Fish skeletal debris from a South Pacific zeolitic clay.

than 500 m. The contribution of pteropods (and heteropods, which have somewhat similar shells) to pelagic sediments is severely limited by the instability of their aragonitic shells. They are found more commonly in Mediterranean and Atlantic than Pacific sediments. This apparently reflects the shallower depth and more rapid rate of sediment accumulation in the former.

F. Fish Debris

Fragments of bone, placoid scales and small teeth (Fig. 6) form a minor constituent of all pelagic sediments, but there has been no reported attempt to identify the fish they represent.

G. Radiolaria

Members of this group of Protozoa occur in all parts of the oceans, and contribute significant amounts of opaline silica to sediments, especially in tropical and temperate regions. Their tests, most of them 50–400 μ in diameter or length, exhibit a great variety of shapes generally based on spherical or helmet-shaped ground-forms (Fig. 7). In attempting to evaluate the literature on Radiolaria, it is important to bear in mind that there are two profoundly different, not closely related, groups of siliceous Radiolaria, and generalizations

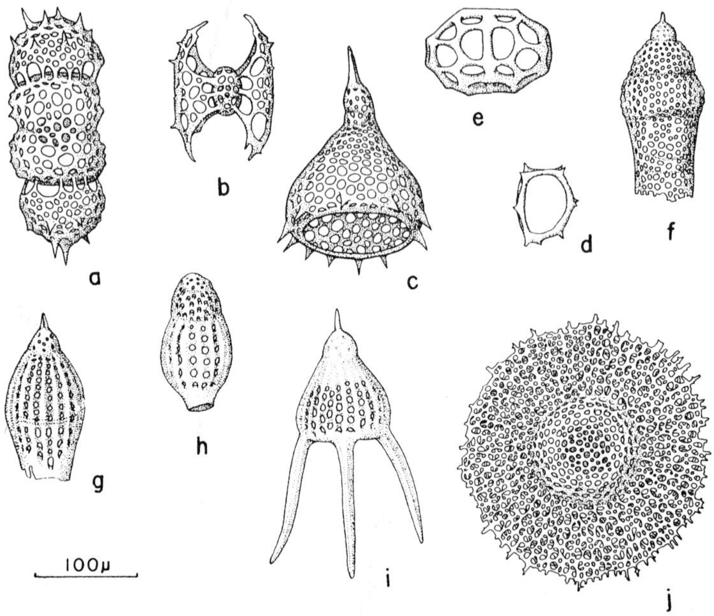

Fig. 7. Radiolaria (Spumellaria and Nassellaria) from Pacific sediments. a–d, Quaternary; e, f, Lower Miocene; g–j, Eocene.

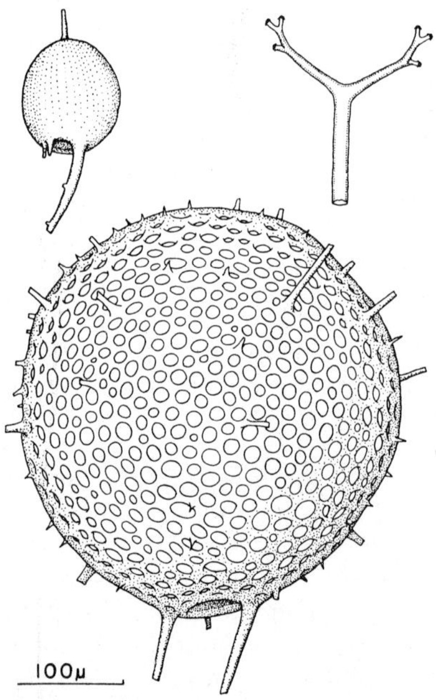

Fig. 8. Tripylean Radiolaria from a Quaternary Pacific sediment. (Scripps core Cap. 10BG1, 22° 03′S, 178° 34′E, 20–22 cm below sediment surface.)

valid for one group are not necessarily applicable to the other. The first of these two groups is the Tripylea (=Phaeodaria), with skeletons apparently formed of an intimate mixture of silica and some organic substance: skeletons of this group are not represented in the fossil record (Deflandre, 1960), though they are preserved in a few Recent sediments deposited under exceptionally favorable conditions (Fig. 8). The second group (comprising Spumellaria and Nassellaria) has skeletons of opaline silica without organic admixture, and it is these which are preserved in sediments. It may be mentioned that the Acantharia, which have skeletons of strontium sulphate and are regarded as Radiolaria by some workers, are not known with certainty to occur in either Recent or ancient sediments.

Because of their relatively large size (up to a few millimeters), the Tripylea are often observed in plankton samples and have therefore been the subject of a number of biogeographic studies (annotated bibliography of Riedel and Holm, 1957). But the Sphaerellaria and Nassellaria, with which we are concerned here, are much more difficult to obtain from plankton samples and their biogeography is consequently sparsely documented. It appears that the majority of them are restricted in life to the upper few hundred meters of the water column, though Haecker (1907) observed some living at depths of 400–5000 m. Popofsky (1908, 1912, 1913) found some species of Sphaerellaria and Nassellaria to be restricted to cold waters and others to warm waters, and the size of some widely distributed species to vary according to the temperature of their habitat.

H. Sponges

Members of this group of sessile Metazoa are widespread on the ocean floor, and contribute siliceous spicules of a variety of forms (Fig. 9) to pelagic sediments. Sponge spicules can generally be distinguished from fragments of other siliceous microfossils by their having a central canal. In deep-sea sediments these spicules usually constitute only a minor proportion of the biogenous silica, though they form a major constituent of some assemblages on shallower banks and continental shelves. The present impossibility of identifying isolated spicules limits their use in paleoceanographic interpretations.

5. Interpretation of Quaternary Microfossil Assemblages

Before attempting any interpretation of microfossil assemblages in Quaternary sediment cores, it is necessary to evaluate the fidelity with which distribution patterns of species in Recent sediments reflect those of the organisms living in the waters above. The most conclusive way to accomplish this is to compare directly the distributions of living planktonic species with the distributions of their skeletal remains at the sediment surface. For most planktonic groups this is impossible at present because of lack of the necessary data, and very few investigations have been made with this object directly in view. On the basis apparently of little evidence, Murray (1897) categorically stated that "the distribution of the dead shells of the pelagic Foraminifera on the floor of the ocean corresponds exactly with the distribution of the living specimens at the surface of the sea . . . ; the distribution of the dead shells on the bottom does not appear to be much if any wider than that of the living specimens at the surface, and this shows that the dead shells must reach the bottom a very short time after the death of the organisms". Schott (1935) examined this matter in the equatorial Atlantic and found Murray's statement to be generally correct if allowance is made for dissolution of some of the more delicate tests. Semina and

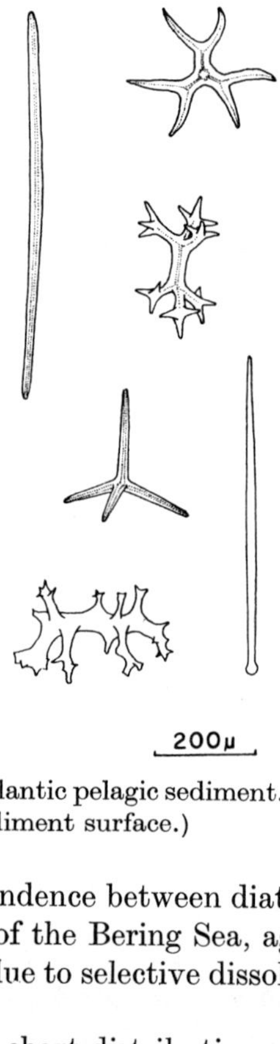

Fig. 9. Sponge spicules from an Atlantic pelagic sediment. (Lamont core A167–44, 25° 40′N, 77° 21′W, 5–10 cm below sediment surface.)

Jousé (1959) found a correspondence between diatom biocoenoses and thanatocoenoses in the western part of the Bering Sea, again allowing for impoverishment of the thanatocoenoses due to selective dissolution of the delicate frustules of some species.

A less direct approach is to chart distributions of species and assemblages in Recent sediments and compare them with the pattern of overlying water masses. If the pattern of faunal assemblages in Recent sediments mirrors that of the more or less discrete water masses above, it appears permissible to conclude that the assemblages of living organisms represented are related to those water masses, and that there has been little lateral displacement of the skeletons during their descent to the sea floor. Kanaya (*in litt.*) is engaged in a study of the diatoms in Quaternary sediment samples from the Pacific floor, and has been able tentatively to distinguish five distinct diatom assemblages in the northeast Pacific area (Fig. 10, I–V). The southern limit of the assemblages characterized by the cold-water diatom *Denticula marina* Semina corresponds rather well to the southern limit of subarctic water, and the fact that the other boundaries between the different thanatocoenoses remain distinct argues against much lateral transport of the sinking frustules by subsurface currents. Lohman (1941) estimated that isolated diatom frustules would require several hundred years to settle through 3000–4000 m of ocean water, and currents are known to exist in intermediate and deep waters. It is, therefore, at first sight remarkable that patterns of diatom assemblages are discernible in Recent pelagic sediments. There appears to be little doubt that a high proportion of at least the smaller microfossils settle through the water column as aggregates, and therefore sink more rapidly than would isolated skeletons. Large numbers

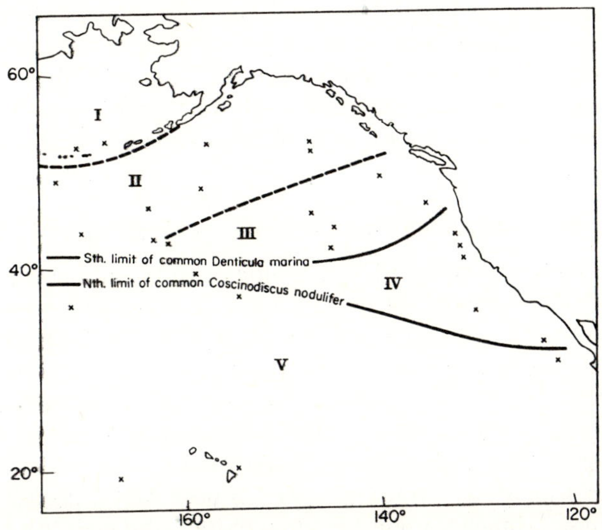

Fig. 10. Diatom assemblages (I–V) tentatively recognized by T. Kanaya in Recent sediments of the NE Pacific. × marks positions of samples forming the basis of assemblage definitions.

of diatoms and silicoflagellates, for example, are observed in the stomachs of planktonic crustaceans and tunicates, and faecal pellets and decaying intestines would provide a source of rapidly settling aggregates.

Having established that the microfossils of Recent sediments reflect conditions in the overlying waters, it is then justifiable to attempt to deduce past oceanic conditions on the basis of changes in microfossil assemblages of sediment cores. Many workers investigating Foraminifera in deep-sea sediment cores have been able to trace successions of alternating assemblages representing warmer and colder conditions in Quaternary sequences: examples are those described by Schott (1935, 1952), Cushman and Henbest (1940), Phleger *et al.* (1953) and Ericson and Wollin (1956) in the Atlantic, by Todd (1958) and Parker

(1958) in the Mediterranean, and by Stubbings (1939) in the Arabian Sea. Similar alternations of diatom assemblages have been described from northwest Pacific sediment cores by Jousé (1960). Such sequences can be related to Quaternary climatic changes, particularly when account is taken also of lithologic changes throughout the cores, as has been done, for example, by Bramlette and Bradley (1940) in the North Atlantic, by Arrhenius (1952) in the Pacific, and by Wiseman (1954) in the equatorial Atlantic. By these means, it is a relatively straightforward matter to trace, qualitatively, changes in certain oceanographic conditions through the Quaternary, and to correlate one core with another in any suitable region; but few investigators have yet obtained sufficient data to determine the actual positions of boundaries between water masses during the various stages of the Quaternary. On the basis of their foraminiferal investigations, Phleger *et al.* (1953) concluded that the northern edge of the North Equatorial Current off the west coast of Africa may have shifted southward through at least 10–15° of latitude during some parts of the Quaternary; and Jousé (1960) found that the southern limit of the boreal diatom assemblage east of the Kuriles, at present situated at about 42–45°N, was displaced northward to about 50°N during the last interglacial period.

6. Micropaleontological Determination of Ages

Passing from a consideration of the pelagic micropaleontology of the Quaternary to that of the Tertiary, we enter a field in which our knowledge is even more fragmentary. The principal reason for this is that fossiliferous Tertiary pelagic sediments are much less frequently encountered than those of Quaternary age; and when they are obtained it is difficult to date them accurately, or to refer them to isochronous levels preparatory to applying them to the interpretation of oceanic conditions at any given time.

The degree to which any group of fossils is potentially useful for age determinations depends primarily upon the areal extent of distribution of individual species, and the length of time between origin and extinction of the species. Species and assemblages of the planktonic micro-organisms here under consideration have relatively large areas of distribution, and the species and genera seem to have evolved at rates comparable with other fossil groups, and, therefore, all are potentially useful as age-indicators. Their application is limited by such factors as inadequacy of the taxonomic systems currently in use, and the small number of described, well-dated assemblages with which samples of unknown age can be compared.

A. Coccolithophorids

The potentialities of these microfossils as indicators of age were pointed out by Bramlette and Riedel (1954), and in recent years our knowledge of Tertiary assemblages has been increased by the investigations of several European workers (including Deflandre and Fert, 1954; Martini, 1958, 1959; Stradner,

1959). Bramlette (*in litt.*) has examined coccolithophorids and related forms from portions of the European Tertiary sequence and their equivalents in other widely separated parts of the world, and has found these microfossils to be useful for rather detailed correlation. The ease with which coccolithophorids can be prepared for preliminary microscopic examination is a factor which favors their routine use. A few milligrams of highly calcareous pelagic sediment need only be puddled on a glass slide with a few drops of water, dried and covered with Canada balsam: in the preparation of less calcareous sediment samples the coccolithophorids may be concentrated by utilizing the different settling velocities of the components of the dispersed sediment. Discoasters appear to have been common in at least the warm waters of the open ocean throughout Tertiary time, and to have become extinct (or virtually so) at about the end of the Tertiary; thus the occurrence of numerous discoasters in a pelagic sediment may be interpreted as indicating a Tertiary age for the sample, or an admixture of reworked Tertiary material.

B. Diatoms

As with several other groups of pelagic microfossils, the paucity of described assemblages of well established ages is the principal obstacle to the use of diatoms as indicators of age. Rheinhold (1937) was one of the first to attempt to correlate Tertiary diatom assemblages, but he was hampered by erroneous age assignments of some of the reference localities which formed the basis of his comparative studies. Both Hanna (1932) and Lohman (1960) have pointed out the striking similarity of Miocene diatom assemblages from the east and west coasts of the United States and have expressed the belief that future investigations will increase the applicability of these microfossils to problems of stratigraphic correlation. Other investigators currently laying the groundwork for future application of diatoms to problems of stratigraphic correlation include Kanaya (1957, 1959), Jousé (1948, 1959), and Sheshukova-Poretskaja (1959).

C. Silicoflagellates

In most pelagic sediments, both species and individuals of this group are too few to permit extensive application as indicators of age. They do, however, offer useful supplementary information on ages. Interpretation of the age-significance of silicoflagellates is aided by several recent articles reviewing these microfossils (Frenguelli, 1940; Deflandre, 1950; Tynan, 1957).

D. Foraminifera

Of all the groups of pelagic microfossils, it is the Foraminifera which have been most extensively applied to stratigraphic problems. For several decades, benthonic representatives of this group received most of the attention of micropaleontologists, but within the last 20–30 years the planktonic forms have come in for more intensive study. From the voluminous literature produced by many authors, it is possible to refer here to only a few papers which indicate the history of the use of planktonic Foraminifera for regional and inter-regional correlation. As an indication of the degree of precision attainable by paleontological correlation, it might be pointed out that within the Tertiary, for example, an inter-regional correlation confidently established within limits of ± 5 million years would be regarded as satisfactory by present-day standards. Grimsdale (1951) showed that the succession of planktonic foraminiferal species in the Tertiary of the Middle East is remarkably similar to that in the region of the Caribbean and Gulf of Mexico. Bronnimann (1950), Beckmann (1953), Bolli (in Loeblich *et al.*, 1957) and Blow (1959) investigated Tertiary planktonic assemblages in the Caribbean region and their relationship to faunas of similar age in other parts of the Americas and in Europe. Hamilton (1953) assigned ages to fossil foraminiferal oozes from Central Pacific seamounts on the basis of planktonic species; and in correlating the New Zealand Upper Cretaceous and Tertiary stages with the international time scale, Hornibrook (1958) based his interpretations largely on planktonic Foraminifera. Monographic regional studies of fossil representatives of planktonic foraminiferal genera, such as those of Subbotina (1953) on Russian globigerinids, hantkeninids and globorotaliids, will enhance their applicability of inter-regional correlation. When a series of Oligocene and Miocene radiolarian assemblages from the tropical Pacific floor required correlation with a sequence of known age on land (Riedel, 1959a), comparison of the associated foraminiferal assemblages with those described from the Caribbean region provided an important part of the basis for the correlation. The pre-eminent position of the Foraminifera in the sphere of Tertiary inter-regional correlation is due largely to their frequent occurrence in a great variety of sediment types in many localities, and to the intensity with which they have been investigated. It seems that the only group which may in the future challenge their position in this field are the coccolithophorids, which apparently present a greater variety of widely distributed species, but which have hitherto received little attention from stratigraphic micropaleontologists.

E. Pteropods

Shells of these organisms can to a very limited extent be applied to problems of age determination. They proved to be of assistance to Rutsch (1934) and Stainforth (1948) in their correlation of Tertiary molluscan assemblages of the Caribbean region with those of Europe.

F. Fish Debris

Since the time of the *Challenger* Expedition (Murray and Renard, 1891), it has been suspected that some of the sharks' teeth dredged in areas of slowly accumulating red clay are Tertiary in age. In the present context, however, discussion is limited to those small skeletal fragments and placoid scales which can be obtained in considerable numbers by sieving a few grams of any sample of pelagic clay. It is not at present possible to use these for stratigraphic correlations, but further investigation to determine their applicability seems desirable because they are practically the only microfossils preserved in many clays lacking siliceous and calcareous remains.

G. *Radiolaria*

Attempts (summarized by Kobayashi and Kimura, 1944) have been made to assign ages to radiolarian assemblages on the basis of the proportion of cyrtoideans to sphaeroideans (two great subdivisions of the Radiolaria). Although it is indeed true that cyrtoideans seem to have developed later than sphaeroideans in geological time, the proportion of species belonging to the two groups in any assemblage can be expected to depend largely on ecological conditions as well as on age. The only reliable method of radiolarian age determination is through comparison of assemblages of species from all available parts of the geological column. This approach has recently resulted in correlations which agree with those based on other microfossil groups (Riedel, 1957, 1959a). Relatively few workers have investigated fossil Radiolaria, and there is no doubt that future descriptions of assemblages of known age will enhance their value for stratigraphic interpretations.

H. *Reworking of Older Microfossils into Younger Sediments*

In the absence of supporting evidence, it is not safe to assume that the microfossil assemblage in any given sample from the surface of the deep-sea floor includes only Recent species. In addition to turbidity currents transporting and mixing sediments in and near areas of relatively high relief, there seem to be abyssal currents capable of eroding and transporting sediments over some areas of low relief on the ocean floor. Evidence of deep-sea erosion and transportation is summarized by Heezen (1959). In view of the fact that late Tertiary sediments may be covered by Quaternary deposits only a meter thick in areas of slow sediment accumulation, it is not surprising that relatively minor agencies of disturbance are able to expose Tertiary sediments on the deep-sea floor where the general relief is of the order of 100 m.

Recognition of reworked Tertiary elements in Recent sediments is hampered by the incompleteness of our knowledge of the Tertiary representatives of pelagic microfossil groups. Where reworked Tertiary microfossils can be recognized in Quaternary sediments, it may also be concluded that inorganic Tertiary sediment components have accompanied the microfossils, and this has considerable bearing on the interpretation of the chemical or mineralogical character of the sediment as a whole. If a sufficient number of closely spaced sediment samples could be collected from around an isolated Tertiary outcrop, it might be possible to determine the strength and direction of the water

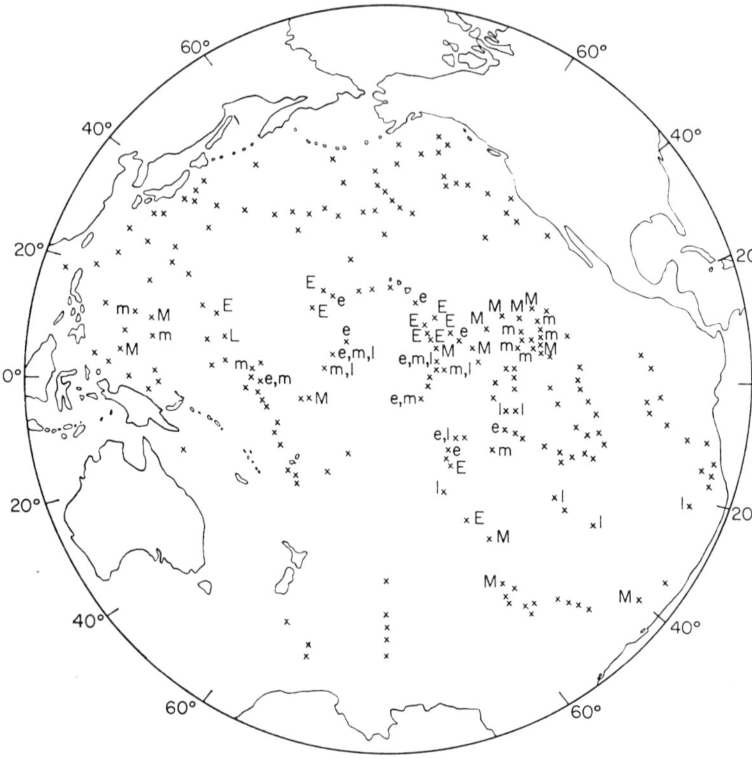

Fig. 11. Tertiary and Quaternary microfossils in surface sediments of the Pacific floor. × marks sample positions, and if not accompanied by another letter the microfossil assemblage is believed to be purely Quaternary. E, M and L indicate admixture of early, middle and late Tertiary microfossils respectively: capitals if Tertiary forms are estimated to constitute more than 20% of the microfossil assemblage, lower case if the Tertiary admixture is estimated to constitute less than 20% of the assemblage. Non-fossiliferous samples and those of doubtful age are not plotted; some available data are omitted in areas of high sampling density. Some of the age assignments on this and the next figure were made by M. N. Bramlette on the basis of coccolithophorids. For a number of samples from the N.W. and S.W. Pacific, we are indebted to the U.S.S.R. Academy of Sciences' Institute of Oceanography, the Japanese Hydrographic Department and the British Museum (Natural History) [*Challenger* samples].

movements which keep the outcrop swept free of Quaternary deposits. Reworked microfossils occurring at many localities in the deep Pacific indicate that this type of disturbance is rather widespread (Fig. 11).

7. Pre-Quaternary Paleoceanography

Sediment cores collected from the deep-sea floor occasionally sample Tertiary strata. The cores at present obtainable are usually not long enough to penetrate a continuous sequence of fossiliferous sediments from the Recent through the Pleistocene and Pliocene to the Miocene or Oligocene: most of the Tertiary samples obtained are disconformably overlain by a relatively thin Quaternary cover. Apart from the Cretaceous and Paleocene on mid-Pacific seamounts (Hamilton, 1956), the oldest known pelagic sediments from the Pacific Ocean basin are two Middle Eocene outcrops on the Tuamotu Ridge (Riedel and Bramlette, 1959). Eocene sediments have been reported in the Atlantic (Heezen *et al.*, 1959), and are believed to crop out also on the floor of the Indian Ocean (Wiseman and Riedel, 1961).

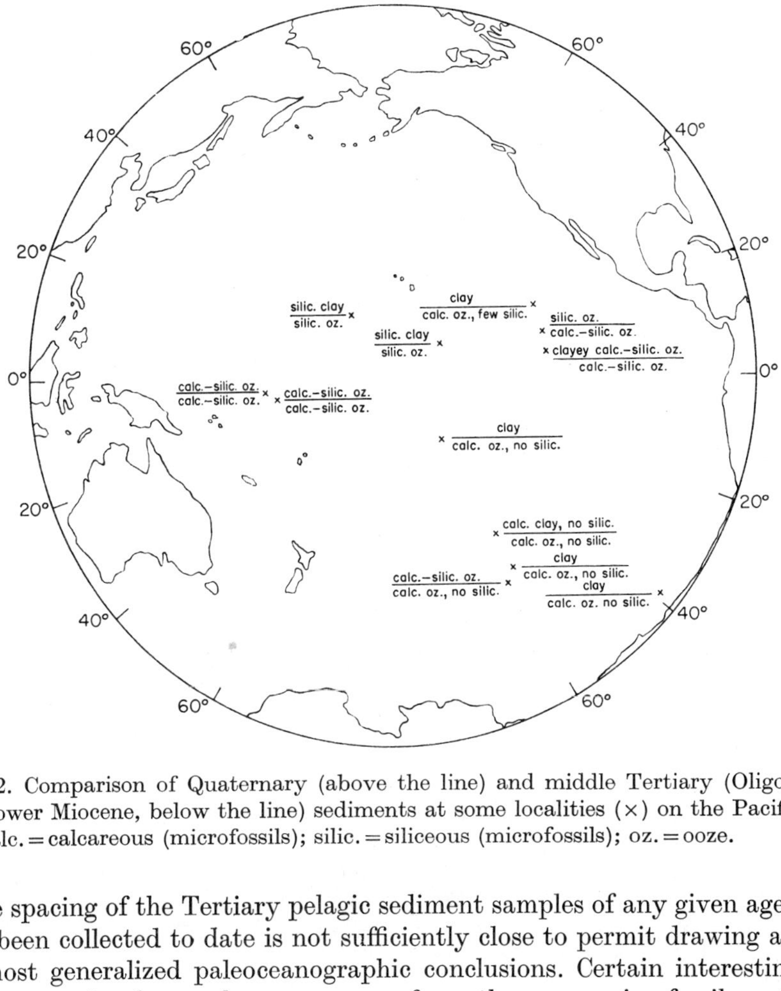

Fig. 12. Comparison of Quaternary (above the line) and middle Tertiary (Oligocene to Lower Miocene, below the line) sediments at some localities (×) on the Pacific floor. calc. = calcareous (microfossils); silic. = siliceous (microfossils); oz. = ooze.

The spacing of the Tertiary pelagic sediment samples of any given age which have been collected to date is not sufficiently close to permit drawing any but the most generalized paleoceanographic conclusions. Certain interesting conclusions can be drawn, however, even from the gross microfossil content of isolated Tertiary samples. Fig. 12 indicates the positions at which middle Tertiary (Oligocene to Lower Miocene) sediments have been encountered in the Pacific Ocean, and the nature of these middle Tertiary sediments at each locality as compared with the type of sediment at present being deposited there. It will be noticed that the contribution of calcareous microfossils at many localities was higher during the middle Tertiary than at present, but this is not paralleled by a higher content of siliceous microfossils in the Tertiary samples. Presence or absence of calcareous microfossils in pelagic sediments depends largely on whether the locality of accumulation is above or below the calcium carbonate compensation depth, but the abundance of siliceous microfossils in such sediments depends principally upon the rate of organic production in the overlying waters (Riedel, 1959). Thus the middle Tertiary data seem to indicate that the calcium carbonate compensation depth was greater at that time than at present, presumably because of higher temperatures of the bottom water (Emiliani and Edwards, 1953), and that the zone of high organic production associated with the equatorial current system, for example, had limits in the middle Tertiary not very different from those at the present day. Long cores of slowly accumulating, non-fossiliferous clay from the north-central Pacific probably penetrate to at least the late Tertiary, and provide evidence that the North Pacific Central Water has maintained its stably stratified, relatively unproductive character throughout the time represented by the deposits.

Many of the most interesting aspects of paleoceanography which can be elucidated by pelagic micropaleontological studies remain practically untouched. When samples are available from many widespread localities on each of several isochronous levels through the Tertiary, investigations of the distribution of species at each level may be expected to reveal the changes in the water mass structure of the oceans throughout that epoch. If the lithologic characters of the ancient sediments are investigated in relation to the associated microfossils, more comprehensive interpretations of past oceanic conditions can be made, as, for example, the rates of contribution of terrigenous minerals, factors controlling preservation and dissolution of biogenous sediment components, and so on. Many species of planktonic organisms have appeared and become extinct during Tertiary time, and studies of pelagic microfossils will very probably shed some light on problems of paleobiogeography and speciation in the open oceans, which are of wide interest to marine biologists. Increased knowledge of the evolutionary history of pelagic microfossils will almost certainly result in improvements in the taxonomic systems of these planktonic groups, which in turn will facilitate their further investigation and make possible more meaningful interpretations of fossil assemblages.

References

Arrhenius, G., 1952. Sediment cores from the East Pacific. *Reps. Swed. Deep-Sea Exped.*, **5**, 227 pp.

Bé, A. W. H., 1959. Ecology of Recent planktonic foraminifera: areal distribution in the western North Atlantic. *Micropaleontology*, **5**, 77–100.

Beckmann, J. P., 1953. Die Foraminiferen der Oceanic Formation (Eocaen-Oligocaen) von Barbados, Kl. Antillen. *Eclogae. Geol. Helvet.*, **46**, 301–412.

Blow, W. H., 1959. Age, correlation and biostratigraphy of the Upper Tocuyo (San Lorenzo) and Pozón formations, Eastern Falcón, Venezuela. *Bull. Amer. Paleont.*, **39** (178), 59–251.

Bradshaw, J. S., 1959. Ecology of living planktonic Foraminifera in the North and Equatorial Pacific Ocean. *Contrib. Cushman Found. Foram. Res.*, **10**, 25–64.

Bramlette, M. N., 1958. Significance of coccolithophorids in calcium-carbonate deposition. *Bull. Geol. Soc. Amer.*, **69**, 121–126.

Bramlette, M. N. and W. H. Bradley, 1940. Geology and biology of North Atlantic deep-sea cores: Lithology and geologic interpretations. *U.S. Geol. Surv. Prof. Paper*, **196-A**, 1–34.

Bramlette, M. N. and W. R. Riedel, 1954. Stratigraphic value of discoasters and some other microfossils related to Recent coccolithophores. *J. Paleont.*, **28**, 385–403.

Bronnimann, P., 1950. The genus *Hantkenina* Cushman in Trinidad and Barbados, B.W.I. *J. Paleont.*, **24**, 397–420.

Cleve, P. T., 1900. Report on the plankton collected by the Swedish expedition to Greenland in 1899. *Kungl. Svenska Vetensk., Akad. Handl.*, **34** (3), 1–21.

Cushman, J. A. and L. G. Henbest, 1940. Geology and biology of North Atlantic deep-sea cores: Foraminifera. *U.S. Geol. Surv. Prof. Paper*, **196-A**, 35–54.

Deflandre, G., 1950. Contribution à l'étude des silicoflagellidés actuels et fossiles. *Microscopie*, **2**, 72 *et seq.*

Deflandre, G., 1960. A propos du développement des recherches sur les radiolaires fossiles. *Rev. Micropaléontologie*, **2**, 212–218.

Deflandre, G. and C. Fert, 1954. Observations sur les coccolithoridés actuels et fossiles en microscopie ordinaire et électronique. *Ann. Paléont.*, **40**, 117–176.

Emiliani, C. and G. Edwards, 1953. Tertiary ocean bottom temperatures. *Nature*, **171**, 887.

Ericson, D. B. and G. Wollin, 1956. Correlation of six cores from the equatorial Atlantic and the Caribbean. *Deep-Sea Res.*, **3**, 104–125.

Frenguelli, J., 1940. Consideraciones sobre los silicoflagelados fósiles. *Rev. Mus. La Plata*, n.s., **2**, sección paleontologia, 37–112.

Gemeinhardt, K., 1930. Silicoflagellatae. In Rabenhorst's *Kryptogamen Flora*, **10** (2), 1–87. Leipzig.

Grimsdale, T. F., 1951. Correlation, age determination and the Tertiary pelagic Foraminifera. *Proc. 3rd World Petroleum Cong.*, Sec. 1, 463–475.

Haecker, V., 1907. Altertümliche Sphärellarien und Cyrtellarien aus grossen Meerestiefen. *Arch. Protistenkunde*, **10**, 114–126.

Hamilton, E. L., 1953. Upper Cretaceous, Tertiary and Recent planktonic Foraminifera from Mid-Pacific flat-topped seamounts. *J. Paleont.*, **27**, 204–237.

Hamilton, E. L., 1956. Sunken islands of the Mid-Pacific mountains. *Mem. Geol. Soc. Amer.*, **64**, 95 pp.

Hanna, G. D., 1932. The diatoms of Sharktooth Hill, Kern County, California. *Proc. Calif. Acad. Sci.*, ser. 4, **20**, 161–263.

Hasle, G. R., 1959. A quantitative study of phytoplankton from the equatorial Pacific. *Deep-Sea Res.*, **6**, 38–59.

Hedgpeth, J. W. and H. S. Ladd (eds.), 1957. Treatise on marine ecology and paleoecology. *Mem. Geol. Soc. Amer.*, **67** (2 vols.), 1296 + 1077 pp.

Heezen, B. C., 1959. Dynamic processes of abyssal sedimentation: Erosion, transportation, and redeposition on the deep-sea floor. *Geophys. J., Roy. Astr. Soc.*, **2**, 142–163.

Heezen, B. C., M. Tharp and M. Ewing, 1959. The floors of the oceans: The North Atlantic. *Geol. Soc. Amer. Spec. Paper* 65, 122 pp.

Hendey, N. I., 1937. The plankton diatoms of the Southern Seas. *Discovery Reps.*, **16**, 151–364.

Hentschel, E., 1936. Allgemeine Biologie des südatlantischen Ozeans. *Wiss. Ergebn. Deut. Atlant. Exped. 'Meteor', 1925–1927*, **11**, 344 pp. + Beilagen 9–42.

Hida, T. S., 1957. Chaetognaths and pteropods as biological indicators in the North Pacific. *U.S. Fish and Wildlife Serv., Spec. Sci. Rep.*, **215**, 13 pp.

Hornibrook, N. de B., 1958. New Zealand Upper Cretaceous and Tertiary foraminiferal zones and some overseas correlations. *Micropaleontology*, **4**, 25–38.

Hustedt, F., 1958. Diatomeen aus der Antarktis und dem Südatlantik. *Wiss. Ergebn. Deut. Antarkt. Exped.*, **2**, 103–191.

Jousé, A. P., 1948. Dotretichnie diatomovye vodorosli. *Botanicheskij Zhur.*, **33**, 345–356.

Jousé, A. P., 1959. Osnovnye etapy razvitija flory morskikh diatomovykh vodoroslej (Diatomeae) na Dalnem Vostoke v tretichnom i chetvertichnom periodakh. *Botanicheskij Zhur.*, **44**, 44–55.

Jousé, A. P., 1960. Les diatomées des dépôts de fond de la partie nord-ouest de l'Océan Pacifique. *Deep-Sea Res.*, **6**, 187–192.

Jousé, A. P., V. P. Petelin and G. B. Udintsev, 1959. K voprosu o proiskhozhdenii diatomovykh ilov s *Ethmodiscus rex* (Wall.) Hendey. *Doklady Akad. Nauk S.S.S.R.*, **124**, 1301–1304.

Kanaya, T., 1957. Eocene diatom assemblages from the Kellogg and "Sidney" shales, Mt. Diablo area, California. *Sci. Rep. Tohoku Univ., Sendai*, ser. 2 (Geology), **28**, 27–124.

Kanaya, T., 1959. Miocene diatom assemblages from the Onnagawa Formation and their distribution in the correlative formations in Northeast Japan. *Sci. Rep. Tohoku Univ., Sendai*, ser. 2 (Geology), **30**, 1–130.

Karohji, K., 1959. Report from the "Oshoro Maru": Diatom associations as observed by underway samplings. *Hokkaido Univ., Bull. Fac. Fisheries*, **9**, 259–267.

Kobayashi, T. and T. Kimura, 1944. A study on the radiolarian rocks. *J. Fac. Sci. Univ. Tokyo*, ser. 2, **7**, 75–178.

Kolbe, R. W., 1957. Diatoms from equatorial Indian Ocean cores. *Reps. Swed. Deep-Sea Exped.*, **9** (1), 1–50.

Loeblich, A. R., Jr., H. Tappan, J. P. Beckmann, H. M. Bolli, E. Montanaro Gallitelli and J. C. Troelsen, 1957. Studies in Foraminifera. *U.S. Nat. Mus. Bull.*, 215, vi + 323 pp.

Lohman, K. E., 1941. Geology and biology of North Atlantic deep-sea cores: Diatomaceae. *U.S. Geol. Surv. Prof. Paper*, **196-B**, 55–86.

Lohman, K. E., 1960. The ubiquitous diatom—a brief survey of the present state of knowledge. *Amer. J. Sci.*, **258A**, 180–191.

Martini, E., 1958. Discoasteriden und verwandte Formen im NW-deutschen Eozän (Coccolithophorida): Taxionomische Untersuchungen. *Senckenbergiana lethaea*, **39**, 353–388.

Martini, E., 1959. Discoasteriden und verwandte Formen im NW-deutschen Eozän (Coccolithophorida): Stratigraphische Auswertung. *Senckenbergiana lethaea*, **40**, 137–157.

Meisenheimer, J., 1905. Pteropoda. *Wiss. Ergebn. Deut. Tief-See Exped. 'Valdivia'*, **9**, 1–314.

Murray, J., 1897. On the distribution of the pelagic Foraminifera at the surface and on the floor of the ocean. *Nat. Sci. London*, **11** (65), 17–27.

Murray, J. and A. F. Renard, 1891. Deep-sea deposits. *Rep. Sci. Results of the Voyage of H.M.S. 'Challenger'*, sec. 3, **1**, xxix + 525 pp.

Parker, F. L., 1958. Eastern Mediterranean Foraminifera. *Reps. Swed. Deep-Sea Exped.*, **8** (4), 217–283.

Patrick, R., 1948. Factors affecting the distribution of diatoms. *Bot. Rev.*, **14**, 473–524.

Phleger, F. B., F. L. Parker and J. F. Peirson, 1953. North Atlantic Foraminifera. *Reps. Swed. Deep-Sea Exped., 1947–1948*, **7** (1), 122 pp.

Popofsky, A., 1908. Die Radiolarien der Antarktis (mit Ausnahme der Tripyleen). *Deut. Südpolar-Exped.*, **10**, 183–306.

Popofsky, A., 1912. Die Sphaerellarien des Warmwassergebietes. *Deut. Südpolar-Exped.*, **13**, 73–159.

Popofsky, A., 1913. Die Nassellarien des Warmwassergebietes. *Deut. Südpolar-Exped.*, **14**, 217–416.

Rheinhold, T., 1937. Fossil diatoms of the Neogene of Java and their zonal distribution. *Ned. Koloniën Geol. Mijnb. Genootsch. Verh., Geol. Ser.*, **12**, 43–133.

Riedel, W. R., 1957. Radiolaria: a preliminary stratigraphy. *Reps. Swed. Deep-Sea Exped.*, **6** (3), 59–96.

Riedel, W. R., 1959. Siliceous organic remains in pelagic sediments. In "Silica in Sediments". *Soc. Econ. Paleontologists and Mineralogists, Spec. Pub.*, **7**, 80–91.

Riedel, W. R., 1959a. Oligocene and Lower Miocene Radiolaria in tropical Pacific sediments. *Micropaleontology*, **5**, 285–302.

Riedel, W. R. and M. N. Bramlette, 1959. Tertiary sediments in the Pacific Ocean basin. *Intern. Oceanog. Cong. Preprints.* Amer. Assoc. Adv. Sci., Washington, 105–107.

Riedel, W. R. and E. A. Holm, 1957. Radiolaria. *Mem. Geol. Soc. Amer.*, **67**, 1069–1072.

Rutsch, R., 1934. Pteropoden und Heteropoden aus dem Miocaen von Trinidad (Britisch Westindien). *Eclogae Geol. Helvet.*, **27**, 299–326.

Schott, W., 1935. Die Foraminiferen in dem äquatorialen Teil des Atlantischen Ozeans. *Wiss. Ergebn. Deut. Atlant. Exped. 'Meteor', 1925–1927*, **3** (3), 43–134.

Schott, W., 1952. On the sequence of deposits in the equatorial Atlantic Ocean. *Handl. Göteborgs Kungl. Vetensk.—och Vitterhets Samh.*, följ. 7, ser. B, **6** (2), 15 pp.

Semina, G. I., 1958. Svjaz fitogeograficheskikh zon v pelagiali severozapadnoj chasti Tikhogo okeana s raspredeleniem vodnikh mass v etom rajone. *Trudy Inst. Okeanol., Akad. Nauk S.S.S.R.*, **27**, 66–76.

Semina, G. I. and A. P. Jousé, 1959. Diatomovye vodorosli v biotsenozakh i tanatotsenozakh zapadnoj chasti Beringova morja. *Trudy Inst. Okeanol. Akad. Nauk S.S.S.R.*, **30**, 52–67.

Sheshukova-Poretskaja, V. S., 1959. K iskopaemoj diatomovoj flore juzhnogo Sakhalina (morskoj neogen). *Vestnik Leningrad Univ.*, ser. Biol., **3** (15), 36–55.

Stainforth, R. M., 1948. Description, correlation, and paleoecology of Tertiary Cipero Marl formation, Trinidad, B.W.I. *Bull. Amer. Assoc. Petrol. Geol.*, **32**, 1292–1330.

Stradner, H., 1959. Die fossilen Discoasteriden Österreichs. *Erdoel-Zeitschr.*, **12**, 472–488.

Stubbings, H. G., 1939. Stratification of biological remains in marine deposits. *Rep. John Murray Exped.*, **3**, 159–192.

Subbotina, N. N., 1953. Globigerinidy, Khantkeninidy i Globorotaliidy. In *Iskopaemye Foraminifery S.S.S.R.* Trudy Vsesojuznogo Neftjanogo Nauchno–Issledovatelskogo Geologo–Razvedochnogo Instituta, n.s., **76**, 295 pp.

Tesch, J. J., 1946. The thecosomatous pteropods: the Atlantic. *Dana-Rep.*, **28**, 82 pp.

Tesch, J. J., 1948. The thecosomatous pteropods: the Indo-Pacific. *Dana-Rep.*, **30**, 45 pp.

Todd, R., 1958. Foraminifera from Western Mediterranean deep-sea cores. *Reps. Swed. Deep-Sea Exped.*, **8** (3), 167–215.

Tynan, E. J., 1957. Silicoflagellates of the Calvert formation (Miocene) of Maryland. *Micropaleontology*, **3**, 127–136.

Wiseman, J. D. H., 1954. The determination and significance of past temperature changes in the upper layer of the equatorial Atlantic Ocean. *Proc. Roy. Soc. London*, **A222**, 296–323.

Wiseman, J. D. H. and W. R. Riedel, 1961. Tertiary sediments from the floor of the Indian Ocean. *Deep-Sea Res.*, **7**, 215–217.

34. THE PLEISTOCENE RECORD

C. Emiliani and R. F. Flint

1. Introduction

The Earth is a geophysical, geochemical and biological system of such complexity that less is known about its structure, composition and dynamics than about those of most other celestial bodies. The complexity of the Earth as a planet rests essentially upon a threefold basis: its mass, large enough to have permitted the differentiation of core, mantle and crust, and the retention of a substantial hydro-atmosphere; its heat balance, permitting surface water to exist in all three phases; and its angular velocity and axis orientation resulting in a diurnal cycle of convenient frequency. A perhaps unavoidable consequence of these things has been the luxuriant development of life.

Motions of a convective nature occur within the hydro-atmosphere, energized almost entirely by the sun. There is some reason to believe that convective motions take place also within the geosphere, energized by the decay of radioactive nuclides and possibly by continuing growth of the core. Thus the Earth seems always to have been in a mildly dynamic state, with a whole array of geophysical, geochemical and biological processes operating concurrently, in both cyclical and non-cyclical patterns, and interacting with each other in complex ways.

The various processes operate through an extreme range of time intervals. Core growth and mantle degassing (two eminently non-cyclical processes) seem to have begun soon after the final accumulation of the Earth, and are likely to be still continuing and to continue for a long time in the future, even though their rates may today be much smaller than in the early times when ^{235}U and ^{40}K were generating relatively large amounts of internal heat. Other non-cyclical processes operating over extreme lengths of time may be the controversial activities of continental drift or polar wandering. Cyclical or quasi-cyclical processes, on the other hand, usually operate over much shorter time intervals. Thus, orogenic cycles may involve tens or hundreds of millions of years; the glacial–interglacial cycle involves only some tens of thousands of years; and the diurnal migration of plankton involves only a day. In addition, there are a multitude of non-cyclical processes of relatively short duration, such as the life of a volcano (a few years to several million years), the life of a marine molluscan species (some ten million years), tectonic motions of various kinds, isostatic phenomena, weathering, non-glacial eustatic changes of sea-level and short term meteorological events. Some of these could be episodes in longer-range cyclical or non-cyclical phenomena, but others are not.

Whereas processes operating over short intervals (up to a few hundred years) are easily studied by direct observation, those operating over longer intervals can be understood only if, in addition to direct observation, their effects through periods proportional to their durations are studied. In this context,

[*MS received March, 1961*]

Pleistocene research assumes an importance that vastly exceeds purely historical interest.

In the view of many, the Pleistocene is the last epoch of geologic time. It has been defined as the time elapsed since the late Cenozoic migration of certain species of northern marine mollusks and Foraminifera into the Mediterranean, and is believed to embrace many hundred thousand years. During this time the history of the Earth has been particularly exciting and eventful. Most of the Tertiary and some of the older mountain chains have been repeatedly uplifted, with vertical displacements estimated in hundreds and, in some areas, thousands of meters (cf. Flint, 1957, pp. 499–503). Volcanic activity has been intense, as evidenced by great lava flows on many lands and by numerous and widespread ash layers in deep-sea sediments (cf. Worzel, 1959). Vast ice sheets have repeatedly covered large continental areas, especially in the Northern Hemisphere. The whole biosphere has been subjected to intense environmental stresses. Evolution, especially among land animals, has been generally fast and, last but not least, man has come into being.

Glaciation is the outstanding feature of the Pleistocene Epoch. Indeed, the definition of the Pliocene–Pleistocene boundary implies climatic deterioration that led to the general development of glaciers. Glaciation may have been made possible by an increase in albedo accompanying mountain building and uplift. The glaciers and pack ice that repeatedly spread over a third of the Earth's surface caused a further large increase of albedo and consequent temperature decrease. Within the last several hundred thousand years the Earth has oscillated between glacial and interglacial conditions; that is, between conditions of lower and higher temperatures. The temperature oscillations have been greatest in the high latitudes, but have not been negligible even at the equator. There is some reason to believe that the major temperature oscillations have been world-wide and generally synchronous on the two polar hemispheres (Emiliani and Geiss, 1959). If so, and if these oscillations are dated absolutely, useful world-wide time markers would be available. Correlation with the thermal record at any locality could then imply absolute dating as well. This is significant because absolute dating, especially of terrestrial material, is at present difficult or impossible to achieve for much of the Pleistocene Epoch.

Before the invention of the piston corer for sampling deep-sea sediments (Kullenberg, 1947), Pleistocene studies centred largely on continental features. From these studies were recognized four or five major glaciations, some of them having second-order oscillations, and separated by interglacial intervals. Dating of glacial and interglacial events was based largely on geologic estimates of dubious validity, until radiocarbon dating was developed. Application of this method to continental material resulted, during the decade 1950–1960, in fixing the last two major glacial maxima at about 18,000 years B.P. (before the present) and about 60,000 years B.P. As the natural limitations of the radiocarbon method preclude the dating of material older than about 70,000 years, most of Pleistocene time is beyond the reach of radiocarbon. In addition, the inherent discontinuity and fragmentary character of the continental record do not permit the dating of earlier Pleistocene events by extrapolating rates obtained from the short time span measurable with radiocarbon. Some attempts have been made to date continental Pleistocene material by the potassium–argon method, but important difficulties, both of sampling and of techniques, are yet to be solved. Deep-sea sediments, which sometimes offer continuous and undisturbed stratigraphic sequences and are more easily datable than continental material by means of radioactive methods having a greater range than radiocarbon, may make possible the establishment of a continuous, dated thermal record covering the whole Pleistocene. Correlation of geophysical, geochemical and biological events with this record should then afford insight into the rates at which many primary processes operate. Although the Pleistocene Epoch represents only a fraction of 1% of the time elapsed since the beginning of the Cambrian Period, probably it is long enough to resolve the activities of many of these processes.

2. Continental Stratigraphy

A. Character of Data

Research in the stratigraphy of the Pleistocene is focused directly or indirectly upon three lines of enquiry: sequence, climate and absolute dating. The goal that can be visualized, but that is still far from being attained, is a stratigraphic sequence correlated from one region and continent to another, and time calibrated at significant points throughout its length. Attached to the sequence would be absolute values of mean annual temperature for representative stations at various latitudes. The maxima and minima for a station, changing with time, would measure roughly the amplitude fluctuations, during the Pleistocene, of temperature, a basic parameter of climate on which other parameters appear to be dependent.

This is a very long-term goal, but progress toward it is already substantial. From evidence of many kinds there has been erected a series of regional stratigraphic sequences to which new discoveries are fitted by correlation. The units within the sequences carry labels that, through widening use, are approaching standardization, although universal standards are still in the future. As much of the early stratigraphic work was done in glaciated regions, the units recognized first were glacial sediments consisting chiefly of drift, and interglacial units consisting of soils, weathering zones and sediments with fossils of temperate- rather than cold-climate aspect. These constitute the chief basis of the sequences developed up to the present.

The sequences follow reasonably well the principles on which the classification of the earlier geologic strata has been built. These principles involve ideally the recognition of time-stratigraphic units, both major and minor, with each unit defined, at least in theory, in terms of the fossils it contains. In parts of Africa, artifacts have been substituted for fossils, with results that, so far, appear reasonable.

In conformity with ancient usage, post-Pliocene strata are grouped into a Quaternary System. By many authorities that System is subdivided into a Pleistocene Series and a Recent, Holocene, or Postglacial Series. Others view the distinction between Pleistocene and Recent as arbitrary, and prefer to consider the Pleistocene strata as representing the entire sequence from the base of the Quaternary System to the sediments that are accumulating today. The later classification is adopted, without prejudice, for the present contribution because it is well adapted to the discussion of deep-sea sediments.

The base of the Pleistocene Series (and also of the Quaternary System) has been defined in various ways. A recent and closely reasoned definition sets the base at the appearance in late Cenozoic sediments occurring in various parts of Italy of a cold-water marine fauna (the Calabrian fauna) differing from the underlying Pliocene fauna. The newer fauna is characterized by the disappearance of certain Pliocene species and by the appearance of a dozen species of North Atlantic mollusks. In northern Italy the Calabrian marine strata grade upward into continental sediments containing a distinctive mammal fauna variously labeled Villafranchian and Upper Villafranchian. In some other areas the mammal-bearing sediments contain a cool-climate continental flora, contrasting with the warmer, Pliocene flora in the underlying rocks. Thus, through fossils of various kinds the base of the Pleistocene is extended by correlation into widely separated regions.

Subdivision of the continental Pleistocene is still based less on fossils than on sediments implying alternation of glacial and interglacial conditions. In both Europe and North America the glacial layers seem to fall naturally into major subdivisions, some of which record a single glaciation, whereas others imply repeated glaciation. Between the glacial units is found evidence, such as zones of deep thorough weathering of glacial drift, or fossils denoting higher non-glacial temperatures, from which long periods of warmer interglacial climate are inferred. In classical North American terminology the major glacial and interglacial subdivisions are *stages*, based only in part on fossils and otherwise on evidence of glacial and interglacial climates. The stages imply at least four, and probably more, major glacial events that seem, through similarity of sequence and wide continuity, to be fairly well established in both Europe and North America. As a result transatlantic correlations have been made with varying degrees of confidence.

Because few sheets of glacial drift possess distinctive characteristics by which they can be identified throughout wide regions, the validity of the sequence of stages is coming to depend more and more on distinctive characteristics within the interglacial sediments that lie between them. The useful characteristics consist chiefly of (1) distinctive fossil plants (commonly represented by pollen) and animals, and (2) distinctive patterns of climatic fluctuations recorded by changes in fossil pollen upward through the sequence.

The areal extent of application of the terms glacial and interglacial is not sharply defined. The terms are applied in the glaciated regions, where the former presence of glacier ice is quite evident. They are applied also in areas of major bodies of outwash sediment deposited beyond the glaciated regions by streams of melt-water. A grand example is the Mississippi River valley, within which outwash sediments are traced downstream through hundreds of miles beyond the limit reached by glacier ice. At a further remove, the terms glacial and interglacial are applied to the much broader areas of loess (wind-blown silt) secondarily derived, in large part from bodies of outwash. Like the glacial drift, the loess can be subdivided into stratigraphic units, separated by ancient soils and more rarely by distinctive non-eolian sediments.

Outwash sediments and loess, indirectly of glacial origin, are developed mainly in the humid and subhumid regions of middle latitudes. In arid regions of middle and low latitudes are found other kinds of evidence of climatic fluctuation. Conspicuous among these are beaches and floor sediments of former Pleistocene lakes, where under existing climates the sites have only shrunken, saline lakes or are completely dry. In western United States, mainly in structural basins, about 120 former lakes have been identified thus far, the largest having an implied area of more than 50,000 km^2 and a maximum depth of more than 300 m. Similar groups of former lakes are known in Asia, Africa, South America and Australia. Such features constitute one of the chief bases for the concept of pluvial and interpluvial climates. According to this concept the lakes imply a combination of greater precipitation and reduced evaporation, compared with today's climate. As early as 1865 the inferred pluvial climates were thought to be contemporaneous with the glacial climates of other regions. In some former lakes two or more alternations of climate are recorded in the stratigraphy. At certain localities a contemporaneous or near-contemporaneous relationship between lake beaches and glacial drift in adjacent mountain valleys has been demonstrated. In cores from a former lake in California, the last high-water phase has been shown by ^{14}C dating to have been contemporary with the last major glacial maximum in central United States (Flint and Gale, 1958). Although similar data are greatly needed from other continents, it seems at present reasonably likely that glacial and pluvial episodes were contemporaneous.

B. Representative Sequences

All stratigraphic columns of regional application are imperfect in that they are broad generalizations from the known evidence. Some of the most interesting, because best supported by detailed information, are discussed below.

Probably the best-supported continental sequence is that of the Netherlands–eastern England region. In that region the column extends from the top of the Pliocene into the upper part of the Pleistocene, with fewer interruptions than in most other regions (Table I). The pre-Cromerian sequence is based in the Netherlands on fossil pollen, which records distinct fluctuations of air temperatures, and in part on terrestrial vertebrates. In eastern England the same part of the sequence is based mainly on marine invertebrates, again with temperature implications. In both areas the Cromerian and post-Cromerian sequence is

TABLE I

Stratigraphic Column for the Netherlands–Eastern England Region

Netherlands (Zagwijn, 1957; van der Vlerk, 1959; see also Zagwijn, 1960, fig. 8)	Eastern England (West, 1958; van der Vlerk, 1959)
UNITS	UNITS
Weichselian	
Eemian	*Ipswichian*
Saalian (3 temperature minima)	Gipping
Hoxnian	*Hoxnian*
Elsterian	Lowestoft
Cromerian	*Cromerian*
Menapian	Weybourne Crag
	Chillesford Clay
Waalian	Norwich Crag
Eburonian	Butley Crag
Tiglian	Newbourne Crag
Pre-Tiglian	Walton Crag
PLIOCENE	PLIOCENE

based on pollen stratigraphy and in part on intervening layers of glacial drift. The Weichselian of the Netherlands has correlatives other than glacial drift in eastern England, and in northwestern England drift of probable Weichselian age is present.

A generalized sequence for the region of the Alps is shown in Table II. Mainly because it was the first sequence with any pretence to completeness to appear in the literature, it has become "classical," and its names have been applied, in some cases with little justification, to stratigraphy outside the Alps.

TABLE II

Stratigraphic Column for Northern Alps Region. (Data from Penck and Brückner, 1909; Eberl, 1930, p. 402)

STAGES
Würm Glacial (at least 2 temperature minima)
R–W Interglacial
Riss Glacial (2 temperature minima)
M–R Interglacial
Mindel Glacial (2 temperature minima)
G–M Interglacial
Günz Glacial (3 temperature minima)
D–G Interglacial
Donau Glacial (more than one temperature minimum)

Much of the evidence on which glacial units are based consists of outwash rather than of sediments deposited by glaciers directly, and most of it has been drawn from a few valleys on the northern flank of the mountains. Data from other parts of the mountains have been gradually embodied in the scheme, but, as the sediments are discontinuous, correlation from one valley to another is based in some instances only on similarity of sequence.

The interglacial or "warm" units are based on both zones of weathering and soils, and, more helpfully, on fossil-bearing sediments in which pollen is outstandingly conspicuous. Some of the fossil-bearing strata occur in such a fashion that although they are unquestionably "warm", their position in the sequence as a whole is in doubt. A possible exception is the fine succession of lacustrine deposits at Leffe, northeast of Milan. The pollen content of this long succession (Lona, 1950; Lona and Follieri, 1957) reveals an alternation of "warm" and "cold" climates, reaching from near the base of the Pleistocene into a "cold" unit considered as possibly equivalent to the Mindel (Table II). The sequence at Leffe is tentatively fitted into the general Alpine succession through local stratigraphic relations (Venzo, 1955).

The units commonly recognized in central North America, the region where the peripheral drift of the great ice sheet is best developed, are fewer than in the regions of Europe discussed above. The column (Table III) includes only four

TABLE III

Stratigraphic Column for Central North America

STAGES	
Wisconsin Glacial	
	Sangamon Interglacial
Illinoian Glacial	
	Yarmouth Interglacial
Kansan Glacial	
	Aftonian Interglacial
Nebraskan Glacial	

glacial and three interglacial stages. Correlation of American with European units is insecure except for upper parts of the Wisconsin stage, which are equivalent, through ^{14}C dating, to upper parts of the Weichsel. Stratigraphic units in other parts of North America have been commonly correlated with this sequence. In the Cordilleran region, however, the use of local names has been the general practice.

C. The Last Major Glaciation

The last major glaciation should be the most useful for study because its record has survived in greatest detail. Actually, however, its stratigraphy has proved difficult to establish, and much of the record is still unclear. The earliest systematic study of it was made by Penck and Brückner (1909), who defined the Würm glacial in the Alps. Those workers recognized within the Würm two chief phases of glaciation, separated in time by an interval of deglaciation. Subsequently, detailed study by F. Brandtner and others (and as yet mainly unpublished) of the sequence of loess sheets and soil zones along the Danube River in Lower Austria revealed a sequence confirming the concept of at least two cold-climate units within the Würm succession in the Alps. The evidence consists basically of three chief loess bodies, representing glacial times in the nearby mountains, and two intervening soils representing times of Alpine deglaciation. The sequence was calibrated in part through ^{14}C dating. Later it was shown that the succession of end moraines and loess bodies in northern Germany, related to the Scandinavian Ice Sheet, could be fitted to the Alpine–Lower Austria scheme by similarity of events, although as yet the fit is confirmed only in small part by ^{14}C. An attempt to show the sequence in graphic form was made by Woldstedt (1958, fig. 1). Although the curves for the last 27,000 years B.P. are fairly well supported, the older part of the record is still very poorly controlled by reliable data. Time calibration of the older part is discussed by de Vries (1958, 1959), Haring *et al.* (1958), and Andersen *et al.* (1960). ^{14}C dates of a few samples from northern Germany, the Netherlands and Britain, taken from positions in the lower part of the sequence and related to fossils that indicate the climatic character at the time, give a reasonably consistent result, although there are still large gaps. The oldest dated sample (de Vries, 1958, GRO-1397) of Würm age has a date of 64,000 ^{14}C years, if a result equivalent to more than 11 half-lives of ^{14}C is accepted. This result would imply that the end of the Riss–Würm interglacial occurred still earlier, an implication with which the results reached by Rosholt *et al.* (1961, 1962) from a study of deep-sea cores are compatible.

The latest major glacial maximum appears to have occurred at around 18,000 years B.P. Thereafter warming and deglaciation occurred, with fluctuations, of which the most notable were the temperature minimum at around 11,000 B.P., following ten or more centuries of markedly higher temperature (Alleröd time); and the Hypsithermal time of relatively high temperature falling between about 9000 and 2600 years B.P.

Much of the record of Wisconsin glaciation in central North America, as recognized thus far, correlates with the later part of the European record, embracing the last 25,000 years or so. The major glacial maximum, fixed in Ohio at 18,000 years B.P. or somewhat later, the Alleröd warmth, the post-Alleröd glacial readvance, and the Hypsithermal are all fairly well documented and were essentially contemporary with the corresponding climatic fluctuations in Europe. Events corresponding to the earlier part of the Würm sequence in Europe, however, are less well known. Research between 1955 and 1962, supported by ^{14}C dates, has made possible a first approximation to a time-calibrated sequence of such events in central North America, summarized by Flint (1963), but alterations can be expected. Not only does the close similarity between the events of the last 25,000 years or so on both sides of the Atlantic suggest the likelihood that earlier events were also similar, but the scanty data published thus far are compatible with the view that the last major glacial age included, in North America, a long sequence of events antedating those of the classical Wisconsin glaciation.

D. Temperatures

Examination of geologic sediments and other features of Pleistocene age has shown that many such things are not in equilibrium with their present climatic environment. Glacial drift is widespread in middle latitudes, even in positions close to sea-level. Terrestrial plants and animals occur as fossils in areas where climatic parameters, such as temperature and precipitation, preclude their existence today. Ancient soils, both buried and at the present surface, and chemical precipitates such as travertines are known to be out of equilibrium with existing climates, chiefly as regards precipitation values. In some regions deflation basins and extensive sand dunes are widespread, now fixed by a continuous cover of vegetation. The shorelines and sediments of former lakes are conspicuous in some regions now arid. Artifacts recording the industries of prehistoric man occur in areas now too dry to support similar cultures. The occurrence of identical species in habitats widely separated by inimical terrain suggests former continuity of habitat.

Disequilibrium features of this nature establish the fact of widespread change of temperature or precipitation or both, within the general span of Pleistocene time. Most of the evidence pertains to temperature rather than to precipitation. Although in a very few instances the data are sufficiently good to indicate the approximate amplitude of change, on the whole they are both widely scattered and imprecise. It is impossible as yet to correlate them from one region to another.

The data are both physical and organic. The physical group consists of changes in the vertical or areal ranges of parameters that are temperature-dependent at least in part. One of the earliest of these parameters to be recognized is fluctuation of the altitudinal zone within which nivation sculptures cirques. In the northern Alps, during the Würm glaciation, the zone was depressed to as much as 1200 m below its present position. A comparable depression (1200 m) is indicated for the Wisconsin glacial maximum in the Rocky Mountains of Colorado, and a depression of at least 1000 m occurred on Mt. Kenya at the equator. Translation of these data into terms of temperature change is not yet possible because the climatic snow-line is dependent also on precipitation. Based on lapse rate alone, without correction, the temperature difference equivalent to a snow-line difference of 1200 m would be about 6.6°C.

The occurrence of glacial drift beyond the present range of glacier ice implies that surface temperatures were no greater than 0°C at the time and place of deposition. Zero temperature at a glacial maximum in Illinois, at 38°N and at an altitude of less than 150 m, compares with around 13°C today. Such temperatures, however, were dependent partly on the presence of the glacier itself and so were not necessarily secular in an independent sense.

The equatorward limit of perennially frozen ground, which is believed to lie close to the mean annual isotherm of −2°C, probably affords a better (although still imperfect) basis for inference as to secular climate. On the assumption that the limit at the time of the last major glacial maximum has been correctly traced across Europe from geologic data (Poser, 1948), mean annual temperatures 8°C to 10°C below existing values can be inferred. The chief uncertainty here lies in the reliability of identification, from relict features, of former frozen ground.

Fluctuations of lakes in basins without outlet are dependent not only on temperature but on precipitation as well. A crude value (probably maximum) for mean annual temperature during the last high-lake episode in the Basin-and-Range region of western United States was found by Meinzer (1922) to be 8°C less than today's values.

As a group, terrestrial plants are thought to be more sensitive to temperature changes than are terrestrial mammals. Consequently special attention has been given to Pleistocene fluctuations of the ranges of fossil plants. In Europe, where most of the studies on record have been made, we find derived mean annual temperatures 4° to 6°C lower than today's at glacial times, and at one locality 2° to 3°C higher than today's at an interglacial time. We find also comparable values expressed as July means.

Three nearly continuous paleobotanical sequences give a picture of temperature fluctuation during earlier parts of the Pleistocene. One is the lacustrine sequence at Leffe, northern Italy (Lona, 1950; Venzo, 1955; Lona and Follieri, 1957), reproduced in its essentials in Fig. 1. The altitude of the locality is about 400 m. Temperature fluctuation equivalent to a change of altitude of more than 2000 m within the full glacial–interglacial range is implied. The apparent mean annual temperature at the coldest times shown by the curves in Fig. 1 was about 9°C lower than that of today (Lona, *in litt.*). The second is the sequence (Szafer, 1954) from Carpathian Poland, with a rich content of megafossils, extending from well back in the Pliocene to a point that is believed to lie near the middle of the Pleistocene. From the Pleistocene part of the sequence, deduced mean annual temperatures range through 13°C (+8°C to −5°C), the lowest value being based on Arctic–Alpine tundra recording an extensive glaciation at the top of the measured sequence. The highest value is close to that of today in the same district. Because of the continental position of the district concerned, the range of temperature fluctuation may well be substantially greater than that of the secular mean. It does not yet seem possible to use the Polish data to extend the curves shown in Fig. 1 further up into the Pleistocene, but the data deserve close study.

Some of these pollen sequences and other sequences having climatic implications for the last several tens of thousands of years were compared by Flint and Brandtner (1961). The agreement among them is fairly good, suggesting general synchrony of temperature fluctuation during the time represented.

The third is the sequence from the Netherlands, developed by Zagwijn (1957), mainly from pollen data. It agrees fairly well with that from Leffe as to

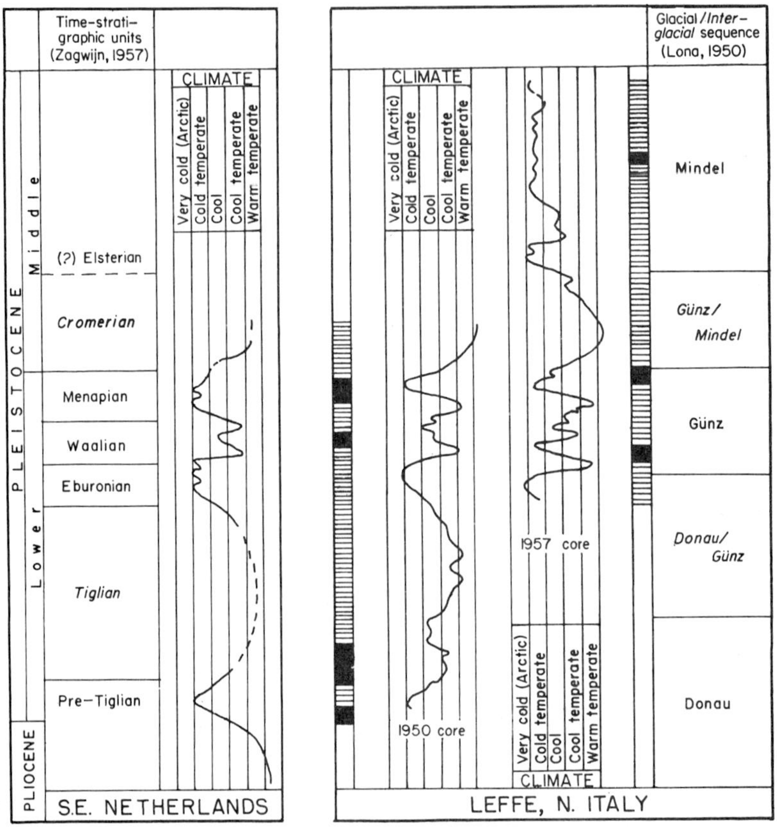

Fig. 1. Pleistocene temperature fluctuations inferred from pollen data. *Left:* Fluctuations between late Pliocene and early middle Pleistocene time, estimated from pollen content in strata in S.E. Netherlands (Zagwijn, 1957, p. 240). *Right:* Fluctuations between earliest (?) Pleistocene and middle Pleistocene time, estimated from pollen content of two cores from points 500 m apart in a single lacustrine sequence near Leffe, Italy (From an unpublished version, kindly furnished by Fausto Lona, of the curves in Lona, 1950, p. 169; Lona and Follieri, 1957, p. 93.)

The Leffe curves were fitted by R. F. Flint to the Netherlands curve by correlating the temperature minima as suggested by Zagwijn (1957, p. 243), who first noticed the similarities. The parts of the Leffe curves between those minima were fitted to the intervening spaces by proportional reduction. Columnar sections show character of sediments: black = peat and lignite; shading = inorganic sediments, chiefly marl and clay.

The vertical scale is not time calibrated in either curve; hence variations in local rates of sedimentation are not compensated for. The troughs and peaks are significant for relative amplitude only.

succession of fluctuations, but as neither column is time-calibrated, spacing of the peaks and troughs of the curves is dependent on rate of sedimentation at each locality, so that curve slopes cannot be compared.

Besides these three sequences there are two paleobotanical sequences in

North America both from core borings. One is a pair of cores from lake sediments beneath Mexico City (Sears *et al.*, 1955). The pollen stratigraphy shows fluctuation of moisture-indicating trees through a range of about 40% of total plants represented, and implies significant climatic variation. However, although it seems likely that the sequence might extend as far back as the last major interglacial age, the fluctuations have not yet been correlated with those derived from other localities. Another sequence is derived from a core in western New Mexico (Clisby *et al.*, 1956, 1957). It shows fluctuation of the proportion among subalpine trees, other woodland and semi-desert plants, and is calibrated by ^{14}C dates in its upper part. This sequence, too, remains uncorrelated with sequences elsewhere. There is a distinct possibility that climate at the locality may have been altered by tectonic movement.

Fluctuation of the evaporation–precipitation ratio is implied also by the results of study of a long core taken from beneath Great Salt Lake in Utah (Eardley and Gvosdetsky, 1960). No pollen is involved, and correlation is only tentative, but it is thought possible that much of the Pleistocene is represented.

In South America pollen stratigraphy is being elaborated by Thomas van der Hammen from cores taken from basin fill near Bogotá, Colombia. Results to date include curves showing fluctuation of temperature and precipitation continuously back into the glacial age preceding the last major interglacial (van der Hammen and Gonzalez, 1960, p. 305). The curve, located close to the equator, resembles curves for localities in Europe, North America and the deep-sea floor.

As to both range of temperature and departures from today's values, then, the record is meager and spotty. However, two generalizations seem justified. First, the ranges derived from physical data are of the same order as those derived from organic data—several degrees centigrade. Secondly, today's values lie within the Pleistocene ranges, usually near their upper limits. But the relation of the temperature fluctuations to the units of the stratigraphic column remains incompletely defined. The view is commonly held that there are major units (the stages of North American usage) and within them secondary units (some of them called substages in North America), but a clear distinction between them has never been agreed on. In European practice glacials and interglacials (essentially the stages of North American usage) are distinguished from stadials and interstadials (essentially substages); here, at least by implication, the difference lies mainly in span of time and amplitude of temperature change represented, rather than in differences among the related fossils, if fossils are present at all. As knowledge grows it may appear that this classification is too rigid; that fluctuations of various amplitudes and durations have occurred; and that a more flexible framework of terminology may be needed.

Whatever the amplitude of fluctuations derived from continental data, the Pleistocene changes of temperature appear to have been rather distinctly localized in time, following a long antecedent period of generally falling temperature. Table IV illustrates the falling trend through a period of some 50 million years. Within the record of that period no evidence of conspicuous

TABLE IV

Trend of Cenozoic Temperature Derived from Fossils

Epoch	Central Europe Mean annual air temperature (°C) inferred from fossil floras (Woldstedt, 1954, p. 5)	Pacific coast, North America. Lat. 40°N. Mean February marine isotherm (°C) inferred from fossil faunas (Derived from Durham, 1950, p. 1259)
Pleistocene	10	(Data inadequate)
Pliocene	14–10	13
Miocene	19–17	20–15
Oligocene	20	22
Eocene	22–20	25

fluctuation has yet been found. The beginning of the Pleistocene, therefore, seems to mark the beginning of new things in the climatic history of the Cenozoic Era. However, not only does the continental record give an inadequate indication of the number and amplitude of the Pleistocene fluctuations, it gives no adequate concept of the duration of the several fluctuations, nor of the Pleistocene time as a whole. For a better concept we must turn to the sediments beneath the sea floor.

E. Chronology

Calibration of Pleistocene temperature changes by means of radio-isotopic dates is a highly important aspect of Pleistocene stratigraphy. Thus far the only radio-isotope that has yielded abundant results is ^{14}C. Because of its short half life, however, ^{14}C seems incapable of dating events more ancient than those of the Weichsel glacial age of western European terminology. The greatest finite ^{14}C date published up to 1960 is about 64,000 years, as mentioned earlier. Pertaining to a Dutch locality, it is believed to represent a short, relatively warm interval following the first cold episode of the Weichsel glacial age (Andersen, de Vries and Zagwijn, 1960). Finite dates between 45,000 and 50,000 years on North American materials have been published. These include part of an interstadial in southern Ontario (de Vries and Dreimanis, 1960) and part of a pluvial in southwestern California (Flint and Gale, 1958). As stated earlier, radiocarbon dates of less than 30,000 years are numerous, and calibrate fairly well climatic fluctuation embraced within the classical Wisconsin glacial of North American terminology.

Radioactive nuclides of the ^{238}U, ^{235}U and ^{232}Th families have been used to date deep-sea sediments back to about 200,000 years ago. This topic will be discussed in a later section.

Preliminary potassium–argon dates of materials significant for Pleistocene chronology have been published (Evernden *et al.*, 1957; Evernden, 1959; Savage *et al.*, 1960; Gentner and Zähringer, 1960; see also von Koenigswald, 1960). Because of the comparatively young age of Pleistocene materials, important difficulties of sampling and analytical techniques exist. If these difficulties can be sufficiently minimized in the future, K–A dating may produce the much-needed absolute time scale for the continental Pleistocene beyond the range of ^{14}C.

3. Marine Stratigraphy

A. Marine-Epicontinental Stratigraphy

The repeated formation of extensive ice sheets on the continents during the glacial ages and their disappearance during the intervening interglacials (Greenland and Antarctic ice sheets excluded) has resulted in repeated glacial-eustatic oscillations of sea-level through much of the Pleistocene. The amplitude of these oscillations is believed to have been at least 130 m and possibly considerably more. Thus, sea-level is believed to have reached at least 100 m below the present level during the last glacial maximum (Main Würm or classical Wisconsin), and to have risen to at least 8 m and possibly 30–35 m above the present level during the last or the previous interglacial age.

In tectonically stable or slowly subsiding areas of the continental margin, where terrigenous sedimentation is appreciable or abundant, the repeated oscillations of sea-level created sequences of alternating marine and continental sediments. Some such sequences have been sampled by means of borings and have been described in some detail. The section of the Lower Versilia (northwest shore of Tuscany) described by Tongiorgi (1936) and Blanc (1937) pictures rather clearly the sea-level changes during the last glacial age (Early and Main Würm) and after. Following the end of the last interglacial age, the sea-level decreased to below 90 m, and continental sand 14-m thick was deposited (Early Würm). The sea-level then rose to -61 m and marine sand with seeds of *Vitis vinifera* was deposited (Early Würm–Main Würm interval). A new lowering of sea-level occurred, during which a layer of continental sand 29-m thick was deposited (Main Würm). This sand contains predominant oak pollen at the base, replaced upward by *Abies* and then by *Pinus mugo* and *Pinus silvestris.* The sequence thus indicates a progressive, marked cooling. Eventually sea-level rose to its present position and additional marine and continental sands, peat and clays were deposited. Carbon-14 analysis of a peat cobble, apparently reworked from the *Pinus* level, gave an age of 18,350 years (Broecker *et al.*, 1956).

A succession of marine transgressions and regressions, presumably controlled by glacial-eustatic movements, was reconstructed by Pfannenstiel (1952) for the coast of Syria and Israel. The sea-level was found to have been lowered to at least -89 m during the Early Würm, in agreement with what was found in the Versilia. In addition, the sea-level appears to have oscillated repeatedly during the time of high sea-level defined as the Tyrrhenian Age with maxima at $+30$–35 m, $+15$ m and $+6$ m, separated by minima of less than $+10$ m and less than $+2$ (all in order of decreasing age). Evidence of sea-level positions during the Early Würm–Main Würm interval and during the Main Würm is apparently missing in the region studied by Pfannenstiel, but sea-level may have risen to +3–4 m about 5000 years ago. Such a high stand of the sea-level in the recent past has been recognized by some authors (Fairbridge, 1958; Hopkins, 1959; Schofield, 1960) but not by others (Shepard and Suess, 1956; Godwin *et al.*, 1958; Graul, 1959; Jelgersma and Pannekoek, 1960).

Sea-level rise during the last 15,000 years can be seen also in the stratigraphy of the Mississippi delta (Fisk and McFarlan, 1955; McFarlan, 1961), but the situation is complicated by the large amount of subsidence and sediment compaction that has occurred (cf. Broecker, 1961).

During the Pleistocene, the Po Valley was filled with marine sediments grading into continental ones upwards and landwards. In limited areas the combined marine and continental Pleistocene sediments attain the remarkable thickness of 2000 m (Perconig, 1956). Preliminary work by Selli (1949), Perconig (1956), Ruggieri and Selli (1950), and others has revealed the repeated occurrence of warm and cold stages. Through most of the Po Valley there is continuity of sedimentation between Pliocene and Pleistocene, the boundary being recognized on the basis of Foraminifera (Perconig, 1956).

Classical Plio-Pleistocene sections of marine epicontinental deposits occur in Calabria, southern Italy. It was in this area that Gignoux (1913) established the Calabrian stage, the base of which is accepted by many as marking the beginning of the Pleistocene Epoch (Intern. Geol. Cong., 1950). In many sections there is continuity of sedimentation between Pliocene and Pleistocene, and the boundary is identified by the sudden appearance of northern species of marine mollusks and Foraminifera (esp. *Cyprina islandica* and *Anomalina baltica*). Oxygen-isotopic analyses of shells of pelagic Foraminifera collected at close stratigraphic intervals from the section at Le Castella, near Crotone, has revealed the occurrence of marked temperature oscillations, present in the Pliocene but becoming especially large in the Pleistocene (Emiliani *et al.*, 1961). In addition, a progressive decrease of the secular maxima and minima was observed, but no sudden, large temperature decrease was noticed across the sharply defined paleontological boundary. Important temperature changes were observed to take place across stratigraphic intervals as short as a meter or two, suggesting that, because of rapidly changing conditions during the Pleistocene, marine-epicontinental deposits of Pleistocene age should be sampled in far greater stratigraphic detail than has generally been done.

Excellent sections of marine-epicontinental sediments exist in East Anglia and the Netherlands. In parts of the Netherlands, marine Pleistocene sediments follow conformably marine Pliocene sediments. The boundary, recognized from the gradual advent of foraminiferal faunas adjusted to cooler conditions, lies as deep as 600 m below sea-level in the area of Haarlem (Pannekoek, 1954, 1956). The foraminiferal faunas have been described in detail by ten Dam and Rheinhold (1942), Pannekoek and van Voorthuysen (1950), van Voorthuysen (1950, 1953), and others. The marine Pleistocene deposits pass upwards and eastwards into estuarine facies, making it difficult to recognize temperature variations other than the cooling that marks the beginning of the Pleistocene.

Marine sediments of lower-middle Pleistocene age, up to a few thousand feet thick, occur in the Los Angeles and Ventura basins of southern California. Thinner marine deposits of the same age occur along the borders of the basins. In the Palos Verdes Hills, southwest of Los Angeles, lower-middle Pleistocene deposits attain a thickness of a little over 100 m (Woodring *et al.*, 1946). In this area three stratigraphic and lithologic units have been recognized: the Lomita Marl, a loose calcarenite lying unconformably upon Upper Miocene shale; the Timms Point Silt, partly overlying the Lomita Marl, partly grading laterally into it; and the San Pedro Sand, partly overlying both the Lomita Marl and the Timms Point Silt, partly grading laterally into them. Attempts to deduce the temperature conditions prevailing when these units were being deposited were made by studying the fossil faunas contained in each unit (Woodring *et al.*, 1946). The results were not very conclusive, probably because temperature varied repeatedly while each unit was being deposited, so that the total faunas consisted of "mixtures" of warmer and colder species, a condition analogous to that in the section at Le Castella, discussed above. Repeated and marked temperature variations were revealed by oxygen-isotopic analyses of calcareous benthonic Foraminifera collected at 15-cm stratigraphic intervals from a 10 m section of the Lomita Marl (Emiliani and Epstein, 1953).

From the above discussion it appears that marine-epicontinental sediments, if studied with sufficient stratigraphic detail, can provide valuable information on the general history of the Pleistocene, in spite of the fact that the marine-epicontinental environment is deeply affected by local variations in rate of terrigenous sedimentation, depth of deposition, current system, salinity, local temperatures, freshwater influx and other parameters.

B. Deep-Sea Stratigraphy

a. Character of the sediment and analytical techniques

Apart from occasional volcanics, ice-rafted material and sand and silt contributed by turbidity currents, terrigenous clastics are absent from pelagic sediments. These sediments are eminently heterogeneous, consisting of water- and wind-borne mineral particles, skeletal remains of pelagic and benthonic organisms (Foraminifera, pteropods, coccoliths, Radiolaria, diatoms, holothurian sclerites, fish teeth and otoliths, etc.), volcanic ash, cosmic spherules and some precipitates (largely iron and manganese compounds). Depending upon the local conditions on the ocean floor, the various sedimentary components or some of them may accumulate and remain *in situ* or may be partly or wholly transported from areas of higher to areas of lower topography by the action of bottom currents and perhaps also by submarine solifluction. We define as "undisturbed" that sediment or sedimentary component which has settled on the ocean floor from the water column above and has remained *in situ*. Because of the great difference in sedimentological behaviour among the various components, a given component may be undisturbed by the above definition, while other components intimately mixed with it may have been largely or even entirely derived from reworking. Generally, the relatively heavy shells of pelagic Foraminifera can settle undisturbed on a given portion of the ocean floor while the finer biogenous components (coccoliths, diatoms, Radiolaria), and especially the much finer mineral particles settling at the same time through the same water column, may be moved away by bottom currents and may be replaced by similar components reworked from adjacent areas. In such cases the reworked components are older than the undisturbed component or components.

The local factors which importantly affect marine-epicontinental sedimentation are integrated in the pelagic environment. The "background noise" is thus greatly reduced and an undisturbed section of deep-sea deposits can yield invaluable information on the conditions prevailing when the sediments were being deposited. The conditions include those on the ocean floor, in the water column above, at the ocean surface, in the atmosphere, on land areas, and possibly on the sun and in the interplanetary space. In addition, the occurrence itself of sedimentary disturbances (turbidity-current deposits, winnowing, etc.) in a column of otherwise undisturbed deep-sea sediments may yield information on the factors causing the disturbances. Thus, the repeated and extensive occurrence of graded and non-graded layers of sand and silt in North Atlantic cores (Ericson *et al.*, 1952, 1955, 1961) which generally contain normal deep-sea clays and oozes in their uppermost portions demonstrates that turbidity currents across the continental slope bordering the North Atlantic basin on the west were far more frequent during the last glacial age than in postglacial time.

Totally undisturbed deep-sea sediments (that is, sediments in which all the components are undisturbed as defined above) are probably a rare occurrence. In cores of *Globigerina* ooze in which the foraminiferal component is undisturbed, the undisturbed character of the finer components (e.g. fine-carbonate fraction and clay) can be considered as proved if parameters measured on these correlate with similar or even different parameters measured on the foraminiferal component. Thus the demonstrated correlation between Mg percentages in the fine-carbonate component and paleotemperatures measured on the foraminiferal component in deep-sea cores from the Caribbean shows that the fine-carbonate component is undisturbed (Wangersky, 1958; Rosholt *et al.*, 1961).

The facility with which information obtained from the various sedimentary components of undisturbed deep-sea cores can be interpreted varies greatly. The foraminiferal record can be interpreted easily by the various methods discussed below, while the record offered by clay mineralogy is still difficult to interpret. This is unfortunate, because red clay accumulates at rates far lower than those of foraminiferal oozes, and can be sampled to far greater ages than the latter. The record offered by coccoliths, diatoms and Radiolaria is still besieged by taxonomic difficulties and also by the facility with which the very light skeletons and skeletal elements of these organisms can be reworked, not only from neighbouring areas of the ocean floor but also from distant lands (cf. Kolbe, 1954, 1955, 1957, 1957a; Riedel, 1957, 1959; Riedel and Bramlette,

1959). Thus information from deep-sea sediments, pertinent to Pleistocene history, has been garnered largely from the foraminiferal component.

Disregarding accessory components, normal *Globigerina* ooze consists essentially of shells of pelagic and benthonic Foraminifera, coccoliths, Radiolaria, diatoms and clay. Generally these various components amount to most of the dry weight of the bulk sediment. *Globigerina* ooze accumulates at rates varying from less than a centimeter to several centimeters per thousand years, depending upon rates of supply of the various components from the water column above, and upon bottom conditions such as currents, morphology and post-depositional solution. Following the invention of the piston corer by Kullenberg (1947), sections of deep-sea sediments as long as 20 m or more were obtained from the ocean floor. In the absence of a reliable method of dating *Globigerina* ooze beyond the range of the ^{231}Pa–^{230}Th method (about 200,000 years), cores showing sedimentary disturbances are difficult to use for the reconstruction of Pleistocene events, because a sedimentary section representing unknown time may have been added or removed by each disturbance. Elisions are most common on the tops and steep sides of deep-sea mounds, whereas additions occur on the floors of depressions. Thus, the most convenient place to sample the ocean bottom for undisturbed sections are the sides of gentle deep-sea mounds.

Research on *Globigerina* ooze cores that seem undisturbed upon megascopic inspection revealed that major unconformities are common in deep-sea sediments, especially in the Atlantic basin. Thus many of the long *Globigerina* ooze cores that have been described in detail show at least one major unconformity, below which older Pleistocene, Tertiary or even Cretaceous sediments occur (cf. Ericson *et al.*, 1961). Indeed, the stratigraphically longest core so far described from the Atlantic and adjacent basins, containing no major unconformities, is a core from the central Caribbean which is only 9 m long. Major unconformities seem to be absent from a suite of five eastern-equatorial-Pacific cores described by Arrhenius (1952). Two of these cores (nos. 58 and 62) may reach back in time about a million years through continuous sedimentary sections. Unfortunately these cores do not seem to be as easy to interpret from the point of view of Pleistocene events as the Atlantic cores, as explained below.

Globigerina ooze does not accumulate below 4800 m in the Atlantic and 4300 m in the Pacific, where post-depositional solution of the carbonate material is already important. The depth at which post-depositional solution becomes appreciable may have varied during the Pleistocene in response to climatic changes, especially in the Atlantic and adjacent seas where significant temperature changes of the bottom water have been noticed (Emiliani, 1955, 1958). During the Tertiary, when the bottom temperature of the oceans was higher, *Globigerina* ooze accumulated in some of the areas where only red clay has been accumulating during the Pleistocene (Revelle *et al.*, 1955).

Globigerina ooze, which accumulates largely on aerobic bottoms, is constantly reworked by bottom organisms. The sedimentary thickness affected by this mixing is believed to range from a few to several centimeters, and corresponds to a time span of a few thousand years. The time resolution of the events recorded by the sediment is thus decreased to a few thousand years, from a theoretical limit of only a few years. The loss of resolution is inversely proportional to the rate of sedimentation, being greatest for the more slowly accumulating sediments. Therefore undisturbed sediments accumulating at higher rates generally yield a record of comparatively greater detail.

Stratigraphic study of *Globigerina* ooze cores involves a variety of methods. Of direct interest to Pleistocene research is the study of those parameters which have been most affected by repeated glaciation of the continents. These parameters are the micropaleontology of the pelagic foraminiferal faunas; the oxygen-isotopic composition of foraminiferal shells; the weight percentage of Foraminifera relative to bulk sediment (possibly only in the Atlantic and adjacent seas); and calcium-carbonate percentages (only occasionally).

The living species of pelagic Foraminifera number about 25, of which 15 are common. Spectrographic and X-ray analysis of 11 species showed that the shell material is nearly pure calcite with only about 0.11% of strontium (Emiliani, 1955b). A small amount of organic material is also present. The common species can be grouped as follows with respect to temperature preferences (data from Phleger *et al.*, 1953; valid for the Atlantic and possibly also for the other oceans; cf. Bradshaw, 1959):

1. Species abundant in low latitudes:
   Globigerina eggeri
   Globigerinoides sacculifera
   Globorotalia menardii menardii
   Globorotalia menardii tumida
   Pulleniatina obliquiloculata

2. Species abundant in middle latitudes:
   Globigerina bulloides
   Globigerina inflata
   Globorotalia hirsuta
   Globorotalia scitula
   Globorotalia truncatulinoides

3. Species abundant in high latitudes:
   Globigerina pachyderma

4. Species abundant in both low and middle latitudes:
   Globigerinella aequilateralis
   Globigerinita glutinata
   Globigerinoides conglobata
   Globigerinoides rubra
   Orbulina universa

The relative abundances of the species listed above, in cores of *Globigerina* ooze, afford a simple method by which past temperature variations at the ocean surface can be estimated. This method was first proposed by Philippi (1910) and was later applied by many authors to a large number of deep-sea cores. In the absence of post-depositional solution (which would make quantitative taxonomic studies impossible because shells of different species dissolve at different rates), the method is sensitive enough to permit the detection of glacial and interglacial stages not only in such areas as the Mediterranean where the amplitude of the temperature change has been large, but also in

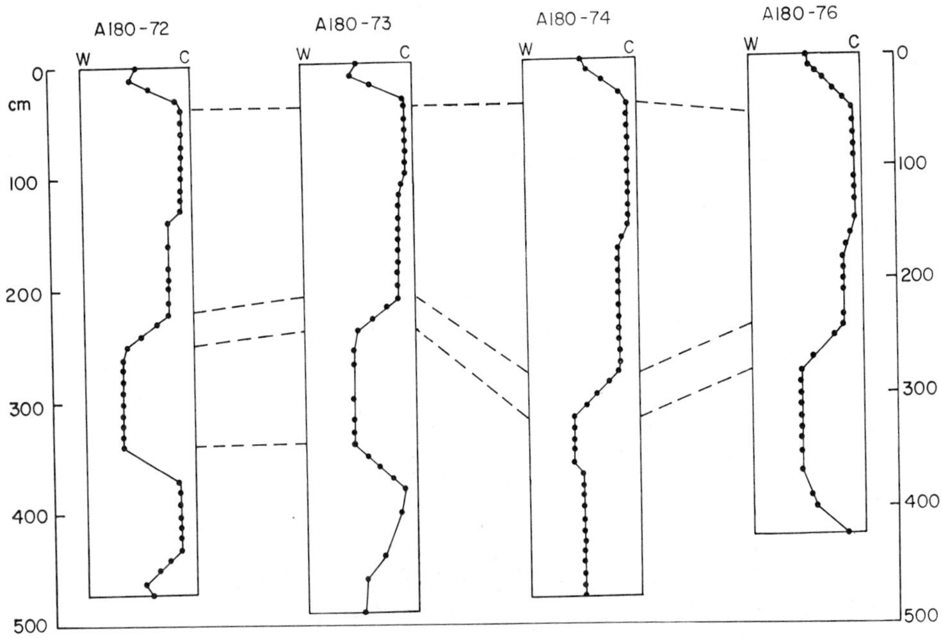

Fig. 2. Temperature variations in a suite of four equatorial Atlantic cores estimated from changes in the relative abundances of pelagic Foraminifera. W means "warm" and C means "cold". (From Ericson and Wollin, 1956.)

the equatorial Atlantic where the amplitude has probably not been larger than 6° or 7°C (Fig. 2).

A much more precise method of analyzing the temperature record contained in cores of *Globigerina* ooze involves mass-spectrometric measurement of the isotopic ratio $^{18}O/^{16}O$ in the calcite of the various species of pelagic Foraminifera. This method, devised by Urey (1947) and developed by Urey, Epstein and co-workers (Urey *et al.*, 1951; Epstein *et al.*, 1951, 1953) is based on the fact that the oxygen isotopes in $CaCO_3$ precipitating slowly in a water solution are fractionated between water and the carbonate ions, and the fractionation factor depends upon temperature. As a result, the $^{18}O/^{16}O$ ratio changes by 0.00023 per degree centigrade within the normal temperature range of biological carbonate deposition, the change being inversely related to temperature. If the oxygen-isotopic composition of the water is known, the temperature existing during the time of carbonate deposition can be determined within one degree centigrade. If the composition is not known but can be reasonably estimated, as in the case of open oceanic water of the not too distant past, reasonably reliable temperature data can still be obtained.

Paleotemperature analysis has been rather extensively applied to the study of Cenozoic deep-sea cores (Emiliani, 1954a, 1955, 1955a, 1956, 1958; Rosholt *et al.*, 1961, 1962). Most of the isotopic work has been done on shells of pelagic Foraminifera, which provide data on the oceanic surface temperature and on its variation through the time represented by the deep-sea cores. No post-depositional recrystallization or ionic diffusion through the solid state of the foraminiferal shell material has been noticed. Since empty shells of pelagic Foraminifera are buried on the ocean floor where temperature is very low, recrystallization and ionic diffusion would tend to increase the $^{18}O/^{16}O$ ratio in the $CaCO_3$. The result would be temperature readings lower than the original temperatures or equal to the bottom temperature for total exchange. It was found, instead, that shells of pelagic Foraminifera of a Middle Oligocene age from a deep-sea core from the subtropical Atlantic Ocean gave temperatures around 29°C, although the shells had been maintained at much lower temperatures for about 30 million years (Emiliani, 1956). Similar results were obtained from two Miocene cores (Emiliani, 1956). The apparent preservation of the temperature record may be due in part to the fact that diffusion through the solid state is not only very slow but is also strongly dependent on temperature (see Urey *et al.*, 1951, for data on calcite). Thus the low temperature at which the foraminiferal shells were maintained may have been important in preserving the temperature record even though the shells are thin and perforated by canals. Organic coatings may also be important.

Because different species of pelagic Foraminifera appear to deposit their $CaCO_3$ at different depths (Emiliani, 1954, 1955, 1958; see also Bandy, 1956; Bandy and Arnal, 1957; Polski, 1959) and/or at different seasons (Emiliani, 1958; Bé, 1960; Phleger, 1960), and because temperature varies rapidly with depth near the ocean surface, and also with time of the year in regions outside the equatorial belt, oxygen-isotopic analysis of deep-sea cores must be performed on monospecific samples and different temperature graphs must be constructed for each different species. The superficial species *Globigerinoides rubra* and *Globigerinoides sacculifera* yield a temperature record of greater amplitude and detail because climatic variations affect more surface water than the water below.

The carbonate component of *Globigerina* ooze may be conveniently divided into two size fractions, one larger than 62 μ and the other smaller. In general, the fraction larger than 62 μ consists almost entirely of shells of pelagic Foraminifera, while the smaller fraction consists of minute shells and shell fragments of pelagic Foraminifera, and especially of coccoliths. In cores of *Globigerina* ooze from the equatorial, tropical and subtropical Atlantic Ocean, and adjacent seas, the weight percentage of the coarser carbonate fraction is in general directly proportional to paleotemperatures (Emiliani, 1955, 1958; Ericson and Wollin, 1956), while the weight percentage of the finer carbonate fraction appears to be inversely proportional (Wangersky, 1958, 1959; Rosholt *et al.*, 1961). Accordingly, the total carbonate percentage may be variously related to temperature, depending not only upon the amount of the non-carbonate component (essentially clay), but also upon the relative amounts of the two carbonate fractions because of their different temperature dependences (Rosholt *et al.*, 1961). A direct relationship between total carbonate percentages and paleotemperatures was observed in a short pilot core from the equatorial Atlantic (Wiseman, 1954, 1959; Emiliani and Mayeda, 1961) and is also noticeable in other Atlantic cores (compare Emiliani, 1955, with Olausson, 1960). Such a relationship might be especially clear in deep-sea cores where post-depositional solution has reduced the amount of the finer carbonate fraction (Wangersky, 1962).

The direct relationship between weight percentages of the coarse carbonate fraction and paleotemperatures should not be taken to indicate a greater productivity of pelagic Foraminifera during warmer ages. In effect, chemical and ^{14}C analyses conducted at close stratigraphic intervals in an equatorial Atlantic core showed that the sedimentation rates of the coarser carbonate fraction, of the finer carbonate fraction and of the non-carbonate component decreased markedly from the last glacial age to the postglacial, the decrease having been greatest for the non-carbonate component, smallest for the coarser carbonate fraction, and intermediate for the finer carbonate fraction (Broecker *et al.*, 1958). The different temperature dependence of the sedimentation rates of the three components explains the observed relationships between their weight percentages and paleotemperatures.

A greater foraminiferal and coccolith productivity during glacial ages is consistent with an overall greater biogenous activity in the oceans, brought about by an increase in the vertical circulation in response to the greater temperature gradient between equator and middle-high latitudes. The greater rate of clay sedimentation in the Atlantic and adjacent basins during glacial ages is explained not only by the faster oceanic circulation (which could carry suspended material farther) but also by the fact that most of the glacial water drained into the Atlantic and adjacent seas. In the equatorial and tropical Pacific, on the other hand, the increase of suspended clay during glacial ages was probably much smaller, explaining the apparent productivity-controlled inverse relationship between carbonate percentages and temperature in the cores described by Arrhenius (1952). Generally, depending upon the relationship among the changing rates of foraminiferal, coccolith, radiolarian, diatom and clay sedimentation, the ratio of Foraminifera to the other components and the ratio of carbonate to non-carbonate fraction can be either directly or inversely proportional to temperature; they can remain constant if the above rates happen to change by the same amounts (cores of type II in Revelle *et al.*, 1955, fig. 11); or can vary independently of temperature (Wangersky, 1958).

Marked variations of the weight percentages of the foraminiferal component and of the carbonate percentages occur also in core sections of Oligocene and Miocene age (Arrhenius, 1952, pls. 2.53 and 2.57; Emiliani, 1956). These variations do not seem to be periodic and do not correlate with the temperatures, which remained more or less constant. Their cause is unknown.

Important, apparently cyclic carbonate variations occur in at least three (nos. 58, 59 and 60) of a suite of five eastern-equatorial-Pacific cores described by Arrhenius (1952). These variations are believed to result from productivity changes in the absence of important changes in the rate of sedimentation of the non-carbonate component, and, therefore, to be inversely related to temperature (Arrhenius, 1952). It is not entirely clear whether this is exactly the case or not, because the carbonate percentage of the modern sediment (shown by two short pilot cores) is actually closer to the carbonate maxima (which should represent glacial conditions) rather than to the minima. In any case, the carbonate cycles seem best interpreted in terms of climatic fluctuations.

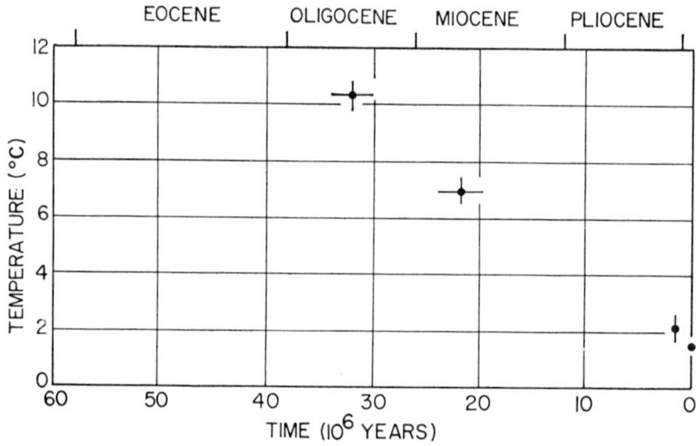

Fig. 3. Temperature decrease of Pacific bottom water measured by oxygen-isotopic analysis of calcareous benthonic Foraminifera. (From Emiliani, 1954a.)

b. The stratigraphic record

As shown in Table IV, a general cooling of the Earth climate occurred during the Tertiary, amounting to about 8°–10°C in middle latitudes. Supporting evidence came from oxygen-isotopic measurements on calcareous benthonic Foraminifera of Tertiary age from deep-sea cores raised in the eastern equatorial Pacific (Emiliani, 1954a). These measurements showed that the bottom temperature of the Pacific Ocean decreased from about 10°C in mid-Oligocene time to about 2°C in the late Pliocene, reflecting a world-wide cooling possibly caused by the Alpine orogenesis (which probably increased the Earth's albedo) and representing cooling of the surface water of the polar seas by approximately the same amount (Fig. 3). A concomitant increase in atmospheric turbulence is suggested by the observation that Pleistocene deep-sea sediments in portions of the Pacific Ocean contain two to three times more wind-borne quartz particles than the underlying Tertiary sediments (Rex and Goldberg, 1958). The low temperature of the Pacific bottom water late in Pliocene time suggests that Antarctica was already fully glaciated, with ice extending to the Antarctic shore.

Oxygen-isotopic analysis of pelagic Foraminifera from equatorial and subtropical Atlantic cores of Middle Oligocene and Lower-Middle Miocene age showed that the ocean-surface temperatures remained rather uniform during the time intervals represented by these cores (perhaps 100,000 years for the Middle Oligocene section and perhaps up to 300,000 years for the Lower-Middle Miocene sections). Only random fluctuations, mostly within 2°C, are apparent (Emiliani, 1956).

Only one core, which probably represents the whole Pleistocene Epoch and penetrates well into the Pliocene, has been analyzed so far by the oxygen-isotopic method. This core, no. 58 of the Swedish Deep-Sea Expedition of

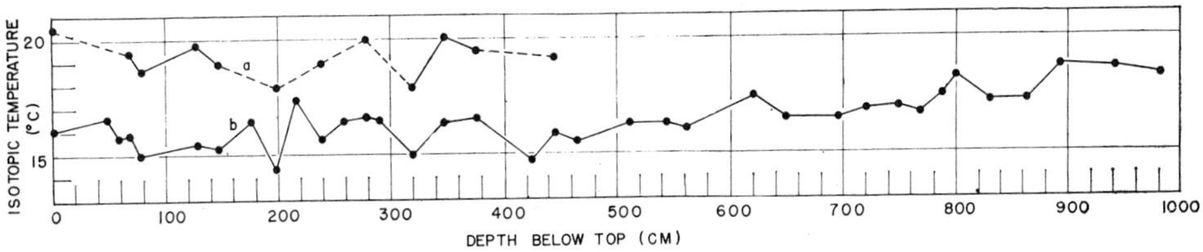

Fig. 4. Plio-Pleistocene temperature decrease of surface water of the eastern equatorial Pacific as shown by oxygen-isotopic analysis of *Pulleniatina obliquiloculata* (curve a) and *Globorotalia menardii tumida* (curve b) from core 58 of the Swedish Deep-Sea Expedition, 1947–1948. (From Emiliani, 1955.)

1947–1948, raised in the eastern equatorial Pacific, shows an appreciable temperature decrease (about 3°C) from the bottom up to the middle portion of the core, followed by small temperature oscillations in the upper half (Fig. 4). These oscillations, however, do not offer a clear picture of climatic changes during the Pleistocene because of the relatively small value of the secular temperature change in the equatorial Pacific, because of the deeper habitats of the foraminiferal species available for isotopic analysis, and because of the local pattern of vertical circulation of the surface water (Emiliani, 1955).

The Pleistocene temperature record is far clearer in cores from the Atlantic and adjacent seas than in Pacific core 58, but unfortunately it is as yet far shorter stratigraphically. In fact, the stratigraphically longest core so far analyzed (Emiliani, 1955) is Lamont core A172-6 from the Venezuelan Basin of the Caribbean, extending back in time an estimated 300,000 years.

Micropaleontological work on this and many other cores from the equatorial, tropical and temperate Atlantic, the Caribbean, the Gulf of Mexico and the Mediterranean (Bramlette and Bradley, 1940; Cushman, 1941; Cushman and Henbest, 1940; Ericson, 1953; Ericson *et al.*, 1955; Ericson and Wollin, 1956;

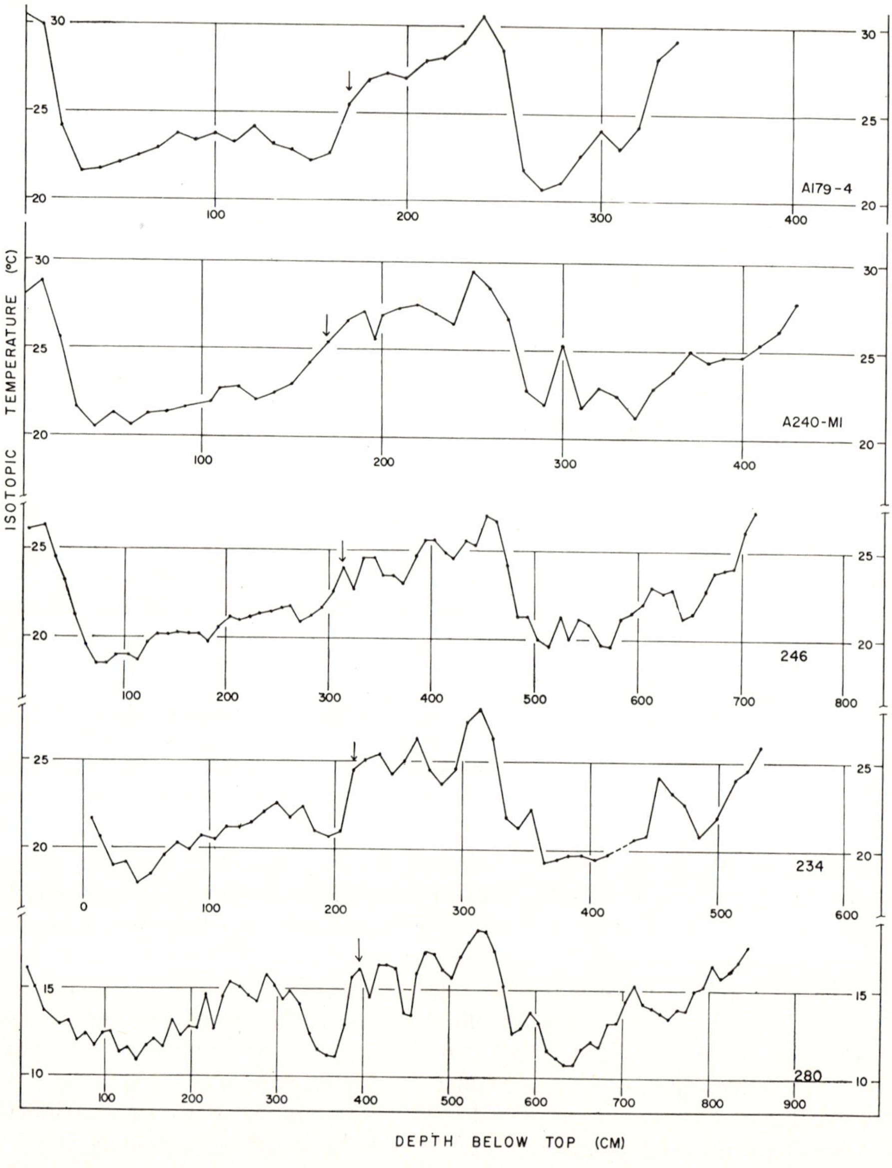

Fig. 5. Paleotemperature curves of five deep-sea cores, determined by means of oxygen-isotopic analysis. Core numbers to the right. Arrows indicate last occurrence of *Globorotalia menardii flexuosa*. Core 280, North Atlantic (34° 57′N, 44° 16′W, 4256 m); core 234, equatorial Atlantic (5° 45′N, 21° 43′W, 3577 m); core 246, equatorial Atlantic (0° 48′N, 31° 28′W, 3210 m); core A240-M1, Caribbean (15° 26′N, 68° 30′W, 4180 m); core A179-4, Caribbean (16° 36′ N, 74° 48′W, 2965 m).

Ericson *et al.*, 1961; Ewing *et al.*, 1958; Parker, 1958; Phleger, 1939, 1942, 1947, 1948, 1951, 1955; Phleger and Hamilton, 1946; Phleger *et al.*, 1953; Schott, 1935, 1952, 1954; Todd, 1958) revealed the occurrence of alternating warmer and colder stages numbering up to nine or ten each in the stratigraphically longest cores. Whenever the climatic record offered by the micropaleontology is of sufficient detail, correlations can be made or attempted among the various cores and between the core stages and the continental-Pleistocene stages. In some cores, the correlations based on changing foraminiferal abundances have been substantiated by correlations based on the changing ratio of left-to-right coiled specimens of *Globorotalia truncatulinoides* (Ericson and Wollin, 1956; see also Chapter 31).

Oxygen-isotopic analysis of deep-sea cores from the Atlantic, the Caribbean and the Mediterranean have yielded temperature records which correlate closely with each other, as shown in Fig. 5 (cf. Emiliani, 1955, 1955a, 1958; Rosholt *et al.*, 1961, 1962), and also with the micropaleontology (Ericson and Wollin, 1956a; Emiliani, 1957). A typical temperature graph (Fig. 6) shows a number of warm units (identified by increasing odd integers in order of increasing age) alternating with a number of cold units (identified by increasing even integers in order of increasing age). One warm unit (no. 3) does not attain a temperature as high as the other warm units, and two cold units (nos. 8 and 12) do not attain temperatures as low as the other cold units. One can thus distinguish major and minor warm and cold units. Within any given region, all major warm units reach about the same temperature extremes, and so do all major cold units. The measured temperature difference between major maxima and minima in cores from low latitudes is about 8° or 9°C, but a correction accounting for the greater concentration of ^{18}O in sea-water during glacial ages reduces this difference to about 6° or 7°C. An average, uncorrected temperature difference of about 12°C is shown by Mediterranean core 189 (Emiliani, 1955a), whereas a difference in surface temperatures of only about 4°C may be inferred from oxygen-isotopic analyses on *Globorotalia menardii tumida* from cores raised in the eastern equatorial Pacific (Emiliani, 1955). Thus it appears that, as expected, temperature oscillations of the surface waters of the oceans during the Pleistocene were smallest in the equatorial Pacific, were larger at corresponding latitudes in the Atlantic, and were still larger in the more northern Mediterranean basin.

Arrhenius (1952) and Olausson (1960a, 1961) published chemical analyses on a number of Pacific cores and included data on the micropaleontology and other features of the sediments. As previously mentioned, carbonate percentages were found to vary markedly and repeatedly in a suite of five cores from the eastern equatorial Pacific, possibly reflecting productivity changes associated with major climatic fluctuations. Correlation among these cores was made using the cumulative amounts of TiO_2 in the sediments, and the last four major carbonate maxima were correlated with the four major glaciations (Arrhenius, 1952). If this correlation is correct, the occurrence of earlier carbonate maxima in two of the cores (nos. 58 and 62) would suggest that

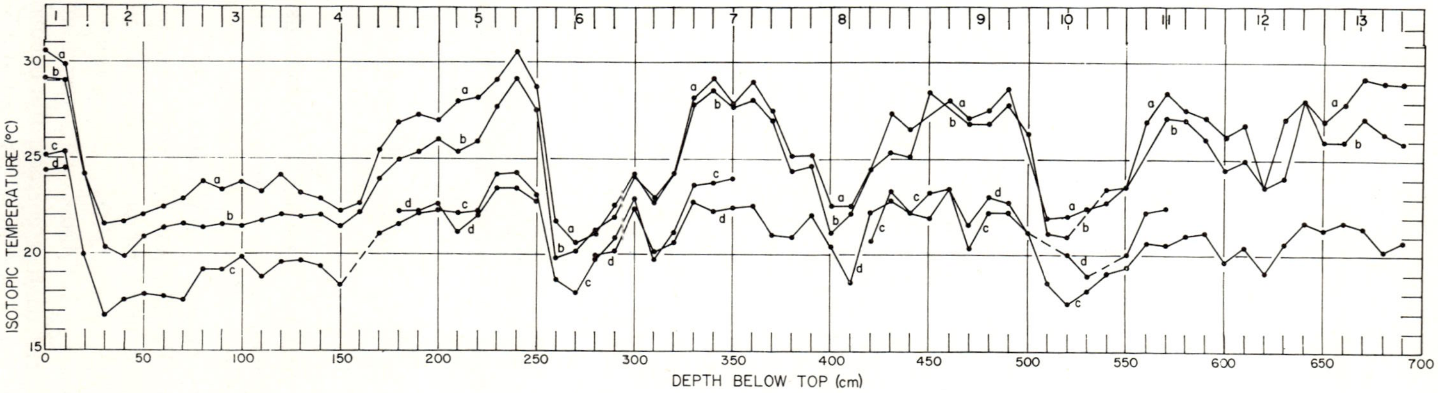

Fig. 6. Core A179-4 (16° 36′N, 74° 48′W, 2965 m). Isotopic temperatures obtained from *Globigerinoides rubra* (a), *Globigerinoides sacculifera* (b), *Globigerina dubia* (c), and *Globorotalia menardii* (d). Numbers above the curves identify temperature units. (From Emiliani, 1955.)

important glaciations occurred before the first major glaciation. Cores 58 and 62 are believed to contain a complete section of Pleistocene sediments and to penetrate into the Pliocene (Arrhenius, 1952). The position of the Plio-Pleistocene boundary in these cores is unknown because a correlation with the Plio-Pleistocene sections of Italy has not yet been made. On paleotemperature evidence alone, the boundary has been placed at 610 cm below the top in core 58 (Emiliani, 1955), but some other lower levels, such as those at 780 or 880 cm below the top, could also be chosen. Ages of 600,000 to 900,000 years can be estimated for these levels (cf. Emiliani, 1955).

So far, only a few cores from the Indian and Pacific Oceans have been analyzed micropaleontologically. The ratio *Globigerina bulloides* to *Globorotalia menardii menardii*, believed to be inversely proportional to temperature, was found to vary markedly in six relatively short cores from the western Indian Ocean (Stubbings, 1939), but the relationship of these variations to Pleistocene ages is obscure. Equally obscure is the relationship between the relative abundances of pelagic Foraminifera in a long core from the central Pacific and Pleistocene events, although a correlation with the Atlantic cores has been attempted (Brotzen and Dinesen, 1959).

Lithological analysis of a core, 194 cm long, from the latitude of 9°S in the eastern equatorial Pacific showed the occurrence of alternating layers of *Globigerina* ooze, red clay and sediments of intermediate type (Hough, 1953). The top 24 cm consist largely of "medium dark brown lutite". Red-clay layers one to a few centimeters thick with interbedded *Globigerina* ooze occur between 24 and 44 cm below the top, between 102 and 114 cm, and from 175 cm to the bottom of the core. These intervals were interpreted as representing glacial ages; and the intervening sections, containing white to medium brown oozes, as representing interglacial ages (Hough, 1953). The correctness of this interpretation is not assured, because pure *Globigerina* ooze occurs interbedded with red clay in the supposedly glacial sections, and the postglacial section is represented by "medium dark brown lutites" which are apparently closer to red clay than to *Globigerina* ooze. The detailed chronology published by Hough (1953) is of uncertain validity. In fact the radium measurements by Urry (1949) on both the *Globigerina*-ooze and red-clay components did not yield the ^{230}Th decay curves hoped for, and allow only the tentative conclusion that the bottom of the core is some hundreds of thousands of years old (cf. Volchok and Kulp, 1957; Rosholt *et al.*, 1961). This difficulty, the obviously large amount of post-depositional solution and the inadequate stratigraphic study of this core, make it impossible to interpret the published stratigraphy in terms of Pleistocene events.

Four cores from north of the Ross Sea, ranging in length from 0.9 to 2.6 m, show the occurrence of coarse glacial-marine sediments in the top 10–20 cm, followed by fine, well sorted sediment a few decimeters thick. This is followed in turn by various layers of alternating coarse to fine glacial-marine sediments (Hough, 1950; Thomas, 1959). As suggested or implied by Thomas (1959), it seems likely that the very fine, well sorted sediments were deposited during glacial times, when sea-ice probably covered a vast area all around Antarctica. Conversely the coarse glacial-marine sediments are likely to have been deposited during interglacial times (as the occurrence of such sediments in the top portions of the cores indicates) when sea-ice was greatly reduced. Radium measurements by Urry (1949) on three of the four cores indicate either greatly varying rates of sedimentation, or radium migration, or both. Important radium migration appears to have taken place in at least two of the cores (nos. N-4 and N-5), where anomalous radium concentrations were observed at some depth in the cores (Urry, 1949, table II; cf. Pettersson, 1951; Kröll, 1954, 1955). Again, the detailed chronology published by Hough (1950) is open to question. The contention that coarse sediments represent interglacial conditions and the very fine sediments glacial conditions is supported by the observation that diatoms and Radiolaria are abundant in the former and absent from the latter (Thomas, 1959).

Altogether, the deep-sea cores from the Atlantic and adjacent basins have yielded a consistent picture of climatic change. The Pacific and Indian Ocean cores have not yielded as good a picture partly because the Pleistocene temperature oscillations may have been considerably smaller than in the Atlantic and adjacent basins, partly because most of these cores have not yet been studied by means of micropaleontological and oxygen-isotopic analysis.

c. Chronology of deep-sea cores and correlation with the continental Pleistocene stratigraphy

^{14}C, ^{230}Th and the ratio $^{231}Pa/^{230}Th$ have been used for absolute dating of deep-sea cores. ^{14}C measurements have been made on the foraminiferal component alone, on bulk core material and on the fine carbonate fraction, using modern shells for the contemporary assay. Measurements on the finer carbonate fraction gave ages up to 10% greater than the ages obtained from the associated foraminiferal component (Rubin and Suess, 1955; Ericson *et al.*, 1956), indicating the occasional presence of some reworked material. Generally, however, the amount of reworked material in the fine carbonate fraction does not seem to be large enough to affect importantly the ages obtained from bulk core material. These, and the ages obtained from the foraminiferal component alone have yielded a consistent chronology for various deep-sea cores, ranging back in time to approximately 40,000 years B.P. (Rubin and Suess, 1955, 1956; Ericson *et al.*, 1956; Broecker and Kulp, 1957; Broecker *et al.*, 1958).

The ^{230}Th and $^{231}Pa/^{230}Th$ methods are based on the fact that thorium and protactinium isotopes produced by ^{238}U and ^{235}U existing in solution in sea-water are rapidly removed by adsorption on particles and deposited on the bottom. The half-lives of ^{230}Th, ^{231}Pa, and of the ratio $^{231}Pa/^{230}Th$ are, respectively, 80,000, 34,300 and 60,100 years. If the concentration of uranium in the ocean remained constant with time, and if the rate of sedimentation of the clay component also remained constant, the activity of ^{230}Th alone would provide a reliable dating method. Urry (Piggot and Urry, 1942, 1942a; Urry, 1949) dated various deep-sea cores from the North Atlantic, the Caribbean and the

Pacific using radium as an index of ^{230}Th. Although radium was later shown to migrate rather freely through deep-sea sediments (Pettersson, 1951; Kröll, 1954, 1955), radium migration is relatively unimportant in sediments with a gross rate of sedimentation greater than about 1 cm/1000 years (Koczy and Bourret, 1958), which is generally the case for *Globigerina* ooze. If the stratigraphy of the Atlantic and Caribbean cores dated by Urry (Bramlette and Bradley, 1940; Cushman and Henbest, 1940; Cushman, 1941) is interpreted in the light of present knowledge, it appears that the last major warm stage was dated from about 110,000 to about 65,000 years B.P., the last major cold stage from 65,000 to 12,000 years ago, and the postglacial from about 12,000 years ago. These figures have been substantiated by recent dating by ^{14}C and ^{231}Pa/^{230}Th methods (cf. Rosholt *et al.*, 1961, 1962). It is unfortunate that stages older than the last major warm stage cannot be clearly recognized in the Cayman core dated by Urry (Piggott and Urry, 1942; Cushman, 1941), a result, probably, of much post-depositional solution of the foraminiferal shells having occurred at the depth from which the core was raised (4800 m).

The difficulty associated with possible changes in rate of supply of uranium to the sea and changes in the rate of clay sedimentation is eliminated by using the ^{231}Pa/^{230}Th method, because such changes would affect equally both parent uranium isotopes. Using this method, Rosholt *et al.* (1961, 1962) dated temperature fluctuations back to about 175,000 years ago in three apparently undisturbed Caribbean cores. The ages obtained from the younger portions of these cores correlate with the ^{14}C chronology within the limits of error of the two methods. ^{231}Pa/^{230}Th dating of a North Atlantic core (Swedish core 280) gave ages that are about 30,000 years greater than the ages obtained from the corresponding levels of the Caribbean cores, and do not correlate with the ^{14}C chronology in the region of common range. This discrepancy is believed to result from the introduction of reworked clay into the Atlantic core (Rosholt *et al.*, 1961).

If the ages of the temperature stages of the deep-sea cores obtained with the various methods mentioned above are compared with the ^{14}C chronology of the continental stratigraphy (Flint, 1957, p. 395; Andersen *et al.*, 1960), it appears that stage 2 (see Fig. 6) represents the Main Würm; stage 3 the Early Würm–Main Würm interval; stage 4 the Early Würm; and stage 5 the last interglacial (cf. Emiliani, 1955, 1958; van der Hammen, 1957). Thus the correlation between core stages 1 to 5 and continental stages from the present to the preceding major interglacial seems to be good. The tentative correlation between older core stages and continental stages suggested by Emiliani (1955) is still considered speculative, as it was when first proposed.

Oxygen-isotopic analysis of calcareous benthonic Foraminifera from the section of core 234 representing the last interglacial showed that bottom temperature followed closely the pattern of surface temperature, but with considerably smaller amplitude (Emiliani, 1958). These and other analyses of calcareous benthonic Foraminifera (Emiliani, 1955, 1958) showed that the maximum glacial–interglacial temperature range of the bottom water was on the average 3°C in the equatorial and temperate Atlantic, and negligible in the equatorial Pacific.

Oxygen-isotopic analyses performed at close stratigraphic intervals on short pilot cores from the Atlantic and the Caribbean generally failed to reveal temperature oscillations related to the units of the last glaciation, with the exception of a minor temperature reversal at 10–15 cm below the tops of the cores possibly related to the Valders unit (Emiliani, 1955, figs. 7–10; Emiliani and Mayeda, 1961). This failure is believed to result from sediment reworking by bottom organisms, affecting a stratigraphic thickness of a few to several centimeters. The implication is that temperature events spaced in time less than a few thousand years from each other cannot be resolved by deep-sea sediments of normal *Globigerina* ooze facies.

The transition between the last glacial age and the postglacial is strikingly clear in the Arctic (Ericson and Wollin, 1959), in the Norwegian Sea (Holtedahl, 1959), and in the northern North Atlantic (Bramlette and Bradley, 1940), where glacial-marine sediments deposited during the last glacial age and containing some or much ice-rafted material pass into postglacial lutites with abundant pelagic Foraminifera. It is also clear at lower latitudes. In the Gulf of Mexico, for instance, a marked sedimentary change occurred 11,000 years ago (Ewing *et al.*, 1958), which implies reduction, by at least one third, in the capacity and competence of the Mississippi River (Emiliani, 1957). The abundance of pelagic Foraminifera in the postglacial sections of cores from the Arctic Ocean and the Norwegian Sea, mentioned above, apparently resulted from an increase in productivity, following reduction or disappearance of sea-ice and/or marked decrease of the terrigenous component. Broecker *et al.* (1960) have marshalled evidence suggesting that most of the temperature increase marking the transition between glacial and postglacial conditions occurred over a period of only about 1500 years, close to 11,000 years ago. This sudden temperature rise might have been caused by decrease of albedo following rapid reduction or removal of sea-ice over the northern North Atlantic, Norwegian Sea and other high-latitude areas (Emiliani and Geiss, 1959).

The evidence offered by deep-sea sediments bears directly on the elaboration of theories of glaciation. No theory of glaciation can be valid unless the facts revealed by the study of deep-sea sediments are also taken into consideration and explained. These facts have already made highly improbable some of the glaciation theories proposed in the past. Thus, Simpson's theory (Simpson, 1934, 1940, 1957) that surface temperatures at low latitudes during glacial ages were alternately lower and higher than during the intervening interglacials is incompatible with the observed or inferred synchronism between glacial ages and the temperature minima of the deep-sea cores from the Atlantic, the Caribbean and the Mediterranean. The groups of theories based on volcanism (cf. Flint, 1957, pp. 504–505, and references therein) and on submarine tectonism along the Greenland–Iceland–Scotland axis (Saks *et al.*, 1955) are perhaps incompatible with the apparent regularity of the temperature variations suggested by the deep-sea cores. Theories based on variations of solar emission are made questionable by Öpik's conclusion, based on elaborate, quantitative analysis (Öpik, 1953), that the reduction of solar emission required to produce the observed ice sheets of the glacial ages would have reduced the equatorial temperatures to 8°C. Theories based on variations of CO_2 content in the atmosphere (Plass, 1956; cf. also Flint, 1957, p. 504) are made improbable by the rapid exchange of CO_2 between atmosphere and ocean, including the deep sea (Craig, 1957; Revelle and Suess, 1957). A recent theory claiming that the Arctic Ocean was open during glacial ages and that it provided the moisture for the northern ice caps (Ewing and Donn, 1956, 1958) has been questioned on theoretical grounds as incapable of maintaining sustained oscillations (Livingstone, 1959) and is apparently contradicted by the pattern of glaciation in Alaska and Siberia (cf. Flint, 1957) and by the stratigraphy of Arctic deep-sea cores (Ericson and Wollin, 1959).

According to a model of glaciation developed by Emiliani and Geiss (1959) and largely based on ideas of earlier authors (esp. Milankovitch, 1938; Köppen and Wegener, 1924; Flint, 1947), glaciation was made possible by Tertiary cooling caused by albedo increase associated with the Alpine orogenesis; the successive Pleistocene glaciations were started by quasi-periodic reductions of summer insolation in northern latitudes (the so-called Milankovitch mechanism, see Zeuner, 1945, 1952, 1959) known to have occurred at intervals of about 40,000 years; the further development of glaciation was largely a self-sustaining process; and complete deglaciations outside of Antarctica and Greenland resulted from surface freezing of the northern North Atlantic and time-delay effects introduced by plastic flow of glaciers, heat absorption by ice melting and crustal warping.

Most of the very numerous theories of glaciation which have been proposed to date have been disproved or made questionable by subsequent discoveries. Of the very few which survive, none exceed the status of theory. The cause of glaciation and the mechanism regulating the repeated advances and retreats of glaciers during the Pleistocene Epoch undoubtedly will become clear in the future, but only through an intensive study of both continental and deep-sea deposits directed to establish standard sections, and only with time calibration by absolute dating methods.

4. Summary

In the foregoing discussion we have summarized what we believe are some of the basic facts in the record of the Pleistocene, gathered from continental and sea-floor sources. The facts concern three chief parameters: (1) stratigraphic sequence, (2) paleotemperatures, and (3) chronology. The sequence displayed beneath both lands and oceans clearly shows both major and minor fluctuations of temperature. The fluctuation is secular, but its amplitude varies with geographic location. Repeated changes of temperature characterize the Pleistocene but have not been discerned in the Tertiary, which is marked by slow, general decrease in temperature.

Radiocarbon dating and the discovery and development of some other methods for absolute dating of Pleistocene materials, invention of the piston core-sampler and the consequent intensive study of relatively long sections of deep-sea sediments, refinements in the techniques for the study of continental deposits (expecially loess and soil profiles), and the pollen-analytical work done on some particularly long sections of lacustrine deposits have greatly advanced Pleistocene science in the decade 1950–1960. Yet, many major advances remain to be made before a clear picture of the Pleistocene Epoch as a whole will emerge. On the one hand, the glacial and periglacial deposits on the continents show four or five, and perhaps more, major glacial advances, each one of which appears to have been double or multiple. On the other hand, the deep-sea cores provide us with a picture of repeated temperature cycles totalling perhaps ten or fifteen during the past half-million years. Correlation between temperature cycles of the deep-sea cores and glacial advances and retreats on the northern continents seems established, at least tentatively, only for the past 100,000 years, i.e. back into the last interglacial. Further, important advances in our knowledge of the Pleistocene Epoch will probably require the collection and study of deep-sea cores representing continuous stratigraphic sections much longer than the ones presently available, together with absolute dating of both deep-sea and continental materials formed in the long Pleistocene time preceding the last interglacial and significantly related to the climatic changes. In addition, much work needs to be done on the Pleistocene of the Southern Hemisphere, both on land and beneath the sea, as well as on continental deposits in equatorial and tropical regions. Techniques for these studies have been already developed or are being developed rapidly, so that a clear, world-wide picture of the history of the Pleistocene may be achieved in the not-too-distant future. The picture, as discussed in the introduction, will be of paramount importance in assessing the rates at which many geophysical, geochemical, geological and biological processes operate.

References

Andersen, Sv. Th., Hl. de Vries and W. H. Zagwijn, 1960. Climatic change and radiocarbon dating in the Weichselian glacial of Denmark and the Netherlands. *Geol. Mijnb.*, **39**, 38–42.

Arrhenius, G., 1952. Sediment cores from the east Pacific. *Reps. Swed. Deep-Sea Exped., 1947–1948*, **5**, No. 1.

Bandy, O. L., 1956. Ecology of Foraminifera in northeastern Gulf of Mexico. *U.S. Geol. Surv. Prof. Paper*, **274**, 179–204.

Bandy, O. L. and R. E. Arnal, 1957. Distribution of Recent Foraminifera off west coast of Central America. *Bull. Amer. Assoc. Petrol. Geol.*, **41**, 2037–2053.

Bé, A. W. H., 1960. Ecology of Recent planktonic Foraminifera: Part 2—Bathymetric and seasonal distribution in the Sargasso Sea off Bermuda. *Micropaleontology*, **6**, 373–392.

Blanc, A. C., 1937. Low levels of the Mediterranean sea during the Pleistocene glaciation. *Q. J. Geol. Soc. London*, **93**, 621–651.

Bradshaw, J. S., 1959. Ecology of living planktonic Foraminifera in north and equatorial Pacific Ocean. *Contrib. Cushman Found. Foram. Res.*, **10**, 25–64.

Bramlette, M. N. and W. H. Bradley, 1940. Geology and biology of North Atlantic deep-sea cores between Newfoundland and Ireland. Part I, Lithology and geologic interpretations. *U.S. Geol. Survey Prof. Paper*, **196-A,** 1–34.

Broecker, W. S., 1961. Radiocarbon dating of Late Quaternary deposits, south Louisiana: a discussion. *Bull. Geol. Soc. Amer.*, **72**, 159–162.

Broecker, W. S., M. Ewing and B. C. Heezen, 1960. Evidence of an abrupt change in climate close to 11,000 years ago. *Amer. J. Sci.*, **258**, 429–443.

Broecker, W. S. and J. L. Kulp, 1957. Lamont natural radiocarbon measurements IV. *Science*, **126**, 1324–1334.

Broecker, W. S., J. L. Kulp and C. S. Tucek, 1956. Lamont natural radiocarbon measurements III. *Science*, **124**, 154–165.

Broecker, W. S., K. K. Turekian and B. C. Heezen, 1958. The relation of deep-sea sedimentation rates to variations in climate. *Amer. J. Sci.*, **256**, 503–517.

Brotzen, F. and A. Dinesen, 1959. On the stratigraphy of some bottom sections from the central Pacific. *Reps. Swed. Deep-Sea Exped., 1947–1948*, **10**, no. 3, 43–55.

Clisby, K. H., F. Foreman and P. B. Sears, 1957. Pleistocene climatic changes in New Mexico, U.S.A. *Veröff. Geobot. Inst. Rübel*, **34**, 21–26.

Clisby, K. H. and P. B. Sears, 1956. San Augustin Plains—Pleistocene climatic changes. *Science*, **124**, 537–539.

Craig, H., 1957. The natural distribution of radiocarbon and the exchange time of carbon dioxide between atmosphere and sea. *Tellus*, **9**, 1–17.

Cushman, J. A., 1941. A study of the Foraminifera contained in cores from the Bartlett Deep. *Amer. J. Sci.*, **239**, 128–147.

Cushman, J. A. and L. G. Henbest, 1940. Geology and biology of North Atlantic deep-sea cores between Newfoundland and Ireland. Part 2, Foraminifera. *U.S. Geol. Surv. Prof. Paper*, **196-A**, 35–50.

de Vries, Hl., 1958. Radiocarbon dates for upper Eem and Würm-interstitial samples. *Eiszeitalter u. Gegenwart*, **9**, 10–17.

de Vries, Hl., 1959. Measurement and use of natural radiocarbon. In Ph. H. Abelson (Ed.), *Researches in Geochemistry*. John Wiley and Sons, New York, 169–189.

de Vries, Hl. and A. Dreimanis, 1960. Finite radiocarbon dates of the Port Talbot interstadial deposits in southern Ontario. *Science*, **131**, 1738–1739.

Durham, J. W., 1950. Cenozoic marine climates of the Pacific coast. *Bull. Geol. Soc. Amer.*, **61**, 1243–1264.

Eardley, A. J. and V. Gvosdetsky, 1960. Analysis of Pleistocene core from Great Salt Lake, Utah. *Bull. Geol. Soc. Amer.*, **71**, 1323–1344.

Eberl, B., 1930. *Die Eiszeitenfolge im nördlichen Alpenvorlande*. Benno Filser, Augsburg.

Emiliani, C., 1954. Depth habitats of some species of pelagic Foraminifera as indicated by oxygen isotope ratios. *Amer. J. Sci.*, **252**, 149–158.

Emiliani, C., 1954a. Temperatures of Pacific bottom waters and polar superficial waters during the Tertiary. *Science*, **119**, 853–855.

Emiliani, C., 1955. Pleistocene temperatures. *J. Geol.*, **63**, 538–578.

Emiliani, C., 1955a. Pleistocene temperature variations in the Mediterranean. *Quaternaria*, **2**, 87–98.

Emiliani, C., 1955b. Mineralogical and chemical composition of the tests of certain pelagic Foraminifera. *Micropaleontology*, **1**, 377–380.

Emiliani, C., 1956. Oligocene and Miocene temperatures of the equatorial and sub-tropical Atlantic Ocean. *J. Geol.*, **64**, 281–288.

Emiliani, C., 1957. Temperature and age analysis of deep-sea cores. *Science*, **125**, 383–387.

Emiliani, C., 1958. Paleotemperature analysis of core 280 and Pleistocene correlations. *J. Geol.*, **66**, 264–275.

Emiliani, C. and S. Epstein, 1953. Temperature variations in the Lower Pleistocene of southern California. *J. Geol.*, **61**, 171–181.

Emiliani, C. and J. Geiss, 1959. On glaciations and their causes. *Geol. Rundschau*, **46** (1957), 576–601.

Emiliani, C. and T. Mayeda, 1961. Carbonate and oxygen isotopic analysis of core 241A. *J. Geol.*, **69**, 729–732.

Emiliani, C., T. Mayeda and R. Selli, 1961. Temperature analysis of the Plio-Pleistocene section at Le Castella, Calabria, southern Italy. *Bull. Geol. Soc. Amer.*, **72**, 679–688.

Epstein, S., R. Buchsbaum, H. Lowenstam and H. C. Urey, 1951. Carbonate–water isotopic temperature scale. *Bull. Geol. Soc. Amer.*, **62**, 417–425.

Epstein, S., R. Buchsbaum, H. Lowenstam and H. C. Urey, 1953. Revised carbonate–water isotopic temperature scale. *Bull. Geol. Soc. Amer.*, **64**, 1315–1325.

Ericson, D. B., 1953. Sediments of the Atlantic Ocean. Columbia Univ., Lamont Geol. Observatory, Tech. Rep. on Submarine Geol., No. 1.

Ericson, D. B., W. S. Broecker, J. L. Kulp and G. Wollin, 1956. Late-Pleistocene climates and deep-sea sediments. *Science*, **124**, 385–389.

Ericson, D. B., M. Ewing and B. C. Heezen, 1952. Turbidity currents and sediments in North Atlantic. *Amer. Assoc. Petrol. Geol.*, **36**, 489–511.

Ericson, D. B., M. Ewing, B. C. Heezen and G. Wollin, 1955. Sediment deposition in deep Atlantic. *Bull. Geol. Soc. Amer. Spec. Paper* 62, 205–220.

Ericson, D. B., M. Ewing, G. Wollin and B. C. Heezen, 1961. Atlantic deep-sea sediment cores. *Bull. Geol. Soc. Amer.*, **72**, 193–286.

Ericson, D. B. and G. Wollin, 1956. Correlation of six cores from the equatorial Atlantic and the Caribbean. *Deep-Sea Res.*, **3**, 104–125.

Ericson, D. B. and G. Wollin, 1956a. Micropaleontological and isotopic determinations of Pleistocene climates. *Micropalentology*, **2**, 257–270.

Ericson, D. B. and G. Wollin, 1959. Micropaleontology and lithology of Arctic sediment cores. In Vol. I of Bushnell, V. C. (Ed.), *Scientific Studies at Fletcher's Ice Island, 1952–1955*, 50–58. U.S. Air Force Cambridge Research Center, Geophysics Research Paper No. 63.

Evernden, J. F., 1959. (Discussion.) *Proc. Geol. Soc. London*, Sess. 1958–59, 17–18.

Evernden, J. F., G. H. Curtis and R. Kistler, 1957. Potassium-argon dating of Pleistocene volcanics. *Quaternaria*, **4**, 13–17.

Ewing, M. and W. L. Donn, 1956. A theory of ice ages. *Science*, **123**, 1061–1066.

Ewing, M. and W. L. Donn, 1958. A theory of ice ages II. *Science*, **127**, 1159–1162.

Ewing, M., D. B. Ericson and B. C. Heezen, 1958. Sediments and topography of the Gulf of Mexico. In Weeks, L. (Ed.), *Habitat of Oil*, 995–1053. Amer. Assoc. Petrol. Geol., Tulsa.

Fairbridge, R. W., 1958. Dating the latest movements of the Quaternary sea level. *Trans. N.Y. Acad. Sci.*, ser. 2, **20**, 471–482.

Fisk, H. N. and E. McFarlan, Jr., 1955. Late Quaternary deltaic deposits of the Mississippi River. *Geol. Soc. Amer. Spec. Paper* 62, 279–302.

Flint, R. F., 1947. *Glacial Geology and the Pleistocene Epoch*. John Wiley and Sons, New York, 589 pp.

Flint, R. F., 1957. *Glacial and Pleistocene Geology*. John Wiley and Sons, New York, 553 pp.

Flint, R. F., 1963. Status of the Pleistocene Wisconsin Stage in central North America. *Science*, **139**, 402–404.

Flint, R. F. and F. J. Brandtner, 1961. Climatic changes since the last interglacial. *Amer. J. Sci.*, **259**, 321–328.

Flint, R. F. and W. A. Gale, 1958. Stratigraphy and radiocarbon dates at Searles Lake, California. *Amer. J. Sci.*, **256**, 689–714.

Flint, R. F. and M. Rubin, 1955. Radiocarbon dates of pre-Mankato events in eastern and central North America. *Science*, **121**, 649–658.

Gentner, W. and J. Zähringer, 1960. The potassium–argon ages of tektites. *J. Geophys. Res.*, **65**, 2492 (abstract).

Gignoux, M., 1913. Les formations marines pliocènes et quaternaires de l'Italie du sud et de la Sicile. *Ann. Univ. Lyon*, n.s., **1**, no. 36.

Godwin, H., R. P. Suggate and E. H. Willis, 1958. Radiocarbon dating of the eustatic rise in ocean-level. *Nature*, **181**, 1518–1519.

Graul, H., 1959. Der Verlauf des glazialeustatischen Meeresspiegelanstieges, berechnet an Hand von C-14 Datierungen. *Deut. Geog. Berlin, 20 bis 25 Mai 1959, Tagungsber. u. wiss. Abhandl.*, 232–242.

Haring, A., A. E. de Vries and Hl. de Vries, 1958. Radiocarbon dating up to 70,000 years by isotopic enrichment. *Science*, **128**, 472–473.

Holtedahl, H., 1959. Geology and paleontology of Norwegian Sea bottom cores. *J. Sediment. Petrol.*, **29**, 16–29.

Hopkins, D. M., 1959. Cenozoic history of the Bering land bridge. *Science*, **129**, 1519–1528.

Hough, J. L. 1950. Pleistocene lithology of Antarctic ocean-bottom sediments. *J. Geol.*, **58**, 254–260.

Hough, J. L., 1953. Pleistocene climatic record in a Pacific ocean core sample. *J. Geol.*, **61**, 252–262.

International Geological Congress, 18th Sess., Great Britain, 1948. Recommendations of Commission appointed to advise on the definition of Plio-Pleistocene boundary. *Rep.*, Part IX, 1950, 6.

Jelgersma, S. and A. J. Pannekoek, 1960. Post-glacial rise of sea-level in the Netherlands (a preliminary report). *Geol. Mijnb.*, **39**, 201–207.

Kay, G. F., 1931. Classification and duration of the Pleistocene period. *Bull. Geol. Soc. Amer.*, **48**, 425–466.

Koczy, F. F. and R. Bourret, 1958. Diffusion of radioactive nuclides in a moving and turbulent medium. Marine Laboratory, University of Miami, Progress Rep. NSF Grant G–3995, 18 pp.

Koenigswald, G. H. R. von, 1960. Tektite studies. *Ned. Akad. Wetenschap., Afd. Wissen Natuurk. Wetensch.*, **63B**, 135–153.

Köppen, W. and A. Wegener, 1924. *Die Klimate der geologischen Vorzeit*. Berlin.

Kolbe, R. W., 1954. Diatoms from equatorial Pacific cores. *Reps. Swed. Deep-Sea Exped., 1947–1948*, **6**, 1–49.

Kolbe, R. W., 1955. Diatoms from equatorial Atlantic cores. *Reps. Swed. Deep-Sea Exped., 1947–1948*, **2**, 149–184.

Kolbe, R. W., 1957. Diatoms from equatorial Indian ocean cores. *Reps. Swed. Deep-Sea Exped., 1947–1948*, **9**, 1–50.

Kolbe, R. W., 1957a. Fresh-water diatoms from Atlantic deep-sea sediments. *Science*, **126**, 1053–1056.

Kröll, V. S., 1954. On the age-determination in deep-sea sediments by radium measurements. *Deep-Sea Res.*, **1**, 211–215.

Kröll, V. S., 1955. The distribution of radium in deep-sea cores. *Reps. Swed. Deep-Sea Exped., 1947–1948*, **10**, *Spec. Investigations*, No. 1, 32 pp.

Kullenberg, B., 1947. The piston core sampler. *Svenska Hydrog.-biol. Komm. Skr.* (ser. 3, Hydrog.), **1**, no. 2, 1–46.

Livingstone, D. A., 1959. Theory of ice ages. *Science*, **129**, 463–465.

Lona, F., 1950. Contributi alla storia della vegetazione e del clima nella Val Padana.—Analisi pollinica del giacimento Villafranchiano di Leffe (Bergamo). *Atti Soc. Ital. Sci. Nat.*, **89**, 123–179.

Lona, F. and M. Follieri, 1957. Successione pollinica della serie superiore (Günz-Mindel) di Leffe (Bergamo). *Verh. 4th Intern. Tagung der Quartärbotaniker 1957, Veröff. Geobot. Institut Rübel, Zürich*, **34**, 86–98.

McFarlan, E., Jr., 1961. Radiocarbon dating of Late Quaternary deposits, south Louisiana. *Bull. Geol. Soc. Amer.*, **72**, 129–158.

Meinzer, O. E., 1922. Map of Pleistocene lakes of the Basin-and-Range Province and its significance. *Bull. Geol. Soc. Amer.*, **33**, 541–552.

Milankovitch, M., 1938. Mathematische Klimalehre und astronomische Theorie der Klimaschwankungen. *Handb. Geophys.*, **9**, 593–698.

Olausson, E., 1960. Description of sediment cores from the North Atlantic. *Reps. Swed. Deep-Sea Exped., 1947–1948*, **7**, 227–286.

Olausson, E., 1960a. Description of sediment cores from the central and western Pacific with the adjacent Indonesian region. *Reps. Swed. Deep-Sea Exped., 1947–1948*, **6**, fasc. 5, 161–214.

Olausson, E., 1961. Remarks on some Cenozoic core sequences from the central Pacific, with a discussion of the role of Coccolithophorids and Foraminifera in carbonate deposits. *Göteborgs, K. Vetensk.-Vit. Samh. Handl.*, Sjätte Följden, ser. B, **8** (10), 35 pp.

Öpik, E. J., 1953. A climatological and astronomical interpretation of the ice ages and of the past variations of terrestrial climate. *Contrib. Armagh Observatory*, **9**.

Pannekoek, A. J., 1954. Tertiary and Quaternary subsidence in the Netherlands. *Geol. Mijnb.*, n.s., **16**, 156–164.

Pannekoek, A. J., 1956. Notice préliminaire sur la structure du bassin sédimentaire Pleistocène des Pays-Bas. Cong. Intern. du Quaternaire, 4th, Rome–Pisa, 1953, Actes, II, 956–960.

Pannekoek, A. J. and A. J. van Voorthuysen, 1950. Some remarks on the marine Lower Pleistocene of the Netherlands. Intern. Geol. Cong., 18th Sess., Great Britain, 1948, *Rep.*, Part IX, 74–77.

Parker, F. L., 1958. Eastern Mediterranean Foraminifera. *Reps. Swed. Deep-Sea Exped., 1947–1948*, **8**, 217–283.

Penck, A. and E. Brückner, 1909. *Die Alpen im Eiszeitalter*. Tauchnitz, Leipzig.

Perconig, E., 1956. Il Quaternario nella Pianura Padana. Cong. Intern. du Quaternaire, 4th, Rome–Pisa, 1953, Actes, II, 481–524.

Pettersson, H., 1951. Radium and deep-sea chronology. *Nature*, **167**, 942.

Pfannenstiel, M., 1952. Das Quartär der Levante. Pt. 1, Die Küste Palästina-Syriens. *Abhandl. Akad. Wiss. Lit. Mainz, Math.-nat. Kl.*, **7**, 373–465.

Philippi, E., 1910. Die Grundproben der deutschen Südpolar-Expedition, 1901–03. II, Geographie und Geologie, **6**, 411–616.

Phleger, F. B., Jr., 1939. Foraminifera of submarine cores from the continental slope. *Bull. Geol. Soc. Amer.*, **50**, 1395 1422.

Phleger, F. B., 1942. Foraminifera of submarine cores from the continental slope, Part 2. *Bull. Geol. Soc. Amer.*, **53**, 1073–1098.

Phleger, F. B., 1947. Foraminifera of three submarine cores from the Tyrrhenian Sea. *Göteborgs K. Vetensk. Vit. Samh. Handl.*, Sjätte Följden, ser. B, **5**, no. 5.

Phleger, F. B., 1948. Foraminifera of a submarine core from the Caribbean Sea. *Göteborgs K. Vetensk. Vit. Samh. Handl.*, Sjätte Följden, ser. B, **5**, no. 14.

Phleger, F. B., 1951. Ecology of Foraminifera, northwest Gulf of Mexico. Part 1, Foraminifera distribution. *Mem. Geol. Soc. Amer.*, **46**, 1–88.

Phleger, F. B., 1955. Foraminiferal faunas in cores offshore from the Mississippi delta. *Deep-Sea Res.*, Suppl. to **3** (Papers in Marine Biology and Oceanography), 45–57.

Phleger, F. B., 1960. *Ecology and Distribution of Recent Foraminifera*. The Johns Hopkins Press, Baltimore, 297 pp.

Phleger, F. B. and W. A. Hamilton, 1946. Foraminifera of two submarine cores from the North Atlantic basin. *Bull. Geol. Soc. Amer.*, **57**, 951–966.

Phleger, F. B., F. L. Parker and J. F. Peirson, 1953. North Atlantic Foraminifera. *Reps. Swed. Deep-Sea Exped., 1947–1948*, **7**, No. 1.

Piggot, C. S. and W. D. Urry, 1942. Radioactivity of ocean sediments. IV, The radium content of sediments of the Cayman Trough. *Amer. J. Sci.*, **240**, 1–12.

Piggot, C. S. and W. D. Urry, 1942a. Time relations in ocean sediments. *Bull. Geol. Soc. Amer.*, **53**, 1187–1210.

Plass, G. N., 1956. The carbon dioxide theory of climatic change. *Tellus*, **8**, 140–154.

Polski, W., 1959. Foraminiferal biofacies off the north Asiatic coast. *J. Paleont.*, **33**, 569–587.

Poser, H., 1948. Boden- und Klimaverhältnisse in Mittel- und Westeuropa während der Würmeiszeit. *Erdkunde*, **2**, 53–68.

Revelle, R., M. Bramlette, G. Arrhenius and E. D. Goldberg, 1955. Pelagic sediments of the Pacific. *Geol. Soc. Amer. Spec. Paper* 62, 221–235.

Revelle, R. and H. E. Suess, 1957. Carbon dioxide exchange between atmosphere and ocean, and the question of an increase of atmospheric CO_2 during the past decades. *Tellus*, **9**, 18–27.

Rex, R. W. and E. D. Goldberg, 1958. Quartz content of pelagic sediments of the Pacific Ocean. *Tellus*, **10**, 153–159.

Reynolds, J. H., 1960. Rare gases in tektites. *Geochim. et Cosmochim. Acta*, **20**, 101–114.

Riedel, W. R., 1957. Radiolaria: a preliminary stratigraphy. *Reps. Swed. Deep-Sea Exped., 1947–1948*, **6**, 59–96.

Riedel, W. R., 1959. Oligocene and Lower Miocene Radiolaria in tropical Pacific sediments. *Micropaleontology*, **5**, 285–302.

Riedel, W. R. and M. N. Bramlette, 1959. Tertiary sediments in the Pacific Ocean basin. *Intern. Oceanog. Cong., New York, 1959, Preprints*, 105–106.

Rosholt, J. N., C. Emiliani, J. Geiss, F. F. Koczy and P. J. Wangersky, 1961. Absolute dating of deep-sea cores by the Pa^{231}/Th^{230} method. *J. Geol.*, **69**, 162–185.

Rosholt, J. N., C. Emiliani, J. Geiss, F. F. Koczy and P. J. Wangersky, 1962. Pa^{231}/Th^{230} dating and O^{18}/O^{16} temperature analysis of core A254–BR–C. *J. Geophys. Res.*, **67**, 2907–2911.

Rubin, M., 1956. U.S. Geological Survey radiocarbon dates III. *Science*, **123**, 442–448.

Rubin, M. and H. E. Suess, 1955. U.S. Geological Survey radiocarbon dates II. *Science*, **121**, 481–488.

Ruggieri, G. and R. Selli, 1950. Il Pliocene e il Postpliocene dell'Emilia. Intern. Geol. Cong., 18th Sess., Great Britain, 1948. *Rep.*, Part IX, 85–93.

Saks, V. N., N. A. Belov and N. N. Lapina, 1955. Our present concepts of the geology of the central Arctic. *Priroda*, **7**, 13–22.

Savage, D. E., G. H. Curtis and J. F. Evernden, 1960. Integration of land-mammal chronology and potassium-argon geochronometry (Abstract). *Bull. Geol. Soc. Amer.*, **71**, 1966.

Schofield, J. C., 1960. Sea-level fluctuations during the last four thousand years. *Nature*, **185**, 836.

Schott, W., 1935. Die Foraminiferen in dem aequatorialen Teil des Atlantischen Ozeans. *Wiss. Ergebn. Deut. Atlant. Exped. 'Meteor', 1925–1927*, **3**, no. 3, 43–134.

Schott, W., 1952. On the sequence of deposits in the equatorial Atlantic Ocean. *Göteborgs K. Vetensk. Vit. Samh. Handl.*, Sjätte Följden, ser. B, **6**, no. 2.

Schott, W., 1954. Über stratigraphische Untersuchungsmethoden in rezenten Tiefsee-sedimenten. *Heidelberg Beitr. z. Mineralog. u. Petrog.*, **4**, 192–197.

Sears, P. B. *et al.*, 1955. Palynology in southern North America. *Bull. Geol. Soc. Amer.*, **66**, 471–474.

Selli, R., 1949. Le conoscenze geologiche sul Quaternario gassifero del Polesine e del Ferrarese settentrionale. *Atti Convegno Naz. Metano, 6th.*

Shepard, F. P. and H. E. Suess, 1956. Rate of postglacial rise of sea level. *Science*, **123**, 1082–1083.

Simpson, G. C., 1934. World climate during the Quaternary period. *Q.J. Roy. Met. Soc.*, **59**, 425–471.

Simpson, G. C., 1940. Possible causes of change in climate and their limitations. *Proc. Linnean Soc. London*, Sess. 152, 1939–1940, Part II, 189–219.

Simpson, G. C., 1957. Further studies in world climate. *Q.J. Roy. Met. Soc.*, **83**, 459–481.

Stubbings, H. G., 1939. Stratification of biological remains in marine deposits. *John Murray Exped., 1933–34, Sci. Reps.*, **3**, 159–192.

Szafer, W., 1954. Pliocene flora from the vicinity of Czorsztyn (West Carpathians) and its relationship to the Pleistocene. *Prace Inst. Geol.*, **11**, 238 pp.

ten Dam, A. and Th. Rheinhold, 1942. Die stratigraphische Gliederung des niederländischen Plio-Pleistozäns nach Foraminiferen. *Med. Geol. Stich.*, Ser. C-V, **1**.

Thomas, C. W., 1959. Lithology and zoology of an Antarctic Ocean bottom core. *Deep-Sea Res.*, **6**, 5–15.

Todd, R., 1958. Foraminifera from western Mediterranean deep-sea cores. *Reps. Swed. Deep-Sea Exped., 1947–1948*, **8**, 167–215.

Tongiorgi, E., 1936. Le variazioni climatiche testimoniate dallo studio paleobotanico della serie flandriana della bassa Versilia presso il lago di Massaciuccoli. *Nuovo Giorn. Bot. Ital.*, **43**, 762–764.

Urey, H. C., 1947. The thermodynamic properties of isotopic substances. *J. Chem. Soc.*, 562–581.

Urey, H. C., H. A. Lowenstam, S. Epstein and C. R. McKinney, 1951. Measurement of paleotemperatures and temperatures of the Upper Cretaceous of England, Denmark and the southeastern United States. *Bull. Geol. Soc. Amer.*, **62**, 399–416.

Urry, W. D., 1949. Radioactivity of ocean sediments. VI. Concentrations of radio-elements in marine sediments of the southern hemisphere. *Amer. J. Sci.*, **247**, 257–275.

van der Hammen, Th., 1957. A new interpretation of the pleniglacial stratigraphical sequence in middle and western Europe. *Geol. Mijnb.*, **19**, 493–498.

van der Hammen, Th. and E. Gonzalez, 1960. Upper Pleistocene and Holocene climate and vegetation of the "Sabana de Bogotá" (Colombia, South America). *Leidse Geolog. Med.*, **25**, 261–315.

van der Vlerk, I. M., 1959. Problems and principles of Tertiary and Quaternary stratigraphy. *Q.J. Geol. Soc. London*, **115**, 49–63.

van Voorthuysen, J. H., 1950. The quantitative distribution of the Plio-Pleistocene foraminifera of a boring at The Hague (Netherlands). *Med. Geol. Stich.*, n.s., **4**, 31–49.

van Voorthuysen, J. H., 1953. Pliocene and lower Pleistocene in a boring near Oosterhout. Foraminifera. *Med. Geol. Stich.*, n.s., **7**, 38–40.

Venzo, S., 1955. Le attuali conoscenze sul Pleistocene Lombardo con particolare riguardo al Bergamasco. *Atti Soc. Ital. Sci. Nat.*, **94**, no. 2, 155–200.

Volchok, H. L. and J. L. Kulp, 1957. The ionium method of age determination. *Geochim. et Cosmochim. Acta*, **11**, 219–246.

Wangersky, P. J., 1958. Mechanisms of marine sedimentation. Yale Univ., Ph.D. Thesis.

Wangersky, P. J., 1959. Mechanisms of marine sedimentation. *Intern. Oceanog. Cong., New York, 1959, Preprints*, 985–986.

Wangersky, P. J., 1962. Sedimentation in three carbonate cores. *J. Geol.*, **70**, 364–375.

West, R. G., 1958. The Pleistocene epoch in East Anglia. *J. Glaciol.*, **3**, 211–216.

Wiseman, J. D. H., 1954. The determination and significance of past temperature changes in the upper layer of the equatorial Atlantic Ocean. *Proc. Roy. Soc. London*, **A222**, 296–323.

Wiseman, J. D. H., 1959. The relation between paleotemperature and carbonate in an equatorial Atlantic pilot core. *J. Geol.*, **67**, 685–690.

Woldstedt, P., 1954. Die Klimakurve des Tertiärs und Quartärs in Mitteleuropa. *Eiszeitalter u. Gegenwart*, **4/5**, 5–9.

Woldstedt, P., 1958. Eine neue Kurve der Würm-Eiszeit. *Eiszeitalter u. Gegenwart*, 151–154.

Woodring, W. P., M. N. Bramlette and W. S. Kew, 1946. Geology and paleontology of Palos Verdes Hills, California. *U.S. Geol. Surv. Prof. Paper*, **207**.

Worzel, J. L., 1959. Extensive deep-sea sub-bottom reflections identified as white ash. *Proc. Nat. Acad. Sci.*, **45**, 349–355.

Zagwijn, W. H., 1957. Vegetation, climate and time-correlations in the early Pleistocene of Europe. *Geol. Mijnb.*, n.s., **19**, 233–244.

Zagwijn, W. H., 1960. Aspects of the Pliocene and early Pliocene vegetation in the Netherlands. *Med. Geol. Stich.*, ser. C-III-1, **5**.

Zeuner, F. E., 1945. *The Pleistocene Period.* Ray Soc., London.

Zeuner, F. E., 1952. *Dating the Past.* Methuen, London.

Zeuner, F. E., 1959. *The Pleistocene Period.* Hutchinson, London.

AUTHOR INDEX

Authors' names of Chapters in this book and the page numbers at which these Chapters begin are printed in **heavy type,** page numbers of citations in the text are in ordinary type and those of bibliographical references listed at the ends of the chapters (including joint authors) are in *italics.*

Where the same author is mentioned in more than one chapter it may be convenient to differentiate by subject between the successive ranges of page numbers by consulting the list of contents at the beginning of the book.

Names associated with known apparatus, equations, laws, principles, etc. are not entered here but in the Subject Index.

SUBJECT INDEX

Titles of Parts and Chapters and the page numbers at which they begin are printed in **heavy type.** Names of ships and of genera and species are printed in *italics.*

Where a single index entry (such as a geographical name) is followed by a string of page numbers, the reader should consult the list of contents at the beginning of the book to differentiate them by subject.

The letters ff after a page number denote either the beginning of a chapter subdivision on the subject in question or that the subject is mentioned again on one or both of the next two following pages.